552
26

# Sedimentology and Sedimentary Basins

D1335530

# Sedimentology and Sedimentary Basins

## From Turbulence to Tectonics

2nd Edition

**Mike Leeder**
*Professor Emeritus and Honarary Research Fellow*
*School of Environmental Sciences,*
*University of East Anglia, Norwich, UK*

LADY MARGARET HALL LIBRARY
OXFORD

WILEY-BLACKWELL
A John Wiley & Sons, Ltd., Publication

This edition first published 2011 © 2011 Mike Leeder
First edition published in 1999 by Blackwell Science Ltd.

Blackwell Publishing was acquired by John Wiley & Sons in February 2007. Blackwell's publishing program has been merged with Wiley's global Scientific, Technical and Medical business to form Wiley-Blackwell.

*Registered office:* John Wiley & Sons Ltd, The Atrium, Southern Gate, Chichester, West Sussex, PO19 8SQ, UK

*Editorial offices:* 9600 Garsington Road, Oxford, OX4 2DQ, UK
The Atrium, Southern Gate, Chichester, West Sussex, PO19 8SQ, UK
111 River Street, Hoboken, NJ 07030-5774, USA

For details of our global editorial offices, for customer services and for information about how to apply for permission to reuse the copyright material in this book please see our website at www.wiley.com/wiley-blackwell

The right of the author to be identified as the author of this work has been asserted in accordance with the Copyright, Designs and Patents Act 1988.

All rights reserved. No part of this publication may be reproduced, stored in a retrieval system, or transmitted, in any form or by any means, electronic, mechanical, photocopying, recording or otherwise, except as permitted by the UK Copyright, Designs and Patents Act 1988, without the prior permission of the publisher.

Wiley also publishes its books in a variety of electronic formats. Some content that appears in print may not be available in electronic books.

Designations used by companies to distinguish their products are often claimed as trademarks. All brand names and product names used in this book are trade names, service marks, trademarks or registered trademarks of their respective owners. The publisher is not associated with any product or vendor mentioned in this book. This publication is designed to provide accurate and authoritative information in regard to the subject matter covered. It is sold on the understanding that the publisher is not engaged in rendering professional services. If professional advice or other expert assistance is required, the services of a competent professional should be sought.

*Library of Congress Cataloguing-in-Publication Data*

Leeder, M. R. (Mike R.)
Sedimentology and sedimentary basins : from turbulence to tectonics / Mike Leeder. – 2nd ed.
p. cm.
Includes index.
Summary: "The sedimentary record on Earth stretches back more than 4.3 billion years and is present in more abbreviated forms on companion planets of the Solar System, like Mars and Venus, and doubtless elsewhere. Reading such planetary archives correctly requires intimate knowledge of modern sedimentary processes acting within the framework provided by tectonics, climate and sea or lake level variations. The subject of sedimentology thus encompasses the origins, transport and deposition of mineral sediment on planetary surfaces. The author addresses the principles of the subject from the viewpoint of modern processes, emphasising a general science narrative approach in the main text, with quantitative background derived in enabling 'cookie' appendices. The book ends with an innovative chapter dealing with how sedimentology is currently informing a variety of cognate disciplines, from the timing and extent tectonic uplift to variations in palaeoclimate. Each chapter concludes with a detailed guide to key further reading leading to a large bibliography of over 2500 entries. The book is designed to reach an audience of senior undergraduate and graduate students and interested academic and industry professionals."– Provided by publisher.

Summary: "The sedimentary record on Earth stretches back more than 4.3 billion years and is present in more abbreviated forms on companion planets of the Solar System, like Mars and Venus, and doubtless elsewhere"– Provided by publisher.

1. Sedimentology. 2. Sedimentary basins. I. Title.
QE471.L375 2010
552'.5–dc22

2010023320

ISBN 978-1-4443-4992-4 (hbk)—978-1-4051-7783-2 (pbk.)

A catalogue record for this book is available from the British Library.
This book is published in the following electronic format: eBook 9781444328479
Set in 9/11.5pt, Sabon by Thomson Digital, Noida, India

# Contents

# Preface

*World is crazier and more of it than we think, Incorrigibly plural.*

Louis MacNeice, 'Snow', *Collected Poems*, Faber

The predecessors to this book, *Sedimentology: Process and Product* (Allen and Unwin, 1982) and *Sedimentology and Sedimentary Basins: from Turbulence to Tectonics* (Blackwell Science 1999) are out of print and partly outdated respectively. I have received much feedback from many persons who have used these books over the years and the current version is intended to try to recapture the spirit of a dynamic and widely applied science. Reasons of space have prevented me from dealing with the subjects of diagenesis and the transformation of sediment to sedimentary rock. I have replaced these with chapters linking sedimentology to climate, sea-level change, tectonics, sedimentary basin architecture and their role in solving interdisciplinary problems. I feel somewhat uneasy about the omission, but it strikes me that the subject of diagenesis has become so based upon the physics of subsurface water flow and the chemistry of low temperature water–rock interactions that the difference of emphasis is too much to encompass within the present text.

Progress over the past decade has been breathtaking. Take some examples: the flow dynamics of opaque mud suspensions can now be monitered by acoustic Doppler probes; knowledge of deep-sea environments has been revolutionized by improved sea-bed imaging; sedimentological reactions to climatic and sea-level change have proved robust and sedimentology contributes vitally to the understanding of the evolution of sedimentary basins, from the birth, life and death of bounding faults to the climatic and palaeontological record contained within them. Further, carbonate sediments and their contained $O_2$ stable isotopes play a key role in establishing ancient oceanic composition (evaporite fluid inclusions), the palaeoaltimetry of high mountains and plateau (calcisols) and the determination of ancient climate (speleothem). All this means that sedimentology is not something that can be done in isolation; the holistic approach is that which I have taken in this book, one based on a thorough understanding of modern processes that I trust will propel the reader into an enthusiasm for the subject and a sense of its place in the wider scheme of earth sciences, specifically in attempts to read the magnificent rock record.

Who do I expect to be reading this book? You will have completed an introductory course in general geology, earth or environmental sciences, and perhaps a more specific basic one in sedimentology or sedimentary geology. You will thus know the basic sediment and sedimentary rock types and also know something of the place of the subject within the broader earth and environmental sciences. You will have enough basic science background to understand, if not feel exactly on top of, Newton's laws, basic thermodynamics and aqueous chemistry. Though mathematically challenged, like many earth scientists including myself, you should at least know where to find out how to manipulate equations to a reducible form. I make no apology for spending a little more time with basic fluid dynamics than with the thermodynamics. This is *not* because I find one more interesting or important than the other—it's just that most high school leavers and graduating university students (even those of physics) do little in the way of fluid mechanics in their syllabi nowadays and it seemed that the theme of 'sedimentological fluid dynamics' is just such a place to set up some sort of foundation. Philosophically you should want to reduce the complicated natural world to an orderly scheme, but at the same time not want to miss out on the romance and poetry of an unclassifiable subject. You will be someone who enjoys talking and arguing with a variety of other earth science specialists.

Just a few final notes are in order.

- More involved derivations of essential physical and chemical concepts are to be found in the end section labelled 'Cookies'. These are meant to be helpful for intellectual health. There is also a short mathematical refresher appendix. I would appreciate it if readers let me know of any mistakes or symbol typos I may have made as I hope to live long enough to make another edition, someday.
- I have tried my best to reference major developments correctly at the end of chapters, to make sure the source of specific case histories can be traced and to respect historical precedence and discoveries. References are given in abbreviated form, but quite sufficient to be of full use in rapid web-based search vehicles like *Web of Science*.
- Many graphs with data points have been generalized to 'clouds' or 'envelopes' of data points—if you wish to get the original data, go back to the cited references.

Thanks to colleagues and friends who either directly, through conversation, or indirectly through me reading their works, have inspired my continued interest in sedimentology and its many applications. I would like in particular to thank long-time collaborators and friends Jan Alexander, Julian Andrews, Jim Best, John Bridge, Rob Gawthorpe and Greg Mack for keeping my mind stretched over past years. I also extend my heartfelt thanks to former faculty colleagues at the fine Universities of Leeds and East Anglia where I have spent my professional life, together with my ex-undergraduate and graduate students, for keeping me on my toes. I am grateful to Dr Jenny Mason for writing the sections on terrestrial carbonates (section 2.9), the role of speleothems in palaeoclimatic studies (section 23.18) and, with Dr James Hodson, for compiling the reference list. Finally, thanks to the whole production team at Wiley-Blackwell and to Ian Francis for his gentlemanly encouragement to complete this project and for putting up with some delays over the past 4 years as I periodically got on with my research and real life instead!

Mike Leeder
Brooke, Norwich
January, 2010

# Acknowledgements

Permission has been granted (or at time of going to press has been requested) to reproduce or alter illustrative materials from the following publishers and holders of copyright:

American Association of Petroleum Geologists: Figs 17.6, 17.10, 21.24
American Geophysical Union: Part 6 Fig. 1; Figs 2.3, 6.5,6.9, 6.15, 66, 8.7, 10.2, 10.14, 14.15, 15.4, 15.7, 16.1, 16.2, 16.6, 16.7, 18.2, 18.5, 18.10, 18.15, 19.9, 21.1, 21.26.
American Mineralogical Society: Figs. 1.3, 1.6.
American Society of Civil Engineers: Figs. 6.8, 11.5, 12.7.
Annual Reviews of Earth and Planetary Science: Fig. 11.6.
Arnold: Fig. 13.2.
Balkema: Fig. 7.27.
Cambridge University Press: Figs. 4.14, 5.8, 5.10–13, 5.15–16, 7.11, 8.1, 8.16, 10.7, 15.1, 15.8.
Chapman & Hall: Figs. 1.4, 19.8
Elsevier: Figs: Part 6 Fig. 2, 1.2, 1.5, 1.8–9, 1.14–1.16, 1.18–19, 2.1, 2.14, 7.6, 7.9, 7.15–16, 7.21, 7.25, 7.32, 7.40, 7.46, 8.6, 11.4, 11.22, 13.4, 17.7–9, 17.12, 19.1, 21.27–28, 23.13.
Geological Society of America: Figs. 1.13, 1.17, 2.2, 2.17, 8.11, 10.5–6, 10.10, 11.3, 12.4, 12.6, 12.9, 14.14, 15.3, 16.13, 17.2–3, 20.1, 20.30, 21.2, 21.14, 22.6–7, 22.11, 22.21.
Geological Society of London: Figs 10.4, 10.13, 11.14–15, 13.15, 21.17, 21.32, 22.12
Harcourt, Brace, Jovanovich: Fig 2.9.
Harper Collins: Figs 7.4, 8.22.
International Association of Sedimentologists/ Sedimentology/Basin Research: Figs. 6.6–7, 6.18, 7.2, 7.7, 7.12–13, 7.22–23, 7.26, 7.31, 7.39, 7.43, 7.45, 7.47–49, 8.3, 8.8, 8.14, 8.17–21, 9.3, 9.6, 9.10–11, 11.11–12, 11.16–19, 11.21, 11.29, 12.10, 13.6–14, 13.17, 14.3, 14.6, 14.8–9, 14.11–13, 14.16, 14.18, 15.10, 15.12–13, 16.4–5, 16.11–12, 17.13, 17.18, 17.27–28, 18.9, 18.14, 18.16, 18.18, 19.10–12, 19.16, 20.2–3, 20.6, 20.14–15, 20.25–27, 20.31–35, 20.38–40, 21.8, 21.10–11, 21.13, 21.16, 21.18–20, 21.25, 21.29–30, 22.13–14, 22.23–27, 22.32–34.
Johns Hopkins Press: Figs 22.9–11.
Journal of Geology: Figs 1.10, 2.7.
McGraw-Hill: Fig. 4.1.
Nature: Figs. 5.20, 7.37, 8.24–26, 10.16, 15.2, 15.5–6, 21.33, 23.2, 23.21, 23.23
Oxford University Press: Figs. 4.9, 5.14.
Pergamon: Figs. 5.21, 5.25.
Princeton University Press: Figs. 4.2, 5.24.
Royal Society of London: Figs. 6.11, 23.3.
Society of Economic Palaeontologists & Mineralogists/Journal of Sedimentary Research: Figs. Part 6 Fig. 2, 3.3, 6.1, 6.16, 7.14, 7.28–30, 7.34, 7.38, 7.44, 10.11, 11.8, 11.24–28, 12.1, 13.16, 14.7, 14.19, 15.11, 16.8–10, 17.4, 17.11, 17.14, 17.16–17, 17.20–25, 18.3, 18.6–8, 18.11, 18.13, 18.20–21, 19.2–3, 19.15, 20.4, 20.8, 20.12–13, 20.16–21, 20.24, 20.28–29, 20.36–37, 20.41, 21.3–7, 21.9, 21.12, 21.15, 21.21, 21.2–23, 22.18, 22.29–31, Cookie 8 Fig. 3.
Soil Science Society of America: Fig. 6.4.
Springer Verlag: Fig. 20.22.
Van Nostrand Reinhold: Figs. 5.2, 5.7, 5.19.
Wiley: Figs. 6.12, 6.14, 11.13, 18.4, 22.8–9, 22.15–17, 22.19–20, 22.22.

# Part 1

# MAKING SEDIMENT

*...the soil which has kept breaking away from the high lands during these ages and these disasters, forms no pile of sediment worth mentioning, as in other regions, but keeps sliding away ceaselessly and disappearing in the deep.*

Plato, *Critias*, Vol. 9, Loeb Classical Library

## Introduction

The noun *sediment* comes to the English language from the Latin root *sedimentum*, meaning settling or sinking down, a form of the verb *sedere*, to sit or settle. In earth and environmental sciences, sediment has a wide context that includes many forms of organic and mineral matter. In Part 1 we look more deeply at the origins of the sediment that occurs on and under the surface of the solid planets and which may be used to infer past environmental conditions and changes. Sediment accumulations may be grandly viewed as the great stratal archive of past surface environments, or more basically as 'dirt'. There has been sediment on the surface of the Earth since the Archaean, with the oldest known sediment grains dating from at least 4.4 Ga (Part 1 Fig. 1). Sediment also mantles the surface of many other planets and their satellites, notably Mars, Venus and Saturn's moon, Titan.

In scientific usage, Earth's sediment is best divided into three end-members:

- *clastic*—originating from pre-existing rock outside a depositional area as transported grains, the commonest being mineral silicate grains, known widely as *siliciclastic sediment*;
- *chemical*—being the result of inorganic or organically mediated chemical precipitation within the depositional area;
- *biological*—derived from skeletal material associated with living tissues.

These simple divisions are robust enough to include even the highly esoteric sediment forms that are turning up in the wider Solar System, like the solid ice particles transported and deposited by liquid methane on Titan. Of course there are unusual, hybrid or mixed origins for some sediment but these can easily be accommodated (e.g. *bioclastic*, *volcaniclastic*). Note that the classification is restricted to grains that were sedimented; there are sedimentary horizons in the stratigraphic record that originated as precipitates below the deposited sediment surface, often bacterially controlled. These were never sedimented as such and are considered as secondary or *diagenetic* sediments that post-dated physical deposition of host primary sediment. Deposited sediment accumulates as successive layers, termed *strata*, and such deposits as a whole are said to be *stratified*. The succession of strata in any given deposit is controlled by environmental factors and their correct interpretation involves a deep understanding of how present and past environments have evolved over time.

The chemical and biochemical processes that produce sediment also give other soluble byproducts; these chemical species control oceanic and atmospheric composition and provide long-term sourcing for base cations that nourish plant life and counteract acid deposition in temperate forested catchments. Chemical earth-surface processes have undoubtedly changed over deep geological time, in response to atmospheric and hydrological changes, whilst biological processes

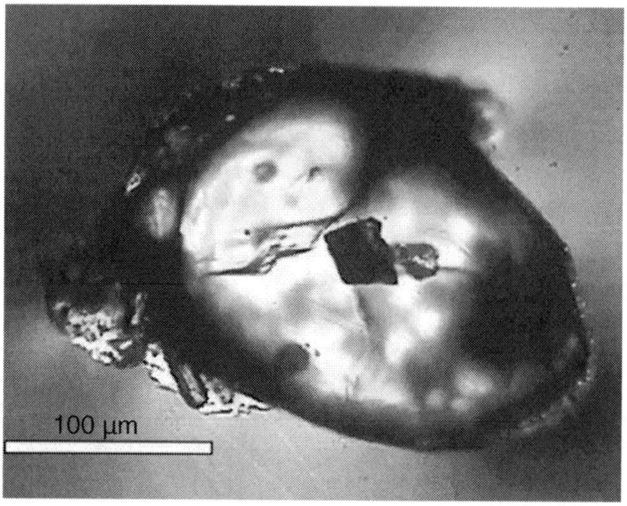


**Part 1 Fig. 1** Image of the famous Jack Hills' zircons from the Cryptic Era of the Hadean Eon (discovery paper of Wilde *et al.*, 2001). The location is ~800 km north of Perth in Western Australia in the Narryer Gneiss Terrane of the Yilgarn Craton, a group of folded and metamorphosed supracrustal rocks thought to have originally contained sedimentary siliciclastic rocks. Detrital zircons with ages greater than 4 billion years old occur here, the oldest being 4.4 Ga (±8 million), the oldest dated material originating on Earth. The source is a metamorphosed conglomerate considered to have an age ~3.0 Ga, so the detrital zircons are sourced from pre-existing rocks, probably subduction zone plutonic igneous intrusions which were then weathered and the resultant sediment deposited as sedimentary rock. The zircons are evidence for the existence of continental-type crust on the surface of the Earth during the Hadean Eon, contrasting with earlier ideas on the earliest phase of Earth's history in which continental crust was thought absent and plate tectonics inoperative until much later. Additionally, oxygen isotopic ratios in the zircons provide evidence for the presence of liquid water on the Earth's surface at this time. The image is a general photograph of a pristine Jack Hills' zircon (Curtin University website). Longest axis of crystal ~250 μm.

have changed hand-in-hand with organic evolution. By this view, sediment production is an accident of weathering and evolution—a waste product. Ever since the Archaean, the planet has 'learnt' how to cope with this waste, just like it has with the waste oxygen produced during plant photosynthesis. There was no predeterminism associated with the processes of sediment production on early Earth or any other planet. Sediment simply fell out (forgive the pun) of the rock cycle in which primary rock is chemically and physically altered. Compare for example sediment on the Moon with the Earth. In the former the sediment is a fragmented remnant from past meteoritic impact events. In the latter sediment is highly varied in its origins, composition, size and physical properties. Its role as an accidental part of the Rock Cycle establishes sedimentology as a fundamental part of Earth System Science. Indeed, as one nice semi-popular review entitled it a decade ago (Stanley & Hardie, 1999), from the point-of-view of calcareous sediment, '*Hypercalcification: Paleontology Links Plate Tectonics and Geochemistry to Sedimentology*'. We shall examine such grand claims later in this book (Chapter 23).

*Chapter 1*

# CLASTIC SEDIMENT AS A CHEMICAL AND PHYSICAL BREAKDOWN PRODUCT

*Few ken to whom this muckle monument stands,*
*Some general or admiral I've nae doot,*
*On the hill-top whaur weather lang syne*
*Has blotted its inscribed palaver oot.*

Hugh MacDiarmid, 'The Monument', 1936, *Complete Poems*, Vol. 1, Carcanet, 1993.

## 1.1 Introduction: clastic sediments—'accidents' of weathering

Terrestrial clastic sedimentary rocks are usually quite different in their composition from the igneous and metamorphic rocks that sourced them. This is because they are derived from an altered *regolith* with a soil profile produced by chemical weathering of pristine bedrock and the source of mineral grains for such sediment. For example, feldspar is the commonest mineral in bedrock of the Earth's continental crust (about 60% of the total) but quartz is usually predominant in clastic sediments and sedimentary rocks. Despite this difference the principle of conservation of mass tells us that for all elements present in the exposed crust and released by weathering, exactly the same levels of abundance must occur in the average total sedimentary mass. Thus the average *chemical* composition of *all* sediments is roughly that of the igneous rock, granodiorite, representing the mean composition of middle to upper continental crust.

It is traditional to divide rock weathering into physical and chemical components, but in reality the two are inextricably interlinked. Water is the chief reactant and plays a dual role since it also transports away both dissolved and solid weathering products. Earth is presently unique in its abundance of water and water vapour, yet Mars also had an earlier watery prehistory. It is easy to take water for granted, the deceptively simple molecule $H_2O$ has remarkable properties of great importance for rock and mineral weathering (**Cookie 1**). These include its solvent and hydration properties, wetting effects due to high saturation and anomalous decrease of density at low liquid temperatures and after freezing. An outline of the near-surface terrestrial hydrological cycle is given in Fig. 1.1.

Want to know more about the structure and properties of water? Turn to Cookie 1.

*Sedimentology and Sedimentary Basins: From Turbulence to Tectonics*, 2nd edition. © Mike Leeder.
Published 2011 by Blackwell Publishing Ltd.

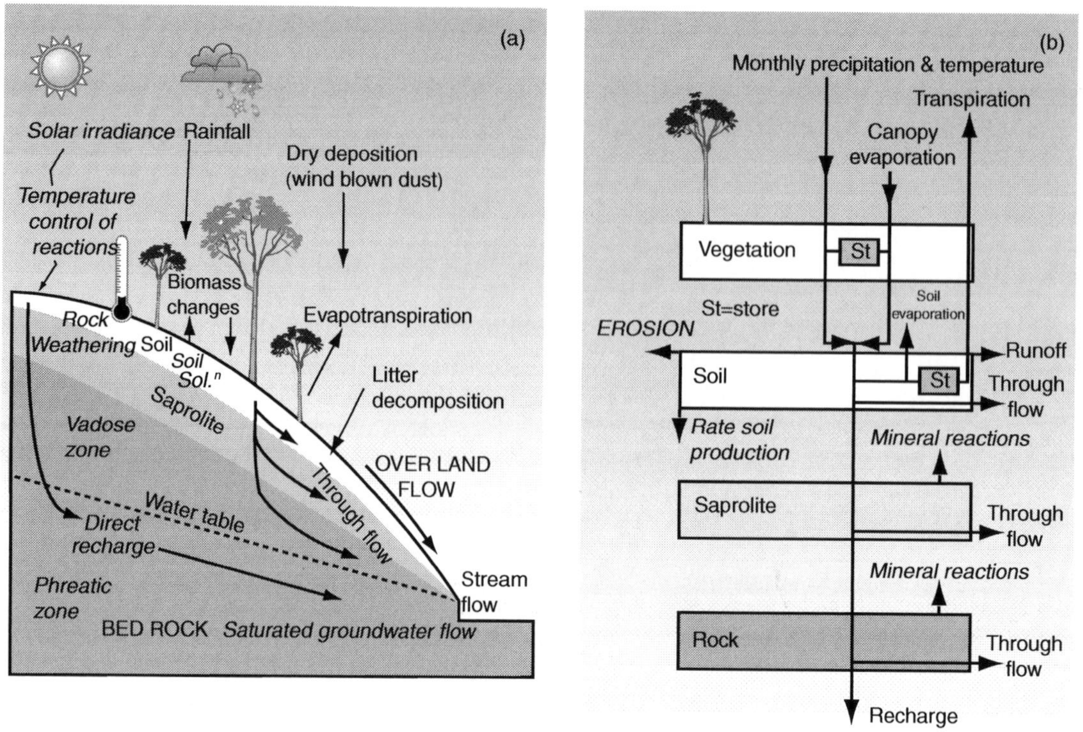


**Fig. 1.1** (a) The main components of the terrestrial hillslope hydrological cycle. (b) Box model for water budget in the surface and near-surface weathering zones.

Chemical weathering involves aqueous reactions with a strong biochemical component since dissolved atmospheric gases are aided by soil-generated gases, dilute acids and organic ligands. Further, the reactions are complex since silicate minerals are involved, with their many constituent anions and cations; also the amount of water and dissolved ions varies in both time and space. Four main mechanisms contribute to chemical weathering: *dissolution*, *oxidation*, *hydrolysis* and *acid hydrolysis*. Reactions usually occur at mineral surfaces in the unsaturated (vadose) zone where, close to the local Earth's surface, rock pores contain atmospheric gases, water, living and dead vegetation and bacteria—all play an important role in weathering. The result is a regolith and soil profile whose characteristics depend upon climate and rock type.

Physical weathering involves the application of differential stresses to rock and mineral discontinuities in the unsaturated zone. These cause fragmentation and are due to erosional unloading, gravity, wind shear, salt crystallization from groundwaters, freeze–thaw and differential thermal expansion.

The combined effects of biochemical and physical weathering produce a weathered regolith profile in bedrock that comprises:

- surface *soil* with significant living and dead organic content;
- chemically altered rock termed *saprolite* that nevertheless preserves its framework coherency without volume loss (*isovolumetric weathering*);
- chemically unaltered but often physically fragmented (exfoliated) *bedrock*.

Interfaces between these layers are in a state of slow downward motion as the landscape reduces. In fact, landscape dating by *cosmogenic isotopes* and other means reveals that a steady-state system often exists, with the material mass removed by erosion being replaced by an equal volume made available from below for further decomposition.

Weathering acts on:

- mineral aggregates at or close to Earth's surface with their many intracrystalline pores and fracture networks;
- imperfect crystals with surface and lattice defects.

Weathering involves:

- bond *breaking*, physically by cracking and chemically by solution (see **Cookie 2**);
- broken bond (danglers) *adoption* by ionic predators;
- electron *removal* from the easily stripped transition metals like Fe and Mn (see **Cookie 3**).

Weathering depends on:

- $H_2O$ *throughput* or hydraulic conductivity by laminar water flow in pore spaces between minerals, microporosity within minerals and cracks across minerals;
- $H_2O$ *dissociation* into the reactive H and OH ions (see **Cookie 1**);
- $CO_2$*concentration* via atmospheric and soil processes;
- temperature, which controls both *reaction rates* (Fig. 1.2) in silicate minerals via the results of the *Arrhenius principle* (see **Cookie 3** and further below) and the rate of aqueous throughput via its control on water's *dynamic viscosity*.

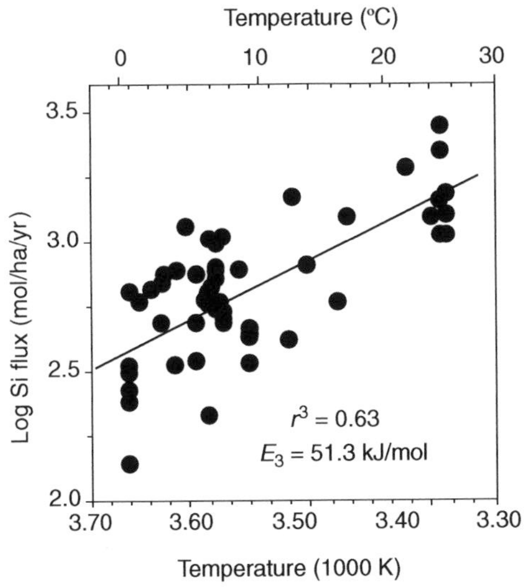


**Fig. 1.2** Arrhenius plot to show that the flux of Si from small, simple-geology granite catchments depends on temperature, once corrections are made to allow for rainfall amount. This is because the rate constant is determined by the Arrhenius effect discussed in the main text and **Cookie 3**. The solid diagonal line corresponds to the activation energy $E_a$. (After White *et al.*, 1999.)

Want to know more about ions, electron transfer and Eh–pH diagrams? Turn to Cookie 2.

Want to know more about proton donors, pH, acid hydrolysis and calcium carbonate weathering? Turn to Cookie 3.

## 1.2 Silicate minerals and chemical weathering

The flux of dissolved elements and altered minerals from continent to ocean is largely controlled by the processes of chemical weathering. Global river sampling indicates that silicate weathering accounts for ~45% of total dissolved load, calcium carbonate weathering ~38% and evaporitic salts ~17%. The recharge of dissolved ions into rivers from the weathering zone (Fig. 1.3) is not only a function of source rock but also of the fluxes from atmospheric deposition, vegetation growth and respiration, and the net reaction of soil and other subsurface interstitial water with minerals in the unsaturated zone. Once liberated by weathering into the hydrological system, some elements behave *conservatively* in that they then proceed down-catchment with little further gain or loss. Other minerals behave in a decidely *non-conservative* way —Ca is a good example since it readily forms mineral precipitates in semiarid soils and in other terrestrial carbonate sinks (section 2.9).

Chemical weathering of silicate minerals plays a major role in the global hydrogeochemical cycle. Since the relative proportion of minerals in clastic sedimentary rocks is different from those in igneous and metamorphic rocks, it is clear that some are more stable than others in the weathering process and some are newly formed; this depends upon the thermodynamics of the reactions involved (**Cookie 4**). Reactions in natural waters include rapid dissolution of ionically bonded minerals like soluble salts (Cookie 1) and acid attack on carbonate minerals (Cookie 2). Both carbonates and evaporites usually dissolve *congruently*,

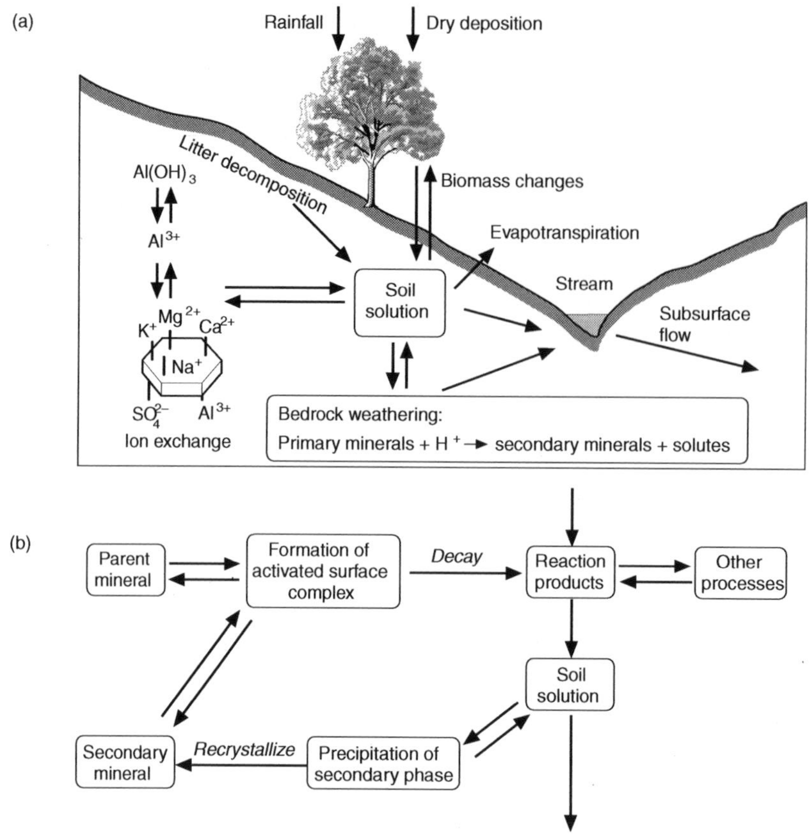


**Fig. 1.3** (a) Some of the processes affecting solute fluxes in catchments. (After Drever & Clow, 1995.) (b) Flow diagram to show characteristics of surface mineral reactions. Note that incongruent behaviour involving mineral precipitation is characteristic of many weathering reactions. Also, non-conservative behaviour means that some dissolved ions may be precipitated *in situ* as other mineral phases, which therefore do not stay in solution to be exported by river water. (After Sverdrup & Warfvinge, 1995.)

i.e. the proportion of elements in solution is the same as the proportion in the former mineral(s). However, in the context of the whole catchment some of the dissolved ions may subsequently become involved in secondary mineral precipitation.

Want to know more about the thermodynamics of weathering reactions? Turn to Cookie 4.

Decomposition of silicate minerals is always *incongruent*—the reactions are very slow and temperature dependent, yielding other solid products in addition to dissolved ones. Reaction pathways vary according to local Eh–pH–temperature conditions. They proceed predominantly by *acid hydrolysis* in which activated surface complexes play a major role (Fig. 1.3) and where small, highly charged protons displace metallic cations in crystal silicate and oxide lattices. Silicic acid is produced from the intermediate metal-bonded silicate complexes and released to solution. $OH^-$ or $HCO_3^-$ ions finally combine with displaced cations to form solutions or local precipitates. Hydrolysis acts along lattice surfaces exposed by discontinuities such as joints, rock cleavage, crystal boundaries, mineral cleavage planes and crystal surface defect sites. Once a dissolved ion is liberated from a crystal lattice by hydrolysis, it may link up with surrounding water molecules and thus stay in solution as a hydrated ion, or, if the force of attraction with water is insufficient to compete with that between the water molecules themselves, the ion will be 'ignored' by the water and will precipitate. Some elements will be exported from the weathering site, perhaps far away. Others will behave non-conservatively and precipitate locally, perhaps forming economic accumulations like in the concentration of Al and Fe in *bauxites* and *laterites* respectively. The type of behaviour in solution is determined by the *ionic potential*, which is the ratio of ionic charge to ionic radius (Fig. 1.4). Cations (metals) whose potential is less than 3 are easily hydrated and highly mobile; anions of potential >12 form soluble complexes. Ions with potentials between 3 and 12 are mostly precipitated as hydroxides (in the absence of protons) and are thus immobile.

The proton concentration of soil water gives rise to pH values in the range 5–9, in which silicon is more soluble than aluminium. Thus prolonged chemical

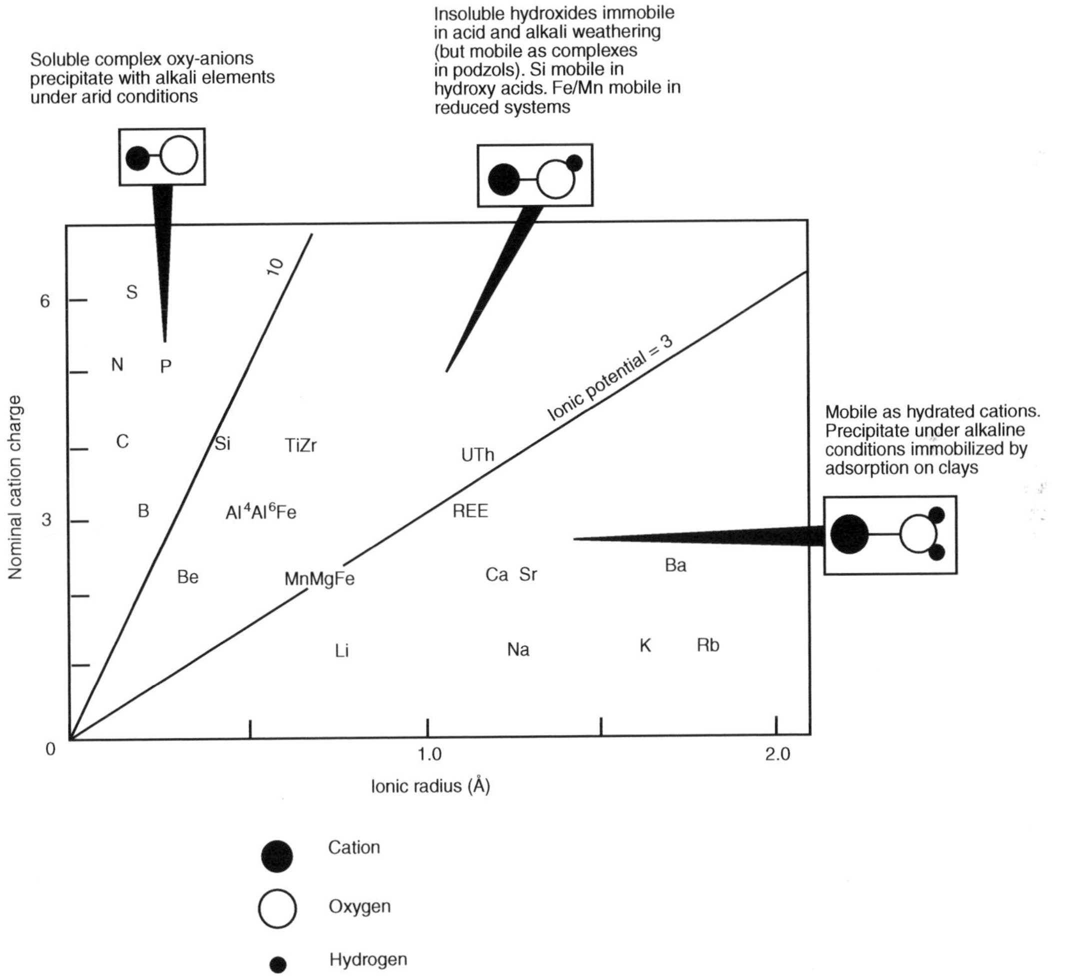


**Fig. 1.4** The relationship between ionic charge and radius for various ions of importance in surface hydrogeochemical and weathering reactions. (After Gill, 1989).

weathering will leach Si from the soil profile, leaving behind a mixture of Al and Fe oxides and clays. These are *lateritic* (aka *ferralitic*) soils, common in well-drained tropical sites. By way of contrast, in highly acid soils Al and Fe are both leached to give a *podzol* with a characteristic light-coloured, silica-rich upper zone. Soils are discussed further in section 6.5.

Quartz, muscovite and K-feldspar dominate amongst clastic mineral components that survive chemical weathering from the weathered *regoliths* of igneous and metamorphic terrains (Fig. 1.5). However, the abundance of these and other primary minerals is highly variable, depending upon a number of factors including sourceland abundance, climate and type of weathering, original grain size, rapidity of sedimentation, and so on. As usual, sedimentology cannot be reduced to mere chemistry or physics. There are also important newly formed minerals produced by weathering.

### Quartz

Quartz makes up ~20% by volume of the exposed continental crust. Its crystals comprise spiral networks of linked silicon–oxygen tetrahedra, making the lattice extremely resistant to chemical attack by aqueous solutions over the acidic and neutral pH range.

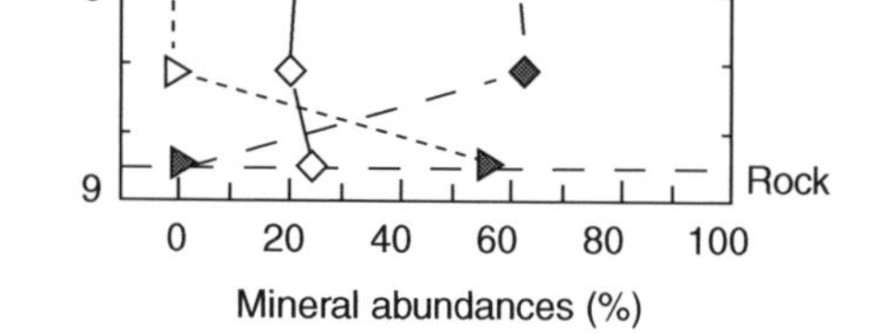

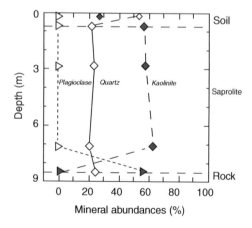


**Fig. 1.5** Selected mineral abundance versus depth for the Rio Blanco Diorite, Costa Rica to illustrate the rapid changes observed in mineralogy at boundaries between bedrock/saprolite and saprolite/soil. (After Riebe *et al.*, 2003.) Note dissolution of plagioclase and neoformation of kaolinite in the regolith.

Solubility increases at pH > 9 when crystal surfaces are subject to increasingly effective *hydroxylation*, particularly in the presence of alkali cations from NaCl and KCl. Hydroxylation is a process that introduces one or more hydroxyl groups ($OH^-$) into a compound or radical, thereby oxidizing it. Polymerization occurs in supersaturated solutions to form silica gels or sols. There is much evidence that quartz dissolution is greatly facilitated (perhaps by several orders of magnitude) by dissolved Al in the porewater environment. The main route by which quartz is made dissolvable, and also fragmentable, appears to be along microcracks and fractures.

### Feldspar

As noted above, feldspar is the most abundant mineral in the Earth's crust and is most notable for being a major player in soil acidification and terrestrial inorganic $CO_2$ uptake (**Cookie 5**). It is an alumino-silicate in which Si–O and Al–O tetrahedra link to form an 'infinite' three-dimensional framework with variable proportions of the alkali cations Na, K and Ca in the interstices. The rate of dissolution is strongly temperature dependent and is also an interesting V-shaped function of pH (Fig. 1.6). At low pH the feldspar weathering reaction is simply a transformation by acid hydrolysis of the Al bonds. The main source of protons is dissolved $CO_2$, both from the atmosphere and the soil as a product of respiration. Such *acid dissolution* is probably the commonest form of soil weathering condition; it results in the precipitation of the white clay mineral, *kaolinite* and the liberation of the alkali and alkaline-earth elements in solution as hydrated ions, carbonate or bicarbonate ions with silica sometimes as byproduct (**Cookie 4,** Equation 4.1). Careful experiments show that the silica probably originates under slow-leach acidic conditions as a gel. Reaction rates are low at or about neutral pH, rising again because of hydroxylation at high pH when Si bonds are susceptible to dissolution in the presence of base cations. K-feldspar is also unstable under acid conditions and yields the products potassium carbonate, kaolinite and silica. There are several other possible products of the feldspar-weathering reaction depending upon local conditions of pH, which control the nature of the dissolved Al species—the most notable is *gibbsite*, ($Al(OH)_3$). At very low pH ($<3$) reactions cause metallic cation leaching and the formation of a

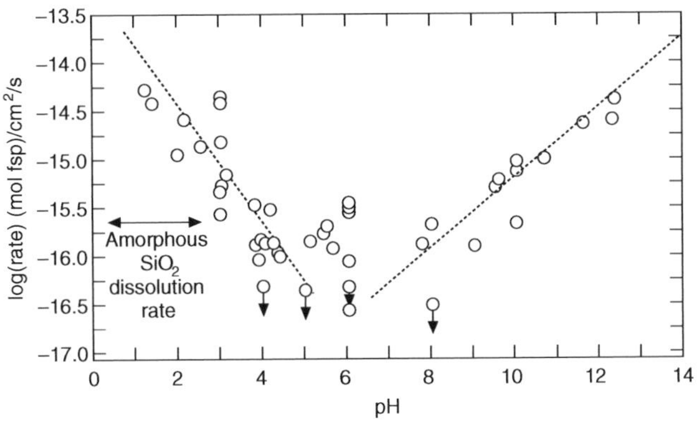


**Fig. 1.6** Experimental sodium-rich feldspar (albite) dissolution rate as a function of weathering solution pH. Note the V-shaped curve common to many silicates. (After Blum & Stillings, 1995.)

thick surface layer rich in Si. Such surface layers do not form at other pH values and the dissolution process is dominated not by diffusion through a surface layer but by direct bond breaking and scavenging of 'danglers' (broken bonds).

Although there are a wide range of apparent activation energies and rate constants recorded in experiments, the generally observed field-weathering order for the various feldspar types is:

anorthite(Ca-plagioclase) > albite(Na-plagioclase) > orthoclase/microcline/sanidine(K-feldspars)

There is some evidence that the dissolution of K-feldspar is inhibited by the presence of Al in solution. Figure 1.7 shows spectacular etch pits and channels formed by Holocene weathering of a K-rich feldspar.

Want to know more about chemical weathering as a geosink for global atmospheric $CO_2$? Turn to Cookie 5.

## Micas

Micas have a distinctive sheet structure in which Mg, Al and Fe cations in octahedral arrangement lie between layers of (Si, Al)$O_5$ tetrahedral sheets. Cation substitutions give rise to negatively charged layers which are neutralized by an interlayer of K in both muscovite (K-rich mica) and biotite (Fe-rich mica) mica. This interlayer is released relatively quickly during weathering. Weathering reactions of the framework ions are slower and chiefly controlled by hydration and hydroxylation at broken metal–oxygen bonds, the process occuring progressively inwards from edge faces. Muscovite exhibits a typical V-shaped dissolution rate curve with respect to pH, with the lowest rates around pH 6. Biotite shows the same dissolution rate trend as muscovite at acid pH, but rates are usually very much higher (five times that of muscovite and up to eight times that of plagioclase) because of the dissolution effects of aqueous oxidation upon ferrous iron, $Fe^2$.

## Chain silicates

These comprise the pyroxene and amphibole groups in which silica tetrahedra are linked in either single (pyroxene) or double (amphibole) staggered chains by oxygen sharing. The chains are linked and strongly held together by metallic cations like Ca, Mg and Fe. Amphiboles have additional $OH^-$ in the rings between opposite chains. Weathering is chiefly by surface reactions that remove the metallic cations and acid hydrolysis by protonation to form surface species bonded as silicon oxide surface species. Rates of dissolution vary inversely with pH, with actual rates varying widely in these often chemically complex groups. As in biotite mica the occurrence of transition-group metals like Fe and Mn means that redox reactions with dissolved oxygen also occur, leading to generally faster rates of overall dissolution. Oxidation of liberated $Fe^2$ leads to surface crusting of

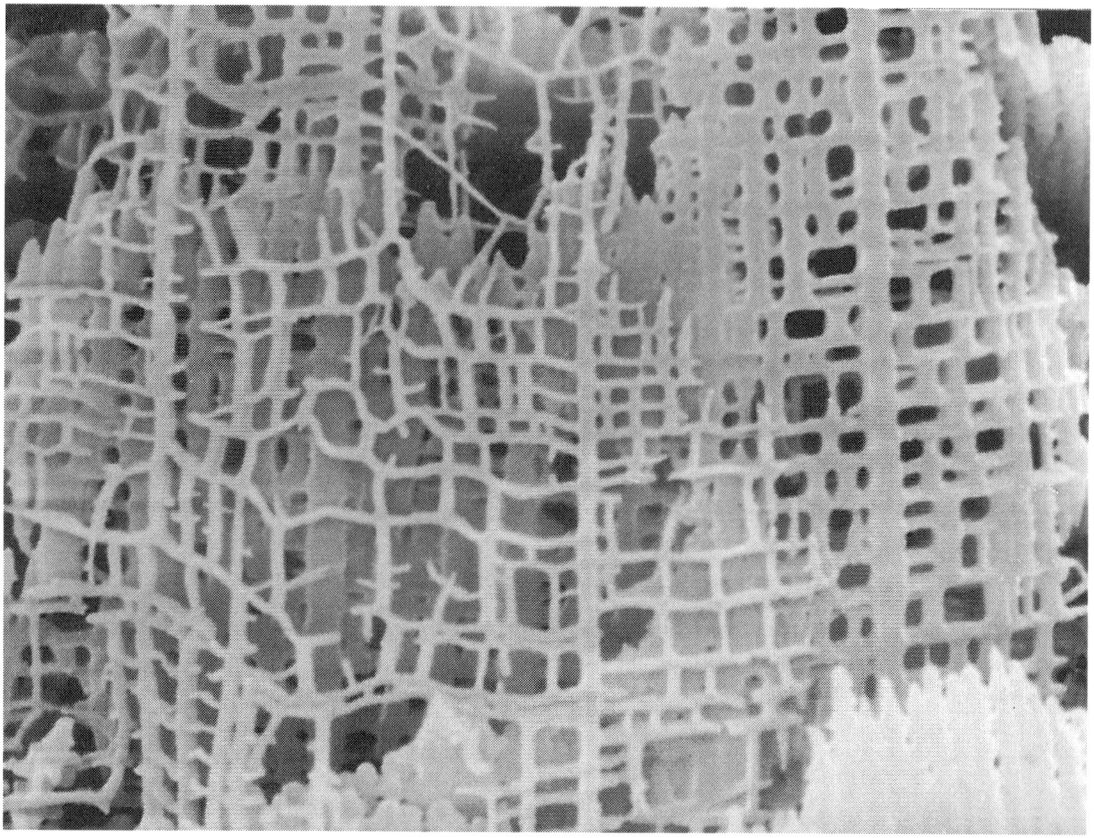

**Fig. 1.7** Scanning electron microscope image of a resin cast of weathered feldspar. The former grain surface of the K-rich alkali feldspar is below the plane of the image and the weathered interior is in focus. The resin network has penetrated into the interior of the crystal along etch pits. Field of view 24 μm. (After Lee & Parsons, 1995, 1998; image courtesy of authors)

amorphous $Fe^3$ precipitates, but it is not thought that these adhere significantly enough to lead to diffusion-controlled reactions.

### Olivine

Olivine has a simple structural arrangement of isolated $SiO_4$ tetrahedra linked by the divalent cations $Mg^2$ and $Fe^2$. It comprises the end-members fayalite (Fe-rich) and forsterite (Mg-rich), with a complete solid solution series in between. Mg is removed by surface leaching and complex formation after protonation. Fe-rich olivines are highly susceptible to oxidative weathering and the formation of hydrated clay minerals. Ferric crusts of *goethite* characterize humid saturated conditions whilst *haematite* occurs under subhumid to arid unsaturated conditions. A major role for microbial oxidation of $Fe^2$ during acid-weathering is suggested by experiments that record lowering of mineral reaction due to surface adsorption of $Fe^3$.

### Clay minerals

Clay minerals are the most important newly formed mineral group in the weathering zone since their eventual erosion and deposition produces copious mud-grade sediments that give valuable information about weathering conditions. We have already encountered *kaolinite*, formed under humid, acid weathering conditions from the alteration of feldspar-rich rocks. *Illite*, a potassium aluminium hydrated silicate, is formed by weathering of feldspars and micas under alkaline weathering conditions where significant leaching of mobile cations such as potassium does not occur. *Smectites* are complex expandable sheet silicates with intracrystalline layers of water and exchangeable cations. They form from the weathering of igneous rocks under alkaline conditions. Gibbsite is simply aluminium hydroxide and forms under intense tropical weathering conditions with high annual precipitation (>2000 mm) when all other cations and silica present in bedrock are leached out.

### Apatite and the P cycle

Apatite is either a stoichometric, fluorinated calcium phosphate (igneous fluorapatite, FAP) or a non-stoichometric sedimentary carbonate fluorapatite (CFA) with variable substitution of $Mg^{2+}$, $Na^{+}$ and $CO_3^{2-}$. The mineral is a common accessory mineral in granitic rocks and forms the majority of sedimentary phosphate rock. Both FAP and CFA define the primary exogenic phosporus sink and serve as the long-term weathering source of P to the biogeosphere where it is an essential nutrient element for cellular life and acts as a control upon net ecosystem production. Weathering is optimal under warm, acidic pH conditions, with FAP being the easiest to dissolve.

### Rock weathering profiles: changes, age and depth

Data from studies of chemical, physical and mineralogical changes with depth in weathered granodiorite are shown in Figs. 1.8 & 1.9. The fresh granitodiorite comprises plagioclase feldspar (32%), quartz (28%), K-feldspar (21%), biotite (13%) and muscovite (7%) with rarer (<2%) amphibole. The main secondary minerals formed during weathering are kaolinite with lesser amounts of halloysite, goethite, haematite, gibbsite and amorphous Fe-hydroxides. Note the rapid loss of alkaline earth elements Na and Ca from fresh to plagioclase-weathered bedrock with K, Fe and Al showing slight relative enrichment. K subsequently declines in saprolite as K-feldspar is incongruently dissolved whilst Fe and Al enrich, the former due to $Fe^{3+}$ precipitation in hydroxides, the latter due to kaolinite precipitation. Si usually behaves conservatively or slightly depletes throughout, apart from in the soil zone where it concentrates and Al/Fe strongly depletes. Cosmogenic isotope dating reveals mean rates of descent of the weathering fronts at speeds of

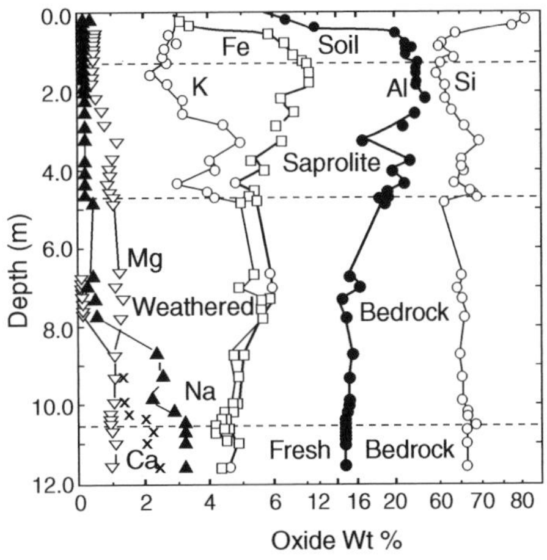


**Fig. 1.8** Elemental distributions with depth in the Panola regolith developed over ~$10^5$ yr on a granodiorite bedrock, Georgia Piedmont province, USA. Weathering was essentially isovolumetric with preservation of primary granitic textures in the saprolite, average saprolite porosities of ~35%, presence of boxwork textures of secondary minerals surrounding primary remnants and etch-pit formation. (After White *et al.*, 2001.)

~4–7 m/$10^6$ yr. These rates are very much slower then experimental feldspar weathering rates and suggest that in old regoliths surface mineral reactivity decreases significantly with time.

### Chemical index of alteration

Although chemical weathering is complex there exists a simple Chemical Index of Alteration (CIA) that is useful to assess its extent. It involves simplifying weathering of the upper crust to a combination of feldspar and volcanic glass, involving only the commonest oxides in the silicate fraction, $Al_2O_3$, CaO, $Na_2O$ and $K_2O$. The CIA is given by $100.Al_2O_3/(Al_2O_3 + CaO + Na_2O + K_2O)$ and varies between 100 and 47. The pristine upper crust has a mean CIA of 47, with 100 signifying wholesale removal of all alkaline earths. CIA values calculated for the suspended load of the major rivers of the world define a chemical weathering path (Fig. 1.10; section 23.3). Rivers draining tectonically active catchments have low CIA values since physical weathering and high mechanical erosion rates predominate here.

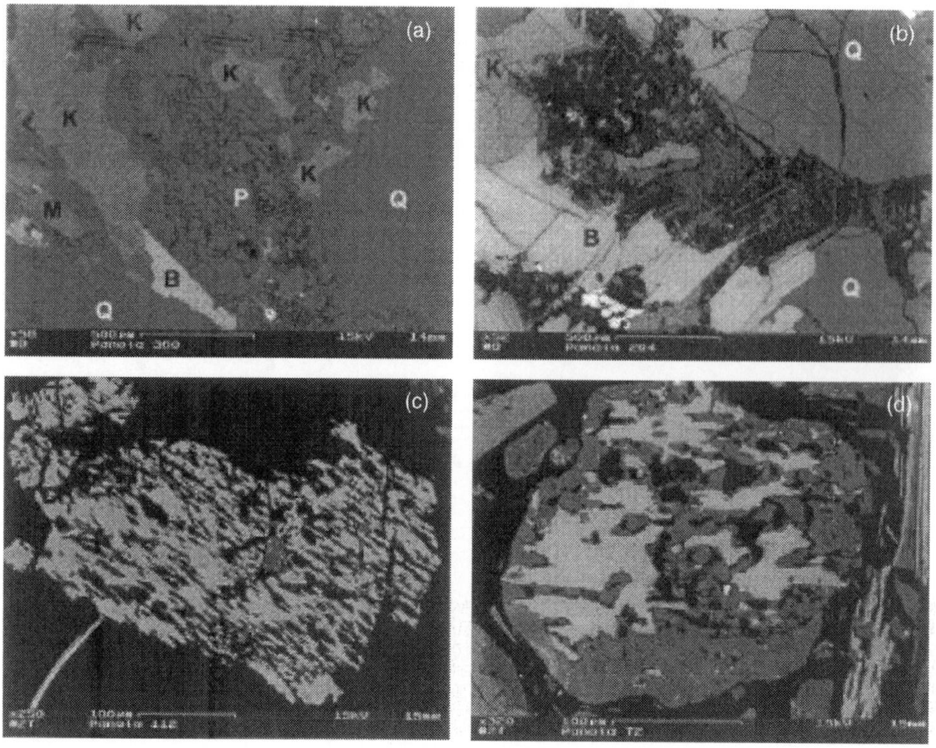


**Fig. 1.9** Scanning electron microscope backscatter-mode images of rock sample thin-sections from various depths in the Panola regolith (P, plagioclase feldspar; K, –potassium feldspar; B, biotite mica; M, muscovite mica; Q, quartz). (a) 9.1 m depth; initial incongruent weathering of plagioclase feldspar in bedrock. Dark areas within the crystal are the clay mineral kaolinite. Note the pristine K-feldspar. (b) 7.5 m depth; the plagioclase crystal (centre) has virtually gone, leaving clay-filled vugs (black areas). The adjacent biotite mica is starting to weather along cleavage planes (dark areas within the biotite). The K-feldspar is again pristine. (c) 2.8 m depth. In this saprolite zone the K-feldspar has dissolved, gaining a skeletal appearance with open vugs (black areas) that originally may have been partially filled with clay. (d) 1.8 m depth just below the soil–saprolite interface. The K-feldspar is partially replaced by clay. Note the relatively unreacted muscovite mica to the right of the image. (After White *et al*., 2001.)

### Al-release from silicates by acid hydrolysis ('acid rain' problem)

Acid hydrolysis reactions in soil zones with base cations (especially $Ca^2$) usually lead to effective neutralization of rainwater, which in unpolluted areas has a pH of around 5.0. However, in many temperate zones with high rainfall and thin soils on Ca-poor substrates, the acidity is not neutralized, leading to formation of characteristic podzol soils (see section 6.6) with leached Al-poor surface horizons. Vastly increased industrial pollutants and emissions ($CO_2$, $SO_2$, NO) have accelerated these processes in many areas, including exports of gases to 'innocent' countries (e.g. Scandinavia), leading to extensive Al-release in acidic waters and widespread environmental damage. The very steep increase of Al solubility in waters of low pH is illustrated by the data of Fig. 1.11. Assessments of acidification depend heavily upon estimations of cation fluxes, sources and sinks through the water–soil–bedrock system.

## 1.3 Solute flux: rates and mechanisms of silicate chemical weathering

In these times of rapid environmental change it is essential to truly understand and to be able to predict the rates and mechanisms of chemical weathering of

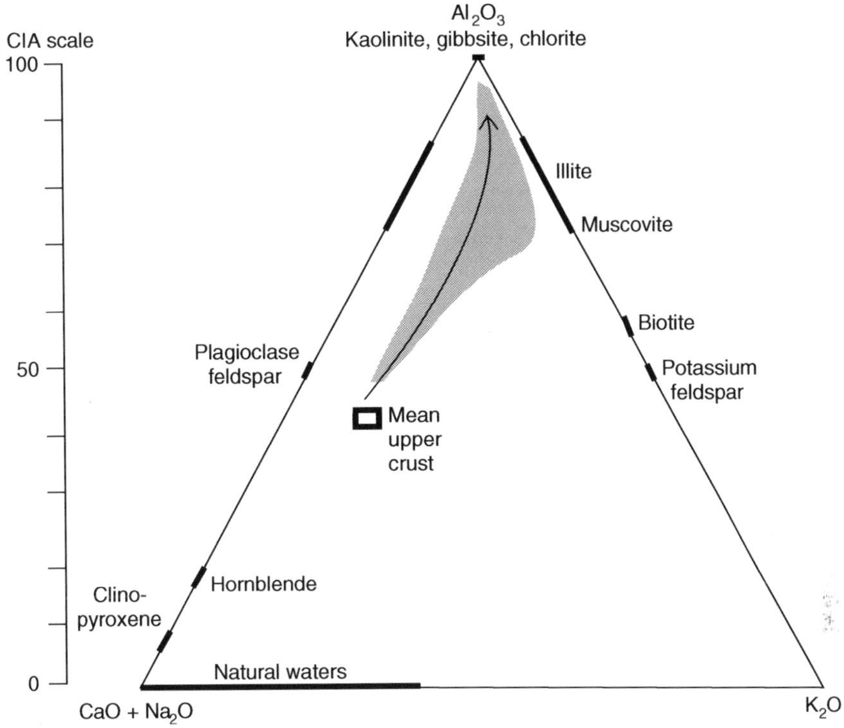


**Fig. 1.10** Triangular diagram $Al_2O_3$–(CaO + $Na_2O$)–$K_2O$, with approximate compositional ranges of natural waters, common silicate minerals and Chemical Index of Alteration scale. (After McClennan, 1993.)

silicate-rich bedrock in response to water throughflow. Very generally the *solute flux* approach to catchment weathering seeks to draw up a balance sheet of all inputs and outputs to the catchment chemical system—it is basically a *conservation of mass* exercise. Its simplest expression states that:

$$\text{cation weathering rate} = \text{leaching} + \text{uptake} - \text{deposition} - \text{base saturation decrease}$$

It is usual to define weathering reactions in terms of a *rate constant*, the amount of mass lost from unit surface of a mineral over unit time. It is defined by *Arrhenius's law* (**Cookie 6**), controlled both by the energy needed to make a reaction happen, the *activation energy*, and by the *temperature* of reaction conditions. The law is easiest to understand kinetically, since any reaction needs sufficiently energetic collisions between reactant ions to cause it to happen; an *energy threshold* must be surmounted. In the case of silicate mineral surfaces in contact with natural waters, the warmer the aqueous phase the more energetic will be the attacking protons and therefore the faster the reaction will be. Not only that, but the energetics increase exponentially as temperature increases. This means that there is a considerable (25-fold) disparity between the energy available for weathering at the poles compared to the tropics.

Want to know more about the rates and mechanisms of chemical weathering? Turn to Cookie 6.

Flow rate of water through the weathering zone is important because much rainfall is recycled back to the atmosphere by evapotranspiration; the amount available for throughflow is dependent on vegetation type and density. Most silicate reactions proceed to a point where the products have concentrations as predicted by the relevant *equilibrium constant*, which for silicates are *very* low (**Cookie 6**). When the rates of forward and backward reactions become equal, no

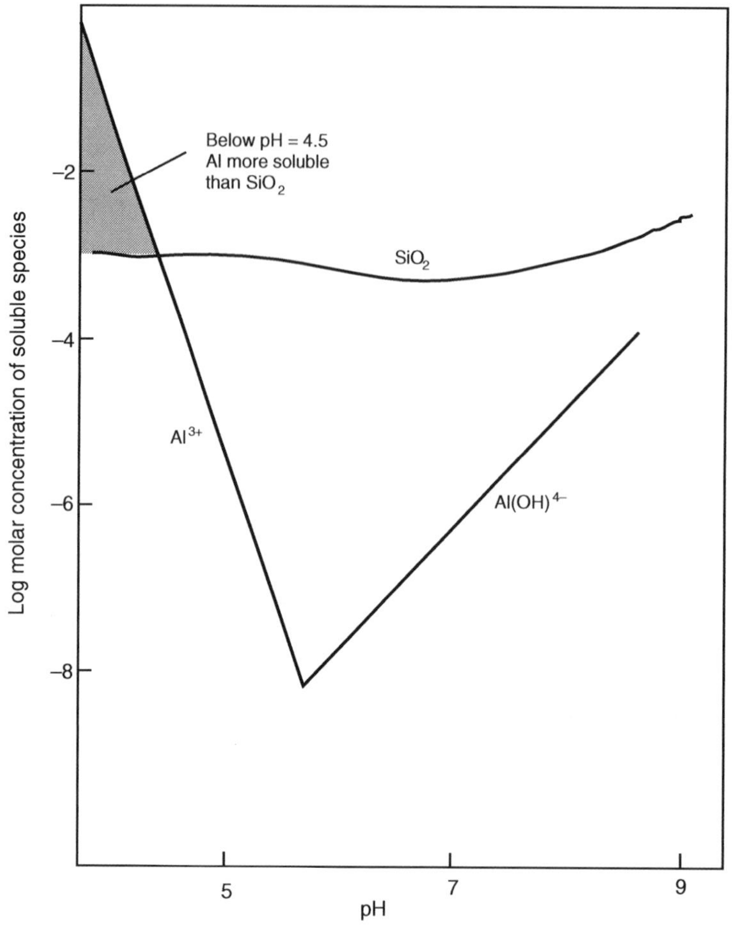


**Fig. 1.11** The solubility of aluminium and silicon as a function of pH. (After Raiswell *et al.*, 1980.)

further weathering occurs. Only by removing saturated porewaters and replacing them with new water can more weathering reactions occur. Optimal conditions for chemical weathering occur in forested orogenic highland areas of the humid tropics, where very thick soil and chemically altered bedrock zones have formed in the Holocene and where soil instability due to slope failure enables periodic exposure of fresh saprolite or bedrock. Lack of physical weathering and erosion on continental plateaux lead to thick residual soils and a marked decrease in total export of ions. Such catchment slope conditions are known as *transport-limited*, for there is always an abundance of weathered product to transport.

There is much experimental work to suggest that weathering rates are linear, controlled by surface reactions between mineral and aqueous phases. Surface in this case refers to fractured and cleaved mineral faces. There is also sound evidence from electron microscopy studies that weathering can occur along uncracked and uncleaved areas of minerals, the process occurring along submicroscopic diffusion paths at sites of lattice dislocations. The chemical attack is seen in the form of etch pits (Fig. 1.7). Rates of surface reactions are usually very slow compared to diffusion rates, and the concentration of products adjacent to the reacting mineral surface are comparable to that of the weathering solution.

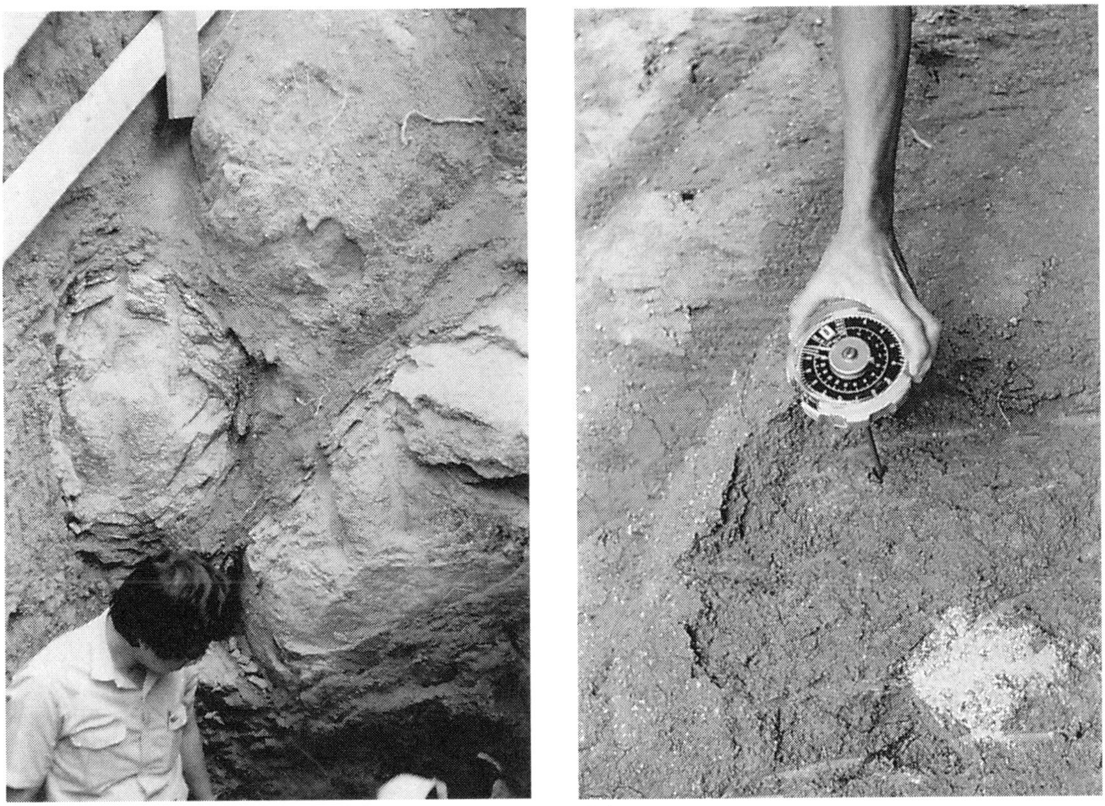

(a) (b)

**Fig. 1.12** (a) Granodiorite corestones showing exfoliated 'onion-skin' morphology in a tropical weathering zone of saprolite. (b) The hand-inserted shear vane is measuring the strength of a saprolite that was once solid granite—testament to the reality of rapid tropical chemical weathering. (Images from Tai Po, Hong Kong, courtesy of Steve Hencher.)

Although experimental and some field studies indicate a constancy of dissolution rates over time, there is also evidence from naturally weathered materials that rates decrease as a result of elemental exhaustion and/or the build-up of a thickening surface layer of weathered material that slows aqueous access during diffusion. A steady state may be set up by removal of the weathered material by erosion (see further below). The local rate of chemical weathering by diffusion will determine the thickness of *weathered rinds*, defined as discoloured and permeable crusts enriched in immobile oxides (e.g. $Fe_2O_3$, $TiO_2$ and $Al_2O_3$) relative to unweathered cores. With calibration to features of known age, it is possible to use weathering rind thicknesses to determine weathering ages. The ultimate signature of progressive chemical weathering of rock masses comes from *corestones*, where rounded remnant nucleii of more pristine rock are surrounded by saprolite (Fig. 1.12).

The rate of chemical alteration of glacial-sourced debris in moraines and soils also shows a well-defined decrease with age, probably because the finely ground carbonate and sulphide mineral components dissolve first and easiest. Silicate weathering increases with distance from an active glacier in the older outwash sediments, particularly if vegetation becomes established. It is thought by some that chemical weathering rates have risen and fallen globally in tune with glacial and interglacial development over the last million or so years, although it should be stressed that there is no *direct* evidence for this. This idea is based upon the greatly increased availability of glacier-ground fine sediments (loess, tills, outwash) that cover higher latitudes and mountain belts at deglaciation. The fresh

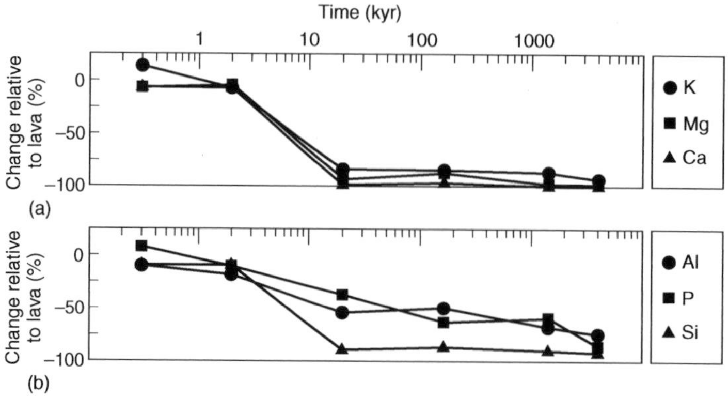


**Fig. 1.13** The time change of elemental proportions in soils developed on Hawaiian basalts of various ages. Note the most rapid rate of change in K, Mg, Ca and Si between 2 and 10 kyr. (From Vitousek *et al.*, 1997; see also Chadwick *et al.*, 1999; Sak *et al.*, 2004.)

mineral surfaces in these are expected to weather very rapidly at first, declining over millenial times scales—we return to this topic in connection with strontium isotope studies relating to global denudation trends (Chapter 10).

Studies spanning $3.5 \times 10^6$ yr on soils atop dated basalt lava flows in Hawaii (humid subtropical forested environment) indicate that the easily weatherable base elements K, Mg and Ca, together with Si, are reduced to less than 10% of their initial values in the top 1 m of the soil profiles within 20 kyr (Fig. 1.13). More resistant Al, present in residual kaolinite, declines much more slowly, reaching such low levels only after 3 Myr. Tables 1.1 & 1.2 summarize work done on the relative rates of dissolution of common catchment rocks and minerals.

It is noteworthy that the relative rates of pure, single-mineral dissolution in laboratory experiments show far greater ranges than those determined from natural rocks in catchments underlain by single lithologies. One problem is that many igneous rocks contain small but chemically significant ammounts of carbonate minerals which may dominate the flux of base cations like Ca. Natural weathering is also more complex, particularly in the flow, concentration and fluxes of both aqueous and solute phases. This is nicely

**Table 1.1** Relative rates of $CO_2$ consumption and solute fluxes in runoff from various catchments based on rock type groups shown. (After Amiotte-Suchet & Probst, 1993.)

| Rock type | Relative rate $CO_2$ consumption | Relative rate solute flux |
|---|---|---|
| Plutonic/metamorphic (granite, gneiss, schist) | 1.0 | 1.0 |
| Felsic volcanics (rhyolite, andesite, trachyte) | 2.3 | ? |
| Basic volcanics (basalt) | 5.0 | ? |
| Sandstones | 1.5 | 1.3 |
| Mudrocks | 6.6 | 2.5 |
| Carbonate rocks | 16.7 | 12.0 |
| Evaporites | 3.1 | 40–80 |

**Table 1.2** Relative rates of dissolution of various minerals in laboratory experiments at pH 5 far from equilibrium. (After Drever & Clow, 1995, and sources cited therein; see also Sverdrup & Warfvinge, 1995, Table 20.)

| Mineral | Normalized rate (mineral rate/ albite rate) |
|---|---|
| Quartz | 0.02 |
| Mica (muscovite) | 0.22 |
| Mica (biotite) | 0.6 |
| K-feldspar (microcline) | 0.6 |
| K-feldspar (sanidine) | 2 |
| Na-plagioclase feldspar (albite) | 1 |
| Na/Ca feldspar (bytownite) | 15 |
| Orthopyroxene (enstatite) | 57 |
| Clinopyroxene (diopside) | 85 |
| Mg-olivine (forsterite) | 250 |
| Dolomite | 360 000 |
| Calcite | 6000 000 |

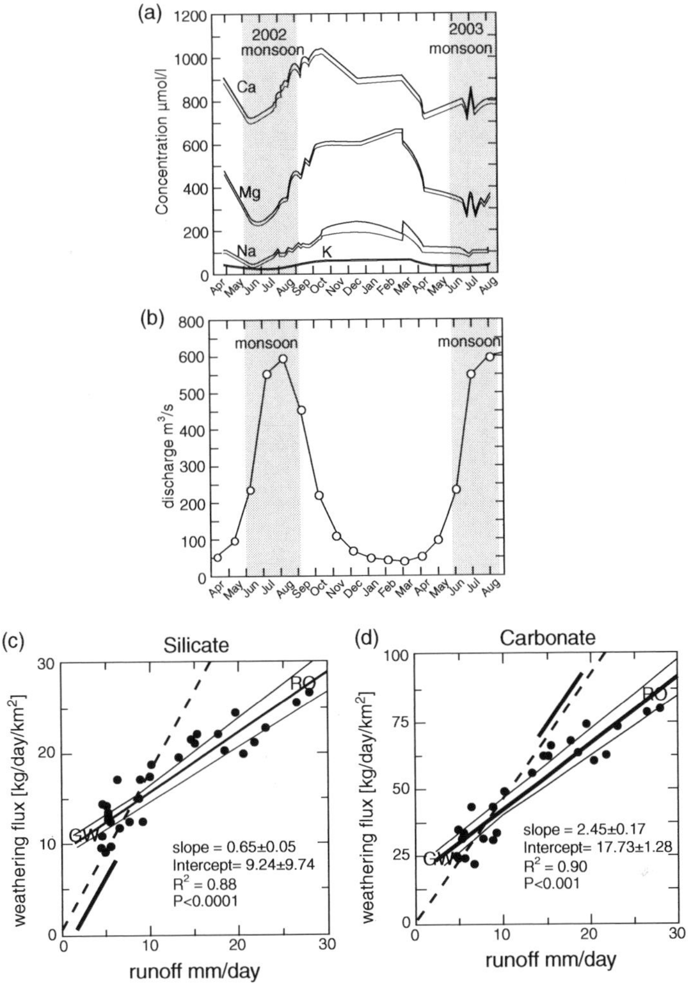


**Fig. 1.14** (a) Selected ionic concentrations through the annual hydrological cycle of the Marsyandi River, Nepal Himalaya. Note highest concentrations occur during dry seasons (b) Monthly discharge of the Marsyandi River averaged over a 10-year period. (c) and (d) The positive dependence of weathering ionic flux on runoff for carbonate and silicate bedrock for a subcatchment of the Marsyandi River system. Note highest fluxes occur during rainy seasons. (After Tipper *et al.*, 2006.)

illustrated by large-magnitude seasonal variations in river-water chemistry and elemental fluxes from Himalayan catchments whose bedrock includes igneous, metasilicate, metacarbonate and sedimentary limestones (Fig. 1.14).

## 1.4 Physical weathering

Erosion causes subsurface rock to be exposed at the surface. During such *exhumation*, elastic volumes are increased by the unloading: deep conditions of higher, but balanced, triaxial stresses are replaced by *in situ* biaxial compressive stresses of order of magnitude (hereafter [O]) 35 MPa, a value in excess of most rocks' strength at the surface. These stresses are vividly illustrated by geologically sudden unloading produced by quarrying or mining with resultant sudden sheeting, arching or catastrophic rock bursting. Together with slow unloading brought about by surface erosion a variety of rock *joints* are produced. These are linear fractures and curvilinear exfoliations parallel to buttressed rock outcrops or valley surfaces. It is now considered impossible for diurnal temperature changes in deserts to cause rock exfoliation since many experiments have failed to reproduce the effects in the laboratory. The effect is ascribed to the spalling-off of partly chemically weathered rock skins from bedrock surfaces, either in saprolite (Fig. 1.12) or on shady parts of rock surfaces. Here, the effects of surface moisture upon minerals is greatest and acts for longer periods.

The various physical imperfections in rock are acted upon by chemical weathering solutions, by biogenic attack and by other physical processes, chief of which are freeze–thaw (frost) and salt weathering. Both of the latter involve the production of stresses through crystallization of solids from solution in tiny rock fractures or near-isolated pores. The net effect, in combination with slow gravity creep down hillslopes (*soil creep*), frost shattering, salt growth and near-surface chemical weathering, is the production of *colluvial mantle* overlying saprolite and bedrock. In this way the landscape reduces, almost by auto-destruction.

*Frost weathering* is particularly important in high mountainous catchments not under permafrost. Up to 165 freeze–thaw cycles per year have been recorded in the French Alps at altitudes of just 2.5 km. A certain amount, sometimes a significant amount, of daily freeze–thaw also occurs in low-latitude and/or high-plateau deserts, where winter rains and dews may provide enough moisture for the freeze–thaw process to be effective. It is partly due to anomalous expansion and decreasing density as water freezes. The accompanying increase in volume of up to 13.5% (at 22 °C) generates up to about 200 MPa of pressure in confined situations. However, the story is inevitably more complicated because of the effect of ice whisker crystal growth, liquid film transfer and stress gradients in tiny to small isolated or tortuous cracks ($<0.5$ mm) and pores. Additional stresses—as much as 10 times those arising from simple expansion above—may arise from ice growth as clusters of parallel ice crystal needles grow normal to the freezing surface. Provided the small crack is supplied with a net input of water, then the stresses arising from crystal growth are limited only by the tensile strength of water, which is drawing water molecules to the ends of the growing crystals through capillary films. This process carries on below normal freezing point in the thin films, being most effective between 4 and 15 °C. It seems that frost weathering is most effective in tiny cracks and crevices of irregular shape in temperate to subarctic climates where repeated thawing and freezing occur on a daily basis, the water moving around by capillary attraction. Between 0.02 and 0.8% by mass of shattered material was generated in experimental 40-yr shatter cycles with a variety of 'hard rocks'. Porous and permeable rocks, where drainage is freer, are not so prone to shattering, but do so nevertheless.

Salt weathering has often been greatly underestimated as a weathering type. It occurs in semiarid to arid climates and in coastal areas of all latitudes where salts are concentrated and where dews, coastal mists, sea spray and ordinary rainfall provide the necessary liquid phases. It would seem to have been particularly important at some stage on the surface of Mars (Chapter 23). There seem to be three ways in which salt expansion may give rise to stresses that lead to rock disintegration: (i) periodic hydration and dehydration, (ii) periodic heating and cooling, and (iii) crystal growth. The last mechanism is probably the most important. As we have seen, crystal-growth stresses are particularly dependent upon rock porosity and the mechanism is most effective in porous sedimentary hosts. Salts vary in their ability to disintegrate rocks by crystal growth, sodium and magnesium sulphates being most effective. Crystal growth stresses occur in tortuous cracks under pressure. They arise at the crystal–rock interface via aqueous films as the crystallizing salts occlude the available space. Open systems, where salts crystallize due to evaporation, cannot give rise to changes of volume and hence cannot give stresses. Finally, a neglected form of salt weathering occurs when sedimentary pyrite is oxidized in fissile calcareous mudrock. The resulting crystal growth of ferrous sulphate and gypsum is highly efficient in disaggregating rock.

## 1.5 Soils as valves and filters for the natural landscape

Soil is usefully envisioned as a valve or filter for the landscape. Imagine it as a graduated semipermeable skin or carapace through which the atmosphere and hydrosphere communicate with the lithosphere. It is graduated in the sense that it divides into more-or-less distinct layers. It is a valve in that it allows some proportion of precipitation to pass through it and hence into saprolite to further the weathering process of bedrock. It is a filter in that the usually well-drained, aerated and oxygenated surface soil layers are the site of a myriad of chemical reactions, many of them biologically mediated, that modify intercepted precipitation. Biological mediation in this case includes:

- elemental uptake and pH-modifying activities due to higher-plant root systems and their adjacent soil volumes (the *rhizosphere*). Rhizosphere pH-diminution is caused by both $CO_2$ production by root respiration and H-ion release to compensate for the excess of cations over anions taken up for plant growth;
- ingestive and metabolic activities of worms;
- metabolizing aerobic soil microbe populations of bacteria (actinomycetes) and symbiotic mycorrhizal fungi on plant roots, algae, protozoa, slime moulds;
- surface encrustation by lichens that cause a substantial increase in silicate mineral dissolution compared to control sites undergoing abiotic weathering.

Concerning the many different soil types, it is the nature of the various constituent layers that forms the basis for soil classifications. There is often an upper accumulation of organic litter of dark hue. Below is a depletive horizon due to mineral alteration, weathering and physical washing-out of materials (*eluviation*). Finally there may be a lower accumulative horizon that represents the deposition of materials from above or the site of deposition of clay-grade material from the throughflowing waters (*illuviation*). In older classifications these three layers are given the codes A, B and C respectively. Below these horizons is saprolite.

An important point is that natural soils develop progressively with time, leading to changes in chemistry and physical characteristics. Residual soils, particularly *duricrusts*, occur in landscapes from which only small amounts of soil breakdown products are removed physically. Such soils are extremely useful to geologists in assessing the nature of past climates and of climate change (Chapters 10 & 23) since they may take many millions of years to form. They are indicative of tectonic stability and frequently occur on ancient erosion surfaces.

In eroding landscapes the extent and age of the soil will depend on the balance between formation and erosion. Studies in landscapes like those of Scandinavia and northern North America, whose young soils were initiated on ground-up mineral grains as Quaternary glaciers retreated in the early Holocene, enable estimates of rates of elemental depletion and soil textural development with time.

We may simplify chemical weathering and soil environments into acid, alkaline and reduced trends (Figs. 1.15 & 1.16). These occur in humid and arid climates and in wetland systems, respectively. We may also simplify soil types into four major groups, ignoring the distinctive but ephemeral character of soil developed upon unconsolidated, glass-rich volcanic ashes.

Acid weathering occurs with high rainfall and a well-developed rhizosphere. This causes hydrolysis and leaching, and results in *ferralitic* and *podzolic* soils.

1 Ferralitic weathering involves Fe–Al enrichment during soil formation and is characteristic of stable, well-drained landscapes in seasonally humid tropical climates encouraging deep, high-pH dissolution of silicate minerals. It has been estimated that such soils cover almost 30% of the continents. Mature soils ($>10^4$ yr old) are dominated by aluminium (defining bauxites) and/or iron oxides and hydroxides (defining ferricretes/laterites), with the minerals goethite, haematite, kaolinite and gibbsite variably present in often very thick weathering profiles. Bauxites form under humid tropical conditions and contain gibbsite and goethite whilst ferricretes/laterites contain haematite and kaolinite formed under subhumid climates. Detailed studies of laterite profiles reveal evidence for both leaching of other cations under alkaline conditions, leading to relative Fe-enrichment, and absolute enrichment of Fe by ionic migration under alternating oxidizing and reducing conditions. The very old ages of some profiles ($10^1$–$10^2$ Ma) leads to opportunities in interpreting past versus present conditions of formation, since climate change has often occurred during their interval of development.

2 Podzolic weathering and *podzol* soil formation (aka *spodosols*) occur in cool, humid-temperate climates under coniferous vegetation that provides acidic leaf litter and also widely in high mountains. Humus accumulates rapidly in an upper horizon since under cool conditions the rate of metabolic breakdown is slow. Moderately to highly acidic conditions ensue as organic breakdown occurs giving *fulvic acids*, which causes Al, Fe and Mn

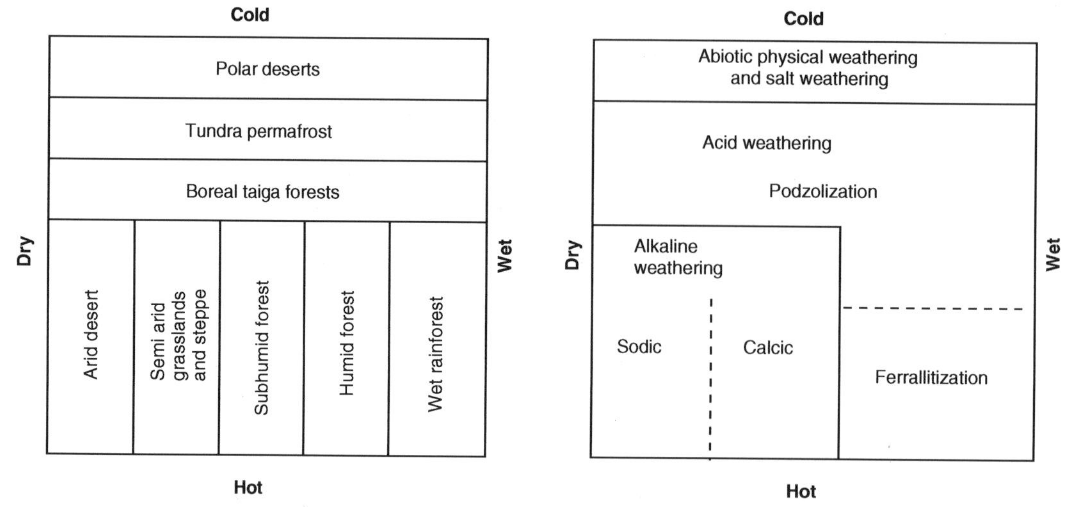


**Fig. 1.15** Simple box classification of climate, vegetation, weathering and soil types developed on well-drained (i.e. oxidising) substrates. This scheme ignores local (sometimes crucial) control by particular lithologies. (Simplified after Chesworth, 1992, and sources therein.)

leaching from a pale middle-depth Si-rich layer and their subsequent eluviation and ultimate reprecipitation in lower soil layers.

Alkali weathering trends occur in arid, semiarid and Mediterranean areas with annual water deficit, and include calcic and sodic soil types.

1 Sodic soils occur in arid climates where there is an upward movement of porewaters leading to evaporative concentration of alkali and alkaline-earth elements such as sulphates, bicarbonates and chlorides. They often occur in interior drainage catchments and reflect ions translocated to the soil profile from considerable distances in through-flowing groundwaters. They are a major problem in irrigated farmlands with poorly planned drainage.

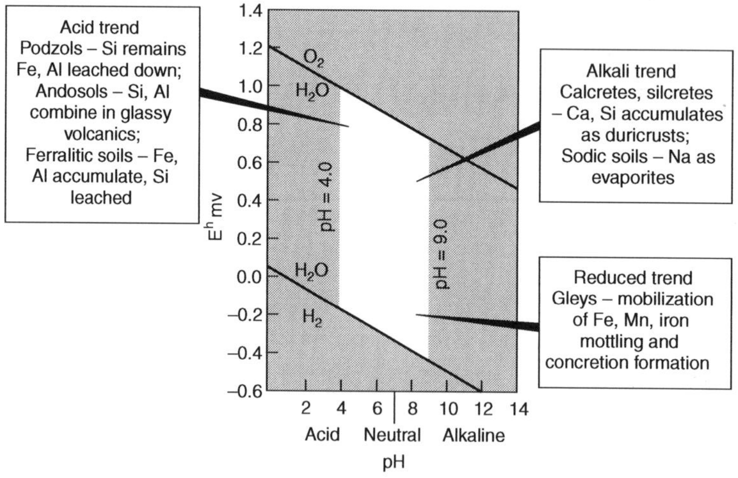


**Fig. 1.16** Broad classification of soil chemical processes in terms of Eh (redox potential) and pH. (After Chesworth, 1992.)

2 Calcic soils (section 2.9) occur widely in semiarid and Mediterranean climates where they result from a balance between downward leaching of $CaCO_3$ and its reprecipitation as calcite in lower soil layers. Stable isotope analysis ($\delta^{13}C$, $\delta^{18}O$) of calcic soils gives valuable evidence for palaeoclimate regime.

*Gley soils* are poorly drained soils that form where the water table is close to the surface and where reducing conditions occur below it. The diagnostic gley horizon has red/brown mottles in a paler matrix. The mottles are coloured by oxidation that occurs when the water table lowers and reduced $Fe^{2+}$ species in the pale horizon can oxidize to $Fe^{3+}$ (**Cookie 2**).

A final point concerns classification and field description of ancient soils (palaeosols) in the sedimentary rock record (Fig. 1.17). Here it is important to be able to assess soil type independently of degree of secondary alteration during burial of what were once primary soil features. Thus physical attributes such as soil agglomerations (*peds*) and expansion/contraction cracks are easily destroyed by compaction (unless mineralized or otherwise infilled) during burial.

## 1.6 Links between soil age, chemical weathering and weathered-rock removal

Chemical weathering and soil formation have been considered separate from physical processes of sediment removal so far in this text; they are in fact strongly coupled. As noted above, it is widely recognized that the rate of chemical weathering from any given soil volume decreases with time. Thickening of soil profiles into saprolite replenishes the supply of fresher mineral surfaces available for chemical reaction. Because of soil-thickening most soils are very much older than the residence times of the minerals within them. However, as soils thicken so their hydraulic conductivity and aqueous throughflow lessens and so the *rate* of thickening decreases. It is the periodic or constant removal of weathered material (both soil and saprolite) by physical transport processes such as *colluvium* (by mass gravity flow) and *alluvium* (by aqueous transport) that provides *new* bedrock material for chemistry to get to work on and produce new soil and saprolite. Chemical erosion rates are thus coupled to physical erosion rates (Fig. 1.18), a topic further considered in Chapter 10.

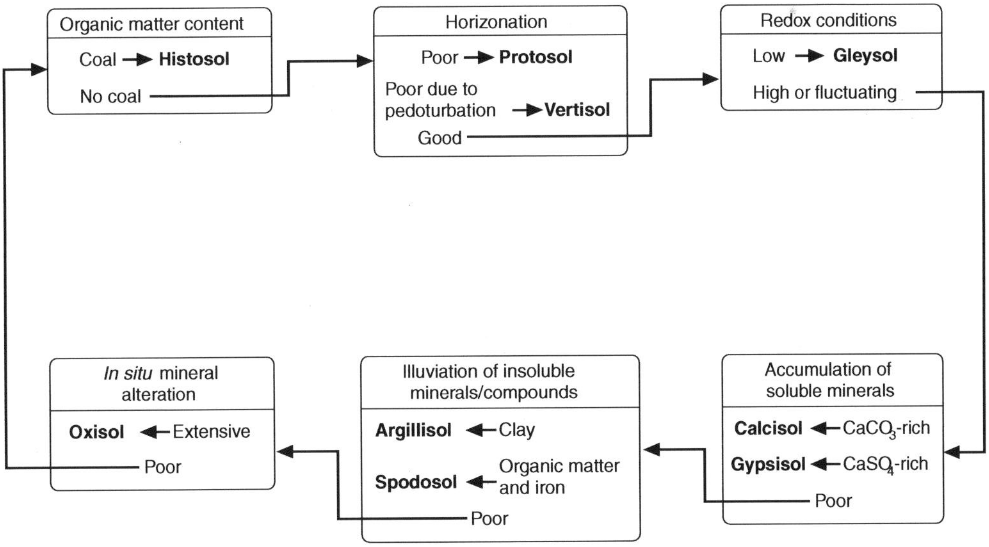


**Fig. 1.17** Simplified flowchart of palaeosol orders to be defined for ancient terrestrial environments by systematic application of six key attributes. (After Mack *et al.*, 1993.)

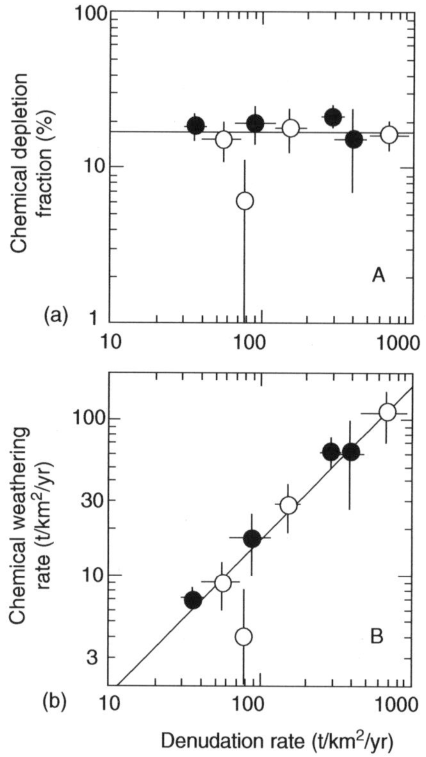


**Fig. 1.18** Chemical depletion fractions and weathering rates from two sites in the Sierra Nevada, California where denudation rates vary substantially. (After Riebe *et al.*, 2003.) (a) The fraction of chemically depleted rock is similar from catchment to catchment. (b) The rate of chemical weathering increases systematically with denudation rate. Lines correspond to relationships based on average chemical depletion fractions of 18%. The coupling shown in (b) suggests weathering rates are supply-limited, i.e. dependant upon fresh exposure of mineral surfaces due to physical denudation processes.

## 1.7 Provenance: siliciclastic sediment-sourcing

In general, provenance means the particular source of something and takes it ultimate meaning from conjunction of the common Latin preposition *pro-*, meaning before and the verb *venire*, to come, i.e. literally, to come before. In sedimentary geology the term denotes the ultimate source of sediment. Thus for the geologist working in the ancient stratigraphic record of uplifted and eroded sedimentary basins (Chapters 22 & 23), siliciclastic sediment may contain the key evidence for crustal evolution in deep geological time. In this section we briefly consider the evidence for sediment-sourcing from the sediment grains themselves; it is vital to recognize that other attributes of sedimentary deposits are equally important, including palaeoenvironmental and palaeocurrent analysis. Major advances in large-scale provenance studies and crustal evolution have come with the measurement of immobile rare-earth elemental concentrations and isotopic compositions of fine sediment fractions (e.g. Sm–Nd (*samarium–neodymium*); Os (*osmium*). Sm–Nd systems behave conservatively during weathering processes and are now widely used to establish mean age and composition of large catchments and to infer rates of continental growth and changing tectonic regimes. For example, a major reorganization of regional drainages in eastern Tibet and southwestern China took place in the Cenozoic as deformation from the growing Himalayas and Tibetan Plateau affected an increasingly wider area. Geochemical and Nd signatures of sedimentary rocks on the northern margin of the South China Sea reveal a major change during the Oligocene when the centre of rifting transferred south and basins on the north margin of the South China Sea experienced rapid subsidence. Further uplift and erosion then exposed Mesozoic and Cenozoic granites that supplied large amounts of granitic detritus. A mid-Miocene change occurred at ca 13 Ma, resulting in less input from local sources (i.e. the fault blocks formed by Mesozoic–Cenozoic tectonics and magmatism) to an increasing contribution of older continental material, mostly from Indochina to the west (see also section 23.3).

Regarding the rock-specific geology of palaeocatchments we must ask, given the destruction visited upon original source rocks by weathering processes, whether an assemblage of coarser sediment grains can tell us anything at all about catchment lithology? The answer is a surprising but qualified yes, particularly when certain mineral grains, properly chemically analysed, may be directly dated by radiogenic techniques. This is despite post-depositional alteration by subsurface waters during burial and diagenesis. The task becomes more difficult though with each tectonically induced cycle of uplift, weathering, erosion and deposition. First-cycle deposits are those derived directly from igneous or metamorphic rocks and are clearly easiest to assess in terms of their provenance. To mix many metaphors, sedimentary geologists can see ancient worlds in their grains of sand, but more darkly further

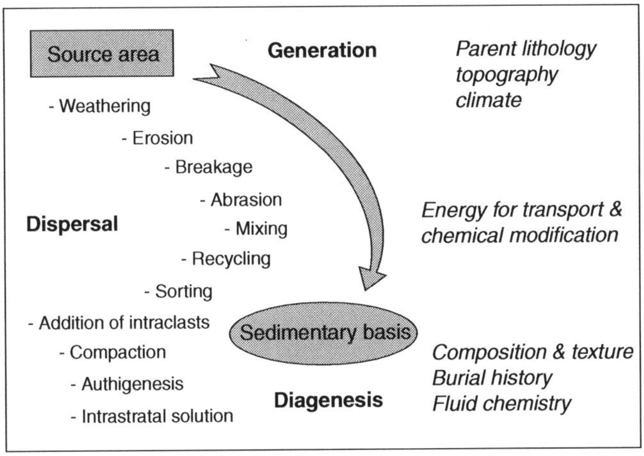


**Fig. 1.19** To illustrate the main pathways in sediment grain evolution (bold) and the principal processes whereby sediment grains might be altered in their journey from sediment source to sediment sink by the main controlling variables (italics). From Weltje & von Eynatten, 2004.)

back in time. A scheme to illustrate the main steps involved in sediment evolution from the point of view of provenance studies is given in Fig. 1.19.

## Quartz

This mineral comprises the most abundant sand- and silt-sized grains in clastic sediments due to a combination of original abundance, hardness, lack of cleavage planes and general chemical durability under normal earth-surface pH conditions. Sections of single crystals may show normal or undulose extinction under the crossed polars of a petrological microscope. Undeformed quartz from volcanic rocks shows normal extinction, but plutonic igneous and metamorphic rocks show either normal or undulose extinction, the latter due to lattice strain. It has been found that metamorphic quartz grains show mean extinction values of $>5°$, whereas plutonic grains show $<5°$, although the overall distributions overlap considerably. Quartz grains may also be single crystals or polycrystalline. Almost all quartz of volcanic origins is of single crystal type. The amount of polycrystalline quartz is least in plutonic igneous rocks, increasing in high-grade metamorphic rocks and is highest in low-grade metamorphic rocks. Also, the average number of crystal units in polycrystalline quartz is greatest in low-grade metamorphic rocks and least in high-grade examples.

The origin of angular quartz silt (4–64 μm size) in ancient marine sedimentary mudrocks is more varied than previously thought. Whilst much sediment of this grade has been considered to have originated as wind-blown dust, detailed imaging and stable oxygen isotope measurements indicate much of it is in fact biogenic, originating as precipitates in algal cysts. It therefore records organic productivity (radiolarian, diatoms, etc., see Chapter 2) of oceanic waters rather than sourceland weathering and detrital sedimentation as silt from planetary wind systems.

## Feldspar

Feldspar also occurs widely in some sedimentary deposits, though its liability to destruction by acid hydrolysis and then break-up during transport along well-developed cleavage planes eventually eliminates many survivors. They are best and most voluminously preserved as first-cycle deposits derived from igneous or metamorphic rocks under limited chemical weathering in semiarid climates. In decreasing order of abundance they include alkali feldspars rich in potassium (orthoclase, microcline), perthite (intergrowths of K-rich and Na-rich feldspars) and the plagioclase series (albite to anorthite) rich in mixtures of sodium and calcium. This ordering follows the stability order of feldspars to chemical weathering discussed previously. Their tendency to dissolve in the subsurface makes further preservation less probable; the stability of alkali feldspars in particular depends on their exact crystal microstructure. Microcline and orthoclase are widely distributed in both metamorphic and acidic plutonic igneous rocks. Plagioclase also occurs widely in metamorphic rocks; in igneous rocks it is increasingly calcic with decreasing silica content of the parent rock. Importantly, sanidine, the high-temperature alkali feldspar rich in K may be radiogenically dated (sometimes as single crystals or as tiny aliquots) by the ratio of its K and/or Ar isotopes, the technique being particularly successful in sediments containing sanidine-bearing volcanic ash.

### Rock (lithic) fragments

These are some of the most useful grains to come across in a sedimentary deposit since they provide direct samples of catchment bedrock. Many igneous rocks have characteristic tectonic modes of origin along active plate margins; good examples would be subduction-related volcanic-arc lavas like andesites and dacites. Volcanic clasts and ash layers are particularly valuable for both provenance and for age-dating the sedimentary sequences in which they lie. The coarser plutonic igneous and schistose/gneissic metamorphic rocks are usually rarer in anything other than first-cycle deposits, breaking down during chemical weathering into individual minerals or breakdown products and fragmenting during sediment transport. Notable exceptions are in glacial lodgement till and oceanic *dropstone* deposits where, enclosed and plucked by ice from their frigid catchments, they often provide extraordinary evidence for glacier ice-provenance and iceberg dispersal paths (Chapters 15 & 21). Fine-grained sedimentary rock fragments in multicycle deposits are dominated by long-lived chert and quartzite lithologies which have extreme mechanical and chemical durability, properties that also encourage their longevity through many sedimentary cycles of uplift, erosion and renewed sedimentation, thus muddying the waters of their provenance.

### Heavy minerals

This large group of non-quartz/feldspar minerals rarely make up more than 1% of a sedimentary deposit. They tend to be concentrated by density during transport and deposition, with the denser forms concentrated in higher energy deposits called *placers*. For study they are separated out magnetically or by density using various high-density liquids to filter them, though since the advent of high-powered microscopy the separation process by hand-picking has become somewhat less laborious and healthier. Opaque heavy mineral grains are usually most abundant, but generally of little use in provenance. They include the iron oxides *haematite* and *magnetite*, titanium-rich *ilmenite* and tin-rich *cassiterite*. The iron sulphide, *pyrite*, is of particular interest in some Archaean sediments since it has lent support to theories for non-oxygenated weathering environments in deep geological time. Non-opaque forms include the hard and ultraresistant *zircon* (Part 1 Fig. 1), with *muscovite* mica, *rutile*, *garnet*, *monazite*, *rutile*, *tourmaline*, *staurolite*, *kyanite* and *sillimanite* common locally. Muscovite, zircon, tourmaline, monazite and garnet may all be radiogenically dated, often nowadays as single crystals by laser ablation. Extremely rarely and locally, provenance studies are enriched by spectacular occurrences of heavy mineral and metal placer deposits of *diamond* and *gold*.

### Phyllosilicates

Amongst these platy minerals, muscovite mica is the most resistant to chemical weathering and is a fine provenance indicator since its high K content means it can be directly dated by Ar-isotope analysis. Clay minerals are generally of little use in provenance studies because of their common origin as weathering products; yet for this reason they are much-loved indicators of palaeoclimate and soil type.

### Organic grains

Charcoal (fusinite) fragments may have an absolutely key role in palaeocatchment studies, for they record the occurrence of wildfires and associated sediment flushing. According to some authors it is the only direct evidence in the geological record for assessing whether levels of oxygen in past atmospheres approximated to those of the present day.

### Wind-blown dust

In Earth's arid regions and on Mars a major sediment production mechanism is by impact-induced abrasion during atmospheric transport. Studies have found that natural dust particles on Earth are more likely to be produced by wind abrasion of weathered sands that have acquired a superficial clay coating, instead of grains with clean surfaces. This coating is removed by the abrasion process and the dust produced by this mechanism in experimental runs has a modal size of 2–5 μm and material < 10 μm comprises up to 90% of the particles produced.

## Further reading

### *General*

A standard chemistry text, with material relevant to some of the chemical principles used in the study of silicate weathering, is by Atkins (1992). Readers lacking college-level chemistry and wanting texts with real geological and environmental relevance would do well to consult the sadly out-of-print Gill (1989), the commendably succinct Andrews *et al.* (2004b) and Chapter 6 of Albarède (2003). The all-time classic on Eh–pH controls is by Garrels & Christ (1965). Stumm (1992) is the fundamental book on chemical weathering and is made of sterner stuff. Drever (1988) discusses a wide range of natural water geochemistry. The collection of papers on the chemistry of rock weathering edited by White & Brantley (1995a) is invaluable, and a bargain. Selby (1993) is very good on physical weathering from a landscape perspective.

### *Specific*

The Jack Hills' zircon discovery paper is Wilde *et al.* (2001), see also Hopkins *et al.* (2008). Plummer (1977) discusses Florida aquifer soil $CO_2$. Estimates for dissolved load to the oceans are by Holland (1978) and Wollast & Mackenzie (1983). Factors controlling detrital mineral abundances are discussed by Suttner *et al.* (1981) and Nesbitt & Young (1989).

References for weathering and dissolution of various minerals are: quartz (Dove, 1995; Hochella & Banfield, 1995), feldspar (Velbel, 1993; Blum & Stillings, 1995; Teng *et al.*, 2001; I. Parsons *et al.*, 2005; Hellmann & Tisserand, 2006), micas (Nagy, 1995; Malmström & Banwart, 1997), chain silicates (Brantley & Chen, 1995; Zakaznova-Herzog *et al.*, 2008), olivine (Hochella & Banfield, 1995; Welch & Banfield, 2002; Zakaznova-Herzog *et al.*, 2008), apatite by Guidry & MacKenzie (2003).

Rates of chemical weathering are in White & Brantley (1995a) and White *et al.* 1999, 2001). For the effects of physical erosion on chemical denudation see Ferrier & Kirchner (2008). There are many experimental studies of chemical weathering rates and mechanisms, notably by Schott & Berner (1985), Furrer & Stumm (1986), Holdren & Speyer (1986), Muir *et al.* (1990), Stumm & Wollast (1990), Casey & Bunker (1991), Dove & Elston (1992) and Shotyk & Metson (1994). Field studies of granitic weathering profiles are by White *et al.* (2001) and of basalt weathering by Stewart *et al.* (2001). Time and space trends in the chemical weathering of glacial outwash sediment are discussed by Anderson *et al.* (2000). The influence of temperature on chemical weathering rates of silicate minerals is discussed by White *et al.* (1999), Dalai *et al.* (2002) and Richards and Kump (2003). Comparative lichen- versus abiotic-weathering studies from Hawaii are in Brady *et al.* (1999), whilst Hinsinger *et al.* (2001) present data on enhanced basalt weathering under higher-plant cover, in the realm of the rhizosphere, a concept due to Darrah (1993).

Weathering rind studies are by Colman & Pierce (1981), Knuepfer (1988), Taylor & Blum (1995), Sak *et al.* (2004). CIA studies are by Nesbitt & Young (1982), McClennan (1993) and Dalai *et al.* (2002). Rates of granitic weathering are by White *et al.* (1999) and basalt weathering by Vitousek *et al.* (1997), Chadwick *et al.* (1999) and Das *et al.* (2005).

Whole catchment studies of chemical fluxes in seasonal rivers draining mixed bedrock types are exemplified by Tipper *et al.* (2006) in the Nepal Himalaya. Problems with modelling chemical weathering processes without a priori knowledge of rock type and mineral precipitation in soil profiles (non-conservative elemental behaviour) are by Goddéris *et al.* (2006) and for the common calcite phases present in many granitoids by White *et al.* (2005). The critical link between rates of chemical weathering and physical erosion is explored by Riebe *et al.* (2003) and the concept of mineral residence time in soil profiles is discussed by Yoo & Mudd (2008). Henderson *et al.* (1994), Foster & Vance (2006), Vance *et al.* (2009) present contradictory views on likely trends in global interglacial/glacial weathering rates. Regolith controls on glaciation and weathering fluxes are by Millot *et al.* (2002) and Roy *et al.* (2004)

Fe–Al-rich palaeosols and their controversial development under changing climatic conditions are discussed by Beavais (1999) and Brown *et al.* (2003), and with the aid of O isotopes by Girard *et al.* (2000). Calcic palaeosols are discussed by Mack *et al.* (1994) and Andrews *et al.* (1998). The contribution of dust to basalt soils is by Kurtz *et al.* (2001).

Interesting experiments on frost weathering are by Lautridou and Seppala (1986). Salt weathering is reviewed by Cooke *et al.* (1993). Molnar *et al.* (2007) discuss active tectonics and fracture formation.

An excellent review of grain provenance is by Weltje & Eynatten (2004), with information on rapid provenancing on the AutoGeoSEM by Paine *et al.* (2005). Ingersoll & Eastmond (2007) give an illuminating case history and a comparison of methodologies. DeCelles *et al.* (1991) is a fine study of progressive source-rock denudation. For K-feldspar stability use in provenance, see Parsons *et al.* (2005). On the algal origin of marine quartz silt see Schieber *et al.* (2000). Nd provenancing is by Yan *et al.* (2007).

# Chapter 2

# CARBONATE, SILICEOUS, IRON-RICH AND EVAPORITE SEDIMENTS

*...changing school, sandstone changed for chalk*
*And ammonites for the flinty husks of sponges,*
*Another lingo to talk*
*And jerseys in other colours.*

Louis MacNeice, 'Autumn Journal', 1938, Faber.

## 2.1 Marine vs. freshwater chemical composition and fluxes

In Chapter 1 the role of weathering of the continental land surface in producing siliciclastic sediment was discussed—the dissolved and particulate byproducts of this weathering are ultimately transferred to the oceans. These vast sinks for dissolved chemical and biochemical weathering products provided means for early estimates of the age of the Earth, assuming that they behaved passively, receiving but not getting rid of elements like sodium. Thus in 1899 the Irishman Joly calculated that the Earth was $9 \times 10^7$ yr old. The present ocean is now thought to be in a state of dynamic equilibrium with a balance between inputs from weathering and mid-ocean-ridge activities and outputs due to chemical and biological precipitation. However, it is a major research issue as to whether there have been secular variations in ocean water chemistry over periods of $10^7$–$10^8$ Myr in the geological past (section 23.1).

The oceans are well-mixed chemically, with mean residence times calculated to be [O] $\sim 10^3$–$10^4$ yr (Fig. 2.1). This broad chemical homogeneity is accompanied in detail by a number of local contrasting physical gradients of temperature and density arising from so-called *'conveyor belts'* of heat energy to-and-from from pole to equator (Chapter 21). These cause some disparities between residence intervals in separate water masses, perhaps by factors up to two orders of magnitude. Table 2.1 reveals that:

1 seawater contains about 300 times more dissolved solids than does fresh water;
2 seawater cations and anions in decreasing order of abundance are Na, $Mg^2$, $Ca^2$, K and $Cl^-$, $SO_4^-$, $HCO_3^-$;
3 freshwater cations and anions in decreasing order of abundance are $Ca^2$, Na, $Mg^2$, and $HCO_3^-$, $SO_4^{2-}$, $Cl^-$;

*Sedimentology and Sedimentary Basins: From Turbulence to Tectonics,* 2nd edition. © Mike Leeder.
Published 2011 by Blackwell Publishing Ltd.

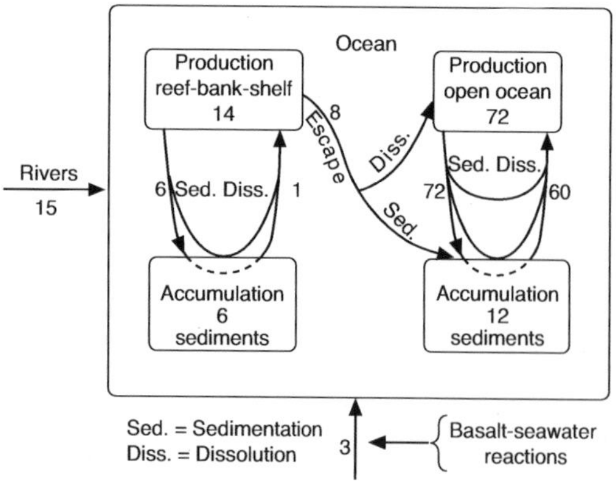


**Fig. 2.1** Global calcium carbonate cycle. Carbon fluxes are in units of $10^{12}$ mol/yr as $(Ca,Mg)CO_3$. (After Mackenzie & Morse, 1992.)

4 river fluxes are dominated by $Ca^{2}$ and $HCO_3^-$, both of which take part in active recycling via carbonate reactions.

Thus seawater is not simply 'concentrated' river-water, because of the different proportions of solutes. Observed seawater composition results from:

- inputs due to composition of continental runoff;
- chemical and biological fractionation by mineral reactions taking place in the water itself, and around the sediment–water interface;
- burial of oceanic waters as 'porewaters' in sediment;
- seawater–basalt reactions along the ocean ridges.

Secular seawater composition may change if the balance of input–output is altered. A notable example is the $^{87}Sr/^{86}Sr$ ratio considered in Chapter 10. Calculation of the magnitude of their various fluxes into and out of seawater leaves significant anomalies, which reflect the importance of hydrothermal basalt–seawater reactions at the world's ocean ridges: Pluto as well as Neptune play a role in the oceanic balance sheet.

## 2.2 The calcium carbonate system in the oceans

The major $CaCO_3$ minerals are calcite, aragonite and dolomite (**Cookie** 7). The composition and origin of marine carbonate sediments plays a crucial role in the analysis of the changing state of ocean chemistry and the evolution of life forms. To understand this it is first necessary to consider the global carbonate cycle (Fig. 2.1). Each year some $15 \times 10^{12}$ mol of bicarbonate are brought to the oceans. Annual production from the oceans is far greater, at about $86 \times 10^{12}$ mol. Some of the difference is made up by carbonate provided at the mid-ocean ridges, but most is explained by post-mortem deep-water oceanic dissolution acting upon the shallow-water production. In total the annual deposition of calcium carbonate from the oceans, as chemical and (mostly) biologically mediated precipitate and skeletons, is only about $18 \times 10^{12}$ mol, with about one-third of this sourced in the *shallow-water*

**Table 2.1** Comparison of the compositions of 'average' river-water and 'average' seawater Note that in seawater, with its high ionic strength, not all of any given anion or cation is available for reaction because of the pheonomena of complexing and ion pairing. The flux of chemical species from the world's rivers into the oceans may be calculated by making use of the measured concentration of elements in river-water and the mean annual discharge of the rivers involved. To this we must correct for the flux of oceanic wind-blown salts to the continents and for the effects of pollutants. Chief amongst the latter is $SO_4$, whose proportion provided by pollutants is estimated to be 20–30%. (Data of Livingstone, 1963; Drever et al., 1988.)

| Ion | Seawater concentration (mol/kg) | Order of abundance | River-water concentration (mol/kg) | Order of abundance | Net river flux—no cyclic salts or pollution ($10^{12}$ mol/yr) | Sea/river |
|---|---|---|---|---|---|---|
| $Na^+$ | 0.47 | 2 | $2.7 \times 10^{-4}$ | 4 | 5.91 | 1740 |
| $K^+$ | $1 \times 10^{-2}$ | 5 | $5.9 \times 10^{-5}$ | 7 | 1.17 | 170 |
| $Ca^{2+}$ | $1 \times 10^{-2}$ | 5 | $3.8 \times 10^{-4}$ | 2 | 12.36 | 26 |
| $Mg^{2+}$ | $5.4 \times 10^{-2}$ | 3 | $1.7 \times 10^{-4}$ | 3 | 4.85 | 318 |
| $Cl^-$ | 0.55 | 1 | $2.2 \times 10^{-4}$ | 4 | 3.27 | 2500 |
| $SO_4^{2-}$ | $3.8 \times 10^{-2}$ | 4 | $1.2 \times 10^{-4}$ | 6 | 3.07 | 317 |
| $HCO_3^-$ | $1.8 \times 10^{-2}$ | 6 | $9.6 \times 10^{-4}$ | 1 | 32.09 | 1.9 |
| pH | 7.9 | | ~7 | | | |
| Ionic strength | 0.65 | | 0.002 | | | |

*carbonate factory* and the remaining majority from the *oceanic carbonate factory*.

Want to know more about basic calcium carbonate minerals? Turn to Cookie 7

Carbonate reactions producing calcium carbonate minerals (**Cookies 7 & 8**) play a key role in maintaining seawater pH. Seawater is well buffered and has a pH in the narrow range 7.8–8.3. Any tendency to increase surface seawater acidity, caused for example by higher atmospheric $CO_2$, is opposed by the other reactions leading to its reduction. Ionic strength plays a key role since the solubility of calcium carbonate in pure water is very much less than in a solution like seawater where other abundant ionic species are present. These ionic species together with polar $H_2O$ molecules tend to cluster around the oppositely charged ions of $Ca_2$ and $CO_3$ preventing the ions coming together to precipitate, i.e. the solubility increases. The greater the charge on the 'pollutant' ions, the greater the effect. Thus solutions like seawater are said to have a high *ionic strength* due to the formation of complex ions and ion pairs. Seawater has an ionic strength of about 0.7 whereas fresh water has values around 0.002.

Want to know more about calcium carbonate equilibria reactions? Turn to Cookie 8

Surface seawater is distinctly supersaturated with respect to aragonite, calcite and dolomite (Table 2.2), yet precipitation of inorganic $CaCO_3$ (**Cookie 9**) is confined to subtropical and tropical locations where it is probably less important worldwide compared to shallow-water biological fixation of $CaCO_3$. The level of carbonate supersaturation decreases with increasing latitude and decreasing surface seawater temperatures (Fig. 2.2). The concentration of $Mg^{2+}$ has a major effect upon which of the polymorphs calcite or aragonite precipitate from seawater. Increasing Mg/Ca ratio progressively retards calcite precipitation whilst at the same time increasing the percentage of $Mg^{2+}$ in the calcite lattice and leaving aragonite unaffected (Cookie 8). Concerning dolomite, the double carbonate $CaMg(CO_3)_2$, modern occurences like those in the Cooring lagoons of South Australia and in ancient peritidal sediments seem to be microbially mediated by $SO_4$-reducing bacteria in very shallow subsurface environments (Cookie 7).

**Table 2.2** Solubility products (*K*) and ion activity products (IAP) for calcite, dolomite and aragonite in surface seawater at 25 °C. Note that surface seawater is supersaturated with respect to all these chief carbonate minerals. (Data of Berner, 1971; Hsü, 1966.)

| Mineral | *K* | IAP | IAP/*K* (=*W*) |
|---|---|---|---|
| Calcite | $4.39 \times 10^{-9}$ | $1.35 \times 10^{-8}$ | 3.4 |
| Aragonite | $6.3 \times 10^{-9}$ | $1.35 \times 10^{-8}$ | 2.1 |
| Dolomite | $\sim 1.00 \times 10^{-17}$ | $1.0 \times 10^{-15}$ | ~100 |

Want to know more about calcium carbonate crystallization? Turn to Cookie 9

Inorganically precipitated $CaCO_3$ occurs in several areas, notably the Bahamas, Arabian Gulf and the enlarged Lake Lisan, Pleistocene ancestor to the Dead Sea. In each case the mineral form is aragonite, in a characteristic needle-like crystal habit with crystal sizes of a few micrometres. In the Dead Sea, events of mass precipitation have been correlated with the appearance of the ghostly *whitings*; large patches of aragonite suspensions that appear suddenly in surface waters. They coincide with an immediate decrease in $HCO_3$ in the water mass, indicating $CaCO_3$ precipitation (Cookie 8). Chemical data are not available for the Arabian Gulf occurrences, but the size and nature of the whitings leave little room for an alternative explanation to inorganic precipitation. Increase in $CO_2$ uptake during periodic diatom 'blooms' has been advanced as the cause for these whitings. A small problem here is the rarity of preserved aragonite in the bottom sediments of the offshore Gulf. Studies in the shallow seas and lagoons of the subtropical Great Bahamas Banks (Fig. 2.3) have established that a significant percentage (~50%) of the fine-grained $CaCO_3$ fraction is inorganically precipitated. As cool Atlantic water passes towards the central bank, it is warmed up, salinity increases to ~1045‰, and carbonate precipitation occurs. The rate of loss of $CaCO_3$ is obtained by dividing the $CaCO_3$ deficit by the mean residence time on the Banks for a particular water sample, originally calculated knowing the degree of incorporation of atomic-bomb-produced $^{14}C$ from

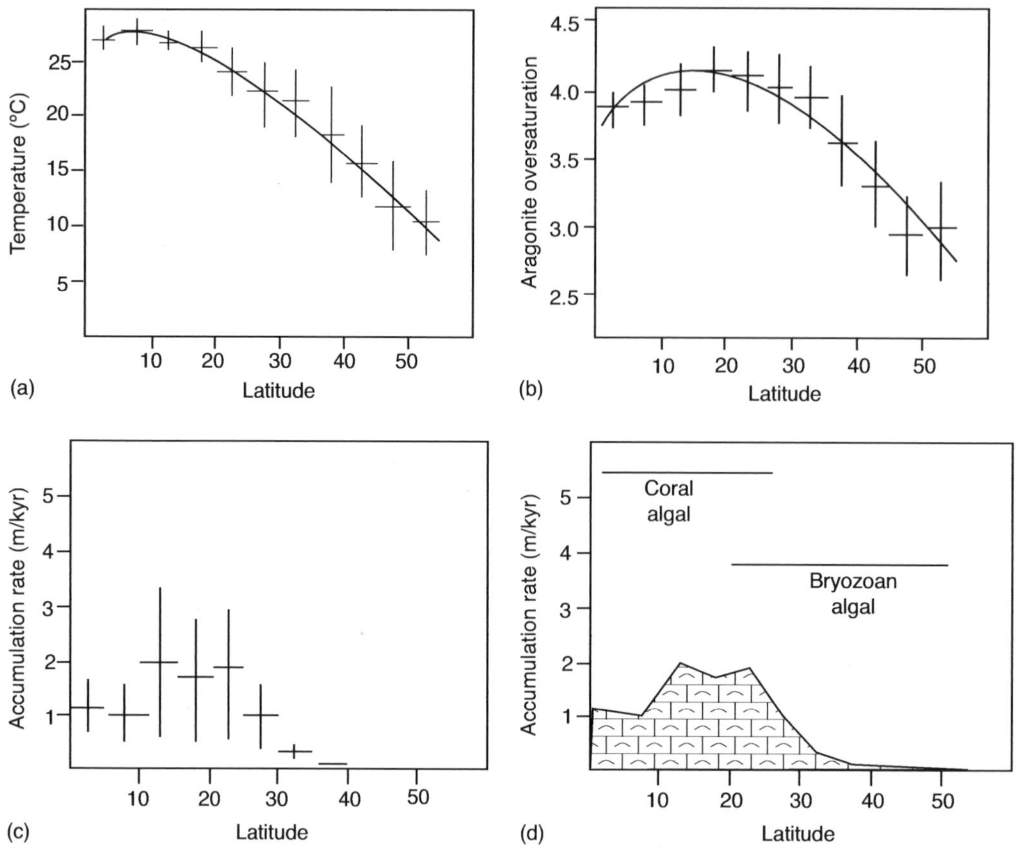


**Fig. 2.2** (a) and (b) Mean sea-surface temperature and aragonite saturation as a function of latitude. Trends defined by best-fitting second-order polynomials drawn through over 200 data points (individual points not shown for clarity). Vertical error bars are for one standard deviation drawn every 5° of latitude (horizontal lines). (c) and (d) Holocene carbonate deposition rates (>200 data points), summarized as mean values every 5° of latitude and one standard deviation. Also shown are the major subtropical and temperate sediment groups. (All after Opdyke & Wilkinson, 1993.)

the atmosphere. Mean precipitation rates of $CaCO_3$ obtained are around 50 mg/cm$^2$/yr.

Finally, there are rare but interesting occurrences of a multihydrated form of calcite, *ikaite* ($CaCO_3 \cdot 6H_2O$), first discovered in 1963. The mineral is a low temperature, high pressure form and precipitates naturally at the Earth's surface only in cold water under conditions of elevated alkalinity and high dissolved phosphate concentrations. At its type locality at the Ikka Fjord, Greenland, spectacular ikaite towers up to 18 m high are created from alkaline submarine freshwater (meteoric) springs. These groundwater seeps are rich in $CO_3^{2-}$ and $HCO_3^-$ ions; the ikaite precipitating on mixing with cold (< 6 °C) marine fjord waters rich in calcium. It is possible that normal calcite deposition is inhibited due to methane oxidation of organic-rich marine sediments through which the seeps pass. Ikaite has also been reported as occurring in many other high-latitude marine sediments. Some of its rarity may be more apparent than real for in core samples the mineral rapidly dissociates to calcite and water at normal surface temperatures. In addition, ikaite also forms as large crystals within sediment. *Glendonite* is a calcite pseudomorph after ikaite formed during diagenesis and may give valuable evidence for previous cold water conditions, as in the

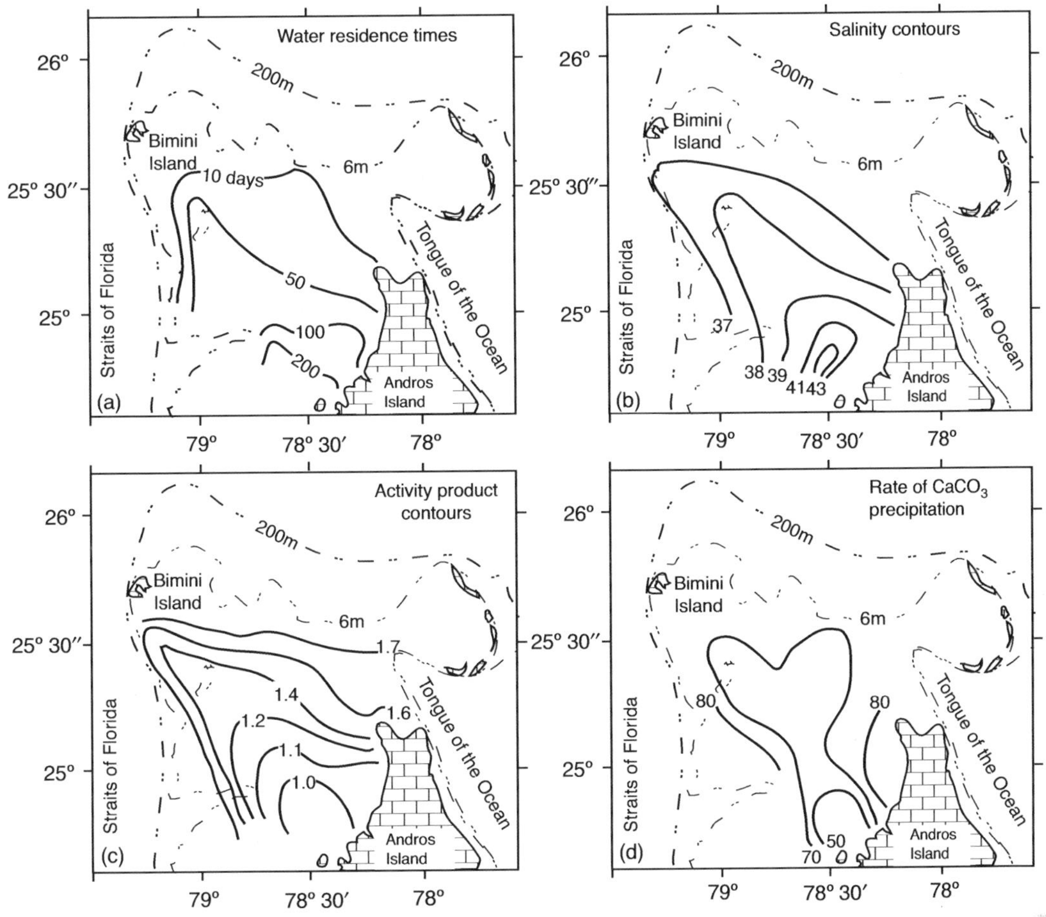


**Fig. 2.3** Bahamas Bank bathymetric maps and chemical data. (After Broeker & Takahashi, 1966.) (a) Mean summer residence times of water. (b) Typical values of salinity (parts per thousand). (c) Activity product contours for calcium carbonate. (d) Calculated rates of $CaCO_3$ precipitation ($mg/cm^2/yr$).

Permian Gondwana glacio-marine deposits of the Sydney basin, Australia and Neoproterozoic Cryogenian deposits of northwest Canada.

## 2.3 Ooid carbonate grains

Carbonate ooids are one of a number of kinds of *coated grain*. They are spherical to slightly ovoid, well-rounded carbonate particles (Figs. 2.4d & 2.5e–h) possessing a detrital nucleus and a concentrically laminated outside (cortex) of fine-grained aragonite and/or high-magnesian calcite often with many thin internal organic layers of algal origin. Diameters range from 0.1 to 2 mm or so. When the cortex is very thin, comprising one or two thin aragonite laminae, the ooid is said to be superficial. Ooids occur in marine sandwave and dune complexes in areas of strong tidal currents or in marine or lacustrine littoral or shallow sublittoral beach deposits. Radiocarbon studies reveal that modern ooids have grown very slowly, with surface layers giving ages as young as 225 yr, and cores as old as 2300 yr.

Many sections through unaltered aragonite ooids (Fig. 2.5f) show a pseudo-uniaxial cross under crossed polars, which gives a negative figure. This means that the predominant alignment of aragonite crystals must

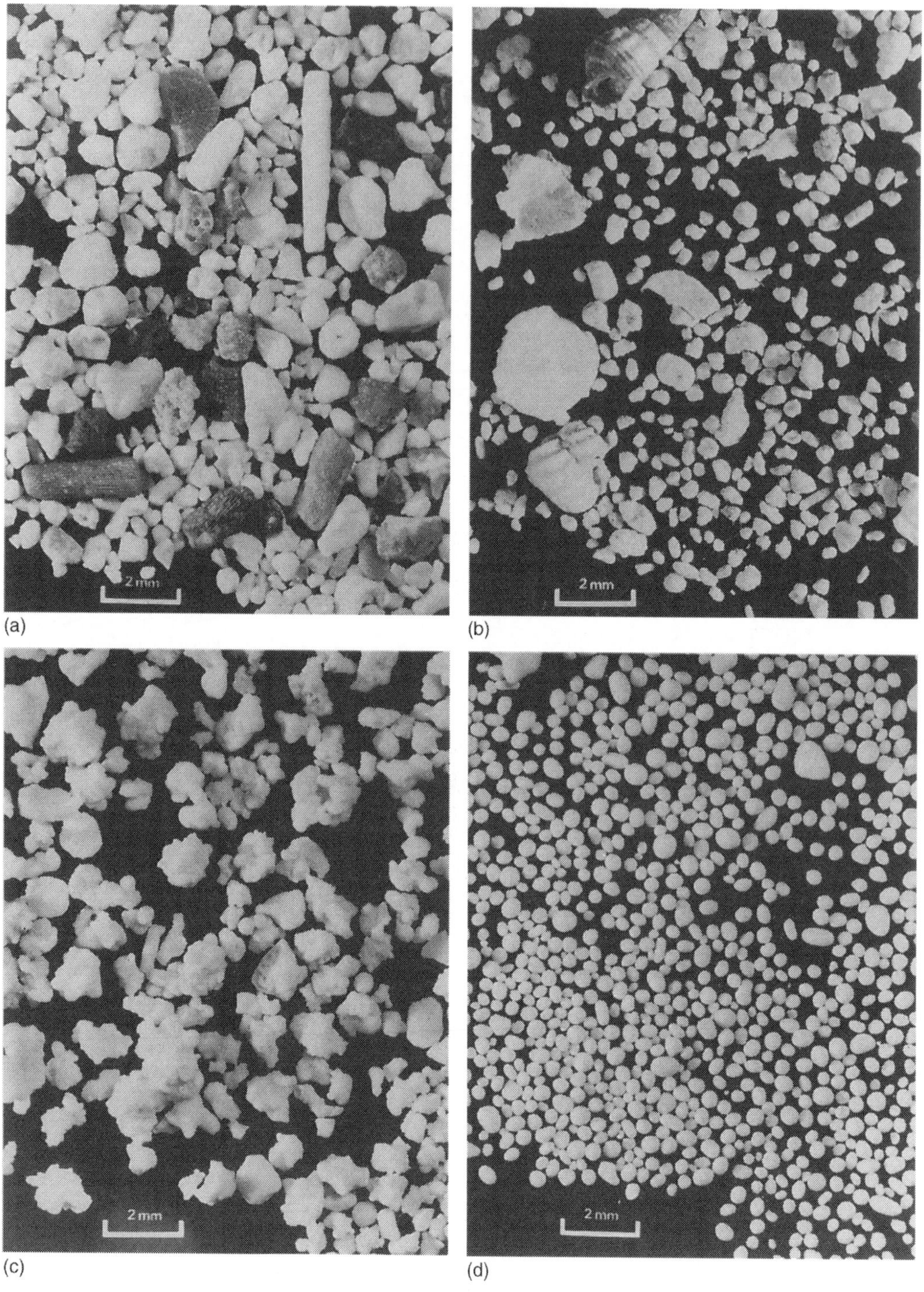


**Fig. 2.4** Various carbonate grains. (a) Skeletal grains, comprising abraded mollusc, foraminifera, echinoid and algal fragments. (b) Skeletal–pelletal sand; note ovoid pellets, abraded forams and gastropods. (c) Grapestone sand with aggregates of micritized skeletal grains and pellets. (d) Oolites, note good sorting and high polish on individual ooids.

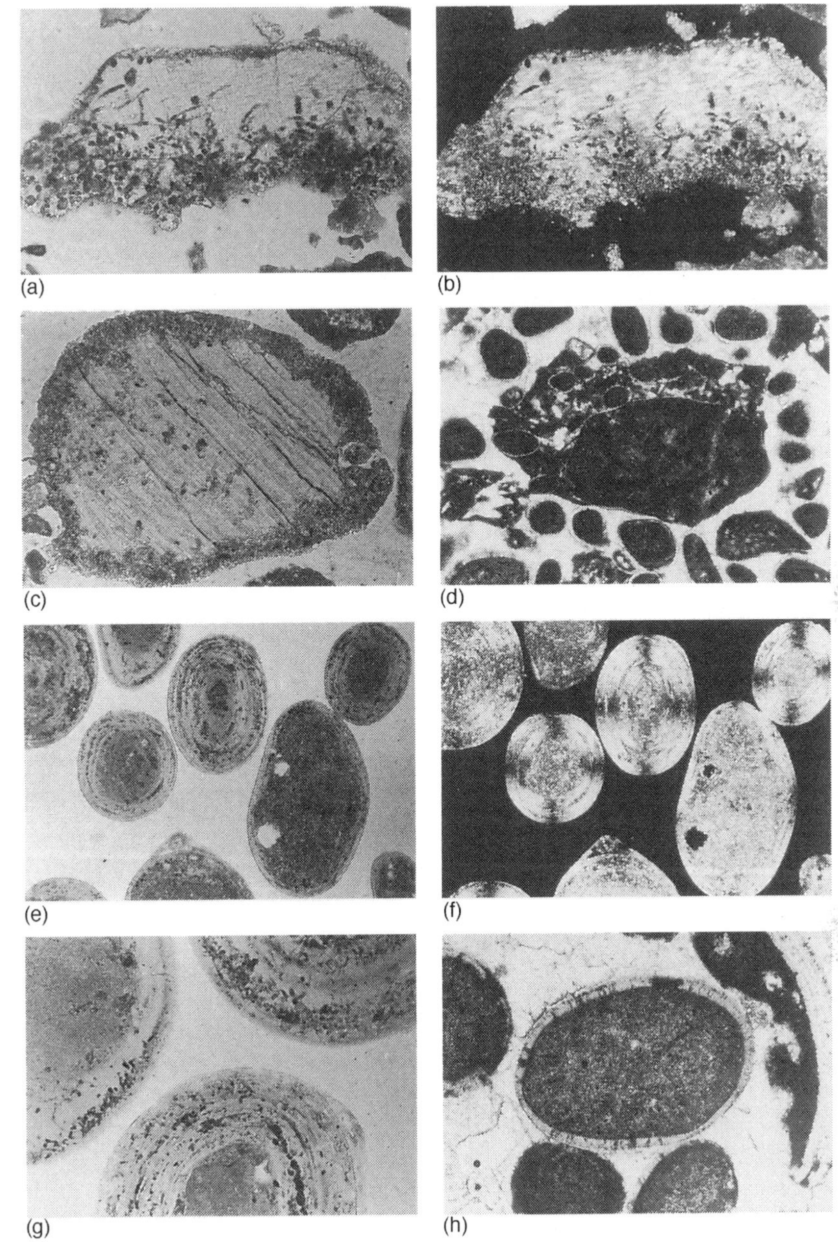

**Fig. 2.5** Carbonate grains in thin section. (a) and (b) Molluscan fragments with well-developed algal borings viewed in plane-polarized light and under crossed nicols. (c) Molluscan fragment with well-developed micrite envelope formed by coalesced and infilled algal borings. (d) Compound intraclast of large amorphous lump and micrite-cemented peloids. (e) and (f) Modern ooids viewed under plane polarized light and under crossed nicols. Note superficial ooid with pellet nucleus and the pseudo-uniaxial crosses. (g) Close-up of ooid to show fine laminations and darker areas of organic mucilage and algal borings. (h) Ancient superficial ooid with pelletal nucleus and radial fabric in the ooid cortex. All scale bars 100 μm. ((a)(c), (e)–(g) from recent carbonate sands of Bimini lagoon, Bahamas. (d) and (h) from Dinantian (Mississippian) of Northumberland basin, England.

be with their *c*-axes tangential to the ooid surface. This is confirmed by X-ray and electron microscope analysis, the latter technique revealing that the aragonite crystals are in the form of rod-like particles 1–2 μm long with flattened end terminations. In the Arabian Gulf, ooids from high-energy environments have a well-developed concentric arrangement of individual aragonite rods, whereas those from more sheltered areas have a radial orientation with looser packing. Some ooids, notably those from the Great Salt Lake (USA), exhibit spectacular mixtures of radial aragonite, tangential aragonite and unorientated fabrics rich in clay minerals.

Regarding the vexed question of ooid genesis, Sorby's original 'snowball' analogue stated that inorganically precipitated aragonite needles simply stuck on to the rolling exterior of an ooid. The lack of evidence for a sticky cohesive matrix on the smooth polished ooid outer surfaces (Fig. 2.4d) has made this 'snowball' mechanism difficult to accept. We may be sure that high-energy environments somehow encourage tangential aragonite growth whilst low-energy environments encourage radial growth. Successful laboratory precipitation of alternating oolitic carbonate and organic laminae around spherules may occur where the concentration of organic material as humates is high and where organic membranes may form around suitable nuclei. Experimentally produced carbonate laminae show radial aragonite growth under 'quiet' laboratory conditions and tangential fabrics after agitation. It is possible that aragonite crystals seed out from the ooid organic matrix to grow radially outwards, as in the growth of any uninterrupted crystal fabric from a solid surface. Such radial fabrics would be modified by turbulence and periodic abrasion into tangential fabrics in higher-energy environments.

The source of ooid aragonite needles remains a great problem since it is difficult to imagine tiny delicate crystals occurring in the bedload layer of high-energy water bodies such as tidal flows over oolite shoals. Paradoxical though it may be, the very process of intense bedload transport may encourage precipitation by a process akin to *collision breeding* of initial aragonite protocrystals. Tangentially orientated ooliths form in laboratory experiments with bicarbonate solutions. Intermittent stirring leads to aragonite needle precipitation around nuclei with the tangential crystals held together by surface bonding forces. Further agitation caused a high polish of the ooid surfaces as aragonite needles develop well-rounded edges. In natural situations ooliths spend much of their time stranded in the inside of moving bedforms such as sandwaves, dunes and ripples. Organic coats may develop here, causing radial aragonite laminae to begin to seed out from the porewaters between the oolite grains. Re-emergence of the grains into a turbulent bedload layer would then cause mechanical alignment of the aragonite rods into a tangential arrangement or would encourage tangential growth. Ooliths in quiet waters would remain with a radial structure. Periodic entrainment and burial would also explain the development of successive concentric layers.

The occurrence of calcitic ooids with well-preserved radial microfabrics in the geological record has posed problems of interpretation. Despite Sorby's acute original deduction that these were primary, the common Bahamian aragonite forms were considered to be the prototypes for most ancient ooids. Thus ancient calcitic ooids were once all considered secondary, resulting from the layer-by-layer replacement of tangential aragonite by radial calcite in concentric organic templates. More recently they have been considered as primary high-magnesian calcite forms, pointing to major changes in global oceanic chemistry that subdued aragonite precipitation (Chapter 23). By analogy with other similar replacements (e.g. echinoderm carbonate), their fine preservation is ascribed to diffusional substitutions of $Fe^2$ for $Mg^2$ during conversion to low-magnesian calcite on shallow burial. There is much evidence that oceanic chemistry was substantially different in the early Precambrian and that it has undergone cyclical changes in Mg/Ca ratios throughout the Phanerozoic (Cookie 7; Chapter 23). Evidence from Archaean carbonate platforms indicate that changes in ocean chemistry such as a significant increase in oxidation state at about 2.2 Ga and a long-term decline in atmospheric $CO_2$ probably had a substantial influence on the chemistry and texture of Proterozoic and Archaean carbonates For example, Neoarchaean carbonates commonly contain centimetre- to metre-thick beds of aragonite and calcite crystals that grew directly on the seafloor from open-marine waters (**Cookie 7**). The abundance of aragonite pseudomorphs in shallow-water environments and the resulting requirement that oceanic pH was not acidic supports atmospheric models of the

Neoarchaean atmosphere that contain significant concentrations of non-$CO_2$ greenhouse gases such as methane, with ferrous iron and reduced manganese acting as inhibitors to deep-water calcite nucleation.

## 2.4 Carbonate grains from marine plants and animals

In a memorable and stimulating phrase, N.P. James considered that carbonate sediments are '*born, not made*' (Chapter 20). Thus fossiliferous limestones are graveyards of former life, the remains of dead creatures proving that massive fixation of calcium carbonate has occurred from the world's ocean over much of geological time. The living biomass of carbon is a very tiny fraction of that preserved in the sedimentary record as limestones and also in mudrock, coal, oil and natural gas. This biomass is a substrate for the carbon cycle and provides the main ions needed for chemical weathering and soil formation, the products of which are enfolded into the oceans as nutrients to provide for further growth. **Plate 1** shows that marine carbonate sediments are very widely distributed, the majority derived from the calcareous hard parts of algae and invertebrates.

A local biotic community gives rise to a characteristic death assemblage of calcareous debris (Fig. 2.4a & b). Identification of calcareous fragments depends upon the degree to which they have physically and chemically broken down—many particles may therefore be classified only to the order level of their appropriate phylum (Fig. 2.6). If the sediment is partly or wholly lithified then thin-section analysis techniques must be used. Particular problems arise if a shell was originally aragonite, since diagenetic dissolution and calcite reprecipitation may totally destroy original shell structure.

The resemblance between a death assemblage and the original faunal community will obviously reflect the degree of post-mortem physical, chemical and biological destruction and redistribution. Many assemblages produce *in situ* carbonate deposits, either as solid reefal build-ups, chiefly of coral, or by the baffling, binding or trapping activities of sea-grasses, benthic algae and other organisms. These give rise to unlithified mounds made up of fine-grained carbonate debris and various skeletal remains (Chapter 20).

Three groupings of marine carbonates may be distinguished.

**Fig. 2.6** Calcareous algae of the genus *Penicillus* with their holdfasts. Vast numbers of these algae live in subtropical carbonate-producing seas like the shallow lagoons of Florida and the Bahamas Banks. Upon death the delicate organic tissues decay and release tiny, blunt-ended aragonite needles of the plant framework on to the lagoon substrate. Some proportion of it survives resuspension and export seawards during storms to accumulate as aragonite mud that is indistinguishable in external form from a biogenically precipitated aragonite.

1 *Oceanic carbonates of pelagic origin.* These are widely distributed (with the exception of the North Pacific, Arctic and Antarctic), closely associated with the mid-ocean-ridge system and areas of oceanic deep-water upwelling (**Plate 1**). They are calcareous oozes comprising the remains of minute coccoliths, foraminifera (forams) and pteropods. These life forms originate in the warm shallow photic zones of the ocean and accumulate after death in maximum water depths of 3.5–5 km, as determined by the local *carbonate compensation depth* (CCD), below which no $CaCO_3$ occurs due to dissolution. The CCD is best imagined as the equivalent of a terrestrial snow-line.
2 *Shelf carbonates of subtropical and tropical origin.* These are shallow-water carbonates of biological origin. Inorganically precipitated $CaCO_3$ is important locally, as discussed previously and in **Cookie 8**. Coral, algal and molluscan species dominate biomass production, with coral reefs alone accounting for over 75% of Holocene shallow-water biological $CaCO_3$ fixation. In addition to shallow-water chemical aragonite precipitation, it is apparent that the breakdown of benthic calcareous algae (Fig. 2.7), can contribute significantly

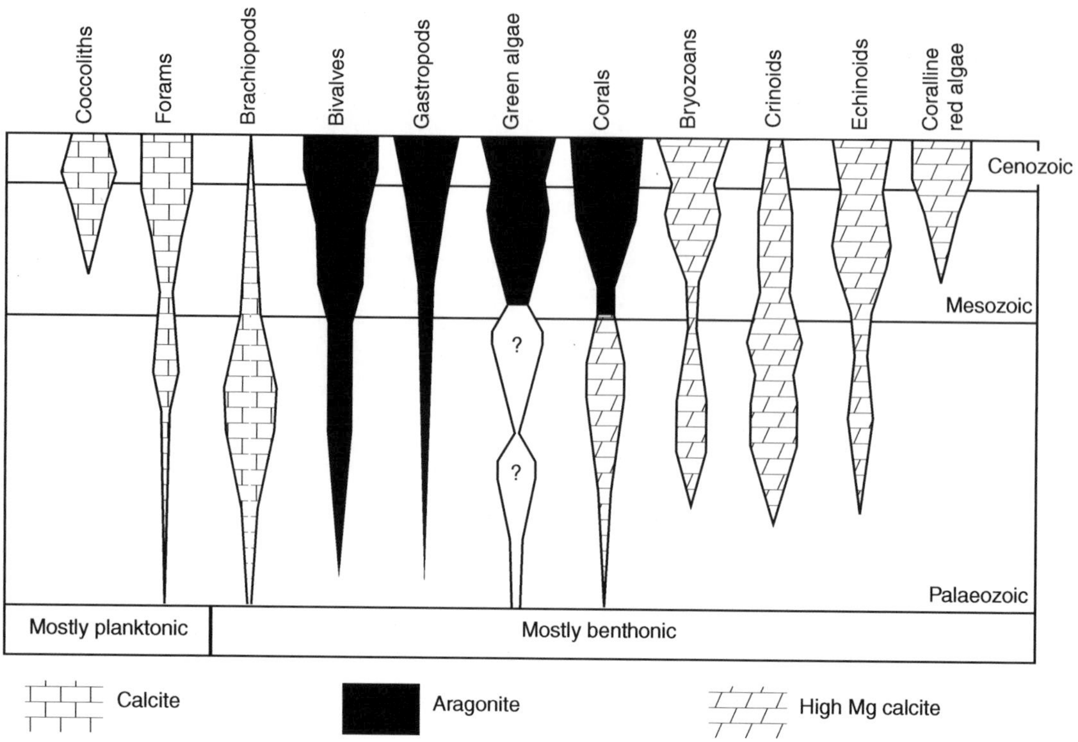


**Fig. 2.7** Variation of carbonate mineralogy with evolution of different calcifying organisms over Phanerozoic time. Width of bars indicative of relative abundance of individual groups as sediment producers. (After Lowenstam, 1963.)

to carbonate mud production. The major controls upon organisms that secrete $CaCO_3$ are water depth, temperature and salinity. Water depth is a variable because of the attenuation of life-giving energetic photons in visible light passing through water, most of which are lost at about 30 m depth. Mean sea-surface temperatures of $>15\,°C$ and normal salinities in the range 32–40‰ encourage the so-called *chlorozoan* (aka *photozoan*) assemblage of zooxanthellae-symbiont schleractinian *reef-building corals*, calcareous green algae and other phyla, together with ooliths (see below). Elevated salinities or excess nutrients cause the corals to disappear, defining the *chloralgal* group.

3 *Shelf carbonates of temperate- to cold-water origin*. These are dominated by contributions from molluscs, forams, red algae and bryozoans, the so-called *foramol* (aka *heterozoan*) assemblage. Deeper shelf and continental rise cold-water coral mound and reef ecosystems discovered in the past decade (Chapter 20) comprise branching corallites of the stony coral *Lophelia pertusa* in environments swept by deep currents rich in nutrients. Brozoan reef mounds were active during Quaternary sea-level lowstands in cool-water, upper slope deposits in water depths of 200–350 m of the Great Australian Bight. In general, accumulation rates of temperate shelf carbonates are slower than their lower latitude equivalents because of lowered metabolic rates and somewhat lower levels of carbonate oversaturation in cool temperate seas. Both biogenic aragonite and calcite make up Holocene sediment, though the former is often completely dissolved in Quaternary equivalents by early diagenesis as a byproduct of bacterially mediated organic decay reactions. It may be that this process has led to underestimation of production rates.

Calcareous organic hard parts may be composed of aragonite, low-magnesian calcite or high-magnesian calcite; some are multimineralic. Most genera have a similar composition and whilst many orders, classes and phyla do also, e.g. modern warm-water corals are all aragonitic, many do not, e.g. Mollusca are exceptionally varied (**Cookie 10**). This differing chemical and mineralogical composition of carbonate hard parts gives rise to a corresponding chemical/mineralogical *preservation potential*. The nature of skeletal carbonate particles formed in a specific environment (e.g. reef, intertidal flat, shelf sand) has obviously changed through geological time in response to evolution (Fig. 2.7). Thus a Silurian reef faunal and floral assemblage is distinct from Carboniferous, Triassic or modern ones.

## 2.5 Carbonate muds, oozes and chalks

An ooze is defined as a deposit of predominantly pelagic origins comprising mud-, silt- or sand-sized organisms such as coccoliths, forams and pteropods. Studies on shelf ramps indicate the importance of ocean current dynamics in determining the relative proportion of the latter two sand-grade grain types. Pteropods thrive in fertile waters of more-or-less constant (*stenohaline*) salinity, whilst some forams can tolerate wider salinity fluctuations. Deposits cored on the Florida ramp reveal alternating pteropod-rich and foram-rich layers whose cyclicity is attributed to interglacial and glacial periods respectively. The accumulation of silt/clay-sized coccolith debris in the oceans is possible because of the high $CaCO_3$ production rates of these plants; coccolith oozes of the shallow mid-ocean ridges are one of the largest sinks for $CaCO_3$ on Earth. In summer sunlight individual coccolith plates are secreted and extruded through the surface of the coccolith algal cell once every 2 h. The results are gigantic 'blooms' of coccoliths, visible from aircraft and satellites (Fig. 2.8), whose constituents sediment as individual plates after death and organic decay. Data from oceanic oozes indicate that during glacial periods calcitic pelagic organisms are replaced by siliceous plankton in regions of lower water temperatures and high productivity, with the silica–calcite 'front' moving generally equatorwards and the proportion of $CaCO_3$ preserved in bottom sediments decreasing and that of organic carbon increasing.

Want to know more about carbonate grains of biological origins? Turn to Cookie 10

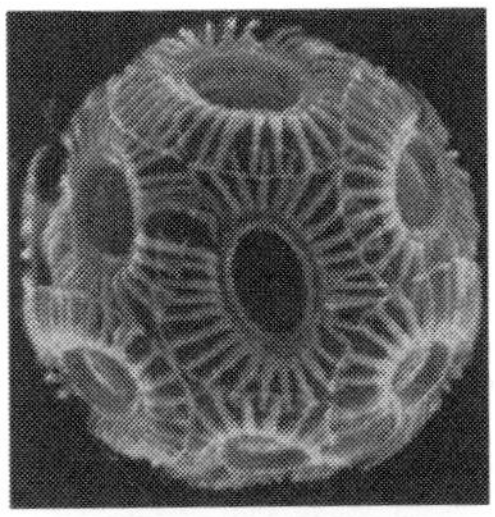

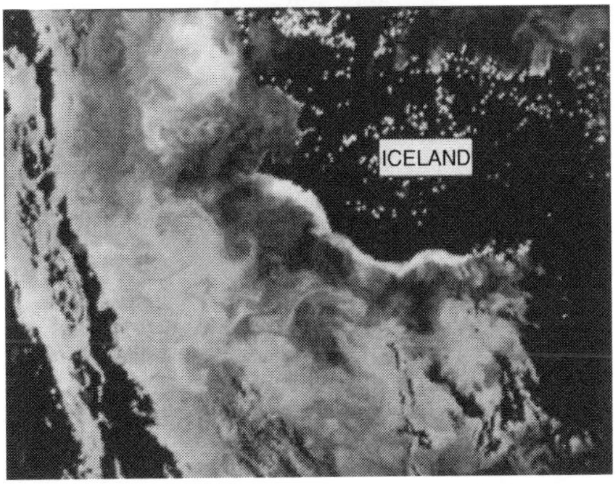


**Fig. 2.8** (a) Single *Emiliania huxleyi* coccolith organism (approximate diameter, 6 μm) to show the constituent platelets which detach from the creature after death to form sediment particles. (Photo, J.R. Young, Natural History Museum website, London.) (b) Satellite image of coccolith bloom south of Iceland, North Atlantic, June 1991. (Photo, S. Groom, P.Holligan, NOC website, Southampton.)

## 2.6 Other carbonate grains of biological origins

Microscopic examination of skeletal fragments collected from quiet-water carbonate environments often reveals a dark rind (Fig. 2.5a–c) of variable thickness and intensity. Individual tube-like cavities may be seen in the less advanced rinds and on the inner margins of the thicker rinds. The tubes may be empty or they may be filled with very fine-grained aragonite or high-magnesian calcite, defining *micrite envelopes*. Dissolution of the shell-fragment carbonate with dilute acid yields a gelatinous residue in which various types of blue-green algae may be identified. The tubes are evidently the result of boring blue-green algae (fungal bores also exist), which rapidly infest all carbonate particles after deposition. The reason for carbonate precipitation in the bores is poorly known but may be connected with the local high pH in the tube following $CO_2$ uptake by adjacent photosynthesizing algae or by bacterial action on the organic residue left in vacated tubes. Algal bores found in temperate-water carbonates do not show carbonate precipitation. Micrite envelopes are also produced by coatings of filamentous endolithic algae. Rapid precipitation of low-magnesian calcite occurs on and within dead filaments, which project from the substrate into the sea.

Rinds may also be due to *calcified bacterial biofilms*. Biofilms in general are thin (10 s to 100 s μm) layers of bacterial populations that adhere to sediment and organism surfaces and which are contained within a matrix of polymeric organic substance which acts as an ambient diffusion barrier so that the bugs can interact in a stable microenvironment. Calcitic micrite precipitation acts to preserve the biofilms—best seen preserved in former tiny crypts within ancient limestones.

Micrite envelopes formed by boring algae may extend inwards to include the whole shell fragment. An *amorphous lump* grain results, with no structure left to testify to its original origin. Such grains belong to the class of carbonate particles known as *intraclasts* (grains derived within the general area of sedimentation) that make up a high proportion of sediment in certain Bahamian and Persian Gulf lagoons. Intraclasts are a highly diverse, polygenetic group of particles, including reworked beachrock fragments, hardground debris, grapestones, older lithified carbonate particles and amorphous lumps.

### Algaliths

Algaliths may be defined as unattached granule to pebble-sized nodules (*oncoids*) of carbonate-secreting algae. *Rhodoliths* are nodules of red coralline algae, and are widely distributed from the tropics to the poles and are a particularly important contributor to temperate carbonate sediments. They comprise laminar crusts, branching or columnar/mamillated forms, the exact morphology depending to a strong degree on energy conditions. *Cyanoliths* are nodules constructed by cyanobacteria whose beautifully delicate sheath-like cells are preserved by calcite precipitation induced by photosynthesis during life. Examples include nodules of the genus *Girvanella* that are very common in the sedimentary record of Palaeozoic to Mesozoic limestones.

### Faecal pellets

In quiet-water lagoons with aragonitic mud substrates, large numbers of molluscs (chiefly gastropods and bivalves), worms and crustaceans continually ingest organic-rich muds, feeding upon the nutrients. The mud is excreted as ovoid faecal pellets (Figs. 2.4b & 2.5d). Internally the pellets are dark, fine-grained, usually structureless and rich in organic matter. Older pellets are fairly hard, probably due to rapid interstitial carbonate precipitation. Winnowing of associated aragonite mud may cause a sand-grade deposit of pellets to be produced, as in lagoons of the Arabian Gulf. Faecal pellets in ancient carbonate rocks may be very difficult to tell apart from abraded amorphous lumps produced by micritization of shell fragments and other processes, including complete micritization of ooliths. For this reason it is best to use the term *peloid* for any structureless ovoid micrite particle unless the genesis may be independently deduced.

### Grain aggregates—grapestones

Grapestones are composite particles comprising cemented aggregates of grains (shell fragments, ooids, pellets) resembling microscopic bunches of grapes (Fig. 2.4c). Intense micritization by infilled algal bores usually obscures any original internal structure of the particles. It is thought that the cementation, by micritic aragonite, and the intense micritization reflect a mode of growth within a subtidal blue-green algal mat (section 2.15) that stabilizes the substrate to resist tidal and wave currents. Periodic mat rip-up during storms then yields grapestone aggregates. Grapestones are thus a type of intraclast, as defined previously.

## 2.7 Organic productivity, sea-level and atmospheric controls of biogenic $CaCO_3$ deposition rates

Productivity of subtropical sediment-forming algae (*Penicillus*, *Halimeda*, *Neogoniolithon*, etc.) is ~500 g/m$^2$ of $CaCO_3$ biomass per year, equivalent to a deposition rate of about 2 mm/yr. Sediment budget studies suggest that algal sediment and chemically precipitated aragonite is overproduced compared to observed depositional fluxes and that very substantial losses must occur to the shelf and deeper ocean, probably as suspended plumes during storms and hurricanes (Chapter 20).

Coral reef growth can vastly exceed lagoonal or forereef rates of deposition—Darwin recognized the importance of high coral growth rates over 150 years ago and featured it in his then-novel 'volcano-drowning' hypothesis for the origin of atolls. Extraordinary rates of up to 6 mm/yr have been observed in the typical reef-building coral genus *Acropora* which has its most rapid growth rates at night, with the algal symbiont zooxanthellae playing little role in this aspect of growth. Coral reefs are by far the greatest sink for $CaCO_3$ on modern shelves and they have been assigned a key role in the recycling of carbon during interglacial as compared with glacial intervals. Highstand interglacial periods have very high 'coral reef' carbonate fixation but also relatively high atmospheric $CO_2$ concentrations. The opposite holds for glacial lowstands. Thus the modern (highstand) fixation of $CaCO_3$ by the world's coral reefs cannot be much less than about $2 \times 10^{13}$ mol/yr for the past 5000 yr. This is approaching the total $Ca^2$ flux into the ocean of about $2.3 \times 10^{13}$ mol/yr. During lowstands the shallow shelf area available for coral reef colonization is much reduced and thus $CaCO_3$ fixation with it. During such times the locus of carbonate deposition must shift to the deep sea. Further, glacial lowstands will show marked dissolution of highstand reefs and other porous carbonates, leading to a net seaward alkalinity flux. Model results indicate that this highstand–lowstand cycle is capable of producing a significant proportion of atmospheric $CO_2$ compositional shifts as measured in fossil ice cores. There is also an important additional contribution due to seafloor and shallow-subsurface $CO_2$ production by organic carbon oxidation.

As noted previously, temperate-water carbonates of the foramol assemblage lack oolith, intraclast and pellet allochems and the contribution from green algae that characterizes warm-water carbonates. Their deposition rates are usually very much lower, as a consequence of the rapid decrease in metabolism and carbonate supersaturation with increasing latitude (Fig. 2.2).

## 2.8 $CaCO_3$ dissolution in the deep ocean and the oceanic $CaCO_3$ compensation mechanism

Oceanic $CaCO_3$ deposits (**Plate 1**) are dominated by coccolith/foram oozes on the crests and flanks of mid-ocean-ridge systems. These are rare below ~5 km water depth in the Atlantic and ~3.5 km in the Pacific due to increased solubility of $CaCO_3$ in the deeper, cooler parts of these oceans. As noted previously the downward limit to $CaCO_3$ preservation is termed the carbonate compensation depth (CCD). It has been calculated that about 80% of the $CaCO_3$ produced in the warm sunlit areas of ocean waters by planktonic organisms is destroyed by deep-water dissolution. Aragonite, the more soluble $CaCO_3$ polymorph, becomes undersaturated long before calcite, at about 500 m depth in the Pacific ocean and 2 km in the Atlantic ocean (Fig. 2.9). Pelagic organisms with aragonite skeletons, like the abundant pteropods, are thus more prone to dissolution than are the calcitic coccoliths. The water depth below which the diminution of carbonate grain size due to dissolution becomes increasingly rapid is known as the sedimentary *lysocline*, though this is a somewhat ill-defined concept as supralysocline dissolution is widely recognized. Calcite becomes undersaturated at depths of between 400 and 3500 m in the Pacific and between 4000 and 5000 m in the Atlantic. These zones where $CaCO_3$ approaches undersaturation for aragonite or calcite are known as *carbonate saturation depths* (CSD). Whether or not calcitic organisms can survive as debris below the CSD is a complex function of the rate of fall of the grain (or agglomeration of grains in faecal pellets) through the water column vs. the rate of $CaCO_3$ dissolution. The rate of dissolution thus depends partly on skeletal size. Solution inhibition by adsorbed molecules may also be important.

Dissolution is caused by increased $CaCO_3$ solubility in cold deep waters which also contain more $CO_2$ in solution than do warm surface waters, e.g. $CO_3^{2-}$-undersaturated Antarctic Bottom Water. Also important is the increase of pressure with depth, which also causes increased $pCO_2$ and thus lower pH. Both effects cause the oversaturation of surface seawater with respect to $CaCO_3$ to decrease with increasing depth. Another crucial effect is the rate of production of total organic carbon in the oceanic mass. Some proportion of its buried flux is oxidized by respiration, producing $CO_2$, either in the oceanic water mass, at the ocean floor or in the few tens of metres below the sediment–water interface; 30–50% of the carbonate that survives oceanic descent is dissolved in bottom-sediment porewaters. In the South Atlantic, gradients in dissolution rates have been measured

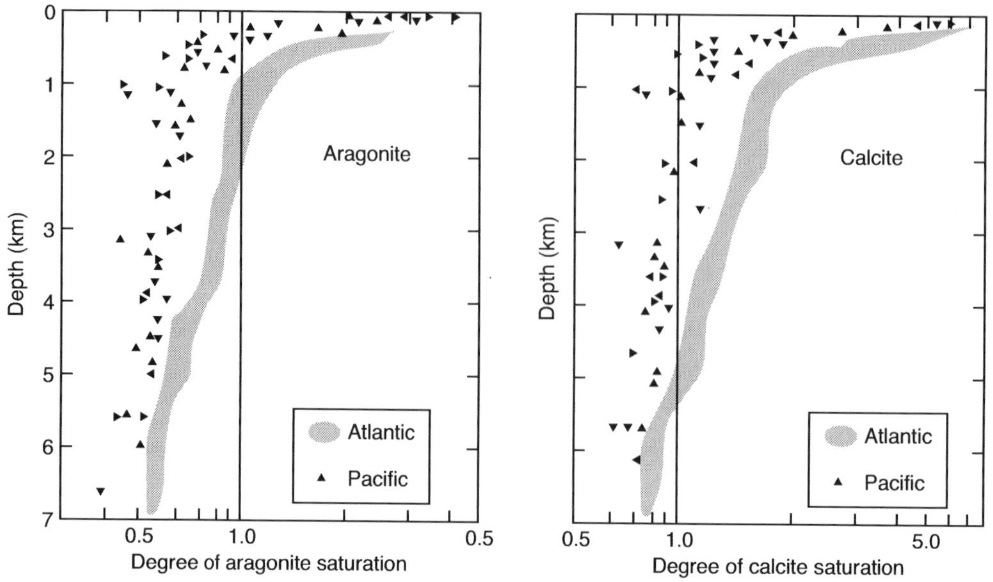


**Fig. 2.9** The degree of aragonite and calcite saturation as a function of depth in the Atlantic and Pacific oceans. Classic data of Broecker (1974).

between eastern and western areas, attributed to higher accumulation rates of organic carbon in more eastern parts of the equatorial upwelling compared with areas further west and beyond. Local deepening of the lysocline by several hundred metres may occur where $CO_3^{2-}$-undersaturated bottom-water flow is restricted by topographic barriers, as in the Angola basin of the southeast Atlantic.

From the previous discussion it can be appreciated that the variation of CSD between oceans is largely a function of pH. Pacific ocean water contains much larger quantities of organic debris than does Atlantic ocean water. Oxidation of this produces additional H ions, some of which are used up in seawater buffer reactions; the small remainder not balanced in this way increase the deep-water acidity. Calcium carbonate is thus dissolved and is less abundant in Pacific than in Atlantic sediments. Adsorbed phosphate and organic ions also act as inhibitors to dissolution; only when these are removed in the deep ocean can dissolution proceed rapidly. During glacial periods the production of organic carbon increases; increased degradation of this produces more $CO_2$, which dissolves more sedimenting $CaCO_3$, which drives the various $CaCO_3$ compensation reactions discussed above to increase total ocean alkalinity. For example, lysocline depth in the South Atlantic during the Last Glacial Maximum was as much as ~1 km shallower than in the Holocene.

Palaeoceanographic studies indicate a major deepening of the CCD, and thus an increase in amount and extent of coccolith calcite, at the Eocene/Oligocene boundary coincident with the initiation of major Antarctic ice cover. This might be thought to be the opposite effect than would be expected from enhanced oceanic circulation and cooler bottom-water production following ice-sheet growth and global cooling. The problem is a complex one—more efficient circulation would also have encouraged higher primary oceanic productivity because of greater nutrient fluxes. The resulting higher organic turnover in the surface would have been accompanied by greater organic decay at depth, leading to lower global oceanic pH and hence a shallower CCD. Possible additional compensating factors are a decrease in atmospheric $CO_2$ and an increase in the continental Ca flux—there is good evidence for greatly increased continental weathering recorded in the Tertiary marine $^{87}Sr/^{86}Sr$ record (Chapter 10). Increased Ca would have 'mopped up' excess acidity due to higher productivity and perhaps overcome Ca-limited growth of calcareous plankton.

## 2.9 The carbonate system on land (provided by J.E. Mason)

$CaCO_3$ sediments are widely distributed on land (for lake carbonate sediment see Chapter 14), occurring as cave speleothem, river tufa, warm-spring travertine and calcisols (Chapter 1). In this section we mainly feature the origin of speleothem carbonates because of their key role in palaeoclimatic studies, an aspect explored further by means of case histories in Chapter 23.

### Speleothem

These are mineral deposits that form in caves, typically in karstified limestone host rocks. The bedrock limestone above a cave is dissolved by downward-percolating meteoric waters containing carbonic acid obtained from atmospheric $CO_2$ and as a product of the decomposition and respiration of plants (Fig. 2.10; Equations 2.9.1–2.9.4). The percolating water drains through the soil and underlying fissured and jointed limestone. This so-called *epikarst* feeds both major conduits and lower transmissivity fissures, which in turn feed zones of dripwater in caves. Dripwater emerges through a stalactite tip or another opening into a cave, the atmosphere of which has a lower $pCO_2$ than that in solution. The water therefore begins to degas $CO_2$ and precipitate $CaCO_3$ via Equation 2.9.4. The precipitate forms either at the point of emergence of the water on the cave roof, forming a stalactite or curtain, or on the cave floor where it may build up to form a stalagmite or flowstone (Fig. 2.10b). Stalagmites (the type of speleothem primarily used for palaeoclimatic study; section 23.18) consist of a series of growth layers, representing successive generations of $CaCO_3$ precipitate. Cave reaction stages are as follows:

1 carbon dioxide dissolves in water

$$CO_{2(g)} \Leftrightarrow CO_{2(aq)} \qquad (2.9.1)$$

2 formation of carbonic acid

$$CO_{2(aq)} + H_2O \Leftrightarrow H_2CO_{3(aq)} \qquad (2.9.2)$$

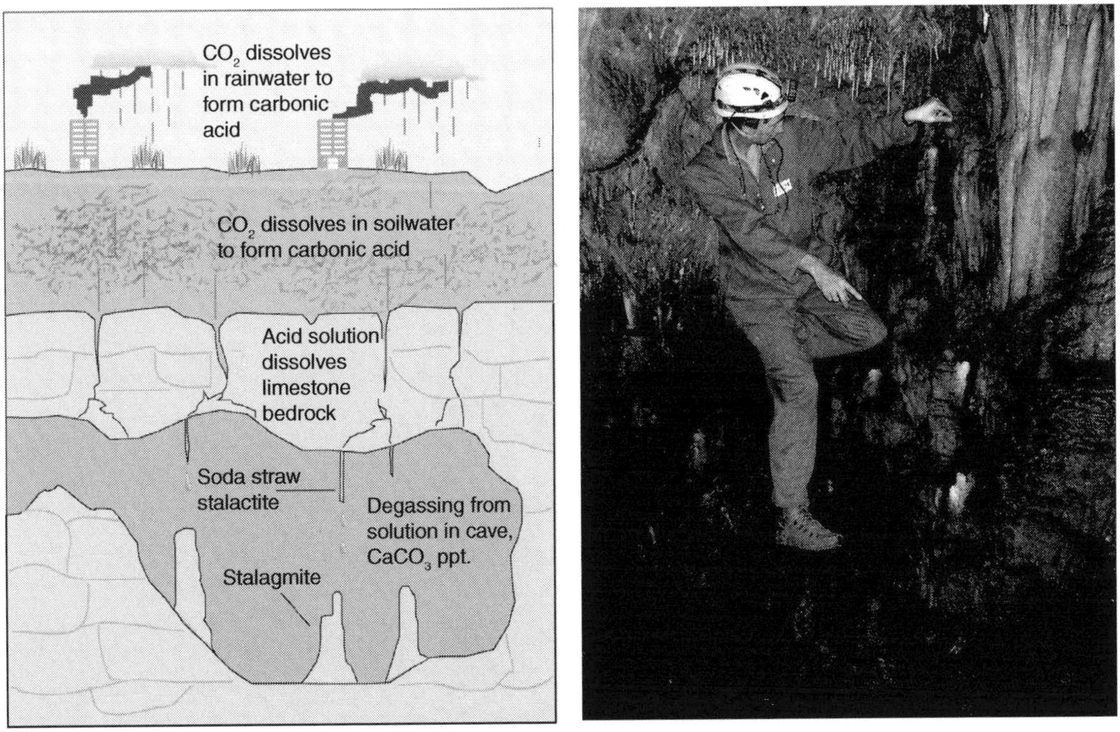


**Fig. 2.10** (a) Diagram to show the nature of the cave environment. (b) Inside a cave to show typical decoration with straw stalagtites, curtain flowstone and stalagmites.

3 dissolution of carbonate in the epikarst

$$CaCO_3 + H_2CO_3 \Leftrightarrow Ca^{2+} + 2HCO_3^- \quad (2.9.3)$$

4 precipitation of carbonate in the cave

$$Ca^{2+} + 2HCO_3^- \Leftrightarrow CaCO_3 + H_2O + CO_2 \quad (2.9.4)$$

### Tufa and travertine

Tufas are terrestrial carbonates that form under open air conditions in streams, rivers and lakes. They precipitate at ambient temperature from waters supersaturated with respect to calcium carbonate due to dissolution of carbonate bedrock. They are, in effect, above-ground equivalents of speleothem. Tufa calcite precipitates when dissolved $CO_2$ in karst groundwater outgases at a spring, or due to further $CO_2$ degassing downstream of springs when rapids, waterfalls (Fig. 2.11A) or other obstructions such as fallen trees or reed beds cause water turbulence. Because tufas form in the light, unlike speleothems, they almost always contain microbial (bacterial and cyanobacterial) and sometimes other algal components; many tufas encrust higher plants that live on the margins of streams, waterfalls and wetlands (Fig. 2.11b). *Travertine* is distinct from tufa, it is a term used to denote hydrothermal (warm to hot water) deposits.

### Calcrete

Calcretes are near-surface terrestrial accumulations of predominately calcium carbonate, which occur in a variety of forms from powdery to nodular to highly indurated. In terms of soil classification they are a form of semiarid- to Mediterranean-climate calcisol (section 1.5). They often feature a clay-rich horizon towards the top of the soil profile, beneath a usually thin organic A horizon: this is the argillic B horizon, which forms as $Ca^{2+}$ ions bind clay minerals with net negative charges together. Leaching of $Ca^{2+}$ further down the profile deflocculates the clays which are eluviated downwards. The calcrete horizon results from the cementation and displacive/replacive introduction of calcium carbonate by vadose and shallow phreatic groundwaters saturated with respect to calcium carbonate. High pH leads to aggressive dissolution of silicate minerals in parent materials. Dissolution of carbonate in wind-blown dust derived from often distant limestone outcrops and deposited

Fig. 2.11 (a) Foreground shows waterfall scene with draping curtain of tufa. (b) Close-up to show a laminated reed-bed tufa.

on soil or bedrock surface has an important role in many areas.

## 2.10 Evaporite salts and their inorganic precipitation as sediment

Evaporites are salts precipitated from brines in basins with limited connections to the ocean, or in closed continental basins. High evaporation rates due to high temperatures and low relative humidity are required for brine formation. Holocene conditions have not created the necessary environments for large-scale marine evaporite precipitation, although several periods during the Phanerozoic have been marked by impressive occurences, e.g. opening of the Atlantic and the Red Sea ocean basins, and desiccation of the Miocene Mediterranean (Chapter 20). The main evaporite minerals are listed in Table 2.3. Here we concentrate on primary marine evaporite sediments.

Seawater is normally undersaturated with respect to all evaporitic salts (Table 2.3). Halite is much more undersaturated than gypsum or anhydrite, so that any evaporation will cause gypsum to precipitate before halite. The most undersaturated salts are the complex series of potassium salts and these represent the final precipitates from highly concentrated brines. It should be noted that progressive evaporation is not a linear process since it is progressively more difficult to vaporize water from brines because of changes in surface tension. Figure 2.12 shows the sequence of salts that precipitate as seawater is progressively concentrated by evaporation in the laboratory. This differs somewhat from those recorded in evaporite deposits in the geological record, which show increased proportions of $CaSO_4$ and decreased proportions of sodium/magnesium sulphates. Magnesium depletion occurs by a combination of dolomitization and clay mineral precipitation. Important changes due to the effects of percolating brines arise during the last stages of brine concentration. Also, influxes of seawater may cause dissolution and reprecipitation.

**Table 2.3** Chemical composition for common evaporate minerals and selected values of IAP (ion activity product) and *K* (solubility product) for halite, gypsum, anhydrite in seawater solutions. (Reaction/concentration data after Berner, 1971.)

| Mineral | Formula | IAP | *K* |
|---|---|---|---|
| Halite | $NaCl$ | 0.12 | 38 |
| Gypsum | $CaSO_4{\cdot}2H_2O$ | $4.6.10^{-6}$ | $2.5.10^{-5}$ |
| Anhydrite | $CaSO_4$ | $4.6.10^{-6}$ | $4.2.10^{-5}$ |
| Sylvite | $KCl$ | | |
| Kieserite | $MgSO_4{\cdot}H_2O$ | | |
| Carnallite | $KMgCl_3{\cdot}6H_2O$ | | |
| Polyhalite | $K_2MgCa_2(SO_4)_4{\cdot}2H_2O$ | | |

The phase diagram for gypsum/anhydrite stability fields (Fig. 2.13; **Cookie 11**) shows why anhydrite has never been observed to precipitate directly from seawater. In highly saline brines, gypsum is still precipitated under earth surface conditions, but as a metastable phase, which during shallow burial may subsequently alter to anhydrite. This dehydration process occurs in the shallow ($\ll 1$ m) subsurface of arid coastal salt flats in very hot climates (*sabkha*; Chapter 20) at temperatures in the range 40–65 °C, encouraged by low relative humidities. There is some textural evidence for primary anhydrite precipitation in such environments and experiments indicate that precipitation is possible in the presence of certain organic molecules that act to inhibit gypsum precipitation.

Want to know more about gypsum/anhydrite reaction? Turn to Cookie 11

The geochemistry and composition of continental brines are much more complicated and unpredictable than marine brines, basically because there is no all-buffering oceanic reservoir to smooth out contrasts between local ionic inputs. Streams and groundwaters will have compositions determined by amount of runoff, catchment bedrock type and local chemical weathering processes (Chapter 1), with important contributions from longer-travelled groundwater brines. Seven major water types are recognized in continental brines (Fig. 2.14): those poor in alkaline-earth elements, i.e. (a) Na–$CO_3$–Cl, (b) Na– $CO_3$–$SO_4$–Cl; those poor in calcium, i.e. (c) Na– $SO_4$–Cl, (d) Na–Mg–$SO_4$–Cl, (e) Mg–$SO_4$–Cl; and those poor in sulphate, i.e. (f) Ca–Na–Cl, (g) Ca–Mg–Na–Cl.

Marine brines approximate to type (d), whilst many subsurface (*connate*) continental brines approximate to type (g). Brine composition obviously depends upon mineral precipitation and brine evolution in space and time (Fig. 2.14). Continental evaporites are distinctive in that $HCO_3^-$ and $CO_3^{2-}$ dominate the

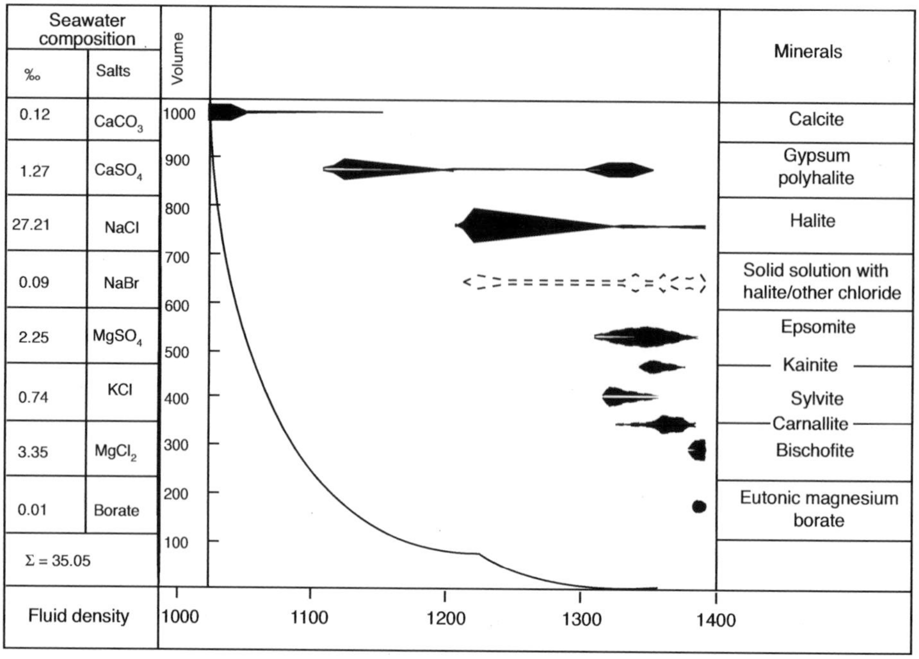


**Fig. 2.12** The effects of seawater evaporation on brine volume, density and type of salt precipitate. (Data of Valyashko, 1972.)

anions and are still available in the late stages of fractionation to form alkaline-earth carbonates (like *trona*, a complex sodium carbonate). In seawater, calcium and carbonate ions are used up early on in calcite, aragonite or dolomite precipitation reactions, leaving fractionated brines poor in $Ca_2$ and $CO_3^{2-}$

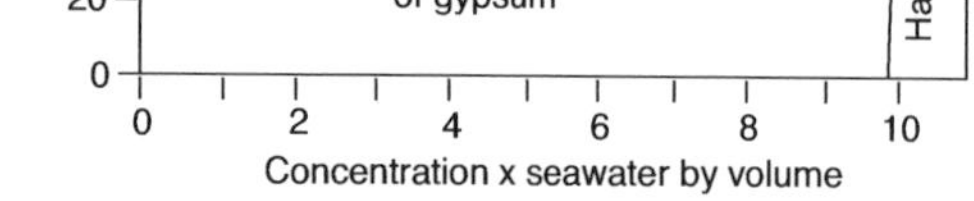

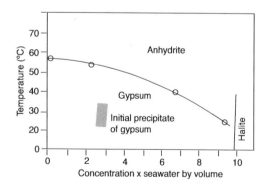


**Fig. 2.13** Phase diagram to show the stability fields of gypsum and anhydrite with respect to brine temperature and seawater concentration. (Data of Hardie, 1967 as recalculated by Blatt *et al.*, 1980.)

and rich in Na, $Cl^-$ and $SO_4^{2-}$. The lack of ancient sodium carbonates in marine evaporite sequences argues strongly for a fairly constant oceanic pH and composition, approximating to the recent, throughout most of geological time.

It has long been recognized that simply concentrating seawater by evaporation is insufficient to produce the great thicknesses of evaporite salts observed in the stratigraphic record; complete evaporation of the world's oceans with a total salt volume of $\sim 2.2 \times 10^7\,km^3$ yields a mean thickness of only 60 m of evaporites. Some ancient evaporite successions of great areal extent may reach over 1 km in thickness. Such evaporite production must have had drastic short-term effects upon global levels of seawater salinity and oxygen isotope composition before a steady state was once more established. For example, the Permian Zechstein evaporites of northwest Europe have a total volume of about $2.4 \times 10^6\,km^3$, some 10% of the oceanic reservoir. However, spectacular short-term events must be set aside the long ($>10^8$ yr) mean residence time of $Cl^-$ in the oceans. It has been calculated that the total mass of Cl in sedimentary

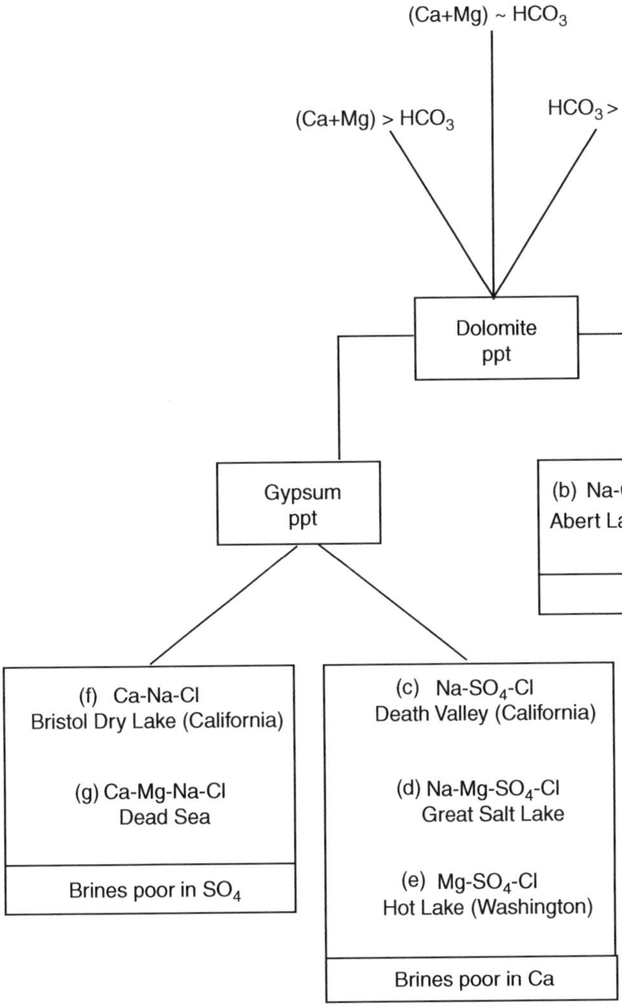

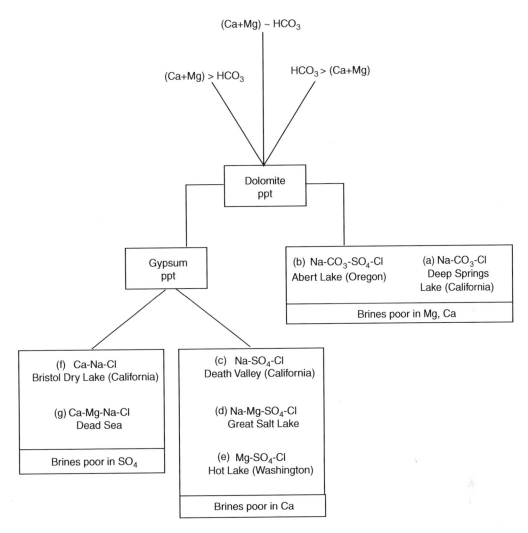


**Fig. 2.14** Flow diagram to show brine types and evolution with the two solid chemical 'divides' due to gypsum and dolomite precipitation. (Data of Harvie *et al.*, 1982.)

rocks and porewaters is roughly equal to that of the modern oceans, and so it is unlikely that the salinity of the oceans could have changed by more than 50% above its present value since the early Precambrian.

Models for evaporite evolution must obey a hydrological accounting system and have the following components: (i) initial brine mass, (ii) rate of influx from 'mother' basin, (iii) rate of evaporative loss, (iv) rate of evaporite solid deposition and (v) rate of seepage or outflow (termed *reflux*) of dense brine (Fig. 2.15). Clearly the system is a complicated one, even neglecting dynamic effects arising from internal stratification and convection. Some light may be shed on the vexed problem of the origin and depth of evaporating brines by making use of bromine geochemistry. Br does not precipitate as a salt, but is incorporated in the halite lattice. Halites precipitated from marine-sourced brines have much greater Br contents than halites from non-marine-sourced brines; a value of less than 65 ppm Br is usually taken to be the upper limit for the latter case. The partition coefficient of Br between crystallizing halite and the residual brine is around 0.1, leading to increased Br in halite as crystallization proceeds. Simple models for the rate

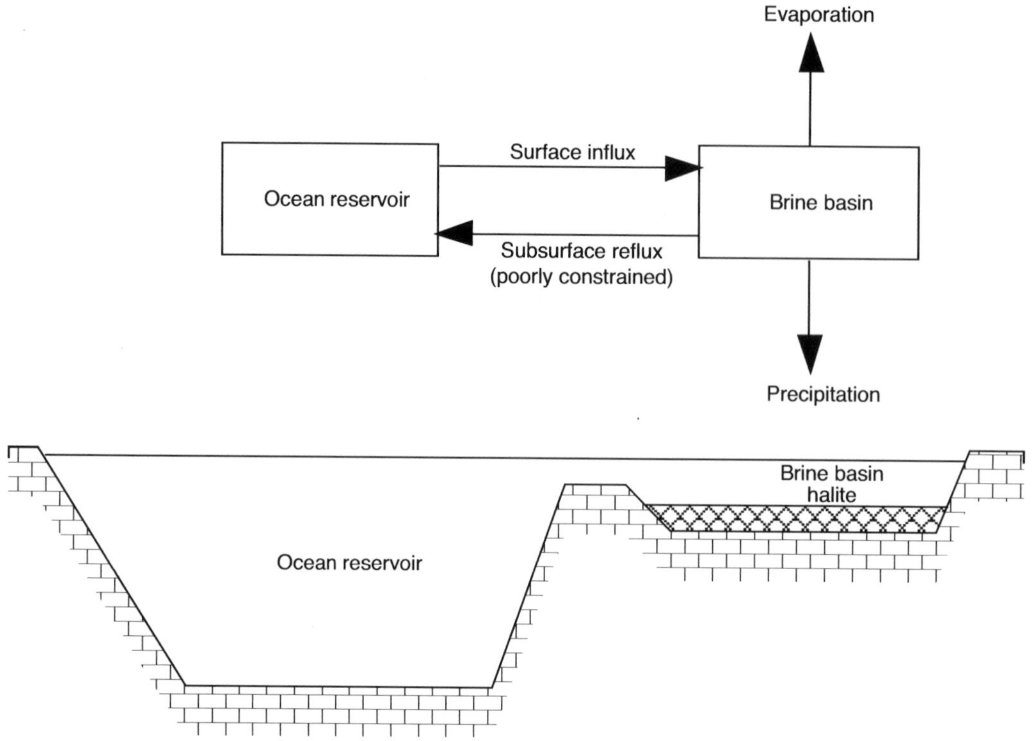


**Fig. 2.15** A simple inputs-outputs model for a brine basin. (Inspired by Tucker & Cann, 1986.) Note that various contributions of marine vs. continental waters to the evaporating brine basin must be determined by techniques such as Sr-isotopic ratios.

of change of bromine with height in halite precipitated from closed basins yield very large estimates of brine depth. Once more realistic open-system conditions are modelled (as functions of rates of influx and reflux), more realistic depths are found. However, in Br tracer studies the role of post-depositional recrystallization in disturbing the original values is important. $^{87}Sr/^{86}Sr$ ratios relative to the known secular Sr-growth curve are more useful chemical tracers to separate brines derived from marine and continental waters in the stratigraphic record. Continental brines derived largely from meteoric water may have high radiogenic Sr contents and this can help decide correct hypotheses for brine origins.

The simplest model for subaqueous evaporite formation is the shallow-water barred basin (Fig. 2.15) where evaporation proceeds in semi-isolation, with replenishment of seawater over a restrictive entrance sill or obstruction. Brines are thus progressively concentrated, and progressive crystallization of the various salts then proceeds. In such models basin brines may stay at particular concentrations for long periods and so abnormally thick sequences of sulphates or chlorides may accumulate. Gypsum crystal fabrics record prolonged preferred directions of brine flow due to slow long-term circulation in such basins (Chapter 20). Cycles of evaporites that approximate reasonably closely to the 'ideal' cycle result when basin brines are evaporated to completion. Lateral changes in evaporite composition are to be expected in barred basins since incoming seawater will precipitate first gypsum and then halite as it spreads over the sill towards the basin shallows. A contrasting model involves closed-basin drawdown of brine bodies with an important contribution to brine volume provided by continental meteoric spring waters. However, Sr ratios do not support this idea, at least for the classic Castile varved evaporites of Texas/New Mexico.

One further point concerns the presence of small-scale evaporite rhythms on a millimetre to centimetre

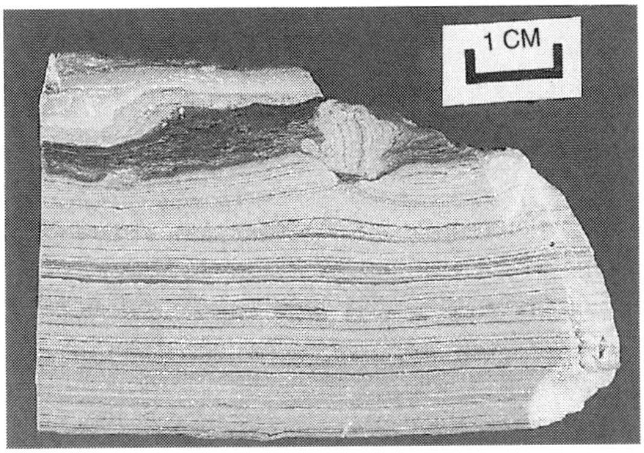


**Fig. 2.16** Subaqueous evaporite precipitates; varved anhydrite–dolomite, late-Jurassic Hith Formation, Saudi Arabia. Note reworked intraclast, proving primary depositional layering, at top right of specimen. Such evaporites are topmost caprock to the legendary Arab Formation oil reservoirs.

scale observed in many ancient evaporites (Fig. 2.16). These may comprise alternating (a) dolomite–anhydrite + clastic clay, (b) clay–dolomite–anhydrite + halite, (c) halite + sylvite–carnallite, (d) aragonite– gypsum and (e) calcite + anhydrite. Inflowing runoff 'freshens' the brine body and this, together with cooler air temperatures, causes either cessation of evaporite precipitation or precipitation of a less undersaturated phase—the runoff also brings in the suspended clastic sediment. The laminae thus represent annual varves by this interpretation, with changes in precipitation, runoff, sediment supply and temperature controlling the detail of the sedimentary rhythms. Type (d) laminae feature in the lake cycles of fresh to saline conditions from Lake Lisan, the precursor to the Dead Sea.

Mention is finally made of the apparent occurrence of global cycles in marine evaporite deposits, perhaps reflecting secular variations in seawater chemistry. Large marine potash evaporite bodies are dominated either by Mg-rich phases like polyhalite and kieserite or by KCl salts like sylvite (see Table 2.3). There is a periodicity to these trends which coincides with those established for carbonate minerals, viz. the occurrence of aragonite vs. calcite seas (**Cookie** 7). In particular, it seems that aragonite seas coincide with periods dominated by $MgSO_4$ evaporites and calcite seas with KCl evaporites (Fig. 2.17). The explanation is sought in large-scale secular seawater compositional changes due to changing rates of seafloor spreading causing mid-ocean-ridge hydrothermal/submarine weathering reactions to provide greater or lesser amounts of magnesium, calcium, potassium and sulphate to the oceans (**Cookie** 7; Chapter 23).

Subaqueous evaporite facies are further discussed in sections 14.5 & 20.11.

## 2.11 Silica and pelagic plankton

An outline of the silica cycle in the oceans is shown in Fig. 2.18. Seawater is vastly undersaturated with respect to amorphous silica (**Cookie 12**) because of efficient removal of Si by siliceous plankton, chiefly diatoms (algal primary producers) and the predatory radiolarians, which construct their skeletons with the amorphous opal-A form. This biogenic extraction is opposed by bacterially mediated dissolution of skeletal material at depth, causing a rapid increase of dissolved silica with depth in the oceans. The total deposition of biogenic silica per year is about $10^{13}$ mol. Some angular quartz silt (4–64 μm size) in ancient sedimentary mudrocks could record the past presence of siliceous diatoms and radiolaria in the Palaeozoic, long before their preservation as recognizable macrofossils in the later Mesozoic (Chapter 1). The main areas of diatom siliceous production correspond to areas of high surface productivity where ocean current divergence or upwelling causes fertile deep waters rich in phosphorus, nitrogen and iron to rise into the warm photic zones (Chapter 21). Radiolaria feed upon the abundant remains of various primary producers. We have already discussed data from oceanic oozes that indicate that during glacial periods calcitic pelagic organisms are replaced by siliceous plankton in regions of lower water temperatures and high productivity. The silica–calcite 'front' moves generally equatorwards, the proportion of $CaCO_3$ preserved in bottom sediments decreases and that of organic carbon increases.

| |
|---|
| Want to know more about silica saturation? Turn to Cookie 12 |

Siliceous diatoms are also a feature of many temperate lake deposits (Fig. 2.19) where they also cause marked near-surface depletion in dissolved silica as the diatom tests are sedimented faster than the dissolved silicon is replaced by runoff. Typically,

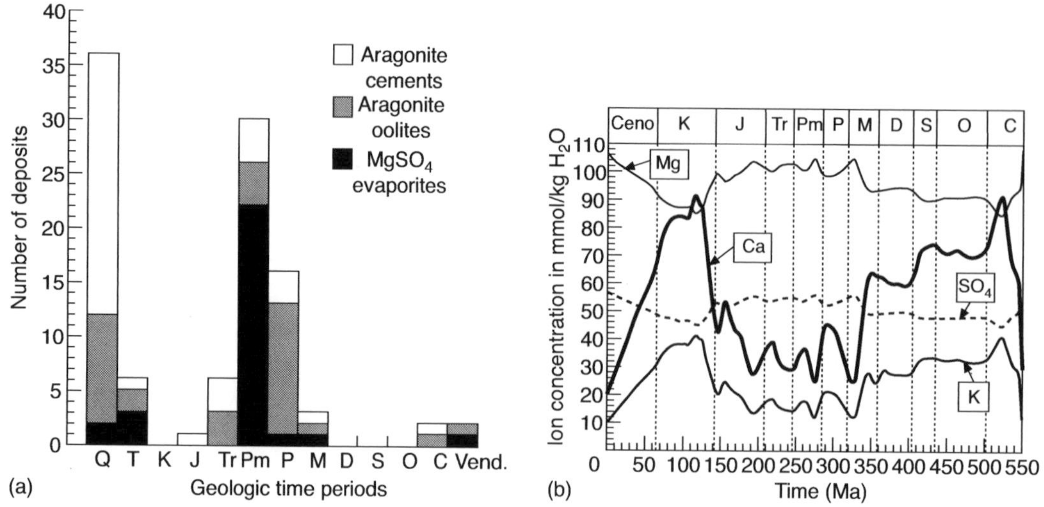


**Fig. 2.17** (a) Time changes in carbonate and evaporite mineralogy. (b) Time changes modelled for variation in major ion chemical composition of the Phanerozoic oceans. (After Hardie, 1996.)

maximum diatom productivity in temperate lakes occurs at the time of the early spring overturn when the surface waters have their maximum concentrations of nutrients. Important fixation of Si by diatoms also occurs in the highly productive flooded wetlands and estuaries of major tropical rivers, notably the Amazon.

## 2.12 Iron minerals and biomineralizers

Iron is present in most igneous and many metamorphic rocks as the ferrous form ($Fe^{2+}$) in silicate minerals like olivine, pyroxene, amphibole and biotite mica. Surface and near-surface oxidative weathering to the ferric form ($Fe^{3+}$) has major significance for earth surface biogeochemical processes, not least because of its implications for atmospheric oxygen abundance and the diversity of Fe oxide, hydroxide and sulphate weathering products. Fe is brought to depositional basins in solution as $Fe^{2+}$, in clay minerals that contain $Fe^2$ in their lattices and as adsorbed species on clay mineral surfaces. Five physico-chemical variables control the nature of reactions involving $Fe^2$ and $Fe^3$ with other dissolved species. These are $E_h$, pH and the activities of dissolved $HS^-$, $HCO_3^-$ and $Fe^2$. *Haematite* ($Fe_2O_3$) and mixed species like *magnetite* ($FeO{\cdot}Fe_2O_3$) are the only iron minerals that can exist in equilibrium with depositional waters above the sediment–water interface. Arid-zone weathering over time encourages red haematitic pigment production in near-surface environments and the formation of *red beds*. The haematite may remain stable as long as organic matter remains absent; its presence encourages reduction of $Fe^3$ to $Fe^2$ and the red pigment disappears, sometimes in spherical greenish areas termed *reduction spots*.

Microbial reduction of $Fe^3$ to $Fe^2$ *and* oxidation of $Fe^2$ to $Fe^3$ are common under anaerobic conditions (**Cookie 13**). The natural oxidation of $Fe^2$ to $Fe^3$ may occur by bacteria or by chemical oxidation, both involving molecular oxygen, *or*, by a group of *purple bacteria* present in all marine muds that utilize the process in the absence of oxygen in order to reduce $CO_2$ to cell material, i.e. the bacteria photosynthesize utilizing energy from the reaction $Fe^2 \Rightarrow Fe^3$. These *anoxygenic phototrophic* bacterial strains are very important because they imply that oxygen-independent iron oxidation was possible long before the evolution of oxygenic photosynthesis.

Want to know more about iron anaerobic reactions? Turn to Cookie 13

A prolific source of minute magnetite grains in sediments (mostly forming in the shallow subsurface

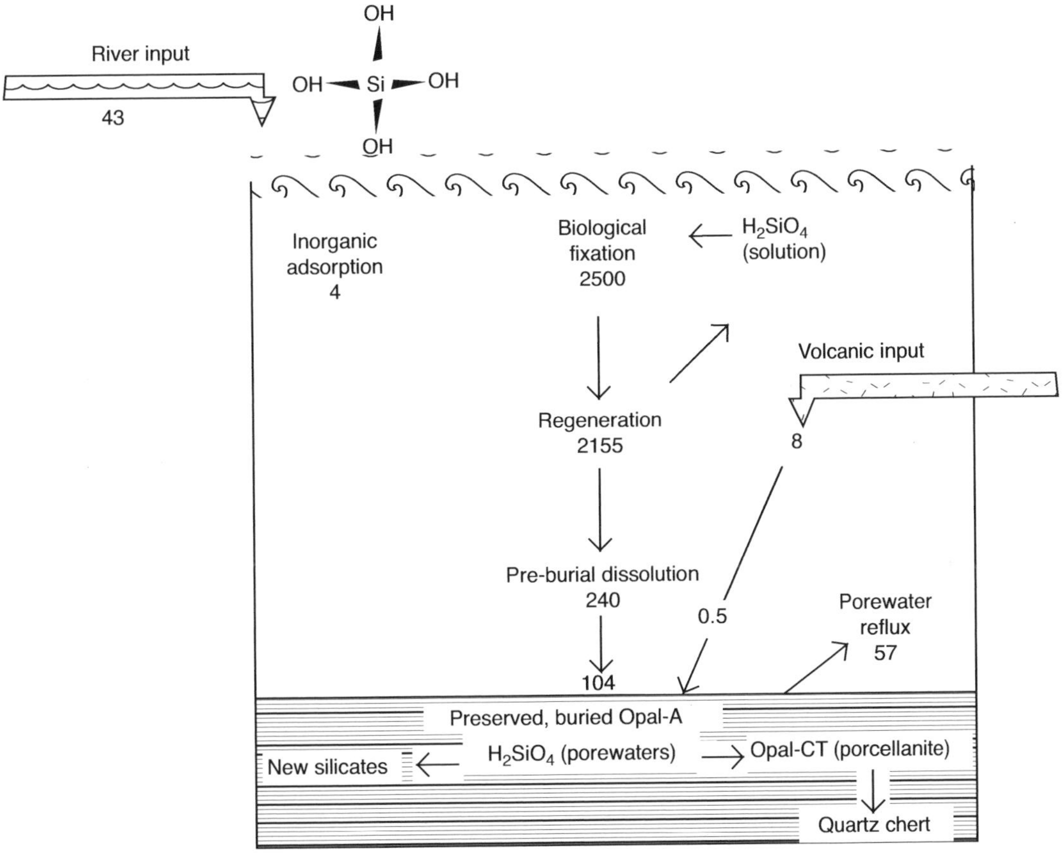


**Fig. 2.18** The oceanic Si biogenic cycle with the magnitudes of dissolved silica as $SiO_2$ in $10^{13}$ g/yr. Note the dominance of biological fixation and regeneration. Annual river input is roughly balanced by diatom burial. (Data of Heath, 1974; Riech and von Rad, 1979.)

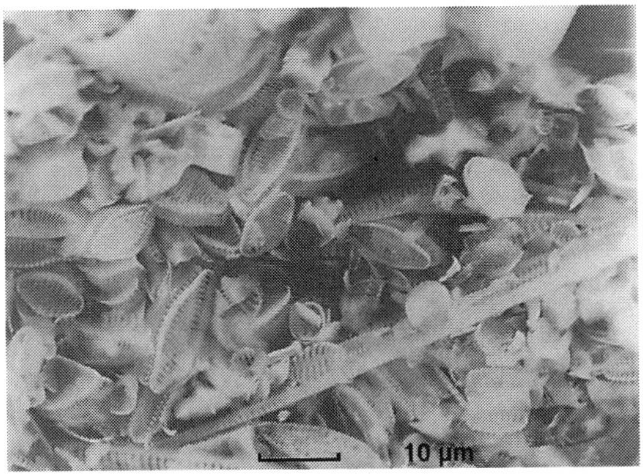


**Fig. 2.19** Freshwater diatoms from mid-Holocene sediments of proto-Lake Hebgen, Montana, USA. Scale bar = 10 µm. Photo courtesy of Jan Alexander.

sulphate-reducing zone) are *magnetotactic bacteria*, in whose cells crystals of magnetite precipitate as part of a biomineralization process. It is these *prokaryotes* that impart the majority of depositional remnant magnetization to sediments. Much of the palaeomagnetic record in ancient sediments depends upon this process; even continental arid-zone sediments that contain predominantly haematite seem able to preserve this bacterial magnetite signature.

*Goethite*, ferric iron oxyhydroxide (FeO·OH), is the major source of detrital iron in sediments. It is derived from weathering and soil reactions, including lateritization (Chapter 1). The dehydration reaction (**Cookie 13**) explains the theoretical instability of yellow to brown limonite/goethite sediments in the subsurface, but crystalline goethite is kinetically stable and occurs widely. Of particular interest are iron ooids that comprise goethite. These are common in the stratigraphic record but very rare in recent environments, being restricted to one occurrence in a shallow marine setting adjacent to a volcanic island. These modern ooids form by chemical precipitation of microcrystalline goethite in waters enriched in Fe, Al and Si by volcanic ash and hydrothermal fluids. Ancient goethite ooids frequently occur in strata that accumulated very slowly (so-called *condensed deposits*). They show microscopic textures like flakey microcrystals arranged tangentially around the ooid cortex. This is considered indicative of primary precipitation as microcrystalline goethite in oxic near-surface layers of marine waters followed by periodic erosion to orientate the goethite microcrystals and to encrust them with marine microfauna.

*Pyrite* ($FeS_2$) and *siderite* ($FeCO_3$) are common diagenetic iron minerals in many marine sediments. Below a thin zone of oxygenated porewaters, anaerobic bacterial reduction of sulphate occurs. The $H_2S$ that forms reacts with iron to form the iron monosulphide FeS. Continued $H_2S$ production encourages sulphur-oxidizing bacteria to form elemental sulphur. This sulphur reacts with the FeS over a period of years to form microscopic aggregates of pyrite crystals termed *framboids*. The most important factor limiting pyrite formation is the availability of organic matter that can be metabolized by bacteria. Both iron and sulphate species are usually present in abundance in marine diagenetic porewaters. Thus the greater the amount of organic material, the greater the amount of pyrite produced. Marine black shales rich in organic material are frequently pyritous in the geological record. By way of contrast, non-marine diagenetic porewaters are generally very low in dissolved $SO_4^{2-}$ and hence pyrite is usually absent. Rapid marine deposition will tend to inhibit pyrite formation since there will be a limited time available for $SO_4^-$ diffusion from the overlying marine reservoir. The $SO_4^{2-}$ available is restricted to that contained within the buried porewaters.

*Siderite* also forms as a diagenetic mineral when a very low dissolved sulphide concentration is coupled with high dissolved carbonate, high $Fe^2/Ca^2$, low $E_h$ and near-neutral pH. These conditions usually restrict siderite formation to non-marine diagenetic environments (low $SO_4^{2-}$) where abundant $Fe^2$ is present (tropical-zone weathering). The mineral is particularly common in deltaic swamp facies where concretions give evidence for continued growth during progressive burial. For siderite to be stable relative to calcite, the iron concentration must be greater than 5% of that of calcium. In seawater it is less than 0.1%. The occurrence of sideritic beds in certain marine mudstones points to somewhat unusual conditions. Associated burrows and fauna often indicate minor depositional gaps. The siderite may have formed later in diagenesis from an oxide precursor when marine $SO_4^2$ was exhausted and when $Ca^2$ ions were also used up. It is probable that microbially mediated reactions control much diagenetic anoxic siderite precipitation, with the iron-reducing *Geobacter metallireducens* organism chiefly responsible.

*Chamosite* is a complex ferrous–aluminium–silicon oxyhydroxide that occurs most characteristically as ooids and as mud-grade crystals with siderite in *minette-type iron ores*. Associated faunas indicate fully marine conditions of formation with periodic agitation and reduced background clastic sedimentation to encourage ooid growth. The environmental requirements for ferrous silicate stability resemble those of siderite modified by low carbonate activity and saturation with respect to some active silica form. They suggest reducing conditions at or just below the sediment–water interface, but this alone does not explain the ooidal accretionary fabric. It is likely that chamosite ooids were originally formed of some early mineral (or minerals) subsequently converted to chamosite during diagenesis. The pioneering sedimentary geologist H.C. Sorby originally postulated in the nineteenth century that the ooliths were formerly calcitic, but detailed fabric studies do not support this conclusion. The mineralogical question revolves around the observation that ooidal ironstones are characterized by the iron-rich clay mineral *berthierine* (commonly altered on burial to chamosite), intimately interlaminated with goethite and cemented by siderite; pyrite is rare to absent. Berthierine precipitation must take place from reducing pore fluids in the absence of bacterially mediated sulphate reduction. The fabric evidence requires episodic exhumation and ooid accretion from the shallow subsurface, suboxic zone to the sea floor. This requires agitated boundary layer conditions to alternate with periods of benthic stasis, a clear signal of wave storm conditions rather than regular tidal processes. The mechanism also causes periodic oxidation and formation of goethite coatings.

Such conditions may commonly arise as a result of marine transgression during initial relative sea-level rise following development of a *sequence boundary* (Chapter 10).

*Glaucony* (aka *Glauconite*) is a mica-structured (phyllosilicate) mineral with a beautiful blue-green colour, hence its name from the Greek *glaukos*. It is a constituent of '*greenstone*' sedimentary iron formations and is typically found as peloidal aggregates or shell cavity infills of very finely xlline grains. It typically occurs as a marine phase in areas of much-reduced sedimentation where it infills shell cavities and replaces faecal pellets. General sedimentary environmental context indicates slow formation under conditions of sediment starvation on *ravinement* sequence boundary surfaces during periods of sea-level rise—it is thus a useful *sequence stratigraphic* indicator (Chapter 10). Glaucony varies tremendously in composition from potassium-poor smectite clays to potassium-rich glauconitic micas with a general trend towards increasing potassium with time. It is thought to form at or just below the sediment–water interface where it is associated with organic matter and generally positive but fluctuating $E_h$ conditions. Rb/Sr and oxygen isotopes indicate development through a series of simultaneous dissolution/crystallization reactions during very early diagenesis in a closed or isochemical system, isolated from the ambient marine environment. The constituent ions of the mineral are derived primarily from terrigenous smectite clay minerals, but considerable potassium is sourced indirectly from seawater, through potassium enrichment of clay precursors. Its gradual, very early diagenesis mode of formation makes radiometric dates derived from potassium/argon decay products in the mineral rather difficult to interpret in some cases.

## 2.13 Desert varnish

Many rock outcrops and exposed clast surfaces in arid climates show a characteristic dark brown to black surface patina. Careful chipping reveals a millimetre to centimetre thick coating with a sharp contact on bedrock. The colour contrast between coating and underlying bedrock often commended itself as a natural engraving plate to native North American pictograph artists (Fig. 2.20). The coating is termed *desert varnish*, consisting of precipitated manganese and ferric oxides and wind-blown dust; in ultrathin section under very high magnification the rather boring-looking patina is seen to be beautifully laminated, with individual darker laminae either Fe- or Mn-rich and the lighter laminae of silicate-rich material, chiefly phyllosilicates. The laminae are extremely thin (just a few microns) and the Mn phases are various, including the hydrated manganese/magnesium oxide, *birnessite*, whilst the Fe phase is predominantly haematite. Detailed high-resolution imaging and precise microanalyses indicate varnish formation by repeated wetting and drying cycles with leaching and oxidation of Fe and Mn from very slow aerosol deposition of soluble $Fe^{2+}$ and $Mn^{2+}$. Post-precipitation modification also occurs, with fracture fillings, recrystallization and secondary barite mineral precipitation. Some workers suspect a microbial influence on varnish precipitation. Radiometric dating shows that desert varnish accumulates incredibly slowly, of order

**Fig. 2.20** Outcrop of naturally pale-coloured hornblende andesite. Native American pictograph artists have chipped away well-developed desert varnish to reveal pristine rock. Three Rivers Petroglyph site, Tularosa, New Mexico, USA. Peak in background is Sierra Blanca (~3600 m).

< 1–40 μm/kyr and this has encouraged its recent use as a sensitive palaeoclimatic indicator; darker Mn/Ba-rich layers are thought to be of wetter pluvial origins whilst lighter Mn/Ba-depleted layers form in more arid intervals.

## 2.14 Phosphates

Phosphates form an important minority of sedimentary rocks. Concentrations of phosphorus in seawater average about 0.07 ppm, derived by weathering of igneous and sedimentary apatites (Chapter 1). Despite these low concentrations, phosphorus is a very important element, being an essential component of all living cells. Increased use of fertilizers and exploitation of natural phosphate reserves have inevitably focused sedimentological attention on the origins of phosphate rock and particularly of the concentration mechanisms involved. Phosphorus turns out to be a key element in tracking the changing magnitude of continental weathering because the weathered flux of phosphorus from the continents dwarfs any primary inputs from the mid-ocean ridges or from atmospheric deposition. Increased P fluxes may correlate with periods of enhanced global weathering, as is suspected for intra-Miocene events at 21–25 Ma, 17–13 Ma, 9.8–10.9 Ma and at the late-Miocene/Pliocene transition at 5 Ma—all major phosphatization episodes in ocean history. Sinks for phosphorus include organic and skeletal matter, iron and manganese oxyhydroxides and the phosphorus-mineral phases francolite–apatite and collophane. Extensive phosphorite deposits are typically associated with upwelling continental margins where organic-rich sediment is deposited beneath highly productive surface waters. Phosphorus is then released into the porewater of the upper sediment column by microbial respiration of the buried organic matter where it precipitates as an early diagenetic phase. It is also commonly associated with sediment-starved transgressive events along shelf margins receiving high fluxes of P-rich weathering products. In field situations it forms what is known as a *lag deposit*, i.e. winnowed fine gravel-grade sediments with phosphate replacing and/or infilling burrows, vertebrate bones, teeth and shell fragments. Such lag deposits may be associated with sequence boundaries (Chapter 10). Primary sediment nucleii are commonly reworked into intraclasts, indicating that the benthic boundary layer, though sediment-starved, was influenced by storm waves.

## 2.15 Primary microbial-induced sediments: algal mats and stromatolites

Filamentous cyanobacterial colonies (chiefly the genera *Schizothrix* and *Scytonema*; **Cookie 10**) create laminated algal mats (Fig. 2.21a) by trapping detrital sediment provided by storms and tidal currents around their mucilaginous sheaths. They are widespread in both carbonate and siliciclastic sediments deposited in intertidal and subtidal environments—examples have been found in the sedimentary record as far back as the early Archaean (3.2 Ga). The latter indicate that their photosynthetic activities were important in gradually establishing an oxygenic atmosphere during early Proterozoic times. Detailed studies and comparisons with modern ecosystems can reveal precise environmental controls on mat morphology. Famous stromatolite localities include Shark Bay (Western Australia), Exhuma Sound (Bahama Banks) and the Trucial Coast (see Chapter 20).

*Stromatolites* are mat forms (Fig. 2.21e & f) lithified both by external precipitation of calcium carbonate around sheaths (*Scytonema*) and by periods of microbial degradation in surface organic layers when sediment-trapping is absent. Since such mineral precipitation is governed by microbiochemical reactions external to the bacterial cells, the colonies are said to be *non-obligate calcifiers*. The *organo-sedimentary structures* that result thus contain distinct millimetre-scale layering, including primary clastic–organic couplets and detrital laminae trapped and bound by *Schizothrix* alternating with photosynthetically and microbially mediated micritic calcite and dolomite precipitates. Algal mats and stromatolites form different morphologies according to ecological niche, including mats, polygons, ridges and a variety of bulbous protuberances (Figs. 2.21 & 2.22). Complex associations of forms exist in some examples and where calcification is abundant and early the term *stromatolitic reef* may be appropriate. The living, sticky filamentous layers enable gravity to be defied locally with internal growth laminae showing characteristic thickening and thinning trends in relation to local light availability, e.g. laminae frequently thin into overhangs and side crypts. Evidence for local

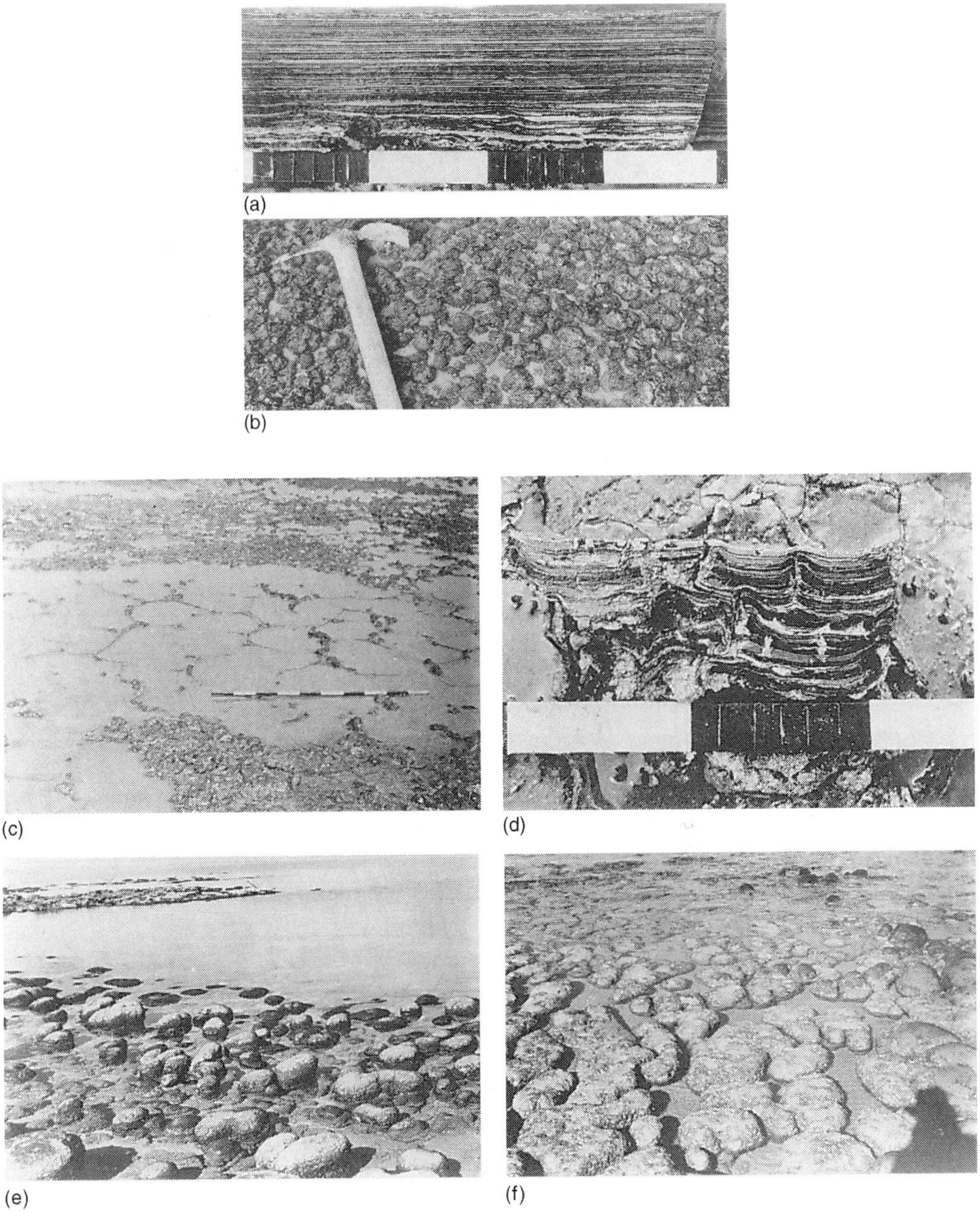


**Fig. 2.21** (a) Smooth, regularly laminated algal mats from the lower intertidal zone of the Trucial Coast, UAE. Note dark algal-rich and light sediment-rich interlaminations; white of scale bar is 10 cm. (b) View of blister morphology algal mat from the mid-upper intertidal zone, same locality as (a). (c) View of large-scale polygonal algal mat with raised rims and areas of blister mat growth; white of scale bar is 10 cm; locality as (a). (d) Section through smooth algal mat disrupted into polygonal mat by desiccation shrinkage cracks. Note raised rims and periodic evidence of crack-healing; white of scale bar is 10 cm; locality as (a). (e) Lithified stromatolite columns showing seaward asymmetry, Shark Bay, Western Australia. (f) Lithified stromatolite ridges in exposed, high-energy intertidal zone, Shark Bay. Ridges are separated by skeletal carbonate sands and show elongation normal to direction of wave propagation. (Photographs (a)–(d) courtesy of R. Till, (e) and (f) courtesy of P.G. Harris.)

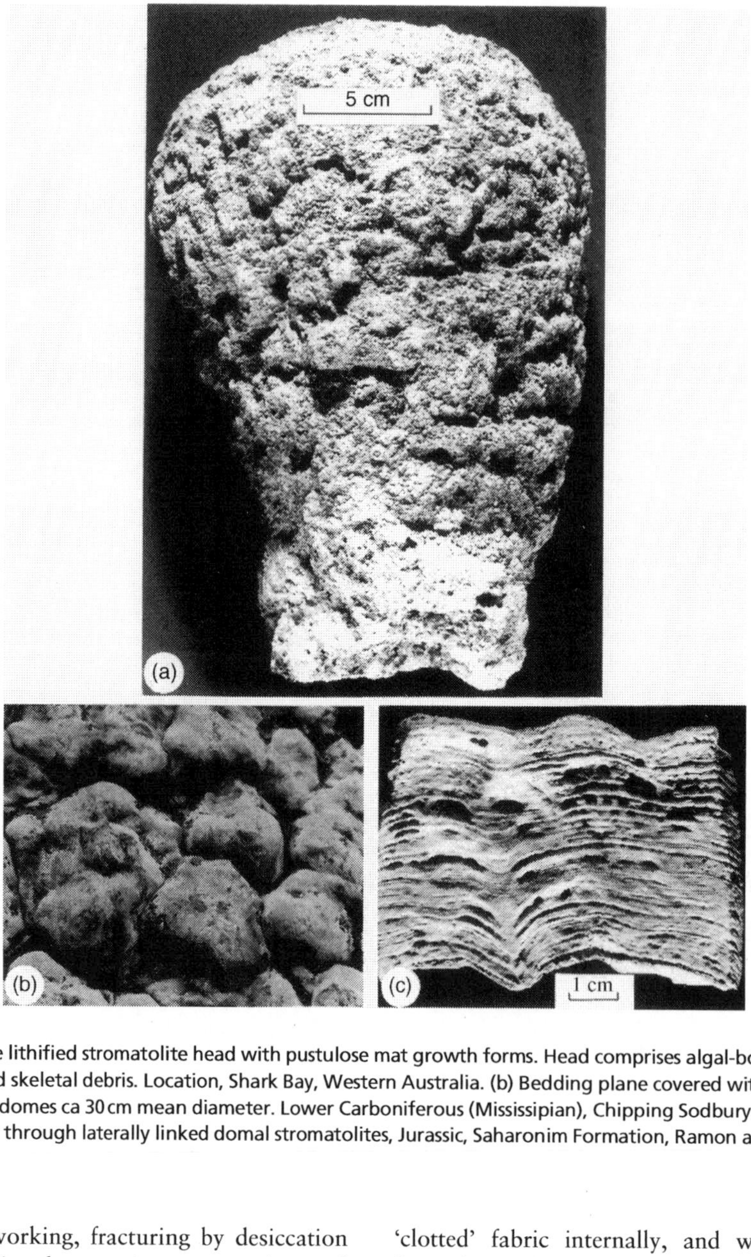


**Fig. 2.22** (a) Single lithified stromatolite head with pustulose mat growth forms. Head comprises algal-bound and aragonite-cemented skeletal debris. Location, Shark Bay, Western Australia. (b) Bedding plane covered with laterally linked stromatolite domes ca 30 cm mean diameter. Lower Carboniferous (Mississipian), Chipping Sodbury, Somerset, England. (c) Section through laterally linked domal stromatolites, Jurassic, Saharonim Formation, Ramon anticline, Israel.

mat erosion, reworking, fracturing by desiccation and post-depositional expansion, contraction and gas-escape structures are visible in many examples. The rather cohesive nature of surface mats affords extra protection to underlying mud/silt substrates from erosion during tidal flow and storms, even when only a very thin *biofilm* of cyans is present.

*Thrombolites* are a variety of stromatolite that do not show internal laminations, rather a sort of 'clotted' fabric internally, and which have been formed entirely by microbially induced precipitation of $CaCO_3$ by non-filamentous cyanobacteria.

## Further reading

### *General*

Krauskopf (1979), though out of print, is still an excellent starting point for elementary aqueous geochemistry and, as in Chapter 1, relevant sections of Atkins (1992), Gill (1989), Andrews *et al.* (2004b) and Albarède (2003) are warmly recommended. Bathurst (1975), though long in tooth, is still delightful. Chapters 11 and 12 in Summerhayes and Thorpe (1996) are excellent on global oceanic geochemistry (Burton) and the carbonate system (Varney) respectively. Tucker & Wright (1990) is an encyclopaedic treatment of all things carbonate. Konhauser (2007) is the main source for microbiological geochemistry. A thick source for microfabric studies is Flügel (2004).

### *Specific*

Changing ocean chemistry and its effects on precipitation of minerals over geological time are discussed by Wilkinson *et al.* (1985), Archer and Maier-Reimer (1994), Hardie (1996), Sumner & Grotzinger (1966, 2004) and Morse *et al.* (2006). Mg contents of echinoderms by Dickson (2004). Mg-based palaeotemperatures in rhodoliths by Kamenos *et al.* (2008) and Halfar *et al.* (2008). See Tripati *et al.* (2005) for Mg/Ca thermometry results. For effects of oceanic Mg/Ca on algal sediment production see Stanley & Hardie (1999) and Ries *et al.* (2006a,b).

Carbonate precipitation in the oceans is discussed by Neev & Emery (1967) Lippmann (1973), Wollast *et al.* (1980), Burton & Walter (1987), Opdyke & Wilkinson (1993), Dove & Hochella (1993), Morse *et al.* (1997) and Choudens-Sanchez & Gonzalez (2009). Ikaite occurrences and glendonite pseudomorphs are discussed by Pauly (1963), Shearman *et al.* (1989), Buchardt *et al.* (1997), De Lurio & Frakes (1999) and Selleck *et al.* (2007).

Aspects of carbonate dissolution in the deep oceans are discussed by Chave & Suess (1970), Morse & Berner (1972), Takahashi (1975), Berner (1976), Thunell (1982), Archer (1991), Archer & Maier-Reimer (1994) and Frenz and Henrich (2007).

The story of Bahamas whitings is discussed in a classic study by Cloud (1962) and later by Broecker & Takahashi (1966), Morse *et al.* (1984), Shinn *et al.* (1989) and Milliman *et al.* (1993). Bahamas stable isotope trends are discussed by Swart *et al.* (2009). The contribution of algae to aragonite muds is discussed by Stockman *et al.* (1967), Bosence *et al.* (1985), Pernetta (1994), Riding (2000) and Arp *et al.* (2002). Aragonite dissolution in late Quaternary temperate carbonates is discussed by James *et al.* (2005a).

AFM and SFM studies of carbonate mineral nucleation are in Dove and Hochella (1993) and Gratz *et al.* (1993).

Dolomite precipitation is discussed by Hsü (1966), Tucker (1982), Mazzullo *et al.* (1995), Vasconcelos *et al.* (1995) and Vasconcelos & McKenzie (1997). Recent interpretation of '1ry' dolomite precipitates as microbially mediated are by García del Cura (2001), Wright & Wacey (2005), Mastandrea *et al.* (2006) and Perri & Tucker (2007).

Iron fertilization to mop up excess atmospheric $CO_2$ is discussed by Martin *et al.* (1990) and Peng & Broecker (1991).

Carbonate ooid genesis is discussed by Shearman *et al.* (1970), Suess & Futterer (1972), Loreau & Purser (1973), Kahle (1974), Sandberg (1975), Halley (1977), Davies *et al.* (1978), Deelman (1978), Ferguson *et al.* (1978), Medwedeff & Wilkinson (1983) and Richter (1983).

Global aspects of skeletal carbonate components are discussed by Opdyke & Walker (1992). Algal rhodoliths and cyanoliths are discussed by (Bosence, (1983a,b) and (Riding, 1983) respectively. Temperate to cold water biogenic coral mounds and reefs are discussed by Williams *et al.* (2006) and bryozoan mounds by James *et al.* (2000).

The connection between algal borings, micrite envelopes and intraclasts were originally noted in a landmark paper by Bathurst (1966); the analogous calcified bacterial biofilms are discussed by Riding (2002).

Carbonate lithification of algal mats to form stromatolites is discussed by Reid *et al.* (2000, 2003), Visscher *et al.* (2000), Arp *et al.* (2002), Perri & Tucker (2007).

Coral reef growth rates (see more refs in Chapter 20) are described by Vago *et al.* (1997) and their role in the carbon cycle by Berger (1982); Opdyke & Walker (1992); Archer & Maier-Reimer (1994). Coral growth inhibition in low-Mg seawater by Ries *et al.* (2006a,b).

Controls on marine evaporite precipitation are discussed by Kinsman (1976), their chemistry by Hardie (1967), Valyashko (1972), Cody & Hull (1980). Continental brines are discussed by Eugster & Hardie (1978) and Harvie *et al.* (1982). Large-scale aspects are covered by Borchert & Muir (1964) and Tucker & Cann (1986). The story of the desiccated Miocene Mediterranean (see also Chapter 20) is told incomparably by Hsü (1972). Sr-ratio studies useful in eliminating groundwater influences during precipitation of the Castil evaporites, New Mexico are in Kirkland *et al.* (2000). Spectacular oriented gypsum crystals recording brine flow are investigated by Babel & Becker (2006).

Aspects of marine siliceous deposition are dealt with by Drever *et al.* (1988) and Bidle & Azam (1999).

Microbial Fe precipitation is discussed by Berner (1970), Lovely *et al.* (1987), Bazylinski *et al.* (1993), Widdel *et al.* (1993), Brown *et al.* (1997), Mortimer & Coleman (1997) and Konhauser (2007). Modern goethite ooids were discovered by Heikoop *et al.* (1996), see also Sturesson *et al.* (2000). A recent paper on goethite ooids is by Collin *et al.* (2005). The problem of chamosite ooids is discussed in a Curtis & Spears' classic (1968) and by Talbot (1973), Kimberley (1979), Bradshaw *et al.* (1980), Taylor & Curtis (1995), Macquaker *et al.* (1996), Donaldson *et al.* (1999), Taylor *et al.* (2002). Glaucony by Odin & Matter (1981), Odin & Dodson (1982), Kelly *et al.* (2001) and Hesselbo & Huggett (2001). Phosphates by McArthur *et al.* (1986), Glenn *et al.* (1994), Dickinson & Wallace (2009) and Follmi *et al.* (2008).

Illuminating recent papers on the origin and palaeoclimatic significance of desert varnish are by Garvie *et al.* (2008) and Liu & Broeker (2008).

Stromatolites are revealed in lovely recent papers by Reid *et al.* (2000, 2003) and Allwood *et al.* 2006. Archaean microbial mats in Noffke *et al.* 2006. A revealing account of various microbial signatures found as films and mats in siliciclastic sediment is by Gerdes *et al.* (2000) and on microbials generally by Riding (2000).

# Chapter 3

# SEDIMENT GRAIN PROPERTIES

*I would have chosen to feel myself rough and elemental*
*like the pebbles that you roll,*
*gnawed through and through by the salt;*
*splinter beyond time, a witness to*
*a cold will that does not fail.*

Eugenio Montale, 'I would have chosen', Transl. George Kay, from *Mediterranean* (1924). Penguin Books, 1969.

## 3.1 General

Sediment has *size*, *shape* and *density* as fundamental granular properties. Grains vary in maximum linear dimensions from minute ($10^{-3}$ mm) specks of wind-blown dust to gigantic (>10 m) boulders, in shape from perfect rhombohedral calcite crystals to near-spherical '*millet-seed*' grains of desert sand, in density from porous coral or algal fragments (ca 1500 kg/m$^3$) to dense (3500 kg/m$^3$) ultrabasic igneous rock. Sediment also has properties in the bulk, those of a pile of sand or a slurry of mixed muddy sand with pebbles and boulders. In these cases we must refer to terms like sediment *packing*, aggregate *porosity* and *permeability*. These define *grain aggregate* properties and yield plenty of information on the erosional and depositional history of sediment.

## 3.2 Grain size

Size is measured by a linear dimension (μm, mm, cm, m). No further discussion would be necessary if all grains were spherical in shape, then grain radius or diameter would be universal and sufficient. To take grain shape into account, one useful parameter is *volume diameter*; the diameter of a sphere having the same volume as the grain in question. Another is to define maximum, intermediate and minimum grain diameters and to consistently use one of them in routine analysis. In this account, grains will be treated mostly as approximate spheres (spheroids), a common shape determined by the abrasion of silicate grains during transport, but the reader should remember that many natural grains, e.g. shell fragments, defy such rational description. A widely adopted scale of grain size suitable for most spheroidal clastic and some calcium carbonate grains is the Udden–Wentworth scale (Table 3.1) in which the various grades are separated by successively halving or doubling about a grain-size centre of 1 mm.

Grain volume and mass vary as the cube of the radius for spheroids. A 10-mm diameter spheroid is five times larger than a 2-mm diameter spheroid, but

*Sedimentology and Sedimentary Basins: From Turbulence to Tectonics*, 2nd edition. © Mike Leeder.
Published 2011 by Blackwell Publishing Ltd.

**Table 3.1** Summary of the Udden–Wentworth size classification for sediment grains, a grade scale in almost universal use amongst sedimentologists.

| | *US Standard sieve mesh* | *Millimeters* | | *Phi (ϕ) units* | *Wentworth size class* |
|---|---|---|---|---|---|
| GRAVEL | *Use wire* | 4096 | | -12 | |
| | *squres* | 1024 | | -10 | boulder |
| | | 256 | 256 | -8 | |
| | | 64 | 64 | -6 | cobble |
| | | 16 | | -4 | pebble |
| | 5 | 4 | 4 | -2 | |
| | 6 | 3.36 | | -1.75 | |
| | 7 | 2.83 | | -1.5 | granule |
| | 8 | 2.38 | | -1.25 | |
| | 10 | 2.00 | 2 | -1.0 | |
| SAND | 12 | 1.68 | | -0.75 | |
| | 14 | 1.41 | | -0.5 | very coarse sand |
| | 16 | 1.19 | | -0.25 | |
| | 18 | 1.00 | 1 | 0.0 | |
| | 20 | 0.84 | | 0.25 | |
| | 25 | 0.71 | | 0.5 | coarse sand |
| | 30 | 0.59 | | 0.75 | |
| | 35 | 0.50 | 1/2 | 1.0 | |
| | 40 | 0.42 | | 1.25 | |
| | 45 | 0.35 | | 1.5 | medium sand |
| | 50 | 0.30 | | 1.75 | |
| | 60 | 0.25 | 1/4 | 2.0 | |
| | 70 | 0.210 | | 2.25 | |
| | 80 | 0.177 | | 2.5 | fine sand |
| | 100 | 0.149 | | 2.75 | |
| | 120 | 0.125 | 1/8 | 3.0 | |
| | 140 | 0.105 | | 3.25 | |
| | 170 | 0.088 | | 3.5 | very fine sand |
| | 200 | 0.074 | | 3.75 | |
| | 230 | 0.0625 | 1/16 | 4.0 | |
| SILT | 270 | 0.053 | | 4.25 | |
| | 325 | 0.044 | | 4.5 | coarse silt |
| | | 0.037 | | 4.75 | |
| | | 0.031 | 1/32 | 5.0 | |
| | | 0.0156 | 1/64 | 6.0 | medium silt |
| | *Use* | 0.0078 | 1/128 | 7.0 | fine silt |
| | *pipette* | 0.0039 | 1/256 | 8.0 | very fine silt |
| CLAY | *or* | 0.0020 | | 9.0 | |
| | *hydro-* | 0.00098 | | 10.0 | clay |
| | *meter* | 0.00049 | | 11.0 | |
| | | 0.00024 | | 12.0 | |
| | | 0.00012 | | 13.0 | |
| | | 0.00006 | | 14.0 | |

$5^3/1^3$ (= 125) times 'larger' in terms of volume or mass of the same material. This is important in the analysis of sediment grain transport by fluid flow, since the grain mass represents the resistance to movement (gravitational *inertial mass*) that must be overcome by moving fluid before grain transport occurs. It also controls the rate of descent by gravitational acceleration of grains settling through fluids.

A number of alternative methods for sizing grains exist, the choice being largely dependant upon the physical state of the sedimentary deposit and the sizes of the grains involved. In the field, use is made of a hand lens and grain-size comparators, these being selected grains or their images within the subdivisions of the Udden–Wentworth scale. Coarser clasts in indurated rock are usually measured by apparent long axes. In the field or laboratory, loose sands and gravels lend themselves to sieving whilst axes of larger loose clasts must be laboriously measured by hand with calipers. In the laboratory unconsolidated silt to sand grades are measured extremely accurately in various automated counter devices in still or falling grain suspensions; by lasers (*Malvern*, *Cilas*, *Coulter* and *Sympatek*), X-ray scans (*Sedigraph*, *Quantachrome*) or electrical-sensing (*Coulter*, *Elzone*). In solid sedimentary rock, sand- and silt-grade grains are measured from thin-sections and the measurements 'corrected' from standard charts so that comparisons can be made with sieved samples. Recent developments include rapid particle recognition analysis in thin rock sections using digital image scanners and autocorrelation algorithms.

## 3.3 Grain-size distributions

Every sediment sample shows a range of grain size (see the example of a sieved desert sand in Table 3.2). Such variation must be characterized statistically so that samples may be compared and interpreted in terms of transport and depositional processes. It is therefore necessary to plot frequency of occurrence of given size fractions in some way.

The simplest plot is that of a histogram, the area of each vertical bar representing the weight per cent of grains present in a given grain diameter interval (Fig. 3.1a). The simple histogram has many advantages in terms of ease of interpretation and perusal of the whole distribution at a glance. It has the disadvantage that it implies discontinuous variation at the breaks between classes. Clearly, the smooth frequency curve is more suitable (Fig. 3.1b & c). Errors arise, however, when reading from the low gradients of the curve at either end of the distribution. It is also necessary to overcome the problem of the wide range of grain sizes that may occur in a sample, frequently >2 orders of magnitude. This is easiest to deal with by plotting data on logarithmic scales (Fig. 3.1d). If a sediment sample is normally distributed then when the distribution is plotted on a log-normal graph it should show a parabolic shape. Overall, many sediment samples only approximate to normal distributions, many are skewed (see below) or have prominent fine or coarse 'tails'; these show a hyperbolic shape on log-normal graphs (Fig. 3.1d).

A cumbersome alternative to logarithmic plots, much in vogue in the last century, is transformation of the mm unit of measurement into logarithmic form as the pseudo-unit known as *phi* ($\phi$). This uses the fact that the subdivisions of the Udden–Wentworth scale define a geometric series, e.g. 8, 4, 2, 1, 0.5, 0.25 mm, etc. This may be manipulated by taking logarithms to the base two, the series becoming $2^3$, $2^2$, $2^1$, $2^0$, $2^{-1}$, $2^{-2}$, etc. $\phi$ is then defined as $-\log_2$ mm; the negative sign avoids assignation of sand grades to negative values. The grain sizes listed then become $-3, -2, -1, 0, 1, 2\phi$, etc. (Table 3.1). Although useful for plotting purposes, the scale is scarily unfamiliar to beginning students for field or laboratory usage, for direct measurement or use in casual conversation.

**Table 3.2** Details of a sieve analysis of a medium-grained, well-sorted, positive-skewed desert sand (data of Bagnold, 1954b). This analysis, from a pioneering and classic account of desert sedimentation, is used as the basis for the various graphical presentations of size distributions shown in Fig. 3.1.

| Sieve mesh (UK) | Aperture (mm) | $\phi$ | Wt (%) | Cumulative wt (%) |
|---|---|---|---|---|
| 12 | 1.58 | −0.65 | 0.005 | 0.005 |
| 16 | 1.17 | −0.23 | 0.043 | 0.048 |
| 20 | 0.915 | +0.13 | 0.338 | 0.386 |
| 24 | 0.755 | +0.40 | 1.855 | 2.241 |
| 30 | 0.592 | +0.75 | 14.120 | 16.361 |
| 40 | 0.414 | +1.13 | 51.776 | 68.137 |
| 50 | 0.318 | +1.65 | 20.300 | 88.437 |
| 60 | 0.261 | +1.92 | 6.080 | 94.517 |
| 80 | 0.191 | +2.40 | 3.860 | 98.377 |
| 100 | 0.114 | +3.13 | 1.105 | 99.482 |
| 150 | 0.099 | +3.33 | 0.404 | 99.886 |
| 200 | 0.073 | +3.79 | 0.082 | 99.968 |
| 300 | 0.054 | +4.21 | 0.024 | 99.992 |

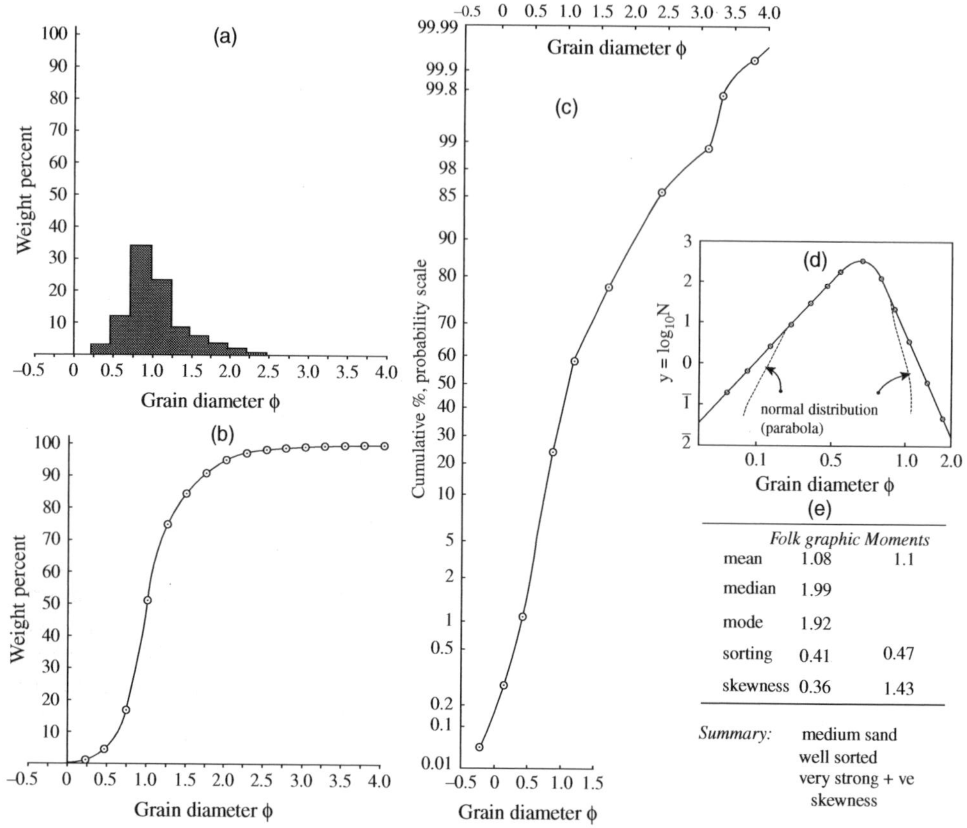


**Fig. 3.1** Some examples of graphical representation of the grain size data given in Table 3.2. (Partly after Bagnold, 1954b.) (a) Histogram. (b) Cumulative plot on arithmetic ordinate. (c) Cumulative plot on probability ordinate. (d) Logarithmic ordinate scale where $N$ = % weight/diameter interval for the sand retained between successive sieves. Note that both (c) and (d) disprove the notion that grain-size distributions follow the normal distribution curve. Indeed the curve of (d) is hyperbolic rather than parabolic (see Bagnold & Barndorff-Nielsen, 1980.) (e) A comparison of Folk graphic grain population statistics with those derived by moment analysis. Note how the graphical method considerably underestimates skewness.

It is absolutely necessary to characterize whole grain populations in terms of the standard common statistical quantities (formulae are given in Maths Appendix) These are:

- *mode*—value of the most commonly occurring particle size, corresponding to the highest point of a frequency curve or the steepest part of a cumulative frequency curve;
- *median*—divides the normal frequency curve into two equal parts and corresponds to the 50% mark on the frequency curve;
- *mean*—symbol $\mu$ is the sum of measurements divided by their number. It is a much superior estimator of the whole distribution than either median or mode. Symmetrical frequency curves have a unique average value that is mode, median and mean;
- *standard deviation*—symbol $\sigma$ (called *sorting* in sediment studies) is the square root of sample variance. $1\sigma$ about the mean includes 68.3% of a normal distribution, $2\sigma$ includes 95.5% and $3\sigma$ includes 99.7%. The larger the sample spread about the mean the larger is $\sigma$ and the worse sorting is, and vice versa;
- *skewness*—a measure of the asymmetry of any frequency distribution. Symmetrical distributions have zero skewness, whilst samples with an excess of fine grains have positive skewness and those with excess coarse grains have negative skewness (Figure 3.2).

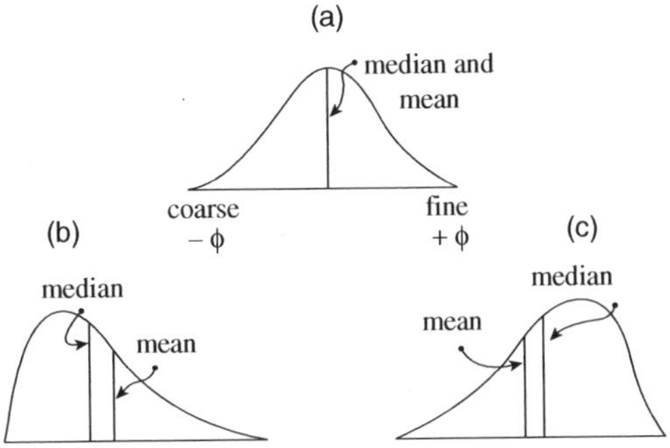


**Fig. 3.2** To illustrate skewness of distributions: (a) symmetrical distribution, (b) positive skewness and (c) negative skewness.

## 3.4 Grain shape and form

Two aspects of grain shape are frequently confused. These are *roundness* and *sphericity* (Fig. 3.3).

Roundness is an estimate of the smoothness of the surface of a grain. It is thus possible to have a well-rounded grain of rod-like shape. Roundness may be quantified as the ratio of the mean radius of curvature of the grain corners to the radius of the largest inscribed circle. This ratio is rather time consuming to measure for a whole population of grains. It is normal practice to estimate mean roundness using a standard set of grain images.

Sphericity is a measure of how closely a grain approximates to a sphere. Various formulations of this have been proposed but the best is the maximum projection sphericity, $\psi_p = (s^2/li)^{1/3}$, where $l$, $i$, and $s$ are the long, intermediate and short axes of a non-spherical grain respectively. The formula takes into account the settling behaviour of a grain in a fluid since it compares the grains maximum projection area (the denominator) to that of the projection area of a sphere of equal volume. In thin-section, sphericity is assessed as *elongation*, being the ratio of width to length of a grain.

Measurements of short ($s$), intermediate ($i$) and long ($l$) orthogonal axes of pebble-sized grains enable them to be classified as one of four end-members; blades ($l > i \gg s$), rods ($l > i \approx s$), spheroids ($l \approx i \approx s$) and discs ($l \approx i \gg s$). Each has distinctive $s/i$ and $i/l$ *Zingg ratios*. Generally during river transport, homogeneous original rocks and minerals without cleavage tend to evolve toward ellipsoidal shapes. However, the process of beach attrition involves the hurling by waves of sand and larger clasts onto the beachface during storms. This seems to encourage the formation of discoidal forms from initially homogeneous rock. However, the shape of pebble-sized clasts can equally owe much to any original fabric present in the rock or mineral, e.g. cleaved, slatey or schistose rocks tend to always form blades or discs.

## 3.5 Bulk properties of grain aggregates

The accumulation of grains in a deposit leads to development of a *packing structure* which determines many bulk grain properties. The fractional volume concentration, $C$, of grains within a freshly deposited

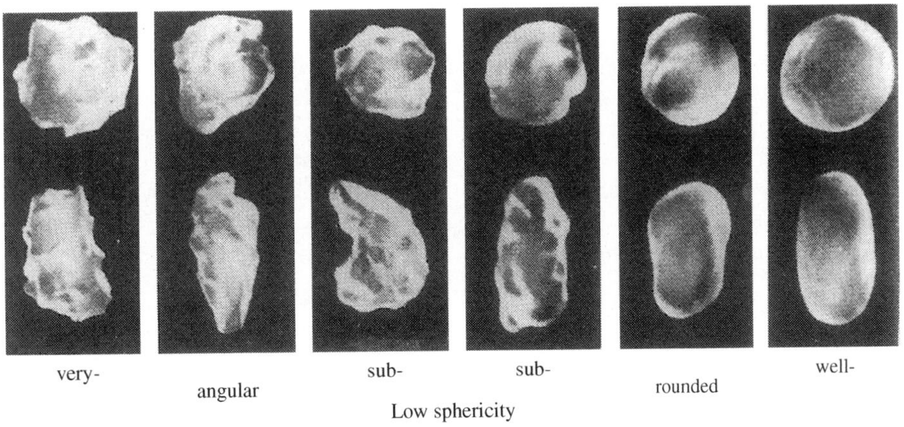


**Fig. 3.3** Photographic images of grains for the visual determination of grain roundness and sphericity.

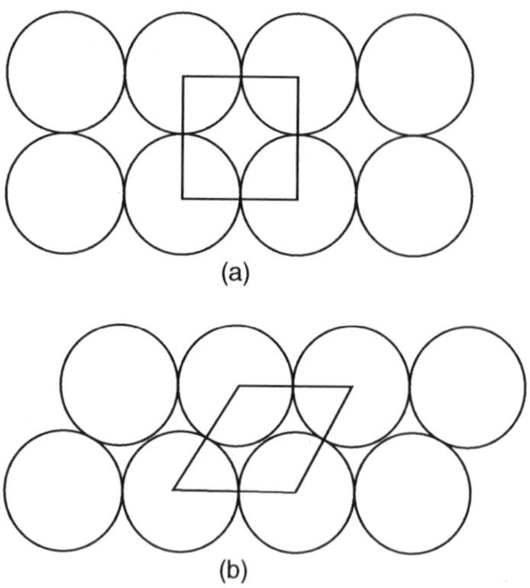


**Fig. 3.4** Vertical sections through piles of spheres to show (a) cubic and (b) rhombohedral packing modes. The two modes give rise to theoretical maximum and minimum values of porosity respectively.

aggregate is the ratio of grain-occupied space to whole unit space. The amount of pore space, *porosity P*, or 'grain-unoccupied' space is then $P = 1 - C$.

Various *packing* modes for grain aggregates exist. Maximum and minimum states for spheroids are *cubic* and *rhombohedral* (Fig. 3.4), with porosities of 48% and 26%. Natural loose granular aggregates of spheroids have porosity values intermediate between these. Note that in equigranular aggregates, porosity is independent of grain size. Grain shape is probably the greatest control on porosity, witness the huge porosities of irregular-shaped aggregates of mollusc fragments, clay minerals or cornflakes. Grain sorting is also a fundamental control, since in poorly sorted aggregates the finer grains 'clog-up' the pores provided by larger grains.

*Deposition rate* affects packing; high rates cause random packing with voids and local cubic packing at the rapidly upward-moving sediment depositional interface. Such a process occurs on grain avalanche slopes during grain flow (Chapter 9), making the surfaces highly unstable should seismic or other disturbances to the grain fabric occur. By way of contrast, high fall velocities cause energetic grain–bed grain collisions and jostling during deposition, tending to eliminate local voids and cubic packing domains (Fig. 3.5).

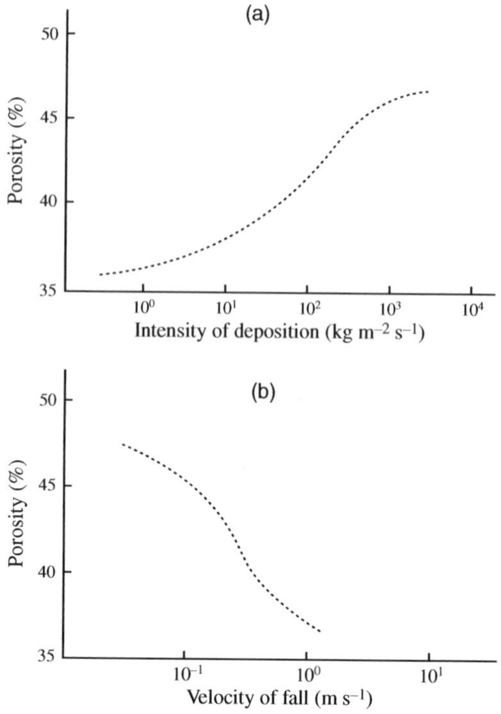


**Fig. 3.5** The porosity of sand aggregates as a function of (a) deposition rate and (b) the fall velocity of individual particles. (Data of Gray, 1968.)

*Permeability* is the rate at which pore fluid can move through the pore system. There may be a relationship between permeability and porosity, or there may be none at all. An analogy is a house, where individual rooms can be compared to pore space yet connections between them depend on the size of doors and corridors which control the flow rate of circulating air. These concepts are relevant to subsurface flow of fluids and gases in water, oil and gas reservoirs. A further control in these situations is that of primary pore elimination by secondary processes such as compaction and mineral precipitation in pore spaces during diagenesis.

## Further reading

A useful book on particle size analysis is by Syvitsky (1991). Precise modern automated measurement of silt-sized particles and its pitfalls are discussed by Bianchi *et al.* (1999) and McCave *et al.* (2006). Rapid particle analysis of digital images by high resolution scanning of thin sections is by Seelos & Sirocko (2005) and autocorrelation algorithms by Rubin (2004). A review of grain-size distributions in terms of mixed elementary populations is by Weltje & Prins (2007). Information on log-hyperbolic grain-size distributions (LHD) is in Bagnold & Barndorff-Nielsen (1980) and Christiansen *et al.* (1984). Accessible applications of LHD to environmental discrimination are provided by Vincent (1996) and Knight *et al.* (2002) and to palaeohydraulics by McCave *et al.* (1995, 2006) and Hall *et al.* (1998, 2001). A hydraulic interpretation of grain-size distributions is by Bridge (1981). Severe warnings on the sole use of grain-size distributions in sediment trend analysis to detail sediment-transport pathways are by Flemming (2007) and Masselink *et al.* (2008). A review of particle shape and some new approaches to this old subject is by Blott & Pye (2008). A nice little note on pebble shape on beaches, the disc-problem, is by Lorang & Komar (1990).

# Part 2

# MOVING FLUID

*So that water might be*
*elephantine and pinpoint,*
*what an industry of air,*
*what transformation of heat.*
*And water goes off–it skulks*
*through the cracks in rocks, jemmying them open…*

Norman MacCaig, 'Water', 1977, *Collected Poems*, Chatto and Windus 1993.

## Introduction

Sedimentological fluid dynamics investigates how fluids transport and deposit sediment over planetary surfaces. It is a branch of geophysical fluid dynamics that arose in the twentieth century from the efforts of a few pioneering mathematical and engineering physicists who were interested in explaining the workings of the natural world. This was amusingly described by J.J. Thomson, discoverer of the electron and isotopes, a student of O. Reynolds whose own efforts in this area we will study closely below, as 'Out-of-doors Physics'. Yet even though the dynamics of pure fluids in the laboratory is not a simple subject, in sedimentology there are added complications. Some examples of natural flows (Part 2 Fig. 1) illustrate this.

- Planetary lower atmospheres are mixture of gases and usually dilute quantities of suspended materials (dust particles and liquid aerosols).
- River water contains suspended solids, sometimes reaching concentrations up to 30% by volume.
- Seawater is an electrolyte with ~3% by weight of dissolved salts; it also carries charged suspensions of particulate organic matter rich in proteins and other long-chain molecules that behave as flexible polymers.
- The meeting of freshwater and seawater in estuary channels leads electrochemical processes to cause the bringing-together of dispersed clay particles into larger clumps, a process known as *flocculation*.
- Turbidity currents are subaqueous turbulent suspensions that are in motion down gravity slopes because of their excess density over ambient freshwater or seawater.
- Sediment avalanches in air or underwater are concentrated aggregates of solids whose flow behaviour is governed by frequent granular collisions; they are termed *granular fluids*.
- Ice is a solid phase yet transports rock fragments; it breaks along crevasses at the glacier surface and flows at depth.

In most fluid cases, the presence of grains increases bulk density and viscosity, both properties that help control the flow characteristics of fluids. But in addition the grains also affect pressure gradients within the carrier fluid and consequently influence the rate of change of momentum. It is the play-off between the effects of grains on mixture density, viscosity, momentum and pressure that lead to the fascinating complexities of sedimentary fluid dynamics.

Since sediment is transported by both air and water it is useful to contrast the characteristics of the two fluids. The obvious physical properties of air are low density, absence of rigidity and very high

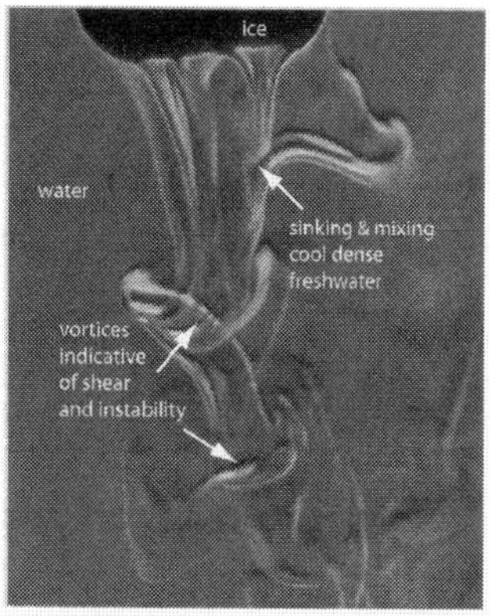


**A.** Water circulation in the oceans is aided by density contrasts due to temperature and salinity variations, illustrated here by the downflow of dense water from melting ice.

**B.** Dust storm front with typical overhanging head composed of lobes and clefts. Prowers Co, Colorado 1937.

**C.** Flash flood showing antidune waves typical of supercritical flow. Arizona, August 1982. Flow top to bottom.

**D.** Pyroclastic flow descending Montserrat volcano, West Indies. Flow moving to left, note various scales of mixing eddies on upper surface shear layer with the atmosphere.

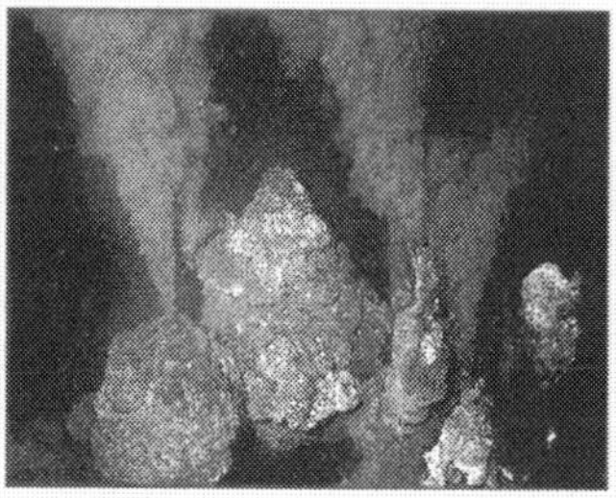

**E.** Black smokers venting metal sulphide at Monolith vent site on the Juan de Fuca ocean ridge.

**F.** Turbulent eddies (at scale ca 10 m here) give unpredictable consequences for velocity or stress distributions.

**G.** W.S. Chepil made quantitative measurements of lift and drag. Here he is pictured adjusting the test section of his wind tunnel in the 1950s. Much research into wind blown transport in the USA was stimulated by the Mid-West 'dust bowl' experiences of the 1930s.

**H.** R.A. Bagnold is pictured here (left) in the 1930 s as a pioneer desert explorer soldier on leave and amateur scientist. On the right he is vigourous and active in his 80 s, widely recognized as a pioneer in the field of physical sediment transport.

**Part 2 Fig. 1** A rather random selection of photographs to illustrate various sorts of fluid flows with sedimentological interest and of two pioneering physical sediment transport scientists.

compressibility. These reflect the low concentration of individual molecules in constant random high-speed motion. The 'spaciousness' of gases explains their low density and high compressibility. Liquids on the other hand are typically [O]$10^3$ denser than gases, with low compressibility and no rigidity. Liquid molecules are relatively close together, though the pattern is generally disordered and unsteady, with the individual molecules still moving randomly at the same high speeds as in gases. The closeness of neighbouring molecules explains the increased density and the difficulty of compression. Liquids and gases share the characteristic that they cannot permanently withstand the action of shearing forces. Thus water in a container will respond immediately to the slightest tilt or bump. Fluid deformation is quite unlike the recoverable deformation of an elastic solid as typified by the action of an elastic spring; but, as we shall see, some substances may overlap in their behaviour between liquid and solid.

A *Newtonian fluid* is defined as a substance that deforms immediately and then continuously when acted upon by a shearing force; the original shape cannot be recovered. *Non-Newtonian fluids* are deviant in the sense that they disobey the rules implicit in this definition and require an initial applied force before they will deform and the subsequent deformation rate may vary—many important examples occur in sedimentology, some of which are highlighted in Table 4.1.

## Chapter 4

# FLUID BASICS

*In the cabin Stephen said, 'Jack, I fear I have been so indiscreet as to ask Mr Wright to dine aboard without consulting you. I particularly wish to hear his view on the action of water flowing the whole length of the horn you so very kindly gave me long ago, upon the nature of turbulence set up by the whorls or convolutions, and upon the effect of the more delicate ascending spirals.'*
*'Not at all, not at all,' said Jack. 'I should very much like to hear him: no man more. Although I have been waterborne most of my days, I am sadly ignorant of hydrostatics except in a pragmatic, rule-of-thumb kind of fashion...*

Patrick O'Brian, from 'The Hundred Days', Harper Collins, 1998 (Stephen Maturin & Jack Aubrey on turbulence along a narwhals horn).

## 4.1 Material properties of fluids

Material in this sense means the *bulk* properties associated with the substance in question; we generally ignore the details of molecular-scale interactions in sedimentological fluid dynamics.

### Density

Density ($\rho$) is mass ($m$) per unit volume with dimensions $ML^{-3}$, in SI units $kg/m^3$. It has an important role in fluid dynamics, controlling dynamic quantities such as the *vectors* fluid momentum and force (momentum flux), and the *scalars* fluid pressures and immersed weight. Water is very much more dense than air and thus for a given velocity ($\mu$) the momentum ($mu$) of unit volume of water is far greater. Similarly the density ratio between the two fluids and sediment grains varies considerably, a feature that has fundamental implications for the buoyant forces exerted by the two fluids on natural mineral grains.

Density is a function of both temperature (Fig. 4.1) and pressure in pure fluids. Under earth-surface conditions, water and air masses are obviously most influenced by temperature changes, the generally negative relationship between density and temperature in both fluids being responsible for the important class of thermohaline currents that dominate deep oceanic circulation (Chapter 21). There are also important effects due to the effects of pressure on the anomalous expansion of water below about 4 °C; this decreases the temperature of maximum density by about 0.021 °C/bar. Thus, whilst a water mass near 4 °C at the surface is at maximum density, it is less dense than slightly colder waters at depth due to this *thermobaric* effect.

Seawater is rich in dissolved ionic species (Chapter 2) and these serve to increase its density over that of freshwater (Fig. 4.2). In coastal mixing zones there is commonly a lateral gradient of density due to mixing of freshwater and seawater masses. These mixing and density gradients set up important buoyancy affects

*Sedimentology and Sedimentary Basins: From Turbulence to Tectonics*, 2nd edition. © Mike Leeder.
Published 2011 by Blackwell Publishing Ltd.

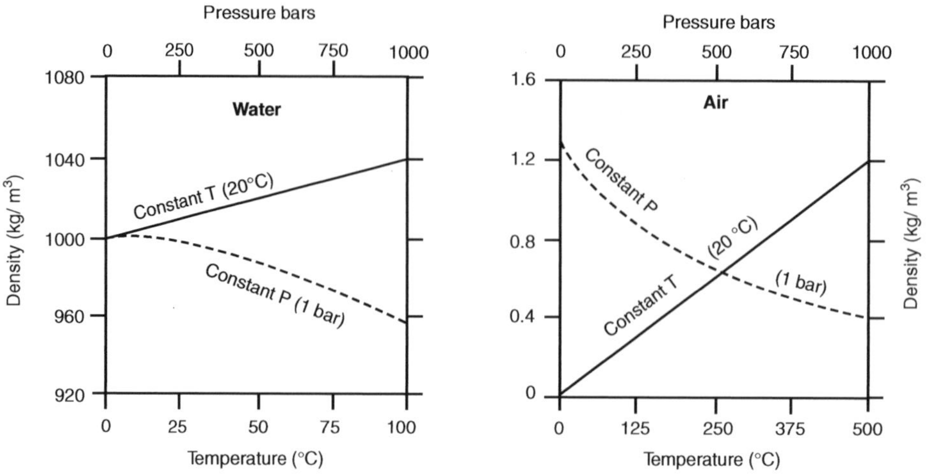


**Fig. 4.1** The variation of water and air density with temperature and pressure. (After Vardy, 1990.)

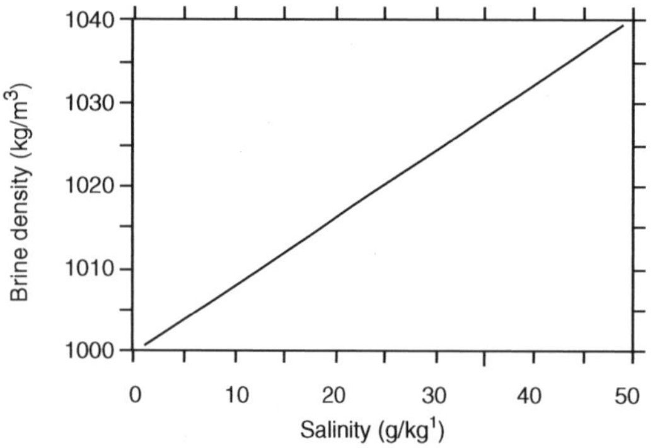


**Fig. 4.2** The variation of brine density with salinity. (After Denny, 1993.)

which control local fluid motions. Transported solid mineral grains give an increased effective bulk density ($\rho_b$) to any fluid volume, $\rho_b = (1 - c)\rho + c\sigma$, where $c$ is the fractional concentration of solids and $\sigma$ is the solid density (Fig. 4.3).

## Viscosity

Molecular or dynamic viscosity ($\mu$) controls the ease of deformation of a fluid. Its effects are illustrated by comparing the force needed to stir fluid substances. It is a struggle to stir treacle just taken from the refrigerator, but as the temperature gradually increases, the stirring becomes easier (some of the temperature increase is due to energy used in shearing the fluid) and requires less work. By comparison, it is always easy to stir water. Newton himself called viscosity (the term is a more modern one, due to Stokes) *defectus lubricitatis* or, in colloquial translation, 'lack of slipperiness'. Viscosity thus acts as a kind of frictional brake that controls the rate of deformation by an applied shearing stress. To set up and maintain relative motion between fluid layers or between fluid and solid layers requires work to be done against the viscous force of resistance.

The dimensions of molecular viscosity are $ML^{-1}T^{-1}$, in SI units kg/m s or N s/m$^2$ (= Pa/s). The role of viscosity is best illustrated by Newton's relationship, $\tau = \mu du/dy$, where $\tau$ is the shearing stress (i.e. the force that you apply whilst stirring), $\mu$ is the molecular viscosity and $du/dy$ is the velocity gradient (or strain rate), i.e. the relative movement between layers that are unit distance apart. Molecular viscosity is thus the proportionality factor that links shear stress to the rate of strain (Fig. 4.4). Note that in a fluid it is the *rate* of deformation, not the actual deformation, that provides the criterion for the stress equilibrium. A succinct definition of viscosity would be: 'the force needed to maintain unit velocity difference between unit areas unit distance apart'. Viscosity is sometimes quoted in units of *poises* (named in honour of Poiseuille who did pioneer work on viscous flow): these are $10^{-1}$ Pa/s. Remember that viscosity is a scalar quantity, i.e. it possesses magnitude but not direction.

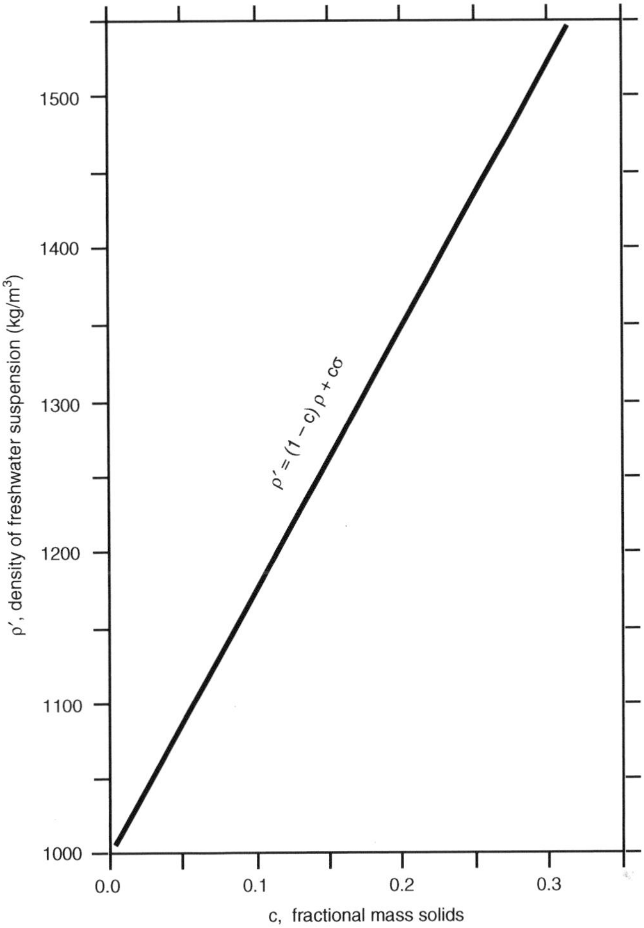


**Fig. 4.3** The variation of freshwater density with volume fraction of added solids of density 2750 kg/m$^3$.

The causes of viscous forces are complex. In gases with a positive velocity gradient flowing past a surface, molecules continuously diffuse across the flow and effect a net transfer of drift momentum from fast- to slow-moving layers and ultimately to the solid surface. By Newton's second law (**Cookie 16**) the rate of destruction of momentum is a force, in this case called viscous *shear* or *drag*. In liquids, in addition to molecular diffusion, there are substantial cohesive attractive forces due to the phenomenon of hydrogen bonding discussed in Chapter 1. Viscosity is a sensitive function of temperature (Fig. 4.5); in gases, viscosity increases with temperature because more kinetic diffusion occurs. By way of contrast, water viscosity decreases with increasing temperature because cohesive molecular forces due to hydrogen bonding decrease.

The reader should be aware that another expression for viscosity is in common use. This is *kinematic viscosity*, $\nu$ (Fig. 4.6), derived by dividing molecular viscosity by density, i.e. $\nu = \mu/\rho$. Its dimensions are $L^2T^{-1}$, in SI units $m^2/s$. The quantity is said to be *kinematic* because only length and time units are involved; a dynamic quantity must include mass. The property is a useful one because it expresses the ratio between a fluid's ability to resist deformation with a measure of its resistance to acceleration.

As with density, viscosity is affected by admixed sediment grains (Fig. 4.7) According to the Einstein–Roscoe equation, $\mu_m = (1 - 3.5c) - 2.5\,\mu$, where $\mu_m$ is

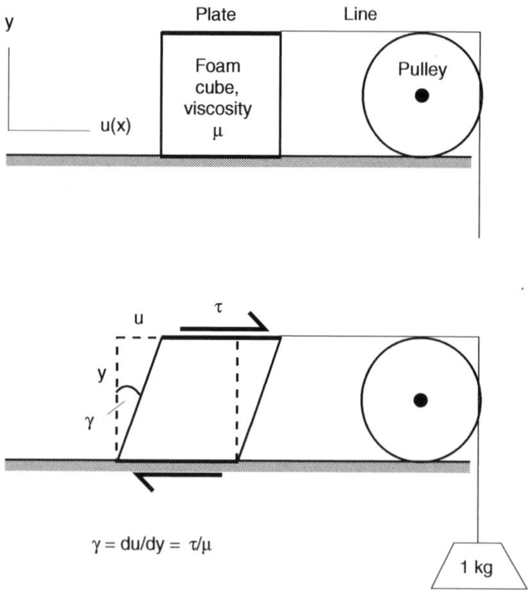


**Fig. 4.4** Analogue model for the action of fluid viscosity in resisting an applied force. The force is exerted on the top unit area of the foam cube. In continuous fluid deformation, as distinct from the equilibrium displacement shown here, the displacement in *x* is the velocity, *u* (as shown).

the apparent viscosity of the fluid–solid mixture and *c* is the fractional concentration by volume. The basic point is that each sediment grain represents a solid surface in a fluid and is thus a potential slip plane that increases the fluids internal resistance to shear. In order to keep the same overall velocity gradient it is necessary to increase the applied shearing stress. In the homely language of our previous thought experiment, we must stir a treacle-peanut mix more energetically than treacle alone—the more peanuts, the greater the effort required.

As defined above, Newtonian fluids are those where the ratio of force to velocity gradient is finite and linear for all values (Fig. 4.8). Major problems arise in the analysis and characterization of viscosity in non-Newtonian fluids where shear rate controls viscosity. *Pseudoplastic* substances are those where μ decreases as rate of shear increases. *Dilatant* substances are those where μ increases as rate of shear increases. *Thixotropic* substances are those where μ decreases with time as shearing forces are applied. In *rheopectic*

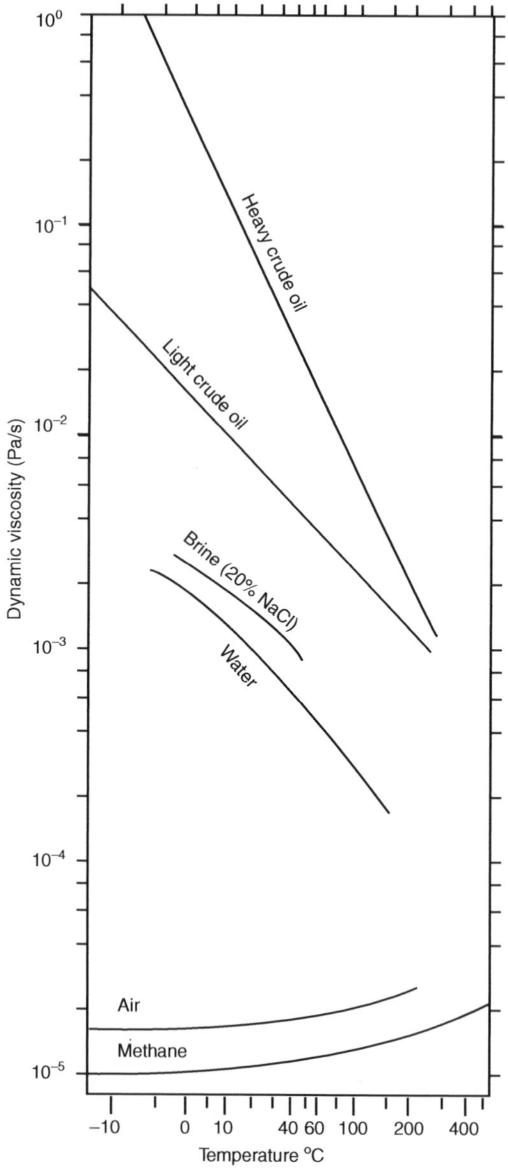


**Fig. 4.5** The variation of dynamic (molecular) viscosity with temperature for some natural fluids.

substances, μ increases with time as shearing forces are applied.

## Plastic behaviour

In contrast to Newtonian fluids, plastic substances show an initial resistance to shear, termed a *yield*

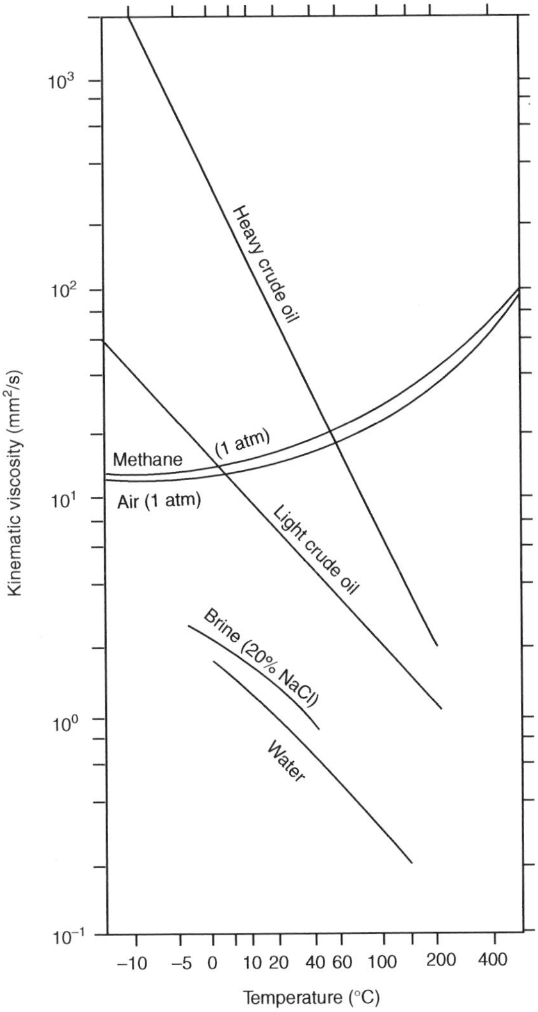


**Fig. 4.6** The variation of kinematic viscosity with temperature for some natural fluids.

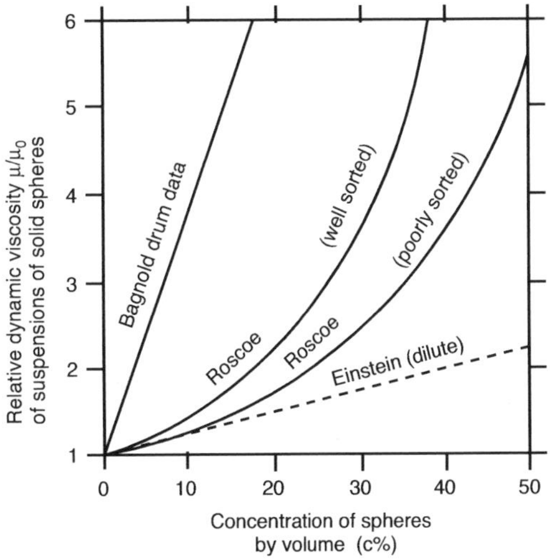


**Fig. 4.7** The positive variation of dynamic viscosity with solids concentration according to Einstein–Roscoe theoretical models and Bagnold (1954a) experimental data on apparent viscosity in sheared suspensions. The vertical axis scale is ratio of fluid viscosity with solids, $\mu$, to pure fluid alone, $\mu_0$. The Roscoe curves are most suited to concentrations observed in most natural flows.

*stress*, followed by subsequent deformation (Fig. 4.8b). *Bingham plastics* show a constant Bingham viscosity whereas non-Bingham plastics show variable viscosity according to shear rate. Many *debris flows* and *lava flows* show Bingham plastic behaviour. The addition of solid contaminants to Newtonian fluids eventually induces plastic behaviour as the limits of concentration (~30% by volume) are approached. The finite yield strength that comes with plastic behaviour allows raised morphological features like levees, flow snouts and flow wrinkles to be preserved during flow and after motion has ceased. In such flows, grain settling is hindered or even impossible. The rate of viscous shear varies across a Bingham flow; marginal parts suffer high strain whilst the flow core moves as unstrained solid *plug flow*.

## 4.2 Fluid kinematics

The study of motion in general is termed *kinematics*. We may directly observe motion of the atmosphere and most of the hydrosphere. Glaciers and ice sheets of the cryosphere move more slowly, as do permafrost upper layers on slopes during summer thaw. Some motions may be regarded as *steady*; they are unchanged over specified time periods. Other motions, as we know from experience of weather or river flood waves, are decidedly *unsteady*, either through gustiness over minutes and seconds or from day to day as weather fronts pass through. How we define unsteadiness at such different timescales is clearly important. For more details about notations for fluid flows, see **Cookie 14**.

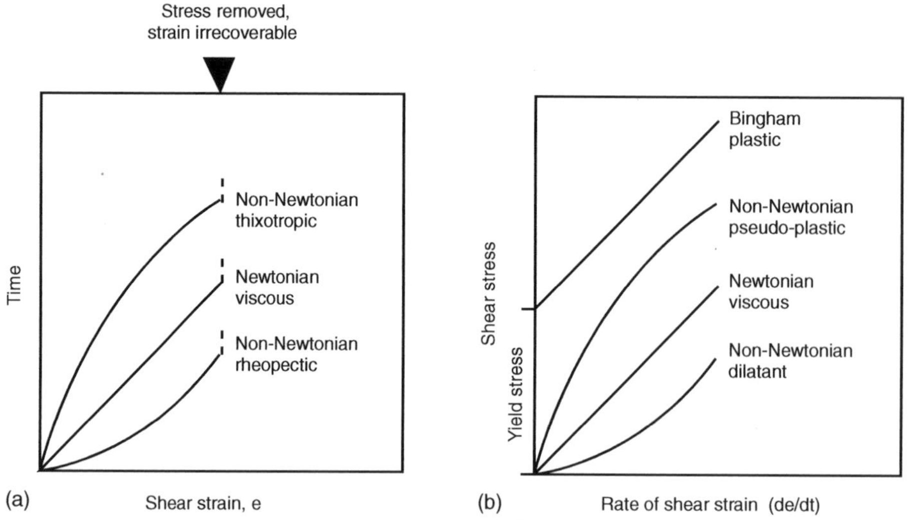


**Fig. 4.8** The behaviour of various types of fluids and plastics as they are strained by an applied force. Natural sedimentological flows exist that show each of these behaviours. (a) Shear strain vs. time. (b) Applied shear stress vs. rate of shear strain. Note that shear strain rate is equivalent to velocity gradient.

## Speed

Faced with the complexity of motion we clearly need a framework and rigorous notation for describing it. The simplest starting point is rate of motion measured as *speed*, defined as increment of distance travelled, $\delta s$, over increment of time, $\delta t$, i.e. $\delta s/\delta t$, length travelled per standard time unit (usually per second; units, L/T) In math terms, speed is a *scalar* quantity, expressing only the magnitude of the motion; it doesn't tell us anything about where a moving object is going.

## Velocity

Many practical analyses of motion need extra information to that provided by speed. Velocity (symbol $\boldsymbol{u}$, units L/T) is the physical quantity of motion we use to express both direction and magnitude of any motion. A quantity such as velocity is known generally as a *vector*; it specifies both distance travelled over unit of time *and* the direction of the movement (Maths Appendix). Vectors will sometimes be written in bold type, like $\boldsymbol{u}$, but you may also see them on the lecture board or other texts and papers underlined, $\underline{u}$, with an arrow, $\vec{u}$ or a circumflex, $\hat{u}$. Any vector may be resolved into three orthogonal (i.e. at 90°) components. We may represent velocity with *vectorial arrows*, the lengths of which are proportional to speed, with the arrow pointing in the direction of movement. With vectorial arrows it is easy to show both time and space variations of velocity, and to calculate the relative velocity of moving objects.

Want to know more about notations for fluid flows? Turn to Cookie 14

## Reference frames

Relative motion is familiar to occupants of stationary vehicles. When a neighbouring vehicle moves forwards or backwards it is initially unclear as to who is moving—it is usually only the absence of an *inertial* reaction (acceleration or deceleration) that can convince us that one is indeed stationary. Now imagine that you are a sedimentologist moving with a sand grain (sitting on it perhaps, like a child's bouncing ball) or riding an advancing turbidity current like a surfer. You would see entirely different patterns of flow in the air or water as they pass by you compared to those observed as fluid passes by a stationary colleague. This excursion into relativity may seem trite, but study of Fig. 4.9 will hopefully soon convince

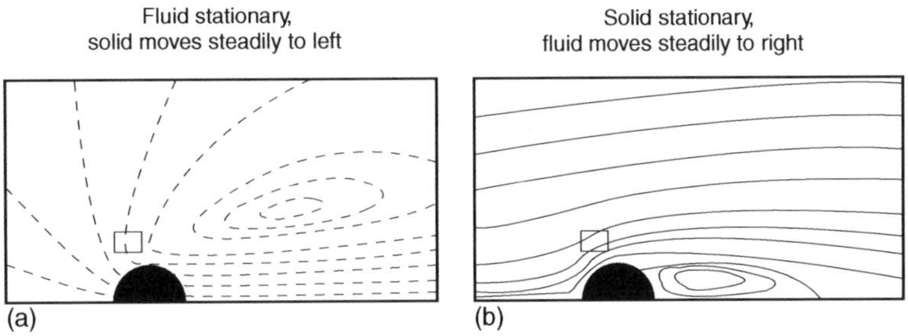


**Fig. 4.9** Sketches to illustrate relative motion of fluid past a solid hemisphere. (After Tritton, 1988.) (a) Case where the fluid is stationary and the solid moves steadily to the left. The flow pattern is 'seen' by an observer moving with the solid in Lagrangian fashion. (b) Case where the solid is stationary and the fluid moves steadily to the right. The flow pattern is that 'seen' by a stationary observer in Eulerian fashion.

the reader that, although the dynamics of the two systems illustrated are similar, the kinematics are not.

Systems where the coordinates $x$, $y$ and $z$ are fixed in space for a stationary observer are known as *Eulerian* (pronounced *oilerian*) systems. All kinematic and dynamic analysis is done with respect to a fixed control volume through which the fluid passes. Velocity measurements at different times are thus gained from different fluid 'particles'. Systems where the reference axes move with the control volume and are seen by a similarly moving observer are known as *Lagrangian* systems. All kinematic and dynamic analysis is thus done with respect to a given, constant mass of fluid.

Most systems benefit from an Eulerian treatment, when the velocity is a function of $x$, $y$, $z$ position in a fluid and time. The mathematics is easier and we consider dynamical results 'at a point', rather than the fate of a single fluid mass as it evolves downstream. However, it must be stressed that sedimentologists cannot ignore the Lagrangian velocity field, as discussed next.

## Steadiness and uniformity

Forces such as those due to viscosity or pressure gradients are set up in flows and these control sediment transport, erosion and deposition. The forces act to change velocity in time and/or space. To understand these forces the twin concepts of *steadiness* and *uniformity* are crucial. A small fluid parcel may be accelerated or decelerated in two ways.

1 A change of velocity may occur at a given point. Imagine you are a stationary observer continuously measuring the velocity, $\boldsymbol{u}$, of river flow from a vantage point on the bank. Should the velocity remain unchanged over the arbitrary measuring interval (within the recording ability of the instrument used), the flow is said to be *steady*, with zero rate of change of velocity with time, i.e. $\delta \boldsymbol{u}/\delta t = 0$. Conversely, when $\delta \boldsymbol{u}/\delta t \neq 0$ the flow will be accelerating or decelerating and is said to be *unsteady* (Fig. 4.10). Thus steadiness is always measured with reference to Eulerian coordinates fixed in the flow boundary at the observation point. Almost all natural flows are, strictly speaking, unsteady, but

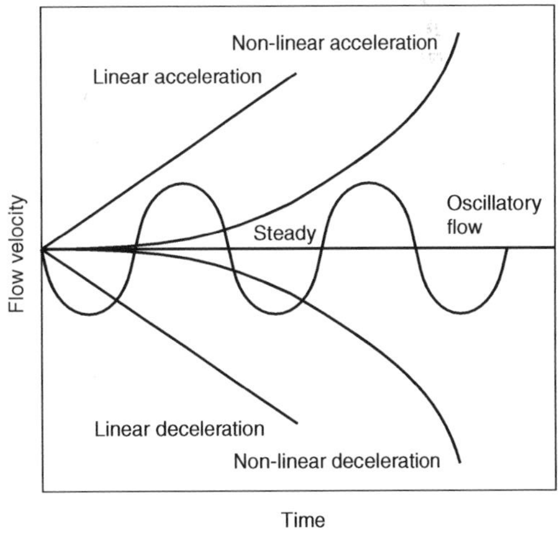


**Fig. 4.10** Graphoid with arbitrary coordinate scales of Eulerian velocity (i.e. measured at a point through time) vs. time to illustrate various types of flow unsteadiness.

this clearly depends upon the time interval over which the flow is being measured. Thus tidal flows are unsteady over periods of minutes or hours whereas most rivers are unsteady over periods in excess of days. Although turbulence (Chapter 5) leads to unsteadiness over periods of seconds, we can speak of a steady turbulent flow if the *mean* velocity remains unchanged over the specified time interval.

2 A change of velocity may occur between given points. If, by use of several flow-monitoring instruments, the observer can make sumultaneous measurements at different places upstream or downstream in a flow and they all indicate the same velocity, then the flow is said to be *uniform* in space, i.e. the rate of change of velocity in space is zero, i.e. $\delta \mathbf{u}/\delta s = 0$ (where $s$ is distance). A change of velocity may occur in space at some point downstream or upstream when $\delta \mathbf{u}/\delta s \neq 0$. This is called *nonuniform* flow. Nonuniform effects commonly arise where constrictions or expansions occur in channels (Fig. 4.11).

In the most general possible case the total velocity change at a point, is $\mathbf{a} = \Delta \mathbf{u}/\Delta t = (u\delta \mathbf{u}/\delta s) + (\delta \mathbf{u}/\delta t)$, where $\boldsymbol{a}$ is acceleration and $s$ is distance. The $\Delta \mathbf{u}/\Delta t$ term is called the substantive (total) rate of change. The first term on the right-hand side (RHS) of the equation is the *advective* acceleration appropriate to nonuniform flows, and records any velocity changes on the way to the observer. The second term is that pertaining at a point to unsteady flows. For uniform, steady flow $\boldsymbol{a} = 0$. An important analysis of turbulent flows, done originally by Reynolds (Chapter 5), makes much use of the total acceleration and,

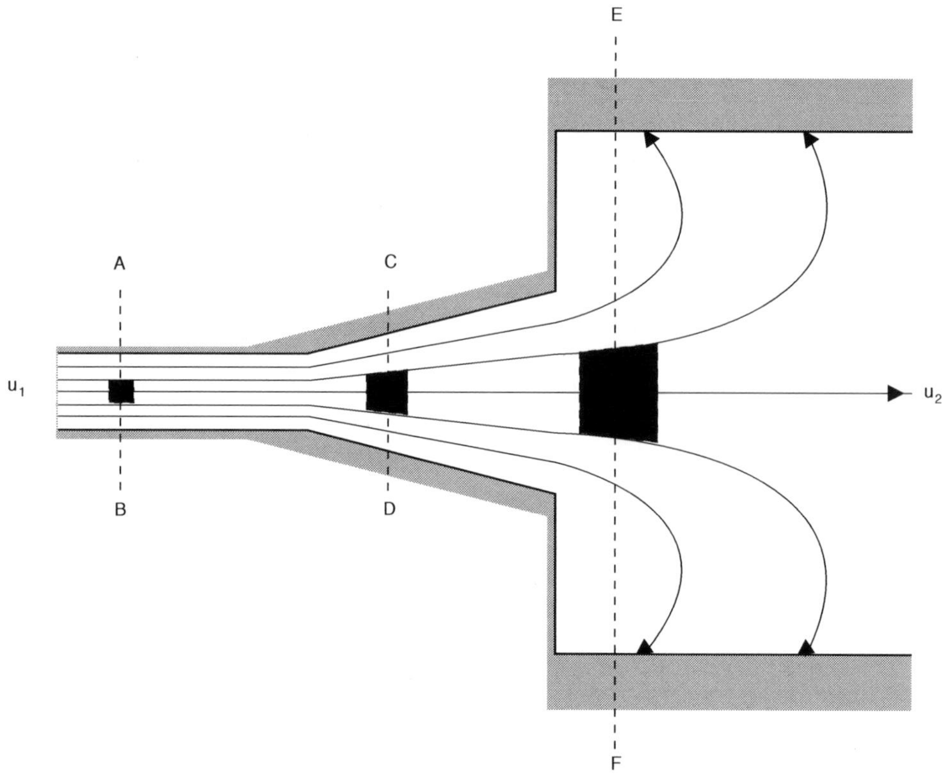


**Fig. 4.11** Sketch of a constant discharge into a straight then flaring channel emptying into a wide reservoir. Lines are streamlines. (a) There is flow uniformity in the parallel-sided section A–B. (b) Flow nonuniformity occurs in the flaring channel section C–D and in the reservoir E–F and where $u_1 > u_2$.

although the expression looks long and cumbersome, it contains a wealth of information about a fluid flow.

## Harmonic motion, angular speed and angular velocity

Harmonic fluid motion deals with the periodic return to a previous level of some surface relative to a fixed point. This is exemplified by water waves, where displacement of the water surface occurs during wave passage (Fig. 4.12). The wave itself has various geometrical terms associated with it, period, $T$, for example, and can be considered mathematically most simply by reference to a sinusoidal curve. Consider curved (rotating) motion in Fig. 4.13; in going from A to B in unit time a particle sweeps out an arc of length $s$, subtending an angle ($\phi$) with the centre of curvature, radius $r$. We can talk about a constant quantity for the travelling particle as $\delta\phi/\delta t$, the *angular speed*, $\omega$, usually measured in radians per second (a radian is defined as $360/2\pi$ degrees). The *linear speed*, $u$, of the rotating particle is the product of angular speed of the particle and its radial distance, $r$, from the centre of curvature, i.e. $u = r\,\omega$.

*Angular velocity* has both magnitude and direction and is thus a vector, denoted $\Omega$ (omega). It has units of radians/s. In order to give angular velocity its vectorial status, the direction is conventionally taken as a normal axis to the plane of the rotating substance, $\Omega$ pointing towards the direction in which a right-handed screw would travel if screwed in by rotating in the same direction as the rotating substance (Fig. 4.13b). For example, in the case of clockwise rotation from *a* to *b* of a fluid eddy in the horizontal xz plane, the axis being in the vertical sense, c, $\Omega$ points downwards and thus has negative sign. Vice-versa for anticlockwise flow. We can denote the position of any rotating particle by means of the position vector, $\boldsymbol{r}$ (Fig. 4.13c). This leads to the important result that the angular velocity vector, $\Omega$, and the linear velocity vector, $\boldsymbol{u}$, of a position vector, $\boldsymbol{r}$, are at right angles to each other (Fig. 4.13). Vector geometry relates the linear velocity vector, $\boldsymbol{u}$, to the *vector product* of the angular velocity vector and the position vector, i.e. $\boldsymbol{u} = \Omega \times \boldsymbol{r}$.

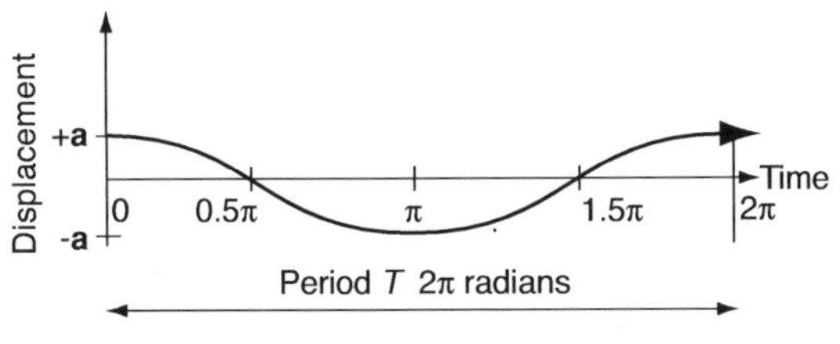


**Fig. 4.12** Harmonic motion. A wave has periodic, often sinusoidal, motion. The example is a curve traced out in time, best imagined as the track to a point on a moving wheel.

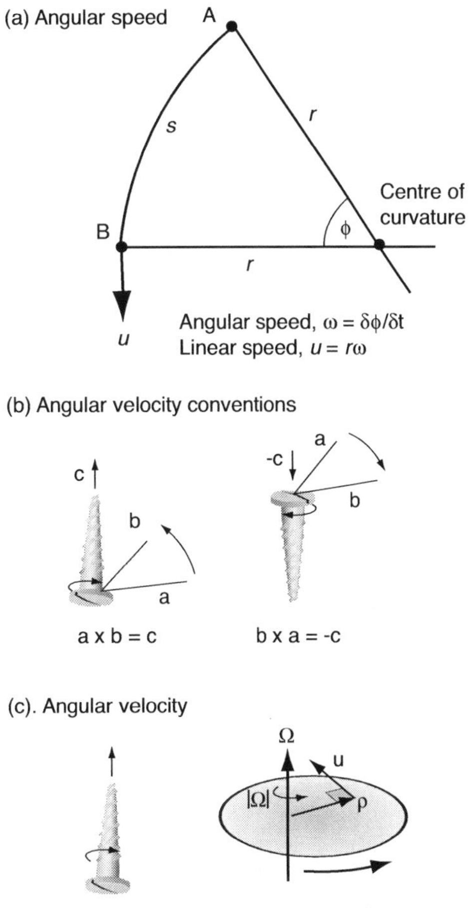


**Fig. 4.13** Definition diagrams for angular motion. See text for explanations.

## Vorticity

Vorticity is related to angular motion and is best envisaged as 'spin', or rotation; it is the tendency for a parcel of fluid or a solid object to rotate. It is sometimes given the symbol, ω, but in oceanographic contexts more usually, ζ, a convention we follow subsequently. Vortical motions occur all around us: the whole solid planet possesses vorticity (appropriately termed *planetary vorticity*), on account of spin about its own axis; atmospheric cyclones and anticyclones rotate; spinning eddies of fluid turbulence are readily observed in rivers and from satellite images in ocean currents. Fluid vorticity is termed *relative* or *shear vorticity* and is due to velocity gradients across fluid elements. Vorticity must be conserved according to the principle of the *Conservation of Absolute Vorticity*. Vorticity has very important consequences for both turbulence and the large-scale flow of ocean currents and motion of the atmosphere, largely through the effect of the Coriolis force.

**Flow patterns—visualization**

Since the mathematical description and analysis of flow can be difficult, particularly for turbulent flows, experiments play a vital role in sedimentological fluid dynamics. Apart from the various techniques used to measure flow velocity, a key aspect is flow visualization, in which tracers of various kinds are introduced into the flow. No dynamical analysis may be confidently begun without some idea of flow pattern (Fig. 4.14).

Some flow visualization techniques can also be used to determine velocity. Visualization is particularly important in the case of separated flows (Fig. 4.14a; most readers will have seen advertisements for cars using wind-tunnel flow visualization, usually with

(a)

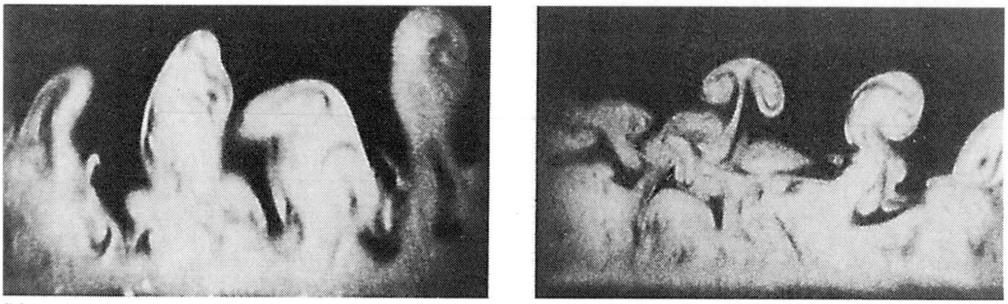

(b)

**Fig. 4.14** Flow visualization. (a) Particles in water flow over a negative step (like the lee-side of a ripple of dune) filmed over long exposure to give pathlines. Flow left to right. Note the splendid flow separation zone. (After van Dyke, 1982, and references cited therein.) (b) and (c) Smoke in the lower part of a wind tunnel boundary layer showing spectacular 'burst' vortices of low-momentum fluid advecting upwards into the outer boundary layer. (From Head & Bandyopadhyay, 1981.)

rear-end separation well shown). However, we must be very careful indeed when interpreting flow visualization images, chiefly in making sure that we have knowledge of the relative motions involved between recorder and flow. Flow patterns may be divided into the following types:

- streaklines—short photographic exposures of continuously introduced tracer;
- pathlines—long exposure of a spot of tracer;
- streamlines—these need not strictly be flow patterns but may be constructed from streaklines and pathlines. Formally, they are lines drawn such that the velocity of every particle on the line is in the direction of the line at that point.

In steady flows all the above are equivalent. They are all different in unsteady flows.

## 4.3 Fluid continuity with constant density

A fundamental principle in fluid flow is that of *conservation of mass*, being the interaction between the physical parameters that determine mass between adjacent fluid streamlines. The transport of mass, $m$, along a streamline involves the parameters velocity, $\boldsymbol{u}$, density, $\rho$, and volume, $V$. These determine the conservation of mass discharge, termed *continuity*. River, sea and ocean environments essentially comprise incompressible fluid. They contain layers, conduits, tributary channels or straits which vary in cross-sectional area, $A$. Now, referring back to Fig. 4.11, consider a steady discharge, $Q$, (units $L^3T^{-1}$) of constant density fluid through fluid elements at AB, CD and EF. Generally, if there is cross-sectional area $A_1$ and mean velocity $\boldsymbol{u}_1$ upstream in AB and area $A_2$ and mean velocity $\boldsymbol{u}_2$ downstream at CD or EF, the product $Q = \boldsymbol{u}A$ must remain constant (you can check that the product $Q$ has dimensions of discharge, or flux, $L^3T^{-1}$). We then have the equality $u_1A_1 = u_2A_2$ so that any change in cross-sectional area is accompanied by an increase or decrease of mean velocity and there is no change in $Q$, i.e. $\delta Q = 0$. Any changes in $\boldsymbol{u}$ naturally result in acceleration or deceleration. This simplest possible statement of the continuity equation may be used in very many natural environments to calculate the effects of decelerating or accelerating flow.

To be applicable, continuity of volume has important conditions attached:

- the fluid is incompressible, so no changes in density due to this cause are allowed;
- fluid temperature is constant, so there is no thermally induced change in density;
- fluid density due to salinity or suspended sediment content also remains unchanged;
- no fluid is added, i.e. there is no *source*, like a submarine spring or oceanic upwelling;
- no fluid is subtracted, i.e. there is no *sink*, like a permeable bounding layer or thirsty fish.

One natural environment where most of these conditions are satisfied is a length of river channel, where cross-sectional area changes downstream, as in Fig. 4.11. Other examples of volume and mass continuity will be met in many later Chapters. See **Cookie 15** for more information on continuity and fluid flow.

Want to know more about continuity and fluid flows without any physics? Turn to Cookie 15

Finally consider *divergence* and *convergence* with respect to sources and sinks. We stated previously that the continuity expression depends upon the lack of sources or sinks linked to the system in question. Two important cases arise in hydrological, oceanographic and meteorological flows. Surface *divergence* of streamlines, most obviously seen when flow is diverging from a point implies that a source is present below the surface, leading to a mass influx and *upwelling*. Surface *convergence* of streamlines to a point implies a sink is present and that *downwelling* is occurring. An added complication for meteorological flows is that vertical motions of fluid in downwelling or upwelling situations also cause changes of temperature and density, which cause feedback relevant to the stability of a moving air mass.

## 4.4 Fluid dynamics

Physical sedimentology cannot be understood without an appreciation of what forces are, how they arise and how they operate upon earth materials. For example, in rapidly moving turbulent fluids, though body forces are always present due to gravity, pressure forces act to cause rapidly changing vectors of motion within the fluid. The key concept of fluid dynamics is thus *force* and this itself comes from the concept of *momentum*, symbol $\boldsymbol{p}$, and how changes may come about to this moving property over time and space. Momentum is the product of the mass, $m$, of any substance (gas, liquid or solid) and its velocity, $\boldsymbol{u}$, hence dimensions are $MLT^{-1}$, the product is a vector and there are no special units for it. The vector is always orientated through a mass in the same direction as its velocity vector, $\boldsymbol{u}$.

The big clue concerning the significance of momentum can be obtained simply from first principles. Mass gives us a measure of the quantity of matter present in a solid or fluid; this helps determine an objects *inertia*, its tendency to carry on in the same line of motion or to resist changes in velocity. This can be the inertia of a stationary object or that of a steadily moving object. In many physical situations the mass of a given volume is constant with time and it is velocity change that determines *conservation of momentum*. From Newton's Laws (**Cookie 16**), any change in momentum over time is due to an equivalent force, $\boldsymbol{F} = \delta p/\delta t$. We expect momentum changes to arise in Nature very frequently; in fluids when an air or water mass changes direction and/or speed due to changes in external conditions; in fluid flow over solid boundaries where a velocity gradient is set up. The further example of colliding solid bodies such as sand grains violently impacting on a desert floor or colliding in a granular fluid brings us to a useful definition of the conservation of linear momentum: '*...the sum of momenta of an isolated system of two bodies that exert forces on one another is a constant, no matter what form the forces take...*'. In other words the collision of bodies or their interaction leads to no change in overall energy (the production of collisional heat energy is included in the balance). This principle of the conservation of momentum forms the basis of Newton's third law (**Cookie 16**).

Want to know more about Newton's Laws? Turn to Cookie 16

As discussed previously, acceleration, $\boldsymbol{a}$, is change of velocity over time and/or space; $\boldsymbol{a} = \delta \boldsymbol{u}/\delta t$, with dimensions of $LT^{-2}$. The standard acceleration due to gravity, $g$, at sea-level is $9.83\ \mathrm{m\,s^{-2}}$. Natural accelerations may be *extreme* compared to this, e.g. turbulent eddies are subject to accelerations of many times gravity, up to [O] $10^3\,\boldsymbol{g}$. A generalized definition of force, $\boldsymbol{F}$, is that it causes an acceleration, $\boldsymbol{a}$, to act upon a mass, $m$. In symbols, $\boldsymbol{F} = m\boldsymbol{a}$. It is clear from this definition that despite a mass being in motion, if there is no acceleration there can be no net force acting, though every moving substance, whether accelerating or not, has momentum. Any acceleration or change of momentum implies that an equivalent force must be acting to cause the change. Weight is a force, given by the product $mg$; units $MLT^{-2}$, with designated units, N, for *Newtons*. This definition means that a substance doesn't need to be moving for gravity to exert a force: gravity acts upon everything, moving or stationary.

In this book we will come across additional natural forces to those due to gravity, including viscous, buoyant, pressure, radial and rotational forces. Each, together with the gravitational and inertial forces (including contact forces due to friction), may contribute to the total force acting on any substance. For the moment we simply say that Newton's second law (**Cookies 16 & 17**) states that the sum, or resultant, of all these forces must equal the observed change of momentum of a substance, or $\boldsymbol{F} = \mathrm{m}\boldsymbol{a}$. In a steady flow of material, however fast or slow, or in a substance at rest, $\boldsymbol{F} = 0$ and for that very important case the arrayed forces must balance out to zero. Major progress in physical dynamics may be made in such cases.

Want to know more about forces in fluid flows? Turn to Cookie 17

As a vector quantity, force acts in the same direction as the acceleration produced by it. Complications arise in turbulent flows where the direction of the force is constantly changing in time and space. In slowly deforming or static material like glacier ice we can more easily speak of the orientation of forces with respect to the distortion they produce: *compressive* forces tending to push adjacent portions of ice together (which is what is happening under a surface load) or *extensional* forces doing the opposite. Across any plane, force vectors pointing towards each other are designated compressional and positive and pointing away from each other extensional and negative. Forces acting on a plane can also have any orientation. The two end members are *normal* and *shear* force orientations, the former normal to the plane in question and the latter parallel. As a vector, force may have any orientation and can always be resolved into components.

One important force affecting fluids and sediment grains in bulk contact is the *friction force*. This is a contact force that acts to resist sliding and depends upon the nature (roughness, physical state), but not the contact area, of opposing surfaces. Newton's first and third laws apply in cases of steady flow over beds of sediment; the applied fluid force is opposed by an equal reactive force from the sediment bed. Generally speaking in sediment erosion, transport and deposition studies we are interested in the bed stresses set up by a natural fluid due to current flow, the passage of surface or internal waves or the combination of current and wave. Unfortunately the natural situation is a deal more complicated than the frictional opposition of a flat sediment bed—the transport of grains is necessarily accompanied by (i) momentum exchange from fluid to grain and (ii) bedform creation (like ripples and dunes) with extra resisting forces due to longitudinal and vertical pressure gradients. So, there is a total loose-boundary resistance, $\tau_b$, to applied fluid forces comprising grain surface drag, $\tau_{sd}$, bedform drag, $\tau_b$, and sediment transport drag, $\tau_{st}$, such that $\tau_b = \tau_{sd} + \tau_b + \tau_{st}$.

## 4.5 Energy, mechanical work and power

Energy, work and power are interrelated scalar quantities that determine the ability of a fluid to transport sediment. When water in a flowing river or air in a desert storm is displaced from one position to another, *mechanical work* must have been done to achieve the displacement. This is equal to the force required, $F$, times the distance moved, $x$, or $F\,x$. *Mechanical* or *flow work* thus has units of force times unit distance, or N m, of dimensions $ML^2T^{-2}$. A single unit of work done is a scalar quantity and is termed a Joule, $J$.

All materials, moving or stationary, may be said to be capable of doing mechanical work: they all possess *energy*. This energy must also be a scalar with units $ML^2T^{-2}$. A moving object or portion of fluid has energy proportional to its mass, $m$, times velocity squared, $u^2$. This is called *kinetic energy* (i.e. energy of movement) and may be shown equal to $0.5mu^2$. A stationary object or piece of fluid has the energy of its weight force, $mg$, times distance, $h$, from the centre of gravity to which it is being attracted. This is *potential energy* (i.e. energy of position) $mgh$. It is usual to define $h$ with respect to some convenient reference level. The conservation of energy principle for fluids is written in the form of Bernouilli's equation (**Cookie 18**)

Energy may be released at variable rate; the time rate of liberation being J $s^{-1}$, a scalar quantity termed *power* with dimensions $ML^2T^{-3}$ and specified units called Watts, $W$. In terms of work, power is the rate at which work is done.

The interrelated concepts of force, energy, work and power are perhaps easiest grasped by reference to fluid flowing down a sloping river channel under the influence of gravity (Fig. 4.15). Fluid movement is created by gravity and the resulting applied force is opposed by the reactive friction force of the channel bed and banks. The flowing fluid has kinetic energy of motion that is provided by the fall in elevation of its surface as a loss of potential energy, though some of the latter is lost in friction. The available power of the river is the time rate of change of this potential energy to kinetic energy minus the frictional losses. The

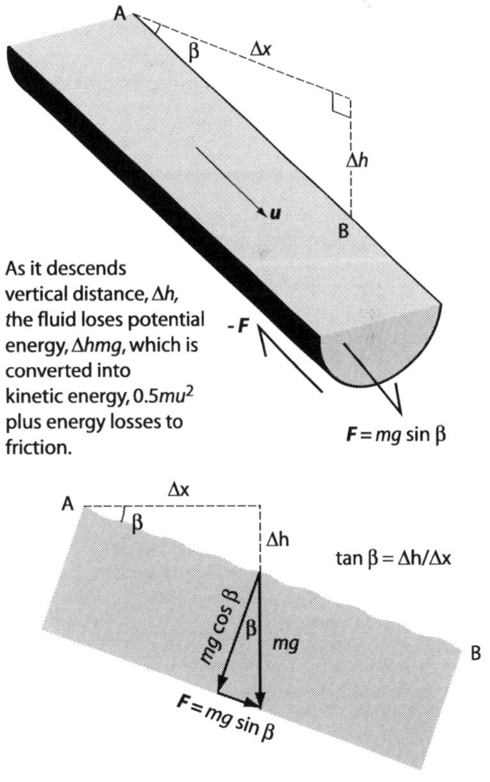


**Fig. 4.15** A mass, $m$, of water or debris flowing at mean velocity, $u$, under influence of gravity down a channel sloping at angle, $\beta$, exerts a force, $mg \sin \beta$, on the perimeter.

**Table 4.1** Earth-surface geophysical flows and their driving mechanisms.

| Geophysical flow | Driving mechanism | Flow type |
|---|---|---|
| Atmospheric gas flows (e.g. planetary and surface winds) | Atmospheric pressure forces driven ultimately by convection and severely affected by Coriolis force | Newtonian |
| Surface water flows in channels (river and delta channels) | Gravity | Newtonian |
| Surface sheet flows (e.g. hillslope runoff, some river floods) | Gravity | Newtonian |
| Aqueous surface waves of translation (e.g. sea surface waves) | Gravity | Newtonian |
| Ocean surface currents | Wind shear, pressure gradient and Coriolis force | Newtonian |
| Tidal flows | Gravity | Newtonian |
| Surface mass sediment flows (solids only) (e.g. grain flows/rock avalanches; rock flows; some powder snow avalanches) | Gravity | Granular behaviour (Bagnoldian) |
| Surface mass flows of sediment and water (e.g. debris, turbidity and other mass flows) | Gravity acting upon density contrasts caused by presence of suspended sediment | Non-Newtonian to Newtonian |
| Deep ocean currents | Gravity acting upon density contrasts caused by salinity differences, differential heating or cooling | Newtonian |
| Lake and ocean vertical mixing (chimneys, overturns) | Seasonal convection due to salinity differences, heating and cooling of surface waters | Newtonian |
| Subaqeous internal solitary waves (e.g. reflected density and sediment flows) | Gravity | Newtonian |
| Near-subsurface (vadose/unsaturated zone) flow in pore spaces | Gravity | Newtonian |
| Deeper subsurface fluid flow in pore spaces and other discontinuities (e.g. burial diagenesis of aqueous solutions, oil, gas) | Gravity causing compaction and chemical reactions and thermal effects causing density contrasts | Newtonian |

energy of flow is available to do mechanical work, artificially in turning a water wheel or naturally in sediment transport.

Highlighted in Table 4.1 are the driving forces that cause the various flows relevant to sedimentological fluid dynamics. We can distinguish momentum- and buoyancy-driven flows. Momentum-driven flow is due to externally applied gravity or pressure forces, which set up velocity gradients. There is no density difference affecting the fluid(s) in motion, although in Nature, as we shall see below, admixed solids may often impose their own density gradients. Buoyant flows depend for their motion on a density contrast with a surrounding (ambient) fluid.

Want to know more about the development of Bernoulli's theorem? Turn to Cookie 18

## Further reading

A fine general physics undergraduate level text is Fishbane *et al.* (1993). A classic little book (now sadly out of print) by Shapiro (1961) has introduced many non-mathematical students to the joys of fluid mechanics. Massey (1979) and Vardy (1990) are clearly written introductory texts in fluid dynamics, but both have the needs of the student engineer primarily in mind. Denny (1993) has many nice ways of presenting basic fluid physics, but in this case with most relevance to the biologist. Furbish (1997) is the best moderate-level text for earth and environmental scientists, being an introduction to fluid motions both on the Earth surface and within its crust. Middleton & Wilcock (1994) is also a very good, sound text on mechanics (solid and fluid) written with advanced Earth scientists in mind. It is excellent, but requires reasonable levels of, and enthusiasm for, higher maths. Allen (1997) provides a valuable introductory text on surface processes for earth and environmental scientists. If you are coming to physical sedimentology from physics or geophysics, then Acheson (1990), Faber (1995) and Tritton (1988) are the books for you. All contain nuggets of clarity, especially the latter, written by a much-respected teacher. More recently, the subject is presented at a very introductory level for earth and environmental science students by Leeder & Perez-Arlucea (2005). Finally, the CD by Homsy *et al.* (2007) cannot be recommended too highly, it covers many topics, has movies, simulations, virtual laboratories and is easy to navigate.

# Chapter 5

# TYPES OF FLUID MOTION

*Pride of play in a flourish of eddies,*
*Bravura of blowballs and silver digressions,*
*Ringing and glittering she swirls and steadies,*
*And moulds each ripple with secret suppressions.*

Hugh MacDiarmid, 'The Point of Honour', 1934, *Complete Poems*, Vol. 1, Carcanet, 1993.

## 5.1 Osborne Reynolds and flow types

As we have seen (Chapter 4), velocity in any flow may change spatially and/or with time. In the late nineteenth century, Osborne Reynolds, first professor of mechanical engineering at the University Manchester, England, set out to investigate the nature of the laws of resistance to flow of fluids in channels and pipes. He approached the problem both theoretically (or 'philosophically' as he put it) and practically and found '...*that the general character of the motion of fluids in contact with solid surfaces depends on the relation between a physical constant of the fluid, and the product of the linear dimensions of the space occupied by the fluid, and the velocity*'.

Designing the apparatus reproduced in Fig. 5.1, Reynolds measured the pressure drop over a length of smooth pipe through which water was passed at various speeds (Fig. 5.2). Pressure drop is due to frictional losses as the fluid moves through a system, converting its potential energy to kinetic energy and finally to heat energy. Reynolds found that the pressure loss per unit pipe length increased in a linear fashion with velocity but that over a transition region the losses began to increase more quickly, at about the square of the velocity. Deducing that the flowing water was also changing its flow pattern, Reynolds confirmed this by introducing a dye streak into a steady flow of water through a transparent tube (Fig. 5.3). At extremely low flow velocities the dye streak extended down the tube as a straight line, and Reynolds described the flow as 'direct' (now known as *laminar* or viscous flow). With increased velocity the dye streak was dispersed in eddies and eventually coloured the whole flow: this was described as 'sinuous' flow (now known as *turbulent* flow).

The fundamental difference in flow types between the two flow regimes is arguably the most important result in the whole field of fluid dynamics. Repetition of the pipe experiments with fluids of different viscosity (the '*physical property*' that Reynolds alluded to previously) and different smooth pipe diameters showed that the critical velocity for the onset of turbulence was not the same for each experiment. Reynolds found, however, that the change from laminar to turbulent flow occurred at a fixed value of a quantity defined as, $\rho du/\mu$ or $du/\nu$, where $u$ is the mean flow speed, $\rho$ is the fluid density, $\mu$ and $\nu$ are the molecular and kinematic viscosity respectively and $d$ is the internal diameter of the pipe. This expression has become known as the *Reynolds number (Re)* in honour of its discoverer. *Re* is a ratio of two forces, viscous

*Sedimentology and Sedimentary Basins: From Turbulence to Tectonics*, 2nd edition. © Mike Leeder.
Published 2011 by Blackwell Publishing Ltd.

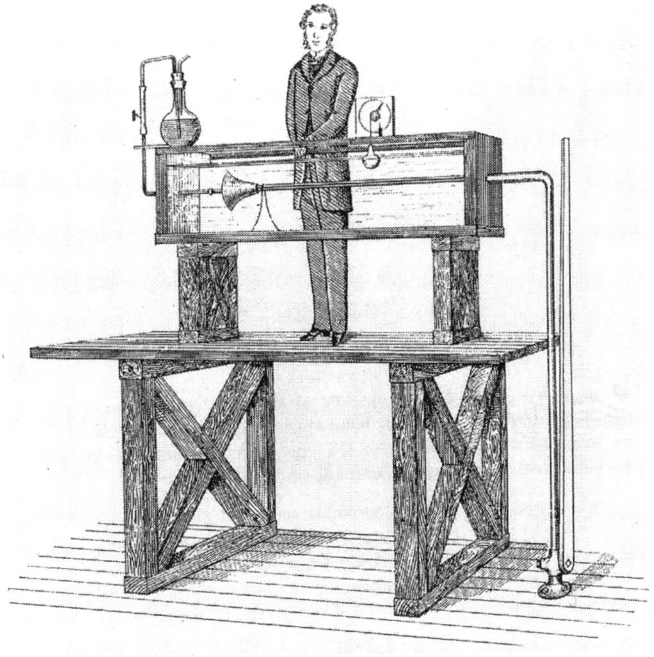

**Fig. 5.1** The original sketch of Reynold's apparatus in his epoch-making 1883 paper.

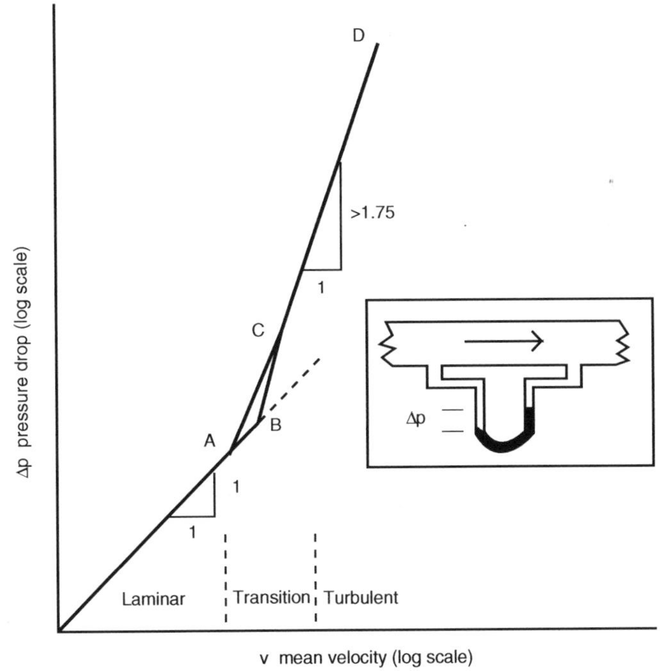


**Fig. 5.2** Graphoid to show the flow energy losses measured as the pressure drop, $\Delta p$, along a smooth pipe (inset) as a function of mean speed ($u$) through the pipe. (After Massey, 1979). The rate of loss increases uniformly from the origin to A in laminar flow. The onset and increased rate of loss at ABC (the transition zone) is highly variable, according to local conditions, but eventually at C the rate of loss stabilizes at some higher constant slope in the turbulent state of flow.

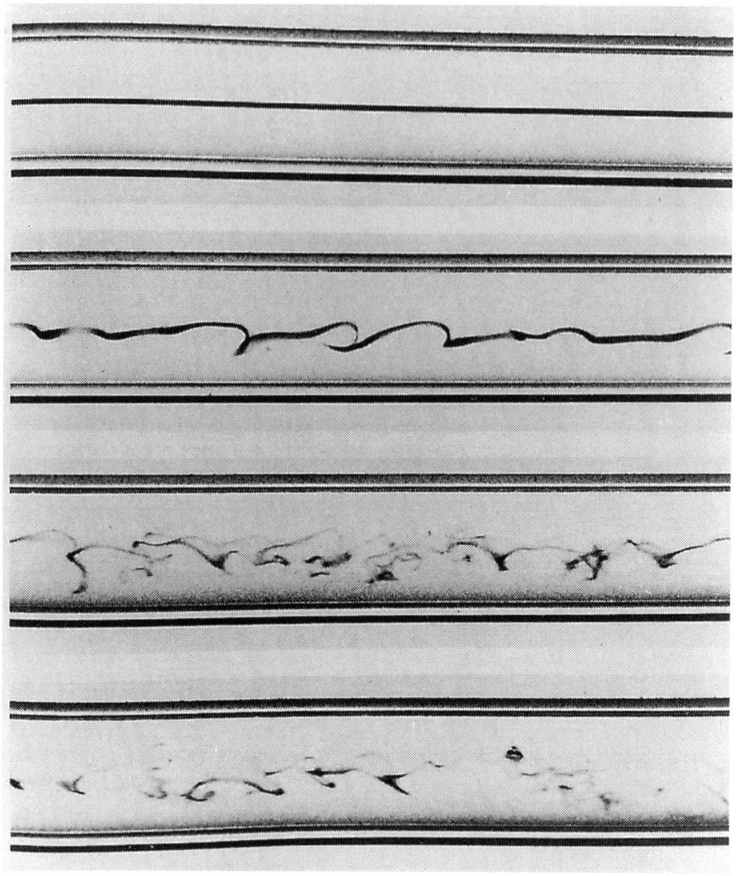

Fig. 5.3 A modern replication of Reynold's experiment. Top to bottom shows laminar, transitional and turbulent flow patterns visualized by a central dye streak deforming in water flowing left to right through a glass tube. (From sources cited in van Dyke, 1982.)

and inertial, acting on any fluid (**Cookie 19**). Viscous forces resist deformation of a fluid; the greater the molecular viscosity, the greater the resistance. Inertial forces represent resistance of the fluid mass to acceleration. With no dimensions itself, *Re* is a pure or dimensionless number (the reader should check this).

When viscous forces dominate, then *Re* is small and the flow is laminar. Energy losses and pressure drop are also low due to the absence of flow mixing. When inertial forces dominate, *Re* will be large and the flow turbulent. Now, energy losses and pressure drop are high as intense turbulent motions do work in mixing-up fluid. For water flows in pipes and channels, the critical region for the laminar–turbulent transition lies between 500 and 2000. We must not assume that laminar flow only occurs in high-viscosity liquids. As Reynolds himself took great pains to emphasize, the flow state is dependent upon *four* parameters of flow, not just one. Thus a very low density or very low velocity of flow has the same reducing effect on *Re* as a very high viscosity. Thus as Shapiro states in his classic introductory text '...*it is more meaningful to speak of a very viscous situation than a very viscous fluid*'. Flow systems with identical *Re* are said to be dynamically similar, a feature made use of in many modelling experiments.

Some comment is necessary on the length scale in the *Re* criterion. For a pipe this is the diameter, but for a river channel or other free-surface liquid flow the mean flow depth is an appropriate scale. For wind and the oceans the length scale presents more problems. In air tunnels this may be taken as the height of the tunnel, but in atmospheric or oceanographic flows the boundary-layer thickness must be chosen. In such very deep (or thick) flows, large calculated *Re* have little physical meaning.

Want to know more about the derivation of the Reynolds number? Turn to Cookie 19

### Grain Reynolds number

The concept of a change in flow behaviour at a critical value of *Re* has been extended to the changing pattern of fluid flow around a sediment grain. This is controlled by variations in the *grain Reynolds number*, $Re_g$, one of whose formula is given as $Re_g = (\sigma - \rho) dU_g/\mu$, where $\sigma$ = grain density, $\rho$ = fluid density, $d$ = grain diameter, $U_g$ is grain velocity and $\mu$ = fluid molecular viscosity. $U_g$ is usually less than the local fluid velocity since the grain's inertia causes it to move more slowly and to an observer moving with the fluid the grain would appear to be moving backwards. Such relative motion is directly confirmed by a technique of modern laser-based flow monitoring equipment, *Phase Doppler Velocimetry*, that can distinguish between motion of fluid and solid phases.

## 5.2 The distribution of velocity in viscous flows: the boundary layer

The movement of fluid past any solid boundary (the basic situation of many sedimentological flows) is seriously affected by friction in a finite zone known as a *boundary layer* and where the fluid is substantially slowed down with respect to fluid further away from the boundary (Fig. 5.4). Given their general form, with velocity increasing away from the bed, the simplest expression for the boundary-layer force per unit area (i.e. stress, $\tau$) is given by Newton's relationship, $\tau = \mu\, du/dy$, where $\tau$ is the shearing stress, $\mu$ is the molecular viscosity and $du/dy$ is the velocity gradient. Figure 5.4 reveals the answer as to why net forces are set up in boundary layers; where a velocity gradient occurs, net forces act across shear planes within the fluid. These forces depend upon velocity gradients in time ($\delta u/\delta t$) *and* space ($\delta u/\delta y$) and the material property of viscosity.

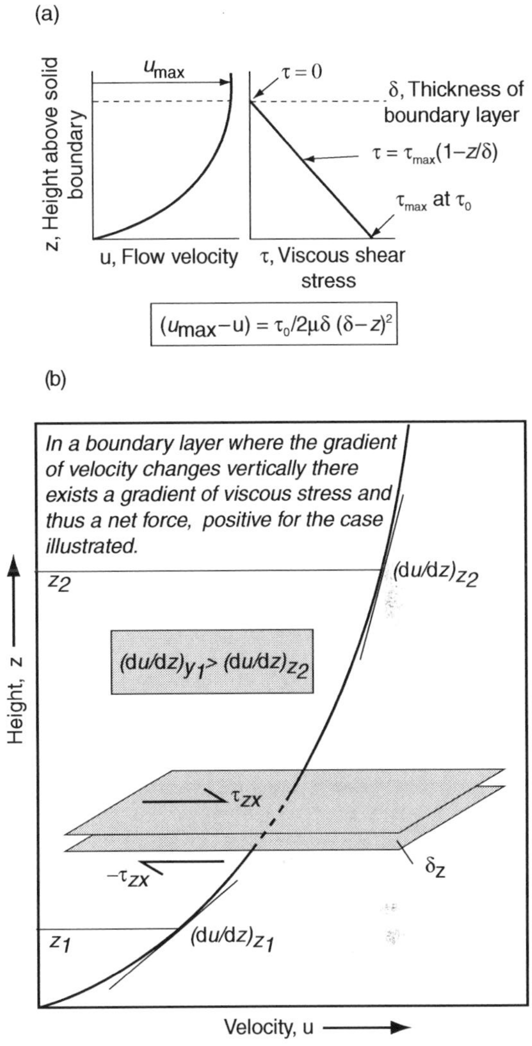


**Fig. 5.4** (a) Velocity and shear stress distribution (to first order only) in a flow. (b) To show definitions of velocity gradients and viscous shear stresses in a boundary layer whose velocity is changing in space across an imaginary infinitesimal shear plane, dz. Such boundary layers are very common in the natural world and the resulting net viscous force reflects the mathematical function of a gradient of stress via the second differential coefficient of velocity with respect to height, *y*, i.e. $F_{viscous} = -d\tau_{zx}/dy = -\mu d^2u/dy^2$.

Boundary layers are familiar from everyday experience, for example in our inclination to duck down closer to the ground when the wind blows exceptionally strongly. They are all-important in sedimentology because much sediment transport and bedform development occurs within them. Close to the wall or flow boundary we can imagine that there is an attached layer of liquid molecules: the *adsorbed layer*. As the liquid flows, the velocity tends to zero at the wall because the adsorbed layer refuses to move because of surface attraction to the solid atoms. This is the *no-slip boundary condition* first proposed by Navier and developed by Reynolds. This viscous retardation gradually dies out away from the wall until at some point in the flow, termed the *free stream*, there is no velocity gradient and hence no stress. It is the exact rate of change of velocity with distance from the solid boundary that is of interest. If it is a linear change, then the shearing stresses set up by viscosity will be everywhere equal. If on the other hand it is nonlinear, then there will be important localization of stresses.

> Want to know more about the derivation of the laminar flow velocity profile? Turn to Cookie 20

Let's investigate the velocity distribution in a laminar flow between two rigid walls, as in a channel. If we measure point velocities across such a flow, then a characteristic parabolic curve results (Fig. 5.5). This parabolic curve may be exactly predicted by simple theory (another result due to Reynolds, **see Cookie 20**) when the pressure and viscous forces acting on the fluid are balanced. It should be noted that the parabolic expression for viscous flow is only strictly applicable to Newtonian fluids at low Reynolds numbers. A case in point would be the *very* slow flow of water along a deep channel. In the boundary layers of Bingham materials like debris flows the velocity gradients and hence strain rates are much higher at the margins of the flow and much lower in the centre of the flow (Fig. 5.5). In fact, the central zone hardly shows differential motion at all, tending to move *en masse* as a plug. Plug behaviour in debris flows gives rise to a number of features of sedimentological interest (Chapter 8).

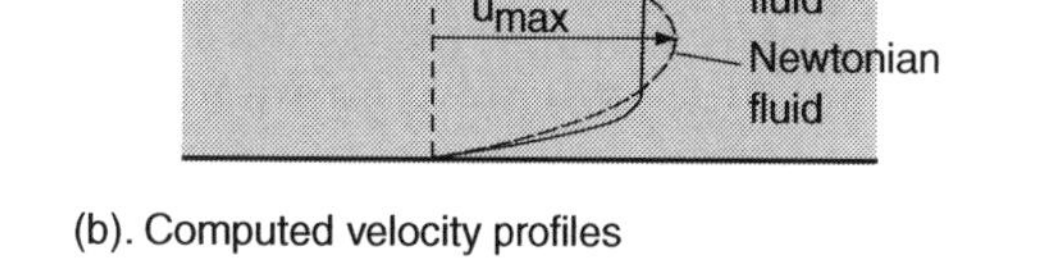

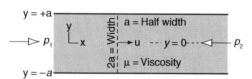

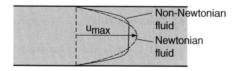


**Fig. 5.5** Definition diagram for the derivation of parabolic Newtonian laminar and plug-like non-Newtonian laminar flow (see **Cookie 20** for derivation).

## 5.3 Turbulent flows

Turbulent flows dominate sediment transport processes—almost all wind and water flows are turbulent. Much of the necessity for understanding turbulence came originally from the field of aeronautics, and it is perhaps no coincidence that serious analysis of turbulence started around the date of *Homo sapien*'s first few uncertain attempts at controlled flight. Sixty years later, photographs of the effects of turbulent atmospheric flows on Earth were taken by persons standing on the Moon, whilst radar and Lander images have more recently revealed Venusian and Martian sand grains, ripples, dunes and sandstorms to Earthlings (section 23.14).

So, what is turbulence all about? Insertion of a sensitive flow-measuring device into a steady turbulent flow for a period of time will result in a fluctuating record of fluid velocity (Fig. 5.6), regardless of the fact that the flow is steady in the longer term. The instantaneous longitudinal streamwise velocity component velocity, $u$, is equal to the mean velocity $\bar{u}$ plus or minus the instantaneous deviation $u'$ from the mean. In symbols: $u = \bar{u} \pm u'$. Thus we can only really talk about a characteristic time-mean flow velocity for turbulent flows, with the mean measured over several tens of seconds or longer depending on the nature and size of the flow. A steady turbulent flow must be defined by repeated measurements yielding equal $\bar{u}$. Also note that since $u'$ may be positive and negative about the mean at different times, over a long time period the mean fluctuation, $\bar{u}'$, must be zero.

In turbulent flow, fluid moves around in what initially seem to be confusing patterns, though in fact there is well-defined coherence or structure to the motions. At one moment a small volume of fluid is close to a solid boundary and then it suddenly crosses

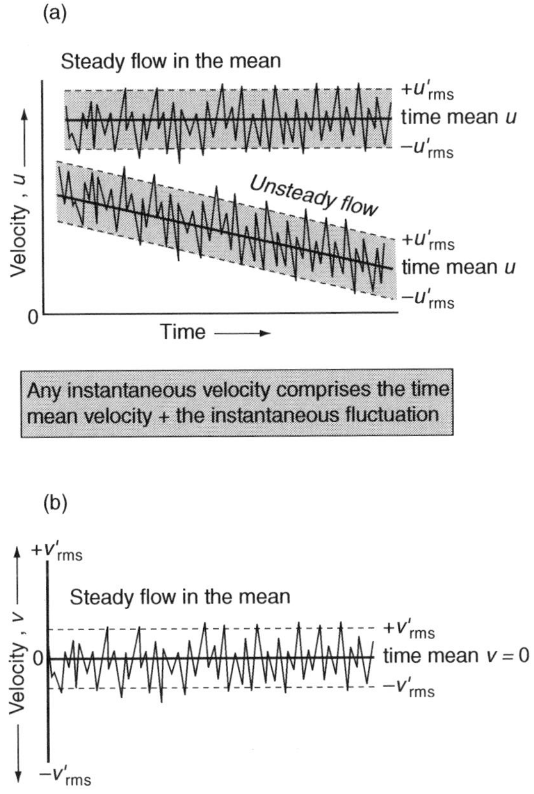


**Fig. 5.6** (a) Turbulent flow velocity time series in $u$, the streamwise velocity component. (b) Turbulent flow velocity time series in $v$, the vertical velocity component.

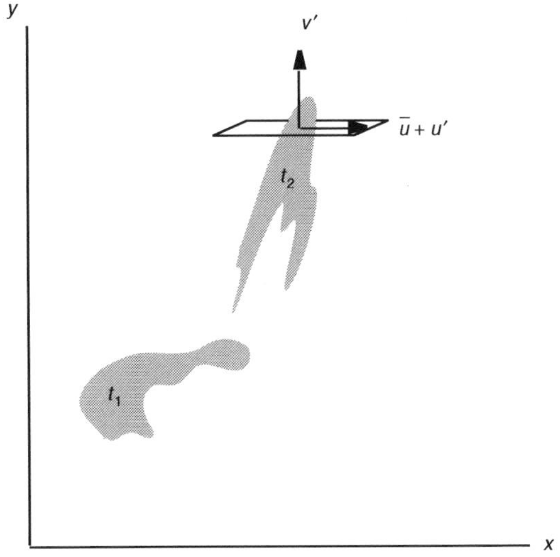


**Fig. 5.7** Sketch to define fluid momentum transfer across unit area of the $xy$ plane by a fluid eddy moving in $xy$ velocity space from time, $t_1$, to time, $t_2$. It carries mean $x$-momentum of $-\rho\bar{u}'\bar{v}'$. (After Massey, 1979.)

the flow. Such fluid *eddies* contrast with the time-mean streamlines that we could define for the mean fluid motion. It is obvious that eddies transfer mass; they must also transfer momentum and redistribute it within the flow (Fig. 5.7). Since transfer of momentum is caused by some force, it is clear that additional accelerations to those in laminar flow must be caused by something other than viscous-induced force. The only candidate is turbulent pressure gradient force. The additional turbulent 'stresses' that cause an apparent increase of energy loss were termed the 'eddy viscosity' by their discoverer Boussinesq in 1877 (**Cookie 21**).

Want to know more about turbulence? Turn to Cookie 21

Turbulent eddies have a 3D form, so in addition to the $x$-component of velocity there will be $y$- and $z$-components as well, $v = \bar{v} \pm v'$ and $w = \bar{w} \pm w'$. For steady 2D flow the mean values of $v$ and $w$ are zero, i.e. there is no *net* upward or spanwise transport of fluid; the up–down/side to side fluctuations all cancel out over time. Fortunately, $v$ and $w$ are usually small in comparison with $u$ for most shear flows of sedimentological interest.

Three-dimensional turbulent fluctuations are caused by fluctuating pressure gradients and the local velocity gradients set up severe local accelerations. These work against the mean velocity gradient to remove energy from the flow; this turbulent energy is ultimately dissipated by the action of viscosity. Thus we now have an explanation of the greatly increased energy losses in Reynolds' pipe flow experiment for turbulent flow as compared with laminar flow. Newton's viscous stress equation must be replaced in turbulent flows by $\tau = (\mu + \eta)\mathrm{d}u/\mathrm{d}y$, where $\eta$ is Boussinesq's *eddy viscosity*. Unlike $\mu$, $\eta$ is a variable quantity dependent upon the size and velocity components of the eddies. Its magnitude is usually much larger than that of $\mu$ and is controlled by local turbulent velocity accelerations—these are termed

*Reynold's accelerations*, often wrongly referred to as Reynolds stresses (**Cookie 21**).

Just as in viscous flows, a boundary layer exists in turbulent flows such that the mean velocity decreases with distance from a solid boundary. When plotted on semilog graph paper (i.e. one axis has log scale), mean streamwise velocity, $u_x$, measurements across the lower third of a boundary layer define straight-lines because velocity is proportional to log height (Fig. 5.8). This is because of a simple maths rule concerning the rate of change of a quantity like velocity being inversely proportional to its distance from a boundary. A physical derivation of this log 'law of the wall' as it is known is given in **Cookie 22**. The rate of increase of fluid velocity with log height is proportional to the slope or rate of change of the velocity distribution curve. This defines the *shear velocity* (aka friction or drag velocity), $u_*$, equivalent to $(\tau/\rho)^{0.5}$, where $\tau$ is fluid shear stress on the boundary and $\rho$ is fluid density. We meet the expression in our derivation of the 'law of the wall'. The reader can check that the expression has the dimensions of velocity.

Want to know more about the derivation of turbulent flow velocity profiles? Turn to Cookies 22 & 23

Very close to a solid boundary the log law breaks down and the velocity there decreases towards the bed in a linear fashion as outlined in **Cookie 23**. This defines the *viscous sublayer*, the one area of a turbulent flow where the eddies of turbulence are reduced to near-zero. Figure 5.9 shows how very thin the viscous sublayer actually is. Boundaries with sediment grains enclosed entirely within the viscous sublayer are said to be *smooth* and are only subject to viscous forces. When grains project through the sublayer they shed off eddies whose size increases directly with that of the particles. The surface is then said to be *transitional* or *rough*, depending on the degree of penetration, and the rate of energy loss by turbulent eddy-shedding is higher. The graph of Fig. 5.9 shows $d$ in relation to flow strength for sand grains in air and water flows, plotted so that the critical $u_*$ is that appropriate to the threshold of grain motion (see Chapter 6). In water, for example, the flow boundary ceases to behave smoothly at the threshold of motion for all grain diameters greater than about 0.6 mm; important consequences for sediment bedform development follow from this fact (see Chapter 7, **Cookie 36**).

In flows passing over rough surfaces composed of coarse sand grains, granules or pebbles, the height of the intersection of the velocity curve with the height ordinate is unchanged with increasing velocity. If the rough surface is immobile, then the height of the focus of all the different velocity curves is approximately equal to 1/30th of the diameter of the surface grains. This $k$ is equivalent to the constant $C_1$ in the derivations of **Cookie 22** and gives physical meaning to the integrations. Stationary grain roughness is generally termed *skin friction*, but is not the only means by which energy is dissipated in a moving fluid. Sediment transport produces a moving bedload layer that further dissipates energy. Bedforms also add a significant component of friction, known as *form drag*, which arises from the generation of eddies due to the phenomenon of flow separation. Finally, energy is lost whenever a channel flow turns a bend or constricts or expands. All these energy losses may be lumped into one term, a generalized *friction coefficient* for a particular flow in a particular context (**Cookie 24**).

Want to know more about friction and mean stresses in turbulent flows ? Turn to Cookie 24

## 5.4 The structure of turbulent shear flows

We now know that Reynolds' 'sinuous motion' in turbulent flow is dominated by 3D eddies with strong vorticity. The pioneering fluid dynamicist Prandtl regarded eddies as 'lumps' of fluid (*Flüssigkeitsballen*) transferring momentum throughout the boundary layer. The eddies gradually develop as the Reynolds number increases from within a stable laminar flow field. We would like to know a little more about the nature of these eddies for they are relevant to sediment transport and bedform theory. For example:

1 Are eddies random and hence unpredictable in time and/or space, or do they have characteristic repeatable structure?
2 What does a 'typical' eddy look like in 2D and 3D?
3 How do eddies transfer momentum?

Although the mathematical development of hypotheses for turbulent behaviour remains limited, direct observation and analysis of turbulent structure is

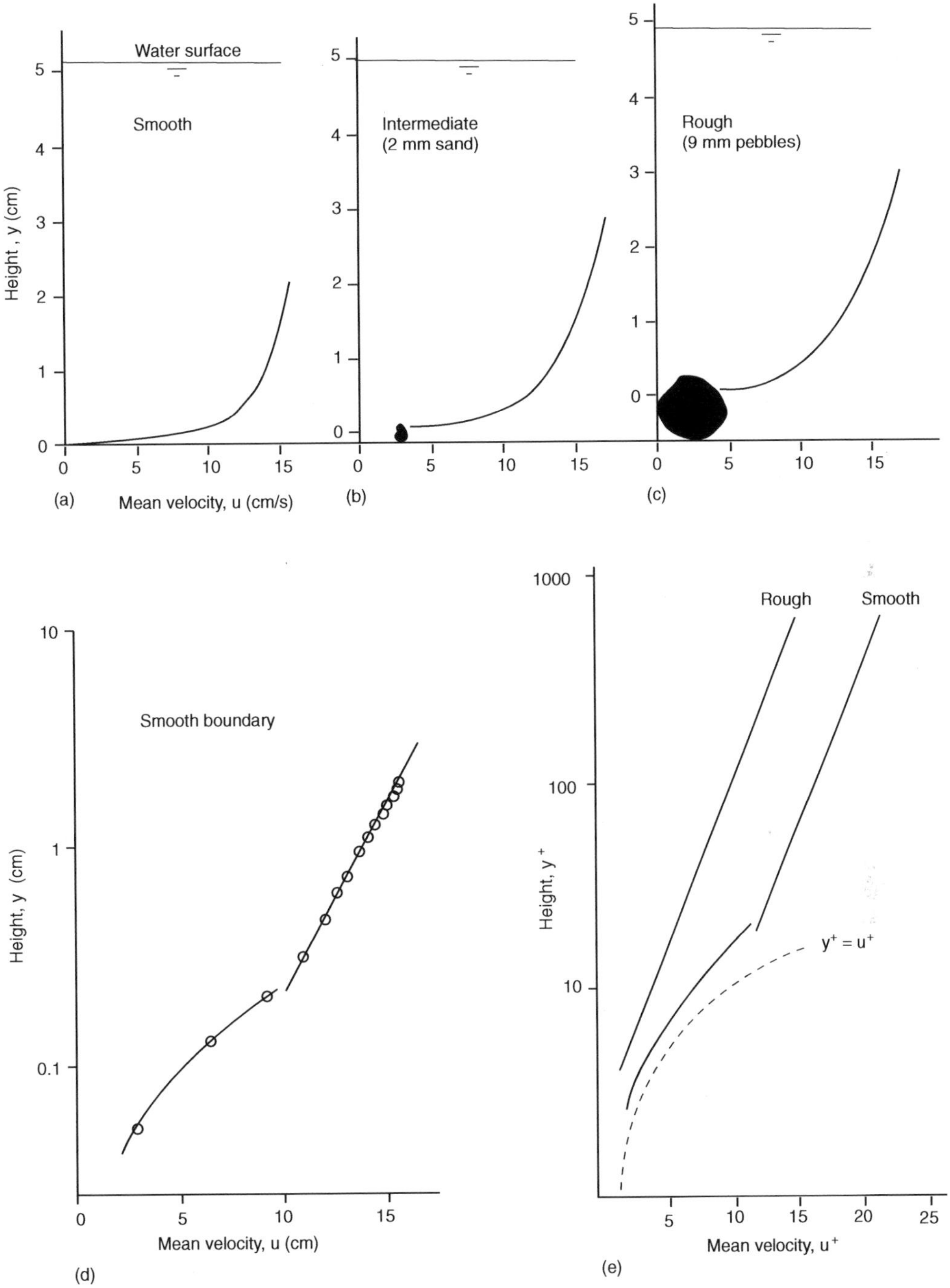


**Fig. 5.8** (a)–(c) Graphs to show the vertical distribution of mean turbulent velocity, $\bar{u}$, measured over smooth, intermediate and rough boundaries. (d) Smooth boundary data of (a) replotted on semilog ordinate scale to show viscous sublayer and log-linear distribution in the remaining flow (note deviations increasing towards surface). (e) Rough boundary data superimposed on curve from (d). Note absence of viscous sublayer. Dimensionless scales are $y^+ = yu_*/\nu$; $u^+ = \bar{u}/u_*$. (Data of Grass, 1971.)

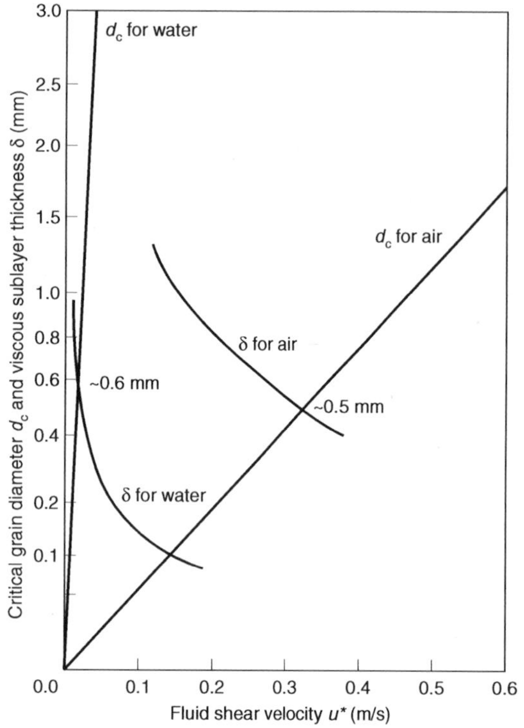


**Fig. 5.9** Plot of fluid shear velocity, $u_*$, against grain diameter, $d$, and viscous sublayer thickness, δ. Intersections of the δ and $d$ curves for air (~0.5 mm) and water (~0.6 mm) define the value of grain size, $d_c$, at which the grains cause disruption of a smooth boundary at the threshold for motion. (After Carson, 1971.)

now possible using sophisticated flow visualization techniques. Coherent eddies are more complex and interesting than Prandtl's 'lumps' of fluid, more similar in fact to Reynolds' 'sinuous' description. The flow visualization results (Figs 5.10 & 5.11) enable the following simple observations.

1 In the *xy* plane:
   (a) Relatively fast 'lumps' of fluid move down from the outer to the inner flow region. These are called *sweep* motions.
   (b) Relatively slow 'lumps' move out from the inner to the outer flow. These are called *burst* motions.

2 In the *xz* plane:
   (a) Close to the bed there exist flow-parallel lanes of relatively slow and fast fluid that alternate across the flow. The low-speed lanes are termed *streaks*.
   (b) The streaks become increasingly less well defined as the flow is ascended.

Three-dimensional reconstructions of eddies (Fig. 5.12), the result of nearly 40 years experimental work into the structure of turbulent flows, show large-scale coherent vortex structures within the boundary layer. They have the shape of languid hairpins which arise at the flow boundary and evolve into groups which extend and grow as coherent structures well beyond the log region. These attached eddies grow in proportion to their distance from the bed in a self-similar fashion; their 'legs' define low-speed inner flow streaks. Streak spacing $\lambda_s$ in the viscous sublayer is given by the experimental relation $\lambda_s u^*/v = 100$, where $u^*$ is the shear velocity and $v$ the kinematic viscosity. Thus with constant $v$, $\lambda$ decreases as $u^*$ increases. The rate of bursting of low-speed streaks increases with $u^*$. It should be noted that the streak pattern is quasicyclic, new streaks forming and re-forming constantly across the flow as groups of hairpins advect pass the bed.

The turbulent boundary layer comprises two rather distinct zones (Fig. 5.13): (i) an inner zone close to the bed with its upper boundary between the transition and logarithmic region of the turbulent boundary layer; and (ii) an outer zone extending up to the flow free surface. Both are dominated by the outward evolution of coherent packets of hairpin vortices.

1 The inner zone is distinguished by:
   (a) being the site of most turbulence production;
   (b) containing low- and high-speed fluid streaks that alternate across the flow;
   (c) occurrence of lift-up and intense vorticity of low-speed streaks in areas of high local shear.

2 The outer zone:
   (a) provides the source of high-speed fluid for the sweep phase near its lower boundary that probably initiates a burst cycle;
   (b) contains large vortices from groups of hairpin eddies that are disseminated through the outer zone and may reach the surfaces as 'boils'.

Use of a simple quadrant diagram (Fig. 5.14) brings out the essential contrasts between burst and sweep turbulent interactions. Turbulent bursts are *quadrant 2* events because they involve injections of slower-than-mean horizontal velocity fluid (i.e. instantaneous $u'$ values are negative) travelling *upwards* (i.e. instantaneous $v'$ velocities are positive); the average product

(a)

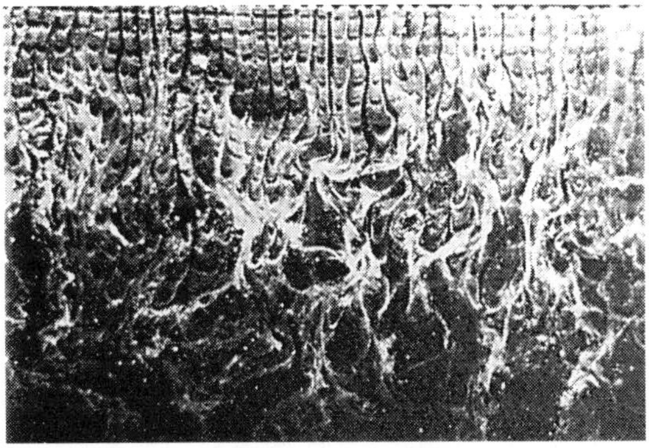

(b)

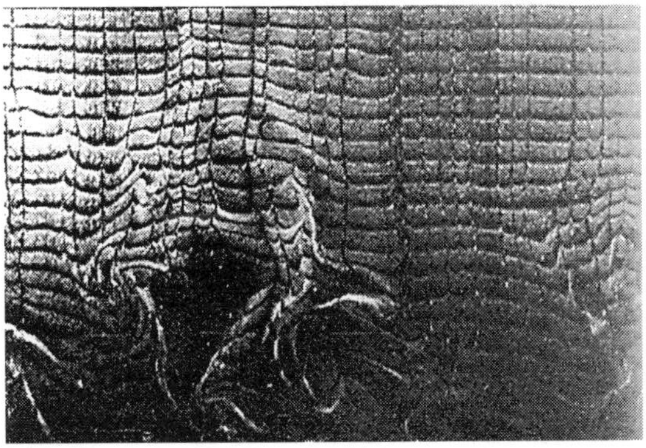

(c)

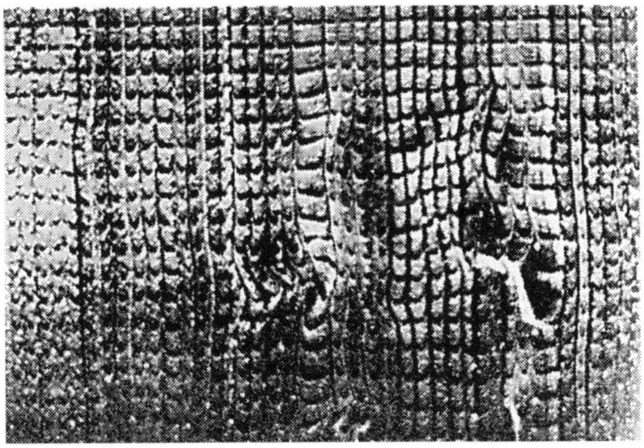

(d)

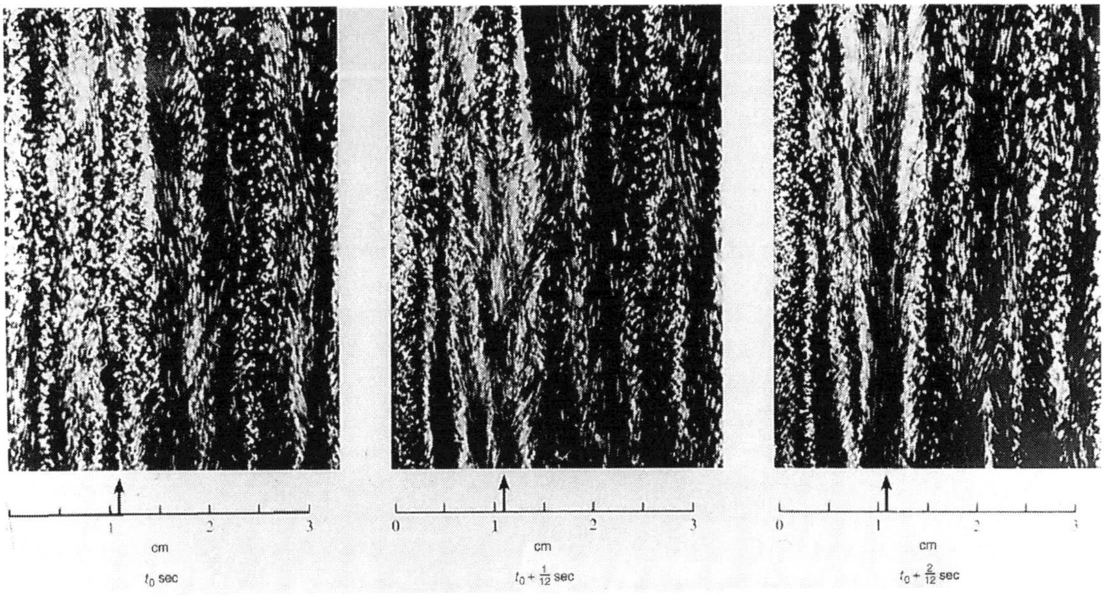


(e)

(a)

(b)

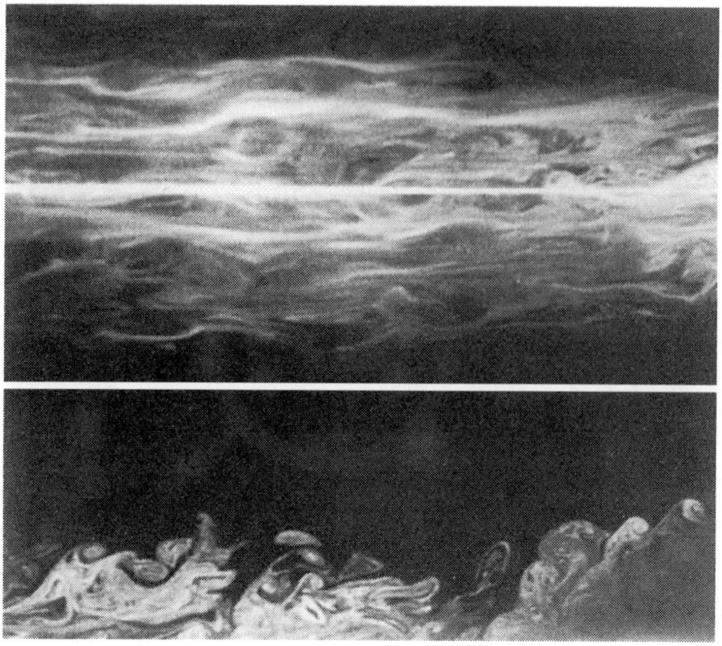

**Fig. 5.11** Turbulent eddies in right-flowing air visualized by smoke injected from below the bed. (a) Instantaneous view from above to show the streaky inner zone structure (the thin straight white line is the beam of light used to visualize the contemporaneous vertical section in (b). (b) A sideways view of (a) to show the beautiful form of turbulent mixing, with periodic ascending low-speed burst elements (white) and descending high-speed outer and transitional layer sweep fluid (black). See Fig. 5.12 for an idealized 3D view of the vertical structure. (After Falco 1977.)

is negative. Similarly for sweep motions, but here the motions are *quadrant 4* events, with faster-than-average downward motion.

Regarding momentum transfer by eddies, high positive contributions to local Reynolds accelerations occur across the whole flow depth when migration of momentum-deficient fluid occurs in bursts. The inrush/sweep phases also give a positive contribution to Reynolds accelerations, but their effect is at a maximum in the area close to the wall (Fig. 5.13). Almost all ($> 70\%$) of the Reynolds accelerations in turbulent flows are due to burst/sweep processes and the majority are produced close to the flow boundaries. Inner-layer fluid deforms in this way in response to critical pressure gradients imposed from the outer flow. The nature of bursts and sweeps also explains why bed-parallel Reynolds accelerations act in the same direction as the viscous bed shear stress.

Of great sedimentological interest are experimental results on burst/sweep phenomena over boundaries

**Fig. 5.10** (a)-(d) Instantaneous photos of $H_2$-bubble blocks in water, taken from above the flow looking downwards toward the bed, the field of view being in the *xz* plane with current flow from top to bottom of each photo. The speck-insulated platinum wire where the bubble blocks are periodically generated is at the top of each photograph. The sequence represents successively higher positions of the wire above the bed. (a) The streaky deformation of the bubble blocks is well shown, each streak identifying a low-speed phase of the turbulent cycle in the viscous sublayer. (b) The streaks have become tangled and less obvious as they pass into the logarithmic portion of the turbulent boundary layer. (c) and (d) Blocks are mostly undisturbed in the outer regions of the flow but larger areas of macroturbulence are prominent. (After Kline *et al.*, 1967.) (e) Viscous sublayer structure visualized from above by means of illuminated 0.1 mm sand grains moving over a smooth black boundary to define pathlines. The sequence of photographs are separated in time by 1/12 s with a 1/30 s exposure and illustrate the development of an inrush or sweep event (arrowed), with slow or stationary grains adjacent. (After Grass 1971.)

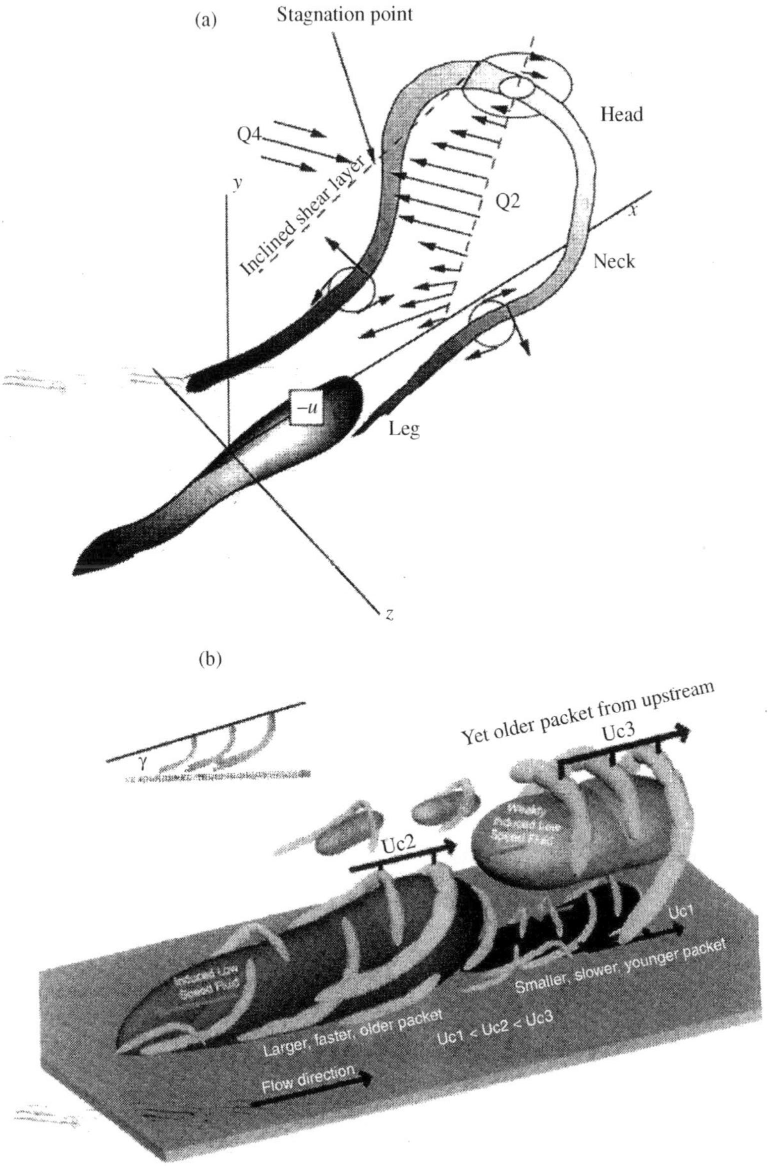


**Fig. 5.12** (a) Sketch to illustrate a bed-attached hairpin vortex together with its induced motion in ambient fluid. (After Adrian *et al.*, 2000.) (b) Conceptual sketch reconstruction of nested packets of hairpin and cane vortices growing up from the bed. The packets align in the streamwise direction and coherently sum to create large zones of nearly uniform streamwise momentum. Large-scale motions in the outer boundary layer (wake) ultimately limit their vertical growth. (After Adrian *et al.*, 2000.)

roughened by sediment grains glued one grain thick over the flow channel bed (Fig. 5.15). For flows of constant Reynolds number, increasing boundary roughness caused increasing mean bed shear stress, as expected. The turbulence intensity data scale directly with $u^*$, independent of roughness conditions for $h/d > 0.2$, where $h$ = height and $d$ = grain diameter, implying that beyond a certain height the intensity

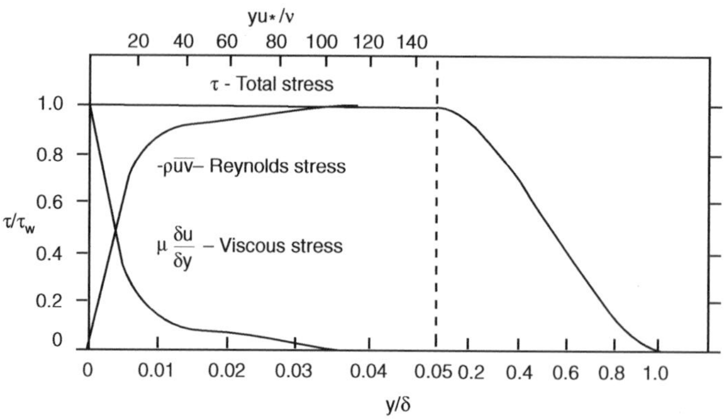


**Fig. 5.13** Distributions of total stress, $\tau$, turbulent Reynolds 'stress', $-\rho\bar{u}'\bar{v}'$, and viscous stress $\mu(\partial u/\partial y)$ across a turbulent boundary layer ($Re = 7.10^4$). Note 30-fold change in the abscissa scale at $y/\delta = 0.05$. The only stress exerted directly at the smooth bed is a viscous one, but *just* away from the bed, turbulence generates an increasingly dominant contribution to the total stress. (After Grass, 1971.)

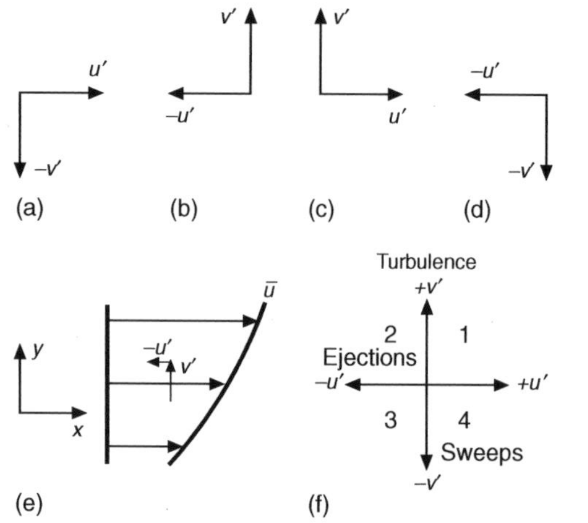


**Fig. 5.14** (a)–(d) Geometric interpretations of turbulent Reynolds 'stress'. (After Tritton, 1988). Patterns (a) and (b) dominate the turbulent boundary layer, giving rise to positive contributions to these 'stresses'. (e) A sketch to show how a Reynolds 'stress' arises from a burst motion in a mean velocity gradient. (f) A quadrant plot to show the four possible types of eddy motion in turbulent flows. Quadrants 1 and 3 contribute little to turbulent 'stress'.

depends solely on boundary distance and shear stress but is independent of the conditions producing the shear stress. Closer to the bed the data separate so that, with increasing boundary roughness, the longitudinal turbulence intensity decreases and the vertical intensity increases. The average Reynolds acceleration measurements are equivalent to a linear mean shear-stress distribution tending towards zero at the flow free surface. We can thus envisage viscous sublayer fluid and the fluid trapped between grain roughness elements as 'passive' reservoirs of low-momentum fluid that is drawn upon during ejection phases. Extreme pressure gradient forces over rough-boundaries causes violent fluid entrainment, with vertical upwelling of fluid from between the roughness elements. Very significantly the streaky pattern of the viscous sublayer observed on the smooth boundary is much less conspicuous in transitional and rough-boundary flows. Faster deceleration of sweep fluid on the rough boundary due to form drag of the grains is thought to cause the decrease of longitudinal turbulent intensity and the increase of vertical turbulent intensity noted above. The turbulent acceleration contribution is also increased by increased roughness close to the wall region.

## 5.5 Shear flow instabilities, flow separation and secondary currents

In many sedimentological contexts, fluid flow may occur within or at the base of another fluid; the enclosing fluid is termed the *ambient* fluid which may itself be moving or stationary. Density currents in air (like volcanic pyroclastic flows and surges) or under water (like turbidity currents, Chater 8) are examples.

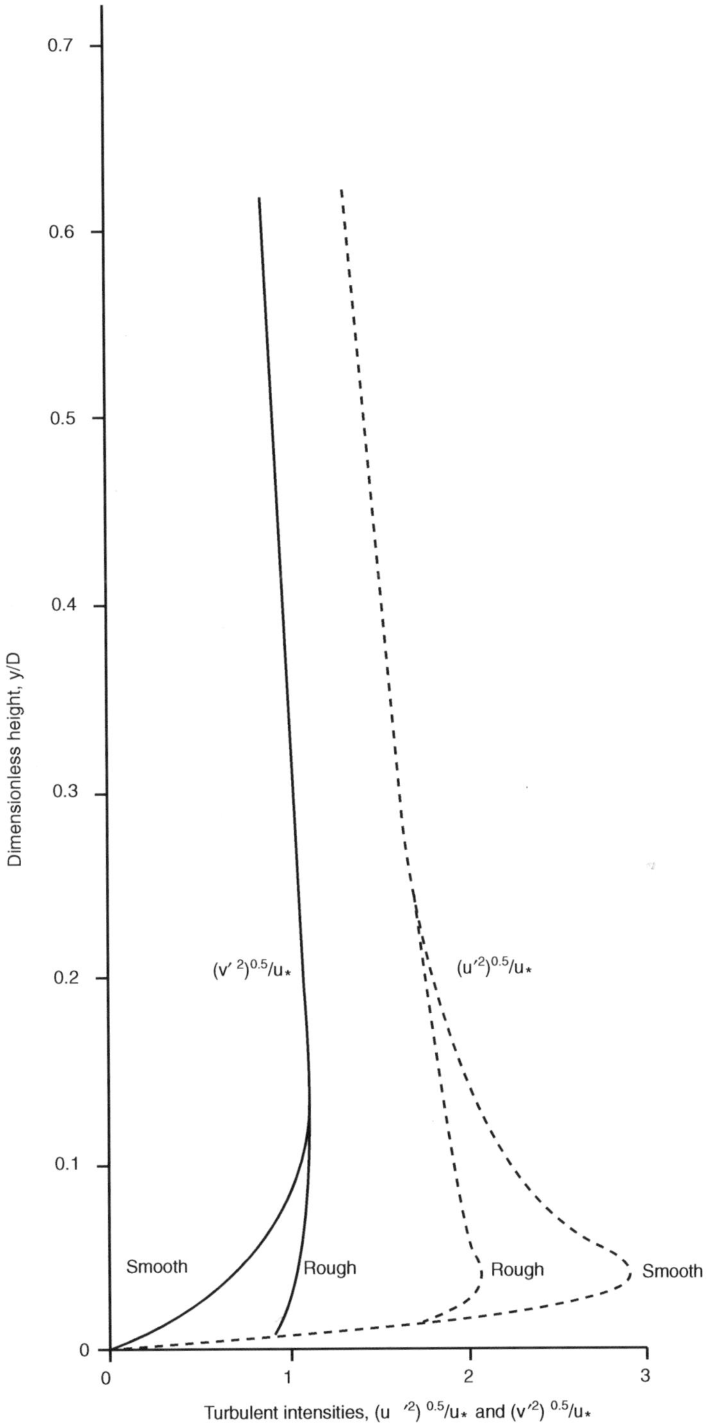


**Fig. 5.15** The effect of a rough boundary upon turbulence production intensity and Reynolds stresses in turbulent flows of similar *Re* (6700). Note the increase of vertical turbulence intensity and the decrease in horizontal intensity from smooth to rough boundaries close to the bed and their independence from roughness further from the bed boundary. (Data of Grass, 1971.)

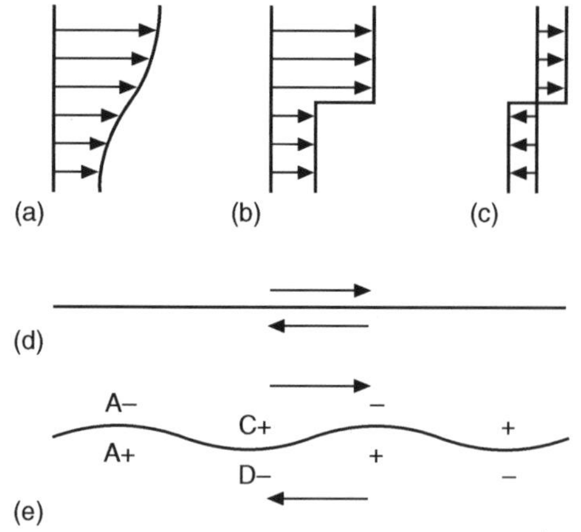


**Fig. 5.16** Sketches to illustrate the cause of Kelvin–Helholtz instability. (After Tritton 1988.) (a)–(c) Various velocity distributions that might cause the instability. (a) An approximation to a turbulent boundary layer. (b) and (c) Fluid layers in relative motion such as density-stratified flows. (d) An initially planar shear layer. (e) Wavy shear layer produced by pressure differences. At locations A– and D– fluid is accelerated and pressure drops (by Bernouilli's theorem, see **Cookie 18**); vice versa for A+ and C+. The system is thus unstable and deforms in response to the vertical pressure gradients until some equilibrium is reached with viscous resistance across the boundary.

In such cases there is shear not only between the moving current and a solid bed, but also between the current and the overlying fluid; we may imagine the situation as like those sketched in Fig. 5.16, with a planar shear layer between two fluids. Now, such a situation becomes unstable if some undulation or irregularity appears along the shear layer, for any acceleration of flow on the part of one fluid will tend to cause a pressure drop and an accentuation of the disturbance. Very soon a striking, more-or-less regular, system of vortices appear (Fig. 5.17) rotating about approximately stationary horizontal axes with respect to the plane of shear. Such vortices are termed *Kelvin–Helmholtz instabilities*. These are highly important mixing mechanisms and the cause of frictional energy losses along pyroclastic flows (Part 2 Fig. 1D), at river tributary junctions (Part 2 Fig. 1F), on the lee side of dunes and on the fronts and tops of both subaqueous (Fig. 5.18) and atmospheric density currents.

The phenomenon of *flow separation* (see Fig. 4.15A) is a consequence of the slowing down of fluid as it moves past certain boundaries. Consider flow over a bed convexity (Fig. 5.19); the mean streamlines converge and then diverge. From the continuity equation (Chapter 4; **Cookie 15**) the flow will speed up over the bump and then slow downstream. Bernoulli's equation (**Cookie 18**) states that fluid pressure should decrease in the accelerated flow section. Pressure and velocity gradients exist that are favourable upstream ($\delta p/\delta x > 0$) and adverse downstream ($\delta p/\delta x < 0$). The pressure gradient has greatest effect on low-speed fluid near the bed since it is more easily retarded there; it tries to push fluid close to the wall upstream (Fig. 5.19). Boundary-layer separation begins at the *separation point*, *S*, whilst *reattachment* occurs some distance

**Fig. 5.17** Beautiful Kelvin–Helmholtz vortices produced at an interface between two immiscible fluids of contrasting density. The long tube was initially horizontal with a stable stratification of a clear, less-dense fluid overlying a dark, denser fluid. The tube is then carefully tilted and the upper, less dense, clear fluid moves upwards with respect to the darker, denser fluid, with resultant shear and vortex formation across the boundary. (From van Dyke, 1982, and sources cited therein.)

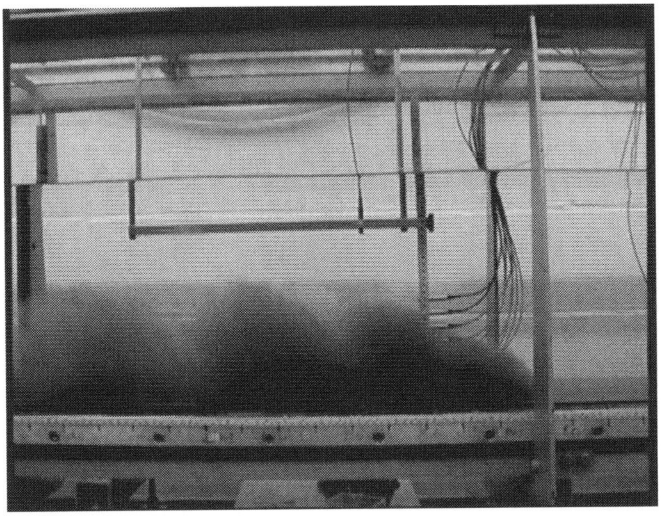

**Fig. 5.18** Digital video image of a train of well-developed Kelvin–Helmholz waves on the upstream-shearing interface of an experimental lock-exchange turbidity current 30 s after release (see Chapter 8 for stuff on turbidity currents). Image taken from a fixed camera approximately 2 m downstream of the lock gate. The flow was generated from a mixed grain-density suspension comprising 50% silicon carbide (3214 kg/m$^3$) and 50% glass ballotini (2650 kg/m$^3$), both with a mean grain size of 70 mm. The flow has a volume concentration of 1% and a bulk density of 1019 kg/m$^3$. Ultrasonic Doppler velocity profiler probes can be seen suspended in the flow. Water depth is 0.30 m and the tank floor marked in 0.05 m intervals. The uppermost probe aligned parallel to the flow is suspended 136 mm above the tank floor. The distance between the uppermost visible probes is 20 mm. (Image and data courtesy of James Hodson.)

downstream. Boundary-layer separation also occurs around spheres or cylinders placed in a flow or when a negative step or small defect occurs on a bed (Chapter 7). Both laminar and turbulent boundary layers can separate, the former more readily. In sedimentology, when flow is occurring over bedforms like ripples or dunes (Chapter 7) we are most interested in turbulent separation. We can define two types of separation zone. *Vortex bubbles* form when a downward ramp or step is skewed $< \sim 45°$ from the flow direction. *Roller bubbles* are skewed $> \sim 45°$ to flow direction. The streamlines of a roller are closed loops and those in a vortex are helical spirals. Important effects arise at the upper junction of the bubbles with the mainstream fluid. Here we have relative motion between two flows, which produces strong vortices of the Kelvin–Helmholtz type along the unstable interface (Chapter 7).

*Secondary flow* is a name given to more-or-less regular patterns of streamlines that diverge from the mean direction of downcurrent or downslope flow. It is common in both straight and curved channels (see Chapter 11 and **Cookie 42**) for an account of the latter). In straight channels the effect takes the form of streamwise vortex cells showing divergence of flow towards the channel wall and a compensatory flow inwards and across the bottom. Several pairs of these longitudinal vortices may occur. The effect is due to an imbalance of turbulent accelerations at the wall compared to the main body of the flow, the velocity gradients setting up forces that cause the observed flow pattern. Such stable secondary vortical flows are known as Taylor–Görtler vortices, responsible for the transport of sand in prominent flow-parallel 'windrows' and for erosion of flow-parallel 'gutter-marks' (Chapter 7). The vortices set up their own transverse pattern of greater and lesser bed shear stresses that cause longitudinal, flow-parallel bedform elements to form.

There are also several other types of natural secondary flows, due to spanwise changes in roughness for example. Here coarser sediment makes up ridges and finer sediment troughs. This size segregation is self-sustaining because the coarser ridges cause higher-than-average vertical turbulent motions, and

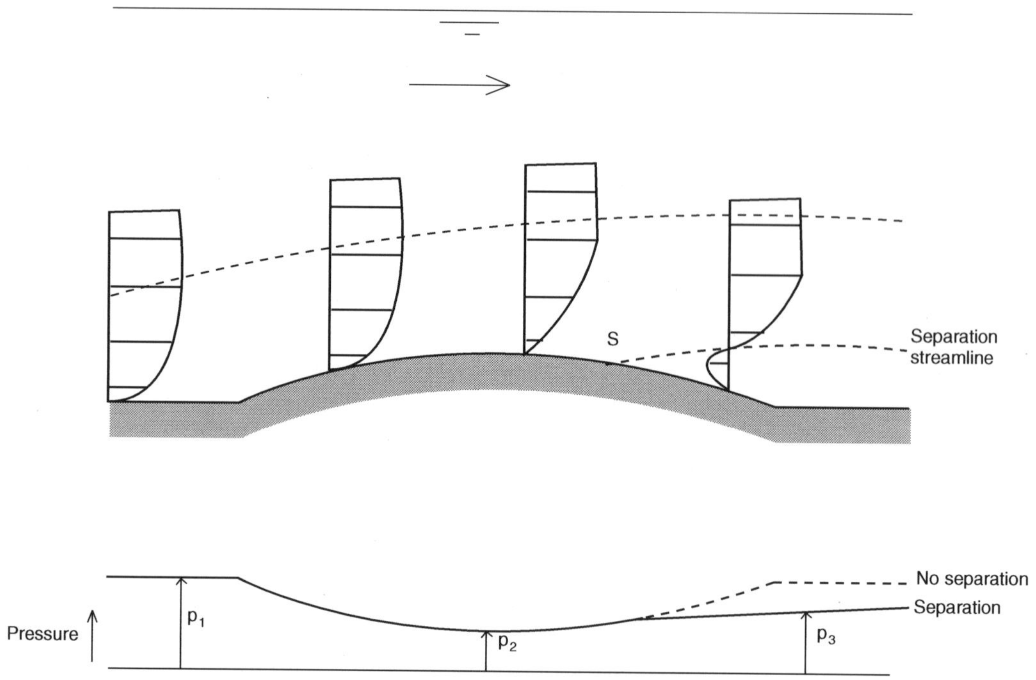


**Fig. 5.19** Diagram to illustrate how flow over a convex surface gives rise to an upstream negative and then a downstream positive pressure gradient, the latter causing flow separation from *S*, the separation point. (After Massey, 1979.)

vice versa for the finer troughs. A stable secondary circulation is thus set up that transfers upwelling fluid over the crests to downwelling fluid in the troughs.

## 5.6 Subcritical and supercritical flows: the Froude number and hydraulic jumps

It is often apparent that the free surface of flowing turbulent water streams can easily be roughened during a strong wind or breeze by the formation of a myriad of small water waves on the surface—a particular nuisance when fishing. Now, when the flow of water is relatively slow and the wind is blowing upstream it can be seen that wind shear causes the waves themselves to propagate upstream. Or the same thing occurs when a pebble is dropped into the flow, the radiating solitary waves migrate upstream. This state of the water flow is said to be *tranquil* or *subcritical*. As the water speed increases there comes a time when the wind, no matter how strong it blows, or a pebble no matter how big, is unable to transmit waves upstream and the water surface in the area of flow appears glass-like, unruffled by waves. Moreover, at this time the general water surface itself may begin to deform into larger scale sinusoidal *standing waves*; the flow is now said to be *rapid* or *supercritical*. Observations such as these led the pioneer naval architect, William Froude (pronounced Frood) to introduce a dimensionless number (derived from the Bernouilli equation, see **Cookie 18**) that could be used to separate the two regimes of flow. Froude reasoned that since any flow has inertia and that the surface waves indicate the flow free surface was deforming due to action of a gravity force, then the ratio of the inertial and gravity forces acting would serve to delimit the flow types. With the water flow having a mean speed, $u$, and the velocity of movement of a surface gravity wave (**Ch. 5.8**) being $\sqrt{gh}$, where $h$ is water depth, their ratio is $u/\sqrt{gh}$ which defines the Froude Number, $Fr$. When $Fr > 1$ (favoured by the combination of fast and shallow flows) then waves cannot travel upstream. Standing or upstream-breaking waves are found to exist when $Fr \geq 0.84$.

*Hydraulic jumps* are familiar when a water jet from a tap strikes a surface—a discontinuity appears as a heightened concentric ring (Fig. 5.20). For a steady

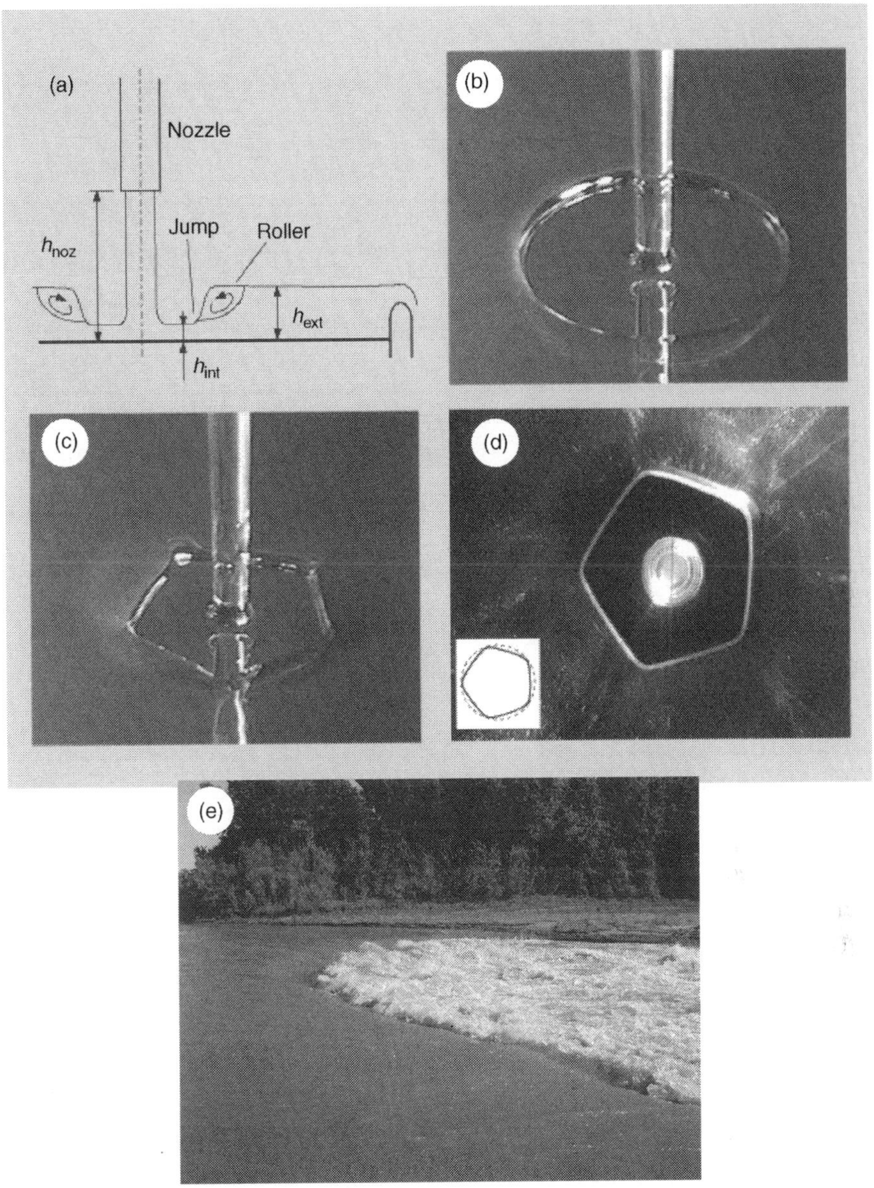


**Fig. 5.20** Hydraulic jumps in the kitchen sink and in the field. (a) Cross-sectional flow structure. The roller vortex overlies a zone of fluid travelling radially outwards and away from the jump. (See MacDonald *et al.* (2009) for more details of flow structure.) (b)–(d) Video images of three experimental states. The fluid is ethylene glycol (antifreeze; about 99% pure) and typical steady flow rates, $Q$, are in the range 30–50 mL/s. The external height, $h_{ext}$, is increased from (b) to (c), causing the familiar circular jump in (b) seen in all water flows to change spontaneously to polygons with many corners (not shown), then to a pentagon (c), viewed from underneath in (d). In (d), the bright circle in the centre is the vertical jet; the corners of the surrounding polygonal jump are clearly evident. The inset shows one solution for a model. (e) Spectacular natural hydraulic jump formed in a ca 50 m wide cut-off channel whose flow is passing to the right over a flute-like negative bed step. Flow is accelerating and supercritical upstream to the left and decelerating and subcritical to the right. The degradation of flow energy and raised lip of the roller vortex is spectacularly vivid. ((a)–(d) after Ellegaard *et al.*, 1998; (e) pers. comm., March 2009, from Jim Best and Bruce Rhoads, University of Illinois.)

flow this stationary, wave-like deformation defines a circular hydraulic jump. For fluids with higher viscosity than water, the jump is polygonal. The jump, a more-or-less vertical discontinuity, occurs at the exact junction between shallow, fast flow and deeper, slower flow at some particular distance from the impacting jet, exactly where supercritical flow is succeeded radially across-flow by subcritical flow. Such conditions in nature occur commonly when a flow boundary changes from high to lower slope and flow deceleration is accompanied by downstream flow thickening. This causes a rapid pressure rise (from Bernouilli's theorem, **Cookie 18**) and flow separation occurs in the form of spectacular macroturbulence in a separation *torus vortex* that indicates the degradation of flow kinetic energy into thermal energy—outward flow of fluid occurs below the torus. Note there is no change in *Re* across a hydraulic jump for decrease in mean flow velocity is compensated for by an increase in flow depth.

## 5.7 Stratified flow generally

Many sedimentologically important flows occur because of instabilities due to local density contrasts. A stratified flow generally is one which exhibits some vertical variation in its density brought about because of heat energy transfer, salinity variations or in our case the effects of suspended solids. If the density increases with height, e.g. due to a desert surface heating an incoming wind, then the situation is unstable and the forward transport of fluid is accompanied by turnover motion that tries to reverse the unstable stratification. This is an example of *forced convection*. Here we are interested in the case of a *stable stratification* in which density decreases vertically, as might be produced in a wind blowing over a cool surface or being heated from above, or in a flow in which the concentration of suspended sediment decreases vertically, like in a turbidity current (see Fig. 5.18). Left to its own devices a stationary stratified fluid or a purely laminar stratified flow would simply slowly lose its density contrast by molecular diffusion. Once put into turbulent motion, however, burst motions in the stratified flow boundary layers will try to lift denser fluid upwards and sweep motions carry less dense fluid downwards. In both cases work must be done by the flow against resisting buoyancy forces. We might imagine in the most general way that the turbulent energy may be either capable of overturning the buoyancy or incapable. In the former case the stratification tends to be destroyed by turbulent mixing, in the latter it remains there and the turbulence itself is dissipated. In both cases an inhibition exists against the tendency for turbulent mixing and a loss of turbulent energy results.

### Richardson criteria for shear stability of density-stratified flow

Turbulent accelerations in the boundary layer of a stratified fluid with a negative-upward gradient of density undergoing shear have to overcome resistance due to buoyancy. Appropriate dimensions of applied inertial force per unit volume and resisting buoyancy force are $\rho \boldsymbol{u}^2/l$ for the former and $-\Delta\rho g$ for the latter, where $\rho$ is mean fluid density, $\boldsymbol{u}$ is mean flow velocity, $l$ is the thickness of the flow and $\Delta\rho$ is the density difference across the flow. Taking the ratio of these, all dimensions cancel and we have the simplest possible form of what is known as the bulk *Richardson number*, *Ri*. The smaller the value the more likely it is that any stratified shear flow will undergo mixing and homogenization. Although this derivation has the correct basic physical principles, it somewhat ignores the physical situation envisaged, i.e. a shear flow with a continuous stable vertical variation of density. The two relevant gradients are those of velocity, and density $d\rho/dy$ Most sedimenting flows are characterized by a negative velocity and density gradients. The two can be combined to give the *gradient Richardson number*, $Ri_g$, written as $Ri = -g(d\rho/dy)/\rho(du/dy)^2$ and controlled by the sign of the density gradient. Negative *Ri* means that there is a destabilizing density gradient and that both boundary layer shear (dominant at low $-Ri$) and buoyancy (dominant at high $-Ri$) generate turbulence. Positive *Ri* means that the density gradient is a stabilizing influence, high values eventually leading to turbulent dissipation

### Stratification and the phenomenon of double diffusion

The dual control of density by temperature and salinity in ocean waters leads to an interesting scenario because adjacent water masses lose thermal contrast much more quickly than they can lose salinity contrast. This is because the molecular diffusion (conduction) of heat is ca $10^2$ faster than the molecular diffusion of salinity. We must imagine a scenario of *metastable stratification* of water layers. Consider for example, an upper salty warm layer with an initial density, $\rho_1$, less than that of a lower cool, fresh layer, $\rho_2$. Such a scenario is to be widely expected in the oceans as a consequence of summer evaporation and warming of surface layers, or to the inflow and outflow of contrasting water masses. The situation leads to enhanced mixing by convection, sometimes called *double-diffusive convection*, at much greater rates than mixing by molecular diffusion. In this process, more rapid heat diffusion across zones of thermal contact cause the stable stratification to break down. In our example, cooling of the saltier layer from below across the boundary of thermal contact causes the cooled saltier fluid and any contained suspended sediment to fall. An intricate pattern of small-scale mixing gradually develops as moving fluid 'fingers' its way downwards. Such double-diffusive instabilities can set up regular layering in the water column, with layer boundaries having high rates of change of temperature and salinity.

## 5.8 Water waves

Waves are widespread phenomena that transfer energy and sometimes mass. Thus physicists postulate the existence of sound, shock and electromagnetic waves, and we observe waves of concentration each time we enter and leave a stationary or slow-moving traffic jam. Surface water waves are more-or-less regular periodic disturbances of the water surface created by surface shear due to blowing wind, the regular or irregular impact of objects, and the sudden motion of the adjacent land due to tectonic faulting. We can make waves of period 1–10 s or so in laboratory tanks using paddles or other oscillatory devices. A great range of waveforms affect the oceans, with periods ranging from $10^{-2}$ to $10^5$ s (Fig. 5.20). Many waves are in motion, travelling from here to there as progressive waves, although the waveform of some are too low frequency to observe directly, like tides. Yet others are standing waves, like the resonant standing oscillations present in many coastal inlets and estuaries. The commonest visible signs of wave motion are the surface waveforms of lakes and oceans. In the oceans, waves are usually superimposed on a flowing tidal or storm current; these tidal and combined flows are particularly important in marine sediment transport. In later chapters we shall discuss waves at density interfaces in connection with the motion of density or turbidity currents and the astonishing solitary waves seen in Nature as tidal bores, tsunamis and reflected density currents.

The chief features and simple physical expressions for the properties of an oscillatory wave are shown in Fig. 5.21. The overall shape of a wave is a curve-like form and this smoothly varying property is a simple

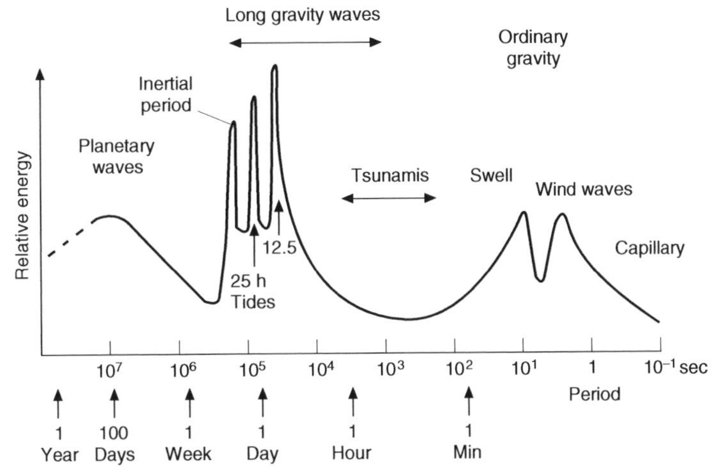


**Fig. 5.21** The relative energy spectrum of various oceanic wave motions. (After Pond & Pickard, 1983.)

mathematical guide to our study of wave physics (**Cookies 25–27**). It is a common mistake to imagine all water waves as heaps and troughs of water moving along a surface. Just one waveform, the *solitary wave* (see below), approximates to this mode. In other oscillatory waves travelling in deep waters, it is just the wave energy that is transferred, without involving *net* forward water motion. The stationary observer, fixing their gaze at a particular point such as a partially submerged marker post, will see the water surface rise and fall up the post as a wave passes by through one whole wavelength. This rise and fall signifies the conversion of wave potential to kinetic energy.

Wind-generated waves are a form of surface gravity wave driven by a balance between fluid inertia and the restoring force of gravity. The simplest approach is to set the shape of the waveform along an $x$–$y$ graph and consider that the periodic upward motion of $y$ will be a function of distance $x$, wave height, wavelength and celerity. Thus we have $y = f(x,H,\lambda,c)$. Attempts to investigate wave motion in a rigorous manner assume that the wave surface displacement may be approximated by curves of various shapes (Fig. 5.22), the simplest of which is a harmonic motion used in linear (Airy) wave theory. We see from this theory (**Cookie 25**) that surface gravity waves travelling over very deep water are *dispersive* in the sense that their

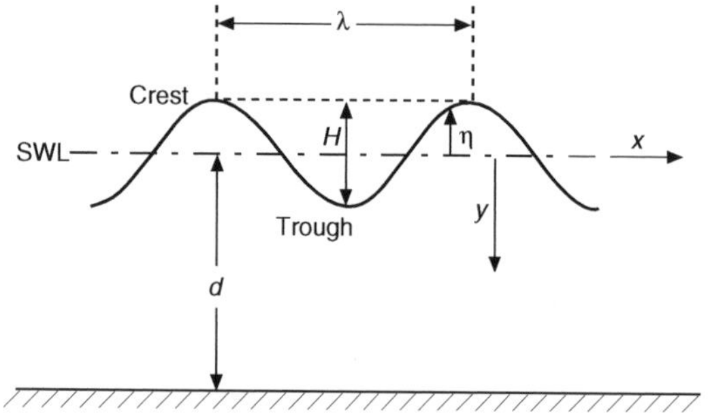


**Fig. 5.22** Standard parameters and definitions for surface water waves. Parameters: $\lambda$ = wavelength from crest to crest or trough to trough; $H$ = wave height; $a$ = amplitude = $H/2$; $\eta$ = displacement of water surface from its still water level. The speed of a wave motion is termed celerity, symbol, $c$. From these parameters we may define: celerity = $c = \lambda/T$ or $\sigma/k$; wave period, $T = \lambda/c$; wave frequency = $\sigma$ = $1/T$ or $2\pi/T$ (radian frequency); wavelength = $cT$; wavenumber = $1/\lambda$ or $2\pi/T$ in radian frequency.

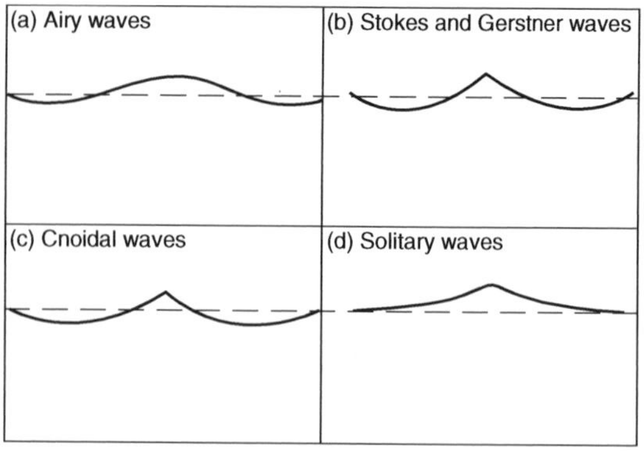


**Fig. 5.23** Profiles of wave types derived by mathematical analysis. (a) Airy waves are simple harmonic (sinusoidal) forms of small amplitude in deep water. (b) Stokes and Gerstner waves are trochoidal deep-water forms. (c) Cnoidal waves are trochoidal shallow- to intermediate-water forms. (d) Solitary waves are single forms and are the only ones that move water mass forwards, as distinct from energy, over shallow water (imagine them as heaps of water moving forwards).

rate of forward motion is directly dependent upon wavelength. Wave height and, of course, water depth (by definition) play no role in determining wave speed. This linear theory of sinusoidal *deep-water waves* predicts that at any fixed point the fluid speed caused by wave motion remains constant whilst the direction of motion rotates with angular velocity $\omega$. Any fixed surface particle must undergo a circular rotation below deep-water waves (Fig. 5.23). At any one instant the water in any vertical plane is all moving uniformly according to position with respect to the wavy surface (Fig. 5.24). These instantaneous sheets of flow lie adjacent to other sheets slightly out of phase and so on. The radius of these *orbital motions* decreases exponentially below the surface according to $H\exp(-2\pi y\lambda)$. Most wave energy (about 95%) is thus concentrated in the half-wavelength or so depth below the mean water surface.

Want to know more about deep-water wave theory ? Turn to Cookie 25

In Nature waves may be quite different in shape and dynamics from that assumed by simple linear

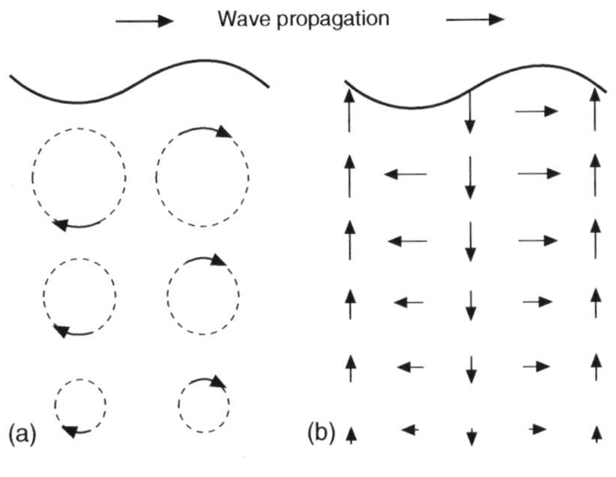


**Fig. 5.24** Two views of orbital motions induced by the passage of waves.(a) Although every water particle rotates about a time-mean circular motion, the circles drawn to illustrate this are *not* stationary vortices shearing past each other. (b) The arrows show instantaneous motion vectors at each arrowhead. (After Vogel, 1994.)

theory of sinusoidal deep-water waves. One reason is that waves appear more irregular, being combinations of different-frequency sinusoidal waves, all superimposed. These can be disentangled using *energy–frequency plots* and *Fourier decomposition*, the energy in a wave being proportional to the square of its height. A second reason is that deep-water wave theory fails when the ratio of water depth to wavelength falls below about 0.5; the deep-water waves pass into *shallow-water waves*, defined when $h < \lambda/20$—they suffer attenuation through bottom friction and significant horizontal motions are induced towards the bottom. The waves take on new forms, with more-pointed crests and flatter troughs. Such waves are obviously of great interest to sedimentologists since most wave-induced sediment transport takes place under them in shallow water. After a transitional period, when wave speed becomes increasingly affected by water depth, shallow-water gravity waves move with a velocity that is proportional to the square root of the water depth, $c = \sqrt{(gh)}$, i.e. independent of wavelength or period. The wave orbits of shallow water waves are elliptical at all depths (Figs 5.25 & 5.26), with increasing ellipticity towards the bottom, culminating at the bed as horizontal straight lines representing to-and-fro motion. Under such conditions the maximum orbital horizontal velocity is given by: $u_{max} = (H/2h)\sqrt{(gh)}$, where $H$ is wave height.

Want to know more about group wave theory ? Turn to Cookie 26

## Surface wave energy and radiation stresses

As noted previously, most wave energy (about 95%) is concentrated in the half-wavelength or so depth below the mean water surface; the rhythmic conversion of potential to kinetic energy and back again maintains wave motion. The displacement of the wave surface from the horizontal provides potential energy that is converted into kinetic energy by the orbital motion of the water. The total wave energy per unit area is given by $E = 0.5\rho g a^2$, where $a$ is wave amplitude ($= 0.5$ wave height $H$). Note carefully the energy dependence on the square of wave amplitude. The energy flux (or wave power) is the rate of energy transmitted in the direction of wave propagation and is given by $w = Ecn$, where $c$ is the local wave velocity, and the coefficients are $n = 0.5$ in deep water and $n = 1$ in shallow water. In deep water the energy flux is related to the wave *group velocity* (**Cookie 26**) rather than to the wave velocity. Because of the forward energy flux, $Ec$, associated with waves approaching the shoreline, there also exists a shoreward-directed momentum flux or stress outside the zone of breaking waves. This is termed *radiation stress*.

Want to know more about shallow-water wave theory and bottom stresses set up by waves? Turn to Cookie 27

## A note on tsunami

Deep-water wave theory fails when water depth falls below about $0.5\lambda$. This can occur even in the deepest oceans, for the tidal wave (see Chapter 18) and for very long (10 s to 100 s km) wavelength *tsunami*. The Indian Ocean tsunami of December 2004 focused world attention on such wave phenomenon. Tsunami is a Japanese term meaning 'harbour wave'. They are generated as the sea floor is suddenly deformed by earthquake motions or landslides; the water motion generated in response to deformation of the solid boundary propagates upwards and radially outwards to generate very long wavelength (100 s km) and long

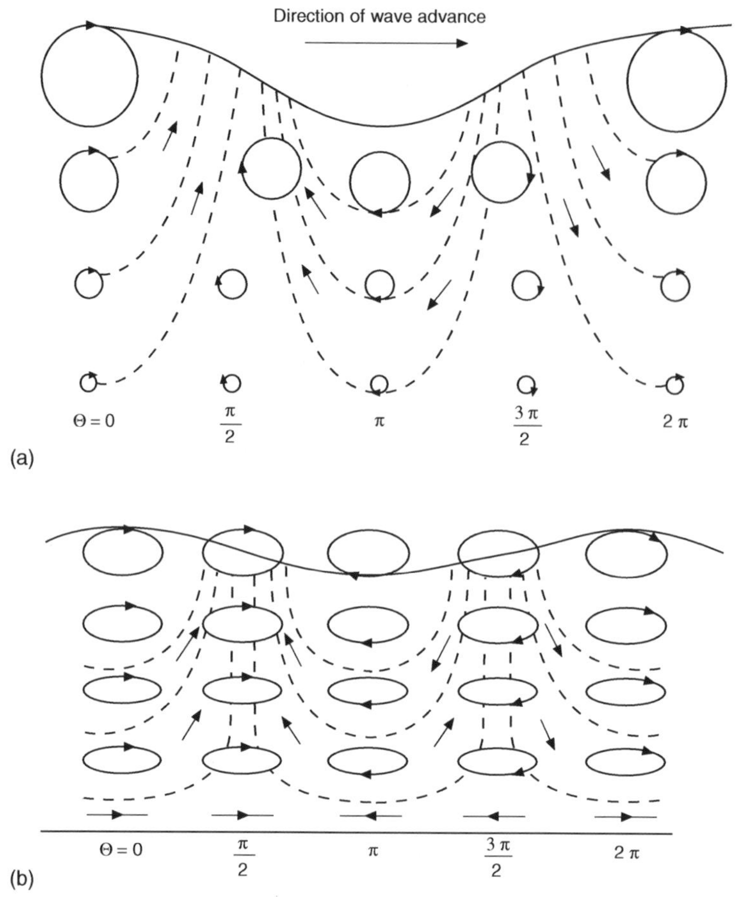


**Fig. 5.25** Orbits, relative orbital velocities and streamlines. (After McLellan, 1965.) (a) For deep-water waves. (b) For shallow-water waves. In the latter case, although the orbits are shown closed, there is in fact a net transfer of water mass in the direction of wave propagation (see Fig. 5.26).

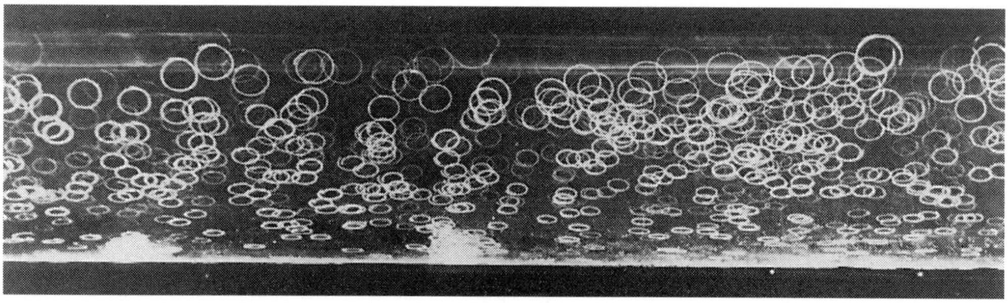

**Fig. 5.26** Flow visualization of suspended grains photographed orbiting under a shallow-water wave traversing one wavelength left to right. Wave amplitude is $0.04\lambda$ and water depth is $0.22\lambda$. The clockwise orbits pass from near-circular at the surface to ellipses of increasing elongation towards the bottom. Some open loops indicate a slow near-surface drift to the right and a near-bed drift to the left. (From van Dyke, 1982, and sources cited therein.)

period (> 60 s) surface wave trains. By very long we mean that wavelength is much greater than oceanic water depth and hence the waves travel at tremendous speed, governed by the shallow-water wave equation $c = gh$. For example, such a wave train in 3000 m water depth gives wave speeds of order 175 m/s or 630 km/h. Tsunami wave height in deep water is quite small, perhaps only a few decimetres. The smooth, low, fast nature of the tsunami wave means wave energy dissipation is very slow, causing very long (could be global) runout from source. As in shallow-water surface gravity waves at coasts, tsunami respond to changes in water depth and so may curve on refraction in shallow water (Chapter 18). Accurate tsunami forecasting depends on the water depth being known very accurately, for example in the oceans a wave may travel very rapidly over shallower water on oceanic plateaux. During run-up in shallow coastal waters, tsunami wave energy must be conserved during very rapid deceleration: the result is substantial vertical amplification of the wave to heights of tens of metres. Massive sediment transport occurs under such conditions, the resulting depositional and erosional phenomena termed *tsunamites* (Chapter 18)

### A note on solitary waves

Especially interesting forms of *solitary waves* or *bores* may occur in shallow water due to sudden disturbances affecting the water column. These are very distinctive *waves of translation*, so-termed because they transport their contained mass of water as a raised heap, as well as transporting the energy they contain (Figs 5.27 & 5.28). These amazing features were first documented by J.S. Russell who came across one in 1834 on the Edinburgh–Glasgow canal in central Scotland. Here are Russells own vivid words, written in 1844:

> *I happened to be engaged in observing the motion of a vessel at a high velocity, when it was suddenly stopped, and a violent and tumultuous agitation among the little undulations which the vessel had formed around it attracted my notice. The water in various masses was observed gathering in a heap of a well-defined form around the centre of the length of the vessel. This accumulated mass, raising at last to a pointed crest, began to rush forward with considerable velocity towards the prow of the boat, and then passed away before it altogether, and, retaining its form, appeared to roll forward alone along the surface of the quiescent fluid, a large, solitary, progressive wave. I immediately left the vessel, and attempted to follow this wave on foot, but finding its motion too rapid, I got instantly on horseback and overtook it in a few minutes, when I found it pursuing its solitary path with a uniform velocity along the surface of the*

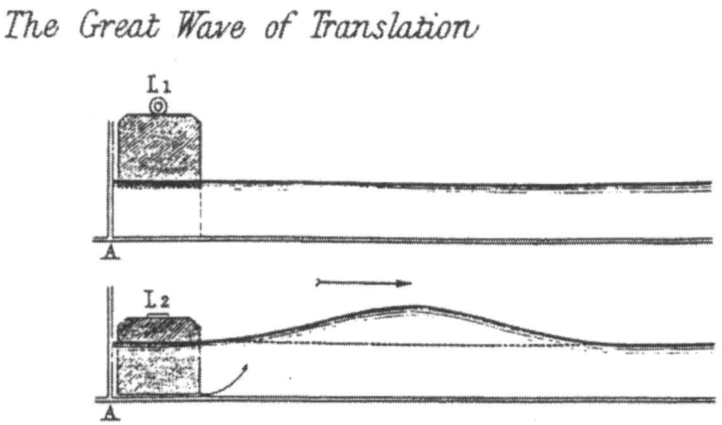


**Fig. 5.27** Solitary waves: Russell's original sketch to illustrate the formation and propagation of a solitary wave. You can achieve the same effect with a simple paddle in a channel, tank or bath. The solitary wave is raised as a 'hump' of water above the general ambient level (Fig. 5.23d). The 'hump' is thus transported as the excess mass above this level, as well as by the kinetic energy it contains by virtue of its forward velocity, *c*.

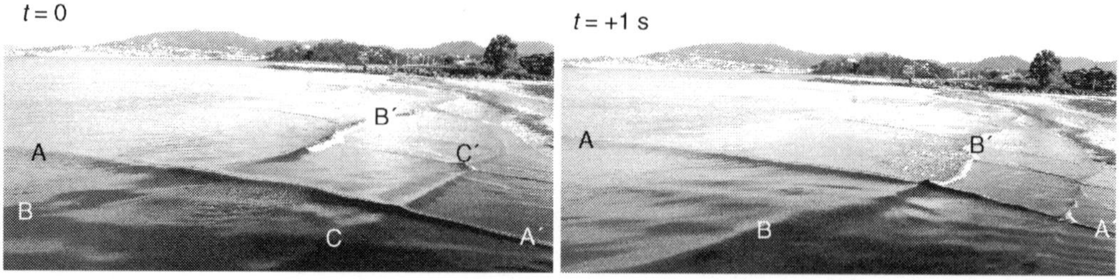


**Fig. 5.28** Photographs of interacting solitary and normal gravity waves, Baiona, Galicia, northwest Spain. Solitary wave A–A′ has just formed as a reflected wave from harbour wall behind and to the left. The views show the wave moving forward (c = ~1 m/s) through incoming shallow-water waves B–B′ and C–C′ with little deformation or diminution.

*fluid. After having followed it for more than a mile, I found it subside gradually, until at length it was lost among the windings of the channel.*

Briefly, a solitary wave is equivalent to the top half of a harmonic wave placed on top of undisturbed fluid, with all the water in the waveform moving with the wave; such bores, unlike surface oscillatory gravity waves, transfer water mass in the direction of their propagation. Somewhat paradoxically we can also speak of trains of solitary waves within which individuals show *dispersion* due to variations in wave amplitude. They propagate without change of shape, any higher amplitude forms overtaking lower forms with the very remarkable property, discovered in the 1980s, that, after collision, the momentarily combining waves separate again, emerging from the interaction with no apparent visible change in either form or velocity (Fig. 5.28). Such solitary waves are called *solitons*. Solitary waves are important agencies of sediment transport in lakes and in the oceans since they can evolve from reflected or refracted turbidity currents (Chapter 8).

### Internal waves

As noted above, the water column is often stratified, with sharply defined sublayers (Fig. 5.29) which may differ in density. The density contrast between layers is small enough (in the range 3–20 kg/m$^3$, or 0.003–0.02) so that the less dense and hence buoyant surface layers feel the drastic effects of *reduced gravity*. Any imposed force causing a displacement and potential energy change across the sharp interface between the fluids below the surface is now opposed by a reduced gravity restoring force, $g' = (\Delta\rho/\rho)g$. The wave propagation speed, $c = gh$ is also reduced in proportion to this reduced gravity, as $c = g'h$, whilst the wave height is very much larger. Internal waves of long period and high amplitude progressively 'leak' their energy to smaller length scales in an *energy 'cascade'*, causing turbulent shear that may ultimately cause the waves to break. This is an important mixing and dissipation mechanism for heat, energy and turbidity in both lakes and the oceans.

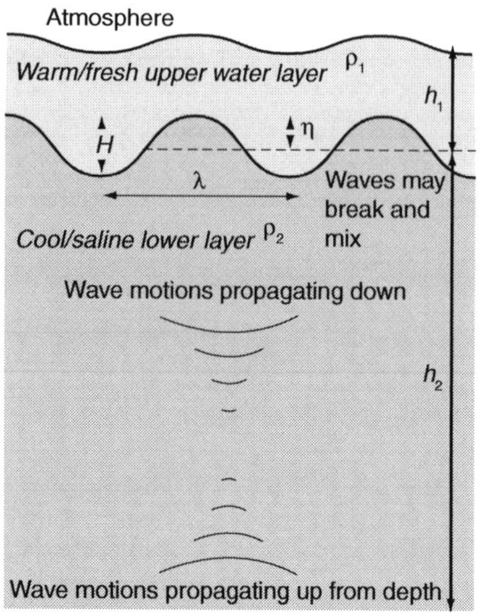


**Fig. 5.29** Diagram to show internal waves at a sharp density interface. Experimental versions of these may be studied at leisure in a tall glass of carefully prepared, café latte.

## 5.9 Tidal flow—long-period waves

The periodic rise and fall of sea level visible around coastlines has long fascinated both scientist and mariner alike. Data from tidal gauges reveal a variety of types of tides with important modifying factors arising from (i) the irregular morphology of the land–sea boundary, (ii) variable water depth, (iii) the Coriolis force produced by Earth's rotation, (iv) inertia of the water mass and (v) frictional effects. The nature of the tidal oscillation depends critically on the natural periods of oscillation of the particular ocean basin. These coincide with the 12-h tide-forming forces in the Atlantic Ocean, giving *semidiurnal tides*. The Gulf of Mexico oscillates in sympathy with the 24-hour forces, giving *diurnal tides*, whilst *mixed tides* result in the Pacific Ocean, which does not oscillate so regularly.

From the account of the origins of the principal lunar diurnal tide in **Cookie 28**, an observer fixed with respect to the Earth might envisage the tidal wave advancing progressively from east to west. In fact, the tides of the world must evolve on a rotating ocean whose water depth and shape are highly variable with latitude and longitude. The result is that discrete rotary and standing waves dominate the oceanic tide such that, at the oceanic margins, as in straits and channels on continental shelves, the tidal wave may advance in almost any direction. Advance of the tidal wave in estuaries that narrow upstream is accompanied by amplification, shortening and steepening of the tidal wave whose ultimate form is that of a spectacular bore. Further, the tide often occurs as a *standing wave*, as off the east coast of most of North America, where tidal currents are zero in the centre of the oscillating water near the shelf edge and reach a maximum at the margins (*nodes*) where the shelf is broadest. They are zero everywhere at high and low water and reach maximum values at half-water stages. The amplitude of the oscillation is greatest when the periods of the local sea and the oceanic tides coincide at about 12 h. It has become conventional to describe tidal ranges according to whether they are *macrotidal* (range $> 4$ m), *mesotidal* (range 2–4 m) or *microtidal* (range $< 2$ m), but it should be borne in mind that tidal range always varies very considerably with location in any one tidal system.

It is usual to regard the relatively thick tidal-wave boundary layer to be fully turbulent. There is also little lag (a few tens of minutes at most) between the local bed shear stress beneath the boundary layer and the sediment transport rate, a state of affairs rather different from surface gravity waves of much shorter period than the tidal wave. Here the maximum in transporting capacity often occurs at the beginning of reversal of the wave motion. The velocity of any tidal current clearly varies in both magnitude and direction over the tidal cycle. Most simply, consider a symmetrical harmonic tide and the variations of velocity with time that result. Any sediment on the bottom will suffer periods of motion, deceleration, rest, acceleration and so on during the course of the tidal cycle. It is an error, though, to consider the tidal wave as purely a symmetrical oscillator since for one thing their is a diurnal inequality in all tides and also because of the occurrence of strong harmonic constituents of the major lunar tide. The vector variation of tidal currents may be usefully summarized by means of a tidal current ellipse (Chapter 19).

Want to know more about tide-raising forces and the bottom stresses set up by currents? Turn to Cookie 28

## Further reading

### *General*

Many turbulent structures are beautifully illustrated in van Dyke (1982) and Samimy *et al.* (2003). Clifford *et al.* (1993) and Ashworth *et al.* (1996) are research monographs with many sedimentologically-relevant papers on turbulence. Once more, Tritton (1988) will set you on the right path for the mathematical physics of turbulence whilst the Homsy *et al.* (2007) CD ROM will inspire you to experiment. Bradshaw (1971) has written with great clarity on turbulence, though the techniques he describes for turbulence measurement have changed much in the past 30 years with the advent of probes like laser, acoustic and phase Dopplers. Turbulence is simply introduced in an environmental context by Leeder & Perèz-Arlucea (2005).

### *Specific*

A nice account of O. Reynold's life and work written for the non-technical reader is by Jackson & Launder (2007) – this contains an amusing account of Stokes and Rayleigh as uneasy and rather uncomprehending referees for his now-famous turbulence papers of 1883 and 1895. The classic non-statistical papers on turbulent flow are the visualizations by Kline *et al.* (1967) and Grass (1970). Visual evidence of links to larger scale turbulent structures were first provided by Falco (1977) and Head & Bandyopadhyay (1981). Adrian *et al.* (2000) is a key later paper that addresses the group hairpin concept for large-scale flow eddies.

Secondary flow in non-circular channels and its role in flow-parallel sediment transport and bedform development is discussed and reviewed by Gerard (1978), Pantin *et al.* (1981) and Colombini (1993), the latter needing advanced fluid dynamic knowledge to be comprehensible.

Hydraulic jumps in gravity currents are clearly discussed by Ellegaard *et al.* (1998), MacDonald *et al.* (2009) and Waltham (2008).

An accessible account of G.G. Stokes contributions to linear wave theory is by Craik (2005).

Solitary waves are nicely summarized on http://www.ma.hw.ac.uk/~chris/scott_russell.html.

# Part 3

# TRANSPORTING SEDIMENT

*As when a man channels water from a dark-welling spring and directs its flow among his plants and garden plots, knocking the dams from its trench with a mattock in his hand: as the water starts to flow it clears all the pebbles from its path, then gathers speed and runs gurgling down over the sloping ground, outstripping the man who guides it.*

Homer, *The Iliad* (c. 10 ka), from Book 21, lines 257–263. Transl. Martin Hammond, Penguin Classics, 1987

## Introduction

The study of sediment transport as the interaction between fluid and sediment grains is a long one, Homer clearly liked it as a metaphor and Roman engineers knew well how to clear sediment from aqueduct pipes using Bernouilli principles. In the early nineteenth century, Scottish geologist Charles Lyell (Part 3 Fig. 1a) established the philosophical basis for investigation of ancient sediments from close observation of the action of what he termed '*present causes*', and also clearly set out the notion that investigation of such causes had to involve the difficult technical problem of dealing with the action of flowing fluids. Thus in an amusing letter written in 1830 to his friend Scrope, a pioneering volcanologist, he says that they might investigate the formation of sand ripples: '*A large and deep trough, with gently slanting sides, might enable us to experiment. get a paddle-wheel which will turn with the hand, and make a ripple ad libitum, and sand and mud of kinds to be deposited. Then we afterwards mix matter in chemical solution. After a due series of failures, blunders, wrong guesses etc, we will establish a firm theory.*'. Another pioneer, the polymath H.C. Sorby (Part 3 Fig. 1b), rediscovered and subsequently developed Lyell's wheel in this regard in the 1840s, culminating in his great paper of 1912. His own voyage of discovery began during a walk taken near his native Sheffield, England, in 1847: '*...when walking from Woodbourne to Orgreave, I was caught in a shower of rain, and whilst sheltering in a quarry near Handsworth my attention was attracted by what I afterwards called "current structures", namely structures produced in stratified rocks by the action of currents present during the time of deposition*'. Subsequent developments in this most technical of areas had to await exploitation of advances in the understanding of fluid dynamics to fully explain the physical processes of sediment transport by wind and water.

Two types of sediment bed are recognized: granular/cohesionless and cohesive. The first type includes all transport systems made up of solid grains that are kept in contact with adjacent grains at flow boundaries purely by gravitational effects. The sediment is therefore loose and flows freely by gravity down steeper slopes or is driven along a desert surface, river bed or tidal stream by flowing fluid. Note carefully that aggregated flocs of mud- and fine silt-grade sediment also behave as granular solids, though on deposition they compact. This field of sediment transport is termed *loose-boundary hydraulics*, distinguishing it from the analysis of pure fluid motion alone.The second type applies most commonly to cohesive mud beds where tiny clay flakes are mutually attracted by electrostatic forces that are large compared to gravitational ones. Once

**Part 3 Fig. 1** (a) Charles Lyell in field gear in his early 40s. He is resting his right arm on a copy of his 'Elements of Geology' published in 1838. This book, what some might today rather condescendingly term a 'training text' and a distillation of his three-volume 'Principles of Geology', introduced process geology to the masses. (b) Henry Clifton Sorby, hand lens at the ready, as a young man in the 1840s. Sorby and Lyell are recorded as having met on some of the former's rare visits to London, from Yorkshire.

deposited the sediment surface may also be coated with a gelatinous biofilm. The bed is thus bound together as an aggregate with a yield strength, only deforming in the bulk by impact of solid tools in a flow, by large raindrops impacting onto exposed tidal flats or by plastic shear deformation.

Sediment transport is a critical aspect of earth surface processes and has a myriad of applied engineering and environmental applications. Since sediment has a vast range of size, the transport of particular grades of sediment locally must be related both to *supply* of different grain-size fractions from upflow or other locations and to the energy available to transport it. It will therefore never be possible to come up with any grand a priori sediment transport theory because both the source/supply and transport phenomenon need to be determined independently. As we shall see it is a different matter when sediment *mass transport* is considered independently of sediment grain size; here a mechanical equilibrium view of transport is possible from simple Newtonian momentum conservation arguments, though again the natural system is so unsteady and nonuniform in its energy supply that only very limited theoretical progress is possible. All sedimentary systems thus have to be classified at the outset into either *supply-limited* or *transport-limited*. There is much room for experimentation and field measurement in this area of study and research.

# Chapter 6

# SEDIMENT IN FLUID AND FLUID FLOW—GENERAL

*Full from the rains, but the flood sediment gone;*
*Under the brace of the glancing current*
*Each pebble shines with a life of its own,*
*Electric, autonomous, world-shaking-divergent.*

Hugh MacDiarmid, 'The Point of Honour', 1934, *Complete Poems*, Vol. 1, Carcanet, 1993

## 6.1 Fall of grains through stationary fluids

Sediment grains falling through static or very slowly moving water and air masses are common in Nature; from silt in river-mouth plumes to shells of pelagic organisms sinking through the ocean. Generally, if we introduce grains of density $\sigma$ into a static liquid of density $\rho$, such that $\sigma > \rho$, then the spheres will initially accelerate until a steady velocity known as *terminal fall velocity* ($U_g$) is reached. A plot of $U_g$ vs. grain diameter, $d$, from experimental data for smooth quartz spheres and for natural grains in water (Fig. 6.1) shows that fall velocity increases with increasing grain diameter but that the rate of increase gets less as the drag coefficient, $C_D$, assumes a constant value of 0.4 for small pebble-size and larger grains. This lessening of fall rate and the constancy of drag coefficient is due to the onset and full development of turbulent flow separation around grains at a critical grain Reynolds number appropriate to ~5 mm grains in water at 20 °C. This is accompanied by shedding of von Karman vortices in the turbulent wake (**Cookie 29**). Accurate prediction of terminal fall velocity was first made by G.G. Stokes and the physical expression is often named *Stokes' law* in his honour—fall velocity is also sometimes known as the *Stokes velocity*. However, the 'law' is only strictly applicable in one highly specialized case—that of the *steady* descent of *single, smooth, insoluble spheres* through *still Newtonian fluid* in *infinitely wide and deep* containers when the *grain Reynolds number* (Chapter 1) is low ($< 0.5$) and hence viscous flow separation does not occur (Fig. 6.2). The reader should note the eight or more restrictions to the Stokes analysis (*italicized* above), perhaps concluding that the Stokes formula is practically useless. Thus natural silicate grains or calcareous biological debris are not spherical, usually fall in a group and the latter are soluble. Nevertheless, we may make use of Stokes' formula as a good approximation in some situations. Results for a more general expression derived from dimensional analysis that applies over a large range of grain sizes for natural grains is shown in Fig. 6.1.

Want to know more about Stokes' law of settling? Turn to Cookie 29

*Sedimentology and Sedimentary Basins: From Turbulence to Tectonics*, 2nd edition. © Mike Leeder.
Published 2011 by Blackwell Publishing Ltd.

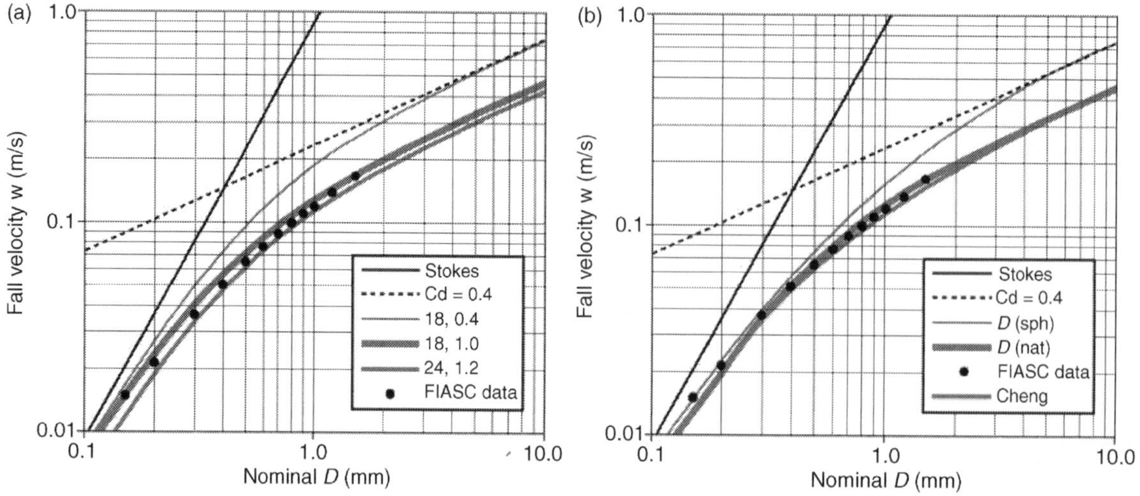


**Fig. 6.1** Predicted relation between fall velocity and diameter for quartz grains in water at 20 °C according to (a) the expression $U_g = (\sigma - \rho)gd^2/C_1\nu + (0.75C_2(\sigma - \rho)gd)^{0.5}$ and (b) previous authors. (See Ferguson & Church, 2004 for data sources and full discussion.) Straight lines in both plots show expected asymptotic trends for smooth spheres (Stokes' law with $C_1 = 18$, and constant drag coefficient $C_2 = 0.4$). Points labelled FIASC are experimental values from the US Federal Inter-Agency Sedimentation Conference. Upper, middle and lower curves in (a) are for spherical, natural and angular grains using $C_1$, $C_2$ shown in legend. Upper and middle curves in (b) are Dietrich's (1982) relation for spheres and natural grains, respectively; lower curve is Cheng's (1997) equation.

In many natural systems, problems arise because of the presence of rough, non-spherical grains and multigrain settling. The sphericity problem must be solved by recourse to specific experiments, particularly when fragments of irregular biological grains (Chapter 2) at high grain Reynolds numbers ($Re_g$) are involved. In multigrain settling, fluid streamlines relative to the individual grains interact. Increased drag results in

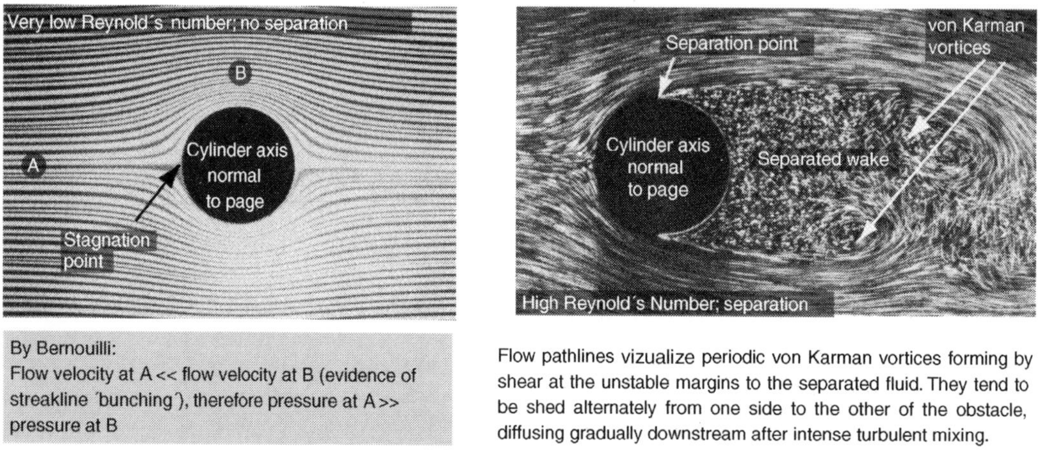


**Fig. 6.2** Flow visualization photographs to show: (a) 'creeping' flow at very low grain Reynolds number ($Re_g < 0.5$) when Stokes Law is appropriate (see Fig. 6.1); (b) the situation at higher $Re_g$ with well-developed flow separation and the alternate shedding of *von Karman vortices* in the flow wake. Images from van Dyke (1982) and sources cited therein.

a decrease of the grain fall velocity relative to that in a grainless fluid, the *Richardson–Zaki effect*. This states that the fall velocity, $U'_g$, of a spherical grain in a dispersion of other falling spherical grains varies as $U'_g = U_g(1 - C)^n$, where $U_g$ is the fall velocity of a single grain in an otherwise grainless fluid, $C$ is the volume concentration of grains in the falling dispersion, and $n$ is an exponent that depends upon flow and grain characteristics, varying for spheres between 2.32 (for $Re_g > 1$) and 4.65 (for $Re_g < 1$). Experiments on natural quartz-density mineral sands give a steady almost linear reduction of $n$ from ~4.4 for 0.2 mm grains to ~3 for 1 mm grains. This tells us that the fall velocity of a grain in a dispersion will be smaller than that in an otherwise grainless fluid and is strongly dependent upon concentration. For fine sediment at a high value of $C$, $U_g'$ may be only a few per cent of $U_g$. This consideration is of great importance in understanding the settling behaviour of sediment-laden flows such as estuarine tide/wave currents and turbidity currents. Further interesting complications arise when grains:

- are soluble so their size and mass decrease (but not necessarily linearly) during fall;
- fall through density-stratified fluid;
- fall through a plastic-like substance whose yield strength varies and may exceed the applied gravitational force due to the grains.

## 6.2 Natural flows carrying particulate material are complex

The majority of natural flows are complex in that they contain particulate matter that influences fluid behaviour under conditions of shear. Four end-members may be defined, though natural 'dirty' flows may show mixed behaviour.

1 In freshwater or seawater transporting cohesionless, non-electrically charged grains of silicate minerals, additional resisting stresses may arise through grain-to-grain interactions, leading to an increase in the apparent viscosity of the two-phase system.

2 Transport of high concentrations of cohesive particles by freshwater may eventually lead to plastic-like behaviour (section 4.1), with significant increase in apparent bulk viscosity directly related to the amount of transported clay and rate of shear. Such behaviour occurs because as the concentration of cohesive particles increases the mixture behaves in a rigid fashion analogous to a colloidal solution such as starch. The fluid is increasingly contained within the rigid aggregate rather than flowing around the individual grains and hence the viscous dissipation of the flow drastically increases.

3 Despite the tendency of solids to increase flow density and viscosity, the presence of dilute concentrations of dense solids in turbulent flows has usually been assumed to have negligible effect upon fluid accelerations; this is termed the *Boussinesq approximation*. In dense saline flows, this may well be satisfied, also for very dilute particle concentrations in turbulent flows. But for grain-rich flows, with their many solid point-masses of excess density, there occurs major enhancement of inertial reaction to extra applied forces. Inertial reaction to the extra pressure-driving force manifests itself in enhanced turbulent accelerations. The extra force component needed to cause the observed accelerations must arise from enhancement of pressure gradients since we know from experiments that these increase even when neutrally buoyant solids are added to shear flows.

4 Seawater flows transporting dilute (0.15–10 g/L), flocculated, organic-rich aggregates of clay minerals have been found to maintain a Newtonian flow structure and show no noticeable gradient of sediment concentration towards the wall. The thickness of the inner wall layer may be enhanced by a factor of 2–5 and the friction velocity decreased by up to 40%. This *drag-reducing* behaviour is the opposite to that expected if the clay particles had caused a small increase in fluid viscosity. Any reduction in fluid drag exerted adjacent to the bed by smooth boundary flows must come about by decreases in some or all of (i) apparent molecular viscosity, (b) velocity gradient in the viscous sublayer and (c) turbulent accelerations. The first effect is the opposite to that expected for added clay since this increases molecular viscosity. The other causes of drag reduction have been widely used to manipulate turbulent boundary layers. Studies with drag-reducing polymers (the *Toms effect*) have revealed that drag reduction is achieved through modification of the near-wall turbulence structure within the buffer region of the flow. This indicates that drag reduction is linked to reduced turbulent momentum exchange with outer regions of the boundary layer.

Many creatures, sharks for example, have evolved roughened platelets and ornaments on their external body parts that reduce drag. These mix up the boundary-layer fluid during motion and prevent flow separation. Humans do this with platelets on the leading edge of aerofoils, spoilers on the back of automobiles, dimpled golf balls, cycling headgear, etc.

## 6.3 Fluids as transporting machines

The fact that moving fluids can do useful work is obvious from their role in powering waterwheels, windmills and turbines. In each case kinetic energy is converted to mechanical energy by the machine in question. Energy losses occur, with each machine operating at a certain efficiency and with work rate = available power × efficiency. Applying these basic principles to a water channel, the concept of useful work is now replaced by natural work. The water will try to transport sediment grains supplied to it. How much sediment can be carried will depend upon the power available to the river, the local power supplied to the channel boundaries and the efficiency of the energy transfer between fluid and grain. As we saw previously (Chapter 5), flow power is made available as fluid potential energy is converted to kinetic energy down a gravity slope; but only some of this power is utilized in sediment transport—this may be termed the *efficiency problem*, one that has proved difficult to solve.

Useful as the concept of flow power is for channelized water flows, there are problems in other fluid systems.

1 The wind has no readily definable upper flow limit so that flow power is impossible to calculate.
2 The availability of sediment to a transporting system is not uniform—it depends upon delivery from hillslope systems so that non-equilibrium effects are often dominant.

Regarding the dynamics of sediment transport by moving fluid, this is due to exchange of momentum between grain and fluid, i.e. forces are set up in the transport process. Three stress components of applied fluid force (see **Cookies 18, 21 & 24**) are involved:

- a fluid shearing stress drives *rolling grain* transport close to the bed;
- a fluid lift force makes grains rise from the bed, whence gravity returns them;
- an upward impulse from turbulent acceleration supports *suspension* in the body of the flow.

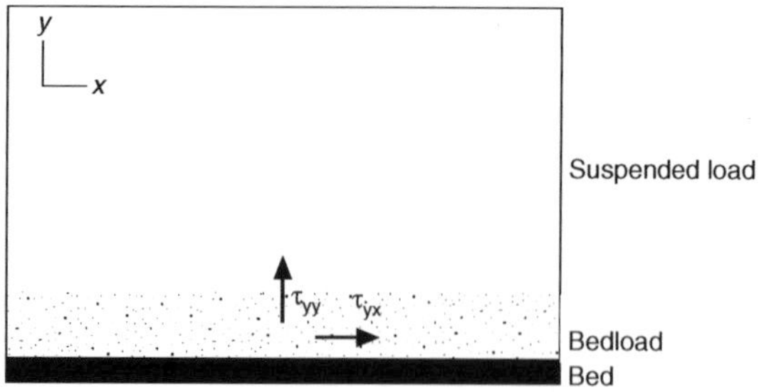


**Fig. 6.3** Simple definition sketch to show the two stresses responsible for bedload and suspended-load transport.

Figure 6.3 illustrates this division of labour into *bedload* and *suspended-load* transport.

Want to know more about fluid forces? Turn to Cookies 18, 21 & 24

Working from fluid dynamic principles, we might expect the following general points to apply.

1 In order to move a layer of initially stationary grains as bedload, it must be lifted over underlying grains as a *dilatation*—work must be done to achieve this.
2 The energy for dilatational work comes from the kinetic energy of the shearing fluid boundary layer.
3 Close to the bed, fluid momentum transferred to moving grains is transferred in turn to other stationary or moving grains so that a dispersion of colliding grains evolves. The efficacy of grain collisions depends upon grain immersed mass, a dynamic friction coefficient (**Cookies 30 & 31**) and fluid viscosity. Major differences occur between wind and water transport.
4 For suspended grains, burst and sweep motion of turbulent shear (Chapter 5) are the source of momentum exchanges to and from bedload.

## 6.4 Initiation of grain motion

As fluid shear velocity, $u_*$, is slowly increased over a levelled bed of similar-sized grains, a critical point, the *threshold for motion*, is reached when the grains begin to move downstream; sediment transport has begun. The *critical threshold shear velocity*, $u_{*c}$, for the grains in question is used to define a *transport stage* expressed as the ratio $u_{*c}/u_*$. A great deal of attention has been paid to the determination of the critical threshold for grain movement; it is an important practical parameter in civil and environmental engineering such as design of canals, irrigation channels, model experiments, erosion of estuarine tidal flats and sediment budget studies. It is also of special interest to sedimentologists interested in establishing estimates for the magnitude of palaeocurrents.

Concerning the magnitude of applied stresses, if $\tau_0$ is the mean bed shear stress, then the mean drag per grain is given by $\tau_0/n$, where $n$ is the number of particles over unit bed area. A lift force also acts on each grain due to the *Bernoulli effect* (**Cookie 18**) as fluid streamlines converge over projecting grains causing velocity to increase and pressure to decrease. At threshold, the magnitude of lift force is comparable to the drag force when the grain is on the bed (Fig. 6.4). It rapidly dies away as a grain rises from the bed whilst the drag force rapidly increases. Both fluid forces try to move bed grains—they are resisted by the grain's normal weight force. A useful way of non-dimensionalizing these threshold forces is to express the ratio of applied shear and lift to normal weight force; grain motion occurs when the fluid force exerts a critical moment about the pivot point of the stationary grain with its neighbour (Fig. 6.5). It is difficult however to theoretically determine the critical applied shear stress for grain motion, despite the initial attractions of using the 'moments of force' approach. This is because of the large number of variables involved, one of the most unpredictable being the degree of exposure of individual grains, no matter how carefully a flat bed is prepared for experiment. It is also difficult to directly estimate the lift-force contribution and to account for variations in mineral density, important in concentrating 'heavy minerals' into economic *placers*. The critical conditions for the initiation of particle motion must therefore be determined experimentally. The simplest approach involves a measure of flow shear velocity vs. grain diameter (Fig. 6.6). Such plots may be used, with suitable caution, in the determination of ancient current strength, a procedure that has been of great use in establishing the palaeovelocities of past ocean currents transporting cohesionless fine sediments in the range 10–60 μm (termed '*sortable silt*').

In order to have greatest generality, experimental results should be applicable to a wide range of fluids and particles. Sedimentologists are most interested in natural mineral grains in air and water, but these systems must be treated as special cases of more general

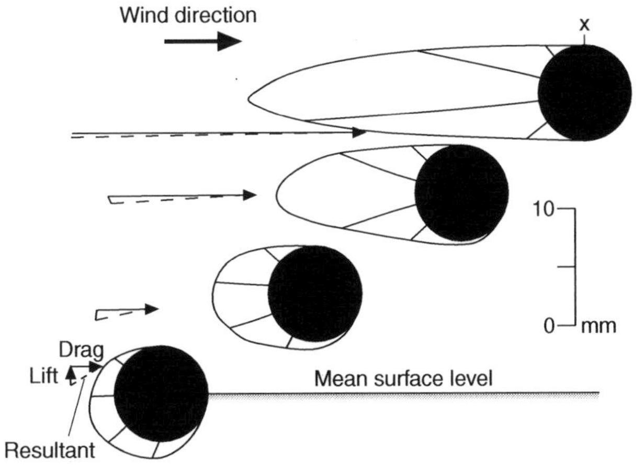


**Fig. 6.4** The distribution of approximate pressure differences measured in experiments between position X on top of a 7.5 mm sphere and other positions on the sphere at various heights in a windstream. The lengths of the lines in the enclosed areas outside the spheres denote the relative differences in air pressures. Both lift and drag (shear) forces act on the sphere, but lift decreases rapidly with height whereas drag increases because of direct pressure from the wind. The air velocity at 20 mm above the surface in the experiments was 7.7 m/s; shear velocity was 0.98 m/s. (Data of Chepil, 1961; for photograph of W.S. Chepil in his wind tunnel, see Part 2, Fig. 1G.)

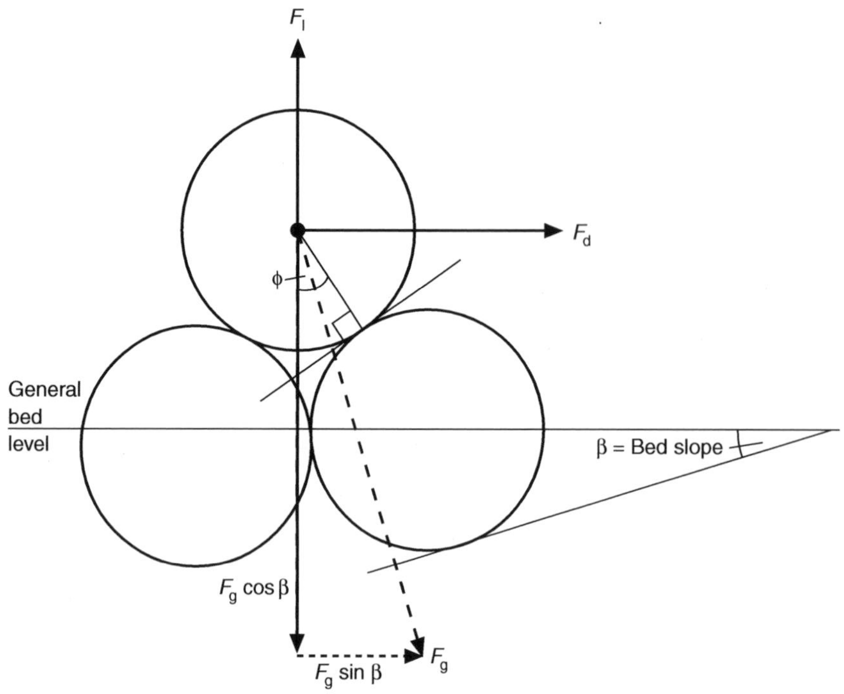


**Fig. 6.5** Force moment-balance diagram for the entrainment of a single sediment grain pivoting at angle $\phi$ about an adjacent like-sized grain on a bed sloping at angle β. $F_d$, $F_l$ and $F_g$ are the drag, lift and gravity forces respectively. At equilibrium, $F_g \sin\beta + F_d = (F_g \cos\beta - F_l)\tan\phi$. (After Bridge and Bennet, 1992.)

application, for example in considering sediment transport on Mars, Venus or Titan (Chapter 23). From first principles, critical shear-stress conditions for motion, $\tau_c$, must be dependent upon gravity, $g$, grain diameter, $d$, or radius, $r$, immersed grain mass, $(\sigma - \rho)$, fluid kinematic viscosity, $\nu$, and applied bed-shear stress, $\tau$, or shear velocity, $u_*$. Thus: $\tau_c = f(g, d, (\sigma - \rho), \nu, \tau)$. The controlling quantities are

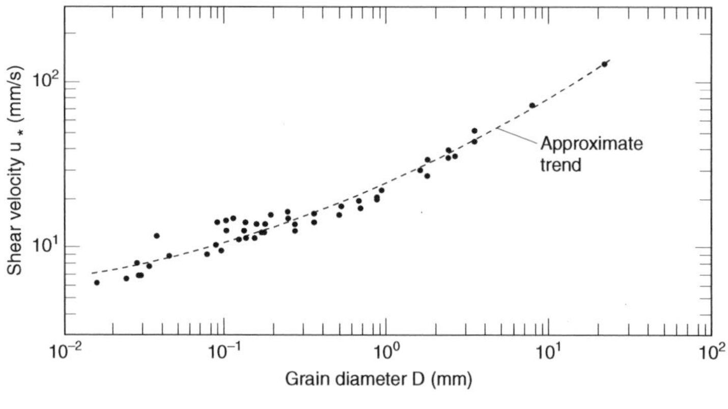


**Fig. 6.6** The variation of threshold shear velocity necessary for the initiation of movement of quartz-density grains in water at 20 °C. (Modified after Miller *et al.*, 1977, by addition of approximate trend line.)

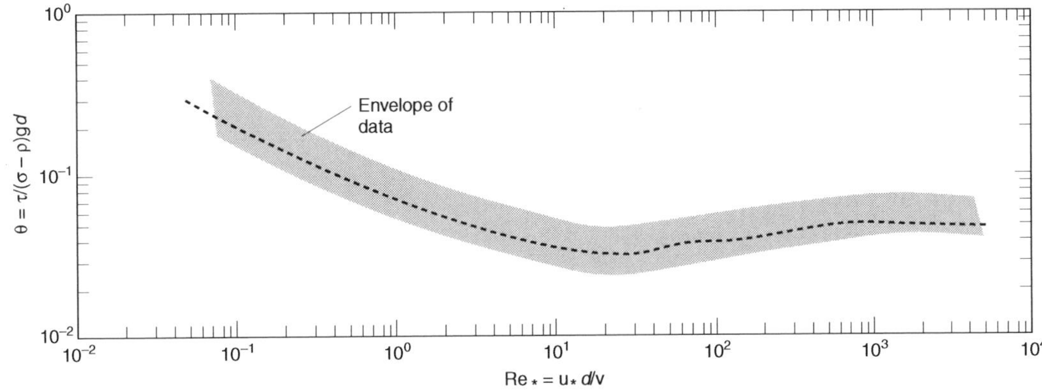


**Fig. 6.7** The variation of threshold non-dimensional shear stress vs. grain Reynolds number for water flow at 20 °C over a wide range of quartz-density grain sizes. (Modified after Miller *et al.*, 1977, by substituting data points with an envelope of data.)

arranged into two dimensionless groups for the purpose of plotting experimental results: $\theta_c = \tau/gd(\sigma - \rho)$ and $Re_g = u_* d/v$, where $\theta_c$ is the critical dimensionless bed shear stress and $Re_g$ is a form of grain Reynolds number. A plot of $\theta_c$ vs. $Re_g$ for liquids (known as a Shields diagram after the German engineer A. Shields) is shown in Fig. 6.7. Scatter is evident in the data because of the many different sets of experimental conditions used in the plot and the difficulty in deciding when threshold is reached. However, $\theta_c$ is an average of ~0.05, for a wide range of grain diameters. For $Re_g \leq {\sim}1$, $\theta_c$ increases steadily to around 0.3. The increase is thought to be due to smooth boundary conditions and presence of a viscous sublayer (Chapter 4). An example of use of the Shields/Bagnold approach in determining palaeohydraulics for flood deposits is given in **Cookie 32**.

Taking a general view, threshold in turbulent flow is best considered overall as the interaction between two variables. The first comprises initial movement characteristic of a given bed material in a fluid of given viscosity and density. Each bed grain is susceptible to a characteristic instantaneous stress, and because of random shape, weight and placement of individual grains this has a probability distribution. The second variable is the applied local instantaneous bed shear stress caused by burst/sweep events (Chapter 5). This also has a probability distribution, dependent on flow conditions. At the onset of grain motion, the most susceptible particles (those with the lowest characteristic critical shear stress) are moved by the highest applied shear stress. Experiments enable histograms to be constructed for each distribution (Fig. 6.8); these show that threshold occurs when the two stress distributions overlap by a certain constant amount. Much of the scatter on the Shields plot evidently reflects different observers deciding on different degrees of overlap—this can be a particular problem with gravel-grade sediments.

A further threshold problem concerns very poorly sorted sediment, perhaps with bimodal or even polymodal grain-size distributions. For a pebbly sand, the motion of pebbles occurs at lower critical applied fluid stress than for pebbles alone. This is because: (i) pebbles project further into the shearing boundary layer and are affected by higher overall shear stress than sand; (ii) direct normal stresses tend to 'push' the pebble forwards and (iii) sand around the pebble margins is scoured due to flow separation eddies. Both sand and pebbles may then move simultaneously and a condition of *equal mobility* is said to exist at threshold. However, the situation rarely reaches equilibrium in perennial streams because in frequent flows close to threshold the bed quickly *armours* once a certain amount of the sand has been scoured away; a surface layer of interlocking and immovable pebbles protecting the underlying sand–pebble mixture (Fig. 6.9). It is only well above threshold, about twice that needed to transport the median bed grain size, $D_{50}$, that equal mobility conditions apply. Below that

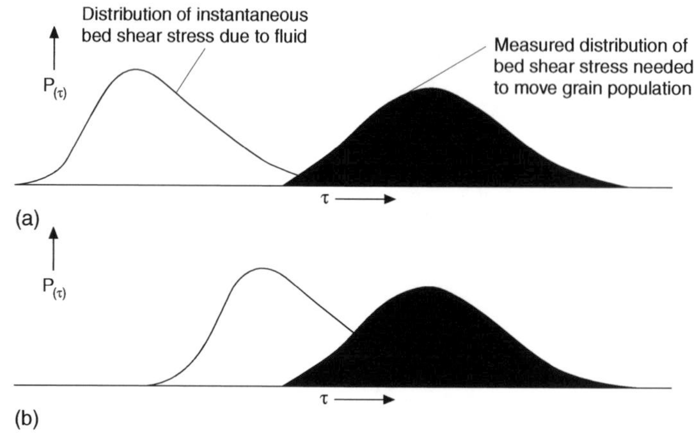


**Fig. 6.8** Graphoids to show that the threshold of grain movement must be defined by the statistical overlap between the distribution of instantaneous applied shear stresses due to the turbulent fluid and the actual distribution of stresses needed to move the particular grain population involved. (a) Small overlap, little motion. (b) Large overlap, general threshold exceeded. (After Grass, 1970.)

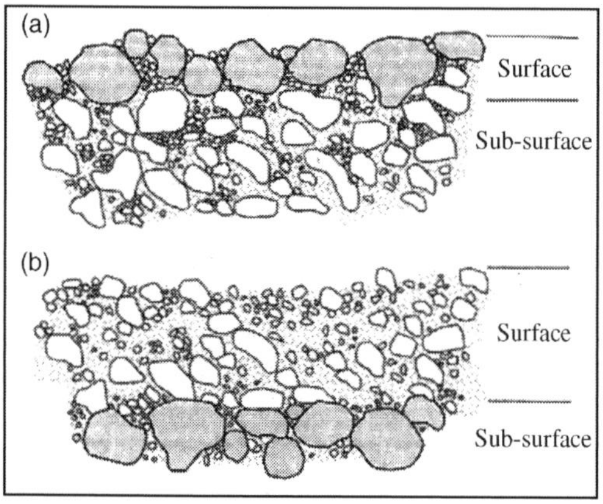


**Fig. 6.9** Notional sections through the beds of (a) armoured and (b) non-armoured streams. (After Almedeij & Diplas, 2005.)

transport is selective. Also, many rivers are not in transport equilibrium, with slopes imposed by prior conditions (glaciation, changed sediment/water balance, incision, etc.) so that downstream decrease in slope (see Chapter 11) reduces flow competence (bed shear stress) whilst at the same time capacity (flow power) increases. Together, the processes of upstream-armouring and downstream bed stress-decline explain the almost universal trend found in rivers for *downstream-fining*, with physical abrasion a relatively inefficient contributor in most situations outside of mountainous channels. The process is necessarily accompanied by channel reach aggradation due to deposition of the coarser bedload as competence decreases, a process vividly apparent at confluence junctions of tributary streams to major trunk channels.

In *ephemeral streams* armouring is impossible, for as discharge decreases, deposition of coarse grains is succeeded by fines and the bed shows no armouring (Fig. 6.9). Come the next ephemeral discharge event this surface finer sediment is instantly available for incorporation into bedload and suspended load, perhaps explaining the very high concentrations of sediment found in the early stages of highly seasonal and ephemeral stream discharge rating curves (see section 6.9).

Finally, thresholds for sediment motion under waves and combined flows have their own particular problems related to peculiarities of the oscillatory flow boundary layer and to interactions between this and any overall flow induced by tides. These problems must be solved by detailed and specific experiments, as must problems due to irregular shapes and low density of biogenic carbonate grains.

## 6.5 Paths of grain motion

Once the threshold for motion is exceeded it is impossible to follow the path of an individual grain,

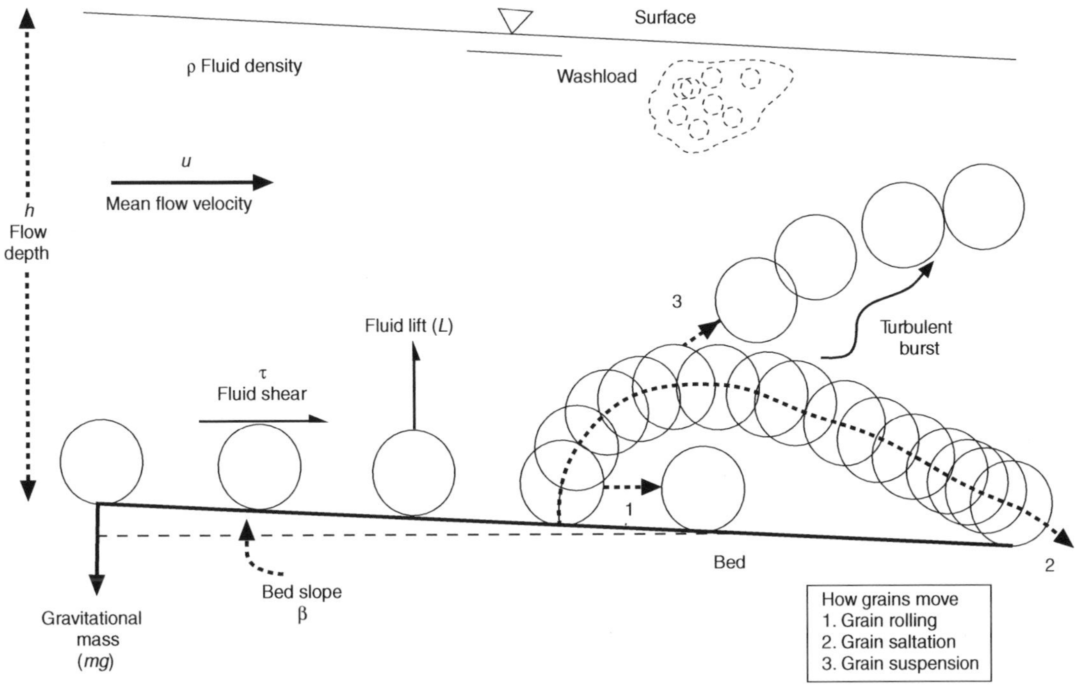


**Fig. 6.10** Diagram to show various sediment transport modes.

particularly in a crowd of other grains; it is only from application of skillful photographic techniques that we have a full idea of the different types of grain paths (Fig. 6.10).

1 *Rolling/sliding*: the grains have continuous contact with the bed. In air, rolling grains are kept in motion as *creep* or *reptation* by impacts of incoming saltating grains.
2 *Saltation* (Latin, *saltare*, to dance or jump): grains rise steeply ($> 45°$) from the bed, ascending to a height of a few grain diameters, then descend in a shallow-angled ($> 10°$) path back to the bed. Saltation may occur in both laminar or turbulent flows. Grain spin during flight causes some extra lift (*Magnus effect*), which probably aids particle ascent and steepens descent. A saltating grain may experience an upward fluid thrust due to turbulent bursting during the descending part of a trajectory—a state of *incipient suspension*.
3 *Suspension* (Latin, 'held up'): grains move within advected bursts of inner layer fluid, moving upwards with bursts and downwards with sweeps. Trajectories are therefore much longer, higher and more irregular than in saltation.

The varying proportions of time spent in the above three modes, the length and height of grain trajectories and mean grain speed are a direct function of transport stage. Characteristic grain paths occur only when grains have no interactions; as transport stage is increased, more and more grains are entrained and *grain–grain interactions and collisions* become certain. The onset of fully developed suspension occurs when upward turbulent momentum exchange close to the bed exceeds the weight force of saltating grains (**see Cookie 33**). However, fine bed sediments ($d < 0.1$ mm) are protected from suspension for a while above threshold as they move initially within the viscous sublayer; this protects them from the action of advective turbulence burst motions.

## 6.6 Categories of transported sediment

1 *Bedload*: (*aka* traction load) includes rolling, saltating and collision-interrupted saltating grains.

The grains receive an applied fluid thrust that maintains saltation and they in their turn transfer momentum to the stationary bed surface by solid–solid contacts (Fig. 6.11; **Cookies 30–32**). Rate of change in this momentum over unit area in unit time must equal the immersed weight of the bedload grains. The definition of bedload is thus both positional and dynamic. The transport of bedload sediment may have marked effects upon the distribution of fluid velocity with height from the bed surface. This is clearly shown by wind transport, where velocity measurements within and just above the bedload zone show that as the wind strength increases above the threshold for movement, there is a clear retardation of air velocity in the centre of gravity of the bedload zone. The retardation reaches about 20% of the pure air flow velocity expected by the Karman–Prandtl equation (Chapter 4) for turbulent flow in the absence of moving solids: it is explicable by transfer of momentum from fluid to solid during saltation. A similar effect should be present in water flows, but although it has been computed theoretically, the effect is difficult to measure because of the thinness of the subaqueous bedload zone.

2 *Suspended load*: includes all grains kept aloft by fluid turbulence (Fig. 6.11; **Cookie 33**) so that the weight force of the suspended grains is balanced by an upward momentum transfer from the fluid eddies. The process is more efficient in water than air because of (i) low density contrast between grain and water and (ii) high molecular viscosity of water.

| For more on suspension theory see Cookie 33 |
|---|

3 *Washload* is a broad term used to describe long-term suspended clay-grade 'fines' present in water flows. *Dustload* is an equivalent term for atmospheric flows which may transport sediment for thousands of kilometres in the planetary boundary layer.

## 6.7 Some contrasts between wind and water flows

There are a number of critical differences between air and water as transport agents. These arise because of differences in fluid material properties, magnitude of stresses exerted and length scales of boundary layers (Table 6.1). The differences are chiefly:

1 Low shearing stresses set up in moving air mean that transported grains are restricted to sand/granule grades.

2 Low buoyancy of grains in air means that bedload is dominated by grain–bed momentum exchange; significant 'splashup' occurs each time a grain collides with a bed of like grains, leading to a chain reaction downwind. This causes a significant *creep/reptation* component of bedload transport. In water, collisions are dampened by viscous retardation after the grain's elastic rebound has occurred—the grain momentarily halts or undergoes a rolling motion (i.e. grain linear momentum is not conserved) before fluid forces make it rise once more.

3 Energetic collisions mean that wind-blown transport is more effective in abrading and rounding grains. Also in bombarding solid surfaces to produce characteristic *ventifacts* such as facetted pebbles (*dreikanter)* and grooved rock surfaces (*yardangs*).

4 The bedload layer is much thicker in air, [O] $10^1$–$10^2$ grain diameters, and adds significant roughness to the flow boundary.

5 Suspension transport of coarser grains by fluid turbulence is more difficult in air because of lower grain buoyancy.

6 Turbulent microstructure of air probably plays little role in initiation of bedforms (Chapter 7).

7 Two types of grain motion threshold occur in air. At the *impact threshold*, grain motion can be started and propagated downwind by simply letting sand grains fall on to the bed. Other grains are bounced into the airstream, which upon falling cause further movement as they impact and so on. Grain motion ceases as soon as the introduction of artificial grains stops. Grains are moved at higher shear rates by the direct action of the wind, defining a *fluid threshold*.

8 The very rapid response of air to heating and cooling causes rapid vertical variations in boundary-layer density and close to the bed has a major effect on a wind's ability to shear a sediment bed. Major reductions of sediment transport are thus to be expected in hot desert conditions (say at +40 °C) compared with those in frigid Antarctic

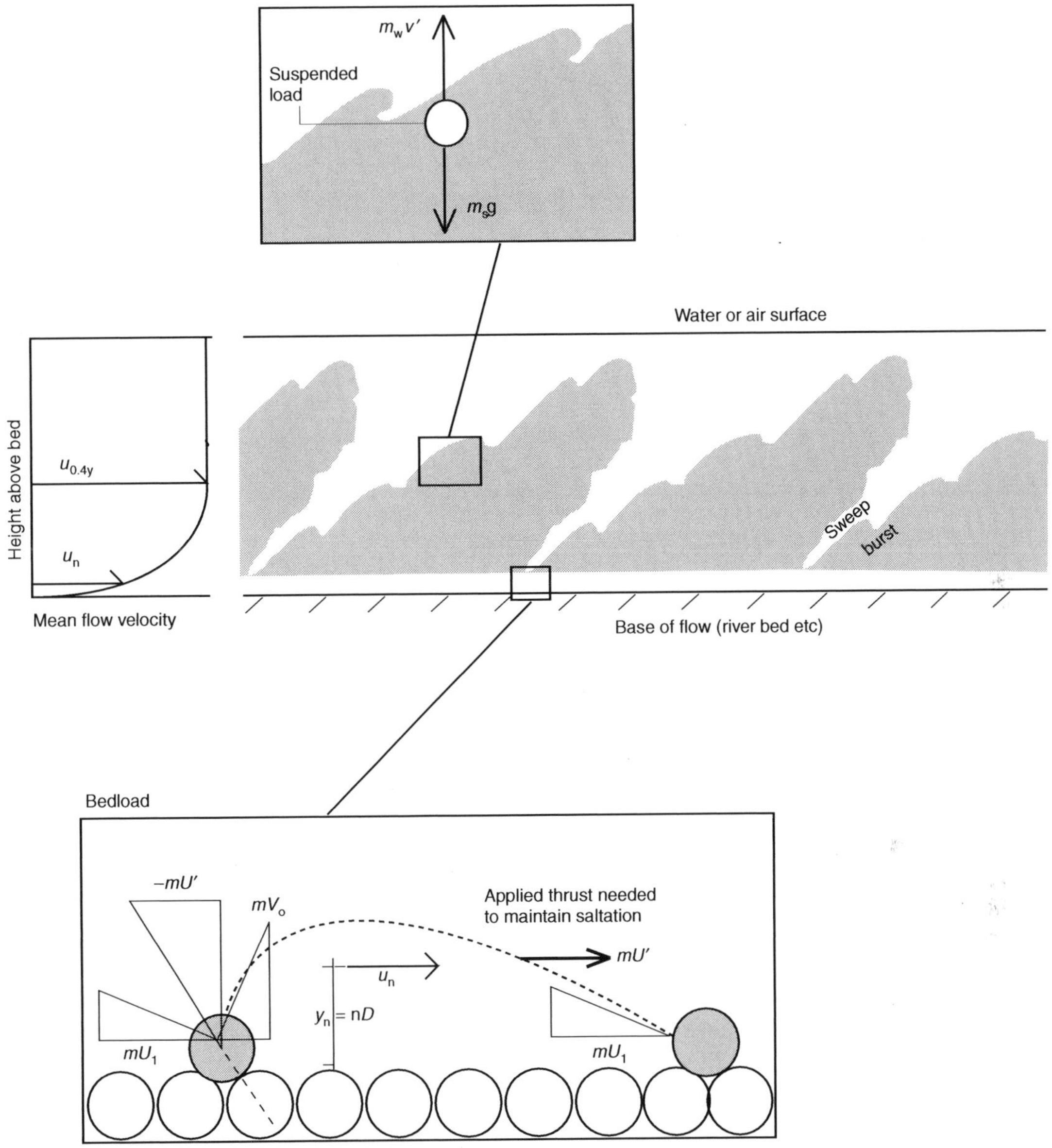


**Fig. 6.11** Diagrams to illustrate the division of transported sediment into bedload and suspended load. (a) and (b) The shaded areas represents bursting inner boundary layer fluid moving upwards in the flow and carrying suspended sediment along with it. (c) Illustratation of the applied fluid thrust that is needed to maintain saltation. Symbols: $m_w$ and $m_s$, water and grain mass respectively; $v'$, vertical turbulent velocity; $u$, streamwise flow velocity; $u_n$ effective velocity at centre of thrust for bedload; $y_n$, height of $u_n$, above bed; $U_1$ incoming grain velocity; $V_0$, outgoing grain velocity; $U'$, grain velocity lost on impact. (After Bagnold, 1973.)

**Table 6.1** Some physical contrasts between air and water and in relevant transport parameters of their flows.

| | Air | Water |
|---|---|---|
| Density (kg/m$^2$) | 1.3 | 1000 |
| Quartz/fluid density ratio | 2039 | 2.65 |
| Buoyant force per unit volume | 12.8 | 9810 |
| Viscosity (N s/m$^2$) | $1.78 \times 10^{-5}$ | $1.00 \times 10^{-3}$ |
| Stokes velocity, $U_S$, for 1 mm diameter sand (m/s) | –8 | –0.15 |
| Representative bed shear stresses exerted (N/m$^2$) | –0.088 for a wind speed of about 6 m/s measured 1 m above the bed (shear velocity $u_* = 0.26$ m/s) | –68 for a comparable shear velocity |
| Critical shear velocity ($u_*$) needed for motion of 0.5 mm sand grain (m/s) | 0.35 | 0.02 |

dry valleys (at −40 °C). Some variation for polar versus tropical surface water streams should also occur, but the effect is not nearly so marked.

9 Wind flow sediment transport is radically reduced by the presence of ground-covering vegetation which may filter, trap and protect otherwise exposed sand and silt surfaces. Similar effects in water flows over loose sediment are provided by the binding and stabilizing effects of cyanobacterial mats and other biofilms such as diatoms (Chapter 5).

10 Vertical convective motions of atmospheres cause vorticity and the production of spiralling dust devils to evolve on land surfaces which radically increase opportunities for suspension transport (see Chapter 23 for application to Mars).

## 6.8 Cohesive sediment transport and erosion

So far we have assumed that sediment grains behave as individual granules and that during transport their form is uninfluenced by chemical composition of the ambient fluid. None of these assumptions is true for fine silt and mud below about 10 μm diameter; below this limit clay minerals aggregate by *flocculation*. Important in understanding estuarine sedimentation patterns (Chapter 16), there is also a specific need for floc research concerning the disposal of pollutants and dredged material. When this is pumped away for dispersal, the particles flocculate as they settle, thereby influencing the composition and behaviour of the natural sea-bed. It may take years for the material to even approximate its original volume, affecting capitalization costs of these projects. In the sewage treatment and water purification industries, 'dirty' water is stirred to enhance flocculation, allowing large aggregates to settle to a sludge cake and be taken away to landfills or used as fertilizer. Here it is important to create large flocs for efficient settling.

The behaviour of clays in transport follows rules established for colloidal suspensions.

1 Clay mineral grains are small enough for short-range atomic attractions (*van der Waals–London forces*) to bring individual clay platelets together, so that a deposited bed of clay has considerable cohesive strength that must be overcome before erosion of the bed can occur by turbulent stresses.

2 Many clay minerals carry a net negative electrical charge on their basal lattice faces caused by ionic substitution. $Mg^{2+}$ may substitute for $Al^{3+}$ or $Al^{3+}$ for $Si^{4+}$. Both give a net unit charge deficiency per substitution. The approach of two clays in face-to-face contact during turbulent transport in freshwater therefore leads to particle repulsion, even with weakly held cationic species in attendance. On the other hand, attractive forces between clay particles in freshwater depend upon the existence of a small net positive charge at the broken edge of a clay platelet where metal–oxygen bonds are broken. When particles are brought very close together during turbulent transport, this edge charge participates in an edge-to-face linkage; the result is a loosely packed aggregate in a 'house-of-cards' arrangement. The probability of this attraction, as opposed to face-to-face repulsion, is relatively low.

3 Seawater is a strong electrolyte, so that negative face charges are neutralized with respect to the ambient fluid through the effects of cations, which cluster tightly adjacent to the faces to form what has

been termed an electric double layer. Because of the very high concentrations of ions (like $Na^+$) in the ambient fluid, there is now little gradient of electrical potential (*zeta potential*) and so approaching clay particle faces (the majority of the 'target' area) no longer repel, allowing van der Waals–London atomic forces to establish attraction. This is the process of *flocculation*, important in estuaries at salinities of only a few parts per thousand, whereby larger aggregates of grains are formed from a myriad of individual platelets. There is an interesting analogy here with blood coagulation and its facilitation by the addition of ions like $Al^{3+}$ in alum. Flocculation becomes important at the head of an estuary where freshwater is diluted by turbulent mixing with seawater. The effect leads directly to the production of *turbidity maxima* and sometimes *fluid mud* (Chapter 16).

4 The bringing together of clay platelets to form flocs is exceedingly complicated. At the smallest scale in near-stationary fluids, collisions between settling grains are due to Brownian motion or the differential settling of vari-sized particles. In laminar boundary layers, collisions are due to differential shear arising from velocity gradients. In turbulent boundary layers, flocs approaching a certain critical size corresponding to the local scale of microturbulence are continuously broken down by turbulent accelerations in the zone of highest turbulent production, only to flocculate once more higher in the boundary layer where both velocity gradients and accelerations are less. Careful experiments show that increasing turbidity at low turbulent shear conditions encourages floc growth but once the fluid shear reaches a certain critical value the flocs are disrupted into smaller sizes. Clay aggregation is important in *hyperconcentrated flows* (see below)

5 Flocs are intimately associated with organic matter, the 'sticky' varieties of which overcome short-range electrostatic repulsion in seawater and encourage agglomeration into irregular large (several millimetres) but weak aggregates.

Because of points 1–5, critical conditions for mud entrainment by erosion due to fluid shear stresses are a complex function of clay type, fluid chemistry, state of fluid flow, organic concentration, etc. The problem is not an academic one, for sedimented muds often concentrate pollutants and contaminants and it may be important to determine the stability of such beds to tidal and wave erosion. For example, the critical erosion rate for muds is a sensitive function of electrolyte concentration, shown by adding various concentrations of $NaNO_3$ to deionized porewaters of pure kaolinite muds. The added salt greatly increases critical erosive stress. Consolidation due to *compaction* is also of great importance, leading to an increase in cohesive strength and a decrease in erodibility with depth so that surface erosion may be followed by bed stability at depth (Fig. 6.12). The nonuniform surface growth and binding effects of diatom and cyanobacterial films and coatings on many marine intertidal muds (clastic *and* carbonate) emphasize the additional role of biological stabilization. In view of all these factors it is impossible to generalize values for critical erosive stress and transport relations of muds. In particular, experimental data pertaining to freshwater muds are hardly likely to apply to marine situations. A number of techniques have been established in attempts to measure *in situ* critical bed shear stresses, such as portable miniflumes and jetted fluid-impact probes.

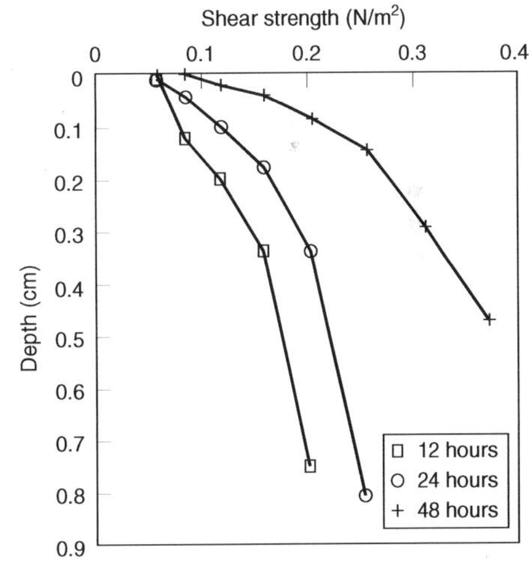


**Fig. 6.12** The increase of mud shear strength with depth for deposited Wadden Sea marine muds that have been experimentally compacted and consolidated for different times. Results mean that the critical bed shear stress needed to erode subsurface sediment increases rapidly from only millimetric depths. This is true apart from the very surface layer which may have increased cohesive strength due to biofilm or cyanobacterial mat. (After Johansen *et al.*, 1997.)

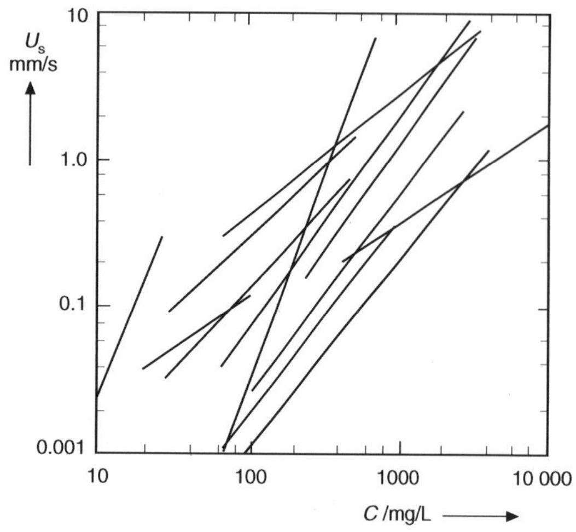


**Fig. 6.13** Data from several estuaries showing the variation of settling velocity and particle concentration for muddy suspensions.

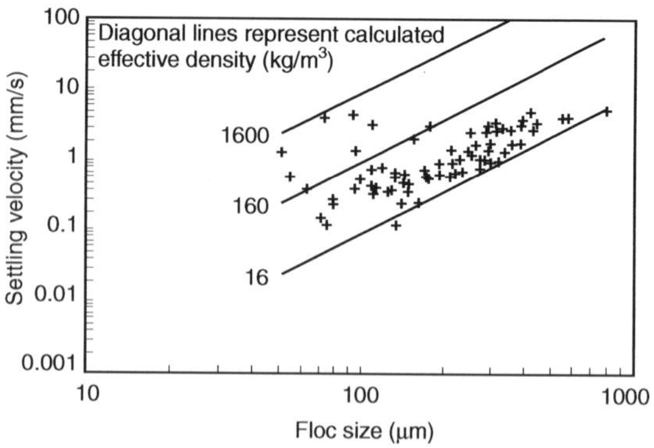


**Fig. 6.14** Settling velocity vs. floc size measured *in situ* in the Elbe estuary. Note the very large range of flocs that have the same settling velocity, and the range of calculated effective (immersed) densities ranging from values close to that of pure organic matter (~30 kg/m³) and solid mineral density (~1600 kg/m³). (After Eisma *et al.*, 1997.)

There are many problems involved in the study of flocculated mud aggregates, not least of which is the difficulty of estimating floc size from seawater samples, the extraction of which may lead to floc disaggregation. Central to understanding the process is the concept of collision-induced flocculation between suspended particles. This explains the positive relationship in dilute concentrations of suspended mud ($<$ 10 g/L) between collision frequency and floc size. In such flows, settling velocity, $U_g$, of flocs is a function of mud concentration, C, and may generally be expressed as, $U_g = kC^m$, where $k$ and $m$ are widely variable empirical constants for particular cases (Fig. 6.13). At low mud concentrations, as we have seen, fluid drag reduction is associated with turbulent boundary layer reorganization and restructuring and is a version of the process of shear-thinning (reduced effective viscosity). For greater concentrations $U_g$ declines because of the effects of hindered settling. Calculations of deposition rates and turbulent diffusion in muddy flows also require knowledge of floc size and density (Fig. 6.14), but these vary widely. As floccing increases in a dense clay suspension undergoing shearing motions, a condition known as *shear-thickening* may ensue in which bulk fluid viscosity increases, causing flow to terminate. This phenomena may be analogous to a physical arrangement of increased colloidal clustering, termed *hydroclustering*, that causes smaller grains and aggregates to be retarded in their efforts to shear past larger aggregates (Fig. 6.15).

## A note on hyperconcentrated freshwater turbulent flows and their deposits

Hyperconcentrated freshwater flows are those turbulent-modified systems that carry $> \sim 10\%$ sediment in suspension and where saline flocculation is not an issue. Laboratory experiments using clay-rich suspensions show transitional behaviour between Newtonian and non-Newtonian flow. This tendency begins at concentrations of only around 2% clay and hence the definition of hyperconcentrated behaviour is likely to be a rather fuzzy affair. As concentrations increase, a stratification develops such that a lower dense sediment-rich layer is trapped in stable mode beneath an upper carapace of more turbulent and dilute flow. Shear at the interface develops Kelvin–Helmholz instabilities and these generate turbulence in addition to wall shear. When hyperconcentrated flows with low ($<$ 2.5%) mud concentration occur over a rippled bed, flow separation and near-bed turbulence are enhanced at first. As concentration increases up to $>$ 10%, stratification and turbulence suppression increasingly dominate in the boundary layer, the velocity gradients decrease and the bedforms themselves become damped due to diminished efficiency of

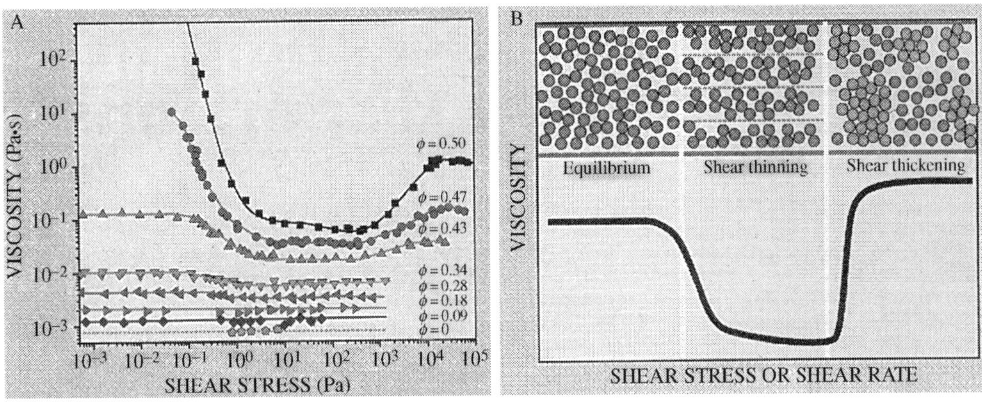


**Fig. 6.15** The viscosity and structure of colloidal dispersions as a function of applied shear stress. (After Wagner & Brady, 2009, and references cited therein.) (a) Viscosity as a function of applied shear stress. $\phi$ is the volume fraction of particles in each dispersion. Note initial yield stress increasing with $\phi$. Beyond this yield stress, viscosity decreases by shear thinning. It then increases with continuing shear for certain high $\phi$. (b) Changes in microstructure of colloidal dispersions with increasing shear. Random structure gives way to ordered microaggregates causing lowered viscosity. At high shear rates hydroclustering occurs and viscosity increases.

flow separation processes. At higher flow strengths the tendency of clay particles to aggregate due to electrostatic attraction into a mud framework is opposed by intense turbulent shear events in the near-bed boundary layer whilst a plug flow with low turbulence develops above (Fig. 6.16).

The deposits of hyperconcentrated flows are intermediate between those of stream flow and debris flow; essentially poorly stratified and clast-supported with variable local inverse and normal grading. In many natural environments they may transform both *from* debris flows by dilution or *to* debris flows by *bulking-up*. The latter process occurs in decelerating mud-rich seawater currents when deposition occurs by *hindered settling* under conditions of reduced shear. This quickly establishes shear-thickening, a mud-floc framework which can support part or all of the sedimenting fraction of granular load. Laminar/plastic behaviour ensues and at reduced gradients, deposition of a matrix-supported layer—the flow has essentially morphed into a debris flow (see Chapter 9).

## 6.9 A warning: nonequilibrium effects dominate natural sediment transport systems

Mechanistic notions of transport equilibrium are ill-suited to studies of natural systems when water and sediment supply 'demons' create disequilibrium. These exist when flows of extreme power rapidly accelerate from almost zero base-flow conditions. For bedload transport the effect is well illustrated by measurements taken during desert flash floods which transport far more sediment at very high efficiency that cannot be easily predicted by theory (Fig. 6.17), in contrast to the situation of perennial streams. This may be the result of non-armoured prior bed conditions (section 6.4). Natural *lags* also arise from river-independent supply (initial storm hillslope sediment runoff, rainsplash effects, saturation, etc.) and the inability of a turbulent flow to react instantaneously to oversupply, particularly of silt-sized and smaller particles. Suspension disequilibrium in *hyperconcentrated flows* (see below) is demonstrated in a tropical system like the Huang-He (Yellow River) which drains easily eroded silt-mantled hillslopes. This river features gigantic suspended load of silts, far in excess of the river's theoretical ability for mechanical transport, chiefly on account of a *settling lag* once the sediment is introduced by hillslope runoff into the main river channels. The effect of this high imposed sediment load has been analysed using the Richardson number (section 5.7) to divide flows into (i) subsaturated stable flows, (ii) supersaturated unstable flow with high deposition rates as turbulent intensity declines and (iii) hyperconcentrated

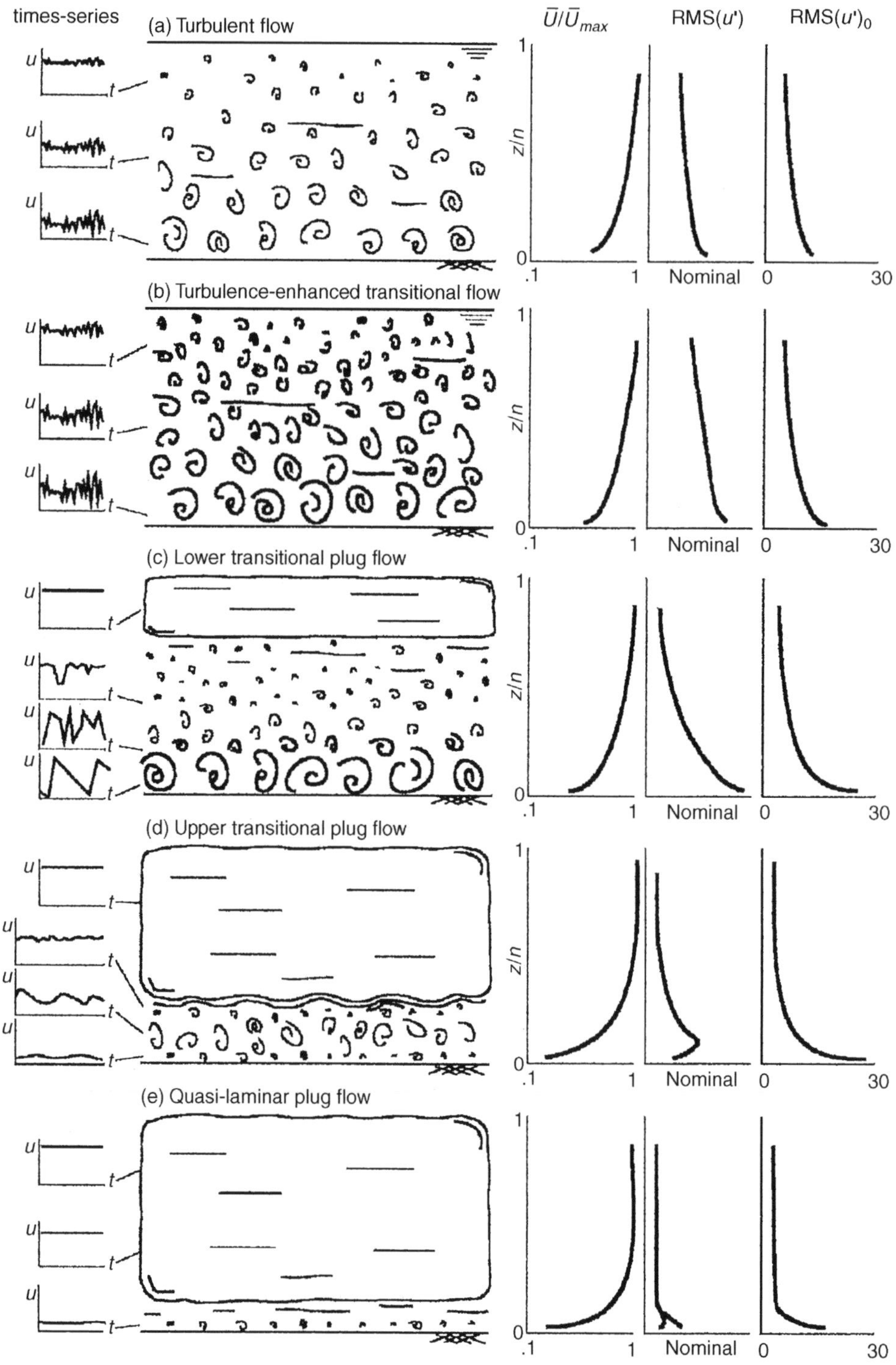
times-series
(a) Turbulent flow
(b) Turbulence-enhanced transitional flow
(c) Lower transitional plug flow
(d) Upper transitional plug flow
(e) Quasi-laminar plug flow
u
t
Ū/Ū max
RMS(u')
RMS(u')0
z/n
0
1
.1
1
Nominal
0
30

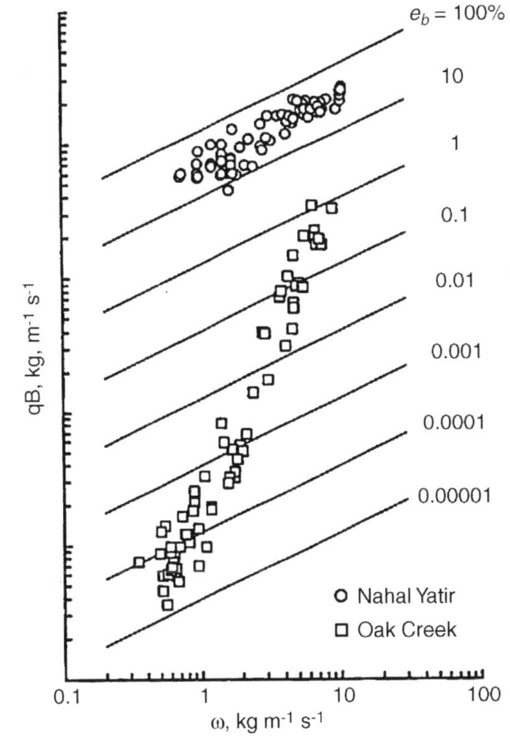


**Fig. 6.17** Bedload transport rate, $i_b$, vs. stream power, ω, for the perennial Oak Creek, USA and the ephemeral Nahal Yatir, Israel. Note that Oak Creek obeys the mechanistic model of Bagnold (1966b, 1973) whilst Nahal Yatir transports at almost constant ultra-high efficiency. The result is dispiriting for those who believe in equilibrium sediment transport theory. (Data of Laronne & Reid, 1993, Milhous, 1973; as presented by Almedeij & Diplas, 2005.)

subsaturated stable flow with hindered settling. At the same time the Reynolds number separates turbulent, transitional or laminar flow types. Many tropical rivers (e.g. Burdekin River, Queensland, Australia) also show marked *sediment hysteresis* effects with peaks in sediment suspension occurring before peaks in water discharge due to initially high surface sediment pickup from initial overland flow and rainsplash erosion (Fig. 6.18).

## 6.10 Steady state, deposition or erosion: the sediment continuity equation and competence vs. capacity

The most fundamental question in sedimentology sounds innocent enough—'Can one predict when sediment grains will deposit or be eroded?' It is only possible to answer this with a great deal of fudging. The starting point is an adaptation of the accounting system developed previously in the context of pure fluid flow itself (Chapter 3)—the continuity equation. We have seen that fluid particles may accelerate or decelerate in response to changes at a point or over time along some distance. If the fluid in question is charged with sediment grains and the transporting machine is in a state of equilibrium, then any local deposition or erosion will cancel out and the system is in a steady state. Should fluid be forced to change velocity, say by a channel width change or by flow over an obstacle, then the motive power available will change and deposition or erosion will result. In the simplest possible way we can thus make a box model of the accounting system: *deposition or erosion* = (*sediment in*) minus (*sediment out*). This concept is expressed slightly more mathematically in **Cookie 34.**

Want to know more about the sediment continuity equation? Turn to Cookie 34

### Competence vs. capacity in sediment transport

This is a simple point to make but it is also a profound one when the sediment transport system is considered. We may talk of the competence of any flow—meaning its ability to transport grains of a certain size—by reference to a threshold for motion graph such as the Shields diagram considered above. The larger the applied stress, the larger the grains that can be transported. However, this does not inform us concerning the amount of sediment the flow can transport, i.e. its capacity. Also, the grains present within a flow of particular competence depend on the supply of

**Fig. 6.16** Sketches for five different flow types in turbulent, transitional and laminar clay suspensions. Wiggles in sketches represent turbulence. The graphoids to the left of the sketches denote characteristic velocity time series at various heights in the flows. The graphoids to the right of the models represent characteristic vertical profiles of dimensionless downstream velocity ($U/U_{max}$), root-mean-square of downstream velocity (RMS($u'$)) and dimensionless turbulence intensity (RMS($u'$)$_o$). (After Baas *et al.*, 2009.)

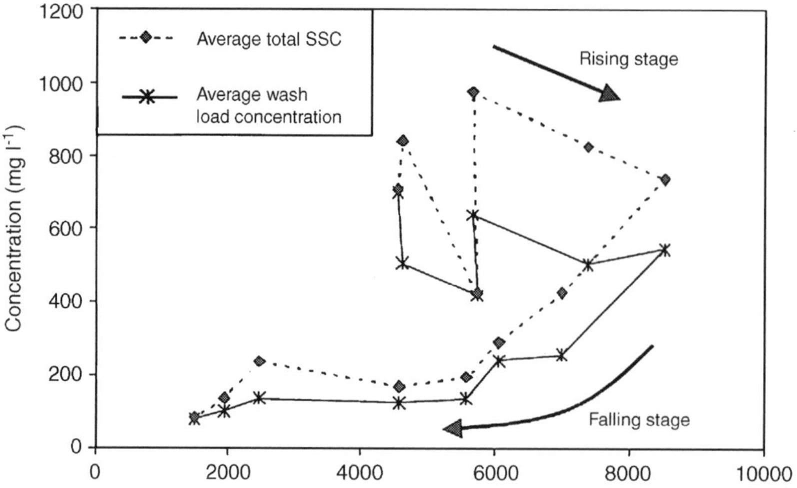


**Fig. 6.18** Mean suspended load concentration (SSC) vs. water discharge ($Q$) for the southern hemisphere autumn floods of the Burdekin River, Queensland, Australia. There is a very clear clockwise hysteresis (non-linearity) whereby the peak of SSC is displaced in time from $Q$. (After Amos *et al.*, 2004.)

material available for the flow to transport. So, inspection of the grain size of a deposit does not necessarily tell us anything about competence of the flow that transported and deposited it. On the other hand, the capacity of a flow relates to how much sediment mass may be transported at a particular flow power. As power declines then capacity falls and deposition ensues. The grain size and grain mass of any sedimentary deposit will thus reflect the combined effects of supply, competence and capacity.

## Further reading

### *General*

Clifford et al. (1993) and Ashworth et al. (1996) are research monographs with many relevant papers on sediment transport. Thorne *et al.* (1987) deals with the research problems of sediment transport in gravel-bed rivers. In the absence of any modern readable text solely dedicated to the principles of sediment transport, a return to Bagnold's 1954b classic (for wind) and to his 1966b (for water) should provide the necessary inspiration. In this context it is worth the reader looking into the selected edition of Bagnold's writings published by the American Society of Civil Engineers (Thorne *et al.*, 1988). Procedures for calculating bed stresses and sediment transport set up under tidal, wave and combined currents are very clearly set out by Soulsby (1997).

### *Specific*

See Dietrich (1982), Cheng (1997) and Ferguson & Church (2004) for development of universal settling velocity expressions. The fall velocity of a grain through groups of grains is discussed in a pioneer paper by Richardson & Zaki (1954). Important experiments on the Richardson–Zaki effect as applied to natural sand grains are by Tomkins *et al.* (2005).

Drag reduction in dilute clay suspensions was discovered by Gust (1976) and confirmed by Best & Leeder (1993b) and Li & Gust (2000).

Bagnold's classic experiments (1954) and his later bedload transport papers (1954b, 1966b, 1973) concern the magnitudes of solid-transmitted stresses. Confirmation of this approach comes in Abrahams & Gao (2006). Suspension dynamics are discussed by Sumer & Oguz (1978), Sumer & Deigaard (1979), Leeder (1983), Wei & Willmarth (1991), Leeder *et al.* (2005). Bagnold's central arguments are confirmed by McEwan *et al.*'s (1999) simulations and experimental data.

Grass (1970) published a pioneer paper on the probabilities involved in the initiation of grain motion, with later experimental work on mixed grain sizes in unidirectional flows by Parker *et al.* (1982), Wilcock & Southard (1989), Paola *et al.* (1992b) and Bridge & Bennett (1992). Buffington & Montgomery (1997) give a very careful and illuminating account of the problem of determining incipient motion in gravel-bed rivers. A note by Almedeji & Diplas (2005) on armoured beds is useful. Downstream-fining is discussed by Ferguson *et al.* (1996) and Rice (1999).

The threshold of motion under oscillatory and combined wave/current flows is investigated experimentally by Komar & Miller (1973), Green (1999), Wallbridge *et al.* (1999), Paphitis *et al.* (2001) and You & Yin (2006; see also Le Roux 2007). The thresholds for biogenic carbonate grains are investigated by Kontrovitz *et al.* (1979), Paphitis *et al.* (2002) and Yordanova & Hohenegger (2007). Heavy mineral transport by Li & Komar (1992).

The concept of 'sortable silt' and instrumental and conceptual techniques for determining past oceanographic current flows is discussed by McCave *et al.* (1995, 2006) and Hall *et al.* (1998, 2001).

Transport by wind is discussed by Bagnold (1941). Grain-bed collisions are investigated by Wang *et al.* (2008). Some peculiarities of the nature of atmospheric flow by Wyngaard (1992) and especially the strong temperature control on shear velocity and sediment transport by McKenna-Neuman (2004). Threshold problems in field conditions are covered by Wiggs *et al.* (2004) and on suspension threshold criteria generally by Cornelis & Gabriels (2004) and Leeder (2007). Surface stability, erosion and sediment trapping induced by vegetation has been investigated by many authors, most recently by Wiggs *et al.* (1995), Lancaster & Baas (1998), Wolfe & Nickling (1996), Kuriyama *et al.* (2005) and in more botanical detail by Levin *et al.* (2008).

Hyperconcentrated flows are discussed by Wan & Wang (1994), Wang (1994), Baas & Best (2002, 2008), Baas *et al.* (2009), Jiongxin (2004), van Maren *et al.* (2009) and Sumner *et al.* (2009).

Concerning erosion and transport of clay particles there are many papers of interest in Mehta (1993) and Burt *et al.* (1997). Raudkivi & Hutchinson (1974) discovered the effects of added electrolyte. For flocs see Mehta (1989). Mud erosion is reviewed by Kusuda *et al.* (1985), Partheniades (1993) and Tolhurst *et al.* (2009). Compaction and dewatering effects on mud erosion are discussed by Gomez & Amos (2005). Experiments on floc size and shear rates are by Manning & Dyer (1999). Shear thinning and thickening in colloidal dispersions in general are discussed by Wagner & Brady (2009). Particle imaging systems for floc research are covered by Lintern & Sills (2006).

Development of the sediment continuity equation (SCE) is by Wilson & Kirkby (1976). Applications of the SCE in the context of river incision and erosion are by Begin (1988) and Leeder & Mack (2007) and a full exploration of the subject is by Paola & Voller (2005).

Nonequilibrium transport and supply-limited sediment hysteresis is discussed by Reid & Frostick (1987) and Amos *et al.* (2004).

# Chapter 7

# BEDFORMS AND SEDIMENTARY STRUCTURES IN FLOWS AND UNDER WAVES

*What a strange, strange boy*
*He sees the cars as sets of waves*
*Sequences of mass and space*

Joni Mitchell, 'A Strange Boy', *Hejira*, Asylum Records

## 7.1 Trinity of interaction: turbulent flow, sediment transport and bedform development

There are many interesting forms and shapes that develop on beds of sediment during sediment transport. These *bedforms* occur at a variety of length scales, from a few grain diameters to veritable hills of moving sand, as in deserts. Bedform migration leaves behind tell-tale structures in the deposited sediment, enabling sedimentologists to 'read' sedimentary deposits in the ancient record; an acute example is provided by recognition of of subaqueous, as distinct from wind-blown, ripple bedforms on the ancient Martian surface. Leaving aside details of bedform morphology and structures for the moment, it is obvious that a fluid flow field will be greatly changed over and around any bedform; complex interactions and feedbacks occur (Fig. 7.1). In the sections below, water-flow, wave, combined-flow and wind-flow bedforms are considered. Additional forms, like large-scale *unit bar* bedforms of river channels that scale with flow width and linear ridges found on many tidally dominated shelves, are considered in later chapters.

## 7.2 Water-flow bedforms

### Current ripples

These are stable above the threshold of grain motion in fine–medium sand and silts (including flocculated muds) at relatively low flow strengths. They do not form in sands $\geq$ 0.7 mm diameter. On artificially smoothed beds, the forms evolve gradually from

*Sedimentology and Sedimentary Basins: From Turbulence to Tectonics*, 2nd edition. © Mike Leeder.
Published 2011 by Blackwell Publishing Ltd.

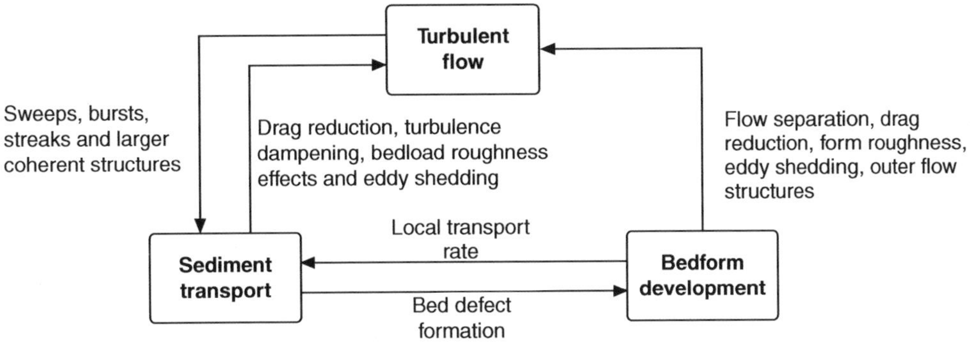


**Fig. 7.1** To show the 'trinity' of feedback between turbulent flow, sediment transport and bedform development. (After Leeder, 1983; Best, 1993.)

chance bed defects formed by impacting groups of turbulent sweep motions (Fig. 7.2). They may also form from initial bed irregularities on a smooth bed, triggered by local flow separation over random defects a few grain diameters high. The forms are asymmetric in the streamwise plane, with a gently sloping upstream or *stoss* side, sometimes with a prominent *crestal platform*, and a more steeply sloping (30°–35°) downstream or *lee* side (see Figs 7.3 & 7.4). Ripple height, $h$, ranges up to 0.04 m, ripple wavelength, $\lambda$, to 0.5 m, with typical ripple indices, $\lambda/h$, of 10–40. Populations of current ripples show a variation of size with individuals continually forming, growing and decaying. There is no clear relationship between ripple size and either flow strength or water depth. However, $\lambda$ does vary with grain size, being $\lambda \approx 1000d$, although the plot shows much scatter. There is a clear separation of size (but not general form) between current ripples and larger dune bedforms (see below).

Ripple crestlines may be straight, sinuous or strongly-curved linguoid or tongue-shaped (Figs 7.3 & 7.5). Careful experiments in laboratory channels reveal that straight- and sinuous-crested forms are metastable, always changing to linguoid ripples given sufficient time at a rate dependant upon the inverse of mean flow velocity. For flows close to the movement threshold this equilibrium time may be long and so the metastability is rather academic. It may be objected that wall effects in the experimental flows inevitably give rise to secondary flows (noted briefly in Section 5.5) and hence to longitudinal, flow-parallel, elements in *any* bedform.

On cohesive or other hard substrates (such as the floor of experimental channels) and on sediment beds where sand supply is limited, disequilibrium current ripples occur as trains and rows of isolated half-moon shapes in plan view, with elongated arms extending downcurrent. These *lunate* forms are termed *barchanoid*, as are larger-scale forms described from deserts (see Section 7.8).

Flow over current ripples involves crestal separation and reattachment slightly downstream of the trough (Figs 7.4 & 7.6). The size of the separation zone is a function of mean flow velocity and ripple height. Grains are moved in bedload up the ripple stoss side until they fall or diffuse from separating flow at the crest to accumulate high up on the steep ripple lee face. Periodically, accumulated grains become unstable as they oversteepen; small grain avalanches result which terminate at the toe of the lee face, propagating upwards to accrete a single lamina onto the lee side. Ripple advance by such lee slope deposition results in the flow attachment point shifting up the back of the downstream ripple where increased erosion occurs because of the very high turbulent stresses generated at the reattachment point. In this way ripples constantly shift downstream preserving their overall equilibrium shapes. They move at a mean velocity dependent upon the magnitude of sediment transport.

Sections cut through current ripples parallel to flow reveal the aforementioned avalanche laminae, defining the sedimentary structure known as *small-scale cross-lamination*. Normal to flow, cross-laminae formed by straight-crested ripples are parallel and

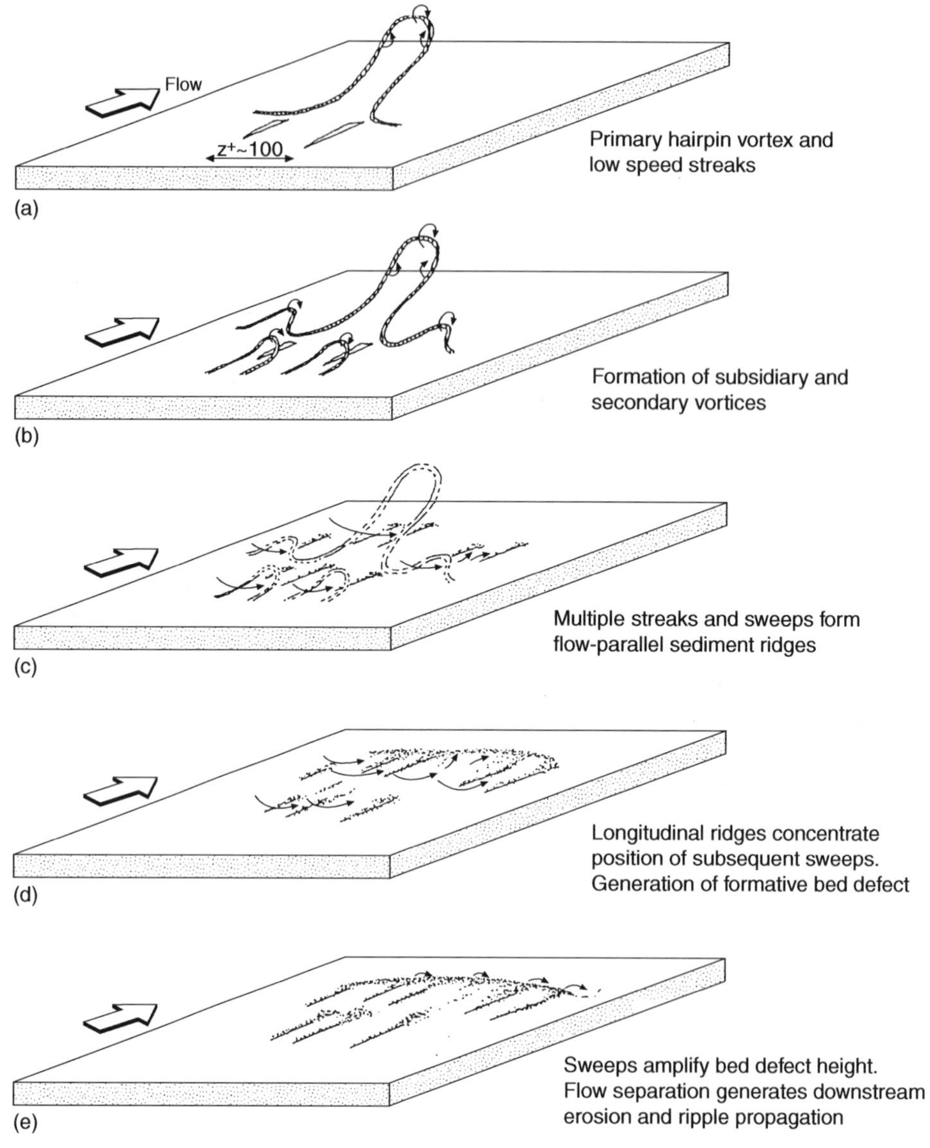
Flow
z+~100
Primary hairpin vortex and
low speed streaks
(a)
Formation of subsidiary and
secondary vortices
(b)
Multiple streaks and sweeps form
flow-parallel sediment ridges
(c)
Longitudinal ridges concentrate
position of subsequent sweeps.
Generation of formative bed defect
(d)
Sweeps amplify bed defect height.
Flow separation generates downstream
erosion and ripple propagation
(e)

**Fig. 7.2** Three-dimensional representation of the interaction between hairpin vortices of turbulence and a sand bed. The interactions lead to production of transverse bed defects which flow separation/reattachment processes amplify to current ripples by enhancing turbulent stresses downstream. (After Best 1992a.)

horizontal, defining *planar cross-lamination*. Sinuous to linguoid ripples produce *trough cross-lamination* (Fig. 7.5) since migration downflow occurs into heel-shaped scour troughs eroded by 3D separation eddies (Fig. 7.5). If there is no net sediment deposition over time then the only cross-lamination produced is that found within individual ripple elements as slightly larger and smaller individuals overtake each other or interact downflow. With net deposition from suspension in a depositing flow, a ripple has a vertical component of motion as well as a horizontal component (Fig. 7.7); sets of cross-lamination are formed, each bounded by erosive surfaces. The thickness of the sets and the amount of erosion is directly proportional

(a)

(b)

**Fig. 7.3** (a) Sinuous and minor linguoid current ripples (typical wavelengths ~ 25 cm) exposed on tidal flat at low tide; flow from right to left; Solway Firth, Scotland. (b) Linguoid current ripples exposed on same tidal flat at low tide; flow from bottom to top. Scale 10 cm longest dimension. Severn estuary, England. (c) Large-scale trough cross stratification; flow towards the observer; hammer shaft = 0.3 m; Old Red Sandstone, Lower Devonian, Welsh Borders. (d) Downward-dipping sets of large-scale cross-stratification, perhaps typical of unit braid bar (see Chapter 11); view is ca 2 m high; Fell Sandstone, Lower Carboniferous, Northumberland, England.

to the variability of ripple height and to lee-side erosion scour; it has little control by net rate of upward movement at low aggradation rates. Internally the set boundaries are seen to 'climb' at an angle to the horizontal; the structure is thus known as *climbing-ripple cross-lamination*. High angles of climb, with little erosion and good preservation of stoss-side laminae, indicate high rates of net deposition.

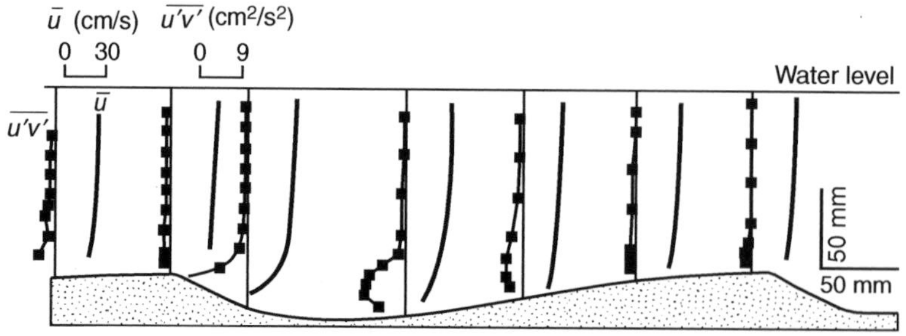


**Fig. 7.4** Classic profiles of mean velocity ($\bar{u}$) and a measure of turbulence intensity ($\overline{u'v'}$) as measured over a fixed and sand-coated experimental ripple model. Note the indication of a roller vortex in the ripple lee and the turbulence generated at flow reattachment. (Data of Sheen, 1964; as reported by Raudkivi, 1976.)

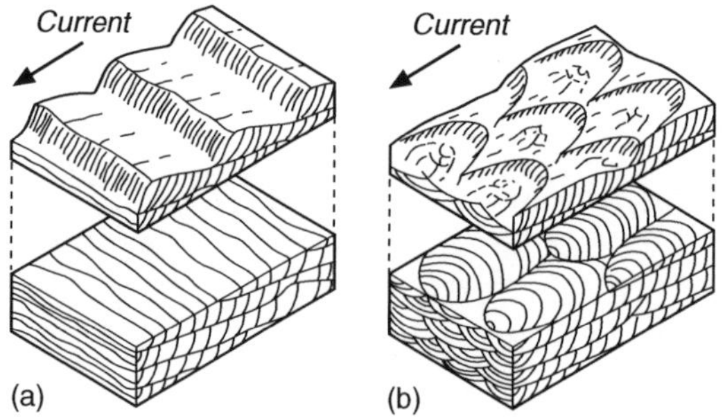


**Fig. 7.5** Diagrams to show that the migration of (a) straight-crested and (b) curved-crested ripple and dune bedforms produces planar and trough cross-stratification respectively. Note the preservation of successive cross-stratification sets produced by migration of one bedform during its lifetime. This may involve a degree of net deposition and when it does so then such sets appear to 'climb' upwards at some angle from the local bed inclination, so that their boundaries are tilted upflow (see Fig. 7.7). (After Allen, 1970.)

As flow strength increases, current ripples in very fine sands or silts are eliminated and replaced by plane beds (see following sections). Characteristic scour-like features accompany the change in fine silty sediments. In fine to medium sands ripples are succeeded by dunes.

### Lower-stage plane beds and cluster bedforms

As noted above, ripples do not form at the threshold for motion in coarse sands when $d \geq 0.7$ mm and a rough boundary exists. Instead an equilibrium *lower-stage plane bed* exists which in well-sorted sands exhibits shallow scours and narrow irregular grooves two to three grain diameters deep over its surface. Net deposition on a lower-stage plane bed should give rise to crude planar laminations, but convincing examples have yet to be described from the sedimentary record.

Rough boundaries on poorly sorted sediment show the effects of lee-side eddies shed by larger grains, examples of *von Karman vortices*, which periodically erupt into the free flow. In pebbly sands and bouldery gravels the coarsest grains ($d > 84\%$) cause lee-side sheltering and upstream hindrance that reduces drag and lift and encourages other grains to deposit both in front and behind. These stationary microforms are termed *cluster bedforms*.

### Dunes

Dunes (Fig. 7.8) are larger bedforms that develop as flow strength over current ripples and lower-stage plane beds is increased. They are similar to current

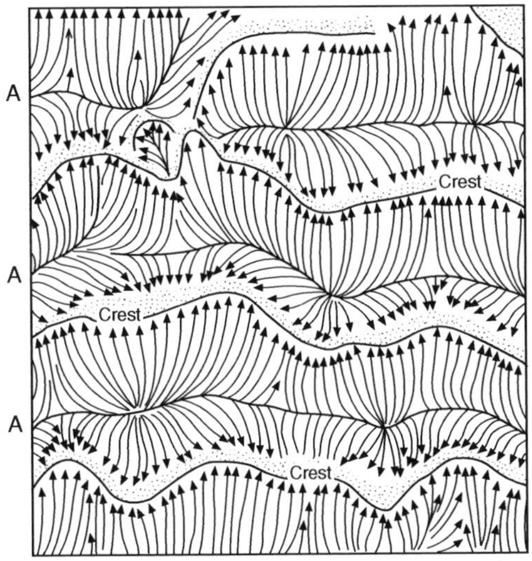


**Fig. 7.6** Classic photograph of the pattern of skin-friction lines for water flow (mean velocity 0.22 m/s, depth 0.095 m) over a current-rippled bed made as a cast from real sand ripples using Plaster-of-Paris according to a technique devised by D. Hopkins of Reading University in the 1960s. Overall flow from bottom to top. The ripple crests mark lines of flow separation. The steep ripple avalanche faces are stippled. (After Allen, 1969.)

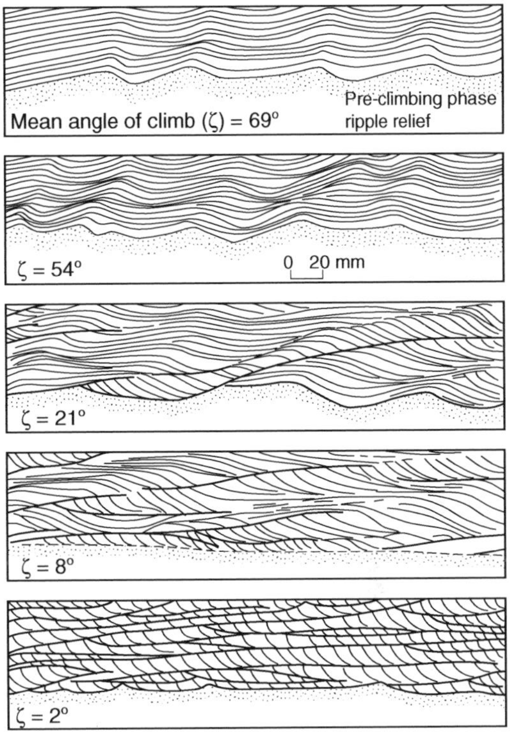


**Fig. 7.7** Experimental climbing ripple cross-lamination vertical profiles sampled and sketched parallel to flow (from left to right). The increasing angle of climb (ζ) of the cross-set boundaries from the bottom to top is caused by an increased rate of net vertical deposition relative to speed of advance of the ripples. (After Allen 1972.)

ripples in general shape but are dynamically distinct; they are emphatically not just large ripples as implied by the alternative term 'megaripples'. Their distinctiveness is indicated by:

- lack of overlap between ripple and dune *form indices* (wavelength:height ratios);
- a positive relationship between dune dimensions and flow depth (Fig. 7.9);
- absence of dunes in sediment finer than [O] 0.1 mm;
- the observation that current ripples may be superimposed upon the backs of dunes (Fig. 7.10a) in an apparently equilibrium relationship; the strong interaction between lee-side flow vortices (see Fig. 7.16) and the outer flow.

Dune wavelengths range from [O] $10^1$–$10^3$ m with heights [O] $10^{-1}$–$10^1$ m or more. In plan view dune crests may be straight to strongly curved. The large size of many dunes makes study of their equilibrium shapes difficult; it is never clear how exposed dunes in tidal channel and river beds have been modified by the waning flows that led to their exposure. This has more recently been revealed by detailed echo-sounding studies during waxing, bankfull and waning discharges in rivers. These show that, like ripples, dunes undergo birth, growth, modification and decay, often with strong interactions between neighbours, including cannibalism, superimposition and splitting.

The flow pattern over dunes is similar to that over current ripples but in addition large-scale advected eddy motions (termed boils or kolks) rich in suspended sediment rise at low angles (20–25°) from the separating flow to periodically erupt at the flow surface. These are caused by a combination of shear instabilities (low frequency wake 'flapping') and vortex interactions (high frequency vortex 'shedding') between

(a)

(b)

**Fig. 7.8** (a) Low-tide view of emergent straight-crested dunes with superimposed linguoid ripples formed during the falling stages of tidal ebb flow; dune wavelength, ca 3 m. (b) Sinuous-crested dunes with well-developed scour pools; dune wavelength ca 5 m. Both from Solway Firth, Scotland.

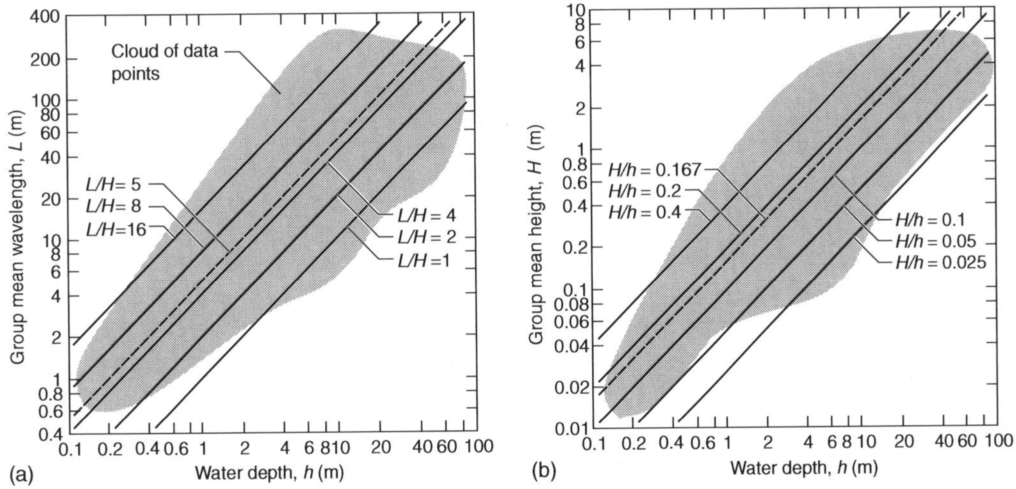


**Fig. 7.9** Graphs to show the correlations between (a) dune group mean wavelength and mean water depth and (b) dune group mean height and mean water depth, both for one-way river flow only. Despite the considerable scatter of data, it is apparent that the scale of dunes increases with boundary-layer thickness, suggesting that dunes owe their origins to flow processes that affect the whole boundary layer. (Data of Allen, 1982.)

the separating flow of the free stream and the lee-side fluid closer to the bed at and downstream from the dune crest (**Plate 3**; Fig. 7.11; and see Fig. 7.17).

Migration of steep, angle-of-repose lee-side dunes give rise to large-scale cross-stratification of high-angle planar or trough type, as for current ripples. Tangential contacts between individual cross-laminae and bounding set surface are encouraged by relatively weak lee-side separation eddies and by a high fallout rate of particles from suspension in the dune lee. *Counterflow ripples* result when grains are swept back up the lower parts of tangential foresets by near-bed flow in the separated flow. The thickness of cross-stratified sets depends upon the same variables noted above for current ripple sets, viz. variability of dune height and depth of lee-side scour (see below). However, echo-sounding measurements in large river channels also reveal that many dune-like forms have quite low-angle lee-side faces. The downflow advance of such forms comes about because of superimposition of smaller dunes that migrate down the low-angle slope. These generate *downdipping sets* of high-angle cross-stratification (smaller dunes) separated by low-angle surfaces marking the lee-side of the larger bedform. It has been somewhat controversial as to whether the small and large dune forms are in true flow equilibrium since such cross-sets may also be cut by erosion of the dune crest and lee into a more gently inclined bar platform by falling-stage or low-stage flows; smaller dunes then nucleate on these and migrate downslope. Such features may be preserved as *reactivation surfaces* and downdipping sets of cross strata within the migrating dune when normal avalanching events begin once more at rising- and high-flow stages. However, extensive echo-soundings in many rivers in recent years reveals that superimposition is ubiquitous and unrelated to hydraulic change.

Dunes in fine bedstock where suspension dominates and those close to their upper flow stability limit, at the transition to upper-stage plane beds (see below), are characterized by streamwise symmetry, lack of avalanche face and absence of lee-side flow separation, though periodic upward advections of macroturbulence rich in suspended sediment occur periodically. These forms are termed *symmetric* or *humpback dunes*. We know little of the internal structures of those formed at high flow, but during reduced flow small dune forms may migrate down the gently dipping lee side of the larger dune-shaped mound to cause successive reactivation surfaces and downward-dipping cross-stratification (Fig. 7.3d).

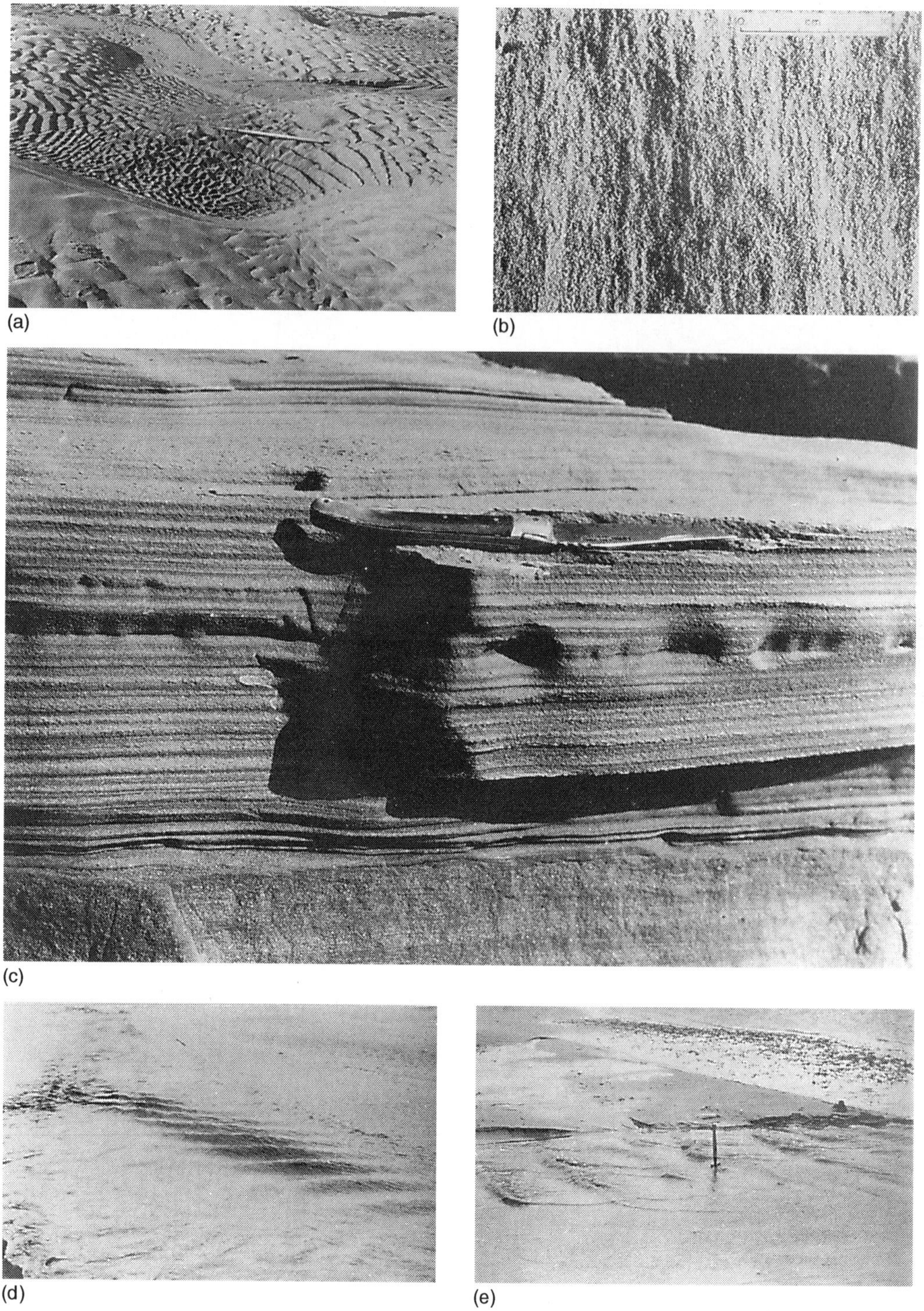


**Fig. 7.10** (a) View at low tide over a sinuous dune crest to show scour pool and ripple fan on back of next dune downflow; scale = 0.5 m; Loughor Estuary, Swansea, Wales. (b) Primary current lineations in fine sands. (After Allen, 1964.) (c) Upper-phase plane beds; knife = 0.15 m long; St Bees Sandstone, Triassic, Cumberland, England. (d) Train of antidunes (wavelength = 0.3 m) in fast shallow tidal channel flow; Barmouth estuary, Wales. (e) Train of upstream-breaking antidunes in tidal channel; flow left to right; shovel handle = 20 cm long; Solway Firth, Scotland.

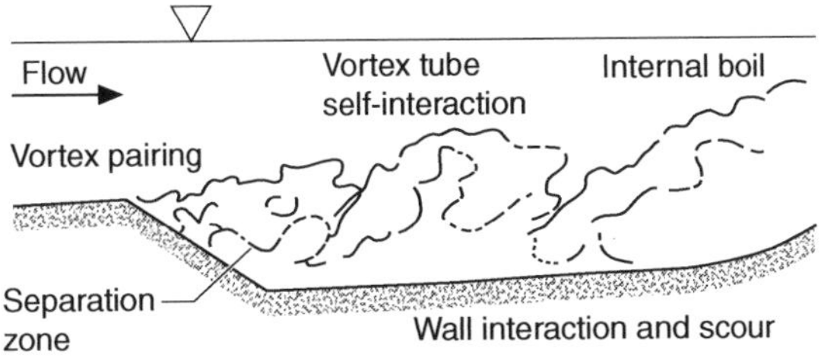


**Fig. 7.11** Vortex development and 'boil' generation associated with the free shear layer of a lee-side flow separation zone over an experimental fixed dune. (After Muller & Gyr, 1986.)

Concerning efforts to establish the height of subaqueous dunes from the thickness of cross-stratified sets, experimental data from a range of flumes and rivers shows that the distribution of dune height can be calculated directly from the distribution of cross-set thickness, and vice versa, independently of aggradation rate. Mean dune height, $h_m$, is approximately 2.9 ($\pm$0.7) times mean cross-set thickness, $h_c$. Since dune height can be related approximately to formative flow depth, $h_f$ (Fig. 7.9), it provides a useful complement to other methods of estimating flow depth from sedimentary information. Most dunes fall within the range of $3 < h_f/hm < 20$. The large scatter is due primarily to two reasons. First, the height of dunes relative to their length and flow depth increases from near zero at the lower boundary of their hydraulic stability field (transition from ripples or lower stage plane beds), to a maximum in the middle of the field, and then to near zero at the transition to upper-stage plane beds. Second, when measurements of dune height and flow depth are made in natural rivers, it is not certain that the dunes were in equilibrium with the prevailing flow conditions. However, it appears that $h_f/hm$ is most commonly between 6 and 10.

### Upper-stage plane beds

As flow strength is further increased over fine to coarse sands, dunes give way to upper-stage plane beds (Fig. 7.10b & c). Careful experimentation establishes that the bed surface actually comprises small-amplitude/long-wavelength *bedwaves*. These are asymmetric in the *xy* plane parallel to flow, range from 0.75 to 11 mm in height, 0.7 to 1.0 m in wavelength and travel with velocities of up to 10 mm/s. Flow measurements suggest that the flow accelerates over the bedform crests and decelerates over the bedform troughs. Upon these bedwaves are superimposed flow-parallel ridges and hollows known as primary current lineation. Primary current lineation is the direct result of viscous sublayer structure (Chapter 5) whereby incoming sweeps, spaced parallel to flow, push grains aside to form the tiny grain ridges separated by broad troughs. It is important to emphasize that primary current lineation is not restricted to the upper-stage plane bed regime but may occur on the stoss side of ripples and dunes.

Migration of bedwaves gives rise to an internal structure of planar to wavy laminations 5–20 grain diameters thick (Fig. 7.12). Lamina preservation and thickness are dependent upon aggradation rate and passage of bedwaves of varying size. Laminae are made visible by lee-side sediment sorting, usually by infiltration of finer grains into an underlying surface. Ancient examples of upper-stage plane beds seem not show preservation of the bedwaves themselves; rather, primary current lineations on more-or-less planar stoss surfaces are the most obvious feature.

### Supercritical flow bedforms: antidunes, transverse ribs, chutes and pools, and related forms

Antidunes are sinusoidal forms with accompanying *in-phase water waves* that typically occur in shallow, fast water flows (Fig. 7.13) when the Froude Number, Fr $\geq$ 0.84. They occur in long trains; the individual waveforms may be stationary or undergo cycles of amplification and destruction in which they periodically steepen, move upstream and break up in a great rush of turbulence, the process then beginning again. Sedimentary structures associated with experimentally produced antidunes are primarily lenticular laminasets with concave-upward erosional bases (troughs) in which laminae generally dip upstream or fill the troughs symmetrically. These laminasets are associated with growth and upstream migration of water-surface waves and antidunes, and with surface-wave breaking and filling of antidune

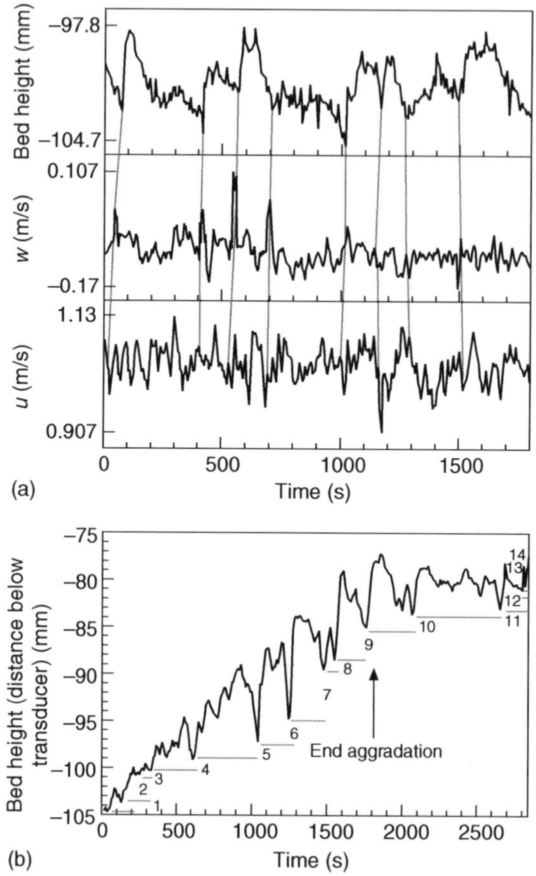


**Fig. 7.12** (a) Time series of elevation and velocity profiles over a sand bed during passage of low-amplitude bedwaves on an upper-phase plane bed. (After Best & Bridge, 1992.) Note the correlations (dotted verticals) of bedwave troughs with high spanwise (*w*) and low streamwise (*u*) velocities, indicating downstream decelerating and laterally accelerating flow conditions there. (b) Time-series plot to show the aggradation of an experimental upper-phase bed with time. The peaks and troughs originate as successive low-amplitude bedwaves pass over the aggrading bed. Numbers 1–12 indicate the laminae preserved in the final deposit that was subsequently cored and sliced. It can be seen that these laminae originated as erosively based features whose preservation potential is controlled by the relative rates of bed aggradation, bedform height and wavelength, and bedform migration rate. (For details see Paola & Borgman, 1991; Best & Bridge, 1992.)

troughs respectively (Fig. 7.13a). In addition, sets of downstream-dipping laminae are produced by rapid migration of asymmetrical bedwaves immediately after wave breaking. Rare convex-upward laminae define the shape of antidunes that developed under stationary water-surface waves. The laminasets and internal laminae extended across the width of the experimental flume, but varied in thickness and inclination, indicating that the antidunes have some degree of three dimensionality. The length and maximum thickness of the lenticular laminasets are approximately half the length and height of formative antidunes, providing a potentially useful tool for palaeohydraulic reconstructions.

Antidune bedforms in coarse, gravelly substrates have been termed *transverse ribs*; they are flow-transverse ridges only a few grain diameters high frequently preserved on bar tops in braided stream settings. Spectacular large-scale gravel antidunes are recorded with wave-lengths up to 19 m, heights up to 1 m, and with steeper upstream than lee slopes. Internally they featured erosive-based lenses of sandy gravel with low-angle downstream-dipping laminations.

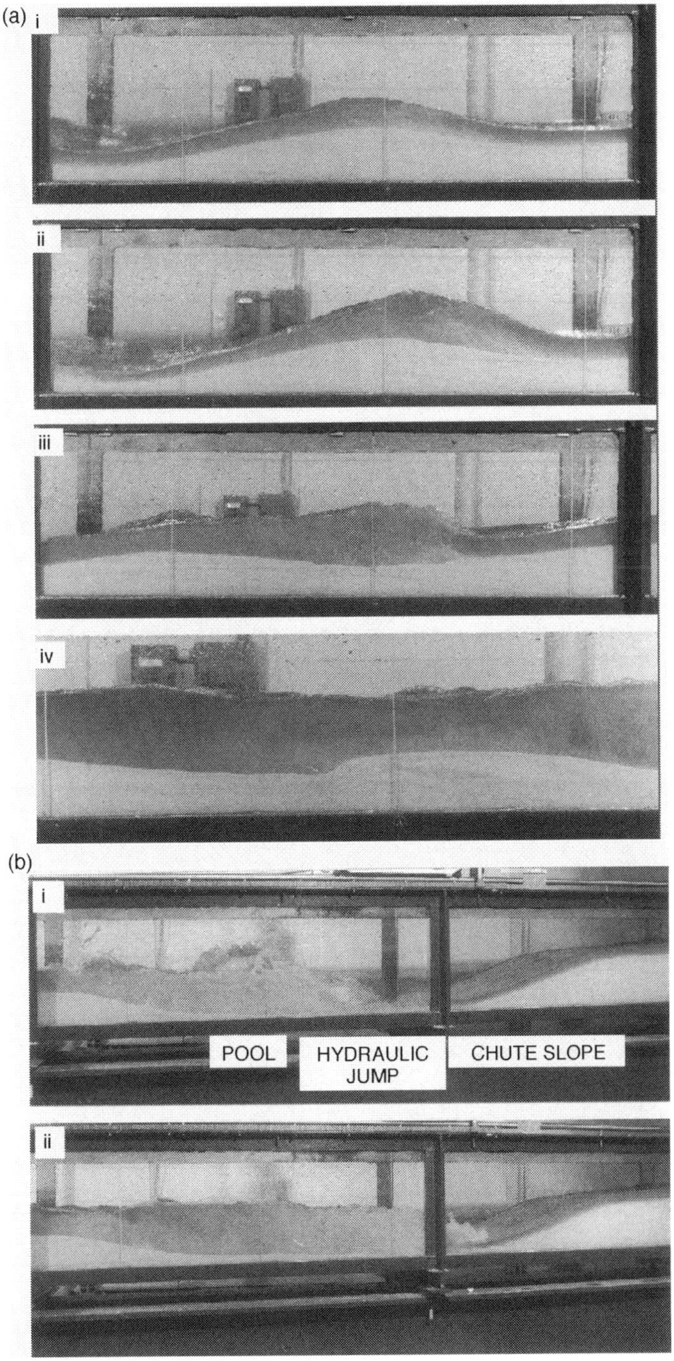


**Fig. 7.13** Supercritical flow bedforms and structures. (After Alexander *et al.*, 2001.) (a) Photographs of experimental test section (1.5 m wide) with water flow from right to left. (i) Standing water-surface wave and developing antidune, (ii) asymmetrical water-surface wave and antidune, with flow separation zone on upstream side of antidune prior to wave breaking, (iii) a breaking wave and (iv) asymmetrical bedform migrating into antidune trough after wave breaking. NB: These photographs are not all of the same antidune. Although the pictures represent a sequence of events, the position of bedforms has changed. (b) Photograph of chute-and-pool showing upstream migration of hydraulic jump (i) and its breaking on the chute slope (ii).

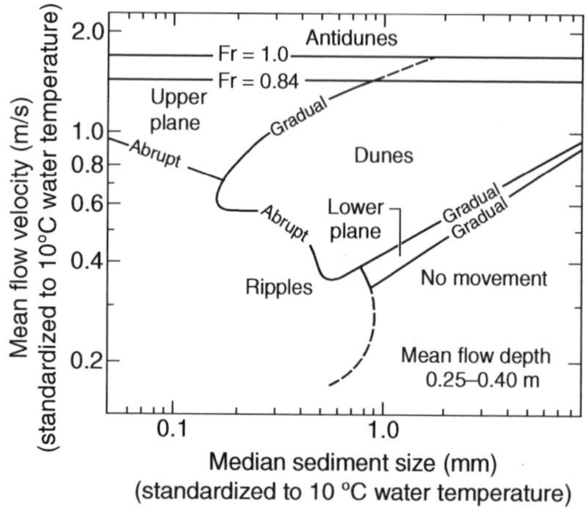


**Fig. 7.14** The simplest and most easily interpreted unidirectional water flow bedform phase diagram is this plot of mean flow velocity vs. mean sediment size for a wide range of flow depths. Variables are standardized to water temperature, viscosity and density conditions at 10 °C. (After Southard & Boguchwal, 1990a, as slightly modified by Ashley, 1990.)

Further increase in Fr leads to formation of *chute-and-pool* bedforms (Fig. 7.13b). The chutes form in shallow supercritical parts of a flow which end abruptly in deeper pools where the flow is subcritical. The upstream boundary of a pool is marked by rapid flow deceleration, the violently breaking water defining a *hydraulic jump*. Sedimentary structures associated with chutes-and-pools are sets of upstream-dipping laminae and structureless sand.

### Bedforms and sediment transport in poorly sorted sediment: bedload sheets and low relief bed waves

In poorly sorted sediment it is common for the finer sand or gravel fraction to be moulded into isolated barchanoid ripples, dunes or dunoids or flow-parallel sand ribbons and for these to then migrate over a more-or-less stationary substrate of immobile coarser gravel that is then left behind as a *lag deposit*. As flow strength increases into the dune stability field, equal mobility (Section 6.4) may cause formation of dunoid-type *low relief bed waves*. The interaction of rolling grains causes particles to group, slow down and trap finer sediments in their interstices. At higher flow strengths within the overall dune stability field, once a critical concentration of bedload grains is reached, the friction between large grains is reduced and most sediment available at any given applied bed shear

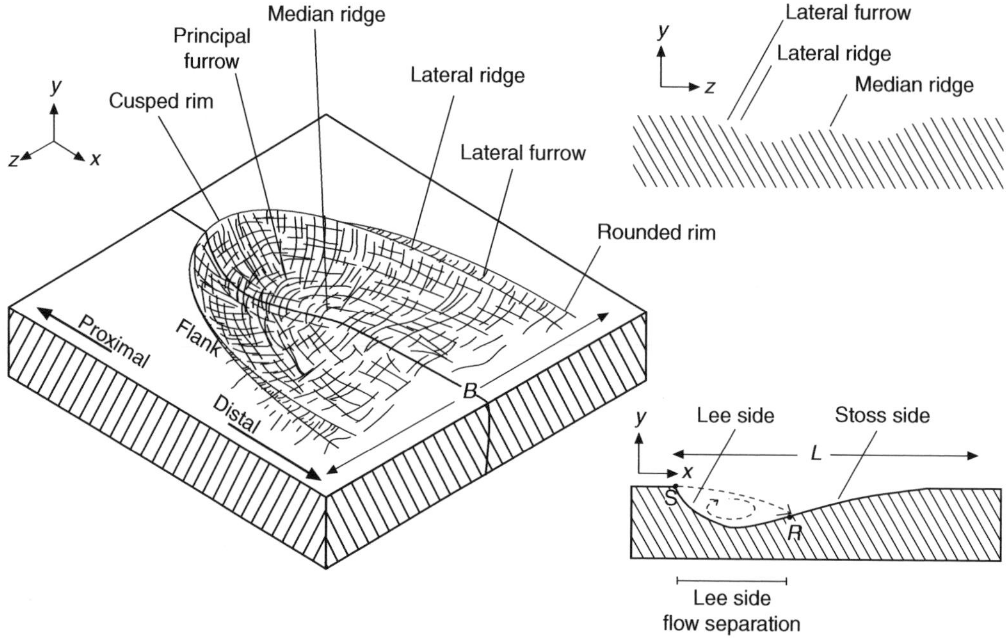


**Fig. 7.15** The morphology of an idealized flute mark cut into a cohesive substrate. (After Allen, 1982.)

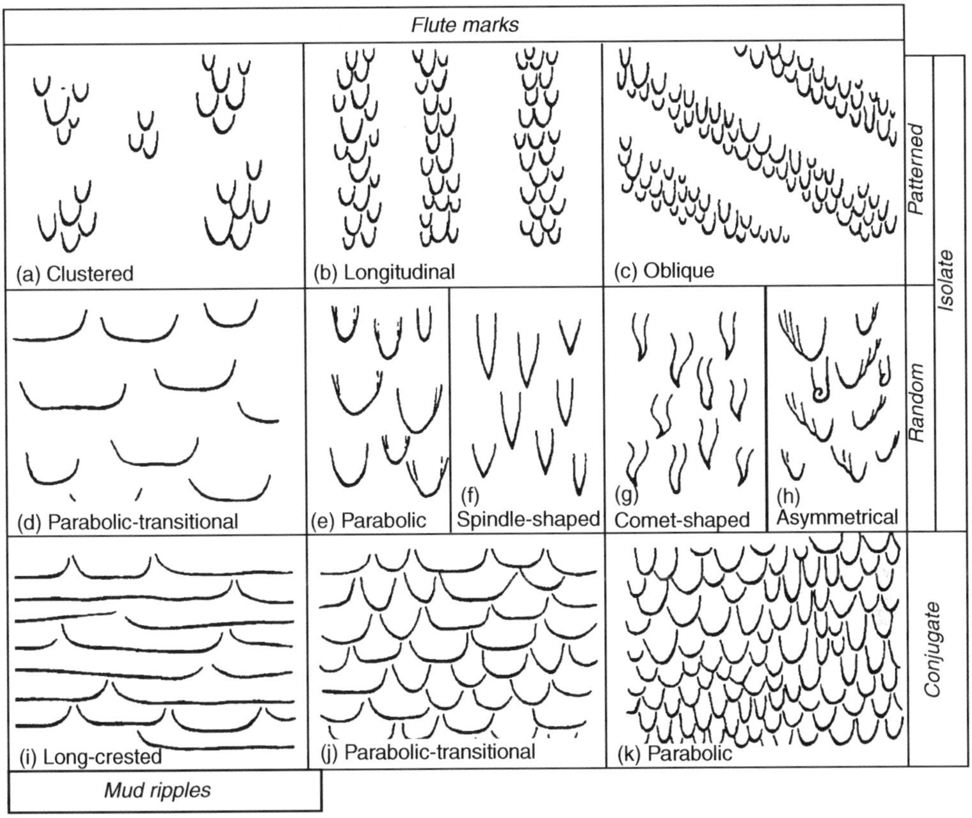


**Fig. 7.16** Some of the main types of flute mark assemblages and morphologies. (After Allen, 1982.)

**Fig. 7.17** View of base of turbidite bed showing a nice array of conjugate parabolic flute casts. Palaeoflow was right to left. Silurian, Horton-in Ribblesdale, Yorkshire, England.

**Fig. 7.18** The upstream part of a spectacular giant (> 6 m wide and 2 m deep) isolated parabolic flute mould (original form of flute) in fine-grained turbidite sandstone bed. Carboniferous, Ross Sandstone, Co. Clare, Ireland.

**Fig. 7.19** Sinuous, long gutter casts on the base of a crevasse-flood sheet sandstone. Middle Jurassic, Scalby Formation, Yorkshire, England.

**Fig. 7.20** The base of a thin fine-grained sandstone showing criss-crossing and 'swirling' casts that probably represent tool marks made by the carriage of plant stems and branches over a partly consolidated mud bed during a river flood incursion on to a muddy floodplain. Lower Carboniferous, Dumfriesshire, Scotland. Scale is 10 cm long.

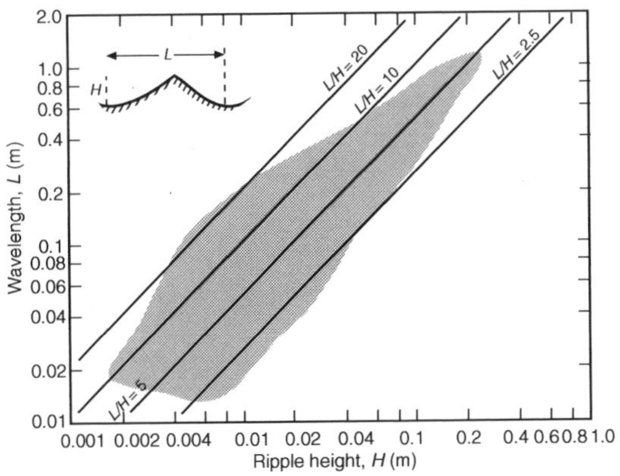


**Fig. 7.21** Wavelength and height of wave ripples; shaded area is an envelope of data points. (Data of Allen, 1982.)

stress may become mobile; the bed evolves into a pattern of low relief, flow transverse alternations of finer and coarser sediment, the latter concentrating and defining the bedform crests of *bedload sheets*. Should these show adjustments to scale with water depth and possess well-developed lee-side separation zones then they are regarded as incipient dunes. Should inadequate supplies of bedload material persist at higher transport stages undersaturated or sediment-starved bedforms persist as isolated forms.

### Bedforms and deposits of hydraulic jumps

Hydraulic-jump-related bedforms (Section 5.6, Fig. 5.20) are thought widespread in subglacial and marginal glaciolacustrine settings, though not directly studied apart from in experiments on sediment-fed but fixed-bed hydraulic jumps where no scour was allowed. These reveal the roller vortex-loop and downstream jet structures of the flow with massive pulsating deposition forming upstream accretion of what are termed *hydraulic-jump unit bars*.

Natural exposures through deposits of this kind where scour did occur are reported from Ontario, Canada, interpreted to be of hyperconcentrated flow origin and emplaced under a regime of rapid flow expansion and loss of transport capacity within a plane-wall jet with an associated hydraulic jump. Massive gravels with unconsolidated sand intraclasts and open-work gravel/gravel–sand couplets were deposited in the zone of flow establishment by hyperconcentrated and supercritical flows, respectively. Immediately downflow, low-angle cross-stratified sand incised by steep-walled scours infilled by diffusely graded sand define the hydraulic jump transition zone of maximum erosion under the roller vortex. These strata record rapid bed aggradation from sediment-laden supercritical flows that were episodically scoured by large vortices generated within the migrating hydraulic jumps. The downstream zone of established flow features medium-scale, planar cross-strata and small-scale cross-lamination related to migrating 2D dunes and current ripples, respectively. The sediment architecture suggests that the plane jet fan was deposited during a relatively short period of time (days, weeks).

## 7.3 Bedform phase diagrams for water flows

Bedform phase diagrams are extremely useful plots of measures of flow strength against sediment mean

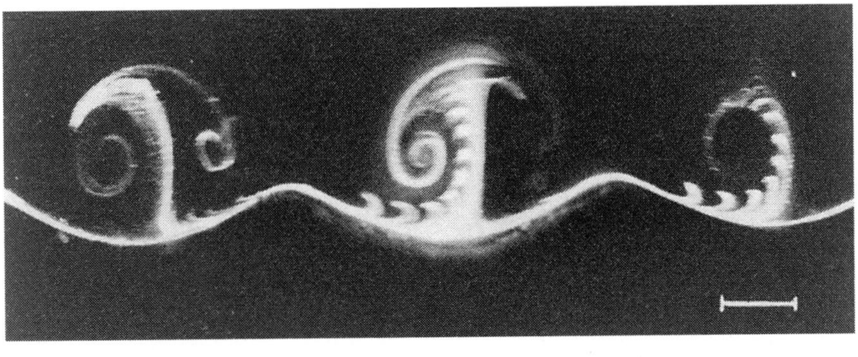

**Fig. 7.22** Flow visualization of steady streaming above a wave ripple model form. Scale = 1 cm. (After Honji *et al.*, 1980.)

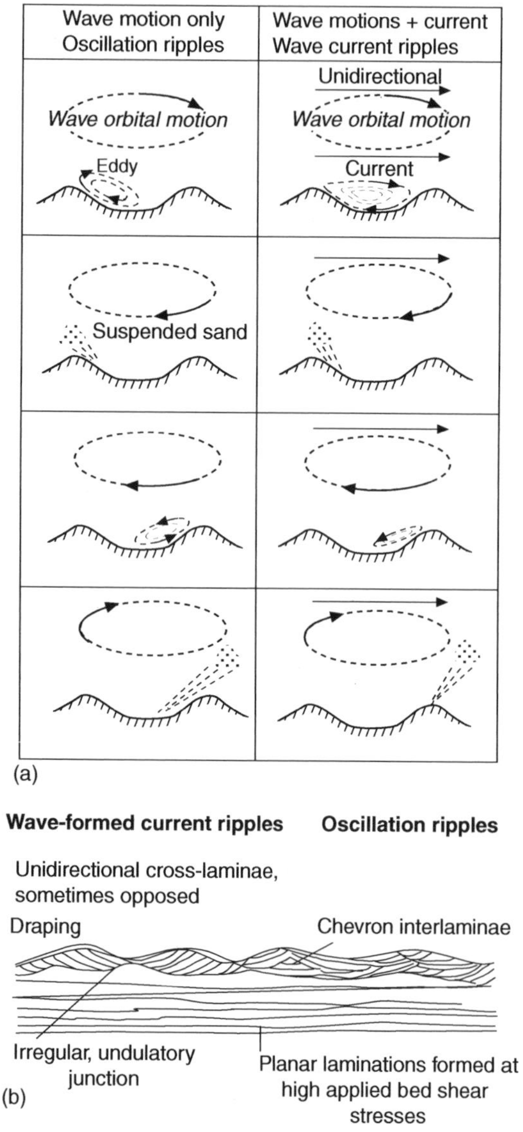


**Fig. 7.23** (a) The relationship between sand transport over a rippled sand bed and the orbital motion of wave action with and without a superimposed one-way current. (After Komar, 1975, from the original data of Inman & Bowen, 1963.) (b) Some characteristic internal features of wave-formed and combined wave-current ripples. (After de Raaf *et al.*, 1977.)

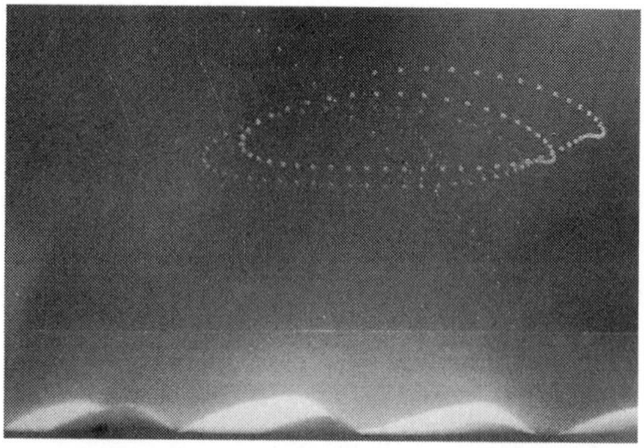

**Fig. 7.24** Splendid stroboscopic image of suspended neutrally buoyant particles to illustrate net forward motion of water (left to right translation of ellipse) in the mid-depth of shallow-water waves flowing over fixed model asymmetric wave-formed ripples. Water depth is 1.9 wavelengths, $T = 2.9$ s. (Photo courtesy of K. Tietze.)

grain size. They define a space where the stability limits of the various bedforms discussed above may be plotted and are therefore to be regarded as obligatory tools of the sedimentological trade, rather analogous to the phase diagrams in physics, physical chemistry and igneous petrogenesis that some readers will be familiar with. The axes of a bedform phase diagram may be dimensional, with units of velocity, stress, power, grain size, or they may be dimensionless like the Shields plot for threshold of sediment motion seen previously in Fig. 6.7.

In days gone by, water flow bedforms were separated into two broad groups. Ripples, lower-phase plane beds and dunes defined a *lower flow regime* with relatively high flow resistance and surface signs of lee-side dune eddy bursts. Upper plane beds, antidunes and chute-and-pool structures defined an *upper flow regime* where flow resistance is relatively low and where surface waves are in phase with any bed undulations. But bedwaves on upper-stage plane beds are out of phase with overlying diverging and converging flow, nullifying this scheme.

A problem with using bedform phase diagrams comes from the fact that friction coefficients (Section 5.3; **Cookie 24**) calculated for current ripples and dunes are two to five times those calculated for upper- and lower-stage plane beds. The increase is due to turbulent energy created and destroyed by lee-side flow separation and reattachment. The extra friction is termed *form drag*. The contrast in friction coefficients means that bedform phase diagrams in which bed shear stress or flow power is used as the

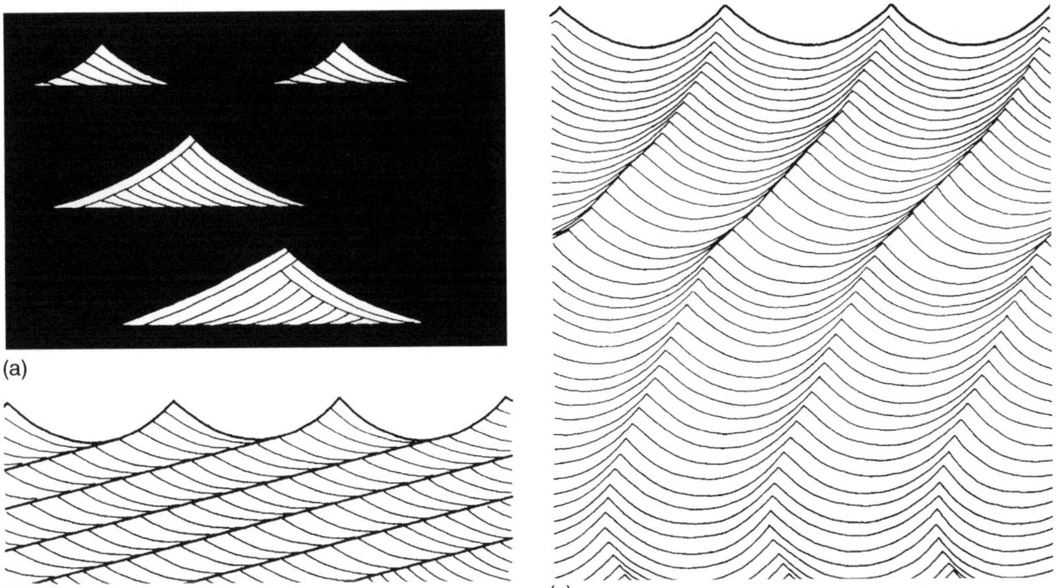


**Fig. 7.25** Sketches of cross-stratified sets formed by wave-related ripples. (a) Internal structures of these form-sets indicate isolated, sand-starved, ripples that periodically grew and migrated over mud (shaded black) substrates, the lower two eventually remaining stationary with chevron laminae forming at their tops. (b) Climbing sets with erosive upper set boundaries that indicate formation by net forward migration and deposition. (c) Climbing sets with nonerosive boundaries indicating very high deposition rates from suspension such as might occur during waning conditions after storms. (After Allen, 1982.)

measure of flow strength may be difficult to interpret. Writing the applied fluid shear, $\tau$, as, $\tau = \rho f u^2/8$, where $f$ is the Darcy–Weisbach friction coefficient (**Cookie 24**), $\rho$ is fluid density and $u$ is mean flow velocity, we see that $\tau$ is a direct function of the friction coefficient; itself a function of bedform type. So the same shear stress could be produced by slower flow over a rough (e.g. dune) boundary or faster flow over a smooth (e.g. upper plane bed) boundary. This effect is partly responsible for the overlap of the dune and upper-phase plane bed fields in plots involving bed shear stress; this may be overcome by plotting bedform phase diagrams as mean flow velocity vs. grain size (Fig. 7.14), but here the flow depth must be kept constant.

> Want to know more about bedform lag, bedform theory and trough cross-stratification? Turn to Cookies 35–37

## Bedforms and deposits of rapidly decelerating flows

In many natural situations a fully charged, sediment-transporting flow may be forced to decelerate rapidly. Examples might be out-of-channel flooding in rivers, downstream deposition after hydraulic jumps and certain turbidity currents. Extreme flow unsteadiness or nonuniformity is a potentially important yet poorly understood control on depositional processes since it has been suspected for some years that high concentrations of settling sediment may modify bedform development. However, there is some controversy about any modifying role of high sediment concentration on both turbulence and bedform development. One group working with quasi-steady, depletive slurry flows found no evidence for modification in rapidly aggrading (up to 4 mm/s) deposits at bedload concentrations of up to 0.35 by volume with parallel lamination formed from upper stage plane beds and migrating bedwaves at bed aggradation rates up to

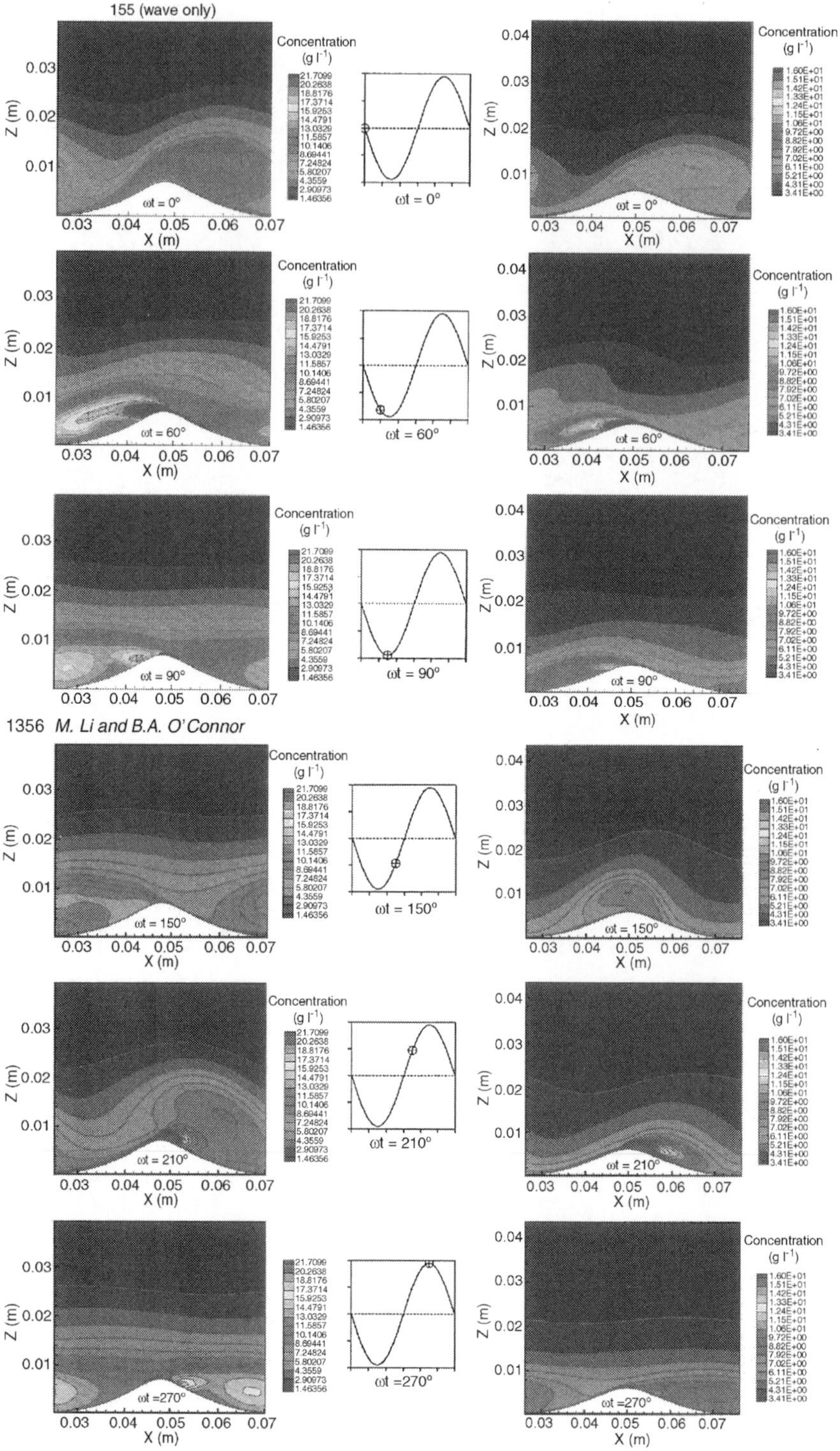


Fig. 7.26 Computational predictions of suspended sediment concentration over fixed bed symmetrical ripples over a single wave period for waves alone and combined flow (wave + following current). (After Li & O'Connor 2007.)

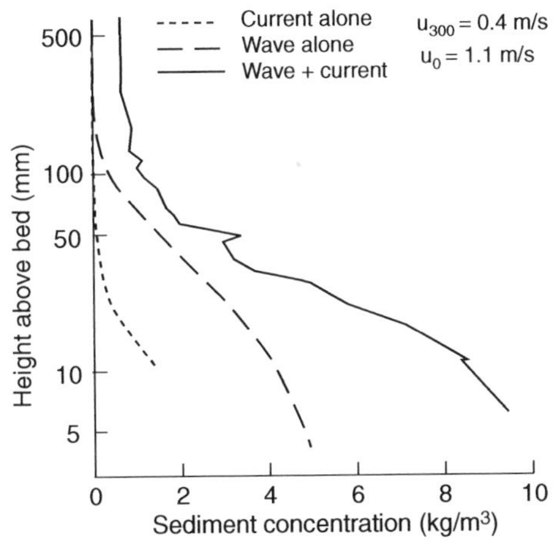


**Fig. 7.27** Experimental profiles of wave-cycle-averaged suspended sediment concentration for current alone, wave alone and wave plus current. It can be seen that the former two are not simply additive to obtain the latter. (After Murray *et al.*, 1991.)

4 mm. A second group conducted experiments in an annular flume and were able to rapidly decelerate (from speeds of up to 3.5 m/s) turbulent flows of fine sand with sustained periods of sediment fallout. In particular, the collapse of high-concentration, moving, thin (< 5 mm) near-bed layers (*laminar sheared layers*) were an important mechanism by which the bed aggraded beneath these unsteady flows. At bed aggradation rates in excess of 0.44 mm/s the sequential collapse of laminar sheared layers produced a structureless, poorly graded and poorly sorted deposit. When bed aggradation rates fell below 0.44 mm/s the collapsing laminar sheared layers were reworked by turbulence to form planar laminae, but these formed in a very different manner to planar laminae attributed to bedwaves on upper phase plane beds in equilibrium steady open-channel flows (see previous discussion). Inverse grading that developed at the base of the deposits of slowly decelerated flows was probably the result of grain sorting in a high-concentration layer that persisted at the base of the flow for many minutes prior to the onset of deposition.

## 7.4 Water flow erosional bedforms on cohesive beds

Once a critical stress is exceeded, an initially flat kaolinite mud bed is eroded by freshwater to give rise to three sorts of bedform.

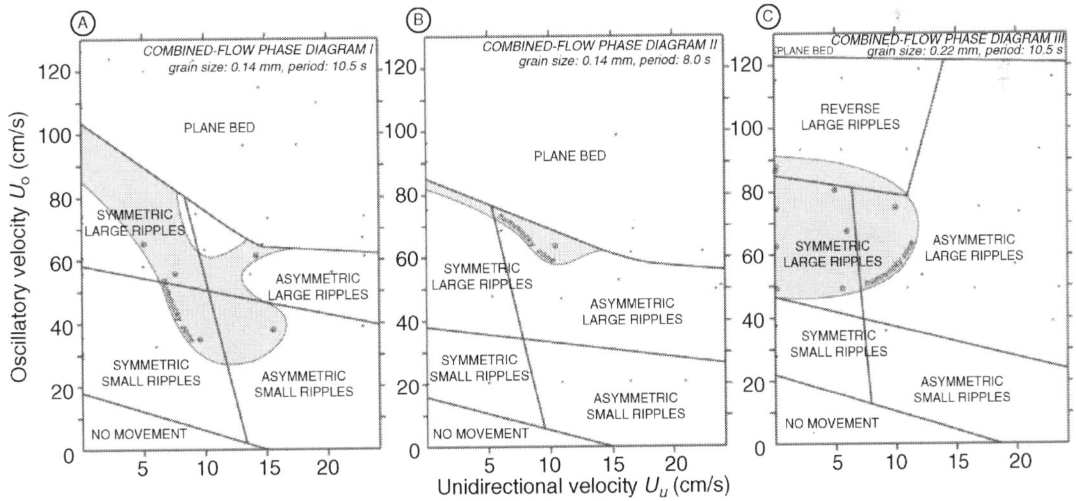


**Fig. 7.28** Combined-flow phase diagrams for fine and very fine sands with oscillatory velocity and unidirectional velocity as axes. Solid circles, runs; open circles, hummocky bed form; shaded area, hummocky zone; solid triangles, additional runs. (After Dumas *et al.*, 2005.) (a) Phase Diagram 1 (median grain diameter, $D_{50} = 0.14$ mm, wave period, 10.5 s). (b) Phase Diagram 2 ($D_{50} = 0.14$ mm, wave period, 8.0 s). (c) Phase Diagram 3 ($D_{50} = 0.22$ mm, wave period, 10.5 s).

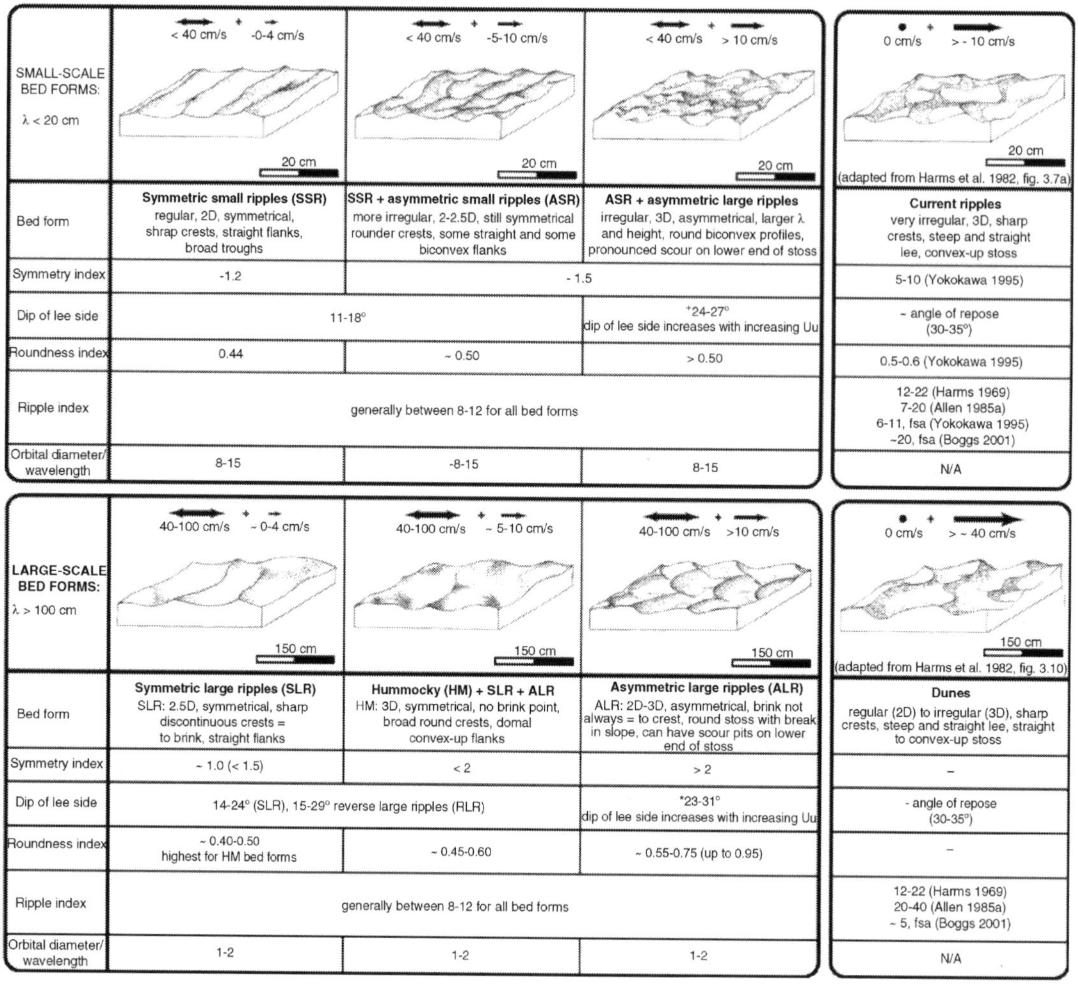


**Fig. 7.29** Morphological attributes of small-scale and large-scale oscillatory-flow, combined-flow, and unidirectional-flow bed forms. (After Dumas *et al.*, 2005.)

1 Small-scale longitudinal grooves and ridges have a typical spanwise spacing of around 0.5–1.0 cm. The ridges are sharp with broad, rounded intervening furrows. Tiny striae produced by the flow enlargement of air bubbles in the mud indicate a bottom flow structure and scaling consistent with their origin by erosion by flow vortices typical of viscous sublayer streaks.

2 At slightly higher flow velocities the longitudinal grooves change into meandering grooves, indicating a transverse instability affecting the sublayer streaks.

The meandering grooves cause deep corkscrew-like erosional marks to form as flow strength is further increased; these gradually develop into characteristic spoon-shaped depressions called *flute marks* (Fig. 7.15-7.17). A large variety of such flute shapes occur, the exact shape depending on initial shape of the defects. The alignment of rows of flutes with intervening flute-free areas may originate from spanwise variations in bed shear stress such as those associated with lobes and clefts at the front of turbidity currents (Chapter 8). Flutes may also commonly form from bed defects such as hollows or impact marks. Flutes range in

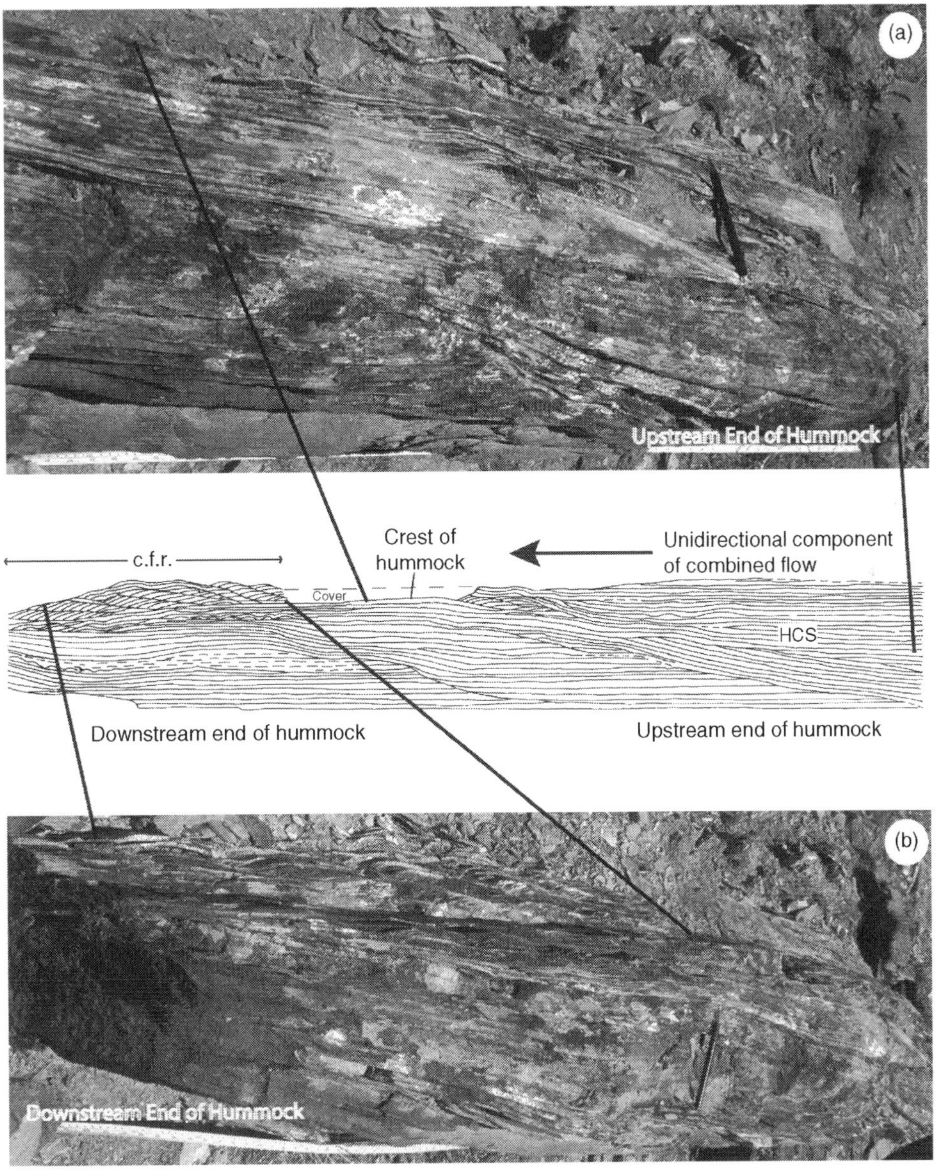


**Fig. 7.30** Detailed sketch of Carboniferous sandstone bed to show large-scale hummocky cross stratification at its base and combined-flow ripples (c.f.r.) on top. Ripple stratification is up to 15 cm thick on the southern (downstream) end of the underlying HCS bedform. Time-equivalent stratification on the upstream end of the bedform shows only a few poorly developed ripples preserved within parallel lamination. The distribution of ripple stratification, all of which shows southward flow, reflects flow acceleration on the upstream end of the hummock and flow expansion on the downstream side. Pencil is 14 cm long. (After Lamb *et al.*, 2008.)

size from several to tens of metres long and wide, aptly termed *megaflutes* (Fig. 7.18), to just a few centimetres. They form as a result of flow separation from the lip of an initial hollow. High turbulent stresses at the point of flow reattachment cause deepening and lengthening of the incipient flute. In a mature flute the deepest part usually lies some distance upstream from the flow reattachment point. Swarms of megaflutes and other large scour-type bedforms of chevron-shape have been imaged on the ocean floor around the outflow regions of certain submarine channels. They occur downflow from large antidune-like bedforms and are thought to record a hydraulic-jump-type transition from supercritical to subcritical flow.

Laboratory experiments cannot reproduce the spectacular large-scale longitudinal grooves called *gutter marks*. These may have a spacing of a metre or more and be up to 20 cm deep (Fig. 7.19) with flutes covering their margins. Intergroove areas may be perfectly flat and show few effects of erosion. Gutter marks are probably caused by the action of boundary-layer secondary flows but there has been little work on the forces and flows responsible.

Finally, detritus carried by a flow may form a great variety of impact marks on cohesive mud beds as the 'tools' bounce or are dragged over the substrate creating *tool marks*. Sometimes these marks may be recognized as caused by a particular tool, as in the case of a saltating orthocone, rolling ammonoid or dragging plant stem (Fig. 7.20). Preserved tool casts may be useful palaeocurrent indicators if the original mark was slightly eroded by flow subsequent to impact, or shows puck-drag features such as downstream-veeing chevron marks.

## 7.5 Water wave bedforms

The to-and-fro bottom motions caused by shallow-water waves (Section 5.6) cause shear stresses to be set up whose effect at some critical wave condition on an initially planar sand bottom is to cause grain movement by rolling and saltation. This causes formation of symmetrical *wave-formed ripples*, whose crests are usually very persistent laterally but which bifurcate in a characteristic manner. The forms vary greatly in size since they are dependent only upon the dimensions of surface waves—commonly 0.009–2.0 m wavelength, 0.003–0.25 m height, with a ripple index 2–20 (Fig. 7.21).

Detailed experiments show that once the threshold for motion is reached, rolling and saltating grains tend to come repeatedly to rest along metastable *rolling-grain ripple* crests that lie normal to the oscillation direction. Ripple crests are small, perhaps $< 20$ grain diameters high, and sweep to and fro with broad, flat or gently curved troughs in which no grain movement occurs. The detailed long-term average pattern of flow is fascinating and beautiful. Close to the bed, fluid and sediment pass from trough to crest and back again in circulating loops. These were first visualized using ink by G.H. Darwin (son of Charles) in 1884 and termed by him 'ink mushrooms'. Reaching further from the bed and sourced in trough or crest depending upon wave stroke are more sluggish arcuate streamlines, termed 'ink trees' by Darwin (Fig. 7.22).

Eventually, rolling-grain ripples evolve to orbital ripples whose size scales with the near-bed water excursion, $\lambda \approx 0.7d$ and $\eta/\lambda \approx 0.15$, where $\lambda$ is ripple wavelength, $d$ is length of near-bed oscillation and $\eta$ is ripple height. To-and-fro flow over orbital ripples causes the formation of flow separation vortices on either side of the well-defined symmetrical crestline. The vortices are able to suspend finer sediment grains as they flip from side to side of the ripple crest during the passage of overlying waves. The resulting ripple forms were originally termed *vortex ripples*, but are more widely known nowadays as *orbital*, *oscillation* or just plain *wave ripples*. They are the common symmetrical wave ripples seen on many sandy tidal flats and beaches.

Wave ripples are initially 2D in plan view, but as oscillatory speed is increased, particularly at longer oscillation periods (2–20 s), they become increasingly 3D. For very fine to fine sands, ripple wavelengths on the sediment bed increase from a few centimetres for periods of $< 2$ s to greater than 1 m for periods $> 10$ s. For a given maximum orbital speed ripple spacing increases with increasing wave period, the ratio of near-bed orbital diameter to ripple wavelength being $\sim$0.5. The near-bed concentration of suspended sediment increases with orbital speed and is strongly dependent upon the vertical advection of sediment associated with the ejection of separation vortices from first one and then the other lee- side of ripple crests during the wave cycle (Figs 7.23a and 7.24). Ultimately oscillation and combined flow ripples are washed out and a plane bed with intense sediment transport is established.

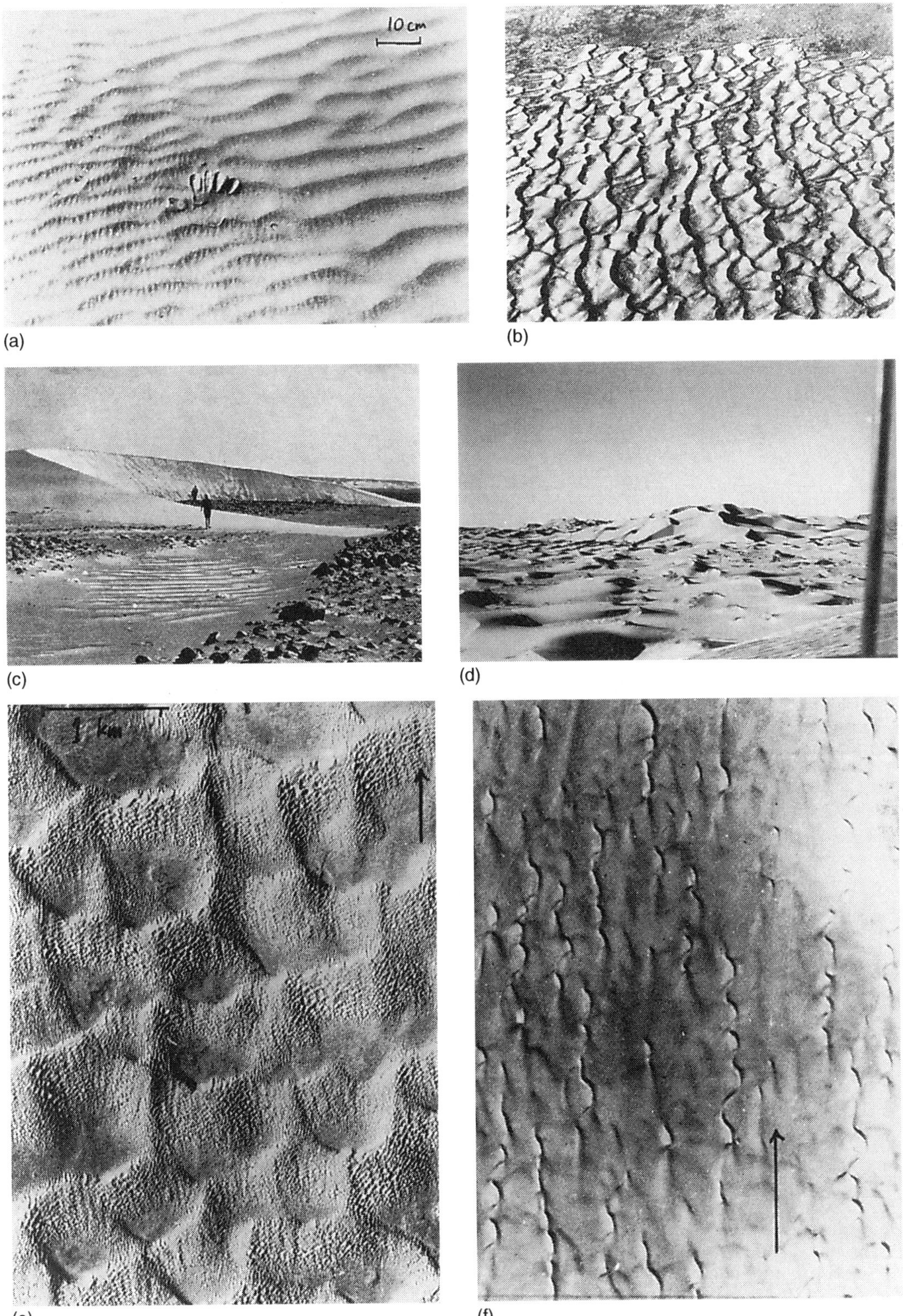
10 cm
(a)
(b)
(c)
(d)
1 km
(e)
(f)

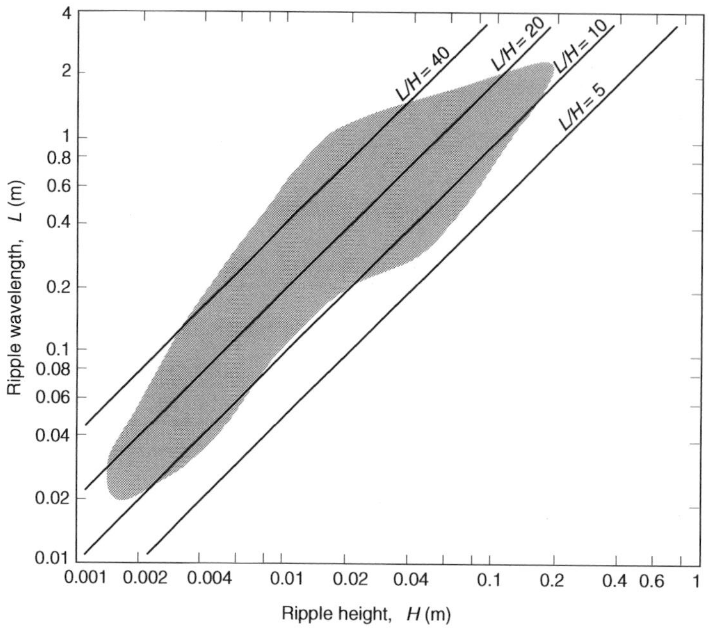


**Fig. 7.32** Graph to show correlation between wavelength and height of aeolian ballistic ripples. Shaded area indicates envelope of data points. (After Allen, 1982, and sources cited therein.)

Regarding the internal structures of growing wave ripples, the alternate crestal vortices scour sand from ripple troughs to deposit it as laminations dipping concordant with, and on either side of the crest causing ripple amplitude to increase. Spanwise sections (Fig. 7.23b) reveal an internal structure of chevron-like laminae accreted onto either side of the ripple crest.

In shallow water, Airy wave theory no longer applies and boundary-layer Reynolds stresses play a role in wave mechanics. There is net foreward transport of water seen (Fig. 7.24) as successive wave orbital motions fail to close in the direction of propagation. The effect causes symmetrical oscillation ripples to migrate very slowly and when accompanied by net sedimentation, internal *translatory ripple* structures result (Fig. 7.25); large-scale examples have provided compelling evidence for major storm events like those that occurred during the nemesis of Snowball Earth. The landward transport of sediment is best appreciated by consideration of the large effects of tsunami in transporting extensive sand sheets into otherwise fine-grained coastal sedimentary environments. Identification and dating of these establishes valuable data for palaeotsunami research, analogous to palaeoseismological studies that also rely on identification of extreme events in the sedimentary record. It is also possible to estimate the magnitude of ancient waves from the observed dimensions of wave-formed ripples. In shallow water the water orbit/ripple wavelength ratio is about 0.5, but the relationship does not

**Fig. 7.31** (a) Ballistic ripples illustrating the grain-size control of wavelength; the smaller ripples on the left are in finer-grained sand and have been affected by a later, gentle wind at 90° to the first, which did not disturb the coarser sand on the right. (After Wilson, 1973). (b) Aerial view of aklé dunes (scale unknown) from Utah, USA. (After Cooke & Warren, 1973.) (c) Barchan dune advancing across a lag gravel pavement; La Joya, south Peru (After Cooke & Warren, 1973.) (d) Draa comprising superimposed aklé dunes; draa height ca 30 m; Erg Occidental, Algeria. (Photo by Ian Davidson.) (e) Aerial view of barchanoid draa and superimposed dunes. (After Wilson 1972a.) (f) Aerial view of 'meandering' seif dunes migrating across a lag of coarse sands; Edeycnubari, Libya (After Wilson 1972a.)

**Fig. 7.33** Figure stands on ballistic wind ripples in fine sand which pass downslope into longer-wavelength granule ridges developed at the margins of Medano Creek, a fluvial arroyo. Great Sand Dunes, Colorado, USA.

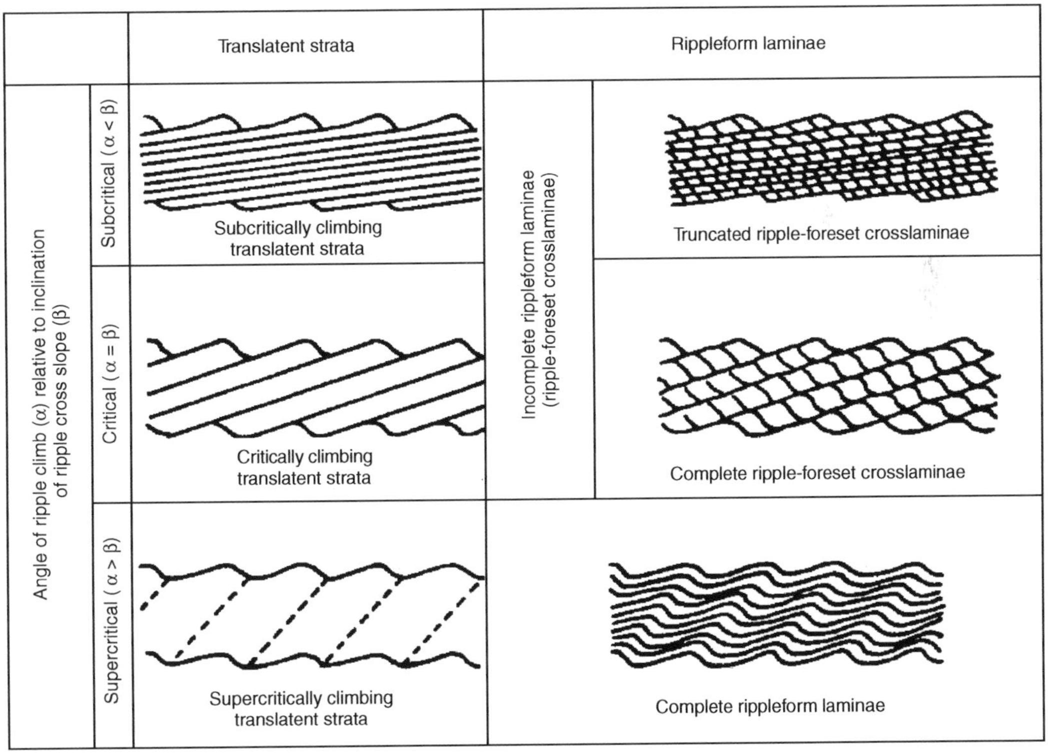


**Fig. 7.34** Types of small-scale sets produced by aeolian climbing ripples in areas of sand experiencing net accumulation with time. The troughs and lower set boundaries are usually in finer sands or silt and may appear as 'pin-stripe' laminae when differentially cemented as rock. (After Hunter, 1977; Fryberger & Schenk, 1988.)

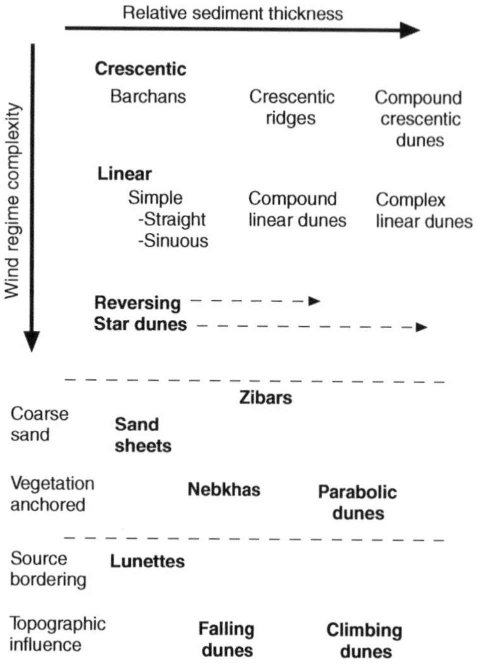


**Fig. 7.35** A useful morphological classification of desert dunes. (After Lancaster, 1995, and sources cited therein.)

hold for deeper-water waves and varies with grain size. Thus it is difficult to calculate ancient wave and water conditions (wave celerity, wavelength and height, water depth) unless there is independent geological evidence for water depth.

## 7.6 Combined flows: wave–current ripples and hummocky cross-stratification

Combined flows are the result of oscillatory wave motion superimposed upon tidal current flow. Ripple forms generated under gentle combined flows rarely show ripple indices greater than ~15. Experiments show the net flow field to cause reinforcement and then opposition of lee-side vortices as the wave cycle progresses (Fig. 7.26). In such flows lee-side vortices interact with the current flow turbulent boundary layer and the near-bed peak in suspended sediment tends to stay lower in the water column. The effects of wave and current on rough boundaries are not simply additive and some sort of interaction is postulated since very high peak vertical turbulent velocity values and increased turbulent stresses occur. It seems that periodic violent vortices are generated during combined flows close to the rough bed, up to three times the values in pure flows alone. The rapid fluctuation of turbulent stresses arising from these vortices during the wave cycle causes near-bed suspended sediment to be advected up into the flow (Fig. 7.27). It is probably this effect that generates turbidity flows and powerful shelf gradient currents during storms. In natural combined flows the local vectors of fluid shear due to tide and wave are not co-linear and constantly change with time. This aspect of nearshore sediment transport and bedform development is discussed in more detail in Chapter 18 & 19.

Combined flows during storms are dominated by great waves whose periods may exceed 10 s and whose surface shapes are strongly 3D. Analogue laboratory experiments using fine to very fine sands, wave periods of 10.5 and 8 s, oscillatory velocities of 0–125 cm/s and unidirectional velocities of 0–25 cm/s, reveal that at low unidirectional velocities ($< 10$ cm/s), addition of an increasing co-linear oscillatory flow caused the bed to evolve from small-scale (wavelength, 20 cm), symmetric, anorbital ripples (i.e. unrelated to orbital excursion magnitude), to large-scale (wavelength ~100 cm), symmetric, orbital ripples and finally to plane bed (Figs 7.28 & 7.29). At higher unidirectional velocities ($> 10$ cm/s), a similar trend was noted, but ripples were more asymmetric. Distinctive features of small-scale asymmetric combined-flow ripples generated are: a 3D planform, round crest, and convex-up sigmoidal profile with local pronounced scour at the toe of the stoss side giving the ripple profile a 'boxy' appearance. Similarly, large-scale asymmetric combined-flow ripples have broad and round crests, convex-up stoss sides, and 'compressed profiles" due to scouring at the toe of the stoss side. Hummocky bedforms (Fig. 7.30) are generated under moderate to high oscillatory velocities and low unidirectional velocities. Hummocks are not observed as a distinct bed state but rather appear to mark transitions in bedform scale and symmetry, being more prevalent at longer oscillatory periods and in finer-grained sediment. Stratification produced by 'synthetically' aggrading hummocky bed profiles closely resembles natural hummocky cross-stratification. With the introduction of only a small unidirectional-flow component, hummocks evolve into downstream-migrating large-scale asymmetric ripples and the

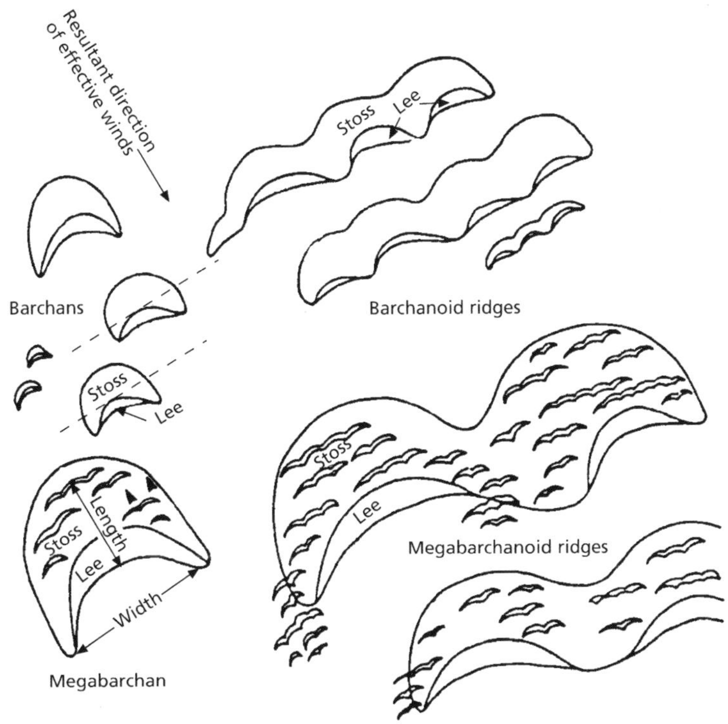


**Fig. 7.36** Sketches in *xz* space (plan view) to show various types of flow-transverse dunes. Sand supply and availability increases from top to bottom. (After Breed, 1977.)

resultant cross-stratification becomes similar to that produced by unidirectional-flow dunes. This suggests that much of the hummocky cross-stratification observed in the sedimentary record is produced by storm-generated long-period oscillatory-dominant combined flows. Inasmuch as long-period, high-energy waves require deep, wide basins to form, hummocky cross-stratification may therefore serve as a useful indicator of deposition in unrestricted, open-water conditions. In coarse sands, instead of hummocks, experiments reveal that large, steep, sharp-crested and 2D orbital wave ripples form instead. These would deposit high-angle (15–25°) cross-stratification with large set thickness (> 5 cm) that might be mistaken for dune deposits.

## 7.7 Bedforms and structures formed by atmospheric flows

Bedforms under atmospheric flows range in size over more than four orders of magnitude, from decimetre-scale ripples to kilometre-scale sand dunes (Fig. 7.31). Pioneer studies plotted grain size against bedform wavelength to define three distinct bedform groups. In ascending order of magnitude, these are *ballistic ripples/ridges*, *dunes* and *draas*. Subsequent research showed overlap between the dune and draa fields. The term draa may still be applied to large-scale composite bedforms made up of a hierarchy of smaller duneforms of various shapes. Recent work stimulated by discovery of Martian desert dunes makes it clear that bedform scale and type depend upon both the availability of sand and the magnitude of the wind regime.

### Ballistic ripples and ridges

These two forms are a continuous series (Fig. 7.32) with wavelengths 0.02–2.0 m and heights from a few millimetres to 0.1 m. Ripple indices generally fall between 8 and 50. Wavelength increases linearly with mean grain size (Fig. 7.33) and also with decreased

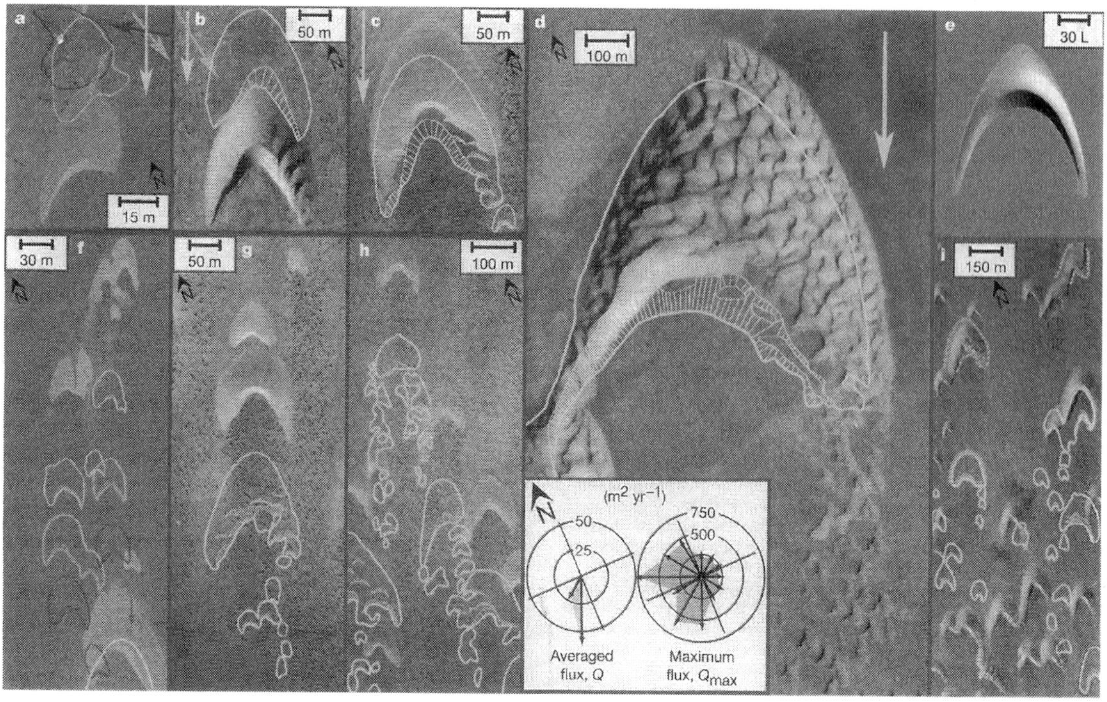


**Fig. 7.37** Response of barchan dunes in North Africa to changes of wind direction, $u$, and to bedform collisions. Contours are from GPS surveys and aerial photos. Inset rose diagrams show (i) mean sand flux (for 1999, $Q = 80\,m^2/yr$) with a narrow unimodal wind regime coincident with mean barchan motion and (ii) maximum flux with its more widely distributed distribution of extreme events. (After Elbelrhiti *et al.*, 2005.) (a)–(d) Response of barchan dunes of various sizes to changes of wind direction. (a) Two outlines of a small dune and its aerial photo at time, $t = 0$; black, $t = -15$ months; white, $t = -12$ months. The slip faces quickly change in response to $\Delta u$ (arrows). (b) Prominent sinuous wave-like minor dunes on the eastern arm of a medium-sized megabarchan that had previously moved from its reference position outline in white ($t = -80$ months) induced by a transient NW wind (arrow). (c) Destabilization of a medium-sized mother megabarchan leading to the shedding-off of barchanette offspring. Reference position at $t = +50$ months with respect to aerial photo. (d) Large megabarchan photographed at $t = -350$ months. The sinuous superimposed minor dunes are permanently driven by daily wind variations. (e) A perfectly symmetrical numerical dune produced by a simulation algorithm. (f–h) response of barchan dunes to collisions. (f) Edging collisions, amalgamations and shedding between small barchans (white, $t = 17$ months; grey, $t = 23$ months; black, $t = 33$ months). (g) Co-linear collision of medium-sized barchans leading to a larger barchan that sheds a series of barchanettes as a wake from its eastern arm ($t = 60$ months). (h) Chain destabilization of medium-sized barchans ($t = 60$ months). The trail of barcahnettes from the arms of a barchan destabilizes the next barchan downflow. (i) Long-term ($t = 350$ months) evolution of a barchan field. Some of the largest dunes persist but many others appear or disappear.

sorting. For the same grain size and sorting, wavelength increases with increasing wind strength. The smaller ripples have slightly sinuous crests that are very persistent normal to the wind flow direction. In flow-wise cross-section the forms are slightly asymmetric, with a gently sloping, convex windward side and more steeply dipping (but usually $< 10°$), concave lee side. A common attribute of all but the smallest forms in very well-sorted sands is the concentration of coarser grains in crestal areas.

There is usually no clear internal structure of cross-laminations in aeolian ripples, unlike their subaqueous counterparts. This is because the ripples migrate less by repeated lee-side avalanches than by saltation bombardment. A particularly common feature of strata deposited by migrating ballistic ripples is a

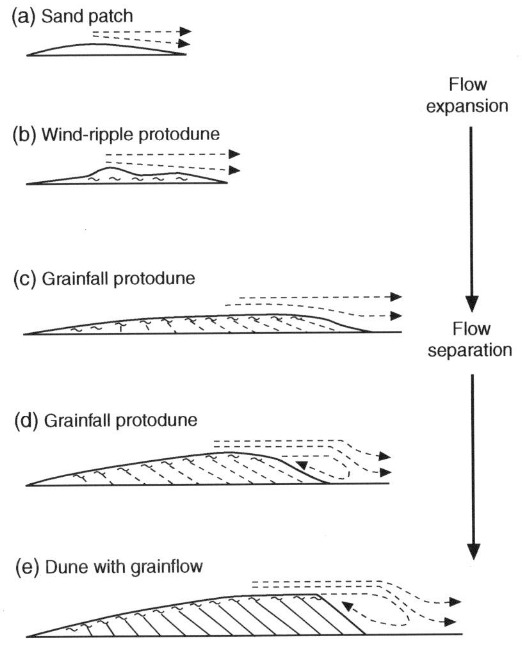


**Fig. 7.38** Sketch sections through dunes in *xy* space (vertical views) to illustrate the successive stages of dune formation from an initial sand patch. Dune dimensions are metres high and tens/hundreds of metres wavelength. (After Kocurek *et al.*, 1992.)

strong reverse grading or bimodality of grain size, a consequence of the concentration of coarser creepload along the ripple crests. The bounding surfaces between the ripple laminae frequently dip slightly upwind, indicating a slow build-up of the sediment bed during ripple migration, a feature analogous to bedset boundaries in subaqueous climbing-ripple structures (Fig. 7.34).

Concerning the origin of aeolian ripples, the superficial resemblance of form to subaqueous ripples hides a fundamental contrast in the mechanism of formation. The higher immersed mass of saltating grains in air means that conditions at the air–bed interface are dominated by grain splashdown effects and not by inner layer turbulent boundary layer streaks—hence the name *ballistic* ripples. Saltating sand grains impact much of their kinetic energy into nudging coarser grains along (up to six times their diameter) by intermittent rolling, sliding or hopping (termed *reptation*) to form the *creepload* part (up to 25%) of the total sediment transport. Ballistic ripples and ridges increase in wavelength with increased grain size and decreased grain sorting since the saltation jump length of fine particles increases markedly with the size of the bed grains with which they collide. In addition, the impact of high-energy saltating grains preferentially ejects smaller grains. This leaves zones of coarser grains that migrate in reptation as patches over downwind plinths of finer grains protected in the lee of the patch. Intense bombardment on the upwind side of the patches thus alternates with lee-side sheltering and a stable ripple form gradually emerges with a characteristic coarser crestal grain size.

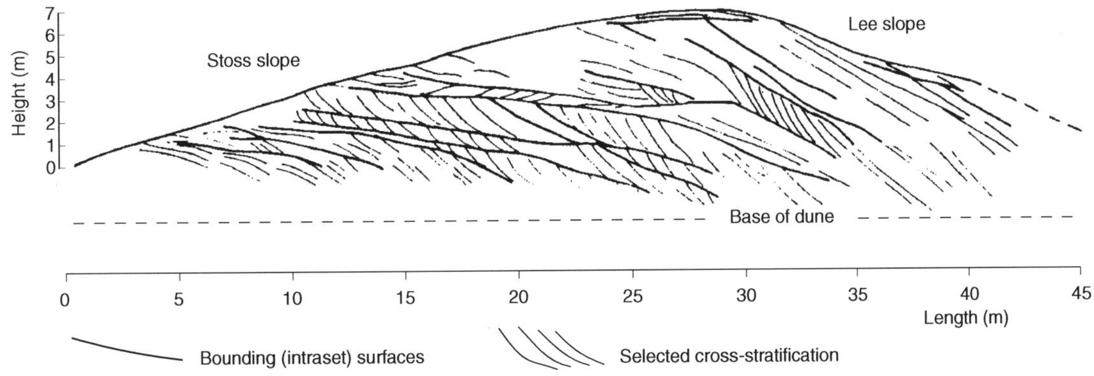


**Fig. 7.39** Internal structure of a barchanoid dune from White Sands, New Mexico, USA. Note that it has grown and evolved its morphology over time (the actual time remains unknown), from a domal dune with no lee-side slipface when smaller dunes migrated over the larger form to define downdipping sets, into a higher and more asymmetric barchanoid dune with a prominent steep lee-side slipface. (After McKee, 1966.)

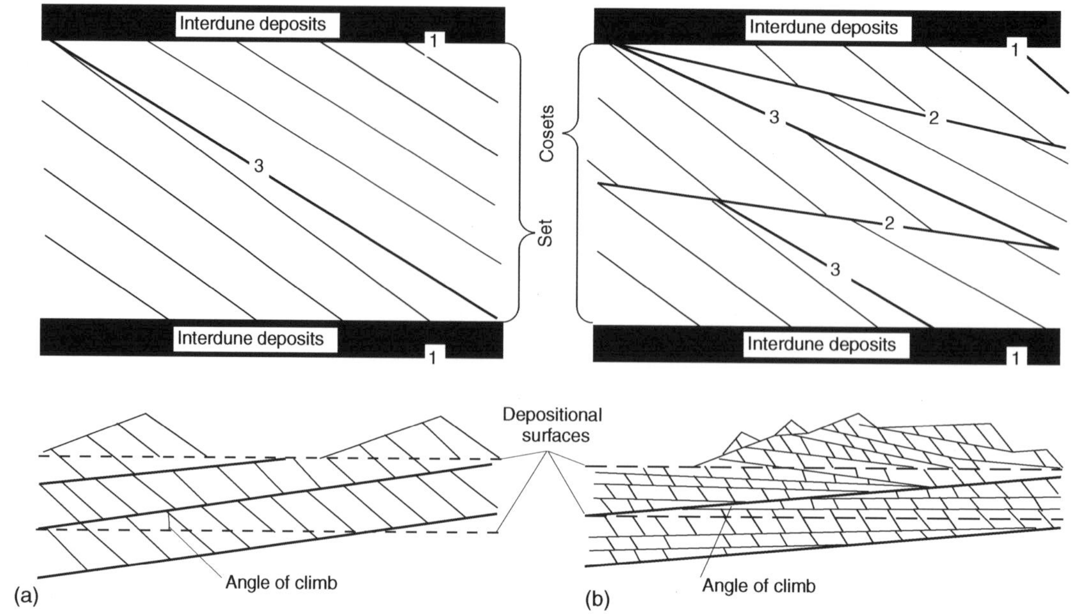


**Fig. 7.40** Sketches to show (a) first- and third-order bounding surfaces in simple dunes migrating and 'climbing' due to net deposition from depositing winds. (b) First-, second- and third-order bounding surfaces and downdipping sets (see Fig. 7.39) in compound draas with parasitic smaller dunes, the larger form migrating and 'climbing' downwind due to net deposition from depositing winds. (After Kocurek, 1988.)

## Dunes in general

Aeolian dunes show more diverse morphologies than do their subaqueous counterparts. A simple division (Fig. 7.35) into linear (flowwise), transverse (spanwise) and complex dunes is useful, though transitional forms occur and many dunes commonly show combinations of elements. There are also a variety of dune types related to the topographic forcing of the wind by isolated hills and escarpments. A second control is exerted by the ratio of available sand sediment supply to the potential transport capacity of the local or regional wind system. Many past and present

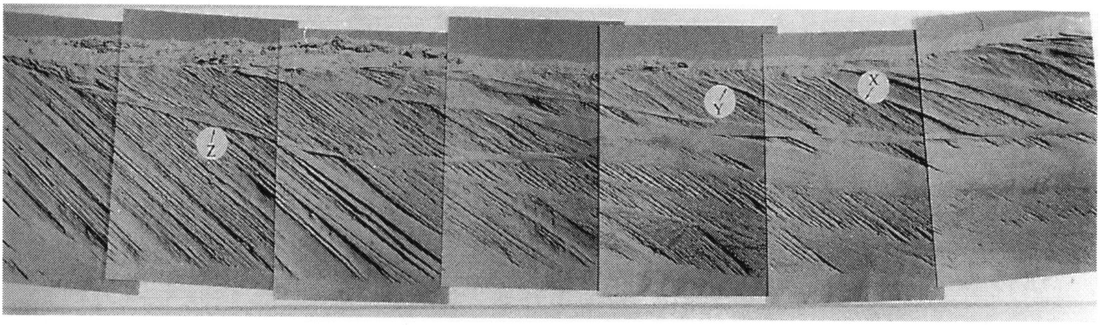


**Fig. 7.41** Photo shows downdipping intraset surfaces (labelled X, Y, Z) associated with the migration of small dunes over a larger flow-transverse draa-like form. Abu Dhabi, UAE. (After Bristow *et al.*, 1996; photo courtesy of C.S. Bristow.)

aeolian systems are or were in a state of supply-limited sand transport. Vegetation and soil moisture play an important secondary role in stabilizing and modifying duneforms. Finally, the reader should be aware that aeolian dunes with large mass are subject to extreme forms of disequilibrium due to climate change—regional wind systems may change and large dunes take time to readjust or are simply present as relics. Also, increased humidity may cause vegetation to stabilize dunes into relict forms. Though not indicative of present-day conditions, such dunes may provide vital evidence for past and future climate change, as for example the generation of intense wind-blown sediment transport and dune advance seen in many areas of the world during the Medieval Warm Period and the subsequent Little Ice Age.

### Transverse dunes

Flow-transverse dunes occur when the predominant wind is unidirectional. There is a continuum of forms related to the availability of sand cover (Fig. 7.36). Crescent-shaped *barchan* dunes, with their 'wings' tapering downwind, are perhaps the most evocative and widely known type, even though they are a minority in most deserts. They are typically 1–20 m high, 25–250 m wavelength and up to 300 m wide, with migration rates of up to 30 m/yr. The speed, $c$, of barchans decreases with overall height, the rate given by $c = aQ/H$, where $a$ is a constant ~2.7, $Q$ is sand bedload flux and $H$ is dune height. A single Libyan barchan studied in the 1930s by R.A. Bagnold and reidentified in the 1980s was found to have migrated at a mean rate of around 7.5 m/yr. Examples from the Ceará coast, northeast Brazil are highly mobile, with average migration rates of 17.5 m/yr for barchans and 10 m/yr for sand sheets.

Barchan dunes occur in areas of reduced sand supply, with individual dunes separated from their neighbours by either solid rock floor or by immobile coarse pebbles (the desert *reg*, a lag deposit). Successive barchans may propagate downwind from their wings. Megabarchans (Fig. 7.37) are those with superimposed smaller barchanoid or sinuous crested dunes. The rarity of isolated barchans in sand seas (ergs) with plentiful supply means that they are unlikely to be preserved in the stratigraphic record. Isolated domal dunes without a prominent slip-face seem to form from the degradation of barchanoid dunes during long periods of gentle wind flows. As expected, such dunes show complex internal cross-stratification patterns.

The nucleation and propagation of barchans and other transverse forms (Fig. 7.38) occurs after sand has collected due to frictional retardation in an initial patch from surrounding bare bedrock or lag deposits. Continued vertical and downwind growth of the trapping patch leads to a convex, mound-like form, a morphology that causes flow streamlines to converge and sediment transport to increase according to the third power of the wind shear velocity (**Cookie 32**). The wind eventually becomes sand-saturated at equilibrium and eventually the downwind margin amplifies so that flow separation occurs and a slipface is produced. This is a critical transition, for the slipface acts as a kinematic and dynamic barrier such that saltating sediment is carried over the incipient dune crest into a sort of temporary suspension, only to fallout rapidly onto the bed slipface, generally within 1 m or so on the lee side. Such *grain fall deposition* leads to local oversteepening instability and causes periodic avalanches or *grain flows* (Chapter 8) down the steep lee slope. Barchan wings form because sediment transport velocities are larger on the margins of a patch of sand where intergranular frictional effects during saltation impact are reduced and rebound is more efficient. Long-term monitoring of megabarchans (Fig. 7.37) shows that the forms are inherently unstable in that intradune 'collisions' and changes in seasonal wind direction destabilize the forms and help create their subsidiary dunes. These latter are surface waves that propagate at a higher speed than the overall mother-barchan and help create new mothers by breaking the elongate horns of the original form. In this way, fields of megabarchans are preserved as multiplying individuals rather than merging into one giant bedform.

Sinuous-crested *aklé dunes* (Fig. 7.31b) are common in areas of plentiful sand supply. These show slipfaces orientated normal to local flow vectors on dune lee sides, giving rise to large internal sets of cross-stratification. Frequent reactivation surfaces and downward-dipping cross-sets result from periodic modifications of dune shape by aberrant winds. Aklé dunes are indicative of minimal effects by longitudinal secondary flows. If an analogy with subaqueous dunes is apposite then they may be related to boundary-layer thickness. with wavelength controlled by the repeat distance of

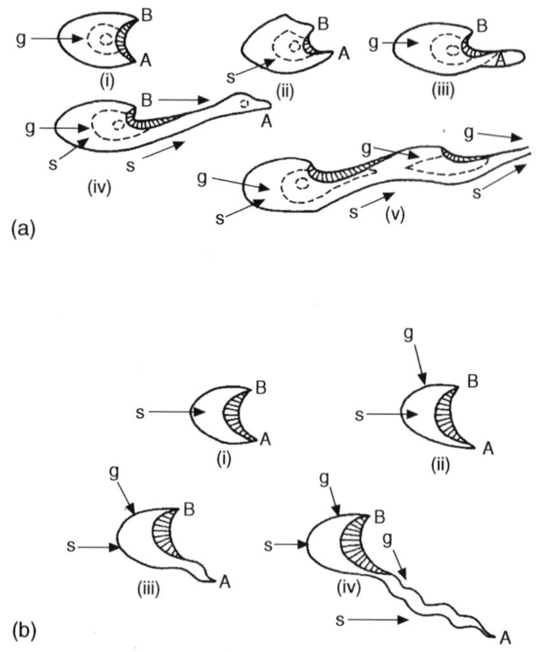


**Fig. 7.42** Bagnold's classic model for the origin of beaded seif dunes from barchans modified by the seasonal alternation of bidirectional strong (s) and gentle (g) winds. (After Bagnold, 1941, as slightly modified by Tsoar, 1983.)

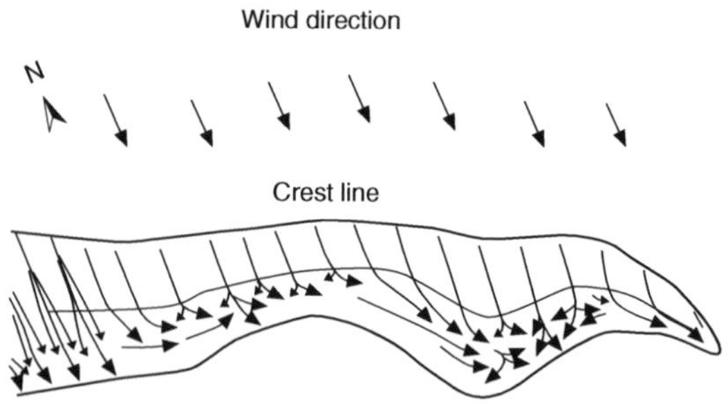


**Fig. 7.43** The pattern of airflow visualized by smoke close to the bed of an obliquely incident wind over a sinuous linear dune. (After Tsoar, 1978.)

large-scale vortical events generated in the lee-side separation zone.

*Draa* bedforms are composite, usually transverse, elements with wavelengths up to 4000 m and heights up to 400 m (Fig. 7.31d). They may be aklé or barchanoid in plan; their great size means they take a long time to form, requiring appreciable original sand cover to provide an adequate original nucleus. Some draas show giant slipfaces up to 50 m high, others feature dune migration down a gently sloping lee side. The resulting internal structure shows various orders of upwind- and downwind-dipping erosional and depositional surfaces (Figs 7.39–7.41). In ancient deposits, identification and mapping of these surfaces in three dimensions is needed to determine the morphology of the draa responsible (Chapter 16).

## Linear dunes

Sharp-crested linear (aka *longitudinal*, or *seif* from the Arabic for sword) dunes may sometimes be traced flowwise for tens of kilometres, with heights up to 70 m or more, widths of sand cover of several hundred metres and lateral spacing between adjacent dunes of

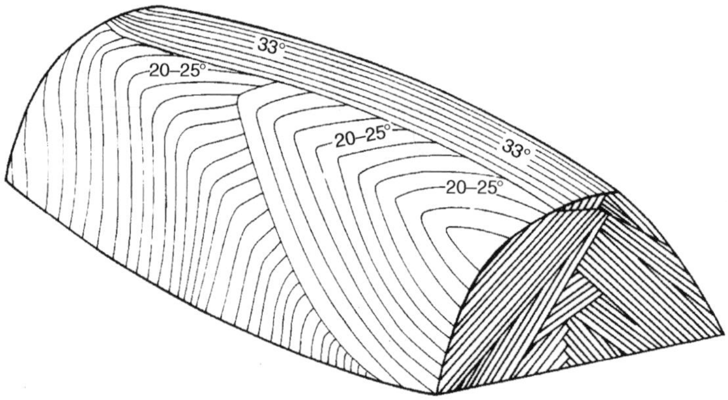


**Fig. 7.44** Three-dimensional internal structure deduced for a typical linear dune of seif type. (After Tsoar 1982.)

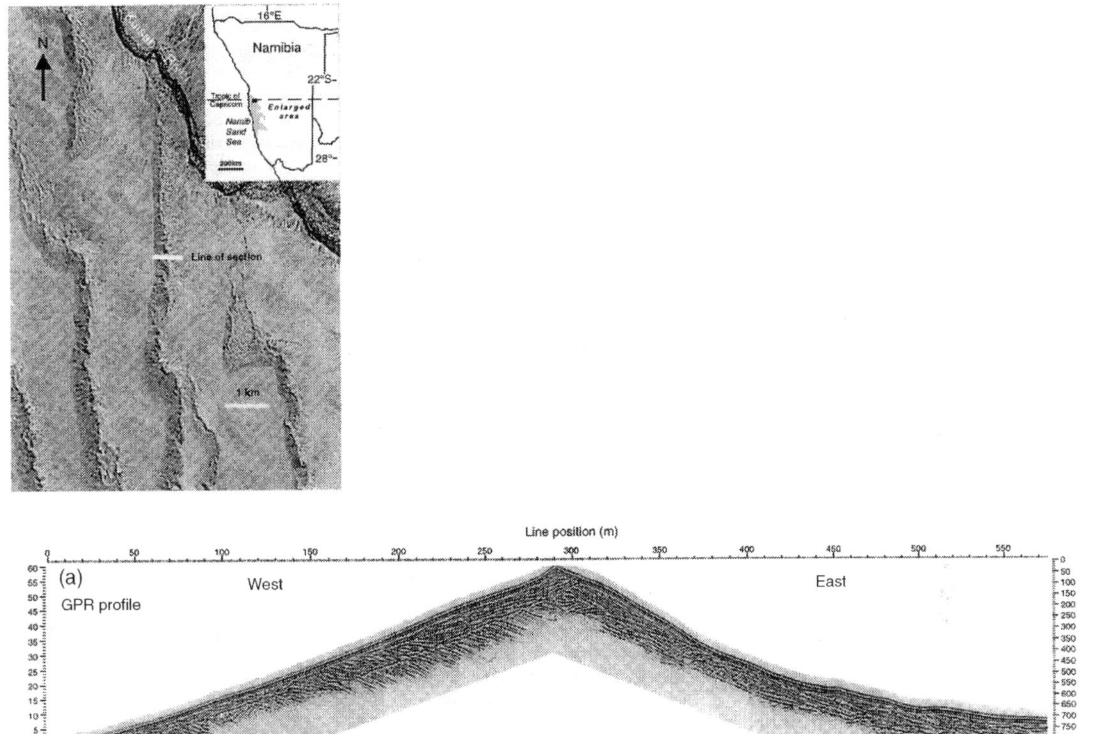

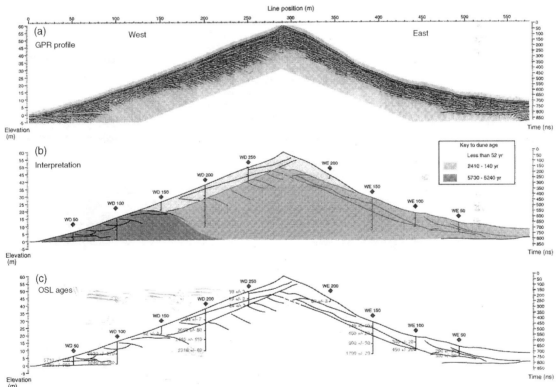


**Fig. 7.45** (a) Linear dunes of the northern Namib desert. (b) Ground-penetrating radar profile across a study dune showing dipping reflections interpreted as sets of cross stratification and bounding surfaces within the dune. On the western dune flank, sets of cross-stratification formed by superimposed dunes truncate older east-dipping sets. (c) Interpretation of dune evolution based upon OSL dating of sand samples retrieved from boreholes. (All after Bristow *et al.*, 2007.)

over 1–2 km. Dune crests sometime coalesce to produce Y-shaped junctions which most commonly fork upwind. Some seif dunes show sharp, smooth crestlines but the majority have a sinuous crestal planform made up of shorter, almost *en echelon*, individual crests (Fig. 7.31f) or have periodic humps. Repeated monitoring of dune populations over multidecadal timescales reveals that the crestlines 'flap' sideways back-and-forth in response to a *bimodal wind pattern* that is a universal feature of linear dunes. The classic explanation for the forms is that they arise when transverse barchanoid dunes are subjected to winds from two directions at acute angles to each other (Fig. 7.42). One barchan wing becomes elongated, later to become the nucleus of a new barchan as the wind re-establishes itself in its former mode. The resultant *beaded seif dune* has its long axis orientated parallel to the resultant of the two wind azimuths. This theory is broadly supported by flow visualization studies on the beaded seif dunes of the Sinai desert (Fig. 7.43), where the oblique incidence of seasonal winds to seif crestlines causes lee-side helical flow spirals to be set up. Internally, seif dunes show a bimodal pattern of large-scale cross-stratification produced by avalanche accretion of alternate sides of the dune during seasonal 'flapping' flows, each with high angles of incidence to the local crest direction (Fig. 7.44).

Detailed GPR and OSL age-dating of Namib linear dunes has shed startling light on longer term longitudinal dune evolution (Fig. 7.45), establishing first

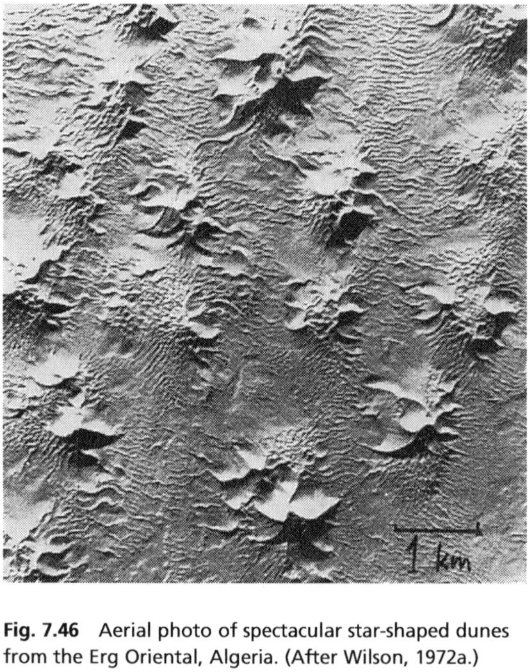


**Fig. 7.46** Aerial photo of spectacular star-shaped dunes from the Erg Oriental, Algeria. (After Wilson, 1972a.)

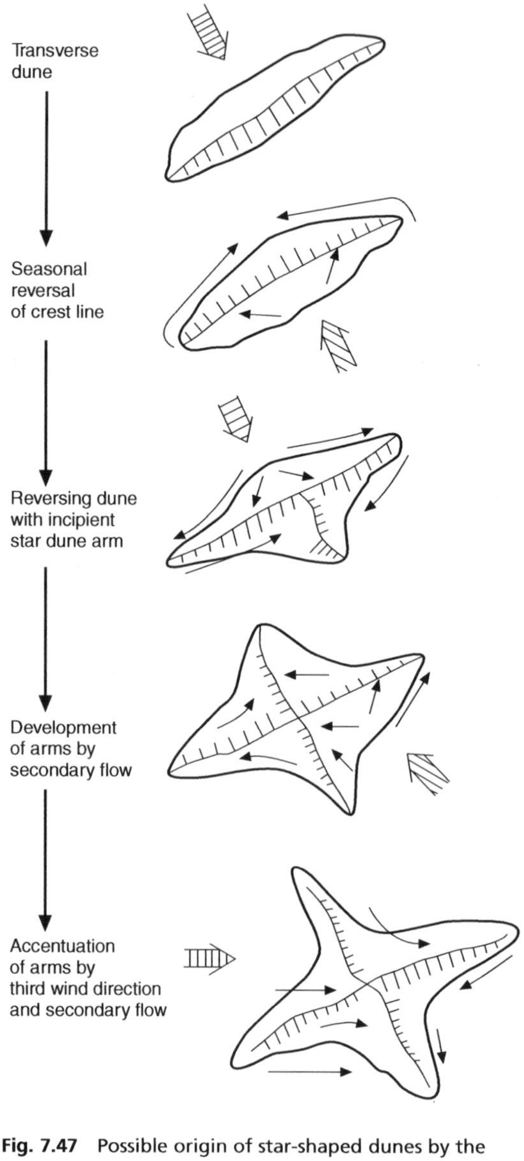


**Fig. 7.47** Possible origin of star-shaped dunes by the development of secondary flow circulations as originally transverse dunes migrate into an area characterized by strong bidirectional winds. (After Lancaster, 1989.)

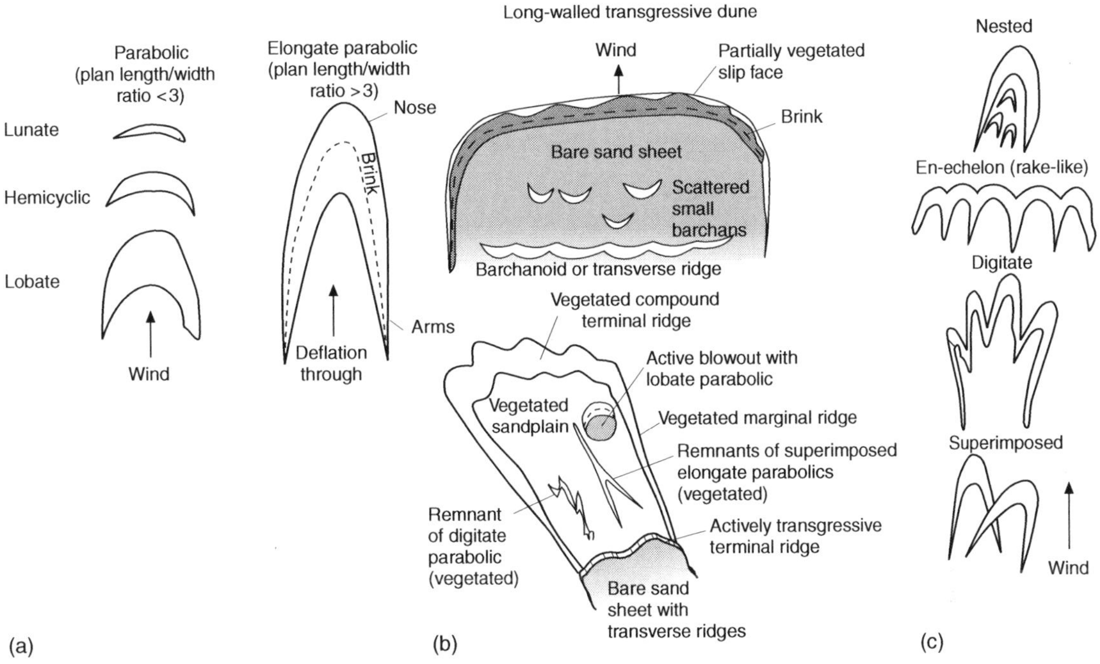


**Fig. 7.48** Types of (a) simple, (b) complex and (c) compound parabolic dunes. (After Pye, 1993.)

that these classic forms are entirely of Holocene age and that previous theories concerning their origin as ancestral Last Glacial Maximum active structures when winds were much stronger are in error. Secondly there was a dune rest period of ca 3 kyr in the moister climate of the mid-Holocene when stabilization by vegetation probably caused dune hibernation. Thirdly, over the past 2 kyr large-scale internal dune cross-sets show net eastward migration of several hundred metres. Finally, in the latest decades, northward migration of superimposed transverse dunes on the main dune flanks has been superimposed upon the older internal structures with young active east-dipping and west-dipping foresets at the crest originating from seasonal 'flapping' in response to the present day bimodal flow. Evidently the south-southwest regional winds have dominated over the easterlies on millenial timescales and have led to net eastern sand flux.

Concerning the origins of linear aeolian dunes in general, the bimodal wind hypothesis seems to best fit observations. However, in the past some authors have proposed that the presence of streamwise secondary flow is important, witness the commonly observed windrows of blown snow or sand over immobile surfaces during strong winds. According to this scheme, linear dunes develop along the axis of the meeting point of pairs of oppositely rotating streamwise vortices. Finer saltating sands are swept inwards in broad lanes where deposition occurs and, given sufficient sand supply, the duneform grows into equilibrium with the flow. Once formed, the dunes reinforce the secondary flow cells. Although superficially attractive it has not been demonstrated how small-scale sand windrows of wavelength an order of metres may grow large-scale dunes. A closer comparison in terms of scale might be made with linear cloud formations (cloud *streets*), whose persistence and wavelength resemble linear dunes.

## Complex dunes

Spectacular star- or pyramid-shaped (aka *rhourd*) dunes with multiple (> 2) slip faces (Fig. 7.46; **Plate 4**) commonly range from 500 to 1000 m wavelength and from 50 to 300 m height. They are the largest of

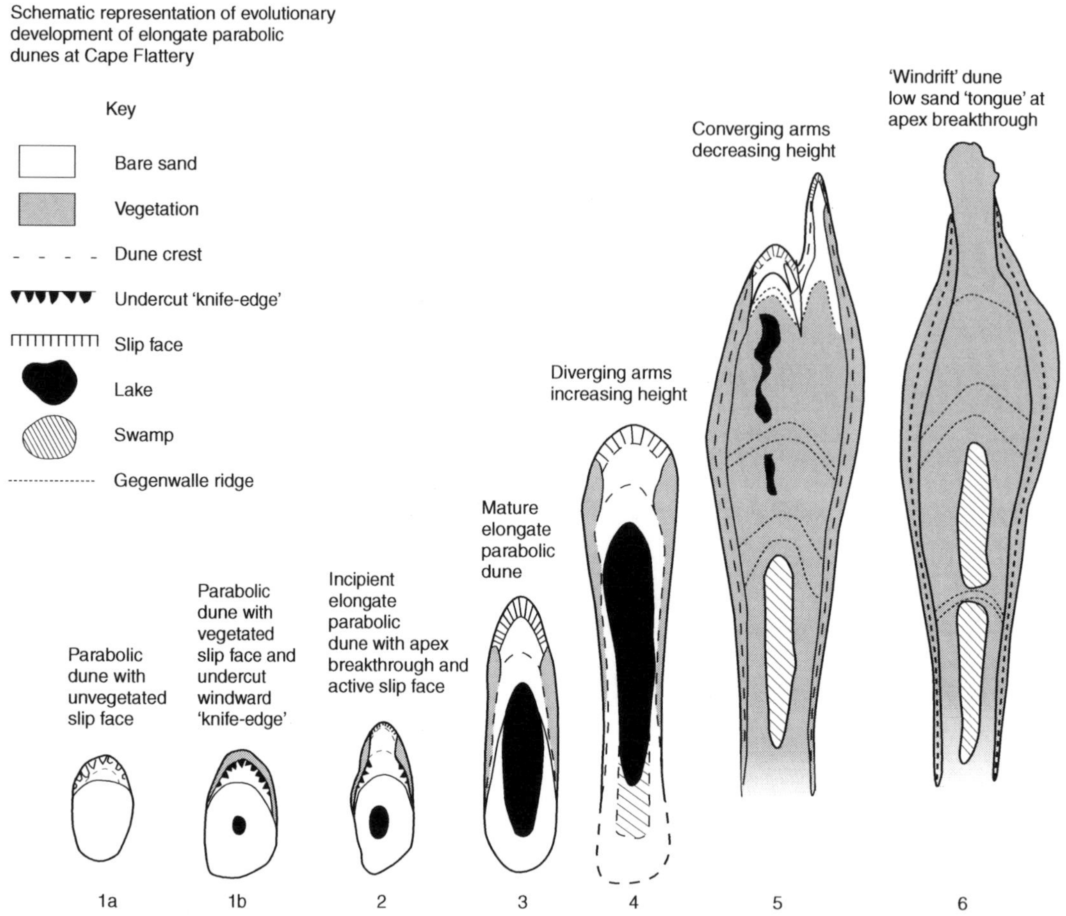


**Fig. 7.49** The supposed evolution of a parabolic dune from northern Queensland, Australia. (After Pye, 1982.)

the aeolian bedforms in terms of sediment mass. The forms have central peaks about which straight to curved crests radiate, the latter reminiscent of vortex lines, into broad plinths. They may be spaced randomly, separated by immobile rock or gravel substrates, or in rows, and seem to arise from the interaction of multidirectional or equal strength bimodal regional winds with less certain contributions from local winds due to convected air masses. The rows of star dunes in the Gran Desierto of Mexico (Chapter 12) are thought to have originated from precursor linear dunes as wind regime changed in the past 3 kyr. The forms show evidence for only very slow lateral migration of their centre of mass; overall they seem to exhibit strongest vertical growth and are thus efficient as local sand sinks. These considerations indicate that flow over the forms is unlikely to result in any strong net transport vector; in fact it is particularly complicated (Fig. 7.47), with the occurrence of both positive and negative helicoidal spiral flow along dune arms dependent upon the local and temporal mean flow directions of winds relative to the dune crests. Details of their internal structure remain poorly known.

## Vegetated parabolic dunes

These curious 3D forms (Figs 7.48 & 7.49) have the opposite orientation with respect to the unidirectional wind from barchans, i.e. the long trailing 'arms' of the dune point upwind. They owe their origins to the interaction between vegetation clumps and thickets and an active source of sand, commonly point-sourced from a central 'blowout' where sand is liberated by intense turbulent shear. Vegetation growth on the relatively inactive 'arms' of the dune stabilizes these as ridges, and hence with time the point source continues to supply sand downwind to the active convexo-concave core of the duneform, causing the 'arms' to elongate downwind. Once the central source of sand becomes exhausted all that is left are the long (sometimes tens of kilometres) flow-parallel 'arms'. Sometimes an elongate lake forms in the central part of the dune.

### Windflow and streamflow bedform theory

Despite the dynamic contrasts between wind and water sediment transport, the process of ripple formation and the final ripple forms are sufficiently similar for us to conclude that some universal law is responsible. This law operates on small moving bed defects to make them larger and more regular; a complex system exists from which regular forms arise (or emerge) due to the convergence of a state known as an *attractor*. Bedform development from an initial random or flat bed is thus an example of *self-organizing* behaviour. A simple simulation model for sediment transport generates dunes of shape and size that depend upon a nonlinear transport law and upon the increased probability of transport over a nonerodible substrate. The model has not been applied to erodible substrates, but doubtless a similar behaviour is to be expected.

## Further reading

*General*

Allen (1982) is a fundamental reference for all bedforms and sedimentary structures. Collinson *et al.* (2006) is more accessible, with many fine examples and clear explanations. Lancaster (1995) is a good source for active desert bedforms. Pye & Tsoar (1990) is also useful. Greeley & Iverson (1985) give stimulating interplanetary perspectives. Read Bagnold (1954b) again for inspiration. P.A. Allen (1997) is good on waves and sediment transport. Sleath (1984) is a fundamental (and difficult) reference on sea-bed mechanics. Haworth (1982) is a good introduction to tidal currents, whilst Soulsby (1997) is a thoroughly good 'cookbook' of great use in making hydraulic calculations in tidal, wave and mixed-regime flows.

*Specific*

Current ripples are researched by Williams & Kemp (1971), Best (1992a), Baas *et al.* (1993), Baas (1994, 1999) and Van Gelder *et al.* (1994). Development of current ripples in very fine silts a and of mudfloc ripples are by Mantz (1978) and Schieber *et al.* (2007) respectively.

Two-dimensional dunes and bedload sheets are considered by Whiting *et al.* (1988) and Bennett & Bridge (1995).

Cluster bedforms are covered by Brayshaw (1984) and Strom & Papanicolaou (2008).

Dunes are usefully reviewed by Best (2005) and researched by Smith & McLean (1977), Muller & Gyr (1982), Saunderson & Lockett (1983), Gabel (1993), Kostaschuk & Church (1993), Bennett & Best (1995, 1996), Kostaschuk & Villard (1996), Kostaschuk (2000) and Parsons *et al.* (2005). Stage fluctuations with respect to 'starved' coarse sand dune morphologies are by Carling *et al.* (2000).

Controls on the thickness of cross-stratified sets formed by both ripples and dunes are by Paola & Borgman (1991), Storms *et al.* (1999), Leclair & Bridge (2001) and Leclair (2002).

Upper stage plane beds are considered by Best & Bridge (1992)

Bedload sheets and low-relief bedforms in poorly sorted coarse-grained bedstock are discussed and reviewed by Kuhnle *et al.* (2006).

Antidunes and other supercritical flow bedforms and deposits are covered by Carling & Shvidchenko (2002), Alexander & Fielding, (1997), Alexander *et al.* (2001), Duller *et al.* (2008).

Hydraulic jump deposits are considered by Russell & Arnott (2003) and Macdonald *et al.* (2009).

Bedform phase diagrams are in Southard (1971).

Rival ideas on rapidly-depositing, sediment-charged flows are presented by Leclair & Arnott (2005) and Sumner *et al.* (2008).

Bedform theory is studied by Baas *et al.* (1993), van der Berg & van Gelder (1993) and Bennett & Best (1996).

Wave ripple sediment transport is studied in pioneering papers by Bagnold (1946), Longuet-Higgins (1953), Southard *et al.* (1990) and Southard (1991). Various forms of dimensionless bedform phase diagrams for orbital ripples are presented by Pedocchi & Garcia (2009a,b).

Sediment transport under water waves is discussed by Osborne & Greenwood (1993) with visualizations and measurements of suspended sediment vortices over individual wave cycles by Villard & Osborne (2002) and Li & O'Connor (2007).

Procedures for calculating bed stresses and sediment transport set-up under tidal, wave and combined currents are very clearly set out by Soulsby (1997).

Detailed accounts of protocols to establish ancient wave parameters, with illuminating case histories, are presented by Allen (1984).

Palaeotsunami studies include a nice paper by Huntington *et al.* (2007).

Palaeowave studies from oscillation ripples include Tanner (1971), Komar (1974), Allen (1984), Clifton & Dingler (1984), Diem (1985) and Adams (2003; see discussion by Le Roux 2004).

Combined flows are studied experimentally by Reineck & Wunderlich (1968), Kemp & Simons (1982), Murray *et al.* (1991), Dumas *et al.* (2005) and in the field by Green *et al.* (1990), Vincent & Green (1990), Kapdasli (1991), Osborne & Greenwood (1993), Park & Vincent (2007) and in beautiful numerical simulations by Li & O'Connor (2007).

Bedform phase diagrams for combined flows are in Myrow & Southard (1991), Dumas *et al.* (2005) and Cummings *et al.* (2009).

Papers relevant to the origin of hummocky cross-stratification are the discovery paper by Harms *et al.* (1975), relevant fluid dynamics studies by Kemp & Simons (1982), Leckie (1988), Southard *et al.* (1990), Murray *et al.* (1991), Panagiotopoulos *et al.* (1994) and especially Dumas *et al.* (2005). Lamb *et al.* (2008) is an interesting take on the generation of storm-modified shelf turbidity currents.

Allen (1982) is a comprehensive source on the origins and characteristics of erosive bedforms in both muds and rock.

Large megaflutes and other erosive bedforms are described and discussed by Morris *et al.* (1998).

Problems of scale and thermal complications of the atmospheric boundary layer are studied by Wyngaard (1992) and Brutsaert (1999).

Aeolian sediment transport and bedforms are discussed generally by Wilson (1972a,b), McKee (1978), Wasson & Hyde, (1983). Ballistic ripples are studied by Seppälä & Linde (1978), Hunter (1977), Anderson & Bunas (1993), Bagnold (1954b), Werner (1995); aeolian dunes by Bagnold (1954b), Hanna (1969), Wilson (1972a), Cooke & Warren (1973), McKee (1978), Kocurek (1981), Pye (1982, 1993), Tsoar (1983), Clemmenson (1987), Wopfner & Twidale (1988), Haynes (1989), Kocurek *et al.* (1992) and Fryberger (1993); Gran Desierto dune evolution by Beveridge *et al.* (2006); Kalahari dunes by Blumel (1998), Stokes *et al.* (1997), Thomas *et al.* (1997) and Lawson *et al.* (2002). Much insight into barchan evolution is provided by the field and theoretical approaches of Elbelrhiti *et al.* (2005). Doubtless future dune evolution studies will follow the pioneering path of Bristow *et al.* (2007) in determining subsurface dune history from a combination of GPR surveys and direct OSL dating of the sand within individual dune units. The riddle of stellate/pyramidal dunes is only partly solved by the laboratory and field observations of Wang *et al.* (2005).

Historic, Holocene and Pleistocene relict dunes and the effects of future climate on dune migration are studied by Thomas *et al.* (2005) and Sridhar *et al.* (2006).

Controls on aeolian bedform growth by variable transport rates at different wind strengths are considered by McKenna-Neumann *et al.* (2000). The issue of sediment supply and transport is addressed by Kocurek & Lancaster (1999). Field sediment transport over dune slip faces is discussed by Nickling *et al.* (2002). Werner (1995) discusses self-organized bedforms.

Chapter 8

# SEDIMENT GRAVITY FLOWS AND THEIR DEPOSITS

*Not just the beck only,*
*Not just the water —*
*The stones flow also, ...*

Norman Nicholson, 'Beck', 1981, Selected Poems, Faber 1982.

## 8.1 Introduction

A sediment gravity flow is one in which a mass of shearing sediment moves downslope against the efforts of frictional resistance, with no motive help from the ambient medium through which it flows. The medium may be air or water and the frictional grain mass may comprise loose cohesionless granular solids, a turbulent suspension or a partly compacted mass of cohesive clay, with every combination or mixture in between. Sediment gravity flows transport vast volumes of sediment from continental shelves to the deeper oceans. They do this after landslide failure of near-surface deposited sediment bodies (after earthquakes, for example) and their collapse downslope into sliding, deforming masses. As well as such *surge-like* events, suspended sediment-rich rivers may enter the ocean with their inertia and bulked-up negative buoyancy carrying them on subaqueously as quasi-continuous *underflows*. In both cases flows may *run-out* vast distances, commonly 1500 km or more from their source, like those fed from volcanic island landslides or even further for Amazon River underflows that feed offshore to the Central Atlantic with its surface plume extending alongshore almost to the Caribbean.

Most workers would agree that three end-member sediment gravity flow types exist.

1 *Granular flows*, regarded by solid-state physicists as 'granular fluids', exhibit grain support by solid-to-solid contacts. They are thus friction-controlled, dominantly by intergranular collisions in rapidly flowing natural materials.
2 *Debris flows* are dense, mostly laminar but sometimes turbulent, non-Newtonian flows. Their cohesive strength defines them as Bingham substances. Granular examples show internal framework support by some combination of (i) fine-matrix clays, (ii) elevated internal pore pressures and (iii) friction reduction due to bulk internal shear along micro-particle clay/grain boundaries.
3 *Turbidity flows* are a variety of *turbulent wall jets*, generally more dilute than grain or debris flows and,

*Sedimentology and Sedimentary Basins: From Turbulence to Tectonics*, 2nd edition. © Mike Leeder.
Published 2011 by Blackwell Publishing Ltd.

crucially, with fluid turbulence both supporting the moving suspension and responsible for driving bed-load along.

Both debris and turbidity flows may be hyperconcentrated, as discussed previously (Chapter 7) but as turbulence declines with increasing sediment concentrations so the latter morph into the former, an example of *gravity flow transformation*.

## 8.2 Granular flows

We observe granular flows every time we tilt or spoon grains from jars of beans, tea, muesli, sugar, etc. In nature, grain flows occur as avalanches. These are on a small scale when groups of grains tumble down the steep slip-face of a dune or ripple after *upper slope failure*. On larger scales are rock-falls down talus slopes on mountainsides or submarine scarps. Originally considered an arcane hobby for physicists (the discipline of *tribology* concerns granular friction, wear and lubrication), they have been much studied and commented on in recent decades because of many practical applications, e.g. to the nonlinear dynamics of granular fluids (the so-called 'physics of muesli' problem) in packing technology, in unexplained 'jamming' of material transfer along sloping ducts and in efforts to understand natural hazards associated with sediment slides, ice and powder snow avalanches and the generation of tsunami after seafloor slope failure. However, these approaches are not new, for O. Reynolds himself, towards the end of his life, tried to develop a universal theory of matter by reference to granular flow. The concept of collisions between moving particles also of course underlies the *kinetic theory* of gases, from which much inspiration has been drawn by more recent workers.

### Grain statics and initial flows

*Amonton's laws* state that static grain friction is due to a force that is directly proportional to the applied load and is independent of the area of grain contact (though not the microscopic nature of the contacts themselves). *Coulomb* extended this to moving friction and stated that moving friction is independent of relative sliding velocity. Consider a volume of dry grains in a container that is tilted at ever increasing angles. A simple force balance (Fig. 8.1) for any given slope, $\phi$, gives the normal force as $P = mg \cos \phi$ and the tangential or shear force as $T = mg \sin \phi$. At some critical value of $\phi$ the grains slide down the tilted surface. In order to shear the grains over the tilted surface at this critical $\phi$, $T$ must exceed a certain limit. The ratio of $T/P$ at this point is a *friction coefficient*, indicating the amount of energy that must be expended to make the grains move. At some critical angle, $\phi_{max}$, some of the grains will flow off the tilted grain surface as an *avalanche* or *grain flow*. The remaining grains are now bounded by a surface resting at a residual angle, $\phi_r$, some 5°–15° less than $\phi_{max}$ (Fig. 8.2) The results are unaffected by the experiment being conducted under water.

Attempting to explain the above phenomenon note first that the downslope movement of a mass of grains must involve an expansion of the mass at failure (Fig. 8.2). This is known as *dilatant expansion* (another discovery of Reynolds) and it requires energy to be expended. Evidently $\phi_{max}$, termed the *angle of initial yield*, must include the expansion involved in moving one grain over another on the zone of potential shear. No such effect is included in $\phi_r$, the *angle of residual shear*. $\phi_{max}$ is strongly dependent upon porosity, which varies between 32° and 40° for tightly to loosely packed sands (Fig. 7.4). This is because for tightest possible (rhombohedral) packing, maximum energy has to be done in lifting all grains over their neighbours. The stress relation at the point of shear is, $T = P \tan \phi_{max}$, and so for a natural grain aggregate to shear, the applied shearing stress must exceed 62–84% of the normal stress due to the static body force. As porosity increases, the difference between $\phi_{max}$ and $\phi_r$ steadily decreases, till at maximum (cubic) porosity of about 46% the two values converge to a limiting value. Here there is no need for any dilatant behaviour and we have this limit as $\phi_{cv}$, the final or *constant-volume friction coefficient* appropriate when initial variations in shape and porosity arising naturally through primary deposition are reduced by remoulding and repacking during shear (Fig. 8.2).

Grain shape is also an important variable in determining $\phi_{max}$ and $\phi_r$; angular gravels on *talus slopes* stand at very much steeper angles (up to 60° for loosely packed deposits) than those of beach, dune and river sands or rounded pebbles.

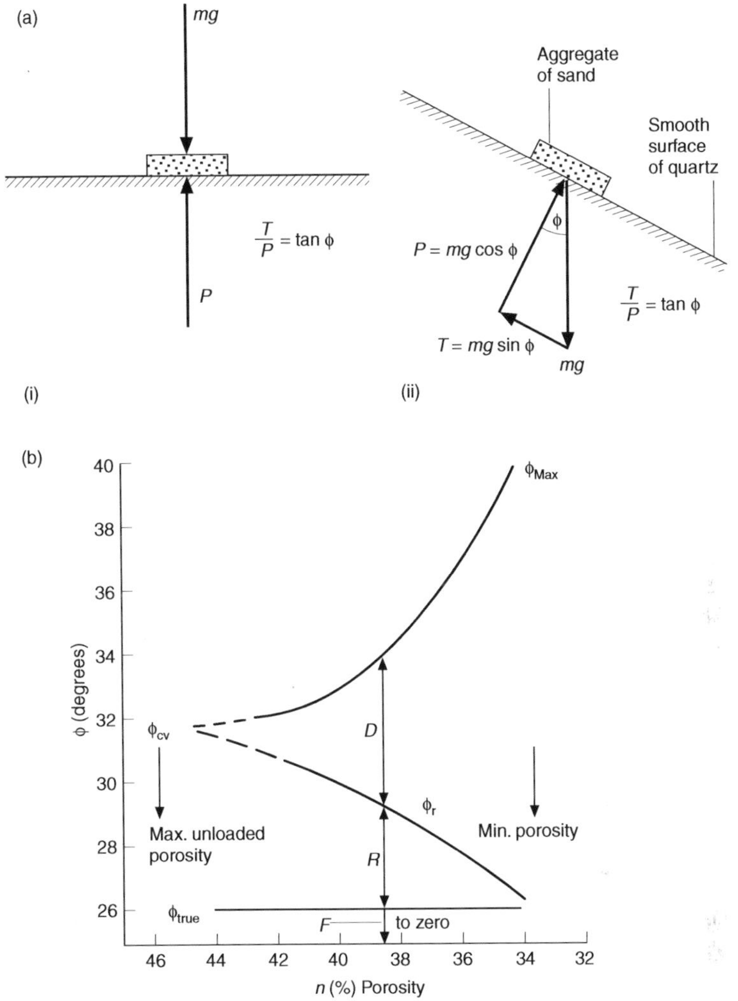


**Fig. 8.1** (a) Definition diagrams for (i) normal and shear stresses and (ii) the ratio between them, across a plane separating an aggregate of sand grains from a smooth silica surface. (b) Shear cell experimental results concerning the relation between variants of the angle of internal static friction, $\phi$ ($\phi_{max}$, $\phi_{cv}$, $\phi_{true}$ and $\phi_r$), and bulk mean porosity, $n$%, for a medium–fine sand. For closely packed sands there is a larger difference between $\phi_{max}$ and $\phi_r$ (the difference, $D$) than for loosely packed sands because the shear stress has to do more work in dilating the grain aggregate. The two curves for $\phi_{max}$ and $\phi_r$ converge to $\phi_{cv}$ at high values of porosity. The difference between $\phi_{true}$ and $\phi_r$ decreases with increased packing due to less energy spent on grain fabric remoulding (the difference, $R$). (After seminal work by Rowe, 1962.)

## Granular flow dynamics and moving friction

Granular flow avalanches are rapidly moving grain aggregates that shear at the boundary between moving grains and loose granular or solid bounding surfaces. At the boundary, granular concentrations and strain rates are very high and where Coulombic friction is modified by intense grain interactions. The degree of interaction is governed by the Bagnold number (**Cookie 30**). We may introduce the essential nature of granular flow by illuminating experiments designed

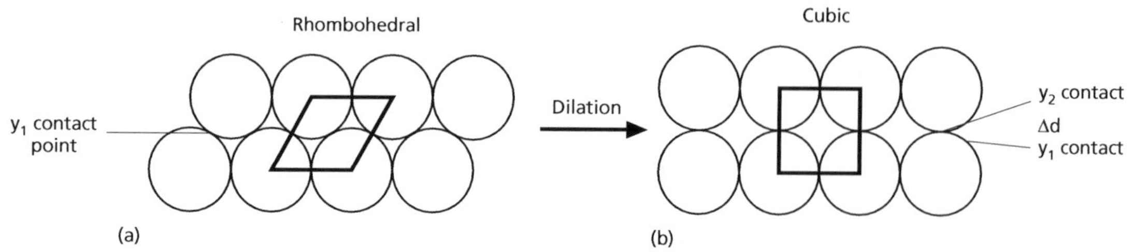


**Fig. 8.2** Sketches to show that (a) bulk shear of a closely packed granular aggregate to (b) wider spacing and looser packing state, must involve a dilation $\Delta d$ and therefore for work to be done in the process. (Based on the work of O. Reynolds.)

to simulate powder snow avalanches using large volume flows of large grains (half million ping-pong balls!) moving as aggregates down a ski-slope in air. These reveal the development of a high, rapidly moving flow head (Fig. 8.3) that leaves a lower, similar velocity body behind it as it accelerates away from a much more slowly moving flow tail. The rapid flow head velocity, > 300 km/h in some natural powder snow examples, and its long runout distance may be connected to the efficient transfer of fluid to solid momentum there and the flow of ambient air through the front of the granular flow.

Want to know more about collision dynamics? Turn to Cookies 30 & 38

It is evident from the ping-pong experiment that the flow dynamics of purely cohesionless (no-mud) avalanches is dominated by solid-to-solid collisions. More sophisticated experiments, done originally by Bagnold in the 1950s, and computer simulations on shearing basal and lateral boundary layers indicate that intergranular ('dispersive') stresses are produced which are proportional to the square of shear rate and grain diameter and linearly to particle density and concentration (**Cookie 30**). By analogy with gaseous kinetic theory, the intensity of grain stresses may be expressed in terms of a granular temperature. Experiments show that the avalanching of granular material at slope angles between the angle of repose and the angle of maximum stability, such as on a sand pile or on the lee (slip) face of a sand dune, typically produces droplet-shaped flows that consist of a well-defined head at the front, as with the ping-pong balls. Particle-image velocimetry results suggest that the presence of a deformable bed (layer of loose or

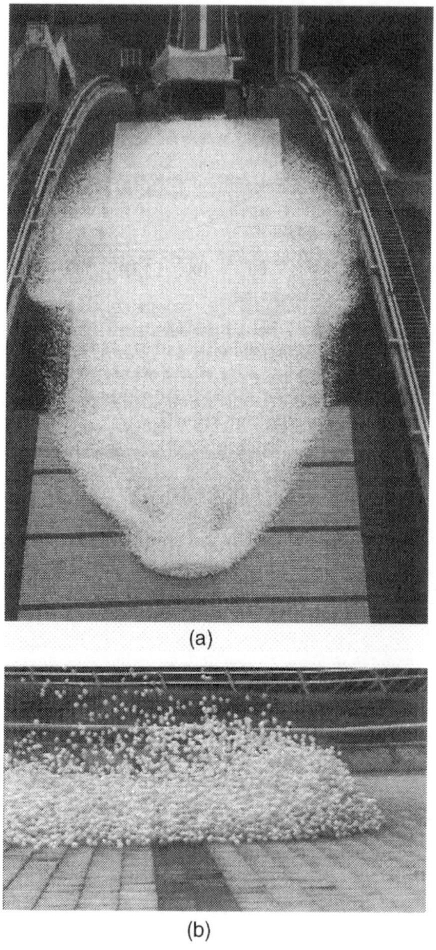


**Fig. 8.3** (a) Front view of a 550 000 ping-pong ball granular flow descending a ski jump. Note the tapering, slug-like frontal snout with two surreal 'eyes'. The horizontal lines are 5 m apart and the lower one is ca 90 m from the release point. (b) Side view of the head. (After McElwaine & Nishimura 2001.)

erodible particles) on the surface is a necessary condition for the occurrence of these flows and that a spontaneous transition occurs in flow behaviour with downstream distance on a uniform slope, from a surface shearing flow to a compressing flow that has the characteristics of a shock wave propagating upslope through the deformable bed. Shock waves also propagate upslope from discontinuities at the base of dune and ripple lee surfaces

In subaqueous grain-flow avalanches the interstitial water dampens the exchange of momentum experienced during collisions. Friction reduction in small subaerial grain flows is probably minimal, but larger subaqueous examples and powder snow avalanches may entrain appreciable amounts of fine-grained muddy sediment or wet snow that may act to reduce friction and buoyancy contrast by elevating pore pressures, causing the grain flow to become a debris flow (see below). Also, in snow avalanches the grain collisions may be destructive, leading to particle melting, refreezing and growth of granular ice duing flow.

A feature of granular flow deposits most commonly seen in delta-front avalanche or colluvial talus avalanche deposits is the frequent occurrence of open-framework gravels and coarse sands with reverse grading, i.e. where grain size increases upwards in a single deposit. The process is illustrated from a kitchen table experiment in Fig. 8.4) and the topic is aptly known as the *Muesli effect*. Two hypotheses for this problem of granular physics have been proposed. The first is that *dispersive stresses* are set up by granular collisions (**Cookies 30 & 38**). This has always been assumed to be greatest close to the shear plane so that large grains there exert a higher stress causing larger particles to move upwards from the shear plane through the flow to equalize the stress gradient. However, a recent theoretical analysis suggests that dispersive pressure cannot account for the upward segregation of large grains. Instead, rapid granular flows are self-organized in such a way that dispersive pressure at any level in a shearing mass of grains is always equal to the applied normal stress. An increase in dispersive pressure causes an immediate expansion of the flow and a consequent decrease in dispersive pressure until it equals the applied normal stress again. The gradient of dispersive pressure is therefore 'hydrostatic' and only grains lighter than the bulk density of the flow are pushed upward, as observed in experiments. The inverse grading of grains observed in a great variety of deposits is not caused by dispersive pressure and must be explained by another mechanism. This is termed *kinetic filtering*, whereby small grains simply filter through the voids below larger jostling grains until they rest close to the shear plane—the larger grains are therefore forced to rise. Kinetic filtering probably plays the dominant role in size-separation phenomena in multisize aggregates.

Marked downslope variations in sorting and grain size develop spontaneously in grain flows, exemplified by the avalanche deposits of **Plates 5 & 6**. It is common for larger grains to be carried further than smaller grains because they have the largest kinetic energy. This leads to lateral (downslope) segregation of grain size, a process helped by later flows that may drag protruding larger grains even further down with them. More interestingly, when the larger grains have higher

**Fig. 8.4** An initial random mix of Riojanas beans and Valencia rice in a glass container is shaken at 3 Hz for 20 s. All the beans rise, magically, to the surface. Physicists use such behaviour to shed light on the properties of granular fluids as analogues for the kinetic theory of gases and solids.

**Fig. 8.5** A typical view of the active lee-side avalanche face of a desert dune. Granular flow lobes originate near the crestal area of the dune, where excess sand falls out from bedload transport of the upwind (stoss) side. The flows are confined as elongate lobe-shaped masses and several generations may be clearly seen descending the steep (ca 30°) slope, their wrinkled surfaces evidence for dilitant shearing and normal grain stresses in jostling sand grains comprising the sliding granular flows. (Photo from the Mali Sahara, courtesy Ian Davidson.)

$\phi$, the mixture spontaneously stratifies as the smaller grains halt first and the larger grains form an upslope-ascending grain layer above them (Fig. 8.5).

### Granular flows, run out and the inadequacy of Coulombic dry friction

Coarse-grained granular flows and those containing fine-grained sediment in a subaqueous environment may be very thick. In high-gradient subaerial and subaqueous environments, coarse sheet-like units termed *megabreccia* are derived from catastrophic rockfall avalanches; a group of phenomena termed '*sturzstroms*' by alpine German-speakers. These rock avalanches travel at phenomenal speeds (up to 100 m/s) and may ride up over substantial topography during their passage. Many owe their initiation to the effects of large earthquakes on steep rocky slopes. The fact that such deposits are often found large distances from their source leads us to further consider the phenomenon of *runout*, defined as the ratio of the vertical fall distance of the flow to the subsequent horizontal travel distance (**Cookie 39**). This ratio defines an apparent friction coefficient. For small, short-path sand avalanches like those on the lee side of a dune or ripple, the runout is very small and the apparent friction very high. Calculated values of the apparent friction for megabreccias are surprisingly low; the larger the initial avalanche volume, the lower the apparent friction and the longer the runout (not uncommonly up to 10 times the fall distance, giving friction coefficients of 0.1). There have been many explanations offered for long-runout phenomenon, including:

1 *flow lubrication* caused by ingestion of basal air under pressure (flymo or hovercraft effect);
2 *fluidization* by ingested air;
3 *fluid-like behaviour* of intraparticular rock dust created by grain collisions;
4 *acoustic fluidization* due to violent grain impacts from high-fall flows causing high-frequency pressure fluctuations that locally exceed normal stresses and reduce friction;
5 reduced granular friction by increased *internal pore pressure*, in direct contradiction to Coulombic principles.

Want to know more about runout? Turn to Cookie 39

Although there is still (and may always be) uncertainty as to the exact internal stress–strain regime inside rock avalanches, theoretical analysis indicates that frictional effects are confined to very thin basal and lateral boundary layers of intense shear with the majority of the flow travelling inside these as a jostling plug. Violent particle collisions with the bed in the thin basal boundary layer support the entire non-shearing grain mass, though the latter must continually provide kinetic energy to the boundary layer where it is dissipated by largely inelastic collisions (the rebound or *coefficient of restitution* of rock is small). Once the kinetic energy of the main mass has been exhausted then the flow comes to a halt. The largest kinetic energy comes from the longest initial avalanche freefalls, for doubling the fall velocity quadruples the kinetic energy available. Additionally, during runout the flow falls over solid rock or relatively smooth substrates of low friction. Computer simulations show that dry friction is increased as shear rate is increased in large rapid avalanches; possibly caused by non-elastic effects at points of intense collisional contact. They indicate that additional effects must cause reduction of granular friction in a granular flow boundary layer, probably including enhanced pore pressures maintained by ingested water, clay-grade fines and the ability of water/organic/clay mixtures to somehow lubricate contact surfaces at the grain microscale.

## 8.3 Debris flows

Debris flows (Figs 8.6–8.8) are slurry-like flowing aggregates of diverse grain size, concentration, velocity and internal dynamics—they are non-Newtonian systems with a plastic-like yield strength which means they often stop on tilted-surfaces of critical minimum slope and form a rigid deposit. A useful everyday analogue for consistency and behaviour is that of wet concrete. Their destructive ability arises because of the combination of extreme mobility and large momentum; velocities $> 10$ m/s are not uncommon and bulk densities are in the range 1800–2300 kg/m$^3$. This lethal combination (they cause more casualties worldwide, on average per year, than any other natural hazard) arises because the presence of densely packed coarse solids, perhaps $> 60\%$ by volume occurs mixed with a smaller proportion of watery clay/silt matrix (though often $\ll 10\%$ by volume). In water-saturated granular examples the above combination is conducive to the build-up, behind the flow head, of *excess pore pressure* that reduces shear strength so that despite clast angularity, overall frictional retardation is low. Fresh debris-flow deposits consolidate under their own weight. Modelling and experimental measurements demonstrate that changes in fluid pressure and effective stress evolve upward from the base of a deposit, and show that hydraulic diffusivities of muddy slurries containing about 5 to 50 wt% mud are remarkably similar, about $10^{-6}$–$10^{-7}$ m$^2$/s. By comparison, cohesionless sandy-gravel debris flow deposits containing $< 2$ wt% mud have higher hydraulic diffusivities, $\sim 10^{-4}$ m$^2$/s. Significant dissipation of fluid pressure is restricted to post-depositional consolidation and deposition results from frictional effects concentrated along flow margins where high pore-fluid pressures are absent. Sustained high pore-fluid pressure following deposition fosters deposit remobilization, which can mute or obliterate stratigraphic evidence for multiple events. A thick deposit of homogeneous, poorly sorted debris can result from mingling of soft deposits and recurrent surges rather

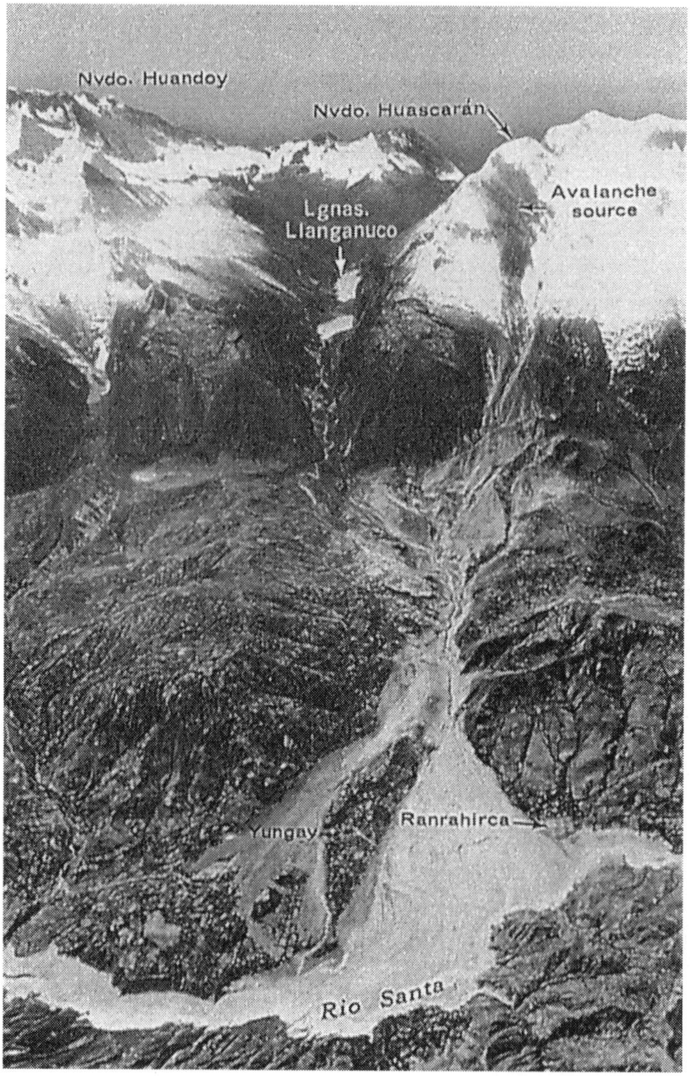


**Fig. 8.6** The Nevados Huascaran avalanche and debris flow, Peruvian Andes. A large earthquake triggered the collapse of part of the 6654 m high mountain, causing 50–100 × $10^6$ $m^3$ of rock to drop 4 km and flow out 16 km to the Rio Santa,where it was transformed into a gigantic debris flow. (From Plafker & Ericksen, 1978.)

than from a single flow wave if deposit consolidation time greatly exceeds typical sediment emplacement times. In addition, water:clay mixtures serve to lubricate grain contact surfaces at a microscale, as in grain flows. Regarding their 'runout', debris flows usually have greater runout, volume for volume, than grain flows. Debris flows are usually laminar in character (Fig. 8.8a) but more watery examples may be mildly turbulent (Fig. 8.8b); the former show typical non-Newtonian plug-flow profiles, with the vast majority of internal deformation taking place along flow-margin boundary layer/shear zones.

Subaerial debris flows are common in most climatic regimes and are usually initiated as debris slides or avalanches after heavy rainfall from steep debris-laden catchment slopes or vegetated slopes with well-developed soil or saprolite (Fig. 8.7). Coastal cliffs and land escarpments cut into mudrocks are common sites

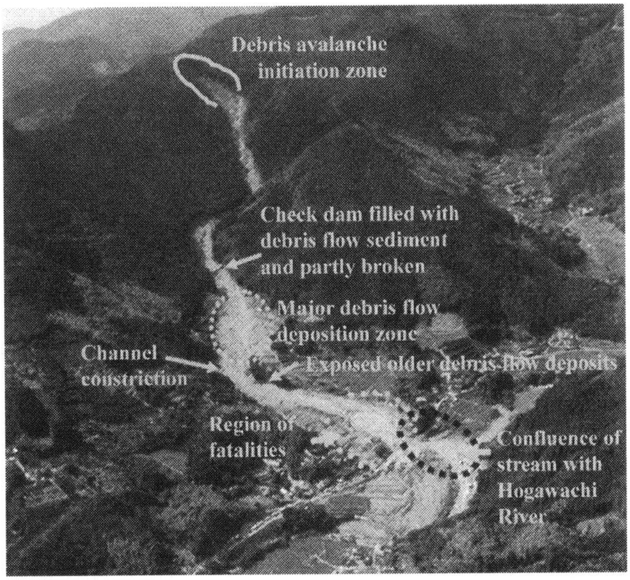


**Fig. 8.7** Trace of flow path of debris avalanche from source via piedmont deposition and junction with river, Kyushu, Japan (from Sidle & Chigira, 2004.)

of debris flows. Slope failure seems to occur due to bulk friction reduction as pore pressure increases in weathered regolith during heavy precipitation. Natural forest fires act as powerful triggers for conditions necessary to trigger debris flows after subsequent rain storms. Debris flows are of particular importance in volcanic areas when torrential rains or snowfield melting that frequently accompany or follow eruptions lead to widespread and catastrophic flows (termed *lahars*, an Indonesian word) sourced from unconsolidated ash on volcanic slopes. These are usually more destructive than the direct products of the volcanic eruptions themselves. The transformation of hyperconcentrated flows into lahars and debris flows is common in such environments where there is an abundant supply of loose surface sediment rich in silt and clay fines. Incorporation of these into storm-induced hyperconcentrated flows by surface wash and channel bank erosion causes 'bulking', a term used to describe an increase of sediment concentration from a few percent to many tens of percent and a change from Newtonian to non-Newtonian behaviour. Many avalanches/landslides also transform themselves, and the rivers into which they flow, into debris flows as they run out (Fig. 8.6).

For information on debris flow runup and rheology see Cookie 40.

Less is known about the initiation of subaqueous debris flows but submarine slope failures, avalanches and slides seem to provide most of them (see also section 9.3 & Chapter 21). Such failures reflect great sensitivity to changes in sediment pore pressure due to loading or storm wave action. Impressive examples of debris flow initiation and downslope movement of distinct flow lobes from gullies occurred during the 2005 tropical storm season in the Gulf of Mexico offshore the Mississippi delta front when hurricanes Katrina and Rita caused 10–15 m waves. Debris flows also develop from slides, slumps and submarine escarpment collapse caused by earthquake shocks. The high slopes around volcano-constructed islands (e.g. Hawaii, Canary Islands) are particularly prone to failure.

### Debrites: recognizing the deposits of debris flows

Debrites (admittedly an ugly term) are characteristically very poorly sorted (Fig. 8.9) and have few signs of

Fig. 8.8 Field photos of debris flows, Mt Thomas, New Zealand. (a) Laminar flow of mean velocity ~ 0.2 m/s, Re = ~30. Channel at constriction is ca 2 m wide. (b) High-velocity, low-turbulent debris flow. Cobbles and boulders at flow margins define levees. Mean velocity ~5 m/s, supercritical flow (see standing waves opposite persons) and Re = ~3 × $10^3$. (From Pierson, 1981.)

internal structures because of the preponderance of plug flow which allows no shear fabric or sorting to develop. Traces of shear cryptofabric are revealed by automated image anlysis of experimental cohesionless deposits—these occur along internal bulk shear planes and in high shear areas such as basal and marginal zones. Many debrites occupy pre-existing water-cut channels which also feature overspill deposits—only the more turbulent and fast moving can cut channels themselves. The non-Newtonian property of *matrix strength* causes overspill to 'seize-up' by basal friction into marginal levee-like ridges, a feature also seen in the overhanging nature of wrinkled flow snouts. A mysterious phenomenon observed in experiments is the violent expulsion of coarse clasts from moving frontal areas. Debris flow surging also occurs in experiments and in natural flows and is an integral part of flow dynamics (Fig. 8.10); it may sometimes produce recognizable stratification and asymmetric folding due to overriding and thrusting of plugs along internal shear planes (see also flow slides, section 9.3). Normal-graded debris-flow deposits are attributed to

**Fig. 8.9** View of vertical wall cut through saturated experimental debris-flow deposit to show the massive character and local development of inverse gading in the $> 0.8$ cm pebble population. (From Major, 1997. Photo courtesy of Jon Major.)

the cumulative effects of sustained deposition from a travelling debris floodwave (Fig. 8.11). Fluidization pipes (section 9.2) indicative of high water content and enhanced pore pressure prior to deposition, occur in relatively fine-grained flow deposits.

Long-runout debris flows into water, especially lahars, show marked downstream changes in internal fabric and structures as they decelerate, mix with ambient waters and spread laterally. For example, a remarkably extensive debrite called BIG'95 dated to 11.5 ka in the western Mediterranean affected 2200 $km^2$ of the Ebro continental slope and rise with a deposit volume of some 26 $km^3$, which is up to 150 m thick. High-resolution imaging and coring reveals that the deposits comprise relatively coarse material, mostly remoulded during flow and finer more cohesive material which moved as independent, partially buoyant, rigid blocks. These kept their internal coherence but were broken up as they were pushed and dragged up to 15 km by the Bingham-like mobile

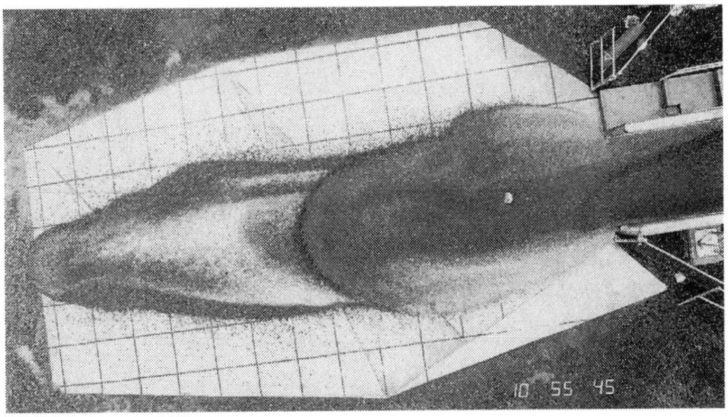


**Fig. 8.10** View from above of an experimental debris flow with well-defined traces of lobe surge fronts. Flows exit a 30° sloping channel and deposits on a 1 m square gridded platform. The flow comprised a water-saturated sand–gravel mixture very low in silt or mud content ($< 1\%$). Note the elongate shape and flange-like levees. Successive surge lobes (one seen) give rise to an incremental deposit whose internal structure may reveal little of the actual surge-like behaviour (From Major, 1997. Photo courtesy of Jon Major.)

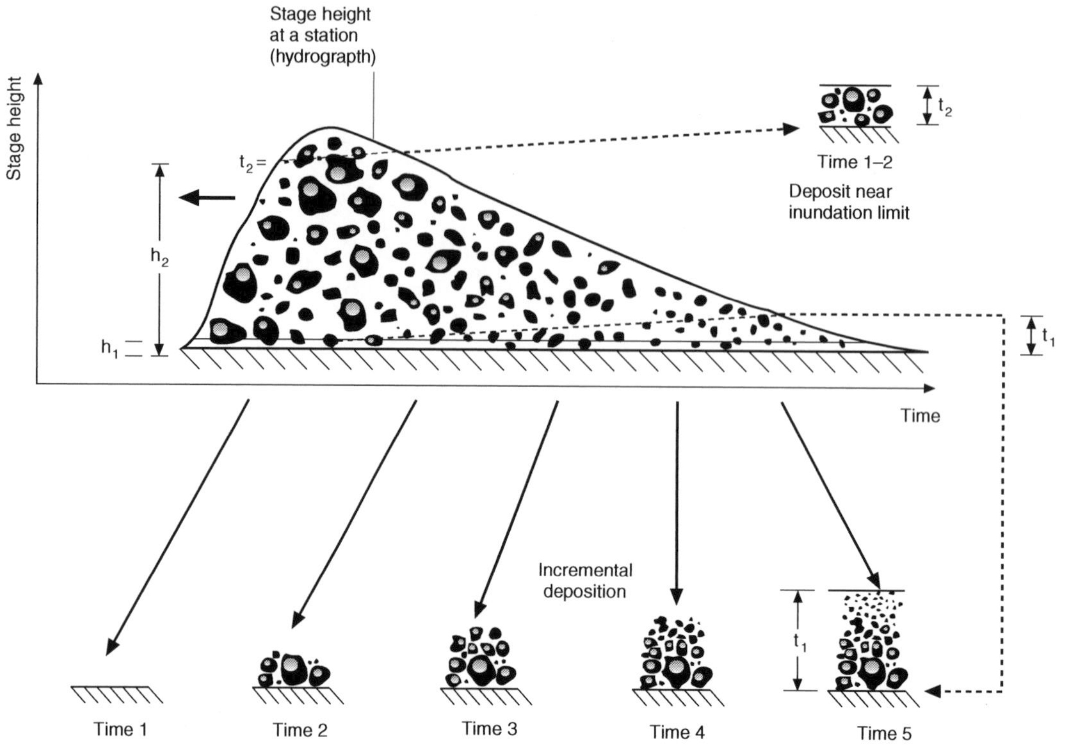


**Fig. 8.11** Debris flow as a sediment wave. Schematic diagram of stage height vs. time (debrigraph) for mass flow at a point, and production of normal-graded and ungraded deposits. (After Vallance & Scott, 1997.)

material. The more mobile material kept on flowing, reaching almost 2000 m water depth in the Valencia Channel after travelling ~110 km. Numerical results demonstrate that observed runout is physically possible with quite low values [O] 1° of Coulomb frictional angle (**Cookie 40**) and [O] 800 Pa for yield strength.

### Flow transformation of debris flows to turbidity currents

Sediment *slides* and *slumps* caused by slope failure transform and disaggregate by liquefaction, fluidization (see section 9.3) and downslope surface shear into debris flow (Fig. 8.12). The latter cannot transform into the turbulent suspensions that define turbidity currents (see below) without entrainment of ambient fluid and this may not be possible if their irrotational fronts have significant strength. Instead, experiments show that debris flow transforms into turbidity flow in a rather inefficient way along their upper edges by entrainment stresses due to turbulent separation and reattachment; the turbid mixture may then overtake the parent debris flow. However, careful experiments also show that basal *aquaplaning* under the head region causes acceleration, partial lift-off and vertical head growth, so dramatically increasing upper surface mixing. Flow transformations from debris flows are facilitated during sea-level lowstand adjacent to deltas, along shelf margins, grounding ice masses or iceberg 'graveyards'. All these environments experience high deposition rates so that their sediment beds are loosely packed and thus unstable in the face of applied shear stresses which can initiate increased pore pressures. There is also a suspected role for *methane gas hydrates*—these are ice-like solids that grow from reactions involving buried organic matter.

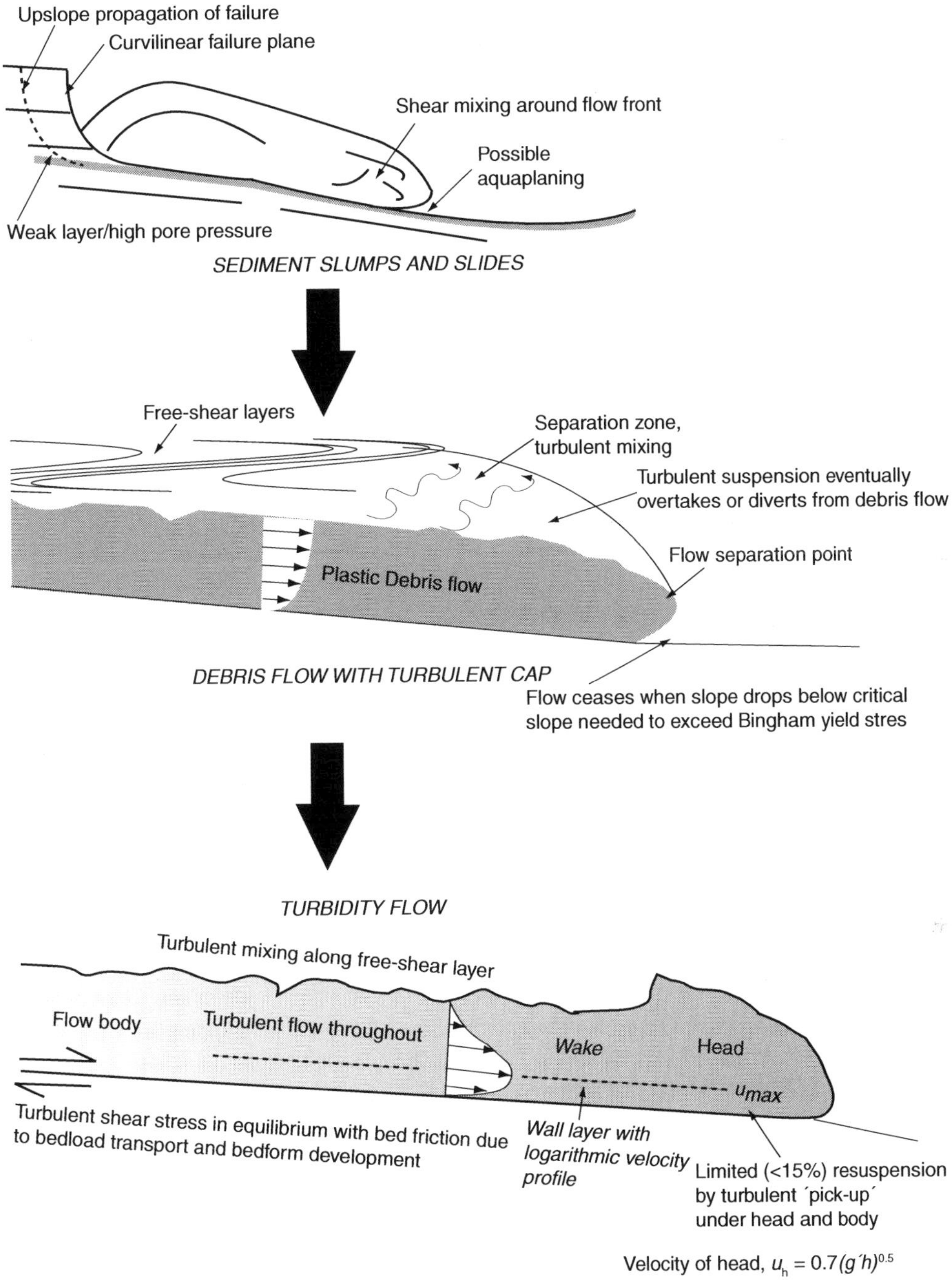


**Fig. 8.12** Various possible flow transformations from subaqueous sediment slumps and slides via debris flows to turbidity flows.

They are prone to disassociation and methane-emission during periods of lowered sea level causing regional mass failure planes in buried shelf sediments.

## 8.4 Turbidity flows

Turbidity flows are a subgroup of gravity flows (see Fig. 8.13) whose changed density, $\Delta\rho$, compared to ambient fluid is due to grains suspended by turbulence. They are not restricted to subaqueous conditions for dense suspensions may flow downslope in air, witness the frequent occurrence of powerful powder-snow avalanches. The fluid dynamics of turbulent suspensions is a highly complicated field because the suspended particles feed back to the turbulent characteristics of the flow, with effects on viscosity, turbulence generation and stratification. This leads to the consideration of what has been termed autosuspension (**Cookie 41**).Turbidity currents originate in a number of ways, including the following.

1 Transformations from debris flows (see previous section).
2 Direct *underflow* of suspension-charged river water in *hyperpycnal plumes* or quasi-continuous *sediment underflows*. These produce sustained turbidity flows, recorded during snowmelt floods in steep-sided basins like fjords (**Plate 9**), in front of estuary channels, delta channels and in river tributaries whose feeder channels have extremely high suspended loads. The ability of a river discharge to underflow is a function of salinity, temperature, dissolved load, stratification of suspended load and estuarine mixing. Specifically, the critical concentration of suspended sediment for hyperpycnal flow formation is reduced by near-bed estuarine mixing and by flocculation, which can reduce the critical concentration by several orders of magnitude since only the lower portion of formerly riverine discharge descends as underflow. Underflows are proven to be much commoner during periods of global sea-level lowstand when rivers issue oceanwards almost at the shelf edge.
3 Along tectonically active coasts with very narrow shelfs, heads of *submarine canyons* may rise close inshore where they can collect sediment provided by coastal *longshore drift*. The process is most efficient at sea-level highstand, during and following storms and tends to lead to sand-dominated sustained turbidity currents.
4 *Shelf turbidity currents* arise during *storm surges* when a seaward negative hydrostatic pressure gradient is produced by water set-up due to the combined effects of high flood tide and onshore windshear. This gradient causes near-bottom return flow of suspended sediment stirred up by the storm waters far out onto the shelf (see further details in Chapters 18 & 19).

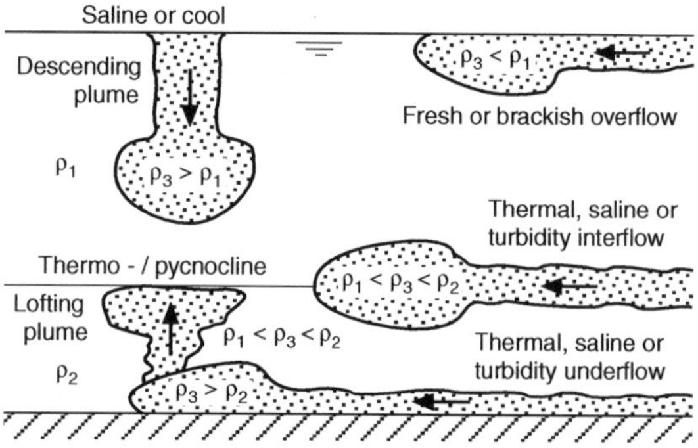


**Fig. 8.13** Types of density currents.

> For a short summary of the concept of autosuspension, see Cookie 41

There are also a number of particular features of turbidity currents that need attention.

1 Double shear boundaries (Fig. 8.14) at both sediment-flow–stationary-bed and sediment-flow–ambient-fluid interfaces. Frictional effects may generally be neglected at the upper interface in ambient air, but in subaqeous flows the double boundary layers cause significant mixing, turbulence generation and frictional drag. Fluid dynamicists refer to such wall-attached flows generally as *wall jets*.
2 The accelerative state of the flow, arising from the *sediment continuity equation*. This defines whether the flow is steady or unsteady, uniform or nonuniform. If unsteady then we need to know whether the unsteadiness is waning or waxing. If it is nonuniform, whether it is depletive or accumulative. We can then refer the flow to its correct position in a *Branney–Kneller acceleration matrix* (Fig. 8.15).
3 The time-evolution of any flow. As a flow travels on its long or short, slow or fast journey into a receiving basin, be it lake or ocean, it may perforce flow

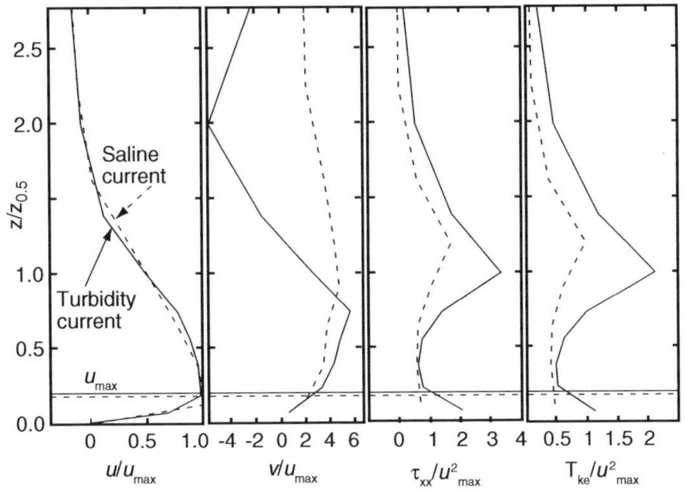


**Fig. 8.14** The characteristic double boundary layer of a wall-jet flow. Data for turbidity currents and saline currents of exactly the same excess density. Mean streamwise and vertical velocities ($u$, $v$), streamwise turbulent stress ($\tau_{xx}$) and turbulent kinetic energy ($T_{ke}$) exhibit contrasts between experimental saline and turbidity flow. Dimensionless height is with reference to $z_{0.5}$, the height at which $u$ reaches value $0.5\,u_{max}$. (After Gray *et al.*, 2005.)

down variable slopes, along variablewidth channels and be either depositing and losing sediment mass to the bed, or eroding and gaining sediment mass from the bed. Any flow may therefore continuously or periodically change or evolve downslope from source to sink. We can call these various possibilities *flow transformations*. They are currently a hot research topic.

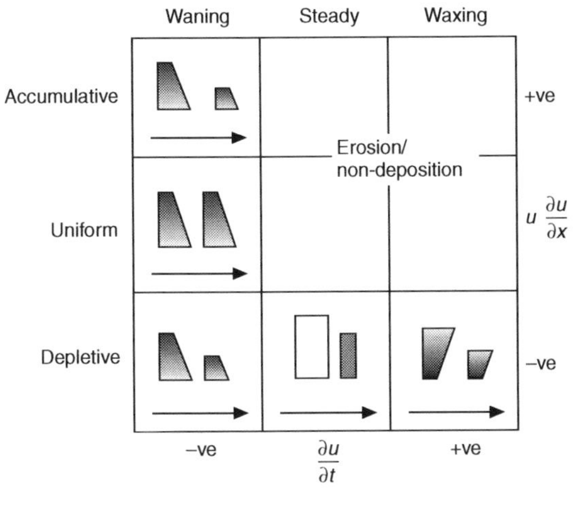


**Fig. 8.15** Branney–Kneller plot: an acceleration matrix for turbidite beds, with illustrative bed sequences for each field of behaviour showing downstream and vertical (time) changes in the relative grain size of the deposits. (After Kneller & Branney 1995; Kneller 1996.)

4 The strange behaviour of turbidity flows as they traverse meandering channels on submarine fans. The nature of secondary flows set up by these currents is thought by some, rather controversially, to contrast with those set up in river meanders (Chapter 11) and will be discussed further in Chapter 21.

### Turbidity surges

Release of finite volumes of sediment-charged denser fluid from restraining barriers into a channel filled with less-dense ambient fluid causes a turbidity surge to move along the floor as a wall jet. The flow shows well-developed *head* and *body* regions. Under zero-slope conditions the head is usually 1.5–2 times thicker than the body, with the ratio approaching unity as the depth of the ambient fluid approaches the depth of the density flow. We may regard surges simply as shallow-water waves (Chapter 5) so that flow velocity, $u_h \propto \sqrt{(gh)}$, where $u_h$ is forward head velocity and $h$ is mean head height. The proportionality coefficient depends directly upon the density difference, $\Delta\rho$, via the action of *reduced gravity* $g' = (\Delta\rho/\rho)g$. For

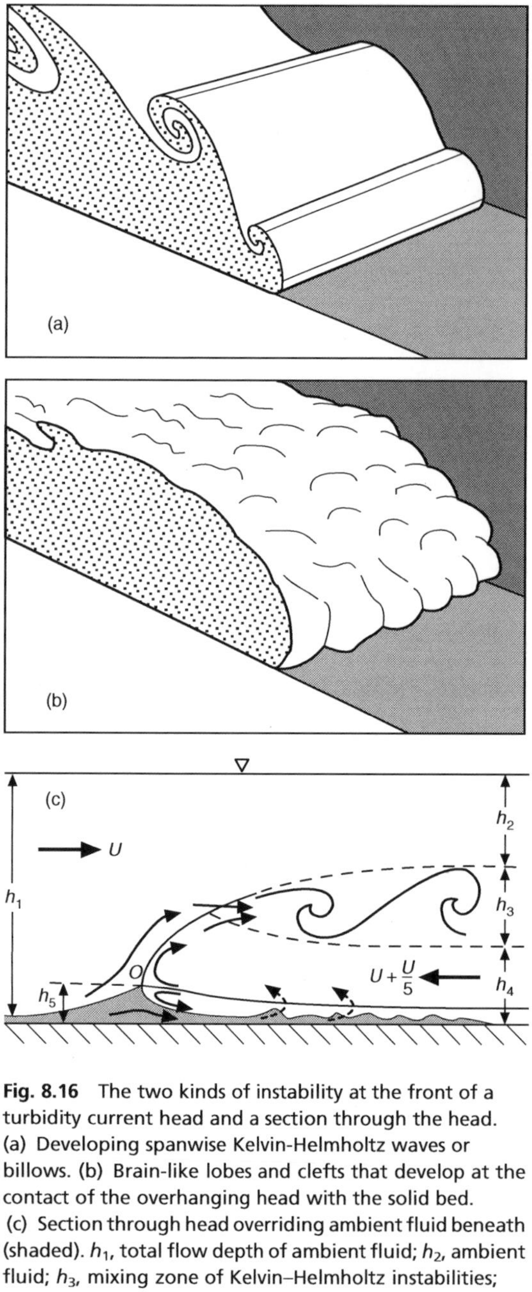


**Fig. 8.16** The two kinds of instability at the front of a turbidity current head and a section through the head. (a) Developing spanwise Kelvin-Helmholtz waves or billows. (b) Brain-like lobes and clefts that develop at the contact of the overhanging head with the solid bed. (c) Section through head overriding ambient fluid beneath (shaded). $h_1$, total flow depth of ambient fluid; $h_2$, ambient fluid; $h_3$, mixing zone of Kelvin–Helmholtz instabilities; $h_4$, height of main body of flow; $h_5$, height of frontal overhang; $U$, mean forward velocity; $O$, frontal separation point. (After Simpson, 1987.)

$Re > 1200$, experiments show that a density current head flows with a mean velocity, $u_h = 0.7\sqrt{(g'h)}$.

Close examination of the head (Figs 8.16 & 8.17) shows it to be an array of bulbous lobes and trumpet-shaped clefts. The overhang height of lobes is up to 20% of head height for $Re > 1000$. Ambient fluid mixes into the flow body under the overhanging lobes and through the clefts. Most mixing occurs by breaking Kelvin–Helmholtz waves behind the head. Internally the head region shows strongly diverging, predominantly upward mean flow vectors, which help to provide the turbulent fluid stresses needed for sediment suspension. Continued forward motion of the head at constant velocity requires a transfer of denser fluid (buoyancy flux) from the faster-moving tail into the head in order to compensate for boundary friction, fluid mixing and loss of denser fluid. Surge-like flows must therefore decelerate because the supply of denser fluid from behind the head is finite and the buoyancy force driving the flow is insufficient to overcome frictional energy losses. The head thus shrinks until it is completely dissipated.

Concerning erosion and deposition under turbidity flow surges, it is commonly thought that passage of the head is accompanied by erosion, with production of erosive bedforms like flutes and grooves. Extra sediment added to the flow will cause the head of the flow to accelerate. Careful biostratigraphic and sediment budget studies of Quaternary turbidites reveal that 12% or so of turbidite volume may be composed of reworked materials. This converts to an average of ~4.3 cm of erosion over the total areal extent of deposits.

Experiments with initially channelized surge-type density/turbidity flows show that as these enter wide reservoirs or basins, rapid dissipation occurs as their heads spread radially; resulting deposits define laterally extensive sediment lobes.

### Sustained turbidity currents

*Sustained turbidity currents* arise from prolonged underflows provided by sediment-laden river flux or the progressive upslope development (*retrogressive failure*) of submarine slope sediments (see Chapters 9 & 19). A steady state is brought about by near-constant input over time. In such *quasi-steady flows* head velocity is approximately 60% of the tail velocity in the slope range 5° to 50°, leading to the head

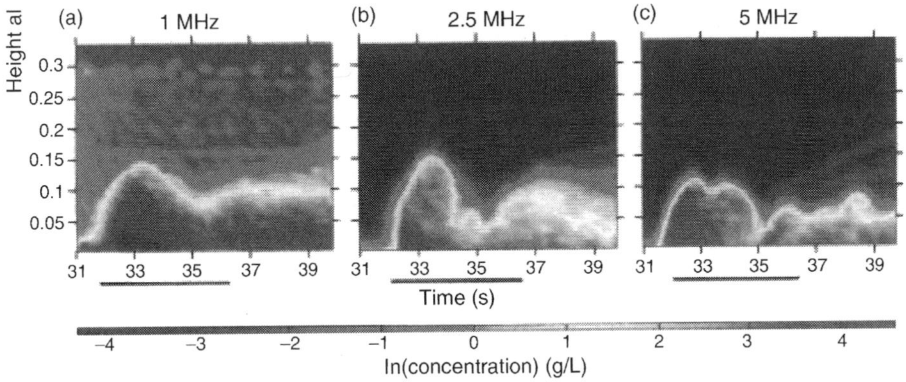


**Fig. 8.17** (a)–(c) Acoustic backscatter system (ABS) images of suspended sediment concentration to show the structure of a passing turbidity current head and proximal body with their signature Kelvin-Helmholtz vortices. Increasing frequencies from 1 to 5 MHz in the ABS receivers gives progressively higher sensitivity imaging of silt-grade suspended sediment. Note evidence for internal flow layering in the head, shown especially clearly in the 2.5 MHz frame. Brightest line bounds higher sediment concentrations (> ln = 0) below. (After Alexander *et al.*, 2008.)

increasing in size as it travels downslope. Entrainment of ambient fluid also causes head growth, increasingly so at higher slopes; the momentum transferred from the current to this new fluid acts as a retarding force to counteract the buoyancy force due to the slope. This *steady velocity/growing head* behaviour is also a characteristic of starting thermal plumes but has not been investigated from the point of view of turbidity current deposition and erosion. It might be expected to play an important role in causing flow overspill onto levees from submarine fan channels. An interesting situation occurs when sustained subcritical turbidity currents pass over a negative slope break at the end of a sloping submarine channel. The slope decrease causes longitudinal flow deceleration (nonuniform, depletive mode) and enhancement of vertical turbulent accelerations. The mean streamwise slowing is transferred into turbulence production at the slope break; this causes increased sediment transport, a decrease of deposit mass initially downstream and thereafter increased sand transport basinwards. If the channelized flow is critical (Fr > [O] 1), then a hydraulic jump may occur (find more out about these in tc's)

Little is known about the behaviour of natural quasi-continuous turbidity current underflows. In examples studied entering glacial lakes, flows exit delta distributary mouths, reach a plunge-line and then flow unchannelized down gradually declining delta-front slopes. Whole flow-field monitoring reveals very distinct unsteady 'pulsing' of the flow at any particular location, a feature indicative of frequent flow diversions or possibly meanderings from the descending plunge-line as flows gradually decelerate in a nonuniform, depletive fashion.

Experimental sustained-type turbidity current underflows exiting an inclined channel across a break in slope into a wide reservoir (Fig. 8.18) simulate natural flows from river channel to delta front and shelf margin submarine channels. They show initial head-expansion on entering the wide tank and then a *steady wall jet* phase which is much more important in controlling deposition than the head phase. This is in marked contrast to surge flows. Steeper channel slopes produce greater flow velocities and turbulence intensities, but these effects diminished markedly with distance from the channel mouth. Flow velocity vectors show a persistent central core of faster velocity and narrow vector dispersion (Fig. 8.19), with slower flow and larger dispersion at the jet margins. Suspended sediment concentrations are higher within flow heads with vertical concentration gradients and dense basal layers in the flow body. The deposits comprise a thick central ridge of similar order width to the channel mouth, with abrupt margins and a surrounding, very thin, fan-like sheet (Fig. 8.20).

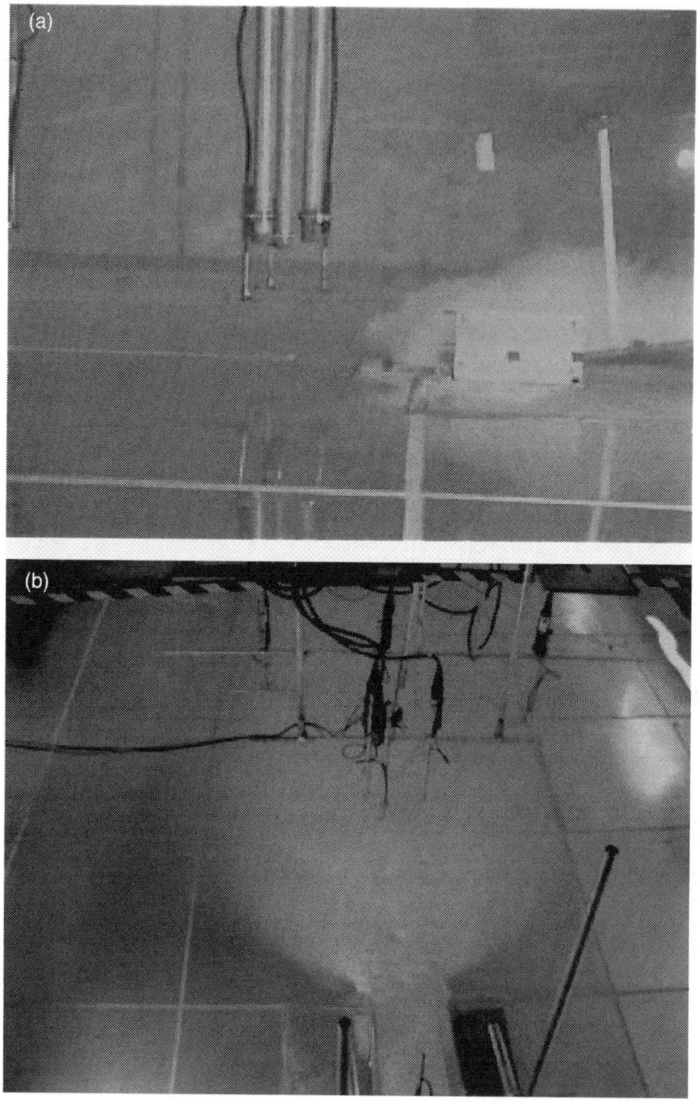


**Fig. 8.18** (a) Images of quasi-continuous turbidity currents entering large reservoir from sloping inlet channel. Quasi-continuous turbidity current head with nose at the channel mouth viewed through flume sidewall. Also in view are ADV meters. Input sediment concentration 1.21 vol.%, discharge 1.74 $m^3$/sec $\times 10^{-3}$, channel slope 9°, mean grain size 59 mm. (b) View down the axis of the flume after end of a run, showing the deposit shape and gantry mounted equipment. The pock marks on the deposit are where samples have been extracted. (After Alexander *et al*., 2008.)

Ridges are coarser grained and better sorted than the original sediment, with grain size fining downflow. The ridges suggest that in quasi-steady turbidity currents, vertical turbulent momentum exchange is more significant for sediment dynamics than spanwise momentum exchange due to lateral expansion. The streamwise elongate geometry of the ridges contrasts with the fan-like modes of surge-type flows. They suggest that deposition in natural turbidity currents will be predominantly streamwise, with little lateral

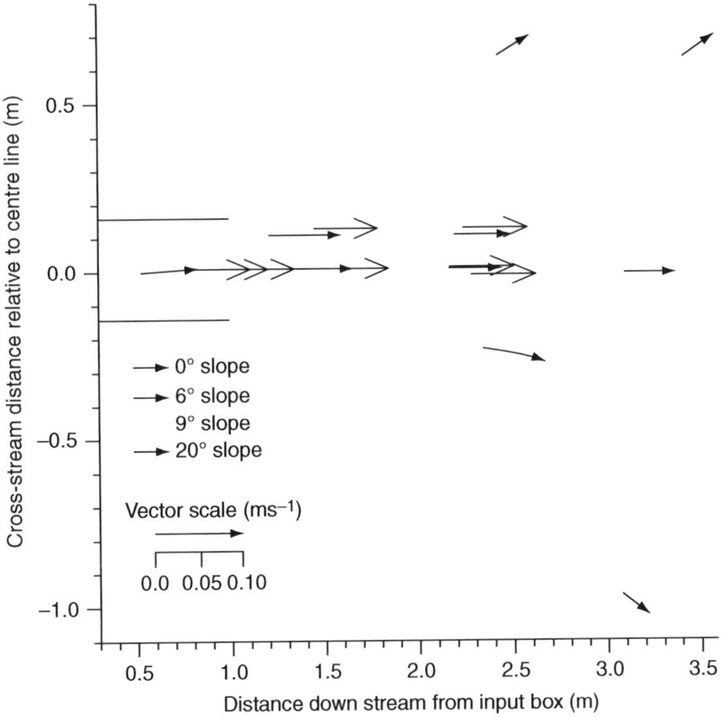


**Fig. 8.19** As for experiments imaged in Fig. 8.18, ADV (acoustic Doppler velocity) data for four slope runs plotted as vectors on a map of the experimental tank floor. The diagram shows the consistency of flow pattern whose directions and magnitude define a marked central core of high velocity with little directional deviation from axial streamwise which is very similar for all runs. The magnitudes of the vectors have a similar pattern of decline with distance from the channel mouth, with those originating from steeper slopes only marginally larger. (After Alexander *et al.*, 2008.)

dissipation, a feature deduced from whole-basin studies of ancient turbidites (see Chapter 21).

## Reflection and refraction of turbidity currents

At positive slope changes, such as when gradients increase around topographic obstacles, flows may partially or completely run up and overshoot, be partially or wholly blocked or be diverted or reflected. The process of run-up and full or partial reflection up ramps (*sloshing*) is particularly interesting (Fig. 8.21). Reflection is accompanied by the transformation of the turbidity flow into a series of translating symmetrical waves which have the properties of *solitary waves* or *bores* (section 5.8). They travel back in the up-source direction, undercutting the slowly moving nether regions of the still-moving forward current. Such *internal bores* have little vorticity and smooth forms. Deposits from the flow reversal process show distinctive current reversals with mud laminae sometimes having time to form in the interval between forward and reversed flow. Flow *over* obstacles may create distinctive *lee waves* whose properties lead to the formation of regular large-scale bedforms in turbidity flow deposits.

## Buoyancy reversal and flow lofting from freshwater turbidity currents in the oceans

This occurs when the bulk density of a freshwater turbidity current is greater than seawater and when deposition of sediment occurs along its flow path. As

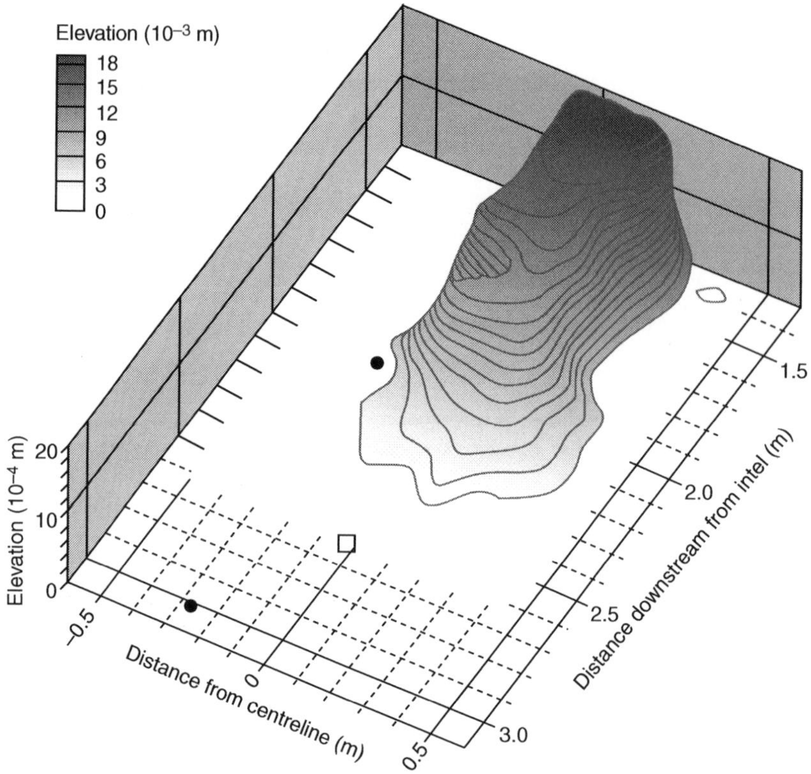


**Fig. 8.20** Three-dimensional contour plot of the deposit from experiments outlined in Fig. 8.19 for the 9° input channel slope. Only the area of the horizontal tank with significant deposition is shown. Note the significant exaggeration of the vertical axis and the finger-like nature of the deposit.

flow bulk density reduces below that of average seawater, 1030–1040 kg/m$^3$, the current loses its forward momentum and starts to rise up through the water column as a freshwater plume, still with a considerable amount of suspended silt- to mud-grade sediment. Experiments show that it does this initially from the top of the current in the upper mixing zone where suspended sediment is progressively lost towards the bed. Such lofting behaviour serves to aerate oceanic water layers and contribute to the suspended sediment-rich layers of the ocean termed *nepheloid layers*. The process may happen on a vast scale adjacent to melting continental ice sheets, particularly during the occurrence of Heinrich events when iceberg discharge rates weere at maximum. Sediment cores in areas like the Labrador Sea reveal deposits of graded muds with ice-rafted debris that may represent the combined products of deposition from quasi-continuous underflow turbidity currents, lofting into the ambient lower ocean and iceberg meltout in the overlying oceanic water column.

## Turbidites: their character and how to spot them in the sedimentary record

Deposits from turbidity currents are called *turbidites*. Individual beds range in thickness from a few millimetres up to 10 m or more; *megaturbidite* is an acceptable term for extraordinarily thick (many metres) examples of regional extent that record exceptional current generation. Turbidites range in grain size from coarse gravel to fine silty muds. Geologically ancient examples are a great challenge to interpret because of the many variables that must be reconstructed or parameterized before a rational choice of explanations is arrived at. Younger turbidites whose parent currents flowed

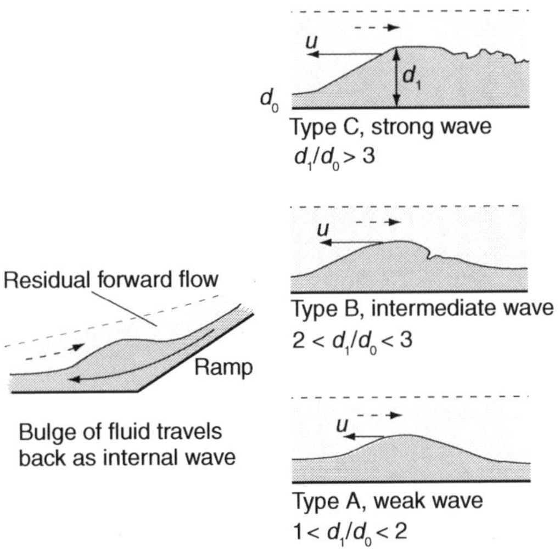


**Fig. 8.21** Density flow reflection. (a) Forward turbidity flow meets opposing topographic ramp slope and reflects back (b)–(d) under residual forward flow fluid as an internal solitary wave. (After Edwards *et al.*, 1994.)

over sediment beds of similar slope and extent to those still present in a given basin are easier to interpret, but their still remain serious problems. This is because the detailed character of turbidites varies according to:

1 (un)steadiness, (non)uniformity of the parent flow as previously discussed with reference to rapidly decelerating sediment-laden experimental flows;
2 mechanism of emplacement, e.g. surge vs. quasi-continuous underflow;
3 entry point and distance from source;
4 position on complex depositional surfaces with slope changes such as submarine fans traversed by channels, levees, crevasses, erosional submarine topography, growth structures and abyssal plains with low-amplitude relief;
5 evolution of grain size in time and space from source to sink.

Unfortunately, discussions of turbidite character have often been dominated by the concept of 'ideal' sequences of sedimentary structures, called *Bouma sequences* after their originator, but modified in great detail by many subsequent workers. The Bouma sequence was once implicitly thought to record unsteady (decelerating) flow, but flow nonuniformity must also be taken into account in turbidite interpretation; the exaggerated role of loss of *competence* by a turbidity current to transport sediment must be accompanied by a consideration of loss of sediment-carrying *capacity*, a fundamental issue termed the *Hiscott problem*. The several types of turbidites summarized in Fig. 8.22 provide very important evidence as to the nature of the turbidity currents themselves; the carrying capacity of flows may be affected by unsteady and/or nonuniform behaviour according to the *Branney–Kneller acceleration matrix* (Chapter 8.4) and also to the rate of flow deceleration.

In the original 'ideal' sequence of structures in turbidite beds (Fig. 8.22), the following divisions A–E occur (listed from base to top); these are still useful as simple descriptors, based on interpretations from experimental studies.

A Massive unit whose lack of structure and general grading is thought to represent very rapid deposition from a dispersed higher-density portion of a current with much-reduced internal turbulence. Transport may be by near-bed laminar shear layers but very rapid aggradation during deposition prevents visible grading or internal structure development. Reverse grading takes place when deceleration rates are low. Frequent dewatering pipes and dish-and-pillar structures attest to frequent fluidization (Chapter 9) during deposition.

B Parallel-laminated unit. This reflects reduced but still high deposition rates from unit A and is either indicative of 'normal' turbulent boundary layer deposition at high bed shear stresses from low-amplitude bedwaves on upper-phase plane beds (Chapter 7) or by lamina-by-lamina deposition from concentrated shear layers that modify the turbulence. More widely spaced and inversely graded stratification (Fig. 8.23) may represent sedimentation of successive granular 'traction carpets' kept aloft by dispersive granular stresses, large-scale variations in turbulent burst/sweep cycles, current surges or possibly bedwaves larger than those normally present on upper plane beds.

C Cross-laminated fine to very fine sand unit often with climbing ripple stratification recording current ripple migration under conditions of high net bed aggradation.

D Silts and interlaminated silts and mud, sometimes with starved ripples and mudfloc/silt ripples. Records turbulence—or Kelvin–Helmholz-modified suspension fallout

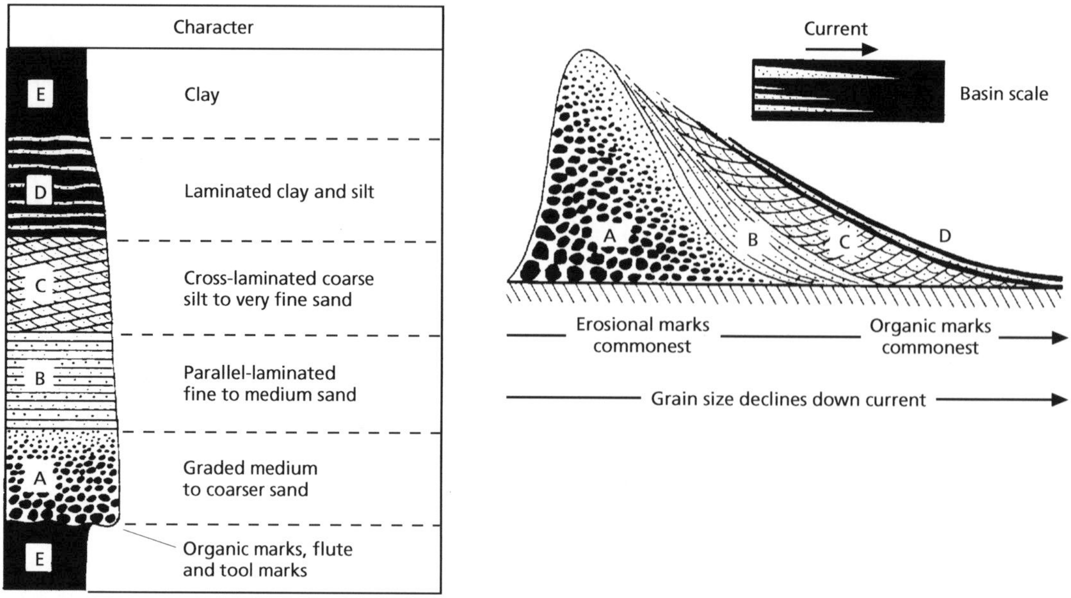


**Fig. 8.22** The traditional view of the depositional sequence for an 'ideal' turbidite deposited by waning flow, and the hypothetical variation of such a turbidite sandstone downcurrent. (After Allen, 1985; following Walker as well as Bouma.)

E Homogeneous, structureless mud. Final suspension fallout but also possible production of a stratified fluid mud layer.

The thickness and occurrence of these various 'ideal' subdivisions may vary downcurrent for the many reasons discussed above; sometimes there is a trend for fining and progressive loss of the A–D divisions until only mud is deposited—this is the classic proximal–distal turbidite model. The situation is particularly complicated for the deposits of turbidity currents suspected to have been generated or strongly modified by large-amplitude storm waves. Studies of Pleistocene and modern turbidites reveal that *mud turbidites* are considerably commoner, thicker and more widespread than was previously thought. These turbidity current deposits were probably due to ponding, reflection and lofting processes. The great mobility of mud-grade currents arises from a combination of drag reduction, hindered settling and stratified flow (Chapter 6) development.

Concerning shelf turbidites, turbidity currents generated by nearshore gradient currents during storm set-up must flow through a field of oscillatory flow caused by wind-generated waves. Evidence from ancient sandstone event beds in Carboniferous sediments in Colorado, USA, outboard of wave-dominated units show reverse-to-normal grading and sequences of sedimentary structures that indicate deposition from waxing to waning hyperpycnal flows. These are thought to have been generated directly from highly concentrated river plumes, which accelerated and decelerated in response to rising and falling tidal discharge.

## 8.5 Turbidite evidence for downslope transformation from turbidity to debris flows

A major discovery in recent years arose through the ability to take deep, closely spaced and precisely located ocean cores through turbidite layers along accurately levelled seabed and to date and correlate depositional events within them. Following the discovery of long runout of sediment slides and avalanches from continental margins it has become apparent that not only did some debris flows never transform, but also transformations have occurred

Fig. 8.23 Examples of enigmatic inversely graded structures in fine- to coarse-grained sandstones from turbidites of Ordovician age, Cloridorme Formation, Quebec, Canada. Centimetric scale. (After Hiscott, 1994. Photo courtesy of R. Hiscott.)

from debris to turbidity currents and vice versa. Such inferences are based on the occurrence downslope of *debrites* within or laterally equivalent to well-correlated turbidites, again revealed by painstaking fieldwork. As noted above, turbidites are generally relatively well-sorted and may show grading and bedload current structures. They also tend to gradually reduce in thickness when traced bed-for-bed over tens to hundreds of kilometers, a feature explicable by the gradual flow-wise decay of a turbulent flow. Debrites on the other hand are relatively poorly sorted, show little grading and few if any current structures indicative of turbulent boundary layer development. In addition the individual beds tend to thin or end abruptly as befits the onset of relatively sudden frictional stick-slip due to applied stress in the debris flow falling below yield stress. These observations require a mostly turbulent flow to transform by bulking-up into a more concentrated, mostly laminar debris flow. The evidence indicates that they do this when downflow slopes drastically reduce, as in the central Atlantic (Figs 8.24–8.26) where there is a fivefold decrease from the Agadir Canyon (0.05°) to the Agadir Basin (0.01°). Such slope reduction presumably induced mass depositional fallout from a current previously in suspension and bedload equilibrium, a state of steady but nonuniform depletive flow.Sedimentation from the outer Gaussian boundary layer into the zone of reduced velocity gradient and former log-layer would lead to hindered settling and transformation to a laminar, high viscosity and hyperconcentrated flow—mass flocculation perhaps encouraged by reduced turbulent shear stresses. It is not impossible that the whole process followed upstream energy degradation by hydraulic jumping, triggered in the area of slope reduction. After the jump zone, flow recovery and tranquil flow might have allowed reconstitution of laminar flow and bottom-up deposition along a sedimenting front until the debris flow 'froze' as autocompaction exceeded local debris-flow yield strength. Meanwhile the cloud of turbulent flow passed overhead and deposition of overlying and downflow turbidites begins to bury the debrite layer.

## Further reading

*General*

Simpson (1997) is an excellent general introduction to the experimental fluid dynamics of gravity currents. Though we await a good introductory text on sediment-transporting gravity flows, the Special Publication edited by McCaffrey *et al.* (2001) has many high-quality papers of general and particular interest to grain, debris and turbidity flows. Those in the volume edited by Hutter & Pudasaini (2005) are more fundamental in their treatment of the theoretical physics. Mulder & Alexander (2001) is a thoughtful qualitative review of the wide spectrum of subaqueous sediment density flows and their deposits.

*Specific*

Grain stresses in static aggregates are discussed by Rowe (1962), Jaeger *et al.* (1996) and Fineberg (1996). Häner & Spencer (1998) provide a nice semi-popular article on 'rubbing & scrubbing' friction.

The physics of grain flow and granular materials is discussed by Bagnold (1954a, 1966a), Barker &

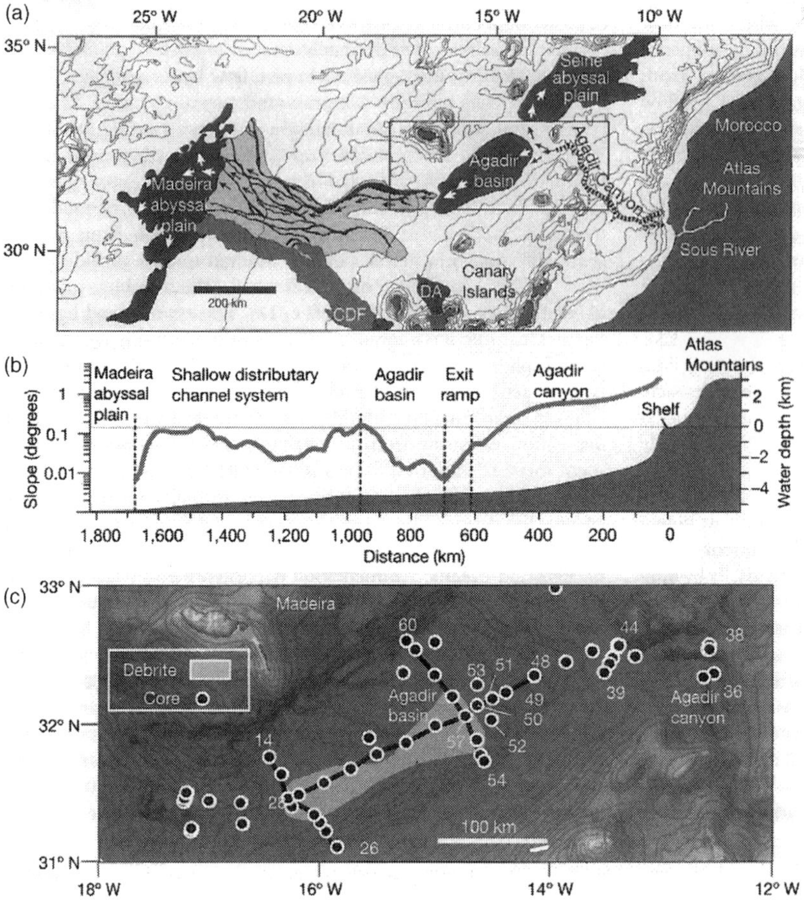


**Fig. 8.24** Agadir turbidity current transformation. (a) Location map and generalized bathymetry for northwest African continental margin. The Canary debris flow (CDF, shaded brown) and debris avalanches (DA) are also shown. Arrows indicate the flow paths of events that deposited bed 5. (b) Change in seafloor gradient along flowpath of deposit 5. (c) Location of cores and debrite in Agadir basin (location boxed in (a)). (After Talling *et al.*, 2007.)

Grimson (1990), Maddox (1990), Jaeger *et al.* (1996), Fineberg (1996), Makse *et al.* (1997, 1998) and Straub (1997). More recent accounts of the initial Bagnoldian advance are by Straub (2001), Iverson & Denliger (2001) and McClung (2001). A fundamental physical review from the constitutive equation point-of-view is by Forterre & Pouliquen (2008).

Snow avalanches are discussed by Hopfinger (1983) and McClung (2001), with the spectacular ping-pong experiments in McElwaine & Nishimura (2001). See also avalanche web sites like Swiss Avalanche Institute, www.slf.ch or www.avalanche.org.

Megabreccias are discussed by Burchfiel (1966), Yarnold (1993) and Friedman (1998), and sturzstroms by Hsu (1975).

Dynamics of runout in grain flows are considered by Kent (1966), Hsu (1975), Melosh (1979, 1987), Savage (1979), Yarnold (1993) and Friedman (1998). Computer simuations of grain flow friction are in Campbell (1989), Cleary & Campbell (1993) and Campbell *et al.* (1995). Avalanche velocimetry studies are by Tischer *et al.* (2001).

Grain sorting during granular flow (the 'muesli problem') is discussed by Rosato *et al.* (1987), Middleton (1970, kinetic filterer), Sallenger (1979,

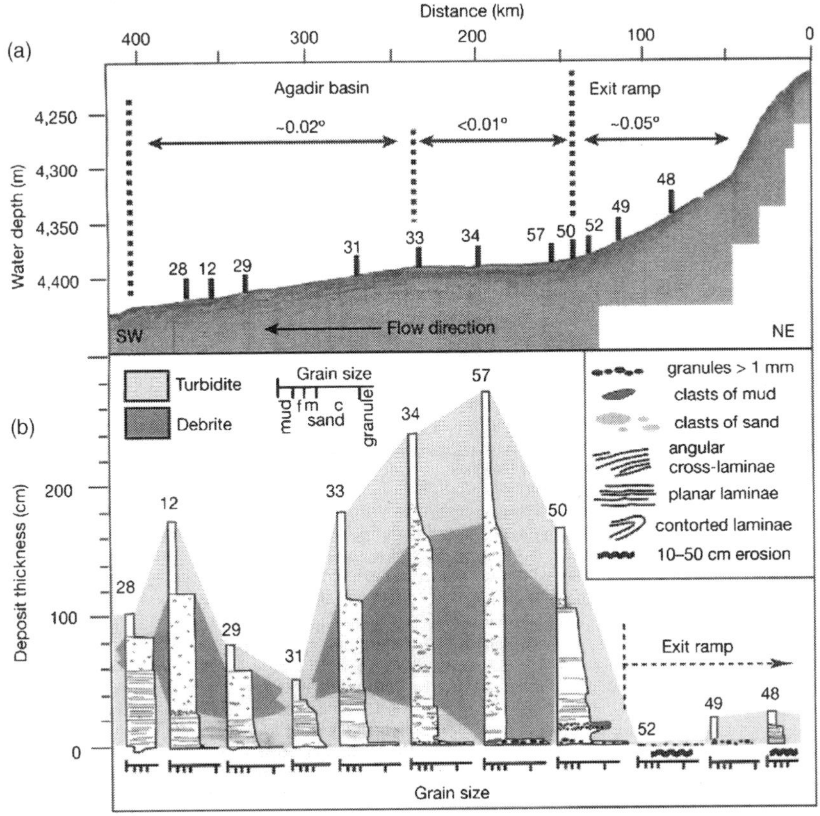


Fig. 8.25 Agadir turbidity current transformation. (a) Changes in seafloor gradient. (b) and (c) The shape and internal structures of bed 5. Numbers are core locations of Fig. 8.25c. (After Talling *et al.*, 2007.)

dynamic filterer), Jaeger *et al.* (1996), Knight *et al.* (1993), Umbanhowar *et al.*, (1996), Makse *et al.* (1997, 1998), Pouliquen *et al.* (1997) and in a magisterial way by Legros (2002). Sorting concepts from the above are applied to lee-side avalanching of subaqueous dunes and deltas by Kleinhans (2005). Runout transformations from avalanching grain flows to turbidity flows and their deposits are considered by Postma & Roep (1985) and Falk & Dorsey (1998).

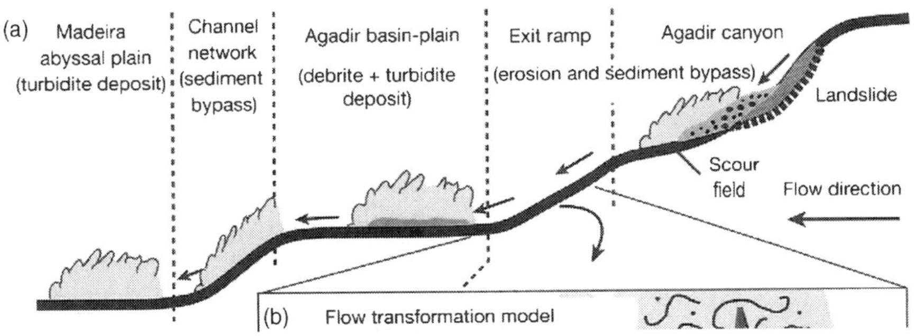


Fig. 8.26 Agadir turbidity current transformation. (a) Evolution of the whole flow event that deposited bed 5. (b, c) Two alternative mechanisms for generating the debrite. (After Talling *et al.*, 2007.)

Debris flows are discussed by Johnson (1970) in a pioneering concept book on physical processes that still reads like a classic. Papers and reviews are by Plafker & Ericksen (1978), Pierson (1981), Pierson & Scott (1985), Herd, (1986), von Heune *et al.* (1989), (Best, 1992b), Meyer *et al.* (1992), Cronin *et al.*, (1996), Iverson (1997), Iverson *et al.* (1997), Vallance & Scott (1997), Whipple (1997) and Major (1997, 2000). The important role of enhanced pore pressure in debris flow is highlighted in the insightful papers of Major & Iverson (1999), Major (2000) and Iverson & Vallance (2001). The role of clay-colloid fines in aiding flow mobility is partly addressed by Bardou *et al.* (2007). Basal aquaplaning under debris flow heads is investigated by Mohrig *et al.* (1998). Cryptofabrics are analysed by Naruse & Masuda (2006). More recent 'newsy' examples of flow destructive capabilities in subaerial and subaqueous events are by Larsen *et al.* (2001), Sidle & Chigira (2004), Lagmay *et al.* (2006) and Walsh *et al.* (2006). Major runout of submarine debrites and inferences on their flow evolution are beautifully documented by Gee *et al.* (1999, 2001) and BIG'95 is modelled by Lastras *et al.* (2005).

Density/turbidity flows are discussed generally by Simpson (1987), Fernando (1991), Monaghan (2007) and, with respect to oceanic mixing, Ivey *et al.* (2008).

Sea-level controls on the frequency and magnitude of turbidity currents are by Brunner *et al.* (1999), Rothwell *et al.* (2000), Reeder *et al.* (2000) and Normark *et al.* (2006) The flow transformation of initial sediment slides to liquefied debris flows and finally to turbidity currents is the subject of classic experiments by Hampton (1972) and later studies of variable concentration and strength of debris flows by Felix & Peakall (2006); an interesting application is to the Grand Banks turbidity current (Piper *et al.*, 1999b) and to other lowstand shelf-margin deposits by Jenner *et al.* (2007). A thoughtful and accessible review and develoment of ideas on flow transformations across hydraulic jumps is by Waltham (2004). Turbidity flows are studied by Middleton (1966a–c), Hampton (1972), Kersey & Hsu (1976), Britter & Linden (1980), Lowe (1982), Piper *et al.* (1985, 1999b), Prior *et al.* (1987), Prior & Bornhold (1988, 1990), Stow & Wetzel (1990), Rothwell *et al.* (1992, 1998), Sparks *et al.* (1993), Mulder & Syvitski (1995) and Mulder *et al.* (1997). Aspects of the velocity structure and turbulence characterstics of turbidity flows are by Kneller *et al.* (1999) with a long useful review by Kneller & Buckee (2000). Ancient megaturbidites are discussed by Ricci Lucchi & Valmori (1980), Pickering & Hiscott (1985, 1995), Awaallah & Hiscott (2004) and Amy & Talling (2006). Quaternary examples taken in cores and correlated over wide areas are by Rothwell *et al.* (2000) and Reeder *et al.* (2000).

Modelling and simulations can be found in Felix (2002). Underflow generation (hyperpycnal conditions) is discussed by Mulder & Syvitski (1995), Felix *et al.* (2006) and ingeniously, invoking finger convection from particulate overflows (hypopycnal conditions), by Parsons *et al.* (2001). Modelling of direct turbid underflows of river water is by Kassem & Imram (2001) and observations of underflow pulsing by Best *et al.* (2005).

The effects of obstacles on turbidity currents is discussed by van Andel & Komar (1969), Beghin *et al.* (1981), Pickering & Hiscott (1985), Pantin & Leeder (1987), Muck & Underwood (1990) Kneller *et al.* (1991), Pickering *et al.* (1992), Lucchi & Camerlenghi (1993), Weaver & Thomson (1993), Edwards *et al.* (1994), Haughton (1994), Sinclair (1994), Alexander & Morris (1994), Dade *et al.* (1994) and Kneller & McCaffrey (1999).

Autosuspension is studied by Bagnold (1962), Pantin (1979), Bouma (1962) and Parker *et al.* (1986), and enhanced suspension of fine-grained particles by Gladstone *et al.* (1998).

Turbidite sequences are studuied by Bouma (1962), Walker (1965), Stow & Shanmugam (1980), Porebski *et al.* (1991), Kneller & Branney (1995) and Kneller (1996). An important experimental study by Sumner *et al.* (2008) of decelerating sediment-charged channel flow has much of relevance to turbidite structural development. Contrasting pithy reviews of turbidite interpretations and sediment gravity currents in general are by Shanmugam (2000) and Mulder & Alexander (2001) respectively. The Hiscott competence problem is discussed by Hiscott (1994) and Leeder *et al.* (2005). Deposits thought to have been influenced by storm wave action are nicely discussed by Lamb *et al.* (2008).

Unsteady whole flow-field behaviour of underflows is documented by Best *et al.* (2005), whilst laboratory sustained turbidity currents forming elongate sediment ridges are discussed by Alexander *et al.* (2007).

The lofting/reversed buoyancy story began with laboratory experiments; the key paper is Sparks *et al.* (1993). A plausible story of massive sediment deposition due to lofting flows is by Hesse *et al.* (2004).

The story of downflow transformation from turbidity to debris flow in young oceanic deposit cores may be followed in McCave & Jones (1988) and Talling *et al.* (2007a) and for older turbidite/debrites in the rock record by McCaffrey & Kneller (2001), Haughton *et al.* (2003), Talling *et al.* (2004, 2007a,b), Talling (2007), Amy *et al.* (2005), Amy & Talling (2006) and Ito (2008). The latter papers contain meticulous field observations on streamwise turbidite thickness and grain-size distributions which require serious explanation. Experiments that justify the Talling–Amy hypothesis of debritization of turbidity currents are in Sumner *et al.* (2009).

# Chapter 9

# LIQUEFACTION, FLUIDIZATION AND SLIDING SEDIMENT DEFORMATION

*The mud, thick as molasses, dripped back into the water with a slow flab flab, and the pole sucked lusciously. It was very beautiful, but it all stank so: yet to his surprise he found he rather enjoyed the rotting smells of the estuary.*

Lawrence Durrell, Mountolive p. 12, (part of The Alexandria Quartet), Faber and Faber, 1958.

## 9.1 Liquefaction

We have seen (section 8.2) that a mass of deposited grains is stable if its slope is less than some critical angle $\phi_{max}$. This slope is much reduced when small amounts (0.5%) of mud are added, flow occuring when the mass exceeds the Bingham yield stress $\tau_B$. It is also drastically reduced when elevated pore pressures develop in the mass of sediment. Concerning the general mechanics of *slope failure*, the applied shear stress at failure, $\tau_f$, on any slope must exceed the shear strength of the material in question. For shear failure to occur, the *Mohr–Coulomb criterion* must be equalled or exceeded, $\tau_f = c' + (\sigma - p)\tan\phi$, where $c'$ is cohesive strength, $\phi$ is friction angle and $p$ is pore pressure. The interesting term $(\sigma - p)$ is the Terzaghi *effective normal stress*, given by the *bulk normal stress*, $\sigma$, minus the pore pressure. The important point here is that the Terzaghi stress is much lowered if pore-pressure is increased by processes such as fabric collapse, vibrational liquefaction and subsurface pore fluid seepage.

In all saturated deposits *liquefaction* may easily occur, even on zero slope. Liquefaction is when a stationary saturated granular mass, either at or below the surface, is changed *in situ* to a fluid-like state. Examples attesting to this change of state are common after large earthquakes; here are the words of an eyewitness observer to the great New Madrid earthquake of 1811–12 in the lower Mississippi Valley:

> *'Great amounts of liquid spurted into the air, it rushed out in all quarters. . .ejected to the height from ten to fifteen feet, and in a black shower, mixed with sand. . .The whole surface of the country remained covered with holes, which resembled so many craters of volcanoes. . .'*

*Sedimentology and Sedimentary Basins: From Turbulence to Tectonics*, 2nd edition. © Mike Leeder.
Published 2011 by Blackwell Publishing Ltd.

**Fig. 9.1** Visible signs of subsurface liquefaction and piping flow—line of sand volcanoes (cones 1–2 m diameter) erupted from a liquefied layer at depth into a field during the 1980 Imperial Valley, California earthquake. (US Geological Survey).

Although earthquakes undoubtedly impose an effective trigger mechanism for liquefaction (Fig. 9.1), they are by no means the only one.

There are two main methods of liquefaction.

1 Temporary collapse may occur due to grain-to-grain contact sliding that reduces the volume of loosely packed saturated sands. This is called *contractant* or *reverse dilatant* behaviour and causes an increase in pore fluid pressure. When sudden shock occurs, the grains are shaken apart and momentarily suspended in their own porewater, causing the high pore pressures and reduction or elimination of intergranular friction and hence complete loss of strength. In effect the granular solid is transformed into a Newtonian fluid. The energy for granular sliding or disaggregation is provided by high accelerations experienced during (i) cyclic shock caused by earthquake ground motions, (ii) pressure changes due to the passage of large storm waves, (iii) sudden arrival or departure of a tidal flow, mass flow or turbidity current. 'Vibracoring' devices owe their efficiency to the liquefaction of a narrow zone of sediment around the tip of the core barrel provided by an industrial vibrating tool. Laboratory simulations of liquefaction due to earthquake-induced accelerations use shaking tables.

   Should liquified sediment be resting on a slope, however slight, then *liquefied slope failure* along a discrete basal and curvilinear shear plane will occur, the liquefied flow maybe transforming eventually into a debris or turbidity flow. On a horizontal bed, liquefaction cannot persist and gravity causes the sand grains to settle back once more into grain contact in a new, tighter packing. This resettlement causes net upward displacement of pore fluid of volume proportional to the difference in pre- and post-liquefaction porosity.

2 Upward displacement of fluid may be sufficient in itself to cause overlying grains to remain in suspension by a process known as *fluidization* or *seepage*

*liquefaction*. Here adjacent sediment grains are kept apart by an upward force due to the movement of pore fluid along a pressure gradient. This gradient is usually vertical, with $p$ decreasing upwards. The fluidized suspension may be *en masse* or just restricted to selected pipe-like conduits (as described by the acute observer of the New Madrid earthquake), often arranged in circular or polygonal forms. Readers may be familiar with industrial and technological uses of fluidization, notably the passage of hot gases through powdered coal beds to optimize burn efficiency. In this case gas fluidization is often accompanied by bubble formation, which is uncommon in liquid fluidization. Fluidization in its most simple sense is analogous to turbulent sediment suspension in that it requires the velocity of moving pore fluid, $u_f$, to exceed the grain fall velocity, $U_S$. Moreover, in order for *en masse* fluidization to occur, the upward-moving fluid must exert a normal stress $P_f$ equal to the immersed weight of any overlying sediment. This stress is, $P_f = \Delta\rho gCh$, where $\Delta\rho$ is effective grain density, $g$ is gravity, $h$ is layer thickness and $C$ is fractional grain concentration. In the *piping flow* noted above, upward-moving fluid is concentrated into discrete jets, which may separate areas of adjacent unfluidized sediment.

## 9.2 Sedimentary structures formed by and during liquefaction

Piping flow from a consolidating fluidized bed causes formation of *dewatering pipes* or *sediment dykes* (Fig. 9.2). These may be arranged in crudely polygonal arrays and range from a few millimetres to several metres in height. Laminations in the liquefied bed serve to outline the shearing effects escaping water has on the liquefied sand. Permeability barriers may be violently punctured during the process (Fig. 9.3). Grains suspended by the escaping water may be transported up to the bed surface to form *sand volcanoes* with diameters of up to 1 m and cone angles of up to 16° (Figs 9.1, 9.4 & 9.5). These are preserved only in quiet subaqueous environments. They are best known in the rock record on the top surfaces of turbidite beds and subaqueous slumps. Rare elongate examples provide clear evidence for residual current motion that served to direct the outgoing ejaculations of sand slurry (Fig. 9.5).

*Dish-and-pillar* structures (Fig. 9.6) are also due to water escape mechanisms. Dishes are thin, subhorizontal, flat to concave-upwards clayey laminations in silt and sand units. Pillars are vertical to near-vertical cross-cutting columns and sheets of structureless sands. Both structures are clearly post-depositional since they cut primary sedimentary structures. During dewatering less-permeable horizons act as partial barriers to upward flow and they force the flow to become horizontal until upward escape is possible. As the water seeps outwards, fine sediment grains such as clay flakes are filtered out and concentrated in pore spaces. The resulting clay-enriched laminae form the dishes, which may later be deformed at their margins by upward flow. *Pillars* form during more forceful water escape—they are essentially tiny dewatering pipes.

Should liquefaction affect actively moving subaqueous dune bedforms, then an interesting arrangement presents itself; the flowing water now exerts a shear stress upon a liquefied bed. Such a simple shear system will cause progressively less shear to be transmitted to the deeper parts of the liquefied dune. Also, since the liquefied bed solidifies from the base upwards, there will be progressively less time for shear to operate in these lower areas. The net result is to cause a cross-stratification plane within the dune to become sheared over into a parabolic curve, defining the structure known as *overturned cross-stratification*, sometimes termed *omelette structure*. Liquefaction also induces gravitational collapse of any relief features present on depositional surfaces. Subaqueous bedforms are particularly vulnerable, collapse of dunes and ripples causing formation of generally slumped and deformed cross-strata.

Further structures arise from gravitational instabilities during liquefaction. Generally, when a light fluid lies beneath a denser one, gravitational instability causes the lighter fluid to intrude into the denser in an effort to overcome the gravitational instability. Such *Rayleigh–Taylor instability* gives rise to pipe- or ridge-like intrusions termed *diapirs*. The two main arrangements that give rise to diapirism are mud–sand and salt–sediment interlayers. In the former case, large-scale differential loading such as that experienced at the front of prograding deltas gives rise to diapirs known as *clay lumps*, which may reach the depositional surface from depths of up to 150 m (Fig. 9.7). The density inversion needed to form clay

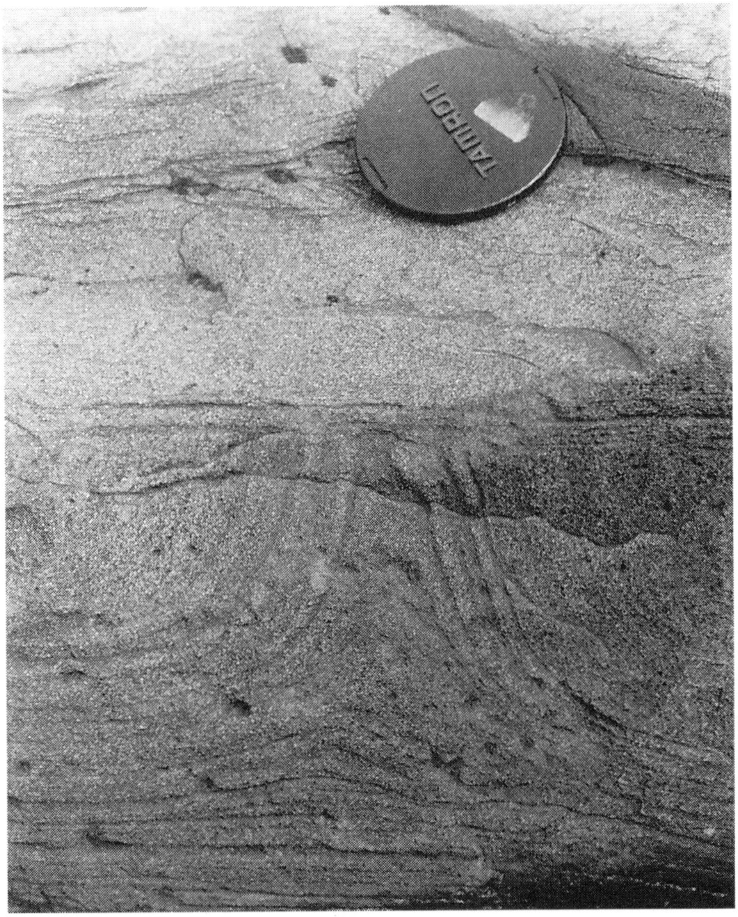


**Fig. 9.2** Vertical section through fluidization pipe in fine sandstone. The upturned laminae were originally horizontal and have been dragged-up by shear accompanying water expulsion during a piping flow event. The deformed laminae may mark the site of a former sand volcano but this has been removed by erosion since the pipe is succeeded by horizontal laminae as deposition resumed. Lens cap rests on the later laminae. Upper Carboniferous of Co. Clare, Ireland.

diapirs disappears at depth because of compaction-induced dewatering of the clay sediments. Diapiric injection of liquefied mud into sand beds or collapse of lava into saturated substrates gives rise to narrow *flame structures* adjacent to broader downbulges, defining *load casts* (Fig. 9.8). The shape of the structures and their wavelengths are controlled by density, viscosity and layer-thickness parameters. In some instances sand pillows sink into liquefied mud to form detached *sandstone balls* with characteristic internal deformed laminations. Rarely, the lamination within detached sandstone balls records the advance of sinking current ripples over a liquefied mud patch.

*Convolute laminations* are common structures comprising muddy, fine-grained sands and silts. They exhibit narrow vertical upturned laminae, often truncated at the top surface, separated by broader synclinal downfolds with wavelengths of a few centimetres or decimetres. They are most commonly developed in sediments that have been rapidly deposited, such as climbing ripple cross-laminations in distal turbidite and river flood facies. There is some evidence that the convolutions develop because of fabric readjustment following the gravitational collapse and flattening of a rippled depositional surface during liquefaction. Associated *water escape pipes* are common.

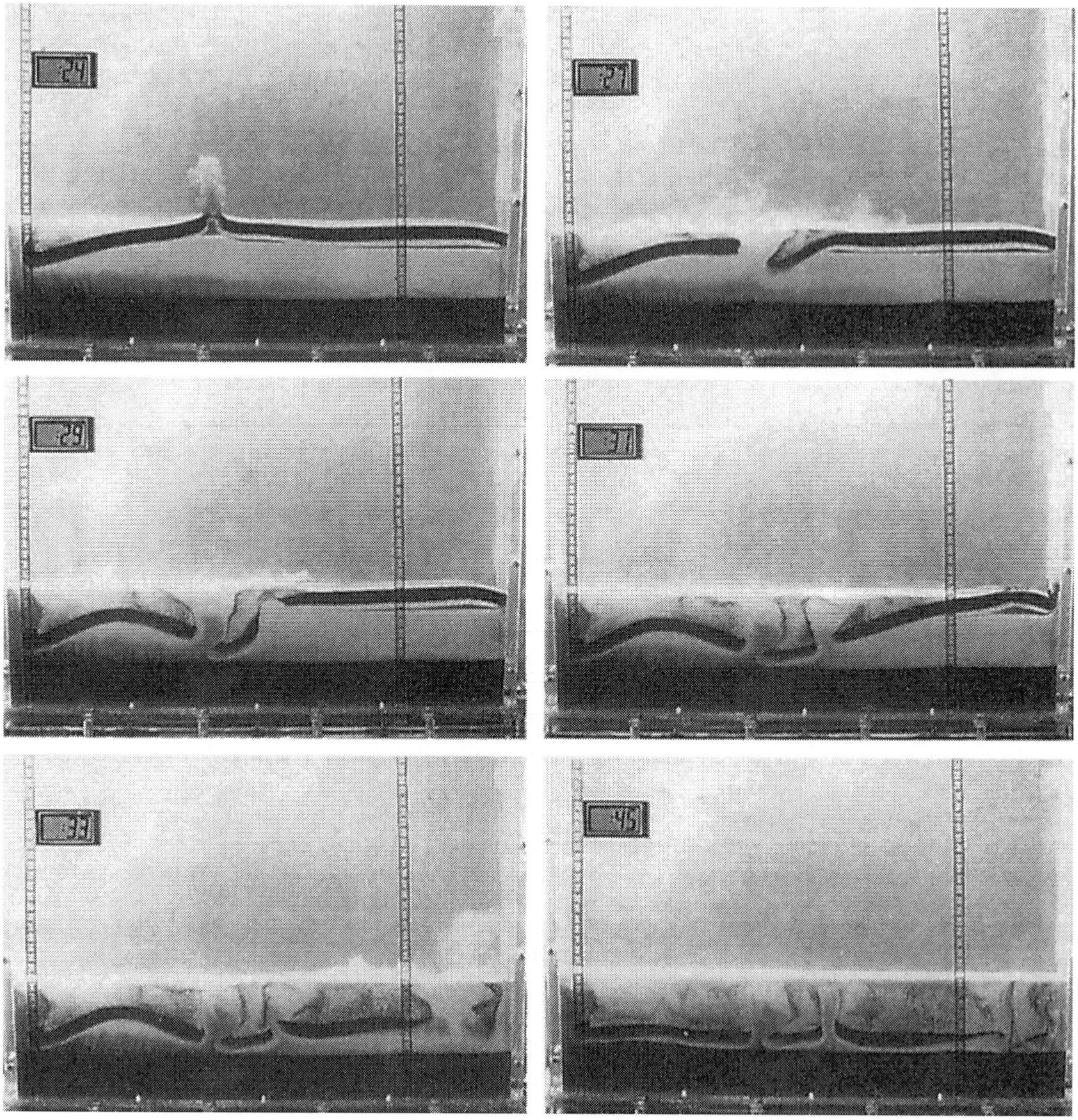

**Fig. 9.3** Photographic sequence (numbers in top left of each image are time in seconds) to show 180–250 mm silicon carbide grains (dark coloured layer) overlying 74–125 mm clear glass spheres through which a flow of water is passed via a perforated baseplate. Sequence shows blow-out of a void produced by fluidization of the lower layer followed by piping flow and collapse of the upper layer. (After Nichols *et al.*, 1994. Photo courtesy of R. Nichols.)

We only briefly consider salt diapirs here. The rate of flow of rock salt by *creep* increases with burial until a critical point is reached at which the upper surface of the salt bed deforms and amplifies by the process known as *halokinesis*. The production of one rising diapir or salt pillow often triggers an adjacent structure, and so on. One result is a fairly regular separation distance between salt diapirs in particular areas. Examples are known of salt diapirs that have risen vertically a distance of 5–6 km, sometimes reaching the surface to flow as salt glaciers, as in the Cambrian salt diapirs of southern Iran and the Persian Gulf. Spectacular diapirs of Permian salt in the Gulf of Mexico and the North Sea are noteworthy because

**Fig. 9.4** A fossil sand volcano preserved on the top of a sediment slump. Cone has diameter ca 3 m; author and Storm for scale) with a marked central vent and parasitic flank cone partially visible to left. Upper Carboniferous, Co Clare, Ireland. (Photo courtesey of Jim Best.)

of their deformation of the contemporary seabed. This has interesting consequences for the routing of turbidity currents and accumulation of slope turbidites.

## 9.3 Submarine landslides, growth faults and slumps

Landslides are mass movements of sediment above a plane of shear failure. They are common in both subaerial and submarine situations; we shall concentrate on the latter here. It is useful to distinguish between end-members; a *slide* is still in some sort of stratigraphic order and close to its initial failure plane whilst a *slump*, although occurring above basal/lateral shear planes and containing recognizable original stratification, is more highly deformed internally by viscous shear and probably further travelled. Slide blocks initially shear in brittle fashion close to their headwall failure zones and large-scale striations or grooves are recorded on these shear planes from detailed sonar records in several modern examples. Further downflow the slide develops an internal boundary layer and a plastic/fluid no-slip condition prevents brittle failure features, encouraging longer runout. Gently sloping depositional surfaces in cohesive mud deposits are particularly prone to rotational failure; the slip surfaces define *listric faults* that approximate to arcs (Fig. 9.9) in plan view and with a characteristic downward-curving aspect in vertical section. These may occur on a gigantic scale, as in submarine (particularly volcanic) escarpments, and may be triggered by major earthquake shocks. Catastrophic continental margin collapses and slides are believed to have occurred over deeply buried planes of weaknesses (*décollements*) located above *gas hydrate* horizons. On a smaller scale any muddy intertidal point bar and cutbank surfaces show such rotational slides to perfection. In such environments, failure of very gently sloping muddy tidal flat surfaces may even occur if the receding tide can reduce the effective strength by increasing the pore pressure. More important volumetrically are the rotational slides that occur offshore from major deltas. These slides, moving slowly along the fault planes, develop as deltaic clastic wedges prograde out over muddy delta-front deposits in which overpressuring and reduction of effective stress can occur. The active listric fault is termed a *growth fault* because beds on the downthrow sides

Fig. 9.5 Exquisite small, elongate, asymmetric sand volcanoes on the top of a fine-sand turbidite. The forms indicate that after rapid deposition of the turbidite a fluidization event allowed small-scale piping flow upward to the depositional surface whilst there was still some forward flow of the turbidity current that caused the asymmetry. Very slow settling of silt and mud in the distal remants of the current, or some other oceanographic current, then buried and preserved the forms for posterity. By Kirkandrews, Galloway, Scotland.

of such faults are thickened where greater sediment deposition occurs in response to ponding. Associated with listric faults are *roll-over anticlines* which are are important hydrocarbon traps in delta slope settings.

Gently dipping rotational slides on continental margins and delta fronts give rise to large-volume sediment slumps. Some slumps are clearly triggered by cyclic earthquake shocks acting upon clay sediments of high water content in which porewaters cannot easily escape (undrained condition). The resulting bulbous-nosed slumps with pull-apart structures and imbrications may slide on their basal fault planes (Fig. 9.10) and run-out for large distances on gentle slopes. Many slumps do not mix into debris flows or disintegrate into turbidity currents. Instead, they come to rest at the base of a slope where they have high preservation potential. As increasing internal friction causes slumps to reach the limit of their ability to slide, they suffer deceleration and *frontal arrest*, with a postulated compressional shock wave running back into the still-advancing flow tail that is thought to cause folding-deformation, liquefaction and pillar fluidisation of both substrate and slump mass (Fig. 9.11). Moving slumps may quickly give rise to true debris flows and, ultimately, turbidity flows if the ambient waters can be mixed into the slump

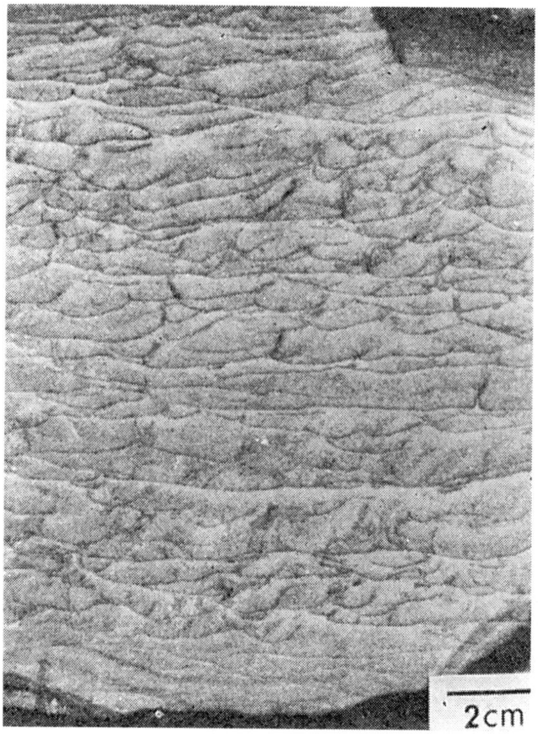


Fig. 9.6 Dish and pillar structures defined by clay-rich laminae. Jack Fork Group, Oklahoma, USA. (After Lowe & Lopiccolo, 1974.)

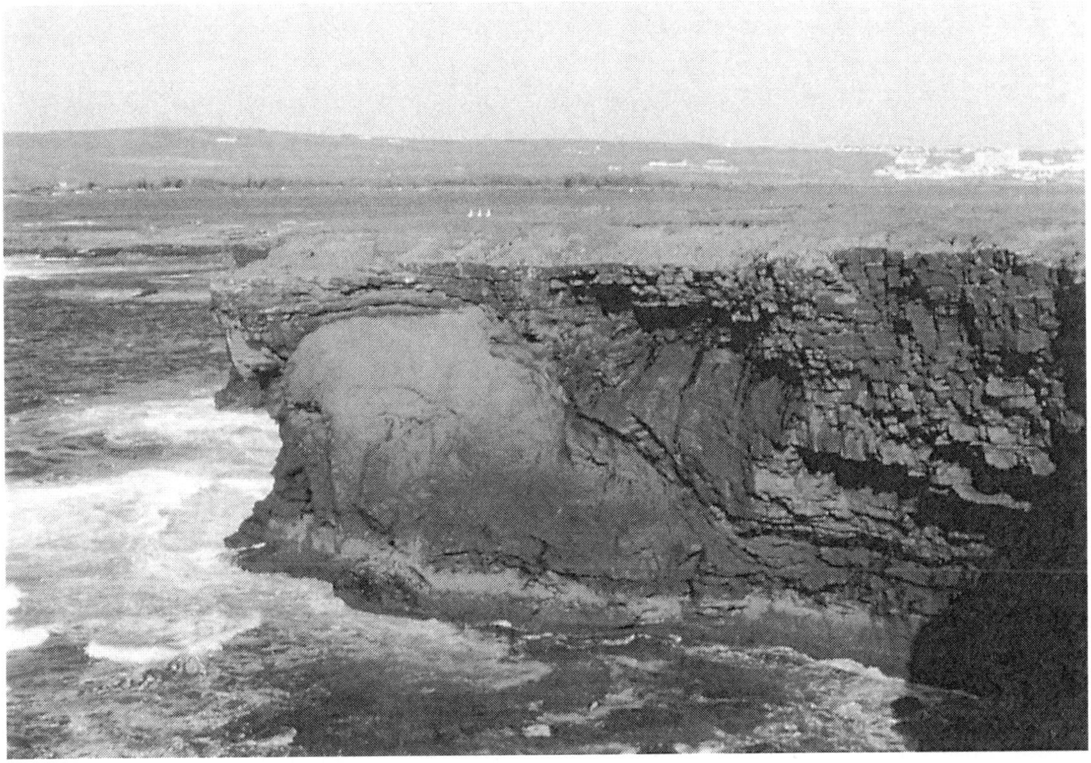

Fig. 9.7 Impressive exposure of a ca 35 m high mudlump intruding prodelta and delta-front sediments. Distant figures in the Irish mist are for scale. Upper Carboniferous, Co Clare, Ireland.

sediment. How exactly slump transformations occur, as with the other gravity flow transformations we have studied in Part 3, is poorly understood. Sonar images of modern slumps suggest slump disaggregation and breakup is important and very rapid; study of arrested slump masses show that slump flow unsteadiness and nonuniformity seem to control density evolution by fluidization and mixing causing transformation to a three-phase flow; the parent slump, and sibling debrite and turbidite deposits (Fig. 9.12).

## 9.4 Desiccation and synaeresis shrinkage structures

It is a matter of common observation that exposure of wet cohesive sediment to the atmosphere causes the formation of polygonal *desiccation cracks* as porewater in the sediment dries out and the sediment volume contracts. Deep cracks often show the *plumose markings* seen on rock joint surfaces that also form due to elastic material rupture, but of course without volume change. Cracking propagates upwards and outwards from the base of the mud layer towards the free surface in the desiccating mud. The downward-tapering cracks, usually preserved in the rock record as casts on sandstone bases, exist on a variety of scales; the thicker the desiccated layer, the deeper and wider the crack systems. Desiccation cracks are rectangular on sloping surfaces such as exposed lake margins. Desiccation of thin mud drapes or layers may produce *mud curls*. These have a low preservation potential, tending to become reworked to mudflake intraclasts by subsequent water current action. Rarely, wind-blown sand may preserve mud curls *in situ*.

Shrinkage cracks may also form under water following porewater expulsion by chemically induced synaeresis. This causes clay aggregate contraction and volume decrease. The subaqueous shrinkage cracks thus produced may be single, elongate 'eye-shaped'

**Fig. 9.8** Flame structures of upward-intruded river floodplain mudrock and accompanying convex-downwards load casts. The latter are in highly deformed, what was originally a heterolithic alternation of sandstone and mudrock. It is possible that this example originated as a seismite, triggered by a large earthquake. Camp Rice Formation, Plio-Pleistocene of southern Rio Grande rift, Rincon, nr Hatch, New Mexico. Greg Mack for scale.

**Fig. 9.9** Brace of listric growth faults in delta-front sediments; note marked thickening of the lighter sandstone beds in the fault hangingwalls. Breathitt Group, Kentucky, USA. Large tree partially obscuring scene is about 10 m high.

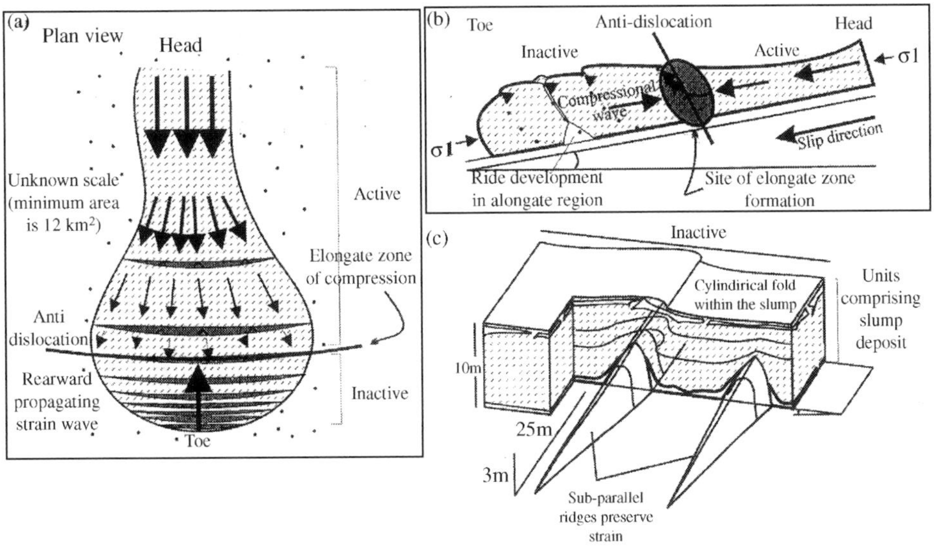


**Fig. 9.10** (a) Plan view of schematic slump unit undergoing frontal arrest. (b) Cross-section of (a). (c) The formation of underlying ridge sand bodies formed by the deformation fronts shown in (a) and (b). (After Strachan, 2002.)

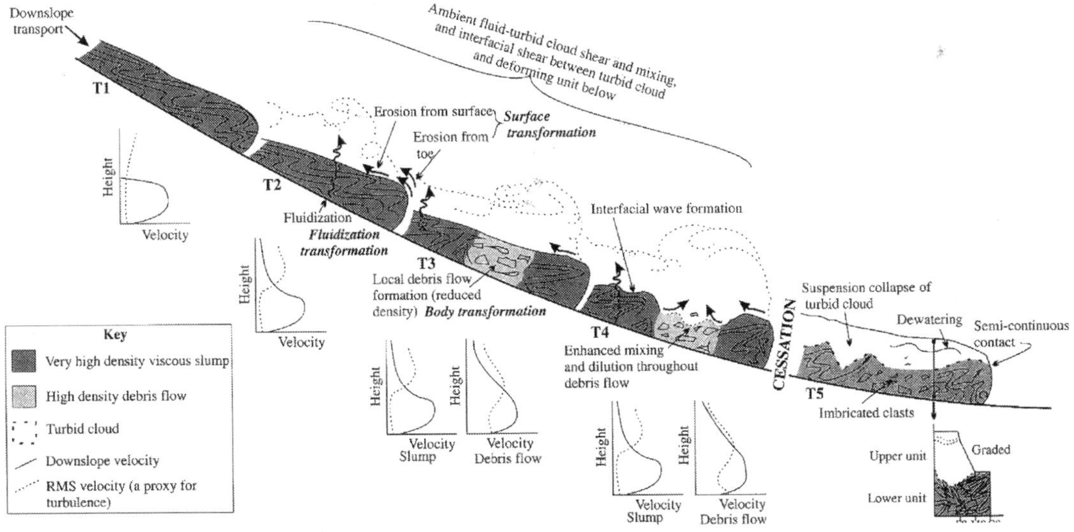


**Fig. 9.11** Schematic drawing of the Little Manly Slump flow-transformation model. T1 to T4 represent progressive development of transformation with time. T5 shows deposition of debrite/turbidite units after runout following slump cessation. (After Strachan, 2008.)

**Fig. 9.12** View of a prominent, partly disembodied slump (with nice fold noses visible in places) on low-angle basal slide plane at about chest-level of the two figures who stand on a horizon buckled by internal compressional strain caused by passage of the overlying slump mass. Slide masses are overlain by parallel bedded turbidites, storm deposits and sand volcanoes (see Fig. 9.4). Ross Slump, Upper Carboniferous, Co Clare, Ireland.

features or trilete cracks radiating from a central point with no connection to adjacent cracks. The single cracks often show preferred orientations. Delicate infill of subaqueous shrinkage cracks by sand or silt leads to preservation, but compactional effects often cause the preserved fill to be considerably deformed when seen in sections normal to the former depositional surfaces. The delicate nature of the preservation process means that subaqueous cracks are most common in sheltered shallow-water lake environments.

## Further reading

### *General*

Discussions of sediment strength are found in all soil mechanics texts; the reader is especially recommended to peruse Lambe & Whitman (1969) for a clear introductory account.

### *Specific*

See Owen (1996) for shaking table experiments on liquefied sands. See Davidson *et al.* (1977) for classic experiments on fluidization. Experimental liquefaction structures are in Nichols *et al.* (1994), Nichols (1995) and Owen, (1996). Field dish structures are studied by Lowe & Lopiccollo, (1974) and Lowe (1975). A theoretical treatment of cross-stratified omelette structures is by Allen & Banks (1972). Field loading structures are studied by Needham (1978), and growth faults by Crans *et al.* (1980) and Weber & Daukoru (1975).

The internal structure of slides and slumps from the sedimentary record and their transformations into gravity flows are nicely described in suitable detail by Strachan (2002, 2008). Transformations in young slumps are discussed by Masson *et al.* (1994), Piper *et al.* (1999a,b) and Gee *et al.* (2006).

# Part 4

# MAJOR EXTERNAL CONTROLS ON SEDIMENTATION AND SEDIMENTARY ENVIRONMENTS

*Here in the wide*
*inter-tidal lull*
*The estuary suffers*
*a Pleistocene age of change.*
*A night's storm*
*recontours a continent;*
*Sand-slide and canyon*
*dissect the exposed plateau.*

Norman Nicholson: 'Tide Out', from *Sea to the West*. Faber & Faber Ltd, 1981.

## Introduction

Environment in the sedimentary context involves the physical, chemical and biological characteristics which control sedimentation at any point in space and which make it distinctive. It is useful to divide environmental variables into compositional factors and scalar or vector quantities (**Part 4 Fig. 1**).

- *Suprasurface variables* relate to stationary or moving atmosphere or hydrosphere. Examples are: density, viscosity, pressure, temperature, lapse rate, length scale (water depth), speed or velocity, momentum, force, energy, power, elemental composition, redox potential, pH.
- *Surface variables* are attributes of solid (land, seafloor, ice) or liquid (river, lake, ocean, magma) surfaces at interfaces with the ambient suprasurface environment. Examples are: elevation, gradient, strength, permeability, porosity, hardness, surface tension, surface form roughness, waviness, albedo, temperature gradient, relative velocity.

Most variables vary with time, some on such very short timescales that a temporal mean value is required to define the variable at a point (e.g. turbulent velocity fluctuations) and others over very long timescales. There may also be additional spatial variations defining gradients, like scalar temperature or elevation gradients or the vectorial gradient in velocity that determines acceleration.

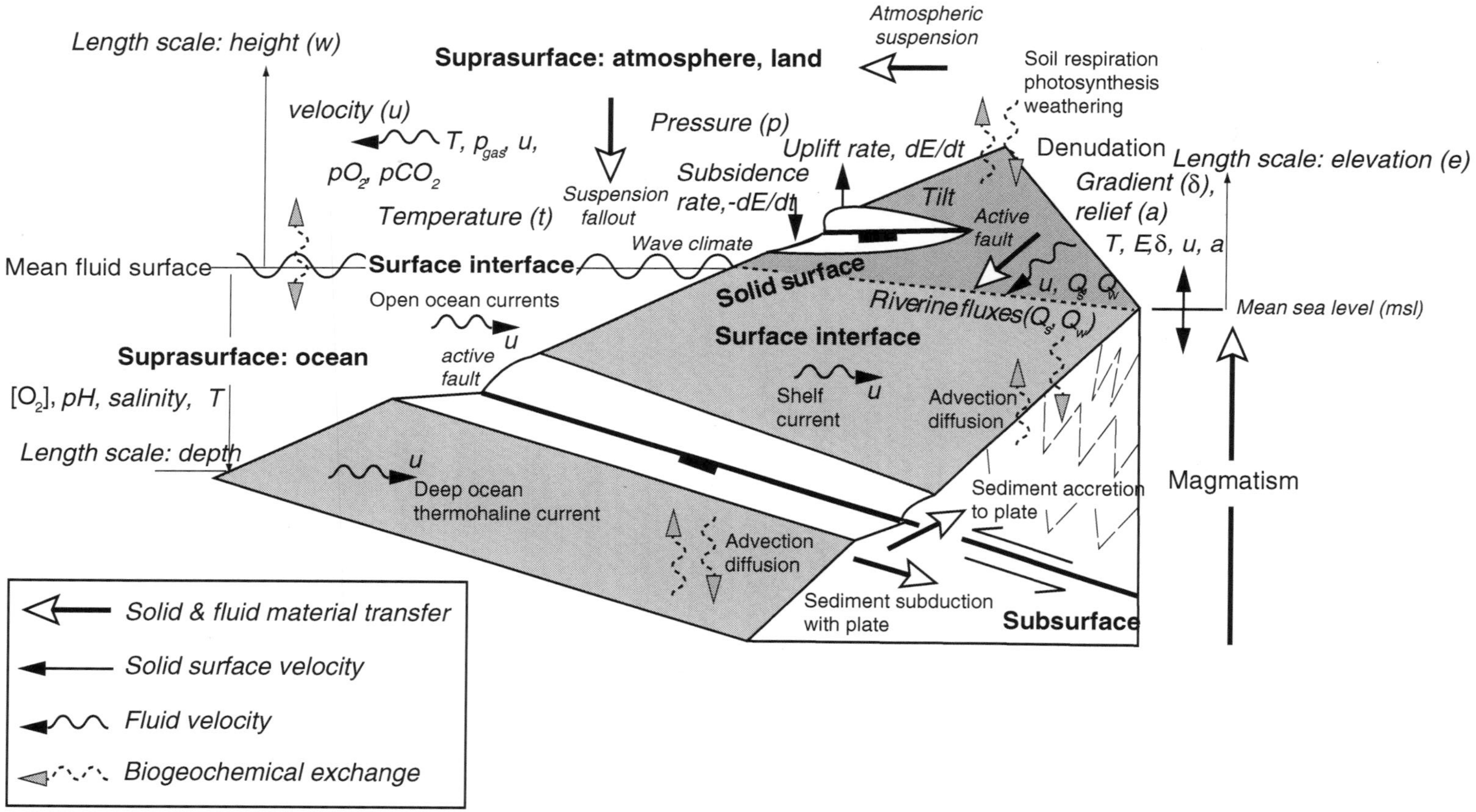


**Part 4 Fig. 1** Sketch to show a selection of suprasurface and surface environmental parameters and their magnitudes that may be determined or constrained by studies of sediment or sedimentary rock.

Palaeoenvironmental analysis makes use of the sedimentary record to make and test theories and models. This is most revealing when accompanied by accurate determination of geological age—it is the spectacular success achieved in radiometric and palaeomagnetic dating of sedimentary deposits in recent decades that has enabled full exploitation of the depositional archive. A major goal in palaeogeography and tectonics is the determination of palaeogradient, both the direction and magnitude of the slopes involved. This is because slope is a major physical variable that controls the power available to transport sediment. Palaeogradient may either be on the large scale, e.g. delimiting the slope away from mountain belt catchments into sedimentary basins, or the regional to local scale, e.g. determining the slope associated with individual tectonic structures like faults or folds. Large-scale gradients are established from provenance studies (Chapter 2) and regional palaeocurrent analysis. Local gradients come from evidence for tilting of once horizontal marine terraces, migrating stream courses and longitudinal valley gradients on river terraces. Provenance studies have been revolutionized by the ability to radiogenically date single mineral grains.

Some of the most important palaeoenvironmental signals come from knowledge of accurate time series of sedimentary successions. Sedimentary cyclicity has recently been thoroughly revisited in the younger Tertiary with the aid of a priori knowledge of time-variance in Milankovich-forcing parameters such as precession and obliquity. Once established, lithological cycles yield data on the magnitude of sea-level fluctuation and their detailed climatic and vegetation records have been further used as correlation aids. Clear signals are also obtained concerning rates and magnitudes of sea-level change from raised marine terraces constructed during global highstands (**Part 4 Fig. 2**). Once accurately mapped and correlated with

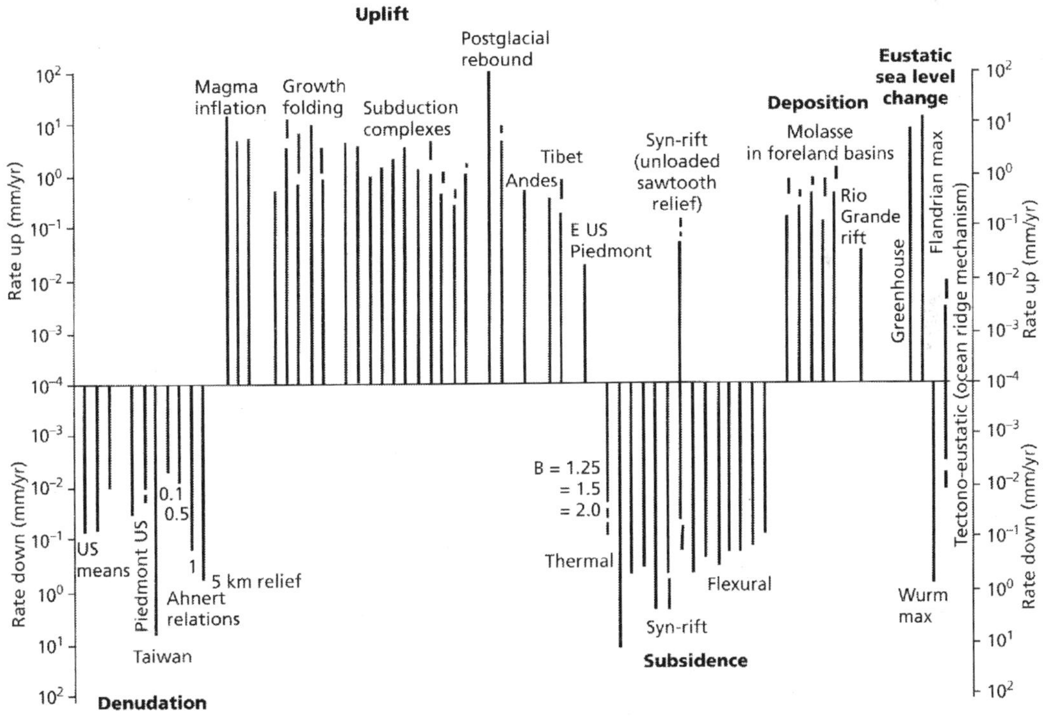


**Part 4 Fig. 2** The 12 orders of magnitude of velocity of the Earth's surface caused by tectonic subsidence, uplift, denudation, deposition and eustatic sea-level changes. Note that subsidence and uplift rates are generally of greater magnitude than denudation rates and that uplift rate is generally greater than deposition rate. Note the high rates of eustatic sea-level change produced by glacial melting and the rapid uplift rates produced by glacial rebound. By contrast, note the low values of eustatic sea-level change produced by mid-ocean-ridge inflation/deflation mechanisms.

the global isotopic record of change from deep-sea cores they may yield a valuable chronology for the timing and magnitude of local or regional tectonic uplift.

It must be stressed that determination of palaeoenvironments is constantly open to the possibility of misinterpretation or of alternative interpretation. The challenge of the subject lies in the ability of its practitioners, on the one hand to keep on top of technical aspects and ahead in the techniques game, on the other to carefully evaluate the possibility that certain signals might require radically different explanations in the way of parallel or alternative hypotheses. It is an inevitable consequence of the vagaries of geological preservation that many palaeoenvironmental signals will get fainter and estimates of their timing more inaccurate the further back in time. The proliferation of new techniques leading to new data on Quaternary palaeoenvironments leads to strong hopes that the older sedimentary archive will become progressively opened up more fully. In some cases this will be led by the application of new radiometric dating techniques, such as the Re–Os geochronometer recently applied to black shales. Perhaps the greatest danger in palaeoenvironmental analysis comes from too great a reliance on correlation of events and causes when chronology is poor.

# Chapter 10

# MAJOR EXTERNAL CONTROLS ON SEDIMENTATION

*Memory is beautiful*
*As a stone, simple*
*As a sample of mountain,*
*A handful of hill.*

Norman Nicholson: Silecroft Shore, from *Rock Face*. Faber & Faber Ltd, 1948.

## 10.1 Climate

Climate has a major effect on sedimentological processes through its control of surface temperature, vegetation, the water cycle, weathering rate, runoff and the existence of either greenhouse or icehouse climate states. Since climate is 'average weather' it needs to be defined for a particular area by some long-term (multidecadal) mean value of temperature and precipitation. This averaging interval must be chosen with great care: we have learnt in the past decade that climate changes and the causes of these are exceedingly subtle and complex, involving interactions between atmosphere, oceans and land. This is because the circulation and dynamics of the oceans are largely driven by atmospheric motions—the transfer of heat energy between atmosphere and ocean is thus said to be *coupled*. Earth's recent and distant history makes it clear that climate has undergone major and sometimes abrupt shifts, leading to important events like the shutdown of life under kilometres of ice, major and sometimes frighteningly rapid ups and downs of sea level and large-scale perturbations to geochemical cycling, especially carbon.

Each climatic variable—temperature, light and precipitation—plays a vital role in earth-surface processes.

1 Temperature determines reaction rates in aqueous solutions, primary (photosynthetic) organic production and changes of aqueous state by evaporation, condensation and freezing; spatial temperature gradient enables energy transfer at the Earth's surface; primary solar radiation is reflected, absorbed and reradiated. Clearly, temperature changes in time and space are remarkably complicated.
2 Light level controls rates of photosynthesis and hence primary production levels. Although direct light level follows a predictable yearly pattern related to the changing declination of the Sun, it is

*Sedimentology and Sedimentary Basins: From Turbulence to Tectonics*, 2nd edition. © Mike Leeder.
Published 2011 by Blackwell Publishing Ltd.

strongly affected by degree of cloud cover and by atmospheric pollutants like volcanic and desert dust—even more important on Mars.

3 Precipitation falling as rain or snow may also accumulate as ice. Water flux controls rates of aqueous chemical weathering, runoff and sediment transport. During the growing season when light is at a maximum, temperature is the major control on terrestrial primary production.

More specific to sedimentology, the climate of a region controls:

- the rate of chemical weathering of silicate minerals via the Arrhenius rate law (Chapter 2) and the throughflow of precipitation;
- the amount of surface runoff left over after uptake by vegetation and subsurface seepage—this runoff causes erosion and sediment transport;
- the yearly distribution of runoff, which plays a major role in determining stream channel and floodplain sedimentary processes through its control on the hydrograph, and also on the timing and extent of sediment, nutrient and freshwater export to and across the shelves into the ocean basins;
- the nature of surface sedimentary environments, and subsurface soil development, including the propensity for chemical precipitation under conditions of high evaporation.

## 10.2 Global climates: a summary

Equatorial latitudes are dominated by high insolation, low pressures, light to variable winds, high mean annual temperatures, low daily temperature changes, high water vapour saturation pressures and copious convective precipitation, particularly in summer months (Fig. 10.1). A yearly rhythm of wet and dry seasons follows north and south migration over the equator of the intertropical convergence zone (ITCZ).

The dependable Trade Winds dominate latitudes 15–35°N and S of the ITCZ. Over the hot deserts, air masses are dry and skies clear, with high radiative heat transfer and consequent high daily temperature variations from very hot to cold. Over the oceans, evaporation rates into the initially unsaturated, advecting air masses are very high. Hurricanes and typhoons result from instabilities set up during convective oceanic heating. The Indian monsoon is amplified by a deep convection effect as the High Himalaya topographically blocks incoming onshore moisture-laden southeast trade winds, thus initiating torrential

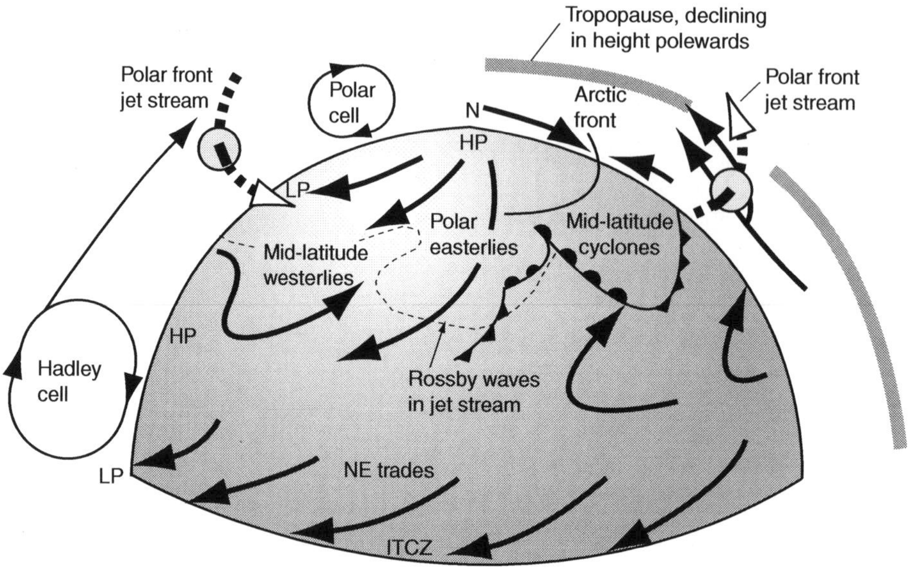


Fig. 10.1 General atmospheric circulation, involving convective cells with Coriolis turning, jet stream, Rossby waves and associated frontal systems: ITCZ, intertropical convergence zone; HP, high pressure; LP, low pressure.

summer monsoonal precipitation over the southern ranges and foothills. Monsoonal rainfall also occurs in East Africa, Southeast Asia, northern Australia and northern Mexico/southwestern USA.

The subtropical high-pressure belt at about latitudes 30–40°N and S is characterized by descending unsaturated air masses and generally light winds. Mediterranean climates result with hot dry summers and cool wetter winters.

The mid-latitude to temperate-latitude maritime low-pressure zones (40–60°N and S) are dominated by the movement of frontal systems that form at the polar–subtropical transition. These sweep warmer saturated air north and cooler unsaturated air south as wave-like intrusions termed *Rossby waves*. Plentiful precipitation results in the cool to cold spring, winter and autumn seasons. Oceanic currents like the warm north-flowing Gulf Stream lead to highly important contributions to air temperatures by latent heat released in the rainstorms associated with frontal systems blowing over them. Continental areas at these latitudes suffer much larger temperature extremes and generally lower precipitation.

The polar anticyclones preside over a stable regime of cold to very cold descending dry air masses with very high albedos over snow- and icefields under cloudless or pervasive 'thin' cloudy skies in summer giving high radiative heat losses to the atmosphere.

### Climate, mountains and plateaux

It has long been appreciated that tectonic landforms have a major effect on local and regional climate. This is because they act as barriers which moist wind systems have to surmount or pass around (topographic 'blocking' conditions; the role of the High Himalaya has been noted above). The act of surmounting leads to reductions of both temperature and moisture, the latter since warm air can hold more water in the vapour state than cold air. This saturation effect leads to the dumping of *orographic rain* on the prevailing windward slopes and summits of ranges and relatively dry conditions (the 'rain shadow') in the lee (Fig. 10.2). Since weathering and erosion rates correlate positively with available precipitation, any uplifting tectonic range will show preferential denudation on windward margins. Mountain ranges also project into the troposphere and hence local air temperatures decrease according to the value of the local lapse rate. This coupling of tectonically created relief and climate creates a profound environmental control on all aspects of erosion, sediment transport and deposition (Chapters 22 and 23).

### Climate change and radiation: any change in the solar constant?

We have little direct evidence that the solar constant changes in response to short-term solar behaviour like the quasi-decadal sunspot cycle but there is circumstantial evidence that over periods of several hundred years that the decreased activity of sunspots may be reflected in lower solar energy output since such periods are associated with severe global cooling. Such periods of low activity are termed *Maunder minima*. An example is the Little Ice Age, a frigid interlude between the late-medieval warm period and the nineteenth century.

### Climate change and orbitally induced radiation changes

Variation in the orbital path of Earth around the Sun and in Earth's own rotation induce longer-term ($10^4$–$10^5$ yr) changes in the relative solar energy flux to particular parts of the planetary surface. Note the term 'relative' because small changes in orbital parameters lead to no net increase or decrease in solar radiation received: the changes simply tend to apportion the radiation at different times of the solar cycle in particular hemispheres. It is this cyclical preferred apportionment that is thought to lead to longer term climate change and the accumulation or melting of great ice-sheets. The determination of the time series to these physical changes and in the seasonal distribution of incoming energy represents one of the great scientific breakthroughs of the twentieth century—its indirect climatic effects have been carefully ascertained by sophisticated geochemical studies of Quaternary marine fossils.

Following the lead of the nineteenth-century amateur scientist J. Croll, M. Milankovitch in the 1920s and 1930s calculated how variations in the three orbital parameters; eccentricity, wobble and tilt would lead to different amounts of radiation being received at different latitudes (Fig. 10.3). The key result relevant to global climatic change was found in the variation of radiation received at temperate

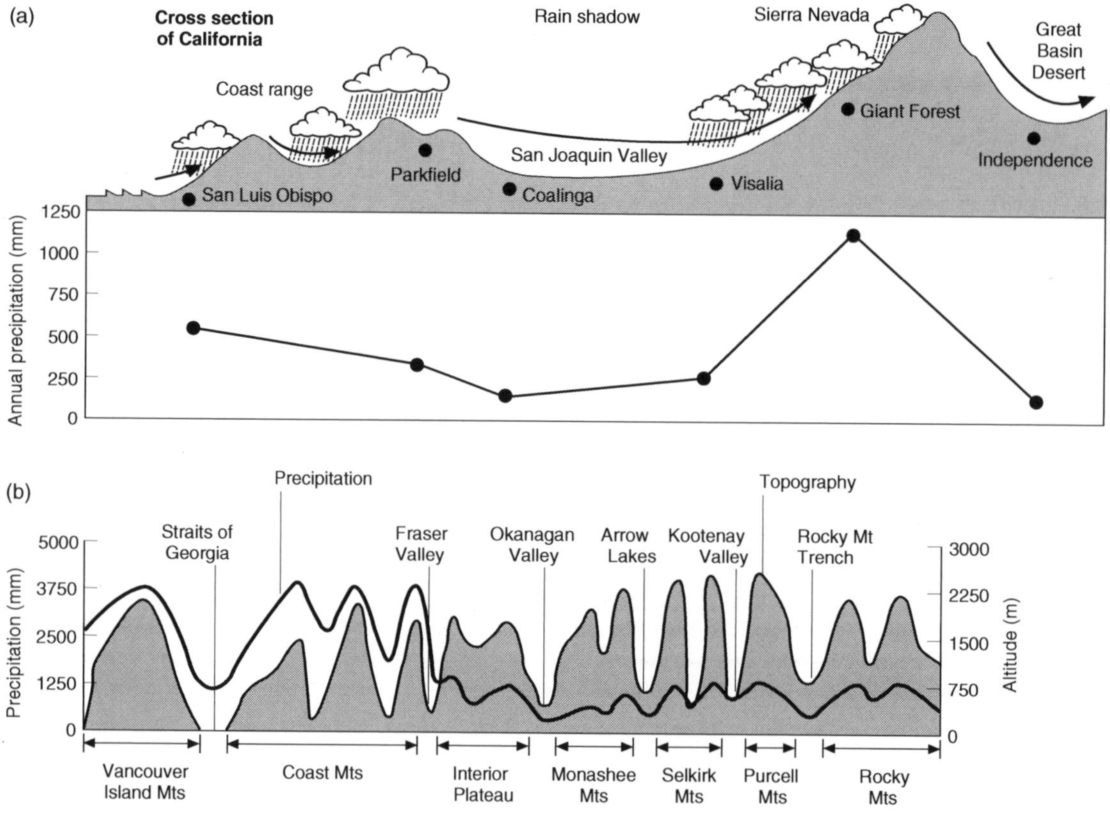


**Fig. 10.2** Elevation and precipitation profiles to illustrate orographic precipitation and rain shadow effects for the Coast Range and Sierran mountains of California. (After Barros & Lettenmaier, 1994).

latitudes during summer. During winter we know that the polar and high latitudes are cold enough to form snow and ice; it is the survival into summer and hence year-upon-year of these seasonal features that will determine whether the ice-sheets can expand below the Arctic or Antarctic circles. Thus any orbitally induced changes that encourage summer cooling by decreasing incoming radiation should lead to climate change sufficient to trigger an Ice Age.

One orbital mechanism is based on the fact that Earth's rotation around the Sun is elliptical and not circular, a fact known since the work of Kepler in the seventeenth century. The very nature of an elliptical path, with the radiating Sun at one focus, means that there are seasonal variations in the amount of radiation received by Earth at aphelion and perihelion. This is based a priori on the premise that the intensity of solar radiation is reduced with distance from the Sun.

At present, Earth is nearest the Sun, by about $4.6 \times 10^6$ km, on 2–3 January (perihelion) and furthest away on 5–6 July (aphelion); note that these dates are not the same as the times of solstice. As a consequence of this elliptical rule the solar radiation received by Earth varies by about $\pm 3.5\%$ from the mean value. Although these figures are not appreciable compared to the other orbital effects noted below, exact calculations of the gravitational effects of the other planets in the Solar System on Earth's elliptical orbit led to the later theory (due to Leverrier in 1843) of time-variable eccentricity, whereby the yearly orbit becomes more and less eccentric on the rather long timescales of around $10^5$ and $4.10^5$ yr. At the present time and for the foreseeable future (Fig. 10.3) we are in a period of average to low eccentricity; at times of highest eccentricity it is calculated (originally by Croll) that the change in incoming radiation may be $>5\%$.

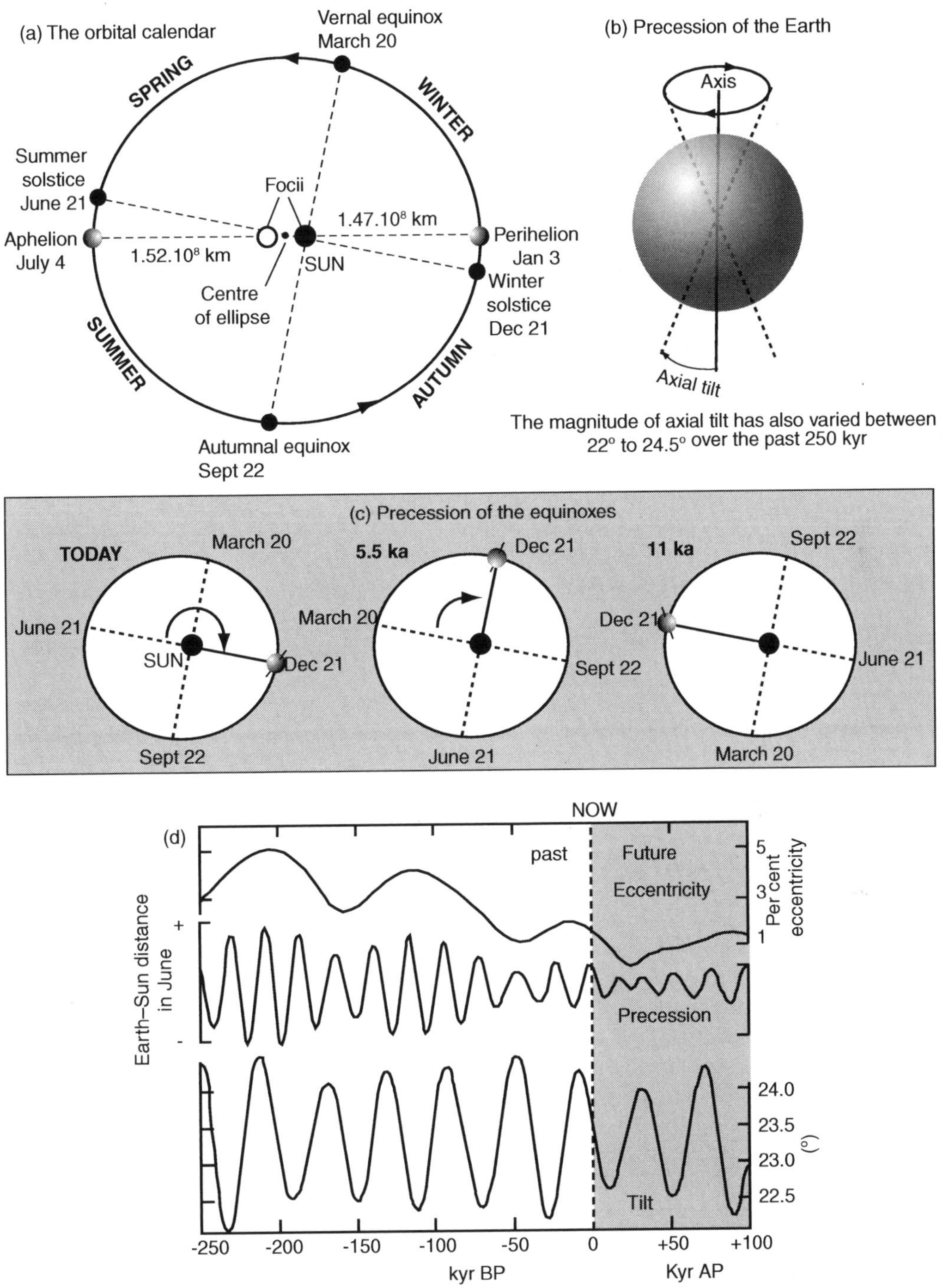


**Fig. 10.3** Milankovich mechanisms. (Largely after Imbrie & Imbrie, 1979, and sources cited therein.) (a) Orbital calendar today. (b) Precession and axial tilt. (c) Precession over a half cycle. (d) Computed long-term changes in eccentricity, tilt and precession for the past 0.25 myr and for the future 0.1 myr. (Data of A. Berger.)

A second orbital mechanism is based on the regular 'wobble' of Earth's inclined spin axis relative to the plane of rotation around the Sun or to some point fixed in space. This wobble is due to the gravitational attraction of the Sun and Moon upon Earth's own equatorial bulge. The practical effect of this leads to *precession of the equinoxes*, a phenomenon discovered in about 120 BC by Hipparchos of Alexandria, whose own observations of star clusters taken at fixed yearly times and positions compared to those observed by earlier Egyptian and Babylonian astronomers (going back to about 4000 BC) led him to note the gradual shift of familiar star clusters around the Earth's ecliptic, the plane of the solar orbit, at times of solstice. One complete wobble involves a circuit of the spin axis about a circle, causing the northern and southern hemispheres to change their times of closest and furthest approach, at perihelion and aphelion respectively, approximately every $2.2 \times 10^4$ yr. This is the explanation for the fact noted above that the solstices (times of maximum tilt of the Earth away from and towards the Sun) do not have to coincide with aphelion and perihelion. In terms of solar radiation received at Earth's surface, the Lambert-Bouguer law makes it clear that the effects of precession are greatest at the equator, decreasing towards the poles. Minimum levels of radiation for either hemisphere away from the polar circles occur when aphelion corresponds to winter. Today we are close to the situation of southern hemisphere summer at perihelion: about 11 ka any Palaeolithic astronomers would have experienced warm northern hemisphere summers at perihelion.

The final orbital mechanism depends upon the angle of the inclined spin axis changing relative to the ecliptic. Calculations and observations indicate that this is currently changing by $\sim 0.7.10^{-4}$ °/yr. Over a $4.10^4$ yr period the axis varies about extreme values of $\sim 21.8°$ and $\sim 24.4°$, with the current value $\sim 23.44°$. In terms of solar radiation received, again according to the Lambert–Bouguer law, minimum levels are to be expected in winter when tilt is maximum, but the effect makes no difference to high polar latitudes since these are in darkness anyway. The effect has greatest influence in moderate to high latitudes.

In what was arguably as big an earth sciences scientific discovery as that of plate tectonics, the Hays–Imbrie–Shackleton determination of orbital control of past climate by the three Milankovitch mechanisms working in unison was termed 'pacemaker of the Ice Ages'. The pacemaker was determined by accurate and precise mass spectrometer measurements of tiny oxygen isotope compositional variations in marine oceanic foraminifera forced by past variations in global sea temperature and, by inference, on the abundance of land-ice and sea-level elevation for the past several millions of years (Fig. 10.4). We still do not know exactly why the eccentricity signal 'kicked in' at about 800 ka and triggered the first major Ice Age. Prior to this it is the axial tilt modulation that shows up most prominently in the isotope record whereas the precessional record is clear at all times.

### Short-term changes in reradiation and the 'greenhouse effect'

As is well known current concerns about global warming arise mainly from the increased *greenhouse effect* due to increased levels of atmospheric $CO_2$ above the current value of around 350 ppm, produced mainly by burning fossil fuels. Of equal interest to students of future and past climate change are variations in atmospheric water vapour content, another major greenhouse contributor. These variations are strongly dependent upon temperature, so that any increase in atmospheric $H_2O$ due to greater evaporation consequent on general global warming will further accentuate the trend by *positive feedback*. Counter to this trend would be the increased levels of incoming short-wave solar insolation reflected back into space by increased levels of cloud cover. This *negative feedback* would itself be reduced in magnitude by reductions in the extent of high-albedo ice and snow cover. Reduction in atmospheric $H_2O$ would tend to reduce surface temperatures by increasing the direct cooling of the oceans by increased escape of infrared radiation to space. Again, counter to this trend would be the increase in short-wave solar insolation not reflected back into space because of decreased levels of cloud cover. The greenhouse effect is also sensitive to short-term variations in atmospheric dust and aerosol gases of volcanic origins. Although the interactions are complicated, the net effect is to cause an increase in reflected and absorbed/reradiated solar energy in the upper atmosphere and thus a general cooling for a number of years after powerful Plinean-type eruptions.

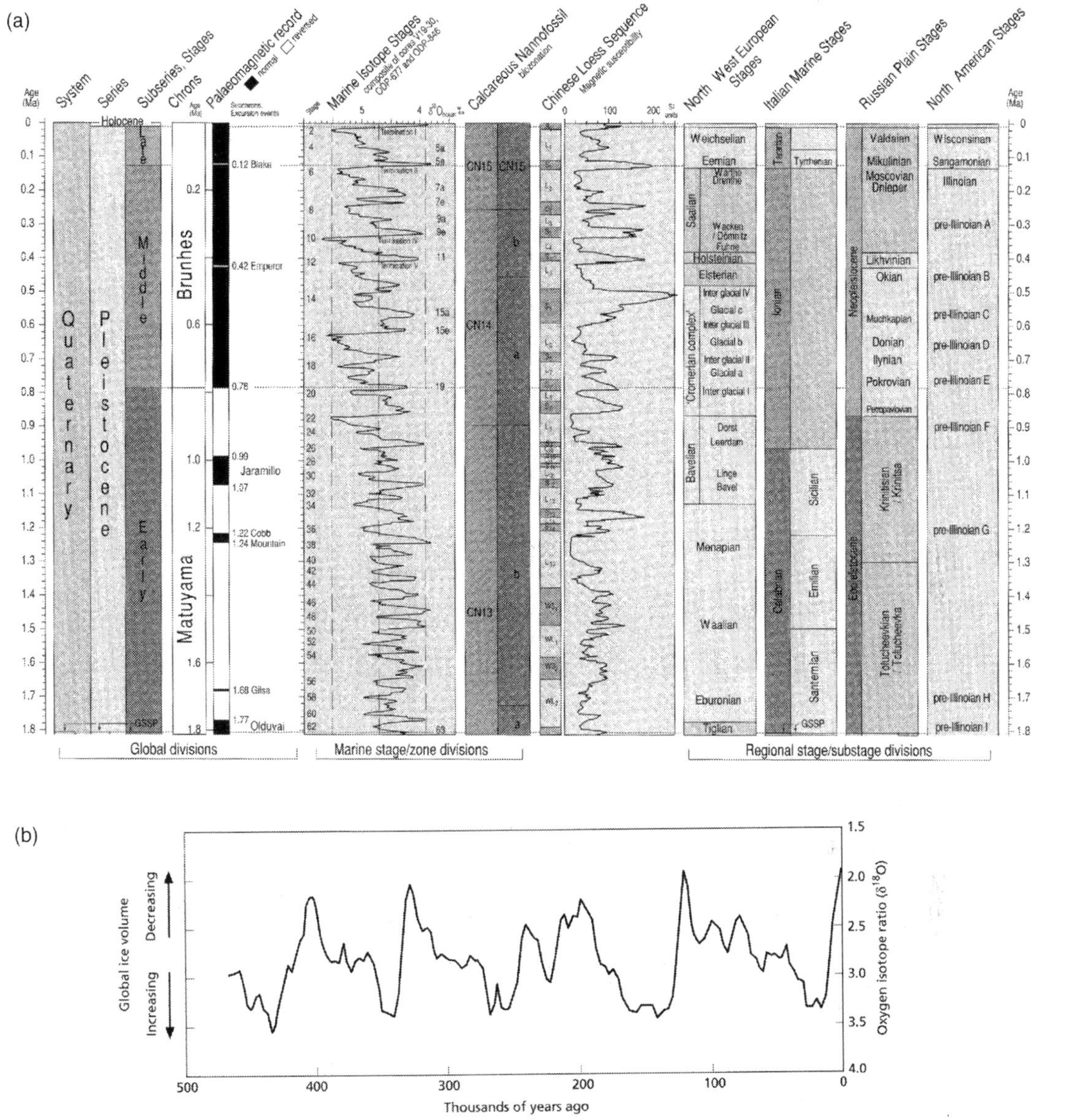


**Fig. 10.4** (a) Global Quaternary events related to marine isotope stages. (From Head & Gibbard, 2005.) (b) The variation of $\delta^{18}O$ vs. time for the past 500 kyr. Note, more positive $\delta^{18}O$ inicates *enrichment* in this heavy stable isotope as more of the lighter isotope $\delta^{16}O$ is sequestered in the polar ice caps. These data, a twentieth century Rosetta Stone, finally unlocked what was memorably termed the 'orbital pacemaker' of the Ice Ages by the discoverers of the effect (Hays *et al.*, 1976).

## Sedimentological evidence for palaeoclimate

The analysis of palaeoclimate can make use of a host of techniques and features preserved in rocks and minerals. Primary evidence for latitudinal range comes from palaeomagnetism. With such constraints we can use other sedimentological features (with some caution of course) to help palaeogeographical reconstructions.

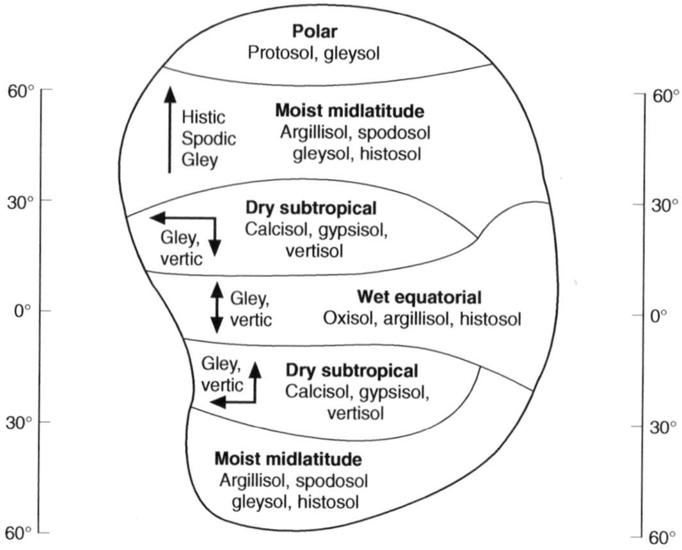


**Fig. 10.5** Generalized latitudinal (zonal) distribution of palaeosol types useful in assessing palaeoclimate. (After Mack & James, 1994.)

- As noted above, oxygen stable isotope fractionation (section 23.18) from calcite-shelled foraminifera give evidence for variations in contemporary ocean temperatures and proxy evidence for sea level and 'icehouse Earth' conditions (Fig. 10.5).
- Certain deposits have intrinsic climatic messages. Arid tropical locations are indicated by the presence of sabkha evaporite successions (Chapter 20). Zonal climates control the broad palaeosol classes: a useful and simple scheme is given in Fig. 10.5. For example, the continued presence of semiarid calcisols through Plio-Pleistocene sedimentary sequences in the southwestern USA indicates that climate has remained essentially semiarid, unchanged since about 5 Ma. By way of contrast, the occurrence of thick oxisols in mid-latitude volcanic sequences of Miocene Germany and Eocene Ireland indicates former 'greenhouse' climatic optima in these now more frigid and desolate places. *Palaeoergs* (ancient sand seas) delimit the extent and vigour of trade wind belts and loess deposits indicate sustained dust deposition from expansion and contraction of synoptic wind regimes.
- Fluid inclusions in evaporites and the Mg-content of calcite yield evidence for past temperatures and salinity variations from normal seawater (Chapter 23). Trace elements and stable isotopes from biogenically or physically precipitated sedimentary deposits and soils reveal critical information concerning salinity, elemental balances and the distribution of mean annual ambient temperatures. Noteworthy modern advances are the determination of $pCO_2$ from the C-isotope composition of calcic palaeosols, and palaeotemperatures and palaeoprecipitation from the O-isotope of terrestrial carbonates, especially speleothem (**Ch. 23.18**) but also tufa. Additional use of oxygen isotope fractionation is in calcisols and lake carbonates as palaeoclimatic and palaeoaltimetric indicators, the latter of exceptional value.
- Checks on models for elevated palaeoatmospheric $pO_2$ come from critical ignition experiments on natural plant materials and the proxy signals of sedimentary charcoal particles. Checks on hypotheses for the nature of Archaean Earth's atmosphere have relied on the presence of oxidative or reducing Fe-products preserved in palaeosols and in the role of P-removal and Ni-enrichment during anoxygenic microbial precipitation of banded iron formations (section 23.2).
- Sapropels and black shales reveal the redox state of oceanic and shelf basin bottom waters (Chapters 19 & 21). Oceanic anoxic events occur due to a combination of changes in surface water dynamics,

biological productivity, $pCO_2$ from methane hydrate degassing and/or coal metamorphism and elevated chemical erosion rates.

- Presence of contour current deposits and the sortable-silt palaeovelocity index indicates variations in the vigour of thermohaline circulation (section 6.4 and Chapter 21).

Listed below are more obvious large-scale attributes useful in zonal climatic reconstructions in the sedimentary record.

1 Equatorial and subequatorial latitudes on the continents, where rainfall is seasonally high, vegetation is lush and reaction rates fast, have distinctive intensely leached Fe- and Al-rich soil types. Runoff is very high, especially from orogenic uplands subject to monsoons, with abundant fluviodeltaic sediments and outflows and underflows into adjacent seas and oceans of suspended-sediment-rich plumes and turbidity currents. On low-relief stable cratons, sediment fluxes from the deeply weathered regolith may be negligible, and runoff dominated by organic fluxes. In the seas and oceans, very high benthic organic productivity gives a productive 'carbonate factory' mostly away from areas of siliciclastic input.

2 Low-latitude Trade Wind belts over continents are arid with little vegetation, so chemical weathering is negligible, and physical breakdown predominates. Aeolian processes dominate in great sand seas (*ergs*), with steady synoptic seasonal winds. Evaporite mineral precipitation is widespread, from groundwaters, playas and coastal lagoons. Sabkha anhydrite is indicative of extreme aridity and very high summer temperatures ($> 40°$). In shallow marine environments carbonate production is again high whilst adjacent oceans with offshore or shore-parallel wind regimes experience strong upwelling and high primary productivity, with bottom-water anoxia encouraging *anoxic conditions* (Chapter 22).

3 Semiarid climates with scrub vegetation have low water tables but sufficient summer convective or weak monsoonal precipitation to develop characteristic calcisols, vertisols and *in situ* oxidative weathering of shallow buried sediment to produce 'red beds'. River regimes are characteristically 'flashy', with a great propensity for delivery of hyperconcentrated and debris flows.

4 Mid- to temperate-latitude maritime climates develop deciduous to coniferous woodlands under which acidic surface waters cause development of strongly leached podzol-type soils. Glaciers exert a tremendous influence during icehouse epochs.

5 Polar climates show negligible chemical weathering. Areas under ice and in the permafrost zone feature a wide range of glacial and periglacial facies and structures.

## 10.3 Sea-level changes

The elevation of the sea surface is a fundamental control on sedimentation because any change will cause coastal, shelf and oceanic environments to be brought closer to, or further from, sources of sediment supply. Further, as shelves are exposed during periods of lowered sea level, rivers traversing them often cut deep incised valleys and deliver sediment to the shelf edge. Any such seaward-directed shoreline shift is termed a *regression*. Such shifts may be accomplished purely by sediment deposition, like in a seaward-moving (*prograding*) deltaic shoreline constructed during a period of stable sea level. Regression due to falling relative sea level is called *forced regression*. Landward shoreline shifts are termed *transgressions* (aka *marine flooding*) with a relative sea-level rise causation termed a *forced transgression*. Landward shoreline retreat caused by transgression is sometimes termed *retrogradation* or *backstepping*.

Confirmation of the Austrian geologist Suess's original idea (from 1888) that absolute global sea level has not been constant but has varied through geological time was one of the major developments in twentieth-century geology. Conditions of maximum and minimum sea level are now referred to as *highstand* and *lowstand* respectively, with periods of rise and fall separating them. Such absolute global changes are termed *eustatic* and must be distinguished from the myriad of local or regional relative changes induced by faulting, thermal subsidence, volcanism, intraplate stresses due to subduction, thermal uplift, sediment compaction, subsurface fluid withdrawal and the near- to far-field effects of glacial ice loading and unloading. Whilst local or regional relative sea-level change may be very important, for example the relative movements generated along seismically-active faults or folds or uplift by post-glacial rebound, it is the elucidation of global sea-level change that represents the major breakthrough in stratigraphic and sedimentary studies. Global changes may be estimated in two ways.

1 From evidence in the sedimentary and/or fossil record for shallowing or deepening or of shoreline shift to landward or seaward. Once established that a sea-level change has occurred locally, then checks by fieldwork and study of the literature can be used to establish whether the change was global. Care is needed in this exercise since the period of change must be exactly replicated by fossil or radiometric dating between localities. The method has been used for some time in the geological literature and serves to crudely constrain the older geological record (> 10 Ma) of sea-level change for Palaeozoic and Mesozoic times.
2 By oxygen isotope fractionation evidence indicative of the abstraction of water into and out of polar ice-caps. This comes from marine benthic formaniferas collected from ocean floor sediment cores and by ice-coring. Needless to say, the method is considerably more accurate than conventional stratigraphic techniques. The most significant highstand and lowstand points on the marine-derived O-isotope time series are defined as Marine Isotope Stages, the former being periods of $^{18}O$ depletion, designated by odd numbers (MIS 2, 4, 6, etc.) and the latter as periods of enrichment, designated by even numbers (MIS 1, 3, 5, etc).

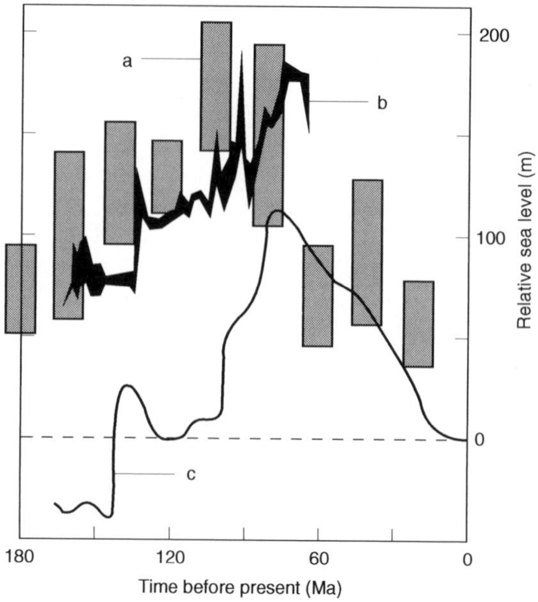


**Fig. 10.6** (a)–(c) Three estimates of long-term sea-level change from the Mesozoic and Cenozoic stratigraphic record. Note that despite the wide range of estimates for magnitude of sea-level variation the trends to and from the late Cretaceous highstand are at least comparable. (Compilation of Heller *et al.*, 1996.)

## Rates and magnitude of sea-level change

Mesozoic sea-level changes occurred in a greenhouse epoch, appearing as low-frequency ups and downs, typically [O] 0.5–3 Myr duration, superimposed on change at very low frequencies [O] $10^2$ Myr (Fig. 10.6). Estimates of the magnitude of low *greenhouse* frequencies are in the range of a few tens of metres. Estimates of the magnitude of the very low frequency changes are very much more difficult to call because of the effects of post-Mesozoic tectonic uplift and subsidence, but may range from 100 to 300 m above present-day sea level. From these data we infer that rates of absolute Mesozoic sea-level change were very low, in the range 0.001–0.05 mm/yr. Such rates are usually less than relative sea-level changes brought about by tectonics. Under such conditions the correlation of apparent sea-level changes from different continental margins must be done with very great care indeed, a practice sadly not achieved in many studies.

Upper Cenozoic sea-level changes occurred in an icehouse epoch, appearing in ocean-sediment or ice-core O-isotope time-series as high-frequency ups and downs at Milankovitch-band periodicities (Fig. 10.7). Prior to the onset of major ice-sheet growth and non-Greenland/Antarctic ice caps and continental glaciations at about 800 ka, the predominant signal back to about the late Miocene period is of a 40 kyr periodic fluctuation of about 20–30 m magnitude produced by obliquity-forced oscillations of West Antarctic and possibly Greenland ice sheets. From about ~800 ka to present, both 20 and 40 kyr fluctuations are superimposed upon (and reinforce) a highly asymmetrical ~100 kyr periodic fluctuation of ~125 m magnitude. The fluctuation is asymmetrical in that the rising curve to sea-level highstand is very much steeper than the falling limb to lowstand. From these data, calculated mean rates of absolute upper Cenozoic sea-level change were very high, in the range 1–10 mm/yr. Such rates are very much greater than sea-level changes brought about by most tectonic mechanisms, other

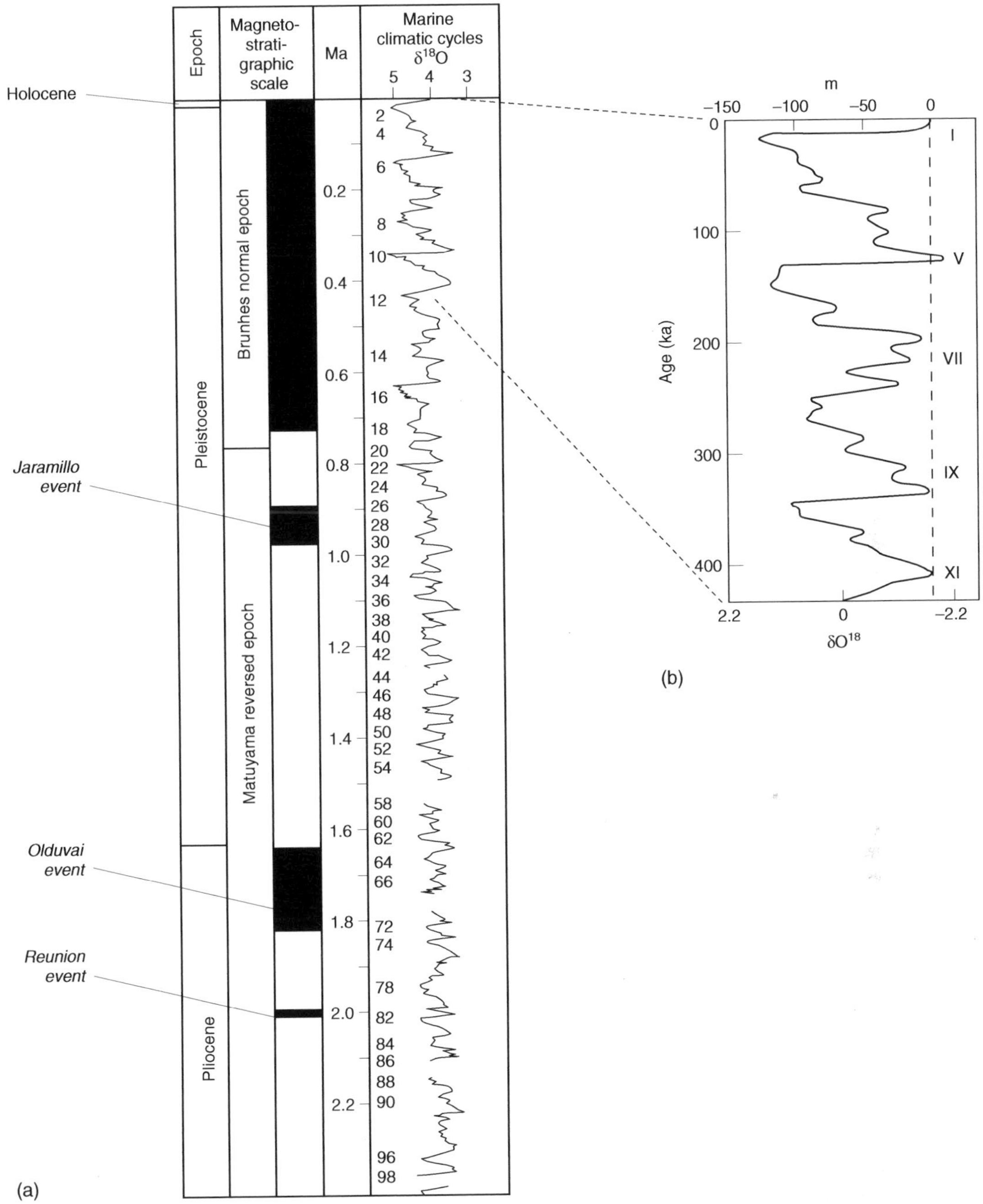


**Fig. 10.7** (a) Time series of late Neogene to Holocene oxygen isotopic fractionation. (Data in Harland *et al.*, 1990.) (b) Details of approximate sea-level variation for the past 400 kyr. (From Imbrie *et al.*, 1984.)

than slip rates along particularly active faults. The highest rates are also more rapid than any counter-acting process of sedimentation or of organic coral reef framework growth.

Concerning the duration of interglacial highstands, these vary from a few thousand to tens of thousands of years. Both MIS 11 and 5 were rather long, at ~30 and ~15 kyr respectively. Of major importance is the discovery that in MIS 5, very rapid *suborbital* sea-level changes, both ups and downs, have occurred. Some of these have been spotted by sedimentological evidence (section 23.17).

The transition from Last Glacial Maximum low-stand to Holocene highstand was marked by at least two major accelerations in sea-level rise up to peak rates of 50 mm/yr at ~19 and ~14.5 ka. Each of these 'meltwater pulses' added the equivalent of 1.5–3 Greenland Ice Sheets to the oceans over a period of 100–500 years. Periods of decelerating to zero rise also occurred during the Younger Dryas (~13 ka) and 8.2 ka 'cold events'. The sea-level record from stable tectonic sites indicates that the post-glacial Holocene highstand seems to have been achieved by ~5 ka, although there is much argument in the literature about this, much of it erroneous from theoretical geodesy.

Global sea level is currently rising at a rate of $\sim 1.5 \pm 0.4$ mm/yr, with about one-third due to thermal expansion of upper ocean waters consequent upon global warming (*steric* effect) and the remainder due to continental ice meltout and return of freshwater to the oceans (*eustatic* effect). Future sea levels could in theory rise by +70 m if both the Greenland and Antarctic ice sheets melted completely. Some model studies show that this is unlikely to happen if global temperature rise remains $< 5\,°C$, since the East Antarctic ice sheet will grow by accumulation under predicted conditions. The Greenland ice sheet is most vulnerable to terminal decline; this would lead to a computed +7 m global sea level rise—an event of extreme interest for future generations of sedimentologists and extreme concern for more normal people.

Finally it is vital to stress that in areas outboard of former ice-sheet margins, late-Quaternary eustatic sea-level changes are convolved with the effects of post-glacial rebound as viscous lower crustal and mantle flow returns material to sites of former ice-loading. Even during periods of eustatic sea-level rise, such sites may themselves be rising and so a rather interesting sedimentary signature will evolve with time. More controversial is the extent to which far-field sites are influenced by such post-glacial return flow. For example, extravagant theoretical geodetic claims that the Mediterranean area has suffered significant post-glacial subsidence as material flows north under Eurasia are not supported by field evidence.

The situation regarding sea-level changes further back in geological time is much less clear. It seems reasonable to suppose that periods with proven ice-house conditions (Permo-Carboniferous, late Ordovician, late Precambrian) would have shown similar high-frequency signals to those seen more recently. Certainly that is the message from studies of cyclical deposits of Permo-Carboniferous age on the ancient Pangaean megaplate, where estimates of sea-level change are comparable in magnitude to those determined for the late Pleistocene, ~ 80–140 m. But for the majority of Earth history, sedimentation took place under greenhouse conditions, with slow eustasy dominant in the background. However, although major ice-caps were never present, palaeobotanical evidence from high latitudes suggests that smaller caps probably were and it is these that would have waxed and waned at axial tilt frequencies—perhaps generating metres to tens of metres of sea-level oscillations—through most of geological time.

### Origins of global sea-level change: slow vs. fast eustasy

Slow-acting eustasy works on the 'bath-tub principle' in that although the total amount of global seawater is conserved, ocean basin volume may change. This is most likely to be due to large-scale plate tectonic processes, since we know that in the long term the Earth is neither increasing nor decreasing in diameter. Given these boundary conditions the cycle of plate tectonic interactions must conserve lithospheric mass and the dilation or shrinkage of the ocean basins or the uplift and/or subsidence of the continents must involve global sea-level change as a serendipity effect from a very complicated sequence of feedback mechanisms.

1 Swelling or shrinking of mid-ocean ridges is seen by some authors in proxy evidence for faster or slower Mesozoic spreading rates recorded in the ocean crust of the Pacific. (Fig. 10.2). Such a mechanism,

if it be real, cannot be directly established for rocks older than the oldest ocean crust, about 180 Ma (early Jurassic).

2 Eruption of large volumes of basalts from mantle plumes into the ocean basins, during so-called megaplume, episodes would reduce their volume. Such bursts of plume activity are postulated for several intervals of Earth history, particularly the middle Cretaceous. They may be related to subducted lithosphere-melting events in the D" region of the core–mantle boundary, the so-called 'slab graveyard'. Contrary to this view is the argument that such plume eruptions are essentially random, or even more sceptically, that plumes do not exist at all.

3 Superplate break-up and oceanic destruction have been implicated by their coincidence with changing global sea levels in the Jurassic and Cretaceous. This is due to replacement of old, cool and dense oceanic lithosphere by young, warm and less dense oceanic lithosphere, and since ocean depth is a function of cooling age, younger Mesozoic oceans would have been shallower on average than older oceans. Hence global mean sea level would have increased during progressive rifting, reached a maximum and then declined as thermal cooling counteracted the shallowing tendency (Fig. 10.8).

Dynamic topographical support of the continents by a combination of sublithospheric mantle convection and mantle advection inboard induced by oceanic lithosphere subduction. Direct tests of the idea are difficult to make. Indirect tests achieve some explanations of ca 100 Myr cratonic 'cycles' but require rather naïve assumptions concerning palaeoplate tectonics and basin-scale modelling.

Fast-acting eustasy has a rather more satisfactory explanation in that seawater volume is not conserved and water abstracted to form polar ice-caps is eventually melted to return freshwater to the oceans. This causes global seawater volume to fluctuate on Milankovitch orbital forcing scales by a maximum of ~5% of total ocean volume. For the past eight ~100 kyr cycles, the asymmetry in the sea-level time series noted above (Fig. 10.4) means that continental ice-caps melted more quickly than they formed, not something that is easy to understand without involvement of large-scale oceanic convective warming at polar ice-cap margins.

## Sequence stratigraphy

Sedimentary deposits are *layered* (aka *stratified*), interleaved and generally arranged in 3D patterns that

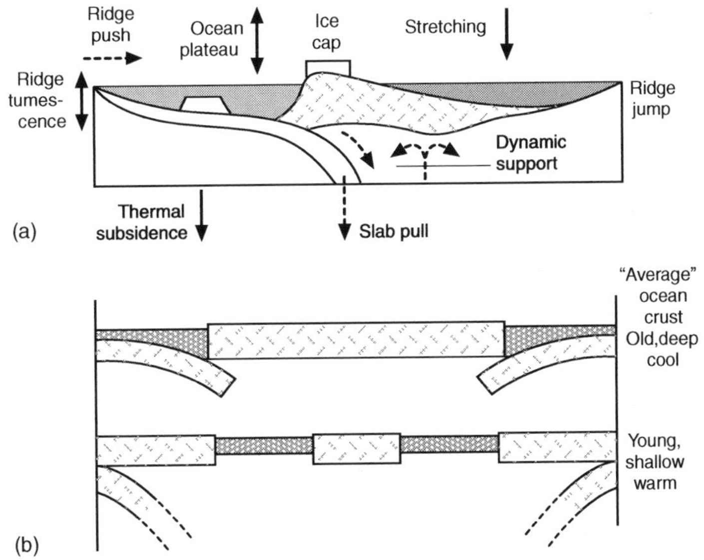


**Fig. 10.8** (a) The main tectonic controls acting to determine mean oceanic and continental volume. (b) The relative oceanic shallowing induced by break-up of a former supercontinent between young, warm oceanic lithosphere.

reflect the shifting environments of their formation. *Stratigraphy* is the discipline that deals with the arrangement of sedimentary layers by mapping and correlating them using a wide range of techniques—natural outcrops, seismic reflection profiles, cored wells, fossils and palaeomagnetism, to name but a few. Like the foundations, walls, floors and roofs of buildings, depositional environments may be divided into mosaics called *architectural elements*. When mapped out they define the basic *lithostratigraphy* or arrangement of distinctive layers. Lithostratigraphic units form a hierarchy of layers in 3D space mapped by the geologist between exposures or wells. The situation faced by a subsurface mapper using seismic reflection traces is somewhat analogous. Here the layercake must be split up based on the succession of distinctive reflector horizons to define *seismic stratigraphic* units. Frequent use is made of the nature of the top and bottom contacts between distinctive layers to shed light on the nature of processes responsible for deposition of the units in question (Fig. 10.9). At this stage the geologist may have no knowledge of the nature of the rocks in question far below the surface, although inspired guesses are based on the pattern and amplitude/response pattern of the seismic signal. Once a well or a number of wells have been drilled through such seismic stratigraphic units, then the seismic units may be transferred into lithostratigraphic ones.

Local, regional or global changes in sea level enable sedimentary deposits to be subdivided according to the phase of the sea-level curve in which they were deposited or in which erosion occurred (Fig. 10.10). Those sediments laid down between successive lowstands are termed a *sequence*. Sequences are easiest to recognize where the sequence boundary is also a surface of erosion or nondeposition, made clear by a distinct mappable surface and/or on the basis of a time gap indicated by a mature palaeosol or from fossil evidence. Erosion during lowstands is most obvious where the continental shelf is exposed to the atmosphere, giving the opportunity for soil development, the cutting of new drainage channels and the downcutting of pre-existing channels into the former marine sediments of the highstand shelf. More controversial is the situation at lowstand on the former coastal plain. For situations where the slope of the newly exposed continental shelf is significantly greater than that of the coastal plain, it is predicted that a 'wave' of channel incision and soil formation will work its way up the river channels at a rate determined by the slope difference and the erosive power of the river. It is also expected that the rate and depth of incision will decrease with time and with distance upstream. As the channels incise, so the likelihood of floodplain accretion will decrease and soils will begin to develop on terraces and interfluve highs. For situations where slopes of the coastal plain and shelf are similar or where the former is less than the latter, then it is less likely that incision will occur.

As sea level begins to rise from lowstand, then the formerly eroding shelf or coastal plain experiences a forced marine transgression and progressive water deepening until highstand is reached (Fig. 10.10). The transgressive deposits of this stage of the cycle and those of the succeeding highstand and regressive phase overlie the lower sequence boundary and are in turn bounded by the next lowstand erosion surface. Each of the subdeposits of lowstand, rising, highstand or falling sea level are termed *systems tracts*.

There are a few final points concerning the application of sedimentology to sequence stratigraphy.

1 The concepts of sequence stratigraphy are at once most useful but also most difficult to apply in offshore and oceanic environments below lowstand sea level. One clue is that rivers debouching their sediment loads closer to the shelf edge, avoiding reworking by tide and wave, will result in more frequent and voluminous turbidity flows. Under such conditions continental margins, submarine fans and their feeder channels should show rapid lateral growth until rising sea level to highstand brings the process to an end with the fan now draped by fine-grained sediments of hemipelagic types.
2 The term *parasequence* describes the depositional products of a progradational regression that is terminated by a relatively abrupt relative sea-level rise.
3 Sequences can be divided into bins or *orders* since slow-acting eustasy and tectonic alterations to sea level act up to $\sim 10^8$ yr and orbital parameters may control sequence development over 20, 40, 100, 400 kyr. There seems to be consensus in referring to sequence duration in terms of units of powers of 10, with 100 Myr cycles being first-order, 10 Myr cycles second-order, and so on.
4 The term *accommodation space* is used to denote the amount of 3D space available to deposit

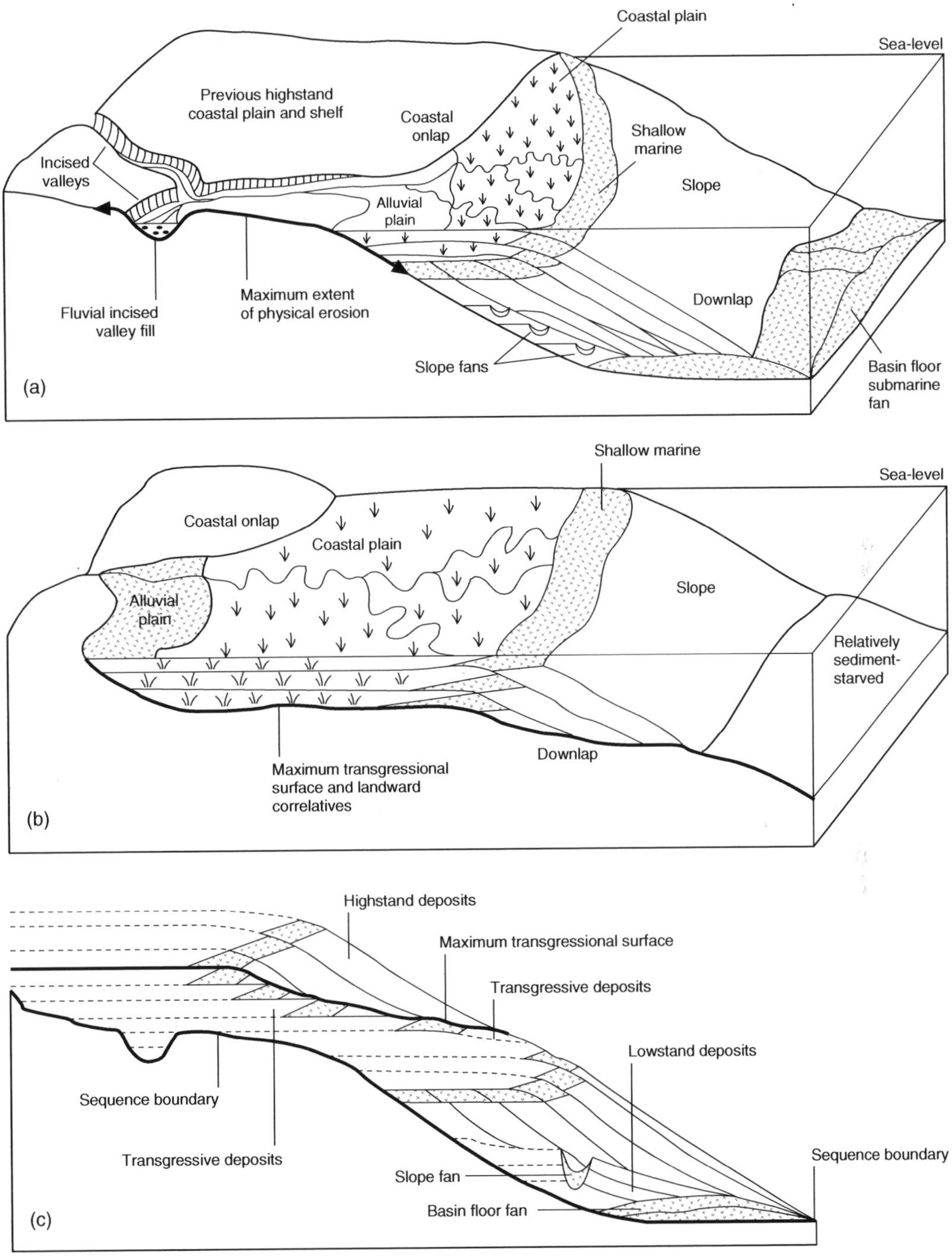


**Fig. 10.9** Diagrams to illustrate stratal geometries, lowstand and highstand sequences. (Loosely based on the discovery paper by van Waggoner *et al.*, 1988; mostly after Myers & Milton, 1996.) (a) Three-dimensional sketch to show components of lowstand deposits on a subsiding coastal margin with a wide coastal plain and a narrow shelf. (b) Three-dimensional sketch to show components of highstand deposits on a similar shelf to (a). (c) Two-dimensional section to show overall architectural summary scheme for (a) and (b).

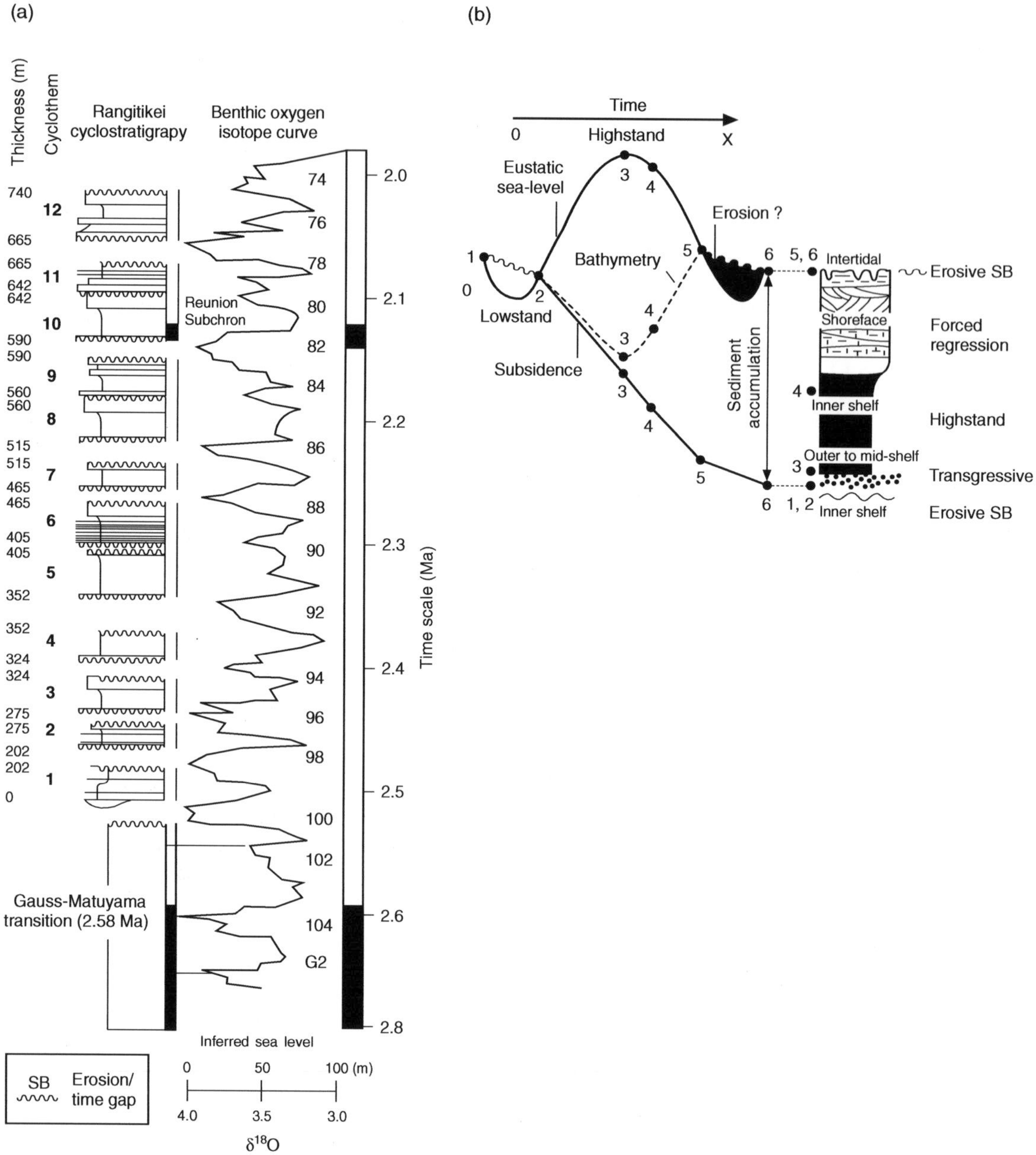


**Fig. 10.10** (a) Correlation of shallow marine Plio-Pleistocene sedimentary cycles from New Zealand and sea-level change from the oceanic oxygen isotope record. (b) Summary model to show how one shallow marine cycle of sedimentation is produced by a combination of subsidence, eustatic sea-level change and sedimentation. SB, sequence boundary. (All after Naish, 1997; see also Naish *et al.* (2009) for Plio-Pleistocene obliquity-forced oscillations of West Antarctic ice sheet and global eustatic implications.)

sediment below mean sea or lake level. The deposition of sediment on land has no such limits since it is governed by the extent of depositional fluid boundary layers.

5 Carbonate sediments are particularly suited to sequence stratigraphic analysis since many carbonate sedimentary environments are closely controlled by water depth and so their recognition in outcrops or wells enables often sophisticated analysis of palaeobathymetry (Fig. 10.11), e.g. oolite shoals, peritidal shrinkage and expansion features, supratidal evaporates, and schleractinian coral reef front buildup (Chapter 20).

Many detailed aspects of sea-level related cycles in specific sedimentary environments will be presented in Parts 5 and 6.

## 10.4 Tectonics

The large-scale pattern of sediment distribution and the partitioning of the Earth's surface into areas of uplift, subsidence, erosion and deposition depend largely upon tectonics. Long-term sediment preservation occurs in subsiding sedimentary basins so it is particularly important to understand basic tectonic mechanisms of subsidence and patterns of adjacent uplift in these systems. Sedimentary basins come in all shapes and sizes with a variety of tectonic origins (Chapter 22). These sediment repositories owe their origins to crustal subsidence relative to surrounding, often uplifted, areas. Basins may be active for hundreds of millions of years, yet within a few million become sourcelands themselves through structural deformation and uplift: a process known as *inversion*. The sediments that accumulated in the basin are then recycled into some adjacent or distant structure.

Basins form due to deformation processes acting upon a compositionally and mechanically layered lithosphere which has variations in material strength due to compositional and thermal structure. The most significant strength change relevant to the surface deformation of continents occurs at the crustal *brittle-to-ductile transition*. Above this, brittle deformation occurs by faulting; below it, deformation is of pure-shear type, taken up by quasi-continuous plastic creep at a rate determined by imposed local or far-field tectonic stresses and mineralogical composition. The depth of the brittle-to-ductile transition is extremely sensitive to thermal gradient: increased gradients cause the transition to migrate upwards. In recent years workers have also recognized that the strength of both the upper mantle and the lower crust also depend upon water content and that generally the former must be considered weak. The following notes on basin origins and architecture are somewhat amplified in Chapter 22.

In active *rift basins*, displacement occurs along normal faults as the crust is stretched during the *syn-rift* phase of basin evolution. The result is rotation of rigid crustal blocks about horizontal axes (Fig. 10.12). Relative uplift of the immediate footwall block to faults is subordinate (about 10–30%) to subsidence of the hanging wall block. Rifts (aka *graben*) so-produced may take the form of long, linear juxtaposed *half-graben* (e.g. Baikal, Rio Grande, Corinth, Rhine, Suez, Awash rifts), or comprise broad zones with many such structures (e.g. Basin and Range). Magmatism often accompanies crustal extension, either due to decompressive partial melting of the asthenosphere as it is forced to rise under the thinning lithosphere or as more voluminous volcanism from an upwelling asthenospheric plume.

As active extensional tectonics ceases a rift province undergoes general subsidence as lithosphere and asthenosphere, previously heated or stretched during extensional tectonics, contracts and slowly subsides (Fig. 10.13). This is the post-rift (aka *thermal sag*) phase of subsidence that dominates the coastal plains, shelves and slopes of passive continental margins.

Basins at or adjacent to actively converging plate margins result when major thrust faults rupture so that the fault hangingwall rocks rises up and over the footwall during a *syn-thrust* phase whose horizontal magnitude may be tens of metres in larger 'quakes (Fig 10.14). As thrusting develops, a monoclinal fold forms above the fault tip, the latter eventually rupturing through to break surface. Tell-tale signs of the process are produced in syn-tectonic deposition around the developing structure. The uplifted hangingwall crust is a new load on the footwall which subsequently bends down over a larger area (Fig. 10.15). Deformation of this type occurs not only in *foreland basins* adjacent to developing mountain ranges but also as oceanic lithosphere is bent into trenches during subduction.

*Strike-slip basins* are due to oblique plate or crustal block motions along transcurrent faults and are able to produce zones of uplift and subsidence because of

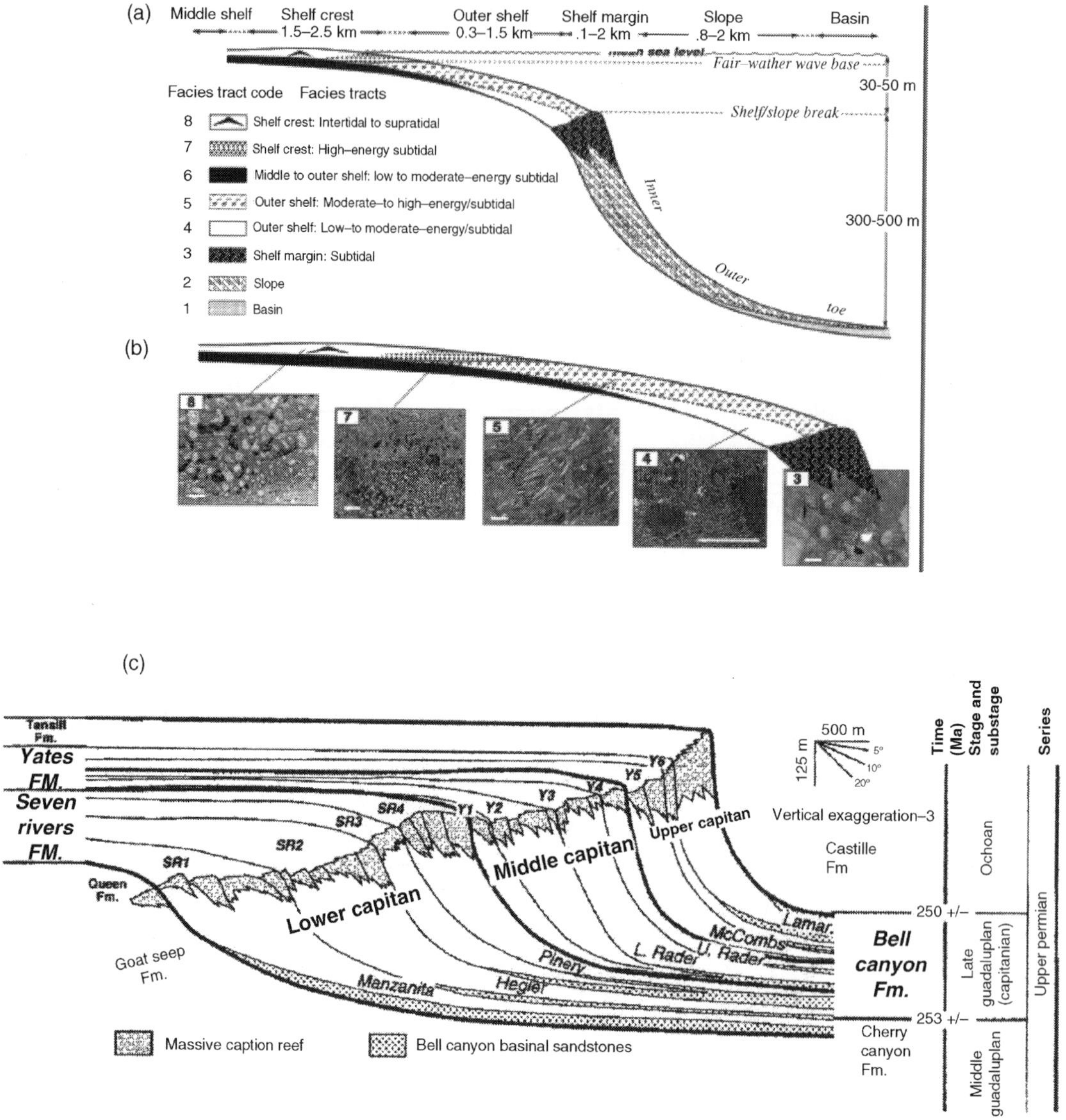


**Fig. 10.11** Sequence stratigraphic analysis in carbonate sediments—examples from the famous Capitan reef complex, New Mexico. (a) Generalized 2D cycle showing vertical and lateral position and width ranges of major environments related to palaeobathymetric profile. (b) Expanded shelf part of the cycle with photographs of key facies showing a general decrease in interpreted depositional energy downdip. Scale bar is 1 cm for all photographs. Numbers correspond to facies-tract legend. (c) Simplified cross-section along McKittrick Canyon. Heavy lines are composite sequence boundaries. Thin lines are high-frequency sequence (HFS) boundaries. Formations included in this study are the Seven Rivers and Yates on the shelf, the lower and middle Capitan at the shelf margin, and the Bell Canyon in the basin. (All after Tinker, 1998.)

the kinematic properties of fault bends, splays and overlaps (Fig. 10.16).

Large-scale (up to $10^4$ km$^2$) *cratonic basins* feature abundantly in the geological record and are somewhat enigmatic in that they appear to be unrelated to crustal extension or shortening. One idea is that they arise through dynamic response to deep subduction in an overriding plate that causes transient patterns of

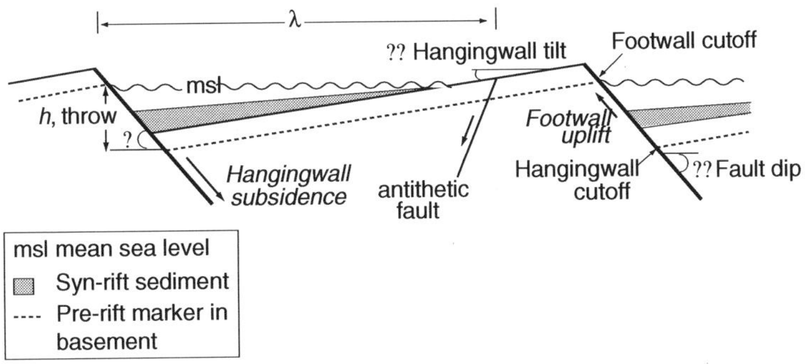


**Fig. 10.12** Tilt blocks and their kinematics: definition diagram.

mantle flow and density changes that induce subsidence or uplift over $10^7$ Myr or so.

## 10.5 Sediment yield, denudation rate and the sedimentary record

The magnitude of sediment yield is expressed as mass lost per time increment, whilst denudation as mass lost per unit surface area per unit time. Local denudation rates vary widely due to involvement of many environmental variables. An exciting development over the past decade has been the recognition that nonuniform denudation rates induced by variable precipitation over a developing mountain belt may significantly affect tectonic exhumation rate, metamorphic grade and lithospheric rheology along active structures like thrust faults. This emphasizes the intimate link that exists between surface and subsurface geological processes.

Catchment- to regional-scale denudation rates may be directly computed from river gauging station data on particulate and dissolved sediment load, particularly when long-term records enable sampling of extreme events, or from selected short-term data that have sampled one or more such extreme events. However, direct sampling of denuded products in most catchments hardly goes back more than 120 years and may not be able to sample enough extreme events. The most convenient way of bypassing this problem involves using single catchments where an outlet leads to a confined depositional basin whose sediment volumetrics may be calculated (Fig. 10.17). Use is frequently made of some datum below a deposit like a regional erosion surface (peneplain, pediplain, marine erosion surface) or a volcanic outflow.

The advent of mineralogical dating techniques, firstly of apatite fission tracks (AFT) and, more recently, cosmogenic nuclides, has revolutionized the scope for computation of volumetric estimates of catchment denudation. For example, they have determined an acceleration in erosion rates in the Himalaya since about 4 Ma which is attributed to the onset of glaciation. It is instructive to compare denudation data over different timescales using different techniques; noteworthy early examples from the Appalachians were followed by data from the Idaho Rockies indicating disparities in mountain erosion rates between short ($10^2$ yr) and long ($10^3$–$10^6$ yr) timescales due to the inability of short-term gauging data to sample more extreme, rare sediment delivery events. In the easily erodible Taiwan orogen, similarities of short- (30 yr) and long-term ($10^6$ yr) erosion rates indicate that rapid uplift, frequent tropical storms and landslides lead to apparent equilibrium steady state between denudation and rock uplift. By way of contrast, in the slowly denuding craton of Galicia, northwest Spain, comparison of long-term estimates constrained by AFT with directly measured data on suspended and dissolved sediment flux for normal precipitation and exceptionally wet years (Fig. 10.17) show that sediment yields for normal years are currently about ×4 greater than the long-term AFT average, with the discrepancy rising to about ×12 for wet years.

### Empirical relations for denudation rates

Given the sheer complexity of the processes involved in determining sediment yield or denudation rate for a particular catchment, statistical links have been

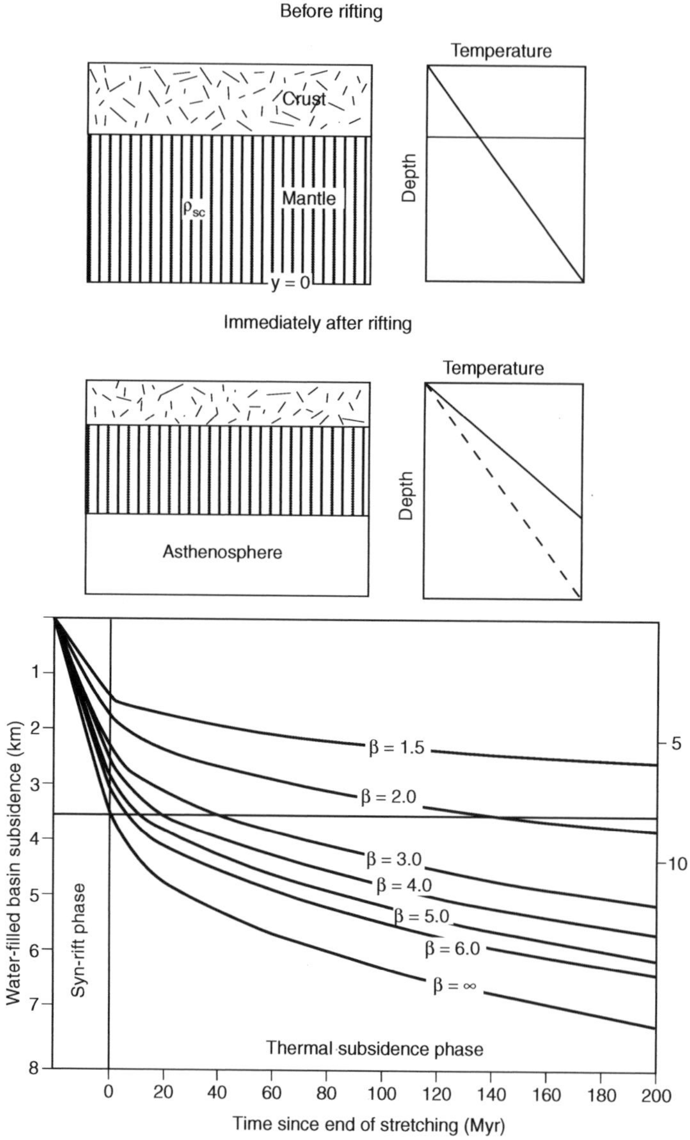


**Fig. 10.13** (a) Simple uniform extension by a factor β and the resulting geotherms. (After classic paper by McKenzie, 1978.) (b) The exponential decay with time of thermal subsidence caused by simple uniform stretching of a lithosphere of initial thickness 125 km, crustal thickness 31.2 km. (From Dewey, 1982.)

sought between them and a number of geological and geomorphological variables thought to be the most important controls. Early efforts established a correlation of denudation rate and mean catchment relief, a useful guideline correlation confirmed by several more recent multivariate analyses. Such correlations exist because of the obvious link between elevation and orographic rainfall on the one hand and elevation and relief on the other. Also important is the fact that many young tropical mountainous terrains (Taiwan, New Zealand) form as accretionary prisms above subduction zones undergoing very rapid uplift: the combination of easily erodible rocks, steep slopes and high rainfall leads to ideal conditions for rapid

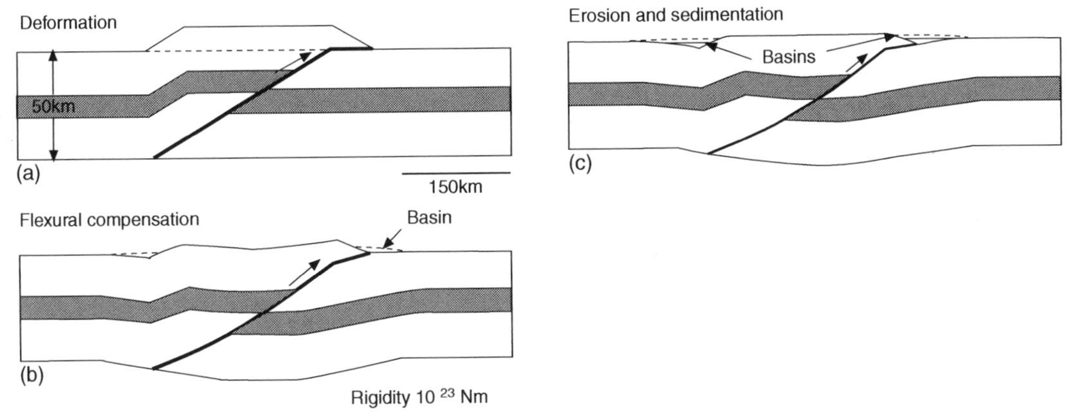


**Fig. 10.14** Cartoon to show how shortening across a deep thrust fault causes a regional load which must flexure the whole brittle lithosphere and cause deeper compensatory viscoplastic flow. (After Flemings & Jordan, 1990.)

erosion. Not only does high rainfall tend to cause more rapid weathering, it also reduces effective stresses within rock masses so encouraging slope-failure, landsliding and generation of avalanche and debris flows. Failure is also frequently triggered in such active tectonic regimes by co-seismic ground accelerations.

Although all of the above processes control the export of rock from catchment slope to drainage channel, it is the stream power of the channel itself that is the true rate-controlling step, for no matter how rapid the process of landsliding, it is the ability of a river to transport dumped debris that will control rate of removal from the catchment. The link between rapid uplift, elevation, relief and precipitation means that stream power will be high in the channels of mountain rivers draining such terrains. This is seen in the predominance of erosion from mountainous catchments worldwide and particularly along active plate margins in central and east Asia.

## Modelling denudation

The water cycle and its control by vegetation is the most analytical approach to modelling erosion rates, a concept we owe to geomorphologist M.J. Kirkby. It is also the most predictive and climate-sensitive through rate-controlling steps that include the Arrhenius effect of temperature on silicate weathering rate and first-order control of vegetation biomass by temperature and rainfall. CSEP/Medalus hydrological models generated by Kirkby and co-workers generate an excess of sediment that is free to be transported out of a catchment by the runoff that is spared from precipitation used in weathering and photosynthesis. Through a gradient term the model predicts stream power available to transport sediment from catchment slopes.

## Global-scale denudation

Attempts to directly determine mean annual global sediment yields and denudation rates must rely on integration of estimates taken over all the world's catchments—one estimate for total world yield is between 1.6 and 2.0 $10^9$ t/yr (Fig. 10.18). There are various problems associated with such efforts, chiefly the lack of accurate time-series data for many large catchments and the effects of dams. Hydrologically based estimates of past global-scale denudation rates were first systematically addressed by S.A. Schumm whose data on the correlation of mean annual precipitation with denudation rate in small- to medium-sized catchments in the mid-West USA led him to deduce that precipitation controls vegetation, which in turn controls rate of physical erosion and sediment flux by limiting hydrological runoff. Since evolution has changed the role and scope for vegetational influence, he identified the advent of Gramineae (grasses), Angiosperms (flowering plants), Gymnosperms (seed plants like conifers, cycads and relatives) and algal colonization of the land as underpinning a speculative group of time series of relative denudation rate with mean annual precipitation (Fig. 10.18). These

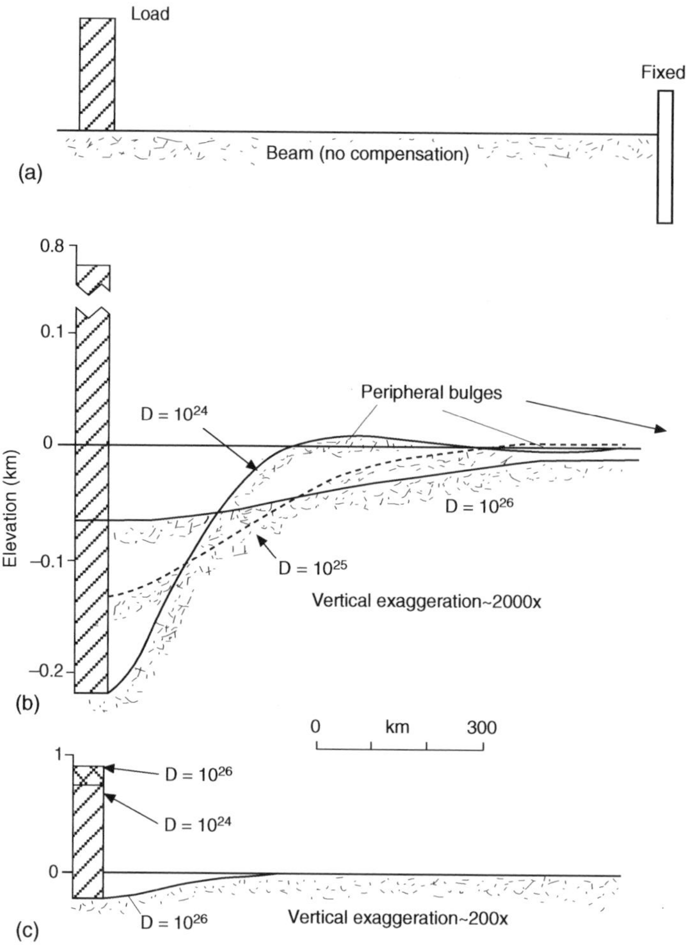


**Fig. 10.15** Sketches to illustrate flexural bending of an elastic beam (fixed at the right in the far-field) as an analogy with a lithospheric plate resting on the asthenosphere and loaded by a thrust. (After Beaumont 1981; Jordan 1995.) (a) Initial state or the case where the beam supports the imposed load. (b) Complete bending subsidence and peripeheral uplift for a constant load of width 50 km and thickness 1 km. Note broken scale at positive elevations > 0.1 km. (c) True scale section of (b) to emphasize the point that complete subsidence due to crustal thrust loading is only a small fraction (about 10% for a flexural rigidity, *D*, of $10^{26}$) of the elevation of the applied load, i.e. thrust-bounded mountains are very much higher than their adjacent basins.

featured a massive increase in rates before Upper Palaeozoic time. Although the approach has remained inspirational there are various problems connected with applying a hydrological data set from a small region to a global context over the vastness of geological time. It remains to update Schumm's speculations in the light of subsequent developments, such as:

- data on modern mean annual global sediment fluxes that, although crude, enable estimates as hindcasts for the geological past (Fig. 10.18);
- global marine radiogenic strontium time-series (see below; Fig. 10.18) that enable relative Phanerozoic denudation rates to be estimated;
- better understanding of the various controls on denudation rate, including the important role of tropical storms, active tectonic uplift, elevation and relief;
- H. Falcon-Lang has pointed out to me that more specific understanding of the timing and effects of key floral evolutionary changes with sedimentological

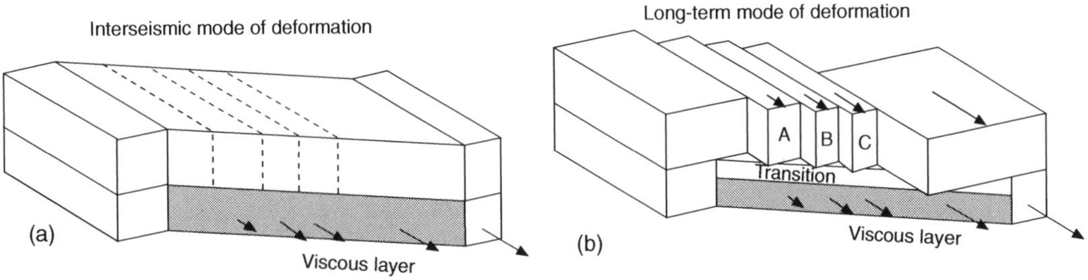


**Fig. 10.16** How strike-slip faults work in a zone of lithosphere subject to simple shear. Between earthquakes the blocks bounded by faults remain locked as they are subject to basal viscous stress and they accumulate elastic strain. During earthquakes the blocks move horizontally relative to each other. The total long-term velocity of the rigid blocks is the same as that of the underlying flowing viscous layer. (After Bourne *et al.*, 1998.)

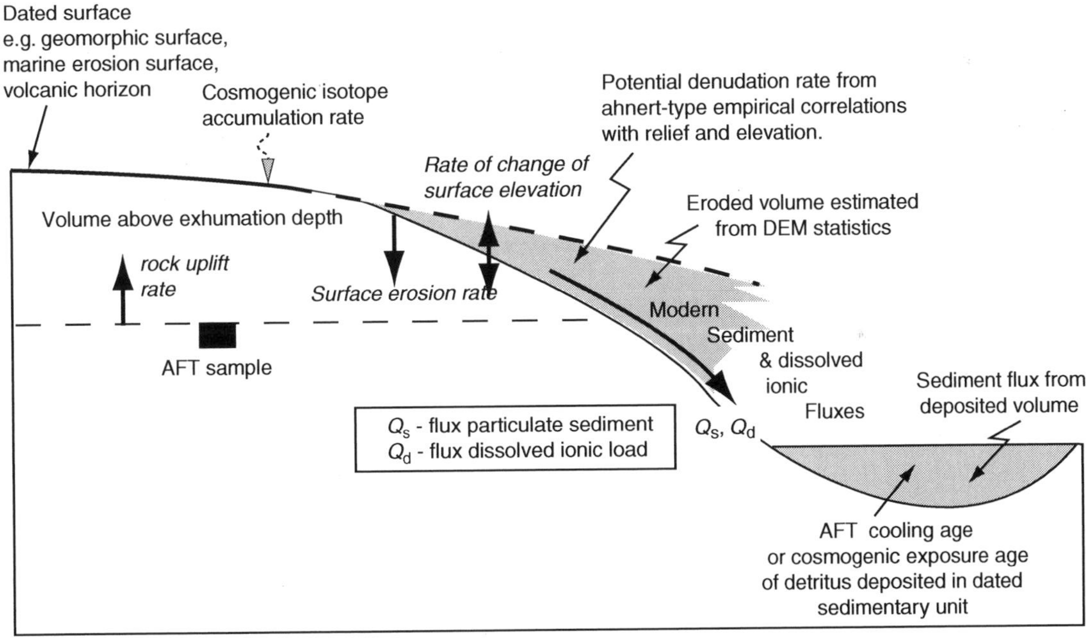


Galician data

| *Long term estimates of denudation rate (mm/kyr):-* | | *Modern sediment flux estimates of denudation rates (mm/kyr):-* | |
|---|---|---|---|
| AFTD from catchment bedrock | 13 | Qs, flux particulate sediment (normal year) | 10 |
| Potential from relief | 27 | *Ditto*, exceptionally wet year | 60 |
| Potential from elevation | 16 | Qd, flux dissolved ionic load (normal year) | 44 |
| Landscape volume loss | 10–15 | *Ditto*, exceptionally wet year | 97 |

**Fig. 10.17** Sketch to illustrate methodologies commonly used to determine catchment or regional scale denudation rates, with measured comparative estimates from Galician (northwest Spain) catchments for some of these techniques. AFTD, apatite fission track data; DEM, digital elevation model. (Data after Perez-Arlucea *et al.*, 2005)

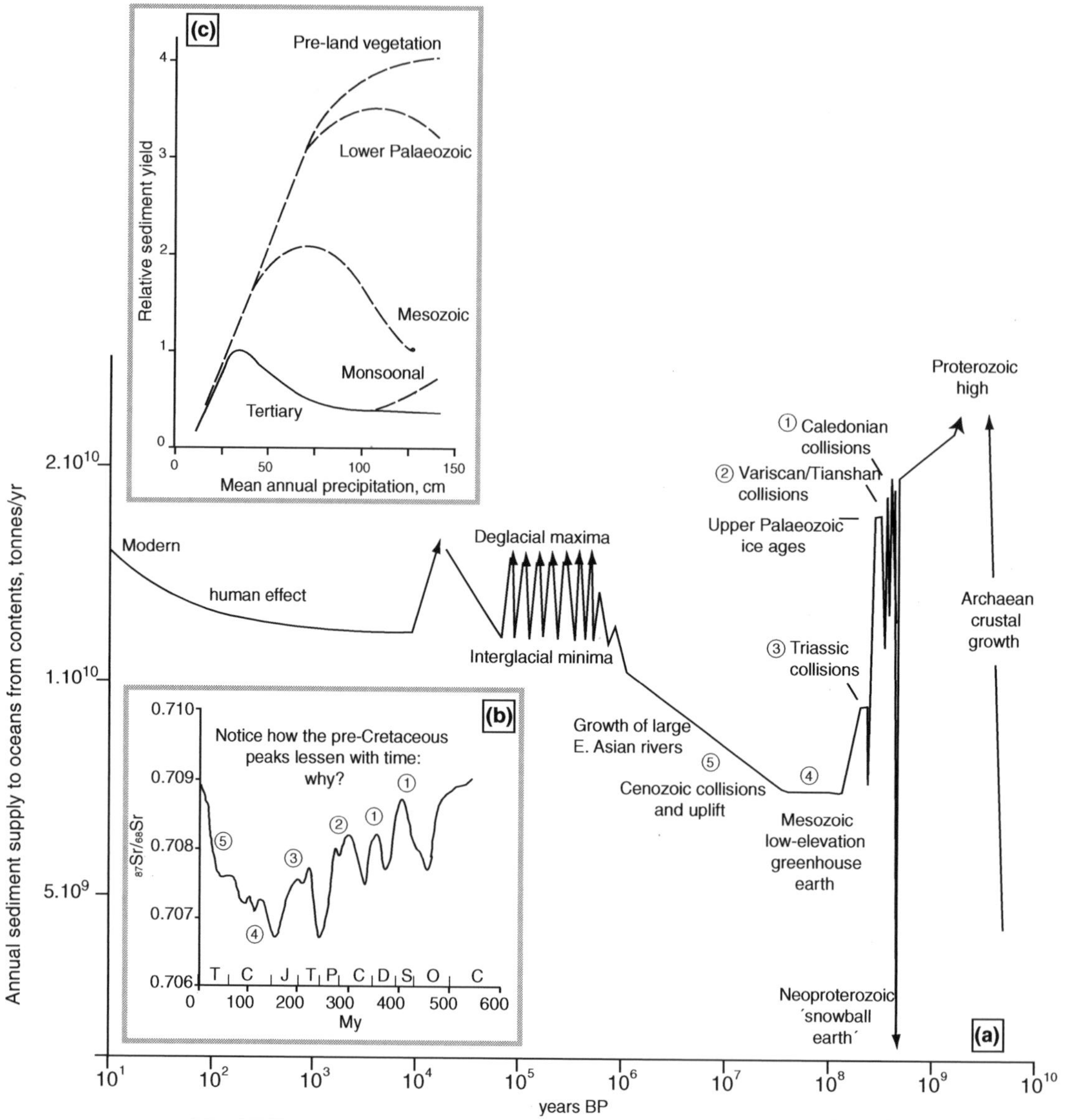


**Fig. 10.18** (a) Speculative time series of denudation rates expressed as tonnes per year of sediment supply to the oceans over geological time. Note the logarithmic abscissa scale. The modern flux is guessed to be about 50% of the pre-Himalayan/Tibetan uplift flux, an assumption based upon current 30–50% of total global sediment load into oceans coming from 'young' East Asian rivers. Arrowheads point to generally unknown upward or downward limits. Events 1–3 and 5 are high sediment supply transients associated with major orogeny (*cybertectonic* control). Event 4 is a postulated global Mesozoic low period of sediment supply. The hypothesis of Snowball Earth requires almost zero supply. (b) The Phanerozoic marine Sr-isotope curve, with events 1–5 as in (a). The reason for the secular changes towards progressively lower $^{87}Sr/^{86}Sr$ peaks with time is unknown, or at least uncommented on in the literature: they may be due to progressive dilution of a more radiogenic Proterozoic ancestral signature or to decreasing chemical weathering contributions relative to stable mantle outwelling. (After Veizer *et al.*, 1999.) (c) Schumm's (Schumm, 1968a) speculative plot of climate/vegetational control of relative sediment supply (see critical discussion in text).

implications are (i) Proterozoic rise of microbial and lichen terrestrial films culminating in (ii) the mid-Ordovician rise of mosses, hornworts and liverworts, (iii) mid-Silurian rise of vascular plants culminating with seed plants (gymnosperms) and large forest trees (= dryland colonists) in the late-Devonian and (iv) Cretaceous rise of angiosperms, especially grasses, into the early Tertiary in response to global cooling and aridification.

The issue of past continental erosion rates has recently been at the forefront of efforts to model the long-term global carbon budget. Thus, as part of the long development of GEOCARB cycling models, R.A. Berner attempted to use a parameter derived from observed $^{87}Sr/^{86}Sr$ marine ratios (see below) to estimate rates of silicate weathering. A more direct approach to mountain uplift and weathering was later adopted whereby estimates for continental weathering were incorporated into long-term C-cycling models. The method chosen was the computed abundance of siliciclastic rocks in the geological record. Although more direct, the method is grossly inaccurate for most of the pre-Tertiary record because of extremely large (and unexplored) errors arising from (i) incorrect stratigraphic identificationand (ii) incompleteness due to non-conservative behaviour in the rock cycle, namely deep burial, erosion, deformation and metamorphic transformations of original sediment deposits.

Marine strontium isotopes play a key role in determining relative denudation rate since $^{87}Sr/^{86}Sr$ ratios depend upon inputs from mid-ocean ridge hydrothermal output (lower ratios) and continental weathering solutes in rivers (higher ratios) (Fig 10.19). Sr has rather a long residence time in the oceans (several hundred thousand years) and so $^{87}Sr/^{86}Sr$ cannot be used to assess short-term climatic signals at Milankovich periodicities. Longer term fluctuations over the past 200 Myr have been ascribed by some to variable seafloor spreading rates (see Chapter 23) but this cannot have been the cause of the most recent rise in $^{87}Sr/^{86}Sr$ recorded in the ocean sedimentary record since about 35 Ma (Fig. 10.18). This increase has been ascribed to either (i) increased Quaternary continental denudation fluxes due to climate working on increased secular global elevations and relief over the past 30–50 Myr and/or (ii) erosion of highly radiogenic calc-silicate metamorphic and igneous sourcelands of the High Himalaya, also due to uplift and erosion since ca 40 Myr. Both effects are likely to have contributed and have increased due to increased

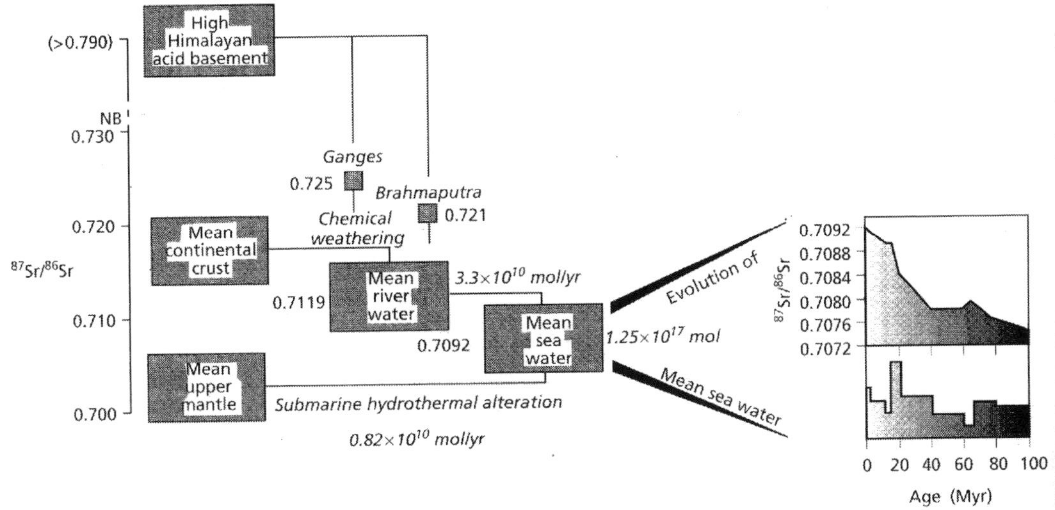


**Fig. 10.19** The major $^{87}Sr$ and $^{86}Sr$ fluxes and sources, and the change and rate of change of the mean oceanic $^{87}Sr/^{86}Sr$ ratio over the past 100 Myr. The oceanic mean composition is very strongly affected by weathered input from high $^{87}Sr/^{86}Sr$ acid leucogranites of the Himalaya. (Various data from Krishnaswami *et al.*, 1992; Richter *et al.*, 1992.)

physical erosion during Pleistocene glaciations, the flux-magnitude term probably being the stronger influence. It is remarkable and highly significant that a small, ca 5%, increase in total global silicate alkalinity flux occasioned by enhanced Cenozoic uplift and basement erosion was capable of triggering major global climate change. Supporting evidence for enhanced Pleistocene denudation rates comes from the increase of sedimentary mass recorded over the past 2–4 Myr in sedimentary basins worldwide and recorded by combined cosmogenic and AFT unroofing studies from the Himalaya in particular. Of great topical interest is the question of whether Sr-budgets changed globally during glacial/interglacial intervals. There is no direct evidence of cyclicity from $^{87}Sr/^{86}Sr$ ratios measured in foram samples from deep-sea cores, rather the data show a remorseless linear increase over the past 0.5 Myr. But a problem arises because the rate of this increase is far to slow compared to modern fluxes of riverine $^{87}Sr$ into the oceans. It is thought by some that this Holocene flux is too high compared to the mean global flux due to a multi-millenial-scale transient imposed by chemical weathering of freshly provided deglacial debris. Over the longer term, the global Phanerozoic secular $^{87}Sr/^{86}Sr$ curve is now well constrained (Fig. 10.18) and its periodic peak excursions to greater $^{87}Sr/^{86}Sr$ ratios are routinely explained by increased mean global erosion rates due to the effects of plate reorganization, orogeny and mountain uplift.

### Sedimentary and stratigraphic process models

Quantitative sedimentary and stratigraphic models attempt to forecast or to hindcast the details of sedimentary successions. To be successful they must robustly withstand field testing so that they include surface environmental variables such as tectonic subsidence, spatial gradient in subsidence rate, sediment flux and response to sea-level fluctuation (Fig. 10.20). Examples exist for carbonate, fluvial and coastal sediments. Details of some of these will be found in specific chapters in Parts 5 & 6. For siliciclastic sediments there is still a great chasm between sedimentary models and more fundamental hydrology-based approaches developed by geomorphologists. Existing clastic sedimentary models are kinematic rather than dynamic since they are artificially separated from the hydrology, whereas in fact they should be driven by the hydrological budget and, ultimately, climate. It is a mistake to assume that the hydrological or climatic cycle in a catchment acts kinematically: it is in fact dynamic, with both physical and thermodynamic energy transfers and transformations taking place constantly within the system. Thus catchment processes create the entire basin landscape from a number

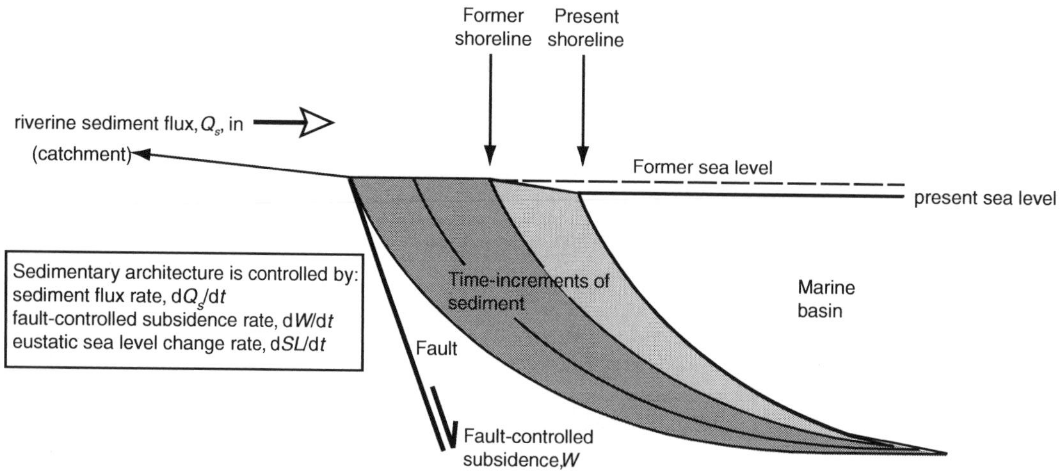


**Fig. 10.20** Definition diagram for modelling sedimentary architecture controlled by: sediment flux rate, $dQ_s/dt$; fault-controlled subsidence rate, $dW/dt$; sea-level change rate, $dSL/dt$. (Based on Ritchie *et al*., 2004a).

of prior conditions, rather analogous to 'Nature vs. Nurture' concepts for individual human development, whereby the genetic make-up of an individual (nature-providing) is acted upon by external circumstances (nurture-modifying). Tectonics and lithological make-up are the given catchment genes whereas hydro-climatic variables, including vegetation, and sea-level change nurture and modify the sedimentary infill. Much work remains to be done in developing sedimentological models based upon energetics and taking full account of the complex feedback between variables discussed in this section, such as climate, vegetation, sediment and water yields, sea-level fluctuations and tectonics.

## Further reading

### *Climate*

Lockwood (1979), sadly out of print, is an excellent, though idiosyncratic, introduction to modern climatology. A more recent text is Linacre & Geerts (1997), with a refreshing antipodean bias. Scorer (1997) is more physical. Boos & Kuang (2010) shed new light on Indian monsoon dynamics and cast severe doubt on any role for its amplification by the Tibetan Plateau. Lamb (1995) is incomparable for historic climate change. Wright *et al.* (1993) is indispensable for post-18 ka climate change. Older palaeoclimate is comprehensively covered by Frakes *et al.* (1992). A host of good papers on the sedimentological consequences of orbital forcing are to be found in de Boer & Smith (1994a). Many interesting papers are also in the volume edited by Head & Gibbard (2005).

### *Sea level*

For high frequency late Quaternary sea-level changes see pioneer paper by Chappell & Shackleton (1986), with later papers by Raynaud *et al.* (2005) and Rabineau *et al.* (2006). Neogene cyclostratigraphy is presented in seminal papers by Hilgen (1991, 1994), Lourens *et al.* (1996) and Naish (1997). For obliquity-forced Plio-Pleistocene eustatic fluctuations due to oscillations of West Antarctic ice sheet and global eustatic implications see Naish *et al.* (2009). For suborbital changes see Chen *et al.* (1991), Zhu *et al.* (1993), Carew & Mylroie (1995), Hearty & Kindler (1995), Stirling *et al.* (1998), McCulloch & Esat (2000), Muhs *et al.* (2002), Jedoui *et al.* (2003), Blanchon *et al.* (2009) and Andrews *et al.* (2007). For Holocene events see Rohling & Pälike (2005) and Yu *et al.* (2007). For ice-sheet melting see Cuffey & Marshall (2000) and Alley *et al.* (2005). For ocean warming see Domingues *et al.* (2008). For ice-core data see *EPICA* community members (2004). For critical stratigraphic field evidence against far-field geodetic theory see Pirazzoli (2005).

Pitman (1978) is the classic paper on global tectonic eustasy. Miall (1997) is the most scholarly account of the whole business of sequences, eustasy and much else. (Hallam, 1997) is also a good scholarly read. Vail *et al.* (1977) is more partial and is of interest in pushing a corporate scheme of science, since much criticized, rightly so in my view (see Miall & Miall, 2001). Emery & Myers (1996) is useful but 'cookbookish'. Case histories and much to ponder over appear in the twin peaks of Weimer & Posamentier (1993) and Loucks & Sarg (1993). Poulsen *et al.* (1998) is a careful account of the 3D complexities of 'real world' stratigraphy. The very important science of 'cyclo-stratigraphy' for Neogene times is reviewed by de Boer & Smith (1994b). Sequence stratigraphy is very clearly explained in the Open University text edited by Coe (2003). For examples and discussion of global sea-level changes see Gurnis (1993), Burgess *et al.* (1997) and Dewey & Pitman (1998). The global sea-level scenario for Cretaceous times is dealt with by Heller *et al.* (1996).

### *Tectonics*

The reader is pointed to Allen & Allen (2004) for an introduction to mechanisms of basin subsidence and for details of the treatment of sediment compaction, loading and lithospheric flexure. Turcotte & Schubert (2002) is the fundamental source on geodynamics, though not for the mathematically challenged.

### *Erosion and denudation*

Carson & Kirkby (1972) is the fundamental text, sadly now out of print. $^{10}$Be and denudation rates are studied by Morris (1991), Lee *et al.* (1993) and McKean *et al.* (1993), and AFT applications by Gleadow *et al.* (1986) and van der Beek *et al.*,

1994). Mass-balance studies are by Pavich (1985, 1986), Boettcher & Milliken (1994) and Pazzaglia & Brandon (1996). Large-scale studies of denudation rates are by Langbein & Schumm (1958), Schumm (1968a), Ahnert (1970), Scott & Williams (1978), Milliman & Meade (1983), Milliman & Syvitski (1992), Jansson (1988), Summerfield & Hulton (1994), Ludwig & Probst, (1996), Hovius (1997), Hovius *et al.* (1997, 1998), Pinet & Souriau (1988) and Poulos & Collins (2002).

Basinal studies of denudation and sediment flux: the inverse approach by Beaty (1970), Copeland & Harrison (1990), Ibbeken & Schleyer (1991) Collier *et al.* (1995), Pazzaglia & Brandon (1996) and Dadson *et al.* (2003).

Modelling sediment supply, vegetation and climate change: implications for basin stratigraphy by Perlmutter & Mathews (1989) and Frakes *et al.* (1992). CSEP models are by Kirkby & Neale (1987), DePloey *et al.* (1991), Kirkby (1995) and Kirkby & Cox (1995). Other yield studies are in Schumm (1968a), Burkham (1972), Graf *et al.* (1991), Prentice *et al.* (1992), Blum (1993), Tzedakis (1993), Hereford (1993), Huckleberry (1994), Nott & Roberts (1996), Kirkby *et al.* (1998), Kirkby (1999), Bogaart & Van Balen (2002) and Bogaart *et al.* (2002, 2003).

Marine Sr-isotope ratio and continental erosion rates are studied by Palmer & Elderfield (1985), Edmond (1992), Richter *et al.* (1992) Rea (1993), Bickle *et al.* (1995), McCauley & DePaolo (1997), Quade *et al.* (1997), Blum *et al.* (1998), Galy *et al.* (1999), Peizhen *et al.* (2001) and Vance *et al.* (2003). Longer term Sr time series studies are by Berner (1994), Bruckschen *et al.* (1999), Jacobsen & Kaufman (1999), Prokoph & Veizer (1999), Veizer *et al.* (1999) and McArthur *et al.* (2001).

Sedimentary and stratigraphic process models: carbonates by Aurell *et al.* (1995), Barnett *et al.* (2002) and Boylan *et al.* (2002); fluvial sediments by Mackey & Bridge (1995) and Karssenberg & Bridge (2008); coastal siliciclastic environments by Ritchie *et al.* (2004a,b).

# Part 5

# CONTINENTAL SEDIMENTARY ENVIRONMENTS

*The bar at the mouth of the Aigouille lay before him at last, a broad strip of sand strewn with bleached tree-trunks: for except at times of flood most of the river stayed in its lagoons and marshes, the rest reaching the sea by a channel no wider than a man could leap.*

Patrick O'Brian, from *The Ionian Mission*, Harper Collins, 1981 (Stephen Maturin just landed in France)

## Introduction

Several of the five great continental sedimentary environments—river, alluvial fan, lake, hot desert and glacier ice—may be closely linked in both space and time. In each of them climate plays the dominant role in determining water flux, ice discharge and sedimentation rates. Although river channels are adjusted to carrying the most frequent discharges, they cannot deal with extreme events that might only recur on multidecadal–multicentennial timescales. Thus river floods are the main events governing deposition and erosion on floodplains and for the tendency of river channels to suddenly or gradually change course. Similarly, lakes are extremely sensitive to changes in runoff and evaporation because these factors will determine water depth, salinity and perhaps the very existence of the standing water body in the medium to long term. The decadal or so influx of floodwater into Lake Eyre, central Australia, controlled by La Niña forcing, is a fine example, as is Lake Chad (Part 5 Fig. 1) and the now entirely dried-up mid-Holocene lake beds present in many areas of the modern day Sahara Desert. All these are *endorheic* lake basins, defined as internally drained, with no outlet to the ocean. Many serve as chief source areas for wind-blown sediment transport in the great trade wind deserts, indeed in the fringes of modern active desert systems there often exist relict areas of once active wind-blown sediment transport, now stabilized by vegetation.

So, in many ways we may best view the continental sedimentary system as a vast interconnected whole, with each subenvironment linked to both past and present environmental variables. We are dealing here with inheritance, with legacies handed down by previous climatic conditions, with baggage if you like. We are also talking teleconnections, with climatic fluctuations like El Niño–Southern Oscillation (ENSO), other decadal or multidecadal oscillations and variations in monsoonal strength adjusted to orbital time periods. All these have a profound effect upon sedimentary systems. Regarding the legacy, the evidence for this is all about us, particularly in areas where there is a direct glacial legacy from the late Pleistocene

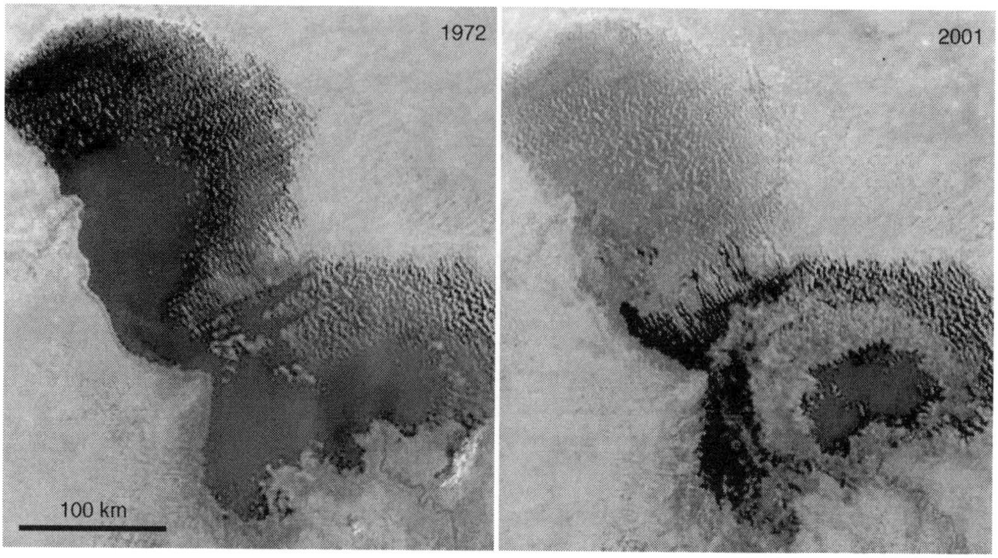


**Part 5 Fig. 1** These NASA images reveal the interactions over decadal time spans between arid-zone sand sea, lake and river in the Lake Chad area on the SW margins of the Sahara desert. Lake expansion and contraction cycles due to climate change have major implications for the sedimentary record.

(Part 5 Fig. 2). Here, paraglacial contributions to the modern sedimentary systems of rivers may dominate transport and channel characteristics. Looking back into the sedimentary record of previous greenhouse and icehouse worlds is therefore a great challenge of interpretation and we should learn much from the past to inform us of possible future trends in continental reaction to climate change by doing so.

Yet neither is the continental sedimentary landscape immune from the effects of sea-level change, for during periods of glacio-eustatic lowstand, vast areas of continental shelf are added to the land area of the planet and rivers advance far across them to the shelf edge. The effects of increased shelf gradients and increased discharge from seasonal melting of ice-choked catchments may cause profound changes to upstream channels, including incision and changes to channel morphology. In other areas away from the direct influence of ice the climate may change quite independently, becoming more or less arid depending upon the regional regime.

This brings us to the final link, that of ecology and especially the control by climate of vegetation. Run-off, erosion, sediment yield and surface sediment stabilization are all closely related to the local biomass and it seems unlikely that the sedimentary system can be described without an adequate appreciation of this factor.

(a)

**Part 5 Fig. 2** Sea-cliff sections from near Start Point, Devon, Southwest England, outwith the furthest southern advance of Pleistocene land-ice over the northwest European continental shelf reveal the often startling evidence of global sea level and local climate change. Here the overview and close-up show a MIS 5e abraded shoreline platform cut into Devonian schists about 5 m above modern high water level (see Chapter 23.15). The top part of the cliff section features massive, structureless, colluvial fan breccia deposits that probably date from a period of late Pleistocene/early Holocene periglacial development prior to post-Last Glacial Maximum global sea-level rise.

# Chapter 11

# RIVERS

*I do not know much about gods; but I think that the river*
*Is a strong brown god—sullen, untamed and intractable,...*

T.S. Eliot, 'The Dry Salvages', *Four Quartets*, Faber, 1941

## 11.1 Introduction

A major part of the precipitation that falls on Earth's surface eventually finds itself flowing as channelized runoff. The river system itself stretches from upland catchment to ocean, lake or inland basin terminus and can be thought of as an extended conduit for the dispersal of catchment weathering products. Rivers control the supply of sediment and water to almost all other environments so their sedimentary record can tell us much about geological and geomorphological evolution, including tectonic slope changes, source-land geology, climate and human modifications to land use. *Alluvium* is the general term used to describe clastic sediment deposited in any part of a river system. Generally, tectonic subsidence creates the large-scale space for river deposition, with changes of gradient and differential tilting allowing either deposition or erosion, depending on the circumstances. For example, the lateral and vertical growth of active faults and folds perturbs local or regional surface gradients (Chapters 22 & 23). Climate on the other hand provides the water needed for vegetation and soil development, the runoff that allows hillslope sediment transport and thus the eventual sediment and water flux into a river system—changing climate or human modification to land use perturbs these variables. Erosion is commonplace in the Quaternary and older alluvial record of tectonically active basins. One reason is global or local sea-level fall or lake draw-down causing exposure and erosion (Chapter 10). On coastal plains, river channels may override subsidence by incising in response to lowered sea level across a gradient change at the coastal plain–shelf break. Another reason relevant to continental basins is that climate change can alter the balance of a river's hydrological and sediment supply variables; in extreme cases rivers may dry up for thousands of years and their former channels and floodplains become active sites of wind-blown deposition, witness the 'ghost' rivers of the Sahara. Finally, changing land use engineered by humans, for example from virgin prairie and forest to croplands in the upper Mississippi basin, can cause orders of magnitude higher floodplain sedimentation rates.

## 11.2 River networks, hydrographs, patterns and long profiles

We shall see later that many sedimentary models for rivers are kinematic rather than dynamic since they are artificially separated from *catchment hydrology*. A key concept here is that of the annual river *hydrograph*, a simple time series of water discharge (Fig. 11.1). However, the time series of sediment

*Sedimentology and Sedimentary Basins: From Turbulence to Tectonics*, 2nd edition. © Mike Leeder.
Published 2011 by Blackwell Publishing Ltd.

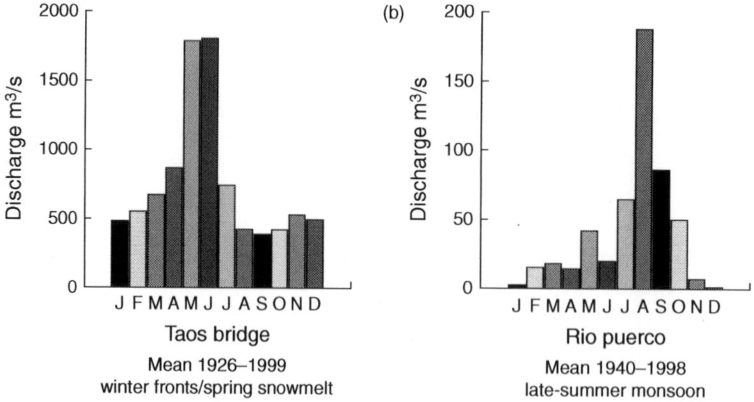


**Fig. 11.1** Annual hydrographs for the Rio Grande, New Mexico to show (a) Typical high mountain northern catchment spring snowmelt discharge pattern from a perennial reach. (b) Summer discharge from ephemeral southern reaches fed by Mexican monsoon. (Data of US Geological Survey.)

discharge may or may not show a correlation to water discharge. Take the example of summer monsoonal rainfall; this falls rapidly over whole catchments and leads to peaks in both water and sediment discharge in river channels. By way of contrast, spring snowmelt-dominated hydrographs (Fig. 11.1a) leads to little sediment flux from catchment hillslopes and hence the channelized water picks-up sediment from the bed of the channel, tending to cause net erosion. It is thus easy to appreciate that sedimentary processes are driven by the nature of the hydrological budget from within a catchment. This is a dynamic process, with both physical and thermodynamic energy transfers and transformations taking place constantly within the system.

River channels vary greatly in size, over more than four orders of magnitude, from mere ditches to the $\gg$20 km wide lower reaches of the Brahmaputra and Ganges. The overall drainage network can be characterized in various ways, but most logically through *stream relative magnitude* whereby tributaries successively add to the downstream relative magnitude of the trunk channel (Fig. 11.2). The number of channels in any system defines the *drainage density*, that is, the

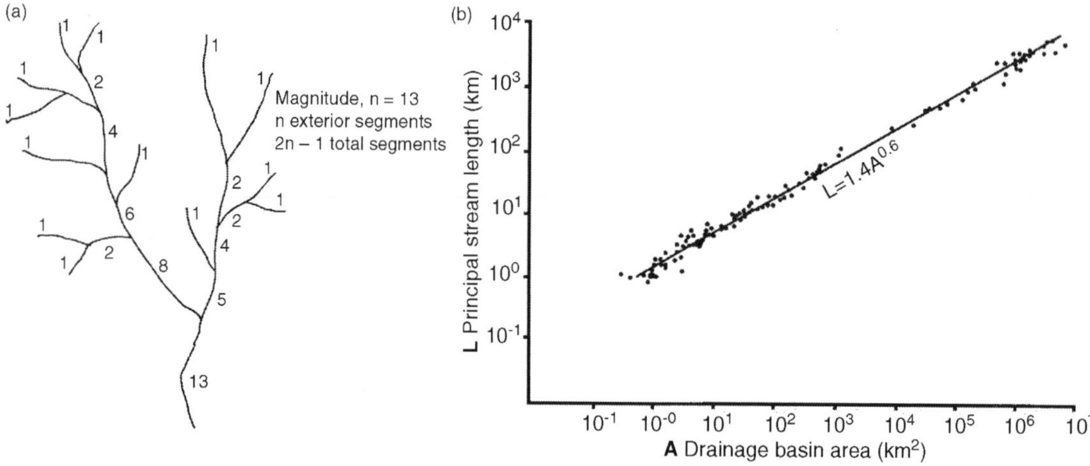


**Fig. 11.2** (a) Definition of stream magnitude. (b) The relationship between drainage basin area and drainage basin length established by Hack (1957) and Leopold *et al.* (1964) with some additional data points.

ratio of total stream channel length to drainage area. This is highest in semiarid catchments where gullying is unimpeded by vegetation. As the length, $L$, of the principal channel increases so does the area, $A$, of the drainage catchment, defined by an empirical fractal relationship termed *Hack's law* which states $L = 1.4A^{0.6}$. Stream length is often controlled by tectonics and thus drainage basin size may be limited according to the nature of the tectonic slope provided.

The actual magnitude of any channel at bankfull flow depends on its width, $w$, and depth, $h$. These basic measures of channel size help to determine the extent of coarse-grained channel deposits. The bigger the channel, the more water it can carry through itself, so we must also characterize channels according to the magnitude of the *mean annual discharge* in $m^3$/sec. Since the mean flow velocity, $u$, in any channel is $Q/wh$, we have $Q = whu$. Expressing width, depth and mean velocity of flow as functions of the mean discharge, we can derive the basic expressions of hydraulic geometry: $w = aQ^d$, $h = bQ^e$ and $u = cQ^f$, where $abc = 1$ and $d + e + f = 1$. The magnitudes of the exponents and constants vary according to different stream types and climatic conditions. The ratio $w/h$ is a particularly key relationship for it appears quite generally that high $w/h$ ratios are characteristic of low-sinuosity channels.

The remarkably smooth longitudinal profile of most channels once they emerge from their bedrock valleys reflects a long-term ability to overcome original and any subsequently imposed irregularities in topography; this is done by the combined effects of erosion and deposition. River profiles are usually concave; all initial or imposed convexities have to decay to concave slopes with time. This can be readily understood by noting that the downstream increase in discharge associated with all stream networks (Fig. 11.3) must be accompanied by a downstream decrease in slope, $S$, if equilibrium, i.e. neither erosion nor deposition, is to be maintained. If slope stayed constant or increased, then erosion would result in a lowering of the bed and the production of a local discontinuity. Profile concavity may be described by an equation of the form, $H = H_0 e^{-kL}$, where $H$ is local profile height with respect to an initial height $H_0$, $k$ is an erosional or depositional constant and $L$ is distance downstream.

Major changes in the gradients of channel networks may arise due to active tectonics, such as when a river crosses particularly large thrust faults or when a normal fault grows laterally into a drainage network. Himalayan examples of the former (Fig. 11.4) reinforce the observation that the production of a smooth profile requires massive redistribution of materials, something that may take many millions of years to achieve once tectonic activity stops. The rate at which a particular channel can make adjustments to its profile will clearly depend on the magnitude of stream power available to the system in question.

The downstream concavity of profile in most aggrading rivers (ignoring the effects of tributaries) leads to a diminution of bed shear stress and flow competence—hence to a general downstream fining of grain size through deposition, although abrasion due to transport also contributes. This is despite the fact that most rivers increase their discharge and depth downstream as tributaries join them. Downstream fining is particularly marked in mixed gravel–sand systems where there are marked changes in river long profile curvature. Such gradient changes are most marked in tectonically active basins at faulted mountain fronts where gravel 'trapping' is commonly observed. Gravel trapping generally is a complicated result of the attainment or otherwise of 'equal mobility' of gravel–sand mixtures (see Chapter 7). Since equal mobility depends upon bed shear stresses attaining more than about twice the value to transport the median grain size of the mixture, then downstream decrease of bed shear stress will inevitably result in gravel deposition and selective sand transport.

## 11.3 Channel form

Channels possess form as well as magnitude and are best described (Fig. 11.5) at some constant river discharge stage (e.g. bankfull discharge) by a combination of the following forms.

- Channel *sinuosity*, $P$—a planform description of channel deviation from a straight path. As discharge and channel cross-sectional size increase, so does the planform wavelength, $\lambda$, of any meanders involved. Measurements of many natural bends reveal that $\lambda \approx 11w$. A second useful scaling relationship is that between meander wavelength and mean annual discharge, $\lambda \approx 106\sqrt{Q}$.
- Channel *braiding*—the degree of channel subdivision by macroscale bedforms (see below) and accreting islands around which channel reaches

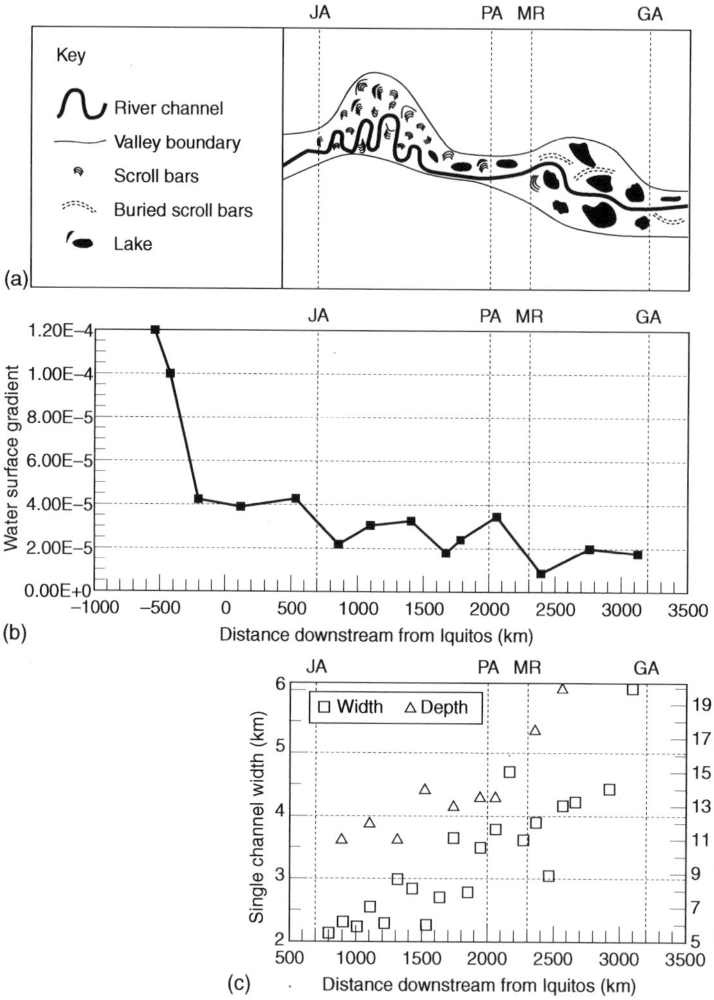


**Fig. 11.3** Amazon River downstream changes. (After Mertes *et al.*, 1996.) (a) Indicative planform changes downstream from Iquitos bends; not to scale. (b) General downstream decrease in water surface gradient. (c) Downstream increase of mean low-water depth and single-channel low-water width.

diverge and converge. However, many braided rivers so-defined are only braided at discharges lower than bankfull, since at high stages many braid-bars are submerged.

- *Anastomosing* river reaches are more permanently interconnected channels subdivided into smaller channels and separated by floodplain, with each channel containing its own channel and point bars. Unlike braided reaches, the cross-valley position of individual channel branches and adjacent vegetated islands are relatively stable. Such examples are true multichannel rivers—but note that they are distinct from distributive channels forming at the downstream deltaic terminus (Chapter 17) of a trunk channel into lake, ocean or interior drainage basin.

The causes of braiding versus meandering behaviour of river channels remains physically obscure (Fig. 11.6). There is a much-quoted slope dependence of braiding involving discriminators of the kind $S = aQ^{-b}$. Some workers include a grain-size term such as $S = aQ^{-b}D^{c}$, or a combined stream power: grain size criterion. These cannot represent the whole

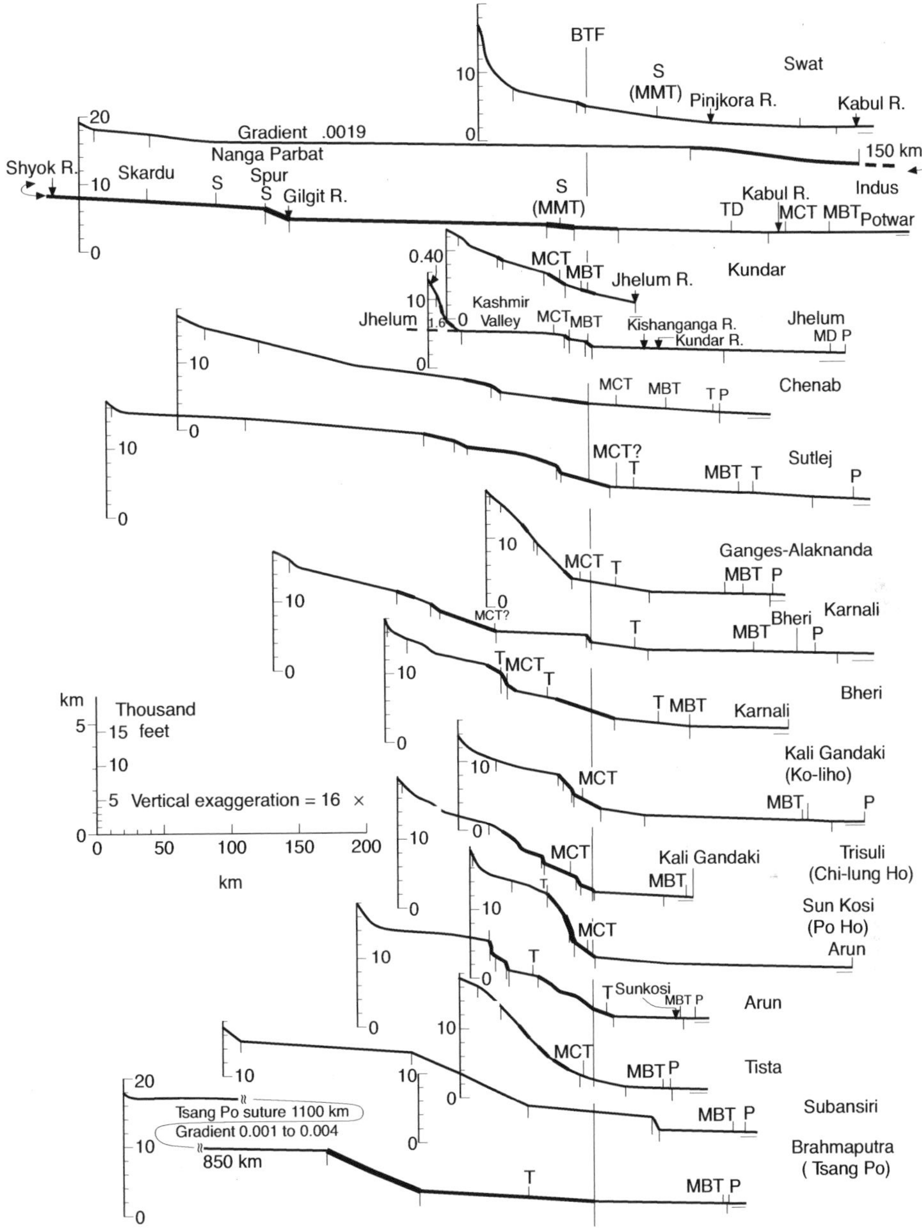


**Fig. 11.4** Longitudinal profiles of the major Himalayan rivers (arranged west to east, top to bottom) to show common zones of high gradients (thickened lines) and curvature convexities associated with great thrust fronts like the Main Central Thrust (MCT) that separates the Greater and Lesser Himalayas. The existence of such curvatures shows that river disequilibrium is common across actively uplifting geological structures. The trans-Himalayan Indus River in the west shows no such development. The Main Boundary Thrust (MBT) separates the Front Ranges of the Lesser Himalaya from the Indo-Gangetic Plain foreland basin. MMT, Main Mantle Thrust; BTF, Basement Thrust Front; S, suture zone; T, thrust; P, entrance point of river into Indo-Gangetic plains. (After Seeber & Gornitz, 1983.)

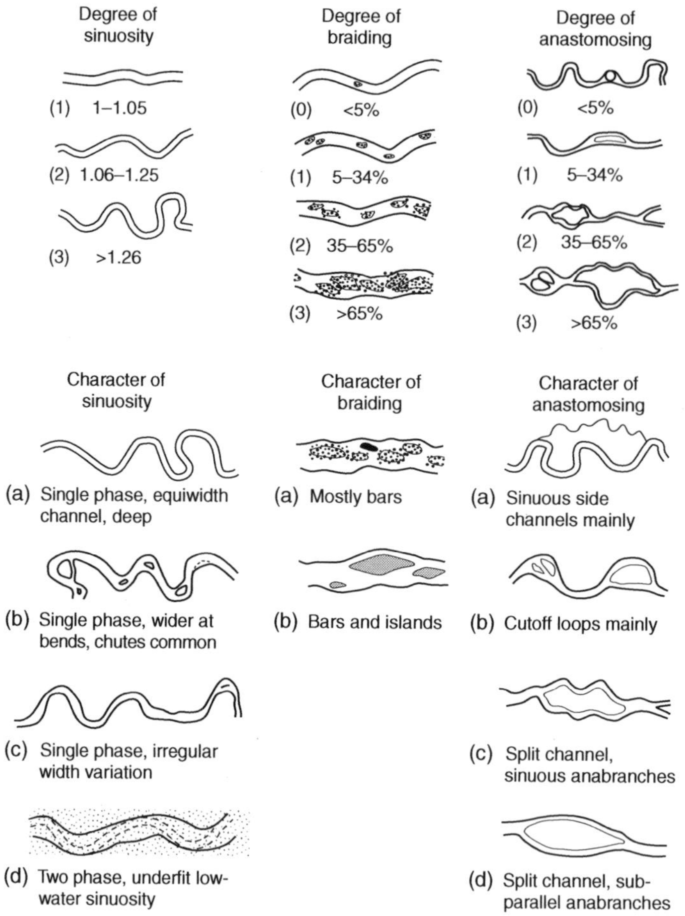


**Fig. 11.5** Indicative sketches to show the range of river channel types. (Adapted after Brice, 1984.)

story since sediment supply characteristics clearly play an important role—high supply invariably leads to braiding. The main point is that braided rivers fed by ample sediment supply are sufficiently wide so that flow can converge and diverge around barforms; indeed, simple but powerful cellular models of braiding achieve exactly this form with these boundary conditions.

A variable that does influence channel form is bank stability: unerodible banks will contain a perfectly straight channel, whilst totally cohesionless banks will be free to widen indefinitely until the decrease of depth and applied bed shear stress at the bank margins means the water can no longer do any eroding. Stable channel perimeters (banks and bed) will clearly contain more cohesive sediment, such as peat, clay or silty clay, than sand. If the river is transporting an appreciable volume of suspended fines, then the floodplain should, *ipso facto*, also contain dominantly fine-grained sediment. A parameter, $M$, that expresses the amount of silt–clay in the channel perimeter shows a high degree of correlation in USA Great Plains rivers between $M$ and both $w/h$ and sinuosity, $P$, with $w/h \approx 225M$ and $P \approx M^{-0.25}$. However, these are not generally applicable.

No rigid classification of any single channel on anything longer than a reach level (a dozen or so channel widths) is universally practicable since many rivers show downstream combinations of sinuosity and braiding. However, many do not and some

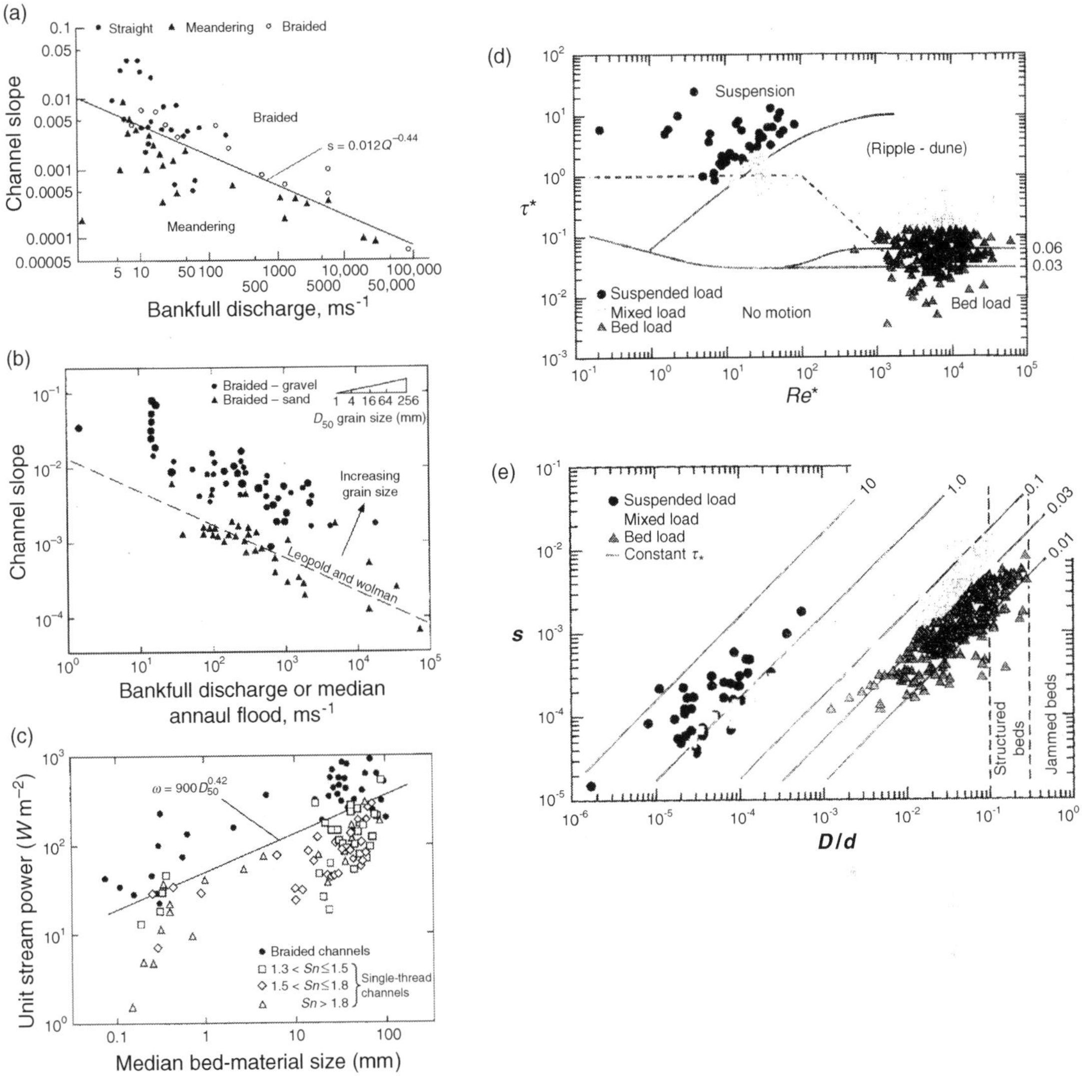


**Fig. 11.6** Empirical data used to discriminate channel patterns and channel transport modes. (a) Channel slope and bankfull discharge. (After Leopold & Wolman, 1957). (b) Channel slope, bankfull discharge and mean grain size. (After Ferguson, 1987.) (c) Stream power per unit bed area and bed material size. (After Van den Berg, 1995.) (d) Channel transport modes as dimensionless bed shear stress and grain Reynolds number. (After Dade and Friend, 1998; Church, 2006.) (e) Channel transport modes as channel gradient and relative roughness. (After Church, 2006.)

rational discriminatory classification may be worthwhile. On the face of it, the magnitude of the energy available to a stream channel and the grain size of the sediment supplied to it might be considered to be primary dynamic variables. Two contrasting examples of discriminant plots using various forms of these variables are shown in (Fig 11.6b & c). In one, available stream power controls channel planform for any given grain size; in the other a clear division into suspended, mixed and bedload channels is achieved on a dimensionless Shields plot of the kind introduced in Chapter 6.

Finally, much of the literature on ancient river deposits seeks too rigid a classification of ancient rivers and too narrow interpretations of the reasons for particular channel form; the tectonic and palaeohydrological inferences thus made are frequently in error. Thus braided rivers do not *have* to occur on steeper slopes, in coarser sediments and in catchments with 'flashy' discharge regimes. They may be made experimentally with invariant discharges. Neither do they comb their floodplains constantly, leaving no floodbasin fines for preservation in the stratigraphic record. Gravel-bed meandering rivers with decent floodplains are not uncommon.

Anastomosing channel patterns have been suggested to reflect the combination of highly resistent cohesive and ultrastable perimeter banks and very low gradients that may arise in areas subject to rapid rise of base level. They are therefore the product of a river regime which has very low stream power and a preponderance of deposition and suspended-load transport of fine-grained sediment. Since the multichannel branches are often quite sinuous, and the channels do not migrate significantly laterally long-term, it appears that they must have evolved slowly from local flood-break river diversions termed *crevasses*—these evolved into permanent courses carrying some fraction of the total river discharge. They are therefore thought of as the product of *partial avulsions* (see below) issuing from an original channel that was not anastomosing.

## 11.4 Channel sediment transport processes, bedforms and internal structures

### Unit bars and dunes—macroforms and mesoforms

The largest river channel bedforms are termed *macroforms* or *unit bars*. These scale with channel width and make up the active portions of all bars. Secondary helical flow is the common denominator to each curved reach that bounds the active unit bar flank (s); bar growth is accompanied by both streamwise and a measure of spanwise *lateral accretion*. The former positions of barforms may be studied from aerial photographs since migration of bar and/or channel during or after flood-induced erosion and deposition leads to partial preservation of former barform boundaries as topographic features (swale-and-ridge topography) in plan view. These are the surface expressions of periodic lateral accretion of sediment on to the bar, particularly due to flow expansion downstream (Fig. 11.7). Unit bars are also asymmetrical in alongstream section, and their steepest, downstream sides may be at the angle of repose or (more commonly) less. In plan view, unit bars are lobate (linguoid) and the front part of their crest is transverse to flow. Bar heights are commonly more than half flow depth.

Unit bars often feature superimposed bedforms like dunes (Fig. 11.7) that scale with flow depth. These are termed *mesoforms*. As noted in Chapter 7, current ripples and upper stage plane beds show no such correlation and are termed *microforms*. Although unit bars are normally larger than dunes their heights may be comparable. This overlap is the reason why it may be difficult to distinguish the deposits of dunes and unit bars, particularly in cores. The deposits of unit bars are mostly low-angle (less than angle-of-repose), downstream-inclined stratasets (large-scale inclined strata) that are internally composed of cross strata formed by superimposed, down-climbing bedforms such as dunes and ripples. Such sets of large-scale inclined strata formed by unit bars are commonly decimeters to meters thick, and more laterally extensive that cross sets formed by dunes. The low angle of the large-scale inclined strata is related to the low-angle lee slope of the unit bar. However, unit-bar lee faces locally and periodically reach the angle-of-repose, such that parts of unit bars are composed of angle-of-repose cross strata. The thickness of these sets of cross strata ranges from decimeters to meters, thus overlapping with the scale of medium-scale cross strata formed by dunes. It is important to distinguish cross strata formed by dunes from those formed by unit bars in order to correctly interpret bedforms from river deposits. This distinction is also important if cross-set thickness is being related to bedform height and flow depth. Angle-of-repose cross strata formed by unit bars can be distinguished in cores by the presence of relatively thick and fine-grained bottomsets, and, in some cases, by the relatively large set thickness. Cores through two modern braided channel belts indicate that only 1–2% of the channel deposits are composed of angle-of repose cross strata formed by unit bars, which is much less than commonly believed.

### River confluences

Important hydraulic effects occur at *river confluences*. At these joining zones, experimental and field studies

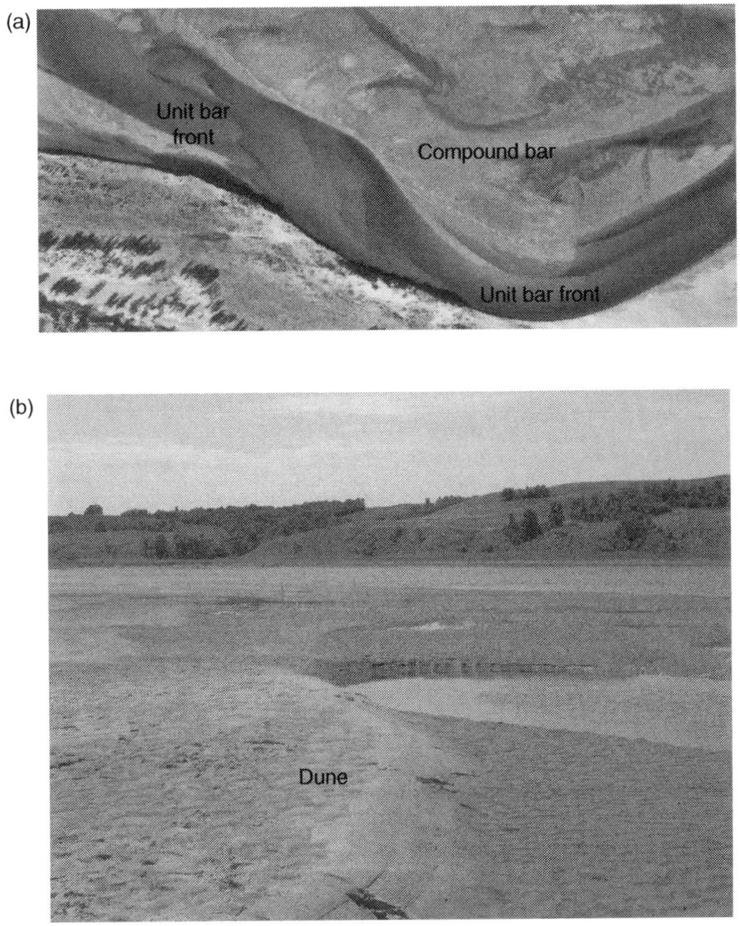


**Fig. 11.7** Unit bars from the South Saskatchewan River, Canada. (a) Aerial photo showing unit bars with superimposed dunes in a channel adjacent to a compound bar. Unit bars are about 200 m long, and the superimposed dunes are in the order of 10 m long. (b) Emergent unit bar with superimposed ripples and dunes. (Photos courtesy of Johnny Bridge.)

reveal the existence of strong vertical-axis vortical structures of Kelvin–Helmholtz type (see Plate 2), which scour deeply into the underlying substrate (Fig. 11.8). The depth of scour may reach several times that of the bankfull depths of the contributing tributaries, so that in the case of major rivers, like those of the Jamuna and Ganges in Bangladesh, the highly mobile scours may reach up to 30 m below low-water level. Bedload sediment supplied from tributaries enters such scour pools via steeply dipping deltas whose deposits will include large-scale planar sets of cross-stratification, perhaps overlain by the deposits of subsequent channel bars. Downstream of the joining zone in the now united discharge, a zone of flow separation occurs into which large separation eddies transfer sediment to mould a separation bar, particularly evident at high river stage. There are other neglected effects of tributary junctions in river dynamics, chiefly their behaviour during unequal flood discharges in joining channels when 'backing-up' of the smaller leads to out-of-channel flooding upstream.

At *channel diffluences*, the calibre and magnitude of sediment load that is transferred from major to minor tributary depends not only on the apportionment of stream power but for sinuous channels also on the position of the channel bifurcation with respect to upstream bends in the main undivided channel. If branching occurs on the outer bank of an upstream bend then because of the way that point bars sort their incoming sediment (see below) relatively coarse sediment is provided that may be immobile apart from at very high flow stages.

## Channel bends and point bars

The basic dynamics of flow in channel bends (**Cookie 42**) usually leads to erosion on the outside parts of bends and deposition of a *point bar* on the inside. Decreasing bed shear stress from the deep toe to the shallow top of the point bar also leads to the development of distinct bedform suites on the point-bar surface and, in sections of the bend where the flow is fully developed, an upwards fining of grain size (Fig. 11.9). The upstream parts of point bars often show a coarsening upwards onto the bar top where

See Cookie 42 for details about flow in channel bends

high power flow from the upstream helical flow cell impinges on the bar head; this effect rapidly diminishes as full establishment of local helical flow

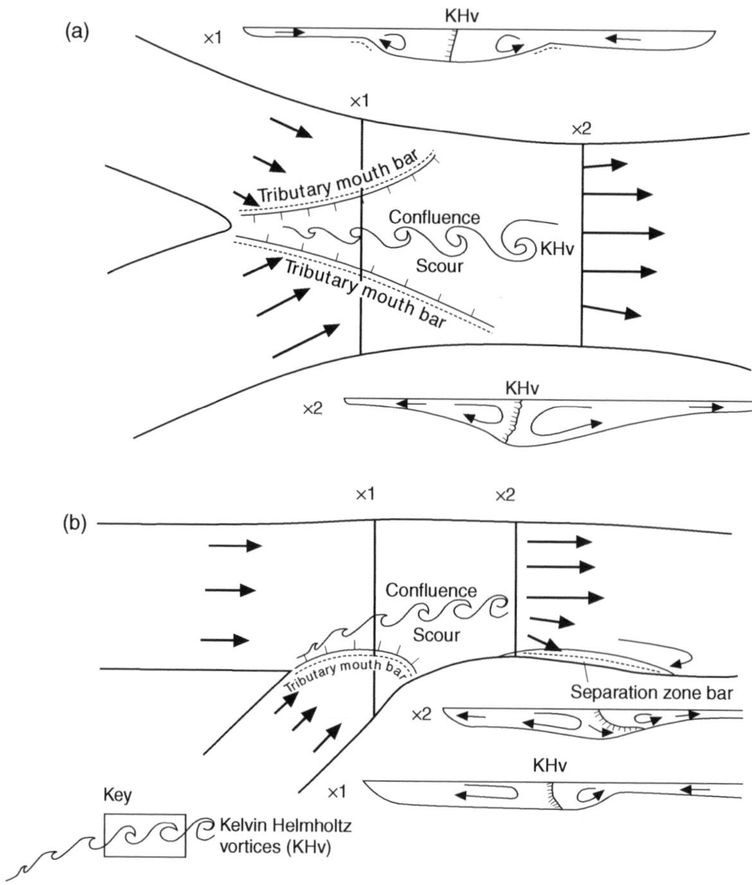


**Fig. 11.8** (a) & (b) General diagrams to illustrate flow, scour and bedform patterns at river confluences. Arrows are vectorial indicators of mean flow velocity. (After Best, 1987; Best & Roy, 1991.) (c) Turbulent mixing and scour at river channel tributary junctions highlighted when tributaries of contrasting suspended sediment content meet, in these cases the Negro/Solimoes (Brazil) and Meghna/Padma (India) junctions. Both the Negro and Meghna rivers have low suspended sediment and high organic content and therefore show dark colours compared to the suspended-rich Silimoes and Padma in these images. (Negro photo courtesy of Marta Pèrez-Arlucea; Meghna photo courtesy of Jim Best.)

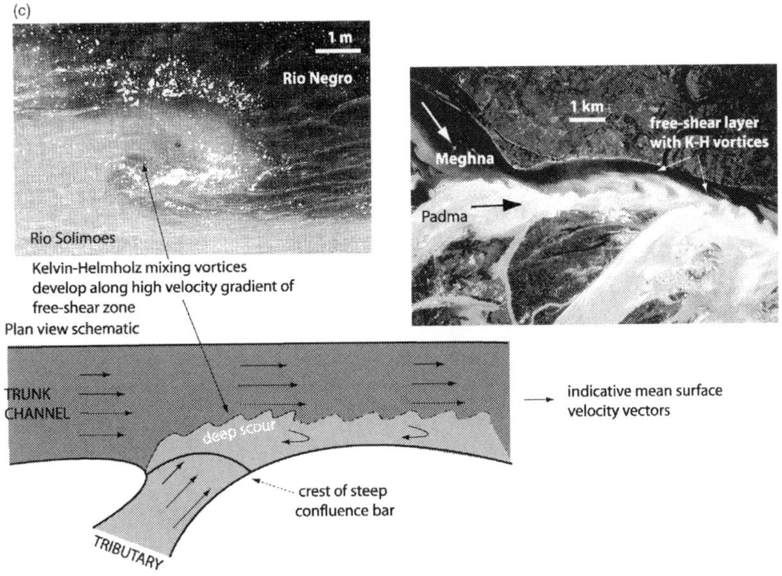


**Fig. 11.8** *(Continued)*

occurs (Fig. 11.10). The cross-stream shallowing and secondary flow (Fig. 11.11) causes a general sediment fining around the middle to downstream ends of point bars with current ripples, dunes and upper plane beds as common bedforms on sandy bars.

Erosion along outer cut banks depends upon bank structure and properties—whether noncohesive, cohesive or composite. Bank undercutting leads to gravitational collapse along planes formed by soil shrinkage. Clay banks may fail along rotational slides. In both cases, failure and collapse may occur preferentially during falling river stage in response to changing porewater levels and pore pressures. In single-channel rivers of very low flow power whose sediment load is mostly silt- and mud-grade carried in suspension, flow occurs through a more-or-less fixed sinuous system that lacks the erosive strength to cause bend migration. Outer *and* inner bank accretion is then common, as is the repeated failure of bank slopes along rotational slides—seasonal deposition of mud aggregates then produces inner and outer bank oblique-accreted deposits.

The combined processes of outer bank erosion and inner bank accretion over the surface of a point bar in moderate- to high-power mixed-load channels leads to whole-channel migration—bends grow, migrate and rotate over time. Periodic bend migration and point-bar accretion in response to major floods lead to the production of a characteristic arcuate topography of unit scroll bars and swale depressions on the point-bar surface (Fig. 11.9). These are marked within the point bar by characteristic sigmoidal *lateral accretion surfaces* whose upper parts are particularly well defined by rapid alternations of thin sands and organic-rich layers. These are key surfaces to look for in the sedimentary record of river channels; the often low-angle dipping surfaces having dips normal to the local trend of palaeocurrents obtained from sedimentary structures. Amplification of channel bends frequently leads to the formation of *meander cutoffs* and the familiar *oxbow lakes*. The process known as *neck cutoff* occurs in highly sinuous meanders where the upstream and downstream limits to a meander loop have evolved so that they become tangent and finally merge. Alternatively, channel flows may take short-cut routes across or around swale depressions on any width of meander neck during flood discharges, a process known as *chute cutoff*.

### Braided reaches and channel bars

As noted above, braided rivers are characterized by relatively high stream power, rapid rates of lateral bank erosion and deposition and frequent shifts in the

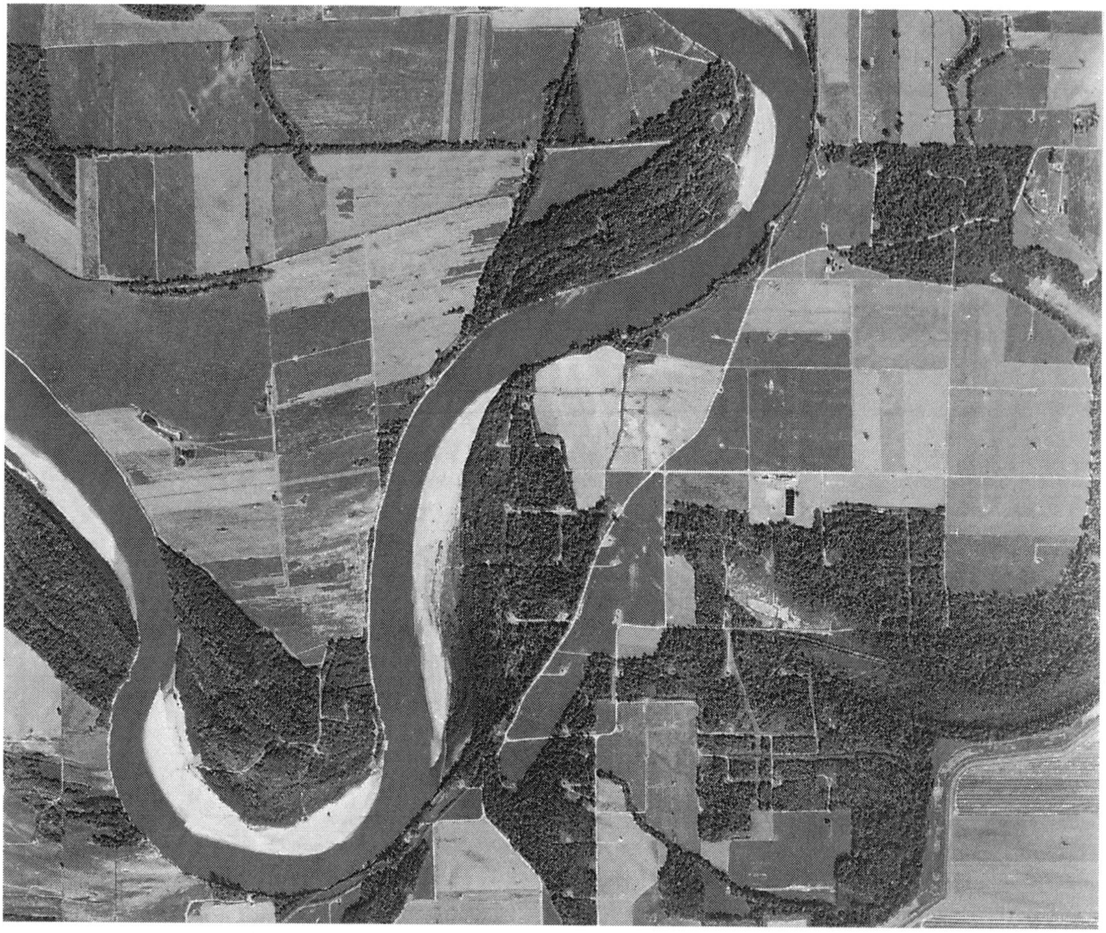

**Fig. 11.9** Vertical aerial photo to show the Bozeman (centre), Helm (lower left) and part of the Pearl meander bends of the Wabash River near Grayville, Illinois, USA. Flow is from top to middle left; photo taken at low flow; scale 1:30 000. The exposed sand–gravel sediment on the active unit bars are the white crescentic areas on the inside of each bend. Note the progressive colonization of the inactive point-bar top by riparian trees and shrubs, the faint traces of scroll bar surfaces, abandoned meander loop (right centre) and the well-developed active scroll bar on the downstream portion of the Helm bend. (Photo dates from 1970s from US Department of Agriculture; for further details see Jackson, 1976.)

position and degree of activity of the braided tributaries. Braid bars are usually compound macroforms that have a range of planform shapes that are strongly stage-dependent (Fig. 11.12) and which scale with channel width, $w$. They comprise chiefly: (i) composite lateral, alternate or side bars which occur adjacent to alternate right and left banks downstream, have a spacing $\sim 2\pi w$ and which migrate downchannel over time; (ii) composite mid-channel bars with superimposed unit bars and/or mesoforms—these split the channel flow and are the most typical braid bars. They are best regarded as mobile, double-sided point bars (Fig. 11.9). However, because of the following, the flow dynamics are more complicated than those of point bars.

1 Hydraulic effects due to channel splitting (*diffluence*) at the upstream end of the bar and rejoining (*confluence*) on the downstream end (Fig. 11.13). The confluence is often the site of deep scour, filled in by the avalanche deposits of relatively straight-crested bedforms of Gilbert-delta type orientated at high angles to the mean downstream flow vector.
2 The frequently unequal channel width and hence magnitude of flow power and secondary flow on either side of the composite bar.

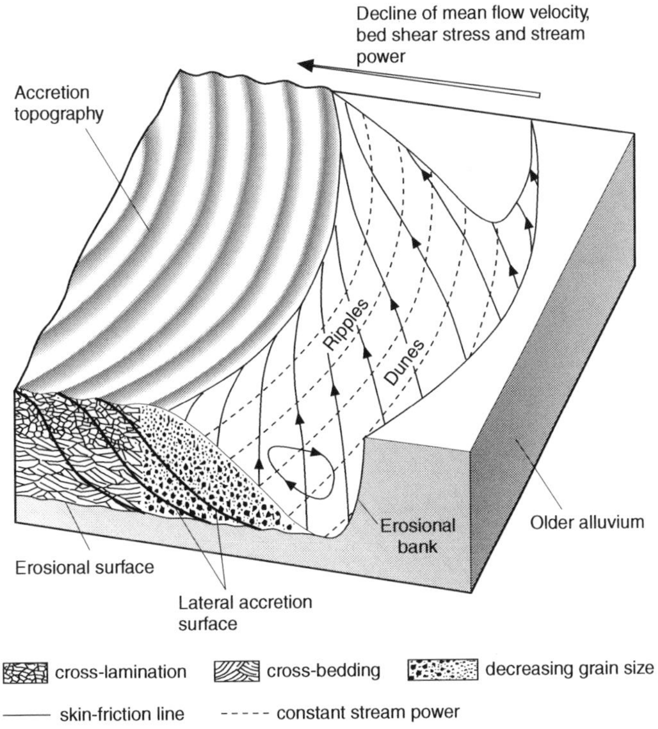


**Fig. 11.10** Helical flow in a meander bend causing regular changes in grain size, bedforms and sedimentary structures in the deposits of an accretionary point bar. (After several authors, including van Bendegom, Rozovski, Allen, Bridge, Jackson).

3 Sudden or gradual migration or shifting of individual channel branches (referred to as within-channel belt avulsion by many authors, but more correctly as *channel diversion*; see further below) that occurs in response to flow, bend and bar evolution.
4 Potentially extreme effects that can arise due to falling stage dissection of the bar by cross-bar channels.
5 The effects of grain roughness in very shallow cobble- to gravel-bed rivers that causes nucleation from cluster bedforms and accretion downstream to the tail of a bar from the bar head.
6 Experimental studies on correctly scaled models of aggrading channels throw light on the process of channel belt accretion, particularly the preservation potential of channel abandonment fines. These indicate that channel diversions into which fines may subsequently accumulate in 'deadwater niches' most commonly occur due to (i) reduction in channel capacity or 'choking' of channels and (ii) tightening of channel bends and inner bar accretion causing flow breaching at the outer bank.

In the simplest possible manner, braided-reach sediments resulting from downstream bar migration over time will feature a lower confluence-scour and scour-fill cross-stratified deposit overlain by very much more complex products of the migrating and accreting bar macroform itself, with its many superimposed laterally accreted unit-bar macroforms (Fig. 11.14). For an example of this complexity consider first sand-bed braided channels, illustrated by the results of long-term studies of flow, sediment transport and bar development in the great Jamuna (aka Brahmaputra) River, Bangladesh. Here, large compound bars are stable except where they pass through major nodes of flow convergence or channel confluences where flow is able to rework the forms. The overall planform morphology of the 55-km long Jamuna study reach (Fig 11.15) is low sinuosity and moderately braided (braiding index 2–4). The channel belt is dominated by two main channels that bifurcate at the bar-heads of large (>15 km long) mid-channel bars and recombine at their bar tails to form major zones of flow convergence and scour. Analysis of

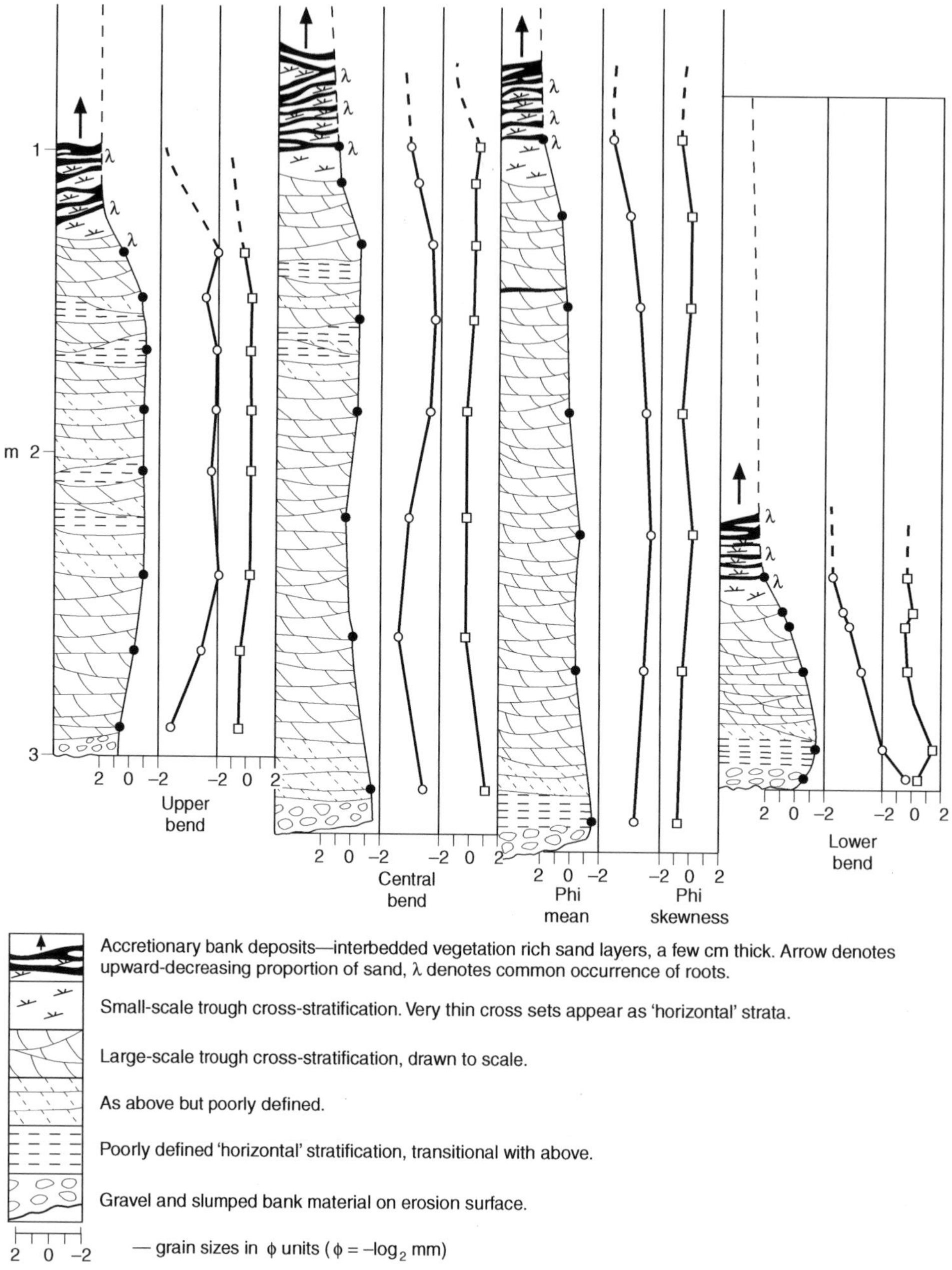


**Fig. 11.11** Logs from vibracores to illustrate sedimentary successions developed at various positions around a meander bend. (After Bridge & Jarvis, 1982.)

satellite images and surveys reveal that channel-bar migration and reworking (Figs 11.16 & 11.17) occurs through: (i) erosion of the bar-head and deposition at the bar tail; (ii) upstream accretion onto the bar-head; (iii) lateral accretion onto existing bars by unit bars; (iv) mid-channel deposition downstream of zones of significant bank erosion; (v) rapid deposition downstream of a major flow convergence. Migration rates

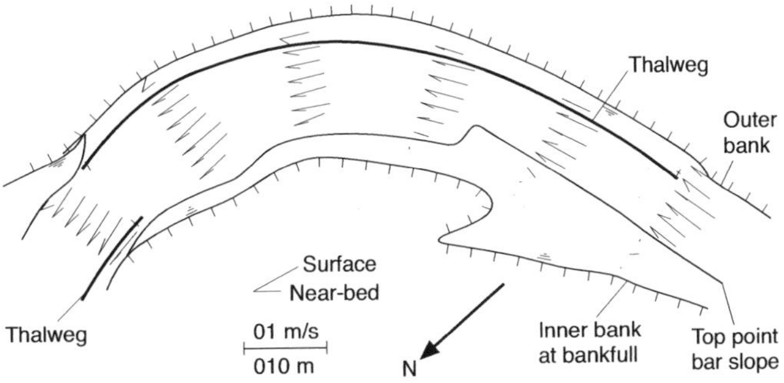


**Fig. 11.12** Stepover-thalweg on upstream part of point bar and measured bankfull flow patterns in the meander bend, River South Esk, Scotland. Flow clockwise. Secondary helical flow cell well developed in the first four radial transects. (After Bridge & Jarvis, 1982.)

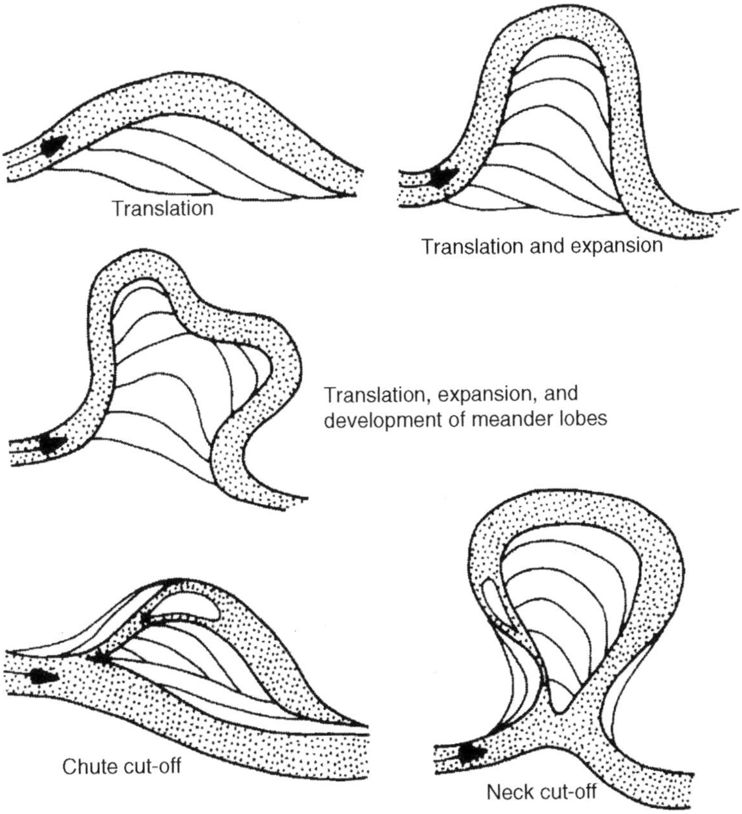


**Fig. 11.13** Typical modes of channel migration and cutoff within channel meanders shown by accretion along former inner banklines (thin curved lines). (After Bridge, 2003.)

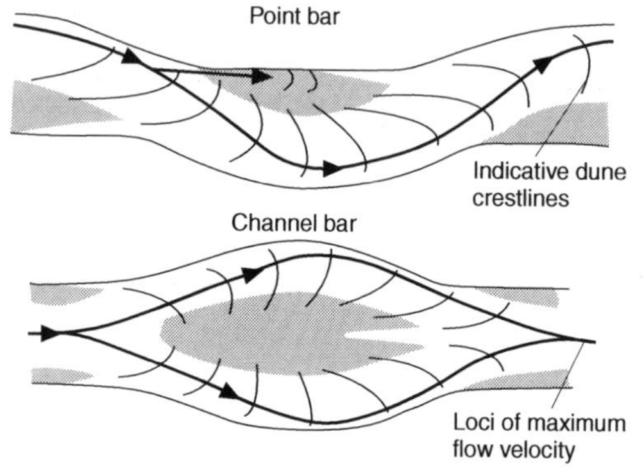


**Fig. 11.14** Sketches to illustrate the simple concept of a channel bar being a 'double' point bar. (After Bridge, 1993.)

of the compound mid-channel bars are typically < 500 m per year. Study of the evolution of a new bar (Figs 11.16 & 11.17) reveals that mid-channel bar growth occurs downstream of a major flow convergence and large-scale sediment input from up to 600 m of bank erosion immediately upstream of the zone of bar deposition. The bar grows through amalgamation of large dunes that form a central bar nucleus. Bar-top aggradation continues through both dune superimposition and development of a 3-m-high, 'accretionary dune front' due to amalgamation of smaller (< 1 m high) dunes in shallower flow on the bar top. At low flow, the bar widens through lateral accretion as dunes migrate around and onto the margins of the bar, forming two or more protruding 'limbs' at the bar tail—these provide a zone of low flow velocity for fine-grained deposition. As bar evolution continues, one anabranch becomes dominant, is enlarged and supplies sediment for deposition within the anabranch. This deposition deflects flow across the bar tail and constructs a broad depositional front attached to the bar tail. Emergence of bars along this lobate depositional front gives the reach a morphology that resembles an alternate bar. The morphological evolution of the sand braid-bar documented in the Jamuna River suggests that the depositional facies of kilometre-scale, sand braid-bars will be dominated by cross-stratification formed by dunes and sets of cross-strata produced by slipface accretion at bar margins.

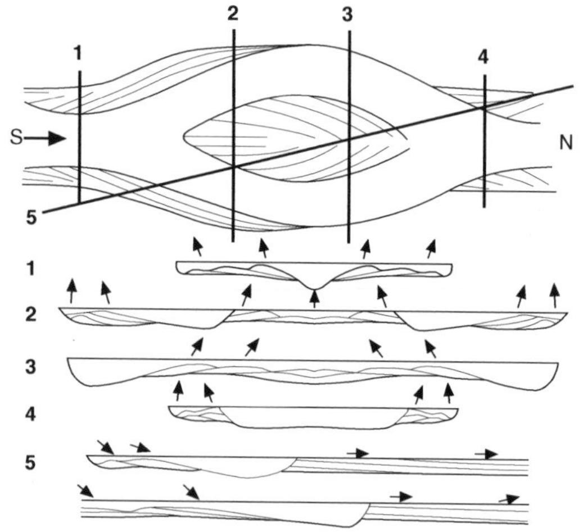


**Fig. 11.15** Sketches to show unit bar-margin accretion surfaces and palaeocurrent orientations for various section orientations across a simple braided channel reach. (After Bridge, 1993.)

Most *gravel-bed* or *mixed sand/gravel rivers* are coarser grained and shallower than sand-bed rivers so they have very much higher relative roughness, $D/h$, where $D$ is mean grain size and $h$ is water depth. This means that gravel bedforms are usually macroforms or very low sandy mesoforms. Despite this, the main principles of diffluence/confluence reaches, mid-channel bar nucleation, growth by lateral accretion, chute cutoff during flood discharges and cross-bar flow during falling stage are shared in common with sand-bed braided channels. Consider the results of an integrative study involving bar monitoring, GPR surveys and coring of an active reach of the Savaganirktok River, Alaska. This is a mixed braided/anastomosing active channel belt that is 2–4 km wide as defined by the extent of flow during bankfull flood events (Figs 11.18 & 11.19). It comprises active and partially abandoned channels, simple unit and compound bars and bar assemblages. Bar lengths scale with width of adjacent channels and have heights comparable to formative channel depth. The main (first-order) channels and bars are cut by smaller (second-order) cross-bar channels that may have their own associated second-order bars. Abandoned bar assemblages make up areas of floodplain adjacent to the main active channels. The main channels are 50–250 m across, and cross-bar channels have widths of 5–40 m. The maximum bankfull depth of the main channels is 3.9 m, and the median bankfull depth is around 1–2 m. The channel geometries observed are similar to those in other sand-bed and gravel-bed rivers,

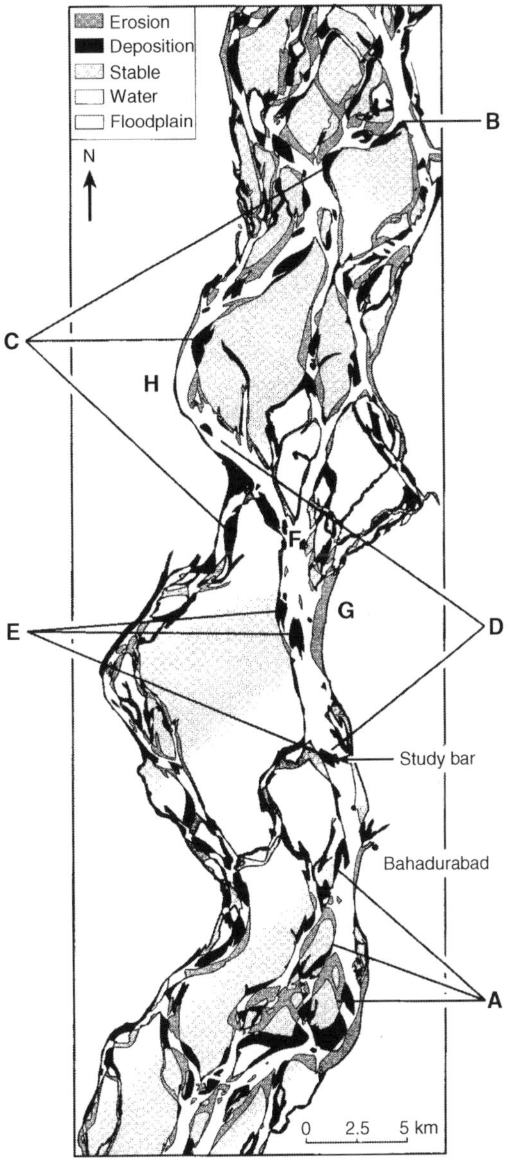


**Fig. 11.16** Channel change along the Jamuna River during the 1994–95 monsoon season. Bar outlines are traced from 1:50 000 colour SPOT images for successive low-flow periods on 3 March 1994 and 24 March 1995 (water level varied by 0.08 m between these dates). Flow direction is north to south. Labels A–G are as follows: A, bar migration and reworking by erosion of the bar-head and deposition at the bar-tail; B, upstream accretion on to the bar-head; C, bar growth from lateral accretion on to existing bars; D, mid-channel deposition downstream of zones of significant bank erosion; E, mid-channel deposition through rapid deposition downstream of a major flow convergence; F, study bar site of Fig. 11.17 located 9 km downstream of a major node of flow convergence in the east channel; G, between March 1994 and March 1995, up to 600 m of bank erosion occurred along a 5-km stretch of the post-confluence eastern margin which may partly explain the initial trigger for growth of the study bar. (After Ashworth *et al.*, 2000.)

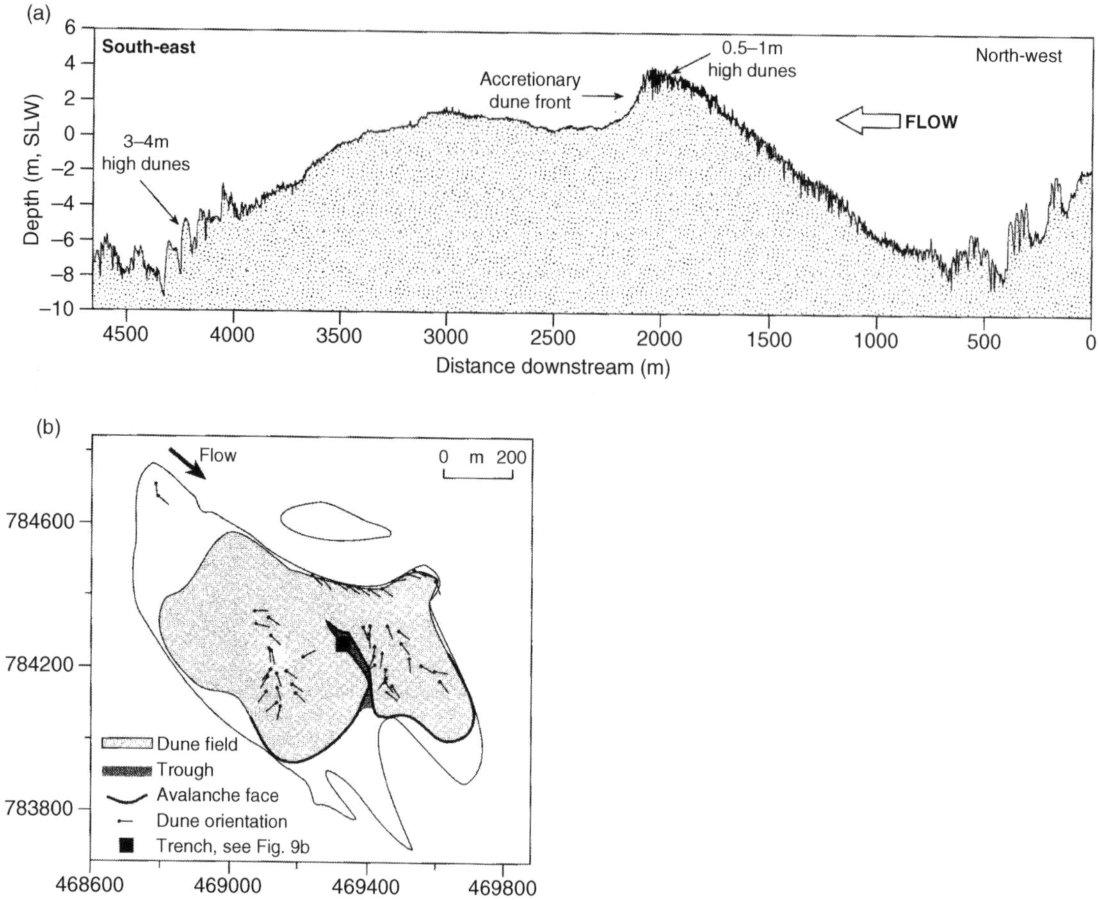


**Fig. 11.17** Jamuna River bar morphology surveys. (After Ashworth *et al*., 2000.) (a) Longitudinal survey line over the centre of the study bar-top on 12 August 1994. Flow depth is 5.83 m above SLW. Note the reduction in dune height from the thalweg to bar-top and the 'accretionary dune front'. (b) Planimetric map of the study bar at low flow on 13 March 1995, showing the position of the avalanche face of the accretionary dune front and the extensive coverage of dunes on the bar top. The bar planform is defined by the surveyed water's edge (stage = + 1.06 m SLW).

including: asymmetrical, triangular cross-sections in curved channel reaches and symmetrical concave-upward cross-sections in straight reaches; the deepest parts of the channels are near the apices of bends and in confluences.

Studies of gravel-bar sedimentation in *ephemeral braided streams* characterized by extremely intense flood discharges reveal that such event-controlled deposits increase in thickness with flood magnitude, are deeply erosive into previous deposits leading to partial preservation, and are ungraded and clast-supported with sand-granule matrix infill.

### Anastomosing channel reaches

As noted briefly above, anastomosed reaches commonly, but not invariably, record rapidly rising base levels, either a sea- or lake-level rise or a local rise due to some aggradation caused by valley-damming transverse fans or tributaries. The depositional facies of

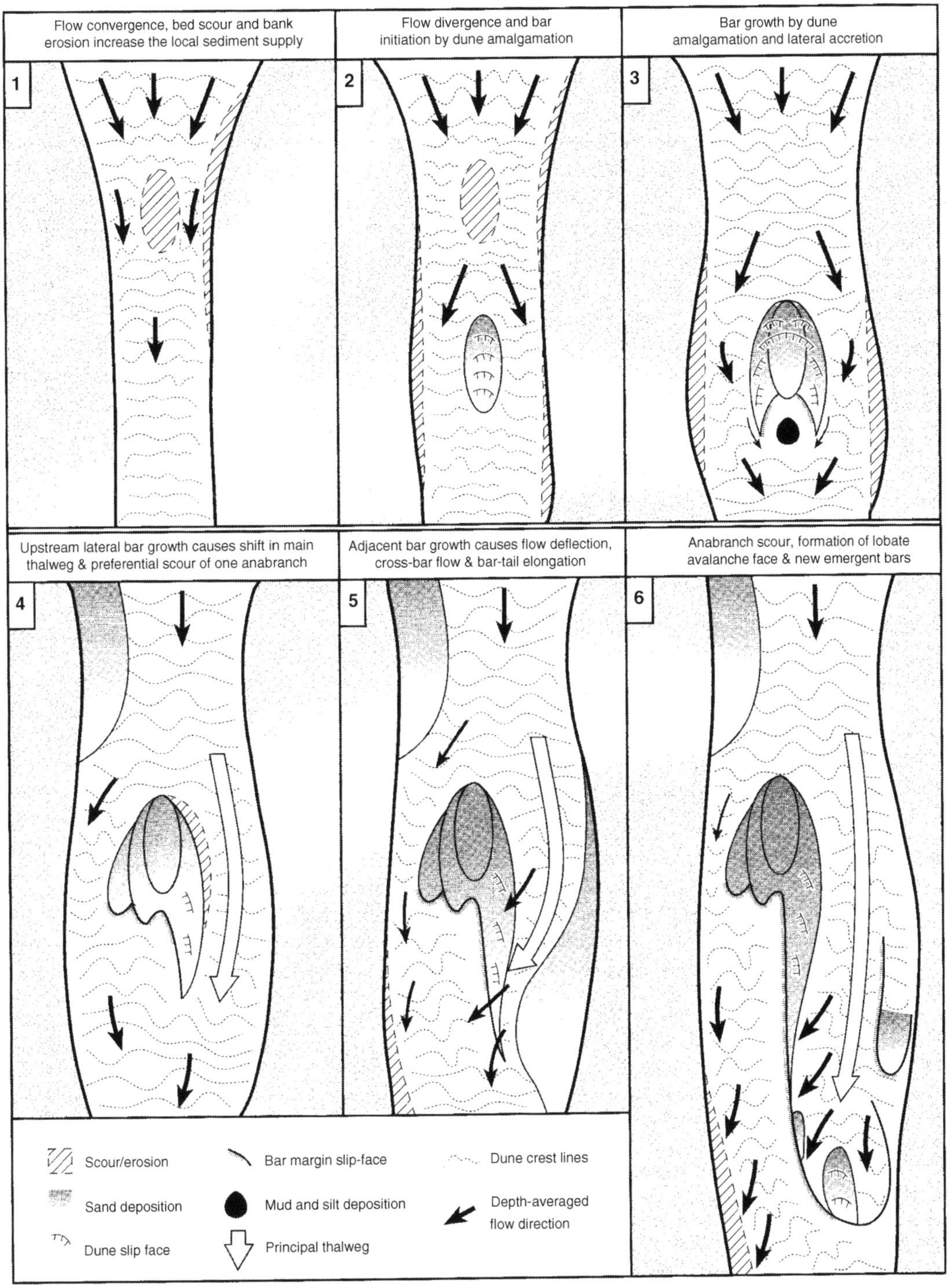


**Fig. 11.18** Summary of the key stages in the evolution of a kilometre-scale, mid-channel sand bar by asymmetric lateral accretion of unit bars with superimposed dunes. (After Ashworth *et al.*, 2000; dune orientations and flow directions at the bed are inferred from the data presented in Roden (1998) and McLelland *et al.* (1999).)

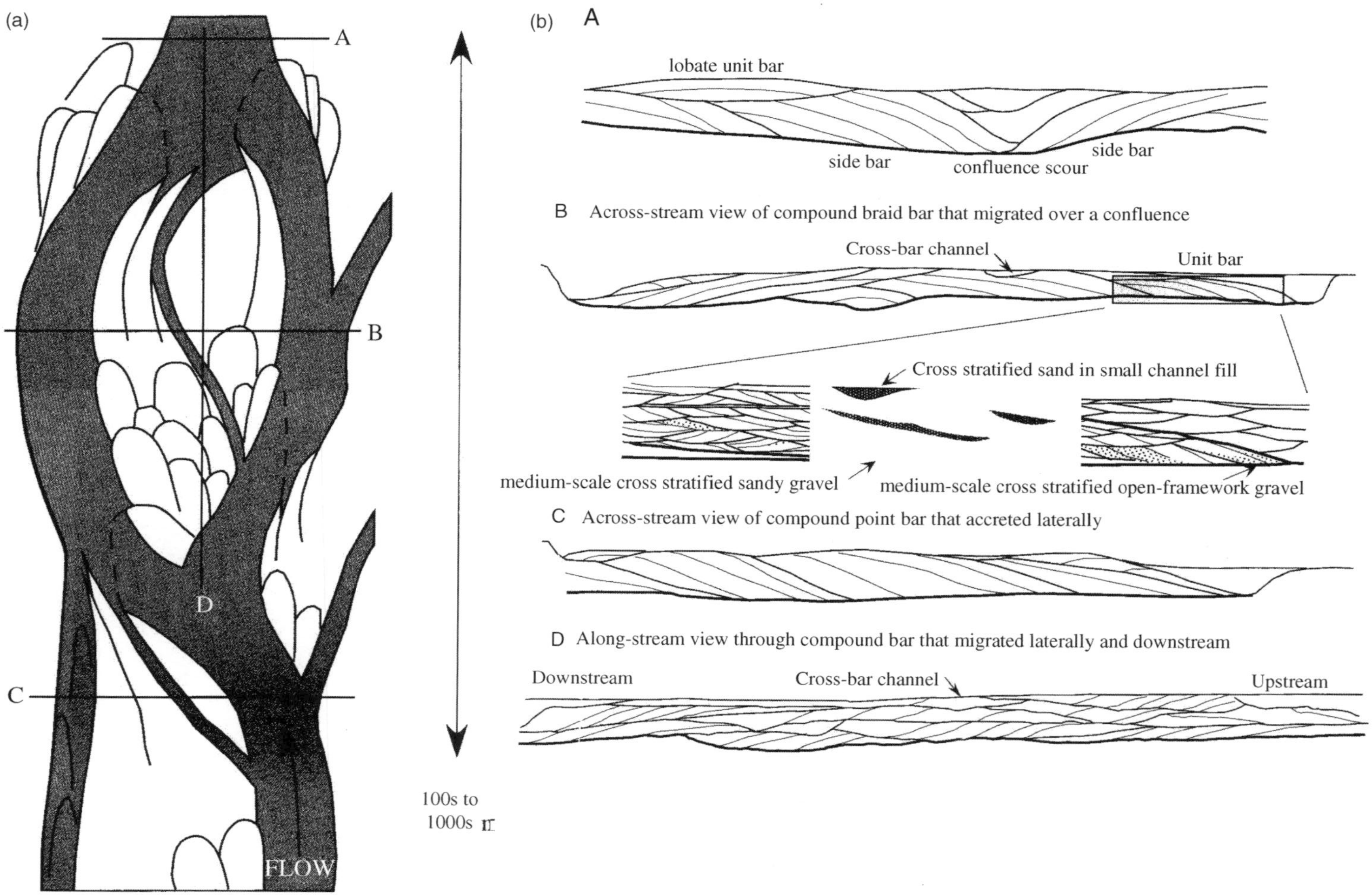


**Fig. 11.19** (a) Depositional model of gravelly braided-river deposits. Map showing idealized channels, compound bars and simple (unit) bars in active and abandoned channels. Cross-sections A to D correspond to those shown in (b). (b) Cross-sections showing large-scale inclined strata (associated with compound and unit bars) from deposits in the active part of the channel belt. Vertical exaggeration is 2:1. Thin lines represent large-scale strata, medium weight lines represent bases of large-scale sets, and thick lines represent bases of compound sets. Large-scale strata generally dip at $< 12°$, but may be up to the angle of repose. Light stipple represents open-framework gravel, dark stipple represents sand and no stipple represents sandy gravel. (All after Lunt *et al.*, 2004.)

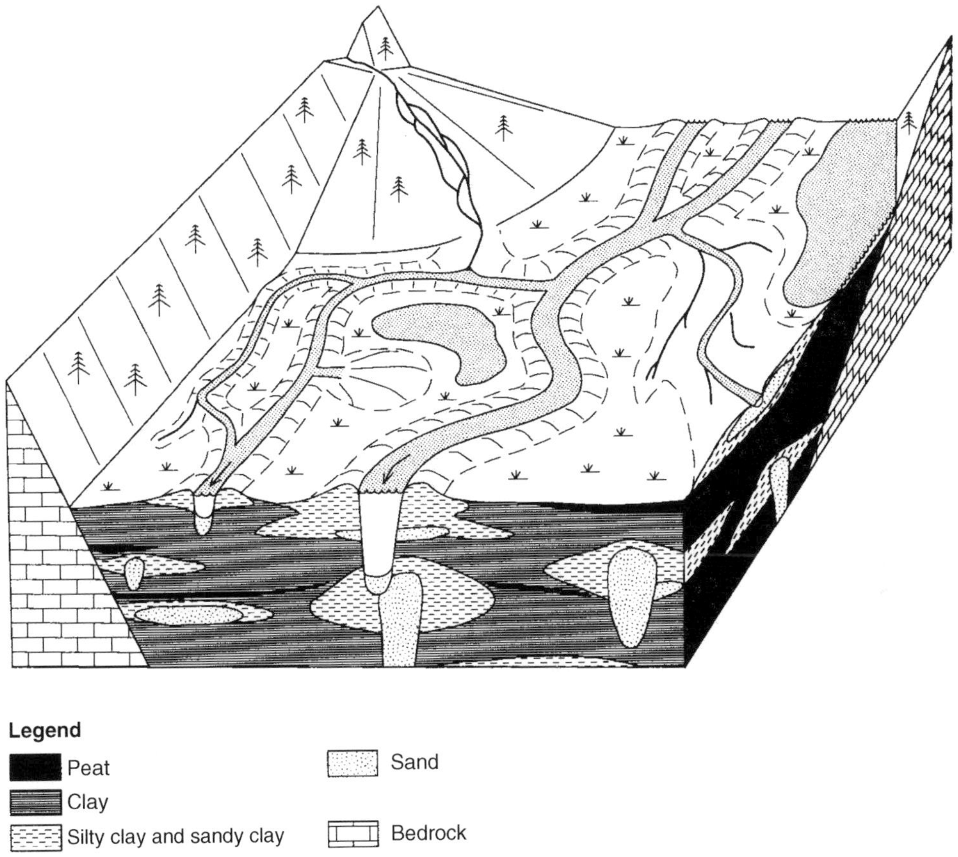


**Fig. 11.20** Diagram to show setting, surface morphology and subsurface architecture of the anastomosing Upper Columbia River, Canada. (After Makaske, 1998.)

such straight to meandering channel networks are dominated by vertical and lateral accretion of channel sediment between stable leveed banks (Fig. 11.20). Slow vertical mud accretion also characterizes the arid-zone and anastomosing river plains of the Channel Country, central Australia, where the channel positions remain fixed.

## 11.5 The floodplain

Adjacent to river channels are swathes of periodically flooded wetlands whose ecological and hydrological well-being has often been neglected by floodplain cultivation and habitation. Out-of-channel sedimentation occurs on *levees*, *crevasse splays* and in the marshy *wetlands* and shallow lakes of flood basins. Floodplain sediments have a special interest since they are the sites of upward accretion and preservation of soils, habitats, habitations and organic remains. They are therefore both *sinks*, sites of *sediment storage*, and *sources* where *in situ* mineral precipitation and carbon fixation occur. The alternation of channel and floodplain sediments in stratigraphic sections emphasizes that the processes of alluvial sedimentation depend on a dynamic interaction in time and space between the two environments. In recent years there have been new data on this interaction, particularly on depositional sequences resulting from channel changes and the role of floodplains as sequesters of downstream-moving sediment.

*Levees* are convexo-concave ridges along river perimeters created by vertical accretion of the banks due to rapid out-of-channel deposition of suspended sediment from the channel during sheet-like overbank flooding. Their elevation above the main channel is controlled by maximum flood stage. They are well-drained compared to surrounding marshy wetlands and in suitable climates bear an often dense canopy of riparian tree and shrub vegetation, especially willows and the like. Most suspended sand-grade sediment is dumped on the levee front, top and back by advection from decaying channel turbulence as suspended-sediment-charged floodwater decelerates basinwards, the process helped by the baffling effects of levee vegetation. Levee morphology and internal stratigraphy generally reflects upward accretion over time but the details depend critically on the absolute power and degree of confinement of flow in the adjacent floodplain that determines depositional fluxes onto the front, top and back levee slopes. For example, in narrow floodplains bounded by valley walls, flood stream power in the adjacent floodplain will tend to be high and levee sediments are characterized by erosion surfaces and flood routing channels adjacent to the levee margins that restrict levee lateral growth. In more open valley or lower power floodplains, levees are wider and comprise uniform, fine-grained deposits with occasional floodbasin-thinning and -fining sand lenses which grade to pronounced distal floodplain muds.

*Crevasses* are point-sourced, initially channelized invasions of water and sediment that cut through levees and spread into the floodplain as delta-shaped splay lobes marked by numerous distributary channels that shallow and widen down-splay. They form as breakout events from channels when levee sediment strength is reduced by increased internal pore pressures as river stage rises; initial floodwater breaks through the levee by bank failure and scour into a deepening channel. Crevassing is frequent in channel reaches undergoing bed aggradation, for example the floodplains of the braided Niobrara River have become flooded by rising groundwater base levels, leading to more than eight crevasse splays to form in four years in the 1990s (Fig. 11.21). Such an

**Fig. 11.21** Oblique aerial photograph looking southwest across the east Niobrara crevasse splay taken in the summer of 1996 shows the splay spreading away from the crevasse with sandy channels extending into distal vegetated areas. Note the abandoned channel on the left side of the photograph, which can be seen on older aerial photographs, and a small splay to the south of the east splay crevasse. At the time the photograph was taken, part of the flow from the crevasse ran into the old abandoned channel and then into a backswamp area in the foreground. The width of the main channel at the top of the photograph is around 250 m, and flow in the main channel is from left to right. (After Bristow *et al.*, 1999.)

enhancement of crevassing may give tell-tale evidence for the effects of base-level rise due to marine transgression in the stratigraphic record. The Niobrara splays are sand-dominated and characterized by bedload deposition within channels and with the development of slipfaces where splays prograde into standing bodies of water. Sedimentary structures include horizontal lamination, ripple lamination and sets of cross-stratication, with a slight tendency for splays to coarsen up, but individual beds within the splays often fine upwards. Contrasting examples of prograding splays producing coarsening-upwards sequences are described below (section 11.6) from the South Saskatchewan River, Canada.

Floodplain deposition rates and sediment grain size are generally found to decrease towards distant floodbasins (see below). Initial flood deposits tapping just-bankfull channel waters tend to be finer than later levee-overtopping events as flood stage rises further and bedload is advected by turbulence into downvalley flow in extreme floods. These flood level processes tend to cause coarsening upwards depositional units to be common in levee and proximal floodpbasin deposits. It is also important to note that at early stages of flooding, inundation by floodwater is sensitive to microrelief on the floodplain, with low areas like old channel reaches, crevasse splay channels (see below) and backswamps more prone to flooding and accretion of sediment (Fig. 11.22). Data on the magnitude of deposition during flooding events are sparse, but careful trapping experiments, comparisons of upstream and downstream gauging station records, and radiometric dating of floodplain cores reveal that a surprisingly high (30–70%) proportion of upstream suspended load may be deposited on the floodplain reach during flooding. On emergence during flood drawdown, freshly deposited levee sediment is progressively bioturbated and penetrated by roots.

The net effect of repeated flooding is the production of an alluvial ridge, whose topography of levees and active and abandoned meander loops may stand far above the general floodplain level. The observed fall-off in mean net deposition rate $r$ at any distance $z$ from the edge of the channel over the levee to the floodplain margin is most simply given by power-law expressions like, $r = a(z + 1)^{-b}$, where $a$ is the maximum net deposition rate at the edge of the channel belt, and $b$ is an exponent that describes the rapidity with which the rate of deposition decreases with distance from the meander belt. The constants vary according to factors such as climate, river size, timing of flood and sediment load. Such expressions fail to predict the observed impact of abandoned channels and other depressions on the floodplain, which very markedly increase deposition rates since they tend to trap and route floodwaters once they have left the levee flanks.

Flood deposits themselves may be dated using various techniques, an important exercise in developing hazard analysis in floodplains by *palaeoflood analysis*. The link between channel change, climate and flooding levels then enables conclusions to be made concerning the causes of palaeofloods. Cycles of flooding and discharge variations occur on an approximately decadal timescale in the west of North America and also in Bolivian Amazonia due to the oscillating El Niño–Southern Ocean (ENSO) effect, causing variations in Pacific-sourced winter storm frequency and magnitude. Sediment cores in the northern Gulf of Mexico reveal evidence for palaeo-Mississippi megafloods at 500–1200 yr recurrence intervals in the upper Holocene, perhaps in response to transport of warmer than normal air masses northwards to the Midwest. The regularity of the famous Nile floods was interrupted several times in the period AD 900–1500 (the Medieval Warm Period) by successive decades of very high then very low flood levels that may be related to goings-on in the North Atlantic Oscillation. We know the facts of this case by a unique palaeoflood record, that of the 'Nilometers' first introduced by the ancient Egyptians to measure Nile water levels (Fig. 11.23). These were like modern flood gauges and measured height in cubits above low water. The annual record of these Dynastic Ethiopian-sourced flood peaks have been lost but fortunately later Nilometers and their records built by Arab and other dynasties have survived (they were used for tax purposes, the higher the flood up to a certain level, the more buoyant the economic prospects and the higher the tax) and give us invaluable evidence for past hydro-climatic variations.

Since river floods are periodic, newly deposited sediment is at once acted upon by soil-forming processes. Rapid vegetation growth in the backswamps of humid climatic regimes leads to the formation of

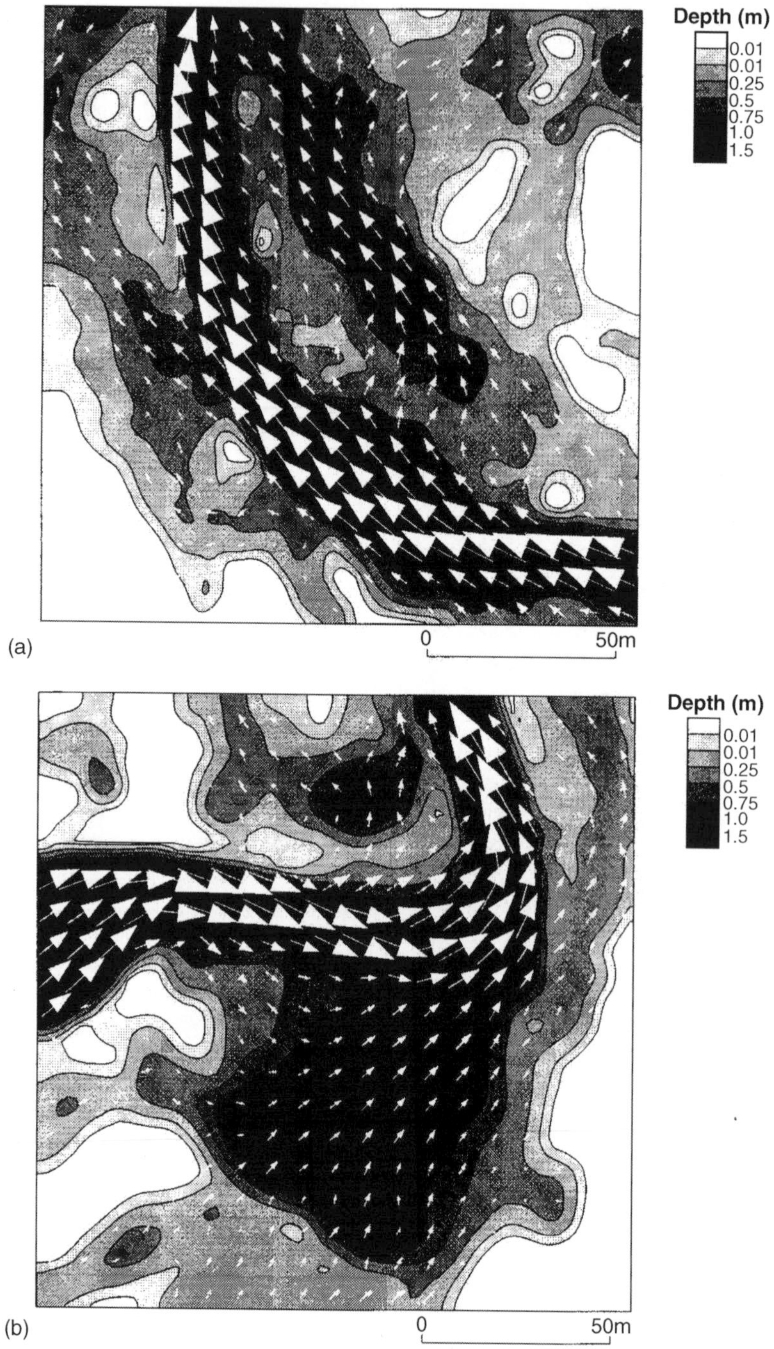


**Fig. 11.22** Computed flood routing maps for reaches of the River Culm, Devon, UK to illustrate the effects of floodplain topography on advective flood discharge. (After Nicholas *et al.*, 2006.) (a) Inundation for a range of discharges up to the peak of a 40-year return period flood. (b) Modelled patterns of flow depth and discharge at a discharge of 60 $m^3/s$. Vector lengths are proportional to square root of unit discharge. Maximum unit discharge is 4.5 $m^2/s$.

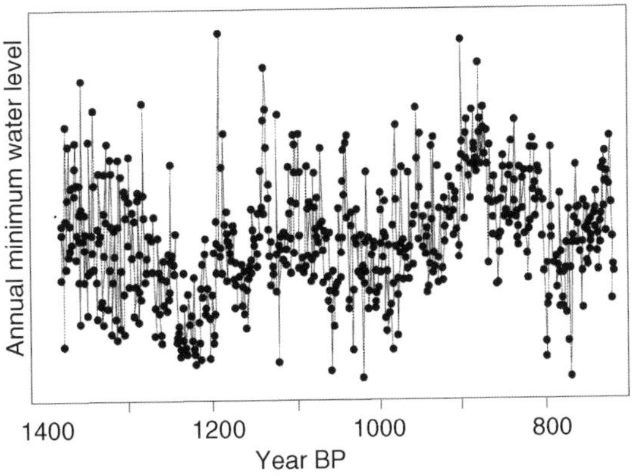


Fig. 11.23 (a) The Umayyad period Nilometer on Roda Island, designed and built by the Turkestani astronomer known as Alfraganus. (b) World's oldest time series from the Roda Nilometer. Such records provide important evidence to evaluate palaeoclimate proxies over medium term timescales. (Both from www.waterhistory.org)

peat beds separated by thin partings of flood-derived muds and silts. Both soil formation and peat formation will be encouraged by slow sedimentation. Well-marked horizonated soils are generally poorly developed in areas of active floodplains because of the rapidity of sediment deposition and by seasonal cycles of soil wetting and drying. Simple models for soil development are possible because there may be an inverse relationship between the extent of pedogenesis and the rate of sedimentation, though the poor drainage in backswamps far away from channels can inhibit soil weathering processes. The concept of a *residence interval* is a useful one in this respect. This is the time during which pedogenesis can act upon a volume of alluvium initially at or close to the floodplain surface. Well-horizonated soil formation may also be inhibited by high water tables and in semiarid climates by carbonate precipitation as groundwater calcrete.

The flux of nutrients, cations and anions from floodplain soils has great relevance to the marine inorganic carbon cycle. For example, it has recently been found that over the past 50 or so years there has been a ca 40% increase in the bicarbonate concentrations per mean unit discharge of water from the drainage catchment of the Mississippi River. This is confidently ascribed to anthropogenic changes in the catchment, notably expansion of liming practices on soya bean and other intensive crops.

## 11.6 Channel belts, alluvial ridges and avulsion

As discussed above, prolonged occupation of a river course leads to production of a *channel belt* produced by successive generations of channel reach migration and bend cutoffs. The width of the belt depends on the ability of a channel to laterally widen by erosion. Data from the Rhine–Meuse and Mississippi rivers suggests that significant downstream channel-belt narrowing occurs as a response to decreased substrate erodibility, especially due to increased peat abundance. When substrate is uniformly more easily erodible then no such trend emerges.

In addition to within-belt processes of reach cutoff, there is evidence that periodic movement of the *whole* channel belt (not just a single reach or bend cutoff) to another position on the floodplain may occur, a phenomenon termed *avulsion*. The process occurs in both meandering and braided rivers and is recorded by abandoned channel belts preserved on floodplains or buried partly or wholly beneath them, the latter as revealed by shallow coring or geophysical surveys. Formerly active channel belts of Holocene/late Pleistocene age may be mapped and dated using radiocarbon analyses of wood or peat in the earliest channel fill deposits that formed after avulsion. These data enable the point and time of avulsion to be constrained for each event so that a picture can be built up of any spatial or time trends in the avulsion process. By far the most detailed database built up in this way comes from the work of H. Berendson's Utrecht group in the Rhine–Meuse floodplains of the Netherlands. There is some evidence for increased avulsion frequency here at a time of rapidly changing sea level (pre-5 ka). But comparing this data with scantier records elsewhere it appears that the frequency of avulsion can vary widely (10–1000 yr); extreme contrasting examples being the milleniall-scale Rhine–Meuse mean interavulsion period and multidecadal shifts of the Kosi River, Bihar state, India over its fan (Chapter 12) and shifts of the nineteenth century Rio Grande channel in New Mexico, USA (Fig. 11.24). Avulsions in these latter examples take place over one or a few flood seasons (perhaps the term 'combing' is more appropriate for these frequent avulsions) compared with the well-studied and ongoing >150 yr old avulsion in the River South Saskatchewan, Canada (see below).

Regarding possible causes of avulsions it must be said at the outset that there is no first-hand detailed observational data on the hydraulics of the process, other than that certain historic avulsions (including the 1855 Yellow River and 2008 Kosi River avulsions, see Chapter 12) have occurred during very major flood episodes. It seems most logical to simply regard avulsion as the end-process of crevasse flooding, whereby levee failure and crevasse channel cutting carries on to enable wholesale escape of the river flood discharge into a new course that is subsequently sustained. This process requires energy to complete it, i.e. excess stream power of the crevasse channel compared to the main channel. It also requires the bed of the new course to be topographically lower than the old channel bed in order that total transfer of the discharge can be facilitated. Several authors have invoked rational reasons for this to happen which depend on *superelevation* of the channel belt above the floodplain, a process that requires net aggradation in the channel over long periods of time. Such superelevation is spectacularly recorded in the Yellow River since an avulsion in 1855.

In one scheme, the probability of avulsion during large flood discharges acts on superelevated channel belts with high transverse-to-valley/downvalley slope gradient ratios that provide the extra increment of flow power needed to excavate a permanent excape conduit. However some authors have stated that this does not work for examples like the Holocene Mississippi Valley and that other substrate, hydrological or even random variables are probably more important, especially the tendency for reoccupation of sandy former meander belts. Another simply appeals to avulsion probability increasing with magnitude of channel belt superelevation above the floodplain. However, the empirical record of Holocene avulsions in the world's rivers is too short (~5 kyr) and the physical data at the time of avulsion too fragmented for any definite model to be preferred. Experimental studies of avulsion in steep fan-like alluvium have invoked the role of increased channel-bed aggradation rate as a positive control on the increased frequency of avulsion. However, the same effect is not seen in scaled experiments with 'normal' gradient braided channels. Finally, it is also known from tectonically active areas that vertical crustal movements must play a role in both initiating and controlling avulsion. Thus diversions may be established because of fault subsidence.

Gradual avulsion leaves very characteristic deposits in the stratigraphic record of a floodplain. This is best seen in the lower Saskatchewan River, Canada (Figs 11.25 & 11.26) where an avulsion in the 1870s has led to the production of a vast complex of splays, wetlands and channels in the Cumberland Marshes. The diversions are gradual and ongoing, with a general progradational aspect well seen in the coarsening-upwards sequences produced by splay-infill of lakes and floodplains. Not all beginning-avulsions actually succeed in diverting the course of a channel, for data from the Rhine–Meuse reveals that certain channel escapes that started off with good intentions never managed to completely divert the main flow, petering out within tens of kilometres in a matrix of floodplain fines; thus either qualifying as *failed-avulsions* or *megasplays*.

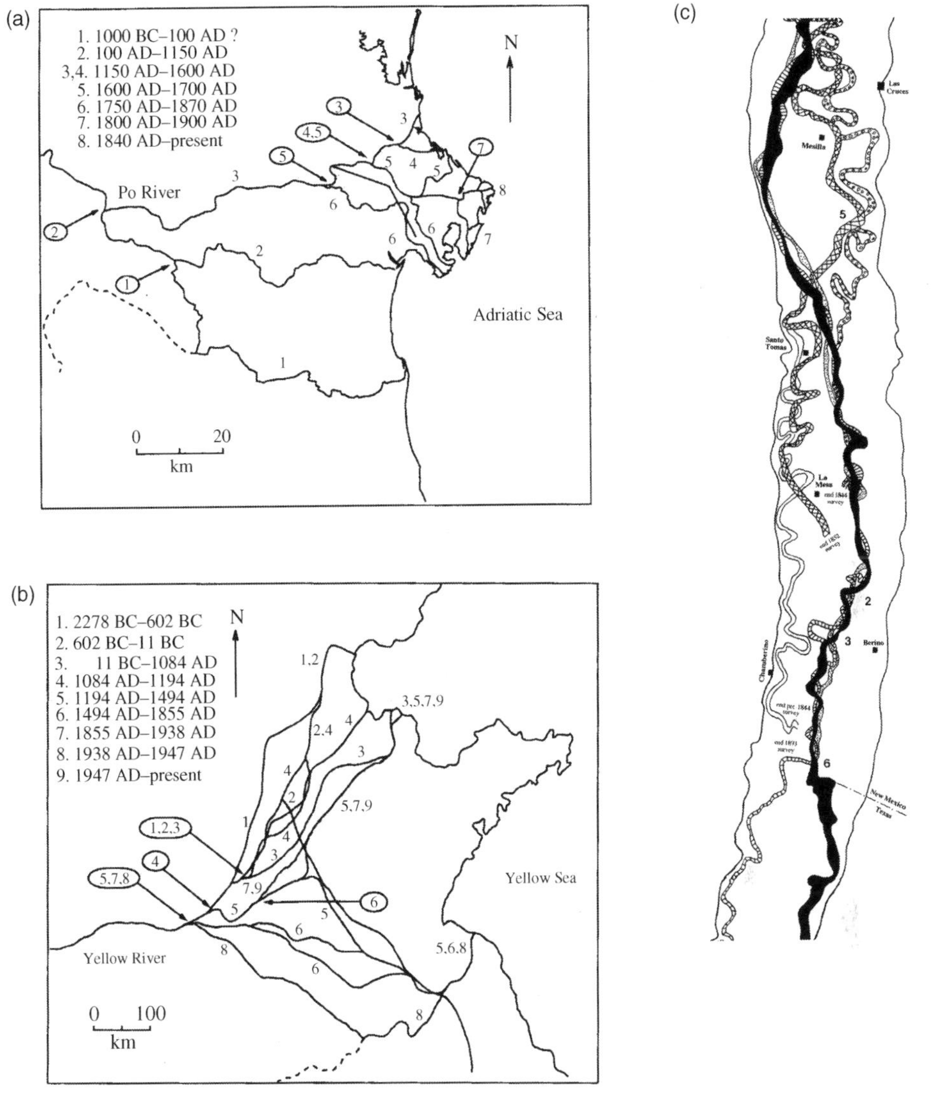


**Fig. 11.24** Avulsion sequences. (a) Po River, Italy. (b) Yellow River, China. (Both after Bridge & Mackey 1993). For both these rivers before the twentieth century note the rather long periods between avulsions (hundreds of years). (c) A 70-km length of the Rio Grande, southern New Mexico, USA. (After Mack & Leeder, 1998.) For this wild river note the frequent channel changes over just 70 years of historical record until upstream damming in 1916. Key to channel positions: black, 1912; lines, 1903; stipple, 1893; cross hatch, 1852; circles, 1844; white, pre-1844.

## 11.7 River channel changes, adjustable variables and equilibrium

A river may adjust the following variables in response to independently imposed climatic or tectonic changes (i.e. changes to runoff/discharge and slope over which the river itself has absolutely no control): cross-sectional area (*wd*), cross-sectional ratio (*w*/*d*), bed configuration, bed material grain size, planform shape (sinuosity) and size (meander wavelength), and channel bed slope. If equilibrium forms truly exist in Nature, then they will include that Holy Grail of equilibrium geomorphology, the *graded stream*. A graded stream has been defined as '. . .one in which, over a period of years, slope, velocity, depth, width, roughness and channel morphology mutually adjust to provide the power and efficiency necessary to transport the load supplied from the drainage basin *without aggradation or degradation of the channels*'. Note that equilibrium applies only to the sediment load, i.e. to a condition of (sediment in) □ (sediment out), or $\nabla . Q_s = 0$.

As we have seen, a river channel of a particular type has several attributes like depth, width and slope that are related in a complex way to imposed sediment load and water discharge. Clues as to the origins of channel type come from changes over time—channels are extremely sensitive to perturbations in slope, sediment load and water discharge. These perturbations may be imposed by climate change, base-level change and tectonics. For example, the hydraulic geometry equations imply that the magnitude of water discharge and the nature of sediment load should radically affect channel sinuosity. Many river systems around the world record major changes in channel magnitude and geometry since the Last Glacial Maximum (LGM), commonly exhibiting a trend from large, braided, aggrading channels to large and then smaller, meandering, incised channels. These changes have occurred due to large decreases in sediment supply in response to a general decrease in runoff and increase in vegetation in the past 20 000 yr. Increased temperature and humidity after the last Ice Age caused

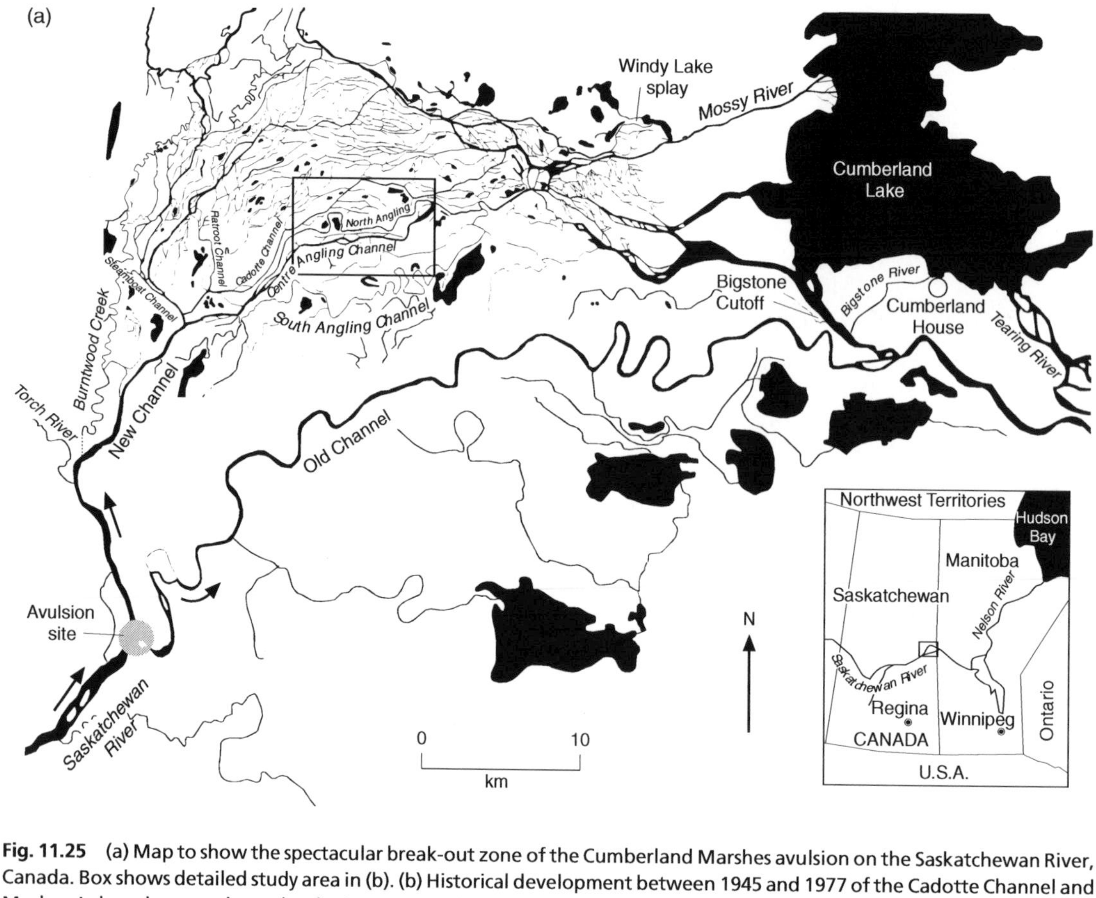


**Fig. 11.25** (a) Map to show the spectacular break-out zone of the Cumberland Marshes avulsion on the Saskatchewan River, Canada. Box shows detailed study area in (b). (b) Historical development between 1945 and 1977 of the Cadotte Channel and Muskeg Lake splay complexes developing against the North Angling Channel alluvial ridge. (After Pérez-Arlucea & Smith, 1999.)

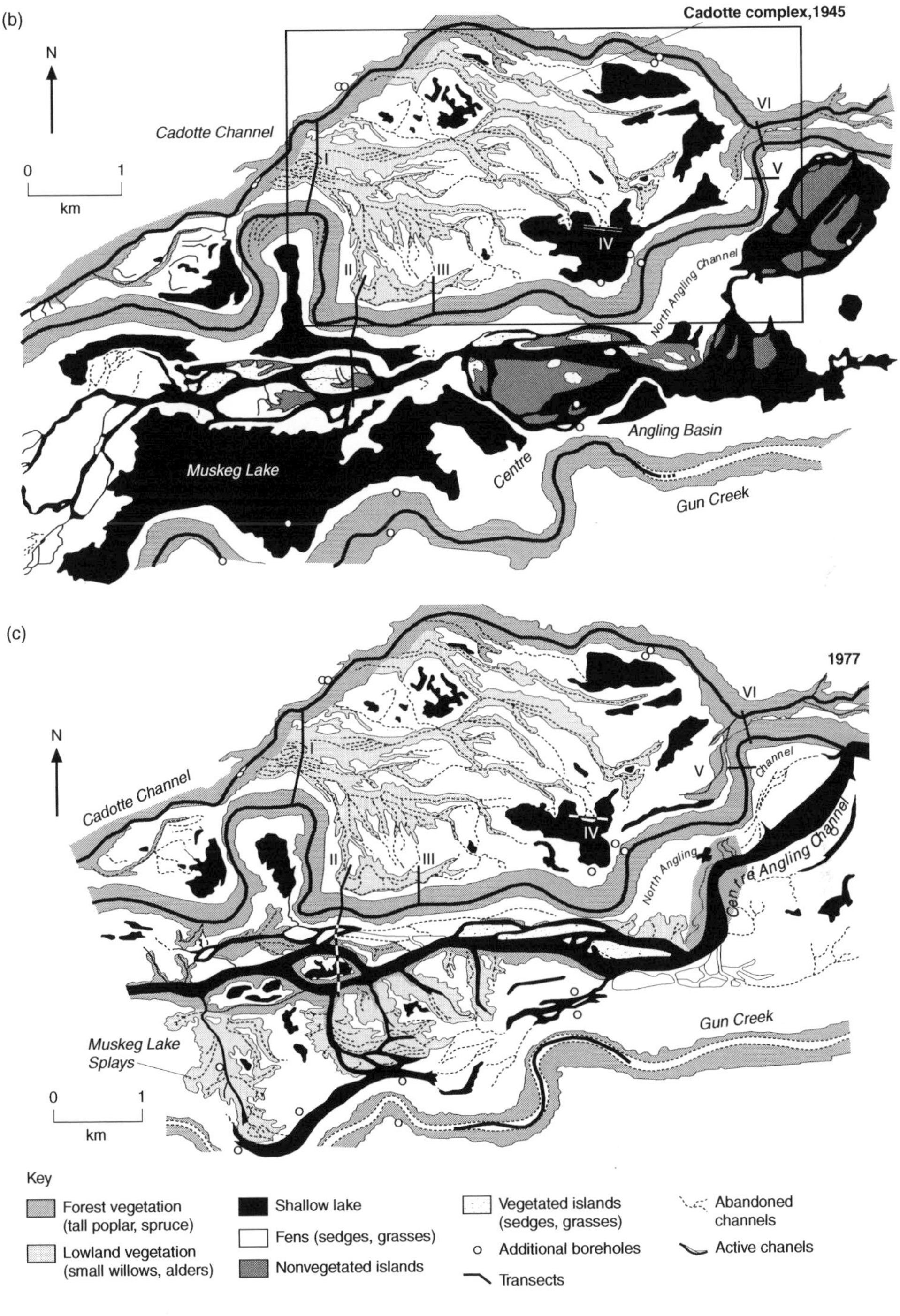


**Fig. 11.25** (*Continued*)

vegetation growth and substantially reduced the amount of coarse sediment liberated from drainage basins (Chapter 15).

Downstream and vertical (time) channel changes from meandering to anastomosing have been carefully documented from tens of thousands of boreholes in the Rhine–Meuse delta, where the cause has been ascribed to a period of rapid sea-level rise about 4–5 ka causing higher groundwater levels and hence aggradation reaching the depositional area (Fig. 11.27). An interesting thing about Fig. 11.27 is the downstream distributive nature of the channel change from meandering to anastomosing, with many blind crevasses and the trunk channel decreasing in magnitude also. These bear some resemblance to the Cumberland Marshes avulsion sequences described previously and are even reminiscent of terminal fans (Chapter 12).

## 11.8 Alluvial architecture: product of complex responses

The stratigraphic *architecture* of river deposits in sedimentary basins defines the 3D distribution of the various river environments we have discussed above. Explanations for the architecture must be sought in the host of palaeoenvironmental variables that contribute to a particular river's history and any past environmental change. Basin type is established by

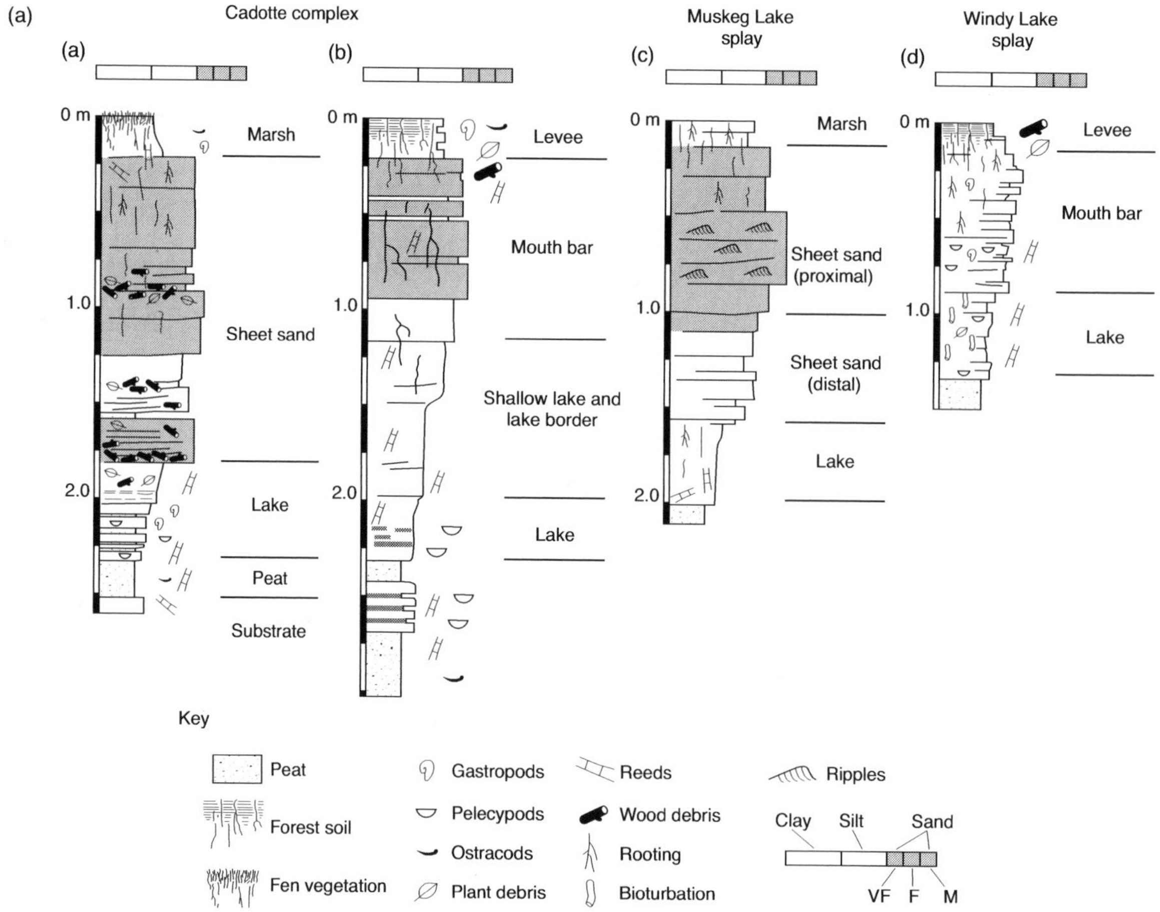


**Fig. 11.26** (a) Logs from vibracores through the Cadotte Channel and Muskeg Lake crevasse-splay complexes to show the various coarsening-upwards sequences. (b) Sketches to show scheme for crevasse-splay evolution, Cumberland Marshes. (After Pérez-Arlucea & Smith, 1999.)

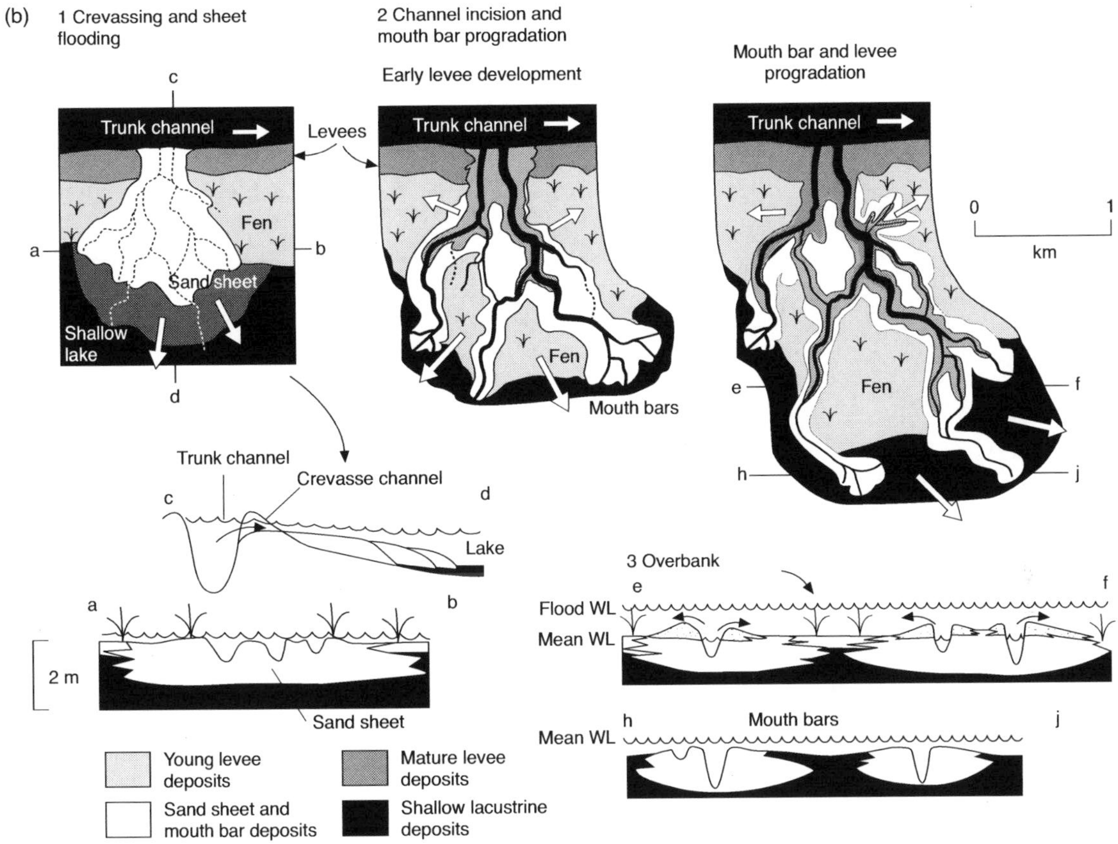


**Fig. 11.26** (*Continued*)

tectonics and provides the large-scale framework for fluvial deposition. River systems in fault-bounded asymmetric basins such as rifts, pull-aparts or thrust-top basins are likely to be strongly influenced by tectonic effects. Occasionally the sedimentary record preserves the deposits of fundamental drainage system outlets that once drained whole orogenic belts and which now provide a record of drainage and structural evolution; such a notable event is preserved within the Sis conglomerate which was a transfer zone trunk palaeovalley of the Sis fluvial system, a drainage system established within the south Pyrenees during Late Palaeocene times. In other larger sedimentary basins, river sedimentation patterns are unlikely to be primarily controlled by active tectonics, either because subsidence is steady and slow (e.g. thermal sagging) and/or because the magnitude of periodic tectonic deformation is too slow to influence channel behaviour.

The advent of sequence stratigraphy has focused attention on relative base-level change to explain incision/aggradation of river and delta channels. On many shallow-water continental shelves lowstand incision to form *incised valleys* due to enhanced erosion over the exposed shelf slope may be observed, for example widely over the Mississippi and Alabama shelves of the northern Gulf of Mexico (Chapter 19). In the ancient record the recognition of incised valley fills formed at lowstands must involve recognition of a regional, unconformity-bounded valley of significantly greater magnitude than individual channel dimensions. This will contain smaller channel-scale sediment bodies deposited during rising sea level and highstand. Such a situation is revealed by core data

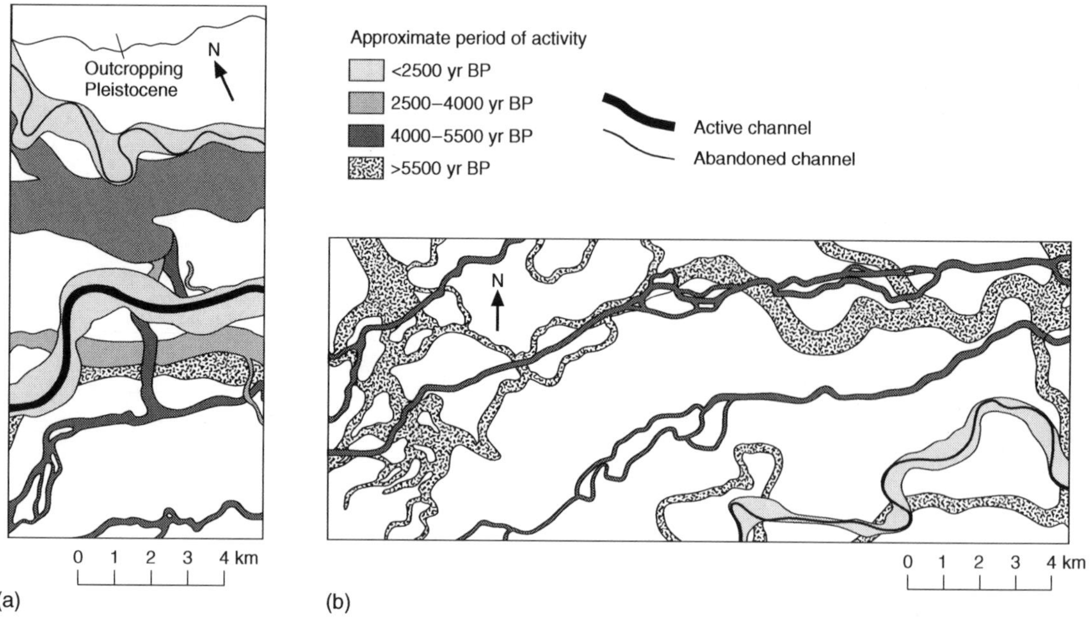


**Fig. 11.27** Palaeochannel pattern (flow east to west) of the Holocene Rhine–Meuse in the region of Utrecht as revealed by an extremely high sampling density of shallow cores. (After Törnqvist *et al.*, 1993.)

of alluviation during sea-level rise and subsequently during highstand. Exceptionally good data from the Rhine–Meuse fluvial complex, with a very clear landward onlap of Holocene alluvium from the position of highstand reached about 5 ka. This has resulted in a tapering wedge of alluvium containing both channel belt complexes and floodplain fines and organic accumulations (Fig. 11.28). Mature to supermature palaeosols are of particular use in recognizing incised valleys, for they form on exposed interfluves that receive little or no sediment deposition over many $10^3$–$10^4$ years, i.e. the soil residence interval in alluvium is very long. Good examples are described from the Dunvegan Formation of British Columbia, Canada.

The fluvial system is, however, complex (in the chaotic sense) since many possible causes for incision exist and there is much nonlinearity due to feedbacks of one sort or another. The state of any river system may thus be seen as the interplay between an 'intrinsic' set of variables (like discharge of water and sediment) controlled by the climate and characteristics of its whole catchment, set against external variables like tectonic and eustatic changes in sea level. For example, base-level change does not affect river channel behaviour in a simple linear fashion, since it takes some time for the wave of incision initiated by base level falling below a previous shoreline and on to a gentle continental slope to pass upstream. The rate of upstream progression of the modified gradient change (the *knickpoint*) and the distance of travel of the upstream limit to incision will vary according to the initial difference in gradients between river/delta plain and shelf, and the transport capacity of the river channel in question, more specifically its ability to erode and transport away the sediment at the knickpoint. In symbols, a simple linear diffusional approach yields, $\partial y/\partial t = \kappa \partial y^2/\partial^2 x$, where $x$ and $y$ are the horizontal and vertical coordinates respectively, $t$ is time and $\kappa$ is a sediment transport coefficient with dimensions $LT^{-1}$. The position of the knickpoint is defined as where the term on the right-hand side of the expression is at a maximum.

Another factor concerns the river's ability to alter its planform rather than necessarily incising or aggrading in response to imposed gradient changes. Thus sinuous river channels have an extra degree of freedom with which to respond to gradient changes: they can change their sinuosity to counteract the imposed gradient change. This is because a high-

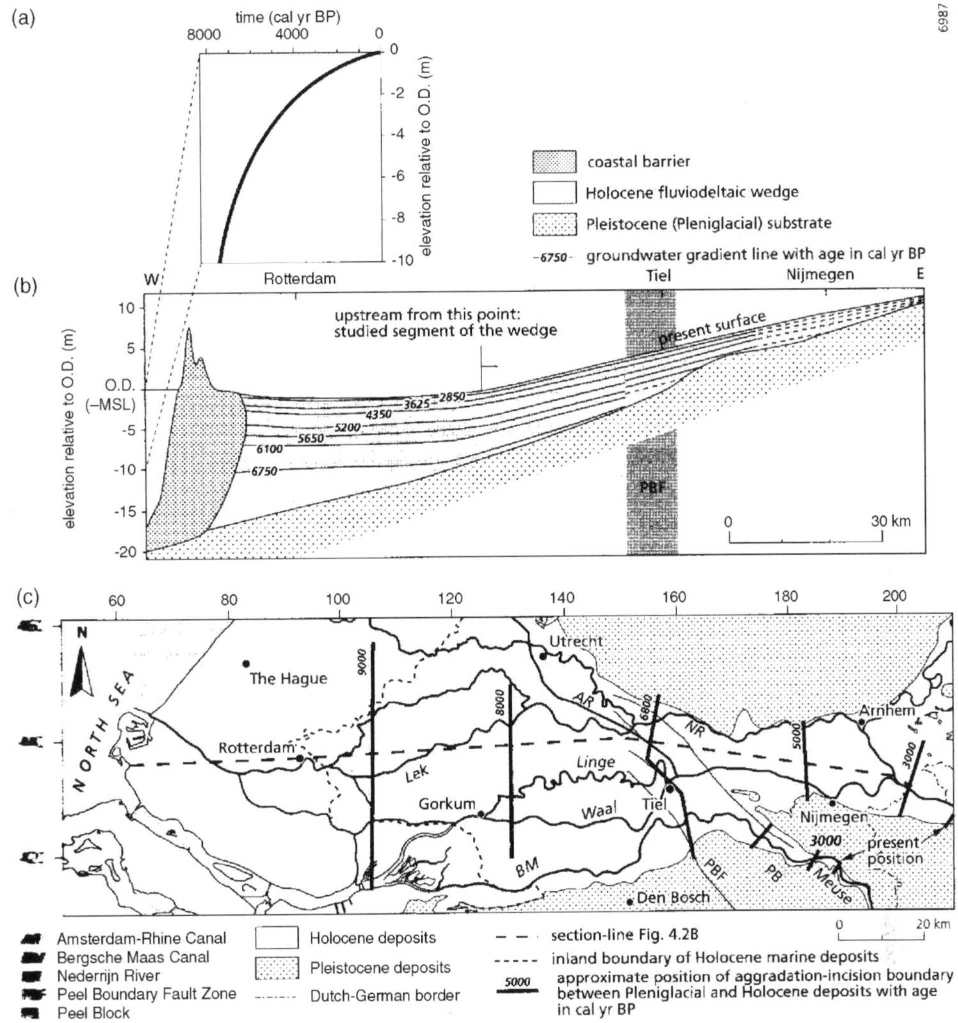


**Fig. 11.28** Fluviodeltaic wedge of the Holocene Rhine-Meuse delta. (After Stouthamer & Berendson, 2403, 2404; Cohen *et al.*, 2002.) (a) Relative sea-level rise. (b) Longitudinal section to show palaeo-groundwater gradient lines. (c) Upstream Holocene migration of the aggradation–incision boundary that marks the upslope limit of the fluviodeltaic wedge, represented in (b) by the intersection of the groundwater gradient lines and the Pleistocene substrate. Between 8–6.8 ka the upstream shift of the aggradation–incision boundary decreased due to the upthrown Peel Block (PBF).

sinuosity channel has a lower mean channel gradient than a low-sinuosity channel flowing down the same valley floor.

Channel aggradation–incision cycles may also have a climatic origin, in that weathering/erosion changes in the drainage catchment cause greater or lesser amounts of sediment and/or water to discharge into the alluvial and deltaic plains. Cycles of aggradation or degradation thus result independently of or out-of-phase with sea-level change. For example, in both the Rio Grande and the Colorado River there are Holocene cycles of alluviation and incision unrelated to sea-level change. Similarly, in the Mississippi delta the maximum phase of incision only affects the river up to Baton Rouge; in alluvial valleys to the north, aggradation occurred at the glacial peak and beginning of deglaciation due to the massive amounts of sediment and meltwater coming down the river from the north. Regional cycles of stream incision (termed *arroyo-cutting*), followed by aggradation (of 40–50 yr duration) have occurred in the American southwest over the past 100 yr or more. Channel incision seems to have occurred during periods of high winter frontal activity. These fronts cause much channel discharge but not so much surface erosion from interfluves as high-intensity summer thunderstorms. The winter discharges are therefore dilute and undersaturated, capable of picking up more sediment from stream beds, causing incision. Aggradation occurs when winter precipitation is less common and floodplain construction by overbank deposition occurs with high sediment yield produced during runoff after summer thunderstorms.

## 11.9 Alluvial architecture: scale, controls and time

On a timescale of $10^3$–$10^4$ years, alluvial architecture is controlled by the two-dimensional migratory response of channels to tectonic gradient changes, climatic fluctuations causing sediment supply and discharge changes, base-level changes causing incision and aggradation, and the effects of compaction and pedogenesis upon channel stability and migration.

For example, the alluvial architecture and soil characteristics of Holocene Mississippi River deposits in the southern Lower Mississippi Valley provide evidence for significant changes in response to sea-level rise. Holocene deposits are subdivided into Lower and Upper Holocene units. The former >5 ka) consist of lacustrine and poorly drained backswamp muds that contain authigenic siderite, pyrite and vivianite, and show little evidence of soil formation. Muds encase crevasse-splay and floodplain–channel sand bodies ( < 1 km wide), and collectively these deposits represent a mosaic of shallow lakes, poorly drained backswamps and multichannel streams, similar to modern examples in the Atchafalaya Basin (see Chapter 17). Upper Holocene deposits ( < 5000 ka) are represented by large Mississippi River meander-belt sand bodies that are up to 15 km wide and 30 m thick. Natural-levee silts and sands and well drained backswamp muds are present between meander-belt sands. Upper Holocene deposits contain abundant soil features, The presence of isolated sand bodies surrounded by mud and the scarcity of soil features suggest that lower Holocene sediments reflect a period of rapid floodplain aggradation during which crevassing, lacustrine sedimentation and avulsion dominated floodplain construction. No evidence of large meandering Mississippi River channels represented by buried, thick tabular sands occurs, and discharge in Lower Mississippi Valley flow was probably conveyed by a network of small, multichannel floodplain streams. Upper Holocene sediments record a dramatic change ca 5000 yr BP from rapid to slower floodplain aggradation, accompanied by extensive lateral channel migration, overbank deposition and soil formation. Similarities in the timing of changes in floodplain processes and fluvial style and decreasing rates of Holocene sediment accumulation strongly suggest that decelerating Holocene sea-level rise in the Gulf of Mexico affected floodplain development at least 300 km inland from the present-day coast. Similarities between the floodplain history of the Mississippi River and those of modern and ancient rivers elsewhere further suggest that avulsion, rather than simple overbank deposition, contributes to the construction of fine-grained floodplains to a greater degree than generally recognized.

One challenge that remains unfulfilled is an account of the evolution of continental drainage through time, especially involving identification of the great rivers of the past. By great here we mean rivers that drain huge areas of continents, like the Mississippi, or highest elevation mountain ranges, like the Ganges–Brahmaputra, and which form large-volume depocentres in the oceans. Ancient equivalents will have had large discharges and would have transported much sediment so that their trunk channels were deep (several tens of metres) and wide (hundreds to thousands of metres). Such great channels may leave tell-tale signs in the sedimentary structures of their channel belt deposits, such as thick lateral accretion surfaces, giant macroform bedforms or thick channel confluence scours. However, care is needed when assessing palaeochannel magnitude, for without such specific structures, thick sedimentary deposits within channel belts may be recording gradual aggradation during periods of overall tectonic subsidence or within deeply incised valleys excavated during periods of lowered sea level. In particular it may be quite difficult to recognize regionally significant sequence boundaries in stacked sequences of braidplain alluvium, witness the concept of *cryptic sequence boundaries*; the use of palaeosols on interfluves has already been mentioned in this regard.

A final point concerns a very obvious feature of many alluvial deposits—the vertical and lateral alternation between coarser-grained channel and finer-grained floodplain sediments. The nature of these alternations, specifically their separation distances and the fraction of the channel deposit sediment bodies that freely connect with each other through an erosive contact, the *connectedness ratio*, has many economic and environmental applications. Thus the so-called *net-to-gross ratio* of permeable reservoir or aquifer channel units to impermeable floodplain units exerts a fundamental control on subsurface fluid flow and on the exploitation and magnitude of water and hydrocarbon reserves. Given the large number of variables that might control these reservoir or aquifer properties, many authors have tried to come up with quantitative models that might somehow simulate alluvial architecture (Fig. 11.29). These seek to evolve a subsurface architecture that results from migration and avulsion of channel belts over a river floodplain undergoing subsidence with time. They succeed to some extent in giving a conceptual view of how a given system can evolve (they are a very nice teaching tool) but the sheer number of input variables means that possible outcomes are almost infinite so that it is difficult to determine which modelling output has any realistic application to particular ancient river architecture. This is illustrated by the paucity of field tests of quantitative alluvial architectural models; very few authors have attempted to get field data from well-dated and exposed alluvial systems and parameterize the results of published models. Another problem is that rivers are essentially a dynamic and

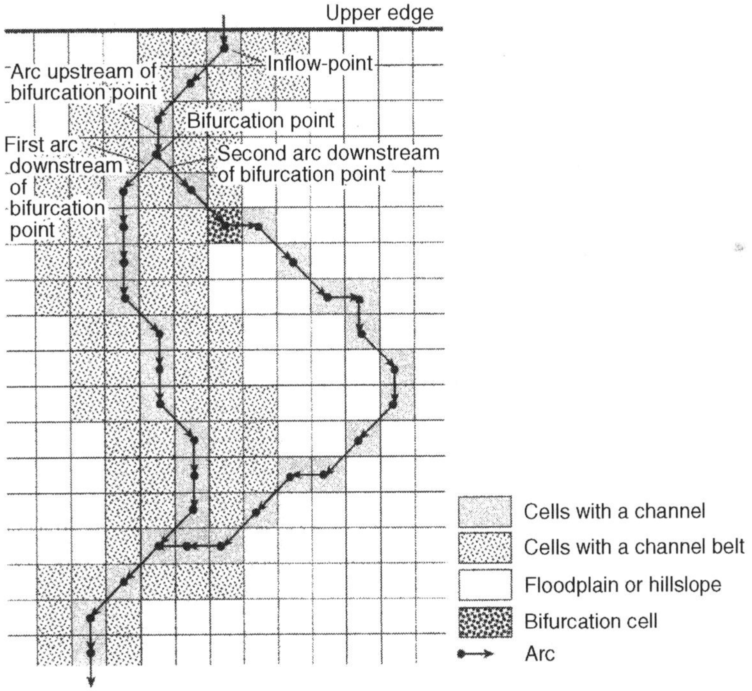


**Fig. 11.29** In a new generation of computational alluvial architecture models, the effects of base-level rise and fall, subsidence and discharge ratios are all incorporated. This sketch (after Karssenberg & Bridge, 2008) shows the basis for a cellular representation of channels, channel belts and floodplain or hill slope areas in the model. Only the upper part of the modelling area is shown. The situation is shown just after a bifurcation of a channel to the right on the figure. In this case, the bifurcating channel rejoins the other channel downvalley, but it may also flow downhill to the lower edge of the modelling area.

hydrologically driven system whilst existing architectural models are kinematic and artificially separated from hydrologically based processes.

## Further reading

### *General*

Bridge (2003) is scholarly, critical and sceptical in every good way. Miall (1996) is massive, descriptive and intelligently informative. Best & Bristow (1993) is excellent on braided rivers. Ashworth *et al.* (1996) is more technical, but none the worse for that, on channel flow structures. Knighton (1998) is a nice geomorphological account of fluvial forms and process whilst Anderson *et al.* (1996) has much of relevance to floodplain processes. Purseglove (1989) is delightful.

### *Specific*

The influence of humans on land use and sedimentation rates is studied by Knox (2006).

Quaternary channel incision is covered by Burkham (1972), Graf *et al.* (1991), Hereford (1993), Huckleberry (1994) and Blum (1993).

Downstream-fining s studied by Wilcock & McArdell (1993), Sambrook-Smith & Ferguson (1995) and Ferguson *et al.* (1996).

Causes of braiding or meandering behaviour are reviewed by Schumm 1960, 1963), Parker (1976), Ferguson (1987), Bridge (1993), Murray & Paola (1994), van den Berg (1995) and Peakall *et al.* (2007).

A pioneering classification of rivers into bedload, mixed load and suspended load was proposed by Schumm (1963). This has been given more recent dynamic teeth by Dade & Friend (1998) and Church (2006).

Anastomosing channels are considered in Smith & Smith (1980), Smith (1983), Törnqvist (1993), Makaske (1998, 2002, 2007) and Gibling *et al.* (1998).

River confluences are studied by (Best,1987, 1988), Roy & Bergeron (1988), Bristow *et al.* (1993), Best & Ashworth (1997) and Siegenthaler & Huggenberger (1993); river diffluences and bifurcations by Frings & Kleinhans (2008); point bars and meander bends by (Jackson (1975), Bridge & Jarvis (1982) and Bridge *et al.* (1995); meander cutoffs by Constantine & Dunne (2008).

Mud-dominated sinuous streams are studied by Taylor & Woodyer (1978), Page *et al.* (2003), Brooks (2003) and Wakelin-King & Webb (2007); gravel-bed meandering channels by Pyrce & Ashmore (2005) and Kostick & Aigner (2007); sand-bed braid bars and processes by Best (1988), Bridge 1985, 1993), Bridge *et al.* (1986) Bridge & Gabel (1992), Bristow *et al.* (1993), Ashworth *et al.* (2000) and Sambrook-Smith *et al.* (2006).

GPR studies of gravel bed bars and processes are by Gawthorpe *et al.* (1993), Siegenthaler & Huggenberger (1993), Lunt *et al.* (2007) and Hickin *et al.* (2009). Open-framework gravels and unit bars are studied by Carling & Glaister (1987), Carling (1999) and Lunt *et al.* (2004, 2007). Interesting scaled experiments on braidplain accretion and within channel diversions are by Leddy *et al.* (1993), Peakall *et al.* (2007) and Ashworth *et al.* (2007). A direct comparison between experiment and field is in Ashworth *et al.* (1999).

Ephemeral stream processes are studied by Laronne & Reid (1993) and Laronne & Shlomi (2007).

Major rivers with extreme discharge variation, notably the Burdekin, North Queensland, are discussed by Fielding *et al.* (1997, 1999).

Floodplain processess and sediment storage are considered by Bridge & Leeder (1979), Smith *et al.* (1989), Gomez *et al.* (1995), Smith & Pérez-Arlucea (1994), Pérez-Arlucea & Smith (1999), Mertes *et al.* (1996), Allison *et al.* (1998), Sellin (1964), Pizzuto (1987), Marriott (1996), Gretener & Stromquist (1987), Asselmann & Middelkoop (1995), Allison *et al.* (1998), Ten Brinke *et al.* (1998), Makaske *et al.* (2002), Törnqvist & Bridge (2002) and Nicholas *et al.* (2006); levees by Fisk (1944, 1947), Coleman (1969), Smith (1983, 1986), Iseya & Ikuda (1989) and Makaske *et al.* (2002). Combined coring, GPS and morphological data are uniquely from Ferguson & Brierley (1999), whilst deposition/accretion rates and levee growth are nicely analysed by Filgueira-Rivera *et al.* (2007); crevasse splays by Smith & Pérez-Arlucea (1994) and Bristow *et al.* (1999); floodplain nutrient and carbon fluxes by Raymond *et al.* (2008) and Davies-Vollum & Smith (2008).

Palaeoflood deposits are studied by Graf *et al.* (1991), Macklin *et al.* (1992), Baker (1994), O'Connor *et al.* (1994), Ely *et al.* (1996), Jansen & Brierley (2004) and Pizzuto *et al.* (2008). For Bolivian Amazonia see Aalto *et al.* (2003). Evidence from the marine record is covered in Brown *et al.* (1999). Coarse-grained tropical flood deposits are studied by Alexander & Fielding (2006).

Nilometer records are from De Putter *et al.* 1998, Hassan (2007) and www.waterhistory.org.

Floodplain soils are studied by Kraus & Aslan (1993), Sanz *et al.* (1995), McCarthy *et al.* (1997) and Aslan & Autin (1998).

Channel belts, alluvial ridges and channel avulsion are discussed in general and critical ways by Törnqvist & Bridge (2002, 2006), Bridge (2003) and Aslan *et al.* (2005). Slingerland & Smith (2004) present interesting dynamic models for channel escape. The most detailed account of a major avulsion, including startling data on Yellow River channel superelevation is by Qian (1990). Case studies from the Rhine–Meuse are by Törnqvist (1993), Stouthamer & Berendson (2000, 2001), Törnqvist and Bridge (2002), Cohen *et al.* (2002), Gouw & Berendson (2007), Makaske *et al.* (2007) and Gouw (2008). South Saskatchewan by Smith *et al.* (1989), Smith & Pérez-Arlucea (1994), Smith *et al.* (1998), Pérez-Arlucea & Smith (1999) and Morozova & Smith (2000). Mississippi by Autin *et al.* (1991). Rio Grande by Mack & Leeder (1998). Columbia River by Makaske *et al.* (2002). The role of peat compaction on avulsion and channel architecture is nicely analysed by Rajchl & Uličný (2005).

Conceptual thoughts and experimental avulsion studies are by Bryant *et al.* (1995), Heller & Paola (1996), Ashworth *et al.* (1999, 2004) and Mohrig *et al.* (2000). Tilt-related avulsions and channel migrations by Alexander & Leeder (1987), Blair & McPherson (1994a,b), Peakall (1998) and Peakall *et al.* (2000a).

Graded stream concepts are discussed by Leopold & Bull (1979).

Quaternary river channel change is put into climatic and sea level context in a masterful review by Blum & Törnqvist (2000). General texts are by Schumm (1977), Bull (1991), Miall (1991), Autin *et al.* (1991), Schumm (1993) and Blum (1990). Specifically: in Europe by Starkel (1983), Vanderburghe (1995), Törnqvist (1993) and Törnqvist *et al.* (1993); North America by Knox (1983), Schumm & Brackenridge (1987) and Aslan & Autin (1999); South Australia by Schumm (1968b) and Page & Nanson (1996); North Africa by Adamson *et al.* (1980).

Incised valleys and their fills are considered in an outstanding study by Plint & Wadsworth (2003), also by Greene *et al.* (2007) and, on valley dimensions, by Gibling (2006). The role of palaeosols in recognizing interfluves of incised valleys is covered by Wright & Marriott (1993), Shanley & McCabe (1994), Ethridge *et al.* (1998) and, specifically for the Dunvegan Formation, by McCarthy *et al.* (1999) and McCarthy & Plint (2003).

Influence of riparian vegetation on channel processes, bar evolution and ecology of floodplains by Fielding *et al.* (1997), Steiger *et al.* (2005), Rodrigues *et al.* (2006) and buried peats by Rajchl & Uličný (2005).

The Rhine–Meuse database of Holocene alluvial development referenced by numerous papers above is essential reading. Width/thickness shotgun plots of palaeochannel bodies and valley fills are in Gibling (2006). Scientific analysis of ancient point-bar deposits, with economic and palaeontological implications, is in Wood (1989), Willis (1993), Diaz-Molina (1993) and Daams *et al.* (1996). Three-dimensional seismic studies of subsurface river architecture are beautifully laid out by Maynard (2006) and Wood (2007). Miall (1994) is good on 3D reconstructions of in-channel bars from the famous Castlegate Sandstone of Utah. Marzo *et al.* (1988) document an impressive valley-fill sequence in the Castissent Sandstone of the Spanish Pyrenees. The alluvial history of the great north-west European rivers (Thames, Rhine, Meuse, Channel Rivers) over the past 3 Myr is impressively documented by Gibbard (1988) with newer treats on the Channel/Manche River megafloods by Gupta *et al.* (2007). Outstanding studies of fluvial sediments within the sub-Himalayan Siwalik foreland basin are provided by Willis & Behrensmeyer (1994), Khan *et al.* (1997) and Zaleha (1997a,b). Mack & Leeder (1999) provide an account of the interaction between lateral alluvial fans and the axial ancestral Rio Grande in the Rio Grande rift. Particularly careful analyses of alluvial architecture via detailed facies analysis are to be found in Ramos *et al.* (1986) and Clemente & Pérez-Arlucea (1993). The problem of cryptic sequence boundaries in stacked channel sandbodies is raised by Miall (2001).

Great rivers past and present are discussed by Potter (1978), Mossop *et al.* (1983), Rainbird *et al.* (1997), Miall (2006) and Potter & Hamblin 2006.

The notable Sis palaeovalley in the southern Pyrennees is explored by Vincent (2001).

Trace the trend to more sophisticated or more specialized alluvial basin architectural models through successive papers by Leeder (1978), Allen (1978), Paola *et al.* (1992a), Bridge & Leeder (1979), Bridge & Mackey (1993) and Karssenberg & Bridge (2008).

# Chapter 12

# SUBAERIAL FANS: ALLUVIAL AND COLLUVIAL

*The water seethed and rose round Achilleus in a fearful wave, and beat down on his shield as it broke over him. He could not stand firm on his feet, but caught hold of a great full-grown elm-tree. The tree came away by the roots and tore open the whole of the bank: it blocked the lovely stream with a mass of branches, and dammed the entire river, crashing full length into it. Achilleus jumped up out of the whirling water and, terrified, made a dash for to go flying across the plain with all the speed of his legs...but the river streamed on in pursuit, crashing loud behind him.*

Homer, 'The Iliad' (ca 10 ka), from Book 21, line 216–250. Transl. Martin Hammond, Penguin Classics, 1987

## 12.1 Introduction

Subaerial alluvial fans are accumulations of sediment deposited from flows issuing from upland catchments via a *point-source*—deposition usually occurs downstream in channels or less commonly as laterally unrestricted sheet flow (Fig. 12.1). The fan builds up over time due to sudden avulsive shifts of its feeder channel, usually switching positions at sites close to its catchment exit point in a process known as *nodal avulsion*. Such sudden shifts mean that human settlements on large fan surfaces are vulnerable to catastrophic flooding, witness the events in Bihar State, India where the capricious Kosi River avulsed in 2008 (see section 12.5). From the exactness of Homer's description in the extract from the Chapter motto it may also have been an avulsion that caused the great flood from the River Skamandros that pursued Achilleus below Troy.

In radial section, fan channel profiles are straight to concave, in cross-section invariably convex upwards. The fan toe grades gently into a basin-floor environment like a range-front parallel (axial) river or a lake. So-called 'toe-trimmed' fans result when channel or wave erosion of the lower fan occurs and a cliff profile is formed. It is common to find linear zones of adjacent coalesced fans, particularly along faulted mountain fronts; these are part of the general *bajada* environment (from the Spanish for 'down', or 'base-of-slope' in this context).

*Colluvium* is mobilized hillslope sediment. Colluvial fans and *talus aprons* are not fed by significant drainage catchments but receive their sediment from a diverse range of upslope processes. These include rockfall and granular flows down enlarged joints or lines of weakness that become *bedrock chutes*, soil creep (especially *solifluction*, Chapter 15), debris flow, localized water flows and powerful snowflow

*Sedimentology and Sedimentary Basins: From Turbulence to Tectonics*, 2nd edition. © Mike Leeder.
Published 2011 by Blackwell Publishing Ltd.

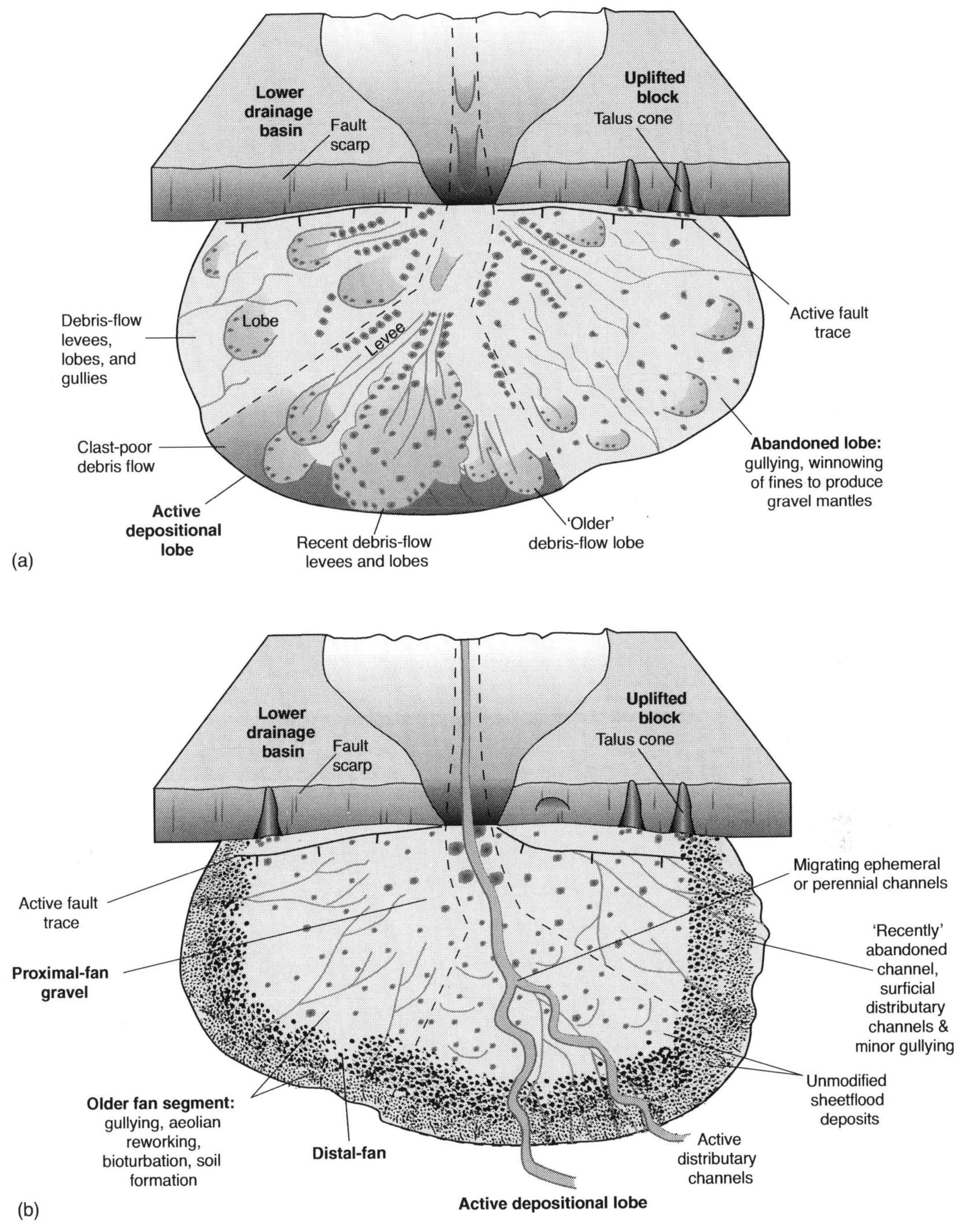


**Fig. 12.1** Block diagrams to illustrate (a) debris-flow dominated and (b) stream-flow dominated alluvial fans adjacent to uplands bounded by active faults. (After Blair & McPherson, 1994a.)

avalanches that may entrain sediment. Like alluvial fans, their colluvial cousins are also commonly found adjacent to active faults, with increments of fan accumulation often triggered by seismic events that create rockfall avalanches; these may fall directly onto adjacent alluvial fan surfaces (Fig. 12.2). Colluvial sediment bodies are also common as slope-mantling talus cones along the sides of major steep valleys or fjords, or at the margins of glacier ice.

In a geological context the most important environment where alluvial or colluvial fan deposition occurs is in faulted sedimentary basins where periodic normal, reverse or strike-slip fault movement enables local or regional subsidence and high fan-sediment preservation potential (Chapter 22).

## 12.2 Controls on the size (area) and gradient of fans

Fan size has numerous implications for basin-margin stratigraphic architecture. The chief controls upon fan area have been investigated by many field studies, and all other things being equal, drainage area is found to be the major influence on alluvial fan size through its control of runoff magnitude and sediment discharge (Fig. 12.3). This is expressed in relations of the kind, $A_f = cA_d{}^b$, where $A_f$ is fan area and $A_d$ is drainage catchment area; coefficient, $b$, and exponent, $c$, are gained from regression analysis with $b$ approximating to 0.9 in many cases; $b$ and $c$ are variable because other factors, such as climate and catchment geology, intervene to cause considerable scatter when data from

**Fig. 12.2** Colluvial talus cones comprising coarsening-downslope limestone clasts, with cone apices issuing from chute-like karst embayments along an active normal fault limestone scarp. Much of the debris is thought to have fallen after an earthquake in 1981. Cone overlies older stream-flow dominated alluvial fan deposits sourced from a drainage catchment to the right of the photo; deposits visible in the quarry cut in the right foreground. Skinos, Greece.

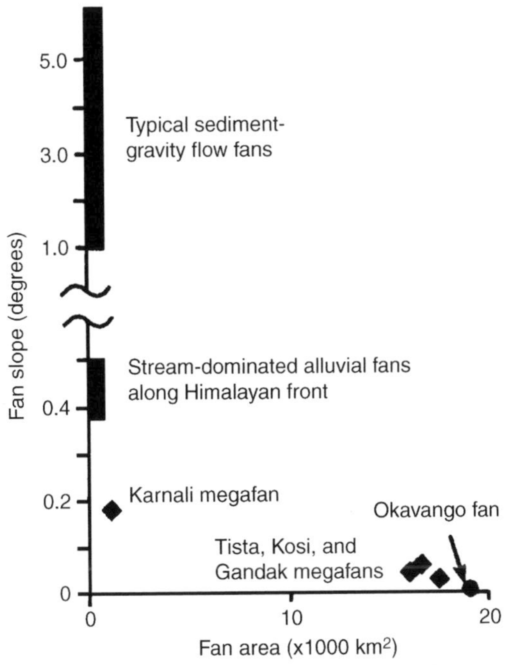


**Fig. 12.3** Cross-plot showing surface area of alluvial fans versus fan slope. (After DeCelles & Cavazza, 1999, and sources cited therein.)

different field areas are compared. There is no such area/drainage area relationship for colluvial fans and cones. In general, mean alluvial fan gradient also decreases with increasing fan size.

Some authors have regarded alluvial fans as essentially small-scale features dominated by hyperconcentrated or debris-flow deposition. They separate them from larger alluvial fans (termed *fluvial megafans*) such as those of the Kosi, Ghagra and Gandak (Indian SubHimalayas) and Okavango (Botswana). These have very large catchments and well-defined fluvial channels and hence predominant streamflow deposits. In reality there is likely to be a complete gradation in catchment area, fan size and transport mechanism (Fig. 12.3) so that size, as usual, doesn't matter. A key point concerns the evolutionary aspects of megafan growth in relation to tectonic processes acting in the drainage catchment. Fault and fold propagation has a major effect here, being capable of fan uplift, tilting, channel incision and the focusing of catchment outlets. The focusing effects are spectacularly seen in the Nepalese Himalaya hinterland, where formerly (prior to ca 10 Ma) a dozen or so drainages emerged from outlets along the mountain front to feed smaller piedmont fans issuing into the Indo-Gangetic plain. Subsequent propagation of the Main Boundary Thrust has focused the entire drainage of a 750 km long sector into just three megafan outlets (Fig. 12.4). Such major changes in basin-margin fan size are likely to dominate the stratigraphic architecture of many ancient foreland basins.

## 12.3 Physical processes on alluvial fans

Channelized flows emerging from a drainage catchment via a rock-bounded valley onto a fan are free to diverge, widen and shallow, break up into smaller channels, though usually one main branch carries most of the discharge, or to infiltrate. Downslope stream power thus dissipates according to the dictates of the sediment continuity equation (**Cookie 34**). Deposition inevitably ensues and a lobe-shaped morphological feature results (Fig. 12.1). Analogies may be made with submarine fan lobes at the ends of submarine channels (Chapter 21) and with distributary mouth bars formed by effluent discharge at the mouths of delta channels (Chapter 17). Nodal avulsion events noted previously are caused by depositional blockage or build-up along channels on any one segment and subsequent 'break-out' down into an area of higher gradient, where the build-up process begins again.

The presence or absence of a fanhead trench is one of the most important features that control the subsequent nature and distribution of alluvial fan sediments. A trench will tend to isolate large areas of the proximal or even medial fan surface from deposition, namely that part of the system upstream of the intersection point (where the fan channel stops being incised). Fanhead trenching, together with channel avulsion and channel plugging by debris flows, are the chief causes and controls on the development of subsidiary fan lobes and of composite fans. In many examples the site of active deposition is periodically shifted over the fan surface in response to phases of late Quaternary climate change, leaving large areas inactive and prone to weathering, rock varnish development and soil formation. The transition from glacial maximum to interglacial climates or vice versa was a prime time for maximum sediment production in many areas of the world, whilst subsequent

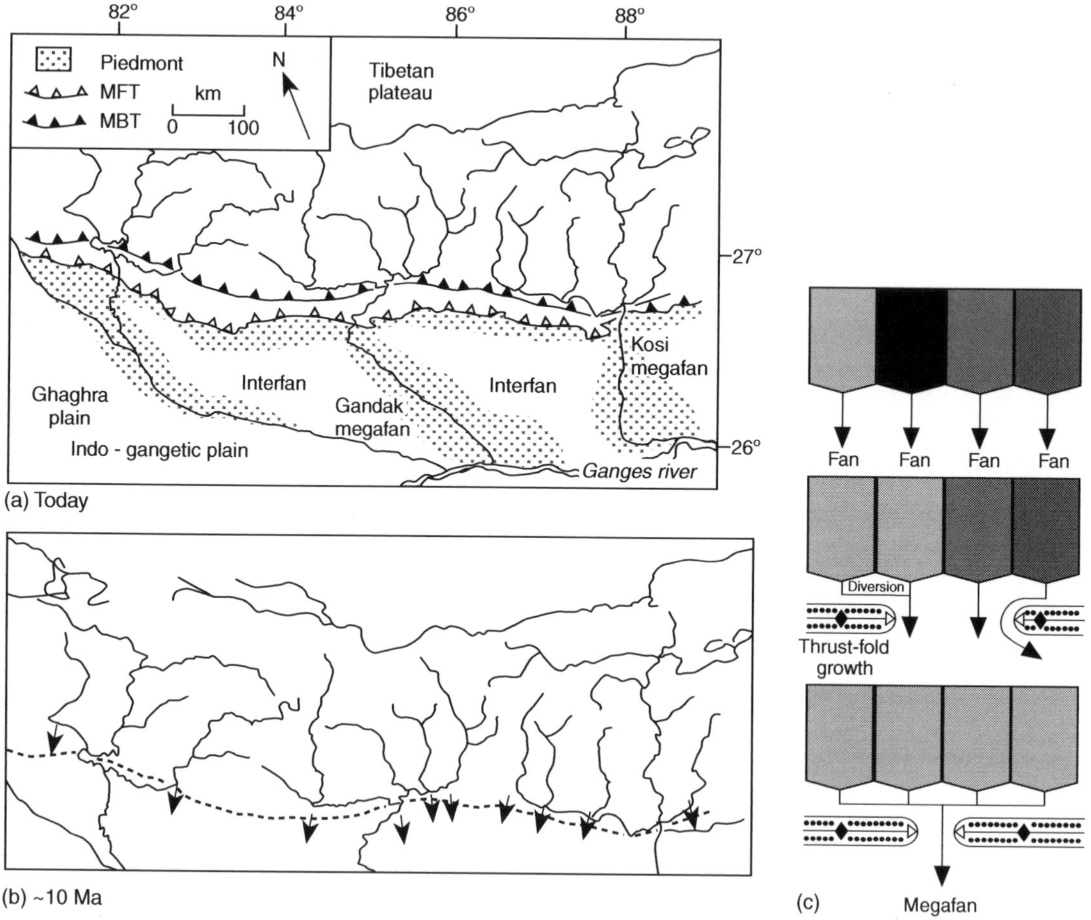


**Fig. 12.4** Maps of the great fluvial megafans issuing into the Indo-Gangetic plains from catchments in the Nepal Himalaya. (After Gupta, 1997.) (a) The present-day configuration with grid-iron drainage catchments feeding into just three main drainage outlets and hence into megafans. Note the way that each tributary drainage turns through a high angle just upstream of the Main Boundary Thrust (MBT; solid triangles) to join the trunk outlet. (b) The situation immediately before initiation of the MBT, with many separate drainage outlets on to fluvial-dominated fans. (c) Sketch to illustrate how the lateral growth of thrust-related folds has focused the drainages. MFT, main frontal thrust.

interglacial growth of vegetation in catchments or climate change to a more stormy rainfall regime has led to incision. Periods of tectonic tilting or fault propagation may also cause local fan incision, as can fan toe-trimming by axial river channels. Spectacular examples of *telesecopic fans* are seen in many mountain valleys (Fig. 12.5) where a trunk river channel incises its former deposits so that lateral alluvial fans must play 'catch-up', leading to construction of successively lower fan surfaces away from the original fan apex.

Major depositional episodes on the lower fan may correspond to periods of entrenchment at the apex, the amount of deposition being controlled by the balance between local rainfall amount and the availability of detritus temporarily stored in the catchment. Thus some high-rainfall periods may see little sediment transfer whilst others may see very substantial transfer—the 'liberation' of stored sediment in the fan drainage basin occurs above a geomorphic threshold that must be exceeded before substantial deposition can occur on the lower fan. A spectacular example has

**Fig. 12.5** Spectacular three-tier telescopic alluvial fans issuing from a faulted mountain front on to the incised floodplain of the Golmud River, northern Tibet. Youngest active fan lobe gravels and sands are mostly unvegetated and splay out on to the modern floodplain.

been documented in the White Mountains alluvial fans that descend into the Owens Valley rift of California and Nevada, western USA. Debris-flow activity here can be shown to be highly sporadic, dependent upon the amount of sediment stored by mass flow in the catchment valleys and upon the probability of a random summer storm affecting all or part of the catchment (convective storms are notoriously localized). It is difficult to overstate the tremendous runout distances (up to 4 km) and transport power (blocks up to 10 m diameter) of such debris flows. From the volume of known historic debris flows, the whole volume of sediment in the Milner Creek Fan (age about 750 kyr) could have been contributed by about three flows per thousand years.

## 12.4 Debris-flow-dominated alluvial fans

Debris- flow-dominated fans can occur in all climatic regimes. Several studies of adjacent fans with identical climate show that source-rock characteristics (e.g. matrix-rich saprolite and mudrocks *versus* more permeable, granular-weathering mantles) play an important role in both providing very fine-grained sediment fractions to encourage cohesive debris flows and in causing enhanced initial debris failure and runoff from colluvium-laden hillslopes. Debris flows initiate on steep colluvial slopes in the fan catchment, often during locally intense thunderstorms. Saturation of the low-permeability, clay-bearing colluvium during such precipitation generates increased pore pressure that promotes slope failure. Such sliding masses quickly transform into debris flows that travel downslope to the feeder channel of the catchment drainage net, and then onto the fan.

Morphologically, debris-flow-dominated alluvial fans comprise a main rockhead valley, which may pass, but not always, into a fanhead channel, canyon or trench and hence into a largely inactive distributive network of fan channels and mid- to lower-fan lobes (Fig. 12.1). These fan channels may have been cut by previous fluvial processes or by more dilute turbulent debris-flow events but are reoccupied by successive generations of cohesive debris flows. The active lobe is separated from inactive lobes by interlobe and interchannel areas. Clasts on inactive fan lobes commonly develop a coating of desert varnish or the sediment develops an incipient soil profile with windblown dust deposition over periods of order $10^3$–$10^4$ yr.

Catchment-valley deposition occurs by scree-fall, colluvial mass flow and debris flow. Fanhead trenches range in depth from a few metres up to tens of metres. They focus most of the falling-stage streamflow discharges and contain localized ribbon-shaped accumulations of poorly sorted, angular and coarse gravels, cut into debrisflow deposits with the clasts being mostly grain-supported. Internal stratification is poorly to moderately developed. At the intersection point of the fan trench with the general alluvial fan surface, the confined flow changes into a distributive network of shallow channels which source active fan lobes, the major locus of deposition on the downstream fan. Both debris-flow deposits and water-lain deposits occur over the mid-fan area. The former comprise interdigitated sheets with nonerosive basal contacts, or occupy prior channels cut by water-flow action. The detailed morphology of the fan surface depends critically upon the effective viscosity of the debris flows themselves. Low-viscosity flows tend to move further down-fan where they spread and radiate, thus smoothing prior topographic regularities and filling in channel traces. High-viscosity flows by way of contrast are unable to move from the upper fan, where they remain as rough, leveed landscape features, each flow influencing the paths of its successors. It is thus clear that whatever controls debris-flow viscosity (sediment type, weathering, precipitation) will also control fan steepness and size.

Traced down-fan, deposits may show decrease in mean grain size, bed thickness and channel depth, and increase in sediment sorting. Because of the short transport distances involved, there is usually little discernible down-fan change of grain shape. Important changes in flow mechanism are responsible for some of these trends; e.g. a down-fan change from thick debris-flow lobes with levees to more fluid mudflows and sheet floods on the lower fan apron. In the coarse deposits of the upper fan, much runoff may infiltrate into the subsurface, causing the rapid deposition of a gravel lobe as a *sieve deposit* with detrital clays infiltrating the interstices of the open gravel framework beneath. Other debris-flow-dominant fans may show little downfan change in deposit character. Thus examples from western Death Valley, California show stacked, clast-rich and matrix-supported debris-flow lobes of cobble to pebble gravel in beds 0.12–1.5 m thick. These dominate the fan from apex to toe, accounting for 75–98% of most exposures. Interstratified with the debris flows are less abundant, thinner (0.05–0.3 cm) and more discontinuous beds of clast-supported and imbricated, pebble–cobble gravels deposited by overland flows and gully flows. These formed by water-winnowing of debris flows during recessional flood stages of debris-flow events.

## 12.5 Stream-flow-dominated alluvial fans

These receive ephemeral, perennial or seasonal stream flow that often originates from fault-bounded ranges subject to rare convective thunderstorms, high seasonal (orographic) rainfall or substantial spring snowmelt. The sedimentary processes on smaller ephemeral-flow fans in arid to semiarid climates is primarily the result of large floods produced by infrequent but high-intensity rainfall in mountainous catchments. For example, palaeohydraulic reconstructions from the Rio Grande rift (**Cookie 31**) of a severe flood in 2006 indicate peak mean flow velocities of several metres per second and discharges approaching $10^3\,m^3/s$. In western Death Valley, fans with active lobes are dominated by deposition from fan sheet floods that were often supercritical, depositing gravels and sands in bedforms such as antidunes and chute-and-pool, together with channel barforms. Individual deposits are planar to wedge-shaped trough-beds with upstream-dipping backsets that dominate apex to mid-fan areas. Pebble–cobble gravel lags are sometimes present above scours cut into the sheet flood and backset deposits. They consist of coarse gravel concentrated through fine-fraction winnowing during recessional flood stage or from non-catastrophic discharge during the long intervals between major flash floods. Although common at the surface, giving rise to a 'braided-stream' appearance, in most of the subsurface fan such deposits are stratigraphically limited, present as thin, continuous to discontinuous beds or lenses that bound thicker sheet flood units. Distal fan environments downstream from incised feeder channels may feature single-event fan lobes that fine rapidly basinwards and show evidence for periodic flood-induced avulsions.

The avulsion and migration of stream channels dominates the sequences produced on larger perennial and seasonal fans. More equable discharge regimes such as spring snowmelt encourages steady channel migration and the formation of channel-bar and point-bar deposits. Some streamflow fans may be very large, with low gradients, an example being the Kosi and other fans noted previously issuing from the Nepal Himalaya fan in northern India. Others in extensional rift basins like the Rio Grande rift of southwest USA completely overlap in size with debris-flow-dominated fans. Fluvial fans often show a proximal–distal change from coarse, cobble-to-boulder grade alluvium deposited in apical braided channels to finer sediment of meandering channel and overbank flood deposits. They too may often show entrenched fanhead channels. In the largest fans, coalescence and fluvial invasion by adjacent stream systems is common. Additionally the fans may be so large and gently sloping that they source their own drainage systems on the temporarily inactive fan surfaces. Probable climate control on megafan development is well shown along the Yamuna–Ganga megafan, north of Delhi, India. Here the modern channel of the Ganga river has deeply incised a formerly active megafan surface of area $1.5 \times 10^4\,km^2$, defining a valley up to 30 m deep and 10–20 km wide. Evidently the megafan surface was produced by significant late Pleistocene deposition from a network of now-abandoned channels under a rather different climatic regime when sediment and water fluxes were very much higher than today.

Fluvial fans feed downfan into trunk or axial rivers and are dominated themselves by the migratory behaviour of their channels. Indeed, the capricious behaviour of the Kosi River channel led to its being named after the abused heroine of a Hindi legend. Up to 2008 the active channel belt of the Kosi lay to the extreme west of the fan accumulation (Fig. 12.6), having migrated clockwise though successive avulsions over the past several hundred years (Fig. 12.7). Slopes decrease markedly from proximal to distal, accompanied by a braided-to-meandering channel change and grain-size fining over the 160 km or so of the fan length. Groundwater-fed streams reworking abandoned fan-surface deposits become increasingly common distally. In 2008 after extreme monsoonal discharge (Fig. 12.8), the Kosi river avulsed from its nodal point to roughly the centre of the fan (between positions 4 and 5 on Fig. 12.7), creating havoc and destruction with much loss of life in farmed and settled areas that had been stable for 200 yr or more. This nodal avulsive behaviour of the Kosi river and the presence of numerous historic channel courses on the megafan surface brings us to a key point concerning nomenclature. None of these channels are *distributive* in correct usage of this term—to be thus means that a channel discharge is *split* between *active* distributaries. This may occur on coastal river deltas (Chapter 16), but not universally so.

Downfan-fining trends are evident in the subsurface deposits of the Bologna alluvial fan bajada that extends 15–20 km basinwards of the Appennine Thrust Front in northern Italy. This area is noteworthy for the influence of both climatic and tectonic controls on fan architecture. It is also one of very few examples where correlations have been made between catchment valley processes of downcutting and terrace formation and the deposition of fan sediment. Avulsive behaviour, but not of a sweeping migratory kind, is also recorded on the Assiniboine Fan (with its spectacular channel levees) of Manitoba, Canada and on the spectacular abandoned Pleistocene fans of northern Oman.

The Okavango megafan of north-west Botswana is extraordinary. Rivers rising in the humid Mozambique Highlands feed discharge down to the hyperarid Kalahari Desert. The main discharge enters a tectonic rift through a 'panhandle'-like gorge, whereupon periodic combing motions of the initially sinuous streams have created a vast wetland fan traversed by channels. The very low proportions of suspended load do not prevent the occurrence of meandering or straight channel forms since the channel margins are encased in thick cohesive peat levees of the papyrus plant community, which stabilize the channels. Periodic abandonment in the distal fan is accompanied by natural peat burnings, which cause the levees to crumble and the abandoned channel sands to stand out in relief.

## 12.6 Recognition of ancient alluvial fans and talus cones

The majority of ancient alluvial fan deposits are located at basin-margin positions adjacent to once-active normal, thrust or strike-slip faults. In view of the prevalence of fluvial megafans in the modern and Miocene Himalayan foreland, the South Pyrenean foreland thrust basin and in the Upper Cretaceous–lower Tertiary stratigraphic record of the North American Mesozoic Cordilleran foreland basin it seems reasonable to agree with the suggestion that

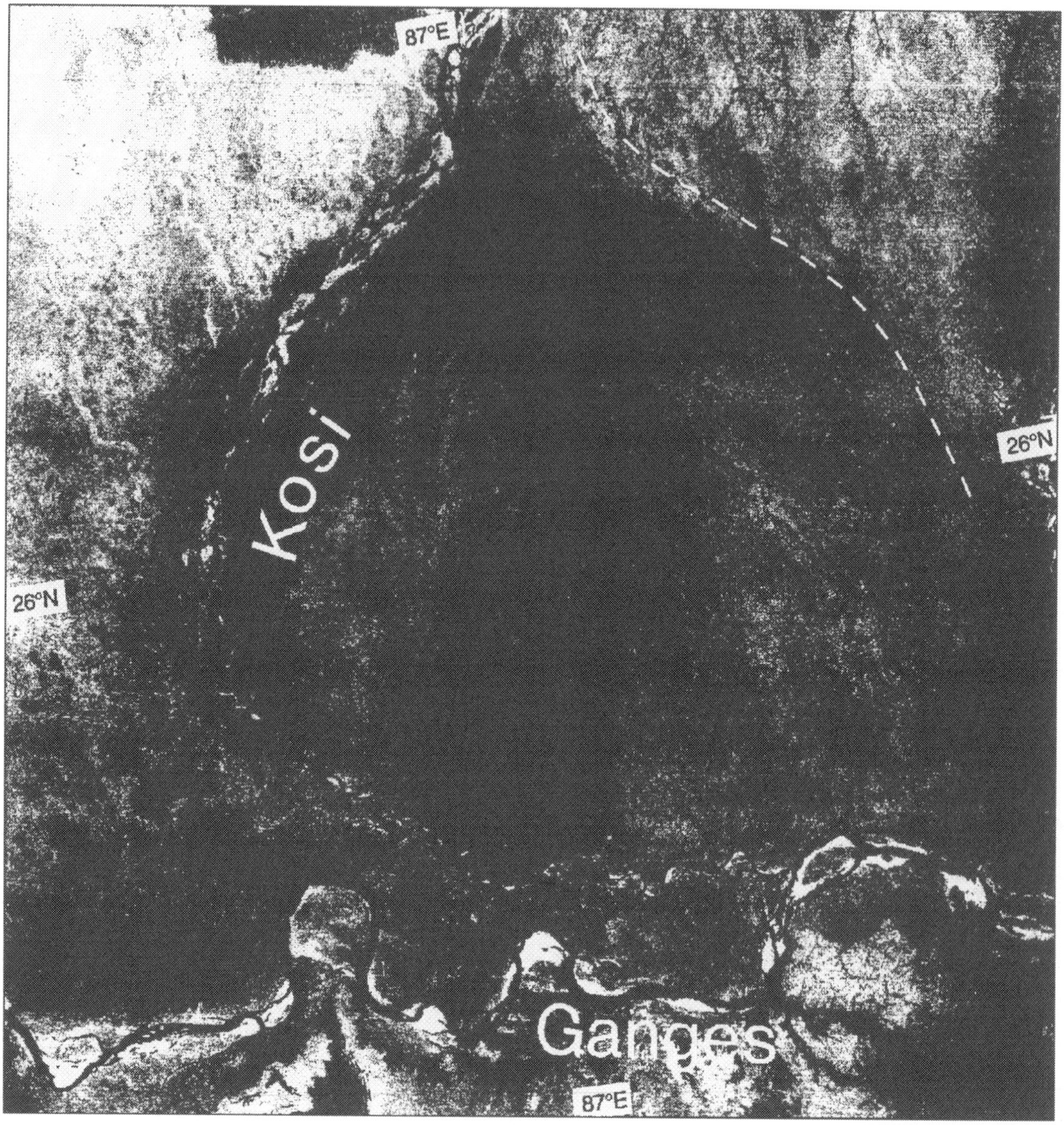


**Fig. 12.6** Landsat image of Kosi megafan that pre-dates the 2008 avulsion (see Fig. 12.8) of the main channel eastwards. Maximum width of Kosi fan across parallel 26 °N is ca 140 km. (After DeCelles & Cavazza, 1999.)

thrust-basin deposits may contain the volumetrically largest gravel accumulations in the geological record (Fig. 12.9). Individual ancient fans or coalesced bajada may be recognized by a combination of palaeocurrent, grain-size, clast-provenance and facies trends. The purist will insist that the radiating form of any contemporary fan-shaped body be established a priori from palaeocurrent or channel trend mapping. The nature of the fan may be established by a consideration of the predominance of streamflow vs. hyperconcentrated- and debris-flow features. Downflow decrease of channel size may provide important evidence for the existence of a *terminal fan*, one in which both water and sediment budgets are

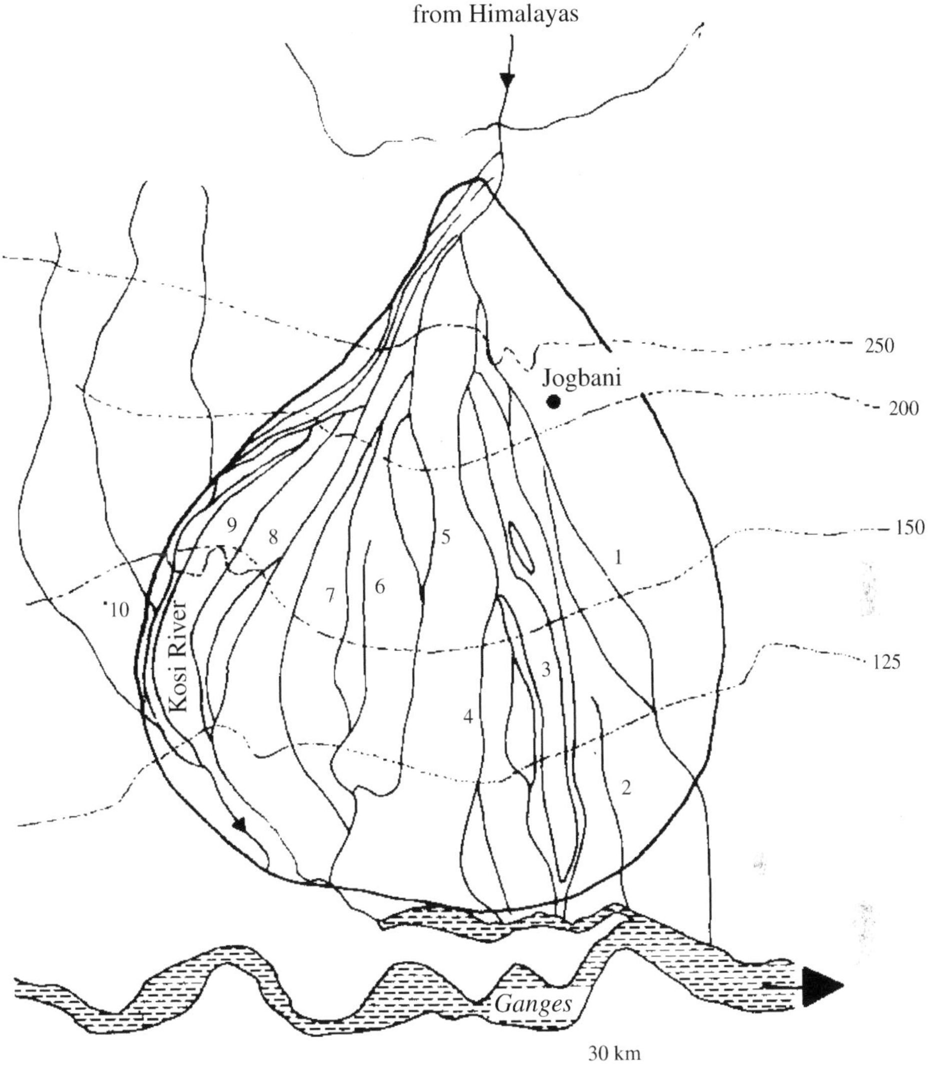


**Fig. 12.7** The numerous positions of the Kosi River on the Kosi fluvial megafan over the past 278 years. Note the apparently systematic combing from east to west. The dotted contours are given in feet above the level of the Ganges floodplain (After a classic paper by Gole & Chitale, 1966.)

conserved, i.e. nothing is exported beyond the fan confines. Many modern debris-flow fans are terminal and conservative in this scheme and fine examples of streamflow fans are inferred from the margins of the south Pyrenean Ebro foreland basin(Fig. 12.9), the Cenozoic of Bolivian Altiplano thrust basins and from several other ancient basins, including the Devonian of England, Scotland and Ireland. However, it is fair to point out that no convincing Holocene or recent example of a large terminal fan has yet been described, leading to marked scepticism concerning the concept on the part of some researchers.

Upward changes in grain size and depositional features and reverse stratigraphic trends in clast analysis (clasts increasing in age upwards) reflect cycles of hinterland erosion and fan growth. The influence of tectonics upon fan cycles may be carefully investigated with reference to field and theoretical models of the

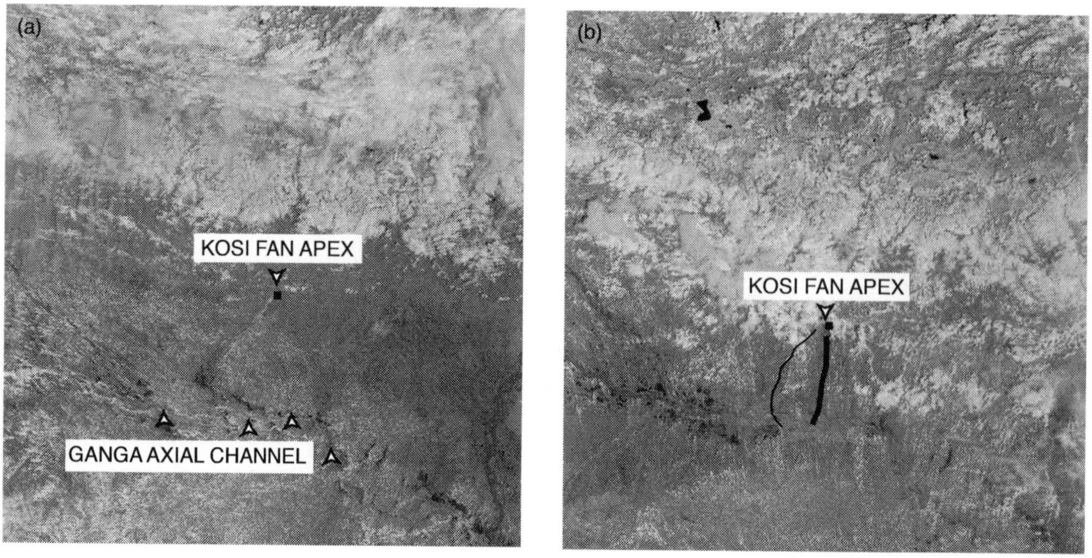


**Fig. 12.8** Satellite images of the Kosi 2008 avulsion, Bihar State, India. (a) Before avulsion. (b) After avulsion. The systematic westwards avulsion sequence indicated by Fig. 12.7 is broken by the 2008 avulsion back eastwards (heavy line) to approximately the position of channels 3 and 4 dating back to the eighteenth century, a truly remarkable illustration of avulsion seeking out maximum gradient advantage.

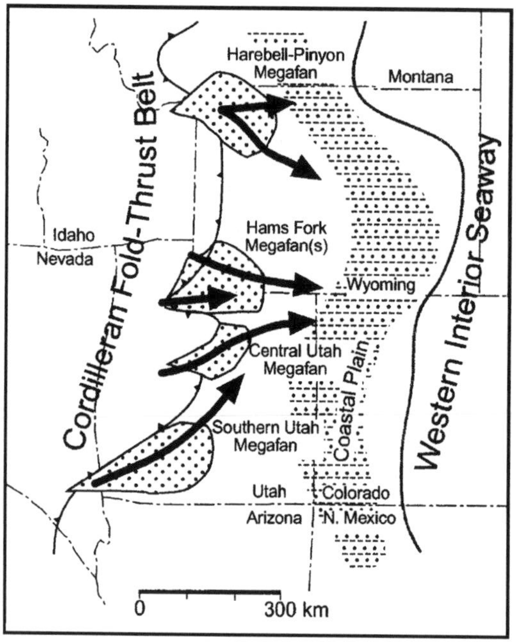


**Fig. 12.9** Palaeogeographical sketch for late Cretaceous–Paleogene times to show fluvial megafans draining the Cordilleran fold-thrust belt and adjacent foreland basin system in the western USA. (After DeCelles & Cavazza, 1999, and sources cited therein.)

effects of subsidence rate and fault propagation-folding upon fan dynamics. Climatic and eustatic base-level effects may be at least as severe as tectonics. For example, in intermontane Himalayan valleys (termed *duns*), thrust-bounded alluvial fans were influenced both by tectonics and glacial climate fluctuations. Surface morphology shows oldest fan surfaces entrenched and onlapped by second-order fans, which are in turn cut by a major throughflowing incised valley showing a pair of terrace levels. Dating by optically stimulated thermoluminescence (OSL) determines that the period of incision separating the younger and older fan deposits (70–80 ka) coincided with enhanced southwest monsoon precipitation. The subsequent development of the second-order fans and their progradation until 20 ka suggest erosional unloading of the thrust hangingwall during a tectonically quiescent phase. Toe cutting, deposition of axial river and lacustrine facies, and fan retreat around 45 ka, indicate fanward shift of the axial river due to thrust-related tilting. Cessation of fan deposition around 20 ka and the onset of through-fan entrenchment suggest reduced sediment supply but relatively high stream power during the Last Glacial Maximum (LGM) and subsequently during intensification of the southwest monsoon. Another example of the powerful application of OSL dating to constrain the timing of deposition in individual fan-lobe deposits from Death Valley, California shows that deposition occurred predominantly during phases of climate change, both from wet-to-dry and vice versa during the past 25 ka.

Stratigraphic models for the sedimentary sequences built up over time by alluvial fans are largely based upon the hypothetical behaviour of a prograding or retrograding fan system. A prograding fan is expected to give rise to large-scale coarsening-upwards sequences as the depositional processes become increasingly proximal. Progradation may be caused by increasing intensity of basin-margin faulting or by increasingly humid climate giving higher runoff and sediment transport rates. A retrograding fan system should give rise to a fining-upwards sequence by the reverse of the above. Initial fan growth along a new scarp or fault scarp should also give rise to a coarsening-upwards sequence, as will advance of active subsidiary fans after deep fanhead trenching. Features of Plio-Pleistocene relict footwall-derived fans in the Rio Grande rift, New Mexico, USA are 3–10 m thick cycles of sediment representing: (i) development of calcic palaeosols; (ii) truncation of soils by channels up to 4 m deep; (iii) infill of channels by streamflow gravels and sands deposited as sheets and channel bars; (iv) deposition of hyperconcentrated-flow pebbly sandstones. The cycles are attributed to climatic changes over periods of order 150 kyr.

Talus cones are usually represented by smaller and steeper accumulations than alluvial fans, though catastrophic runout of megarockfall avalanches may involve travel for tens of kilometres (see account of the Nevadan Huescos in Chapter 8). A particular feature of talus cones formed by granular gravity flow is open-framework gravels with inverse grading and increasing clast size downslope, quite distinct from the usual fining trends found in many alluvial fans.

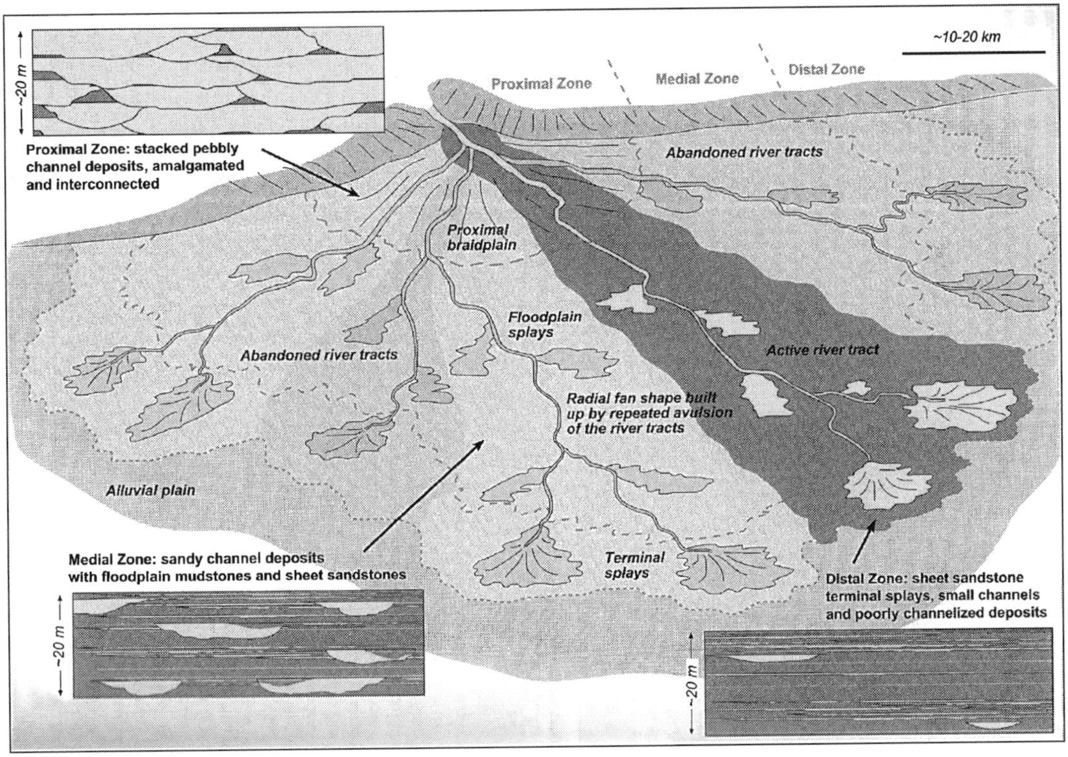


**Fig. 12.10** Inferred distribution of sedimentary environments on an ancient terminal fan complex, gained from many outcrop panels in the late Tertiary Ebro foreland basin, northeast Spain. (After Nichols, 2008; Nichols & Fisher, 2008.) As highlighted by the authors, note that the various channel belts are *not* distributaries since they are not conjointly active.

A final point concerning alluvial fans is the frequent occurrence of *relict shoreline deposits* that record previous times of lake expansion and the conversion of a usually dry alluvial fan margin to that of an active fan delta (see Chapter 17). Particularly fine examples occur in Lake Lisan, Israel and in the western USA associated with pluvial Lakes Lahontan, Bonneville and other local lakes like those in Death Valley, California and Dixie Valley, Nevada (see Chapter 14).

## Further reading

Classificatory and fan magnitude matters are considered by Stanistreet & McCarthy (1993), Blair & McPherson (1994a, b), Harvey (1997, 2005), Kim (1995), McCarthy & Candle (1995) and North & Warwick (2007).

Colluvium plus a general review of physical colluvial fan processes in high latitudes are considered by Blikra & Nemec (1998) and, from Turkey, by Nemec & Kazanci (1999).

Tectonic influences are studied by Hooke (1972), Leeder & Gawthorpe (1987), Muto (1989), Leeder & Jackson (1993) Jackson & Leeder (1994), Amorosi *et al.* (1996), Jackson *et al.* (1996) and Gupta (1997).

Quaternary fan changes are studied by Lustig (1965), Melton (1965), Bull (1979, 1991), Mayer *et al.* (1984), Ono (1990), Harvey (1990), Derbyshire & Owen (1990), Mack & Leeder (1998), Sohn *et al.* (2007) and Ori *et al.* (2008).

Fanhead trenching is discussed by many authors, including Beaty (1970, 1990) on White Mountains fans; telescopic fans by Bowman (1978) and Colombo (2005); debris-flow-dominated fans by Hooke (1967), Beaty (1990), Harvey (1990), Kochel (1990), Whipple & Dunne (1992) and Blair (1999b).

The role of catchment rock types on debris-flow versus streamflow dominance is considered by Blair (1999a–c), Mather & Hartley (2005), Nichols & Thompson (2005), Wagreich & Strauss (2005).

A pioneering paper on fan architecture is by Heward (1978).

Streamflow-dominated alluvial fans and megafans are studied by Gole & Chitale (1966), Hirst & Nichols (1986), Gohain & Parkash (1990), Kochel (1990), Maizels (1990), Amorosi *et al.* (1996), Mukerji (1990), Gupta (1997), Guzzetti *et al.* (1997), Mack & Leeder (1999), Friend (1978), Wells & Dorr (1987), Rannie *et al.* (1989), McCarthy *et al.* (1991), Stanistreet *et al.* (1993), Blair (1999a), Shukla *et al.* (2001), Gábris & Nagy (2005), Weissman *et al.* (2005) and Nichols & Fisher (2007).

The role of sheet floods on distal terminal fans in tectonically active settings covered by Hampton & Horton (2007); toe-cutting of fans by Mack & Leeder (1999) and Leeder & Mack (2001); major floods and their effects on fluvial fans by Mather & Hartley (2005), Magirl *et al.* (2007) and Mack *et al.* (2008).

Recognition of ancient alluvial fan is discussed by Galloway (1980), DeCelles *et al.* (1991), Crews & Ethridge (1993), Nemec & Postma (1993), Yoshida (1994), Amorosi *et al.* (1996), Daams *et al.* (1996) and Mack & Leeder (1999).

Climatic and tectonic influences on Himalayan fans are considered by Suresh *et al.* (2007); terminal fans by Friend (1978), Nichols & Hirst (1998), Nichols (1987, 2005, 2008) and North & Warwick (2007); Torridonian megafans by Williams (2001).

Discussions of analogues between Himalayan-sourced megafans and the North American Cordillera foreland basin can be found in De Celles & Cavazza (1999).

Outstanding examples of outcrop to core-based architectural studies in the sub-Pyrenean foreland basin megafans are presented by Donselaar & Schmidt (2005).

For Pleistocene alluvial fans that were once pluvial-stage fan deltas see Blair (1999a) and Harvey (2005).

# Chapter 13

# AEOLIAN SEDIMENTS IN LOW-LATITUDE DESERTS

*The bulk of the grains flowed as a dense fog, rising no higher than five feet from the ground. Over it we could see each other quite clearly, head and shoulders only, as in a swimming bath. Up above the great fine-grained crests of the dunes were on the move. Cornices dissolved as we looked, swaying along the curving surfaces in heavy dark folds, as if the mane of some huge animal was being ruffled and reset in a new direction by a gale.*

R.A. Bagnold, *Libyan Sands*, Hodder and Stoughton, 1935

## 13.1 Introduction

*Aeolus* was the Ancient Greek god of winds; it was he who gave the vagrant Odysseus his bag of wind to power his ship back to Ithaka. So, we call wind-blown sediment and associated sedimentary processes, *aeolian*. A desert is generally defined as having limited available precipitation and a general deficiency of water for weathering, runoff and accumulation. The technical definition depends on the ratio between actual precipitation ($P$) and the ability of solar energy and vegetation to return moisture to the atmosphere by evaporation and evapotranspiration ($ET$p). Thus a desert is an arid zone that has a $P/ET$p ratio of $< 0.2$. There is no sharp defining boundary, rather a gradation of conditions from extremely arid to semiarid and subhumid (Fig. 13.1). A desert and its hinterland are providers, accumulators and exporters of sediment (see section below on system state), often teleconnecting vast areas on the scale of trade-wind meteorological systems. Deserts may be bordered or traversed by permanent rivers (e.g. Nile in Egypt, Niger in Chad) or lie adjacent to a coast (Oman, Namib, Atacama). Most importantly, most deserts have seen immense Quaternary climate change so that the majority can look back to previous 'pluvial' times when more abundant runoff as river flow provided the sediment now mixed around by the wind system. However, low-latitude deserts have only about 20% of their surface area covered by active sand dunes.

The major deserts of the world with active sand flow occur in the central and southern parts of subtropical high-pressure cells where descending dry, warm air forms Trade Wind belts (Chapter 10). A further category of desert occurs in the continental lee or rain shadow of uplifted mountain belts such as the western USA, Central Asia and Patagonia. In these semiarid and arid zones, sediment was generated in upland areas during pluvial (higher rainfall) periods and transferred by stream flow into depositional basins. En route the sediment was sorted before further more drastic sorting by the blowing wind, which also moulded it into aeolian bedforms. These coalesce to form sand seas, or *ergs* (an Arabic name, widely

*Sedimentology and Sedimentary Basins: From Turbulence to Tectonics*, 2nd edition. © Mike Leeder.
Published 2011 by Blackwell Publishing Ltd.

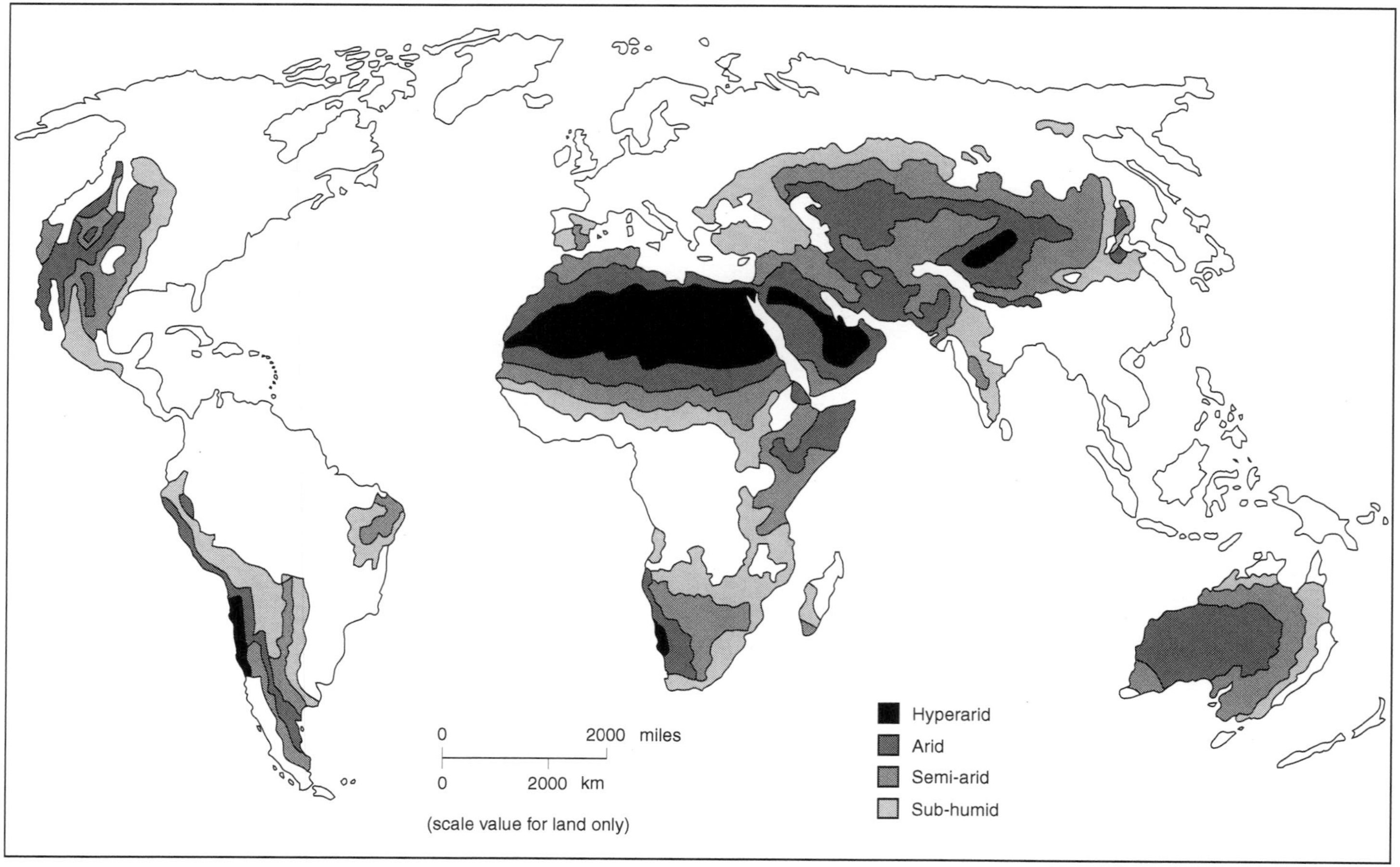


**Fig. 13.1** The world distribution of deserts. Hyperarid zones have *P/ET*p $< 0.03$; arid zones $< 0.2$–$0.03$; semiarid zones $< 0.5$–$0.2$; subhumid zones $< 0.75$–$0.5$. There are Trade Wind deserts (North Africa, Arabia, central Australia), rain-shadow/continental deserts (Central Asia, western North America, sub-Andean South America) and Trade Wind/upwelling-coastline deserts (Atacama, Namibia). (From UNESCO, 1977.)

adopted). In all ergs, sand-grade sediment and finer is very selectively taken into the bedload transport part of the local or regional wind system. *Loess* sheets result in surrounding areas due to deposition of formerly suspended silt-grade dust. Both the Namib and Oman ergs are noteworthy for major sediment sourcing by coastal beaches. That of the Namib is located to the south and fed by narrow deflation corridors that stretch from the coast through over 100 km of intervening bedrock. The sediment source in Oman includes important deflation and reworking of marine shelf-carbonate deposits during late Pleistocene global sea level lowstands.

Vast areas of late-Pleistocene 'fixed' or inactive ergs (Fig. 13.2) attest to much-expanded Trade Wind belts during the Last Glacial Maximum (LGM), which could also be called the Last Desert Maximum. A further major source of data concerning the relative vigour of past atmospheric circulation comes from the ice-core records of Greenland and Antarctica. These show that non-volcanic dust levels were an order of magnitude higher in glacial periods compared to interglacials. Thus the highest peak of dust in the Vostok core was in the LGM between 20 and 21 ka. At this time it is thought that the low-latitude wind systems of the world were more vigourous but in addition there was a large reservoir of increased sediment availability for the wind system since all continental shelves were exposed to deflation during global lowstand.

Widespread Neogene loess such as that in central and east Asia also emphasizes the vigorous nature of full-glacial planetary winds when regional aridity and reduced vegetation cover enabled massive sediment export. This led to accumulation in long-term sinks like the Chinese Loess Plateau whose thick successions are of great use in palaeoclimate studies. Arid glacial-stage loess layers deposited by strong northwest winter monsoon winds are separated by regional soil horizons (analogous to the sandy erg supersurfaces described below) that formed when moist southeast monsoon winds dominated local climate. In another example, the North African Trade Winds were active much further south during the LGM. The 'fixed' ergs in the northern savannah fringes, now stranded and stabilized by vegetation, attest to the long time interval needed to destroy an erg. They lead to the expectation that erg expansion–contraction sediments and loess sheets ought to be widely preserved in the geological record of ancient sedimentary basins where they will have noteworthy palaeoclimatic implications.

## 13.2 Aeolian system state

The introductory remarks make it obvious that in any aeolian environment, the *state* of the local or regional system may be defined by three components: sediment supply, sediment availability and the transport capacity of the wind. *Sediment supply* is the gross source of sediment for the aeolian system. It is partly generated during ephemeral runoff peaks during humid periods when enhanced discharge enables fan deltas or enigmatic terminal fans to discharge into sedimentary basins. It may also be enhanced by shelf reworking during global marine lowstands. Supply may be calculated over time from sedimentation rate and the frequency of floods. *Sediment availability* varies according to the calibre of sediment supplied locally, the degree of exposure of wind-transportable grains and the extent of vegetation cover. Numerous factors have an impact on the susceptibility of grains on a surface to transport, but these are cumulatively manifested by the actual transport rate, which serves as a proxy for sediment availability. *Transport capacity* is the potential sediment transport rate of the wind (**Cookies 32 & 33**): it is usually more than the actual sediment transport rate because supply of available sediment is insufficient to fully charge the wind—this defines the state of *transport undersaturation*. Because the three aspects of sediment state can be given as volumetric rates, they are directly comparable.

## 13.3 Physical processes and erg formation

In Chapter 7 we discussed the various aeolian bedforms. Here we consider ergs and their relations to continental wind systems. In the great Trade Wind deserts, such as the North African Sahara and central Australia, there is a close correspondence between dominant wind flow and sand transport. Thus meteorological observations, bedform orientations and the trend of erosional wind-abraded ridges known as *yardangs* (Fig. 13.3) enable sand-flow distributions to be mapped out regionally. Ideally, a sand-flow map should show resultant directions as flowlines *and* resultant magnitudes as contours, analogous to

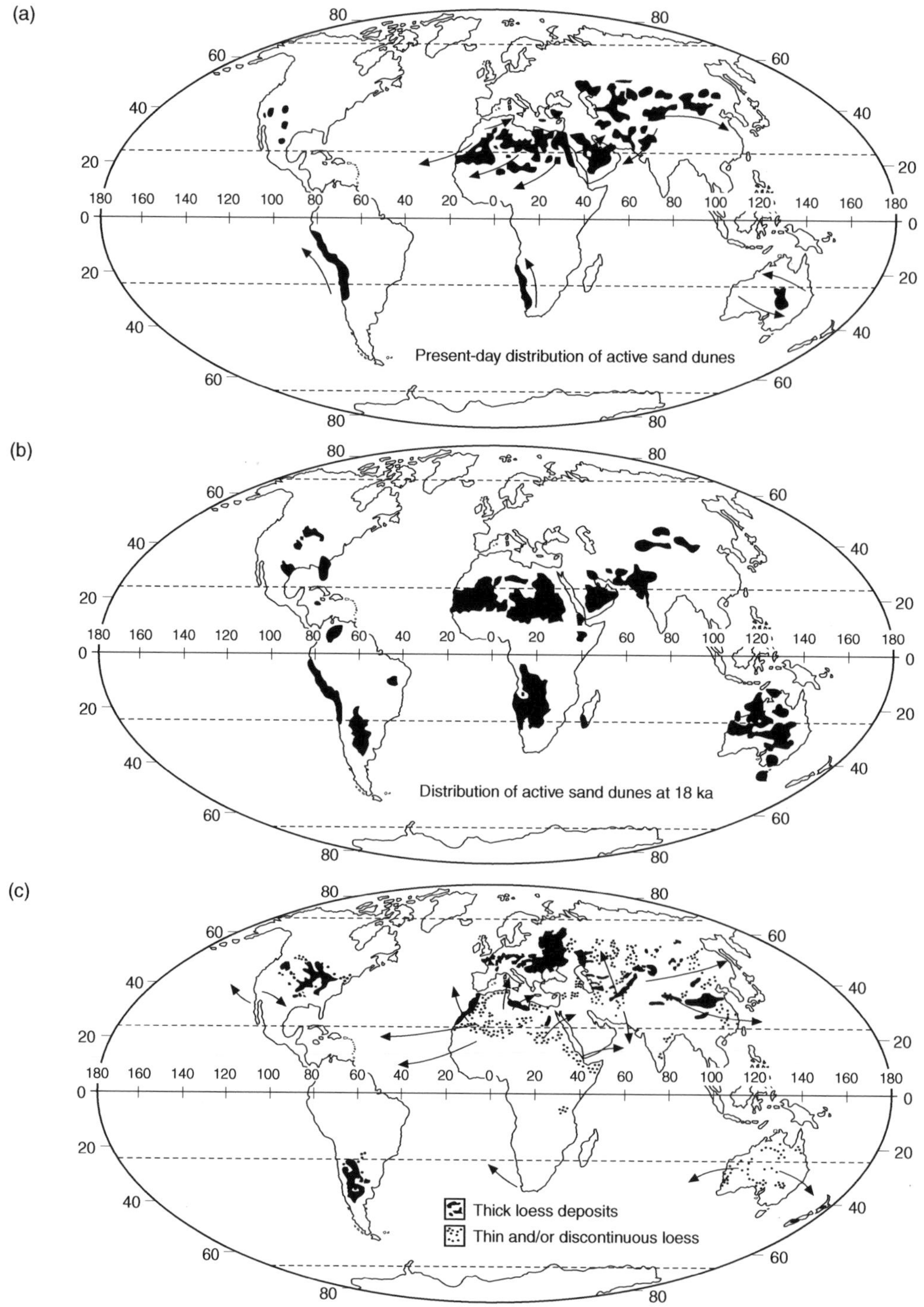


**Fig. 13.2** Distribution of: (a) present-day active sand dunes and dune-forming mean wind regime; (b) Last Glacial Maximum active sand dunes; (c) modern distribution of mostly LGM loess deposits and trajectories of modern aerosol dust tracks. (Mostly after Williams *et al.*, 1993, and sources cited therein.)

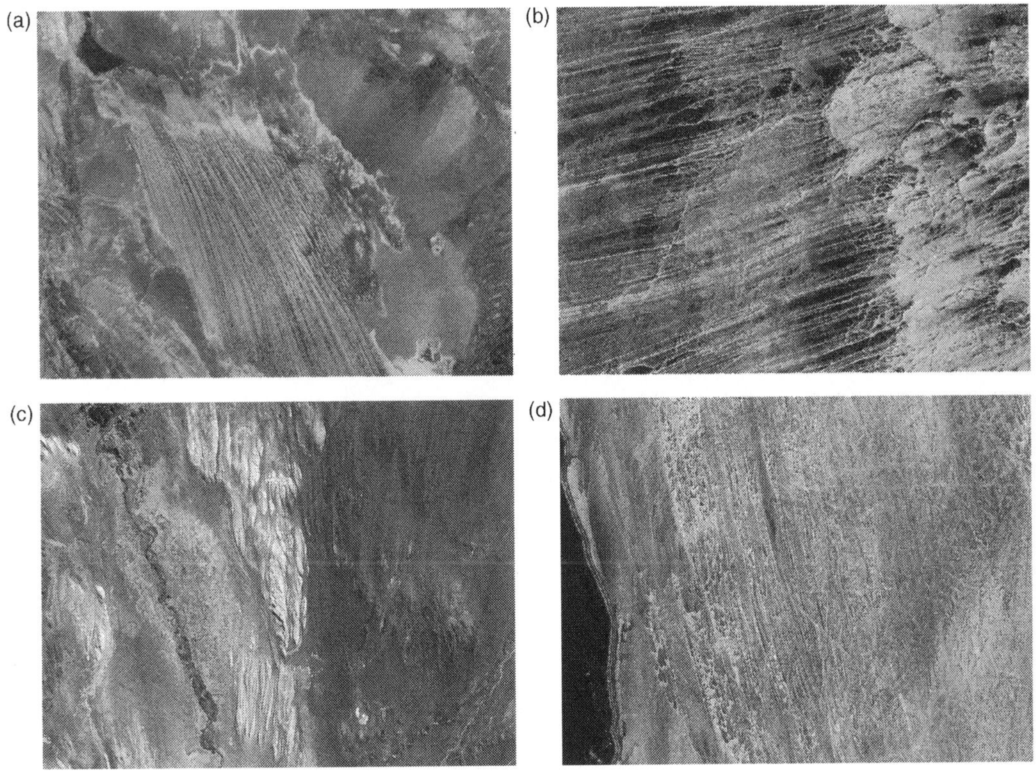


**Fig. 13.3** Spectacular Landsat 7 images of megayardangs of the kind used in establishing regional inter-erg sand transport paths. (a) Lut Desert, Iran, scene 110 km across; (b) central Sahara, scene 44 km across; (c) Peruvian Desert, scene 28 km across; (d) northern Namib, Namibia, scene 21 km across. The remarkable linearity and regularity of yardang spacing must be informing us about an aspect of atmospheric boundary layer structure that has received little attention from researchers. (Images from NASA Zulu, via Goudie, 2008, courtesy of Andrew Goudie.)

a combined wind direction and pressure map. Information currently available is generally inadequate to achieve this aim. Sand-flow maps are also analogous to drainage maps in that they show divides separating distinct 'drainage' basins: peaks in fixed high-pressure areas and saddles in between them. Unlike water drainage, there is little direct relation between sand flow and topography since winds and their sandy bedload may blow uphill.

The flowlines for North Africa (Fig. 13.3) extend from erg to erg, implying very long transport distances downwind, giving ample time for aeolian abrasion and transport processes to work. Evidence for this erg-to-erg transport is provided by satellite photographs showing yardangs in between erg areas along sand-flow lines. All the sand-flow lines arise within the desert itself, with the main clockwise circulatory cell roughly corresponding to the subtropical high-pressure zone. The most vigourous wind regimes and therefore the largest dust deflation of dried-up lake beds (see below) occur where the trade winds are concentrated in a low-level jet that is topographically channelled between the Tibesti and Ennedi mountains. The flowlines eventually lead to the sea, giving rise to a great plume of Saharan dust extending out for thousands of kilometres into the Atlantic ocean. A significant part reaches as far as the Lesser Antilles, providing a steady rain of fine silt- to clay-grade

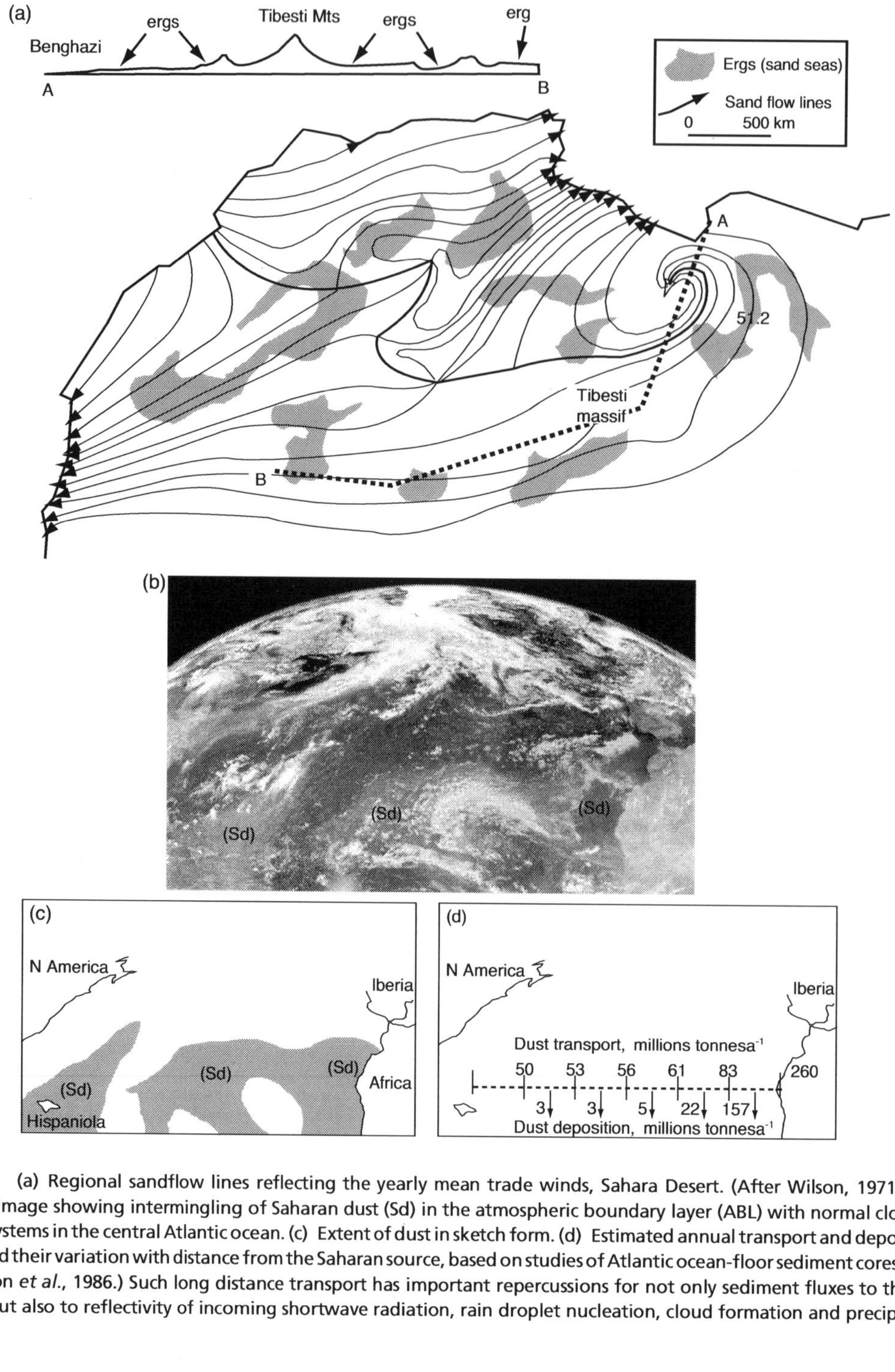


**Fig. 13.4** (a) Regional sandflow lines reflecting the yearly mean trade winds, Sahara Desert. (After Wilson, 1971.) (b) A satellite image showing intermingling of Saharan dust (Sd) in the atmospheric boundary layer (ABL) with normal cloud and frontal systems in the central Atlantic ocean. (c) Extent of dust in sketch form. (d) Estimated annual transport and depositional fluxes and their variation with distance from the Saharan source, based on studies of Atlantic ocean-floor sediment cores. (After Middleton *et al.*, 1986.) Such long distance transport has important repercussions for not only sediment fluxes to the deep oceans but also to reflectivity of incoming shortwave radiation, rain droplet nucleation, cloud formation and precipitation.

particles into the deep ocean (Fig. 13.4). It has been estimated that some $2.6 \times 10^8$ tonnes of Saharan mineral dust per year are removed in this way. The magnitude of this sediment flux varied throughout the Quaternary in response to cycles of climate change causing trade winds to fluctuate greatly in strength. Present-day fluxes are low compared to those at the LGM—the evidence comes from deep-sea cores, which enable dust deposition rates to be calculated as far back as the late Cretaceous. Mineral dust in the size range 0.1–1 μm is a potent shortwave radiation-scattering aerosol and may have an important cooling role to play in climate change.

The other spectacular example of Trade Wind orientated desert dunes occurs in the great ergs of central Australia, where a migrating high-pressure cell provides the dominant control of anticlockwise sand movement along longitudinal dune systems. The pattern of sand accumulation is much simpler than in North Africa, since it is largely undisturbed by topographic obstacles.

Individual desert ergs are confined to basins, whatever their absolute height. They terminate at any pronounced slope break. Ergs can form only when the wind is both fully charged with bedload and decelerating in either time or space according to the sediment mass continuity equation (**Cookie 34**). They can form at sand-flow centres, at saddles and in local areas controlled by topography. Some possible basinal configurations for erg formation are shown in Fig. 13.5.

Deposition and deflation are controlled not only by the regional wind system and the sediment system state, but also in a complex manner by more local bedform hierarchies present on the erg. Despite the continental scale of low-latitude desert sand flow, major local influences upon wind flow are exerted by topography and by mesoscale circulation at wavelengths of 10–100 km. Winds have to flow around as well as over some high-relief areas, much as turbidity currents interact with submarine topography (Chapter 8). The resulting accelerations, decelerations, lee waves and helical flow paths create much interesting complexity whose effects have rarely been appreciated in the geological record. For example, it is likely that smaller-scale ergs not linked to the Trade Wind system and hence to external sand supply will be much more prone to climatically induced interruptions to sedimentation.

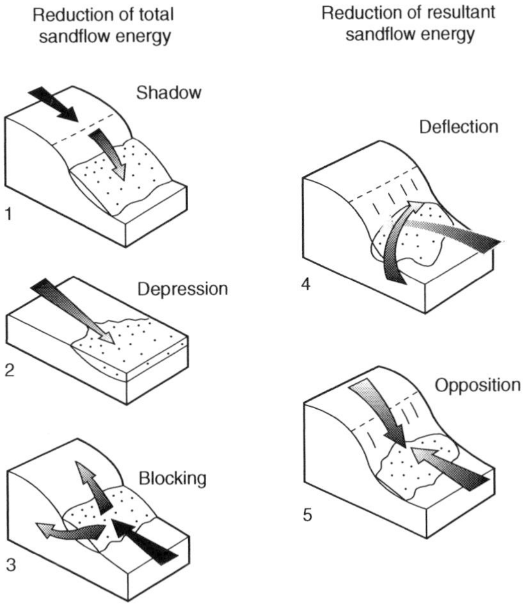


**Fig. 13.5** Ergs, topography and sand deposition. At a regional scale, sand-bearing winds decelerate in time and/or space according to the sediment continuity equation (see **Cookie 34**) and deposit their sandy bedload to form an erg sand accumulation. Sketches show various geomorphological situations leading to deposition. (After Fryberger & Ahlbrandt, 1979.)

## 13.4 Erg margins and interbedform areas

The margins of many ergs are marked by relatively thin sand accumulations, so-called *sand sheets* (Fig. 13.6). There is insufficient fine sand supply here, and maybe other unfavourable conditions, e.g. high water table, surface stabilization, coarse lag desert pavements, for significant dunes to form. Sand sheets form a transitional facies between aeolian dunes and non-aeolian deposits. They originate by gentle wind deceleration in the lee of small surface irregularities, and exhibit small dunes, aeolian ripple remnants, granule ridges, surface lag deposits and internal low-angle erosion surfaces and climbing ripple laminae (Fig. 13.6). Sand sheets may also result when an erg margin is bordered by a high water-table area with permamant or ephemeral streams that rework erg bedforms, as in Medano Creek bordering the Great Sand Dunes of Colorado, USA. In many ergs, areas of interdune or interdraa deposition occur that

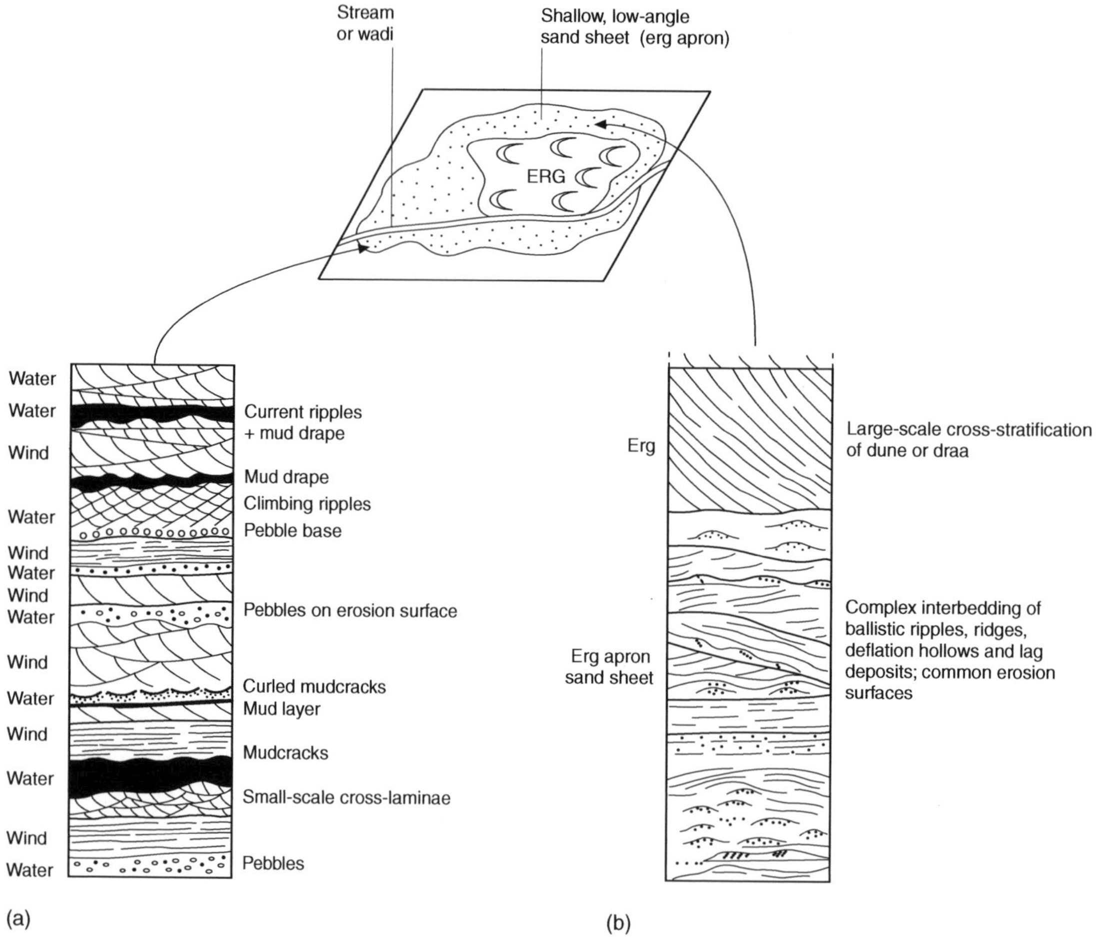


**Fig. 13.6** Vertical sequences liable to be generated at the margin of ergs. (After Fryberger *et al.*, 1979.) (a) Alternating 'wet/dry' conditions such as might be found adjacent to an ephemeral stream. (b) 'Dry' conditions along an erg apron-margin that was formerly 'wet' to source the coarser lags.

have many similarities to the deposits of sand sheets. Sometimes, as in the northern Namib Desert, ephemeral streams actually flow through the erg for a few days a year, their high discharges enabling the river to pass through interdunal depressions to reach the ocean.

The exact nature of interdune areas depends upon the availability of sand, moisture content of the depression floor brought about by changing water-table levels, river flooding, saline groundwater invasion or marine intrusion in coastal ergs. Sparse sand availability under 'dry' conditions leads to dune or draa migration over the areas of immobile sediments and the nonpreservation of interbedform sediment. In this case the interbedform areas are merely transport paths for sand between adjacent dunes. Wet interdunes show ample evidence for deposition and commonly result from slow passive rise of the local water-table surface or more rapid flooding from adjacent ephemeral rivers. Thus careful analysis of interbedform deposits may provide valuable evidence for the general palaeogeographical and palaeoclimatic conditions that existed during erg evolution. For example, deep floods from the Hoanib ephemeral braided river in northern Namibia pond laterally to considerable depths (metres to >10 m) in adjacent interdunes,

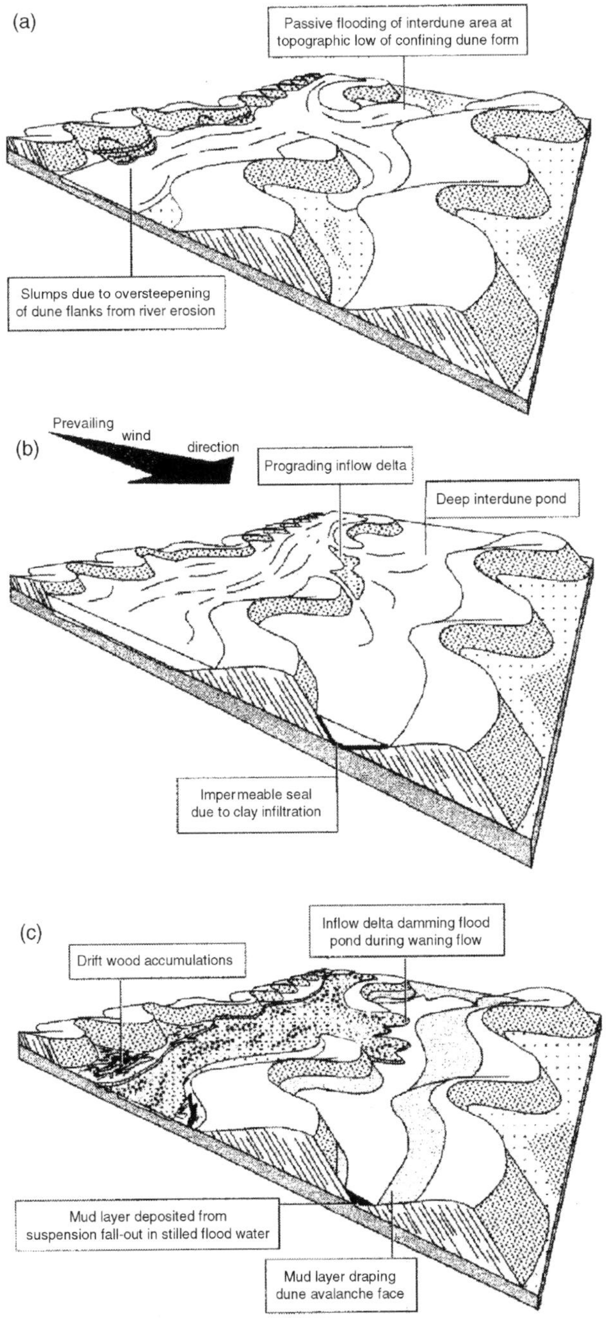


**Fig. 13.7** A summary of the schematic stages in the evolution of a flood pond generated by fluvial interdune incursion. (After Stanistreet & Sollhofen, 2002.)

depositing mud layers 1–50 cm thick on avalanche and stoss faces of bounding dunes. The flooding is slow and hardly disturbs aeolian stratification. Floodwater clay infiltrates and settles as an impermeable seal, with a *flood pond* on top, perched above regional groundwater level. Flood ponds evaporate slowly for long periods (>3 yr). Early emergence desiccates higher parts of a mud layer. Subsequent floods can refill a predecessor pond, benefiting from the existing impervious seal. Potential preservation of such mud layers is lower on the upwind dune face, but high on the leeside avalanche face after burial by subsequent dune reactivation and migration (Figs. 13.7 & 13.8).

More rapid runoff by ephemeral floods causes dune margin erosion and formation of sheet-like channelized deposits of small subaqueous dunes, upper plane beds and current ripples. These may show decelerating flow sequences and be covered with thin mud drapes as shallow ponding occurs, thereafter cracked up by desiccation, sometimes to be incorporated in

**Fig. 13.8** Field photographs showing (a) oblique windward view (to the SW) of inflow delta (ca 100 m wide) at a large lake flood pond, 3 weeks after the April 2000 Hoanib floods. The Hoanib channel is towards the right outside the photo. (b) Burial of a mud-lined, nearly dry flood pond by a downwind-prograding dune avalanche front. (Both after Stanistreet & Sollhofen, 2002.)

wind-blown sands of the next arid period. Sabkhas occur where a saline water table intersects the inter-dune troughs. Blown sand driving across moist areas causes build-up of highly porous *adhesion-accreted sands*. Evaporation leads to evaporite precipitation as crusts and subsurface nodules. Crusty surfaces do not trap sand along transport paths; rather the hard surfaces accelerate transport on to the next dune, draa or erg. In some areas temporary playa lakes may form in interdune depressions following extensive rainfall. Spectacular examples of inter-dune playas between stabilized dunes occur around Lake Chad near Timbuktoo. Oscillation ripples and algal carbonates may develop when such lakes remain for appreciable periods of time. Dry inter-dune depressions develop soil horizons, aeolian ripple remnants and other features similar to those of sand sheets. The generally poorly sorted and fine-grained character of interdune deposits contrasts markedly with those of the main erg bedforms. They form important permeability barriers in both aquifers and hydrocarbon reservoirs, and have thus received much recent study.

## 13.5 Erg and draa evolution and sedimentary architecture

The detailed subsurface architecture of ergs is based on analogues to large-scale fluvial bedforms, conceptual models from modern desert surface processes and especially from detailed studies of well-exposed outcrops of ancient aeolian deposits, particularly in the western USA and southwest Africa (Figs. 13.9 & 13.10). It is also essential to consider erg reaction to climate change.

The migration of dunes and draas and their contained interdune depressions is thought to give

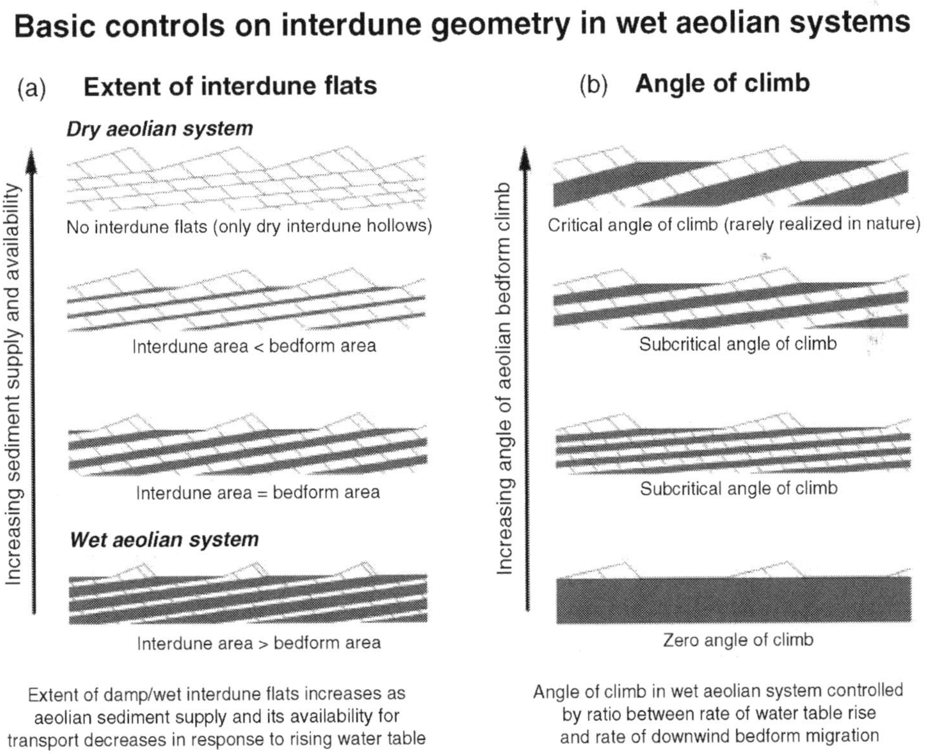


**Fig. 13.9** Depositional models illustrating basic controls on the accumulation and sedimentary architecture of 'wet' aeolian systems. (After Mountney & Jagger, 2004; based in part on Kocurek & Havholm, 1993.) (a) The effects of relative dune–interdune size (proportion of damp surface covered by aeolian dunes). (b) The effects of angle of climb (the ratio between the rate of downwind dune migration and the rate of relative water-table rise).

rise to a hierarchy of cross-stratified sets, bounding planes and interbedform deposits. The particular details of these features depend upon a variety of factors, including dune type, sand supply, depth of water table and capillary fringe, vegetational development, proximity to riverine, sabkha or marine environments and the subsidence rate of the basin floor that is accomodating the transported aeolian sediment.

In migrating dunes and draas the upward growth and downwind migration of bedforms gives rise to internal sets of cross-stratification separated by low-angle (sometimes only fractions of a degree) truncation or accretion planes analogous to those documented from unit bars in braided river channels (Chapter 11). These dipping surfaces separate individual cross-stratified bedsets and are termed *bounding surfaces* and may dip upwind or downwind. Simple bedforms generate bounding surfaces that dip upwind; luminescence dating of Holocene dunes shows individual bedforms may accrete and grow to equlibrium very rapidly as sediment availability is enhanced by environmental change. Migration of dunes over the downwind depositional parts of migrating compound draa without slipfaces generate downwind-dipping surfaces. Furthermore, if

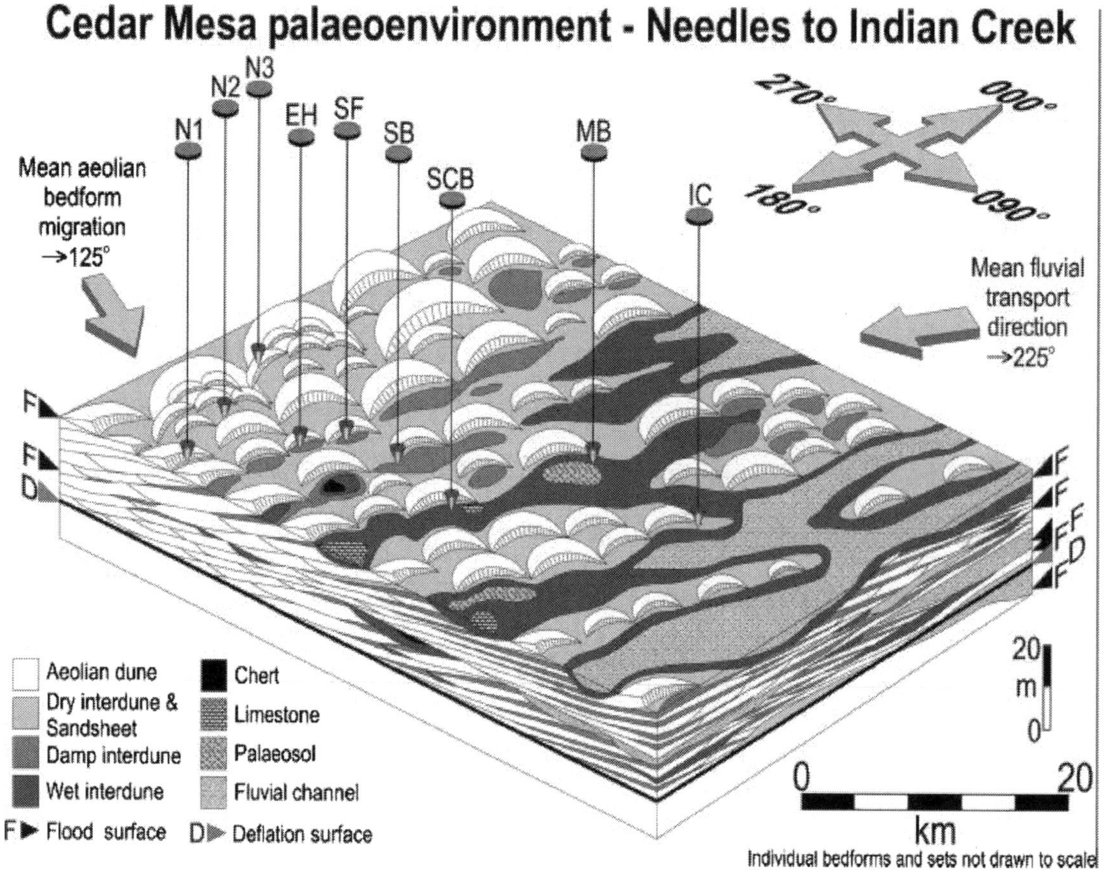


**Fig. 13.10** Palaeoenvironmental reconstruction of the Cedar Mesa Sandstone, Permian erg margin system, southeast Utah, USA. Note how fluvial incursion into the erg margin occurs preferentially along interconnected interdune corridors and how the extent of these incursions is limited by the increased size and frequency of aeolian dunes in the aeolian-dominated part of the erg margin. In this region, isolated, wet interdune ponds occurred during periods of elevated water table. (After Mountney & Jagger, 2004.)

simple or compound bedforms are separated from downwind neighbours by interbedform depressions, the bedform front will migrate over the deposits of these largely immobile features, preserving them to a greater or lesser extent. Preservation is aided if the locality is a sand sink, i.e. net deposition is occurring, for then the net addition of sediment causes surface build-up and generation of the 'climbing' strata sequences of simple large dunes. Ultimate preservation can be guaranteed only if the area is also undergoing tectonic subsidence. The enclosed nature of interdraa depressions means that the preserved interdraa deposits will be of finite extent normal to mean wind direction. Interesting data from Cretaceous draa in well-exposed deposits in the Namibian Etjo basin indicate that the maximum preserved thickness of individual aeolian sets varies systematically across the basin, from 52 m in the basin depocentre to only 8 m at the basin margin. The set architecture indicates that this spatial variation is primarily the result of decreased angles of bedform climb at the basin margin, rather than the presence of smaller bedforms. Similarly, a temporal reduction in the angle-of-climb, rather than a reduction in bedform size, is considered to be responsible for an upward decrease in preserved set thickness. Reductions in bedform climb angle may be due to progressive loss of depositional space as the erg gradually filled a topographic basin depression.

## 13.6 Erg construction, stasis and destruction: climate and sea-level controls

Aeolian erg sediments are valuable archives of climate change. As noted previously, huge areas of stabilized, partly vegetated dunes exist around the margins of modern active ergs. These once-active bedforms are subject to soil development and sediment redistribution into interdune troughs by runoff and mass flow. They are eloquent testimony to the ability of climate change to form regional bounding surfaces, termed *supersurfaces*. Such surfaces are of much greater extent than those within individual draas or dunes discussed in the previous section; they bound *all* elements of the once-active erg. If the aeolian system becomes operative once more and a sand flux into the area is provided, then such regional surfaces will become preserved, bounding a new sequence of aeolian sediments.

The process of episodic stratigraphic accumulation is very well illustrated in the Gran Desierto of northern Mexico (Fig. 13.11). This area is particularly interesting as an analogue for ancient deposits because it is located in the Salton Trough, an area of rapid subsidence associated with the transtensional regime of the San Andreas associated faults. Though the pattern of dunes determined from remote-sensing images is both spatially diverse and complex, it is made up of multigenerational groups of simple dune types (Fig. 13.12). Optically stimulated luminescence age-dating indicates that these represent relatively short-lived aeolian constructional events since 25 ka, comprising: (i) late Pleistocene relict linear dunes; (ii) degraded crescentic dunes formed at ~12 ka; (iii) early Holocene western crescentic dunes; (iv) eastern crescentic dunes emplaced at ~7 ka; and (v) star dunes formed during the past 3 ka. Palaeowind reconstructions, based upon the rule of gross bedform-normal transport, are largely in agreement with regional proxy data. The sediment state over time is one in which the sediment supply for constructional events is derived from previously stored sediment (Ancestral Colorado River sediment), contemporaneous influx from the lower Colorado River valley and coastal influx. Constructional events are triggered by climatic shifts to greater aridity, changes in the wind regime, and the development of a sediment supply. The rate of geomorphic change within the Gran Desierto is significantly greater than the high rate of subsidence and burial of the accumulation surface upon which it rests. The rapidity of dune growth is supported in other areas by OSL dating, for example, in the northern Rub Al Khali of Arabia. There are important lessons to be drawn from these data for the interpretation of ancient erg deposits.

The Wahiba Sand Sea of Oman (Fig. 13.13) is noteworthy and unique because it has been seismically imaged, cored and dated using optical luminescence and other proxy records (Fig. 13.14). It is situated at the northern margin of the area presently affected by Indian Summer Monsoon Circulation and it records environmental changes associated with this major climatic boundary over the past 160 000 yr. The internal stratigraphy and evolution of the sand sea records the influence of sea-level changes on the sedimentary architecture and composition of the dune deposits. During the last two glacial periods, low global sea level was associated with a high input of bioclastic grains, reflecting the significance of subaerially exposed shelf areas as one of the main sources of aeolian sediment. The onset of aeolian sediment transport and deposition is related to the breakdown of stabilizing vegetation during arid periods related to sealevel lowstands. The preservation of aeolian sediments by the formation of supersurfaces and associated palaeosoils took place during times of increased wetness and elevated groundwater tables. This interplay of constructive and destructive periods greatly influenced the sedimentary architecture. Oscillations of wet and dry periods between 160 000–130 000 and 120 000–105 000 yr ago are attributed to the evolution of a wet aeolian system. Younger periods of aeolian deposition around and after the LGM were characterized by dry aeolian conditions. No soil horizons developed during these times.

Longer term records are also available from the southern margins of the expanded Pleistocene Thar Desert, India where it seems that LGM and subsequent fluvial incision and enhancement of local sediment availability led to widespread aeolian dune deposition over a formerly fluvial depositional landscape.

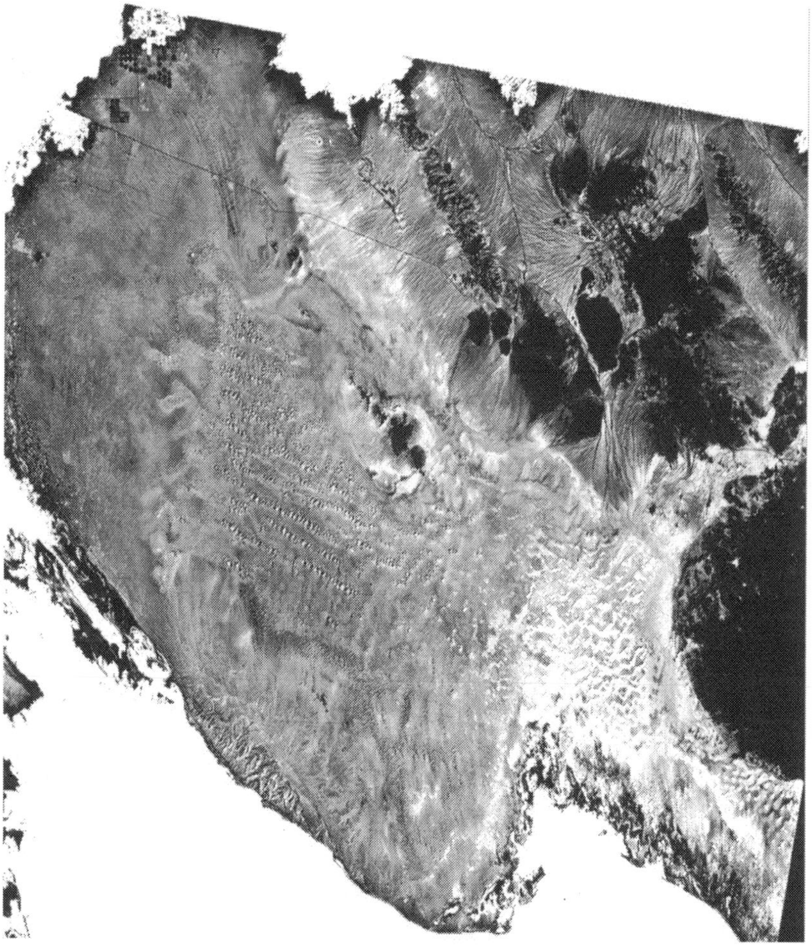

**Fig. 13.11** Landsat 7 image of the Grand Desierto dune field showing the spatial distribution of the groups of dune patterns. (After Beveridge *et al.*, 2006.) See Fig. 13.12 for historical development of various dune types.

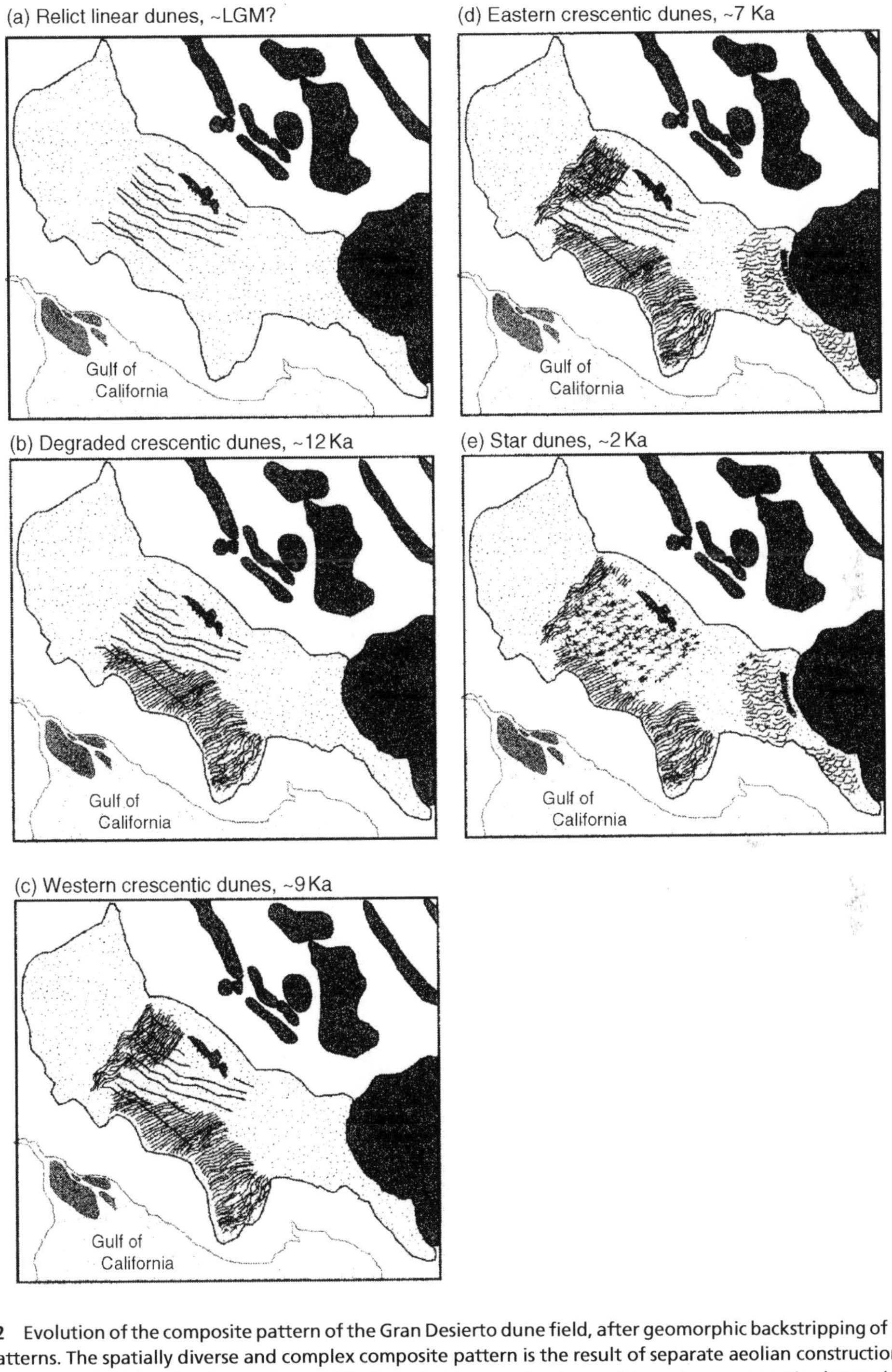


**Fig. 13.12** Evolution of the composite pattern of the Gran Desierto dune field, after geomorphic backstripping of individual simple patterns. The spatially diverse and complex composite pattern is the result of separate aeolian constructional events that gave rise to the (a) relict linear dunes, (b) degraded crescentic dunes, (c) western crescentic dunes, (d) eastern crescentic dunes, and (e) star dunes. (After Beveridge *et al.*, 2006).

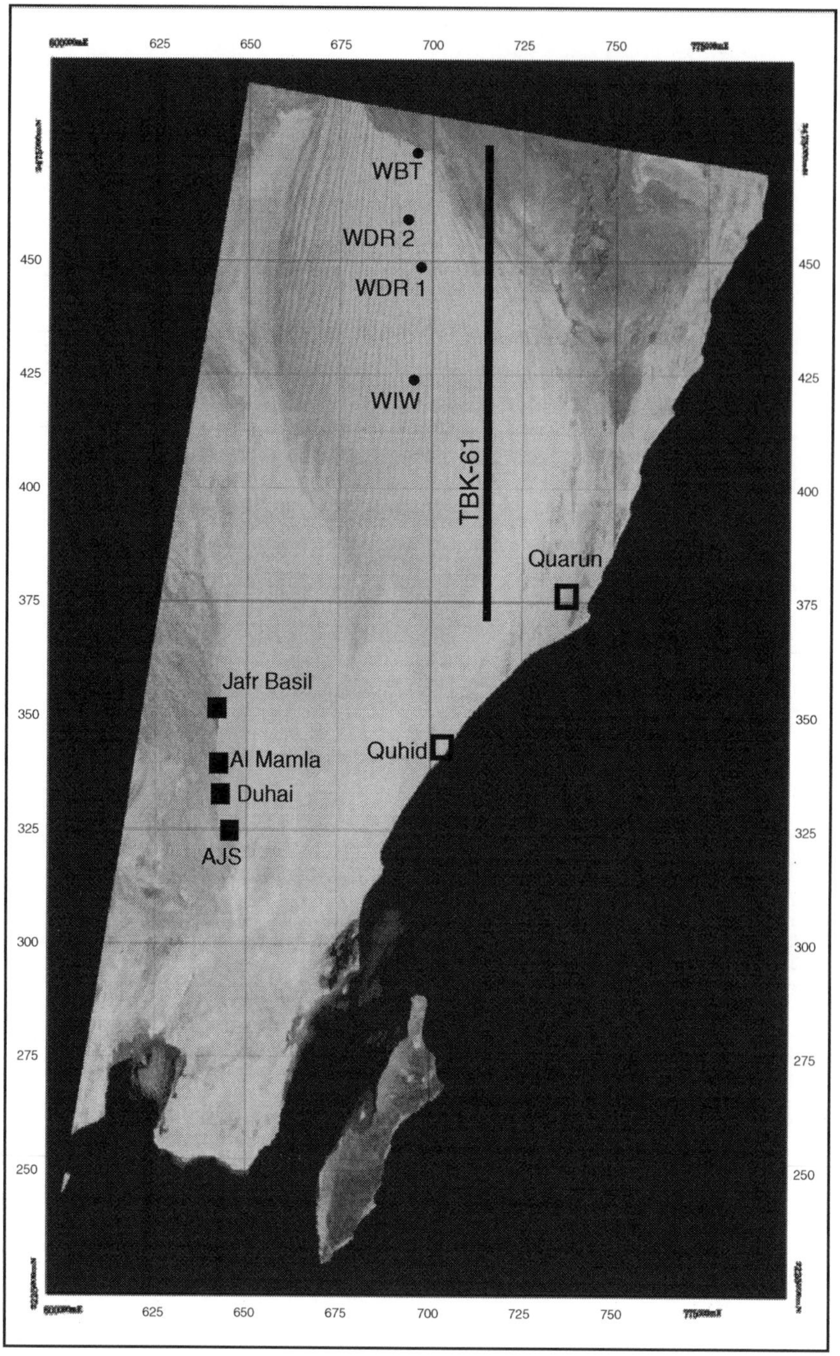


**Fig. 13.13** Landsat satellite image of Oman Wahiba erg with localities of outcrops and drill sites and seismic line TBK-61. UTM coordinate grid with 25 km spacing for scale. (After Radies *et al.*, 2004.).

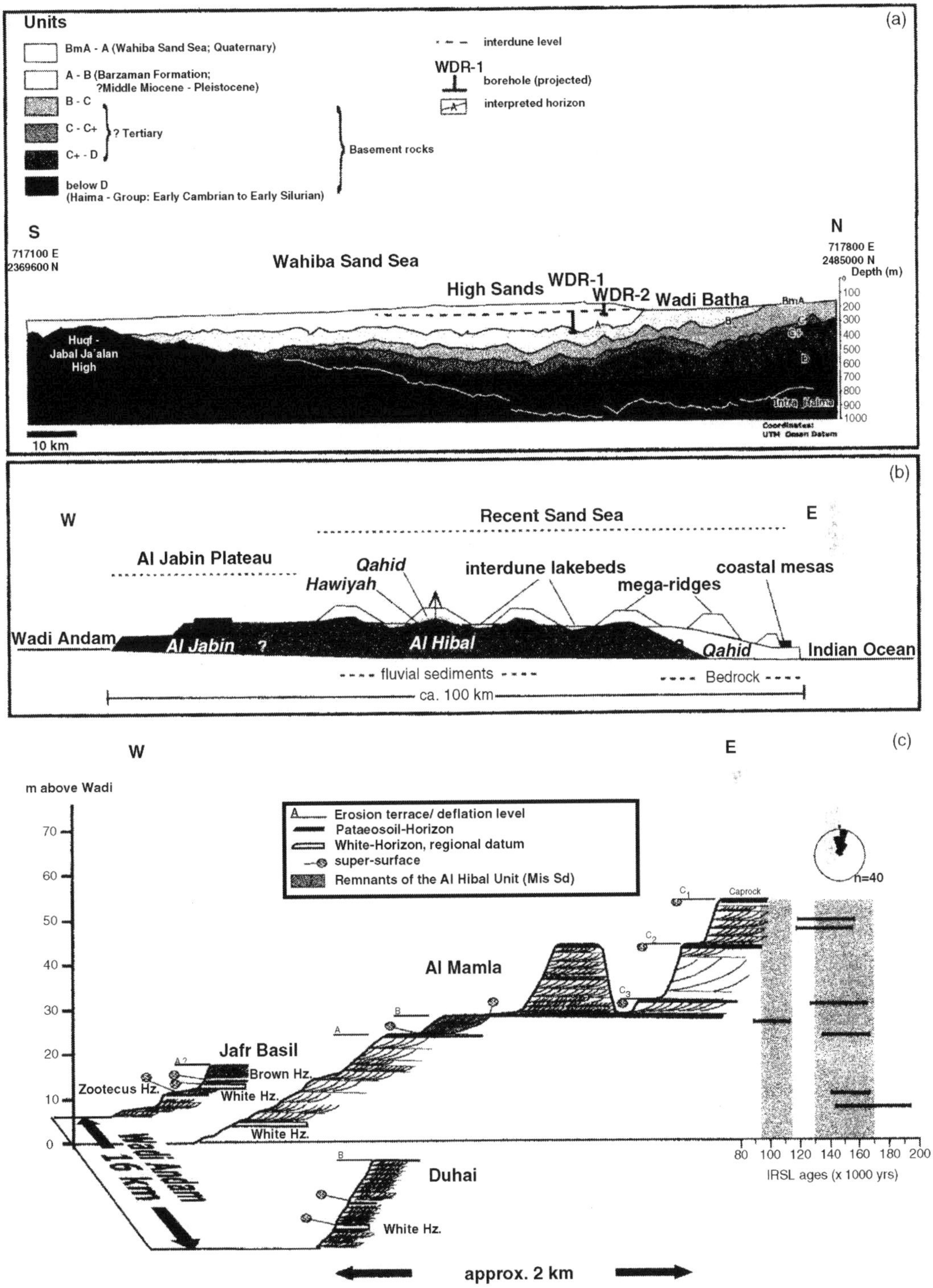
(a)
Units
BmA - A (Wahiba Sand Sea; Quaternary)
A - B (Barzaman Formation; ?Middle Miocene - Pleistocene)
B - C
C - C+
C+ - D
? Tertiary
below D (Haima - Group: Early Cambrian to Early Silurian)
Basement rocks
interdune level
WDR-1 borehole (projected)
interpreted horizon
S
717100 E
2369600 N
Wahiba Sand Sea
High Sands
WDR-1
WDR-2
Wadi Batha
N
717800 E
2485000 N
Depth (m)
Huqf - Jabal Ja'alan High
10 km
Coordinates: UTM Oman Datum
(b)
W
Recent Sand Sea
E
Al Jabin Plateau
Qahid
Hawiyah
interdune lakebeds
coastal mesas
mega-ridges
Wadi Andam
Al Jabin
Al Hibal
Qahid
Indian Ocean
fluvial sediments
Bedrock
ca. 100 km
(c)
W
E
m above Wadi
Erosion terrace/ deflation level
Pataeosoil-Horizon
White-Horizon, regional datum
super-surface
Remnants of the Al Hibal Unit (Mis Sd)
n=40
Caprock
Al Mamla
Jafr Basil
Brown Hz.
Zootecus Hz.
White Hz.
White Hz.
Wadi Andam 16 km
Duhai
White Hz.
IRSL ages (x 1000 yrs)
approx. 2 km

Many dried-up, formerly freshwater, lake beds occur within the Trade Wind deserts. For example, the lakes of central Australia became progressively more saline about 25 000 yr ago due to climatic aridity when gypsum was deposited together with clays. Complete desiccation of the lakes led to wind reworking of the finely powdered gypsum and clay into elliptical lunette and parabolic dunes now seen on the downwind lake margins (Chapter 7). The desiccated lakes of the South-Central Sahara are the source of globally significant dust fluxes, a major contribution being diatomaceous deposits of the Bodélé former lake basin in Niger and Chad, memorably described as the 'dustiest locality on Earth'.

Dating by OSL reveals that the linear dunes of the interior southwest Kalahari Desert were active in two distinct phases, between 30–23 and 16–9 ka during periods of relative aridity and possibly increased mean wind strength. More localized activity also occurred at times in the Holocene, including in the last millenium, like those discussed previously from the northern Namib Desert. Small desiccated lakes within lunette or parablic dunes (Chapter 7) are known as *pans* in the wider Kalahari region. Dating by OSL reveals that the upwind lunette dunes that border these kilometre-scale features have been active at many times in the Holocene in response to mainly local conditions after lake desiccation as local deepening of groundwater levels occurred and aerodynamic conditions changed.

## 13.7 Ancient desert facies

Knowledge that modern ergs of continental scale are adjusted to the motions of the Trade Wind system means that ancient erg deposits may be used in conjunction with continental reassemblies to reconstruct ancient wind systems. Below are two examples as a starting point to appreciate the wealth of data available in the sedimentary record.

### Icehouse erg behaviour: northwest europe and Western USA

Located in the southern and northern North Sea basin (Fig. 13.15) and its extension into central Europe, the *Rotliegendes* forms an important regional natural gas reservoir in the British, Dutch and German sectors. Northern and southern ergs were separated by adjacent uplands (London–Brabant High and Mid-North Sea High) that were areas of sediment supply provided by ephemeral streams and wind deflation. Detailed well-log studies and studies of onshore outcrops in northeast England reveal a series of complex fossilized erg sand bodies up to 500 m thick with marginal facies of fluvial wadi sediments and a basin centre facies of lacustrine clays and playa evaporites. The onshore exposures around the erg margins give convincing evidence for the existence of compound longitudinal draas separated by interdune areas. Regional palaeocurrent studies demonstrate deposition by a clockwise-rotating atmospheric cell. Cored intervals and dipmeter data cannot establish dune type with any certainty because of correlation problems, but dry and wet interdune horizons have been clearly identified. Regional stabilization supersurfaces may be identified from closely spaced well log data. Drying-upwards cycles have been recognized from many wells and these have been used to infer numerous (five or more) periods of large-scale erg contraction and expansion. It is possible that these cycles may reflect Milankovitch-driven responses to mid–late Permian climatic changes, since this was a major icehouse period of Earth history. The final extinction of the great Rotliegendes erg occurred as global Permian sea level rose upon final melting of the Gondwanan glaciers. A distinctive water-reworked horizon, the *Weissliegend*, with soft-sediment deformation in underlying dunes, marks the event in onshore exposures.

Equivalent horizons in the Permian of Utah, western USA (Cedar Mesa Sandstone) represents the product of at least 12 separate aeolian erg sequences, each bounded by regionally extensive deflationary

**Fig. 13.14** Oman erg architecture. (After Radies *et al.*, 2004, and sources cited therein.) (a) Interpreted seismic line TBK-61 (see Fig. 13.13) across the Wahiba Basin redrawn after reprocessed sections. Vertical scale as depth (m) to reference horizon. Wadi Batha is at ca 240 m above mean sea level. (b) Schematic east–west section (west of Qahid) through the aeolian units (italics) of the Wahiba Sand Sea, not to scale. (c) Geomorphological and sedimentological sections for the Jafr Basil, Al Mamla and Duhai localities along the western edge of the Al Jabin Plateau. The White Horizon palaeosol acts as a regional stratigraphical datum.

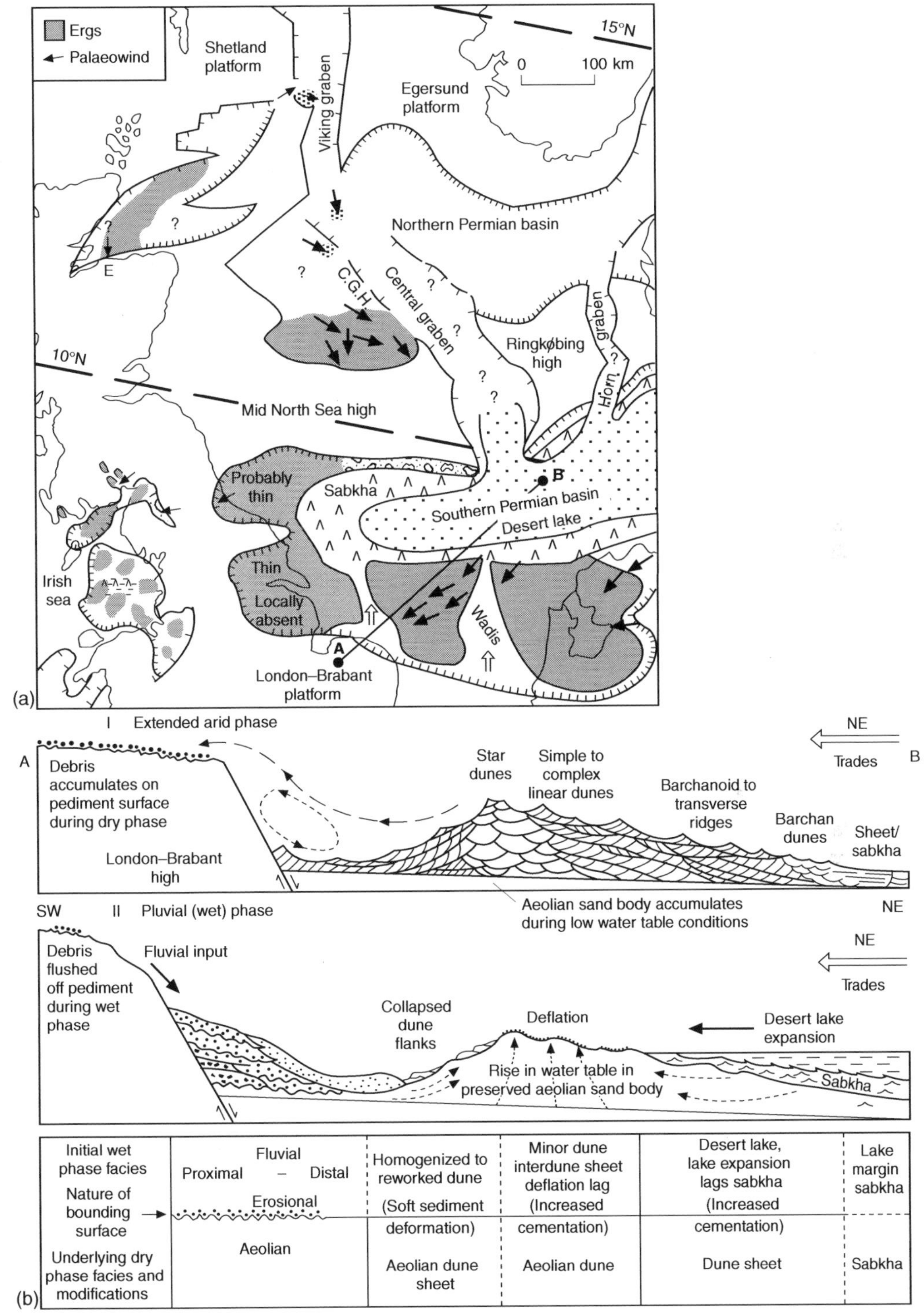


**Fig. 13.15** (a) The extent of Permian ergs in the North Sea basin. (After Glennie, 1982.) (b) An interesting climatic hypothesis for the formation of repeated bounding supersurfaces, (After George and Berry, 1993.)

supersurfaces. Facies analysis indicates that preserved sequences each record a systematic sedimentary evolution (Fig. 13.16). There is a transition upwards from damp sandsheet, ephemeral lake and palaeosol deposition, through dry-climate sandsheet deposition, to the development of thin, chaotically arranged aeolian dune sets and finally to equilibrium erg construction by 300–400 m wavelength dunes migrating over slipfaceless draa. Each aeolian sequence is capped by a regionally extensive deflation supersurface characterized by abundant calcified roots and bioturbation when the aeolian system state was characterized by low sediment availability and high potential transport vectors. Sequence generation is attributed to both cyclical changes in climate and related changes in sea level of probable glacio-eustatic origin with lowstand/arid sand-sea construction and highstand/humid deflation and destruction. There is also an important role for the generation during arid phases of large draa with wide kilometre-scale spacing and high topography (25 m) which survive climate change episodes and whose interdune depressions create a template for barchan bedform nucleation during a subsequent arid sequence (Fig. 13.16).

### Ancient Jurassic ergs of the Western USA

Owing to the late Tertiary downcutting activities of the Colorado and other rivers, no area in the world has such breathtaking exposures of Mesozoic aeolian sediments as the Colorado Plateau of the western USA. These sediments record the continental sedimentary evolution of the western part of Pangea, the supercontinent that straddled Earth's equator in the Late Palaeozoic to Early Mesozoic. Noteworthy studies have set out to test conceptual models like those noted above for erg evolution and architecture. Regional unconformable surfaces attest to periods of widespread erg stabilization, with regional bounding surfaces traced out over hundreds of kilometres, for example in the Middle Jurassic Page Sandstone of Utah and Arizona. Facies architecture developed in zones of intertonguing between ergs and adjacent river systems, in the Lower Jurassic Moenave/Wingate and Kayenta/Navajo Formations show that the position of the river/erg boundary appears to have suffered repeated major lateral translations, on a scale of 1–100 km, perhaps in response to climate change affecting sand flow and river sediment dynamics. The result is a series of 'drying-upwards' and 'wettening-upwards' cycles passing from fluviatile to erg facies and back again.

The palaeoclimatic (and economic) implications of these discoveries are profound. If the Jurassic was indeed a hothouse era, then the major changes in erg dynamics implied by environmental analysis are difficult to explain, particularly since palaeocurrent data and climate modelling indicate quite stable or slowly rotating wind regimes during the periods in question. The timing of the climatic cyclicity is low frequency and not immediately explicable other than by perhaps the long-term eccentricity contribution (ca 400 kyr) from Milankovitch-type orbital forcing mechanisms. It may be that a major long-term cyclical switch between Trade Winds (arid) and monsoonal winds (seasonally moist) is being recorded, tracked by motions of the contemporary equatorial zone of convergence. Comparisons with the icehouse Permian erg cycles of the area and with northwest Europe might reveal some fundamental truths about the longer-term behaviour of planetary Trade Wind belts and their dynamic ergs. On the other hand, it seems clear that at least some regional supersurfaces, notably those of the Middle Jurassic Page Sandstone, correlate with basinwards marine transgressions whose effect was to cut off sand supply and raise the regional water table, causing erg stabilization and abandonment. Rapid marine transgression over remnant dune forms in the Middle Jurassic Entrada Formation of New Mexico led to modification of dune margins by subaqueous mass flow deposits analogous to the *Weissliegende* of the northwest European Permian desert basin noted above.

Cross-equatorial, westerly winds are key features of tropical circulation in monsoonal regions. Although prominent in numerical climate models of the late Palaeozoic to Mesozoic Pangaean supercontinent, such flow is confirmed by migration directions of ancient dunes determined from their palaeocurrents. The prevailing north-westerly surface palaeowinds recorded by dune-deposited sandstones are explicable as cross-equatorial westerlies produced by a steep pressure gradient spanning the supercontinent equator during northern winter and southern summer. Mountains along the western coast of Pangaea are thought not only to have enhanced wind strength, but

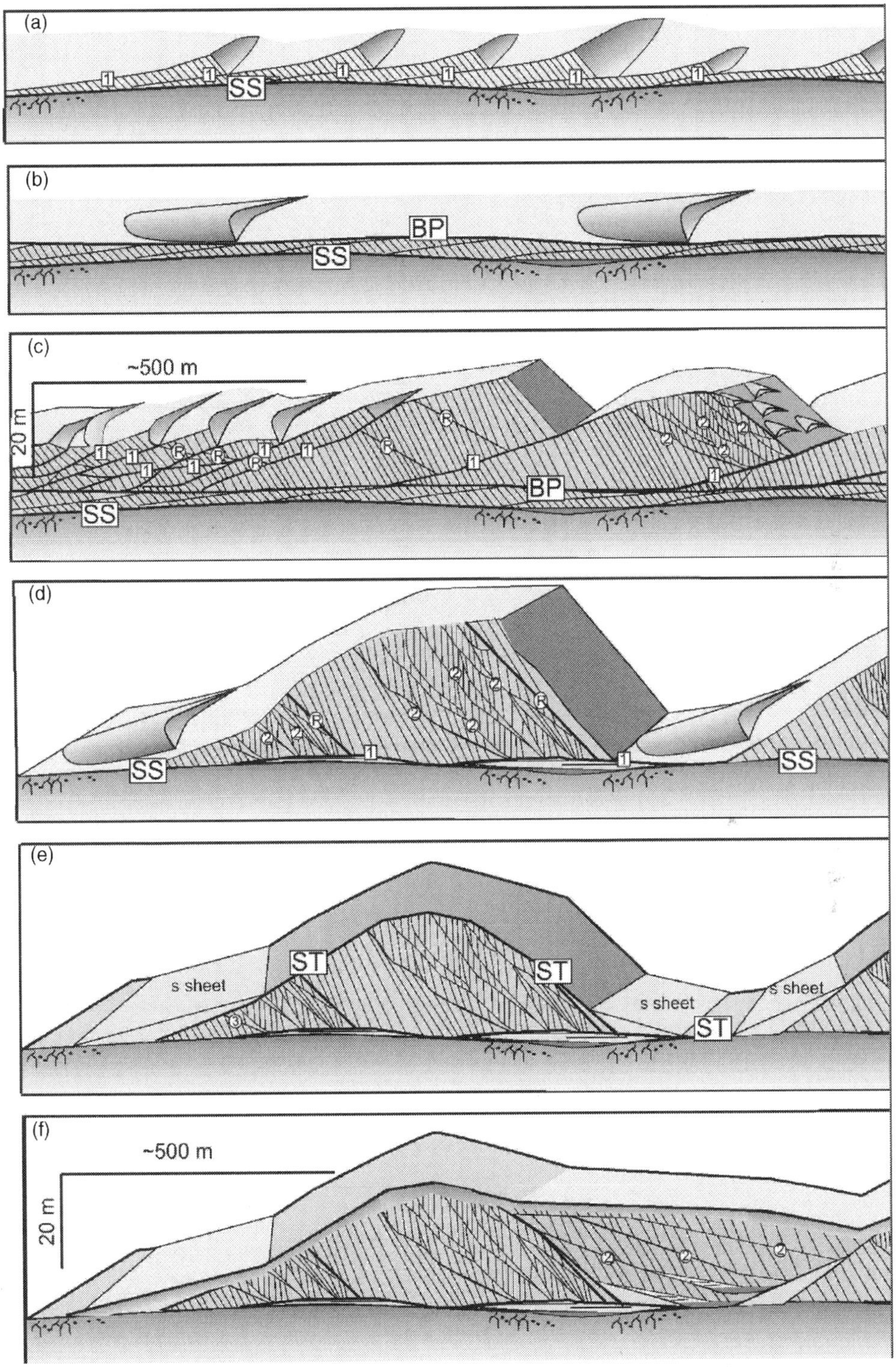
(a)
SS
(b)
BP
SS
(c)
~500 m
20 m
BP
SS
(d)
SS
SS
(e)
ST
ST
s sheet
s sheet
s sheet
ST
(f)
~500 m
20 m

they also cast a rain shadow that allowed active dunes to extend very close to the palaeoequator. (Fig. 13.17)

## Further reading

### *General*

Cooke *et al.* (1993), Thomas (1989), Williams *et al.* (1993), Abrahams & Parsons (1994) and Kocurek (1996) are readable summaries of many features of desert geomorphology and sedimentology. McKee (1978) contains superb satellite photos of erg features, many in colour. Ancient desert deposits are featured in North & Prosser (1993a) and Pye & Lancaster (1993). The general concept of sediment state for aeolian systems is introduced by Kocurek & Lancaster (1999) and applied to the Pleistocene– Holocene Kelso dune field of the Mojave desert. Most useful review of wind erosion in deserts is by Goudie (2008).

### *Specific*

Fixed ergs are considered by Talbot & Williams (1979), Talbot (1980, 1985); regional sandflow by Wilson (1971), Mainguet & Canon (1976), McKee (1978), Mainguet (1978, 1983); Australian ergs by Brookfield (1970); dust in general by Goudie & Middleton (2006); North African dust by Coudé-Gaussen (1984), Lever & McCave (1983), Li *et al.* (1996), Tegen *et al.* (1996), Engelstadter *et al.* (2006); meteorological conditions by Atkinson & Zhang (1996).

Sand sheets are studied by Fryberger *et al.* (1979), Kocurek & Nielsen (1986) and Lancaster (1995); interdune and sand sheet erg margin deposits by Glennie (1970), Kocurek (1981), Crabaugh & Kocurek (1993), Talbot (1980), Mountney & Jagger (2004). The Namibian Hoanib river interdune ponds and ancient analogues are beautifully described by Stanistreet & Stollhofen (2002). Coeval dune migration and high water-table sediment accretion in interdune areas of ancient draa is well documented by Mountney & Thompson (2002).

Rapid accretion and growth of Holocene dunes in discrete time intervals are considered by Goudie *et al.* (2000) and Beveridge *et al.* (2006); bounding surfaces, 'supersurfaces' and interbedform deposits by Stokes (1968), Glennie (1970), Brookfield (1977), Kocurek (1981, 1988), Loope (1984), Kocurek *et al.* (1991), Lancaster (1992), Crabaugh & Kocurek (1993), Fryberger (1993), Langford & Chan (1993) and Blakey *et al.* (1996). Reconstruction of the draa bounding surfaces of the Cretaceous Etjo erg in Namibia is by Mountney & Howell (2000).

Climate change and ergs are studied by Talbot & Williams (1979), Talbot (1985), Blount & Lancaster (1990), Kocurek *et al.* (1991), Lancaster (1992), Williams *et al.* (1993), Williams (1994), Bowler (1977) and Giraudi (2005). The notable Omani Wahabi erg architecture determined from coring, seismic and dating is by Radies *et al.* (2004). The sediment state of the Gran Desierto, Mexico is from Beveridge *et al.* (2006). Kalahari dunes are considered by Blumel (1998), Stokes *et al.* (1997), Thomas *et al.* (1997) and Lawson *et al.* (2002); Thar Desert margin is from Juyal *et al.* (2006).

Permian Rotliegendes of northwest Europe is considered by Glennie (1986), Steele (1983), Clemmensen (1989), Chrintz & Clemmensen (1993), George & Berry (1993), Heward (1994)

**Fig. 13.16** Contrasting (or complementary?) models for aeolian aggradation. (After Langford *et al.*, 2008.) (a)–(c) The growing and climbing dune/draa model for aeolian aggradation. (a) Dunes aggrade through climb on first-order (1) bounding surfaces (SS). (b) Short-lived bypass surface (BP) cut by non-climbing migration of dunes. These may interrupt climb and make recognition of first-order surfaces difficult. (c) Aggradation of complex dunes through climb on first-order surfaces. Stratigraphy is made more complex by downwind-dipping surfaces. Reactivation surfaces (R) (third-order surfaces of Brookfield, 1977) mark reworking of the slipface of the dune. Second-order surfaces (2) mark superimposed barchanoid forms descending the slipface of the draa. (d)–(f) Draa topography model for aeolian aggradation. (d) Growth and migration of draas (megadunes). Annotation is the same as above. Note the discontinuous first-order surfaces (1). (e) Draa stabilization. Formation of a stabilization surface (ST) that drapes the draa. Sand sheets accumulate on both flanks of the stabilized megadune. (f) Interdraa fills with strata separated by downwind-descending surfaces that mark accretion from the side of the draa (2).

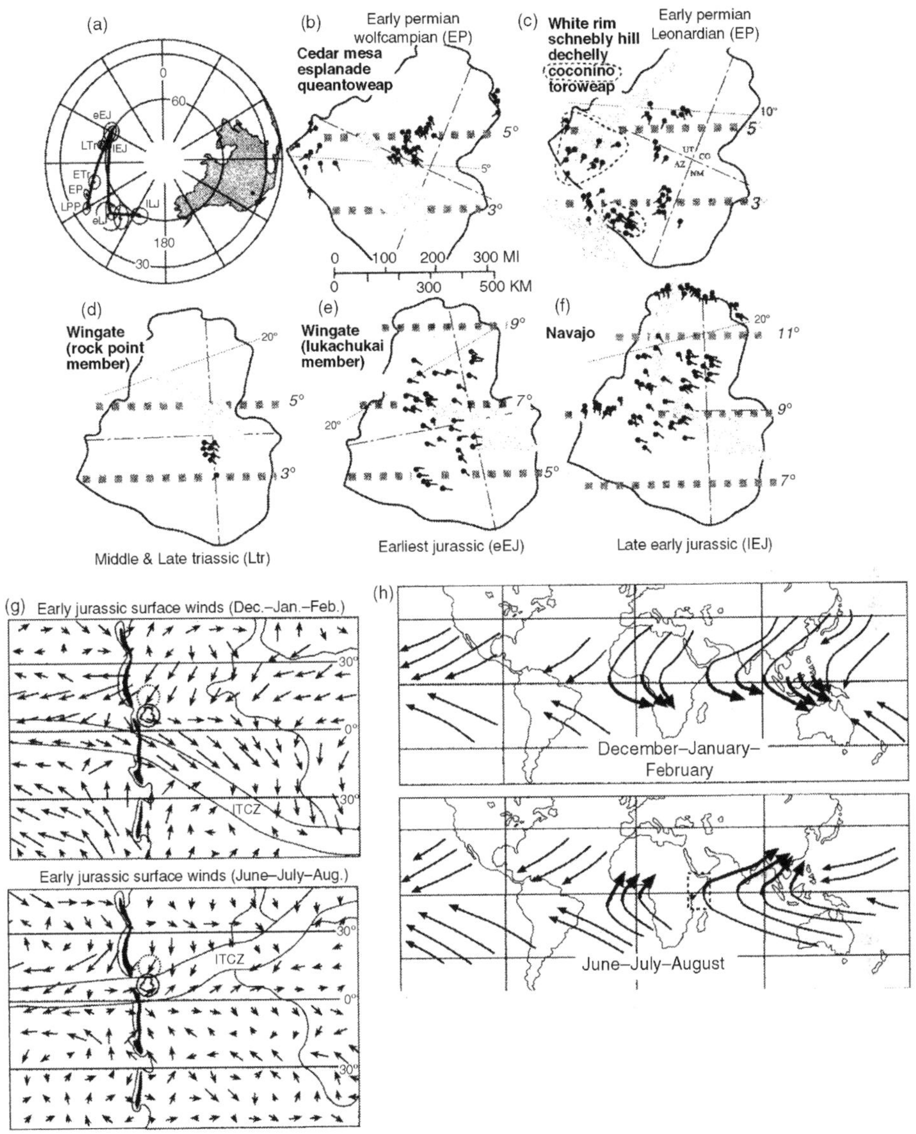


**Fig. 13.17** Palaeolatitudes and palaeowinds of Colorado Plateau aeolian sandstones. (After Loope *et al.*, 2004, and sources cited therein). (a) Palaeomagnetic pole positions for Colorado Plateau strata deposited between Late Pennsylvanian (LPP) and late Late Jurassic (lLJ) time. Circles represent the area of 95% confidence around each pole. (b)–(f) Palaeowind data. Lines projecting from dots represent downwind and the resultant dip direction of at least 20 cross-bed sets. Bold dashed lines represent the palaeolatitudes corresponding to the palaeopoles in (a). Fine dotted lines are palaeolatitudes. (g) Modern surface winds. Within the tropics, westerly winds (bold arrows) are restricted to monsoon systems. Rectangle shows position of East African Low Level Jet; star marks location of Mombasa, Kenya. Note that, in several locations, trade winds are reoriented to westerlies before crossing the equator. (h) Simulated surface winds from Early Jurassic global climate model. Dashed circle shows position of Colorado Plateau; solid circle shows the position of the Plateau that reflects recent palaeomagnetic data. No other palaeogeography has been adjusted. In upper (DJF) panel, note shift of winds from northeasterly to northwesterly in area of Colorado Plateau and the position of mountains just to the west of the Plateau (light shading). In lower panel, note migration of Intertropical Convergence Zone (ITCZ) over Plateau during June, July and August.

and Sweet (1999); *Weissliegende* facies by Glennie & Buller (1983), see also Benan & Kocurek (2000) for analogous Jurassic examples in Permian and Jurassic of the Colorado Plateau. Permian sequences analysis in western USA related to glacio-eustatic/climate changes is beautifully illustrated by Mountney (2006) and Mountney & Jagger (2004); see also the sideways look by the stimulating Langford *et al.* (2008).

Ancient ergs of the western USA are studied by Kocurek (1981), Clemmenson *et al.* (1989), Clemmenson (1989), Crabaugh & Kocurek (1993), Herries (1993), Havholm *et al.* (1993), Blakey *et al.* (1996) and Langford *et al.* (2008); ancient ergs of Pangea and palaeowind reconstructions by Bigarella (1973), Parrish & Peterson (1988) and, with emphasis on monsoonal circulations, by Loope *et al.* (2004).

# Chapter 14

# LAKES

*A lake allows an average father, walking slowly,*
*To circumvent it in an afternoon,*
*And any healthy mother to halloo the children*
*Back to her bedtime from their games across:*
*(Anything bigger than that, like Michigan or Baikal,*
*Though potable, is an 'estranging sea').*

W.H. Auden, 'Lakes', *Selected Poems*, Penguin Books, 1958.

## 14.1 Introduction

Lakes cover about 2% of the Earth's surface—they are important sinks for water and sediment, both largely brought in by river flow, although groundwater is also important locally. They form when spring flow, runoff or river flow is interrupted or intercepted, usually when a depression below local base level creates opportunity for water accumulation. Most large, deep and steeply shelving lake basins form as a result of tectonic subsidence (as in the East African and Baikal rift) or glacial erosion (as in the Great Lakes, USA/Canada). Large shallow lakes occur in crustal sags (Lakes Chad, Victoria). Smaller examples occur widely in coastal, deltaic and fluvial wetlands. Huge ancient lakes have been reconstructed from thrust-loaded foreland basins such as Eocene Lakes Gosuite, Uinta and Flagstaff of the western USA and in interconnected rift basins such as Pleistocene pluvial lakes Bonneville and Lahontan in the Basin and Range province. Lakes also occur in volcanic calderas, in meteorite impact craters (e.g. Lake Bosumtwi, Ghana) and in many glacial margin environments.

Looking at the surface of a lake, one is unaware of the complex processes that go on in the apparently tranquil waters beneath (Fig. 14.1). Depth, bottom slope and surface area represent balances between basin-forming processes and the hydrological variables of outward and inward flow and evaporation. Climate thus emerges as an important modulator of lake dynamics. Solar radiation provides one means of energy transfer through its control of surface water temperature and hence density. Sunlight provides photosynthetic opportunities for primary production at the base of the food chain upon which all else in the biomass depends. Skeletal sediment fragments arise from carbonate-secreting green algae (*charophytes*) and ostracods and from siliceous diatoms. Clastic sediment particles and dissolved ions enter the water body mostly at point sources such as river deltas, although wind-blown detritus may be important locally. Plumes and jets of sediment-laden water then interact with internal density changes and basin floor slopes. Wind shear on the lake surface provides kinetic energy to set up gradient currents, becoming more important as surface area increases. These act to resuspend and transport bottom sediment introduced previously by rivers or provided by organic cycling.

Although most lakes are freshwater bodies, saline perennial and ephemeral lakes are also common in

*Sedimentology and Sedimentary Basins: From Turbulence to Tectonics*, 2nd edition. © Mike Leeder.
Published 2011 by Blackwell Publishing Ltd.

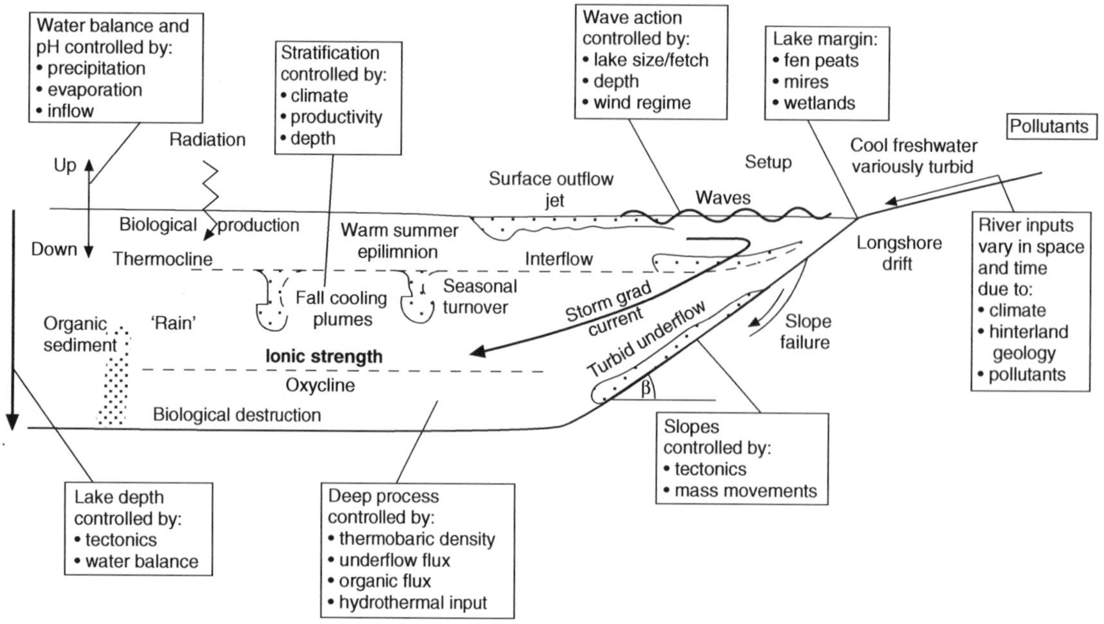


**Fig. 14.1** Aspects of the whole lake system. The sedimentology of lakes is a delicate balance between: (i) clastic sediment input by rivers, redistribution offshore by waves, wave-produced gradient currents, downslope mass movements and turbidity currents (ii) *in situ* biological production and chemical precipitation. These processes are seasonal and further modulated by short- to long-term changes in lake depth which cause transgressive and regressive sedimentary sequences to form over time.

lower latitudes. Ephemeral lakes may form in continental interior basins after abnormal rainfall events—witness the astonishing Lake Eyre, South Australia. Chemical classifications of lakes are based upon ionic concentrations through measurable conductivity, a feature closely related to alkalinity. It is sometimes useful to distinguish closed lakes from open lakes. The former are models of conservative behaviour, perfect *endorheic* sinks which lose no water or sediment supplied to them. The latter are imperfect sinks which may either have a surface outlet controlling maximum lake level, best illustrated by the Great Lakes system, or which lose water and dissolved ions through groundwater flow.

## 14.2 Lake stratification

Lakes may develop distinct layers that differ in their density, chemical composition and biochemical processes; they are said to be *stratified*. Thermal density stratification is due to surface heating caused by solar radiation (Fig. 14.2). Take the case of a temperate-climate lake. In summertime, given sufficient water depth (of order 10 m), it will show well-marked thermal stratification, with an upper, warm layer called the *epilimnion* and deeper, cold water of the *hypolimnion*, the two separated by water showing a gradient of temperature, the *metalimnion*. *Thermocline* is the term used to define an imaginary planar surface of maximum temperature gradient. Most heat is trapped in the epilimnion until, in autumn, cooling from the water surface downwards causes density inversions and mixing of the epilimnion with the deep hypolimnion. In early spring, melting of any surface winter ice also causes wholesale sinking of cold surface water, giving rise to the *spring overturn*. The water of a moderately deep lake will now all be at a temperature of about 4 °C. As spring progresses the topmost waters will gradually warm and be mixed downwards by wind shear. As heating continues, the isothermal, warm surface water will become buoyant enough to resist wholesale mixing and will remain above the cold

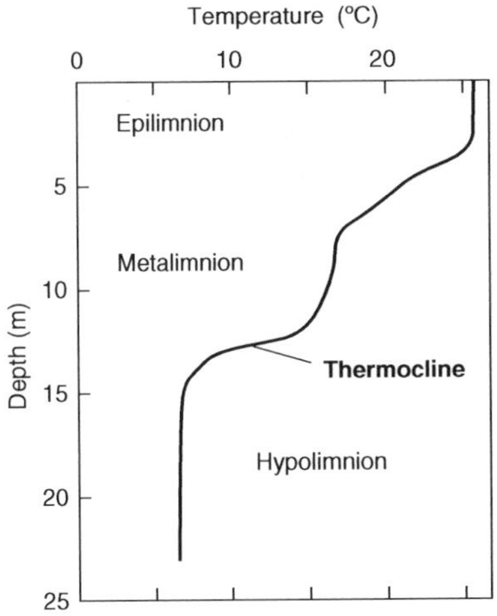


**Fig. 14.2** Sketch to illustrate summer temperature stratification of a temperate lake. (After Wetzel, 1983.)

deep water. Such biennial overturns contribute to the production of distinct laminations in bottom sediment. By way of contrast, in deep tropical lakes stratification is permanent. In all cases, stratification may be strongly disrupted by mixing due to surface waves and more importantly by the passage of internal-waves, termed *seiches* (see below). Density current interflows may also occur along the thermocline.

There are a large number of other terms used by limnologists to describe variations in lake circulation and stratification. In lakes deep enough to form a hypolimnion, some of these are:

- *amictic*—lakes permanently isolated from the atmosphere by ice cover;
- *cold monomictic*—water temperatures never rise above 4 °C, with one period of stratification in the summer;
- *cool dimictic*—lake water overturns twice yearly (described above) in spring and summer;
- *warm monomictic*—water temperature never falls below 4 °C, freely circulating in the winter and stratifying in the summer;
- *oligomictic*—water temperature always well above 4 °C (i.e. tropical conditions) with stable stratification and a small temperature gradient; water rarely circulates;
- *polymictic*—lakes with frequent or continuous circulation due to strong wind regime and/or large short-term temperature variations;
- *meromictic*—lakes with a chemocline separating a near-permanent bottom layer that never mixes.

Concerning meromictic lakes, chemical density stratification occurs when: (i) a layer of salty water is stable beneath upper layers of less-saline fluid, the two separated by a *pycnocline* of salinity gradient. Such conditions occur in some coastal lakes fed by marine seepage and in solar ponds. (ii) An oxygenated layer undergoing mixing and diffusion with the atmosphere is separated from a lower deoxygenated layer along a chemocline. Many lakes worldwide suffer human-induced conditions of seasonal low oxygen bottom waters (*hypoxia*). This is due to the enhanced oxidation of organic matter sinking from highly productive surface water enriched by effluent pollution from agrochemicals rich in nitrate and phosphate.

## 14.3 Clastic input by rivers and the effect of turbidity currents

Water and sediment input to lakes occurs at river deltas. Only a small proportion of surface runoff and suspended sediment enters a lake as a surface plume. The generally higher density of cooler inflowing water means that underflow is common. Proximity to source is a fundamental control on the nature of lake sedimentary deposits. Successively finer sediment will be deposited outwards from the point source, although this regular pattern is affected by surface currents due to direct wind shear. Channelized runoff from steep tectonic or glacial slopes encourages formation of Gilbert-type fan deltas (Chapter 17) whose steep slopes are prone to sediment failure and downslope mass movement. Indeed, the formation, runout and mixing of these flows with lake-water dominates many lake sediment distribution patterns. Base-of-slope fans are preferentially formed during lowstands when subaerial fan deltas become incised and bypassed.

Quasi-continuous turbidity currents in lakes are caused partly by underflowing hyperpycnal river-waters (Fig. 14.3 & Plate 9) and partly by flow transformations from slides, slumps and debris flows. Turbidity current development is hindered by turbulent dissipation of incoming flows in very shallow, well-mixed lakes with gently sloping margins. The density contrasts that drive subaqueous currents vary with temperature and sediment concentration. Detailed monitoring provides evidence of the variability

**Fig. 14.3** The steady underflow of cold, sediment-rich meltwater and interaction with the deep portion of a wind-driven gradient countercurrent produces an intensely turbulent bottom current that ponds against and then spills over an intrabasinal sill in Peyto Lake, Alberta, Canada. (After Chikita *et al.*, 1996.) See also Plate 9.

of density underflows on a daily basis and their interactions with wind-forced currents. In thermally stratified lakes the density of the inflowing water may be greater than that of the lake epilimnion but less than that of the hypolimnion, so that the density current moves along the top of the metalimnion as an *interflow*. Thus, high concentrations of suspended sediment occur at this level, which may then be dispersed over the lake by wind-driven circulation and internal waves. Inflows denser than the hypolimnion flow along the lake floor as *underflows/wall jets*. The underflows bring oxygenated water into the deep hypolimnion and may prevent permanent stagnation in deep lakes. It is possible that the phenomenon of turbidity current lofting (Chapter 8) might also occur once the sediment load has been largely dissipated.

## 14.4 Wind-forced physical processes

Away from the effects of river influx, water movement in lakes is controlled entirely by wind-driven progressive waves and gradient currents, for even the world's largest lakes are too small to exhibit more than minute (<2 cm) tidal oscillations. Wind-driven surface waves effectively mix the upper levels of lake-water and give rise to wave currents along shallow lake margins. These processes are accompanied by the seasonal density overturns described previously. The size and effectiveness of lake waves depend upon the square root of the fetch of incident winds and therefore on the length scale of the lake itself. The energy associated with travelling waves is dissipated along the shoreline as the waves break, the nearshore sediment distribution being largely controlled by these effects. Internal waves may also form at the epilimnion–metalimnion interface.

A steady wind causes a mass transport of surface water by wind shear (Fig. 14.4), though the direction of water movement may be oblique to the direction of wind flow due to Coriolis effects. In very large and deep lakes, this deviation can reach a maximum at the surface of 45°, decreasing down from the surface and thus defining the so-called *Ekman spiral* (**Cookie 43**). The Ekman effect becomes progressively less as the the lake decreases in area and depth. A measurable tilting of the lake-water surface results from wind shear, the downwind part being higher than the upwind part. Similar effects in very large lakes may arise due to differences in barometric pressure. It can be shown that a static equilibrium is possible if the wind stress is balanced by a surface elevation gradient of magnitude $u_*/gh$, where $u_*$ is wind shear velocity and $h$ is water depth. The effect is therefore less in deeper lakes. The surface slope may have values of between $10^{-7}$ (1 cm in 100 km) and $10^{-6}$ (10 cm in 100 km). Although these gradients are small, they generate currents in the whole lake or in the surface layer if the lake is stratified. Such currents have the ability to interact with turbidity undercurrents in an interesting way (Fig. 14.3). The application or disappearance of

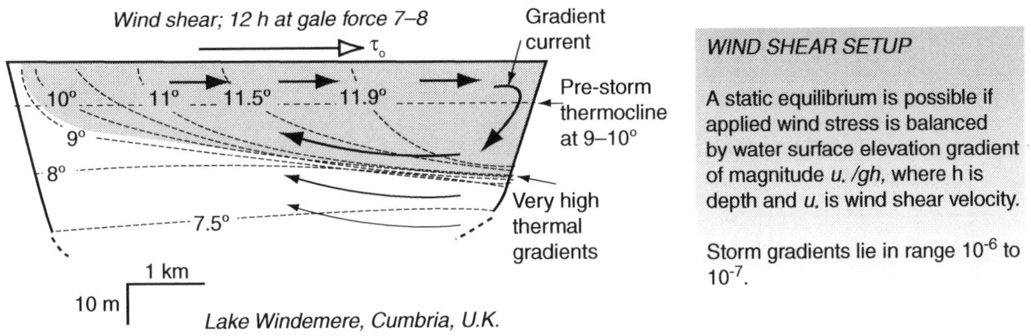


**Fig. 14.4** The results of a 12 h duration autumn wind of force 7–8 across Lake Windemere, Cumbria, England, to illustrate surface wind-shear currents, deep gradient current and epilimnion mixing. (After Mortimer, in Wetzel, 1983.)

wind stress causes lake-surface and internal wave oscillations known as *seiches*, which may further mix surface waters and especially subsurface stratified layers and cause erosion and entrainment along shorelines. The amplitudes and intensity of internal-wave seiches are much greater than surface waves (Section 5.11) and they have been shown to be capable of significant sediment resuspension when they break or surge over bottom topography.

For more info on Ekman effects see Cookie 43

Linear lake bodies may show reasonably fast currents caused by direct wind stressing in shallow marginal areas (up to 0.3 cm/s) with a pattern of surface eddies accompanied by slower compensatory flow in the lake bottom ( < 3 cm/s). Such surface currents can transport very fine sands and silts as bedload and will thoroughly mix suspended load introduced by surface river plumes. Instrumented moorings down to depths of ca 100 m in southern Lake Michigan show that local resuspension of bottom sediment occurs frequently during winter storms after the breakdown of the thermocline. Resuspension is most frequent close to the shore, with advection of resuspended sediment to deeper sites by the general lake circulation. In reality surface currents in large lakes can be exceedingly complex, taking the form of vertical-axis eddies called *gyres*; these are well known from the Great Lakes of North America (Fig. 14.5). Wind-produced currents have been held responsible for deposition of thick sediment drifts in Lake Cardiel, Argentina, though these so-called contourites are misnamed for they cannot have been deposited by true thermohaline currents as in the oceans (Chapter 21).

## 14.5 Temperate lake chemical processes and cycles

The chemistry of lake water is delicately adjusted to the input of solutes from the lake drainage basin and to the amount of evaporation and vertical mixing. The input ions are in turn mediated in an important way by biological cycling. The nature of the input strongly reflects the type and degree of rock weathering in the lake catchment. The total ionic salinity of lake-water is usually dominated by the cations $Ca^{2}$, $Mg^{2}$, Na and K and the anions $CO_3^{2-}$, $SO_4^{2-}$ and $Cl^-$. Many other ions occur of course (e.g. ions containing P, S, Fe, N and Si) but, although essential for biological productivity and metabolism, they usually have very low dissolved concentrations.

The *carbon cycle* in lakes is dominated by the inorganic and organic precipitation of $CaCO_3$, and the fixation of carbon by photosynthetic and heterotrophic algae. Nitrogen is an important constituent of cellular tissue and is a major freshwater nutrient—most nitrogen is fixed by microorganisms but the nitrogen cycle is highly complex and will not be discussed in this text. However, the process is important in understanding the fixation of iron, copper and phosphates by substances like polypetides excreted by blue-green algae. Phosphorus in freshwater lakes occurs as organic phosphate, as particulate phosphates in organisms and as solid mineral phases.

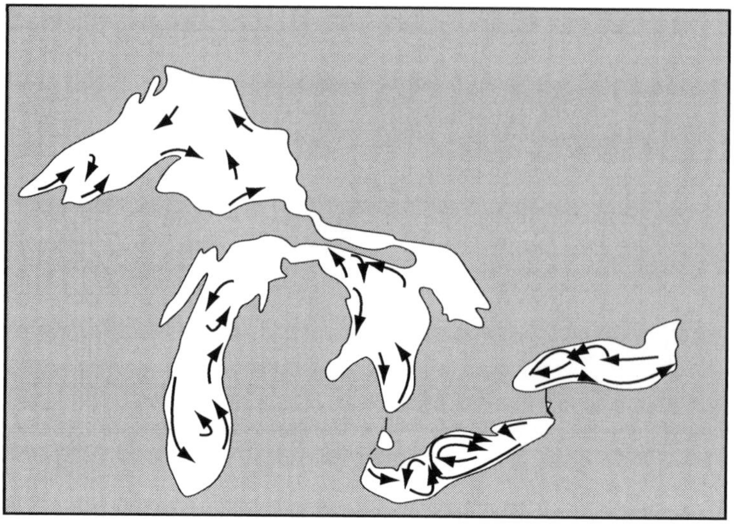

**Fig. 14.5** Mean summer surface circulation in the Great Lakes of North America. There is a general anticlockwise circulation. This can be used in conjunction with the more important winter circulation to obtain mean annual vorticity, of great practical use in applied environmental engineering and pollution control. (after Beletsky *et al.*, 1999.)

Soluble orthophosphate is of great importance as a nutrient, and is also very reactive with cations such as $Fe^2$ and $Ca^2$. There is often a marked increase in total phosphorus in the oxygen-poor hypolimnion of well-stratified lakes. Regarding the *silica cycle* in lakes, the abundance of the dissolved form of silicon (provided by rivers in the form of silicic acids) is chiefly modulated by the uptake of the element by photosynthesizing diatoms in the spring and summer, followed by the return of the silica to the lake floor as diatom carcasses in winter.

Most temperate-zone lakes are dominated by chemically and biologically mediated fluxes of calcium and bicarbonate ions between source (river or groundwater) and sink (lake-floor sediments). Marked annual changes occur in $Ca^2$ and $CO_3{}^{2-}$, but the average composition of lake-water is similar to the unevaporated composition of 'typical' river-waters derived from mixed geological terrains. Calcium is a critically important element for cell growth and metabolism and is thus required for a lake to be a productive ecosystem. Chemical data from many temperate hard-water lakes show that surface waters are only slightly supersaturated with respect to calcium carbonate during the winter and that supersaturation decreases after the spring overturn when undersaturated bottom waters are brought up to the surface and the lake-water is effectively homogenized. Maximum supersaturation, with low-magnesian calcite precipitation, occurs in the summer caused by warming and $CO_2$ removal by charophytes and phytoplankton blooms (Fig. 14.6). $Ca^2$, $CO_3{}^{2-}$ and a number of other trace ions and a great deal of carbon in cellular biomass are thus sedimented out of the lake system via shell and skeletal tissues on an annual basis: this sequestered carbonate forms *marl-lake* sediment common in many temperate cool dimictic lakes. A certain proportion of the solid-phase carbonate may be returned to the hypolimnion by winter dissolution at the lake floor. Generally, lake carbonate sediments are quite distinctive—especially charophytic algal sediments and tufas of the shallow lake margin which have particular use in the elucidation of palaeoclimate and changing shoreline position over time.

## 14.6 Saline lake chemical processes and cycles

A saline lake is one with greater than 5000 ppm of solutes, a playa being a seasonally exposed, evaporitic lake floor (Fig. 14.7). Saline lakes are usually alkaline and occur most commonly in semiarid, closed-drainage

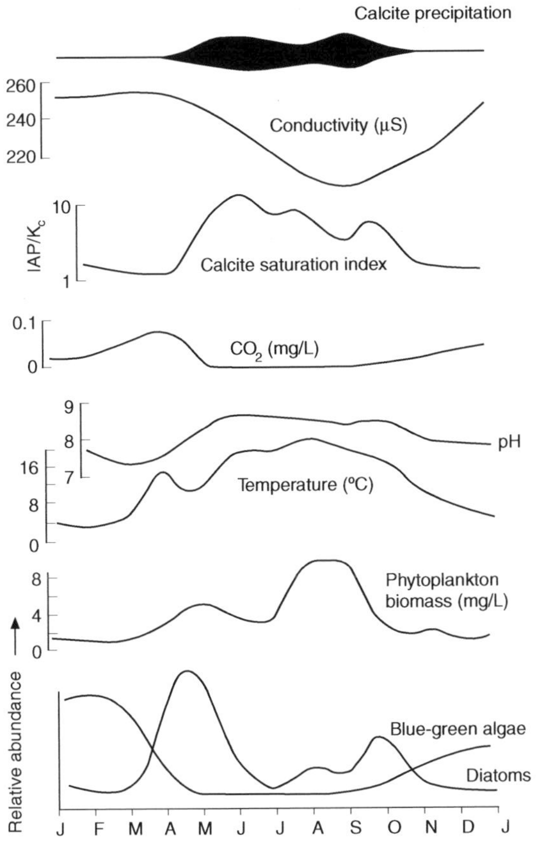


**Fig. 14.6** Time series of several parameters observed and measured in the seasonal cycle of the epilimnion waters of Lake Zurich, Switzerland. (After Kelts & Hsu, 1978.)

(endorheic), fault-bounded basins: the perennial Great Salt Lake of Utah and the playa of Death Valley, California (Plate 10) are perhaps the most spectacular and best-known examples.

Shallow saline lake sediments are characterized by evaporite/dolomite–clastic couplets produced in the ephemeral and perennial lake by cycles of storm runoff from the surrounding alluvial flats, followed by evaporite precipitation. Layered halite rock often results, in which clastic laminae, solution or deflation surfaces separate vertically elongate crystals of halite, which have nucleated from the shallow lake floor or from foundered mats of hopper crystals that grew on the brine surface. Spectacular laminated and particulate gypsum rocks and selenite swallow-tail growths are recorded in the evolving Quaternary coastal saline lakes (salinas) of South Australia. The brines in these lakes were/are fed by marine seepage through coastal dunes. Laguna Mar Chiquita, a highly variable closed saline lake located in the Pampean plains of central Argentina, is presently the largest saline lake in South America (ca $6000\,km^2$). Recent variations in its hydrological budget have produced dry and wet intervals that resulted in distinctive lake level fluctuations. Lake level drops and concurrent increases in salinity promoted the development of gypsum–calcite–halite layers and a marked decrease in primary productivity. The deposits of these dry stages are evaporite-bearing sediments with a low organic matter content. Conversely, highstands like that of the past 200 yr are recorded as diatomaceous organic-rich muds (Fig. 14.8).

Deposits of deeper saline lakes are known from the stratigraphic record, such as those of Pleistocene pluvial lake Bonneville in Utah where cores show periodic deep-water conditions such as those during MIS 2 when diatom-rich, aragonitic marls and sapropels accumulated. Bedded halite deposits formed during lowstands at glacial to interglacial transitions. In the Miocene Teruel Basin, northeast Spain, basin centre gypsum is believed to have precipitated in a deep lake, in which water stratification became unstable with progressive shoaling (Fig. 14.9). Rhythmites, composed of alternating laminae of pelletal gypsum and very fine lenticular gypsum crystals mixed with siliceous microorganisms, formed in addition to gypsum turbidites, intraformational gypsum breccias and slump structures. The pelletal laminae originated from the faecal activity of animals (crustaceans?) ingesting gypsum crystallites in the lake water during episodes of maximum evaporation, whereas the laminae of very fine lenticular gypsum mixed with microorganisms accumulated during episodes of relative dilution.

Playas collect runoff from surrounding footwall and hangingwall uplands (Fig. 14.10), but precipitation is often minimal and a negative water balance exists for most of the year. Clastic detritus is trapped on alluvial fans that fringe such basins so that only solute-laden waters and springs issue into the saline lake. Replenishment flow into the playa may be by direct rainfall, via groundwater springs charged by the topographic relief of surrounding uplands or from ephemeral stream flow with evaporation causing evaporite crusts to appear at the surface. During

**Fig. 14.7** A typical playa, seen here on a Landsat 7 image of Pilot Valley area, eastern Nevada, USA. (From Liutkus & Wright, 2008.)

seasonal highstands, cyanobacterial mats may dominate the playa margins and lake bottom though these do not often add to the deposited sediment since they are destroyed during periods of desiccation. A concentric arrangement of evaporite precipitates is usually seen, with the mineral phases present being controlled by the nature of the weathering reactions in surrounding hinterlands. Subsurface growth of both evaporite salts in the basin centre and dolomite (as dolocrete) in the flanking highs may occur. In some tectonically active rifts associated with volcanism, as in the East African rift system, there may be a significant input of salts (containing Na, Ca and Cl) from juvenile magmatic sources. Some enclosed playas are still open hydrologically because they may still lose subsurface water and ions to adjacent basins. Saline waters are usually dominated by Na–Ca–Cl–$SO_4$, although great variation can occur. Many present-day saline playas of the western USA were once deep perennial lakes during glacial maximum times. Death Valley, for example, was up to 90 m deep between 10 and 35 ka, precipitating calcite, some halite and with significant clastic mud deposition. This contrasts with the playa-like state of the past 10 kyr dominated by Na–Ca–$SO_4$ precipitates. Other lakes show well-developed laminated black-shale facies of algal origins during lake highstands. In Europe, major changes in lake hydrology occurred during the Holocene in response to climatic fluctuations, with early Holocene saline deposits of microbial dolomite, mid-Holocene varves of dolomite and clastic laminae to late Holocene freshwater systems dominated by detritus (Fig. 14.11). Many modern desert margins also show evidence for major Holocene climatic fluctuations that periodically allowed shallow lakes to spread widely and occupy interdune areas of stabilized sandy ergs.

Deep permanent saline lakes such as the Dead Sea show appreciable river-water influx, which balances net evaporation. Most of the solute influx, however,

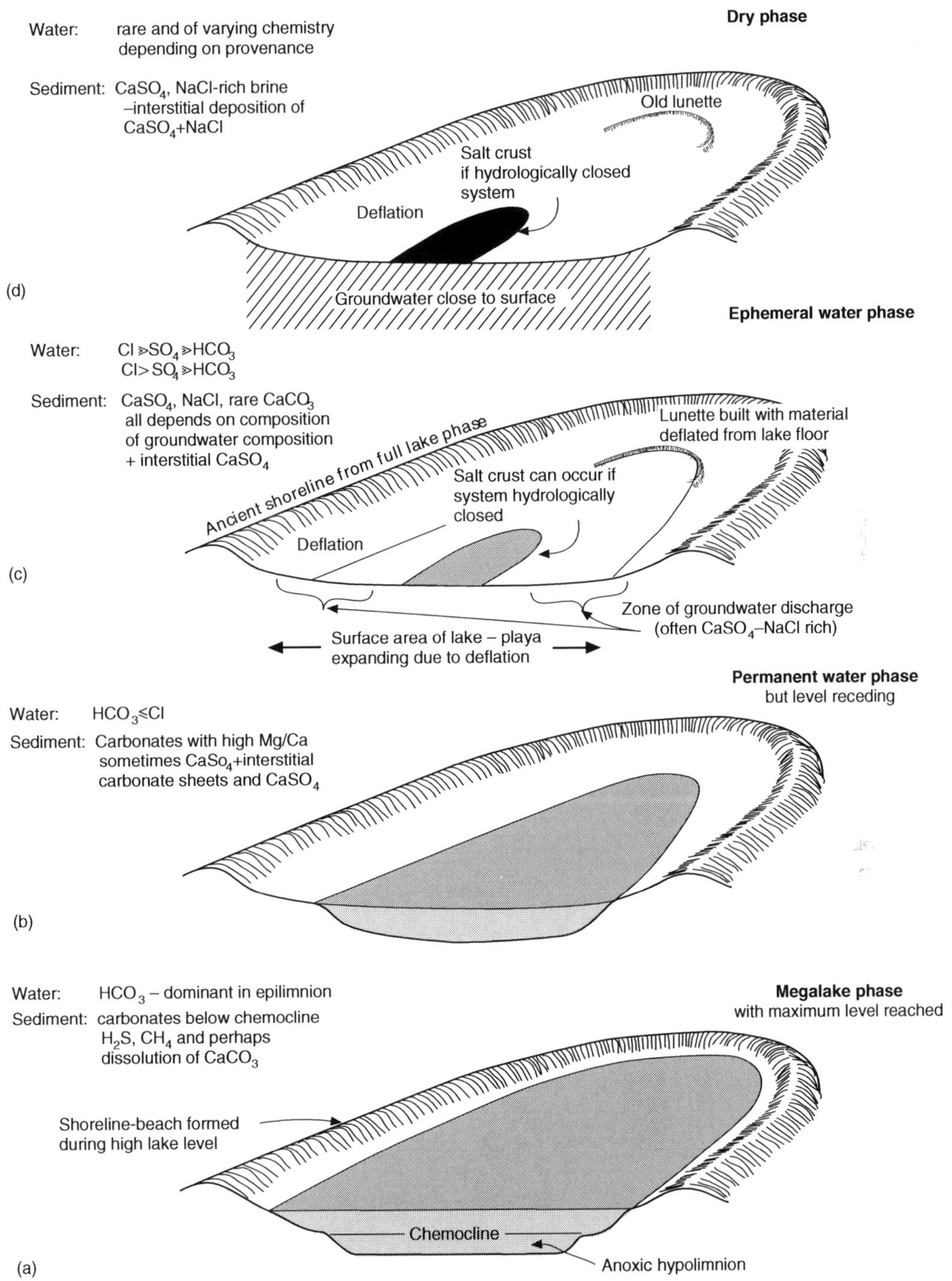


**Fig. 14.8** Succession of stages in the transformation of a pluvial-period permanent lake to a dry-period salina or playa. (after DeDekker, 1988.)

comes from small marginal saline springs around the periphery rather than from the River Jordan itself subsequent to diversion, water theft and training in the past 30 yr. The lake-waters are very saline (> 300 000 ppm) and of an unusual Na–Mg–(Ca)–Cl type, with low sulphate and bicarbonate. The deep (~300 m) main basin has historically been density-stratified because of salinity differences, both aragonite

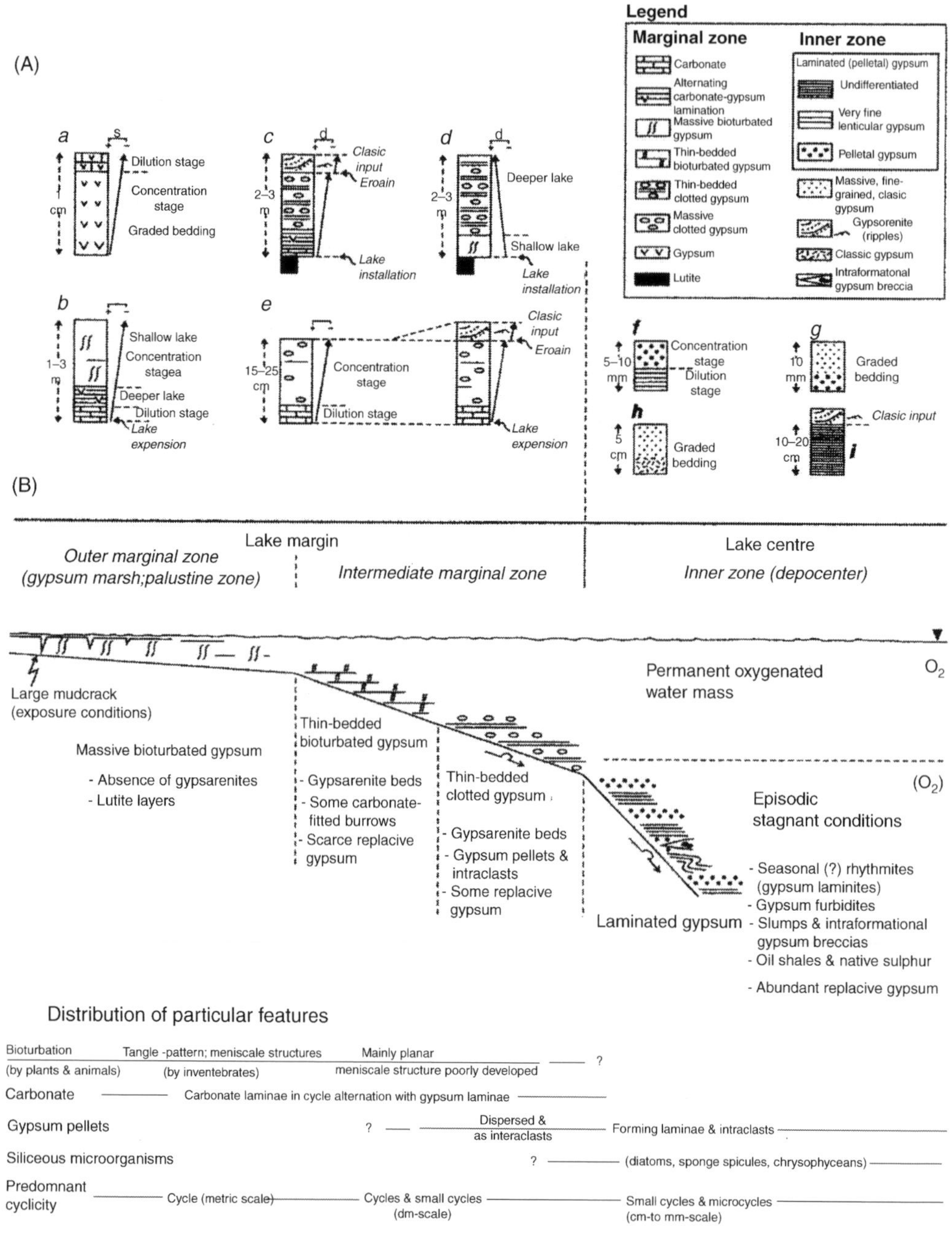

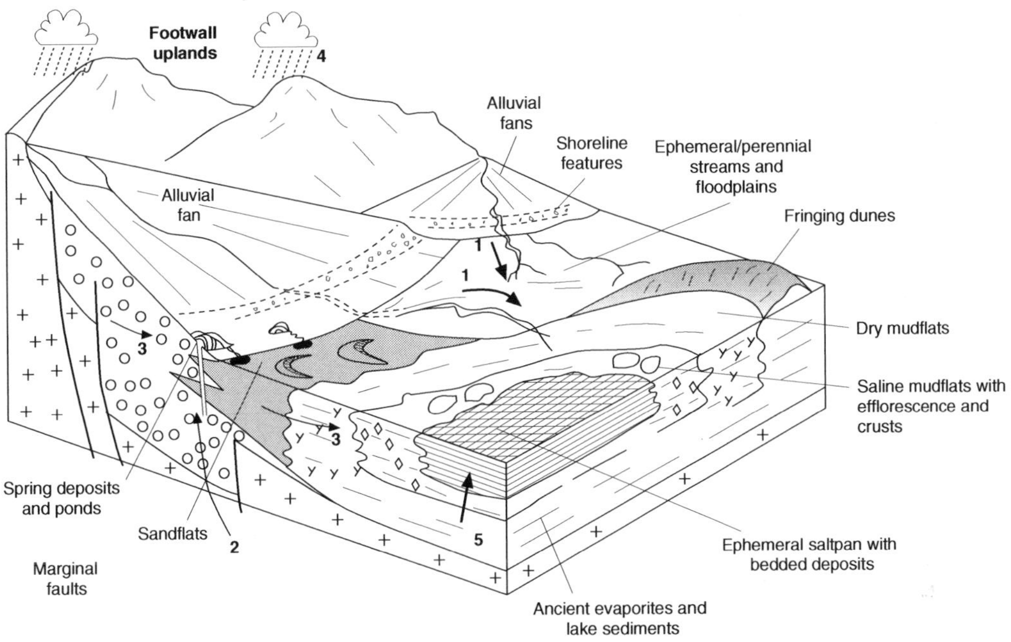


**Fig. 14.10** To illustrate depositional environments and sources of groundwater in typical playa basins: 1, surface runoff; 2, hydrothermal outflows, 3, groundwater; 4, direct rain recharge; 5, subsurface brine reflux. (After Hardie *et al.*, 1978; Rosen, 1994.)

and gypsum precipitating from surface waters. Natural halite precipitation also occurred in the deep lake about 1500 yr ago when the evaporation/inflow ratio reached into the halite supersaturation field. The Dead Sea is now well mixed and its level dropping since an overturn in 1979, with coarsely crystalline halite, aragonite and gypsum being precipitated and preserved on the lake floor. Seismic and field data from exposed Holocene strandline fan delta deposits (Chapter 17) reveal evidence for many phases of climatically driven lake-level rises and falls over the past 5 Myr.

## 14.7 Biological processes and cycles

Nutrients brought into lakes by runoff are utilized by a variety of biological communities. Playa lake-

**Fig. 14.9** Interpretation of lacustrine sedimentation in the gypsum subunit of the Miocene Libros Gypsum unit, southern Teruel Basin, NE Spain. (After Orti *et al.*, 2003.) (A) Selected metre-scale cycles and small cycles in the lake margin (marginal zone):(a) gypsum–carbonate microcycle, with graded bedding; (b) gypsum cycle with carbonate layer and alternating carbonate–gypsum laminations at the base, and massive bioturbated gypsum at the top; (c) gypsum cycle with carbonate layer and alternating carbonate–gypsum laminations at the base, thin-bedded clotted gypsum, and gypsarenite at the top; (d) gypsum cycle with massive bioturbated gypsum at the base and thin-bedded clotted gypsum at the top; (e) thick gypsum cycle with carbonate at the base and thin-bedded clotted gypsum at the top, with or without uppermost gypsarenite layer; s, salinity; d, depositional depth. Selected small cycles and microcycles in the lake centre (depocentre): (f) lamina of very fine lenticular gypsum at the base, and pelletal lamina at the top; (g) pelletal lamina at the base, and lamina of fine-grained gypsum at the top, with graded bedding; (h) lamina of clastic, millimetre-sized gypsum at the base, and lamina of fine-grained gypsum at the top, with graded bedding; (i) laminated gypsum at the base, and gypsarenite at the top. (B) Lacustrine subenvironments of the saline lake system, and distribution of lithofacies and associated features.

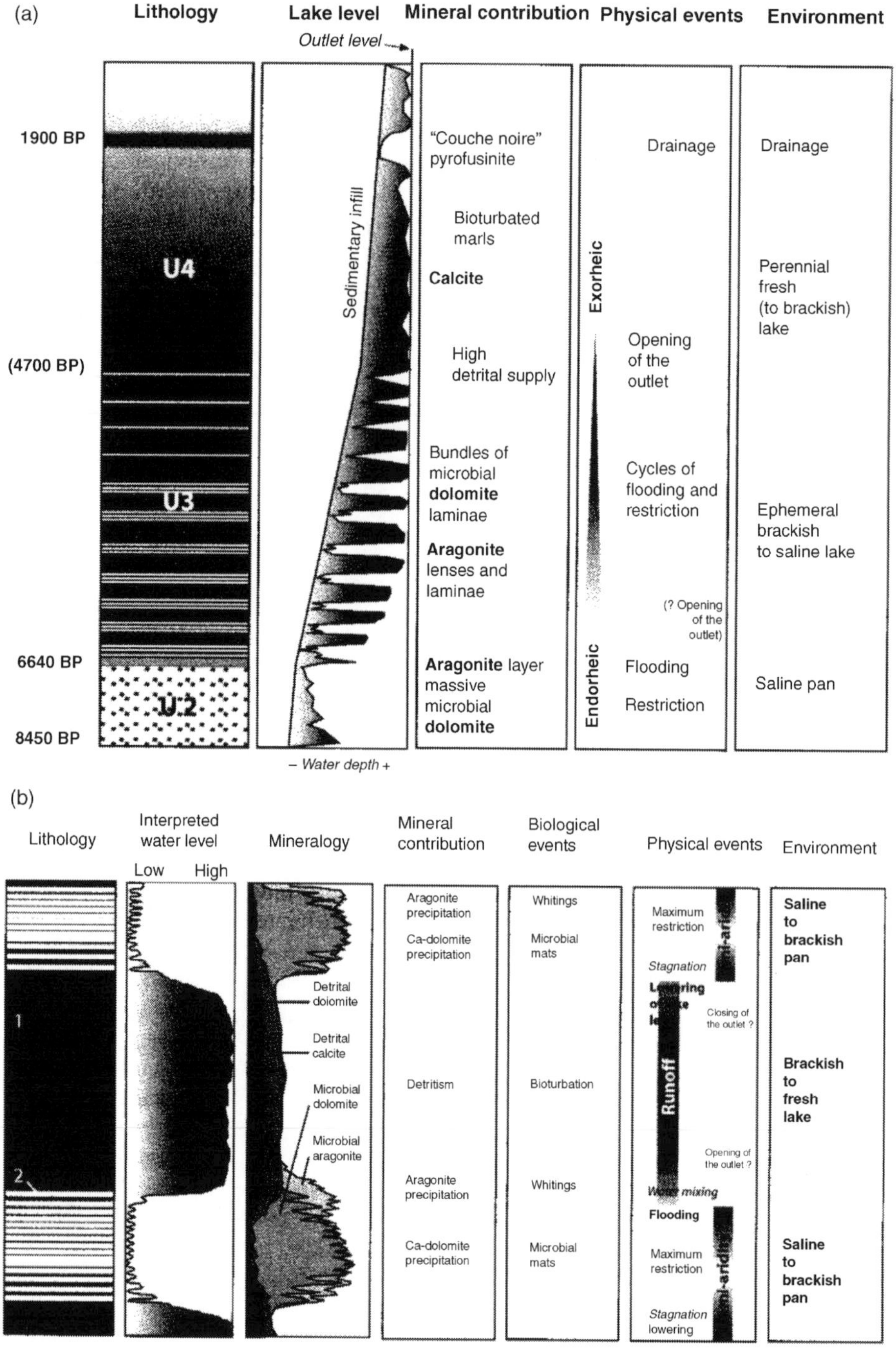


**Fig. 14.11** (a) A scenario suggested for the evolution of the Sarliéve palaeolake during the Holocene according to the sedimentary record, maximum thickness of ~6 m. U2: massive dolomicrite; U3: bundles of dolomite laminae interbedded with homogeneous marls; U4: homogeneous to bioturbated marls. (After Bréhéret, *et al.* 2008.) (b) A scenario suggested for the deposition of a small sequence of Unit 2 (at a decimetre scale), composed of (1) homogeneous marls and (2) bundle of dolomite laminae.

margins fed by saline springs may show extensive stands of *saltgrass* communities encrusted by *rhizoliths*, calcite encrustations precipitated around plant stems during shallow lake highstands. Photosynthetic phytoplankton thrive in the well-lit epilimnion and provide the basic rung of the lake food chain. Diatoms are often highly important in nutrient-poor (*oligotrophic*) lakes like Baikal. Below a certain depth, freshwater lakes with low degrees of vertical mixing (for various reasons) are prone to chemical stratification. Thus most dissolved oxygen is used up in the bacterially mediated oxidation of detrital organic matter, phytoplankton, zooplankton and higher members of the food chain that have settled out after death on to the lake floor from the productive upper waters. It can be appreciated that the abundant fertility of upper lake levels may cause the 'death' of the lower lake—witness the explosion of productivity felt in many lakes polluted by abundant human-introduced phosphates and nitrogenous compounds. Permanently chemically stratified (meromictic) lakes are able to preserve almost all incoming organic debris, causing the formation of dark-coloured laminated organic-rich muds. However, seasonal fluctuations in the depth of oxygenation may cause the production of distinctive organic-rich and organic-poor laminae to be preserved below a certain depth. Early winter mixing of the upper lake-waters of meromictic Lake Crawford caused oxygenation and mass mortality of anaerobic bacteria, which sedimented to form an organic-rich layer on the deep lake floor (Fig. 14.12). These thin ($\sim$1 mm) layers become enriched in iron monosulphides after reactions between dissolved or particulate $Fe^2$ and the sulphur-rich bacterial carcasses (the photosynthetic bacteria themselves oxidize sulphides to elemental sulphur during life). This illustrates the great importance of bacteria in the biogeochemical cycling of lakes.

## 14.8 Modern temperate lakes and their sedimentary facies

The facies pattern of a cool dimictic lake is well illustrated by Lake Brienz, Switzerland, a 14 km long and 261 m deep lake in the Swiss Alps. Sediment deposition is entirely clastic, the detritus being introduced by rivers that enter the lake from opposite ends. As noted above, the fluvial sediment is transported and deposited in the seasonally stratified lake by overflows, interflows and underflows, depending on the density difference between the river- and lake-waters. High-density turbidity currents form underflows and deposit thick ($<$ 1.5 m) graded sand layers. These deposits occur only once or twice per century in response to catastrophic flooding. Low-density underflows occur annually during periods of high river discharge and deposit centimetre-thick, faintly graded sand layers. Fine sediment introduced by overflows and interflows is mixed over the whole lake surface by circulation. It settles continuously during summer thermal stratification to form half of a varve couplet. At turnover in the autumn as the lake is wholly mixed the remaining sediment trapped in the thermocline settles out and forms the light-coloured winter half of the varve that has fairly uniform thickness over the whole basin. The turbidite deposits grade laterally into thin dark laminae similar to the summer part of the varve noted above. The mechanisms of turbidite deposition and clastic summer varve formation are thus related, the two layers having a common sediment source but a differing level of introduction during periods of thermal stratification.

In contrast to the clastic facies of Lake Brienz, the sediments currently being laid down in Lake Zurich, Switzerland, are mostly biogenic and chemical, since flood-control dams have almost stopped fluvial sediment input into the lake since about 1900. The varved sediments laid down here form in response to an annual chemical and biological cycle. The varves are present below 50 m depth but are destroyed on slopes by slow creep. Close analogies may be made with Neogene lacustrine chalks penetrated by deep-sea drilling in the Black Sea.

## 14.9 Lakes in the East African rifts

Lakes Malawi (aka Nyasa) and Tanganyika typify the large, stratified, tropical freshwater lakes developed in the East African rift system (Fig. 14.13). In Lake Malawi (area 45 000 km$^2$, maximum depth 730 m) up to 4.5 km of sediment has accumulated in the lake basin in the past 5 Myr or so. In Lake Tanganyika (area 23 000 km$^2$, maximum depth 1470 m) more than 4 km has accumulated in the past 1 Myr. In each half-graben the bathymetric framework is dominated by the asymmetric basin form and the effects of fault segmentation (Fig. 14.13; Chapters 22 & 23). The magnitude and architecture of the structural basins are a consequence

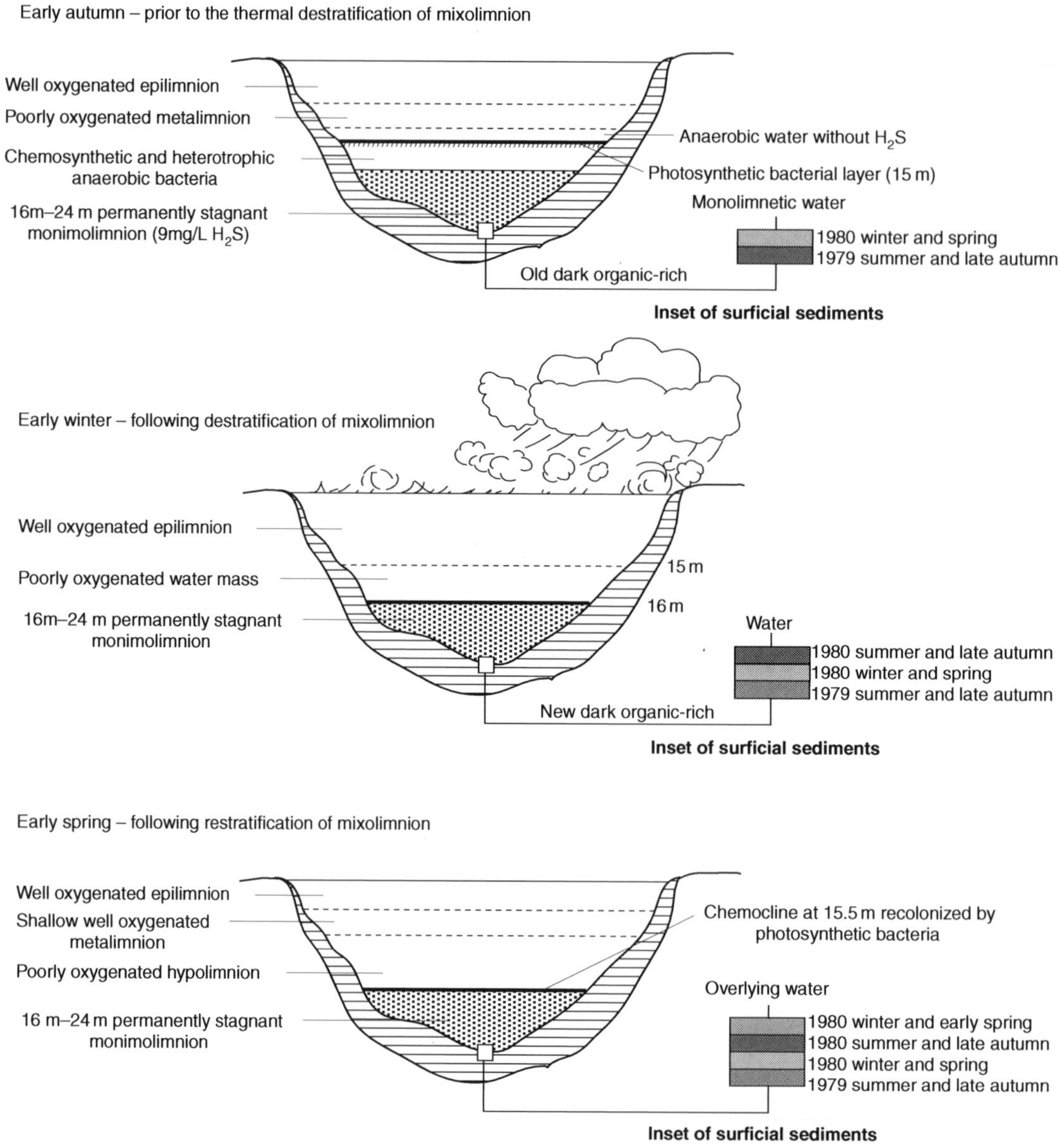


**Fig. 14.12** Sketch sections to illustrate seasonal changes in meromict Lake Crawford, specifically the mortality of photosynthetic bacteria at the chemocline and the formation of dark organic-rich varve laminations. (After Dickman, 1985).

of the extension and faulting of old, thick cratonic lithosphere with continuous fault lengths of >100 km. Lakewards of active faults, the steep subaqueous slopes are predominantly bypass margins, feeding turbidity currents into the lake bottom (Fig. 14.14). Along shoaling hangingwall ramps, lake surface currents are able to rework the coastal sediment much more, with the production of dunefields and a consequent paucity of turbidity currents. Most sediment enters the lakes via axial rivers in Lake Tanganyika, with numerous transverse deltas more significant in Lake Malawi. Most sediment is subsequently redeposited into the deeper water by mass and turbidity flows. Geophysical studies indicate the occurrence of Quaternary lowstand sublacustrine fans offshore from the modern fan deltas with channel and levee deposits. In Lake Malawi the southern lake floor sediments are predominantly fine-grained hemipelagic muds, diatom oozes and near-surface Fe-oolite deposits, the latter probably forming during a lowstand when the lake was well mixed. There is evidence from both lakes that major ($> 150$ m) rapid ($\sim 350$ yr) fluctuations in lake level have occurred in the Holocene. Lowstand lake deposits might resemble those developed in the shallower saline lakes of the eastern and northern rifts (Fig. 14.14).

Lake Turkana (northern Kenya; Fig. 14.13) is in notable contrast to the deep, permanently stratified rift lakes like Malawi and Tanganyika. It has an area of about 5000 $km^2$ and an average depth of around 35 m. The climate here is hotter and more arid than in the western rifts. The lake is saline (2.5‰), alkaline (pH 9.2) and oxidizing (70–100% saturation). It is well-mixed both chemically and physically due to a combination of the shallow depth and the high degree of wave-induced turbulent mixing. Detrital sediment enters from lateral and axial drainages (chiefly the Omo River in the north) to form bottom-hugging expanding plumes during spectacular flood discharges. The lake deltas show well-developed distributary, floodplain, lagoonal and beach environments. Earlier Holocene beach levels are recorded as high as 80 m above present lake level. The varve-like sediments found offshore contain very little organic material owing to a high rate of influx of clastic sediments from the Omo River delta and the well-mixed and oxidizing lake conditions. The fine-grained lake bottom muds contain much montmorillonite that is believed to have neoformed in the Mg-rich interstitial waters. Some contribution to the high alkalinity and $Mg^2$ concentrations in the lake may come from submarine hydrothermal springs associated with shallow level magmatic intrusions. Lake Bogoria in the Kenya rift valley also has substantial hydrothermal water input and is noteworthy because, even though quite shallow ($< 12$ m), it is meromictic.

## 14.10 Lake Baikal

Baikal is the world's largest-volume (23 000 $km^3$) and deepest (1640 m) lake. It is located in the Baikal rift within numerous segmented half-grabens bounded by major crustal-scale normal faults similar in magnitude and extent to those of East Africa (Fig. 14.15). The Lake Baikal Depression itself comprises three down-to-the-west, *en-echelon* basins about 350 km long and 35–40 km wide. The oligotrophic (nutrient-poor) lake-waters are so deep that normal dimictic circulation and overturn processes cannot operate because the hydrostatic pressure gradient prevents vertical mixing of 4 °C cooled surface waters below about 300 m. Deep mixing and aeration are by a combination of coastal-forced downwelling and thermobaric instability (directly observed) and probably by hyperpycnal turbidity current intrusions (inferred). The thickness of the sedimentary deposits under the modern lake may be $>7$ km. High-resolution seismic-reflection and coring reveals deposition in three broad sedimentary environments: (i) turbidite depositional systems, by far the most widespread, characterizing most of the margins and floors of the main basins of the lake; (ii) large deltas of major drainages; and (iii) tectonically or topographically isolated ridges and banks. Holocene sedimentation rates based on radiocarbon ages vary by more than an order of magnitude among these environments, from less than about 0.03 mm/yr on ridges and banks to more than about 0.3 mm/yr on basin floors. Extrapolating these rates, with a correction for compaction, yields tentative estimates of about 25 and 11 Ma for the inception of rifting in the Central and North basins, respectively, and less than 6 Ma for the 200-m sediment depth on Academician Ridge. Clastic sediment enters the delta from drainage catchments in the eastern hanging wall, like that of the Selenga delta. This has the distinctive form of a classic prograding Gilbert-type delta (Chapter 17), but its history appears to represent a complex combination of tectonism and sedimentation. The central part of the delta is underlain by prograding, shallow-water sequences, now several hundred metres below the lake surface. These deposits and much of the delta slope are mantled by fine-grained, deep-water, hemipelagic deposits whose base is estimated to be about 650 000 years old. Modern coarse-grained sediment bypasses the delta slope through fault-controlled canyons that feed large,

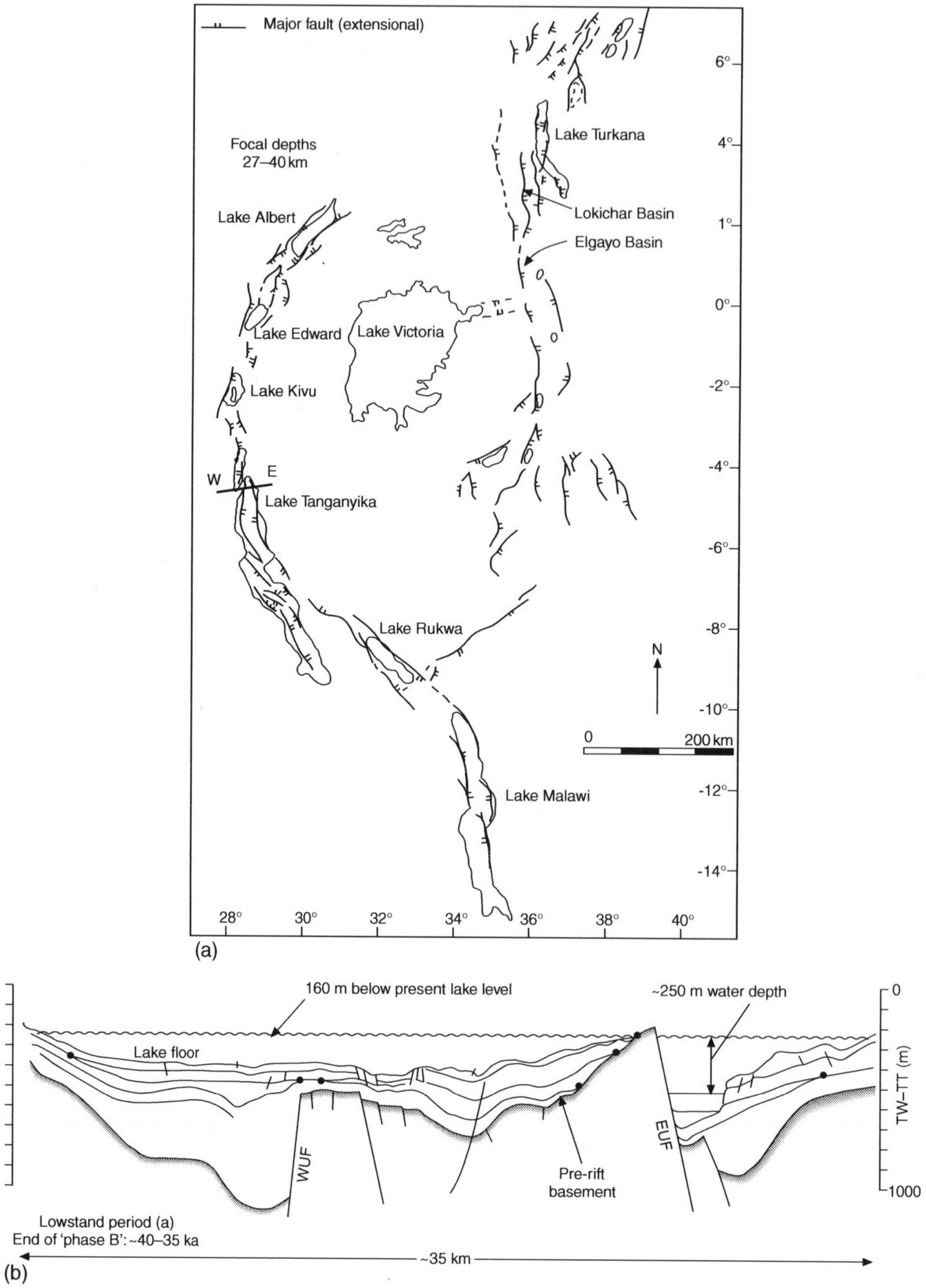


**Fig. 14.13** (a) Regional tectonic map of the East African rift system and (b) a west–east cross-section of the northern Lake Tanganyika rift to illustrate typical asymmetric lake bottom slopes produced by intrabasinal footwall uplift and hangingwall subsidence. Lake level drawn for a −160 m lake lowstand about 35 ka. EUF, WUF—Eastern and Western Ubwar faults; TW-TT, two-way travel. Lake-floor sediments are about maximum 300 m thick extending back to about 1 Ma. (After Lezzar *et al.*, 1996.)

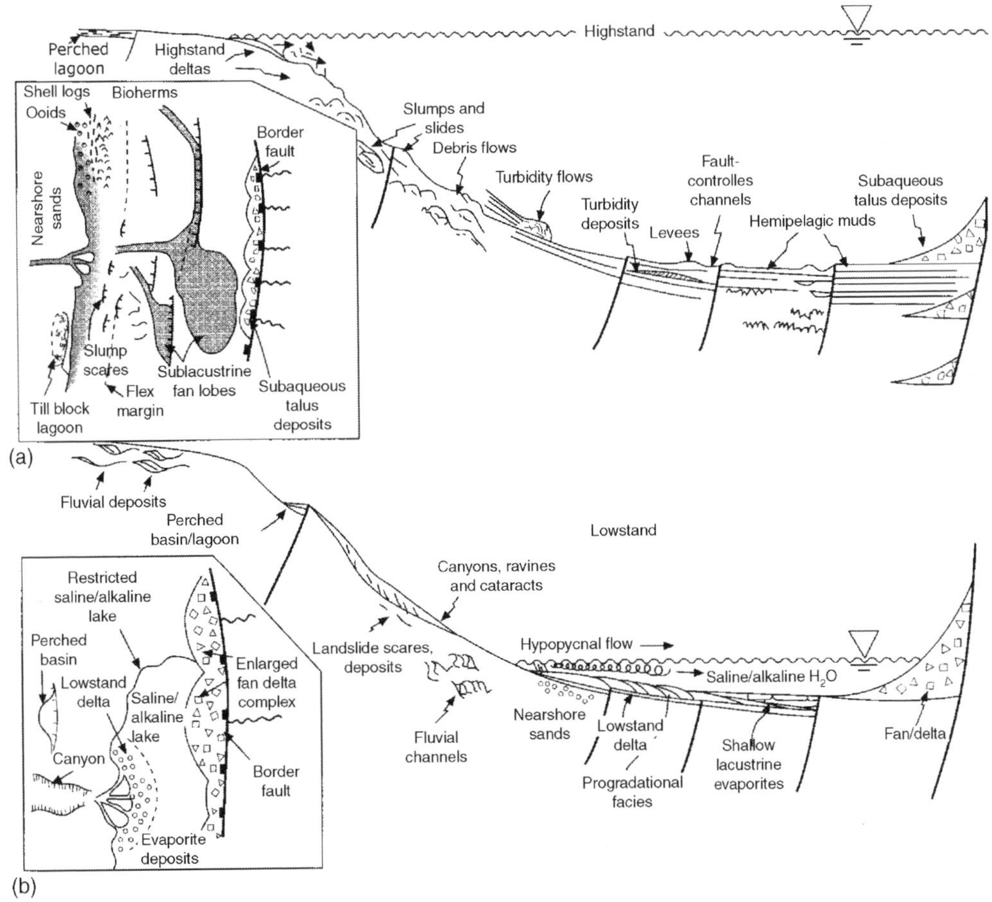


**Fig. 14.14** Sections to illustrate (a) highstand and (b) lowstand deposits of 'deep' rift lakes. (After Scholz *et al.*, 1990.)

subaqueous fans at the ends of the South and Central basins. These relations, along with abundant other evidence of recent faulting and the great depths of the Central and South basins, suggest that these two rift basins have experienced a period of unusually rapid subsidence over the past 650 000 years. Many syndepositional faults cut the offshore deltaic sediments and form structural ponding barriers to downslope turbidity flow. Shallow structural highs like Academician Ridge received little clastic sediment and preserve laminated biogenic sediment rich in diatom remains. Up to 60% of the bottom sediment here is made up of diatoms, with marked decreases observed during periods of full-glacial climate. An exciting discovery has been the location of submarine hot-spring vents in the deep lake floor with thriving communities of organisms.

## 14.11 The succession of facies as lakes evolve

To the student of modern lakes it seems inevitable that lakes should evolve through time (*lake ontogeny*) as they infill from their margins. Such changes might

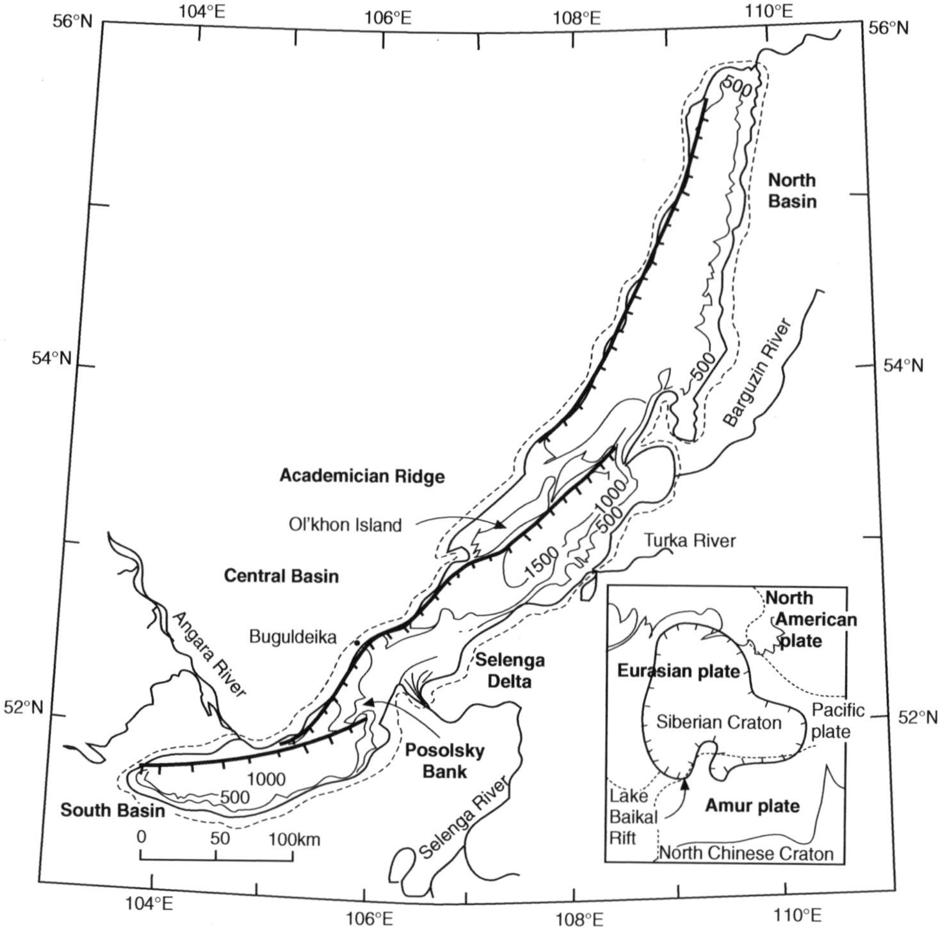


**Fig. 14.15** Bathymetry, inflowing rivers and the three major, left-stepping normal faults that border Lake Baikal. Note ridges and highs located on stepover zones between faults and major river inflow from the southern hangingwall uplands. (After Scholz *et al.*, 1993.)

involve the succession of facies laid down in stratified freshwater lakes, from the products of the lake hypolimnion to the epilimnion, and hence to the clastic input sites at the lake margins, and finally to the products of wetlands and bogs around the lake margins. At its crudest, we might expect a coarsening-upwards trend in grain size and a trend from algal-dominant to higher-plant-dominated organic facies. This sedimentary trend might also record the geochemical change from relatively high-pH to low-pH conditions as the alkaline hypolimnion gives way to the acidic groundwater conditions afforded by encroaching freshwater *Sphagnum* peats. Saline lakes will show a characteristic infill pattern, with subaqueous evaporites passing upwards into particulate forms with wind reworking and dissolution surfaces. However, despite such lake-infill trends, the sedimentologist is aware that changing climate and rates of

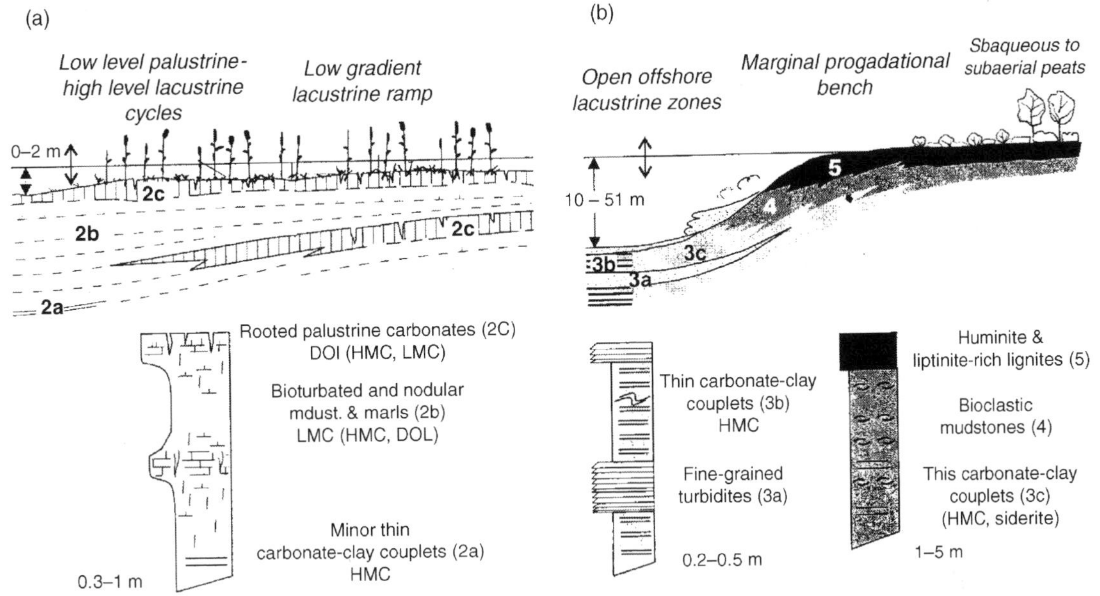


**Fig. 14.16** Contrasting shallow and deeper lake sedimentary environments.(a) Shallow lacustrine-dominated ramp model for deposition of the cyclically arranged mudstones and carbonates. (b) Depositional model of the deeper, lacustrine-dominated evolutionary stage. Both models display the ideal sequential arrangement of the most widespread facies, which are displayed here with their main mineralogical characteristics. (After Saez & Cabrera, 2002.)

subsidence play a vital longer term role in determining whether a lake body expands or contracts. Some lakes in tectonic depressions show lake expansion and deepening trends with time due to increased extent and rate of subsidence.

Climatic fluctuations exert a medium to short time control ($10^3$–$10^5$ yr) upon lake level, enabling sedimentary cycles of various types to evolve within the framework provided by lake infill and crustal subsidence (Fig. 14.16); the concepts of sequence stratigraphy are thus readily applied to lakes. The competing claims of 'climate vs. tectonic' mechanisms for sedimentary sequences need careful sorting in the light of field evidence. Good examples are provided from sedimentary sequences of so-called 'pluvial' lakes in many areas (Fig. 14.17). Here major and often very rapid fluctuations in lake level have occurred during and since the last glacial cycle, witness the large numbers of ancient lakes and swamps that were widely distributed over the North African deserts from about 10 to 5 kyr BP and at various previous times in the repeated cycles of Quaternary climate changes. The chronology of lake shrinkage and expansion established from lake sediment cores is vital in efforts to establish local, regional or global controls upon climate recorded by such events. The development of sequence boundaries and highstand progradational parasequences in lake margin fan deltas also provide a neat way of assessing past lake levels. Good examples are provided from the margins of Lake Lisan, the Pleistocene precursor to the present-day saline Dead Sea.

## 14.12 Ancient lake facies

It has been forcefully pointed out that although sedimentary and limnological processes in modern lakes seem rather complicated, the sedimentary record of lakes is relatively simple, comprising three basic environments, termed fluviolacustrine margin, fluctuating deeper water and evaporitic. Although climate is often cited as *the* primary control on lake environ-

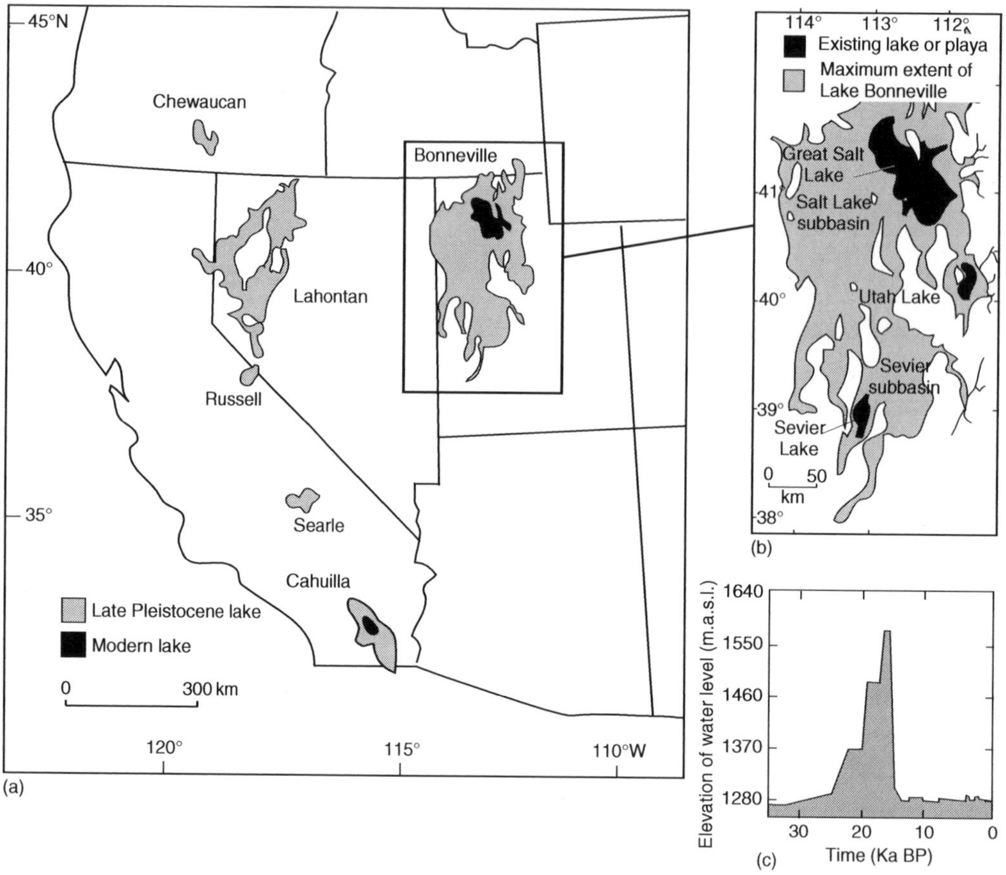


**Fig. 14.17** (a) Map to show the enormous extent of pluvial lakes in the western USA at the time of the Last Glacial Maximum. (b) The extent of Lake Bonneville. (c) Variation of Bonneville lake-level elevation with time. (After Sack, 1994; Benson & Thompson, 1987; Spencer *et al.*, 1984.)

ments, some modern lake characteristics do show little correlation with measures of climatic humidity. On the other hand, lake depth is a fundamental lake paradigm that controls many physical, biological and chemical processes. The key here is the climate/tectonic control on water balance achieved through lake basin outflow and evaporation, providing a useful classification of lake basins into *overfilled* to spill-point, *equilibrium balanced-fill* and *underfilled* due to evaporation. Both tectonics and climate obviously influence the nature of any particular lake basin through time.

## Triassic Lakes Newark (USA) and Ischigualasto–Villa Unión (northwest Argentina)

A 5 km thick succession of numerous lacustrine cycles deposited over some 25 Myr occurs in the Newark graben, northeast USA. The lacustrine facies overlie alluvial fan and fluviatile sequences derived from both the hangingwall and footwall uplands surrounding the lake. The lacustrine sediments usually comprise a threefold division. Lake transgressive sands are succeeded by highstand microlaminated black shales with fish faunas deposited in the hypolimnion of a relatively deep (?70–100 m) stratified lake. The majority of the rest of the cycle comprises mudrocks with occasional desiccation cracks and animal tracks indicative of periodic emergence during falling stage and lowstand. The characteristic cycles are ascribed to basinwide rises and falls of lake level at a periodicity appropriate to the basic precessional Milankovitch timescale of 21 kyr. The presumed precessional cycles are further arranged in compound cycles thought to represent the longer eccentricity timescales ~100 and ~400 kyr. The youngest rocks present in the rifts indicate a long-term, probably climatically driven, trend towards increased aridity, for they contain numerous cycles indicative of periodic playa lake development.

Contemporary with Lake Newark is the Ischigualasto–Villa Unión Lake Basin of northwest Argentina, a half-graben with early stages marked by stacked parasequences bound by lacustrine flooding surfaces related to climatically induced lake-level fluctuations superimposed on variable rates of subsidence on the controlling rift-border fault zone. The youngest sequences shows a switch in sediment supply related to reduced footwall uplift, the possible presence of a relay ramp and/or supply from a captured antecedent drainage network. (Fig. 14.18).

### Devonian Lake Orcadia

Lake sediments make up much of the Middle Devonian Old Red Sandstone in the post-orogenic extensional Orcadian basin of northeastern Scotland. The main Caithness Flagstone Group consists of many sedimentary cycles containing laminites. Some of these show 0.1–1 mm laminae in doublets and triplets up to 10 mm thick of alternating micritic carbonate, organic-rich and/or clastic siltstones, sometimes with phosphates. Where carbonate laminae are dominant, the organic laminae are reduced to 0.1 mm streaks, as in the famous fish-bearing Achanarras Limestone. Non-carbonate laminites show alternations of siltstone and mudstone or of coarse and fine siltstones. Subaqueous and subaerial shrinkage-crack casts are abundantly present on the bases of many coarse laminae. The tops of thicker laminae show preserved symmetrical wave-formed ripples. Thicker, sharp-based and sometimes erosive sandstones cap coarsening-upwards cycles involving the above lithologies. There is a general upward trend from laminites to coarser sandstone lithologies. All these features are readily interpreted as due to the interplay between a large stratified lake and streams draining into the lake margin. The fluvial facies that increasingly dominate the succession in Caithness indicate a major fluviolacustrine regression with time. The lake basin was probably permanently stratified, never evaporitic, and to judge from the regional tectonics, probably intermontane in the fashion of Lake Titicaca or certain southern Tibetan lakes like Nam Co. Basinal subsidence was certainly rapid, since over 5 km of sediment accumulated in only 10 Myr.

### Eocene Lakes Gosuite and Uinta

The famous Green River Formation (maximum thickness 950 m) of Wyoming, Utah and Colorado, USA, is one of the most-studied sequences of lacustrine deposits in the world. The basins in which the Formation developed formed part of the 'broken foreland' basins (Chapters 10 & 22) that evolved during the Sevier orogeny in the late Cretaceous to early Tertiary. It provides an interesting contrast with the Orcadian example noted above. The formation contains the world's largest reserves of trona ($Na_2CO_3$) and its oil shale facies are potentially the world's largest single hydrocarbon reserve. Detailed study of the Wilkins Peak and Laney Members in Wyoming reveals sedimentary facies arranged in cycles and thought to have been deposited in a spectrum of lacustrine environments ranging from central playa and freshwater lakes to fringing alluvial fans (Fig. 14.19). Thus dolomitic flat-pebble conglomerates are thought to have been formed by strandline reworking of algally bound sediments subjected to desiccation shrinkage. Wave-rippled and cross-laminated sandstones and desiccated mudrocks are thought to be of lake shoreline origin, with desiccated mudrocks with siltstone laminae representing playa mudflats. The oil shale lithologies comprise organic-rich dolomitic laminites and oil shale breccias, with common desiccation cracks. The organic laminae were probably gelatinous algal–fungal oozes that accumulated on a shallow, periodically exposed lake floor. High rates of organic productivity are inferred, with the playa flats acting as efficient sediment traps preventing the introduction of much clastic material. Bedded trona deposits contain thin partings of dolomitic mudstone interpreted to result from increased evaporative concen-

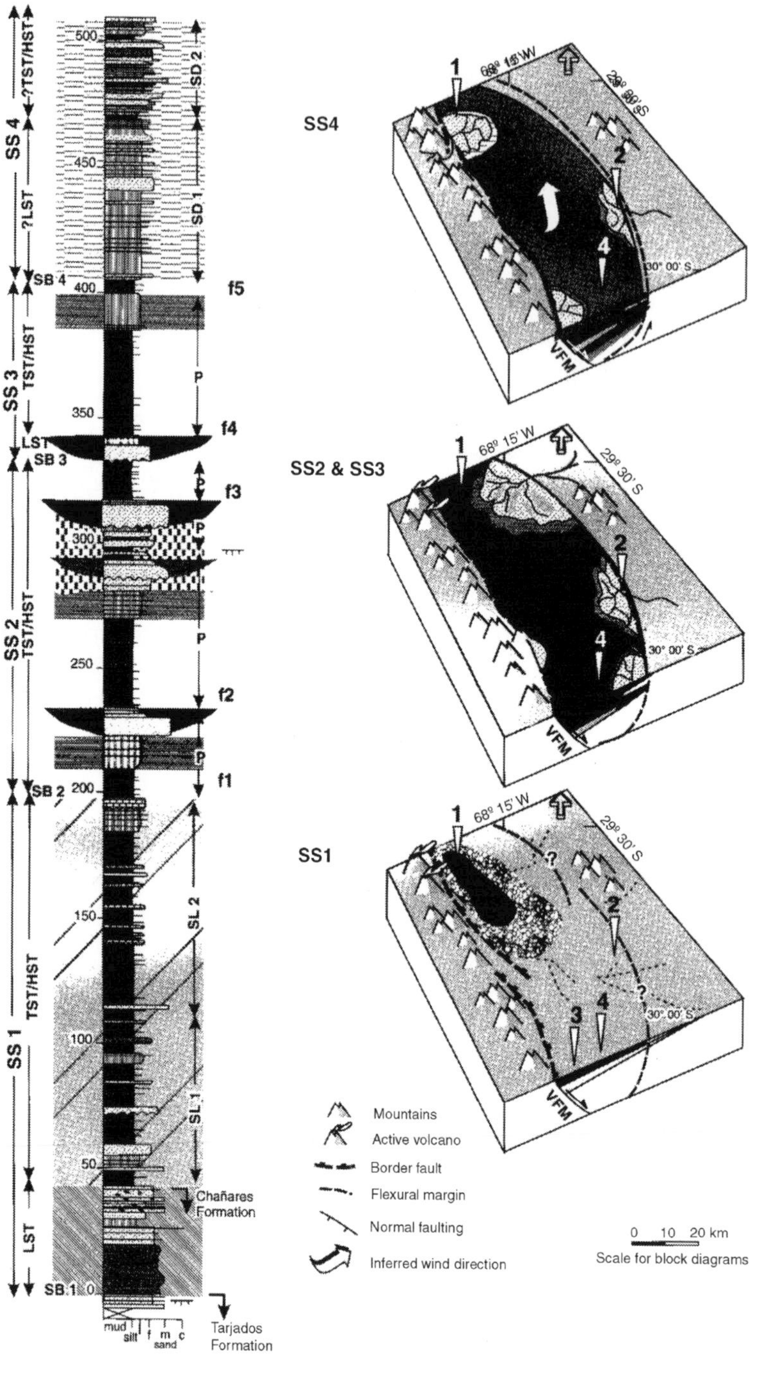


**Fig. 14.18** Evolution of Triassic palaeolLake Ischigualasto/Villa Unión, Argentina. (a) Sequence stratigraphy superimposed on summary lithological succession. (b) Block diagrams illustrating the palaeoenvironmental and palaeogeographical evolution of the studied lacustrine basin fill. LST, lowstand system tract; TST, transgressive system tract; HST, highstand system tract; SB, sequence, boundary; f, lacustrine flooding surface; P, parasequence. (After Melchor, 2007.)

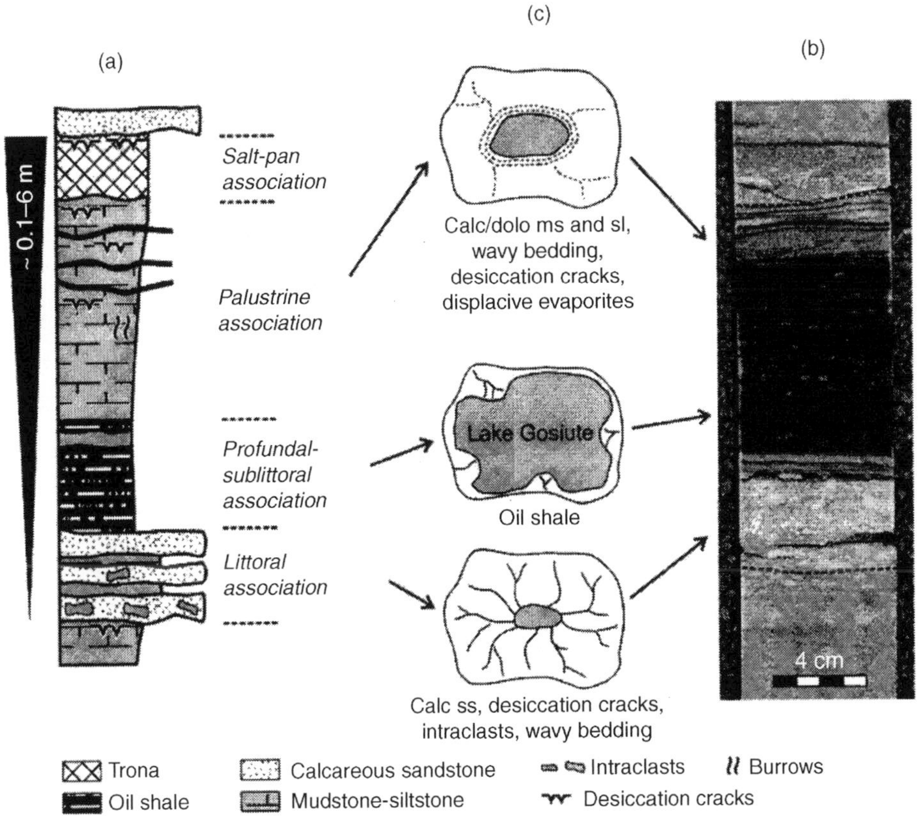


**Fig. 14.19** (a) Diagram of a typical Lake Gosuite expansion–contraction sediment cycle in the Wilkins Peak Member, highlighting common sedimentary structures and the vertical distribution of lithofacies associations. (b) Photo of one complete cycle in a sliced rock core. (c) Sketch palaeogeographies. (All after Pietras & Carroll, 2006.)

tration of the playa lake that gave rise to the oil shale facies. The Wilkins Peak cycles have flora and fauna recording deposition under a subtropical, humid climate, but with major changes in the water balance causing alternating periods of lake expansion and contraction (Fig. 14.19). According to recent accurate radiometric dates on volcanic ash from within the sequence the cycles may be too short to be 'standard' precession-controlled events. Bioturbated calcareous mudstones present in the Laney Member contain freshwater molluscs and ostracods. There is an upward trend in the Laney Member cycles from playa to shallow freshwater lake environments. Cross-stratified sandstones with frequent channel forms are indicative of alluvial fan and braidplain environments with Gilbert delta facies recorded occasionally in the Laney Member. The clastic facies intertongue with lacustrine facies and are sourced from syntectonic highs within and adjacent to the lake basins.

### Tertiary Lake Madrid

The Tertiary deposits of the Madrid Basin, Spain, have well-developed lacustrine facies. In particular, spectacular upward transitions occur from well-drained lake margin alluvial plains, with dolomitized calcisols, to shallow evaporating carbonate pond (*paludal*) deposits with gypsum crusts (Fig. 14.20).

Fig. 14.20 Close-up of the transition from palaeosols to pond facies in the Tertiary of the Madrid Basin. The lower section shows pedified mudrock with dolomitized calcisols. The upper section features ledges of more resistant weathering gypsum crusts. (Photo courtesy of J.P. Calvo.)

## Further reading

### *General*

Many papers of interest for both modern and ancient lake studies are to be found in the volumes edited by Matter & Tucker (1978), Lerman (1978), Fleet *et al.* (1988), Renaut & Last (1994) and Rosen (1994). Talbot & Allen (1996) is a well-balanced review of lake processes and sedimentary products. A fundamental reference on the physics and chemistry of lakes is Hutchinson (1957), and an excellent general text on limnology is Wetzel (1983). East Africa lakes are described in Talling & Talling (1965), and Amazonia in Rai & Hill (1980). Volcano caldera lakes are considered by Sáez *et al.* (2007).

### *Specific*

Lake varves are studied by Sturm & Matter (1978), Pickrill & Irwin (1983), Bréhéret *et al.* (2008); underflows by Weirich (1986), Chikita *et al.* (1996), Wunderlich (1971) and Sturm & Matter (1978); contourites by Zavala *et al.* (2006); wind shear by Csanady (1978) and Hawley & Lee (1999); mean gyre-like circulations in the Great Lakes by Beletsky *et al.* (1999); so-called lake contourite drifts by Gilli *et al.* (2005); internal-wave seiches by Wiegand & Chamberlain (1987) and Gloor *et al.* (1994); resuspension by internal-wave seiches by Horppila & Niemistö (2008).

The chemistry of lake-water is studied by Neev & Emery (1967), Neev (1978), Tuschall & Brezonik (1980), Hardie *et al.* (1978), Kelts & Hsu (1978), Drummond *et al.* (1996), Calvo *et al.* (1989), Platt (1989), Alonso-Zarza *et al.* (1992), Newton (1994), Colson & Cojan (1996), Rosen (1994), Li *et al.* (1997), Csato *et al.* (1997) and Yechieli *et al.* (1998); biological processes and cycles by Kelts & Hsu (1978), Dickman (1985), Colman *et al.* (1995) and Blas *et al.* (2000); rhizoliths by Liutkus & Wright (2008); charophyte stable isotopes by Andrews *et al.* (2006).

Modern temperate lakes and their sedimentary facies studied by Sturm & Matter (1978), Kelts & Hsu (1978); lakes in the East African rifts by Owen *et al.* (1982), Ebinger *et al.* (1987), Scholz *et al.* (1990), Scholz & Finney (1994), Johnson *et al.* (1995), Lezzar *et al.* (1996), Cohen *et al.* (1997), Yuretich (1979), Yuretich & Cerling (1983), Frostick & Reid (1986) and Renaut & Tiercelin (1994); Lake Baikal by Crane *et al.* (1991), Hutchinson *et al.* (1992), Scholz *et al.* (1993), Craig (1994), Colman *et al.* (1995, 2003) and Baikal Drilling Project Members (1997), with deep-water renewal by Schmid *et al.* (2008).

Saline lakes and playas are studied by Shearman (1970) Arthurton (1973), Warren (1982), Blas *et al.* (2000), Orti *et al.* (2003) and Bréhéret *et al.* (2008), with Laguna Mar Chiquita by Piovano *et al.* (2002), and acid saline lakes as interesting analogues to Martian surface hydrology by Benison *et al.* (2007)

Changes to Holocene Great Lakes water levels are considered by Lewis *et al.* (2008a); Pleistocene pluvial Lake Bonneville in Utah by Dean *et al.* (2002); Pleistocene Lake Lisan by Haase-Schram *et al.* (2004) and Bartov *et al.* (2007); desert erg margin lakes formed during pluvial periods by Lewis *et al.* (2008b).

Lake sediment architecture is considered by Kutzbach & Street-Perrot (1985), Gasse *et al.* (1990), Oviatt *et al.* (1994), Drummond *et al.* (1996), Cohen *et al.* (1997), Seltzer *et al.* (1998) and Bréhéret *et al.* (2008).

Increasing subsidence causing lake expansion through time is considered by Saez & Cabrera (2002), and concepts to simplify lake classifications and relate them to both changing tectonics and climate by Carroll & Bohacs (1999).

Ancient lake facies: Newark Supergroup by Van Houten (1962, 1964), Olsen (1986), Schlische & Olsen (1990), Schlische (1992), El-Tabakh *et al.* (1997); Ischigualasto–Villa Unión Lake Basin by Melchore (2007); Devonian Lake Orcadia by Trewin (1986); Green River Formation by Eugster & Hardie (1975), Surdam & Stanley (1979), with radiometric dates and lake cycle duration by Pietras & Carroll (2006) and sequence stratigraphic study with insight into lake evolution by Keighley *et al.* (2003); Eocene Lake Flagstaff, Utah by Bowen *et al.* (2008); Tertiary Lake Madrid by Sanz *et al.* (1995).

# Chapter 15

# ICE

*...and then the earth namely the air and then the earth in the great cold the great dark the air and the earth abode of stones in the great cold alas alas in the year of their Lord six hundred and something the air the earth the sea the earth abode of stones in the great deeps the great cold on sea on land and in the air...*

Samuel Beckett [from Lucky's soliloquy], 'Waiting for Godot', 1953, *Complete Dramatic Works*, Faber

## 15.1 Introduction

Some 10% of the Earth's surface area is covered by ice, representing about 80% of surface freshwater. A further area of 20% is affected by permafrost. This 30% of frigid Earth surface is our *cryosphere*, an environment quite unlike any other. Of the ice-lands, the East and West Antarctic ice-sheets have about 86% by area, containing enough water volume to give ~65 m and 5.5 m of sea-level change respectively, whilst Greenland has about 11% by area and a potential 7.3 m of sea-level change. The many other valley and piedmont glaciers issuing from mountain belts make up the remaining 3% by area. Yet earlier in Quaternary times a staggering 30% of the Earth's surface was ice-covered—vast areas of North America and Europe by two giants, the Laurentide and Scandinavian ice-sheets and elsewhere by extensions of extant ice-sheets and and even larger areas subject to permafrost. At least 11 epochs during Earth history had very extensive and long-lasting *icehouse-mode* climates, the longest and best-documented being the Gondwanan Permo-Carboniferous glaciation (Fig. 15.1).

As discussed in Chapter 10, periodicity in glaciation is best explained by Milankovitch mechanisms. However, the Milankovitch effect is small and cannot trigger an Ice Age. Even in our present interglacial conditions, the Earth's oceanic heat engine is thoroughly dominated by the exchange of heat from deep, cold, polar water masses to warm surface currents that feed air masses with latent heat from condensation. It is these warm winds, not subtle radiation changes, that prevent temperate latitudes from developing permanent ice-cover. The presence of continents over the polar regions and a deep oceanic meridional circulation (Chapter 21) are obviously a major prerequisite for glaciation, variations in solar energy input and reflection/trapping efficiency being secondary controls that modulate the primary cause.

The major environments of glacier ice are:

- *Ice-sheets* and their *outlet glaciers*—amalgamated and relatively fast-moving *ice-streams* and coastal *ice-shelves*. The Greenland ice rests on elevated Precambrian craton above sea level and reaches the ocean via ~30 main outlet glaciers that have cut fjords in coastal mountains. The West Antarctic ice-sheet rests on extended continental crust below sea level and is buttressed and stabilized by the large Ross and Ronne ice shelves. It is divided into five large ice streams and with its base below sea level,

*Sedimentology and Sedimentary Basins: From Turbulence to Tectonics*, 2nd edition. © Mike Leeder.
Published 2011 by Blackwell Publishing Ltd.

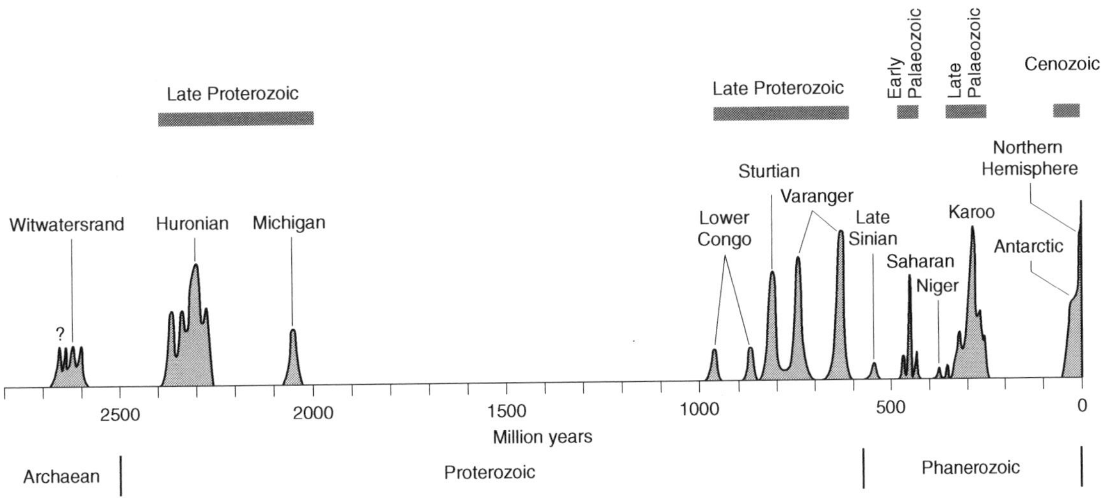


**Fig. 15.1** Time series for glaciations and their approximate magnitude during Earth's history. (After Hambrey, 1994.)

becoming deeper inland, it is most susceptible to ocean warming and loss of the shelves. The East Antarctic ice-sheet has ice thickness >4 km and primarily rests above sea level on thick continental crust. It is drained by outlet glaciers cut through the Transantarctic Mountains, large ice streams draining into the Amery and Filchner ice shelves and numerous smaller glaciers. A variety of techniques, including surface elevation (radar and most recently by ICESat laser altimetry), changing gravity fields and flux estimates (Fig. 15.2) indicate that many outlet glaciers of Greenland and Antarctic ice-caps are accelerating and losing mass at startling rates. Southeast and northwest Greenland glaciers flowing >100 m/yr at rates of ice thickness reduction of 0.84 m/yr, and coastal West Antarctica at rates of >9 m/yr. Dynamic thinning has also propagated far into the interior of each ice-sheet as ice-shelves thin by ocean-driven melt.

- *Valley glaciers* (Plate 12) and their marine outlets called *tidewater glaciers*.
- *Piedmont glaciers*—divergent, fan-like ice masses formed after a valley glacier becomes unconfined.

In all glaciers a dynamic equilibrium exists between snow input, ice formation, ice motion and terminal melting. Depositional products are dominated by a variety of subglacial sediments and by meltout products of glacial outwash plains on land and glacimarine and glacilacustrine environments under water. The preservation potential of glacial environments will generally be low unless the depositional and erosional features are produced in areas of subsidence. Further, problems of chronology still hinder interpretation of even classic areas of the onshore Pleistocene glacial sedimentary record, notably the remnant glacial deposits around the margins of the former northwest European extension of the Scandinavian/Scottish ice-domes. For this reason the depositional products of the glacimarine environments on rifting or thermally subsiding margins, with their wide ring of ice-shelves and calving icebergs, are probably the most likely to be preserved and yield evidence for the origin and evolution of polar glaciations through time. In many ways the sedimentary record of this exciting aspect of Cenozoic Earth history is more useful than that of the more obvious record from ice-cores themselves; the record is far longer and more direct in many ways. Unfortunately these glacimarine environments are also the least known; recent offshore geophysical remote sensing and drilling is gradually righting the balance and this chapter will attempt to draw attention to such developments.

## 15.2 Physical processes of ice flow

Ice as a sediment transporting system is poorly understood and so radically different from any other we have hitherto encountered in this book (it is difficult to do flume experiments with ice!) that a short discussion of the basic principles seems necessary. In the interior

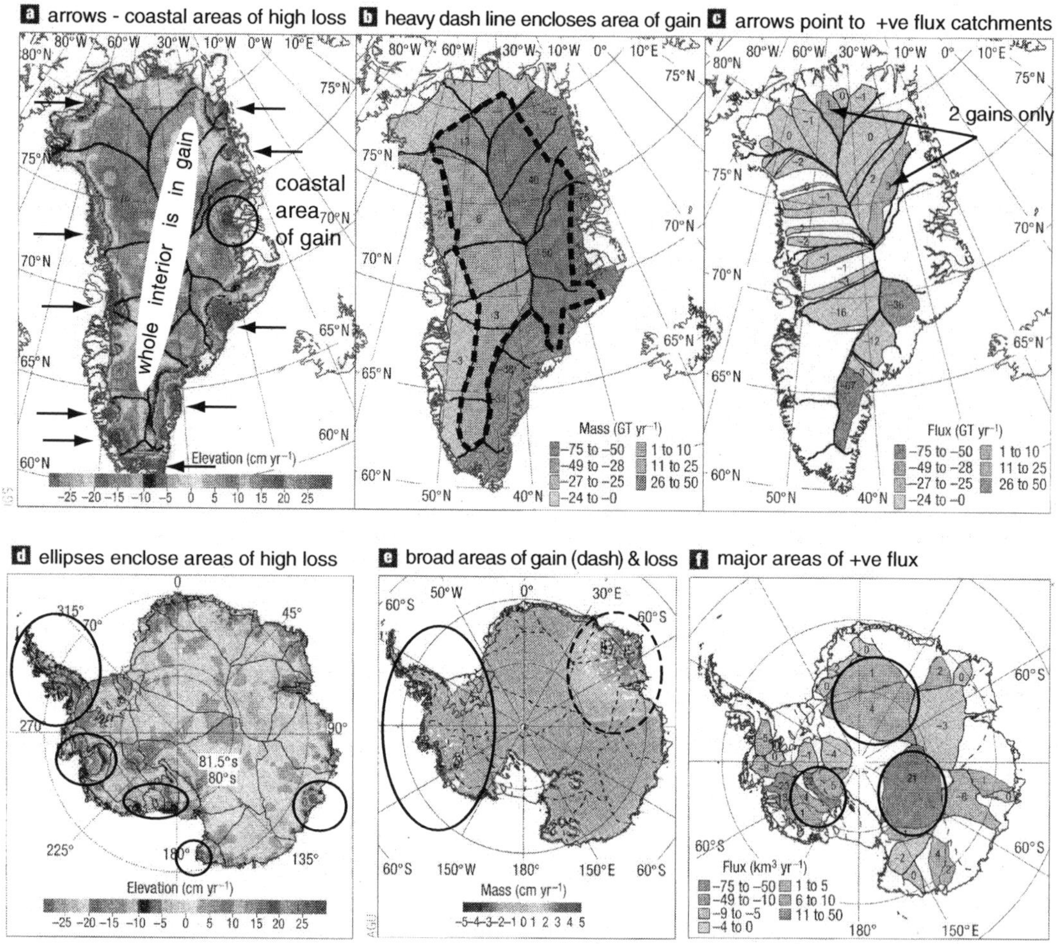


**Fig. 15.2** Estimates of ice-sheet mass balance based on surface elevation, changing gravity fields and flux estimates for Greenland and Antarctica. (After Bell, 2008, and references cited therein; see also the most recent ICESat data in Pritchard *et al.*, 2009). (a) Greenland surface elevation changes in cm/yr. (b) Greenland catchment-based estimate of mass change from gravity data. (c) Greenland flux estimates from space geodesy. (d) Antarctic surface elevation changes in cm/yr. (e) Antarctic mass changes from gravity data. (f) Antarctic flux estimates from space geodesy.

of ice-sheets we are dealing with very slow (a few m/yr) bulk sliding by creep of the crystalline solid phase due to its own body force under the influence of gravity. The direction of movement is set in response to regional pressure gradients caused by the 3D distribution of ice mass in the case of ice-sheets or bedrock slope in the case of descending valley glaciers. The radial flow rate of a mound-like ice-sheet increases with a rate dependent function and spatial pressure gradient depends upon ice thinning towards its circumferential terminus. The ice-sheet pays little attention to local or even regional bedrock relief, i.e. ice-sheets may commonly move 'uphill' with respect to underlying bedrock. Valley glaciers, on the other hand, move downvalley in response to the downslope component of gravitational force acting on the ice mass. The mean surface laminar flow of glacier ice is usually ~10–200 m/yr for valley glaciers, ~200–1400 m/yr for terminal ice-streams and up to ~12 000 m/yr for large outlet glaciers. Although usually slow and steady, spectacular glacier surges occur periodically when ice velocity increases by an order of magnitude and more.

Ice is composed of an aggregate of roughly equigranular crystals whose crystal size increases with time and/or depth. When stressed by burial in a glacier or insertion in a test rig in the laboratory, each crystal deforms easiest internally along glide planes parallel to the basal planes of the hexagonal crystal lattice. Since crystals in ice are not usually aligned along common axes, the polycrystalline aggregate rearranges, and recrystallization takes place during strain or flow. Natural ice crystals in the actively deforming layer of a glacier also contain a myriad of gaseous, liquid and solid impurities and inclusions. It is not surprising therefore that natural ice deformation is rather complex, with time-dependent behaviour seen during the application of continued stress. Primary, secondary and tertiary creep regimes may be identified, with secondary creep (a sort of steady state) dominant in glaciers which are usually responding to load and slope-induced stresses in the range 2–10 kPa. For applied stresses of this magnitude in the laboratory, the shear strain rate of secondarily creeping glacier ice is given by Glen's law, $\dot{e} = k\tau^n$, where $n$ is an exponent ranging between 1.5 and 4.2, $k$ is an experimental constant and $\tau$ is the shear stress. (Compare this to the Newtonian law for fluids when n = 1 and $1/k$ is the molecular viscosity.) It has been stated that $n$ is most reliably estimated as 3, whilst $k$ is partly a temperature-dependent (Arrhenius) function controlled by the energy required to activate creep, a function of crystal size, shape and inclusion content. In view of all this complexity, the reader should note dissenting evidence from Antarctic studies where a linear flow law for strain rate, of the form $\dot{e} = k\tau$, with $k \approx 10^{-15}$ s/Pa best fitted borehole deformation data for −28 °C. The shear stress arising from valley glacier ice sliding over a plane inclined bed may be approximately given by the familiar tractive stress equation, $\tau = \rho g h \sin \alpha$, where $\rho$ is the density of ice, $h$ is the thickness of ice and $\alpha$ is the mean valley floor (or ice surface) slope.

## 15.3 Glacier flow, basal lubrication and surges

Our previous discussion of the mechanical flow of ice left out the most vital part of the whole process, namely that of the interaction of the basal ice layer in a real glacier with (i) solid bedrock substrate of variable hydraulic condictivity, (ii) sediment substrate and (iii) glacial waters. It is the triple influence of the three materials that is really the key to the whole process. Two fundamental end-member types of flowing glacial ice (Fig. 15.3) have been proposed:

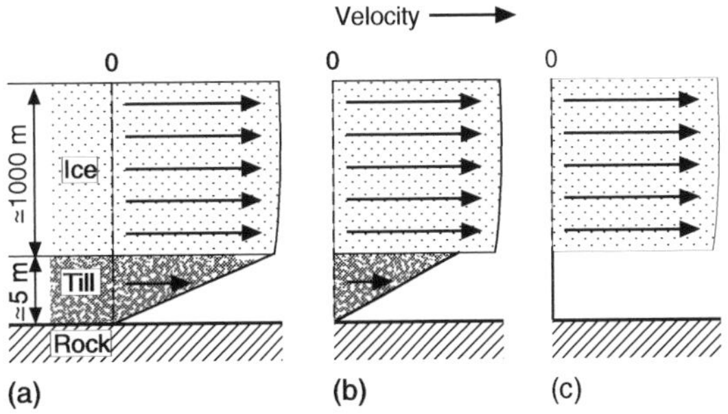


**Fig. 15.3** Velocity profiles possible for sliding glacier ice in contact with a basal till overlying solid bedrock. (a) Predominantly till deformational strain, as proposed for Antarctic ice streams. (b) Till deformation plus basal sliding, appropriate for near-polar glacial termini. (c) Basal ice sliding only, probably rare in nature. (After Alley *et al.*, 1986.)

1 'Cold' and dry ice beds occur when polar ice is relatively thin and therefore lies below its pressure melting point at depth. On Spitsbergen, for example, glaciers less than about 100 m thick are cold-based. In such states, a condition of 'no slip' exists at the ice–bed interface, and there is a general absence of englacial or subglacial drainage. Forward motion of such ice is therefore by internal creep alone. Glacial debris is transported within the ice, with substrate erosion due to plucking and grinding effective only at the summits of protuberances on the bed.
2 'Warm' and wet ice beds lie close to the pressure melting point at the glacier sole (ignoring the conditions throughout the rest of the ice column), and the glacier slides over its bed on a *décollement* or *slip* plane on soft, fluid-rich, highly porous, deformable sediment. On Spitsbergen glaciers >100 m thick are warm-based. At least one ice stream in Antarctica moves with a *stick-slip* sliding mode, releasing seismic energy as it does so, giving sudden spurts of ice motion in the order of metres over a timescale of about a minute.

Most glaciers, including polar ice-streams and temperate valley glaciers, exist with warm beds which, pressure gradients permitting, show quite rapid ice-flow velocities. Remote sensing can reveal both longitudinal and transverse variations in valley glacier surface flow rates (Fig. 15.4). Glacier flow is unsteady and nonuniform on a variety of timescales due mainly to variations in the rate of basal sliding vs. internal ice deformation caused by variation of water content. Slow winter flow occurs because meltwater is in short supply and glacial drainage is minimal. Flow accelerates in spring and summer as more water becomes available, peaking when large supraglacial lakes form, as in Greenland, which may very rapidly drain away to cause surging. Glaciers may also suddenly surge after years of steady slow flow and over a few months move orders of magnitude faster than in preceding and subsequent months. The process suggests that some deformation threshold is crossed. Given the roles that basal fluid pressure and deforming sediment have upon glacier behaviour, it seems likely that changing near-bed water content plays a crucial role. Antarctic ice-sheet streams are located in areas of converging ice flow and have typical flow velocities of thousands of metres per year (Fig. 15.5), an order of magnitude greater than typical velocities of nonstreaming portions of the ice-sheet. Deformability of basal till sediment is witnessed in active and deposited subglacial till by the presence of tight internal shear folds; evidently both ice and basal sediment are in motion, though by the evidence of strain in the till, presumably at different speeds. It is instructive to remember that the weight of kilometers of Antarctic ice is held up by huge pore-fluid pressures in a thin (0.5–6 m) basal deforming layer of till. Radar/seismic surveys establish that the ice–bed interface in sliding regions of the Rutford ice-stream, West Antarctica is much wetter than in regions where the bed is deforming and that the water probably occurs in numerous small water bodies or cavities at the ice–bed interface.

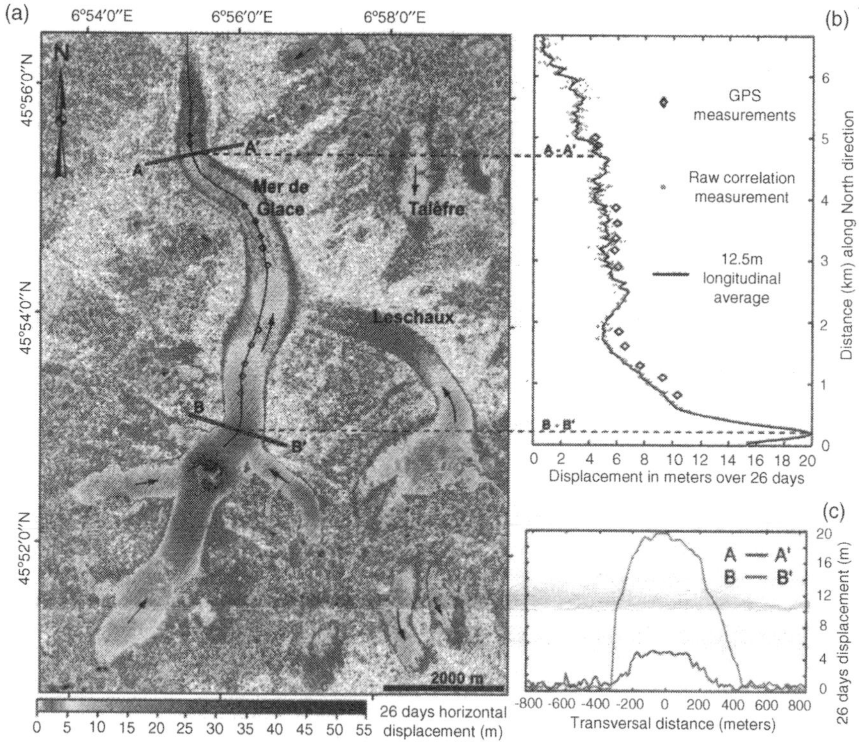


**Fig. 15.4** (a) Valley glacier flow as revealed by 26-day sequential satellite images. (b–c) Flow-wise (to the north at top of image) and span-wise variations respectively in valley-glacier ice velocities as revealed by GPS and satellite images, the former slightly faster as the measuring period was taken slightly earlier in the summer during the 2003 heatwave. (After Leprince *et al.*, 2008.)

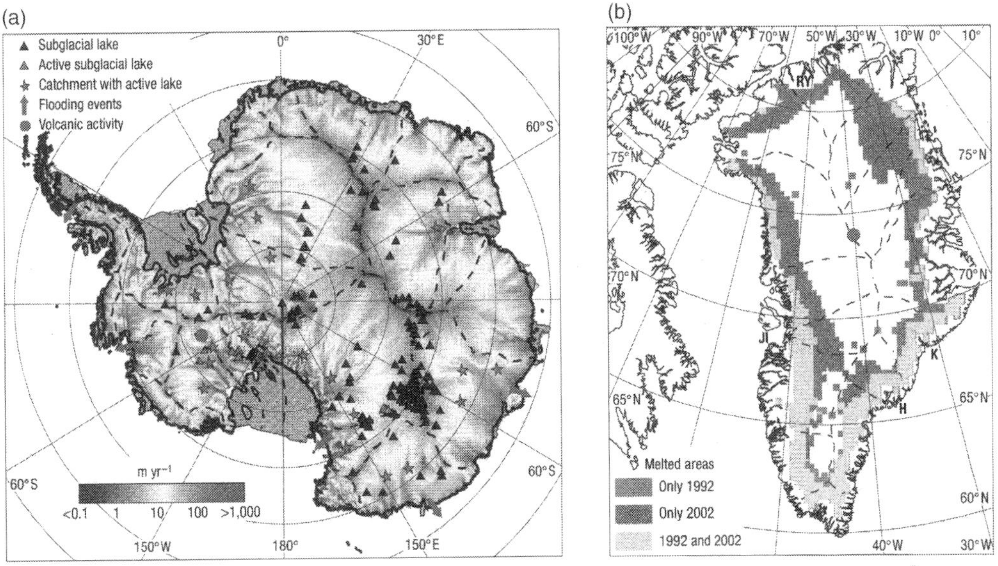


**Fig. 15.5** Distributions of water for the polar ice-sheets. (After Bell, 2008, and references cited therein.) (a) Antarctica, with known subglacial lakes (triangles) on top of ice velocities (see grey scale). Grey triangles are active lakes that transfer water. Arrows record outburst floods. (b) Greenland surface melt extent in 1992 and 2002 overlain on major ice catchment boundaries. The four outlet glaciers Ryder (RY), Jakobshavn Isbrae (JI), Kangerdluqssuaq (K) and Helheim (H) are located. Dots show elevated heatflow.

It seems that up to 90% of total glacier movement must occur by basal sliding, the rest by internal deformation. The concept of effective stress (Section 9.1) is relevant again here, for the shear strength, $\tau_s$, of subglacial sediment must be exceeded by that of the driving bed shear stress for ice flow, $\tau_0$, if deformation is to occur at all. Ignoring cohesive strength, assuming that resistance is due to solid friction ($\phi$) and that strength is much reduced by high porewater pressures, we may write $\tau_0 \geq \tau_s$, or $\rho gh \sin \alpha \geq \Delta P \tan\phi$, where $\Delta P$ is the excess of lithostatic pressure above porewater pressure. The reduced strength allows the driving force provided by the tractive force of the glacier—actually quite small for most glaciers due to the low slopes involved, and about 20 kPa for the Antarctic ice-cap—to cause deformation and steady forward motion. Direct subglacial measurements of rates of till deformation indicate values of 'viscosity' for deforming till of between $3 \times 10^9$ and $3 \times 10^{10}$ Pa s, with yield stresses of about 50–60 kPa. But we know little about the *in situ* properties of deforming till; it may even be likely that normal concepts of yield stress and flow law behaviour apply. In any case, it seems clear that both the glacier ice *and* the deforming subglacial till must move along.

The process of basal sliding must also involve enhanced creep around drag-creating obstacles, pressure melting around obstacles and direct lubrication by abundant basal meltwater. The latter comes from surface summer meltwaters let into the sole of valley glaciers and the Greenland ice-sheet by fracture systems. Under thick insulating polar ice-sheets that have little surface water the basal fluid is due to melting by geothermal heat transfer and ice melted by pressure at the glacier sole. Water flow occurs through both networks of tubular conduits cut into deforming till and in linked cavity fracture networks directly imaged in boreholes and from the evidence of transmissivity tests. The former are high-discharge/low-pressure features whereas the latter are the converse. The evolving fracture networks are generated by internal strain, are in a state of dynamic change as strain evolves locally and largely control permeability and hence the passage of water at all depths once the larger conduits are infilled to their discharge capacity by infill from supraglacial meltwater in subpolar and temperate glaciers. The larger conduits pressurize and drive water into the fracture conduits thus easing friction and increasing ice deformation.

The role of lakes under the East Antarctica ice-sheet on glacier flow, like Lake Vostok and over 140 others so far discovered, has been much investigated in recent years. After the initial discovery of Vostok in the 1990s it was thought that such curious phenomena were isolated systems, but satellite radar, seismic images and modelled flow paths make it clear that the lakes define parts of common subglacial catchment cascades, with periodic subglacial floods transferring water from one lake to another along the drainage paths, just like in any subaerial catchment (Fig 15.5). Satellite imaging of deformation associated with lake contraction and expansion shows that ice-flow velocity increases by an order of magnitude downstream from areas of subglacial lakes in beginning ice-streams like Recovery Glacier, Donning Maud Land. It is thought that lake water freezing onto the ice sole reduces bed friction and adds thermal energy to the advected ice, thus warming it, reducing its effective viscosity and allowing faster flow. It is evident that the outlets of most ice-streams are located above subglacial catchment outlets, a very significant discovery that has many ramifications for the behaviour of ancient ice-sheets. Over time it is likely that the outlets for subglacial drainage may flood and disperse sediment widely under and at the ice-sheet front; one such event has so far been recorded by ICESat in Antarctica, that of the Byrd Glacier in East Antarcica betwen 2005 and 2007. Some authors consider that some Pleistocene fluvioglacial events and drumlin fields may reflect major ice-sheet catchment dewatering events. Flow rates are also likely to be influenced by the removal of buttressed elements like ice-shelves and inland retreat of the grounding line, as seen in the Amundsen Sea sector of the West Antarctica ice-sheet and in eastern Greenland (Fig. 15.6).

The Bering Glacier of Alaska is a well-known example where surges occur quite regularly, every 20 yr or so, the most recent initiated in 1993. As in flooding rivers, surging reflects an imbalance between supply from the gathering area upstream and the geometry of the channel downstream. In addition, resistance to flow is strongly dependent upon water content at the glacier sole. The icy kinematic floodwave that quickly travels through the system causes tectonic thickening, intense crevassing, extreme local velocity gradients, abundant discharges of meltwater and the intense mixing of lateral and medial moraines. As the surge dissipates and the glacier resumes its sluggish phase once more, the advanced snout (though surges do not *always* reach the snout) decays back to its equilibrium position. Distinctive surge moraines then mark the previous maximum position of the glacier front. Surging in Svalbard glaciers is partly controlled by bedrock geology, for when ice overlies erodible mudrocks then basal conditions are influenced by the high content of englacial debris. In one investigated example, the surge front itself coincides with the change from a warm ice-bed upstream to a cold bed over permafrost downstream, with basal sediment and water ejected along a frontal thrust fault system in the compressional glacier forebulge snout downflow from the surge front (Fig. 15.7).

## 15.4 Sediment transport, erosion and deposition by flowing ice

The debris found within glacial ice may come from the glacier top, sides (Plate 12) or sole. In the former cases a variety of angular particles are supplied by the very effective freeze–thaw mechanisms that act on surrounding bedrock slopes. It is supplied to the glacier margin by colluvial scree-fan build-up and subsequent rockfall avalanches and debris flows. Successive bands of debris within temperate glaciers record seasonal cycles of snow accumulation and debris fall from rock outcrops. Some proportion of the supraglacial debris finds its way to the sole via crevasses (Fig. 15.8) and intraglacial tunnels, where it is modified in shape and size during ice motion. Polar ice characteristically contains a basal zone of layered *regelation ice*, heavily charged with debris caused by upstream pressure melting and downstream refreezing around obstacles such as gravel and boulders. Thin regelation layers also occur in warm glaciers but these are lost downflow as the basal sliding process takes over. Subglacial erosion is effected by both sediment-charged subglacial water and by plucking, abrasion, crushing and fracturing. Subglacial transport processes cause frequent grain-to-grain interactions resulting in abrasion and a degree of rounding. Also formed are striated facets and the formation of the abundant rock ‘flour’ that dominates the suspended load of glacial outwash streams. Thrust planes may arise at the glacier snout

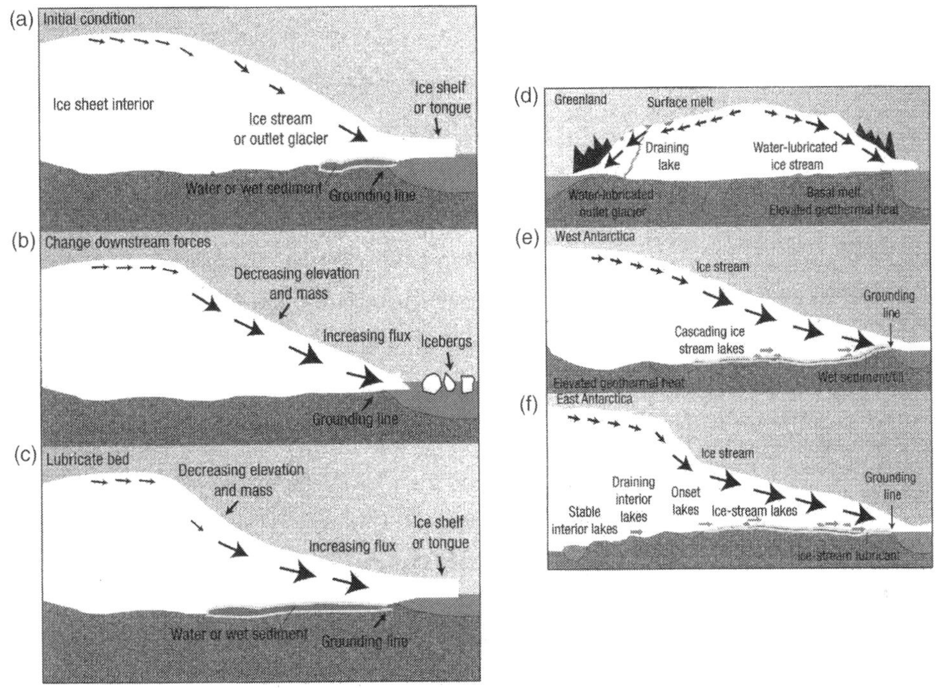


**Fig. 15.6** (a)–(c) Perturbations to ice strain rates. (After Bell, 2008, and references cited therein.) (a) Ice-shelf or floating tongue exerts a compressive force that slows the flow of an ice-stream. (b) Removal of the shelf or tongue causes an increase in velocity, as does landward retreat of the grounding line and loss of ice elevation and mass. (c) Basal lubrication by entrained surface meltwater (Greenland) or subglacial pressure-melted water (Antarctic) and/or water-saturated sediment. (d)–(f) Water and the major ice sheets. (d) Greenland case with the warmest ice and surface meltwater source plus localized geothermal heating as water sources. (e) West Antarctic case with basal melt in the ice-sheet interior and ice-stream tributaries plus localized geothermal heating as water sources for downstream cascading lakes. (f) East Antarctic case with stable interior and interconnecting and draining ice-stream lakes.

(Figs 15.7 & 15.8) in response to compressional stresses between the slow-moving or stationary snout and the fast-moving upstream ice, particularly during surges. These thrusts carry *subglacial debris* to high positions in the ice-front, where it can both melt out and flow.

Deposition (lodgement) of glacial debris at the active ice-sole occurs in response to the cohesive smearing of clay-rich sediment on to bedrock or pre-existing sediments. The process is aided by melting around the upstream parts of bed roughness elements and deposition of the released materials downstream. Thick layers of *lodgement till* with shear fabric thus gradually build up (Fig. 15.8a). Clast orientation within lodgement till and spectacular shear foliation of the ice itself (Fig. 15.9) point to the effectiveness of the shearing–smearing process. Clasts may be orientated with their long *a*-axis parallel to flow, as observed in many grain flows and debris flows, but with a varying proportion of imbricate intermediate *b*-axis orientations, as seen in sediments deposited by bedload rolling mechanisms.

## 15.5 Glacigenic sediment: nomenclature and classification

Much care is needed in the description, nomenclature and interpretation of glacigenic sediments, particularly those that are coarse, poorly sorted and matrix-rich Such muddy, sandy and gravelly deposits are given the quite general and nongenetic term *diamict(ite)*, the suffix for lithified examples in the geological record.

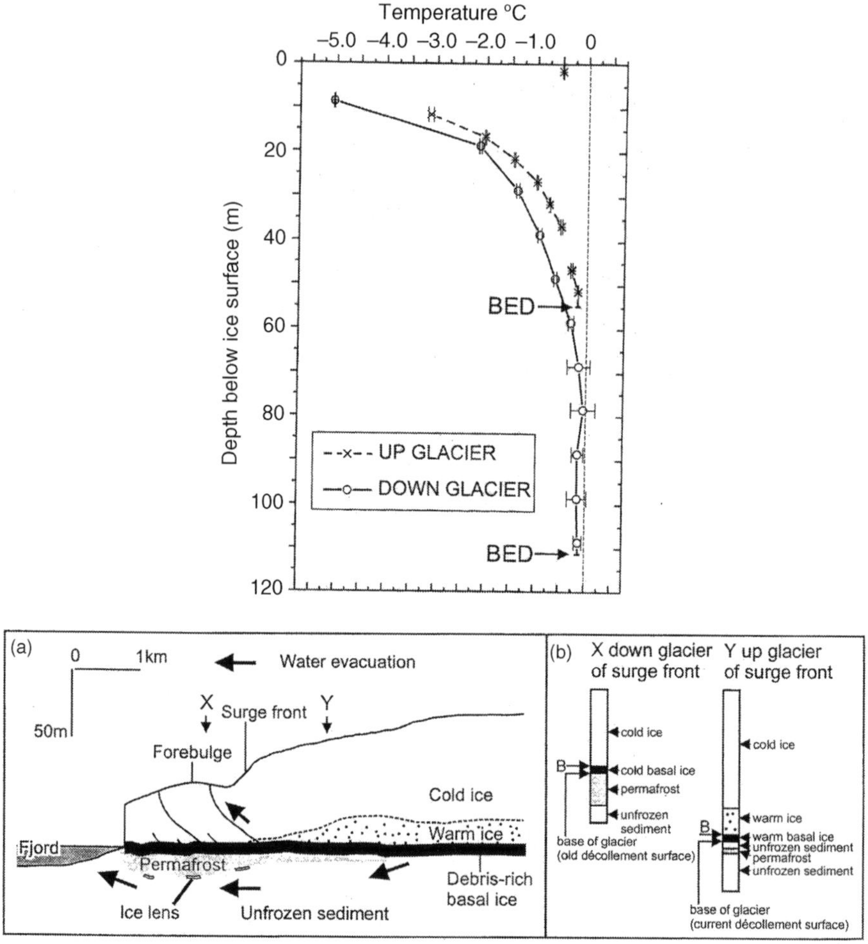


**Fig. 15.7** (a) Temperature profiles through warm- and cold-based ice, Bakaninbreen Glacier (Svalbard) measured in boreholes drilled upstream and downstream from the surge front respectively. (b) Conceptual model for Bakaninbreen glacial surge propagation as warm-based section of thick ice slides more rapidly over a deforming layer of water saturated sediments than the glacier snout which is moving more slowly over frozen (permafrost) sediment. Strain is accommodated by thrusting and shear zone production in the forebulge where water can also escape as well as in groundwater. (All after Murray *et al*., 2000.)

This is not for pedantic or semantic reasons, but because of the great controversy (and difficulty!) often associated with the identification of ancient glacial deposits. Thus a diamictite could have an origin as a true subglacial primary lodgement till (see below) or it could have been redeposited by a debris flow related or quite unrelated to any glacier ice. Many extra observations of both the deposit and its relationships with adjacent deposits are needed before a glacigenic origin can be confidently stated.

The following terms are in general modern use and increasingly accepted as standard by workers in the glacial field.

- *Primary till* is unmodified diamictite deposited directly from glacier ice on land or under the sea. The latter is also known as waterlain till. Deposition is most commonly due to meltout from stationary ice or lodgement from the moving glacier or iceberg base.
- *Secondary till* is diamictite that is modified after deposition (but not by significant addition of

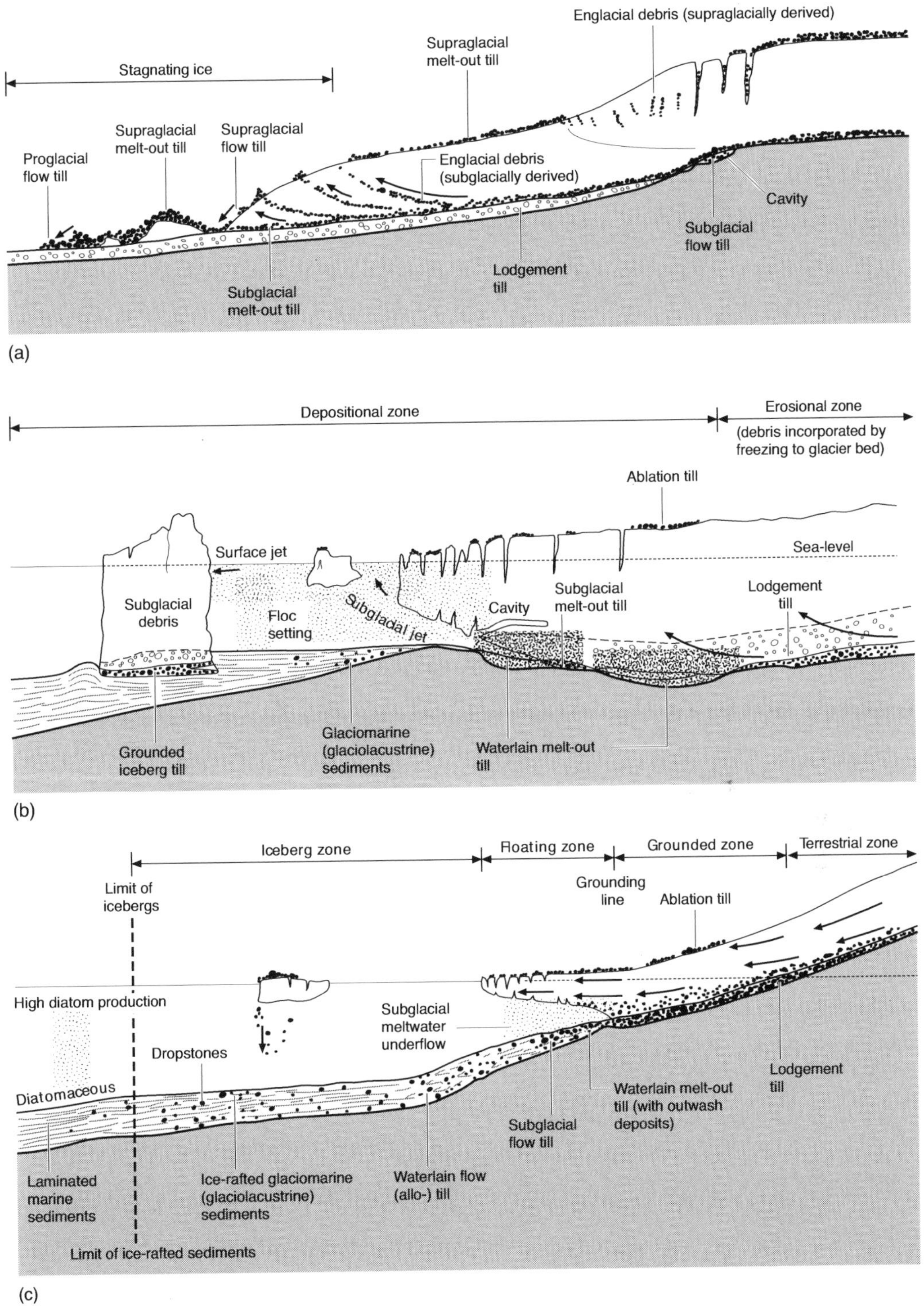


**Fig. 15.8** Flow-wise sketches of glacier termini (not to scale) to show modes of till, outwash and subaqueous proglacial deposition in (a) land glacier, (b) grounded glacier and (c) floating glacier. (After Hambrey & Harland, 1981.)

Fig. 15.9 Basal glacier ice of the Variegated Glacier, Alaska to show intensely sheared and folded pure ice (white) and debris-rich (ca 10%) ice (grey). Foliation planes are about parallel to glacier bed. Ice axe is 85 cm long. (From Sharp *et al.*, 1994; photo courtesy of M. Sharp.)

meltwater), most commonly by gravity flow, slumping or debris flow.

An important point concerns the physical state of lodgement tills, which are frequently quite dense and overcompacted. This was not their condition during glacial flow (at least away from the terminus) since we have seen above that most ice flows on a highly porous and overpressured water-rich sediment base. The till fabric and physical state must evolve rapidly as the snout is approached, with massive dewatering and compaction.

## 15.6 Quaternary and modern glacial environments and facies

A broad spectrum of depositional environments and sediment types exists in association with the movement and melting of glacier ice. We may usefully divide depositional environments and products (see summary in Fig. 15.8) into:

- ice-produced, by direct lodgement from subglacial flow or by passive meltout;
- ice-produced, by subsequent gravity flows from the glacial terminus;
- water- or debris-produced, within the ice or at the subaerial ice contact zone—confined or unconfined ice-stream outlet, tunnel flow, ice-contact (terminoglacial) fans and delta deposits;
- glacimarine, after ice or meltwater comes into contact with marine waters—a great variety of till, density current (plumes and jets), tunnel jets with high sediment inertia and meltout deposits;
- glacilacustrine, after ice or meltwater comes into contact with lacustrine waters—a great variety, as for glacimarine.

In terms of geological preservation potential, the last two are most favoured because the products, particularly those on the shelves and in the oceans, may be spread over vast areas. Picking up this glacigenic signal amongst the deposits of 'normal' shelf and oceanic deposits is obviously highly important for reconstructions of pre-Holocene ice dynamics and distributions (see Chapter 21 for Heinrich events).

## 15.7 Ice-produced glacigenic erosion and depositional facies on land and in the periglacial realm

Glacigenic erosion may produce *grooves*, *channels* and *roches moutonnées* with up to tens of metres of relief eroded in bedrock surfaces; they may be used to map the former margins of rapidly moving ice-streams, as in exposures under the pre-recent Ross Sea Ice Stream. Deposits are dominated by very poorly sorted, often clay matrix-rich, gravelly *tills* (an older term is *boulder clay*) whose clasts are variously angular, striated and of extremely mixed provenance. As noted previously a number of distinct genetic varieties of till are known. *Lodgement tills* result from deposition at the base of temperate ice by plastering effects due to regelation and bulk freeze-on after supercooling of water-rich debris bands. There is often evidence of an upper, more porous, deformed layer with a good development of shear fabric overlying a more massive, denser layer with spaced shear planes. Larger clasts may resist the shear stress of the glacier ice by their greater stability if they 'keel down' into the lower nonshearing layer. Such lodged boulders may resist the shearing stress set up by the glacier bed and act as nuclei for the downstream accretion of finer sediment to form positive-relief, ridge-like mounds (Fig. 15.10) a few metres or tens of metres long (these are called *flutes* in the glaciology literature). Assemblages of

**Fig. 15.10** Former subglacial debris flutes exposed after glacial retreat, Briksdalsbreen, Norway. (From Hart, 2006.)

*subglacial flutes*, saturated till and push moraines, are typical of a deforming bed glacier. The relatively low presence (14%) of core clasts in some study sites suggests that most clasts moved within the deforming layer. Sometimes patches of more competent till nucleate core-less flutes or other clasts may stack upflow to form stoss-stacked flutes. The height of flutes (up to a metre or so) depends on the size of the core stones. Accretion has been postulated to have been brought about by a spanwise component of motion of saturated till into a zone of changed pressure gradients in the lee of the obstructing pebble or boulder. The spanwise motion has recently been confirmed by analysis of pebble orientations.

On a larger scale, but possibly in a similar way to flute ridges, till deposited from ice-sheets or streams may be moulded into larger scale streamlined *drumlin* bedforms. These elongate mounded forms typically have lengths and widths of tens to hundreds of metres. They form most commonly when ice passes over permeable sand and gravel substrates whose high porosity and permeability enable reduction of pore pressure and enhanced frictional resistance. The drumlins are thus to be viewed as obstructional ridges that deform as ice flows more rapidly around adjacent fine-grained substrates with reduced shear strength and low permeability. Drumlin fields may often coincide with passage of ice that has accelerated in the beginnings of ice-stream passage from impermeable to permeable substrates, causing loss of porewater pressure and massive rapid deposition.

*Meltout* or *ablation tills* (Fig. 15.8a) result from the seasonal melting of debris-laden ice in temperate glaciers. Such tills show no development of internal structures or fabrics and may locally overlie lodgement tills. *Flow tills* sometimes showing internal flow banding are extremely common at the glacier front and result from thick *supraglacial debris accumulations* derived from meltout of debris bands along *englacial thrusts*; these occur at glacier snouts due to longitudinal compression resulting from the downflow transition between warm- and cold-based glacier beds. Such poorly sorted deposits then move down the local gravity slope at the glacier snout or lateral margins as coherent slumps or as debris flows that may coalesce in ice embayments to form steep, proximal ice-contact (terminoglacial) fans. Flow tills may come to overlie meltout and lodgement tills to form a characteristic tripartite subdivision. Spectacular valley-wide diamicts of debris flow origin are also sometimes produced during deglaciation.

*Moraines* are a polyglot sedimentary depositional feature that may form from active, stagnant or rejuvanated ice masses. High, laterally extensive *terminal moraine* ridges with tens to hundreds of metres of relief are produced by glacial dumping of both ice-deposited tills (unstratified moraine) and displaced water-deposited outwash deposits (stratified moraines). They may mark longer periods when forward ice motion is exactly balanced by meltout. Some feature spectacular *glacitectonic structures* such as thrust faults and debris bands, thrust-tip folds and nappe formation, with diamictons, outwash deposits and interglacial fluvial terraces overrun by later glacial advances. Alternating beds of massive silty clay diamicton and laterally extensive sorted sand and/or silt in erosional hummocky morainic terrain in south-central Alberta, Canada, share common palaeoflow directions and suggest a subglacial origin. Sheet stream flow for the finer units requires that water was stored beneath the ice prior to drainage. Consequently, some of the diamicton may have been deposited by rain-out through the shallow stored water and, hence, each diamicton bed may represent a combination of direct melt-out onto the bed and rain-out through the shallow water column. Inferred cycles of water accumulation and drainage may have been annual. Many other moraines record motion then stranding of stagnant ice when the dead-ice core disintegrates and mass movement processes rework the sediment cover repeatedly, leading to a gradual lowering of the ice-cored terrain. This evolves to a partially ice-cored terrain where the former coherent ice mass is disintegrated by solution cavities into isolated dead-ice blocks capped by multiple resedimented, slumped deposits. The dead-ice process eventually defines *moraine mounds*; fields of hummocky, rounded, rectilinear to elongate diamicton of several metres relief much influenced by arcuate englacial thrusts that have a strongly 3D nature in plan view, rather like listric faults. Rounded mounds are developed around the stranded masses of melting glacier ice, whilst elongate mounds are glaciofluvial in origin that may also show extensional faulting linked to subsequent melting of ice ridges. Interpretation of such features in the sedimentary record is hindered by structures formed by deformation and relief collapse that occurs over buried ice masses that gradually meltout. Rare examples of preserved thrust-debris bands surrounded by glaciofluvial deposits are recorded from the Quaternary of Poland. In many previously glacial valleys, ice-marginal deposits may nucleate laterally against a bedrock valley wall in a quasi-periodic manner to produce *valley-side morainic mound complexes*.

We finally briefly mention the widespread *periglacial realm* in which frozen ground conditions may reach hundreds of metres below the surface. Diagnostic near-surface features are produced by brief summer thaws and deep winter freezes above a permanently frozen substrate. Hillslope veneers of colluvium may slowly creep downslope, most rapidly on south-facing slopes, during spring and summer thaw, a process known as *solifluction*. *Ice-wedge casts* and various forms of patterned ground that open in summer thaws and freeze with ice wedges in winter are particularly characteristic of periglacial conditions. *Rock-glacier* flows and deposits are ice-cored remnants (an example in Antarctica has the oldest known ice, at about 8 Ma) covered with an often thick layer of avalanche, debris flow or collapse talus; these creep slowly downslope from cirques or ice fronts. More permanent permafrost thawing is increasing in the Arctic and sub-Arctic due to global warming. One consequence of thawing permafrost is the development of *thermokarst*, physical depression of the land surface because of reduced support of overlying soil. This occurs in patches and causes lake development on lower ground. On sloping ground, slab erosion leads to soliflucted shallow sheet slides and gullying.

## 15.8 Glaciofluvial processes on land at and within the ice-front

Subglacial outlet drainage may occur along distinct channels or tunnels that may be alongside or within the ice or at its base. Deep tunnel valleys may be cut below ice-sheets overlying relatively easily erodible sediments. Fluvial depositional facies at the base or sides of glaciers or in the ice-contact zone include *debris fans, colluvial cones* and *kame* terraces. Kames form by subaqueous deposition adjacent to ice masses in contact with bedrock surfaces. They commonly have relatively flat-topped surfaces with slumped margins marking the former position of the melting ice boundary.

Summer meltwater discharged from the glacier front is free to erode, transport and deposit sediment as in any other fluvial system. Although outwash discharges are often heavily charged with suspended rock flour, the abundance of gravel- to boulder-grade debris in the bedload determines the intensely braided form of most outwash plains (Plate 9). These are known by the Icelandic term, *sandur*, and may exhibit single or coalesced fan-like forms. Downstream trends include decreasing grain size and increasing predominance of point and lateral bars. Peculiar effects are caused by both seasonal freezing of channel beds and by melting of buried ice masses. In the latter case, large-scale cross-beds and coarsening-upwards sequences are thought to result from the transport of outwash, as deltas prograde into local depressions (*kettleholes*) formed by the melting of buried ice masses. Imbricate gravels in unit bars are much more common in proximal areas, with particles transported and deposited in layers parallel to the surface of the bar–channel network. Some authors distinguish *terminoglacial* (aka *ice-contact*) *fans* from sandur plains, these being smaller, steeper fan-like bodies fed initially by proximal debris flows but passing downfan into braided outwash environments.

We cannot leave the subject of glacier-front fluvial processes without briefly mentioning two kinds of mass-flooding events: *glacial lake breakout* and *jökulhlaups*.

During deglaciation vast volumes of meltwater may be held back in huge glacial lakes by ice fronts. One such was Lake Agassiz-Ojibway in what is today east-central Canada, ice-dammed to the north by a remant Laurentide ice-cap surrounding a proto-Hudsons Bay. About 8.5 ka, breakout floods are thought to have lifted the ice-dam south of the bay when ice thickness reduced to about the depth of lake water and led to outrush of a swarm of broken ice as bergs whose swirling keels thoroughly churned the floor of the bay. Upcurrent, under the lifted ice, sediment was moulded by rapid flow into a field of sandwaves. It has been claimed that the vast freshwater outburst triggered the 8.2 ka *cool event* in the North Atlantic, but the time lag of several hundred years seems excessive given likely mixing rates in ocean surface environments. More accessible to direct investigation are the channelized and bedformed 'scablands' left behind in northwest USA by enormous early Holocene Lake Missoula breakout floods—an event that has significantly affected our attitude towards extreme events and catastrophism in the geological record. The most recent discovery of Quaternary megafloods involves the catastrophic escape of Anglian-age (ca 400 ka) meltwaters ponded in the southern North Sea due to a breach in the Weald-Artois chalk ridge. High-resolution sonar bathymetry reveals channel scabland-type megaflood deposits occurring within excavated shelf bedrock-floored valley systems ~400 km along the length of the English Channel.

Jökulhlaups are also major floods but this time due to the bursting of subglacial meltwater reservoirs produced by volcanic activity and related geothermal energy release and sometimes with a major addition of pyroclastic materials. Their *locus typicus* is Iceland with the last spectacular example in 1996. Jökulhlaups contribute significantly to the architecture and build-up of sandur outwash plains.

## 15.9 Glacimarine environments

Proximal glacimarine environments (Fig. 15.8b & c) include ice-shelves and ice-stream/valley-glacier terminii at fjord and valley heads in both temperate, subpolar and polar climates. Rapid outflow of ice-streams is seen not only in elevated ice fluxes but also by the outflow of major subglacial water drainage discharge that leads to production of offshore glacial troughs and large submarine *glacimarine sediment fans*. Such associations serve to identify ancient ice-stream locii such as that responsible for the deposition of the 280 000 km$^2$ Bear Island Fan in the Norwegian–Greenland Sea during the Last Glacial Maximum. Here, in these proximal to distal *ground-line* to *far-field glacimarine environments*, deposition rates may be very high, up to several centimetres per year. Proximal environments grade out into the more distal shelf and shelf edge, where deposition rates decline exponentially as glacial debris forms less and less of the local sediment flux.

A floating ice mass is produced when a mass of glacier ice (sheet, stream or valley ice) reaches the coastline and interacts with the marine environment at the limit of ice–bed contact, the *ground-line* (Fig. 15.8c). Since the freezing point of seawater is −1.8 °C, polar conditions are below or close to freezing whilst temperate conditions are always higher. In subpolar ice terminii, the temperature of ambient seawater approximates to or is a little greater than freezing in winter and greater in summer months. Temperate offshore glacimarine environments are dominated by meltwater runoff whilst polar examples are dominated by iceberg discharge and rarer ice-shelf calving episodes. Subpolar examples are intermediate between these end members and show mixed meltwater and iceberg influences.

The complicated dynamics of an ice-shelf may be approximated by considering a wedge-shaped tongue of moving ice that tapers seawards. A mass balance is set up between ice and sediment delivered to the front, contact melting at the ice–seawater interface, ice-wedge flotation and ice taken away by iceberg calving. Lodgement till may continue to be delivered to the ice terminus where, under stable conditions of sea level, it may oversteepen and be resedimented as flow till and further down the local marine basinal slope as debris flows on *ground-line debris fans*. Subaqueous stratified flow till with dropstones derived from meltout of the ice roof of the tapering wedge of the ice shelf occurs above *in situ* lodgement till deposited by a former glacier surge or during a period of lowered sea level. These deposits thin seawards as the layer of basal grounded-ice debris itself dies out under a tapering ice-shelf. Delivery of copious silt and mud at the shelf edge during glacial sea-level lowstands leads to the periodic generation of long-runout *glacigenic mudflows*. Recent exciting finds from drillholes and cores 600 m below the Ross ice-shelf, Western Antarctica ice-sheet demonstrate well-developed 40 kyr obliquity-forced sedimentary cycles of Plio-Pleistocene age marking glacial/interglacial oscillations of the ice-sheet. Interglacial conditions when planetary temperatures were ~3 °C warmer than today, with atmospheric $CO_2$ at 400 ppmv, led to ice-sheet collapse resulting in a switch from grounded-ice/ice-shelf sediments, to open-water diatomites in the Ross embayment.

In temperate terminii, the discrete exit of meltwaters from subglacial tunnels occurs as jets and evolves into widespread plumes; underflows are rare because of the high sediment concentrations needed to form them, but interflows are commonly observed. The meltwaters may have cut deep tunnel valleys or deposited eskers in the forward part of the glacier—their deposits are characteristically coarse. At the ice terminus, bedload sediment from the meltwater is rapidly deposited as mound-shaped *ground-line fluviodeltaic fans*. The fans may exhibit imbricated gravel bedforms in their proximal parts, fining downcurrent and exhibiting well-defined delta-front foresets. Should conditions allow, the subaqueous grounding-line fans may grow to sea level and produce subaerial

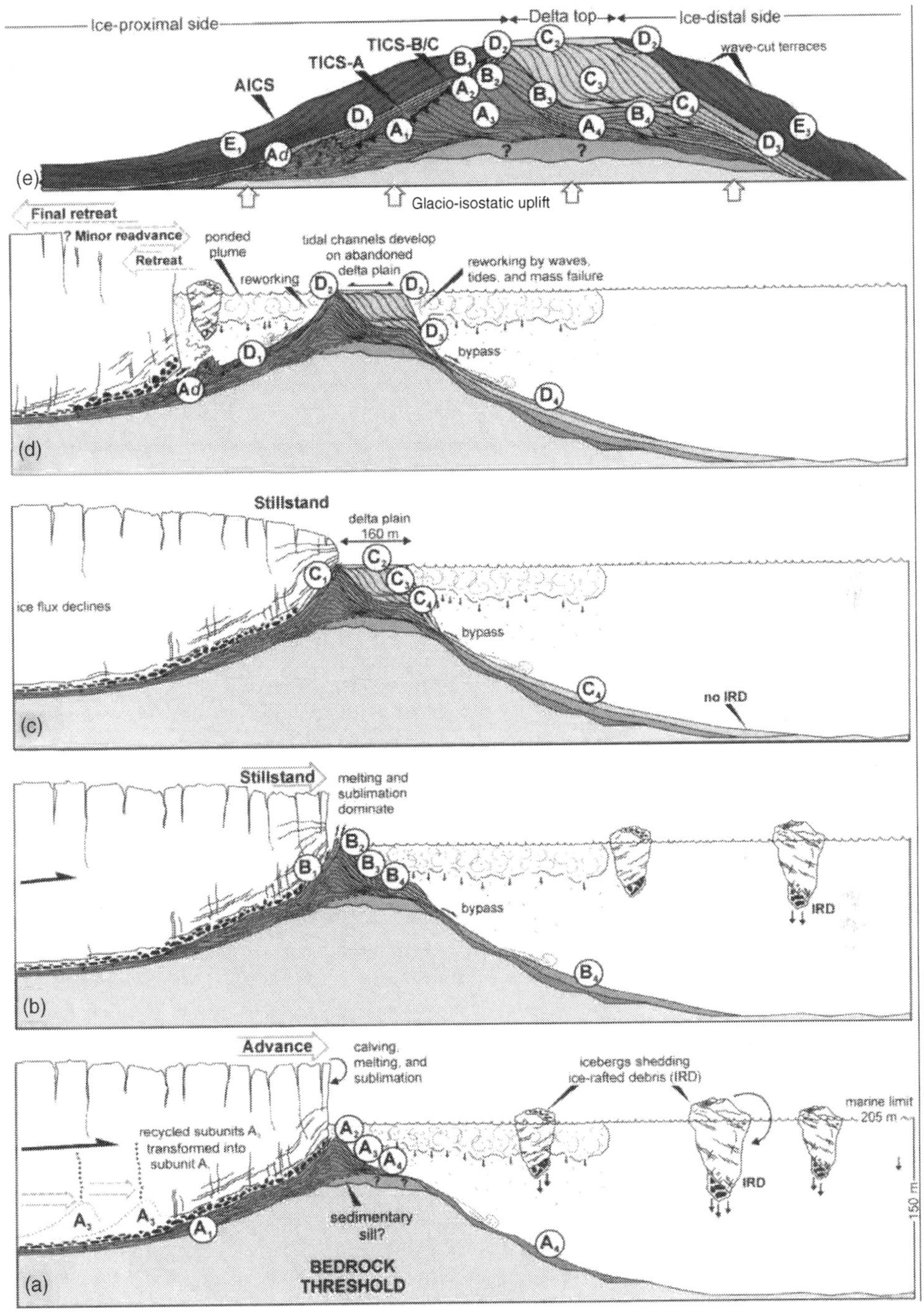
Ice-proximal side
Delta top
Ice-distal side
wave-cut terraces
TICS-B/C
TICS-A
AICS
Glacio-isostatic uplift
(e)
Final retreat
? Minor readvance
Retreat
ponded plume
reworking
tidal channels develop on abandoned delta plain
reworking by waves, tides, and mass failure
bypass
(d)
Stillstand
delta plain 160 m
ice flux declines
bypass
no IRD
(c)
Stillstand
melting and sublimation dominate
bypass
IRD
(b)
Advance
calving, melting, and sublimation
icebergs shedding ice-rafted debris (IRD)
marine limit 205 m
IRD
150 m
sedimentary sill?
BEDROCK THRESHOLD
(a)

ice-contact deltas. Very energetic jet flow at high discharge stages is postulated to cause sets of avalanche faces rather in the style of Gilbert-type subaqueous fan deltas (Chapter 16). After flow expansion and deceleration the jets then provide surface or stratified plume-like extrusions of buoyant meltwater laden with suspended sediment. Temperate valley glaciers will obviously liberate much more sediment from their copious meltwaters into buoyant brackish overflows than will polar ice-sheets. Concentrations as high as 800 mg/L have been recorded from Alaskan glacially sourced plumes. The suspended load settles out rapidly, fining seawards as the plume decelerates, cooled from below by mixing and conduction. Deposition rates decrease markedly due to plume dissipation. Finely laminated sedimentary couplets of coarse and fine grains (not varves) may form in the zone due to the interaction of the buoyant outflow and the cycle of local tidal processes, should these be significant. Seawards, this zone is dominated by buoyant outflows and also affected by iceberg calving, which may cause basal scour and the addition of extra sediment from meltout. The detailed stratigraphy and sedimentary evolution of a glacimarine terminal moraine formed during the Younger Dryas readvance of the Scandinavian ice-sheet is illustrated in Fig. 15.11.

In subpolar terminii, seasonal meltwater and sediment supply is predominantly from glacial surface meltwater or from season-round direct contact melting of submerged ice with basal metre-scale debris sheets and from iceberg meltout. At the calving Maar Glacier front, Antarctic Peninsula, bergs are very small because the glacier front is intensely fractured by englacial thrusting so the metres-scale blocks quickly melt and dump their included sediment load within 1 km or so of the ice-cliff. Meltwater and surface heating causes water column stratification in summer months with a well marked halocline at depths of a few tens of metres. Turbid layers within the column are mixed with the ambient seawater by wave and tidal currents, whilst a more permanent bottom turbid layer located in bathymetric lows in near-terminus locations is due to calving turbulence bottom resuspension and iceberg sediment fallout.

Sediment rafted along by large *icebergs* may become widely dispersed over the adjacent shelf. Very thick bergs (> 600 m) are produced by calving from the fast-moving outlet glaciers fed from the Greenland ice-cap. Not only do these produce copious sediment as their debris-rich basal layers melt, but their deep keels extensively scour and mix up existing shelf sediment. Sediment coring reveals thick sequences of massive diamicton over wide areas of the East Greenland Shelf, whilst seafloor imaging reveals ample evidence of iceberg scours. By way of contrast, formation of multiannual, shorefast sea ice ('sikussak') during cold latest Pleistocene and Holocene periods (Younger Dryas, Little Ice Age) restrains iceberg dispersal within many fjords or bays so that laminated glacimarine sediments are deposited there between massive or stratifed glacimarine diamictons.

Many examples of Pleistocene glacial-epoch iceberg scours and deposits may be found on temperate shelves. By way of contrast, abundant icebergs calved from the great floating ice-shelves of the Antarctic tend to be much thinner and less rich in debris, reflecting the seaward-tapering nature of the ice-apron and the extensive *in situ* meltout that takes place below the shelf prior to calving. In one fairly recent major calving event, in 1986, the calved mass was a staggering 13 000 km$^2$ in area and fragments may have reached 38°S, the latitude of Buenos Aires. It should be said that polar mountain glaciers will deliver more sediment-rich bergs than the ice-sheets themselves, but these are in a minority volumetrically in Antarctic conditions. The high preservation potential

**Fig. 15.11** Reconstruction of the depositional history of Mona moraine, southern Norway with the letter code referring to distinct stratigraphic units. (a) A submarine ice-contact fan is formed on the bedrock threshold at Mona during the southward advance of a temperate glacier with a calving tidewater front. (b) The glacier grinds to a halt and the submarine fan aggrades to the sea surface, which terminates calving and dispersal of ice-rafted debris, IRD. (c) An ice-contact Gilbert-type delta develops, with short, subaerial distributary plain; the avalanching delta slope extends down on to the steep 'bypass' slope of the bedrock threshold, which leads to the deposition of turbiditic gravelly sand in the foot zone of the seafloor threshold. (d) The delta is abandoned due to a rapid ice-front retreat by calving, and the moraine becomes an island surrounded by sea, up to 100 m deep; tidal channels develop across the former delta plain and reworking by waves begins, accompanied by slope failures. (e) The moraine emerges due to regional glacio-isostatic uplift, while its slopes accumulate regressive foreshore deposits, are swept by longshore currents and are terraced by wave erosion. AICS, apparent ice-contact surface; TICS, true ice-contact surface (indicated for units A–C). (After Lønne *et al.*, 2001.)

of iceberg scours and their associated plough-produced lateral ridge 'berms' is emphasized by a notable occurrence in Pleistocene glacilacustrine sediments from ice-damned palaeo-Lake Ontario, Canada.

The distinction between sequences of iceberg meltout diamictons on a continental shelf, as distinct from diamictons formed by nearshore meltout adjacent to the grounding-line, is an important one. Shelf diamictons will be widespread, intercalated within 'normal' shelf sequences as distance from the shoreline increases and should contain a very wide variety of debris sourced from various calving ice-streams. These features, together with the occurrence of seabed scours, might lead to the identification of Greenland-type grounded-ice outlets rather than the presence of an Antarctic-type ice-shelf.

In recent years more attention has been paid to the formation of ice-sheet marginal shelf and shelf-basin glacigenic *sediment drifts*. These are distal 300–1200 m deepwater sediment build-ups in the form of mounds, terraces and accreted lenses that form when very fine-grained hemipelagic sediment of predominantly glacial, but also planktonic origin, are deposited by oceanic thermohaline and shelf-margin currents. Offshore Antarctic Peninsula basin cores show rhythmic alternations of bioturbated clay, silt and diatom laminae that reflect seasonal and longer periodic fluctuations in ocean current forcing and local wind, wave and tidal forces with teleconnections to intensity-controlling processes like ENSO and the Antarctic Oscillation. Periods of enhanced productivity that gave rise to thicker laminated diatom oozes also occurred due to more restricted circulation and lower bottom oxygen levels when glacial activity produced more enclosed estuarine conditions in the mid-Holocene.

The three-dimensional distribution of sediments and their deposition rates in glacimarine environments from coast to ocean is dependent upon the control by meltout plume flux and trajectories, iceberg flux and tracks, local relative base-level change by glacio-eustasy and by regional crustal response to glacial loading and unloading. Off the northeast margin of Canada, unexpected spatial and temporal differences in sediment deposition probably reflect changes in sediment transport trajectories associated with differences in ice extent (glacial or sea-ice) and atmospheric/oceanic circulations. For example, during Heinrich events 1, 2 and 4 (late Pleistocene to early Holocene periods of enhanced ice, meltwater and iceberg flux into the northeast Atlantic, see Chapter 21) sediment was delivered to the slope and into the Labrador Sea, with isopachs showing major delivery along the axis of the Labrador Sea, probably associated with the North Atlantic Mid-Ocean Channel (NAMOC). With deglaciation of the Baffin and Labrador shelves, sediment plumes and icebergs could now track along these shelves, resulting in relatively low rates of sediment accumulation in the Labrador Sea and more ice-distal North Atlantic sites. Concerning eustasy versus isostasy the response time of ice-sheets to global warming is rapid compared to the response of the asthenosphere to glacial unloading. The two effects are thus out of phase, with maximum subsidence occurring well after ice-sheet retreat. At first melting causes marine transgression over former shoreline and land glacial facies, but as the crust responds to unloading the late-glacial and transgressive facies are uplifted and incised. The sequence stratigraphy of ground-line glacial deposits is therefore highly complicated, but nevertheless revealing.

## 15.10 Glacilacustrine environments

Many depositional processes in the glacimarine environment are also present in glacilacustrine settings, much simplified by the lack of tides and the lower effectiveness of waves and absence of other ocean currents. The other difference is the greater density of cold suspended-sediment-rich meltwater over the (often) stratified lake-waters. Jet- or plume-like underflows are therefore much commoner in glacilacustrine environments which are the classic site for production of lake *varves*, annually laminated couplets of silt and clay that form due to seasonally controlled changes in meltwater discharge. Spring and summer meltout gives coarser siltier multilaminae from turbidity underflows whilst late-autumn to winter shutdown and overturn (Chapter 14) gives final clay settlement from suspension fallout enhanced by any pelagic biogenic production. At any one point in space varve thickness is controlled partly by palaeostream discharge but also by diversions in underflows caused by point-source trajectory variations. Autumn to winter clay/biogenic laminae tend to vary little in thickness.

*Eskers* associated with temperate glaciers are thought to have formed at or fairly close to a retreating glaciolacustrine terminus. They are either simple linear ridge-like accumulation of stratified sands or gravels or a multiridge and channel system that was deposited by a single or multithread channel wholly or partly confined by a lake and the glacier ice terminus, the deposits resting usually directly on till or bedrock. They are commonly a few metres to tens of metres high, tens to hundreds of metres wide and kilometres to tens of kilometres long. Narrow eskers up to ca 100 m width may have formed in front of tunnel-like drainage conduits filled with coarser, often bouldery or cobbly sediments. Esker deposits of englacial streams are 'let down' to the local bedrock surface during ice melting and, as a result, may show marginal slump and fault structures related to this movement. However, most eskers are thought to have formed subaqueously from subglacial tunnel streams as they debouched from a glacier terminus into lake bodies as jet-like fans. These would have had high sediment inertia and so would not form classic fan-like bodies but rather linear downstream-fining ridges and have clear basal erosion surfaces cut into older lodgement till or outwash sediment surfaces formed at the proximal channel mouth. Analogous experimental linear ridges form as channelized turbidity currents of high sediment inertia debouch into wide reservoirs (see Chapter 7). The finer, sand-grade deposits of eskers are dominated by current-produced sedimentary structures that developed under conditions of high rates of sediment bed aggradation, particularly climbing ripple cross-laminations with high angles of climb. Esker beads are thought to be caused by gentler flow acceleration and deceleration over the mouth-bar like beads at the front of enlarged subaqueous subglacial streams, giving rise to characteristic down-bead fining.

Evidence of repeated seasonal to annual deposition has been observed in Finnish eskers located in the retreating re-entrant between ice lobes (Fig. 15.12). Exposures reveal time-transgressive, retrograding and overlapping depositional sequences consisting of deposits from two successive melt seasons, including: (i) massive to stratified coarse gravels—summer ice-tunnel deposits; (ii) trough and ripple cross-stratified fine-grained deposits—autumn to winter deposits; and (iii) sandy stratified beds—spring deposits. The depositional environment of each lithofacies association involves a transition from subglacial or submarginal tunnel to a subaqueous re-entrant environment, which then evolves to a proglacial glaciolacustrine environment.

Proglacial lakes are fed by calving icebergs and by seasonal outwash from proglacial drainage which may debouch into the lake to form coarse-grained, steeply dipping delta-front foresets of Gilbert type (first recognized by G.K. Gilbert in the late nineteenth century from the margins of glacial Lake Bonneville (Chapter 14). The topsets of such deltas are valuable palaeowater-level indicators and subsequent lake drainage or deepening may be recorded by younger declining or aggrading delta cosets. The proximal deposits may form the core of more low-angle subaqueous medial to distal fans on which is deposited sediment from gravity currents, including debris and turbidity flows. Finer-grained material finds its way out into the lake body proper by processes of overflow (rare), interflow or underflow, depending upon the relative density of the incoming suspended-sediment-rich water and the ambient lake-waters. The seasonal melt–freeze process gives rise to varved deposits, which may show complex internal rhythms caused by interference between sediment supplied by overlapping density currents. Some coarse-grained proximal lake-fan deposits record the gradual retreat of feeding glacier fronts in the form of fine-grained sediment drapes rich in silts and climbing-ripple cross-laminations. Final lake drainage is marked by transition to subaerial braidplain outwash sediments.

## 15.11 Glacial facies in the pre-Quaternary geological record: case of Cenozoic Antarctica

Of most immediate interest is the pre-Quaternary history of glaciation in the Antarctic, for before the first ocean drilling in the Southern Ocean by Glomar Challenger in the 1970s it was widely thought that polar glaciation was a purely Quaternary phenomenon. The early drilling results found glacigenic deposits at least as far back as ~25 Ma and evidence from O-isotopes of a marked palaeotemperature fall at both the Eocene–Oligocene boundary (~34 Ma) and in the middle Miocene (~15 Ma). Subsequent activity worldwide produced a global sea-level and temperature curve with which to place the polar glacial episodes in context and a broad agreement that the first continent-wide Antarctic glaciation began in the early Oligocene, ~33 Ma, with an Antarctic ice-cap as large or larger than today present since at least ~24.5 Ma.

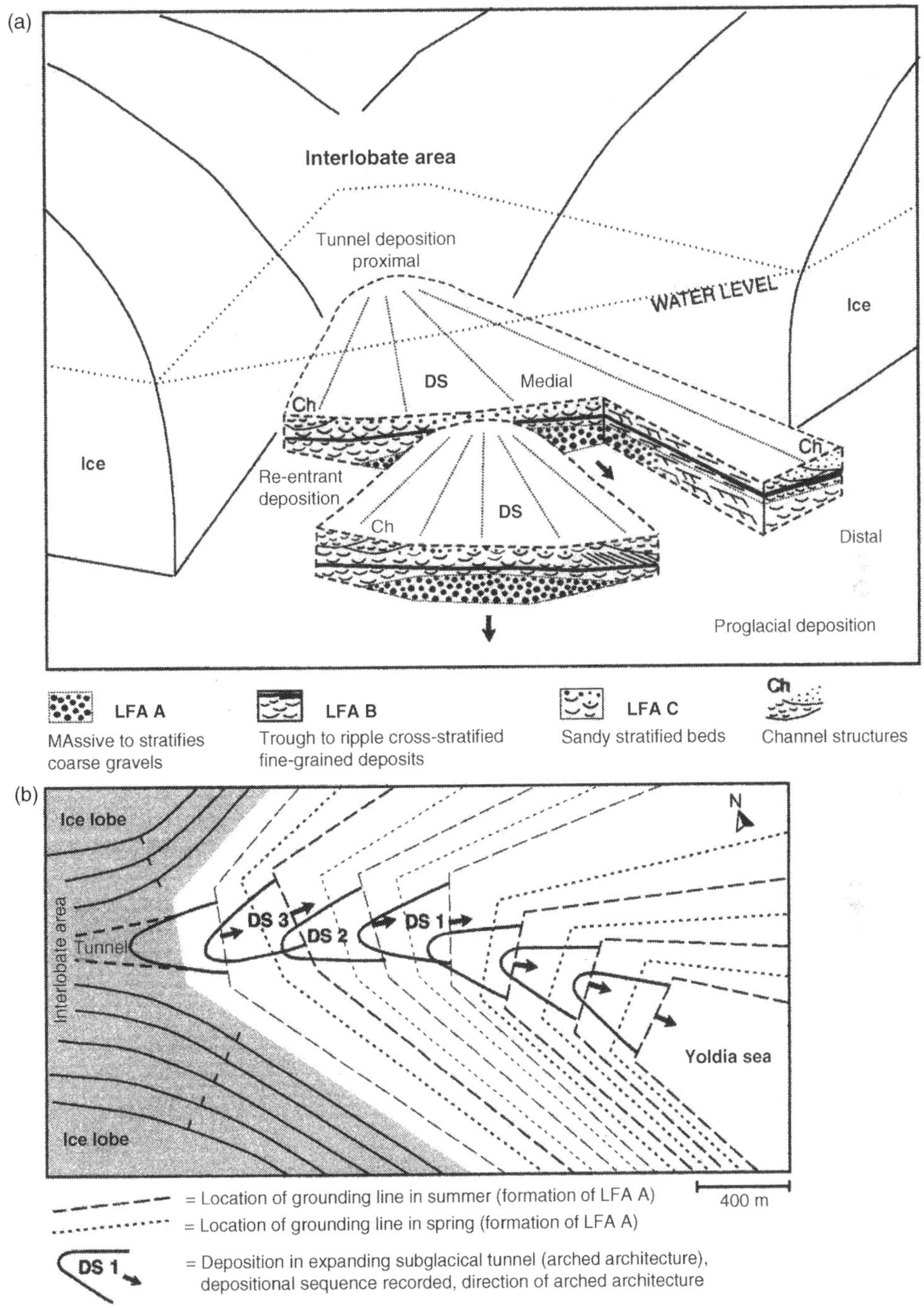


**Fig. 15.12** (a) Block diagram showing development of esker depositional sequences and related depositional environments from southwest Finland. The most proximal bedding planes of each depositional sequence dip about 10° up-ice. Note that massive to stratified gravels (LFA A) are deposited earlier inside the tunnel and extend 300–400 m from ice-proximal contact shown in the diagram. (b) Diagram demonstrates the time-transgressive pattern of depositional sequences within the esker segment. (After Mäkinen, 2003.)

Drilling from sea-ice off Cape Roberts, Victoria Land Basin, part of the actively extending West Antarctic Rift System (WARS), has revealed coastal deposits from 34 to 17 Ma which, though with a fragmentary chronology, give evidence for coastal margin deposition throughout, with wave-dominated marine sediments alternating with diamictites deposited by the advance and retreat of piedmont ice (Fig. 15.13). There was dramatic increase of glacial influence after about 33 Ma, with 10–30% of cyclical glacial/interglacial deposits from 33 to 17 Ma (Oligocene–lower Miocene) and a progressive shift between 33 and 25 Ma from the products of chemical to physical weathering over that time period as recorded in mudrocks by a decline in the CIA index (Chapter 2) and an increase in abundance of illite/chlorite. The cycles record alternating lowstand subglacial till deposition beneath wet-based grounded ice, subglacial channels and grounding-line fans, with highstand interglacial wave-dominated clastic shoreline sediments fed by proglacial rivers draining from more distant glacial termini. Chronology from radiometrically dated tephra indicate the cycles represent 40 kyr obliquity cycles of Milankovitch frequencies and sea-level variations of a few tens of metres.

More recently the Cape Roberts record has been brought closer to present by the Antarctic Geological Drilling Program (ANDRILL) whose first deeper water (~860 m) well cored ca 1300 m of sedimentary and igneous record back from the present to 13 Ma beneath McMurdo Ice Shelf in the WARS. The record accumulated 100 km offshore in 200–1000 m water depths and comprised diamictites, sandstones, mudrock, diatomites and extrusive igneous rocks. These are interpreted from their structures and fabrics as the deposits of ice-proximal massive and stratified tills, turbidity currents, suspension fallout muds and high-productivity phytoplanktonic ooze. During previous glacial periods the ice-sheet terminus lay far to the north in the Ross Sea, whereas during interglacials the area was part of an ice-shelf, like today, or an open-water shelf.

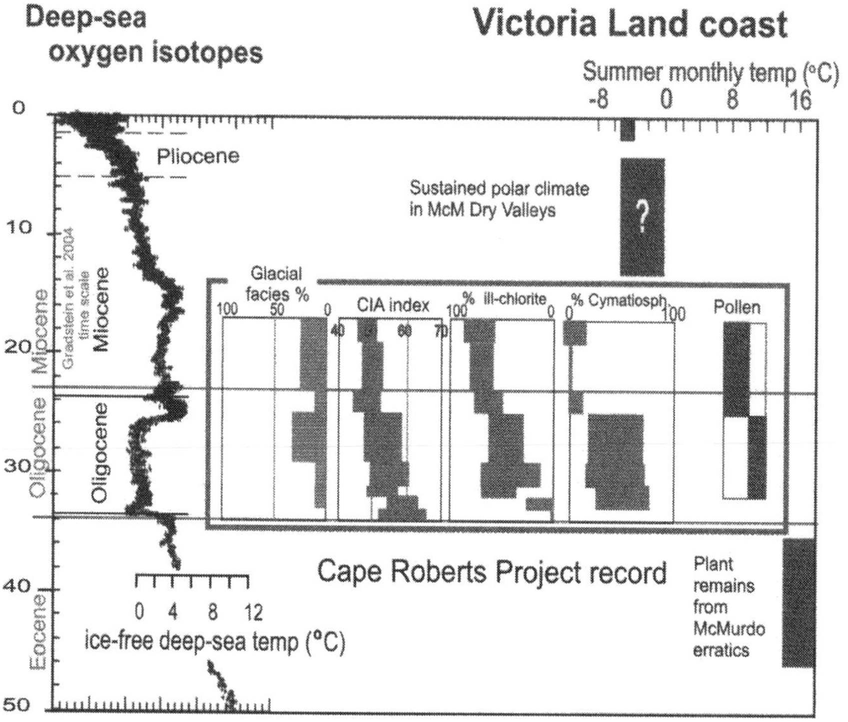


**Fig. 15.13** Trends in time and temperature data from the Cape Roberts area (boxed material) in the context of deep-ocean oxygen isotope curve. (After Barrett, 2007, and references cited therein.)

The record preserves more than 60 cycles of ice-sheet advance and retreat over 13.5 Myr, with the composition of the diamictites indicating ice derivation from the Byrd and Skelton outlet glaciers sourced in the Transantarctic Mountains. Four phases of late Cenozoic climatic fluctuation can be interpreted from the record: (i) an early to late Miocene (13.5–10 Ma) cold period with much subglacial tillite and minor interstratified interglacial mudrocks; (ii) a relatively warm late Miocene (9–6 Ma) period of subpolar ice-sheets that laid down much subaqueous glacimarine outwash, with open-water, ice-distal environments during interglacials and ice-grounding during glacials; (iii) a dynamic interaction during the Pliocene (5–2 Ma) between interglacial open water, with lower Pliocene high productivity diatomites up to 80 m thick recording extended periods of open shelf and favourable currents in the Ross embayment, alternating with subglacial and ice-proximal diamictites—these alternations are especially abrupt in the late Pliocene (2.6–2.2 Ma); (iv) a return in the past 0.8 Ma to cold polar glaciations with evidence for ice-shelf calving conditions during interglacials like the present day.

Compared to the Antarctic ice sheet that of Greenland is quite young, and though valley glaciers probably date back to 38 Ma, the ice-sheet itself first formed ~3.2 Ma.

## Further reading

### *General*

Good, concise accounts of glacial environments, with many fine illustrations, are given by Hambrey (1994) and Menzies (1995). Benn & Evans (1998), Hooke (1998) and Paterson (1994) give accessible accounts of ice and glacier mechanics. An accessible recent review of the role of water in ice-sheet mass balance is by Bell (2008). Pritchard *et al.* (2009) give the latest ICESat data on ice-cap thinning. The Quaternary context of glacial processes is well discussed by Williams *et al.* (1993) and by Dawson (1992). Many important papers are to be found in the volumes edited by Jopling & McDonald (1975), Dreimanis (1989), Molnia (1983a), Dowdeswell & Scourse (1990a), Anderson & Ashley (1991), Maltman *et al.* (2000) and Hambrey *et al.* (2007). Paterson (1994) is also good on physical aspects of ice behaviour; see also the volume edited by Colbeck (1980). The 'bible' for studies of how Quaternary glacier studies may be applied to ancient deposits is the massive Hambrey & Harland (1981), whilst numerous papers of interest occur in Anderson & Ashley (1991). A review of waterflow through temperate glaciers is by Fountain & Walder (1998), updated by discoveries in Fountain *et al.* (2005), and with emphasis on ice-sheets by Bell (2008). Periglacial and paraglacial processes and environments are featured in the volume edited by Knight & Harrison (2009).

### *Specific*

Ice deformation is discussed by Weertmann (1957), Doake & Wolff (1985) and Sharp *et al.* (1994). Valley glacier velocities from remote sensed optical data are in Leprince *et al.* (2008). Stick-slip sliding mode for ice streams is discussed by Wiens *et al.* (2008). Linear rheology and diffusion creep by Chandler *et al.* (2008). Surging in Svalbard glaciers is discussed by Murray *et al.* (2000). Several relevant papers are in Maltman *et al.* (2000).

Ice–till–water dynamics is in Alley *et al.* (1986), Blankenship *et al.* (1986), Boulton & Hindmarsh (1987), Walder & Fowler (1994), Murray & Clarke (1995), Smith (1997), Harbor *et al.* (1997) and Fountain & Walder (1998). The primacy of fracture networks in controlling water flow in a valley glacier is demonstrated by Fountain *et al.* (2005). A response of glacier motion to transient water storage is by Bartholomaus *et al.* (2008). The interconnecteness of sub-Antarctic lakes is demonstrated by Wingham *et al.* (2006) and the influence of such lakes on ice-stream initiation in Recovery Glacier, Donning Maud Land is discussed by Bell *et al.* (2007). The Byrd Glacier, East Antarctica surge and subglacial water flow is by Stearns *et al.* (2008). Water control on sliding versus bed deformation in the Rutford ice-stream, West Antarctica is by Murray *et al.* (2008). Recent thermokarst is studied by Gooseff *et al.* (2009).

Sediment transport, erosion and deposition are discussed by Boulton (1972, 1976, 1978, 1979), Boulton & Hindmarsh (1987), Eyles *et al.* (1988), Eyles & Eyles (1989), Dreimanis (1989), Hambrey (1994) and Benn (1995); exposed sub-Antarctic ice-stream erosional features by Anderson *et al.*

(2001); subglacial flutes by Hart (2006). Drumlin theory is discussed by Boulton (1987) and Fowler (2000).

Moraine formation and deformation are considered by Owen (1988), Hart & Boulton (1991) and Ruegg (1991); alternating diamicton/sheet flow deposits by Munro-Stasiuk (2000); dead-ice processes by Kjær & Kröger (2001); preserved thrust-debris bands by Ruszczyńa-Skaszenajch (2001); moraine mounds by Midgley *et al.* (2007); stratified moraine by Russell *et al.* (2007); valley-side moraine mounds by Graham & Hambrey (2007); glacimarine moraines by Lønne *et al.* (2001).

Glaciofluvial and breakout flood processes are considered by Bannerjee & McDonald (1975), Boothroyd & Ashley (1975), Rust (1975), Einarsson *et al.* (1997) and Maizels (1989); termino-glacial fans by Zieliński & van Loon (2000); Lake Missoula megaflood by Baker (1973, 1990, 1994); English Channel megaflood by Gupta *et al.* (2007).

Glacimarine environments are considered by Rust & Romanelli (1975), Gibbard (1980), Hicock *et al.* (1981), Molnia (1983b), Eyles *et al.* (1985), Syvitski *et al.* (1987), McCabe & Eyles (1988), Eyles & McCabe (1989), Syvitski (1989), Cowan & Powell (1990), Anderson *et al.* (1991), Phillips *et al.* (1991), Blankenship (1993), Domack & Ishman (1993), Vogt *et al.* (1993), Dowdeswell *et al.* (1994), Ashley & Smith (2000) and Lønne *et al.* (2001).

Sedimentation from ice-streams of former large ice-sheets is considered by Stokes & Clark (2001) and Siegert (2007).

Plio-Pleistocene obliquity-forced oscillations of West Antarctic ice sheet and global eustatic implications are discussed by Naish *et al.* (2009); variation in Pleistocene/early Holocene Atlantic glacimarine sediment fluxes by Andrews (2000); glacimarine sediment drifts by Willmott *et al.* (2007); palaeofjord warm-based glaciers in Antarctica by Hambrey & McElvey (2000).

Iceberg scours and Heinrich event deposits are studied by Heinrich (1988), Bond *et al.* (1992, 1993), Hesse & Khodabakhsh (1998), Anderson *et al.* (1991) and Dowdeswell *et al.* (1994); suppression of iceberg dispersal by sea-ice by Dowdeswell *et al.* (2000); see Eden & Eyles (2001) for Pleistocene iceberg scours.

Large-scale aspects of glacial stratigraphic architecture are considered by Boulton (1990); glaciolacustrine deposits by Ashley (1975, 1995), Smith & Ashley (1985) and Delaney (2007); subaqueous lacustrine fans by Winsemann *et al.* (2007); eskers by Shreve (1985), Warren & Ashley (1995), Mäkinen (2003), Bennett (2007) and Gale & Hoare (2007); see analogous experimental ridges formed by high-inertia turbidity current jets by Alexander *et al.* (2008).

Glacial facies in the pre-Quaternary geological record: the fascinating story of pre-Quaternary glacial history of Cenozoic Antarctica by Hambrey & McElvey (2000), Barrett (2007), Naish *et al.* (2007, 2009) and Harwood *et al.* (2009) and; other examples from the older record are by Hambrey & Harland (1981), Crowell (1983), Miall (1985), Eyles & Eyles (1989), Eyles *et al.* (1993, 1998) and Levell *et al.* (1988).

# Part 6

# MARINE SEDIMENTARY ENVIRONMENTS

*'That is Cape Fear,' observed Mr Evans, pointing. 'And now you can see the division between the Gulf Stream and the ocean clearly. There, do you see, the line running parallel with our course, about a quarter of a mile away.'.*
*'A noble headland,' said Stephen. 'And a most remarkably clear division: thank you, sir, for pointing it out.'*

Patrick O'Brian, from *The Fortune of War*, Harper Collins, 1997 (Stephen Maturin in the *Leopard* off New England)

## Introduction

Marine waters and their diverse sedimentary environments form part of the coupled ocean–atmosphere heat engine that redistributes the latitudinally (zonal) unequal radiant heat energy received from the Sun (Part 6 Fig. 1). Solar heating drives atmospheric flows that cause direct wind drag of the surface waters. It also produces density differences during heating, cooling, evaporation and precipitation. The nature of the forces acting upon the ocean reservoirs depends upon: (i) *external* forces due to tides, direct wind shear and Coriolis force; (ii) *internal* forces caused by horizontal pressure gradients arising from variations in the height of the oceanic surface, by horizontal and vertical gradients arising from variations in density due to temperature, suspended particles and salinity, and by friction at the ocean boundaries.

It might be thought that the watery 60% of our planet is rather more insensitive to perturbations in climate, sea-level change and tectonics than the continental land surface, but over the past decades it has become clear that the oceans are linked to continental, atmospheric and tectonic events in an intimate and teleconnected way. The most obvious link comes during icehouse times from the massive transfer of oceanic water vapour into high-latitude ice, a distillation that varies in strength according to the various Milankovitch-band controls on the strength of incoming solar shortwave radiation. Even during current highstand the thermohaline circulation of the entire deep ocean is controlled by polar to tropical transfer of deep cool water and its contained suspended sediment and nutrients. Pleistocene lowstand conditions only involved a few per cent change in water depth over most of the ocean mass but exposure of the shelves was sufficient to vastly increase the transfer of sediment and nutrients into the deep ocean, since the shelf sediment sink was no longer working. Such bypass conditions created shelf-edge deltas that dumped their loads of sediment and nutrients directly on to the continental slope in many cases, generating conditions ripe for sediment slumping and gravity flow transfer directly on to the continental rise and ocean floor. These glacial oceans were not only well-ventilated by

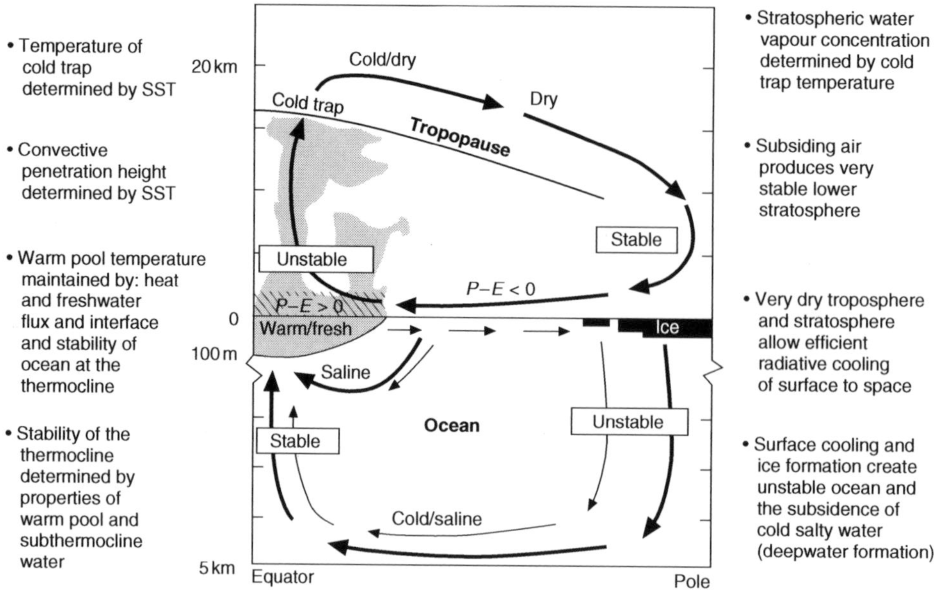
• Temperature of cold trap determined by SST
20 km
• Convective penetration height determined by SST
• Warm pool temperature maintained by: heat and freshwater flux and interface and stability of ocean at the thermocline
0
100 m
• Stability of the thermocline determined by properties of warm pool and subthermocline water
5 km
Cold/dry
Cold trap
Dry
Tropopause
Stable
Unstable
P–E < 0
P–E > 0
Warm/fresh
Ice
Saline
Ocean
Unstable
Stable
Cold/saline
Equator
Pole
• Stratospheric water vapour concentration determined by cold trap temperature
• Subsiding air produces very stable lower stratosphere
• Very dry troposphere and stratosphere allow efficient radiative cooling of surface to space
• Surface cooling and ice formation create unstable ocean and the subsidence of cold salty water (deepwater formation)

**Part 6 Fig. 1** Components of the coupled ocean–atmosphere system: SST, sea-surface temperature. (After Webster, 1994.)

thermohaline flow but they were also benefiting from another continental teleconnection—enhanced fall-out of windblown dust and associated adsorbed nutrients from more energetic trade and circumpolar winds.

There is no doubt that the shallow ocean has the most complicated system of sediment routing, since wave, tide and river input all contribute to the motion of water and the transport of sediment. Thus the tides are strongly unsteady and may be rotary, wave climate is a strongly seasonal factor and may be enhanced by tropical storms and river input depends upon the timing and nature of extreme flood waves generated in continental catchments. There is increasing evidence that this latter process has a profound effect upon the sedimentary architecture of coastal sediments, despite the fact that the smoothing processes of fair-weather wave and normal tidal currents serve to superficially return a coast to normal morphology over a few years. In many cases the river floods too will be controlled by oceanic teleconnection events like the El Niño-southern oscillation (ENSO), north Pacific oscillation (NPO) and Azores–Iceland Atlantic oscillations (AIAO), etc. It is also apparent that coastal and shelf environments will exhibit changed process regimes as sea level varies over time, from lowstand to highstand and back again.

We must also consider the geological impermanence of oceans. Plate motions are adept at shifting connections, switching gateways, changing volume and causing obliteration of entire ocean basins. In interpreting the oceanic sedimentary record, therefore, it pays to be well-informed about ocean tectonic history. It is as well to remember that only 5 Myr ago, at the dawn of the Pliocene Period when primitive hominids were just evolving, a waterfall debouched from the Atlantic ocean through the Straits of Gibralter down the continental slope and into the dried-up bed of the Mediterranean basin, filling it at rates of rise perhaps many metres per year. Such events are witnessed by the oceanic sedimentary record.

A final point concerns the tedious business of nomenclature. Part 6 Fig. 2 gives a useful scheme for the classification of nearshore and coastal oceanic

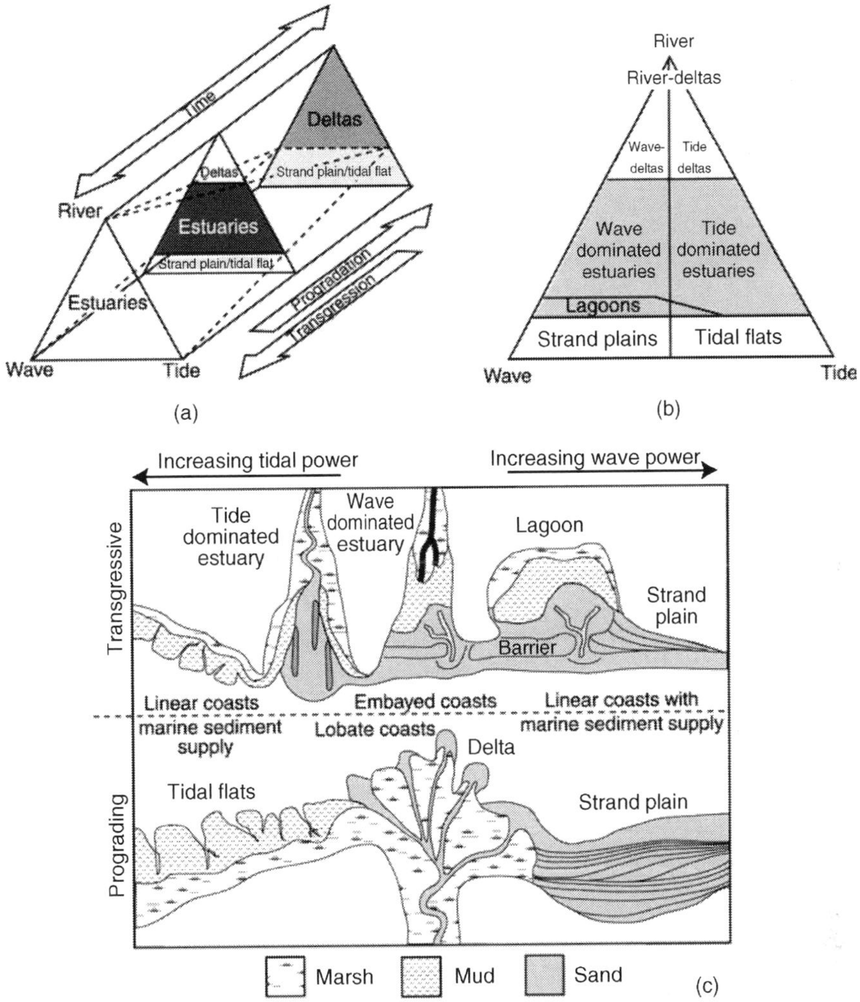


**Part 6 Fig. 2** Various sketches and ternary diagrams that have been proposed to help classify and differentiate coastal siliciclastic environments. (a) This is an evolutionary classification of clastic coastal depositional environments. The long axis of the three-dimensional prism represents relative time, with reference to relative sea level (transgression) and sediment supply (progradation). (After Dalrymple *et al.*, 1992; Boyd *et al.*, 1992; Harris *et al.*, 2002.) (b) The middle triangular section of the prism reflects relative wave/tide power (the *x* axis) versus the rate at which the river delivers sediments to the coast (*y* axis). Deltas occupy the uppermost area of the ternary diagram. Estuaries are located in the intermediate trapezoidal area and coastal strand plains and tidal flats occupy the base of the ternary diagram. C) Plan-view maps for idealized coastal depositional environments, showing the relationships between wave and tidal power, prograding and transgressive environments, and different geomorphic types.

environments which should help the reader to navigate the early chapters of Part 6. The ternary classification of coastal environments according to the relative strength of wave, tide and river power in the figure receives strong support from calculations carried out around the Australian coast.

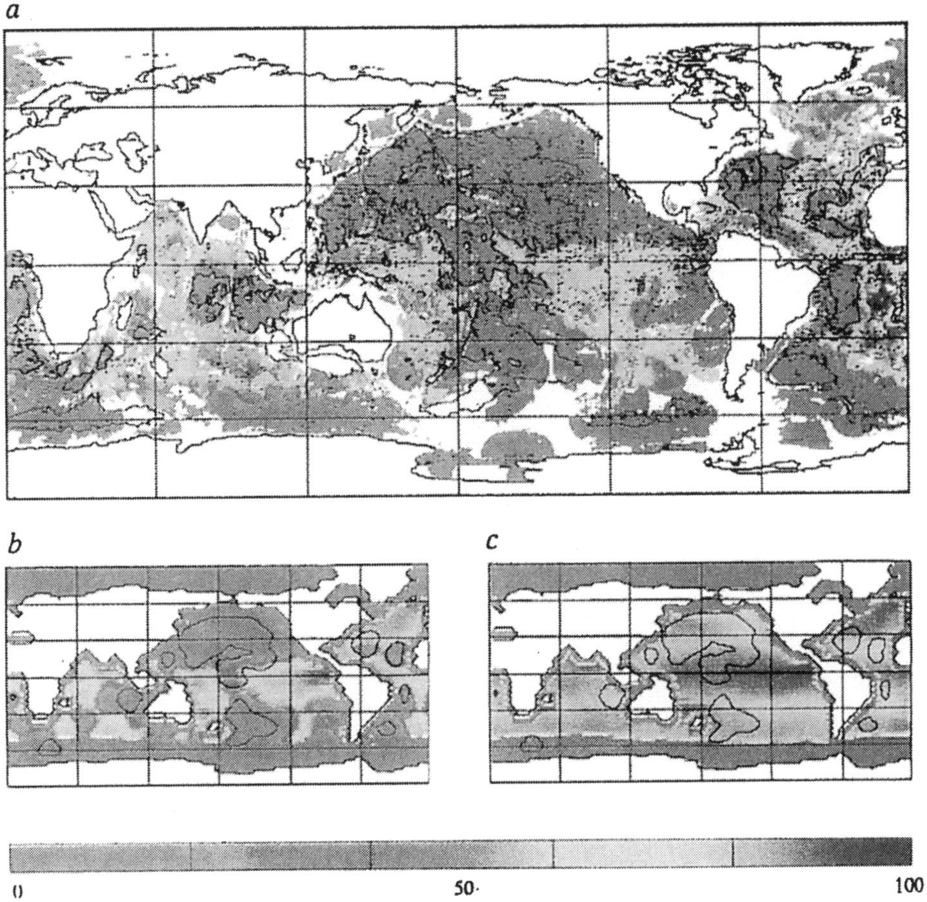


**Plate 1** (a) Distribution of sedimentary calcite (dry weight per cent) in seafloor sediments. (b) Modelled distribution using present day $CaCO_3$ fluxes and a coupled ocean-sediment C-cycle model. (c) Modelled distribution resulting from an increased deep-sea $CaCO_3$ flux (×2.6 present day level) such as might have drawn down the $CO_2$ content of the atmosphere into the glacial ocean. The fact that such abundances of $CaCO_3$ are not seen in glacial-age ocean sediments indicates that another mechanism for drawdown must exist. Contour is the 5 km isobath (Figures and discussion after Archer and Maier-Reimer 1994.)

**Plate 2** Spectacular vertical axis Kelvin-Helmholtz wave eddy at the interface between the Solimos and Negro tributaries to the Amazon. Solimos is loaded with suspended sediment and is brown. Negro water is organic-rich and black. View about 10 m across. (Photo by and courtesy of M. Pèrez-Arlucea.)

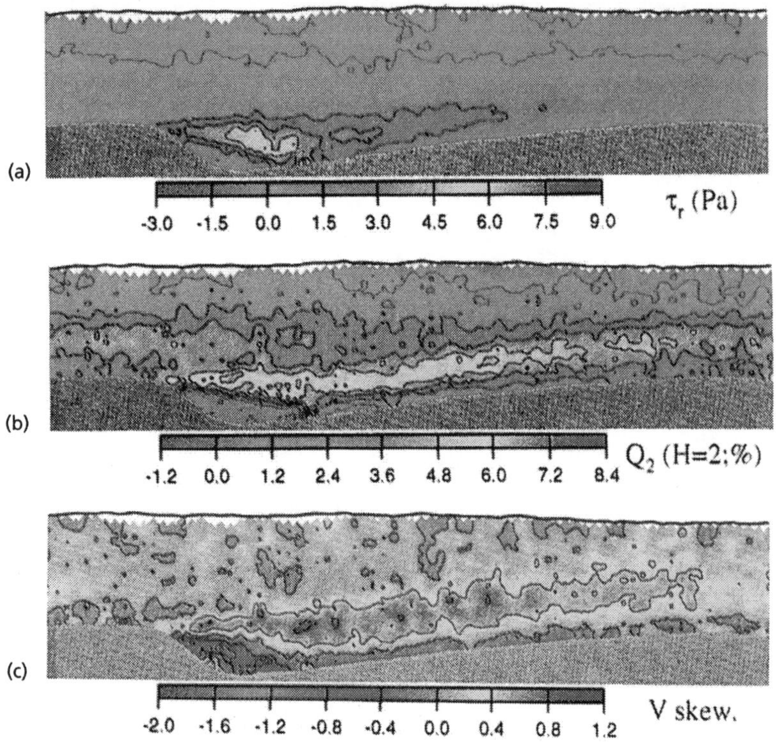


**Plate 3** Contour maps (Bennett and Best 1995) of flow and turbulence parameters measured by Laser-Doppler anemometry over fixed experimental dune-like bedforms (colourscale forms at base of each image). Flow is from left to right in each case. (a) Distribution of Reynolds stress $\rho\overline{u'}\,\overline{v'} = \tau_r$. Note the high stresses extending to the bed associated with the free shear layer of the leeside roller of separated flow and their persistence but gradual decay downstream. (b) Percentage distribution of Quadrant 2 bursting events. Note the persistence of bursting intensity along the downstream extension of the free-shear layer associated with eddies arising from Kelvin–Helmholtz instability (c) Distribution of skewness of the vertical velocity, reinforcing the plot of (b).

**Plate 4** Ground view of large star-shaped dune from the Erg Mehedjibat, 100 km south of In Saleh, Algeria. Dune is approximately 300 m high. (Photo courtesy of R. Dixon.)

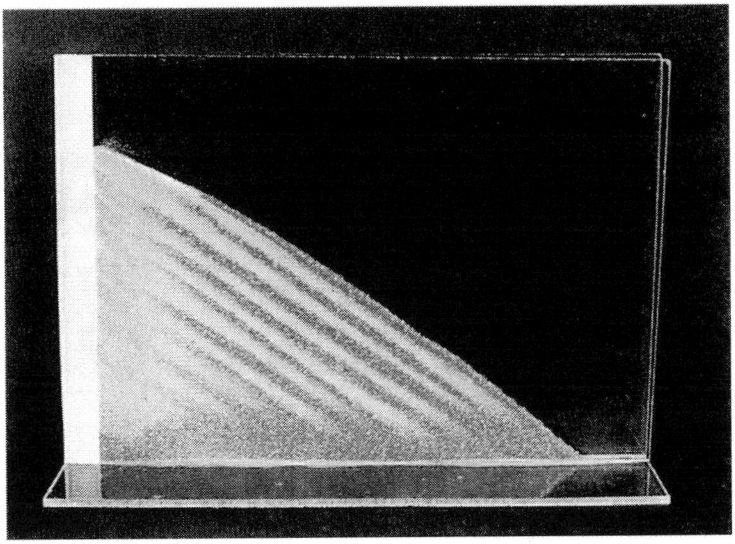

**Plate 5** Experimental (Makse *et al.* 1997) segregation and stratification produced when a mixture of grains is poured into the left hand side of a perspex gate of width 5 mm. Typical result showing the formation of successive layers (about 1.2 cm wide) of fine and coarse grains where the white grains are glass beads of average diameter 0.27 mm, while the larger red grains are sugar crystals of typical size 0.8 mm. (Photo courtesy of H. Makse.)

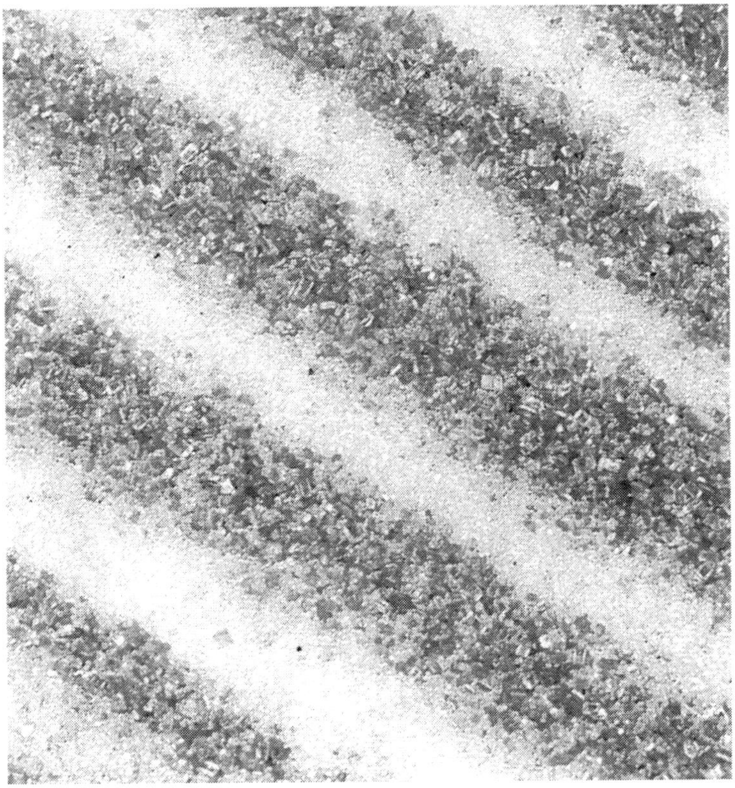

**Plate 6** Close-up of laminations in P.ate 5 showing the coarse/fine couplets. (Photo courtesy of H. Makse.)

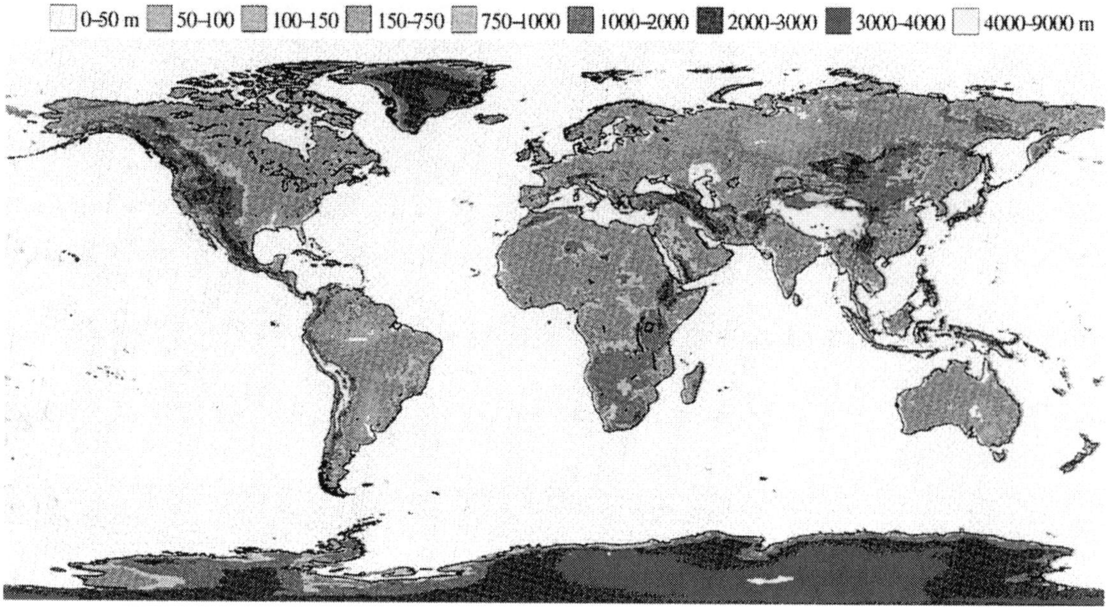


**Plate 7** Map of global continental elevation from DEM of 30′ resolution.

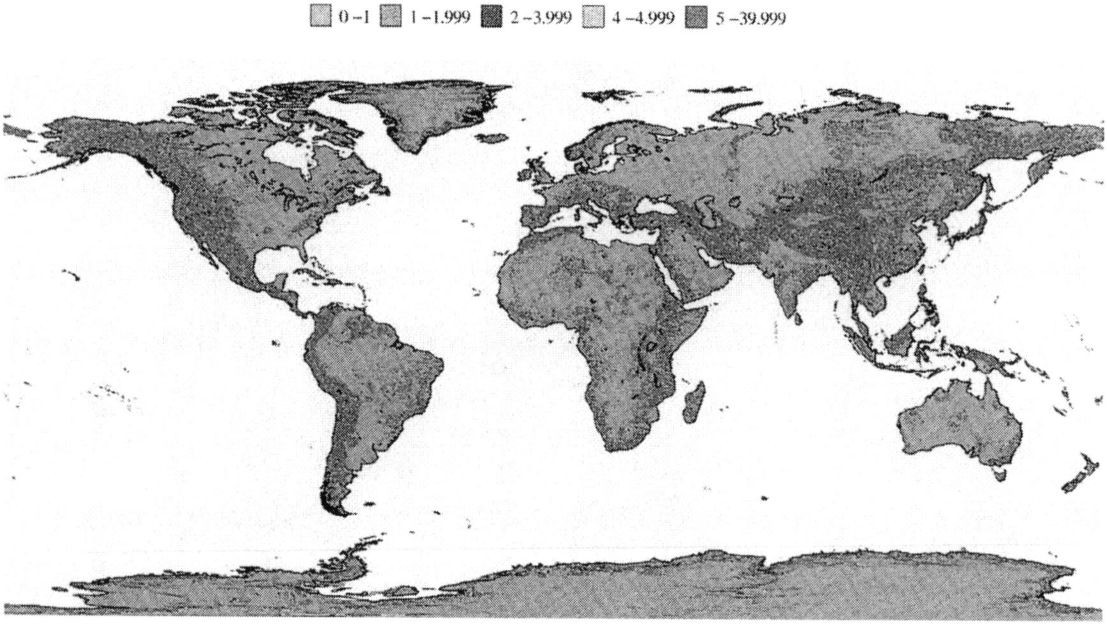


**Plate 8** Map of global continental gradients from DEM of 30′ resolution.

**Plate 9** Glacial meltwater outflows from a sandur braidplain turning rapidly into underflows at well-defined plunge points; glacial Lake Peyto, Alberta, Canada. (Photo courtesy of M. Pérez-Arlucea; see also Chikita *et al*. 1996; and Fig. 14.3)

**Plate 10** NASA multispectral thermal infra-red satellite image of Badwater saline basin, Death Valley, California (Crowley and Hook 1996). Most striking feature is the fringing zone of magenta and salmon-coloured pixels (I, H) representing a sodium sulphate-halite composite phase around the central dark (J) smooth-halite flats. Spectrum K red pixels are gypsum surface crusts. Green (L) and dark blue (M) pixels are siliciclastic mudflats that surround playa. (Photo courtesy S. Crowley, USGS.)

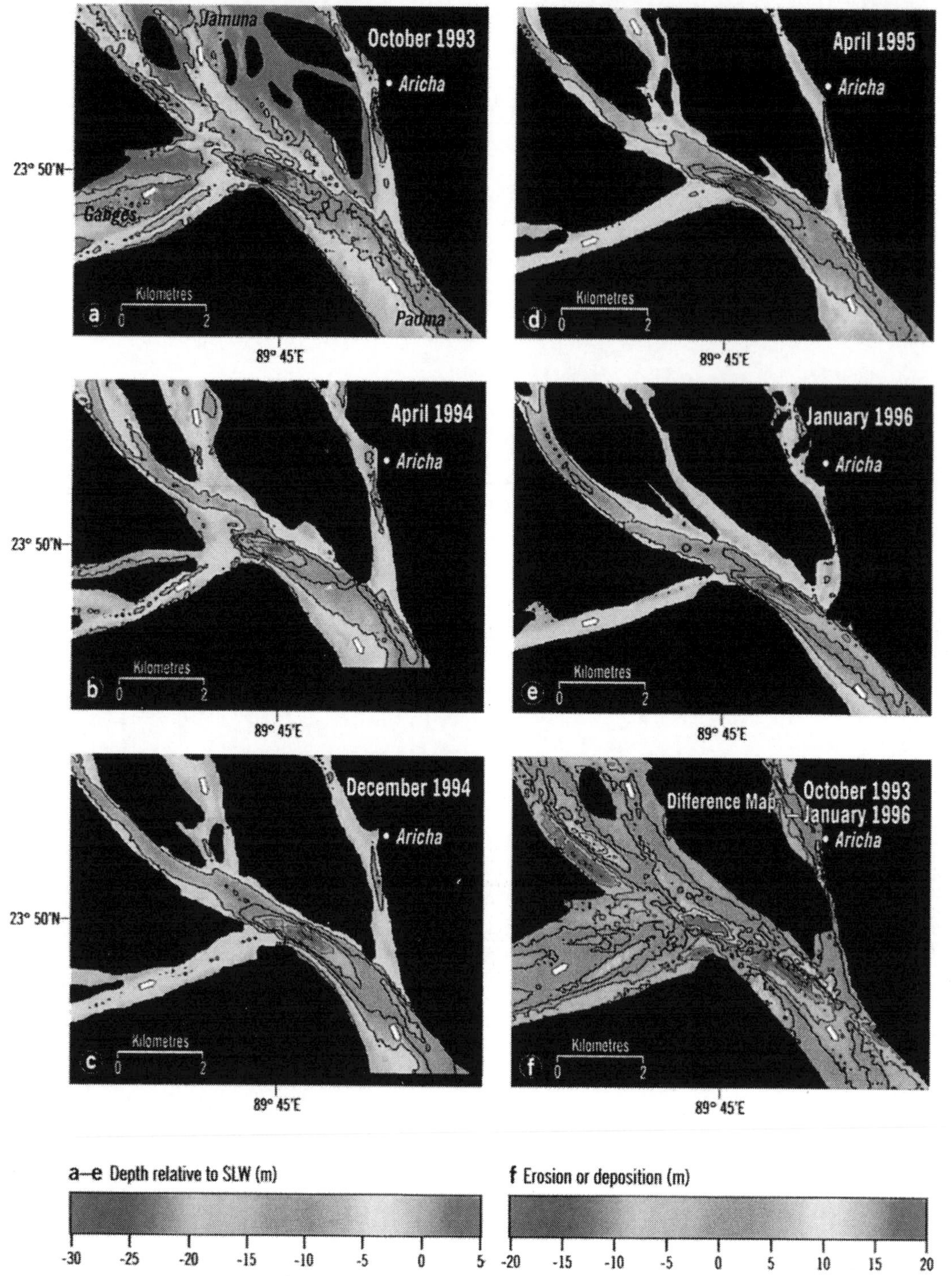


**Plate 11** Bathymetry of the Jamuna–Ganges confluence region, Bangladesh, over a 28-month period, for five surveys between October 1993 and January 1996. Bed heights are expressed relative to a standard low water datum (SLW) at Aricha. Parts (a)–(e) show bed morphology for each survey period; part (f) shows the change in bed level over the total survey period. Note the large fluvial conference scours, which erode up to 27m below present sea level and are highly mobile in the lateral sense. These features mean that apparent ancient incision due to sea level fall must be treated with caution. (After Best & Ashworth, 1997; images courtesy of Jim Best and Phil Ashworth.)

**Plate 12** Near snout of the Glacier, South Island, New Zealand. Note extreme dirtiness of the glacier ice surface which is largely covered with debris from surrounding slopes. (Photograph courtesy of Bruce Yardley.)

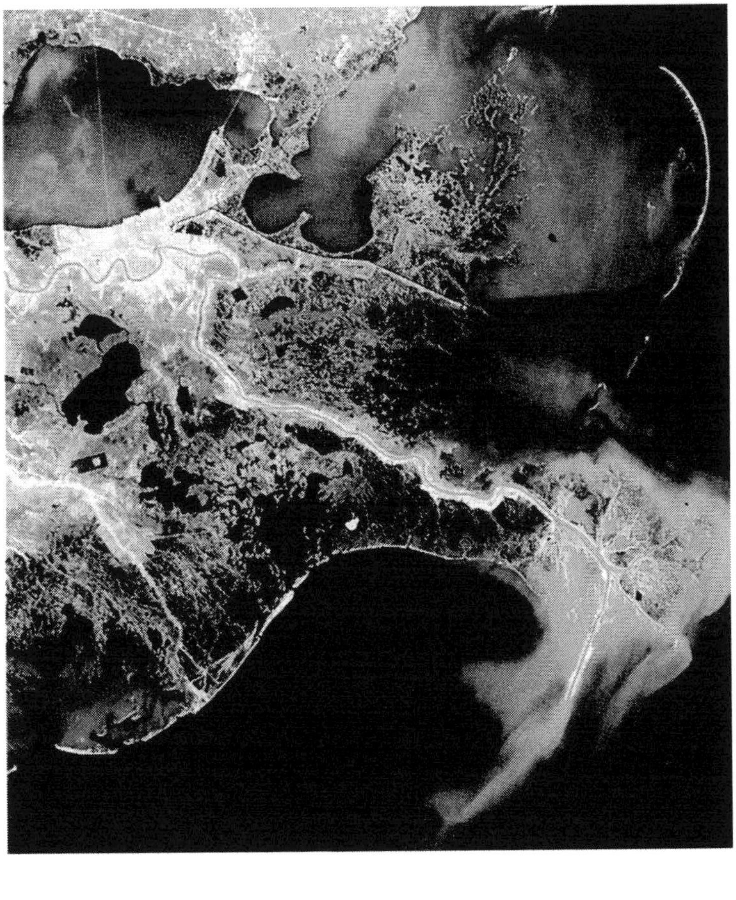

**Plate 13** Infrared colour composite of the Mississippi delta as photographed by the Landsat satellite on 9 April 1976. Vegetation appears red, clear water dark blue and suspended sediment light blue. Note the south-westerly deflection of the sediment plumes by the prevailing shelf currents as they enter the Gulf of Mexico from the modern birdsfoot distributary mouths. This westward transport includes massive introduction of farmland-runoff with nutrients; leading to widespread annual hypoxia all along the Louisiana–Texas Shelf (see Malakoff, 1998) (Landsat images 2443–15 and 2443–15 462 as assembled and discussed by G. T. Moore (1979).) (See Fig. 17.14)

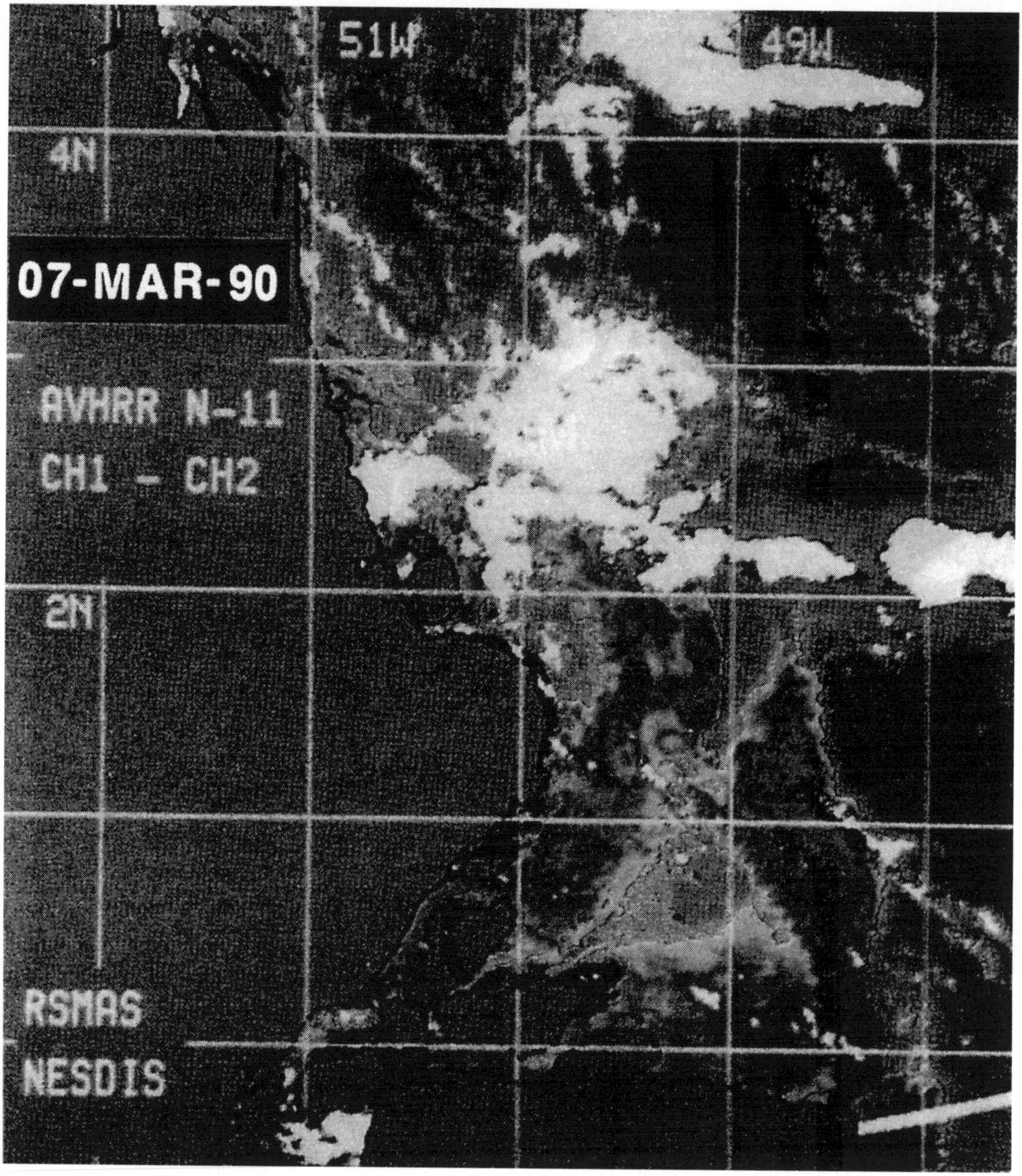


**Plate 14** Satellite image of the turbid plume emanating from the mouth of the Amazon River in March 1990. This image was obtained with the Advanced-Very-High-Resolution-Radiometer (AVHRR) on the NOAA 9 satellite, using a composite image created from channel 1 (visible) and channel 2 (near infrared). The most turbid water is located on the continental shelf along the frontal zone, where tidal resuspension and convergence of the estuarine circulation create extremely high concentrations of suspended sediment. The plume of turbid water extending seaward and northward from the river mouth shows considerable variability in structure, reflecting temporal fluctuations on the order of days to weeks. Note also turbid surface water extending northwards along the coast. (Image used courtesy of C. Nittrouer.)

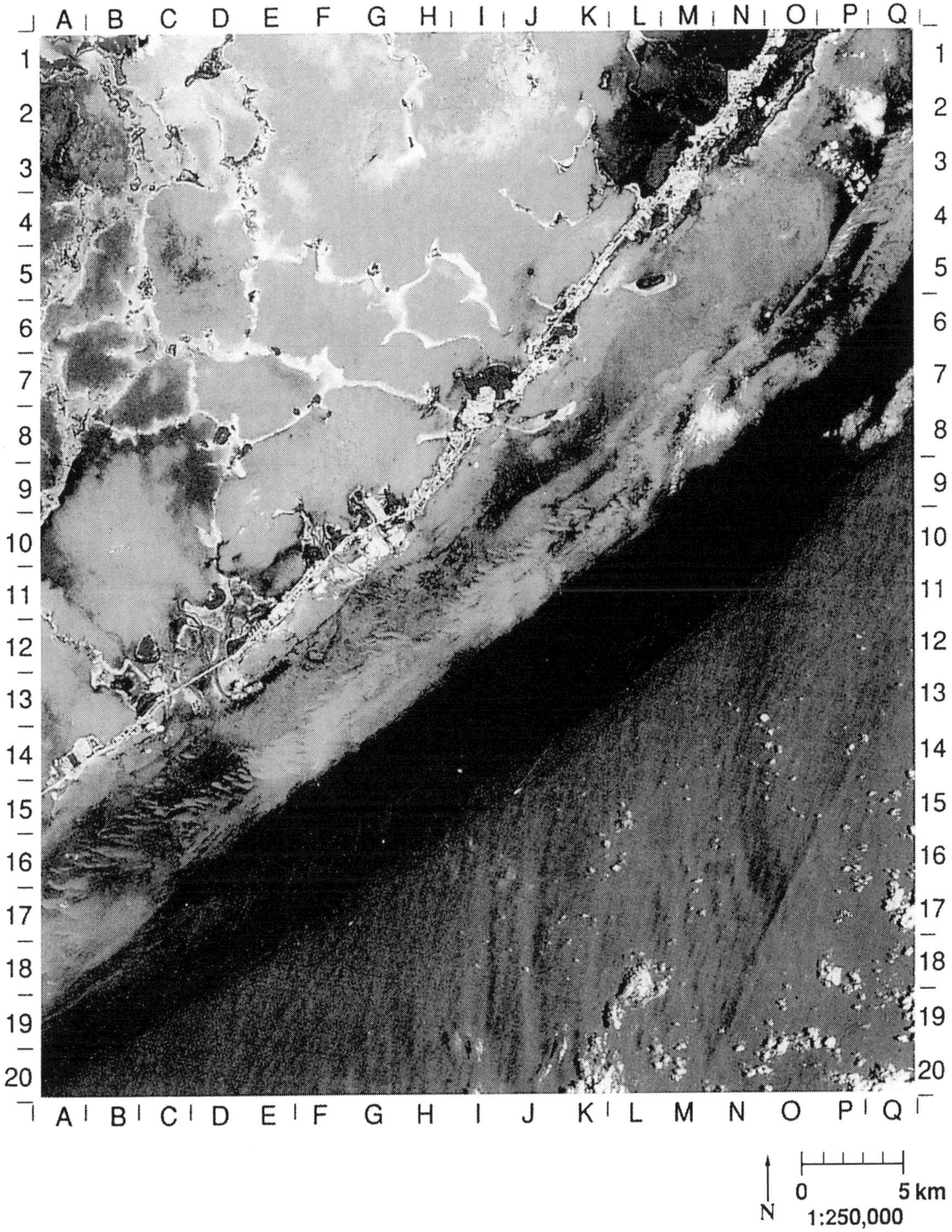


**Plate 15** Landsat TM image of the south-central part of the Florida barrier/lagoon carbonate complex (from Harris and Kowalik 1994). Note: Florida Bay mudbanks and islands; the Keys (barriers) with their ebb and flood tidal deltas; The inner reef with muddy carbonate and local buildups. The outer reef with prominent dunes in skeletal sands. (See Fig. 20.1)

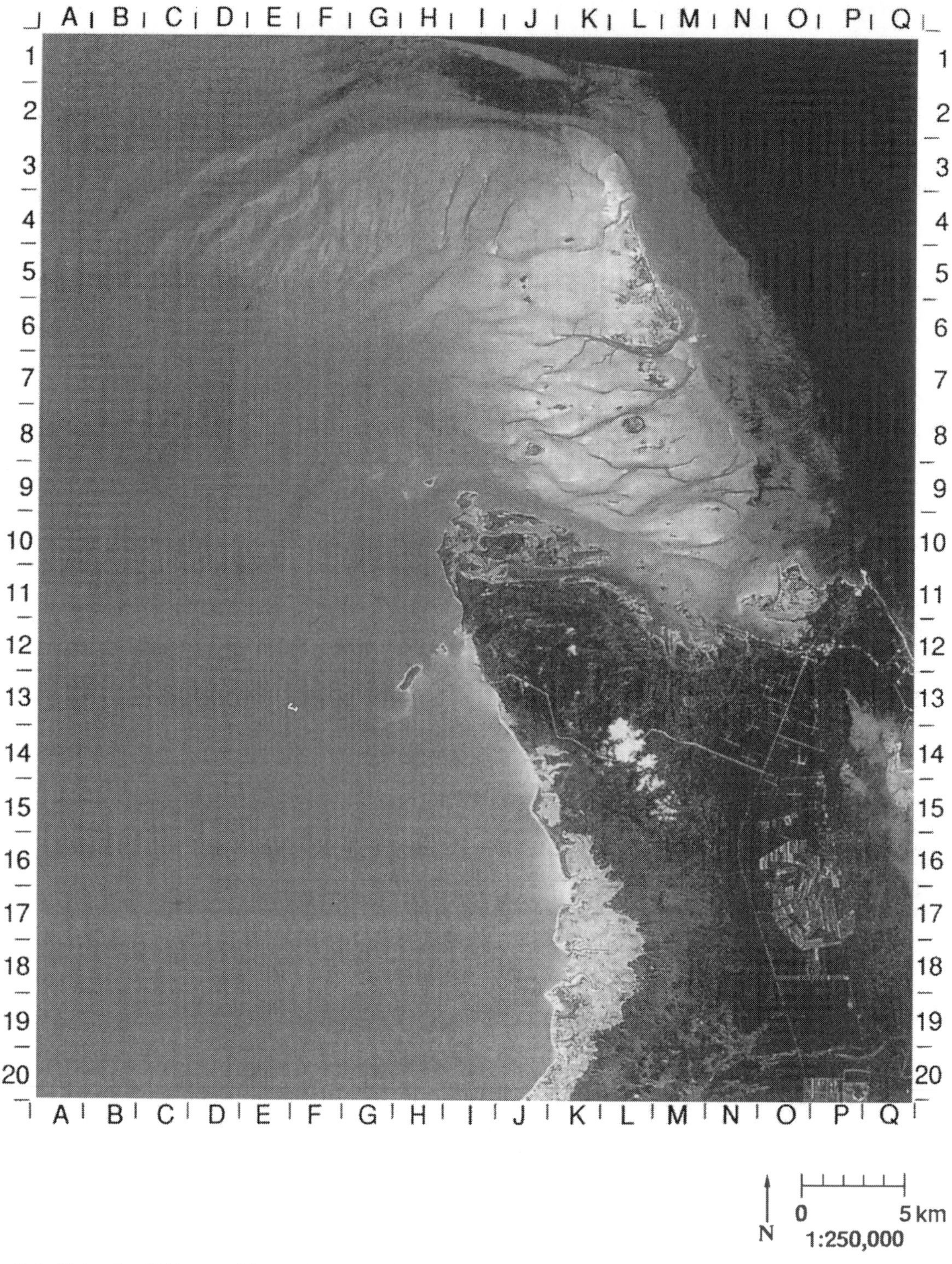


**Plate 16** Landsat TM image of the northern part of the Great Bahama Bank with the northern tip of the Island of Andros, the deepwater tongue of the ocean to the north-west (from Harris and Kowalik 1994). Image shows coastal mudflats and drainage channels of the leeward side of Andros; windward platform bordering the tongue of the ocean; the prominent tidal channels and oolite sand shoals to the leeward of Joulters Cay. (See Figs. 20.1, 20.9 and 20.32).

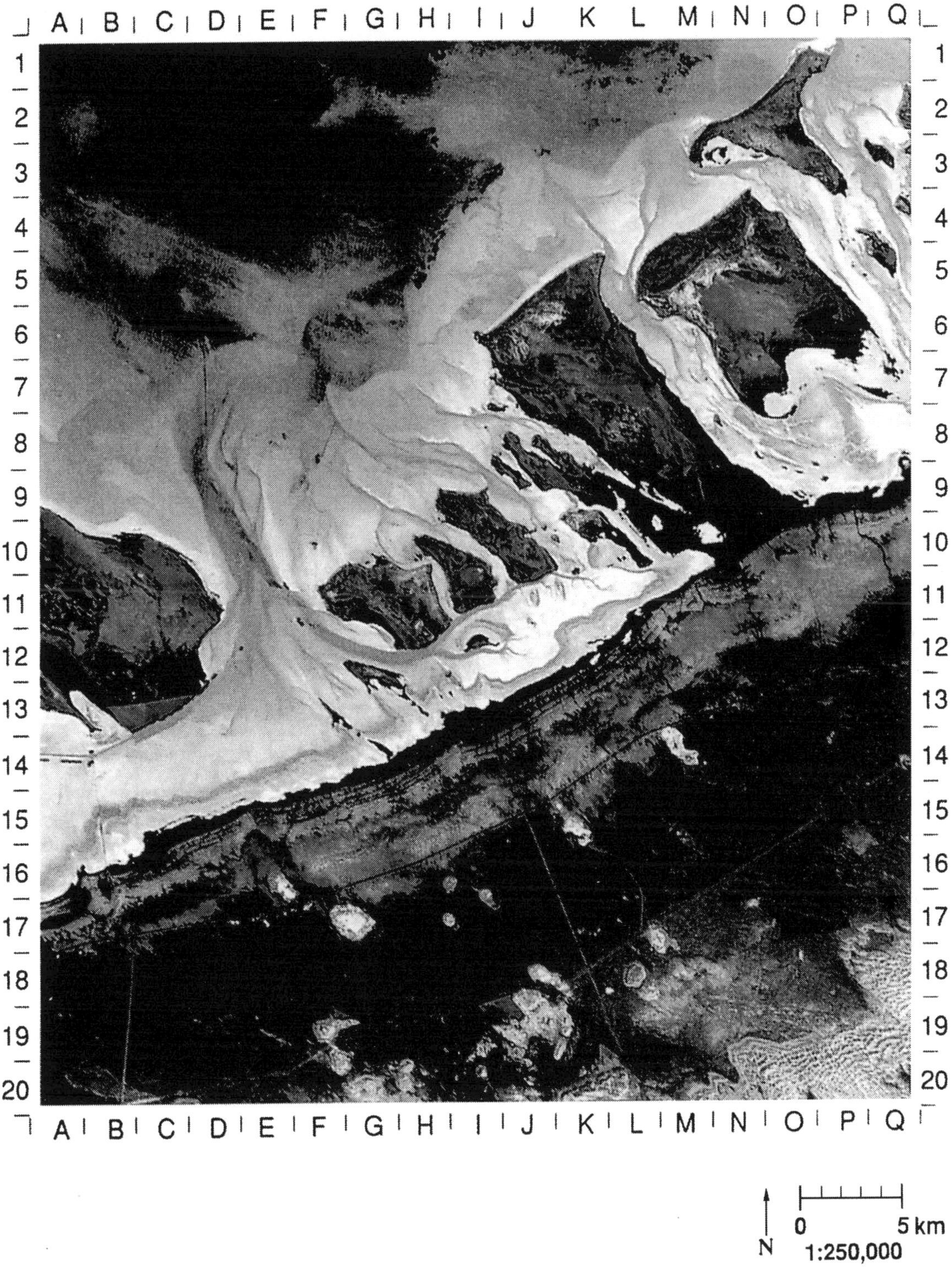


**Plate 17** Landsat TM image of eastern Abu Dhabi area, Trucial Coast (from Harris and Kowalik 1994). Note barrier islands; intervening tidal inlets and ebb tidal deltas; ooid and skeletal sand tidal chanels; dark intertidal stromatolite flats; sabkha with numerous beach ridges; fringe of desert dunes to the SW. (See Section 2.15 and Fig 20.5)

# Chapter 16

# ESTUARIES

*'These, I take it, are sandbanks,' said Stephen.*
*'Just so. And the little figures show the depth at high water and at low: the red is where they are above the surface.'*
*'A perilous maze. I did not know that so much sand could congregate in one place.'*

Patrick O'Brien [Stephen Maturin and Jack Aubrey consulting charts], *Post Captain*, Harper Collins

## 16.1 Introduction

Estuaries are funnel-shaped embayments that narrow upstream, the site of interaction between a freshwater river and marine tidal currents and waves (Fig. 16.1). An estuary is distinct from a delta with its active and inactive distributaries and from a tidal inlet with no incoming river. It is 'a funnel-shaped coastal body of water which has a free connection with open ocean and within which seawater is measurably diluted with a riverine input of freshwater'. Such a definition restricts the term to the dynamic interface between riverwater and seawater as measured by salinity. Many individual delta distributary outlets (Chapter 17) show estuarine characters by this definition. A more sedimentologically appropriate definition goes 'the seaward termination of a single river channel which receives a fluvial and marine sediment flux that is acted upon by tidal, wave and fluvial forces'. Modern estuaries formed in response to early Holocene sea-level rise and are most common along low-lying coastal plains as the successors to drowned lowstand river valleys bounded by pre-Holocene sediment or older bedrock. The occurrence of fluvial incision and subsequent tidal scouring is vital here, for in the absence of confining valley walls a tide- or wave-dominated delta distributary network would form. In view of these points, the abundance of modern estuaries may be unusual compared to some periods of the geological past—the Mesozoic for example, when eustatic sea level changes were of comparitively low magnitude.

## 16.2 Estuarine dynamics

Water and sediment dynamics in estuaries are closely dependent upon the relative magnitude of tidal, river and, to a lesser extent, wave processes. The incoming progressive tidal wave is modified as it travels along a funnel-shaped estuary whose width and depth decrease steadily upstream (**Cookie 44**) and the mixing of fresh- and saltwater causes estuarine circulation in response to density gradients. Sedimentary particles may be of both marine and river origin, with flocculation (section 6.8), floc destruction by turbulent shear and resuspension of faecal pellet material as important controls upon particle size. The concentration of suspended particulate matter depends upon local resuspension of bottom sediment in the upper estuary (tidal pumping effect) and the effects of gravitational

*Sedimentology and Sedimentary Basins: From Turbulence to Tectonics*, 2nd edition. © Mike Leeder.
Published 2011 by Blackwell Publishing Ltd.

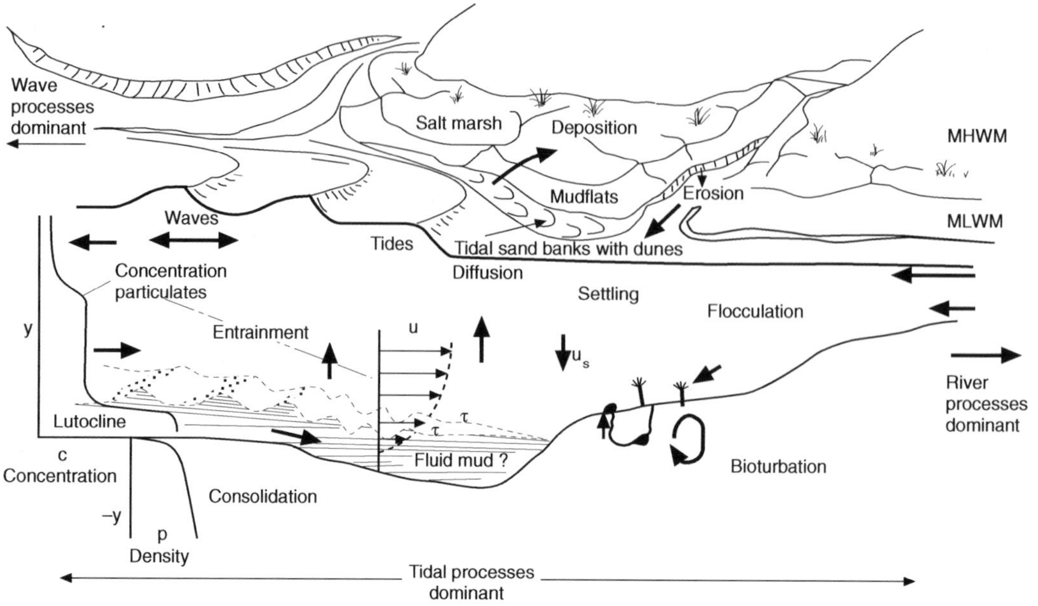


**Fig. 16.1** Sketch to illustrate the major processes affecting estuarine erosion and sedimentation. Variables mentioned are: $y$, height above bed; $-y$, depth below bed; $c$, concentration of sediment; $\rho$, deposit density; $u$, flow velocity; $\tau$, bed shear stress; $u_s$, settling velocity; MHWM, mean high water mark; MLWM, mean low water mark. Discussion of the sedimentation and erosion of fine-grained sediment particles and of stratified flows is to be found in sections 5.7, 6.8 & 7.4, as well as in the present chapter. (Mostly after Dyer, 1989.)

circulation. The higher the amount of suspended clay, the higher the probability of particle collisions, leading to agglomeration and production of sedimenting aggregates whose settling velocity is now considerably enhanced. At the same time, the higher the particle concentration, the lower will be the rate of settling due to the effects of particle hindrance according to the Richardson–Zaki settling equation (Chapter 6). These two effects, agglomeration and hindrance, lead to the formation of distinct layers of suspended material during the period of relatively slack water in estuaries where tidal currents are important (Fig. 16.2). The net accumulation of sediment in the water column due to tidal pumping arises because of inequality in the local magnitude of the ebb and flood tides. If the flood is dominant in the upper estuary, as is often the case, then more sediment enters the upper estuary than leaves, and hence a turbidity maximum occurs.

For more details on tidal amplification in estuaries see Cookie 44

The most fundamental way of considering estuarine dynamics is through the principle of mass conservation, which states in mathematical form that the time rate of change of salinity at a fixed point is caused by two contrasting processes: diffusion and advection. Diffusion is restricted to the flux caused by turbulent mixing whilst advection is the mass flux associated with circulation and various internal breaking waves. Viewed in this way, water dynamics in estuaries may be conveniently represented by four major end-members (Fig. 16.3), which intergrade one with another. It is of vital importance, however, to realize that a single estuary may change its hydrodynamic character from season to season according to changing river, wave and tidal conditions.

*Type A, well-stratified, estuaries* are those river-dominated estuaries where tidal processes are permanently or temporarily at a minimum. The stratified system is dominated by river discharge, with the tidal/river discharge ratio being low, ca $< 20$ (Fig. 16.4). At depth, an upstream-tapering saltwater wedge develops over which the fresh river-water flows as a

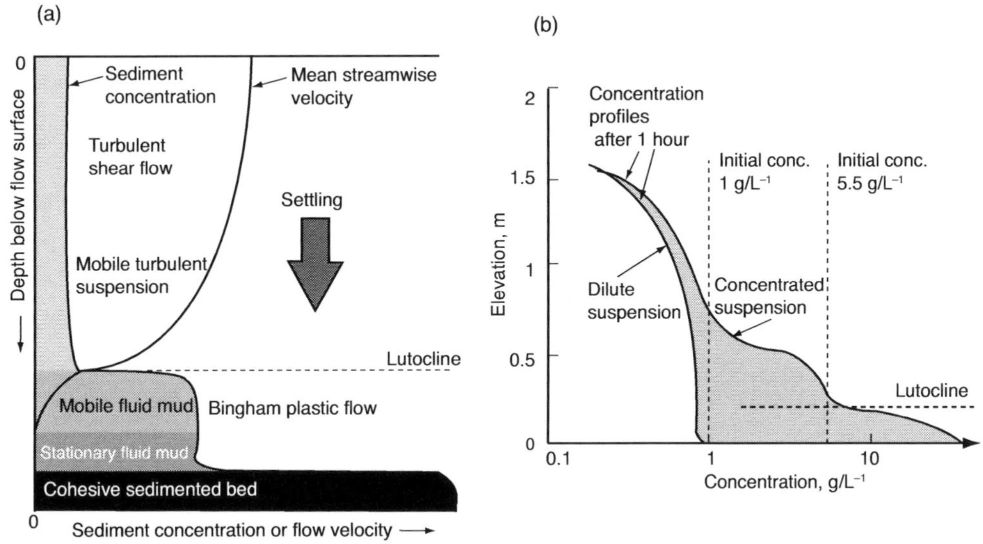


**Fig. 16.2** (a) Graphoid to show idealized suspended sediment concentration and mean velocity profiles relevant to estuarine fluid-mud settling. (b) Experimental suspended sediment concentration profiles with time for dilute and concentrated initial suspensions. Hindered settling and flocculation cause a lutocline to develop. (Data of Ross; as reported in Mehta, 1989.)

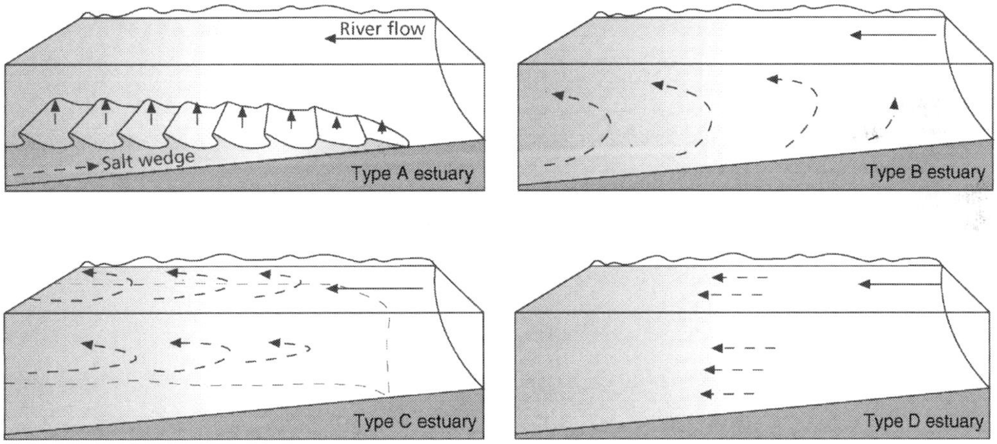


**Fig. 16.3** Diagrams to show the four major types of estuarine circulation originally defined by Pritchard & Carter (1971).

buoyant plume. The scenario is exactly the same in river-dominated delta distributaries such as the modern Mississippi delta (Chapter 17) and is further exemplified by the estuarine-like behaviour of the distributaries of the Fraser River delta, Canada, during periods of high river flow ('freshet') discharge (Fig. 16.5). Internal waves are thought to form at the sharp salt-wedge/river-water interface and these cause limited upward mixing of saltwater with freshwater (advection), but not vice versa. A prominent zone of shoaling at the tip of the salt wedge arises when sediment deposition from bedload occurs in both

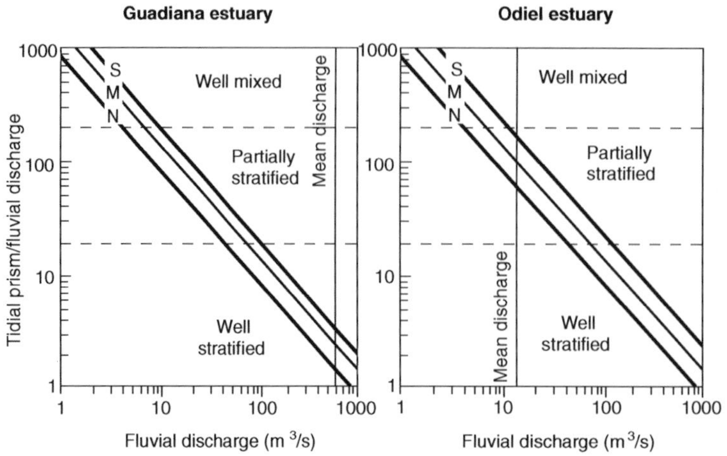


**Fig. 16.4** The division of estuaries into types A (stratified), B (partially stratified) and C (well mixed) according to the ratio of tidal discharge to that of fluvial discharge, with data for mean fluvial discharge and spring (S), mean (M) and neap (N) tides from the Guadiana and Odiel estuaries, southern Spain. (After Borrego *et al*., 1995.)

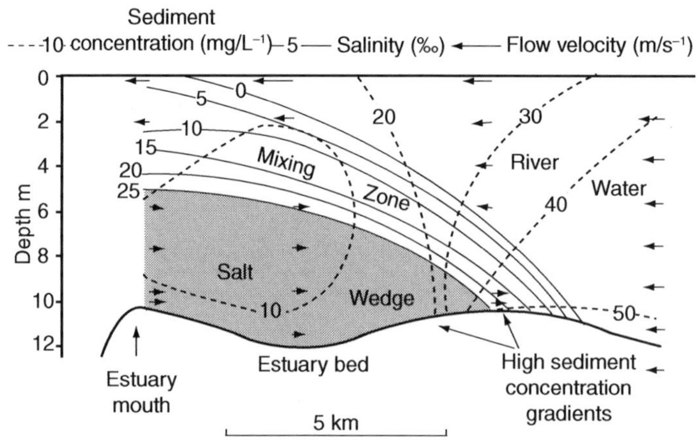


**Fig. 16.5** Salinity, velocity and suspended sediment profiles taken during high tide along a transect of the well-stratified (salt wedge) type A estuary of the Fraser River, Canada. (After Kostaschuk *et al*., 1992.)

freshwater and seawater. This zone of deposition shifts upstream and downstream in response to changes in river discharge and, to a much lesser extent, to tidal oscillation. Thus, deposited fine bedload sediment and flocculated suspended load are periodically flushed out of the system by turbulent shear during high river stage. Further upstream, freshet flows cause rapid dune and bar nucleation with overall down-channel migration.

*Type B, partially stratified, estuaries* are those in which tidal turbulence destroys the upper salt-wedge interface, producing a more gradual salinity gradient from bed to surface water by both advectional and diffusional mechanisms. The tidal/river discharge ratio is ca 20–200. Downstream changes in salinity gradient occur at the mixing zone which moves upwards towards higher salinities. The Coriolis force causes the mixing surface to be slightly tilted so that in the northern hemisphere the tidal flow up large estuaries is nearer the surface and strongest to the left bank (as seen by an observer facing down-river). Sediment dynamics is strongly influenced by the upstream and

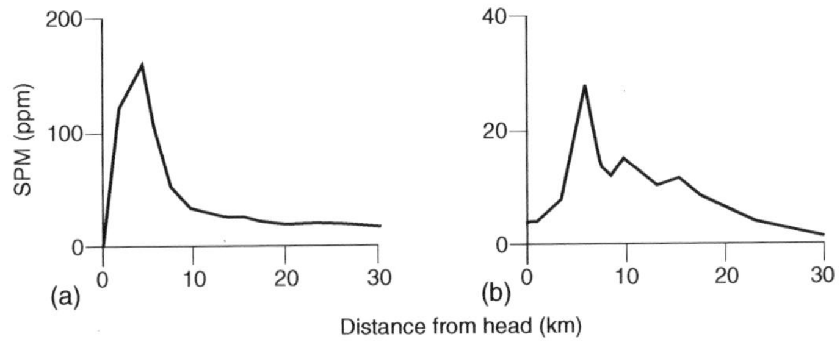


**Fig. 16.6** Typical longitudinal distribution of depth-averaged concentration (ppm) of suspended particulate matter (SPM) in a typical macrotidal, partially mixed estuary. (a) Large neap tide, low runoff; (b) small neap tide, low runoff. Note well-marked turbidity maxima, particularly during the large neap tide. (Data from Tamar estuary,England, Uncles & Stephens, 1989.)

downstream movement of saltwater over the various phases of the tidal cycle. The resulting *turbidity maximum* (Fig. 16.6) is particularly prominent in the upper estuary (~1–5‰ salinity) during spring and large neap ebb and flood tidal phases, and less prominent at slackwater periods due to settling and deposition. Turbidity maxima are affected by the magnitude of freshwater runoff. In strongly seasonal climates a cycle of dry-season upstream migration of the turbidity maximum and locus of maximum deposition is followed by wet-season downstream migration and resuspension by erosion. The turbidity maximum is also acted on by gravity-induced circulations arising from its own excess density. Such effects are responsible for setting up a stratified flow (section 5.7) with a near-bed high density layer whose interface with clearer water above is marked by production of intermittent internal waves which serve to moderate turbulent transport.

*Type C, well-mixed, estuaries* are those in which strong tidal currents completely destroy the salt-wedge/freshwater interface over the entire estuarine cross-section; the tide/river discharge ratio is >200. Longitudinal and lateral advection and lateral diffusion processes dominate. Vertical salinity gradients no longer exist but there is a steady downstream increase in overall salinity. In addition, the rotational effect of the Earth may still cause a pronounced lateral salinity gradient, as in type B estuaries. Sediment dynamics are dominated by strong tidal flow, with estuarine circulation gyres produced by the lateral salinity gradient. Extremely high suspended sediment concentrations may occur close to the bed in the inner reaches of some tidally dominated estuaries. Sediment particles of river origin, some flocculated, will undergo various transport paths, usually of a 'closed loop' kind (Fig. 16.7), in response to settling into the salt layer and subsequent transport by the net upstream tidal flow. Settling of bound aggregates of silt- and sand-sized particles creates large areas of stationary and moving mud suspensions that characterize the outer estuarine reaches of tide-dominant estuaries. For example, major areas of high-density suspensions (up to >200 g/L) are a major feature in the outer Severn estuary, England. These suspensions may be mobile or fixed, the latter grading into areas of more-or-less settled mud. Stationary suspensions up to 3 m thick may be deposited very quickly, cores through such deposits revealing structureless muddy silts with occasional thin sandy laminae. Stationary suspensions show sharp upper surfaces on sonar records. They form during slackwater periods, progressively thickening during the spring-to-neap transition. They are easily eroded by the accelerating phases of spring tidal cycles, to be taken up in suspension once more.

*Type D estuaries* are theoretical end-members of the estuarine continuum in that they show both lateral and vertical homogeneity of salinity. Such conditions apply only in the outer parts of many type B and C estuaries; they are clearly transitional to well-mixed open shelf conditions. Under equilibrium conditions, saline water is diffused upstream to replace that lost by advective mixing. Sediment movement is dominated entirely by tidal motions, again with no internal sediment trap.

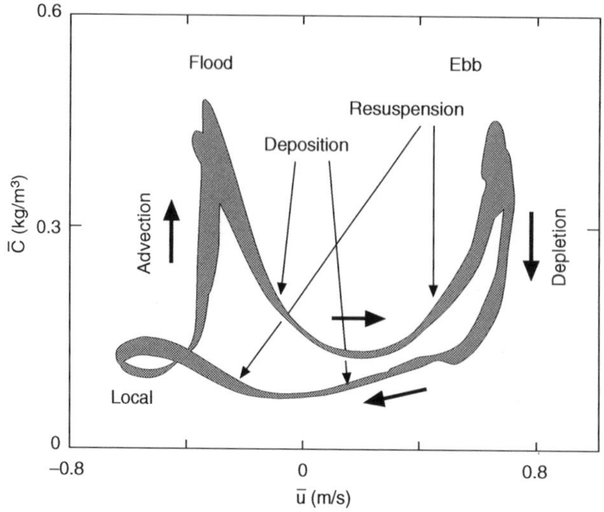


**Fig. 16.7** A remarkable graph to illustrate the repeatable hysteresis (over three tidal cycles, but not all details shown) of suspended sediment concentration ($\bar{C}$) with mean tidal current velocity ($\bar{u}$) at a measuring station in the well-mixed and macrotidal Weser estuary, Germany. Velocities are negative for flood, positive for ebb. Results indicate local advection from downstream at the height of the flood, then deposition during the slack, then resuspension during the ebb, and finally redeposition during the slack. Some net deposition at the site is the end result. (After Grabemann & Krause, 1989.)

## 16.3 Modern estuarine morphology and sedimentary environments

The hydrodynamic review of estuaries stressed the interplay between river and tide so that many estuaries show type A or B characters and hence tend to act as trappers of sediment, particularly of fine grades. Whilst recognizing the efficiency of such estuaries as sediment traps, it should also be pointed out that advective plumes of seaward-directed fine sediment may be driven by both residual tidal and wave currents far out on to the shelf (Chapter 19). Although hydrodynamic estuary classification is of some sedimentological use, behaviour varies according to season and climate. This is best exemplified by estuaries along the macrotidal coastline of northern Australia. These are usually tide-dominated (tidal ranges up to 10 m) type C or D, but during the monsoonal season the tremendous freshwater efflux causes lengthy periods of type B behaviour. A more serious problem with the hydrodynamic classification is that wave energy is ignored. For example, on the wave-dominated and microtidal coastline of southern Africa, beach-barrier cusps and spits partially shield the inner estuary and prevent simple salt-wedge intrusion (Figs 16.8 & 16.9). These beach barriers are periodically destroyed during periodic large river discharges, with the resulting deltas subsequently destroyed by wave action as the barriers reform during long periods of normal river discharge.

Given these complexities, for sedimentological purposes estuaries and their facies belts are best divided morphodynamically into tide-dominated (with seasonal river dominance in some cases) with coast-parallel facies belts, wave-dominated with coast-transverse facies belts and mixed energy with coast-oblique facies belts (Fig. 16.10).

Estuaries with appreciable tidal ranges and low incident wave power are bordered by coast-parallel tidal flats and salt marsh (mangrove swamps in tropical estuaries) with drainage channels. Rapid mud deposition rates occur on intertidal flats during spring tide turbidity maxima, though net rates over time are much reduced by periodic seasonal storm-induced wave erosion and sediment export in outer estuary environments, as in the Seine macrotidal estuary, France. Heavy rainfall at low tide also removes significant sediment from the flats, more than the maximum transport rate at spring ebb in tidal channels of the Solway Firth, Scotland for example, and also marked

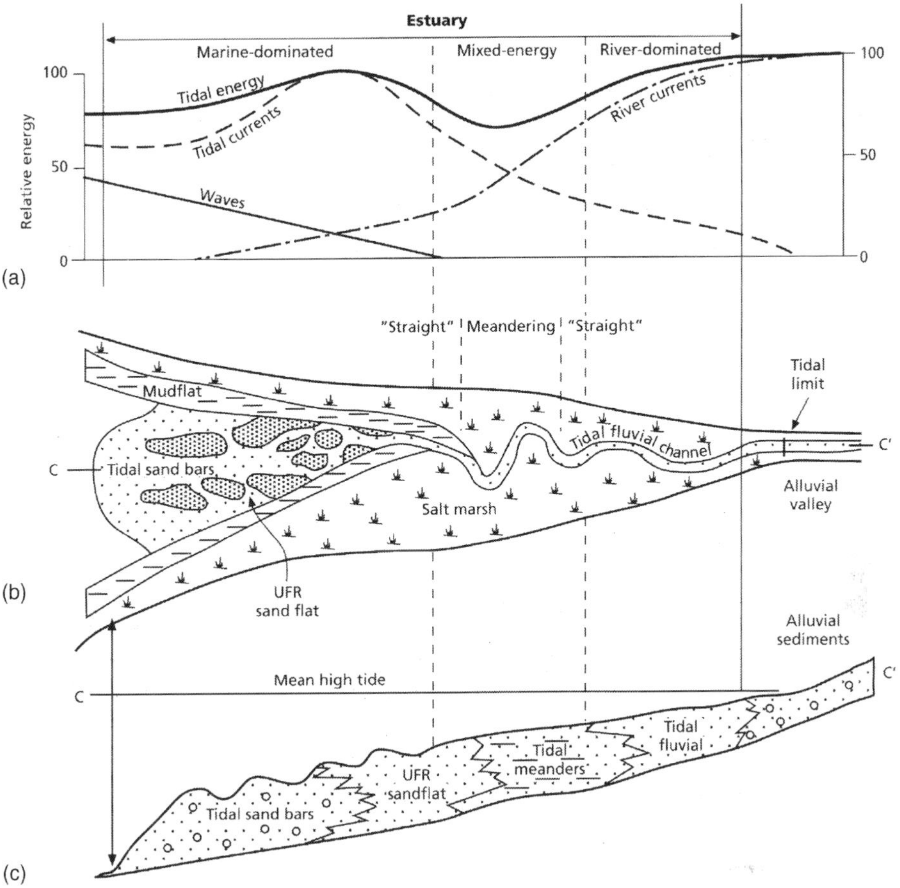


**Fig. 16.8** Seaward facies transitions, grain-size trends and palaeocurrent directions indicating deposition in a tide-dominated estuary as applied to the Eocene deposits of Spitsbergen. Palaeocurrent measurements derived from cross-strata are shown in black, and dip direction of inclined master surfaces in grey. (Estuary model is after Dalrymple *et al.*, 1992, as redrawn by Plink-Björklund, 2005.)

tidal channel erosion. Seaward progradation of funnel-shaped estuarine complexes on a coastal plain cause the deeper-channel estuarine environments to become progressively overlain by fining-upwards tidal-flat sediments (Chapter 20). Studies of main estuarine channels such as the Parker estuary, New England, and of the outer estuary of the Bay of Fundy show well-developed ebb and flood tidal channels and associated shoals with extensive bordering dunefields. Frequently the flood tide dominates in the upstream estuary and the ebb tide dominates in the downstream, as in the Gironde estuary, France where the orientation of dunes is adjusted to this spatial variation in net tidal vector. At the point of equal influence, a suite of symmetrical dunes develop. The magnitude, type, migration rate and orientation of estuarine tidal shoal dunes depends strongly on timing with respect to the neap/spring cycle, quasi-equilibrium forms evolving at and shortly after the spring tidal phase. Large (>2 m high) dunes cannot be altered by the fortnightly tidal variation but must reflect longer-term changes in estuarine dynamics and river floods. Vertical sequences through such sediments reveal alternating ebb- and flood-orientated cross-stratification with clay interbeds (heterolithic stratification; section 18.5) resting on channel-floor erosion surfaces.

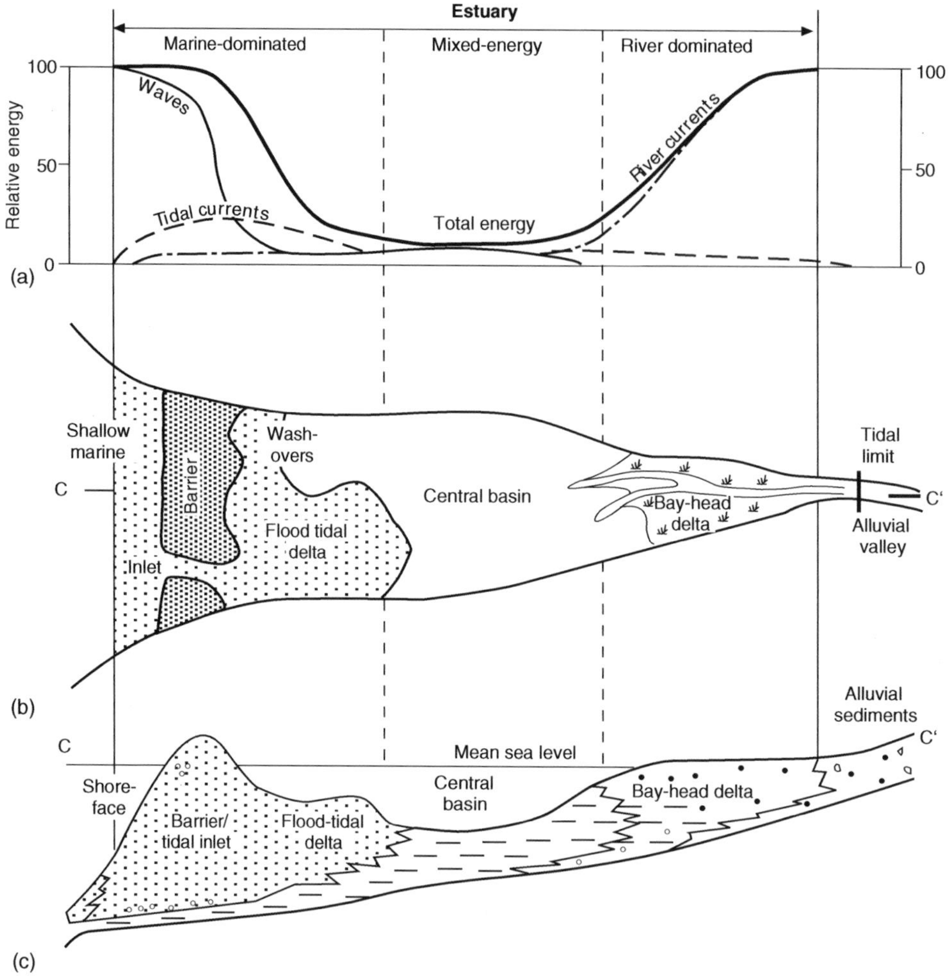


**Fig. 16.9** Sketches and sections to illustrate the salient sedimentary features of partially barred, wave-dominated estuaries. (After Dalrymple *et al.*, 1992.)

Sediment particles are dominantly derived from offshore in many outer well-mixed estuaries. Thus in the Thames estuary, England, fully marine ostracods are found up to 20 km inland from their life habitats and the sediment shows a decreasing mean grain size upstream.

Wave-dominated estuarine systems are partially closed at their mouths by either single or double *beach-barrier spit* systems which form because there is sufficient wave power to generate longshore drift of sand-to-gravel grade sediment from adjacent or distant sources. Entrance and exit of the tidal prism occurs through a *tidal inlet* with both ebb and flood tidal deltas, the latter also feeding sediment on to the barrier shoreface during storms (Chapter 18). Their presence at an estuary mouth in meso- to macrotidal regimes produces a wave-sheltered inner estuary dominated by tidal processes with tidal flats and channels—sometimes the main upper estuarine fluvial channel may feed a *bay-head delta*. Overall the estuarine sediment belts are transverse to the estuary trend. In extreme examples when the tidal prism is small and sediment supply large a frontage barrier may completely seal off the tidal inlet to a formerly

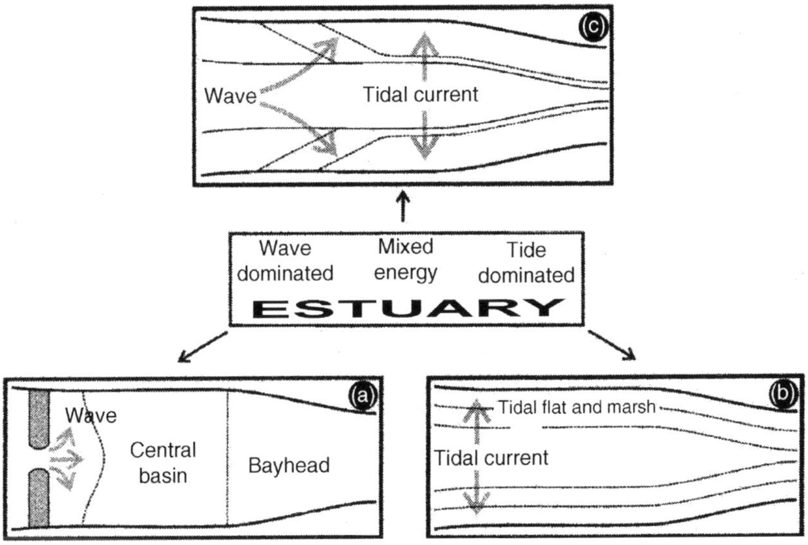


**Fig. 16.10** To show the facies distribution in the transition from (a) wave-dominated estuaries with limited tidal energy through (b) tide-dominated estuaries with insignificant wave energy, to (c) mixed-energy estuaries. The significant penetration of open-sea waves into mixed-energy estuaries, where wave refraction and attenuation occur contemporaneously, allows the facies zonation to be differentiated from the two (a)/(b) end-member estuarine settings. (After Yang *et al.*, 2007.)

marine-connected estuary: a well documented example is that of the barrier-bound, fluvially infilling Ouémé River estuary, Benin, West Africa whose shoreface frontage is fed by copious sand from the 200-km-distant Volta delta (Fig. 16.11).

Many open estuaries, typified by the Gomso Bay estuary, western Korea, have mixed energy regimes that feature up-estuary decreases in wave energy so that physical structures pass from wave-dominated planar lamination and hummocky cross-stratification to tide-dominated heterolithic stratification. Mappable trends in sedimentary facies and ichnofacies appear to be *oblique* to the estuarine margin in the outer and middle bays because of wave refraction, whereas facies belts in the inner bay are parallel to the estuary margin, reflecting tide-dominated conditions.

## 16.4 Estuaries and sequence stratigraphy

The identification of incised lowstand valleys with rising stage and highstand estuarine sediment infill is a key element in the steps of logic that led to the reconstruction of ancient episodes of relative sea-level change (Chapter 10). The association between incised valleys and estuarine environments thus provides an important link in the sequence stratigraphic argument. The high-frequency changes of relative sea level in late-Pleistocene to recent times have enabled estuarine workers to make a decisive and important contribution here (Figs 16.12 & 16.13). For example, one prominent feature of Holocene estuarine systems in wave-dominated microtidal regimes is the development of bay-head deltas that prograde seawards to infill an estuarine incised valley. In macrotidal regimes the lateral sweeping of an energetic tidal channel will cause production of marked erosion surfaces (*tidal ravinement*) cutting both lowstand fluvial and rising sea-level deposits (Fig. 16.14).

Changing sea level in a seaward-deepening and widening estuary is accompanied by very significant changes in tidal resonance and hence tidal range (**Cookie 44**). Lowstand estuaries may be significantly different in terms of the relative strengths of wave, tide and river currents. This is exemplified by the Holocene evolution of the Delaware Bay estuary complex (Fig. 16.13). During relative sea-level rise the developing locus of turbidity maximum, and hence site of maximum mud deposition, migrated steadily

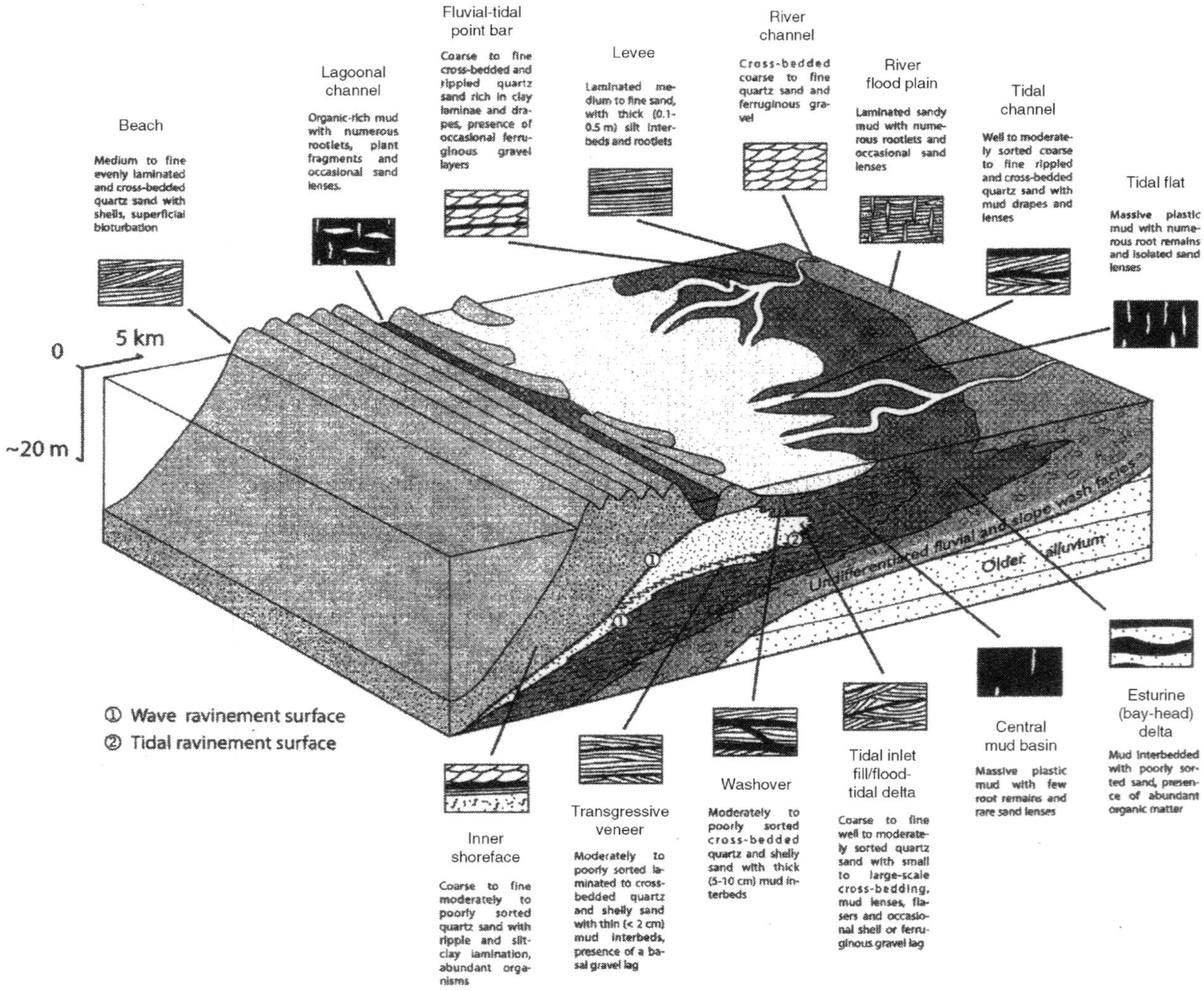
Beach
Medium to fine evenly laminated and cross-bedded quartz sand with shells, superficial bioturbation
Lagoonal channel
Organic-rich mud with numerous rootlets, plant fragments and occasional sand lenses.
Fluvial-tidal point bar
Coarse to fine cross-bedded and rippled quartz sand rich in clay laminae and drapes, presence of occasional ferruginous gravel layers
Levee
Laminated medium to fine sand, with thick (0.1-0.5 m) silt interbeds and rootlets
River channel
Cross-bedded coarse to fine quartz sand and ferruginous gravel
River flood plain
Laminated sandy mud with numerous rootlets and occasional sand lenses
Tidal channel
Well to moderately sorted coarse to fine rippled and cross-bedded quartz sand with mud drapes and lenses
Tidal flat
Massive plastic mud with numerous root remains and isolated sand lenses
0
5 km
~20 m
Undifferentiated fluvial and slope wash facies
Older alluvium
① Wave ravinement surface
② Tidal ravinement surface
Inner shoreface
Coarse to fine moderately to poorly sorted quartz sand with ripple and silt-clay lamination, abundant organisms
Transgressive veneer
Moderately to poorly sorted laminated to cross-bedded quartz and shelly sand with thin (< 2 cm) mud interbeds, presence of a basal gravel lag
Washover
Moderately to poorly sorted cross-bedded quartz and shelly sand with thick (5-10 cm) mud interbeds
Tidal inlet fill/flood-tidal delta
Coarse to fine well to moderately sorted quartz sand with small to large-scale cross-bedding, mud lenses, flasers and occasional shell or ferruginous gravel lag
Central mud basin
Massive plastic mud with few root remains and rare sand lenses
Esturine (bay-head) delta
Mud interbedded with poorly sorted sand, presence of abundant organic matter

**Fig. 16.11** Facies model of the barrier-bound, fluvially infilling Ouémé River estuary. (After Anthony *et al.*, 2002.)

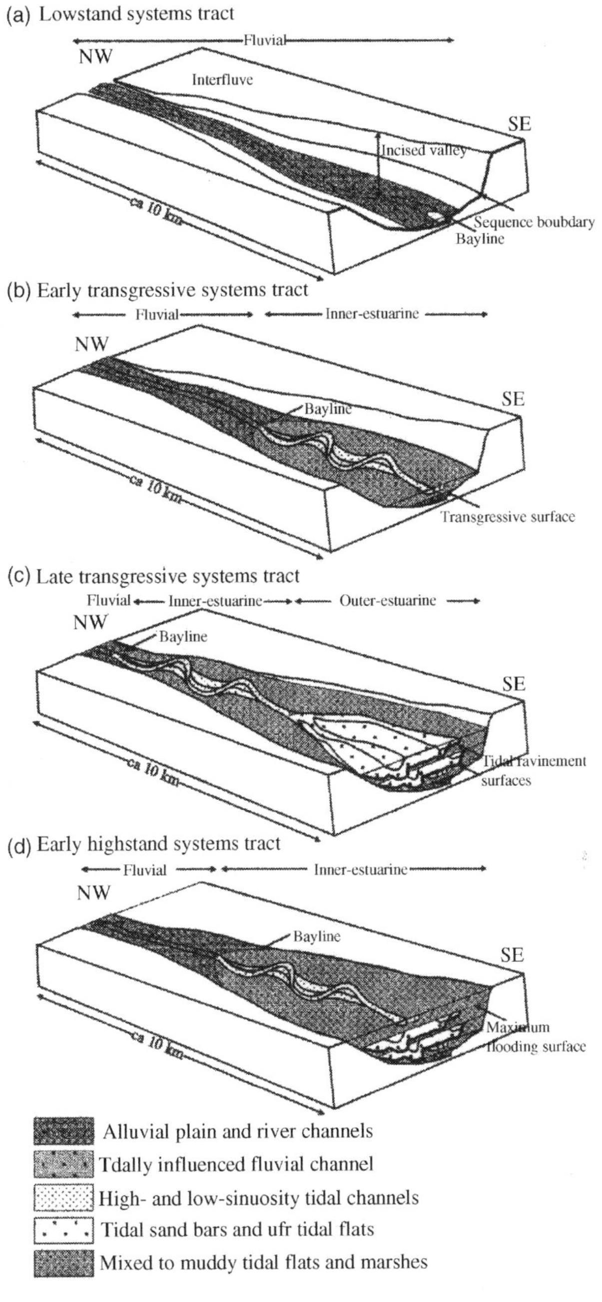


**Fig. 16.12** To illustrate estuarine sequence development. (a) During lowstand, fluvial deposits accumulate on the coastal plain in incised valleys eroded during sea-level falls. River channels are tidally influenced in their seaward ends. (b) & (c) Rising relative sea level drowns the valleys and landward-stepping estuarine deposits accumulate. The transgressive estuarine deposits young landwards as they onlap. (d) During the highstand systems tract (HST) the inner parts of the valleys start filling *in situ* and marshes develop over larger areas as the rate of sea-level rise decreases. Gradually the inner estuarine deposits shift seawards, and the youngest HST deposits cover the oldest transgressive deposits. (After Plink-Björklund, 2005.)

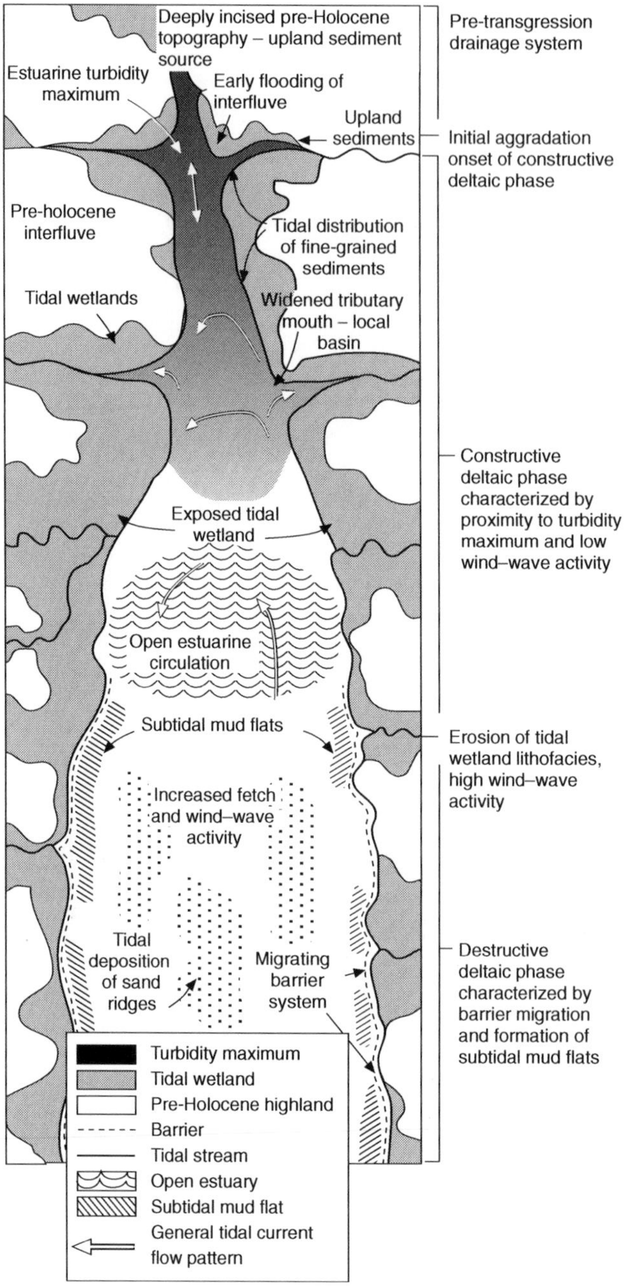


**Fig. 16.13** Sketch to illustrate the morphology of the modern Delaware estuary. The outer estuary was tidally dominant during the early Holocene sea-level rise but is now wave-dominated with the modern turbidity maxima and locus of fine-grained tidal wetland deposition having moved well up the estuary channel. (After Fletcher *et al.*, 1990.)

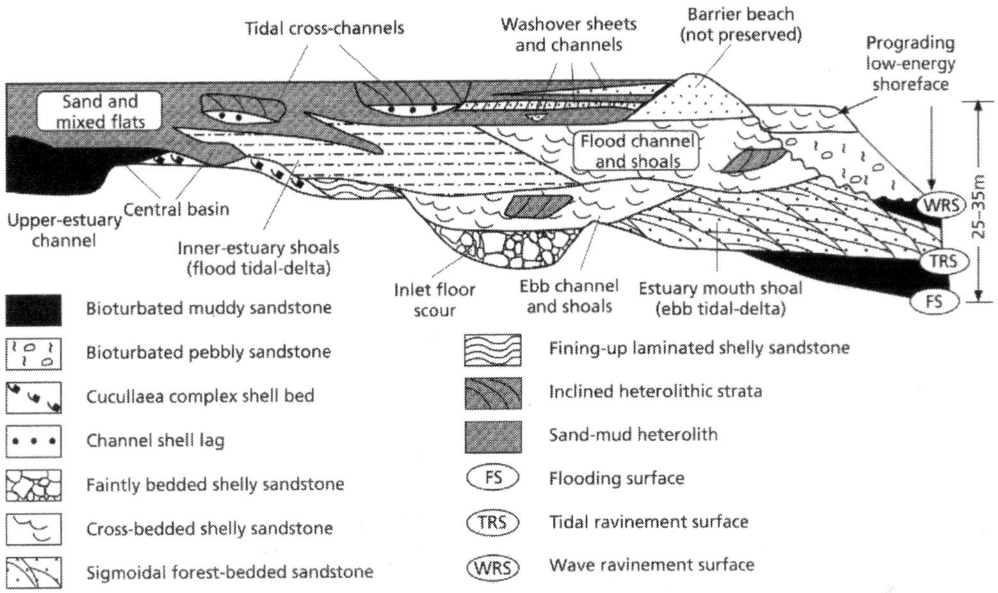


**Fig. 16.14** Sketch section based on Eocene rocks of the Antarctic peninsula to illustrate the lateral relationships between various estuary-fill environments following a transgression and subsequent progradation (After Porebski, 1995.)

upstream. Changes in wave power due to variations in climate or to the effects of estuarine channel migration on offshore water depth and tidal currents led to local episodes of tidal-flat erosion and deposition at highstand. The sediment infill of an estuarine complex is thus time- and space-dependent, as also illustrated by the Cobequid/Salmon River estuary in the Bay of Fundy.

Although the sedimentary architecture of late Quaternary incised valley-fills results from the interplay between sea-level rise, sediment supply and hydrodynamic processes, inundation of fluvial terraces complicates the issue and leads to backstepping parasequences whose timing does not coincide with eustatic sea-level variations. Thus the Trinity incised valley fill, Texas coast, lies under the modern Galveston estuary and formed above an irregular terraced topography. Flooding surfaces formed rapidly across relatively flat fluvial terraces during mid-Holocene decline in overall eustatic rise, up to 6.5 km per century: this led to reorganization of the entire estuarine complex and an associated landward shift in coastal facies.

A final interesting complication occurs in tectonically active areas like coastal south Alaska where coseismic subsidence of estuarine marginal wetlands in the 1964 subduction-related earthquake caused instantaneous transgression by marine flooding. This was followed by slow multidecadal progradation of a thin shallowing-upwards parasequence.

### Ancient estuarine facies

Identification of particular estuarine types in the sedimentary record depends upon recognizing coastal sands of strandlines, barrier spits, tidal inlets and deltas, bay-head deltas, tidal dunefields, intertidal flats and tidal channels (see Chapters 17 and 18). Recognition of a salinity spectrum will, of necessity, depend entirely upon biological inferences and it may therefore be difficult to recognize pre-Phanerozoic estuarine facies. An example of successions inferred to have formed in a Permian mixed wave/tide-dominated estuarine setting is shown in Fig. 16.15.

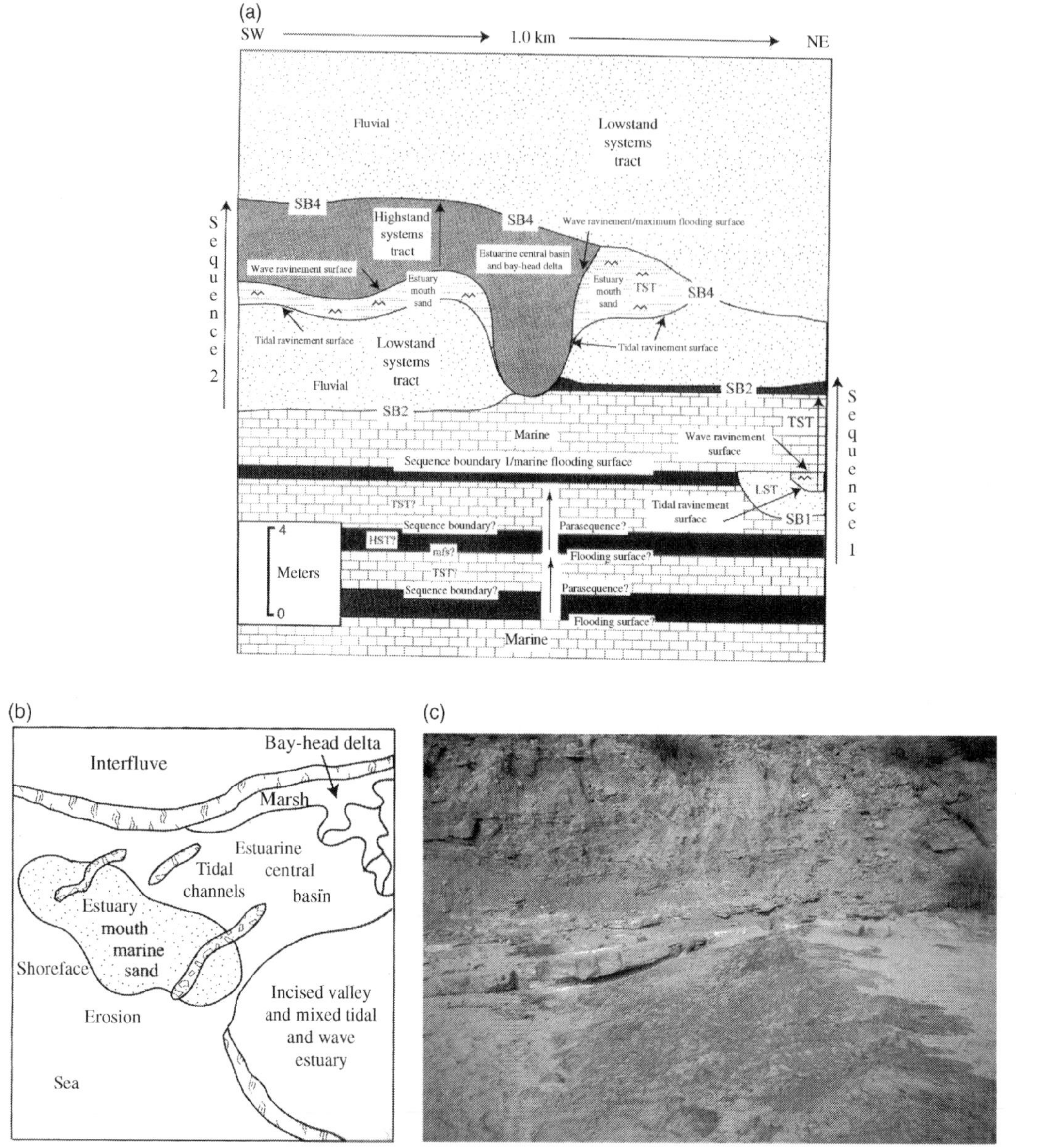


**Fig. 16.15** To show interpretation of ancient estuarine facies in a sequence stratigraphic context based on logged outcrops from the Abo Member of the Hueco Formation (Lower Permian, Wolfcampian), Doña Ana Mountains, New Mexico, USA. (after Mack *et al.*, 2003). (a) Panel with brick pattern as marine limestones, black as grey mudstone and siltstone. Light stipple is mostly fluvial channel, light grey is estuarine central-basin and bay-head delta, light stipple lines with gulls wings is estuary-mouth-margin sandstones with wave-formed ripples. Sequence boundaries (SB1, SB2, SB4) and their lowstand systems tracts (LST) are defined by fluvial facies erosionally overlying marine or estuarine sediments. Subsequent sea-level rise of the transgressive systems tract (TST) of sequences 1 and 2 produced tidal and wave ravinement surfaces and deposited estuary-mouth sands. In sequence 2, deposition of estuarine central-bay and bay-head delta sediments occurred during highstand. Lower part of section comprises interbedded marine fossiliferous carbonate packstone and grey mudstone and siltstone that may represent 5th-order sequences, parasequences or autocycles. (b) Schematic depositional model for (a) inspired by the modern Gironde estuary, France (based on Allen & Posamentier 1993). (c) Photo to show tidal ravinement surface that scours out ~1.5 m thick fluvial channel sandstone to right. Scour is overlain by low-energy estuarine heterolithic strata.

## Further reading

### *General*

Many papers on aspects of estuarine sediment transport may be found in Burt *et al.* (1997). Black *et al.* (1998) also contains much of relevance to estuarine tidal flat processes of erosion and sedimentation. Dyer (1989) provides a useful review of estuarine sediment transport. Sequence stratigraphic aspects of estuarine sediments are found in Dalrymple *et al.* (1994). Harris *et al.* (2002) discuss Australian estuarine location as a quantitative function of wave, tide and river power. Yoshida *et al.* (2007) is a thoughtfull attempt to investigate the idea that variable process regimes during cycles of sea level rise and fall have important repurcussions for sequence stratigraphic studies of coastal deposit architecture.

### *Specific*

Estuarine nomenclature is discussed from the hydrodynamic point of view by Pritchard (1955, 1967), Pritchard & Carter (1971) and Schubel (1971). Sedimentological classifications by Dalrymple *et al.* (1992) and Borrego *et al.* (1995). Yang *et al.* (2007) is the best recent effort to include wave energy into a sedimentological classification. Aspects of estuarine biofacies are in Barbosa & Suguio (1999) and Yang *et al.* (2007).

Flocculation and sedimentation are considered by Schubel (1971), Kranck (1975, 1981), Mehta (1989), Eisma *et al.* (1997), Dyer (2000) and Dyer *et al.* (2000); stratified estuarine flow with internal waves by Dyer *et al.* (2004).

Type A, well-stratified estuaries: Kostaschuk *et al.* (1992); type B, partially stratified estuaries: Allen *et al.* (1975, 1980) and Uncles & Stephens (1989); type C, well-mixed estuaries: Kirby & Parker (1983) and Allen *et al.* (1975).

Variability of estuaries is considered by Woodroffe *et al.* (1985), Cooper (1993), Anthony *et al.* (2002—Oueme estuary) and Yang *et al.* (2007—Gomso Bay estuary).

Macrotidal estuary deposition rates and controls generally are studied by Uncles *et al.* (2002), and specifically by Deloffre *et al.* (2006—Seine estuary, France) and Uncles *et al.* (1998—Ouse estuary, England).

Estuarine channels are studied by Hayes (1971), Langhorne & Read (1986), Dalrymple *et al.* (1990), Berné *et al.* (1993) and Larcombe & Jago (1996); for Fraser estuary channel freshets see Villard & Church (2005); for Gironde sequence stratigraphic model see Allen & Posamentier (1993).

Estuaries and sequence stratigraphy are considered by Fletcher *et al.* (1990), Sondi *et al.* (1995), Dalrymple & Zaitlin (1994), Dalrymple *et al.* (1994) and Nichol *et al.* (1997); Hori *et al.* (2002) discuss the Yangtze incised valley fill; Lobo *et al.* (2003) and Rodriguez *et al.* (2005) discuss the informative Galveston estuary and its incised terraced valley substrate.

Tectonic coseismic subsidence of estuarine margins is studied by Atwater *et al.* (2001).

Ancient estuarine sediments are considered by Hudson (1963), Bosence (1973), Campbell & Oakes (1973), Shanley *et al.* (1992), Richards (1994), Porebski (1995), and Willis (1997). For Permian incised valley/estuarine example see Mack *et al.* (2003). A particularly careful study of palaeo-Amazonia estuarine complexes containing inclined heterolithic cross-stratification is by Hovikoski *et al.* (2008), whilst Plink-Björklund (2005) analyses spectacular exposures of stacked tide-dominated estuarine to offshore cycles from the Eocene of Spitsbergen.

## Chapter 17

# RIVER AND FAN DELTAS

*The brown desert with its monotony of wind-covered dunes had given place now to a remembered relief-map of the delta. The slow loops and tangents of the brown river lay directly below, with small craft drifting about upon it like seeds.*

Lawrence Durrell *Mountolive (The Alexandria Quartet)*, p. 131, Faber and Faber, 1958

### 17.1 Introduction to river deltas

River deltas are coastline features constructed at distributary termini by sediment-laden freshwater discharge. Their exact form, in particular the degree of oceanwards protrusion, depends upon the nature of the interaction of the freshwater discharge with dissipative wave and tide processes in the receiving basin (Plates 13 & 14). Over time an area of subaerial delta wetlands is gradually accreted to an existing shoreline. Upward mire growth dominates the backswamps of this delta plain whose subenvironments include distributary channels, levees, crevasses, marshes, backswamps and lakes. The delta front comprises shoaling distributary mouth bars under effluent discharge, lagoons, tidal channels, barrier beaches and interdistributary bays bounded by active crevasse splays. Offshore, a subaqueous delta slope constructed from sediment dropped from dissipating effluent grades out to the local basin floor. These offshore slopes are highly unstable due to rapid deposition, so much sediment is redeposited downslope by slides, slumps, debris flows and turbidity flows. Eventually the delta slope may grade out into a turbidite-dominated submarine fan. The shape of the long profile of a subaqueous delta is described by $y = ax^b$, where $x$ and $y$ are horizontal and vertical distances from the outlet. Basin water depth emerges as an important control upon the values of the coefficient, $a$, and slope, $b$, terms. The rate of sediment deposition decreases overall with distance from outlet source. Approximating mass movement processes as creep, a simple mathematical analysis yields the result that the delta slope decreases exponentially with distance $x$ from the outlet at any one time.

It would be nugatory to carry on with any more detailed definitions and generalizations since there are a very large number of variables involved in determining the morphology and evolution of particular deltas (Fig. 17.1): as we saw for estuaries, deltas defy any single sedimentary model. Deltas are thus complex sedimentary systems, though there is growing evidence that major river floods may contribute as much, if not more, to sedimentary architecture than the redistributive processes of wave and tide. It would be fair to say that process studies of modern deltas have not been carried on with the same rigour as in river channels (Chapter 11), where interlinked coring, GPR and hydrodynamic measurements have revolutionized our knowledge. A further problem is their sensitivity to sea-level fluctuations and climate variation, for deltas are precious ecosystems, easy to ruin by human greed and folly, especially when upstream sediment sources are cut off by dams, like in the Nile and Yangtze.

*Sedimentology and Sedimentary Basins: From Turbulence to Tectonics*, 2nd edition. © Mike Leeder.
Published 2011 by Blackwell Publishing Ltd.

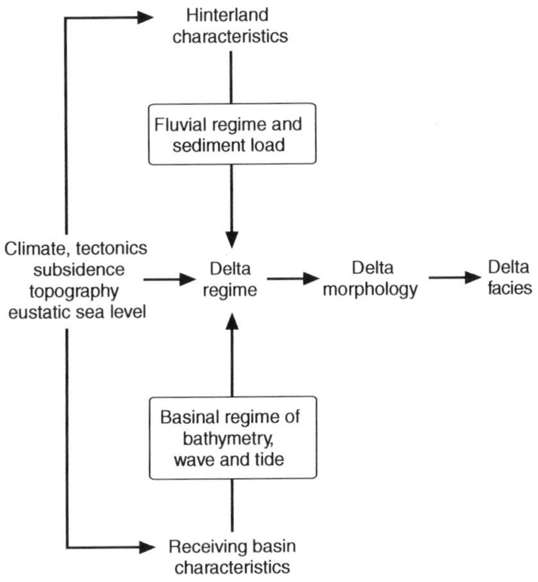


**Fig. 17.1** The various factors affecting delta morphology and sedimentary environments. (After Postma, 1990.)

Tropical deltas are also at the mercy of ferocious tropical storms, witness the frequently overwhelmed Ganges–Brahmaputra delta in the Bay of Bengal and the Mississippi delta in the Gulf of Mexico. A good case could be made for regarding deltas as the single most important clastic sedimentary environment *solely* on the basis of the enormous reserves of coal, oil and natural gas that are located within ancient deltaic sediments. A similar case could have been made thousands of years ago since deltas were amongst the most prized areas of agricultural land in ancient times, witness the great civilizations that grew up based upon deltas such as the Nile and Tigris–Euphrates.

## 17.2 Basic physical processes and sedimentation at the river delta front

The combined discharge of sediment and freshwater issuing from the mouth of a major delta distributary (Plates 13 & 14) occurs as a *jet*, analogous to the expanding flow of fluid issuing from any nozzle or opening. Three factors influence the nature of sediment deposition from a freshwater jet: (i) the inertial and turbulent diffusional interactions between the jet and ambient fluid; (ii) frictional drag exerted on the base of the jet by the delta front slope; and (iii) negative buoyant force due to suspended sediment increasing the jet's density contrast with ambient fluid.

### Homopycnal jets

These have more-or-less the same density as ambient fluid—they are dominated by their own inertia and dissipate by eddy-shedding and turbulent diffusion. A turbulent homopycnal jet (Fig. 17.2) will expand

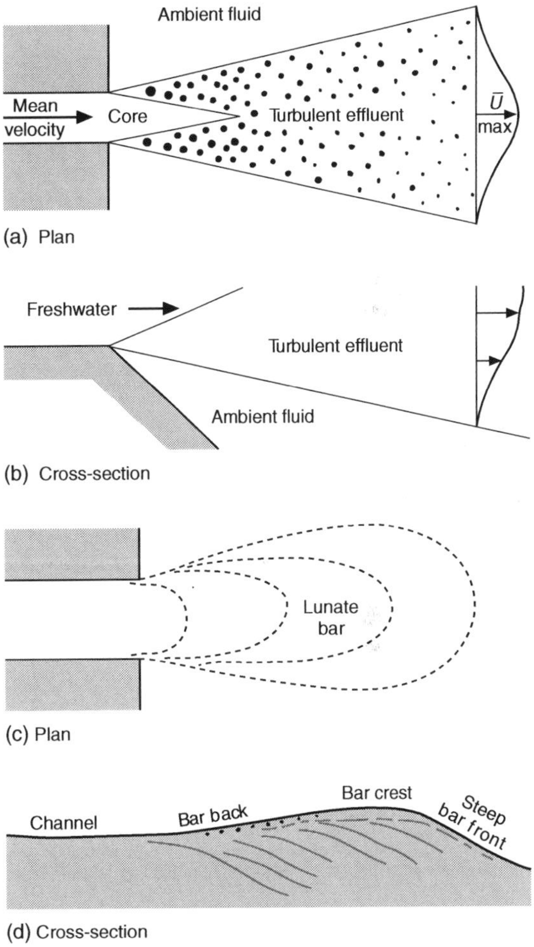


**Fig. 17.2** (a) Plan and (b) section views of inertia-dominated jets issuing into basins that are deep with respect to jet thickness. (c) Plan and (d) section views of resulting distributary mouth bar. Note that seaward fining and steepness of bar front depends upon the grain size of the sediment. Coarse sediment in bedload is deposited rapidly and causes formation of steep, angle-of-repose slopes that define Gilbert-type deltas. (After Wright, 1977.)

linearly with distance from an outlet until it dissipates. Delta fronts dominated by homopycnal flows are commonest in lakes where bedload sediment is deposited very quickly as the jet expands and decelerates. The resulting deposit may feature a steep frontal slope dominated by gravity fall and mass flow deposition. Suspended sediment in homopycnal jets entering lakes does not flocculate and will consequently travel in the jet until diminishing turbulent diffusion allows particle settling. The combined processes of rapid bedload and gradual suspended load deposition causes a marked proximal to distal fining of deposited sediment over the delta front with a comparatively rapid transition from coarser nearshore sandy bedload to offshore suspended load silts and muds.

### Hypopycnal jets

These are negatively buoyant: when such a jet enters a shoreface with gentle slope and shallow water depth, frictional effects arise from bottom drag. This causes rapid initial deceleration as turbulent shear causes bed erosion, channelling into older deposited sediment and eventual deposition of bedload sediment (Figs 17.3 & 17.4). This deposition occurs as a 'mid-ground' distributary mouth bar that emerges and grows by lateral accretion, bordered on its margins by a Y-shaped channel bifurcation rather like a braid bar Chapter 11). Bars of this type are particularly characteristic of subdelta growth in the Mississippi interdistributary bays (section 17.5) and of the Atchafalaya (Fig. 17.4). They are analogous in some ways to riverine crevasse splay lobes (Chapter 11) and similarly feature a myriad of subaerial *terminal distributary channels* and coarsening upwards mouth-bar deposits.

By way of contrast, as hypopycnal jets enter deeper basins with low tidal energy, as we saw in type A estuarine environments (Chapter 16), they spread as narrow expanding plumes above a salt wedge that may extend for a considerable distance up the distributary channel (Figs 17.5 & 17.6). Internal shear waves generated at the salt-wedge/effluent boundary cause vertical turbulent mixing and at higher suspended sediment concentrations an increase in fine-sediment flocculation. During periods of high river flow any salt wedge is expelled from the distributary channel to a position just seaward of the mouth-bar crest where bedload deposition occurs on the bar front as the plume effluent separates from the saltwater.

Massive bar-front deposition and steepening during and after river flooding is well documented from the large 2003 flood in the Rhône delta, France when much of the deposited coarse sediment was derived by deep channel erosion in the immediately upstream distributary channel; subsequent wave action also caused deposition on the mouth bar but little occurred at any time on the bar toe and pro-delta (Fig. 17.7).

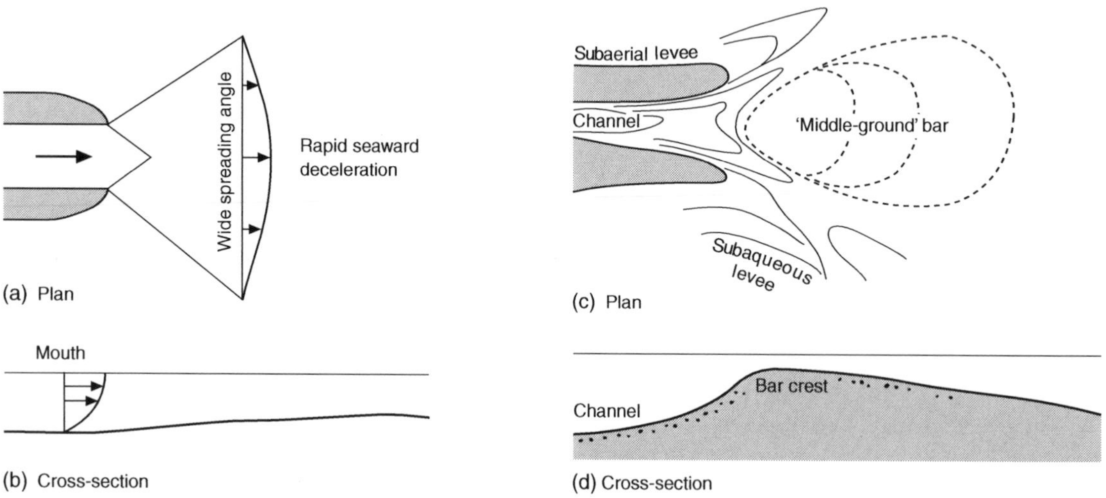


**Fig. 17.3** Plan and section views of friction-dominated jets and their 'middle-ground' mouth bars. (After Wright, 1977.)

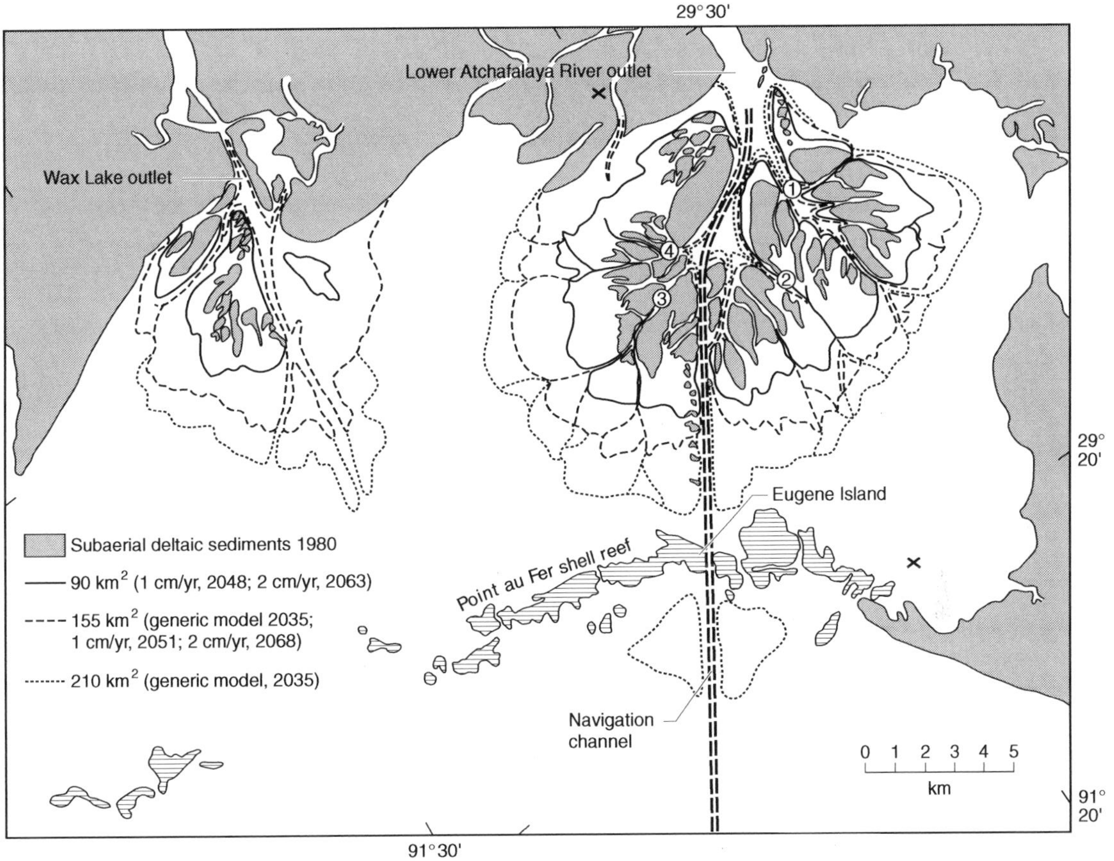


**Fig. 17.4** The shoal-water Atchafalaya subdelta in the Mississippi system was formed by a partial avulsion of the Mississippi River. It represents the fused growth of numerous 'middle-ground' bars located at terminal distributary channels. The growth of the delta has been modelled for the next 100 years from 1980 by slow (1 cm/yr) and rapid (2 cm/yr) relative sea-level rises (after Wells, 1987), though more recent work indicates that there is insufficient sediment flux for progradation to occur (see text for discussion).

Continuing deposition of successively finer sediment occurs on the seaward bar slope out into a basin, chiefly by continued clay flocculation in the turbulent shear layers below and at the margins of the outgoing jet. Studies in the Po delta show this depositional process rapidly removes suspended sediment within a few kilometres of the oufall channel mouth. (Fig. 17.8). Good examples of elongate seaward-fining distributary mouth bars occur at the front of the major Mississippi outfall channels and their progradation leads to production of long 'bar-finger' sandbodies. Buoyant effluent jets (Plates 13 & 14) are also gradually destroyed by basinal wave action but this also suspends finer fractions to be dispersed by background oceanographic currents. In many cases these shelf currents act as barriers to jet penetration and they turn the suspended jet to flow axially normal to the shelf slope and to deposit mud wedges far from the jet outfall (see further notes on wave and tide effects below).

### Hyperpycnal jets

These occur when effluent density exceeds that of basin ambient fluid and the jet underflows as an *attached wall jet*. For marine basins a suspended load of $>\sim 30\,kg/m^3$ must be present to counteract the density of seawater—this may occur during major river flood events, particularly in tropical suspended-sediment-rich rivers fed by seasonal monsoon or storm systems. Perhaps the most spectacular underflowing marine delta system is that of the Huanghe (Yellow River), whose colossal suspended load picked up on its passage through the central China loess belt enables it to often sink without trace into the offshore region as an underflow.

### Effects of waves and tide

These have a modifying effect on simple jet models for delta-front dynamics. In coastal areas of high wave power relative to river discharge, effluent jets may be completely disrupted by wave reworking. The coastlines of wave-dominated deltas tend to be very much more linear in plan view than those of more moderate wave power. Barrier-spit systems are frequently seen fronting the delta plain. Studies and numerical modelling of the active barrier-fronted part of the Red River delta, Vietnam, show that periods of rapid delta-front progradation are followed by barrier formation due to cross-coastal, wave-induced onshore sand transport (Fig. 17.9). The increased fluvial sediment and water discharges periodically create an extensive delta-front platform that is subsequently reworked landwards. In other deltas the effects of wave reworking include fair-weather shoreward transport of sand to form swash bars around broad crescentic mouth bars (Fig. 17.10) and offshore sand transport to form storm sand-sheets during periods of high wave setup and seaward bottom flow. Oblique wave incidence causes beach--spit systems to prograde parallel to the coast and creates delta-front and pro-delta asymmetry. Extensive beach-ridge systems are generated away from the channel mouths, fed by longshore transport of sands, though examples from the Brazos delta, Texas do not conform to a continuous strandplain model (see section 19.5). Lake deltas like Williams River delta, Lake Athabaska, Canada show wave/river interactions with spectacular beach-ridge accretion, partly heightened by wind action, with intervening peatlands. Records produced by GPR show internal erosion surfaces that cut packages of clinoforms and which are thought to have been generated by severe, centennial to millennial-scale, storms (Fig. 17.11).

Tropical storms are a major modifying influence in many subtropical deltas. They (i) dump precipitation on the landward delta surface, (ii) create coastal surges up to 5 m or more above mean high water levels that cause widespread and massive sediment deposition and saltwater invasion of freshwater wetlands and (iii) generate large waves that batter barrier islands and spits causing widespread shoreface erosion and landward washover deposition in bays and lagoons. Tides modify effluent processes by disruption of the salt wedge during tidal ebb and flow. Such distributary mouths are funnel-shaped and most pass seawards into complex zones of emergent tidal current shoals, equivalent to mouth bars in fluvially dominated deltas. But the effect of tides is not a linear 'in-and-out' process for the tidal current vector is quite independent of delta distributary orientation. For example, the mouth bars of the Huanghe (Yellow) River, China interact with a coast parallel tidal system so that effluent river water periodically meets the inshore part of the ebb or flood tide along vertical shear fronts that serve to divert, slow down or even block seaward distributary advance for a few hours twice daily, sending sediment-laden freshwater along temporary coast-parallel routes to deposit elongate shelf mud wedges (Fig. 17.12). Similar effects are seen in the Yangtze, Pearl and Mekong deltas

Strong synoptic wind-driven coastal currents may also have a great effect upon the dispersal of effluent plumes issuing from the distributary outlets and they serve to create prodelta asymmmetry. These are well seen off the Amazon where the strong northwest-flowing Guiana Current transports much silt to sand grade sediment alongshore to the Orinoco delta rather than cross-shelf into the ocean, where it would otherwise construct an offshore platform that might moderate wave action. Similar coast-parallel transport of mud is also a prominent feature of the Po delta, Italy (Fig. 17.8).

## 17.3 Mass movements and slope failure on the subaqueous delta

Seaward delta-front migration in response to net deposition creates opportunities for sediment instability since coarse, rapidly deposited sediment in

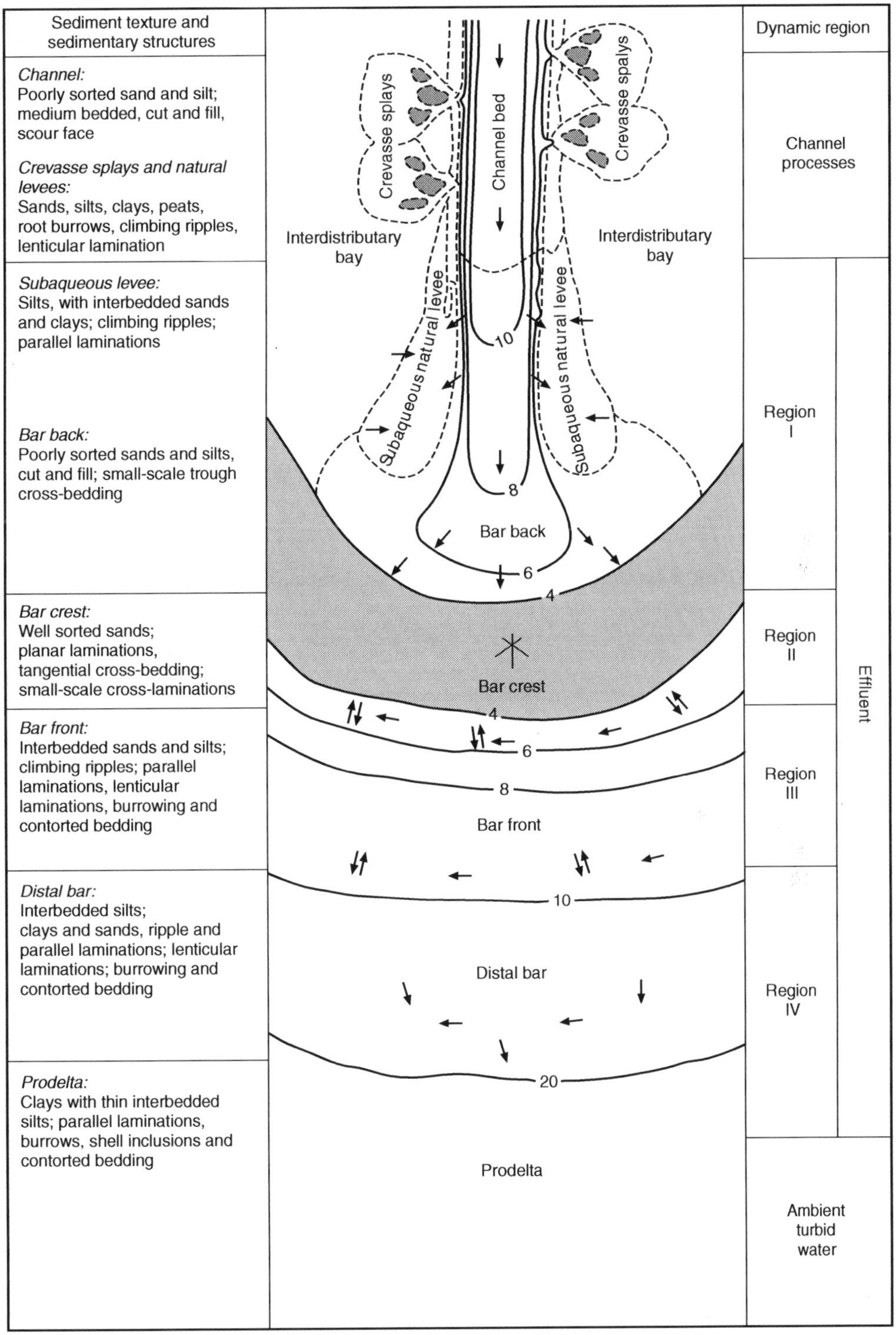


**Fig. 17.5** Sedimentological zonation in a typical buoyancy-dominated mouth bar. (After Wright, 1975.)

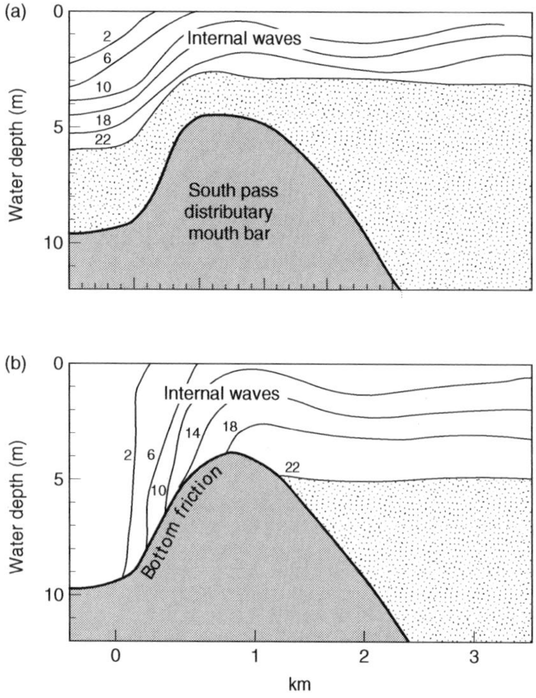


**Fig. 17.6** Longitudinal sections through the South Pass, Mississippi, mouth bar. (a) At low river stage, with pronounced salt wedge. (b) At high stage, with salt wedge displaced seawards. Numbers refer to salinity (‰). (After Wright & Coleman, 1973.)

distributary mouth bars, channels and barrier–beach complexes overlie fine-grained, clay-rich and highly porous sediments of the delta front and prodelta. Intrusive mud lumps record the slow upward flowage of prodelta clays and fine silts in response to loading. Sonar images of mouth-bar surfaces and well-exposed sections through ancient mouth bars (Fig. 17.13) reveal a variety of failure slopes and scarps with development of rotational slides and growth faults. The latter occur on many scales and show characteristic curved morphology in plan view and asymptotic offshore profile. Some rotational slides completely detach from their local position and may travel kilometres offshore, the stratigraphic arrangements within the slide blocks remaining more-or-less intact, although dipping at a high angle to the local horizontal above the basal slide planes. The block rotation of these masses defines local slope basin sinks for turbidity currents so that continued deposition leaves the shallow-water 'orphans' completely surrounded and eventually entombed within offshore deposits.

## 17.4 Organic deposition in river deltas

The formation of wetland mires and peat is the eventual fate of the majority of the lakes, lagoons, abandoned channels and crevasses in a prograding and avulsing delta system. Tropical mires are particularly widespread, e.g. 50–80% of the surface area of the Rajang River delta in Sarawak, and have a high rate of primary plant production, deposits accumulating at rates of several tens of centimetres per year in some areas. Even in high sediment-flux regimes like that of the Holocene Mississippi and Orinoco deltas, thick peats have periodically developed behind reworked shorelines that sheltered abandoned delta lobes whilst in the Williams River delta of Lake Athabaska, peats accumulate in troughs between beach ridges (Fig. 17.11). Tropical peats accumulate as extensive raised mires (peat domes) several metres higher than even the raised levees of the distributary channels that pass through them. In common with raised mires in temperate wet upland areas, the living vegetation gets its supply of water and nutrients largely from rainfall. In lowland tropical deltas the raised mires therefore act as important barriers to channel migration, floodwater runoff and avulsion (a similar role is played by papyrus peats in the inland Okavango delta). The stabilizing effect of raised coastal mires enables them to resist high rates of sea-level rise. Not only that, the largest mires also act as local hinterlands, exporting intercepted rainfall into the 'normal' delta distributaries as humic-rich runoff. This low-density discharge often overlies the denser suspended-sediment-rich runoff, causing sediment bypass to the delta front rather than loss to levees and floodplain. Finally, buried peats have the ability to influence later deltaic events because of their extreme tendency to dewatering and compaction: some authors maintain that coastal relative sea-level rise around deltas is largely controlled by subsidence due to such effects.

## 17.5 River delta case histories

A health warning is necessary here concerning the veracity of inferences made about the predominance

(a)

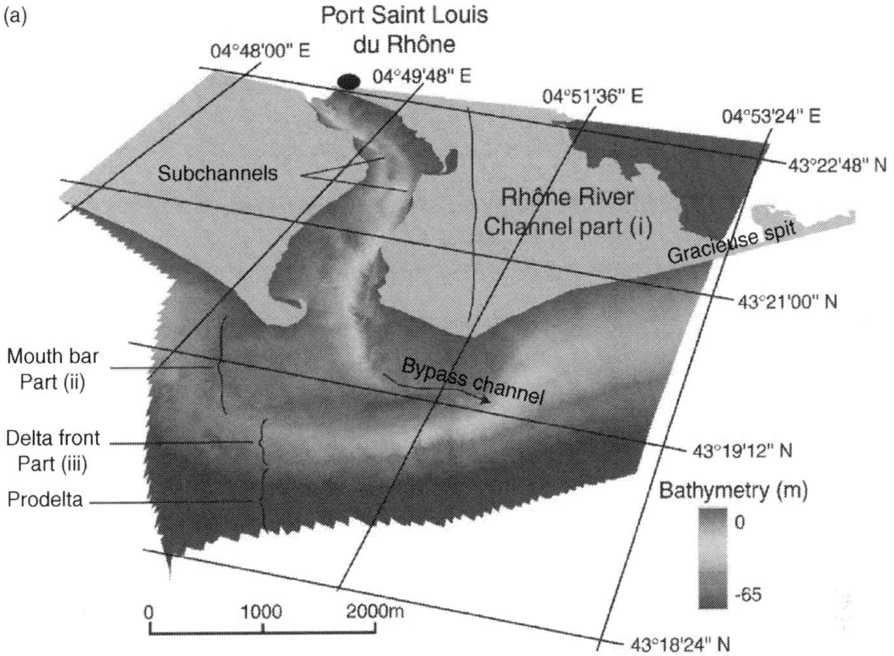
Port Saint Louis
du Rhône
04°48'00" E
04°49'48" E
04°51'36" E
04°53'24" E
43°22'48" N
Subchannels
Rhône River
Channel part (i)
Gracieuse spit
43°21'00" N
Mouth bar
Part (ii)
Bypass channel
Delta front
Part (iii)
43°19'12" N
Prodelta
Bathymetry (m)
0
-65
0
1000
2000m
43°18'24" N

(b)

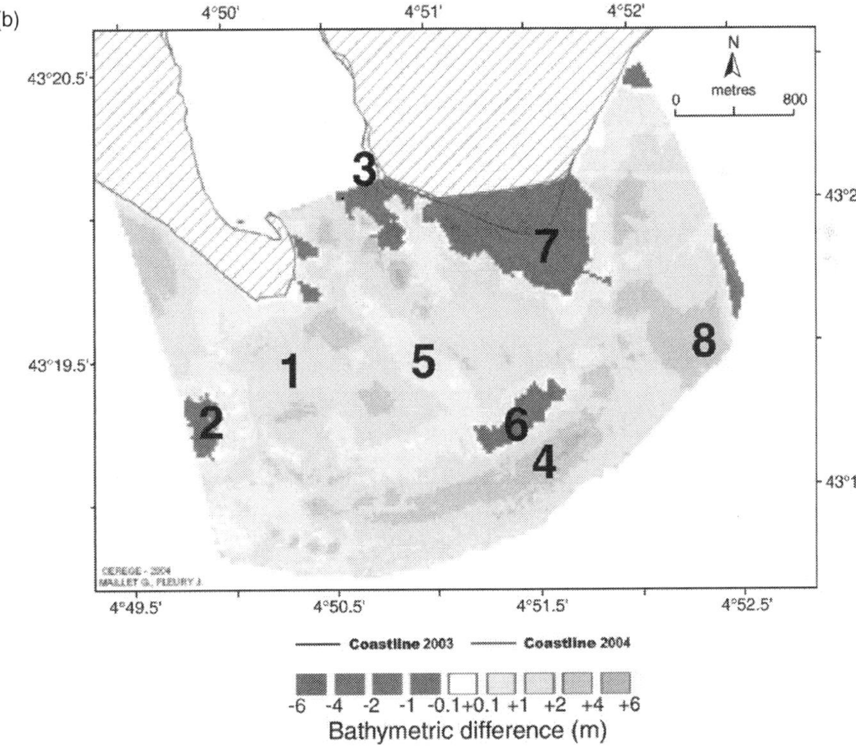
4°50'
4°51'
4°52'
43°20.5'
43°20'
43°19.5'
43°19'
N
metres
0
800
1
2
3
4
5
6
7
8
4°49.5'
4°50.5'
4°51.5'
4°52.5'
Coastline 2003
Coastline 2004
-6 -4 -2 -1 -0.1 +0.1 +1 +2 +4 +6
Bathymetric difference (m)

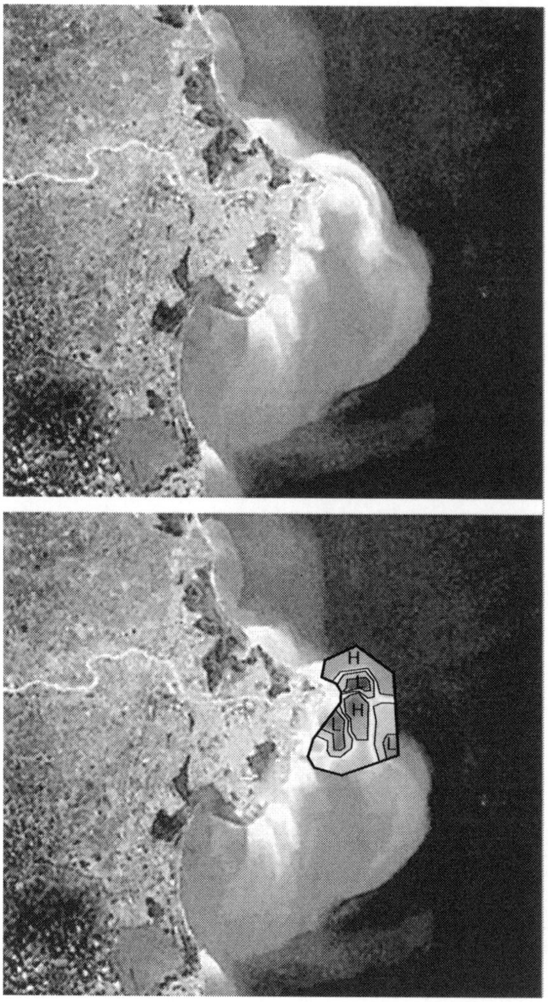


Fig. 17.8 (a) *Modis* image of the Po flood plume in 1985. (b) Contoured values of floc limit calculated for surficial sediments collected in December 2000 and January 2001 are overlain on the image to show depositional patterns. Areas 'H' represent high floc deposition and areas 'L' represent areas where floc deposition was low. (After Milligan *et al*., 2007.)

Fig. 17.7 (a) Rhône delta bathymetry before 2003 flood event. (From Maillet *et al*., 2006.) The study area is divided into three parts: (i) River channel, (ii) mouth bar and (iii) delta front. In the mouth-bar area the incised sub-channel is well-marked and leads to a shoal mouth-bar. Most of the alluvial flow takes place via a bypass channel through the mouth-bar to the east, and via a broad depressed zone to the west. A slump scar is visible on the delta front, linked to a submarine slide occurring before the December 2003 flood. Submarine gullies on the delta front slope extend from the base of the mouth-bar to the top of the prodelta (between −4 m and −30 m isobaths). (b) Bathymetric changes in metres during the 2003 flood event at the Rhône mouth, between 0 and −20 m depth. A significant total volume of 7.8 million $m^3$ (about 4 million tons) was sedimented in this area, corresponding to a mean volume per unit area of about 0.88 $m^3/m^2$. These values are in accordance with upstream fluvial measurements that quantified flood solid discharge as between 3.1 and 5.28 million tons. Key to numbers: 1, spit terminus; 2, erosional surface; 3, erosional channel; 4, final position of main Rhône outlet mouth bar; 5, channelized zone of low accumulation; 6, initial position of mouth bar; 7, extensive area of coastal and nearshore erosion.

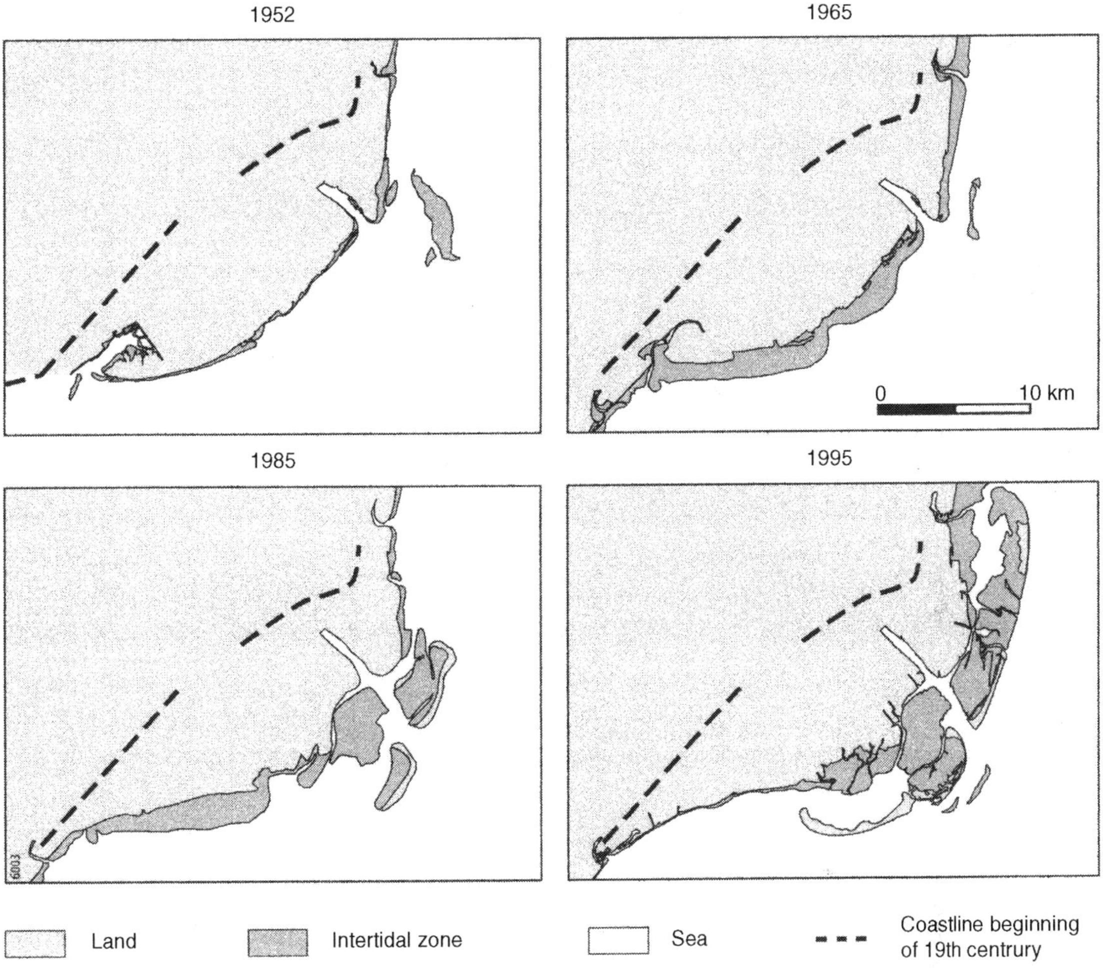


**Fig. 17.9** Evolution of the Red River delta, Vietnam showing development of barrier island frontage after period of rapid delta progradation. (After van Maren, 2005.)

of wave, tide and river processes from morphological studies of modern deltas. It has become evident in the past few years that we have been imprisoned in a misleading conceptual model since the mid-1970s, with insufficient attention paid to the effects of large river floods and local tidal and wave dynamics on internal delta sedimentary architecture. The remarks on individual deltas below may lack this perspective, but will doubtless yield up their secrets as researchers investigate detailed physical processes and their sedimentary products as has been done in many river systems.

## Mississippi delta

River-dominated deltaic environments, with well-developed buoyant forces dominant during jet discharge, are exemplified by the Mississippi delta (Fig. 17.14 & Plate 13) whose front has low tidal range (about 0.3 m). moderately high all-round water discharge, a fine-grained sediment load and moderate degrees of incident wave energy. The latter attenuates markedly in the nearshore area, a direct consequence of fine-grained sediment deposition creating a gently-sloping apron to the nearshore zone. It is upon this

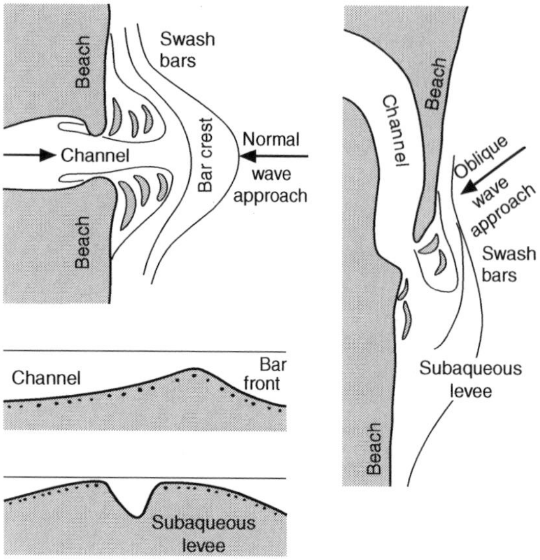


**Fig. 17.10** To illustrate depositional and morphological features of distributary outlets modified by strong wave action. (After Wright & Coleman, 1973.)

apron that deep-water wave energy is expended. The 'exposed' delta plain is dominated by a small number of large distributaries and a host of minor ones. Because of the very low slopes involved, the larger channels tend to be straight. Frequent avulsion occurs as the delta progrades, the channels periodically seeking new routes to the sea with gradient advantages. The 'birdsfoot' morphology of the delta results from these avulsions, with the 'claws' marked by channels and the 'webbed' connections marked by interdistributary bays. The bays are shallow, brackish to marine, and are gradually infilled by minor crevasse deltas with a myriad of terminal distributary channels. Frictional effects dominate in these shallow water bodies and Y-shaped mid-ground mouth bars form as the crevasse channels debouch their discharge (Fig. 17.4). Gradually, by crevasse progradation and overbank flooding, the bays become part of the subaerial marsh of the delta plain. Cores through the prograding marshes and minor channels of the interdistributary bays reveal a variety of coarsening-upwards successions, capped by vegetation colonization surfaces, wetland peats of various kinds and sharp-based fining-upwards successions resulting from deposition in minor channels.

At the delta front the major distributaries widen and pass into well-developed lunate distributary mouth bars. Shoaling from the distributary channel to the mouth bar crest results from the dynamics of salt-wedge intrusion discussed previously. During progradation the channel erodes a portion of the bar crest that is directly downstream from it, causing the thick (50–150 m) coarsening upwards mud-to-sand succession of the mouth bar to be cut by an erosive-based distributary channel sandbody. Deposition of thick successions of muds in the delta front area encourages the development of a wide range of soft-sediment deformation, slump and growth-fault features.

Over the past few thousand years the active parts of the Mississippi delta have undergone periodic avulsive shifts along the coast of Louisiana as successive channels have searched for gradient advantages over their precursors (Fig. 17.15). Transfer of the river to a new location causes abandonment of the previously active delta constructional system (Fig. 17.16). The latest delta lobe is that created by the Atchafalaya avulsion.

Wetland expansion and peat accumulation in the coast-protected environment is followed by eventual marine or bay/lake expansion as very early compactional subsidence causes local submergence and transgression. This process leads to the production of a characteristic abandonment facies at the top of a delta-lobe facies association. Renewed clastic input from further avulsions of the trunk channel may subsequently terminate the shallow bays and lakes. A mechanism thus exists within a switching delta for producing complex deltaic 'cycles', so long as the coastal plain is located in an area of net tectonic subsidence (as is indeed the case in the Gulf of Mexico). Wetland loss by coastal erosion is now a serious matter for concern in the Mississippi area (as in most delta complexes worldwide) but is to some extent balanced by massive sedimentation of muddy sediment during hurricane-induced storm surges, as documented for hurricanes Katrina and Rita in 2005.

The birdsfoot phase of the Holocene evolution of the Mississippi coastal plain is quite atypical of the numerous previous delta lobes that have shifted around the coastal plain due to periodic avulsions. Most of the five or so pre-recent deltas prograded into much shallower water than the modern birdsfoot, whose effluent plumes nowadays almost reach the shelf break. These shoal water deltas now exist as

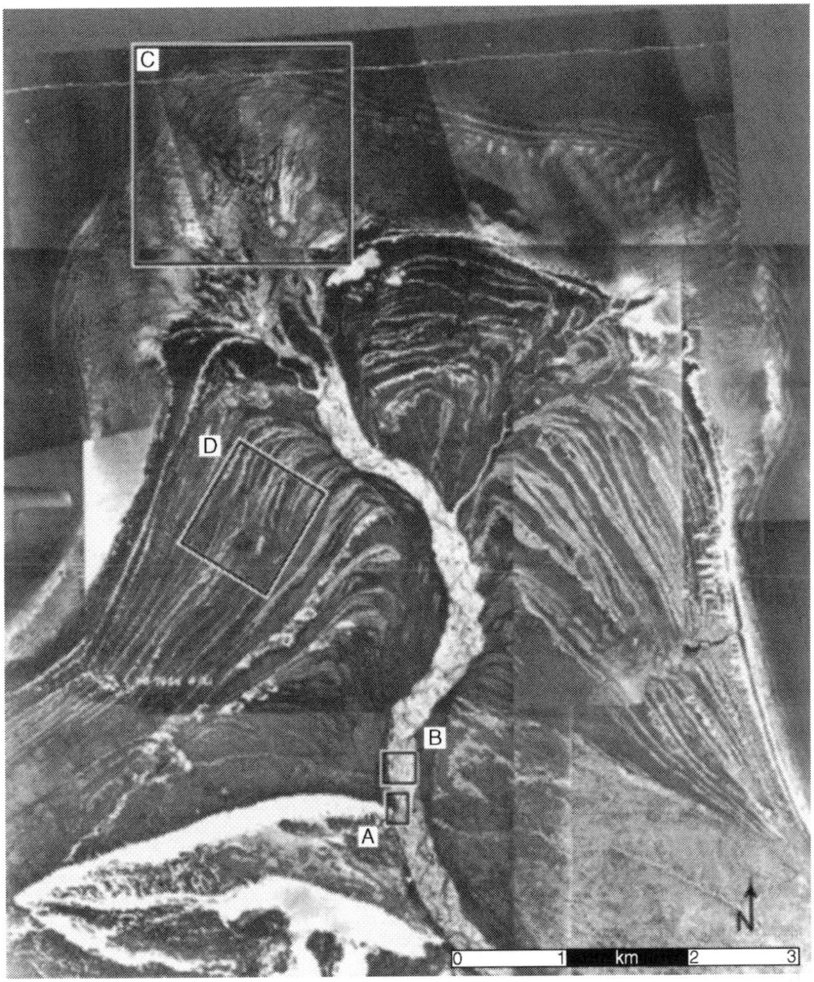


Fig. 17.11 Aerial photo mosaic of the William River delta, Canada showing multiple arcuate-shaped beach ridge sets separated by peatlands. The sand-bed braided William River terminates at the lake as a distributary channel fan complex. Multiple offshore bars extend 1 km farther into the lake from the shore. Boxes refer to detailed images presented in the original paper. (From D.G. Smith *et al.*, 2005.)

eroded and partial remnants whose coarser sand deposits of abandoned mouth bar and channel have been reworked by waves by a sequence of events comprising: (1) flanking barrier/spits; (2) transgressive barrier island arcs with migrating tidal channels; and (3) remnant inner shelf shoals (Fig. 17.17).

## Niger delta

In contrast to the Mississippi delta, the mixed regime (tide/wave-dominated) Niger delta has coastal environments dominated by the occurrence of major coastal barrier islands separated by tidal inlets (Fig. 17.18). The great trunk stream of the braided River Niger breaks up into a host of smaller channels, each of which is tide-dominated. Sandy bedload deposited in ebb tidal-deltas is redistributed alongshore by severe wave action. Delta-front deposits are thus dominated by tidal-inlet channel fills and coastal barrier-sands. Continued delta progradation results in reworking and destruction of these deposits at the expense of the upper delta-plain facies. Investigations of thick

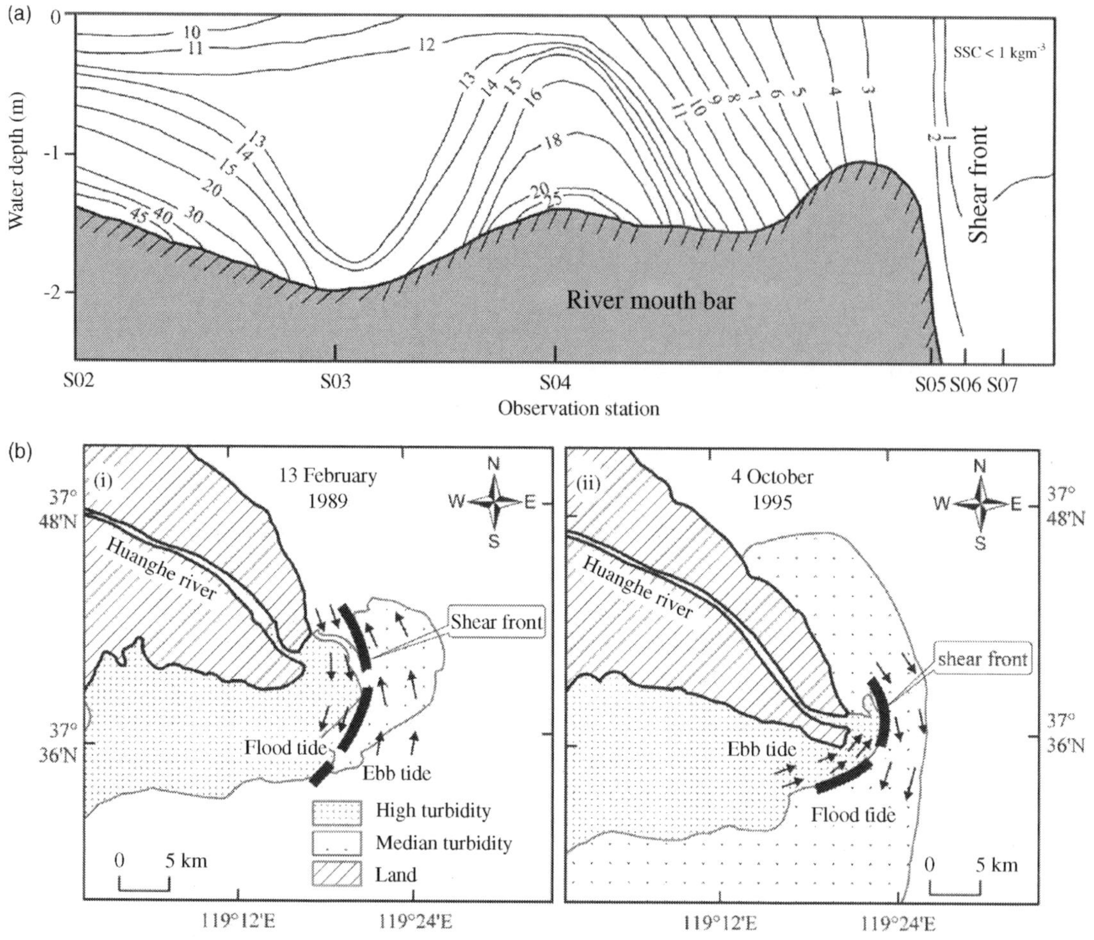


**Fig. 17.12** (a) Vertical profile of suspended sediment concentration (SSC) along the Huanghe River mouth bar during flood tide on 28 July 1984. (b) Tidal currents and shear fronts at the mouth of the Huanghe River inferred from satellite images: (i) inner-outer-ebb shear front; (ii) inner-ebb/outer-flood shear front. (After Fan *et al.*, 2006.)

(9–12 km) Tertiary proto-Niger delta deposits in the subsurface indicate the overwhelming importance of delta-slope failure in controlling delta architecture. Seismic and well penetration indicates that the whole thick succession is broken by a myriad of growth faults, slide blocks, slumps, and associated slope basins, many of which have a tectonic control by basement structures.

## Nile delta

This is the type delta since its naming by Herodotus. It has been somewhat neglected by sedimentologists in the more recent past since attention was shifted to the Mississippi and Niger, both oil-rich. By contrast the Nile (Fig. 17.19) has fewer hydrocarbon reservoirs in its subsurface, but has dominated the rise, fall and rebirth of Egyptian civilization for 7000 yr. The delta is infamous because of the effects of construction of the Aswan High Dam in 1964. Although this assured Lower Egypt of a steady supply of water for irrigation and power, it has led to a vast impoundment of sediment that would otherwise have been added to the floodplains and to the delta front. Very serious coastal erosion is now occurring over much of the Nile coastline as a consequence.

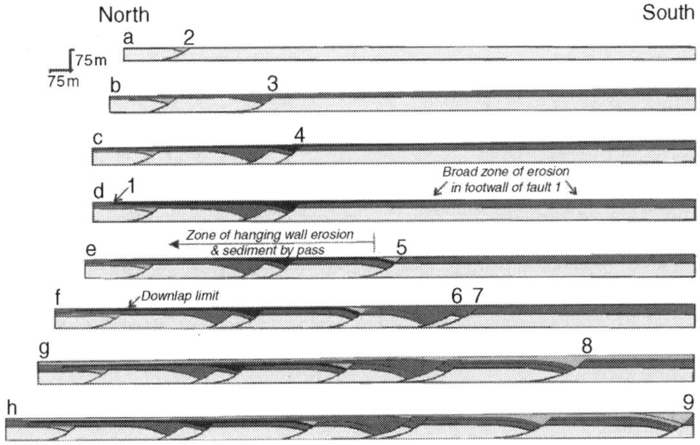


**Fig. 17.13** Reconstruction of the kinematic history of the Cliffs of Moher growth-fault system, Clare, western Ireland. The growth-fault complex affects strata up to 60 m in thickness and extends laterally for ~3 km. Growth faulting was initiated with the onset of sandstone deposition on a succession of silty mudstones that overlie a thin, marine mudrock. A décollement (slip) horizon developed at the top of the latter for the first nine faults, by which time aggradation in the hangingwall exceeded 60 m in thickness. After this time, failure planes developed at higher stratigraphic levels and were associated with smaller scale faults. The fault complex shows a dominantly landward retrogressive movement, in which only one fault was largely active at any one time. There is no evidence of compressional features at the base of the growth faults, thus suggesting open-ended slides, and the faults display both disintegrative and nondisintegrative structure. Thin-bedded, distal mouth-bar facies dominate the hangingwall stratigraphy and, in the final stages of growth-fault movement, erosion of the crests of rollover structures resulted in the highest strata being restricted to the proximity of the fault. These upper erosion surfaces on the fault scarp developed erosive chutes that were cut parallel to flow and are downlapped by the distal hangingwall strata of younger growth faults. (After Wignall & Best, 2004.)

The delta is a wave-dominated system on a microtidal coastline with important east-flowing offshore currents that deflect the effluent jets. The delta itself is unusual for its site in an arid climate, the river collecting its waters mainly from the summer monsoons that drench the Ethiopian Plateau and feed the Blue Nile, rich in suspended sediment. The delta front is dominated by a 500 km long barrier–beach complex, which formerly sheltered extensive back-barrier lakes and lagoons. These are now mostly drained and reclaimed. The coastal barriers are broken by the promontory-like outlets of the two modern subdeltas of the Rosetta and Damietta branches (a total of perhaps six major prehistoric abandoned distributary courses have been recognized; Fig. 17.19). The modern channels are fringed by prominent beach ridges, particularly on their eastern sides, leeward of the prevailing northwesterlies. Post-1964 erosion of several kilometres has occurred in response to lack of sand-grade sediment input to compensate wave erosion and longshore transport.

Great insights into the late-Pleistocene to Holocene history of the delta have been gained in recent years from extensive archaeo-sedimentary studies. It appears that during the LGM (20–18 ka) the Nile ran out to the shelf edge as a sand-dominated incised braidplain. Rapid sea-level rise to 8 ka led to a lowering of river gradients over the present delta and the eventual establishment around 7 ka of stable floodplains that received abundant silts and clays from the flooding river which was no longer in a state of sediment bypass. This deposition formed fertile floodplains that were to provide the springboard for establishment of pre-Dynastic agriculture.

### Tiber delta

This is another 'classic' Mediterranean delta in a microtidal setting with prominent salt-wedge intrusion and hypopycnal outflow—also it is intimately associated with the history of the ancient Roman civilization. The modern delta has evolved from

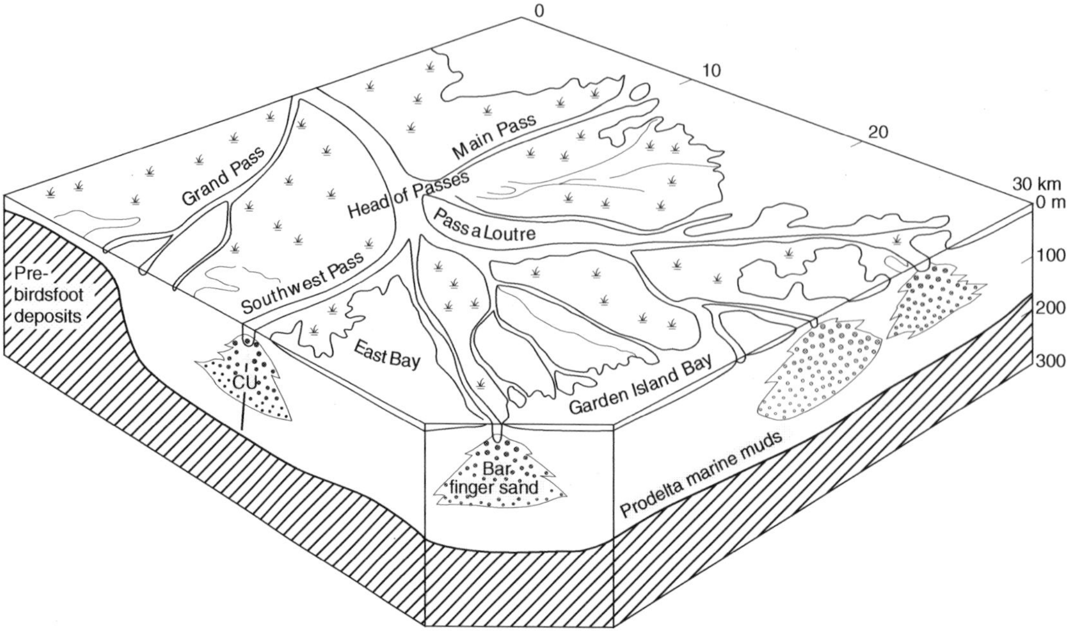


**Fig. 17.14** Block diagram to show the morphology and major facies present around the front of the active modern Mississippi 'birdsfoot' delta. Delta-front dynamics are controlled by homopycnal jetted-effluent with elongate bar-finger sands forming after continued progradation of distributary mouth bars. CU indicates coarsening upwards. (After the classic work of Fisk *et al.*, 1954.) See also Plate 13.

a bay-head delta infilling lagoons during the early Holocene to the wave-dominated highstand feature with reclaimed lagoons that we see today (Fig. 17.20). Progradation has accelerated in the past 500 yr, so much so that Imperial Rome's main port for the import of North African wheat, Ostia Antica, now lies 5 km or so inland.

### Rhône delta

Another wave-dominated Mediterranean delta whose response to a recent very high magnitude river flood has been noted above (Fig. 17.7).

### Po delta

This is a particularly interesting and diverse delta that enters the microtidal Adriatic Sea from a wide coastal plain and a mountainous catchment. The delta is currently south of the Venetian lagoons and this link with a long-lived mercantile centre means that there are many cartographic and literary sources with which to examine late Holocene highstand delta evolution, as well as much shallow seismic and core data. The modern delta system includes an extensive delta plain, a wave-influenced delta front and a wide asymmetric prodelta that is steeper and shorter to the north and merges southwards into an elongate shelf mud wedge (Fig. 17.8). Throughout its highstand history the delta has frequently shifted due to multiple avulsions of the Po River, with rapid progradation, then abandonment, of many delta lobes. A significant feature of the prodelta region in one lobe dating from Medieval times is the presence of silt to very fine sand infilling channels that may record extreme hyperpycnal events.

### Burdekin delta

The Burdekin River, northeast Australia, has constructed a substantial flood-dominated Holocene delta (delta plain area is 1260 km$^2$). It is prograding into a shallow (~10 m) shoreface environment by rapid deposition of mouth bars during river floods generated by tropical cyclones, most of which last for

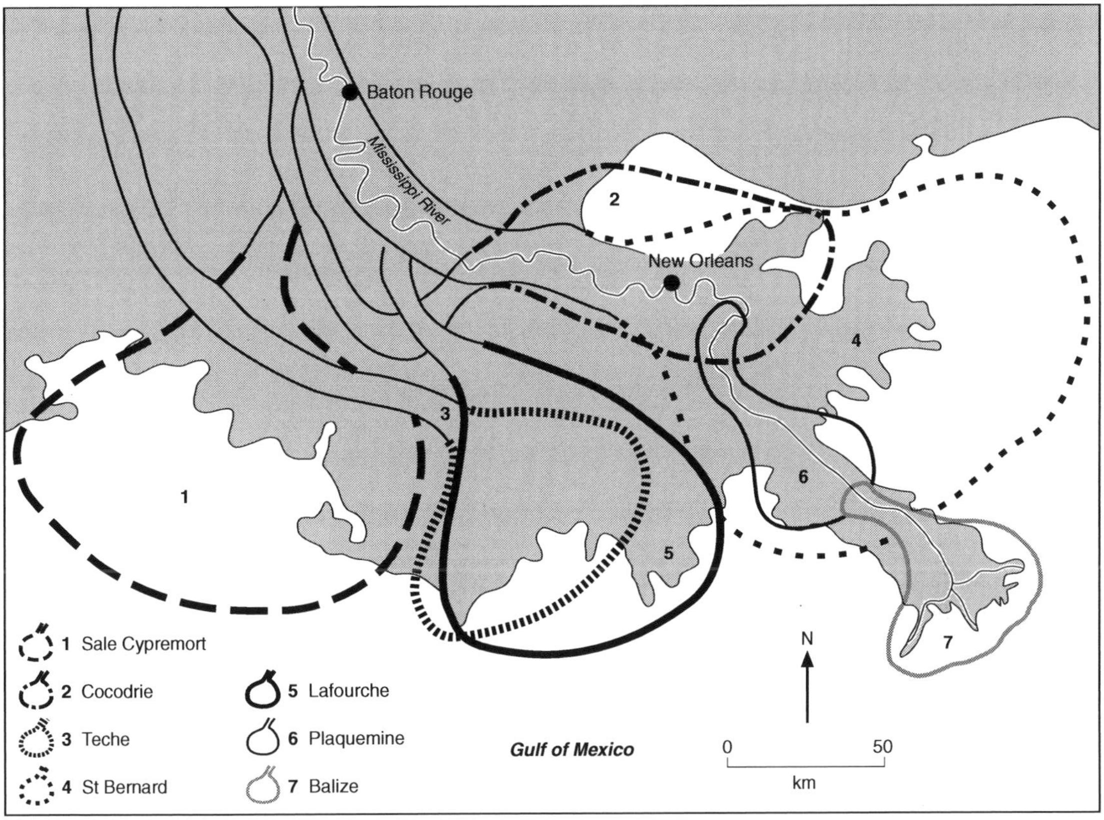


**Fig. 17.15** The seven Holocene delta lobes of the Mississippi delta plain. (After Kolb & van Lopik, 1958, Coleman 1976; for a revised chronology see Törnqvist *et al.*, 1996.)

only a few days (Fig. 17.21). The Holocene sequence was constructed as a series of at least 13 discrete delta lobes which formed after successive avulsions. Each lobe consists of a composite sand body typically 5–8 m thick. Coring reveals a vertical succession of: (i) a basal, coarse-grained transgressive lag overlying a lowstand land surface; (ii) a transgressive coastal mud interval; and (iii) a generally sharp-based sand unit deposited principally in channel and mouth-bar environments with lesser volumes of floodplain and coastal facies. The nature of Burdekin mouth bars differs from those of other river-dominated deltas in that they are often sharp-based and thus difficult to distinguish from distributary channel deposits. The bars have a triangular plan-view subaerial geometry intermediate between the elongate, lozenge-shaped mouth bars typical of river-dominated deltas and the beach-ridge geometries characteristic of wave-dominated deltas. Waves modify sedimentary textures along the foreshore into well sorted fine- to medium-grained sand. Mouth-bar construction takes place over tens of years; GPR reveals that subsurface mouth-bar sand bodies have low-angle seaward-dipping internal bedding surfaces and are bounded by surfaces of similar attitude. Once a mouth bar has become emergent it is stabilized by vegetation and a new bar initiates seawards. In this way, delta 'lobes' are constructed over 100s to 1000s of years before being abandoned following an avulsion of the trunk river channel to another part of the delta.

## Brazos delta

The Brazos delta has been studied by extensive coring, from flood records and by sequential aerial photos (Fig. 17.22). The modern delta originates from the

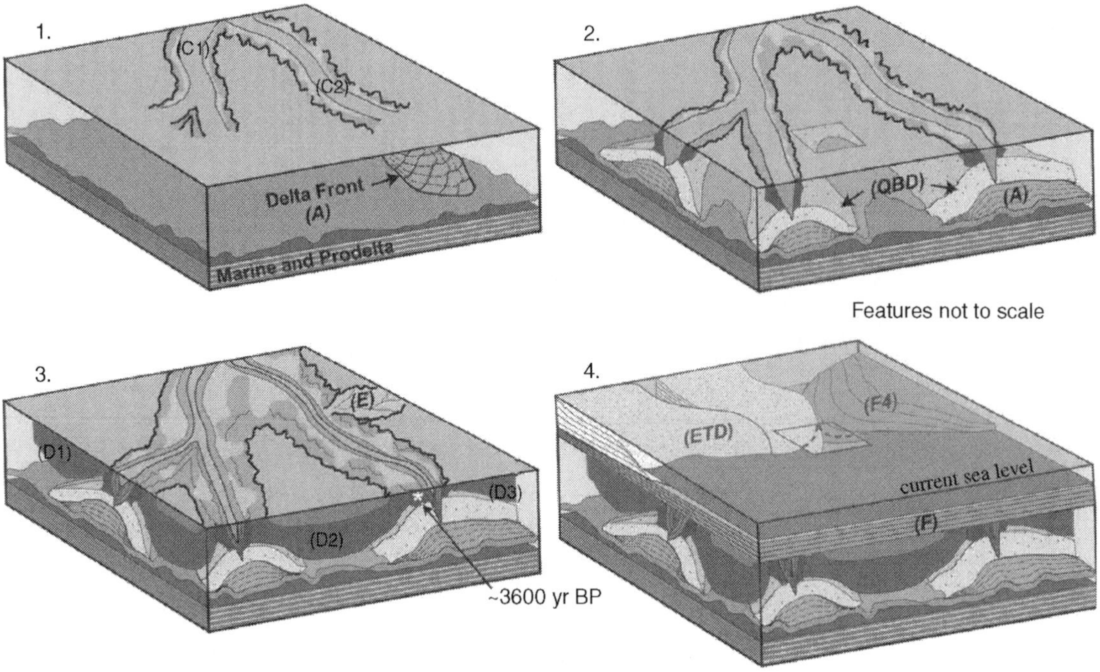


**Fig. 17.16** To illustrate late Holocene development (< 5000 yr BP) in nearshore environment adjacent to Barataria Bay, Mississippi delta region. (1) Initial progradation of delta-front (unit A) and distributary channels (units C1 and C2) into the Barataria Bight, associated with the Bayou des Families delta. (2) Continued deposition of delta-front (unit A) and distributary-mouth bar (unit QBD) deposits through the study area. (3) As progradation passes beyond the study area, interdistributary basins (units D1–D3) and crevasse deposits (unit E) develop, while flows in distributary channels diminish. (4) After abandonment and subsidence, ravinement and marine deposits (F) encroach the area, along with prodelta deposits from the Plaquemines–modern delta (F4). With the development of the barrier islands (not shown), ebb-tidal delta deposits (ETD) associated with the present-day inlets migrate into the study area. (After Flocks *et al.*, 2006.)

1920s, when a planned diversion of the River Brazos brought the new delta some 10 km southwest of the previous delta position. Once regarded as a classic wave-dominated delta with lateral strandplain beach ridges, more thorough consideration has revealed two phases of evolution of the new delta. During the first two decades, sands swept from the old delta were transported westwards by longshore currents, reworked shoreward, and deposited as amalgamated ridges near the mouth of the new delta. This early phase of evolution is typical of the classic wave-dominated delta model. However, depletion of sand supply led to a new morphology characterized by alternating detached ridges and inter-ridge lagoons. The ridges formed initially as mouth bars during and immediately after major river flood discharges in 1941, 1957, 1965 and 1992. During each of these events the channel mouth bar experienced significant growth and a back-bar lagoon was formed. Following each flood, wave action caused a bar to migrate onshore and to the west to become the new shoreline. The old shoreline, in addition to part of the back-bar lagoon, was preserved within the western delta headland. These flood cycles have resulted in the alternating beach-ridge sands and mud-filled troughs that characterize the delta headland. Each of the episodes of rapid delta growth occurred when a period of extensive drought preceded an El Niño-induced flood. The droughts, associated with La Niña events, were instrumental in removing catchment vegetation and preconditioning the drainage basin for erosion. Without a preceding drought, the floods are ineffective in generating sufficient sediment to create a mouth bar/beach ridge. A further striking feature of the Brazos delta is its large flood-enhanced prodelta, nearly 60% of total delta volume.

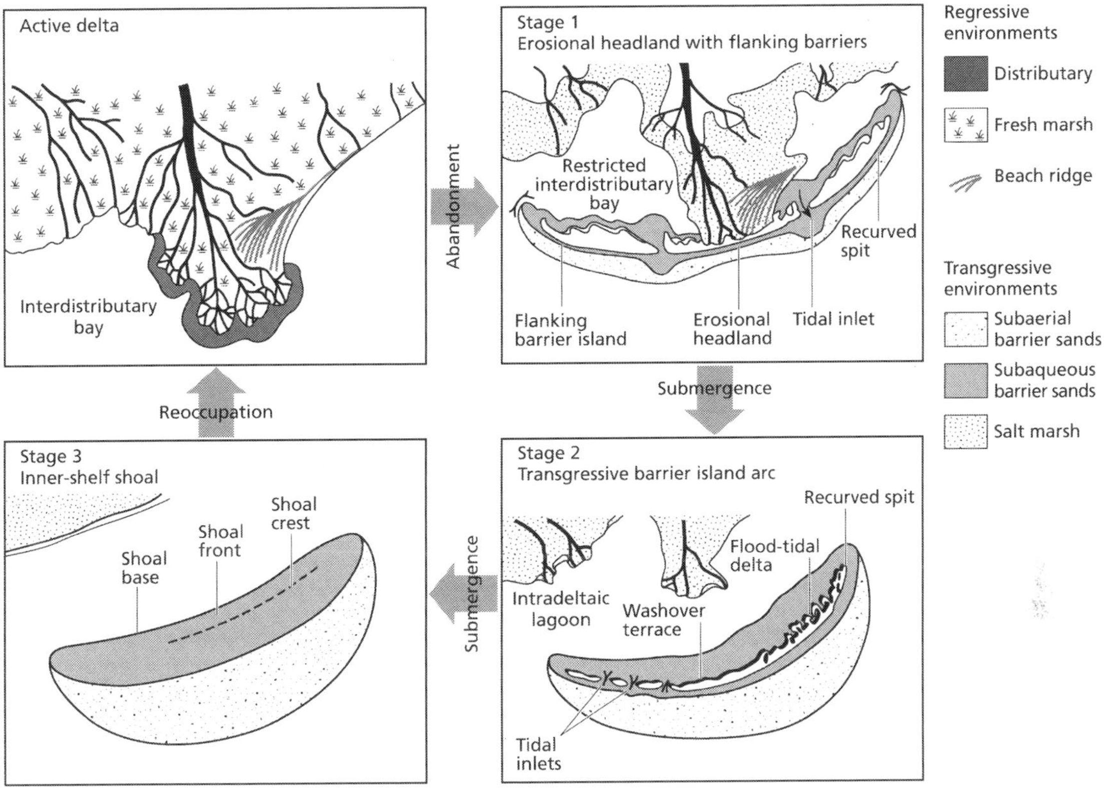


**Fig. 17.17** Evolution of delta-lobe abandonment scenarios appropriate to Mississippi-type shaol-water deltas. (After Penland *et al*., 1988.)

## Delta in a cold climate—McKenzie delta, Arctic Canada

Sedimentation on this high latitude fine-grained delta is greatly influenced by the annual cycle of seasonal processes, including winter freezing of sediments and channels, ice-jamming and flooding in the early spring and declining river stage during the summer and autumn. The subaerial delta plain is characterized by an anastomosing system of high-sinuosity channels and extensive thermokarst (thaw subsidence of permafrost under water) lake development. Deposition occurs on channel levees and in thermokarst lakes during flood events. In the channel-mouth environment, deposition is dominated by landward accretion and aggradation of mouth bars during river- and storm-surge-induced flood events. The subaqeous delta is characterized by a shallow-water platform and a gentle offshore slope. Sediment bypassing of the shallow-water platform is efficient as a result of the presence of incised submarine channels and the predominance of suspension transport of fine-grained sediments by hyperpycnal flows. Sea-ice scouring and sediment deformation are common beyond 10 m water depth where bioturbated muds are the predominant facies. Several indicators of the cold climate can be used as criteria for the interpretation of ancient successions, including thermokarst lake development, freeze–thaw deformation and ice-scour deformation structures. Permafrost inhibits compaction subsidence and, together with the shallow-water setting also limits autocyclic lobe switching. The cold climate can thus influence stratal architecture by favouring the development of regional-scale clinoform sets rather than multiple, smaller scale lobes separated by flooding surfaces.

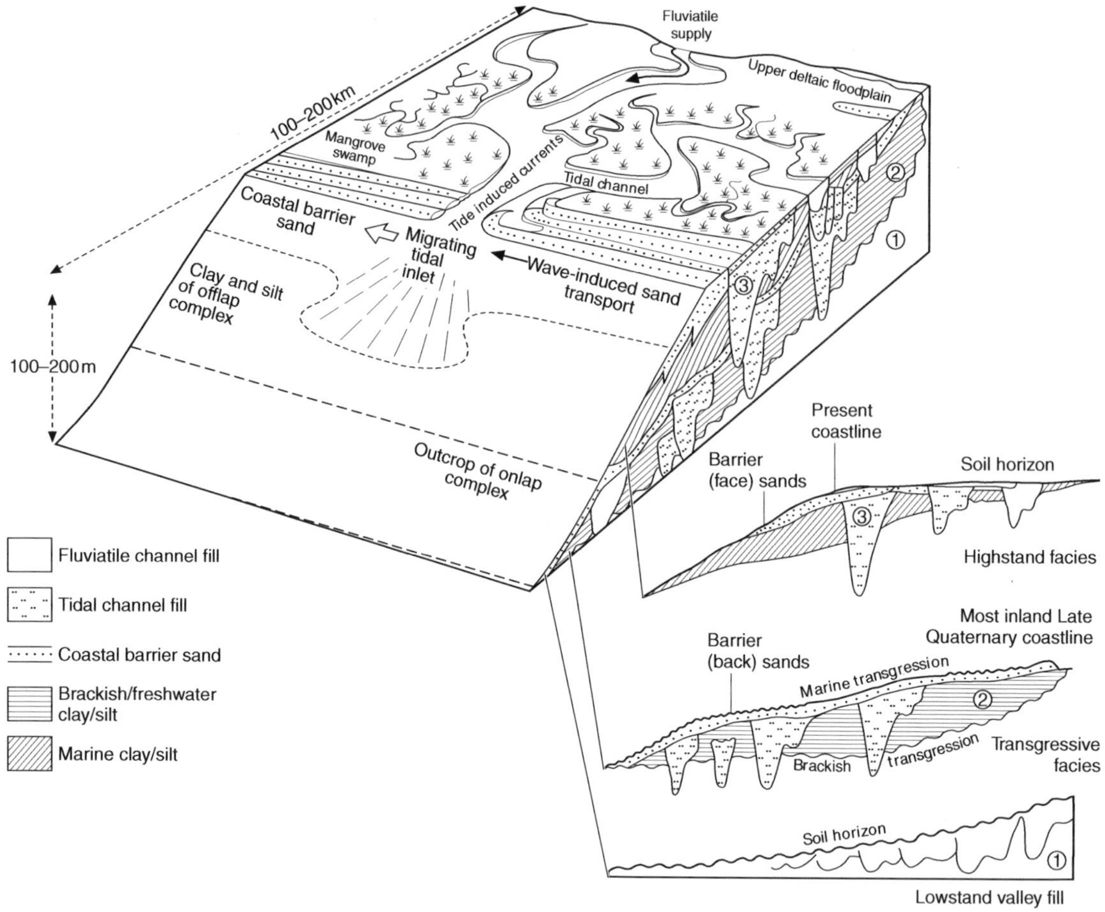


**Fig. 17.18** Lithofacies relations in the mixed-regime, late Quaternary Niger delta. (After classic paper by Oomkens, 1974.)

## Some tide-dominated deltas

Deltas are common on macrotidal coasts where wave action is limited, such as the Ganges–Brahmaputra in the Bay of Bengal, Mahakam, Indonesia, and Han, Korea. They also occur where onshore monsoonal winds are strong, as in the Fly River delta of the Gulf of Papua. Tide-dominated delta-front facies feature a dense network of tidal channels and islands, which pass offshore into coast-normal linear tidal current ridges and shoals. Delta progradation causes gradual emergence of the tidal current ridges so that they become coated with a sequence of intertidal and supratidal fine-grained sediments. The areas between the tidal current ridges eventually become tidal channels and ultimately fluvial channels as progradation continues. In the Fly River delta, for example, the predominant facies in the channels are finely laminated well-sorted sands and silts thought to be deposited during neap tides when wave-induced oscillatory currents are weak. Bioturbation levels are low because of high tidal current velocities (up to 2.4 m/s) and high suspended loads (up to 40 g/L). Meandering tidal distributaries exhibit inclined heterolithic stratification (IHS; lateral accretion deposits) as in the Sukmo distributary channel of the tide-dominated Han River delta, Korea. The IHS occupies the upper 25 m of a 40 m thick, upward-fining channel-bank succession whilst the lower 15 m comprises medium sand with flood-oriented cross stratification. The IHS shows an upward thinning trend in interbedded sand and mud layers. Tidal rhythmites (section 18.5) are

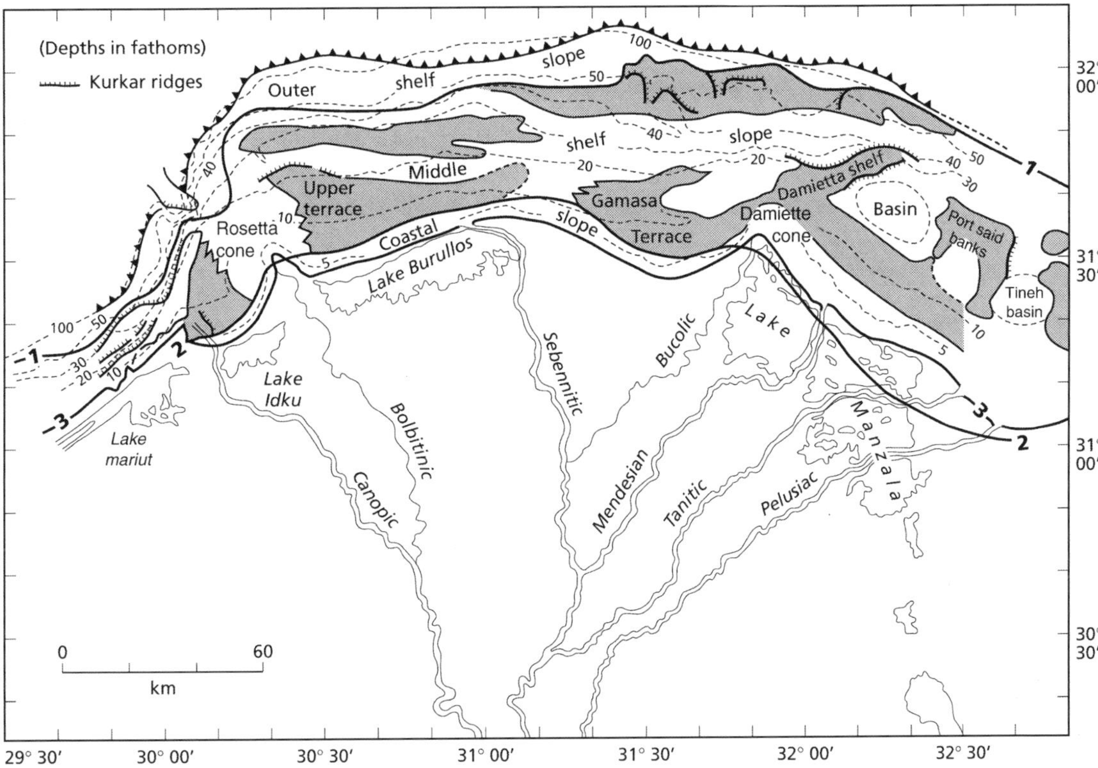


**Fig. 17.19** Map of the Nile delta and shelf showing major morphological features, isobaths and successive (1) early Holocene, (2) historical and (3) modern shorelines. Of the numerous Holocene distributaries only the Bolbitic (Rosetta) and Bucolic (Damietta) are presently active. (After Said, 1981; Scheihing & Gaynor, 1991.)

preserved in the middle and upper intertidal zone at the top of the succession and may also be present in the subtidal zone. Wave action at low-tide produces slightly coarser grained deposits and prevents the formation of rhythmic lamination. Locally generated waves also produce concave-up erosion surfaces at this level.

## 17.6 River deltas and sea-level change

As we see in Fig. 17.1, deltas are traditionally classified by the relative contribution of fluvial, wave and tidal energy flux at their seaward edges: the delta system is seen as the outcome of intrabasinal processes whilst external controls like sea level and sediment supply are held constant. However, studies of Quaternary deltaic deposits show important changes in response to eustatic sea-level falls and rises (Fig. 17.23). Sea-level shift from highstand coastline out to lowstand shelf edge produces bayhead, inner-shelf, mid-shelf, and shelf-margin deltas. Bayhead and inner-shelf deltas tend to form thin units in shallow water (a few metres to tens of metres amplitude, respectively). As they aggrade with rising relative sea level they can generate a thick on-delta succession. Mid-shelf deltas produce units as thick as the mid-shelf water depth, tend to follow a subhorizontal trajectory and generate little or no on-delta deposits—they are also susceptible to shoreface erosional thinning during transgression (ravinement). Shelf-edge deltas at lowstand usually have sparse on-delta deposits, create very thick clinoforms, and feature thick successions of sandy turbidites at the delta front. If sea level falls below the shelf margin, the shelf-edge delta becomes incised by its own channels and large volumes of sand can be delivered on to the slope and the basin floor. During intervals other than lowstand, deltas require a strong fluvial drive to attain shelf transit and as they

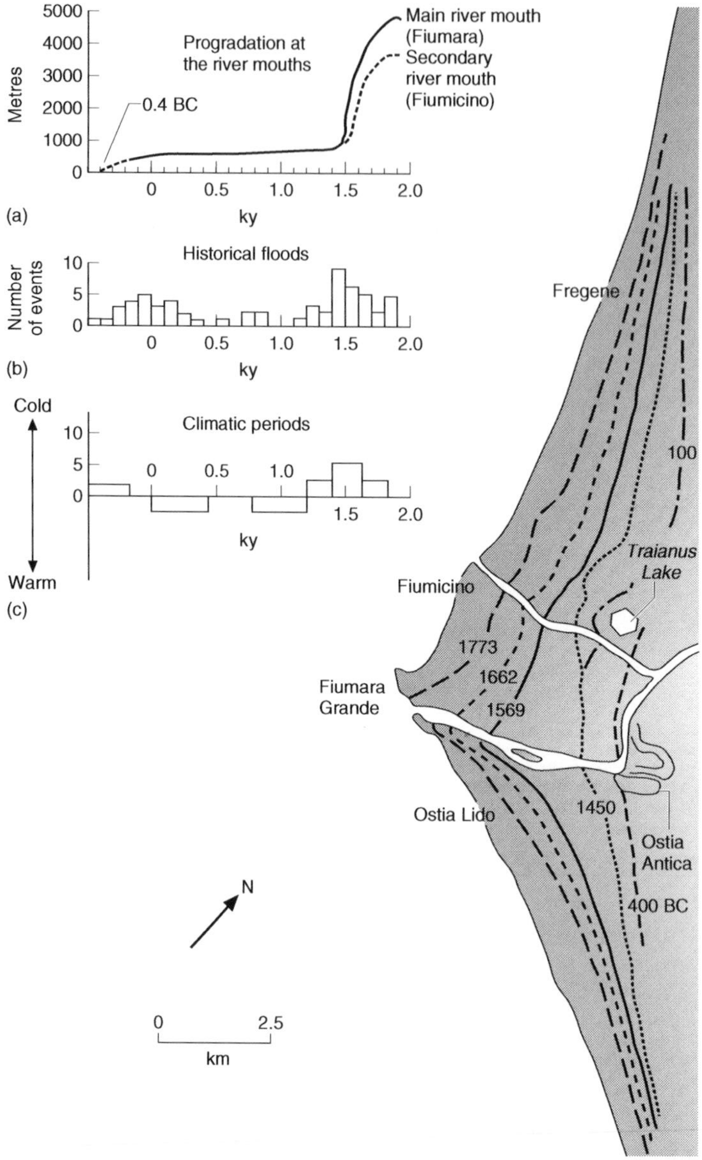


**Fig. 17.20** Map and inset graphs to show the growth of the Tiber delta. Despite obscure records from Rome's 'Dark Ages', note the surge of delta progradation from the late-Renaissance onwards, reflecting both human-induced catchment usages and climate variation. (From Bellotti *et al.*, 1994.)

approach the outer shelf they commonly become wave dominated. Tidal influence can increase on the outermost shelf if relative sea level is falling, if the shelf-break is poorly developed, and if basinal water depth is shallow. Deltas that transit back and forth on the shelf on short timescales ($10^3$–$10^5$ yr) and that are driven largely by sea-level fluctuations are referred to as *accommodation-driven deltas*. Deltas that can reach the shelf edge without sea-level fall are termed *supply-driven deltas*.

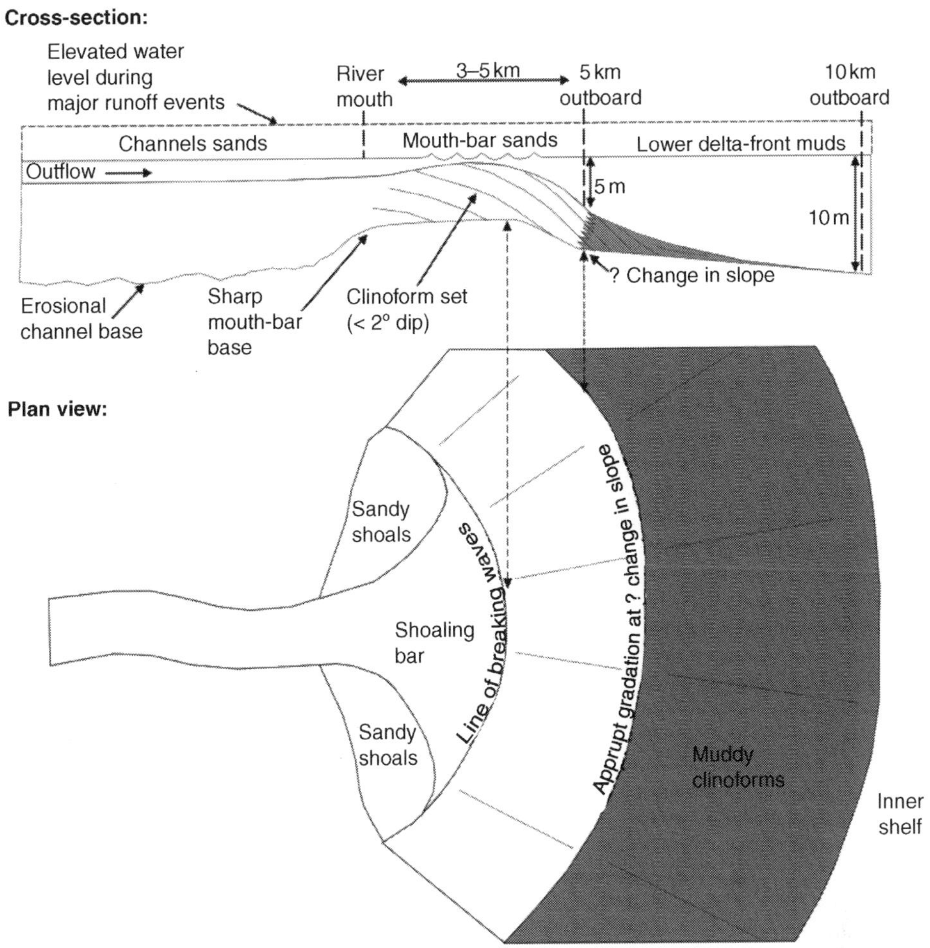


**Fig. 17.21** A depositional scheme for Burdekin delta mouth bars. A bar begins to form following avulsion of the river to a new part of the coast or progradation beyond an earlier mouth-bar system. Sediment is accumulated mainly during major runoff events when the water surface is elevated significantly higher than at low stage flow, and up to several million metric tons of particulate sediment may be delivered to the river mouth in a day or so. Pulsed growth from several such events gives rise to a sand body that is predominantly sharp-based, shows complex or no vertical grain-size variation, fines abruptly in a seaward direction, and internally preserves a low-angle clinoform set. (After Fielding *et al.*, 2006.)

A second aspect of changing sea level is the loss of freshwater wetland environments due to relative sea-level rise (RSLR) induced by a combination of sediment starvation, compaction of subsurface sediment (particularly peat) and groundwater extraction. In many deltas RSLR reaches up to 5–10 times that of a current background eustatic (global) sea-level rise of some 1–2 mm/yr. For example, the city of New Orleans has a mean subsidence rate of 6.5 mm/yr, so that combined with the global eustatic rise this means subsidence relative to changing sea levels of > 7.5 mm/yr. Loss of vegetated freshwater wetland to open-bay seawater means lowered up-delta storm surge protection and loss of mixed salinity ecosystems. Sediment loss leads to irredeemable delta-front erosion during RSLR when the catchment sediment supply teleconnection is broken by upstream dams (Nile, Yangtze, Mississippi). Local losses occur when specific intradistributary wetlands are isolated from channel overbank and crevasse flood sediment supply by

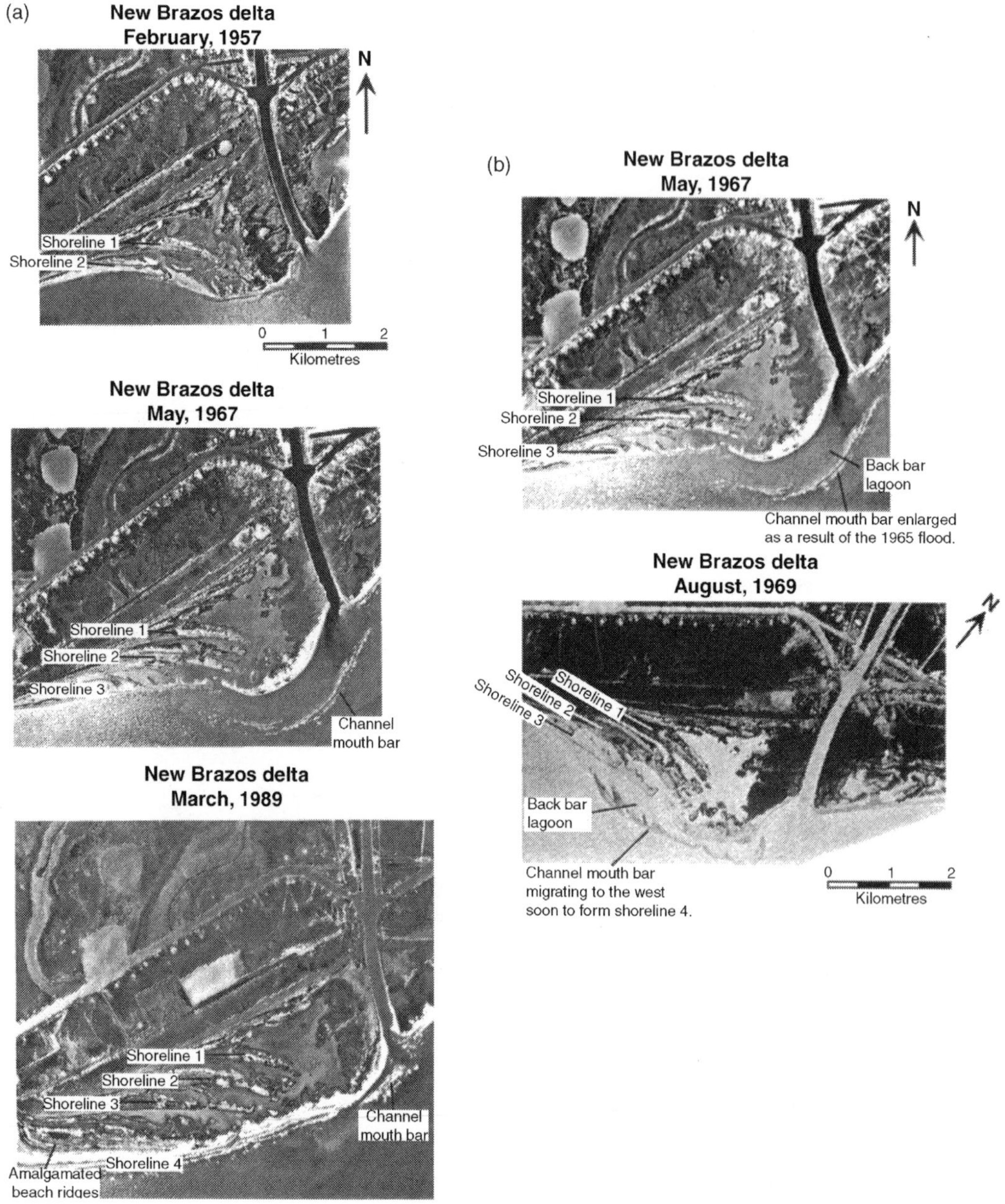


**Fig. 17.22** (a) Scaled and orientated aerial photos of the New Brazos delta illustrating delta progradation. Following each large flood a new ridge/trough pair was preserved in the western delta. (After Rodrigues *et al.*, 2000.) (b) Scaled aerial photos of the New Brazos delta illustrating the progradation event that resulted from the 1965 flood. (Modified by Rodrigues *et al.*, 2000.)

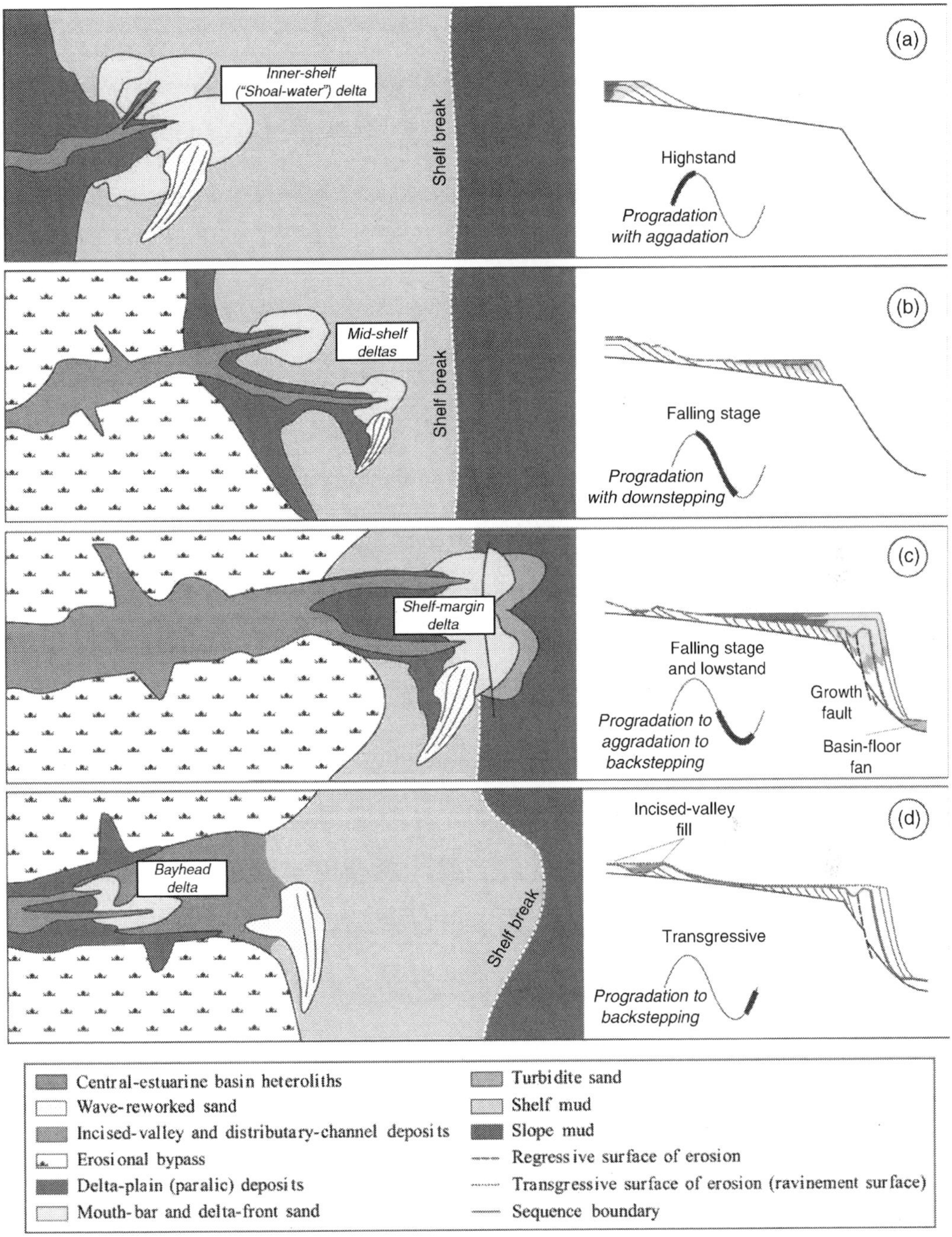


**Fig. 17.23** Classification of shelf deltas in terms of relative sea-level change in a shelf–slope (a)–(d) setting. (After Porebski & Steel, 2006.)

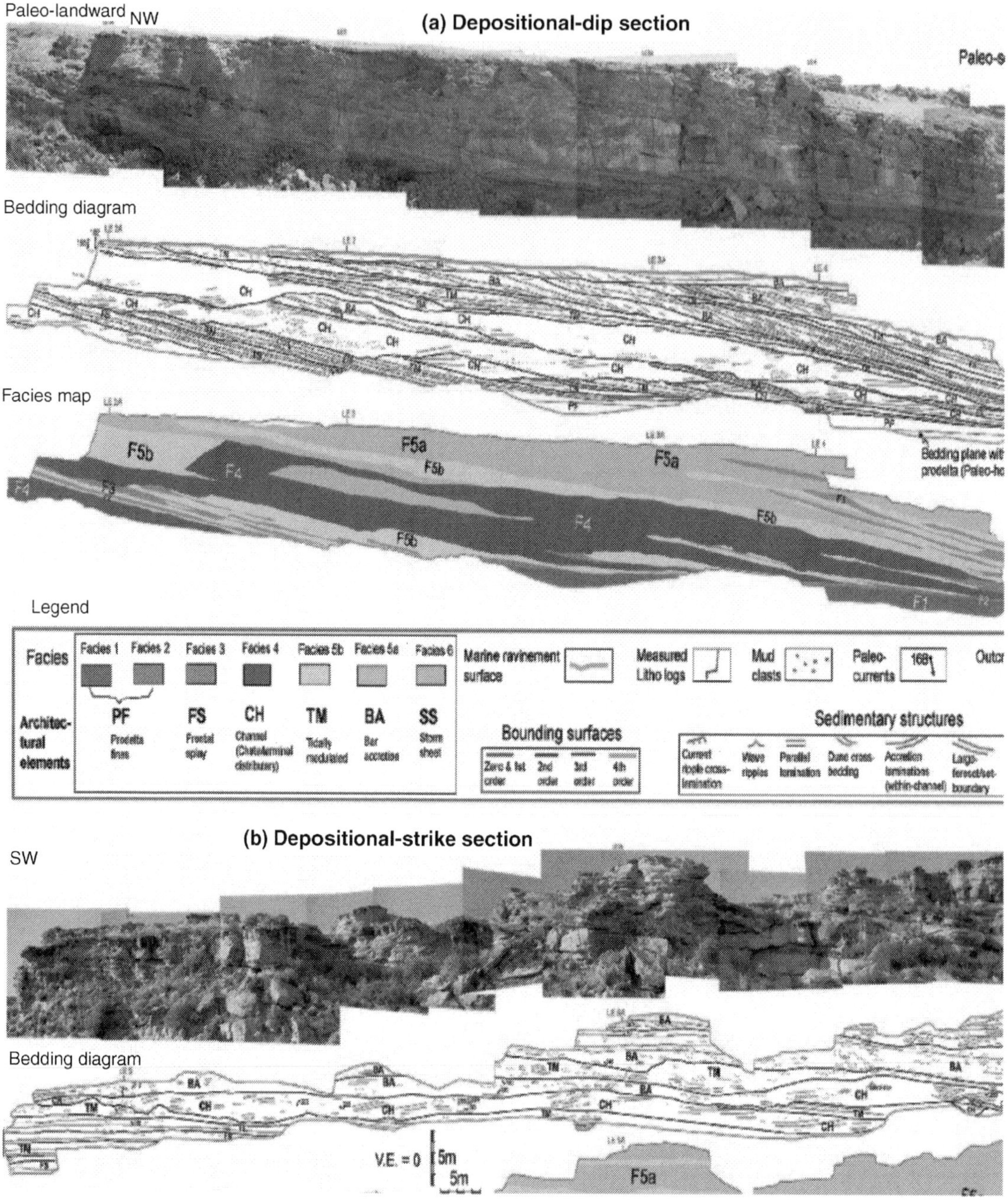


Fig. 17.24 The Cretaceous Wall Creek delta deposits. Outcrop panels (photomosaics with bedding and facies maps) of Sandstone 6 at Raptor Ridge. Detailed identification of sedimentary features, bedding planes, and facies in several depositional dip and strike sections allows reconstruction of bounding-surface hierarchies and architectural elements (marked on bedding diagrams of (a) and (b)). (After Gani & Bhattacharya, 2007.)

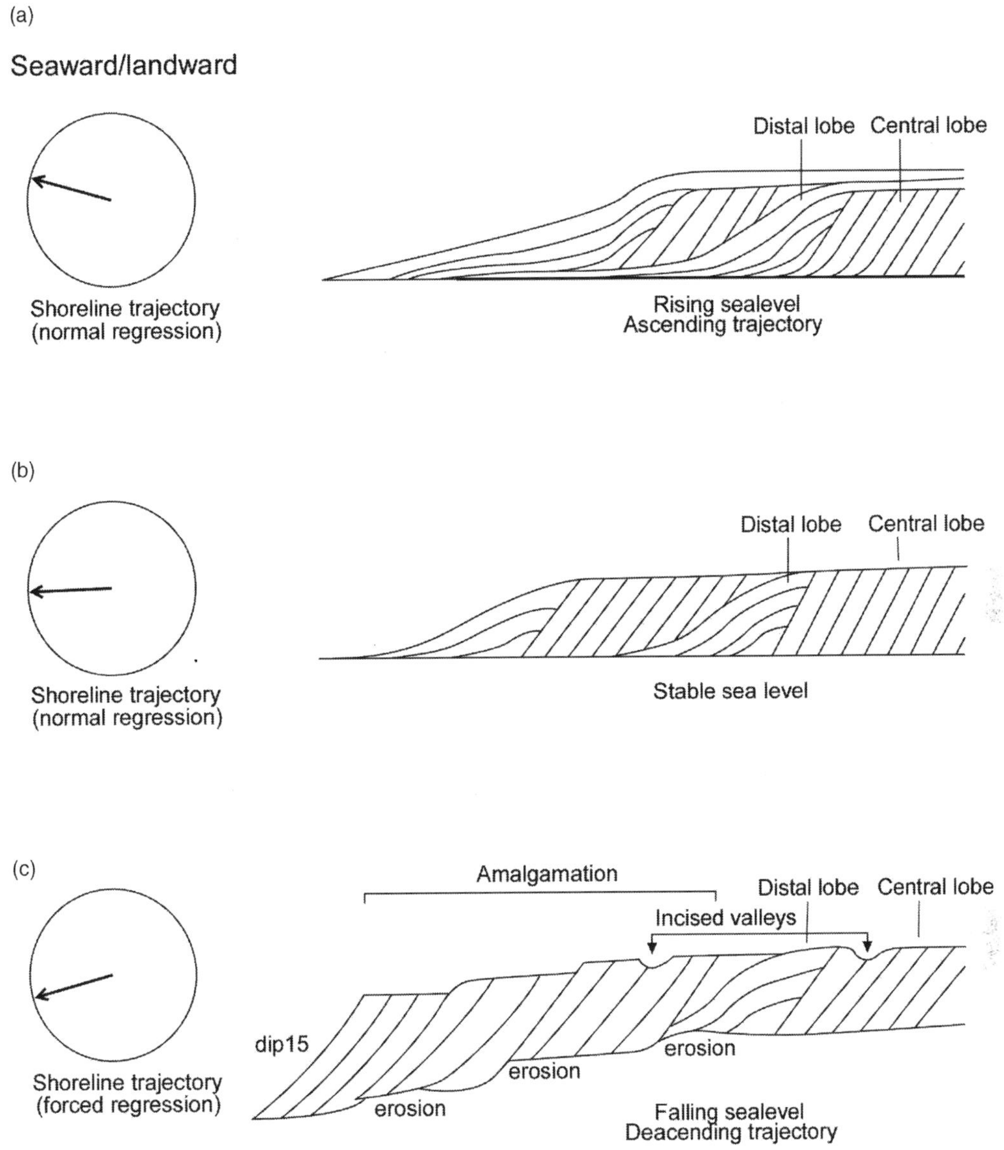


**Fig. 17.25** Illustration of development of the Billund delta complex during rising, stable, and falling sea level. The progradation of the delta is reflected by alternation of parallel clinoformal seismic reflection patterns and sigmiodal seismic reflection pattern, which is interpreted as due to delta-lobe switching. Ascending shoreline trajectory indicates progradation under rising sea level. A stable shoreline trajectory indicates no change in sea level, and a descending shoreline trajectory indicates progradation under falling sea level. Extensive and thickest sands are deposited during falling sea level, which is indicated by a descending shoreline trajectory and deep erosion into the unconsolidated Miocene substratum and incision into the proximal part of the delta complex. (After Hansen & Rasmussen, 2008; based on the concepts of Helland-Hansen & Gjelberg, 1994.)

artificial levees (Mississippi). Efforts to reverse the losses in the Mississippi by timed diversions through floodgates during the rising limb of spring floods have been unable to supply the quantity of sediment required in the time available and which was once supplied more efficiently by natural levee crevasse flooding. Part of this problem is the general reduction of sediment supply to the trunk river channel due to upstream damming of Mississippi tributaries. Other sediment budget studies indicated that by 2100 the delta plain will be effectively abandoned, converted into large open-water bays.

## 17.7 Ancient river delta deposits

As noted above there is a lack of integrated process/facies architectural studies of modern deltaic deposits with studies of several modern deltas suggesting that the role of major river floods has been much neglected. Many of the palaeodeltaic studies in the vast deltaic literature are likely to be in error because of this lack of modern analogue data. The most convincing examples of ancient deltaic systems come from areas of 3D exposures where stratal boundaries are clearly shown, especially the successive *clinoform surfaces* that define the palaeodelta front and the extent of tidal and wave modifications. Examples worthy of close study include the river-dominated Ferron delta (Cretaceous, Utah), the tidally influenced Frewens and Sego deltas (Cretaceous, Wyoming; Cretaceous, Utah) and the wave–influenced Kennilworth delta (Cretaceous, Utah) and Blackhawk delta (Cretaceous, Utah).

For a detailed rock and radar facies analysis of a mixed regime delta consider the Wall Creek delta (Cretaceous, Wyoming) (Fig. 17.24). Five orders of bounding surfaces separate six types of facies in prodelta and delta-front deposits: prodelta fines, frontal splays, subaqueous channels, storm sheets, tidally modulated and mouth-bar accretion deposits. Seasonal to decadal river floods are thought to represent the main building phases of the delta, producing the channels, bars and frontal splay elements. During intervening periods, the delta was reworked by waves, storms, and tides, producing tidally modulated and storm-sheet elements. The plan-view morphology of the delta is a smooth-fronted, arcuate to cuspate shape and gives no clue as to the dominant role of river flooding. Storm-wave-reworked sands are attached to the flanks of the system. Previously published delta models predict incorrectly that the tidal reworking of a river flood-dominated system should result in a more bird-foot shape versus the lobate geometry that is actually mapped. The analysis of internal facies versus external shape matches similar observations on several modern deltas, most notably the Burdekin and Brazos and confirms that external shape is a poor indicator of internal facies arrangement.

The extreme complexity of tidal interactions with distributary outfalls and production of normal-to-shelf-slope mud wedges was highlighted for the modern Huanghe delta and deserves to be better documented in ancient tidally influenced deltas. Another neglected aspect of ancient deltaic architecture is its control by changing relative sea level and the extent to which a delta front is forced to: (i) regress during sea-level fall, (ii) transgress during rising stage, or (iii) remain adjusted to stable relative sea level. This can be estimated from the stacking pattern of outcrop or seismic facies by examination of clinoform geometry as it displays a descending, ascending or horizontal shoreline *cusp trajectory* (Fig. 17.25). Outcrop and subsurface studies reveal that the cleanest sand is often found in resedimented units deposited in narrow bands within structural lows and during periods of forced regression.

## 17.8 Fan deltas

A fan delta is an alluvial fan whose front is an active shoreline, defining an interface with standing water into which it progrades (Fig. 17.26). Marine fan deltas in lower latitudes may exhibit significant biogenic carbonate sedimentation from fringing reefs and patch reefs. Although these coral–algal communities may be periodically disturbed or overwhelmed by flood sediment discharges, the overall clastic sedimentation rate is too low to prevent colonization. Particularly fine examples of such mixed siliciclastic and carbonate environments occur along the modern faulted coastline of the Gulf of Aqaba. Interesting ancient analogues feature in the late Neogene of southeast Spain.

The nature of the beachface to offshore transition (Fig. 17.26) is clearly of primary importance in accounting for variability amongst fan deltas. *Ramp-type fan deltas* have relatively gentle offshore slope with a degree of wave and tide reworking—little sediment bypasses the gently offshore gradients. Internal structure is dominated by low-angle accretion surfaces recording periodic changes in progradation rate associated with

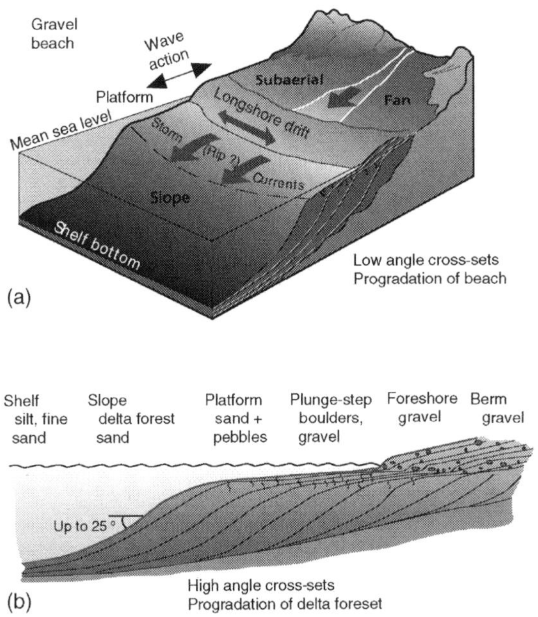


**Fig. 17.26** Sketch to illustrate a wave-dominated fan delta with characteristic low-angle beachface foresets overlying steeper delta-front foresets of 'Gilbert-type'. (After Dabrio *et al.*, 1991b.)

fan-channel avulsions. Such fan deltas are common where tectonic subsidence is low, along inactive faults and particularly on the subsiding hanging walls of extensional basins or the footwall ramps of thrust-related basins (Chapter 22). The most spectacular and instructive ancient example of a shelf-type fan delta is that of the Eocene Montserrat fan delta of Catalonia—great pile of alluvial fan conglomerates that coarsens upwards over 1300 m. Within this sequence, eight individual units (75–250 m thick) fine and thin radially outwards into the basin. Each passes from proximal alluvial fan debris-flow and stream-flow conglomerates to distal fan sandy channel and overbank fines, to beachface, shoreface and mouthbar sands and gravels, and finally to offshore fines. Five of the fan units are separated by thin transgressive marine deposits and the whole complex represents the vertical stacking of wave reworked fan-delta sediments periodically swept landwards by subsidence-induced transgressions followed by renewed progradation. Subtle changes in the character of the fan-delta front during transgression (wave-dominated) and progradation (fluvial-dominated, with coarse mouth-bars, even debris flows) may reflect very marked variations in sediment supply.

*Gilbert-type fan deltas* have an abrupt discontinuity of slope at or near the shoreface such that the subsequent depositional submarine gradient may reach 25° or more. Although the shoreface of the fan delta may be reworked by wave and tide to form significant beachface and shoreface units, a significant sediment flux bypasses the zone during flood events. This sediment is deposited at the top of the delta slope and, together with material provided by frequent submarine failures, is transported down the steep delta front as debris-/grain-flow avalanches or density currents in broad shallow chute channels (Fig. 17.27). Down-

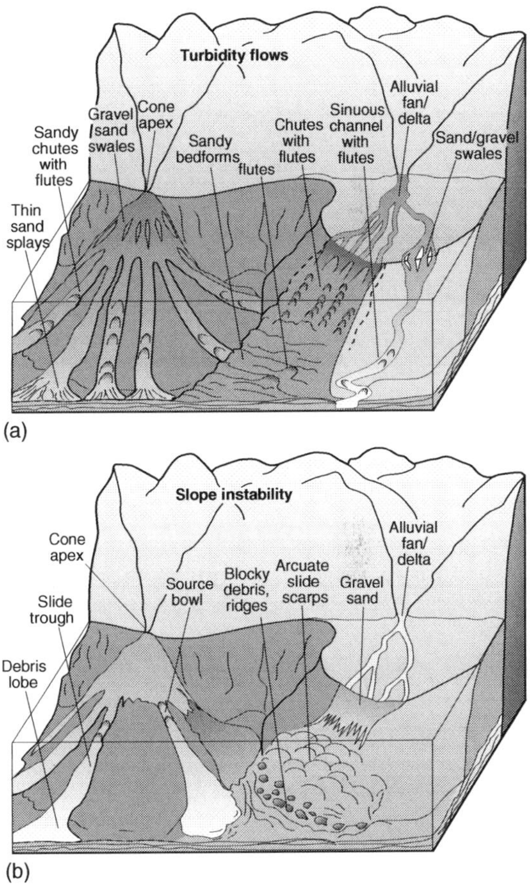


**Fig. 17.27** Three-dimensional views of avalanche, inertia flow, turbidity flow and slope instability features of Gilbert-type fan deltas. In each case the left-hand fan delta has little subaerial expression which is typical only of very young fan deltas or those on very steep slopes fed by low discharge streams, such as on fjord sides. (After Prior & Bornhold, 1990.)

slope the avalanche deposits cease their motion as frictional forces overcome internal driving forces. They accrete on to the pre-existing slope and stand at the angle of repose for the sediment flow in question. At the foot of the delta slope there may be an abrupt hydraulic jump or gradational change to deposits of the basin floor, which may feature resedimented coarse sediment lobes (Fig. 17.28). Sediment entrained into density currents on the delta-slope chutes or generated at the hydraulic jump carries on down across the basin floor where eventual radial spreading and deceleration cause deposition of sand and gravel in lobate bodies interdigitated with fine basin-floor sediments. Narrow basin floors will encourage long runout distances and attainment of very high current velocities.

Recognition of ancient Gilbert-type fan deltas is aided by mapping low-angle topsets of the alluvial-fan channel complex, the connection of these topsets with curvilinear plan foresets and their connection in turn with bottomset and basinal resedimented fan lobes. The thickness of individual Gilbert deltas is controlled

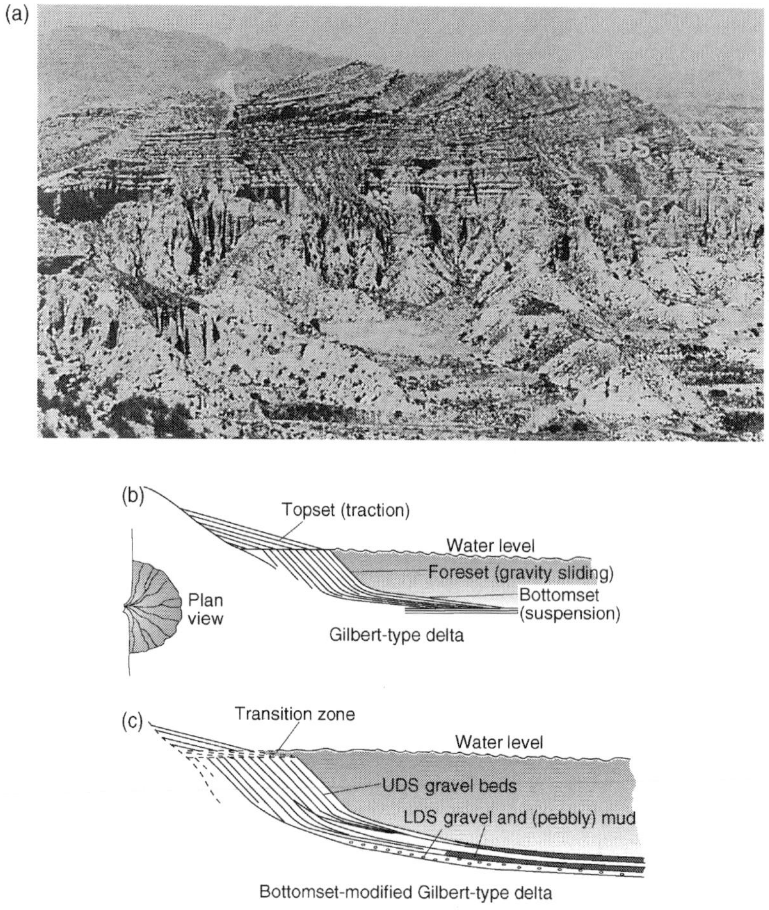


**Fig. 17.28** (a) View of El Hacho, southeast Spain, to show Gilbert-type fan delta deposits. The steep upper foreset bedding (clinoforms) are clearly visible inclined at high angles to the left. These grade out into low-angle bottom sets with thickened gravel beds deposited at the position of hydraulic jumps as high-density turbidity currents sourced from the steep foresets were forced to decelerate quickly. (b) & (c) Sections to illustrate 'normal Gilbert foreset to bottom set transition and thickened gravelly lenses produced by hydraulic jumping. UDS, upper delta slope; LDS, lower delta slope. (All after Postma & Roep, 1985.)

by local basin water depth, ranging from a few to many hundreds of metres. The detailed shapes of the foresets are a response to variations in sediment supply, water depth and relative sea-level change. In deep water and where subaerial-fan channel gradients are very high, there may only be a very small area of subaerial delta exposed, as well seen in some fjords. Spectacular examples of Gilbert fan-delta complexes have been mapped out in the uplifted southern portions of the Corinth graben, central Greece, where foreset complexes of coarse gravels occupy whole mountainsides up to ~ 600 m high.

Because fan-delta foresets are clearly imaged in shallow seismic profiles, they are of great use in elucidating the relationship between relative sea-level change and sedimentary response, particularly the early Holocene sea-level rise. Successive cycles of Gilbert-type fan delta, fluvial and marine transgressive facies reflect changing relative sea level and sediment supply. Convincing evidence for a tectonic subsidence control on repeated high-frequency (~10 kyr) cycles adjacent to the Loreto Fault, Baja California.

## Further reading

*General*

Collections of papers on palaeodeltas are edited by Rahmani & Flores (1984), Lyons & Alpern (1989) and Whateley & Pickering (1989). A notable and stimulating recent effort is by Giosan & Bhattacharya (2005). Harris et al. (2002) discuss Australian delta types and location as a quantitative function of wave, tide and river power. Coast-parallel mud-transport by blocked outfall jets and their role in creating shelf mud 'fingers' is in a nice clear paper by Liu et al. (2009). Yoshida et al. (2007) is a thoughtfull attempt to investigate the idea that variable process regimes during cycles of sea-level rise and fall have important repurcussions for sequence stratigraphic studies of coastal deposit architecture.

*Specific*

A revision of the widespread use of external deltaic form to infer the predominant processes of depositional architecture was stimulated by the work of Alexander *et al.* (1999) and Amos *et al.* (2004) on major discharges in the tropical Burdekin River of Queensland, Australia and proposed by Rodriguez *et al.* (2000) for the Brazos River, Texas, by Fielding *et al.* (2005a, 2006) for the Burdekin river delta and Maillet *et al.* (2006) for the RhÂne delta, and supported by data from the ancient record by Gani & Bhattacharya (2005, 2007) and Lee *et al.* (2007).

Traditional delta classifications, misleading for the above reasons, but widely quoted, are in Wright & Coleman (1975), Galloway (1975), Orton & Reading (1993), Bhattacharya & Giosan (2003).

Physical processes are considered in Bates (1953), Wright & Coleman (1973), Wright (1977, 1985), Matyas (1984), Kenyon & Turcotte (1985), Van Gelder *et al.* (1994), Nittrouer & Kuehl (1995), Adams *et al.* (2001), Li *et al.* (2001) van Maren (2005), Fan *et al.* (2006), Maillet *et al.* (2006), Milligan *et al.* (2007) and Day *et al.* (2007).

Delta-front failure and deformation are studied by Correggiare *et al.* (2001), Wignall & Best (2004) and Maillet *et al.* (2006).

Organic deposition in deltas is described by Anderson (1964), McCabe (1984), Kosters (1989) and Staub & Esterle (1993).

Holocene delta case histories: Amur by Davies *et al.* (2005); Brazos by Rodriguez *et al.* (2000) and Fraticelli (2006); Burdekin by Fielding *et al.* (2005a,b, 2006); Danube by Giosan *et al.* (2005); Fly River, Gulf of Papua by Baker *et al.* (1995); Ganges-Brahmaputra by Kuehl *et al.* (2005); Godavari by Nageswara *et al.* (2005); Han River by Choi *et al.* (2004); Huanghe (Yellow River) by Li *et al.* (2001), Fan *et al.* (2006); Mahakam, Indonesia by Gastaldo *et al.* (1995); McKenzie by Hill *et al.* (2001); Mekong by Ta *et al.* (2005); Mississippi by Fisk (1944), Fisk *et al.* (1954), Coleman & Gagliano (1964), Frazier (1967), Coleman *et al.* (1964), Wright & Coleman (1973), Roberts *et al.* (1980), Penland *et al.* (1988), Kosters (1989), Steinberg (1995), Törnqvist *et al.* (1996), Nadon (1998), Walker *et al.* (1987), Kulp *et al.* (2005), Olariu & Bhattacharya (2006) and Blum & Roberts (2009)—for a nice semi-popular account of wetland loss see Bourne (2004), for effects of Gulf of Mexico hurricanes see Flocks *et al.* (2006), Turner *et al.* (2006) and Day *et al.* (2007), for subsidence controversy see Törnqvist *et al.* (2008); Niger by Oomkens (1974); Nile by Stanley & Warne (1993), Sestini (1989); Orinoco by Aslan *et al.* (2003); Po by Correggiari *et al.* (2001, 2005a,b) and Milligan *et al.*

(2007); Red River by van Maren (2005); Rhine-Meuse by Stouthamer (2005) and Cohen (2005); Rhône by Maillet *et al.* (2006); Tiber by Bellotti *et al.* (1994); Volga by Kroonenberg *et al.* (2005); Williams by D.G. Smith *et al.* (2005).

Ancient deltaic architecture, mostly on sea level controls of sequences, by Helland-Hanson & Gjelberg (1994); Helland-Hansen & Martinsen (1996); Plint & Wadsworth (2003); Bullimore & Helland-Hansen (2004); Anderson *et al.* (2004); Anderson (2005); Amorosi *et al.* (2005); Boyer *et al.* (2005); Correggiari *et al.* (2001, 2005a,b); Stefani & Vincenzi (2005); Gani & Bhattacharya (2007), Lee *et al.* (2007) and Hansen & Rasmussen (2008).

Specific ancient case histories from large outcrops: Ferron delta by Chidsey (2001) and Gani & Bhattacharya (2005); Frewens delta by Willis *et al.* (1999) and Willis (2005); Sego delta by Willis & Gabel (2003); Kennilworth delta by Hampson (2000); Wall Creek delta by Gani & Bhattacharya (2007); Blackhawk delta by Hampson & Howell (2005); Panther Tongue terminal distributary channel mouth bars by Olariu *et al.* (2005).

Shelf-margin deltas and eustatic sea-level changes are considered by Porebski & Steel (2003, 2006), Anderson (2005) and Plink-Björklund & Steel 2005); deltas and relative sea-level rise by Dixon *et al.* (2006), Snedden *et al.* (2007) and Day *et al.* (2007); deltaic sediment compaction by Meckel *et al.* (2007), Törnqvist *et al.* (2008); ancient tidally influenced deltas Willis *et al.* (1999), Willis & Gabel (2001), Pontén & Plink-Björklund (2007).

Fan deltas are studied in useful conference volumes edited by Nemec & Steel (1988), Colella & Prior (1990) and Dabrio *et al.* (1991a); useful general papers are by Surlyk (1978), Ethridge & Westcott (1984), Colella (1988), Massari & Colella (1988), Postma (1990) and Dabrio *et al.* (1991b). Definitions and nomenclature are considered by Nemec & Steel (1988). Mixed siliciclastic and carbonate environments of the Gulf of Aqaba are studied by Hayward, 1985. Interesting ancient analogues feature in late Neogene deposits of southeast Spain (Dabrio & Polo, 1988). The Eocene Montserrat fan delta of Catalonia is described Marzo & Anadon (1988); Gilbert-type fan deltas by Massari & Collela (1988), Dabrio *et al.* (1991b), Horton & Schmitt (1996), Ferentinos *et al.* (1988), Prior & Bornhold (1988, 1990), Postma & Roep (1985), Postma *et al.* (1988), Ori (1989), Syvitski & Farrow (1989), Chough *et al.* (1990) and Falk & Dorsey (1998); ancient Gilbert-type fan delta deposits by Dart *et al.* (1994), Hardy *et al.* (1994), Dorsey *et al.* (1997) and Mortimer *et al.* (2005).

Chapter 18

# LINEAR SILICICLASTIC SHORELINES

*...When the chalk wall falls to the foam and its tall ledges*
*Oppose the pluck*
*And knock of the tide,*
*And the shingle scrambles after the sucking surf, ...*

W.H. Auden, 'Seascape', *Selected Poems*, Penguin Books

## 18.1 Introduction

Linear shorelines are unbroken by estuary embayments or delta protrusions. Detailed morphology is a function of both tidal strength and wave energy, though the quantity and grade of sediment supplied from alongshore or offshore is also important. Tidal strength is expressed as tidal range, with arbitrary microtidal ($<$ 2 m range), mesotidal (2–4 m) and macrotidal ($>$ 4 m) categories. When tidal currents are strong and a plentiful supply of fine sediment is available, a low-gradient wedge of *tidal flat* sediment progrades seawards. Not only is incoming swell wave power greatly reduced by this, but also waves cannot break on any one part of the tidal flat for any length of time during rising or falling tide. The opposite conclusions apply to coasts dominated by high wave power; sediment is usually coarser and shore gradients steeper. Many coastlines have mixed wave/tidal energy levels, some with almost schizophrenic seasonal control by varying river stage, winter storm or monsoon energy fluxes. Linear coastlines may also be subdivided according to whether the coastline is *attached* or *detached*. In the latter case, *barrier islands* or *barrier spits* physically protect an inner, more sheltered coastline that surrounds an estuary, bay, lagoon or tidal flat (Fig. 18.1). The major environments of coastal deposition range from aeolian dunes above extreme mean high water level to the lower shoreface/inner shelf transition which may occur at water depths of $\sim$30 m or so. Specific environments include (i) attached beaches and intertidal flats, (ii) partly attached spits and (iii) detached barriers, tidal inlets and lagoon/tidal-flat complexes. In all these environments there is a single, well-mixed, friction-dominated fully turbulent boundary layer. At the shallow depths of most such shorefaces the Coriolis effect may usually be neglected in fluid force balances. This serves to physically distinguish the shoreface from the inner shelf, where stratified boundary layers are commoner and where the Coriolis effect at higher latitudes causes Ekman transport to become increasingly important in turning geostrophic currents and buoyant plume discharges.

*Sedimentology and Sedimentary Basins: From Turbulence to Tectonics*, 2nd edition. © Mike Leeder.
Published 2011 by Blackwell Publishing Ltd.

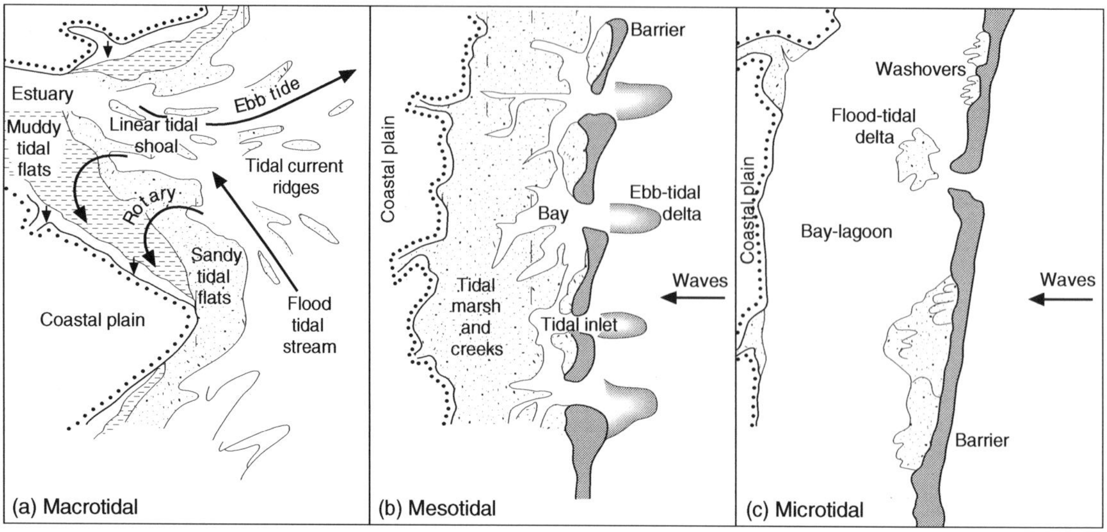


**Fig. 18.1** Morphological sketches of various coastlines, all with moderate wave energy. Sketches ignore time development (contrast with **Part 1 Fig. 1**). (a) Macrotidal coast with tidal range $> 4$ m. Note absence of barriers so an attached coast is featured with complex tides, some separated into ebb and flood routes. Sandy tidal flats on exposed coasts have mixed wave/tide energy. (b) Mesotidal coast with tidal range 2–4 m. Note this is a partly detached coast with barriers, tidal inlets, tidal deltas protecting and guiding tidal flow into more sheltered back-barrier tidal flats and creeks. (c) Microtidal coast with tidal range $< 2$ m. Note few or no tidal inlets and a sheltered bay/lagoon that nevertheless may receive abundant washover sediment generated during storms. (After many authors, starting with Hayes, 1979.)

The distinction between beach and tidal flat can never be sharply defined, but generally beaches occur as narrow, higher gradient, intertidal to supratidal features dominated by wave action in which the sediment coarsens from offshore to onshore. In direct contrast, tidal flats are wider, of lower gradient and dominated by to-and-fro and rotary tidal motions in which the sediment often fines from offshore to onshore. Tidal flats tend to be best developed on open, macrotidal coasts with plentiful mud supply or as part of protected back-barrier complexes on mesotidal coasts. They may also be subject to significant seasonal wave action from winter or monsoon-forced storms. Chenier plains are particular coastal marsh and tidal mudflat environments, broken by very extensive lateral shell ridges up to 50 km long and 3 m high.

Coastal barriers are commonest on mesotidal and microtidal coasts; they require some steady riverine or longshore supply of sand for their sustained development. Those on microtidal coasts are long and linear with sedimentary processes in the back-barrier lagoons dominated by storm washover effects. Barriers on meso- and macrotidal coasts are broken by frequent tidal inlets with flood tidal deltas in lagoons and bays on the back-barrier side and ebb tidal deltas on the seaward side. Fringing tidal flats occur in the back-barrier environment, sheltered by the barrier from most wave action.

## 18.2 Beach processes and sedimentation

### Processes

As the typical sinusoidal 'swell' of the deep ocean passes over the continental shelf towards the coast, waves undergo a transformation as they react to the bottom at values of between about 0.5 and 0.25 of their deep-water wavelength. Wave speed and wavelength decrease whilst wave height increases. Peaked crests and flat troughs develop as the waves become more solitary in behaviour until oversteepening causes

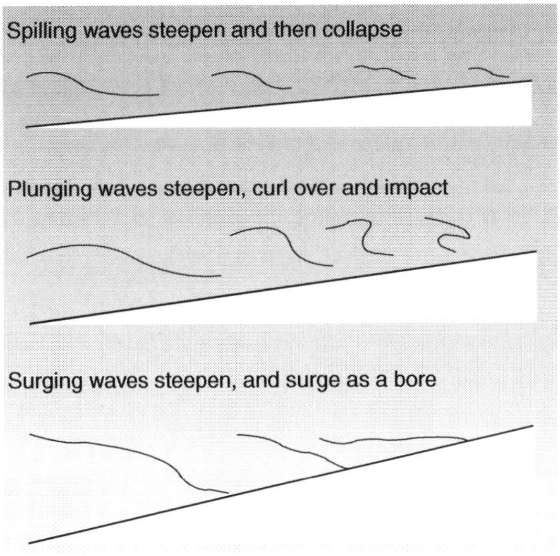

Spilling waves steepen and then collapse
Plunging waves steepen, curl over and impact
Surging waves steepen, and surge as a bore


**Fig. 18.2** Types of breaking waves. (After Galvin, 1968.)

wave breakage. Breaking waves may be divided into spilling, plunging and surging types (Fig. 18.2).

Wave refraction effects occur in response to bottom topography or incidence of wave attack (Fig. 18.3). The behaviour of waves on beaches varies according to the steepness of the beach face. Steep beaches possess a narrow surf zone in which the waves steepen rapidly and show high orbital velocities. Wave collapse is dominated by the plunging mechanism and there is much interaction on the breaking waves by backwash from a previous wave-collapse cycle. Gently sloping beaches show a wide surf zone in which the waves steepen slowly, show low orbital velocities, and surge up the beach with very minor backwash effects.

The nearshore current system may include a remarkable cellular system of circulation comprising rip and longshore currents (Fig. 18.4). The narrow zones of rip currents make up the powerful 'undertow' on many steep beaches, with velocities up to 2 m/s. Rip currents arise because of variations in wave setup along steep beaches. This is the small (cm–m) rise of mean water level above still-water level caused by the presence of shallow-water waves (Fig. 18.5). It originates from that portion of the radiation stress remaining after wave reflection and bottom drag and is balanced close inshore by a pressure gradient due to the sloping water surface. In the breaker zone the setup is greater shoreward of large breaking waves than smaller waves, so that a longshore pressure gradient causes longshore currents to move from areas of high to low breaking waves. These currents turn seawards

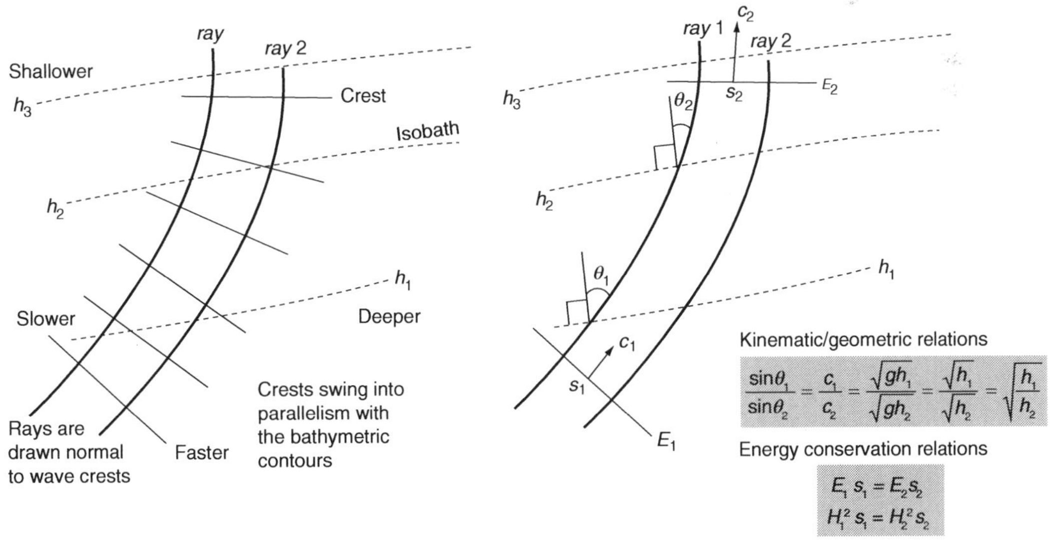


**Fig. 18.3** Wave refraction from deeper to shallower water by shallow water waves whose speed is purely a function of water depth. (After Collins, 1976.)

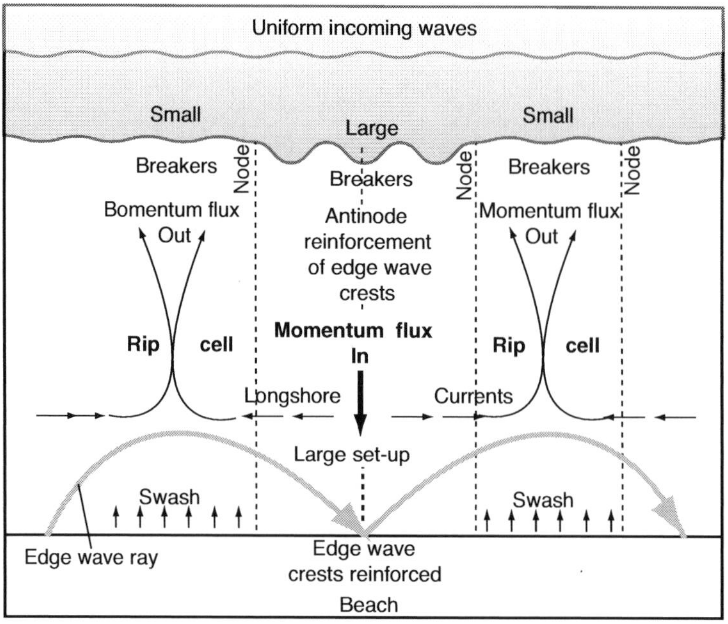


**Fig. 18.4** Rip current cells located in areas of small breakers where incoming waves and standing edge waves are out of phase. (After Shepard & Inman, 1950; Komar, 1975.)

where set-up is lowest and where adjacent currents converge.

What mechanism(s) can produce variations in wave height parallel to the shore in the breaker zone? Wave refraction is one such, and some rip-current cells are closely related to offshore variations in topography. Since rip cells also exist on long straight beaches with little variation in offshore topography, another

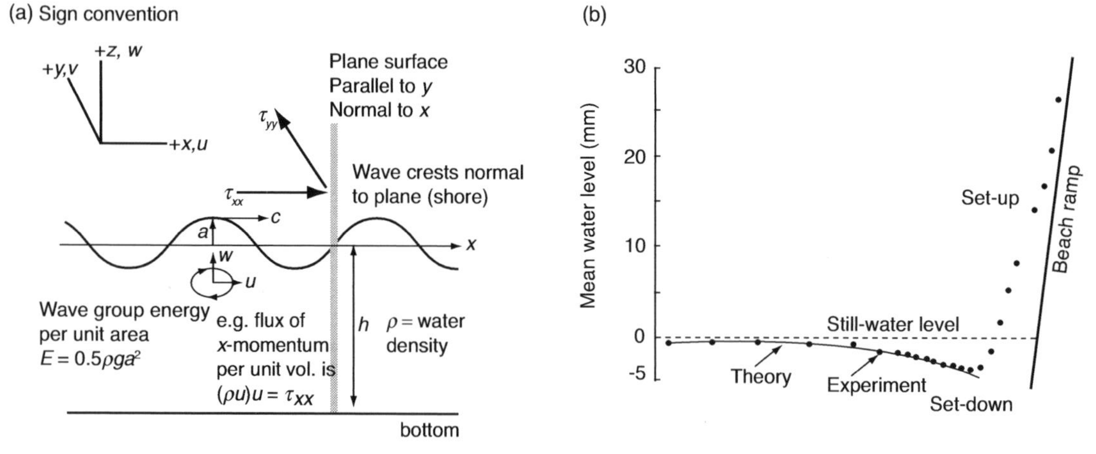


**Fig. 18.5** (a) Definition diagram for the radiation stress, $\tau_{xx}$, exerted on the +ve side of the *xy* plane by wave groups approaching from the left-hand side. The radiation stress is the momentum flux (i.e. pressure) due to the waves. (b) Wave set-up and set-down as produced by radiation stress caused by incoming waves in an experimental tank. (After Bowen *et al.*, 1968.)

mechanism must also act to provide lateral variations in wave height. This is thought to be that of *standing edge waves* which form as trapped waveforms due to refracting wave interactions with strong backflowing wave swash on relatively steep beaches. Edge waves were first detected on natural beaches as short-period waves acting at the first subharmonic of the incident wave frequency, decaying rapidly in amplitude offshore. The addition of incoming waves to edge waves is expected to give marked longshore variations in breaker height, the summed height being greatest where the two wave systems are in phase (Fig. 18.4). It is thought that trapped edge waves may be connected with the formation of the common and intrigueing beach cusps seen on many beaches; these have wavelengths of a few to tens of metres, approximately equal to the known wavelengths of measured edge waves. More recent results concerning the effects of edge waves and 'leaky' mode standing waves (where some proportion of energy is reflected seawards as long waves at infragravity frequency, 0.03–0.003 Hz) indicate that both shoreward and seaward transport of suspended sediment may result depending on conditions. Usually, sediment entrainment under groups of large waves in arriving wavepackets is preferentially transported seawards under the trough of the bound long-period group wave.

Longshore currents are produced by oblique wave attack upon the shoreline and may be superimposed upon the rip cells described previously. Such currents give a lateral thrust to the water and sediment in the surf zone. Long-term longshore transport of sediment depends upon the summed effects of all wave systems that impact upon a coastline. Such long-term transport vectors cancel out seasonal effects and are visually impressed upon the mind of the coastal visitor by piling-up of sediment against groynes, jetties and breakwaters around coastlines and by the migration directions of coastal spits.

### Beachface morphology and sedimentation

Comparison of summer and winter beach profiles reveals major changes. In *summer*, swell waves with low steepness spill and surge on to the beachface, transporting sediment onshore, forming narrow linear ridges called *beach berms*. In *winter*, storm waves with high steepness values plunge on to the beachface and transport sediment offshore forming *offshore bars*. Rip-current cells are commoner in winter on steeper beaches. The beachface slope is governed by the asymmetry of offshore and onshore transport vectors discussed previously. The seaward backwash in summer is generally weaker and longer-lasting than the rapid shoreward movement of water from winter plunging waves, because of percolation and frictional drag on the swash. Sediment is thus continually moved up and down the beach slope until an equilibrium is established. Maximum percolation occurs on the most permeable gravel beaches and these usually have the highest slopes. The coarsening-onshore trend found in almost all beach and nearshore systems (from about surf line inwards) is explained by the fact that the forward orbital motion under shallow-water wave crests is short-lived but powerful compared to the seaward return flow.

Beaches show great sensitivity to wave intensity: the majority of beachface accretion takes place during fair-weather periods. Internal beach structures of the active swash zone include low-angle accretion laminae, gravelly examples showing characteristic open-framework sorting, and seaward-dipping imbrication of discoid pebbles. Wave action at high water mark leaves behind small berm ridges, asymmetric landwards. Storms themselves leave evidence of their occurrence in the form of erosional and truncation surfaces in the beachface and in berm washovers that may be detected by GPR profiling. Tsunami waves have the ability to transport massive quantities of sand landwards, sometimes, as in the 2004 Sumatran tsunami, transporting entire beach systems. Storms generate rip and gradient currents that transport sediment offshore (Chapter 19), giving rise to sharp-based, sheet-like, poorly sorted gravels and sands exhibiting hummocky cross-stratification. Shoreface and foreshore topography reflects the presence of various 'bars', ridges and troughs as well as wave-formed current ripples and dunes (Fig. 18.6). Scuba observations of high-gradient upper shoreface environments down to depths of 6 m or so below a prominent plunge-pool step at low water mark, indicate large-wavelength (up to 2 m) asymmetric (shore-facing) gravel bedforms up to 0.25 m or so in height.

The *shoreface* environment is the transition zone between foreshore/beachface and the inner shelf. It is sensitive to variations in sediment supply, sea level and hydrodynamic processes. The shoreface gradient profile is a surface of 'equilibrium' reflecting wave and

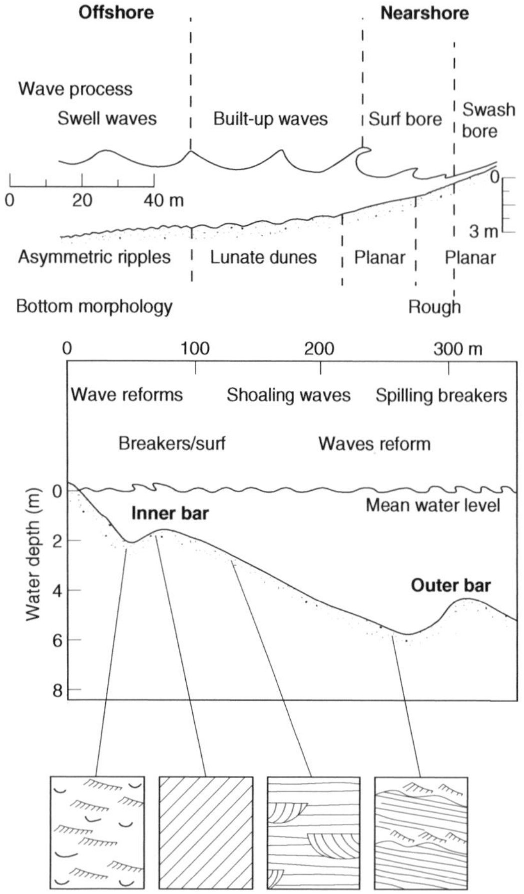


**Fig. 18.6** Classic sketches of beach to shoreface profiles showing the relations of bedforms to waveforms. (After Clifton *et al.*, 1971; Davidson-Arnott & Greenward, 1974.)

current energy regime. Shoreface architecture and Holocene shoreface evolution, whether prograding or retrograding, is highly variable, even in neighbouring transects in the same sedimentary province. This is largely because of variations in sediment supply and wave energy. During transgression, as relative sea level rises, mean storm wave-base marks the point where erosion predominates over deposition—all shallower deposits are eroded and all deeper deposits preserved. The junction point moves over time and defines an erosion or *ravinement surface* or *trajectory*. For example the transgressive surface varies from −6 to −15 m depth along the Texas beach-barrier coast as wave energy changes.

Wide, transgressive, mixed-sediment grade beaches under macrotidal regimes like Waterside Beach, Bay of Fundy feature three main subenvironments: (i) sandy foreshore and shoreface deposits with backshore aeolian sand, moderately seaward-dipping mixed sand and gravel of the beachface and a shallowly seaward-dipping terrace comprising intertidal sand and silty sand and subtidal silty sand and clayey silt deposits; (ii) tidal-creek braidplain and delta sediments are reworked sand and fine gravel removed from the upper and middle intertidal zone and deposited in the lower intertidal and subtidal terrace zone; (iii) wave-deposited gravel and sand bars are surf-zone foreshore and shoreface deposits transported onshore from (ii) by waves as gravel bars up to 5 m high and 800 m long with local mud deposition developed on their landward side. Distinctive muddy gravels feature abundantly in the successions, reflecting high-water settling and infiltration.

Late-Holocene progradational, sandy beach shorelines under macrotidal regime may reach several hundred metres wide when sand supply is abundant; many feature wind-reworked aeolian *foredunes*. These are shore-parallel aeolian dunes formed by bedload deposition within vegetation beyond the upper limit of wave action. Prolonged beach progradation during highstand results in coastal plains kilometres wide, termed *strandplains*, with traces of numerous, formerly shore-parallel *beach ridges* (foredune or high-water berm origins) separated by hollows (Fig. 18.7).

Mapping and subsurface profiling of accretionary coasts enables palaeoshoreline processes, sediment budgets and beach evolution to be assessed. Periodic erosion events often occur between beach-ridge sets having different orientations, with truncation surfaces traced by GPR profiling (Fig. 18.8). Such erosion occurred as the result of a change in synoptic wave and current regimes, caused by either changes in storm frequency and/or intensity or changes in coastal configuration. Studies of Thailand strandplains after the Indian ocean tsunami of December 2004 revealed wholesale transport of modern beach sands on to the Holocene strandplain as a 5–20 cm thick, graded, medium sand to coarse silt sheet by the main 20-m-high wave. Sand deposition was concentrated in vegetated swales between beach ridges. Angular discordances between beach ridge nestings within the 3 km wide strandplains may represent former tsunami

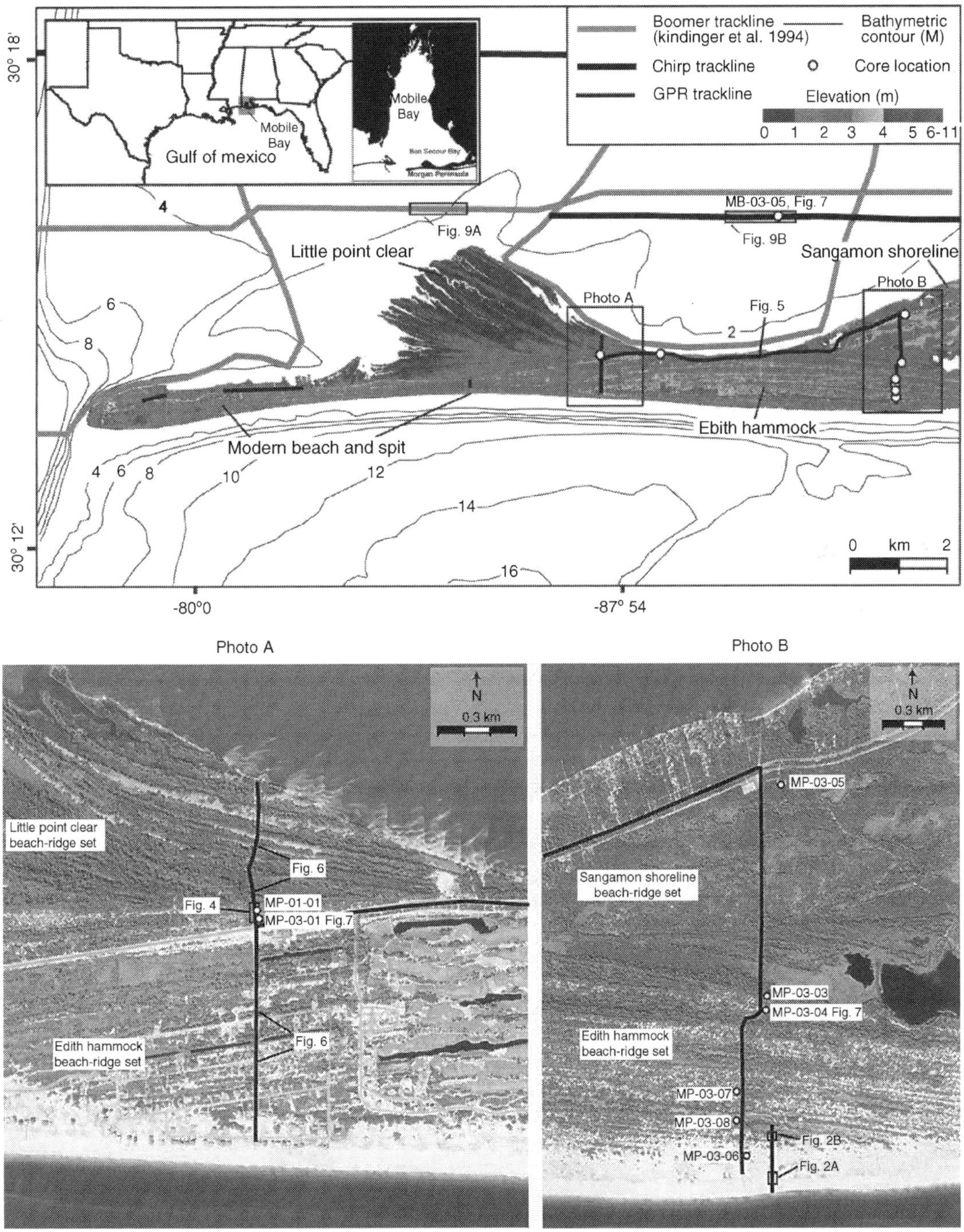


**Fig. 18.7** Beach ridge accretionary coastline, Mobile Bay, Alabama USA. Map of Morgan Peninsula showing LiDAR topography and the locations of data shown in figures in the original paper. The Peninsula is made up of (from oldest to youngest) the Sangamon Shoreline, the Little Point Clear beach-ridge set, the Edith Hammock beach-ridge set, and the modern beach and spit. (After Rodriguez & Meyer 2006.)

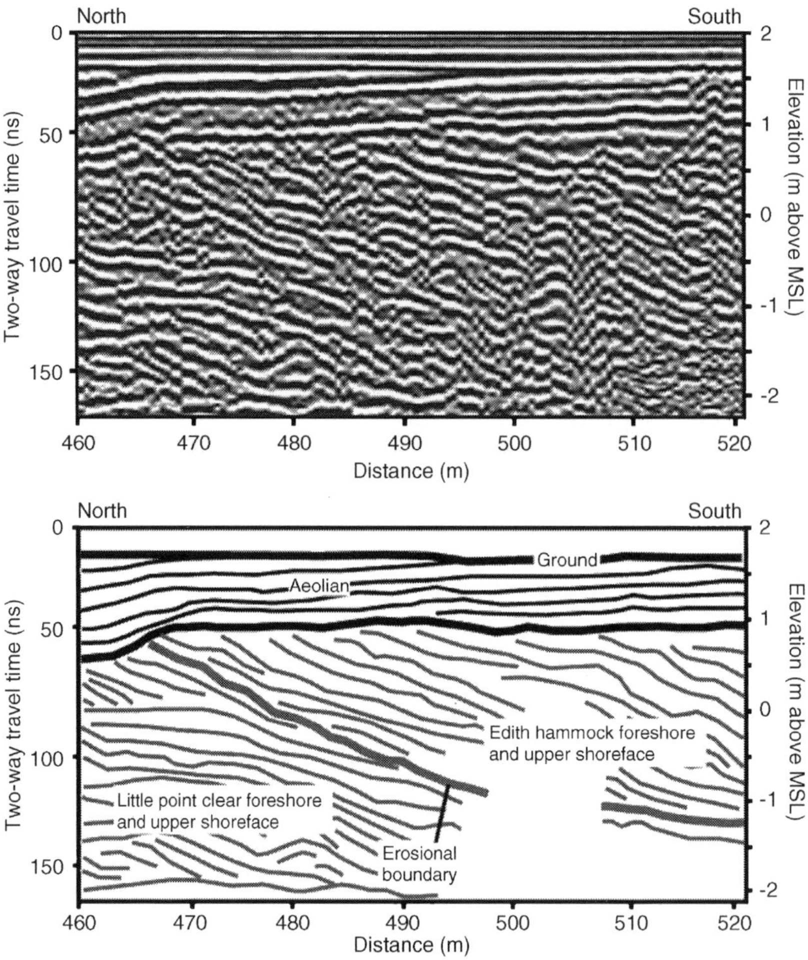


**Fig. 18.8** Uninterpreted and interpreted GPR profile of the transition between the Little Point Clear and Edith Hammock beach-ridge sets. The transition is marked by an erosional surface. Authors profile is labelled Fig. 4 on Fig. 18.7. (After Rodriguez & Meyer, 2006.)

erosion events. Despite such spectacular occurrences, most beach-ridge sets imaged by GPR are zones of uniform shoreline trajectory made up of beach-ridge depositional packages that are bounded by smaller-scale erosional discontinuities cut by storm waves. These units and bounding surfaces make up the complex 'building blocks' of parasequences described from ancient shoreline deposits (see later). An insight into conditions of forced coastal regression and progradation is provided by spectacular tectonically uplifted strandplains up to 10 km wide around Tokyo, Japan which feature successively inset beach-crest erosion surfaces. These, and examples from uplifting post-glacial Hudson's Bay (Fig. 18.9), give interesting analogues for coastal adjustments to falling relative sea level in the older sedimentary record. In the latter case major inset surfaces or erosion are assigned to autocyclic (inherent to sediment system) mechanisms rather than to changes in the isostatic rebound rate.

Coastal gravel-ridge complexes deposited on islands in the Caribbean Sea are recorders of past extreme-wave events that could be have been due to either tsunamis or hurricanes. They consist of locally derived polymodal clasts ranging in size from sand to coarse boulders with morphologies and crest elevations largely controlled by availability of sediments,

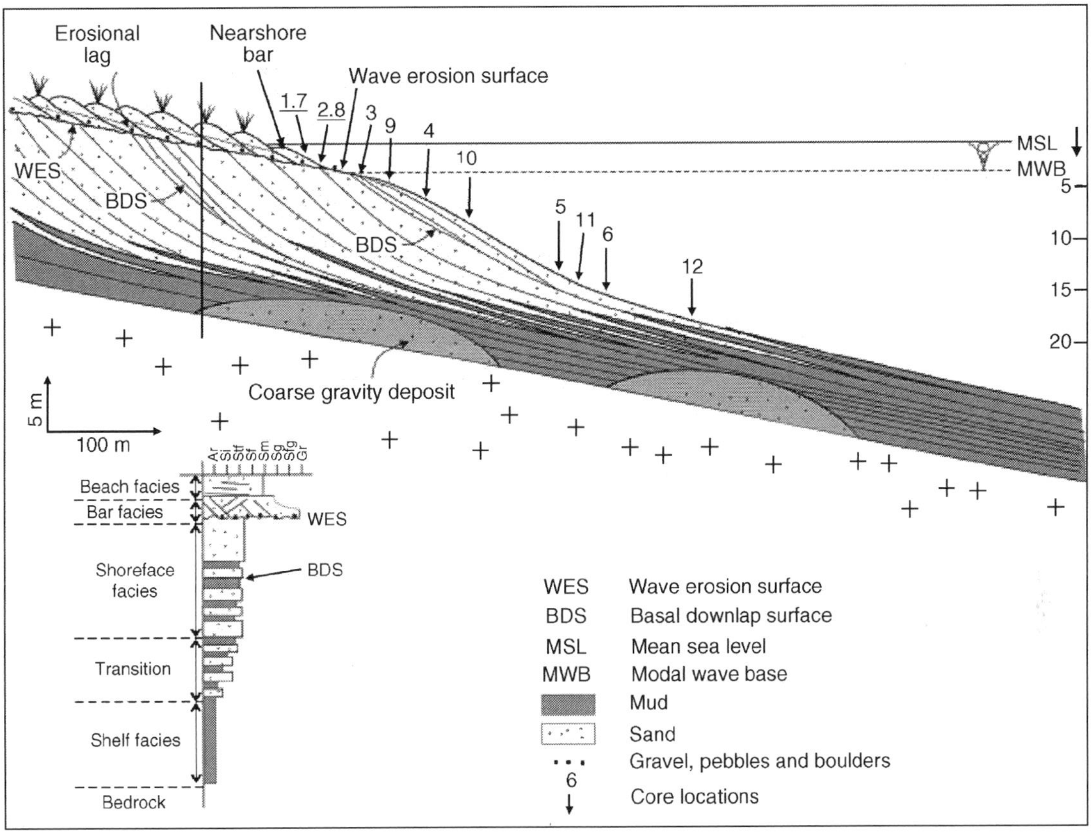


**Fig. 18.9** Scheme based on extensive GPR, coring and map evidence interpreting stratigraphic architecture of a falling-stage (forced regression) strandplain and shoreface sandbody, in this case the relative sea-level fall is due to post-glacial isostatic uplift of the Hudson Bay area. Numbers refer to sediment cores taken by the authors. (After Fraser *et al.*, 2005.)

clast sizes and heights of wave runup. The ridge complexes are internally organized, display textural sorting and show a broad range of ages, including historical events. Some display seaward-dipping beds and ridge-and-swale topography, and some terminate in fans or steep avalanche slopes. They are primarily tropical-storm-constructed features that have accumulated for a few centuries or millennia as a result of multiple high-frequency intense-wave events. Tsunami deposition may account for some of the lateral ridge-complex accretion or boulder fields and isolated blocks that are associated with the ridge complexes.

Larger offshore bars occur on all but the steepest high-energy beaches and are controlled in a complex way by the position of spilling waves during storm conditions. Successive coast-parallel or crescentic bars with wavelengths of tens to hundreds of metres occur on low-gradient shorefaces. The bars tend to increase in height (up to 1.5 m) away from the shore, perhaps individually reflecting the average spilling position of waves of a certain height. The bars (Fig. 18.6) show variably dipping internal sets of tabular cross-stratification directed landwards, and the troughs show small-scale cross-laminations produced by landward-migrating wave-current ripples. Measurements taken during storm conditions indicate that sediment accumulates on bar crests and is eroded from troughs, and that the latter are the site of significant shore-parallel flows. When bars occur in the foreshore area, they are termed *ridge and runnel* topography. During rising and falling tides the runnels come under the influence of slope-controlled flows as they alternately fill up and drain: the resulting current-produced bedforms include dunes and upper plane beds whose orientation is shore-parallel.

Concerning the problem of across-shoreface transport of suspended sediment, field data suggest both time-averaged (mean) cross-shore currents and oscillatory currents are critical in determining net transport direction. In the surf zone of bars there is onshore transport at low wave frequencies (seasonal storm conditions). Outside the surf zone the linkage between the effects of a steady seaward mean current and energy of long-frequency, bounded, wave groups leads to offshore transport. As noted previously, *rip currents* are a distinctive physical mechanism for across-shoreface transport. The very characteristic rip-current cells that occur on many steeper beaches, particularly during storms, produce spaced channels that may dissect offshore bars and contain sharp-based shelly sands and pebbly sands with shell lags. The fan-like terminations to rip-current channels deposit seaward-dipping cross-sets.

## 18.3 Barrier–inlet-spit systems and their deposits

The internal structure of barrier systems has become clearer since the advent of integrative studies involving GPR, coring and radiocarbon dating. Internal structure of prograding highstand barrier-spits is dominated by radar reflectors caused by heavy-mineral concentrates produced on storm erosion surfaces during wave-winnowing. These upper surfaces in coarser sand grades commonly dip at up to 2° whilst the deeper lower shoreface in fine sands to silts shows indistinct reflectors and progressively more gentle slopes. Records of GPR taken at 200 MHz reveal a series of gently sloping, seaward-dipping reflections with slopes similar to the modern beach and spacings on the order of 20–45 cm. Field evidence and model results suggest that thin (1–10 cm) heavy-mineral concentrates, possibly magnetite-rich, or low-porosity layers left by winter storms are separated by thicker (20–40 cm) summer progradational deposits. These results indicate that a record of annual progradation is preserved in the subsurface of the prograding barrier and can be quantified using GPR.

Early theories suggested that barriers resulted from the upbuilding of submerged offshore bars into emergent islands. The absence of offshore facies beneath modern lagoons discredits the theory. We have already seen (Chapter 17) that barriers like those of the Chandeleur Islands overlie marsh and delta-front deposits and mark the reworked rims of abandoned subdeltas of the Mississippi delta complex. Periodically, portions of these islands are flattened or destroyed by hurricanes, but reform by emergence of the remnant submerged sand bars if sand supply is provided. During Hurricane Ivan in 2004 erosion led to 50 m of beach retreat and reduction of relief by 1.5 m with extensive back-barrier sand sheet deposition defining a 100 m wide platform some 0.5 m above mean sea level: detailed studies indicate total barrier mass was essentially conserved, though redistributed. Other barrier islands formed over abandoned deltas isolated from a continuing sand supply may not survive washover of sand into the lagoons and bays on the landward side during storms, e.g. destruction of the Isles Dernieres off coastal Louisiana occurred after Hurricane Andrew in 1992 (Fig. 18.10).

Inherited coastal topography generated during sea-level lowstand plays a key role in barrier evolution. Lowstand palaeochannel networks were separated by

**Fig. 18.10** Raccoon Island, Louisiana, USA, before (top) and after (bottom) the passage of Hurricane Andrew in August 1992. The detached barrier island was originally ca 5 km long but was heavily scoured, shortened and flattened by storm waves and washover transport. (From *Eos*, 1992.)

higher interfluve areas such that during transgression, beaches formed against the eroding interfluves and estuaries along river channel outlets (Fig. 18.11). A variety of barrier, spit, lagoon and tidal-flat environments evolved as transgression continued. Of great importance is a ready supply of sand or fine gravel sediment. One such source may be lowstand fluvial channels, subaerial deltas, delta fronts (particularly glacial outwash deltas) or older barriers and their ebb tidal delta shoals now stranded offshore. On a relatively tectonically stable coastline such as the eastern USA, these might represent MIS 3 or 4 lowstand deposits, now well within modern storm wave base at depths of 30–50 m. A good example is the Merrimack delta off Cape Cod, whose sands are thought to have supplied the Holocene Cape Cod barrier system. These considerations of sediment supply mean that fluctuating rates of Holocene shelf transgression and coastal retreat do not simply reflect the global eustasy curve. Variable inner-shelf gradients and lowstand fluvial deposits serve as local sand sources enabling barrier islands to persist offshore, out of equilibrium with sea-level rise. Once these sand sources became depleted, and/or sea level reaches some critical threshold, barrier shorelines became stranded offshore as banks, and new shorelines formed landward. Fine examples are the stranded offshore Sabine and Heald Banks, offshore Texas coast.

Tidal inlets are present along barriers in macrotidal, mesotidal and, to a lesser extent, microtidal areas. They play an important role, transferring oceanic waters into and out of the back-barrier lagoon or bay. Tidal currents and range are usually enhanced in the inlet compared to the adjacent shelf. They are the main flushing arteries, vital to the ecological well-being of highly prized and productive lagoonal ecosystems.

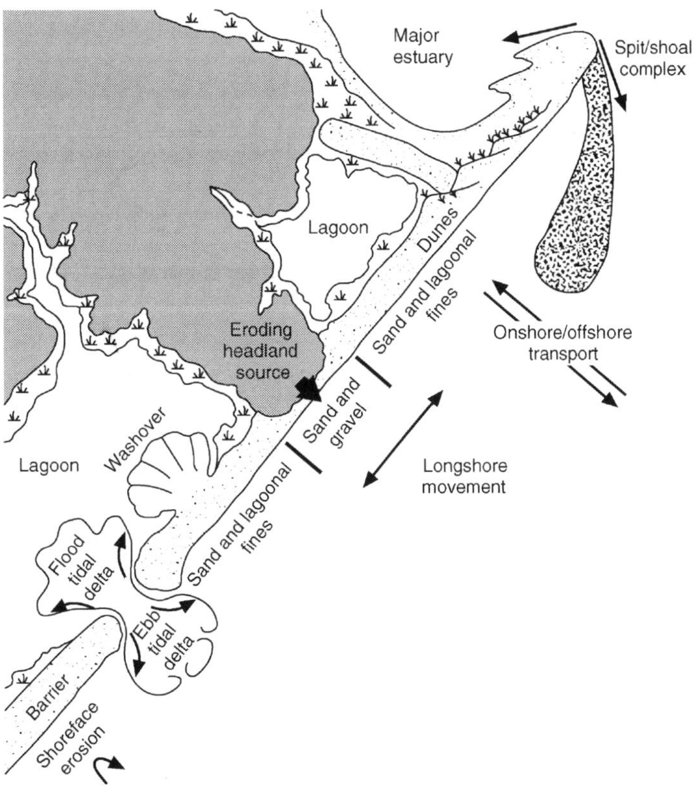


**Fig. 18.11** Sketch of barrier/spit/lagoon/tidal inlet coastal complex showing various sediment sources and sinks. Many inlets will overlie partly buried lowstand fluvial channels, whilst eroding headlands, formerly incised valley sides, supply clastic sediment to the coastal system where it is reworked by tidal currents and waves. (After Kraft *et al.*, 1987.)

The flow passing along the inlet in response to tidal and storm forcing issues as an expanding jet behind and in front of the barrier. In both cases sediment bodies form as flood and ebb tidal deltas respectively (Figs. 18.11 & 18.12). Inlets are often dynamic items, migrating, rotating, opening and closing, or remaining stationary for thousands of years. Their behaviour depends upon local conditions of sediment supply, tidal and wave characteristics, and the changing nature of the tidal prism supplied by the back-barrier catchment. Should this catchment increase in area due to wetland loss, as in many barrier-fringed bays of the Mississippi delta complex, like Barataria Bay, there is an increase in tidal prism which leads to initiation of new inlets, marked deepening of existing inlet and ebb tidal delta progradation seawards. Inlet behaviour also depends upon the inherited topography of the seafloor, for, as noted previously, many prominent stationary modern inlets are found to be coincident with drowned lowstand river and estuary valleys. Coriolis effects are important in larger tidal inlets at higher latitudes, causing the position of the maximum flood and ebb tides to alternate in opposite directions so that one side of inlet channels is flood-dominant whilst the other is ebb-dominant. Inlets play a vital role in the nourishment or starvation of adjacent portions of a barrier. This depends on the degree of bypassing of sediment across the inlet channel. Periodic growth, accretion and then cutoff of ebb tidal deltas control the accretion rate of the barrier itself.

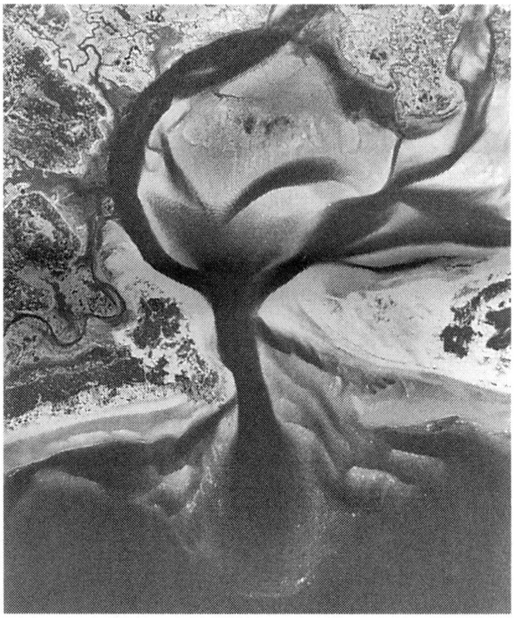

**Fig. 18.12** 1976 vertical aerial photo of the Essex River tidal inlet, Ipswich Bay, Massachusetts, USA showing ebb and flood tidal deltas, attached strandplains and detached barriers, barrier-tip accretion surfaces and coastal tidal wetlands. Local longshore drift is right to left, approximately NW to SE. Mean local wave height is ca 2.0 m, with a mean tidal range of ca 2.7 m. (Data and photo from Fitzgerald, 1993.)

A particularly well-studied example of inlet initiation is in the Cape Cod area of Massachusetts where a major spring tide coincided with a succession of great northeast storms in the winter of 1987. A storm surge overwhelmed the existing barrier opposite Chatham estuary and eroded sufficient sediment so that over a period of days a rapidly widening channel began to transfer water through it. Over a few months a major inlet up to 700 m wide had been cut across the once-continuous barrier, now with tidal currents flowing at over 1 m/s. In this and other equilibrium tidal inlets it is the strong tidal currents that maintain the channel in face of opposition from the deposition of sand by longshore wave-induced currents. It is the delicate balance between longshore supply and tidal currents that determines inlet behaviour, with channels migrating in the direction of longshore supply. The net transport of sediment in an inlet (onshore vs. offshore) depends upon the residual current regime caused primarily by the mean sea-level difference between onshore and the open ocean during the course of the tidal cycle. Tidal amplitude and the phase difference between the two ends of the channel also play a role. In general, the greater the tidal range, the deeper, larger and more powerful is the channel flow in inlets.

Seaward-prograding barrier islands produce coarsening-upwards successions similar to those of attached beaches but with the major addition of tidal inlet and back-barrier lagoon or bay facies. Back-barrier lagoons, bays and wetlands vary tremendously, depending upon climate, degree of tidal flushing by inlets, river inflow and extent of storm/tsunami washover events. All types of barriers in hurricane 'highways' are prone to washover; regional-scale washover occurred along the northern Florida barrier islands caused by hurricanes Ivan in 2004 and Dennis in 2005. Regional overwash, although to a lesser extent, also occurred along southern Florida barrier islands, caused by hurricanes Frances and Jeanne. The washover deposits in the two regions are different in terms of regional extent, basal erosional features and sedimentary structures. Overall, the Ivan–Dennis washover deposits are characterized by horizontal and prograding steep foreset bedding, in contrast to graded bedding in the Frances–Jeanne washover. The different erosional and depositional characteristics are caused by the different overall barrier-island morphologies, vegetation types and densities, and sediment properties. Over eroded dune fields the bases of washover deposits are characterized by extensive, wave-produced, erosion surfaces that truncate older dune deposits. Washover deposits in interior marsh wetlands are characterized by steep tabular bedding, with no basal erosion. The thickest washover deposits, ~2 m, were in back-barrier bays, characterized by steeply inclined, landward-prograding sigmoid bedding. Along the southeast Florida barrier islands, overwash associated with hurricanes Frances and Jeanne penetrated into dense interior mangrove swamps generating both normal and reverse graded beds, the product of intense interaction between overwash and dense mangrove vegetation. Lagoons in microtidal areas are dominated by storm washovers so that the bioturbated lagoonal silts and muds are intercalated with sheets up to 1.5 m thick of parallel-laminated sands derived from storm or tsunami breaching and overtopping of the exposed barrier profile. Washover fans produced in this way may show delta-like landward terminations with internal sets of landward-dipping planar cross-stratification.

The behaviour of regressive barrier systems during periods of stable sea level should be contrasted with that of transgressing barriers. In the latter case little of the barrier facies themselves may be preserved offshore if transgression is slow and sediment supply low. The whole barrier system simply translates onshore (Fig. 18.13) with plentiful landward washover deposits preserved beneath an aggradational beachface. Erosional action at the shoreface zone during storms produces a 'ravinement' erosion surface over the previous beachface slope deposits. Lagoonal environments themselves may narrow and infill as they are forced against higher-gradient land surfaces during transgression and barrier retreat. Older subsurface lagoonal deposits onlapped by the barrier may also be removed by the action of migrating tidal inlets. It is possible that high rates of sediment supply combined with rapid transgression may cause barrier preservation. The behaviour of coastal barrier–beach systems in the context of present-day sea-level rise is a fascinating topic (Fig. 18.13). Early studies proposed that regions subject to rapid sea-level rise with low sand supply would favour stepwise retreat of barriers. Areas with slow sea-level rise and high sand influx might show continuous shoreface retreat and the production of a gently sloping erosional or 'ravinement' surface overlying a truncated shoreface sequence. Barriers would then be free to prograde once more at sea-level highstand. Lessons might be learnt from combined GPR, coring and core studies of the generally transgressive Islas Cíes bay barrier system, northwest Spain (Fig. 18.14). This system shows evidence for periodic aggradation, erosion, washover and progradation related to decadal-scale Atlantic storm intensities and variations in sand supply.

Microtidal lagoons in semiarid climates such as the Texan Laguna Madre show evaporite growth of sabkha types, carbonate precipitation as ooliths and growth of algal mats. Flood tidal deltas in mesotidal lagoons occur on the inner sides of tidal inlets in response to flow expansion and deceleration. The subaqueous delta surface is covered by landward-directed tidal dune bedforms. Some sandflat platforms are wave-dominated but the tidal drainage channels are current-dominated. The remainder of the lagoon in such cases approximates to the physiography of a tidal flat as discussed below. Ebb tidal deltas have a morphology and symmetry that strongly reflect the relative strength of tidal and longshore currents and their ability to transport away sediment. Strongly asymmetric deltas tend to develop when longshore currents are strong.

Concerning the development and migration of coast-attached spit systems (Fig. 18.15), evidence from experiments and late Pleistocene gravel spits indicate that along most of their length the seaward-facing spit shows only limited seaward progradation, since the predominant sediment transport vector is shoreface-parallel. At the spit tip, below a beach platform and a series of subtidal sand bars, gravel and sand are periodically supplied to the deeper water of an estuarine inlet by avalanching. This process gives rise to a very characteristic sequence

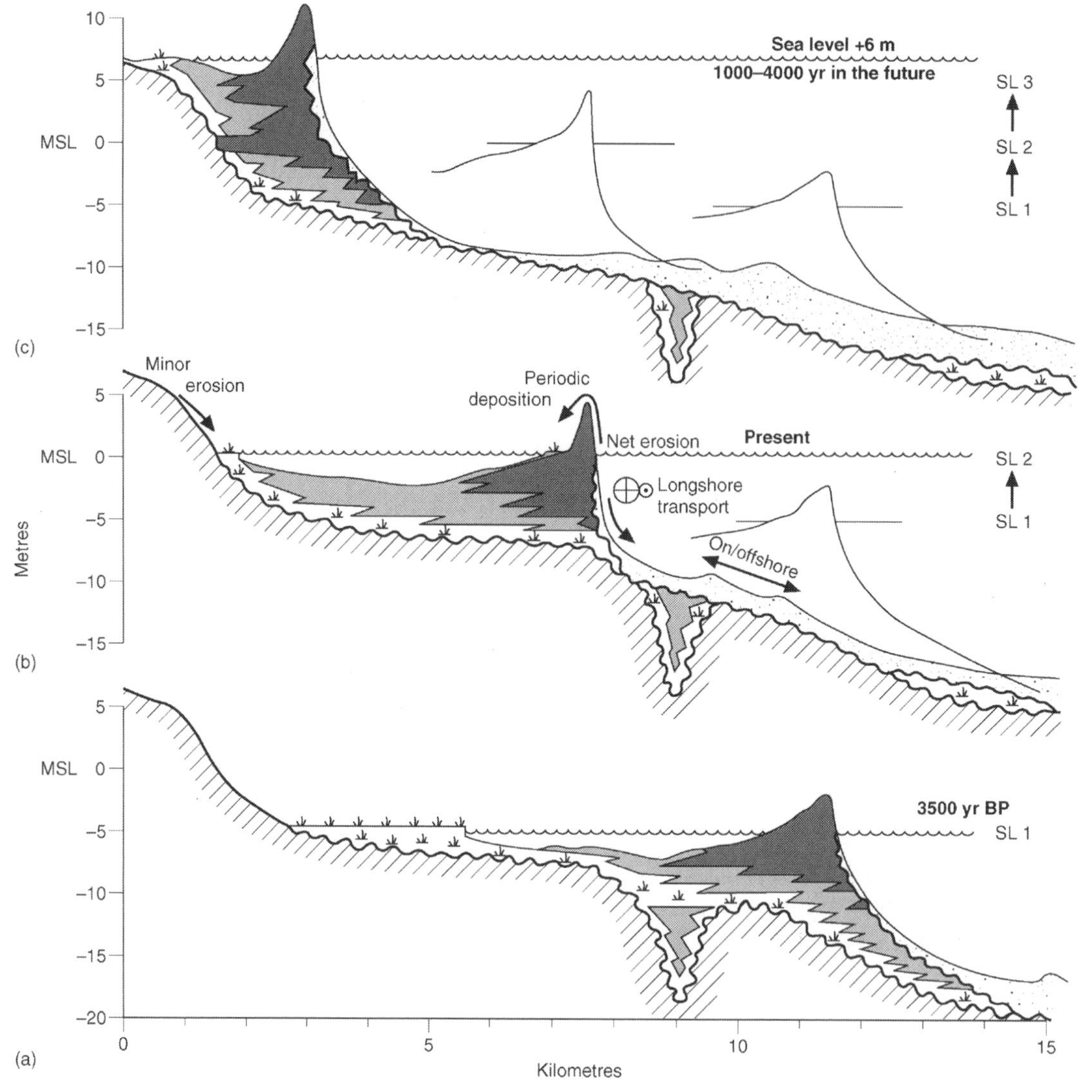


**Fig. 18.13** A series of sketch profiles to illustrate the morphology and lateral relationships of barrier-beach shoreline sedimentary environments in response to changing Holocene to future sea levels. Note the eventual production of an attached strandplain from a detached barrier coast. Model ignores coastal subsidence/uplift. (After Kraft *et al.*, 1987.)

of giant, shoreward-facing and steeply dipping foreset beds underlain by inlet/estuary/bay deposits and overlain by beach and shoreface deposits. However, not all spits follow this model. For example, the 35 km long Skagen-Odde coastal spit complex (Fig. 18.16) at the northern tip of Denmark is a large well-documented system (partly due to its older parts being uplifted and exposed due to glacio-isostatic rebound) that is prograding into the relatively deep waters of the Skagerak between Denmark and Sweden. It began to form 7150 yr BP and from 5500 yr BP to recent times it has prograded $4\,m\,yr^{-1}$ accumulating a prodigious $3.5 \times 10^9\,m^3$ of sand. The sedimentary facies on the accreting leeside of the spit comprise, from below, thick storm sand beds, dune and bar-trough deposits, beach deposits and peat beds. These four units form a coarsening- and shallowing-upward sand-dominated succession, up to

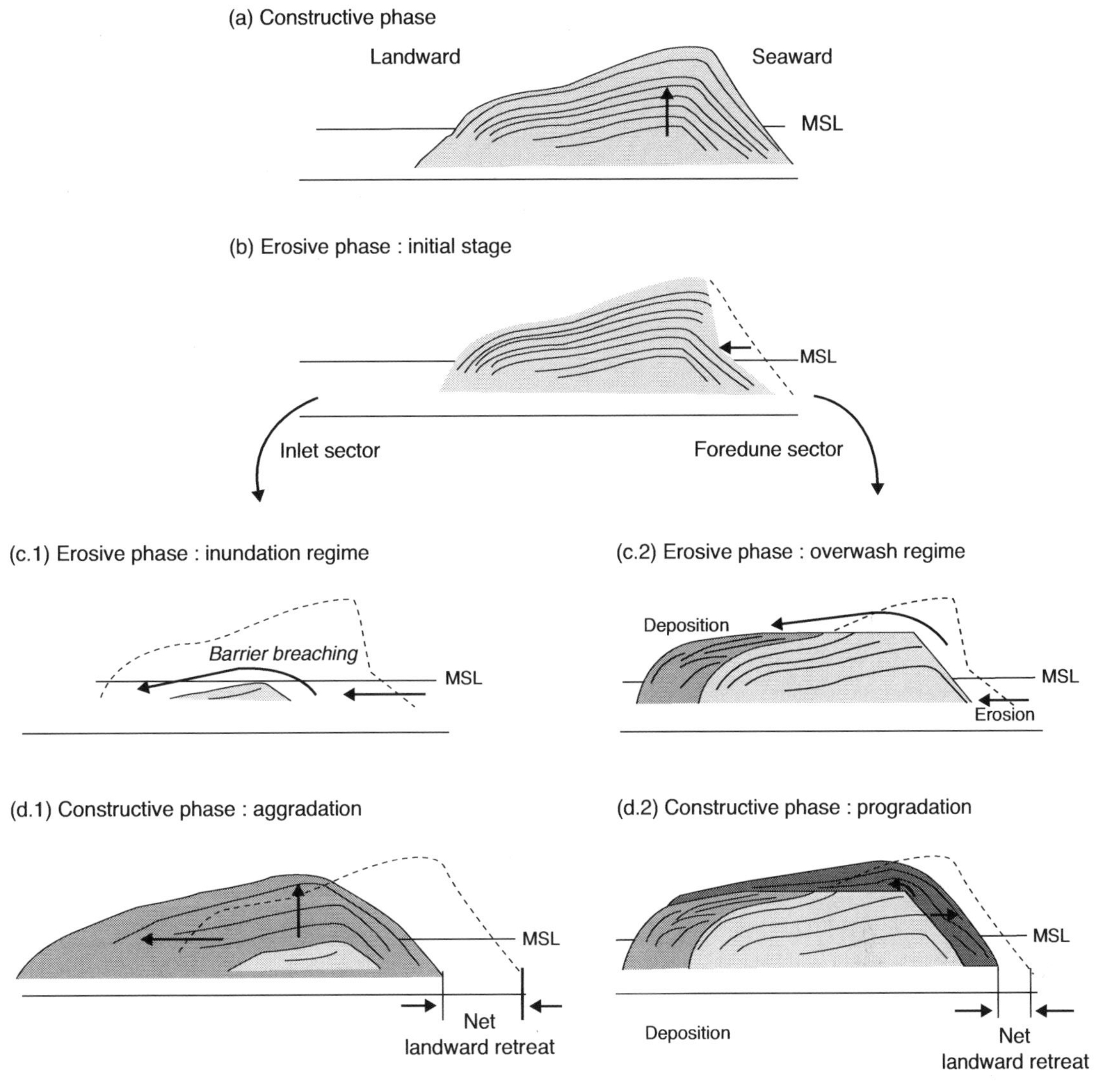


**Fig. 18.14** Evolutionary model of the transgressive Islas Cíes bay barrier, Galicia, Spain. GPR, core and map data show a cyclical behaviour superimposed on net landward retreat from 1956 to 2003. MSL, mean sea level. (After Costas *et al*., 2006.)

32 m thick. The Skagen spit system succession differs from the model noted above by lacking steep avalanche platform foresets. Instead, the main part of the prograding leeward spit succession is composed of thick gently dipping storm sand beds overlain by dune and beach deposits. This is because the Skagen spit system is sourced by finer-grained sediment. Further, it is influenced by much stronger energy with some waves with a fetch of ~800 km, whereas sources for the late Pleistocene spits were composed of coarse-grained sand and gravel and exposed to smaller waves.

## 18.4 Tidal flats, salt marsh and chenier ridges

Tidal flats are generally sinks for fine-grained sediment that originates from delta distributaries, estuarine channels, tidal inlets or in large-scale coastal

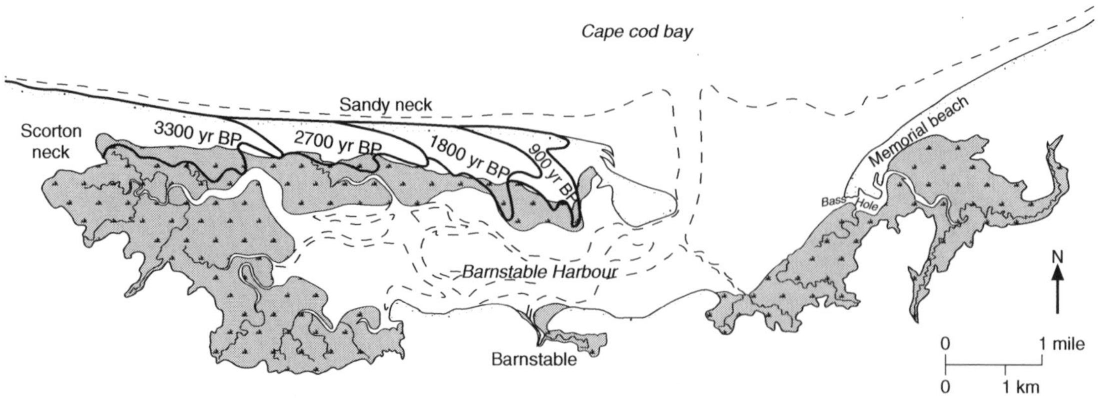


**Fig. 18.15** Map to show the late Holocene easterly growth and development of attached barrier-spit with its sheltered estuary of Sandy neck, Barnstable, northern shoreline of Cape Cod, Massachusetts, USA. Sediments are derived from erosion of Pleistocene cliffs to the NW. (After FitzGerald, 1993.)

embayments, like The Wash, eastern England, or Jade, northern Germany. In The Wash a net supply of fine-grained sediment arrives from offshore North Sea, mostly at spring tides because of a net residual flood tidal current. Fining onshore across tidal flats has generally been attributed to two related processes that encourage silt- and mud-grade sediment to accumulate on the upper tidal flats. First there is the tendency for maximum fine-sediment deposition from suspension to occur at spring-flood-tide high water mark. This deposits fine sediment in a position where subsequent lower tidal levels cannot resuspend it. Second, once deposited during the high flood tide, fine sediment is more difficult to erode at the equivalent point in the ebb because of cohesive forces at the bed, early compaction, biofilm growth and cyanobacterial binding (Fig. 18.17). A marked seasonal contrast in behaviour has been found, notably in the Dollard estuary, Netherlands, where seasonally calm weather causes higher onshore transport of organic-rich flocs on the flood tide and increased threshold for erosion due to algal diatom cover. During stormy weather, waves add to overall bed shear stress, being most effective at high water, thus reversing the calm weather flood-flux during the ebb and producing many small flocs due to disaggregation.

Limited percolation of tidal waters occurs on muddy tidal flats because of low permeability. This encourages surface runoff of both tidal waters and rainwater and the establishment of meandering tidal channel networks (Fig. 18.17). Field measurements reveal a broad rotary pattern as a coast-parallel tide swings on to the tidal flats during the course of the flood tide to become shore-normal. Tidal creeks split the alongshore tidal prism into a number of cells. There is commonly a strong

**Fig. 18.16** (a) Mid-Holocene to recent development of the Skagen Odde coastal complex. 7150 yr BP: Peak Atlantic transgression. Ridges of glacial deposits formed anchor points and sediment sources for the early spit system eroded during peak transgression and recorded in a pebble bed. 5500 yr BP: Former lagoon infilling with sediment and emergent due to ongoing isostatic uplift. A continuous coastline established with effective northeastward sediment transport route developed and the beginning of growth of the Skagen spit system. 4500 yr BP: Eastern lagoon diminished because of sedimentation and uplift as growth of the Skagen spit system continues. Recent: The Skagen spit system continues its growth to present-day size and the old eastern lagoon has changed to a small lake. (b) Schematic depositional model for the Skagen spit system. At the point of the Skagen spit system, subaqueous dunes are driven by currents along the spit coast and obliquely down the spit platform towards deeper water whilst oblique bar trough systems weld to the spit coast. Longshore currents expand laterally and vertically at the spit point symbolized by the widening red arrow. The wind-rose diagram indicates directions of winds that contribute to growth of the spit system. (c) Schematic cross-section A–B displaying the resulting facies architecture. (After Nielsen & Johannessen, 2009.)

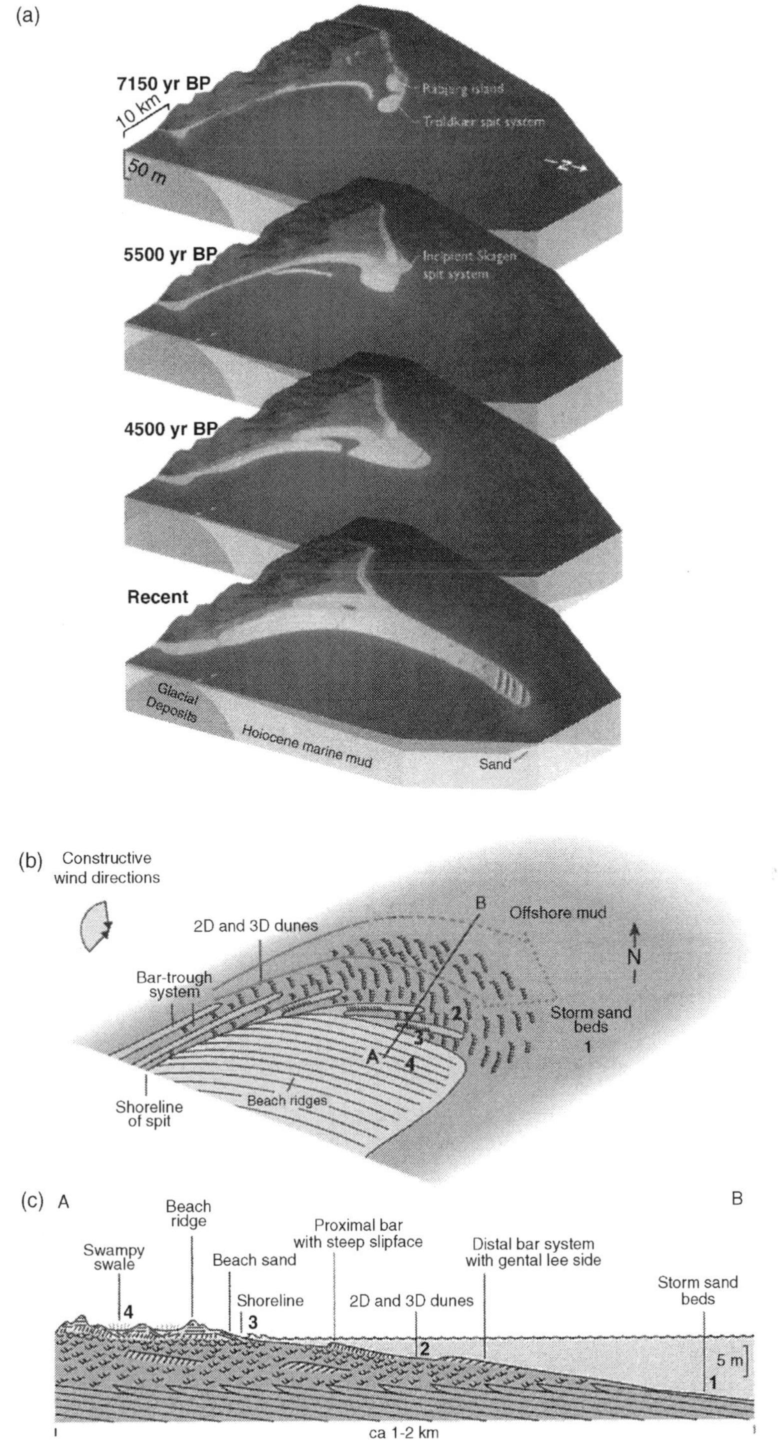
(a)
7150 yr BP
10 km
50 m
Troldkær spit system
N
5500 yr BP
Incipient Skagen spit system
4500 yr BP
Recent
Glacial Deposits
Holocene marine mud
Sand
(b) Constructive wind directions
2D and 3D dunes
B
Offshore mud
N
Bar-trough system
Storm sand beds
1
2
3
A
4
Shoreline of spit
Beach ridges
(c) A
B
Beach ridge
Proximal bar with steep slipface
Swampy swale
Beach sand
Distal bar system with gental lee side
Storm sand beds
Shoreline
2D and 3D dunes
4
3
2
5 m
1
ca 1-2 km

**Fig. 18.17** Low spring tide view on upper tidal flat-to-saltmarsh transition with shallow tidal drainage channels and surface-dessicated cyanobacterial mats. The latter are ~0.5 cm thick comprising *Schizothrix* binding a smelly, muddy fine silt. Flow in the channel is right to left, the direction of local gradient. Width of field of view ~4 m. Thornham marshes, northwest Norfolk, UK.

residual sediment transport towards the shoreline over the tidal flats, with weaker residual transport seawards in the channels (Fig. 18.18). The overall result is a net shoreward transport of sediment and vertical accretion of the tidal-flat surface. Sediment concentrations peak strongly at the beginning of the flood tide and the end of the ebb, reflecting substrate erosion by the shallow, fast gravity-controlled tidal residual and gravity runoff. The latter is a neglected but important seaward-transport process for when copious rainwater falls on to exposed tidal flats, as measured sediment fluxes may exceed those at spring ebb.

Tidal-flat deposits are dominated by the nearshore-to-offshore coarsening trend noted previously (Fig. 18.19). Seaward progradation produces an upward-fining sequence, broken by intertidal and subtidal channels and capped by a rootlet bed or peat accumulation of the salt marsh. The supratidal salt-marsh zone with halophytic (salt-loving) plants passes gradationally outwards at a very low slope (1:100 to 1:800) into a mudflat with a rich infauna. Flow-field measurements show reduction of turbulent kinetic energy near the salt-marsh bed that encourages sediment deposition and reduces erosion. In addition, low levels of vertical turbulent stress in the denser part of submerged *Spartina* canopies increase sediment settling. The evolution of a fringing salt marsh is influenced by the interaction of many factors, including relative sea-level rise, marsh aggradation, nearshore sedimentation, wave climate, and tidal range. Based on the geomorphic history of a fringing salt marsh in Rehoboth Bay, Delaware, USA a response model for a marsh shoreline depends upon the relative rates of marsh and upper tidal flat processes (Fig. 18.20). A marsh shoreline can (i) retreat by erosion, (ii) prograde or (iii) drown in response to local relative sea-level rise (RSLR). Specifically, if the rate of marsh aggradation is equal to or greater than the rate of RSLR, the marsh shoreline progrades. It erodes if the nearshore sedimentation rate is less than the rate of RSLR. If the marsh aggradation rate is less than the rate of RSLR, then the marsh drowns.

In tropical and subtropical upper intertidal and tidal creek levee environments, mangrove swamps show prolific growth and have major effects upon sedimentation by protecting muddy substrates from storm erosion, encouraging settling of silt-grade

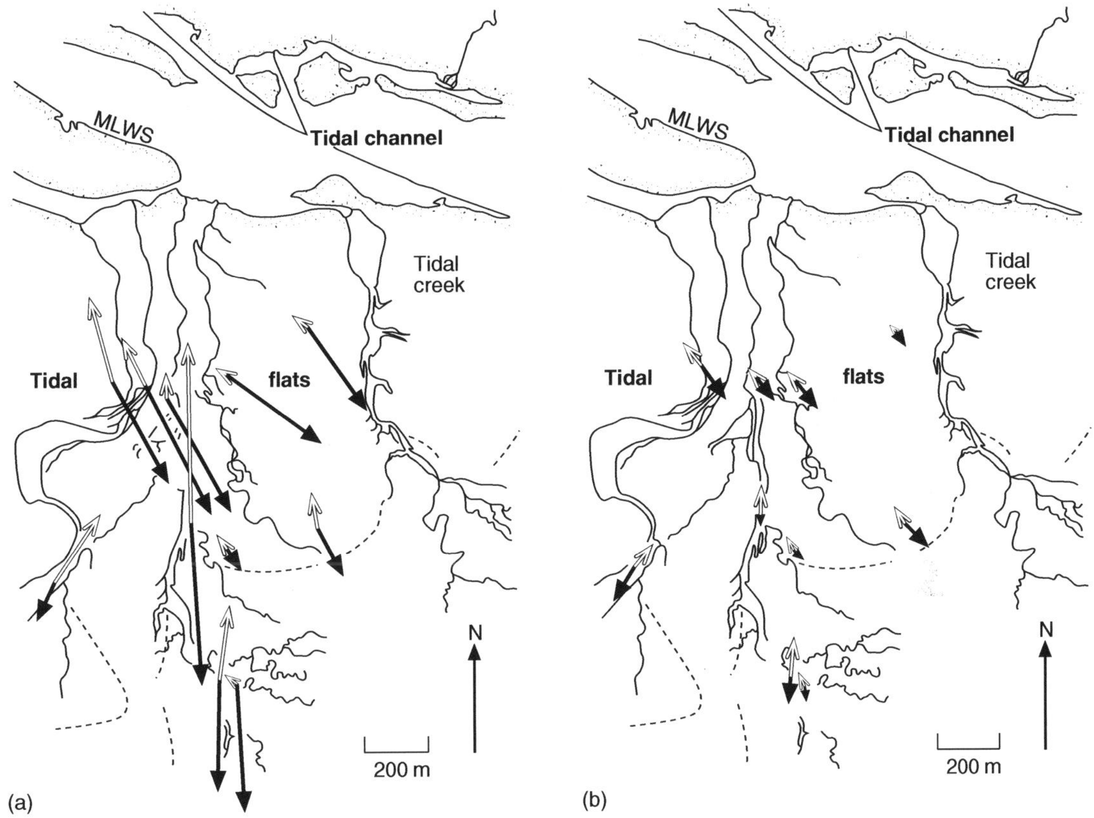


**Fig. 18.18** Maps to show sediment transport pattern for (a) spring and (b) neap tides across tidal flats Black vectors are flood-tide transport residuals, and white arrows are ebb-tide residuals. Note strong onshore asymmetry in sediment flux. MLWS—level of mean low water spring tide. This is the pattern of incoming or outgoing flow once the tide has filled feeder tidal creeks; deeper channels on other tidal flats exert a stronger influence on local tidal flat sedimentation, both by reworking deposited flat sediment and during overtopping at bankfull. (After Carling, 1981.)

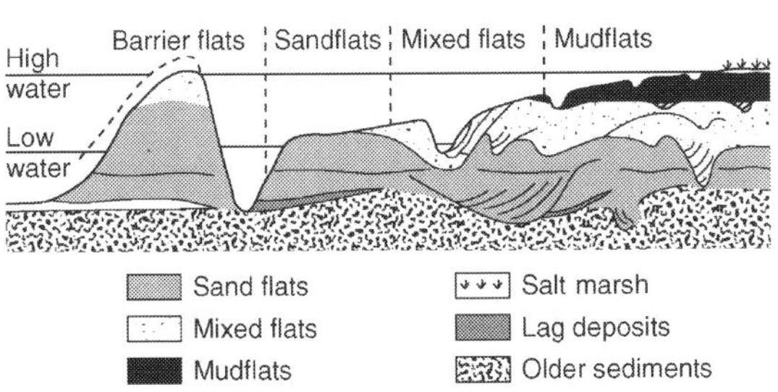


**Fig. 18.19** Classic section through a seaward-progradating tidal flat complex to show the various subenvironments and the generally fining-upwards sequence produced, though broken locally by tidal channel deposits. (After Reineck & Singh, 1973.)

sediments during spring tides. Swamp environments are characterized by a silty substrate, root growth, abundant wood and bark fragments and a distinct fauna of agglutinated foraminifera. The major sedimentary effects of mangrove progradation are to introduce substantial amounts of organic matter into the silty substrate around the root network.

Seaward coarsening of tidal flats from the limit of salt marsh or mangrove swamp gives rise to a mixed sand/mudflat. Again, bioturbation by the abundant infauna is intense. Tidal channels (Figs 18.17-18.20; see also comments in Chapter 16) rework much of the tidal-flat deposits and give rise to inclined lateral accretion deposits of interlaminated silts and muds. Rapid deposition on the point bars discourages infaunas, and hence these deposits are relatively free of

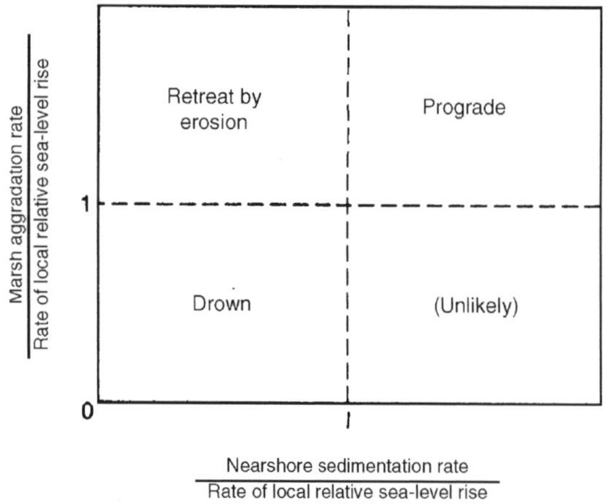


**Fig. 18.20** Three likely responses of a salt-marsh shoreline to local relative sea-level rise. (After Schwimmer & Pizzuto, 2000.)

bioturbation. The sandflats that occur at mean low water mark show a great variety of wave- and current-formed ripple bedforms with complex interference forms caused by gravity runoff effects. Local dunes may result if tidal flows are strong enough in channels. In many areas intertidal channels pass offshore into a subtidal zone of deep channels with major dune bedforms whose migration and accretion are dominated by the periodic ebb and flow of the tidal wave. In channels that transmit both ebb and flood, dunes simply migrate this way and that, with no net displacement. Separate ebb and flood channels (like the tidal inlet channels already discussed) feature dune bedforms with ebb or flood dominance. More rarely, evidence for a mixture of the two vectors may be seen in the occurrence of reactivation surfaces cut into the lee-side deposits of the dominant dune bedform.

Mixed-regime tidal flats with strong seasonal control on predominance of tide vs. wave energy are characteristic of many Southeast Asian coastlines, e.g. open-coast tidal flats of southwest Korea. Here, summer tide-dominance is replaced by winter storm-dominance. Thick, unbioturbated muddy summer deposits reflect low wave energy, weak currents and intense low tide desiccation. During autumn, these are mostly eroded by onshore winds. Winter storm waves dominate sedimentation, generating wave-generated parallel lamination and short-wavelength hummocky cross-stratification. This raises the interesting conundrum that many ancient 'shorefaces' might in fact have been open-coast tidal flats. Distinctive wave-generated tidal bundles recognized on these modern tidal flats may be of use in identifying examples in the sedimentary record.

The classic cheniers of Louisiana, westwards of the modern lobe of the Mississippi delta, and elsewhere reflect alternating fair- and foul-weather influences and upstream supply. Periods of mudflat outbuilding appear to coincide with abundant longshore mud supply from the Mississippi outfalls. Periods of mudflat erosion occur when mud supply is sparse (perhaps following a lobe avulsion) and waves rework the nearshore mudflats and concentrate shell material as chenier storm ridges bordering the mudflats. Chenier sand or shell facies are dominated by storm washover effects that produce landward-dipping, low-angle to planar cross-sets on the landward (washover) side of the biconvex ridges. The nature of the base of the chenier succession varies from a sharp contact with marsh facies on the landward side of the ridge to a gradational contact with shallow-water or mudflat facies on the seaward side.

## 18.5 Ancient clastic shoreline facies

Identification of true shoreline and nearshore facies is of the utmost importance in palaeogeographical reconstructions, since firm limits may then be put on the extent of the ocean during a particular time interval. Additionally, identification of shoreline facies enables deductions to be made as to the magnitude of tides, the relative importance of waves vs. tides and absolute bathymetry. Shallow coring techniques and GPR have revolutionized the interpretation of the coastal sedimentary record, old and young.

High wave-energy shoreface systems are commonly preserved in the sedimentary record as upward-coarsening sandstone lenses which record progressive water shallowing; they are common as the *parasequences* of sequence stratigraphy (Fig. 18.21), recording strandline advance by net deposition under conditions of highstand. Each records a shoreline regression and is capped by a thin transgressive succession over a *flooding surface*. Successive parasequences represent cycles with a period of [O] $10^4$–$10^6$ years. Variability within shoreface parasequences, particularly erosive episodes marked by scours or erosion surfaces, gives information on sea level, sediment supply and wave regime. Detailed GPR and core profiles show erosional or nondepositional episodes are related to reductions in supply or increased wave power in response to changing synoptic wave-climate/storm conditions.

Tidally influenced sediments commonly show the effects of the diurnal and monthly tidal rhythm in the form of *heterolithic bedding* and laminations. Particularly informative are sets of cross-stratification formed by dune migration whose rate of advance is fastest during spring tide and slowest at neap, when fine-sediment fallout from suspension at slackwater creates mud drapes. Also, since tides are periodic phenomenon and tidal-current vectors control rates of sediment transport and deposition, various periodicities, including semidiurnal, daily, semimonthly (neap–spring), monthly, semiannual, annual and lu-

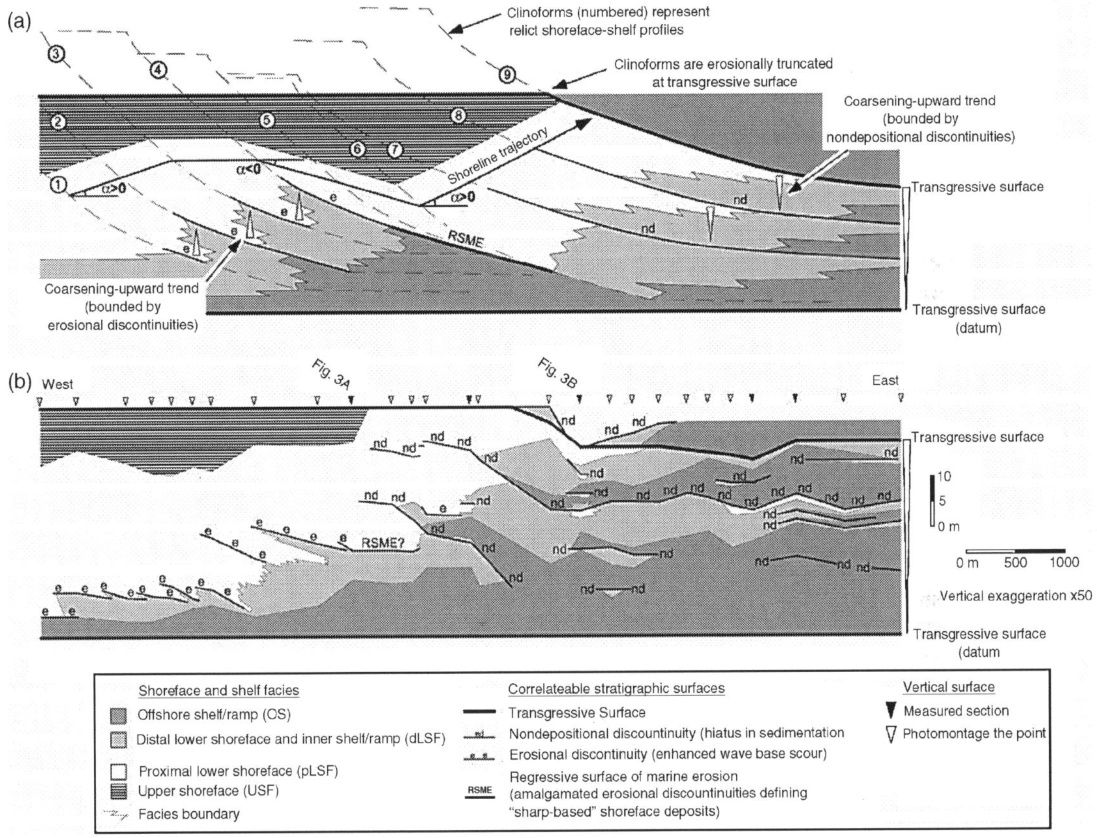


**Fig. 18.21** Shoreface to offshore parasequences from Kenilworth Member, Blackhawk Formation (Upper Cretaceous), continuously exposed in the Book Cliffs, Utah, USA. (a) Schematic panel oriented along depositional dip through a shoreface parasequence. The panel highlights different types of intraparasequence discontinuity surface and variations in facies architecture in between them. Erosional (e) and non-depositional (nd) discontinuity surfaces are marked respectively by seaward and landward shifts in lower-shoreface facies along seaward-dipping clinoform surfaces. Regressive surfaces of marine erosion (RSMEs) are formed by diachronous erosional amalgamation of several erosional discontinuities. (b) Annotated cliff-face panel through the K4 parasequence along the southern face of Middle Mountain and Gunnison Butte, based on measured sections and photomontages. The panel is projected into the plane of regional depositional dip. Transgressive surfaces and selected minor, intraparasequence stratigraphic discontinuities which define clinoforms are labelled. The minor stratigraphic discontinuities are dipping, through-going surfaces that can be traced updip and downdip between facies associations, although they coincide locally with depositional boundaries over part of their length. (After Hampson 2000; Storms & Hampson 2005.)

nar nodal (18.6 yr) periods, have been extracted from ancient cyclic *tidal rhythmites*. If the periodicities are indeed tidal, in many cases inferred net accretion rates would range from decimetre to metre annually with continuous and rapid sedimentation necessary for their formation. Measurements of couplet-thickness periodicities of 13 to 27 have been used to interpret neap–spring tidal sedimentation. The fact that most reported cyclic rhythmites contain less than 28 couplets within a neap–spring cycle has been considered to result from flood- and ebb-velocity asymmetries and/or nondeposition due to low current velocities of subordinate tides and neap tides or simply to non-inundation on the upper intertidal flats during neap tides. However, daily observations on the Yangtse estuary tidal flats show that only short-term mud–-sand deposition during calm weather is controlled by tides. During foul weather the number of couplets preserved per tidal cycle is reduced because of reworking and erosion by waves. Deposition rates thus decrease exponentially as timescales increase because of such periodic erosional intervals.

Although the *mangrove swamp* ecotype extends back in time to the early Cenozoic there are very few documented examples of mangrove-hosted tidal flats in the sedimentary record. This is because of organic matter oxidation and calcareous material dissolution which leaves the silt-rich mangrove unit with only a sparse record of once high-C production.

## Further reading

### *General*

Useful accounts of the physical and geomorphological processes that affect clastic shoreline sedimentation appear in the text by Komar (1998) and in the volumes edited by Hails & Carr (1975), Davis & Ethington (1976), R. A. Davis (1985), Leatherman (1979), Aubrey & Geise (1993), Carter & Woodruffe (1994), Flemming & Bartholema (1995), Black *et al.* (1998), Pye & Allen (2000) and Balson & Collins (2007). Harris *et al.* (2002) discuss Australian coastal morphology as a quantitative function of wave, tide and river power. Yoshida *et al.* (2007) is a thoughtfull attempt to investigate the idea that variable process regimes during cycles of sea-level rise and fall have important repurcussions for sequence stratigraphic studies of coastal deposit architecture. Advanced coastal sediment transport and water wave dynamics are discussed by Nielsen (1992) and Fredsoe & Deigaard (1992). *TELEMAC* modelling of combined tide–wave currents on the Scroby sands, southeast North Sea is by Park & Vincent (2007). A useful analysis of possible coastal change engendered by current and future sea level rise is by Fitzgerald *et al.* (2008).

### *Specific*

For coastal classification see Hayes (1975, 1979), Dalrymple *et al.* (1992), Boyd *et al.* (1992), Harris *et al.* (2002) and Yang *et al.* (2005), and discussions of latter by Chang & Flemming (2006) and Dalrymple *et al.* (2006).

On beach dynamics see Bagnold (1940), Inman & Bagnold (1963), Kemp (1975), Hardisty (1986), Hughes *et al.* (1997) and Masselink *et al.* (2007).

For beach sedimentation see Bluck (1967), Wunderlich (1972), Orford (1975), Davidson-Arnott & Greenwood (1974, 1976); Davidson-Arnott & Pember (1980), Moore *et al.* (1984), Massari & Parea (1988), Hart & Plint (1989), Osborne & Greenwood (1992), Gruszczynski *et al.* (1993), Greenwood & Sherman (1993), Hiroki & Terasaka (2005), Neal *et al.* (2002) and Pascucci *et al.* (2009).

Transgressive muddy-gravel beaches are discussed by Dashtgard *et al.* (2006); uplifted strandplains by Tamura *et al.* (2007) and Fraser *et al.* (2005); shoreface ravinement by Swift (1968) and Rodriguez *et al.* (2001); ravinement transgressive lags by Hwang & Heller (2002).

For Barrier–inlet-spit systems and their deposits see Fisk (1959), Rusnak (1960), Halsey (1979), Oertel (1979, 1988), Nummedal & Penland (1981), Sha & de Boer (1991), Siringan & Anderson (1993), Wellner *et al.* (1993), Aubrey & Geise (1993), Fitzgerald (1993), Oost & de Boer (1994), Biegel & Hoekstra (1995), Fitzgerald & van Heteren (1999), Levin (1995) and Davis & Flemming (1995). Effects of Hurricane Ivan are reported by Stone *et al.* (2005).

For spits see Johannessen & Nielsen (1986), Nielsen *et al.* (1988), Hiroki & Masuda (2000) and Nielsen & Johannessen (2009—Skagen spit).

Changes to Barataria Bay tidal-inlet/ebb-tidal-delta complex are reported by Fitzgerald *et al.* (2004).

Integrated GPR/cartographic/core barrier-beach studies are by Smith *et al.* (1999), Bristow *et al.* (2000), Neal *et al.* (2002), Moore *et al.* (2004), Bristow & Pucillo (2006), Costas *et al.* (2006) (for transgressive barriers) and Rodriguez & Meyer (2006).

For barrier evolution and sea-level change see Kraft (1971), Kumar & Sanders (1974), Kraft & John (1979), Bourgeois (1980), Rampino & Sanders (1981), Kraft *et al.* (1987), Finkelstein & Ferland (1987), Everts (1987), Pilkey & Davis (1987), Ashley *et al.* (1991), Wellner *et al.* (1993), Rodrigues *et al.* (2004) and FitzGerald *et al.* (2008).

Strandplains and back-barrier wetlands with tsunamites and storm-layer washovers are reported in Nanayama *et al.* (2003), Wang & Horwitz (2007), Bondevik (2008), Jankaew *et al.* (2008), Peterson *et al.* (2008) and Morton *et al.* (2008). General discussion of tsunamite semantics is by Shanmugam (2006).

For tidal flats and channels, salt marsh, mangrove swamps and chenier ridges see Reineck (1958, 1967, 1972), Evans (1965), Bridges & Leeder (1976), Carling (1981), Collins *et al.* (1981), de Mowbray & Visser (1984), Ke *et al.* (1996), Schwimmer & Pizzuto (2000), Dyer *et al.* (2000), Lee *et al.* (2006), Neumeier & Amos (2006) and Perry *et al.* (2008); rainwater effects on channel erosion and sediment yield are reported by Bridges & Leeder (1976), and on tidal-flat sediment strength by Tolhurst *et al.* (2006); open coast tidal flats are studied by Yang *et al.* (2005)—see also discussion of this by Chang *et al.* (2006) and Dalrymple *et al.* (2006).

Tidal rhythmites by Visser (1980), Boersma & Terwindt (1981), Sonett *et al.* (1988), De Boer *et al.* (1989), Tessier & Gigot (1989), Kvale *et al.* (1999), Williams (1989), Dalrymple *et al.* (1991), Martino & Sanderson (1993), Chan *et al.* (1994), Archer (1994), Kvale *et al.* (1994), Miller & Eriksson (1997), Cowan *et al.* (1998), Kuecher *et al.* (1990), Tessier (1993), Shi (1991), Brown *et al.* (1990), Lanier *et al.* (1993), Greb & Archer (1995), Archer *et al.* (1995) and Archer & Johnson (1997).

For case histories of ancient clastic shoreline facies Dott & Bourgeois (1982), Hunter & Clifton (1982), Prave *et al.* (1996), Okazaki & Masuda (1995), DeCelles (1987); shoreface-shelf parasequences by Hampson (2000), Hampson & Storms (2003) and Storms & Hampson 2005. Concepts of shoreline migration and sand pinchout are considered by Løseth *et al.* (2006); mangrove swamp facies by Perry *et al.* (2008).

# Chapter 19

# SILICICLASTIC SHELVES

*The roar of the sea is as was*
*heard by me as a child,*
*Without change, without pity,*
*shovelling the sand of the*
*shore.*
*The everlasting swelling, listen*
*to the sound of the swelling...*
*Make up my bed behind the*
*sound of the sea.*

From the Gaelic lament '*An Ataireanhd Ard*' (The High Swelling of the Sea). Traditional

## 19.1 Introduction: shelf sinks and lowstand bypass

The continental shelves are not only half-way houses for sediment that journeys from continent to deep ocean they also act as final resting places, in other words they are partial *sinks*. In order for clastic sediment to arrive on to the shelf it must previously have escaped various nearshore traps such as estuaries, bays, lagoons, deltas and tidal flats. Once on the shelf, a complicated fluid dynamical mixture of tide, wave, wind, ocean and density currents disperse the sediment, allowing some proportion to be transported over the shelf edge into the deep oceans. *In situ* production (*sources*) of calcareous shelly debris may also be important on shelves, e.g. $CaCO_3$ concentrations approach 40% over parts of the northwest European temperate shelf. Shelves extend from the limits of the shoreface (say in the region 4–20 m water depth) out to a prominent shelf-edge break, at the top of the continental slope. The depth of the shelf edge (20–550 m) and the shelf width (2–1500 km) are tremendously variable, depending largely upon tectonic setting. Shelves on Atlantic-type ('passive') continental margins record tens to hundreds of million of years of thermal subsidence and tend to be much wider than those on Andean-type or Pacific margins. The relatively smooth, gentle offshore slope to most shelves is basically a constructional feature moulded by shelf currents and deposited sediment. Shelf subsidence is another factor, since most shelves are underlain by extremely thick sedimentary successions that lie in fault-bounded linear basins and/or in broader thermal downsags superimposed on these relics of continental break-up. These successions were deposited in relatively shallow water ( $<$ 100 m water depth), implying that shelves are prone to continued, gentle subsidence, with rates of sediment supply and production usually able to keep pace with that subsidence.

*Sedimentology and Sedimentary Basins: From Turbulence to Tectonics*, 2nd edition. © Mike Leeder.
Published 2011 by Blackwell Publishing Ltd.

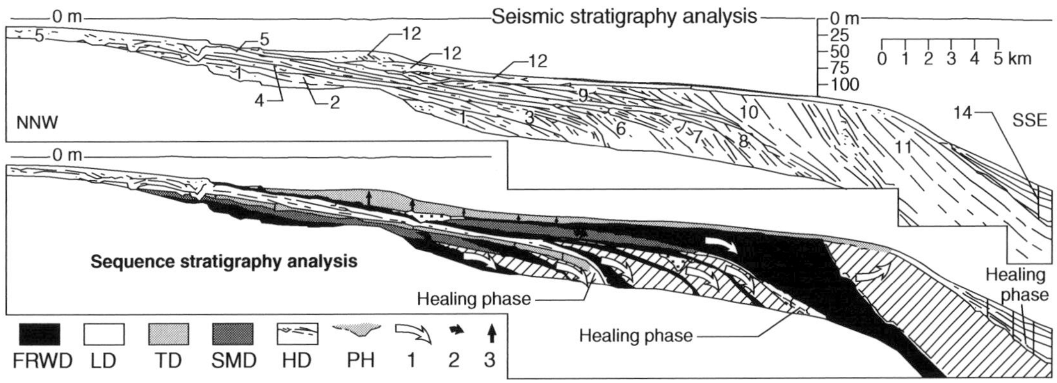


**Fig. 19.1** High resolution seismic reflection profile across the Gulf of Cadiz shelf, southwest Spain. Profile illustrates episodic progradation of the shelf edge in response to high frequency Pleistocene sea-level changes over the past 3 Ma. 1–12, depositional sequences and subsequences; FRWD, forced regressive wedge deposits (sea level falling); LD, lowstand deposits; TD, transgressive deposits; SMD shelf-margin deposits; HD, highstand deposits; PH, incised palaeochannel and infills. Arrow 1, major progradation; arrow 2, minor progradation; arrow 3, aggradation. (Simplified after Somoza *et al.*, 1997.)

Simple shelves outboard a continent and inboard a passive continental margin are the most common type. They show an oceanward-dipping prism of sediments (Fig. 19.1) and have been referred to as *pericontinental* shelves. By way of contrast, some very wide modern shelves like the North Sea, Yellow Sea and Timor–Arafura Seas lie well inboard the continental margin, in the former case because of its status as a failed rift dating back to Mesozoic times. Such *epicontinental*, or *epeiric*, shelves seem to have covered much larger areas in past geological epochs of continental plate evolution, especially in the Mesozoic of northwest Europe, Mediterranean Tethys and the Western Interior Seaway, USA, but also in the upper Palaeozoic of North America and northwest Europe. Narrower shelves exist on destructive or collisional plate margins because shelf sediment 'damming' has occurred behind outboard positive relief features formed by faulting and accretionary prism growth (e.g. Pacific coast of the Americas, Miocene of the central European Paratethys).

Shelves are constructional features with sediment accretion on them during periods of sea-level highstand. During lowstand sediment is accreted close to or at the shelf margin with rivers running out to the shelf edge along incised valleys, terminating in shelf-edge deltas now stranded and defunct in water depths of 100–200 m (Fig. 19.2). It is thus essential to think of shelves at lowstand as *sediment-bypass* systems. During rising sea level, shoreline-retreat deposits (Chapter 18) are most fully preserved in the sedimentary fills to lowstand shelf valleys because such linear depressions often protect sediments from erosion. A variety of sedimentary units may develop during transgression, with valley-fill successions influenced by parameters such as changes in shelf gradient, valley morphology, sea level, climate and hydrodynamic processes. Simple incised valleys are usually fed locally on the old coastal plain or shelf and record complete infill during one lowstand–transgressive–highstand sequence. Compound valleys are usually fed by long-lived catchments that drain into piedmont sourcelands, often structurally controlled. They persisted through multiple sea-level cycles and so contain numerous internal sequence boundaries in addition to that of the basal valley-floor surface.

An idealized shelf incised-valley-fill succession resulting from steady rates of sea-level rise, unvaried mean water discharge and continuous sediment supply (three dubious assumptions) incorporates at least three superimposed transgressive erosion surfaces within it (Fig. 19.3). Since these surfaces form progressively as sea level rises they will define landward-younging features; any surface that varies spatially in age is said to be *diachronous*. The *bay flooding surface* forms by initial flooding of fluvial channels;

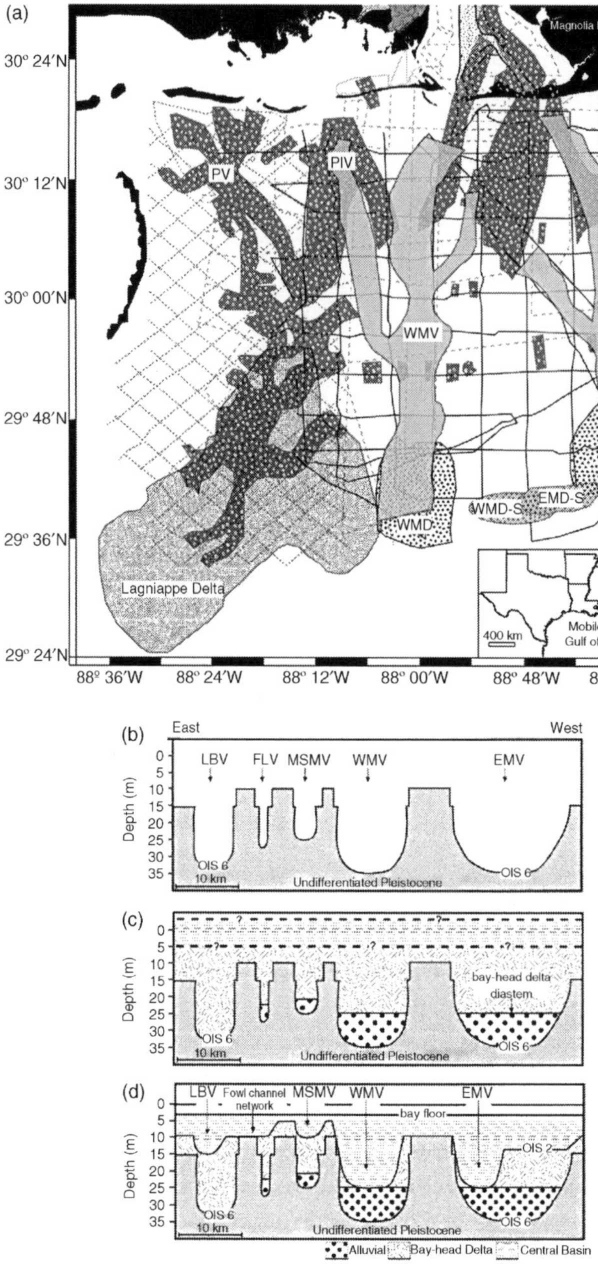


**Fig. 19.2** (a) Incised shelf valleys and shelf-edge deltas mapped from seismic data, offshore Mississippi delta. (b–d) Incised valleys as highstand sediment sinks. Cartoon east–west cross section extending through northern Mississippi Sound and south-central Mobile Bay illustrating the evolution of the area during the late Quaternary. During the lowstand associated with the formation of Sequence Boundary B, the La Batre (LBV), Fowl (FLV), and Mobile (Mississippi Sound Mobile Valley (MSMV), West Mobile Valley (WMV), and East Mobile Valley (EMV)) fluvial systems incised and branched seaward. These valleys were filled with alluvial, bay-head delta, and central basin sediments during the subsequent transgression. During MIS 2 lowstand, the fluvial systems generally reoccupied previous valleys and during the subsequent Stage 2 to 1 transgression filled with central basin sediments. (All after Greene *et al.*, 2007.)

it separates fluvial sediment from overlying estuarine deposits. As the estuary develops (Chapter 16) tidal current erosion near its landward-moving mouth creates a *tidal ravinement surface*. Meanwhile a landwards bayhead channel succession is produced locally by seaward progradation of a *bayhead delta*. It is possible that river flood events through the channels of this delta could also cut erosively into the estuarine sediments forming a regressive *bayhead channel ravinement*. Finally, an extensive *wave ravinement surface* that truncates all underlying estuarine deposits is created by landward retreat of the shoreface during transgression. To create such a surface, wave and current erosion must remove both flood tidal-delta deposits and possibly some central-basin facies. All of these transgressive surfaces are diachronous, and could become amalgamated within a preserved sequence boundary associated with the transgression.

Climatic controls on shelf incision comes from seismic and core data on the east China Sea shelf offshore the Holocene Changjiang (Yangtse) delta. Here, incision was greatest during wet catchment climatic conditions during the initial MIS 2 sea-level fall. However, the subsequent trend toward drier conditions slowed incision to a minimum during maximum lowstand. Furthermore, the low-gradient morphology of the exposed outer shelf diminished the ability of fluvial systems to incise. Instead, fluvial systems migrated laterally, creating a shallow (30 m), wide (300 km) incised valley system on the outer shelf. A similar climatic control is invoked for lowstand incision of the very wide (1000 km) Sunda shelf, Indonesia. Incised valley magnitude (width, depth) on the Gulf of Mexico shelf is more strongly linked to discharge, via the area of upstream catchments, than it is to the gradient of the shelf slope.

Away from incised valley fills generated during transgression, many modern shelves are, to a greater or lesser extent, relict in the sense that pre-Holocene sediment is exposed on many and is being current- and wave-reworked. This is especially true in higher latitudes where much of this relict sediment has a glacial origin that is entirely unrelated to modern shelf energy fluxes. In other cases shelf sediment seems to be in hydrodynamic equilibrium with present-day flows. Shelf architecture, revealed by high-resolution marine seismic surveys, presents a universal picture (extending back well into the middle Tertiary) of lowstand erosion and outbuilding, alternating with transgressive erosion and highstand coastal sediment trapping (Fig. 19.3).

## 19.2 Shelf water dynamics

### General

The dynamics of water and sediment movement on shelves is complicated because of several influences at work, from land-based to ocean-sourced currents (Fig. 19.4). The most important of these are tide, wave and wind, the latter including net mass transport of water by direct wind shear. These all act on incident water and sediment density plumes from river estuaries and delta distributary mouths. Note that the Coriolis force is important as a dynamical consideration on low- to mid- to high-latitude shelves.

Shelves have been classified into tide- and weather-dominated, but most shelves show a mixture of processes over the seasons. The majority of pericontinental shelves are meso- to macrotidal only around their shallower margins; most have a tidal range $< 2$ m. Even the epicontinental North Sea shelf, with its powerful rotary tidal gyres, is macrotidal only over about 50% if its area. It is thus important for geologists to remember that the tidal regime of a whole basin cannot be inferred from evidence gathered from just a small area. Despite all these warnings about complexity, it is considered useful in the context of shelf physical processes to consider a generalized model for shelf physiography and water characteristics that involves an inner-shelf mixed layer where frictional effects of wave and tide are dominant. Then the deepening mid- to outer shelf exhibits increasing differentiation into surface and bottom boundary layers separated by a 'core' zone (Fig. 19.5).

### Shelf tides

In the oceans the twice-daily tidal wave is still of shallow-water (longwave) type since the tidal wavelength is very large (about 10 000 km) compared with the depth of the oceans (say 5 km). The maximum tidal range in the open oceans is about 50 cm and typical open ocean tidal currents are only a few centimetres

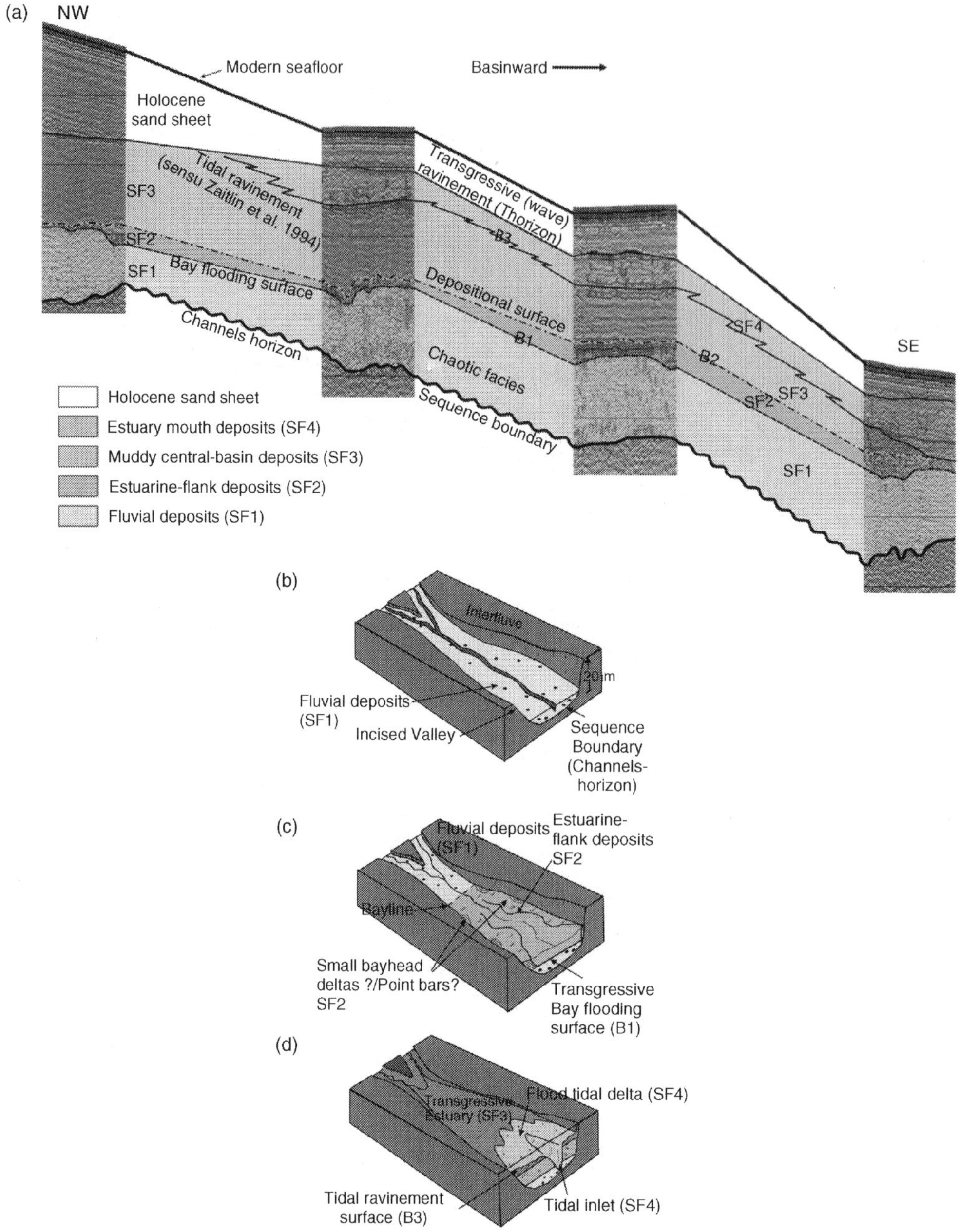


**Fig. 19.3** New Jersey shelf incised-valley fills.(a) Distribution of interpreted seismic facies and bounding seismic horizons along a dip section through the trunk channel of an incised-valley. (b–d) Schematic representations of the evolution of New Jersey incised paleo-valley systems, including sedimentary facies and stratigraphic boundaries: (b) fluvial incisions, with preserved channel lags; (c) aggradational estuarine system, at the initiation of back-filling; (d) passive infilling of the estuary, with developed central-basin muds and estuary-mouth complexes. Not shown is formation of the transgressive ravinement (T horizon), which has reworked and selectively removed portions of these fill deposits. (After Nordfjord *et al.*, 2006, and sources cited therein.)

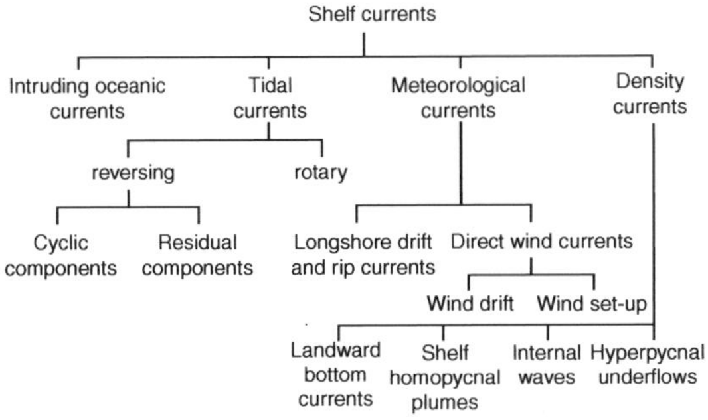


**Fig. 19.4** Components of the shelf-current velocity field. (Slightly modified after Swift, 1972.)

per second. So why does the tidal wave cause so much stronger tidal currents on shelves? This is because the open ocean tidal wave decelerates as it crosses the shallowing waters at the shelf edge. This causes tidal wave refraction of obliquely incident waves into parallelism with the shelf break and partial reflection of normally incident tidal waves. At the same time the wave amplitude of the transmitted tidal wave is enhanced and so is the resulting tidal currents. This is because the energy of a wave must be conserved (neglecting bottom frictional losses) and tidal currents increase because they are dependent on the instantaneous amplitude of the wave (**Cookie 28**). Tidal strength may also vary because of the nature of the connection between the shelf or sea and the open ocean. In the case of the Mediterranean Sea the connection with the Atlantic has become so narrow and restricted that the tide cannot reach any significant range in most of the sea. Locally, in the Straits of Gibraltar, the Straits of Messina and the Venetian Adriatic, tidal currents (but not necessarily tidal range) may be very much amplified when water levels between unrelated tidal gyres or standing waves interact. Another cause of spatially varying tidal strength concerns the resonant effects of the shelf acting upon the open oceanic tide. Resonant effects may greatly increase the oceanic tidal range in nearshore environments and lead to the establishment of dynamic tidal currents and processes (**Cookie 45**).

Want to know more about shelf tidal resonance? Turn to Cookie 45.

The Coriolis force affects tidal streams in semi-enclosed large shelves, like the northwest European shelf, the Yellow Sea and the Gulf of St Lawrence. This is because the tidal gravity wave has a long period and that water on shelves is bounded by solid coastlines so that the tide rotates. Such waves of rotation against solid boundaries are termed *Kelvin waves* (**Cookie 46**). The crest of the tidal Kelvin wave is a radius of any roughly circular basin and is also a co-tidal line along which tidal minima and maxima coincide. Concentric circles drawn about the rotational node are co-range lines of equal tidal displacement. Tidal range is thus increased outwards by the rotary action with further resonant and funnelling amplification taking place at the coastline.

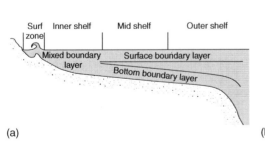

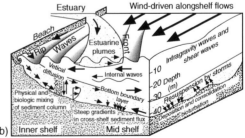


**Fig. 19.5** Shelf boundary layers and sediment transport.(a) Division of shelf waters into mixed, surface and bottom boundary layers. Inner shelf has tide and wave mixing, but is occasionally stable and stratified in season, and is generally said to be *friction-dominated*. Outer shelf is often stratified into an upper layer dominated by geostrophic flow (i.e. Coriolis turning occurs) and a bottom boundary layer with low friction. (b) Diagram to show the major physical processes responsible for cross-shelf flow and sediment transport. (After Nittrouer & Wright, 1994.)

Want to know more about rotary tidal currents? Turn to Cookie 46.

The ellipticity of any tidal current is a function of tidal current type and vector asymmetry. For example, the inequality between ebb and flood on the northwest European continental shelf is largely determined by the M4 harmonic of the main lunar tide. Since sediment transport is a cubic function of current velocity (Chapter 6) it can be appreciated that quite small *residual tidal currents* can cause appreciable net sediment transport in the direction of the residual current. The turbulent stresses of the residual currents will be further enhanced should there be a superimposed wave oscillatory flow close to the bed. A further important consideration arises from the fact that turbulence intensities are higher during decelerating tidal flow than during accelerating tidal flow. This arises from the greater intensity of the burst/sweep process in the unfavourable pressure gradients of such nonuniform and unsteady flows. Increased bed shear stress during deceleration thus causes increased sediment transport compared to that during acceleration, so that the net transport direction of sediment will lie at an angle to the long axis of the tidal ellipse.

A final point concerns the importance of internal tides and other internal waves, particularly upon the outer shelf region. These are common in the summer (fair-weather) months when the outer-shelf water body is at its most density-stratified, with a stable, warm, surface layer overlying a denser layer.

## Wind drift currents

Winter wind systems assume an overriding dominance on most shelves, causing net residual currents arising from wind drift, wind set-up and storm surge. Wind shear causes water and sediment mass transport at an angle to the dominant wind direction because of the Ekman effect arising from the influence of the Coriolis force (**Cookie 44**). For example, northward-blowing, coast-parallel winds with the coast to the left in the northern hemisphere will cause net offshore transport of surface waters and the occurrence of compensatory upwelling.

As discussed in Chapters 14 & 18, wind shear drift causes set-up of coastal waters. Should this coincide with a spring high tide, then major coastal flooding results. The effects are well known in the southern North Sea, Bay of Bengal and Adriatic. Coastal tornadoes hitting the southeastern and southern USA are particularly effective at raising the set-up of coastal waters, sometimes up to 5 m or more above mean high-water level. In semi-enclosed shelves the Coriolis force drives the wind drift current against shorelines where it is further amplified by resonance and funnelling. In the case of the southern North Sea, south-directed wind drift is forced westwards with the southward travelling (anticlockwise) Kelvin tidal wave. In addition to dynamic effects resulting from wind shear, the very low barometric pressures during storms cause a sea-level rise under the storm pressure minimum. The magnitude of this effect is about 1 cm rise per millibar decrease of pressure. So a storm pressure 960 mbar might cause a few tens of centimetres of sea-level rise. Offshore, wave set-up causes a compensatory bottom flow out to sea, driven by the pressure gradient (Fig. 19.6). Such geostrophic or gradient currents (which may be turned by Coriolis forcing) have been proven by measurements during storms to reach over 1 m/s, running for several hours. Such flows are a major means by which sediment is transferred from coast to shelf.

## Shelf density currents

Density currents are very important in distributing and transferring mud-, silt- and sand-grade sediments across the shelf. As discussed previously (Chapters 16 & 17), hypopycnal (positively buoyant) plumes of fresh to brackish water with suspended sediment (but seldom more than $\sim 20\,\mathrm{kg/m^3}$) issue from most estuaries and delta distributary mouths. In higher latitudes, small to moderate buoyancy fluxes are soon turned by the Coriolis force, and they may be trapped along-source in the mid- to inner shelf where they form coastal currents or linear fronts. Large vortices develop along the free shear layer between coastal fronts and offshore circulating shelf waters. Plumes are very sensitive to the effects of coastal upwelling or downwelling currents caused by winds. They may reach some way out into the mid-shelf or right across the shelf break, depending upon their dynamic characteristics, shelf winds and currents, and bottom gradient. Low slopes encourage long passage, whilst the development of vorticity on steeper slopes encourages

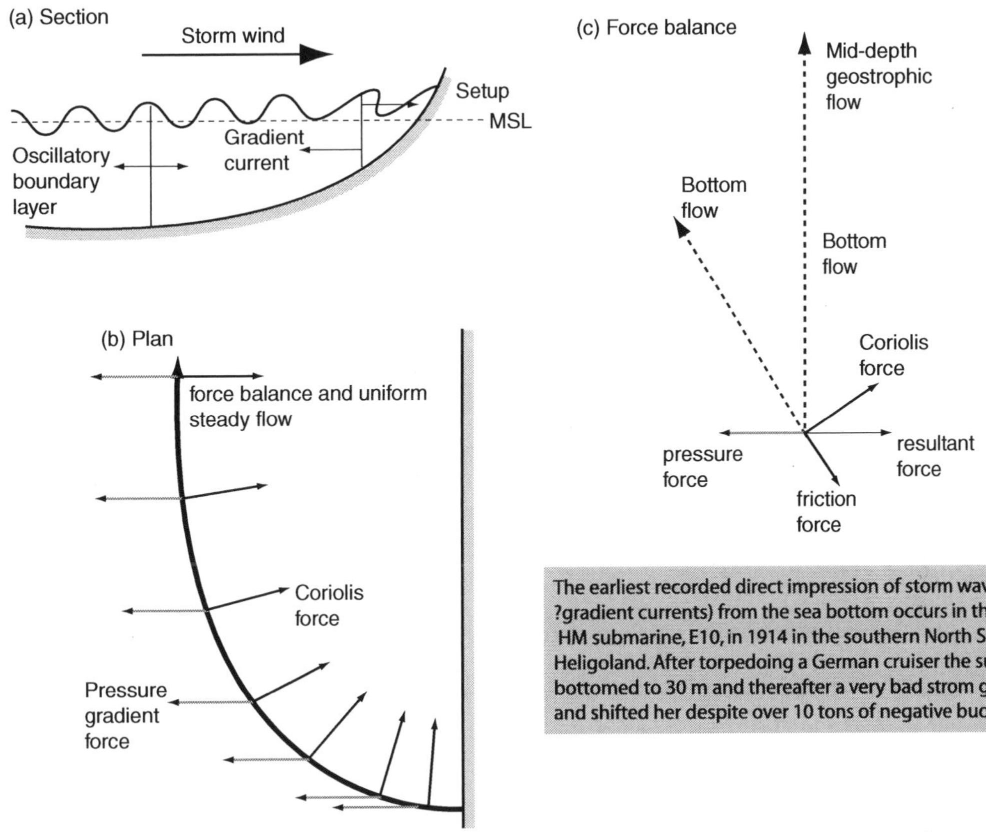


The earliest recorded direct impression of storm waves (and ?gradient currents) from the sea bottom occurs in the log of HM submarine, E10, in 1914 in the southern North Sea, off Heligoland. After torpedoing a German cruiser the sub bottomed to 30 m and thereafter a very bad strom grounded and shifted her despite over 10 tons of negative buoyancy.

**Fig. 19.6** Definition diagrams to show basic pattern of storm-driven shoreface to shelf geostrophic gradient currents in the northern hemisphere where Coriolis forcing is to the right:(a) section), (b) plan and (c) force balance.

turning and termination. The large buoyancy flux of many late spring and summer Arctic rivers causes plumes to extend for up to 500 km offshore, well into the Arctic ocean basin.

## 19.3 Holocene highstand shelf sediments: general

The distribution of sediment of differing grain sizes on highstand shelves depends on a variety of factors, e.g. point vs. line coastal sources, magnitude of freshwater and suspended sediment discharge, existence of advected buoyant plumes, wind/wave/tidal current vectors and regime, and a variety of others, including the general shape of the coastline produced by tectonics. It must be stressed that there is no general shelf grain-size pattern. A further important consideration is sea-level history, especially the influence of lowstand processes, particularly incised river valleys, moraines and so on. These provide environments exposed to, or sheltered from, tide and wave. Many shelves have embayments and gulfs; these are important because generally the flood tidal wave is less dissipated than that of the ebb, and so net up-embayment transport is maintained.

## 19.4 Tide-dominated, low river input, highstand shelves

On tide-dominated shelves with low river input, sediment distribution and bedforms depend to a great extent upon position with respect to tidal current transport paths. These are computed from combined

M2 and M4 tidal current predictions, and show the direction and relative magnitude of *net* bedload transport. Detailed studies on the northwest European tide-dominated shelf, for example (Fig. 19.7), have defined tidal current transport paths by a combination of observations on (i) surface tidal velocities, (ii) elongation and asymmetry of tidal current ellipses, (iii) facing direction of sandwaves and (iv) trends of sand ribbons and grain size. (The reader should compare the methodology of this approach with the principles used to construct sand flow paths in the great Saharan desert ergs; Chapter 13). Note the position of irregular 'bedload partings' that separate opposing sand and gravel pathways. These are associated either with the centres of rotating tidal cells or with local acceleration of the tidal streams in coastline constrictions. There is always a general trend towards decreasing grain size down the tidal current paths, perhaps from coarse sand to mud depending upon the availability of sediment. This trend is due to decreasing net current competence. The concept of bedload partings seems to fall down when residual currents and the nonuniform rotating 3D flows of outer estuaries, embayments and firths are considered. Here the ebb and flood tidal currents commonly separate into distinct pathways, leading to nonuniformity across the whole tidal prism. This is accentuated on shallow shelves by tidal ridge sandbank bedform emergence at low tides.

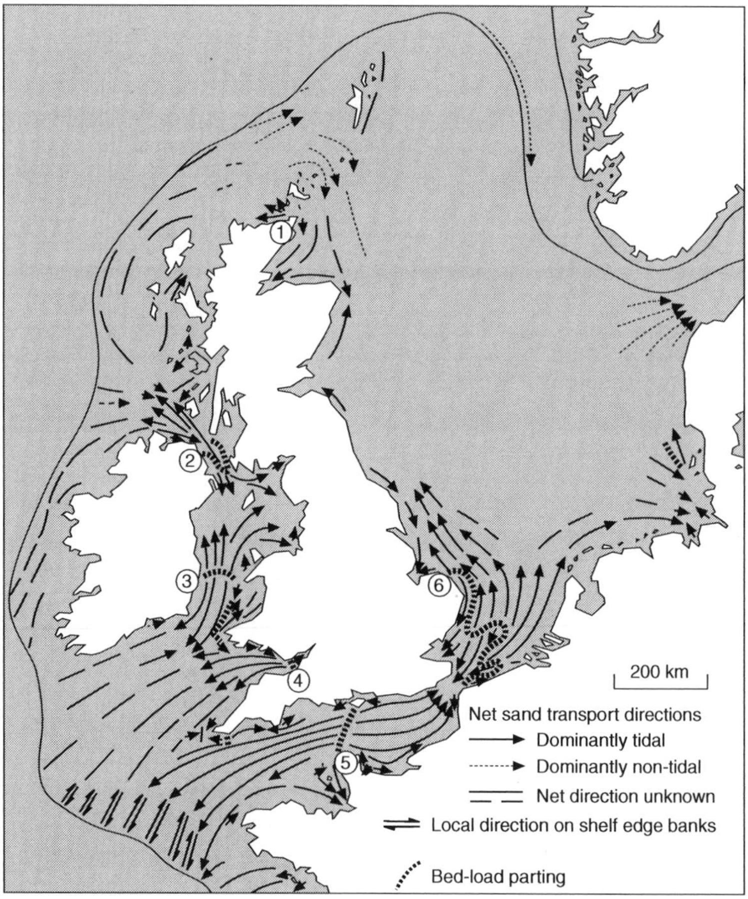


**Fig. 19.7** Map to show generalized tidal current transport paths on the northwest European continental shelf. (After Johnson *et al*., 1981.)

The upstream parts of tidal current paths, with velocities in excess of 1 m/s, may show sand ribbon bedforms (Fig. 19.8) up to 20 km long, 0.2 km wide and a few decimetres thick. These features occur in water depths of 20–100 m on gravel substrates that have a sparse cover of coarse sands. Simple parallel sand ribbons may owe their existence to secondary flows involving pairs of counter-rotating helical vortices (Section 5.5). Dunes 3–15 m high with wavelengths of up to 0.6 km cover large areas (>100 km$^2$) of the higher-energy parts of tidal transport paths (Fig. 19.9). Given a sufficient supply of sand these develop asymmetric forms in areas of marked tidal ellipse asymmetry and as symmetric forms at bedload partings where ellipse asymmetry is absent. During storms, dune heights and wavelengths are significantly reduced. They die out down tidal transport paths as bed shear stress declines, and are absent where wave activity is persistently intense. Little is known concerning the internal structures of tidal shelf dunes, but it may be inferred that they comprise dominantly unimodal large-scale cross-stratification with perhaps smaller-scale sets with opposed orientations. Sandwaves with low-angle lee slopes and superimposed dunes tend to occur in areas of weaker mean tidal currents and are expected to show internal sets of cross-stratification separated by downcurrent-dipping set boundaries.

A very prominent feature of tide-dominated shelves, like those of the southern North Sea (Figs. 19.10 & 19.11), Celtic, eastern Yellow and White Seas, are numerous parallel, large-scale linear tidal *ridges* (sometimes called *banks*). These are the most spectacular bedforms of the continental shelves, reminding us in their scale and persistence of the linear seif dunes and draa of Trade Wind deserts (Chapter 13). The ridges in the North Sea comprise shelly, well-sorted, medium sands up to 40 m high, 2 km wide, 60 km long with spacings of 3–12 km and covered by active dunes. In between the ridges the substrate is usually coarser, sometimes gravelly. Most shelf ridges have their crest lines oriented slightly obliquely (up to 20°, usually clockwise) to the direction of the maximum peak tidal current velocity which reverses during flood and ebb stages of the tidal cycle. They are usually asymmetric, with the steep face inclined at a maximum of about 6° to the horizontal and in the direction of net regional sand transport. The superimposed dunes migrate over the backs of the ridges until they reach the crests, where they turn clockwise to begin their orthogonal descent down the slightly steeper face. Internal structure is revealed by shallow seismic surveys (Fig. 19.11A′), which show evidence for ridge build-up growth and preferential migration. It is clear that the ridges did not develop over pre-Holocene positive or negative relief features developed on the tills and glacial outwash that underlie the area. Asymmetric ridges show internal inclined low-angle foresets parallel to the steeper face, indicative of ridge migration in this seawards direction. Symmetrical ridges show only build-up surfaces internally. Nested ridges in complex tide-wave environments such as the Scroby Sands off Yarmouth, East Anglia, England separate interbank areas as sort of channelized conduits which serve to accentuate ebb-tidal currents whilst the flood-tide is dominant at high water on the bank tops as they are immersed. The ridge systems are in clear equilibrium with the present combined tidal/storm-wave regime, with complex rotary net transport paths that may be computed from tidal current, wave and sediment transport algorithms using a modelling program called *TELEMAC*.

Large fields of apparently inactive ridges occur in deeper waters (ca 130 m) adjacent to the modern shelf break, perhaps most spectacularly developed in the Celtic Sea, although some Yellow Sea tidal ridges are also remnant features of early Holocene transgression. The architecture of the Kaiser Bank in the Celtic Sea (Fig. 19.12) is revealed by high-resolution seismic surveys. Lowest unit 1 has gently dipping (1–8°) strata that are bank-parallel. Unit 2 erosively overlies 1 and forms the majority of the bank. It comprises stacked sets of downcurrent-dipping (7–12°) internal surfaces formed by climbing, sinuous-crested tidal dunes that were up to 20 m high. These deposits are locally incised by Unit 3, an anastomosed channel network that may represent a buried swatchway system. Unit 4 is the upper part of the bank and comprises wave-related deposits mainly preserved on the bank flanks. The outer bank surface is erosional. The bank is believed to have formed during the early Holocene sea-level rise and the evolution from unit 1 to unit 3 indicates an upward increase in tidal energy, mainly characterized by the thickening of dune cross-bed sets in unit 2. The majority of bank growth is inferred to have occurred in water depths of the order of 60 m. This evolution was controlled by relative sea-level rise, which is likely to have caused an episode of *tidal*

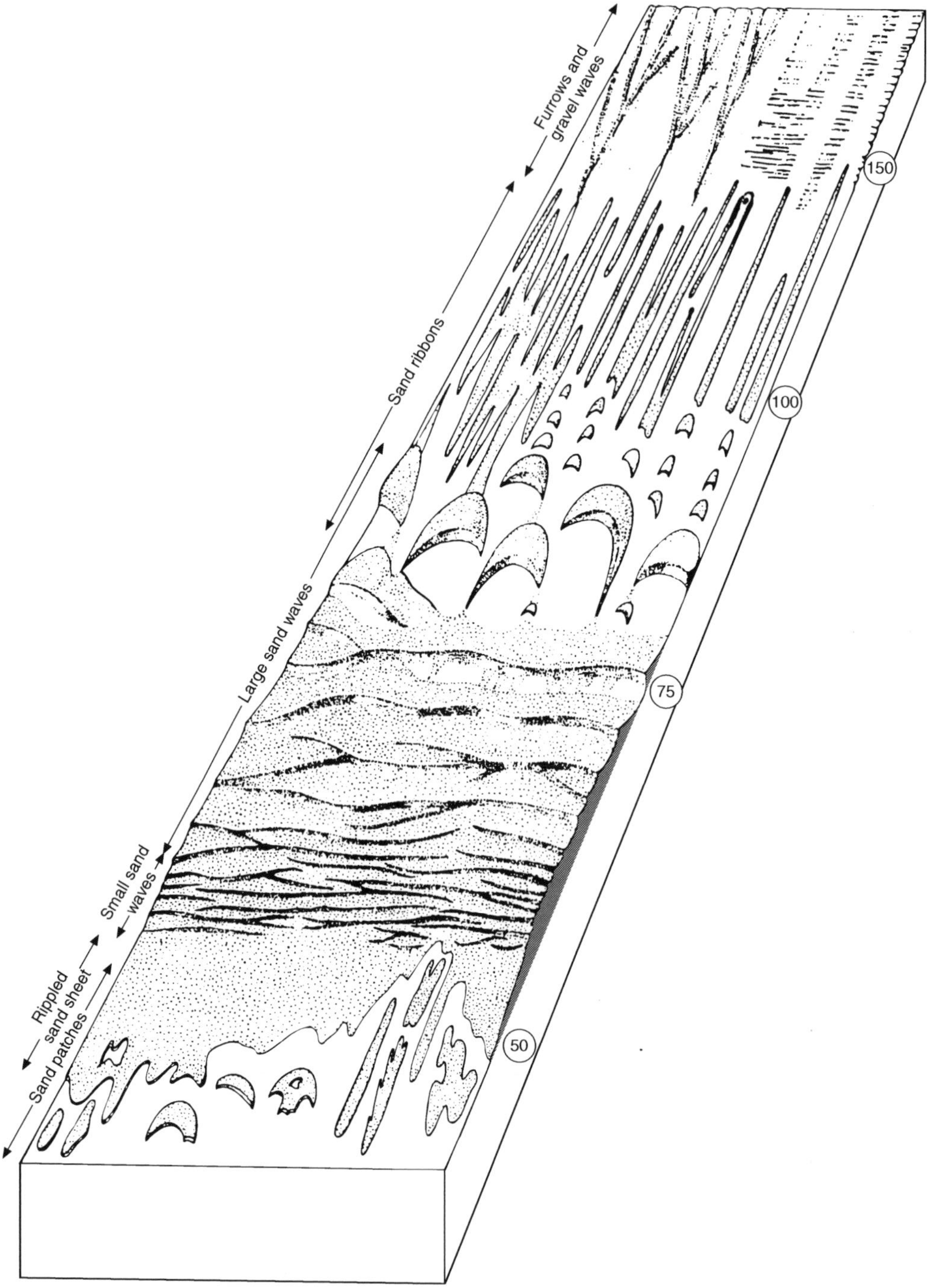


**Fig. 19.8** Block diagram to show the lateral sequence of bedforms and sediment types observed down a typical tidal current transport path, with mean spring-tide peak near-surface tidal current velocities in cm/s. (After Belderson *et al.*, 1982.)

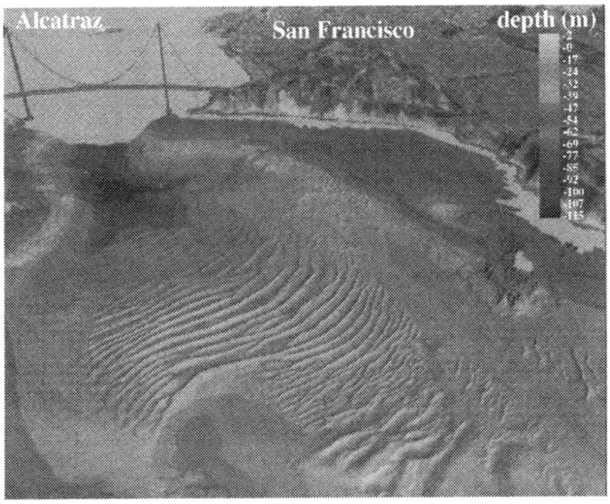


Fig. 19.9 Spectacular oblique view of the dune field at the mouth of San Francisco Bay, viewed from the northwest towards the Golden Gate Bridge (ca 2 km long). The whole field covers an area of ca 4 $km^2$ in 30–106 m water depth. Dune wavelengths and heights are up to 220 m and 10 m respectively. (After Barnard *et al.*, 2006.)

*resonance* (see **Cookie 45**) with associated strong tidal currents that were responsible for the incision of the deep, cross-cutting channels of unit 3. The transition to wave-dominated sedimentation in unit 4 is related to the decay of resonance with continued sea-level rise.

The distal ends of tidal transport paths comprise isolated sand patches and small sandwaves with numerous ripple bedforms and bioturbation features, the paths finally ending in areas of mud deposition. Bioturbated offshore mud deposits with rich infauna can occur only in relatively deep areas of low wave activity, the high deposition rates (3–5 mm/yr) indicating continuous mud fallout from suspension with important storm-produced suspensions contributing significantly. Studies of the mud belt developed in the Helgoland (German) Bight of the North Sea reveal frequent, thin, graded sand and shell layers attributed to storm gradient (geostrophic) currents, which have transported intertidal sands and fauna up to 40 km out into the deep offshore (Figs. 19.13 & 19.14). Measured storm underflow currents reach well over 1 m/s. A well-developed proximal to distal trend is found in these 'tempestite' deposits: from amalgamated, parallel, laminated and hummocky(?) sands with shelly lags in up to 5 m water depth, via interbedded, thinner, sharp-based sands and silts with thin, wave-rippled zones and bioturbated mud caps in 5–15 m water depth, to very thin silts and sands set in thick muds in >15 m depth, some 60 km or so from the shoreline. Thin graded sands are also common off those parts of the Gulf of Mexico coast that are both rich in a sand source and traversed by hurricanes that set up gradient currents.

## 19.5 Tide-dominated, high river input, highstand shelves

On shelves dominated by the advection of sediment-laden surface (hypopycnal) plumes, maximum deposition tends to occur at or about mid-shelf where the plume front is halted as it is redistributed by tidal, wave or oceanic frontal vorticity. At higher latitudes the turning effect of the Coriolis force on the predominant offshore wind regime is crucial. The destination of a plume depends upon the orientation of the effective shelf wind regime. That of the Washington coast of northwest USA encourages oblique passage of river-borne plumes to the mid- to outer shelf, with declining deposition rates all the way. This pattern of dispersal will give rise to lobate mud/silt lobes that fine and thin towards the shelf break. Thus the Amazon buoyant plume has deposited a subaqueous mud-delta wedge over 100 km out from the shoreline into water depths of 70 m or so. In shallow inshore areas, sediments laminated on a millimetre to centimetre scale represent diurnal and spring–neap tidal current variations. The topset wedge ends at a slope break in about 25 m water depth, fluid mud facies occurring landward of this point. On the upper foresets, short-term deposition rates of up to 100 mm/yr have been recorded; not surprisingly there are abundant signs of slope instability in these areas.

The Yellow Sea of Southeast Asia is perhaps the worlds most noteworthy example of a modern tidally dominant shelf fed by massive input of suspended sediment from world-class rivers, the two largest entering on the west being the Huanghe (Yellow) and Changjiang (Yangtze), both with their headwaters in the Tibetan Plateau and heavily influenced by monsoon-fed discharge. Many significant monsoon-fed rivers also enter the eastern Yellow Sea from the western Korean coastline, including the Aprok, Han, Keum and Yongsan. The most enormous area of mud/silt deposition occurs in the northern Yellow Sea (Gulf of Bohai) due to deposition dominated by hyperpycnal underflows and surface plumes. These also

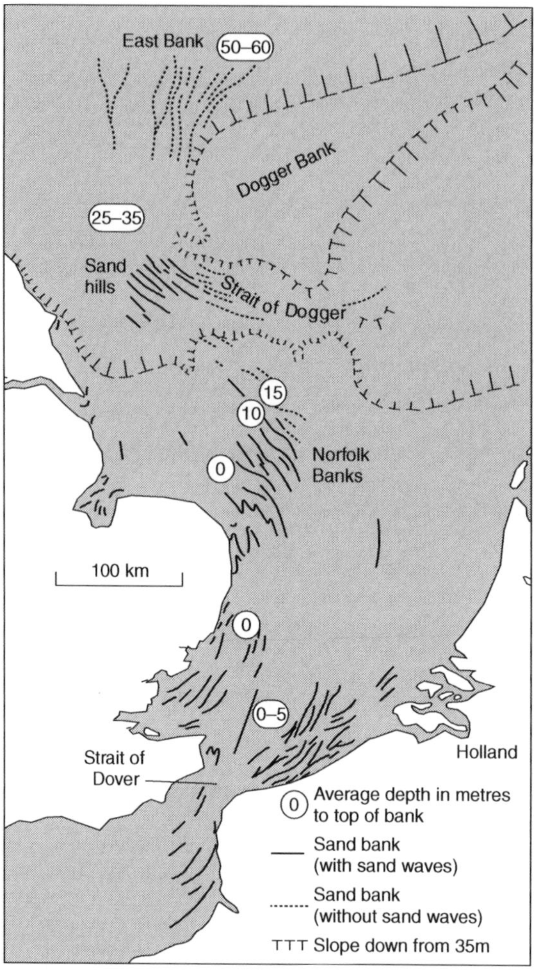


**Fig. 19.10** Areas of tidal sand ridges in the southern and central North Sea. The East Bank, Sand Hills and outermost Norfolk Banks are largely relict features formed during Holocene sea-level rise. (After Houboult, 1968; Kenyon *et al.*, 1981.)

coincide with the surface advection of dilute suspensions up to a concentration of 5 mg/L in the Bohai current. The Huksan Mud Belt (HMB) that fringes the inner shelf of the southeast Yellow Sea has a curious distribution, paralleling the southwestern coast of Korea in water depths ranging from 20 to 100 m and is 20–50 km wide, over 200 km long, and up to 60 m thick (Fig. 19.15). The HMB consists of two mud units that have similar grain texture and clay mineralogy—these can be differentiated on seismic profiles because the lower unit is generally stiffer than the watery upper unit. As elsewhere in the Yellow Sea the area is controlled by strong macrotidal currents, severe winter storms and sporadic summer typhoons. Hydrographic measurements, satellite imagery and clay-mineral provenance data indicate that deposition in the HMB is due to a suspended plume enhanced by winter monsoon winds that originates in the late autumn and winter months from the exit of the Keum River off the west coast of Korea. This winter-season plume carries concentrations of suspended matter one to two orders of magnitude higher than in the summer The suspension-rich plume travels southwest toward the main body of the HMB in a well-constrained corridor. The deposits of the HMB comprise alternating clay-rich and silt-rich laminae, the latter settling

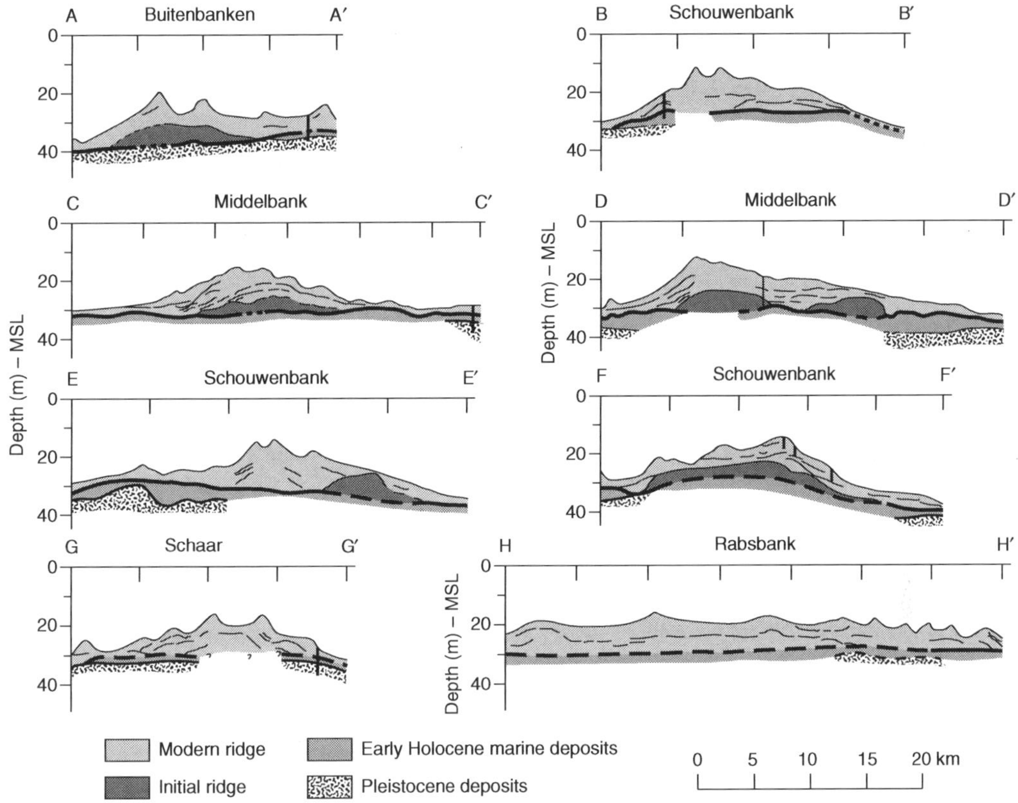


**Fig. 19.11** Seismic profiles and cores reveal the internal structure of these tidal current sand ridges from the Zeeland (Flemish) banks. MSL, mean sea level. (From Laban & Schüttenheim, 1981.)

out from resuspension events after winter wind-generated residual currents. The curious, coast-parallel shape of the HMB is due to the suspended plume being constrained by a pronounced tidal front to the west and, particularly in winter, by a strong thermohaline front to the south, against a branch of the warm saline Kuroshio current. The tidal front marks the eastern boundary to a spectacular train of sandy linear tidal ridges, some of which are remnant late-Pleistocene or early Holocene features. The juxtaposition of these features with the HMB is a startling contrast in grain size and physical shelf processes, a feature that should cause students of ancient tidal shelf deposit architecture much thought. Other areas of fine-grained sediments are relict, recording the former position of the Huanghe south of the Shandong peninsula or are the advected plumes of active riverine input from the Changjiang translated eastwards and southwards over the shelf by strong shelf tidal currents. Very large areas of shelf sands in the eastern Gulf of Bohai, northwest Korea and at the shelf edge fringing the southern Yellow Sea are thought to be relict tidal sand ridges that formed as lowstand deposits and during the post-glacial transgression.

## 19.6 Weather-dominated highstand shelves

Overall, weather-dominated shelves tend to show a general offshore decrease in grain size and Holocene sediment thickness in response to attenuating wave power. This trend is well shown by the East Atlantic USA, Bering, Oregon and southwestern Gulf of Mexico shelves. Mud-grade sediments settling out close to the shelf-edge break are often intermixed with partly reworked transgressive relict sands on the outer

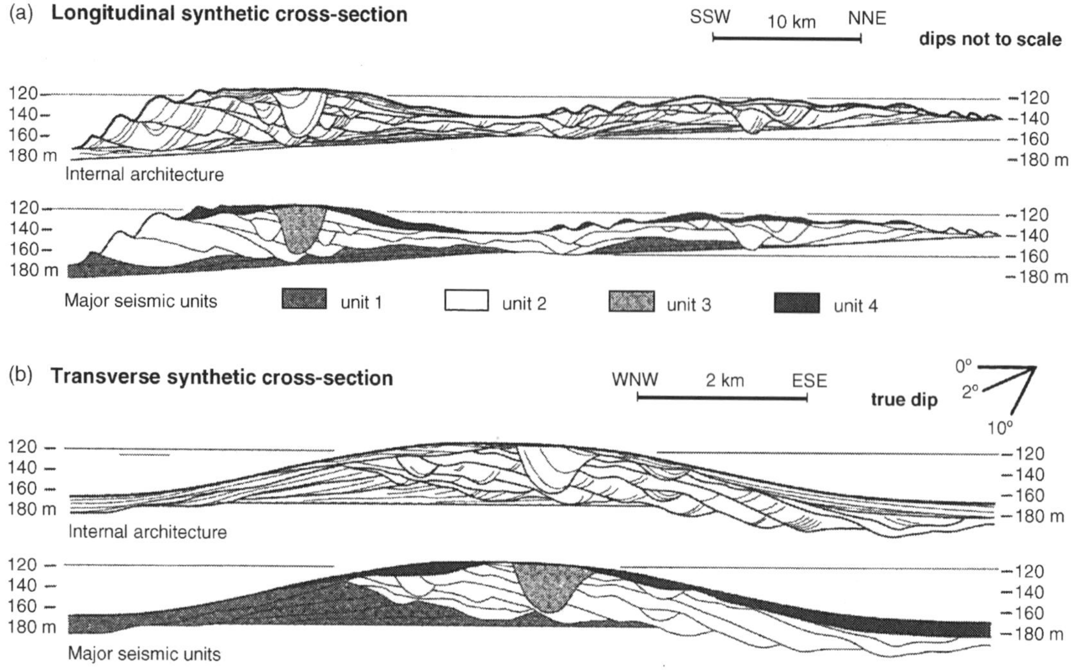


**Fig. 19.12** Synthetic longitudinal(a) and transverse (b) sections showing architecture of the Kaiser Bank, Celtic Sea. Surface slopes are exaggerated (×5–10 for all drawings). Note that in (a), all units are displayed although all may not appear together on a single cross-section along to the bank crest, i.e. they are drawn as a composite section. (After Reynaud *et al.*, 1999.)

shelf. Wave-formed ripples can occur at depths up to 200 m on the Oregon shelf during storms, but fair-weather reworking by burrowing organisms of the rippled sand laminae may obscure these. The predominance of storm waves on exposed offshore shelves means that intermediate- to shallow-water waves possess the necessary wave energy to completely rework many deposits like tsunamites. In deeper, quieter shelf basins there is an important energy teleconnection that comes from distant storm set-up. For example, Hurricane Carla of 1961 has left its legacy on the eastern Gulf of Mexico shelf as a prominent, laterally extensive, graded storm bed up to 0.25 m thick. There is also evidence that shore-normal gradient currents may be significantly turned parallel to bathymetric contours by regional wind drift currents.

Along narrow shelves fronted by steep, mountainous catchments like those of Pacific northwest North America spectacular events known as *oceanic floods* periodically overwhelm the 'normal' weather-dominant shelf environment. These are best described offshore the Eel River, northern California Coast Range, in the mid-shelf. It is important to distinguish these brief oceanic floods from seasonal floods of larger rivers. The steep river catchments have little storage capacity during extreme rainfall events, so that sediment fluxes are briefly huge, leading to rapid flocculation in shelf overflows and underflows. The linked oceanic response to storms means that southerly winds 'pin' the sediment plumes against the coast in $<$ 40 m deep water and are thought to keep the suspensions in a thin wave boundary layer as fluid mud suspension. The modern silt and mud-dominated flood deposits typical of Eel River events show a widespread spatially uniform grain size due to rapid dispersal, quite possibly as pulsating underflows; they have been classified as '*wave-induced turbidity currents*'.

Late Holocene, shelf-transverse lobate mud belts occur on the storm-dominated coast of western France offshore the Gironde estuary (Fig 19.16). The fine sediment-size and shore-normal orientation is unusual for a nearshore location (water depths $<$ 60 m) like this. Such characteristics depend upon several

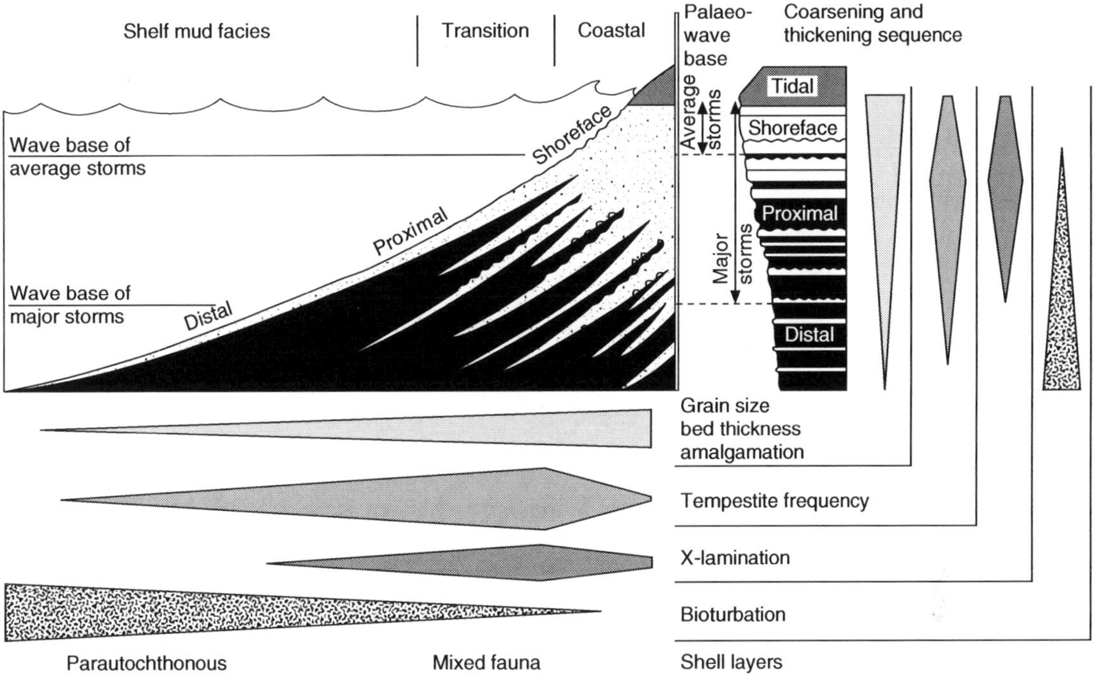


**Fig. 19.13** Summary of the coastal to offshore trends observed in coring transects off the storm-dominated Heligoland Bight, southern North Sea and their use in constructing an idealized vertical sediment succession during coastal-shelf progradation. (After Aigner & Reineck, 1982.)

conditions: (i) a nearby (< 30 km) fluvial source; (ii) two estuarine channels from which fluvial-borne suspended particulate matter is supplied; (iii) a gentle shelf surface with relict depressions (possibly palaeo-valleys); and (iv) a winter hydrological front on the shelf, which prevents the seaward escape of the greatest bulk of estuarine suspended matter. As a result, between 20% and 40% of the annual Gironde

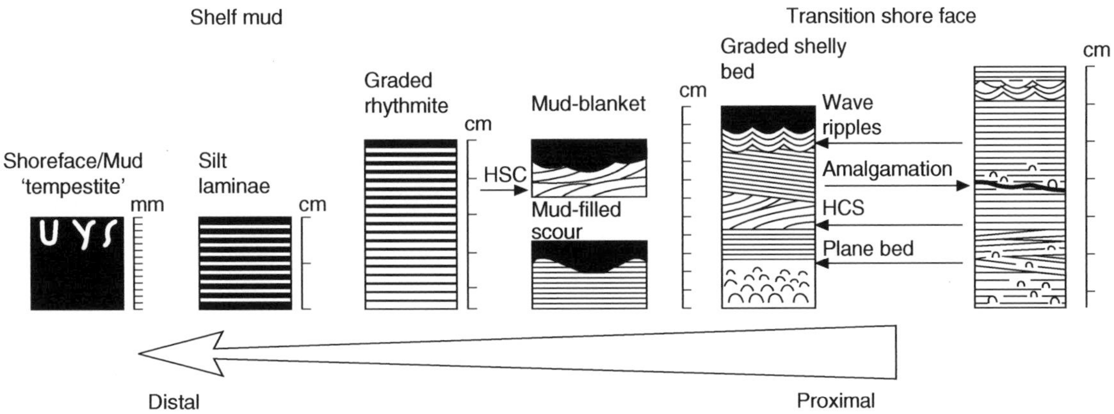


**Fig. 19.14** Simple representation of the sedimentary structures in distal and proximal shelf storm deposits. (After Aigner & Reineck, 1982).

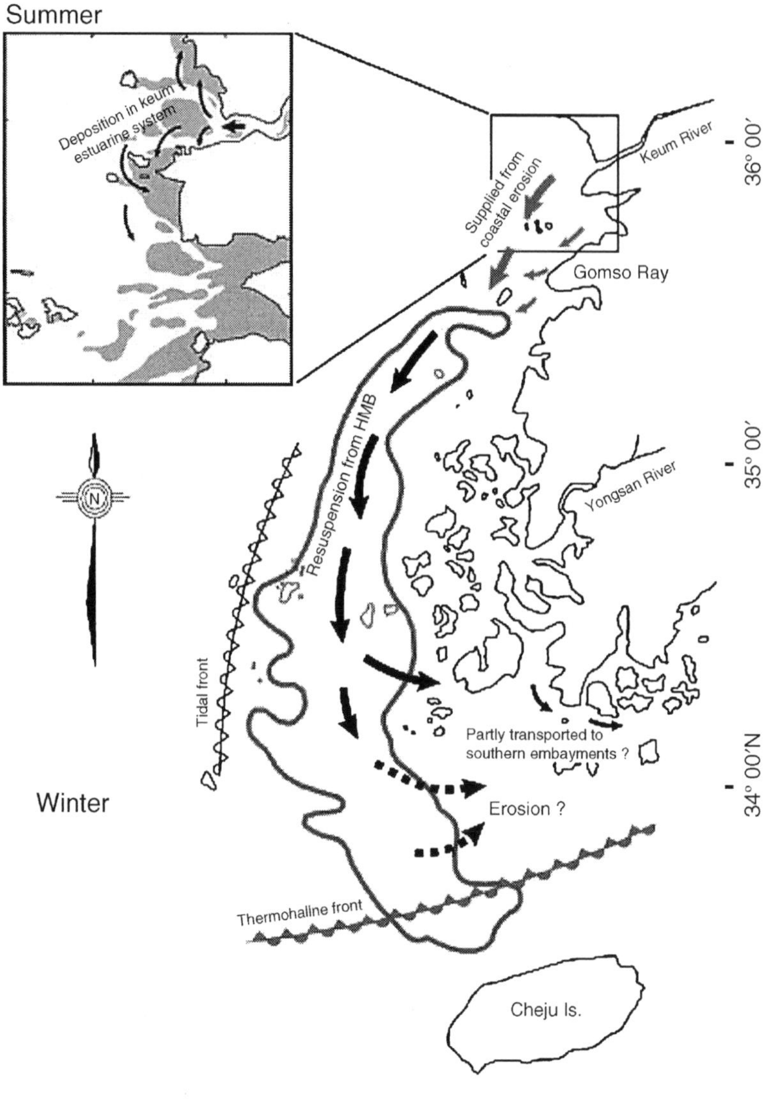


**Fig. 19.15** Diagram illustrating dispersal of suspended matter derived from the Keum River on to the Huksan Mud Belt. Note that the suspended plume is controlled by the presence of tidal and thermohaline fronts and follows the length of the HMB and that some of the suspended material may be transported to the southern coast of Korea. (After Lee & Chu, 2001.)

sediment output is trapped within these mud fields, with sedimentation rates of 1–3 mm/yr.

The Middle Atlantic Bight, the wide (75–180 km) shelf off the eastern USA (Fig. 19.17) between latitudes 35 and 41 °N, stands as a particularly well-investigated weather-dominated shelf of complex morphology. The shelf slopes gently eastwards from the lower shoreface (at a somewhat arbitrary 20 m depth) with the shelf break at between 50 and 150 m water depth. Tides are generally insignificant (apart from in some estuaries and the Nantucket Shoals area around 41 °N) and sediment transport is dominated by the passage of storms. The highest-wave months are September and January–March, with the former bearing the brunt of subtropical-sourced long-period waves and the latter the shorter-period northeasterlies.

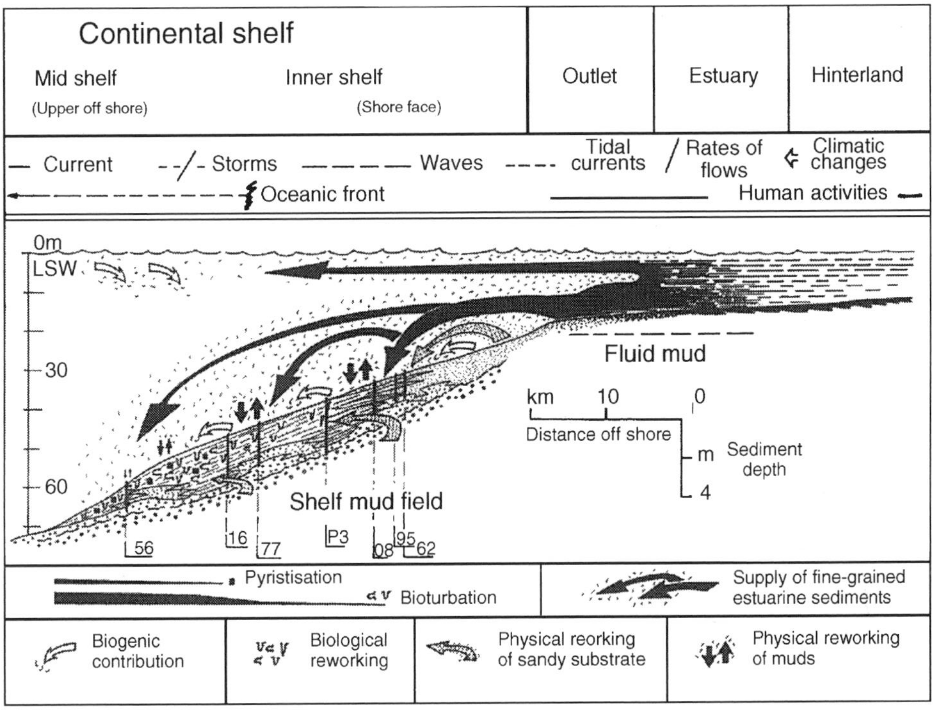


**Fig. 19.16** A synthesis of the different depositional processes responsible for fine-grained sediment deposition on the Aquitaine shelf offshore from the Gironde estuary, France, with locations of some reference cores. (After Lesueur *et al.*, 2002.)

On the inner shelf and lower shoreface there is an upper water regime of asymmetric oscillatory flow landwards, with an important time-mean lower water flow (of geostrophic origins) seawards, the latter enhanced for sediment transport as a combined flow by storm wave orbital motions. All-in-all the surface sediment of the modern Bight is remarkably coarse, with muds and silts being restricted largely to estuaries and back-barrier lagoons. This reflects the paucity of modern sediment flux on to the shelf (compare the Amazon shelf discussed above) and the efficiency of the storm wave transport process acting on sands made available from the pre-Holocene lowstand. The surface of the modern shelf shows abundant evidence that precursor, lowstand and rising sea-level relief features influence and control modern sedimentary processes. This control has been increasingly obvious in recent years with the advent of high-resolution shallow geophysical surveys and extensive coring programmes. Thus traces of incised pre-Holocene river and estuarine channels date from the last glacial lowstand; the landward extension of the smaller of these has already been shown to influence the position of barrier island inlets. Larger channels, such as those of the palaeo-Chesapeake, Hudson and Delaware, may often be traced out as prominent bathymetric valleys all the way to the shelf edge. The pattern of sand distribution on the shelf tracks in a remarkable way the successive positions of the cape and headland high-energy beaches and barriers off New Jersey, Maryland/Delaware and North Carolina that separate the major coastal estuaries. It is envisaged that the predominantly onshore to offshore transport of sand seen in shoreface environments has occurred continuously during transgression, causing the observed modern sand sheet down to depths of 50 m or more to remain under the influence of storm wave processes.

In detail, much of the eastern USA shelf sheet sand (10 m or so mean thickness) is dominated by fields of linear, northeast-trending, ridges up to 10 m high with slopes of a few degrees (usually $< 3°$ nearshore).

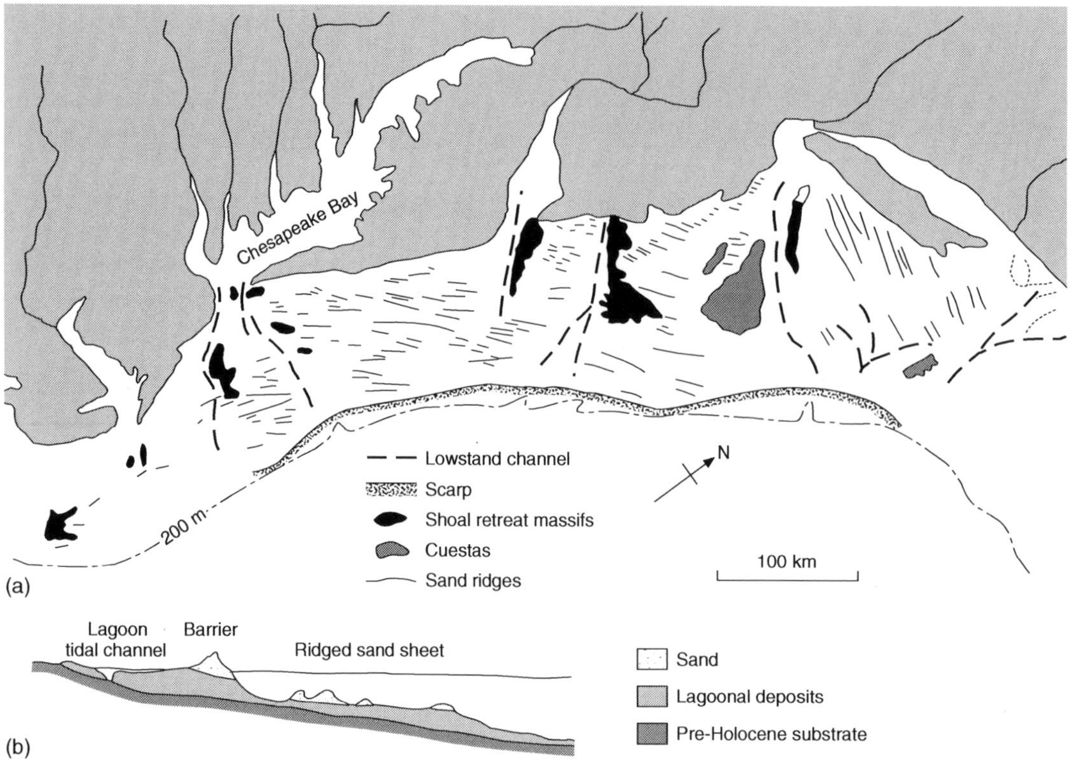


**Fig. 19.17** Middle Atlantic Bight of the eastern USA.(a) Morphology with features of relict and transgressive origins. (b) Section across shelf to show transgressive barrier and progressively abandoned offshore shelf sand ridges. (After Swift *et al.*, 1973; Swift, 1974.)

Shoals are separated by linear hollows bearing gravel over the Pleistocene lowstand surface. Clusters of shoals merge with the modern shoreface in water only 3 m deep. The ridges make small angles ($< 35^{\circ}$) with the modern coastline. Seismic profiles through the more offshore ridges reveal low-angle surfaces that dip to the southeast, the direction of ridge asymmetry. The active shoreface shoals are forming at the present time in response to storm-generated currents running approximately parallel to the shoal crests. Shoal detachment from the shoreface is thought to have occurred periodically during the Flandrian transgression, the detached shoals continuing to evolve at the present day in response to the storm wave surge and water drift currents. This idea is supported by the onshore to inner-shelf transition of ridge morphology observed off the Maryland coast. Here, nearshore ridges are 'attached' to the shoreface as shown by their contours. They are steeper, of lower cross-sectional area and less asymmetric seawards than those detached ridges that lie progressively offshore. Regarding grain size, it is found that the coarsest grain sizes occur in the landward flanks, not on the ridge crests. Long-term studies indicate that the ridges migrate in response to sediment transport aligned south and slightly offshore relative to the ridge crests. It is likely that the ridges owe their origins to the shelf-parallel time-mean geostrophic flows discussed previously. The reasons why such ridges should begin to form as attached spurs on the shoreface is unknown. Also, little is known about the internal structure of the ridges, but it is likely that inclined internal surfaces seen on geophysical records represent storm erosion planes. Successive internal planes may be separated by fine sediment showing small-scale cross-laminations in the wave troughs produced by wave oscillations in normal weather conditions. The reader should note that the Middle Atlantic Bight linear ridges produced by time-mean wave–current flows are quite similar in broad morphology to the tide-dominated ridges noted previously (Section 19.4).

The northern inner shelf (depths 10–30 m) of the Gulf of Mexico offshore Louisiana and Texas is a fascinating locality to study the interactions between major river sediment input, coastal wind-driven current steering, density stratification, tropical storm-induced reworking and the phenomena of *hypoxia*. The shelf regime is characterized by the east to west Louisiana Coastal Current that entrains Mississippi–Atchafalaya homopycnal freshwater plumes with their suspended mineral and organic loads (Chapter 17). The shelf is also affected by both winter storms and major tropical cyclones, the latter of greater significance for longer term sediment reworking and off-shelf transport into shelf-margin canyons. *Hypoxia* is when bottom water dissolved oxygen is too low to support aquatic life and is widely developed in summer months over a vast area of some 20 000 km$^2$. The condition has developed over the past 30 years or so due to increased inputs of nutrients, chiefly nitrogen, from the continental agricultural heartlands of the Mississippi–Atchafalaya River drainage basin. This agro-nutrient spawns massive phytoplankton primary production in shallow waters; in the absence of vigorous mixing of the density-stratified shelf waters by wave or tide, oxidation of this biomass during organic sedimentation uses up all available dissolved oxygen. Hypoxia affects many other shelves influenced by high river discharges, including the eastern North Sea, northwest Black Sea and East China Sea.

## Further reading

### *General*

Despite the explosion of research on shelf flows and sedimentology there are few up-to-date introductory accounts that explain processes *and* sedimentary products, a notable exception being the collection of papers in Nittrouer *et al.* (2007). Swift *et al.* (1991) and de Batist & Jacobs (1996) contain many papers of interest, the latter including a long series of reviews. Nittrouer & Wright (1994) is a more precise and advanced review. For a single case-study of great interest the reader is advised to browse through the collection of papers on the Amazon shelf in Nittrouer & Kuehl (1995). Sediment transport methodology and mechanisms feature in the volume edited by Balson & Collins (2007).

### *Specific*

Many relevant papers are in Stride *et al.* (1982), Suter *et al.* (1987) and Winn *et al.* (1995). Swift *et al.* (2003) is a stimulating account of sequence stratigraphic development and physical shelf processes. Perhaps the clearest account of Milankovitch-forced shelf evolution, erosion and accretion comes from the Cádiz shelf, southern Spain, courtesy of careful studies by Hernández-Molina *et al.* (2000).

Lowstand valleys, shelf bypass to highstand sink conditions are set out in Swift (1968), Nummedal & Swift (1987), Allen (1991), Ashley & Sheridan (1994), Roy (1994), Reynaud (1999), Masselink & Hughes (2003), Greene *et al.* (2007), McHugh *et al.* (2004) and Nordfjord *et al.* (2006). Climatic and upstream discharge controls on shelf sedimentation and incision are discussed in Wellner & Bartek (2003), Mattheus *et al.* (2007), Darmadi *et al.* (2007) and Massari *et al.* (2007); variations in Holocene transgression rates due to antecedent shelf topography and sediment supply by Rodriguez *et al.* (2004).

Tide-dominated, low river input, highstand shelves: shelf tidal currents by Stride (1963), Kenyon & Stride (1970), Belderson *et al.* (1978), Pingree & Griffiths (1979), Johnson *et al.* (1981), Harris & Collins (1991) and Pantin *et al.* (1987); shelf tidal bedforms in Reineck (1963), Houboult (1968), McCave (1971), Johnson *et al.* (1981), Laban & Schüttenhelm (1981), Stride *et al.* (1982), Kenyon *et al.* (1981), Belderson *et al.* (1982), Reynaud *et al.* (1999), Lobo *et al.* (2000), Bastos *et al.* (2003), Barnard *et al.* (2006) and Park *et al.* 2006; TELEMAC routines by Hervouet, J.-M. 2000, and application to modelling of combined tide–wave currents on the Scroby sands, southern North Sea by Park & Vincent (2007).

Tide-dominated, high river input, highstand shelves: shelf plume deposits by Baker & Hickey (1986) and Nittrouer & Kuehl (1995); see Lee & Chough (1989) and Lee & Chu (2001) for the Huksan Mud Belt, Korea.

Weather-dominated highstand shelves are discussed by Sharma *et al.* (1972), Kulm *et al.* (1975) and Shideler (1978); shelf storm deposits by Reineck

(1963), Gadow & Reineck (1969), Aigner & Reineck (1982), Gienapp (1973), Siringan & Anderson (1994) and Morton (1981); storms vs. tsunamites by Weiss & Bahlburg (2006) and Pratt & Bordonaro (2007).

Oceanic flood deposits like Eel River, California are discussed by Leithold & Hope (1999), Wheatcroft & Borgeld (2000), Wheatcroft (2000), Wheatcroft & Drake (2003, Crockett & Nittrouer (2004), Wheatcroft & Sommerfield (2005) and Nittrouer *et al.* (2007). Tropical storm effects on the Mississippi Shelf are documented by Dail *et al.* (2007) and the Middle Atlantic Bight by Swift *et al.* (1973, 1981), Field (1980), Ashley *et al.* (1991, 1993), Wellner *et al.* (1993) and (Levin, 1995). The Gironde/Aquitaine shelf mud belts are beautifully illustrated by Lesueur *et al.* (2002). Dunbar & Barrett (2005) explore hydrodynamic criteria for outward-fining, weather-dominated shelf sedimentation.

Shelf hypoxia with special reference to the Gulf of Mexico is nicely summarized in Boesch *et al.* (2009), with hydrological and chemical data in Goolsby (2000).

Ancient clastic shelf sediments: Anderton (1976) is a classic and pioneering account of an ancient (Dalradian, Scotland) tidally swept shelf environment. Fine accounts of ancient shelf deposits (detailed work on magnificent exposures and hydrocarbon well log correlation) are those of the Cretaceous western North American seaway, a foreland basin stretching from Colorado to Alberta and beyond (e.g. Plint *et al.*, 1986; Plint, 1988; Pozzobon & Walker, 1990; Pattinson & Walker, 1992, Varban & Plint, 2008). Bridges (1982) made the first reconstruction of an ancient rotary tidal system based upon these sediments, deducing the presence of several amphidromic cells acting in the narrow seaway, results extended by Slingerland *et al.* (1996). Spectacular shelf-margin clinoforms occur in the central Tertiary foreland basin of Spitsbergen (e.g. Deibert *et al.* 2003). Another excellent physical reconstruction of an ancient resonating tidal gulf (discussed above) was undertaken by Sztano & de Boer (1995) for the Miocene of the North Hungarian Bay. More recently, great success has been achieved in modelling the tidal regimes of Upper Palaeozoic epicontinental (aka epeiric) seas (Wells *et al.*, 2005, 2007). Boyles & Scott (1982) discussed migrating sandy shelf bars in the Mancos Shale of northwest Colorado, which they compared with those of the wave-dominated Holocene Middle Atlantic Bight. A more recent account of equivalent facies in Wyoming (Mellere & Steel, 1995) concentrates on sea-level influences on detailed sand body geometries. Leckie & Krystink (1989), Duke (1990) and Midtgaard (1996) contribute to the debate on Coriolis-induced turning of gradient storm currents. Brenchley *et al.* (1993) describe coarsening-upwards and thickening-upwards cycles of Ordovician age, which they attribute to cycles of outer- and inner-shelf adjustment to changing sea levels. Shelf tempestite and condensed sequences are explored by Drummond & Sheets (2001). Sinclair (1993) discusses the Tertiary Gres d'Annot shelf storm sequences as part of regressive/transgressive cycles involving a barrier shoreline with inlets and also extruding hypopycnal plumes. A key role for hyperpycnal plumes in dispersing mud out of steeper-angled ($>0.7°$) shelf margins is proposed by Bhattacharya & MacEachern (2009) and for formation of isolated turbidite sandbodies offshore of pro-delta shelf slopes by Pattison (2005) and Pattison *et al.* (2007). Ancient equivalents to the Eel River, California 'oceanic floods' and their distinction from shelf storms sensu stricto are discussed by Campbell *et al.* (2006). Wave-influenced turbidites are deduced from the Carboniferous of Colorado, USA by Lamb *et al.* (2008). Evolution of a sand-dominated gulf-like shelf fed from adjacent strandplains and distributed by storms and strong tidal currents is set out for the Miocene of the North Sea by Galloway (2002). Imaginative and informative use of well logs, natural sections and petrophysical characters in Jurassic shoreface to shelf basin sequence stratigraphy in northern France is by Braaksma *et al.* (2006) and in southern England by Morris *et al.* (2006). A key role for climate in governing rates of sediment supply during sea-level cycles is outlined by Massari *et al.* (2007). Also in the Pliocene Crotone basin a role for tectonic-enhancement of tidal currents along seabed fault scarps is proposed by Zecchin (2005). Quin (2008) analyses the role of high subsidence rate, high sediment supply rate and narrow shelf platform physiography in focusing late Devonian–early Carboniferous tidal and wave reworking of delta-fed shelf sands.

Chapter 20

# CALCIUM-CARBONATE–EVAPORITE SHORELINES, SHELVES AND BASINS

*Full fathom five thy father lies.*
*Of his bones are coral made;*
*Those are pearls that were his eyes;*
*Nothing of him that doth fade*
*But doth suffer a sea-change*
*Into something rich and strange.*

Ariel's second song, from Act 1.2 *The Tempest*, William Shakespeare, 1610

## 20.1 Introduction: calcium carbonate 'nurseries' and their consequences

Despite being subject to the same physical aspects of coastal and shelf processes as siliciclastic sediments—wave, tide and oceanographic currents—carbonate sediments are quite distinctive because of:

- local or *in situ* biogenic or chemical origins *in* the depositional basin;
- local *in situ* sediment production rates are controlled by ecological factors but deposition rates depend on wider physical controls on sediment redistribution;
- hydrodynamic properties of queer-shaped and low-density grains;
- rapid accretion of wave-resistant shallow subtidal to deep shelf reefal 'build-ups';
- organic evolution over geological timescales has controlled many environments (notably reefs and 'build-ups') and grain types;
- a tendency to become lithified after deposition—this is particularly important during sea-level lowstands when resistant aeolian-modified carbonate sediment and karst may persist as islands into subsequent highstands;
- the production of steep lithified/accreted carbonate platform margin slopes.

Nothing in sedimentology is simple, however, for it is common in the stratigraphic record to find carbonate

*Sedimentology and Sedimentary Basins: From Turbulence to Tectonics*, 2nd edition. © Mike Leeder.
Published 2011 by Blackwell Publishing Ltd.

and clastic sediments superimposed vertically or juxtaposed horizontally.

The greatest source for carbonate grains is the warm, shallow, photic zone of the subtidal environment, the *carbonate 'nursery'*, from whence storms and mass flows transfer detritus at declining rates onshore and offshore. *In situ* sediment production rates commonly vary from 0.5 to 5.0 $kg/m^2/yr$. The highest organic productivity of the 'nursery' on recent subtropical and tropical shallow shelves occurs in coral reefs at platform margins, leading to the production of rimmed shelves with abrupt, steep seaward margins and landward lagoons (Fig. 20.1; Plates 15 & 16). Major rimmed offshore banks completely isolated from terrigenous clastic input occur as fragmented continental crustal 'microcontinents' bordered by abyssal plains and deep channels, epitomized by the Bahamas Banks, one of the largest and most studied modern carbonate platforms. Here the distribution of environments and sediment production rates reflects both distribution of *in situ* producers around the bank rim, and of chemical precipitation and sediment redistribution by storm wave and tide around the nucleus of Andros Island, a relatively young feature formed by lithification of former MIS 5 carbonate deposits during the last lowstand. Steep reefal platform margins owe their origins to basement structural controls as evidenced from deep seismic profiles (Fig. 20.1; Plate 16; see also Fig. 20.32) that reveal formation by coalescence of smaller nuclei, probably originally fault-controlled.

Carbonate *ramp* shorelines and shelves (Fig. 20.2) are, as their name implies, gently sloping (less than a degree or so) depositional surfaces that decline away from a shoreline into deeper water without a pronounced ($> 5°$) platform margin break of slope. They have been much investigated in ancient carbonate sediments thought to have developed in regionally subsiding shelf foreland and thermally subsiding basins without fault control—here of course their form can only be inferred. It is not possible to maintain a linear ramp profile if the rate of *in situ* carbonate production and/or seaward redistribution by storms decreases from shoreline to offshore. The effects of nearshore sediment production maxima inevitably lead to the formation of gradient change offshore (Fig. 20.3).

In reality, non-Bahamian platform-type shelves are often seen to be combinations of rimmed shelves and ramps, and the distinction may be scholastic. Take the case of the Queensland shelf, for example. Here the southern and northern shelfward margins are marked by disconnected linear fragments of wall reefs that mark the abrupt shelf edge. Yet all across this sometimes very wide, gently sloping shelf there is a high degree of tidal mixing, such that to all intents and purposes the wide reef-rimmed shelf may be regarded as a carbonate ramp seaway with no discernible chemical or physical barriers across it. The Torres Strait peripheral foreland basin of northeast Australia/Papua New Guinea, at the northernmost extremity of the Great Barrier Reef, is another fine example of such a reef-strewn seaway, with strong tidal currents between the reefs leading to spectacular dunefields. An older example of a ramp–seaway geometry is deduced for the tidally swept temperate carbonate dunefields of the Te Kuiti Group, North Island of New Zealand.

It seems evident from the stratigraphic record that ramps have sometimes been commoner on pre-Cenozoic carbonate shelves. To avoid production of a rimmed shelf margin, shallow-water reefs or buildups must be limited somehow—possible mechanisms are discussed in a later section. In contrast to the northern and southern Great Barrier Reef, the central part of the Queensland shelf has no rimmed margin, it slopes gently offshore as a ramp. Ramp conditions also pertain today in the Arabian Gulf and the West Florida shelf, the latter passing from shoreface to deep water with no pronounced shelf-slope break, although the oceanward ramp slope increases somewhat. It may sometimes be useful to distinguish inner, mid- and outer ramps according to the decreasing degree of wave–current reworking. However, the variation of this and the neglected importance of tidal currents, storm surges, saline underflows, shelf tides and oceanographic currents (section 19.3) within the general hydrodynamic setting of ancient carbonate basins make any generalities as to the absolute water depths appropriate to these zones impossible.

Although carbonate production reaches greatest levels in low latitudes, there is important and distinctive production in the cooler waters of mid-latitude shelves. All shelf sediments contain some proportion of organic-sourced carbonates, but in some areas where siliciclastic supply is limited these may become dominant. Such environments are exemplified by the Cenozoic to recent deposits of the New Zealand shelf

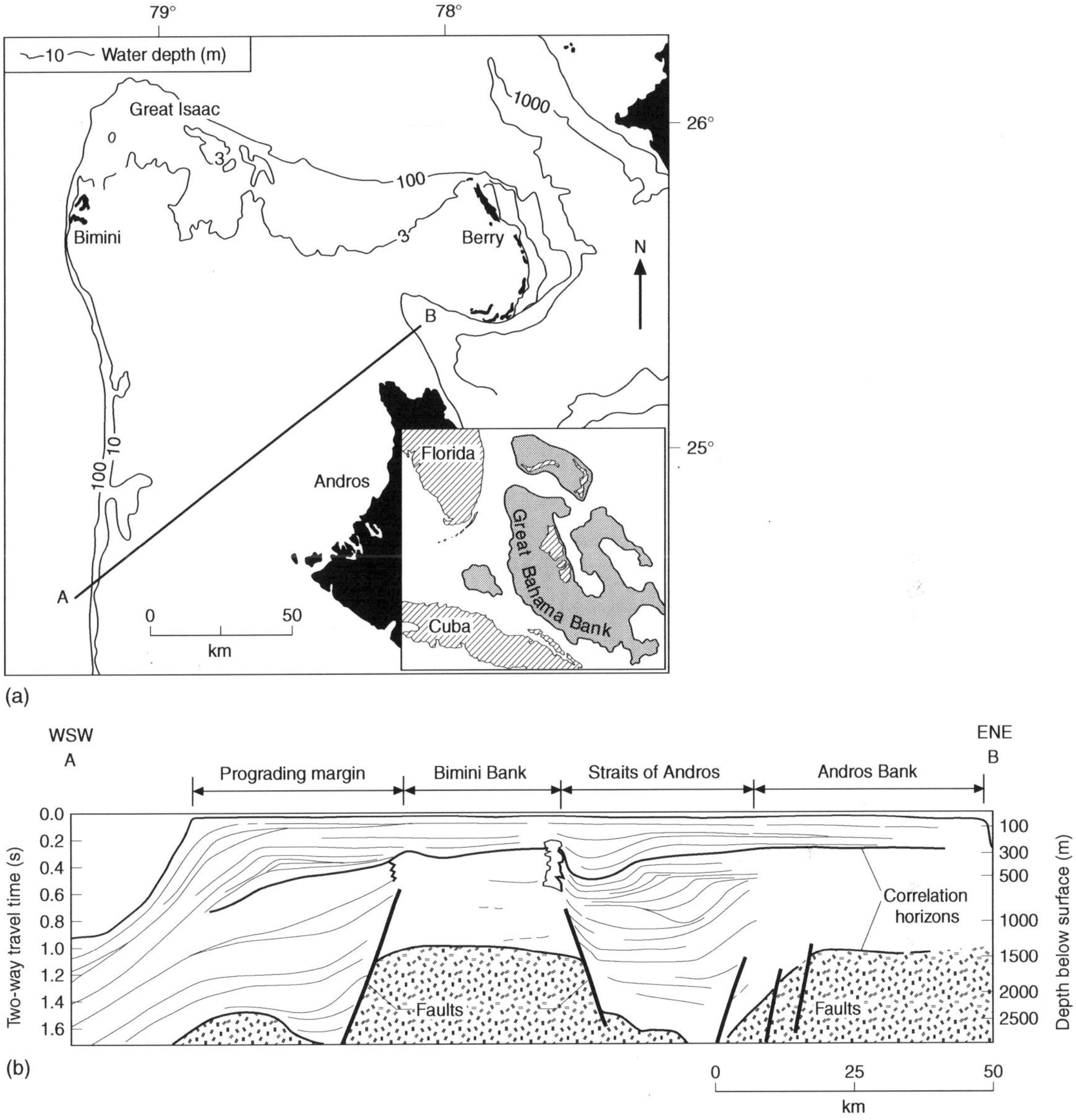


**Fig. 20.1** Natural carbonate laboratory of the Bahamas Banks. (a) Location map, with some detailed study areas mentioned in the text. (b) Section across the Great Bahamas banks to illustrate (i) modern rimmed platform of wide main bank—the Bahamas 'carbonate factory' bounded by a steep western rimmed margin); (ii) growth of western margin through time from ramp to rimmed bank and (iii) the fused, composite nature of subsurface bank architecture, for example the infill of the Straits of Andros by a prograding ramp to steep (25–30°) rimmed shelf through time. (After Eberli & Ginsberg, 1987.)

and particularly the southern Australian shelf, the world's largest modern cool-water carbonate province. At least this is the traditional view, but deep shelf basin and continental rise cold-water coral mound and reef ecosystems have been found to be extensive in the past decade, ongoing exploration efforts indicate they cover an area of at least $2.84 \times 10^5\,\text{km}^2$.

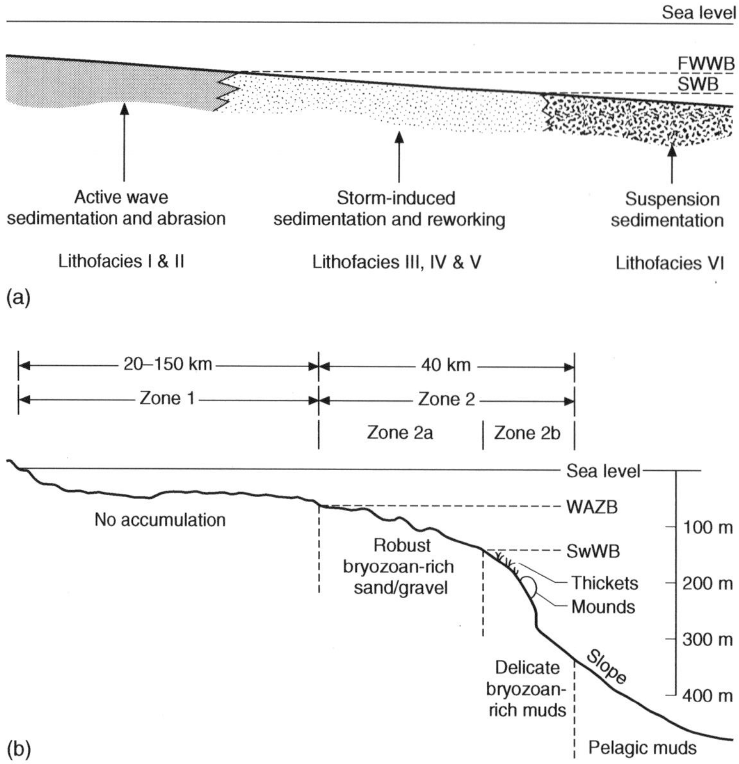


**Fig. 20.2** (a) Reconstructed ramp models for the ancient record, Deschanbault Formation, Ordovician, Quebec, Canada. (After Lavoie, 1995.) Note the rather idealistic linear slope and simple shelf energy regimes. (b) A rather more realistic modern ramp, the Lacapede Shelf, southern Australia. (After James *et al*., 1992.)

## 20.2 Arid carbonate tidal flats, lagoons and evaporite sabkhas

Arid tidal flats, lagoons and sabkhas occur commonly along the southern shores of the Arabian (Persian) Gulf (Fig. 20.4) where they define proximal ramp shorelines sloping seawards at ca 0.5°. The area is dominated by evaporation resulting from a combination of extreme aridity and high annual temperatures—rainfall in the Gulf is a sporadic mean 40–60 mm/yr with evaporation >1500 mm/yr. The major effect of such aridity on the supratidal and high intertidal sediments is greatly increased sediment porewater salinity, which leads to evaporite precipitation and dolomitization. Intertidal sedimentation is dominated by the growth of stromatolitic algal mats, which show well-defined lateral zonation of growth forms due to variations in exposure. Around the At Taf (aka Trucial Coast) the intertidal algal mat zone, up to 2 km wide (Fig. 20.5; section 2.15, Plate 17), is broken up by an irregular network of channels and covered by discontinuous shallow ponds. Along exposed coasts storm beaches feature commonly whilst offshore along unfringed ramps large areas of carbonate-cemented hardgrounds occur in areas of reduced sedimentation, often with sheet-like coatings of live coral. Winter storm winds (Shamal) drive subtidal sediments and water on to the intertidal flats and provide a major proportion of the pelletal sediment bound by the algal mats. Buried algal mat sections reached through pits dug in the prograding sabkha reveal that few of the detailed surface mat forms survive. This low preservation potential is caused by a combination of gypsum precipitation within the buried mat, and compaction and bacterial destruction of organic-rich algal laminae.

The Arabian coastal sabkhas (Figs 20.6 & 20.7) slope gently seawards at about 0.4 m/km and may be

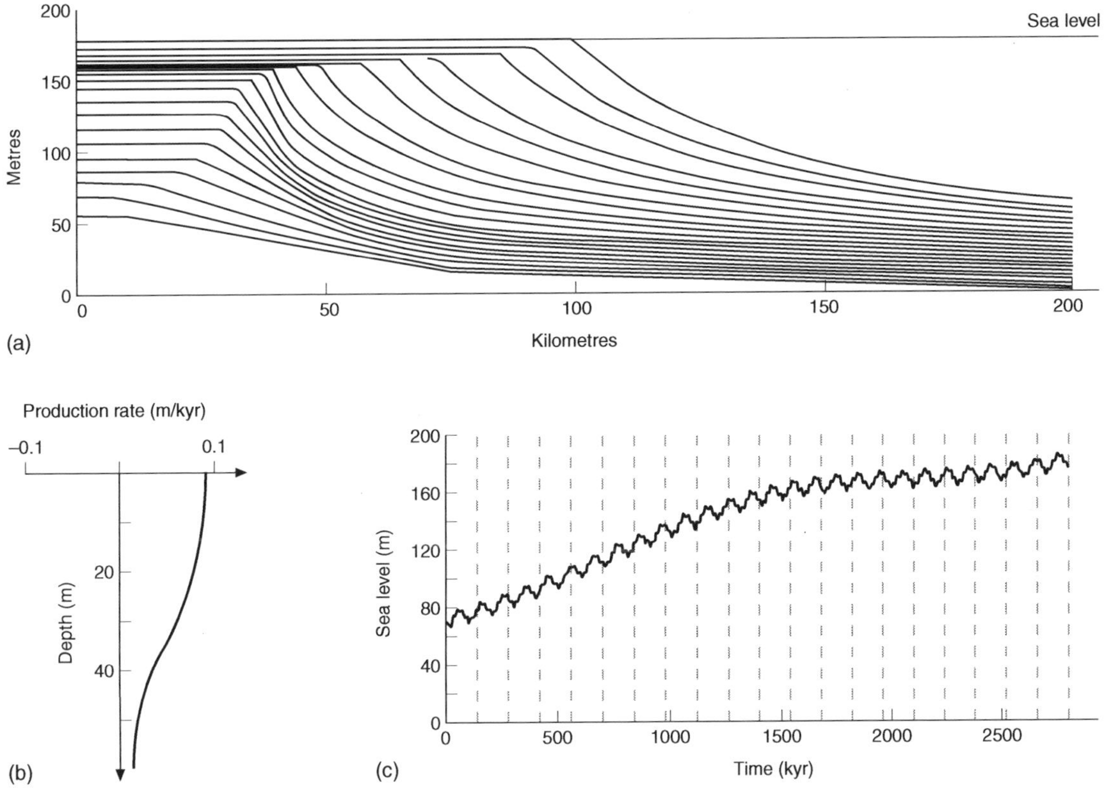


**Fig. 20.3** Modelling the inevitable non-linear development of offshore ramp topography —simplified output from the model of Aurell *et al.* (1995). Notice how closely this evolution concurs with seismic observations of the growth of Great Bahamas platform margin (see Figs 20.1 & 20.32). (a) Evolution of an equilibrium offshore slope (nonlinear) from an initial linear ramp caused by the form of (b) carbonate sedimentation rate and (c) a cyclical sea-level trend combined with linear subsidence and a 'redistribution' function that takes into account offshore sediment transport during storms. Lines in (a) represent 140 kyr time increments; note progradation and aggradation trends.

up to 16 km wide, the largest landward proportion being continental sabkha unaffected by marine flooding. Dolomite and a characteristic suite of evaporitic minerals occur in the shallow subsurface, the most distinctive feature being anhydrite and gypsum with nodular (chicken-mesh) and enterolithic (twisted, gut-like) textures indicative of growth of the evaporite in a carbonate matrix provided by seaward-prograding carbonate lagoons. The supratidal sabkha surface is flooded up to 2 km inland during Shamal wind storms and is covered by halite encrustations formed by both precipitation from shallow brine pools and subsequent efflorescence from evaporating porewaters. The Quaternary sediments below the sabkha surface (Fig. 20.7) reveal Pleistocene lowstand dolomitized sequences and evidence for a major transgressive event that modified a sandy coastal desert zone at about 7000 yr BP. The open coastal embayment so formed was changed into a lagoon/tidal-flat complex when a small (perhaps 1 m) relative sea-level fall caused the emergence of barrier islands and restricted circulation at about 3750 yr BP. Subsequent ramp-like lagoonal and tidal-flat progradation at rates of up to 0.75 m/yr caused the modern sabkha flats to develop. These are now subject to storm processes which periodically renew interstitial porewaters.

Similar fluctuating-salinity environments occur around the deeply indented margins of Shark Bay, Western Australia (Fig. 20.8), though they are less arid and anhydrite-bearing sabkhas do not form. Rainfall

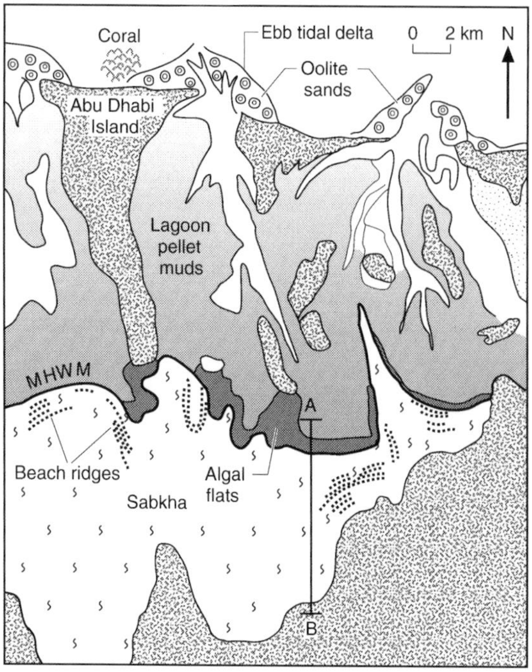


Fig. 20.4 Map to show the distribution of calcium carbonate and evaporite sediments in the Abu Dhabi area along the shoreline of the southern Arabian Gulf. (After Butler, 1970.) A–B is the indicative location of the cross-section in Fig. 20.6. See also the colour satellite image, Plate 17.

at Shark Bay is a variable 230 mm/yr with evaporation at 2200 mm/yr. Low-energy environments (e.g. Nilemah embayment) are dominated by continuous algal mats. Well-laminated sediments with narrow cavities (*fenestrae*) occur beneath areas of smooth mat in the lower intertidal zone, whilst poorly laminated sediments with irregular fenestrae occur beneath areas of pustular mat in the middle to upper intertidal zone. Dominant grain types in the tidal flats are pellets, altered skeletal grains and intraclasts, the last-named being derived as storm rip-up clasts from areas of partially lithified sediment below algal mats in the high intertidal zone. Higher energy environments in Shark Bay are typified by the northwestern margin of the Hutchison embayment (Fig. 20.8). Here lithified algal columns and ridges (section 2.15) form a stromatolitic reef that thickens seawards. The 'reef' is associated with a beach–ridge barrier comprising large-scale, cross-stratified molluscan gravels (*coquinas*). The Shark Bay tidal flats record a history of initial transgression (4000–5000 yr BP) and subsequent coastal progradation. As the supratidal surface expanded, porewater concentrations enabled aragonite and gypsum precipitation, the latter mineral being the major component in upper intertidal and supratidal zones.

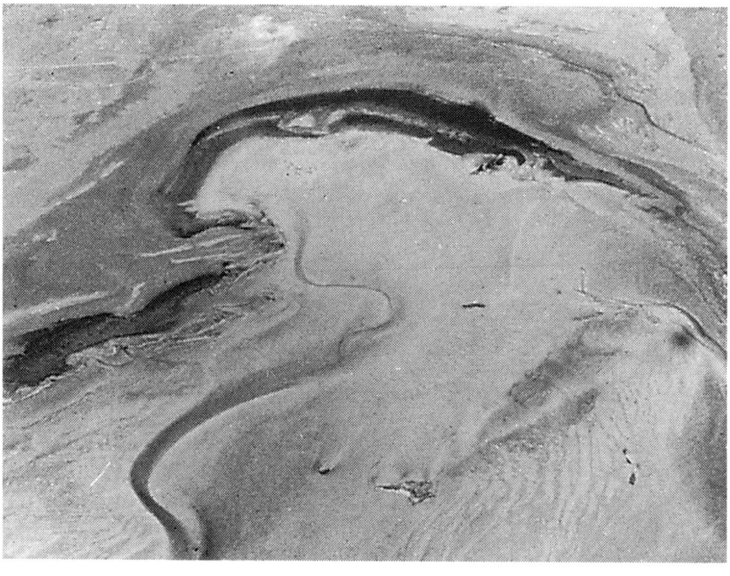

Fig. 20.5 Aerial view looking south along the eastern part of Fig. 20.4. Shows lagoonal/intertidal pellet sandflats, tidal channel, intertidal algal belt (black) and sabkha (S) (Photo courtesy of R. Till).

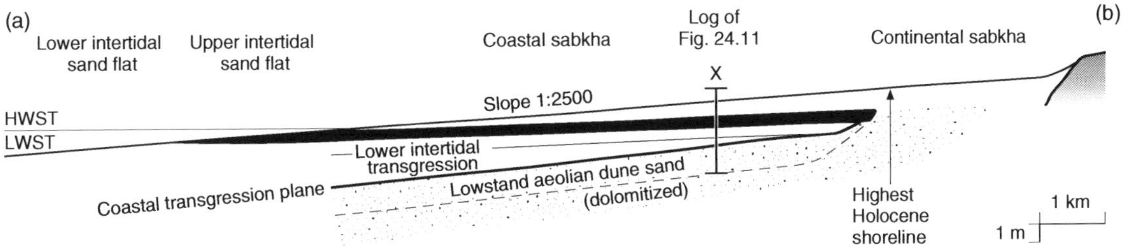


**Fig. 20.6** Generalized north–south cross-section (A–B of Fig. 20.4) of the Abu Dhabi sabkha to show surface slope and shallow subsurface Holocene progradational sediment wedge above the ~7 ka transgressional plane. LWST, low water spring tide; HWST, high water spring tide. (After Evans *et al.*, 1969.)

Offshore carbonate sedimentation occurs in the bay, the general absence of freshwater input and imperfect tidal flushing causes a general landward increase of salinity up to 70‰. Subtidal hypersaline areas that fringe the arid tidal flats are dominated by monotypic shell beds of the small salinity-tolerant bivalve *Fragum hamelini*. The metahaline and oceanic parts of the bay are dominated by spectacular *sea-grass* carbonate build-ups. These have topographic relief on the seafloor, lack an internal skeletal frame, and are composed of *in situ* and locally derived skeletal carbonate from the grass epibiota and sheltered benthos. Skeletal breakdown causes much silt- and mud-grade material to be admixed with the coarser debris of molluscan, algal and foraminiferal origins. The build-ups occur as fringing, patch and barrier types. There is a vertical trend as the sea-grass meadows accrete upwards towards mean tide level from matrix-rich skeletal packstones and wackstones to well-washed skeletal grainstones.

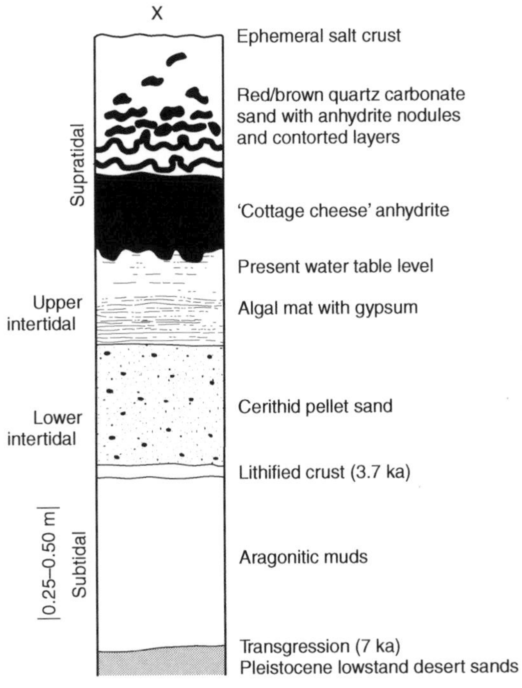


**Fig. 20.7** Sketch log taken at position X in Fig. 21.6 to show regressive highstand sabkha-lagoon sequence overlying lowstand aeolian sands. (After Till, 1978.)

## 20.3 Humid carbonate tidal flats and marshes

The west, leeward, side of Andros Island, Bahamas, fringes Great Bahama Bank lagoon and serves as the type example of non-saline tidal flats and supratidal marshes (Figs 20.1, 20.9 & Plate 16). Similar examples occur around other smaller carbonate platforms worldwide and along the Florida Coast. A tropical maritime climate with winter storms and occasional hurricanes prevails in the area with mean annual rainfall of about 130 cm/yr (range 65–230 cm/yr). This abundant rainwater freshens the supratidal marsh during summer months and prevents development of sabkha-type evaporites. Salinities of the tidal waters usually fall in the range 39–42‰ but may fall as low as 5‰ after heavy rainfall—creating a high-stress (schizohaline) environment and a restricted biota on the flats. The semidiurnal tides have a mean maximum range of 0.5 m, but this is much increased during periodic storm surges. Wave action is not usually important because of the sheltered nature of the

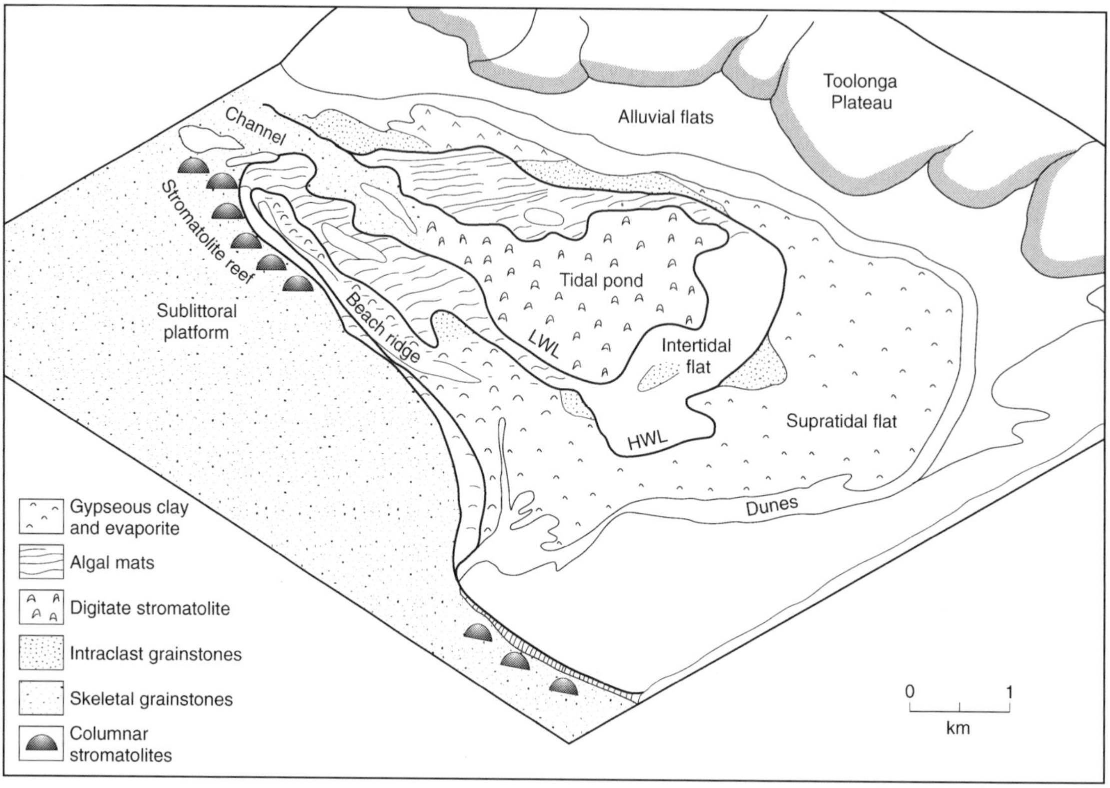


**Fig. 20.8** Sketch map of Hutchison embayment, Shark Bay, Western Australia, to show intertidal flat, lagoon and stromatolite sedimentary environments. LWL, low water level; HWL, high water level. (After Hagan & Logan, 1974.)

lagoon. The tidal-flat sediments are dominantly pelleted carbonate muds, with < 10% of skeletal material, dominantly foraminifera, and extensive algal mats. Three major subenvironments may be defined (Figs 20.10 & 20.11): (i) nearshore marine belt, (ii) tidal-flat complex of channels with levees and tidal ponds, and (iii) supratidal algal marsh.

The nearshore marine belt comprises thoroughly bioturbated, muddy pelletal sands loosely bound by a surface scum of algae. Callianassid (crustacean) burrows are particularly common. The exposed shorelines between channel openings are beach ridges with terraces and washover fans. The latter comprise intraclast gravels and rippled sands showing well-developed internal laminations. Similar facies, with lithified intertidal beachrocks, make up much of the eastern, windward, coastline of Andros Island.

The intertidal flats, partly protected by the beach ridges, are cut by a dense tidal channel network (see Fig. 20.19) with channels 1–100 m wide and 0.2–3 m deep. These meander but show little evidence of lateral migration, in contrast to channels on temperate siliciclastic tidal flats (Chapter 18). The channels contain lag gravels of skeletal debris, intraclasts and fragments of Pleistocene bedrock. The channel banks and stationary point bars are heavily bioturbated by crabs, overgrown by mangroves and covered by complex hemispherical stromatolite heads. Sections through these heads reveal well-preserved domal laminae with abundant uncalcified filaments of the sediment-binding alga *Schizothrix calcicola*. The channel levees occupy 10% of tidal flat area and are only rarely entirely covered by tidal waters. They are coated by thin algal mat with fine undisrupted millimetre-scale lamination. Levee toes show small desiccation cracks.

In between channels are extensive (30% of the total tidal flat area) tidal ponds bounded by algal marshes. The ponds are frequently covered by tidal waters and

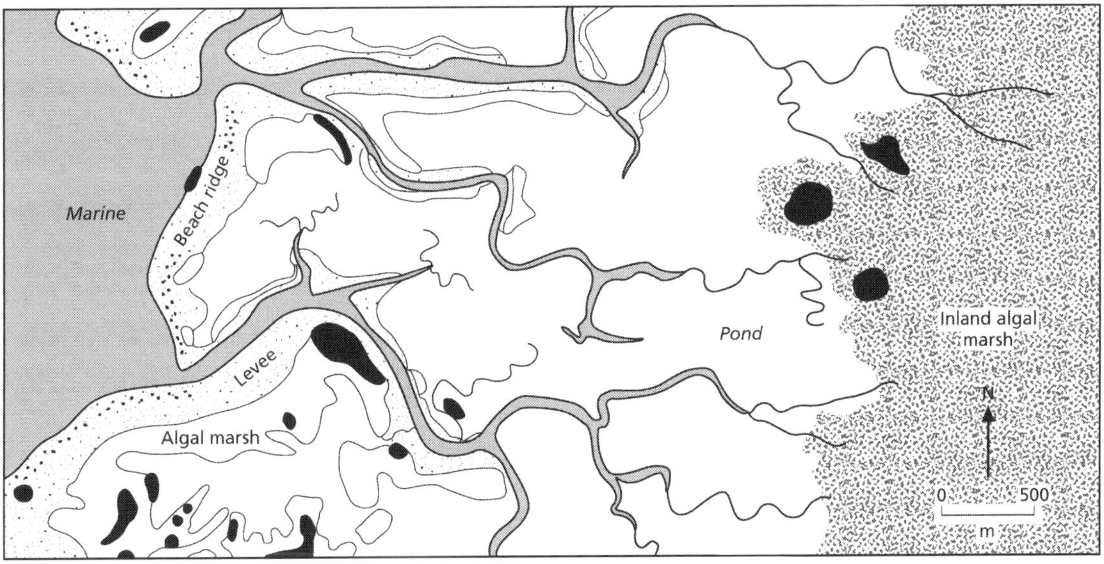


**Fig. 20.9** Map to show the environments of deposition in the Three Creeks area, leeward shore of Andros Island, Bahamas (see also Plate 16). Dark shading indicates cemented surface crusts. (After Hardie & Garrett, 1977.)

their muddy sediment has a thin surface algal mat that is grazed by cerithid gastropods and polychaete worms. Sections reveal unlayered, bioturbated pelletal muds cut by deep (up to 30 cm) desiccation cracks that form during winter and spring low-water periods. The fringing algal marshes comprise a high marsh with continuous algal mats of the freshwater genus *Scytonema* and a low marsh with 'pincushion' *Scytonema* growths. Patchy cementation by high-Mg calcite and aragonite occurs, with sections through the mats revealing a well-developed crinkly lamination with fenestrae (Fig. 20.11). These laminations are thought to be due to sporadic onshore winter storms which suspend and transport lagoonal sediments onshore.

The inland algal marsh lies about 20 cm above the mean high-water level of the channelled tidal flats. A similar zonation of low 'pincushion' marsh to high 'carpet' marsh occurs as noted above from around the tidal ponds. No invertebrates live in this marsh, which

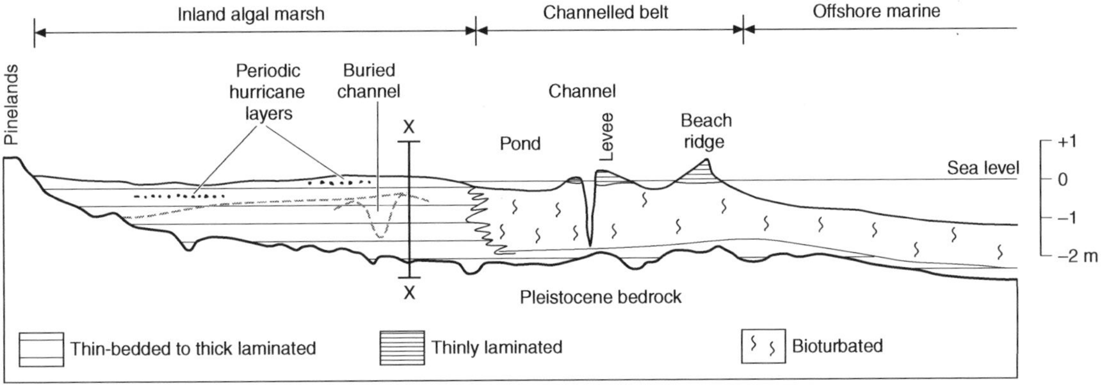


**Fig. 20.10** Generalized section through the Holocene progradational tidal flats shown in Fig. 20.9. (After Hardie & Ginsburg, 1977.)

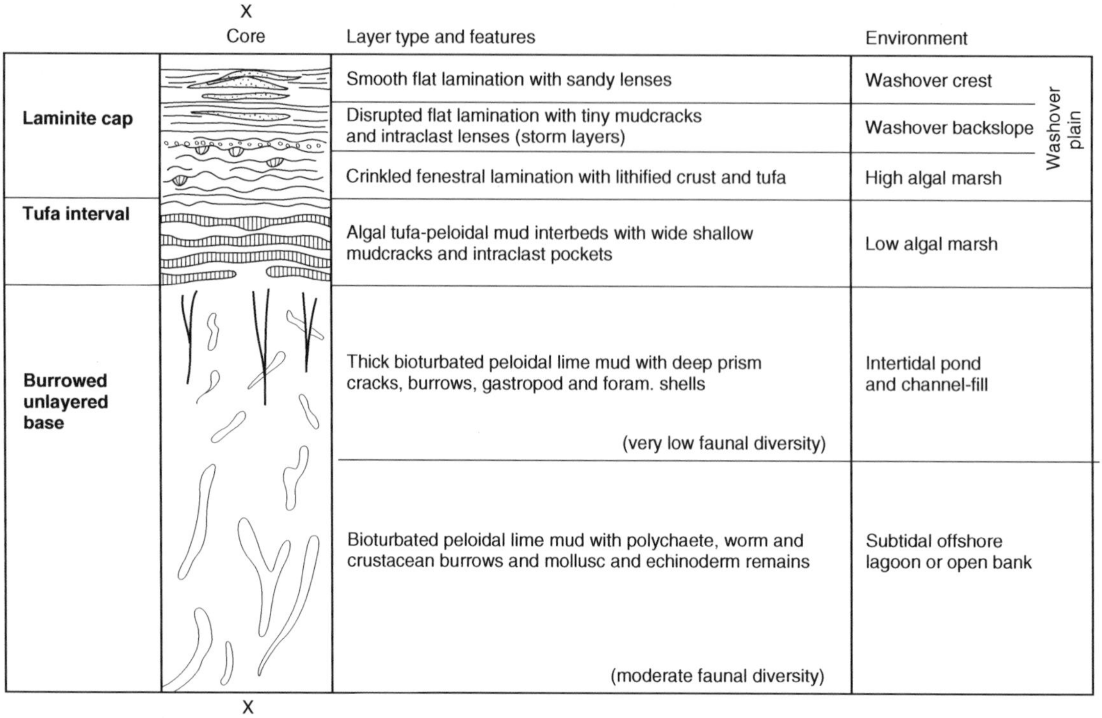
X
Core
Layer type and features
Environment
Laminite cap
Tufa interval
Burrowed unlayered base
Smooth flat lamination with sandy lenses
Washover crest
Disrupted flat lamination with tiny mudcracks and intraclast lenses (storm layers)
Washover backslope
Washover plain
Crinkled fenestral lamination with lithified crust and tufa
High algal marsh
Algal tufa-peloidal mud interbeds with wide shallow mudcracks and intraclast pockets
Low algal marsh
Thick bioturbated peloidal lime mud with deep prism cracks, burrows, gastropod and foram. shells
(very low faunal diversity)
Intertidal pond and channel-fill
Bioturbated peloidal lime mud with polychaete, worm and crustacean burrows and mollusc and echinoderm remains
(moderate faunal diversity)
Subtidal offshore lagoon or open bank
X

**Fig. 20.11** Log at position X of Fig. 20.10, to illustrate general nature of the progradational sequence (after Hardie & Ginsburg, 1977).

may be up to 8 km wide. The *Scytonema* algal mats are frequently lithified by high-Mg calcite, forming a discontinuous algal tufa. Desiccated mats give rise to characteristic polygon heads as the algae attempt to heal over the upturned polygon rims. Sections through the inland marsh reveal up to 1.7 m of laminated sediment with abundant fenestrae. The laminae (1–10 mm thick) represent tropical storm layers, the result of periodic hurricane-driven sheet floods, which carry lagoonal pelletal sediments on to and over the entire tidal flat and marsh, as documented after Hurricane Kate on the Caicos platform. The initially cross-laminated sediment is then bound by renewed *Scytonema* growth and the laminae preserved. The fenestrae are predominantly horizontal sheet cracks with subordinate vertical 'palisade' cracks. They form as primary voids from air pockets and as secondary voids from bacterial breakdown of algal filaments in vertical and horizontal layers and clusters. Lithification of the mats obviously enhances the preservation potential of the fenestrae. Indeed, spar-filled fenestrae (*'birds-eye' fabric*) in ancient carbonate sediments can provide good evidence of high intertidal to supratidal origin in ancient carbonate sediments.

Lateral and vertical sections through the Bahamian tidal-flat complex (Fig. 20.11) contrast markedly with arid tropical sabkhas, being evaporite-free with lithified algal tufa of freshwater-dominated marsh origin. The calcified *Scytonema* filaments and fenestrae in the latter serve to distinguish the stromatolites from the unlithified algal peats found in the intertidal zone of the Arabian Gulf.

## 20.4 Lagoons and bays

Subtropical carbonate lagoons and bays are relatively shallow and quiet-water environments periodically affected by storms and hurricanes. They typify 'detached' shorelines of oceanic or offshore isolated banks, platforms and atolls, as well as linear coastlines. All are partially separated from open marine environments by low offshore islands of lithified Pleistocene limestones (Arabian Gulf), atoll reefs, suprakarstic reefs (Belize, Yucatan), linear reefs (Honduras, Great Barrier), or a combination (South Florida shelf, Bahamas, Cayman Islands). These fringing 'rims' (like siliciclastic barriers or spits) protect the lagoons and bays to a greater or lesser extent from onshore winds and hence the effects of waves. Tidal currents are forced to enter the lagoons via narrow inlets. Efficient tidal exchange may keep lagoons close to oceanic salinity, but in arid tropical areas like the Trucial Coast, salinity may rise as high as 67‰. In humid tropical areas lagoonal and bay water may be considerably freshened by freshwater runoff from tidal flats and the hinterland (West Coast of Andros Island, Florida Bay). Thus, for the most part, shallow coastal lagoons and bays tend to be 'high-stress' environments and a restricted biota may result. In some areas (e.g. Grand Cayma) cores reveal evidence for progressive linkage with oceanic waters during the course of the Holocene transgression, with freshwater pond environments giving way to brackish mangrove swamp peats and finally to a reef-protected fully marine bay/lagoon environment with dense seagrass meadows of *Thalassia testudinum* whose encrusting sponges, bryozoa, and foraminifera mix with benthic green algae and molluscs to define a distinctive sediment assemblage.

Lagoonal sediments often comprise pelleted lime muds, with decreasing amounts of mud as wave action increases in importance in shallower waters. The shallow, wave-stirred Trucial Coast lagoons, for example, are floored by pellet sands excreted by crabs, cerithid gastropods and polychaete worms. The current-scoured outer Florida lagoon (Fig. 20.12) is floored by a winnowed lag of skeletal debris. Aragonite mud is predominantly of algal origin in the inner Florida and Honduras lagoons, coccoliths being an important contributor in the latter area. Controversy still rages as to the origin of Bahamian bank-top lagoonal aragonite muds (Chapter 2), but the very

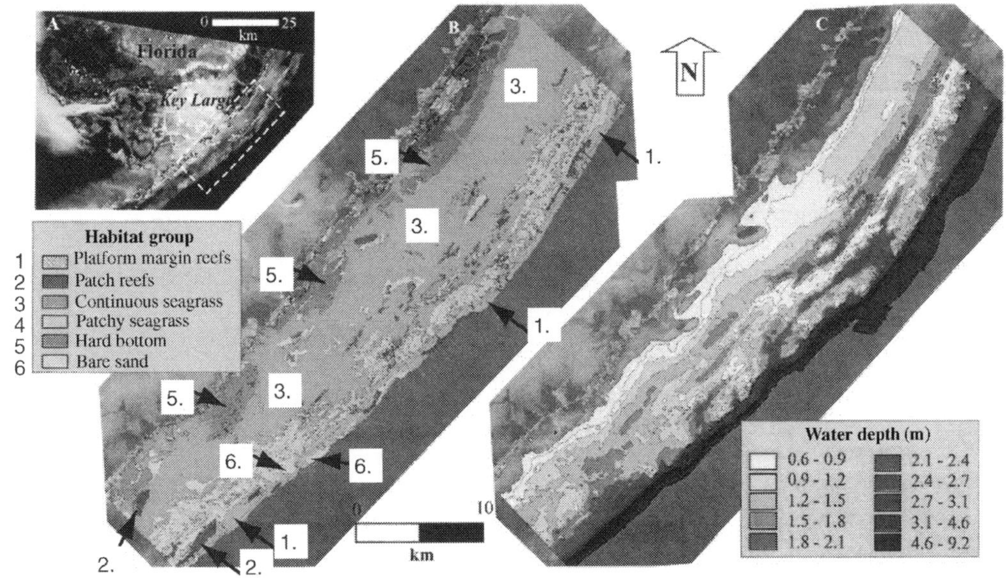


**Fig. 20.12** The South Florida shelf ramp margin. (a) Landsat image. The study region is outlined by dashed box. (b) Major sedimentary environments. (c) Water depth. Rankey (2004) found that in detail there was no unique correlation of sediment substrate with water depth and that other factors were probably important, one being the history of Holocene sediment accretion since highstand. (After Rankey, 2004.)

detailed exploration of aragonite whitings, and chemical and morphological evidence on the Great Bahamas Bank, indicate that most is probably chemical in origin and not resuspended algal material. Minor skeletal debris usually comprises foraminifera and molluscs. Atoll lagoons like those in Polynesia may be deeper than 'linear' lagoons, with green algae and sea-grass beds together with fine foraminiferal sands grading out to white carbonate muds between 20 and 40 m depth. The 2004 Indian Ocean tsunami deposits in Maldive atoll lagoons record overwash but no return flow and hence deposition of across-lagoon tapering wedges of decimetre-thick shallow subtidal sands which are progressively bioturbated post-tsunami. In fact, most lagoons support a thriving infauna, particularly crabs, which effectively destroy any primary laminations. *Thalassia* sea-grass stands occur in Florida Bay and in Bahamas and Arabian lagoons. In the inner part of Florida Bay these act as sediment baffles and have produced numerous elongate build-ups and islands that show vertical and lateral accretion trends with time (Plate 15). Patch reefs occur in many lagoons surrounded by a halos of coarse, reef-derived bioclastic grains.

A particularly interesting anaerobic carbonate environment characterizes shallow ponds and bays in lagoons and on bounding reef platforms in Polynesian atolls. Here, modern lacustrine to brackish stromatolites termed *kopara*, several decimetres thick, cover the bottom of most of these shallow ($<2$ m deep) features and the salinity of the water fluctuates between fresh and fully marine. A millimetre-scale lamination develops in the surficial sediments, produced by an alternation of microbial organic-rich and $CaCO_3$-rich laminae. Such features may be analogous to certain laminated, platy organic-rich limestones found in the stratigraphic record.

Variations in Holocene Floridan lagoonal sediment thickness are caused by differential topography of the lithified and karstified Pleistocene bedrock that underlies it. In Yucatan and Belize offshore platforms such antecedent karstic topography creates bathymetric depressions that are substantially deeper than normal and where muddy carbonate skeletal sands (packstones, wackstones) accumulates under more sheltered conditions (Fig. 20.13). Hurricanes may have temporary smoothing effects on depositional topography of bays and lagoons and lead to the infill and abandonment of open burrows like those of callianassid shrimps. The long-term trend in bays/lagoons like those of Florida is for coastal progradation to occur behind the barrier reef as islands coalesce and are replaced by coastal mangrove swamps and a supratidal zone with shallow lakes. Carbonate sediment budget studies in Bahamian lagoons reveal a substantial overproduction compared with accurately dated deposition rates. This is accounted for by storm and tidal current export of bank-top lagoonal sediment, mostly chemically precipitated aragonite, to periplatform areas.

Strandline carbonate sequences of Holocene age facing high-energy, open marine environments are rare, probably because of the predominance of offshore reef build-ups. An interesting example of a late Pleistocene wave-dominated beach–ridge plain has been documented from the Yucatan Peninsula of Mexico. Here a 7-m-thick sequence of shoreface to beachface grainstones, including common ooids in the upper shoreface, accumulated in a seaward-prograding coastal plain that was up to 4 km wide and 150 km long at its maximum extent. The interesting and perplexing aspect of this arrangement is that the entire high-energy coastal-plain sequence prograded behind a contemporary offshore barrier reef that obviously had no discernible 'protective' influence from incoming wave energy, a situation also seen landwards of the Great Barrier Reef.

## 20.5 Tidal delta and margin-spillover carbonate tidal sands

For reasons outlined in Chapter 2 there is close correlation between strong tidal currents and oolite formation. Tidal currents are amplified as they pass through gaps in offshore islands or reefal barriers on to rimmed Bahamian shelf or into Trucial Coast lagoons. Along these tidal inlets and in the zone of diverging flow bankwards/lagoonwards, active oolite and skeletal carbonate sand shoals form. By way of contrast, the high-energy carbonate sand belt of the southwestern Florida Keys (around and to the west of Marquesas Keys) lacks ooliths. Here skeletal (algal/coral) sands are moulded into dune-like bedforms, which accrete laterally (westwards) under the influence of strong tidal currents (Plate 15). It was only in the last interglacial that oolite shoals formed here—seen now in the famous Miami Oolite deposited by tidal currents on prograding barrier/inlet systems.

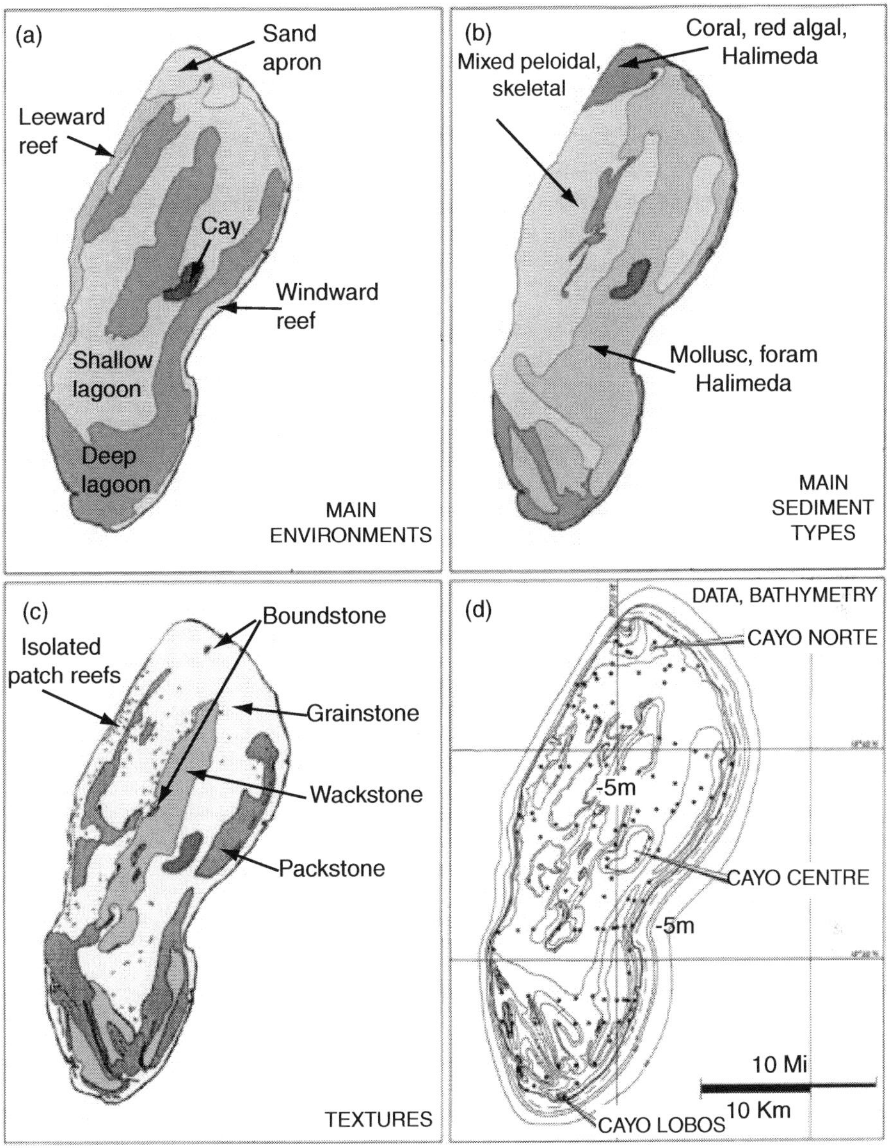


**Fig. 20.13** To illustrate the control on carbonate sediment type by bathymetry inherited from antecedent topography. Maps of Banco Chinchorro, Yucatan shelf ramp, Mexico showing (a) environments, (b) composition, (c) texture of surface sediments, and (d) bathymetry (in metres) and 99 sample locations. The fore-reef facies belt is very narrow and therefore does not show up in the figure. Patch reefs are shown only on the texture map. Note the occurrence of matrix-supported sediments (wackstone) in elongate bathymetric depressions. (After Gischler & Lomando, 1999.)

On the modern Bahamas Bank, active oolite shoals form at many localities around the perimeter. The shoals take many forms depending upon local tidal currents that are governed and steered by platform margin topography, reefs and islands. Although tropical storms may modify shoal morphology it is now thought that ambient tidal activity mainly controls shoal morphology. This is exemplified in the Lily Bank area of the Little Bahamas Bank where detailed remote sensing, bathymetric and hydrodynamic measurements reveal active oolite shoals of area ca 100 km$^2$ bankward of the Abaco Island chain of isolated reefs and Pleistocene islands (Figs 20.14 & 20.15). Two striking bedstates occur under maximum tidal currents of magnitude 0.5–1.0 m/s: (i) ca 1 km wavelength, 3–4 m relief asymmetric linear bars with crests normal to incoming or outgoing flood tides, and with superimposed dunes often parallel to the bar crest; (ii) parabolic bars up to 4 m high with steep leeside slipfaces and bar-crest superimposed dunes. Tidal flows over the former are not compartmentalized into flood or ebb, i.e. each phase of the tide flows over the whole bedform uniformly. Flow over the parabolic bars varies during ebb and flood routes in an extraordinary way. Both bar types show an extremely interesting internal architectural pattern of well-sorted oolite sands on their shallow, high-energy crests with muddier, finer-grained and less well-sorted skeletal and ooid sands in their deeper, low-energy troughs. The active oolite shoals disappear bankwards, replaced by a stabilized oolite and grapestone sand covered with a thin subtidal algal mat and *Thalassia* stands.

The formation of oolite accumulations by tidal inlet processes has also been made clearer by studies along the northern rim of the Little Bahamas Bank in recent years. Here separate flood and ebb tidal paths through inlets between islands generate a zone of active oolite shoal formation reflecting a time-average anticlockwise tidal current gyre (Fig. 20.16). This hydrodynamic pattern around the shoal (called the 'spin cycle'), allowing the sands to remain in motion without being transported out of the ooid 'factory'.

In the intershoal channels of southern Exuma Islands, giant, up to 2 m high, variously shaped stromatolites, some analogous in external form at least to those of Shark Bay (section 20.2) exist *within* active dunefields. They nucleated on to highs on underlying Pleistocene lowstand karst. The growths are periodically constructed by sediment-binding activities of filamentous green algae, cyanobacteria and diatoms. Destruction results from the activities of boring sponges and molluscs, endolithic algae and grazing fish. Growth and decay alternate in a complex series of events as the stromatolites are alternately exposed and buried by dune migration.

Lateral sediment changes from bank edge to lagoon are well illustrated by the Joulters Cay Shoal. Here the site of active ooid sands is located as a windward fringe, 4 m thick, with an extensive bankward spread of altered ooids mixed with skeletal grains and aragonitic muds. The extensively bioturbated muds are stabilized by grasses and are up to 10 m thick. There is an upward trend towards less mud within the inactive interior shoal. Numerous horizons within the interior shoal show penecontemporaneous cementation in areas of stabilized bottom covered by algal scum.

Distinctive oolite shoals also occur in the Schooners Cay area at the north end of Exuma Sound. Here the shoals take the form of linear tidal ridges whose long axes are parallel to the dominant flood tidal currents. Individual ridges are up to 8 km long and 750 m wide with amplitudes of about 5 m. Spillover lobes occur with their long axes orientated subparallel to the ridge long axes. They indicate a component of on-bank flow that is reflected in the asymmetry of the ridges, whose steeper sides are directed bankwards. Ripples and dunes superimposed upon the ridges are also orientated bankwards. Flow in the channels separating ridges is dominantly parallel to the ridge long axes. These linear ridges are very similar to those described from clastic tidally dominated shelves (Chapter 19) and are expected to show the same internal structures, i.e. low-angle cross-sets dipping obliquely to perpendicularly with respect to the ridge long axis. The ridges overlie burrowed muddy pelletal sands. Penecontemporaneous cementation may occur in the channel floors between the active oolite ridges.

Cemented carbonate aeolian dune deposits, often sourced from last interglacial oolite/skeletal tidal shoals when sea levels were 4–6 m higher than today, are common along arid to semiarid, wind-exposed, present-day coastlines bordered by productive carbonate ramps. Such lithified carbonate dunes are distinctively colonized by root traces (*rhizoliths*) and are termed *aeolianites*.

## 20.6 Open-shelf carbonate ramps

Carbonate sedimentation on shallow-marine shelf ramps is not simple and the analysis of limestones thought to have ramp origins in the sedimentary

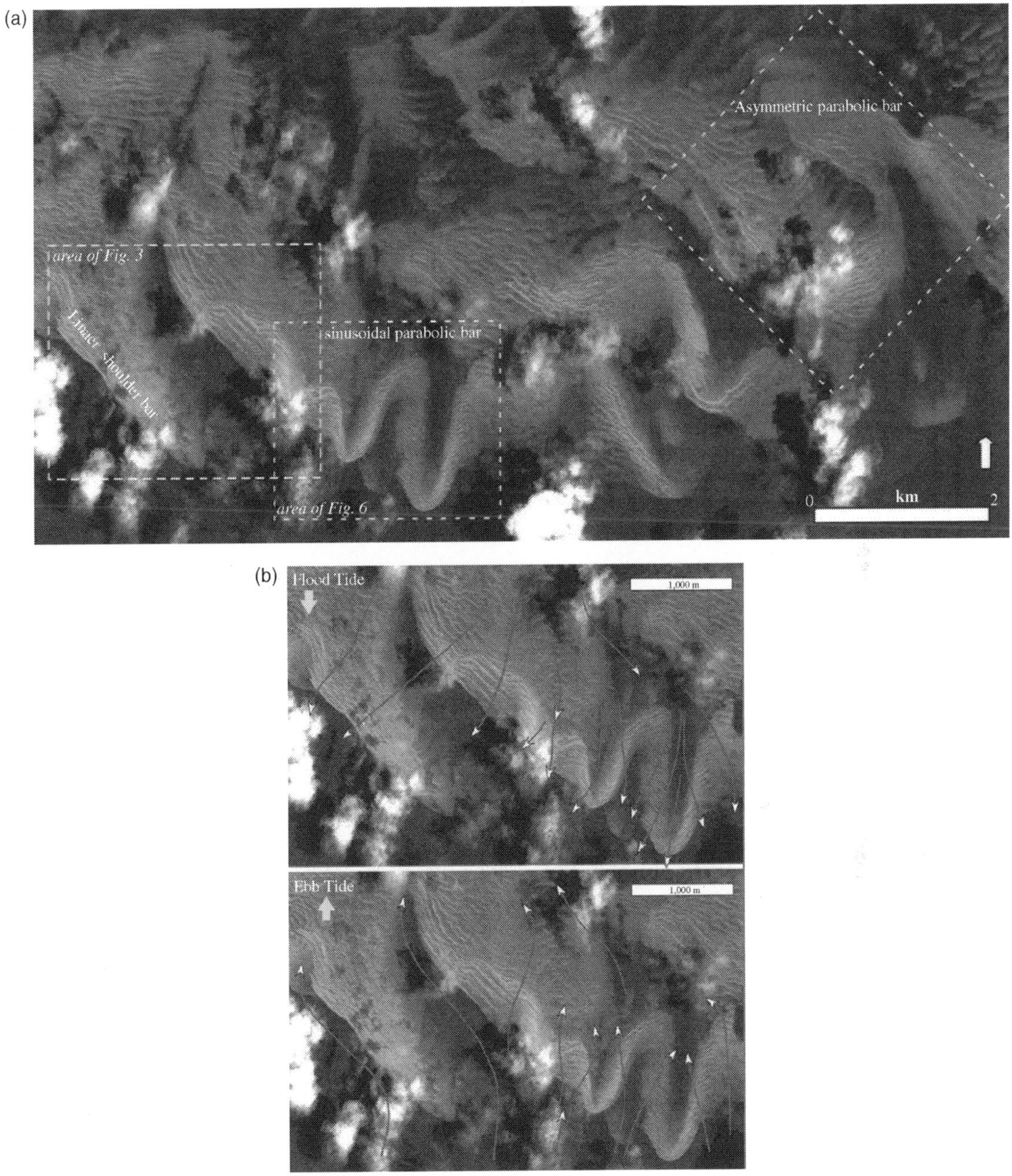


**Fig. 20.14** Bahamian oolite shoals. (a) High-resolution remote sensing image of part of active shoal, Lily Bank. The focus areas for the Rankey *et al.* (2006) study are in the 'linear shoulder bar' and 'sinusoidal parabolic bar' boxes. (b) Schematic model for interpreted flow paths for flood tide (upper) and ebb tide (lower), based on measured flow characteristics and bedform geometry. The lines and arrows schematically illustrate general flow directions, not flow velocities or volumes. As discussed in the text, note that the parabolic features appear to be in 'equilibrium' with both ebb and flood tide, although flow paths will not be perfectly reversed (equal and opposite paths and velocity) in many locations. (After Rankey *et al.*, 2006. Remote sensing image is copyright SpaceImaging.com.)

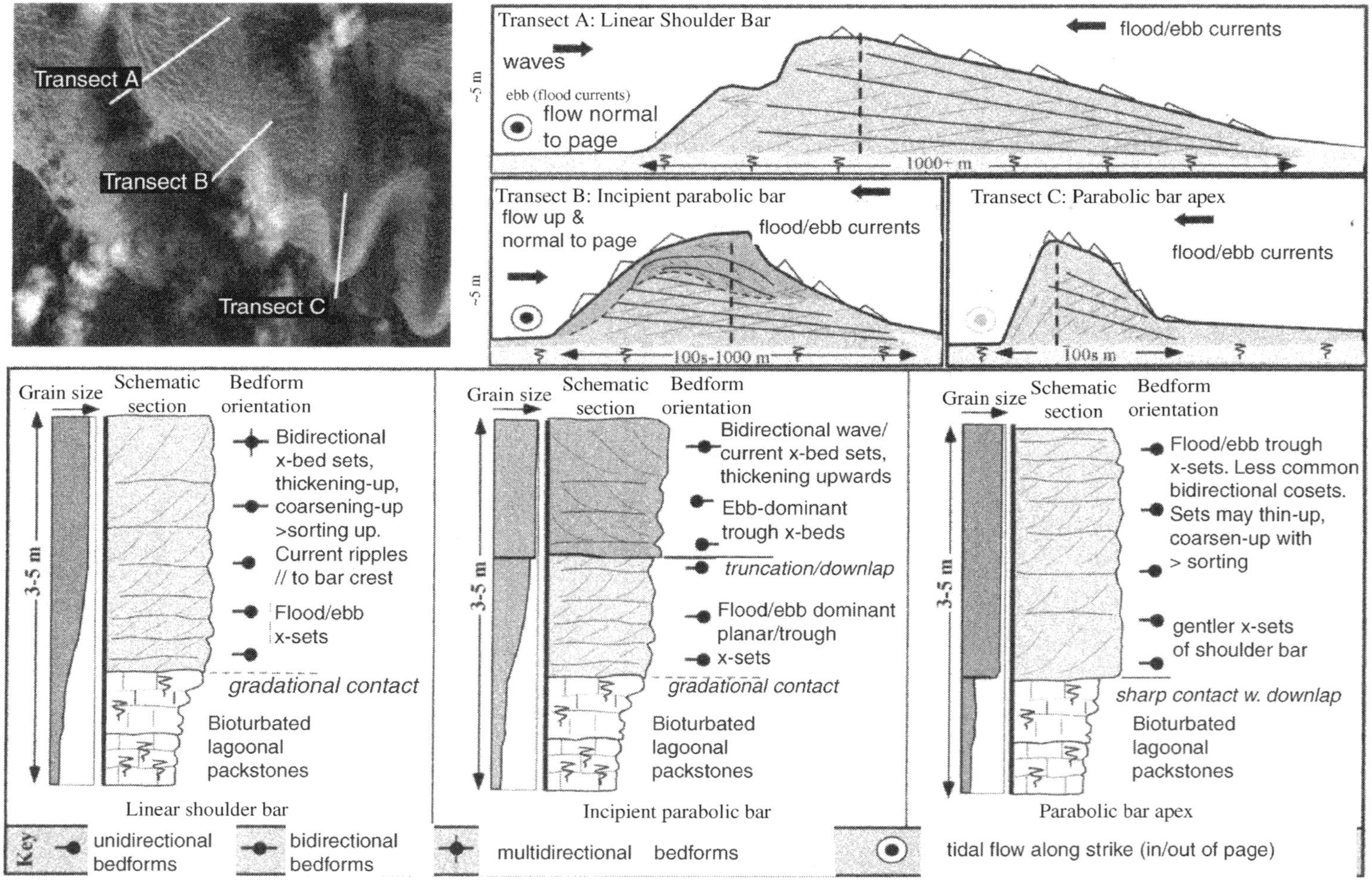


**Fig. 20.15** Panels to illustrate possible internal architecture of Lily Bank ooid shoals shown in Fig. 20.14. Transects well capture processes and resultant stratigraphical geometries that might be expected across linear shoulder bars (a), incipient parabolic bars (b) and on the apex of a parabolic bar (c). The dashed lines represent locations of the schematic stratigraphical sections, which illustrate expected vertical trends in grain-size, sedimentary structures and bedform orientation. (After Rankey *et al.*, 2006.)

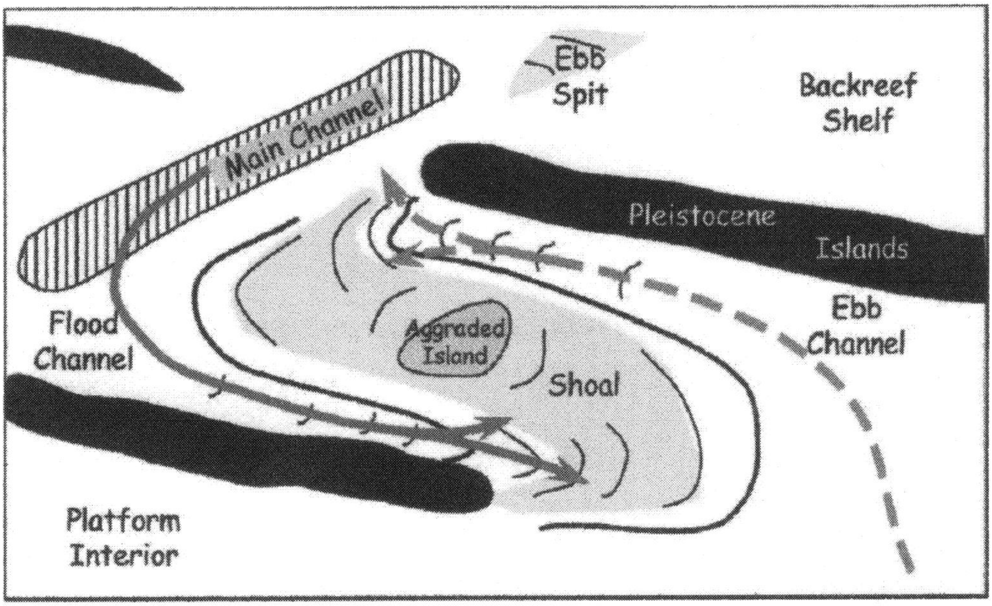


**Fig. 20.16** The closed 'spin loop' of a tidal delta oolite factory, Little Bahamas Bank. Sketch of the dominant flood (solid) and ebb (dashed) tidal flows, illustrating an approximation of the flow paths. The shallowest parts of the shoal are in the central area (low residual flow velocities) and at the ends of the Pleistocene islands (loss of flow restriction, expansion, deposition). (After Reeder & Rankey, 2008.)

record (Fig. 20.2a) must address a gamut of issues that do not just recognize offshore-deepening and declining wave energy as the main control on sediment production and sedimentation rates. The very same physical components of the siliciclastic shelf field apply (Fig. 19.4), with the added importance of ecologically important effects of ocean currents, nutrient levels, saline underflows and eutrophication. Shallow-marine carbonate sediments are classified into (i) cool-water *heterozoan* assemblages dominated by molluscs and bryozoa and (ii) warm-water *photozoan* assemblages of zooanxellic corals, green algae and ooids. In transition zone seas like the Mediterranean a distinct red algal sediment contribution is recognized.

### Warm ramps

As noted above, Great Bahamas Bank carbonate platform evolved from an early ramp-like geometry (Fig. 20.1, see also Fig. 20.32). The Miocene and Lower Pliocene of the leeward flank of Great Bahama Bank provides an example of the poorly known depositional setting of the outer part of distally steepened carbonate ramps. The contrast between its sedimentary patterns and the well-known Upper Pliocene–Quaternary slope facies associations of the flat-topped Great Bahama Bank shows the strong control that the morphology of a carbonate platform exerts on the depositional architecture of the adjacent slope and base-of-slope successions. Deep, ~1.2 km, cores off the central windward margin of the Great Bahamas Bank platform reveal cyclic alternations of light- and dark-grey wackestones/packstones with interbedded calciturbidite packages and minor slumps. Light-grey layers containing shallow-water bioclasts were formed when the ramp exported material, whereas the dark-grey layers are dominantly pelagic. Calciturbidites are arranged into mounded lobes with feeder channels. Internal bedding of the lobes shows a north-directed shingling as a result of the asymmetrical growth of these bodies. Calciturbidite packages occur below and above sequence boundaries, indicating that turbidite shedding occurred during third-order sea-level highstands and lowstands. Cyclicity was driven by high-frequency sea-level changes. Highstand turbidites contain shallow-water components, such as green algal debris and epiphytic foraminifera, whereas lowstand turbidites are dominated by abraded bioclastic detritus. Gravity flow depocentres shifted from an outer ramp position during the early Miocene to a basin floor setting during the late Miocene to early Pliocene. This change was triggered by an intensification of the strength of bottom currents during the Tortonian, which was also responsible for shaping the convex morphology of the outer ramp.

The Arabian Gulf is a carbonate ramp in a foreland basin that flanks the great Zagros thrust-fold belt of Iran. Further, it lies inshore of Arabian Sea upwelling and as a consequence is rich in nutrients and poor in flanking reefal build- ups, though patch reefs are locally common. Over most of its area the shelf waters show salinities between normal seawater values and 42%. In shallow coastal areas (5–30 m deep) skeletal grainstones comprising well-rounded and well-sorted molluscan, foraminiferal, algal and (localized) coral debris are accumulating. In deeper offshore areas (>30 m) sorting becomes poorer and skeletal fragments are more angular, their sharp fracture surfaces perhaps being caused by *in situ* mechanical breakdown. Increasing admixtures of silt- and mud-grade low-Mg calcite occur in deeper areas, giving rise to packstone and wackstone fabrics and, ultimately, marls. The fines are thought to be derived from wind-blown carbonate dusts. Although 'whitings' of precipitated aragonite mud occur periodically in the Gulf, no trace of this aragonite has been recorded in the offshore sediments. Much of the Gulf shelf is covered by a thin, lithified subtidal hardground that supports a specialized epifauna adapted to hard substrate life. This cemented horizon indicates low offshore productivity and sedimentation rates.

The northeast Brazilian shelf ramp stands as a prime example of a high tidal/wave energy environment with mixed siliciclastic and carbonate sediments, occupying almost half the Brazilian shelf area, one of the longest carbonate depositional environments in the world. In over half the area seawards of a relatively narrow siliciclastic zone both the middle (20–40 m water depth) and outer (>40 m water depth) shelf are dominated by carbonate sediments (>75%). In certain areas carbonate sediments form a more continuous blanket from the inner to the outer shelf. The ramp offshore the State of Rio Grande de Norte is influenced by strong oceanic and wind-driven currents, 1–5-m-high waves and a mesotidal regime. Large-scale bedforms consist of: (i) bioclastic (mainly coralline algae and *Halimeda*) sand ribbons (5–10 km long, 50–600 m wide) parallel to the shoreline in 7–17 m water depths; and (ii) offshore (18–20 m depth) very large transverse siliciclastic dunes (3–4 km long on average, 840 m spacing and 3–8 m high), with troughs that grade rapidly into carbonate sands and gravels. Wave ripples are superposed on all large-scale bedforms, and indicate an onshore shelf sediment transport normal to the main sediment transport direction. The occurrence of these large-scale bedforms is primarily determined by northwesterly flowing residual oceanic and tidal currents, resulting mainly in coast-parallel transport. The sand ribbons occur in an area of coarse, low-density and easily transportable bioclastic sands and gravels compared with the siliciclastic dune sands. These latter are thought to originate from seafloor erosion of sands during past sea-level lowstands.

The Yucatan shelf ramp (Fig. 20.13) has an inner zone 130–190 km wide extending down to depths of 60 m where a zone of relict lowstand build-ups occurs along the shelf break with ooids, peloids and lithoclasts. The modern sedimentary cover comprises a thin layer of molluscan debris, everywhere < 1 m thick. At greater depths these nonskeletal sands are increasingly diluted with the tests of winnowed pelagic foraminifera. This pattern of relict outer-shelf facies and contemporary inner-shelf molluscan debris emphasizes the extreme importance of the shallow subtidal carbonate 'nursery' as a sediment producer.

## Cool ramps

The southern and western coasts of Australia present a fascinating array of shelf-ramp environments ranging from cool to tropical in water temperatures and with a variety of cool- and warm-water ocean current systems in contention (Fig. 20.17). It is also an object lesson for understanding the role of sea-level change in carbonate sedimentology.

Ramp complexity is emphasized by the northwest shelf of Western Australia (Fig. 20.18), today a highstand swell-wave ramp notable for the contribution of both heterozoan and photozoan elements to the inner ramp carbonate factory, also because of coastal clastic

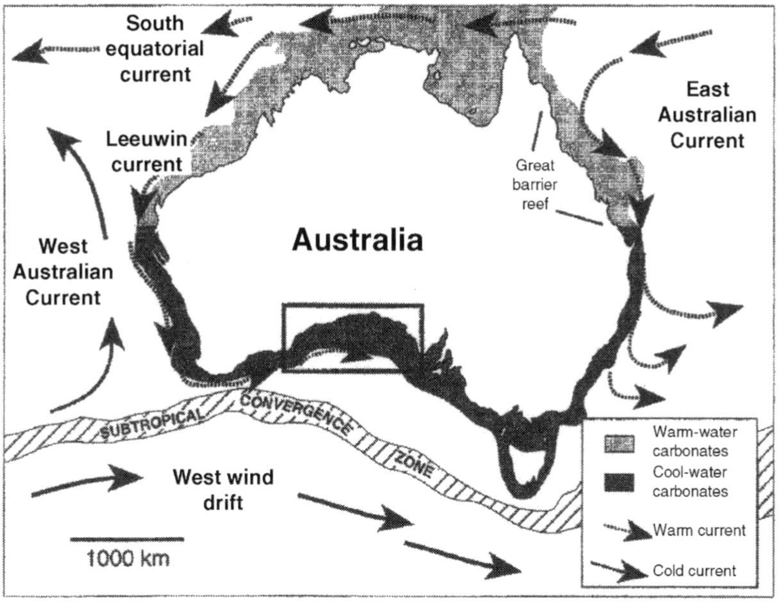


**Fig. 20.17** Map of Australia showing areas of cool-water and warm-water carbonate deposition on the continental margin and major oceanic current patterns. (After James *et al.*, 2001.)

input and lack of nearshore protective barriers. The modern inner- and mid-shelf ramp boundaries are drawn at fair-weather wave base (50 m) and storm wave base (120 m) and comprise skeletal and lithoclastic sands and gravels with much reworked relict grains, including fragments of *Halimeda*, coral, and gastropod at water depths between 40 and 70 m. Further notable features are the occurrence of seasonal saline underflows from coastal lagoons on to the mid-ramp and presence at depth of the south-flowing, low-salinity Leeuwin Current. The underflows seriously reduce mid-ramp productivity whilst the cool current increases deep outer ramp productivity and encourages development of a linear pelagic foram-rich sand ridge at depths of ~140 m. The outer ramp also features huge accumulations of relict aragonite carbonate muds generated by precipitation from seawater during the last glacial lowstand.

Ramps in the transition zone from cool to warm waters occur along the southwest Australia shelf and illustrate the role of changing sea level upon the predominance of upwelling or downwelling and its influence upon ramp carbonate production. Thus the margin is strongly influenced by the poleward-flowing, warm, nutrient-poor Leeuwin Current, which promotes overall downwelling, and the strong summer equatorward-flowing West Australian Current sourced from the Circum Antarctic Current, which generate local seasonal upwelling. Southern cool-ramp slopes feature luxuriant stands of seagrasses and macrophytes growing on coralline-encrusted hardgrounds and rooted in sediments rich in coralline algae and larger, symbiont-bearing forams together with abundant cool-water elements such as bryozoans, molluscs and small forams. Central ramp slopes off Shark Bay are very poorly productive due to saline outflow and low nutrient supply and the ramp sediments are mostly relict. This portion is thus termed a 'subtropical starved ramp' in the sense that although mid-ramp bottom-water temperatures are tropical, the biota is largely subtropical and neither the mid-ramp nor the outer ramp is a site of active modern carbonate sediment production. Biodegraded sediments and clasts, on the other hand, indicate that carbonate production was active in the recent past. Northern areas are influenced by the downwelling Leeuwin Current and this nutrient-poor water bounds the fringing Ningaloo Reef, separated from the coast by a lagoon 0.2–7.0 km wide. During lowstand it is likely that the Leeuwin Current was pushed to the

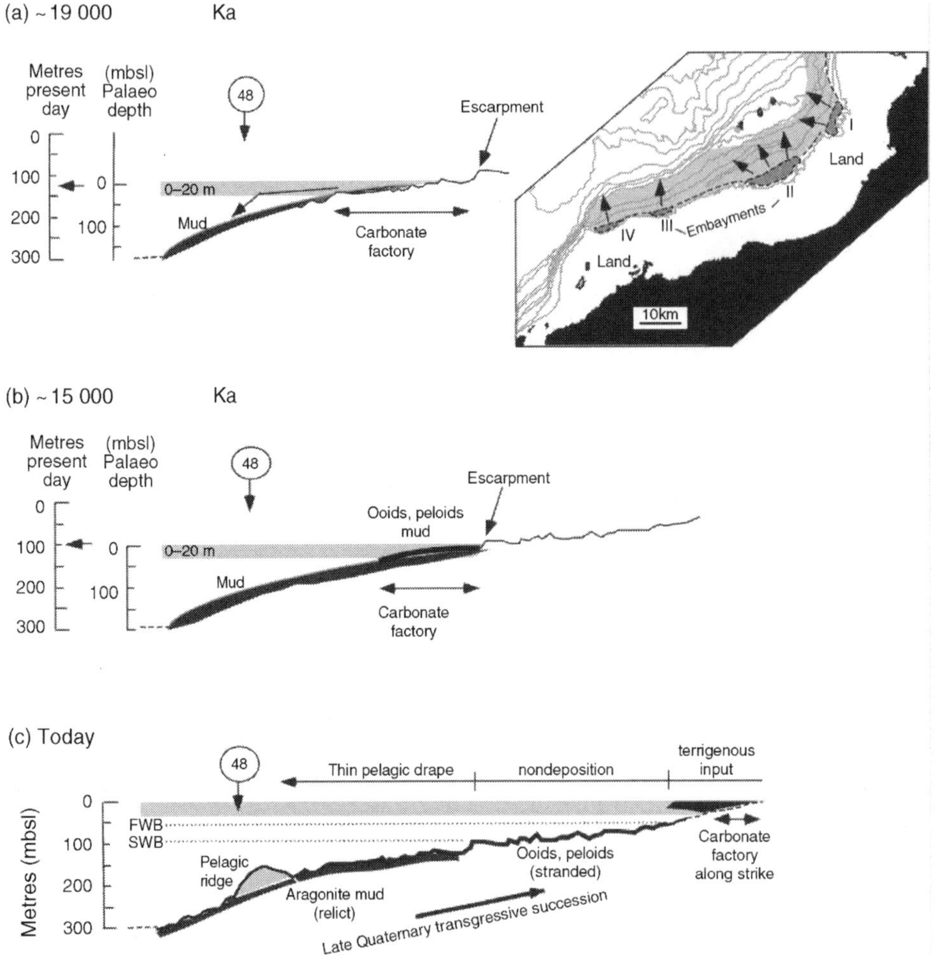


**Fig. 20.18** Summary of late Pleistocene–Holocene sedimentation on the northwest Australian shelf.(a) A narrow shelf developed seaward of −125 m during the Last Glacial Maximum, and broadened locally into four palaeoembayments (I–IV). Seaward export (arrows) and area of micrite distribution (light grey) are indicated relative to four palaeoembayments (dark grey). (b) Earliest post-glacial sea-level rise initiated ooid and peloid growth, with possible continuation of micrite production and/or reworking of existing mud by strong currents and waves. (c) Modern conditions: a relatively narrow inner-ramp carbonate factory occurs in areas not dominated by terrigenous input; the high-energy mid-ramp lies between the fair-weather (FWB) and storm (SWB) wave bases, yielding negligible mud accumulation; pelagic ooze accumulates on the outer ramp and slope. (After Dix *et al.*, 2005.)

north with a concomittant increase in upwelling all along this coast leading to the succession of cool over warm carbonates—a distinctive sedimentary rhthym.

Cool-water carbonate ramp deposits on the huge southern Australia shelf comprise a mixture of Holocene biofragments and late Pleistocene relict and stranded particles. Swell-driven seafloor sediment disturbance to depths of 100 m in conjunction with reduced rates of sedimentation that characterize the cool-water carbonate realm, have resulted in mixing of sediment currently being produced with exhumed relict and stranded sediments generated during previous sea-level stands. The stranded relict sediments are interpreted to have accumulated in an areally restricted photic zone and upon an overall narrow shelf covered in cool, nutrient-rich waters. They originally accumulated in low-energy, inboard environments but were modified in the surf zone during subsequent sea-level fluctuations. This resulted in the formation of intraclasts: abraded, infilled, and cemented biofragments of local origin. The widespread distribution of intraclasts and of composite grain assemblages derived from multiple sedimentary environments that display highly variable degrees of seafloor alteration may be signatures of cool-water carbonate assemblages deposited under fluctuating sea-level conditions.

Patterns of Holocene sedimentation on the main South Australia ramp are linked to modern oceanographic parameters in this high-energy setting characterized by overall downwelling (Fig. 20.19). The inner shelf is the main carbonate nursery, an area of abundant macrophytes and seagrasses, active carbonate sediment production and accumulation, and little relict sediment. The middle portion is a shelf desert, with active sediment winnowing and mostly relict sediment. The outer shelf and upper slope is variably productive, characterized by prolific calcareous epibenthic growth on hard substrate subaqueous 'islands' shedding particles into surrounding sands and muds (Fig. 20.20). Prolific rhodoliths occur on the northwest inner shelf, where shallow summer waters are the warmest in the Great Australian Bight (GAB) (Fig. 20.21),). These warm, saline, nutrient-depleted waters drift eastward across the shelf, suppressing heterozoan carbonate production on the central and eastern mid-shelf. This arrested production in the eastern GAB is countered locally by summer coastal upwelling along western Eyre Peninsula, with bryozoan-rich sediment extending well inboard on to the mid-shelf. The outer shelf and upper slope is an area of prolific bryozoan growth, probably linked to upwelling, except in the central GAB, a region of year-round downwelling (Fig. 20.21) where the area is one of off-shelf fine sediment transport and carbonate mud deposition.

The Lacepede Shelf, South Australia, is a gently sloping, ramp-like feature (Fig. 20.2b) that extends from the siliciclastic beach ridges bordering the Coorong Lagoons down to a shelf break at about 200 m. Although the northwest margins of the Coorong Lagoons mark the site of the large Murray River, little clastic sediment reaches the shelf from the now-restricted lower reaches of this sluggish system. The outer shelf from about 100 m depth slopes more steeply oceanwards into a canyoned slope and hence into the adjacent abyssal plain. The wide shallow shelf is mostly at depths of 40–60 m and is dominated by a wave regime of high (>2.5 m), long-period (>12 s) swell waves. During the summer the shelf water mass is stratified with surface waters of temperature 18 °C overlying a thermocline at 30–80 m and bottom waters of about 13 °C. The shelf waters with mean

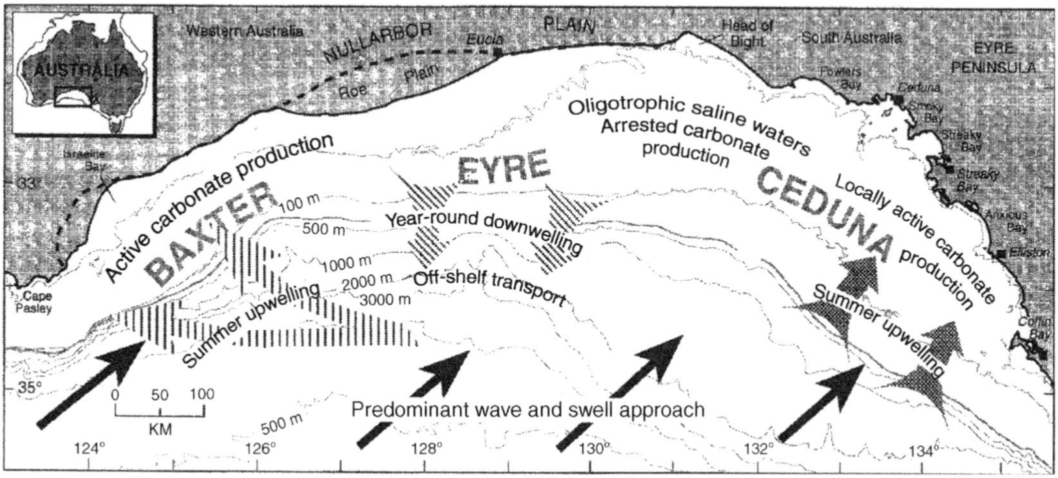


**Fig. 20.19** Map of Great Australian Bight (GAB) illustrating the major oceanographic controls on sedimentation. (After James *et al.*, 2001.)

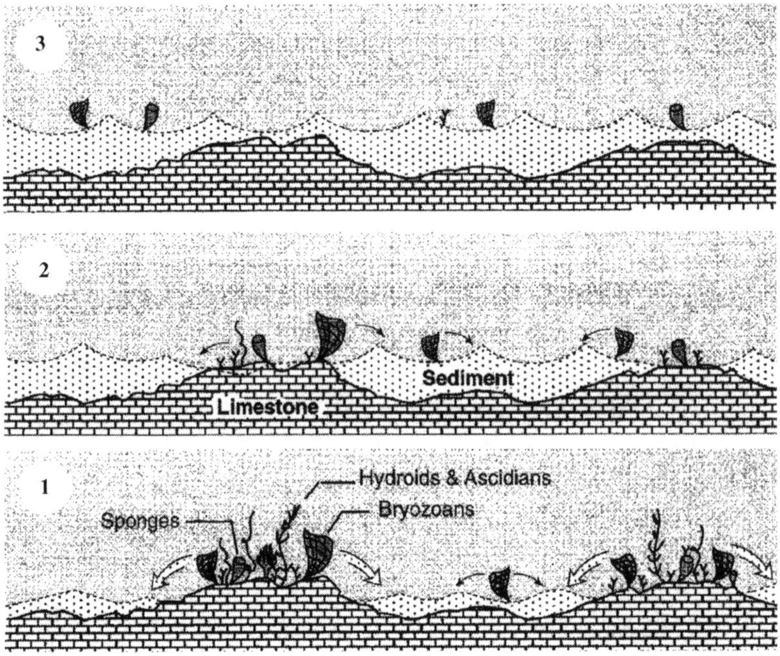


Fig. 20.20 Sketch illustrating nature of the carbonate sediment factory on the middle to outer shelf and upper slope of the Great Australian Bight (GAB) and its evolution with time. Prolific growth of skeletal invertebrates on rocky surfaces produces sediment that buries the hard substrates upon which they grow and so the whole ecosystem self-destructs. (After James *et al.*, 2001.)

temperatures of around 18 °C are well-mixed during the winter down to depths of 100 m. Sediment distribution contrasts markedly in several ways with low-latitude shelf ramps and platforms. Sediment types include various proportions of terrigenous clastic sands and silts, relict skeletal carbonates of pre-Holocene strandlines and modern skeletal carbonate materials, chiefly bryozoan and mollusc sands and gravels. These latter are never dominant in shallow-water areas, usually representing at most 45% of the total and mainly comprising abraded mollusc debris sourced from storm-wave-reworked infaunal bivalves. The main carbonate production comes from an outer-shelf 'bryozoan nursery' at depths >80 m but still well above the swell wave base. Here the bryozoan-rich carbonates are coarse detritus sourced from production (including calcareous algae) on harder substrates lithified by lowstand carbonate precipitation. The deposits are frequently moved by winter swell waves and sorted into rippled patches. Below about 150 m the sediment becomes muddier and generally finer, reflecting the growth of more fragile bryozoan species. The location of most carbonate production on the outer shelf has important implications for ancient cool-water facies models for wave-dominated regimes. Also important is the close proximity of this production to the edge of the slope, with the expectation that significant losses of carbonate will occur downslope.

## 20.7 Platform margin reefs and carbonate build-ups

### Names matter

Modern carbonate platforms are frequently rimmed by reefal carbonate build-ups, predominantly comprising coral but also encrusting coralline algae. These control to a greater (e.g. Bahamas) or lesser (e.g. Queensland) extent, the resultant distribution of carbonate facies on the interior platform because of their modifying influence on the physical processes of wind, wave and tide. Due to organic evolution the

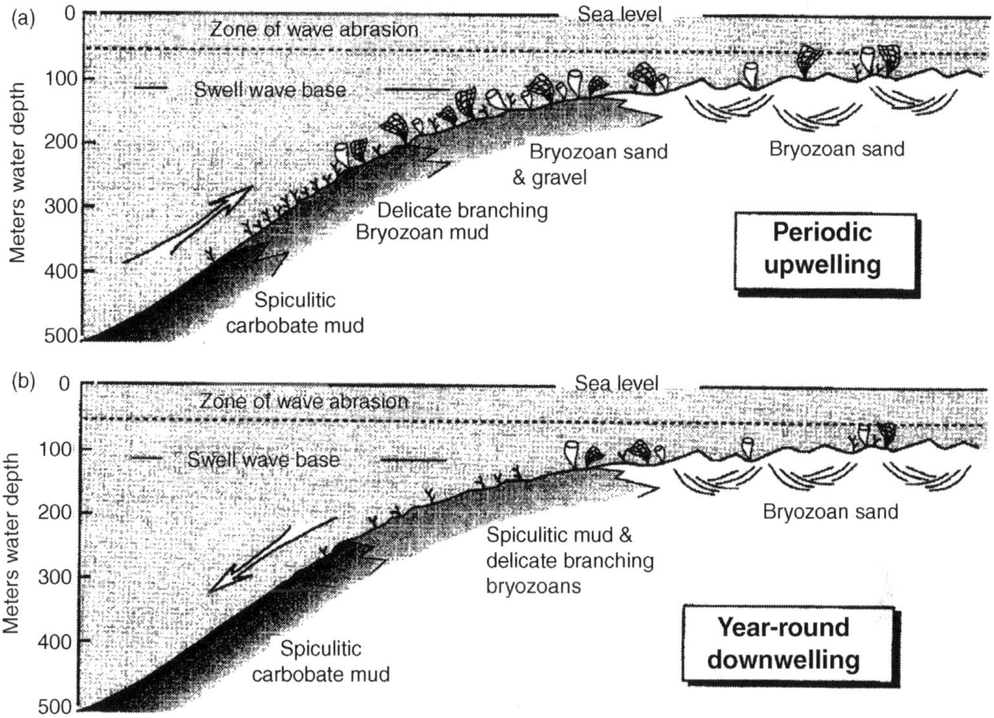


**Fig. 20.21** Summary profiles of different sectors of the Great Australian Bight (GAB). (a) The outer shelf/upper slope on a high-energy, cool-water carbonate shelf with seasonal upwelling, based on the Baxter and Ceduna sectors. (b) Summary of a similar profile on cool-water carbonate shelf with year-round downwelling, based on the Eyre sector. (After James *et al.*, 2001.)

sedimentary record reveals that the composition, depth distribution and influence of build-ups have changed over time. The terms below are used, somewhat variably, by authorities on modern and ancient build-ups and reefs. The term reef itself defies concise definition—this author heard the late Robin Bathurst, that great carbonate specialist, annoyed by a semantic discussion of this point at one of his conferences, leap to his feet and splutter words to the effect that '*. . .a reef is, simply, a lithified structure that can sink a ship*'!

- *Calcimicrobes*—calcified microbial fossils (a 'dustbin' term for a large group of ancient problematica, including suspected cyanobacteria, etc.) with a very long geological history that play an important role in strengthening the basic coral framework of modern and ancient reefs, also in cementing nonframework build-ups such as Miocene *Halimeda* mounds.
- *Carbonate boundstone*—a reefal fabric in a limestone characterized by few metazoan fossil remains but with features indicative of 'anti-gravitational' carbonate trapping, cementation or binding often associated with calcimicrobes.
- *Carbonate build-up*—general term for mostly organic bodies of locally formed and laterally restricted carbonate sediment that possesses topographic relief.
- *Carbonate mound*—particular equidimensional or ellipsoidal build-up in shallow warm or deep cold water.
- *Carbonate pinnacle*—particular conical or steep-sided upward-tapering mound.
- *Patch reef*—small isolated subaqueous build-up in shallow water.
- *Knoll reef*—small isolated subaqueous build-up in deeper water.

- *Atoll*—conical organic accumulation above subsiding oceanic basement with a ring-like surface plan, rimmed by active coral reefs surrounding a lagoon.
- *Barrier island reef*—linear-to-curvilinear belt projecting to sea level (often made up of individual wall reefs) of organic accumulation, steep to seawards, situated somewhat offshore and separated from the coast by a lagoon or broad shelf.
- *Shelf-edge reef*—submerged in 15–60 m water, relict or actively growing sometimes deeper-water corals, often buttressed and grooved to windward.
- *Fringe reef*—belt of organic accumulation built out directly from the shoreline.

A classic subdivision of build-ups at platform/shelf margins is provided by Fig. 20.22. The very existence of a carbonate build-up depends upon a local carbonate production rate that exceeds that of surrounding areas and in which excess production is conserved, although some surplus production may be exported locally or regionally. The higher production measured over the Holocene mounds off the Florida Keys for example (Plate 15) is due to the localized growth of standing crops of the green algae *Neogoniolithon* and *Halimeda* and the coral *Porites*. Coring reveals that the gravel-grade sediments of the modern mound are derived from *in situ* breakdown at a rate of about 2 mm/yr. Early microbially mediated cementation just below the sediment–water interface must also play a role in preventing sediment export by tide and wave and in stabilizing many noncoral framework build-ups such as Palaeozoic phylloid algae and the Cretaceous-to-modern green alga *Halimeda*.

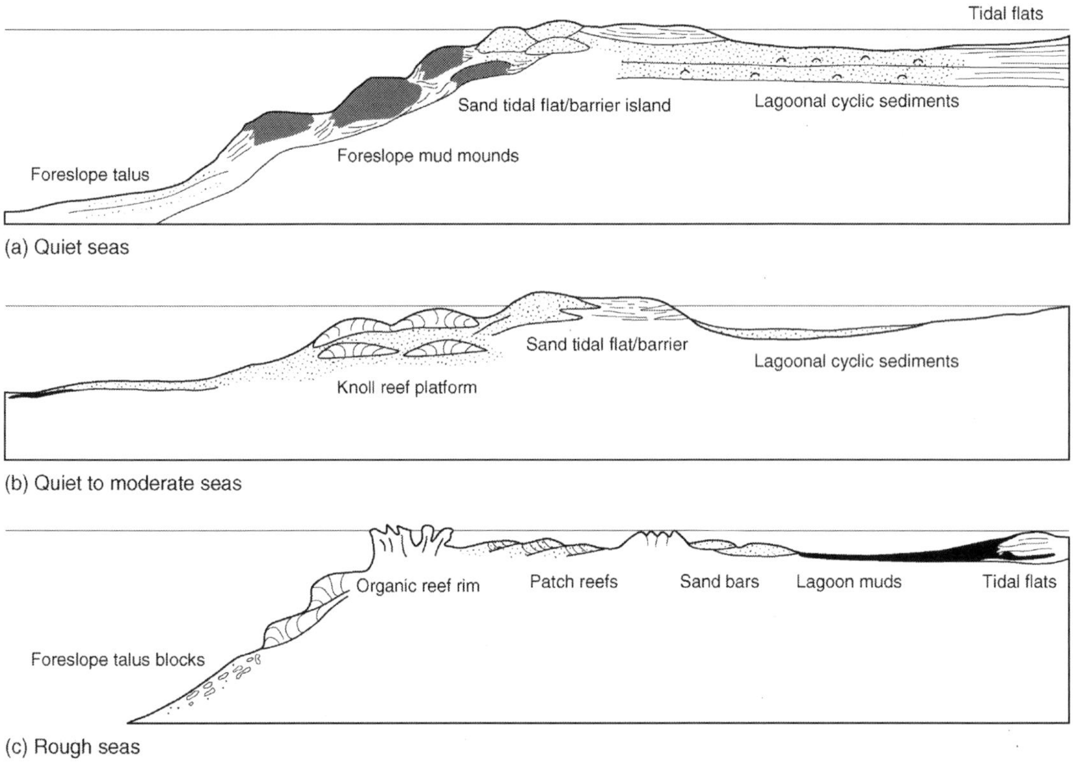


**Fig. 20.22** Classic end-member views defined by Wilson (1975) of build-ups at modern and ancient carbonate platform margins.(a) Downslope and basinal mound accumulations. (b) Knoll-reef platforms of moderate-energy but periodically hurricane-swept tropical margins like those of Florida. (c) Robust framework-reef-fringed platforms, atolls and islands.

### Low wave-energy build-ups and deep cold-water build-ups

No subtidal ramp is *dominated* by framework-poor build-ups at the present day, but mounds of *Halimeda* are common in waters 20–50 m deep on many carbonate platform margins (e.g. Queensland). The *Thalassia*-bound build-ups called *mud mounds* have various proportions of carbonate mud-matrix in the quieter 'lagoonal' environments of Shark Bay (section 20.2) and Florida Bay (Fig. 20.12). Larger-scale Belize examples are in more challenging storm-frequented situations. Deep-water build-ups on the flanks of the Bahamas platform provide important modern analogues to ancient lime mud mounds. These are relatively low-energy ramp margin features (Fig. 20.22) of microbiologically precipitated carbonate mud, *in situ* skeletons and organic detritus. The cementing, binding, trapping and/or baffling organisms may be calcimicrobes, as in many Proterozoic and Palaeozoic reefs, calcified sponges and primitive coral-like ancestors in the earliest, Lower Cambrian, shallow-water metazoan reefs, bryozoans (Ordovician–Pleistocene), platy algae (upper Carboniferous), crinoids (Silurian–Carboniferous), rudist bivalves (Cretaceous), the green alga *Halimeda* (Cretaceous– recent) and marine grasses (Tertiary–recent).

Proof of topographic relief on ancient build-ups revolves around recognition of contemporary, non-tectonic dips using boundstone markers, internal sediment-filled inclined cavities and talus spreads tonguing-out from build-up flanks. Sea-level fluctuations may impress karstic phenomena upon the build-up framework, sometimes accentuating evidence for build-up slopes. Studies of the famous Triassic Latemar platform margin in the Dolomites (section 20.9) reveal that the narrow reefal build-up comprised wave-resistant boundstones containing the microscopic problematica *Tubiphytes* and a variety of organic crustal carbonate growths and cements, but very little skeletal framework support.

Notable examples of ancient finer grained build-ups evidently formed below storm wave base (depths >100 m). For example, spectacular exhumed Middle Devonian build-ups in the Algerian Sahara (Figs 20.23 & 20.24) take the form of mounds, linear fringes, ridges or barriers, constructed in water depths of order 100–200 m. Other notable Devonian examples of both occur in the legendary Canning Basin of Western Australia, noteworthy because of the change from coral framework build-ups to microbial build-ups after the Frasnian mass extinction eliminated the framework ecotype. A particularly interesting example, painstakingly documented and modelled from the Middle Triassic of the Italian Dolomites (Fig. 20.25) establishes the 3D growth of kilometres-wide mounded tropical build-up which did not have the capacity to grow into continuously wave-swept environments because of its small size and the absence of organisms around at the time to construct a wave-resistant energy barrier.

**Fig. 20.23** Field photograph of linear group of seven exhumed Devonian mud mound build-ups in the Algerian Sahara, each 25–30 m high (Photo courtsey of A.J. Wendt).

Recent discoveries have been made of truly deep, cold water, nonframework, mostly unlithified, build-ups. Thus thousands of spectacular cold-water coral mounds occur in up to 1 km water depth in the northeast Atlantic (Rockall Trough, Porcupine Seabight), closely associated with thermohaline flow. They occur on a variety of scales from a few tens of metres to kilometres wide and a few metres to up 200 m relief above the seafloor. Many of the Porcupine mounds are relict, probably Pleistocene in age, being covered by dead coral rubble or buried by sediment, but a few are thriving as thickets of the cold-water corals *Lophelia pertusa* and *Madrepora oculata* that live and grow without photosynthetic algal symbionts. Coring of these show alternating coral layers with coral debris in a matrix of micrite composed of coccolith remains. The Darwin mounds in Rockall (Fig. 20.26) seem to be growing on a pockmarked seabed swept by thermohaline currents and with

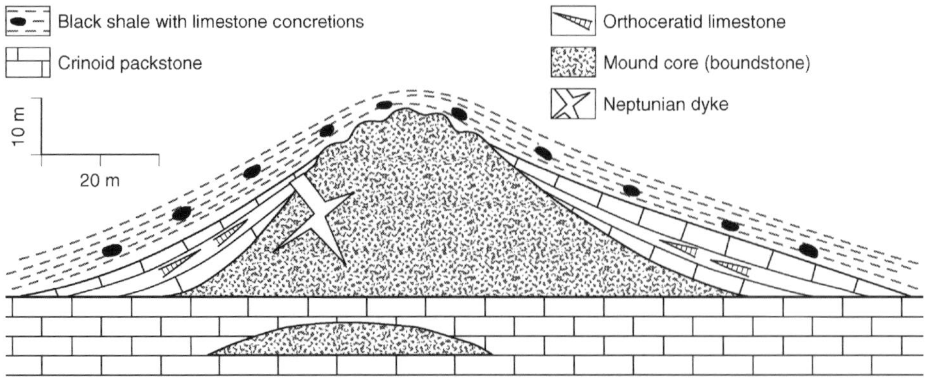


**Fig. 20.24** Restored section through the Algerian Devonian mud-mounds featured in Fig. 22.23. Note the overlying black shales suggesting nutrient-induced termination of the build-ups. (From Wendt *et al.*, 1997.)

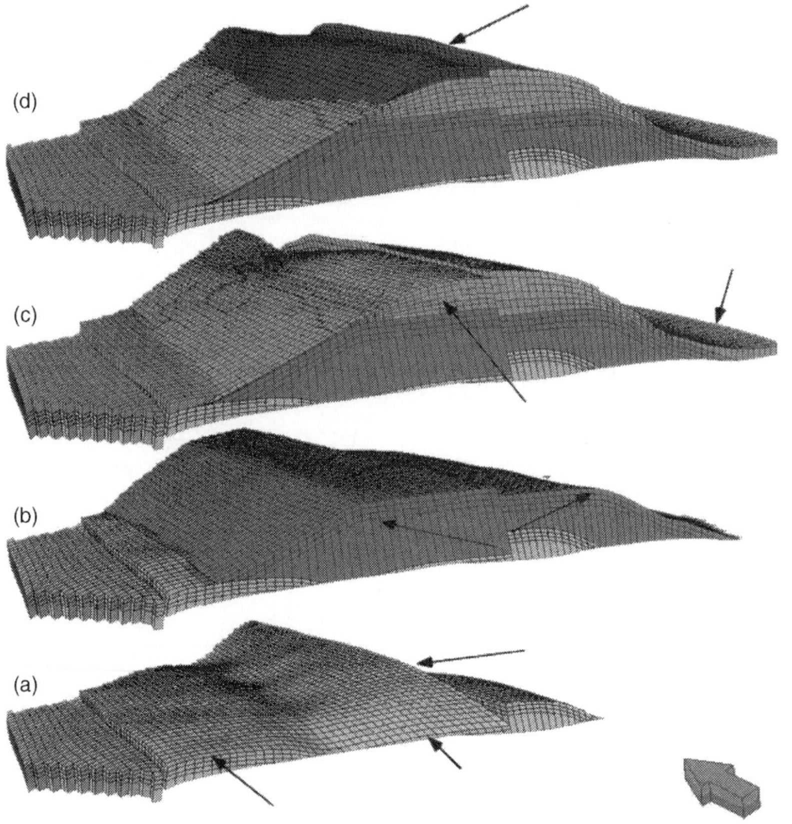


**Fig. 20.25** Modelled growth of a mounded Triassic build-up from field observations. The four main units, with cell dimensions of about 50 × 50 m in an oblique aerial view from the southwest. (a) The Contrin Formation develops as a platform in the western part grading into a mounded platform, now dissected by a fault. (b) Progradation measured from the top of the mounded platform was about 750–800 m and led to an almost flat-topped carbonate platform. A slightly mounded top (vertical relief 50 m) is modelled for this stage. (c) The retrogradational interval is modelled with a gradual reaccentuation of mound architecture that was finally capped by a mounded interval (d). (After Blendinger *et al.*, 2004.)

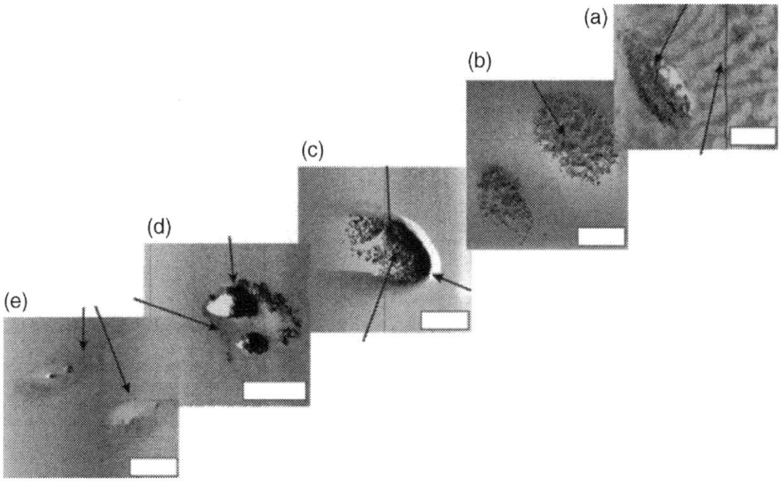


**Fig. 20.26** Exceptional images of deep-water coral mounds, Darwin Mounds Rockall, northeast Atlantic. Selected 100 kHz (a), (b) and (e) and 410 kHz (c) and (d) side-scan sonar images that provide evidence of antecedence from pockmark morphology of the Darwin Mounds. Light tones are low backscatter, dark tones are high backscatter. (After Wheeler *et al.*, 2008.)

a sand substrate made available for initial coral colonization by sand volcano eruptions. The coral colonies thrive in the nutrient-rich thermohaline flows and trap a good deal of fine sand within their biogenic structures. They pass downcurrent into rippled and sand wave sands. Deep cool-water mounds are also known from the Cadiz basin, offshore southern Spain.

Bryozoan-rich biogenic mounds present on 200–350 m deep prograding carbonate slopes of the central Great Australian Bight developed during sea-level lowstands throughout Pliocene–Pleistocene time. It is inferred that the upper limit of growth was swell wave base and the lower limit by impingement of an oligotrophic water mass. Mound accretion characteristically began with delicate branching bryozoan floatstone that increases in bryozoan abundance and diversity upward over a thickness of 5–10 m, culminating in thin intervals of grainstone characterized by reduced diversity and locally abraded fossils. Mound growth during glacial periods is interpreted to have resulted from increased nutrient supply and enhanced primary productivity. Such elevated trophic resources were both regional and local, and thought to be focused in this area by cessation of Leeuwin Current flow. These mounds are analogous to spectacular late Cretaceous examples from the Danish Chalk.

### Framebuilt reefs in shallow warm waters

High-energy shoalwater margins comprise framebuilt reef rims that grow up to or close to mean sea level (but prolifically from 15 m to as deep as 60 m). Mesozoic to Holocene examples are dominantly scleractinian coral associations and occur where sunlight radiation levels provide maximum benefit to the coral–algal symbionts; perforce they must also grow in the zone of greatest wave energy but in relatively nutrient-poor waters, for reasons outlined further below. They form barrier, fringing or shelf-edge reefs preferentially on the windward coastlines of carbonate platforms and are zoned ecologically in parallel belts, with their hexacoral growth forms (platy, domal or branching) precisely reflecting light intensity, sediment concentration, exposure level and energy levels. Microbial crusts and coralline algae play an important role in protecting and strengthening the primary coral framework in some examples.

Framework reefs usually show steep seaward slopes with spur-and-groove morphology (Fig. 20.27) and abundant reef talus. The form and regularity of spur-and-groove suggests a hydrodynamic origin, perhaps analogous to the self-similar beach cusps seen on wave-dominated shorelines. Older reef sections show evidence for frontal accretion as the reef moves

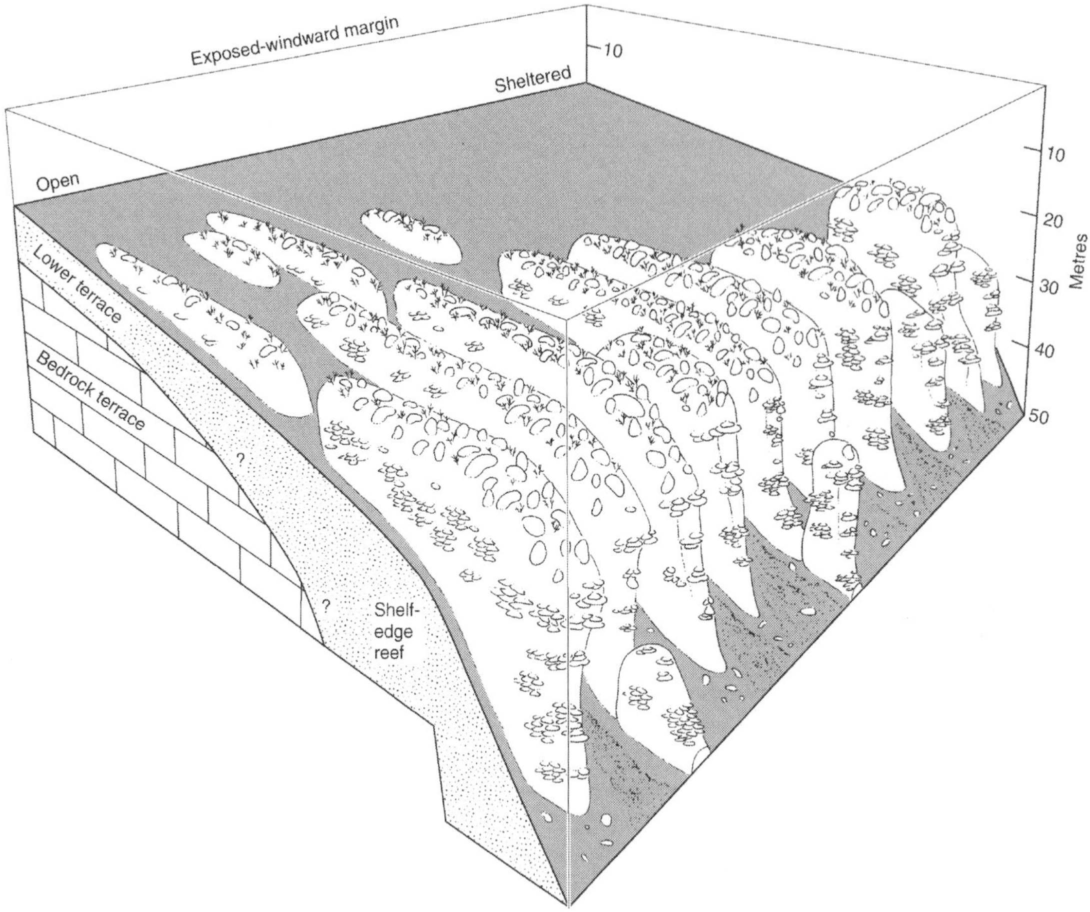


**Fig. 20.27** Drawing to show the highly elongate buttressed spurs and embayed talus spreads typical of domed growths of *Acropora* palmate corals situated on exposed high-energy reef margins. The spatial trend to more sheltered conditions is also shown, with the elongate spurs replaced by more equidimensional 'crowned' growths at depths >20 m. (Drawing courtesy of P. Blanchon; see Blanchon & Jones, 1997.)

uncertainly seawards over their substrate of talus. Along the Grand Cayman shelf-edge the seaward reef wall at depths of 15–60 m is variably buttressed and channelled according to the degree of exposure to hurricane-induced storm wave (and presumably gradient) currents The Great Barrier Reef (GBR) of the Queensland coast of Australia has well-developed linear wall reefs at the steep-gradient shelf edges in the northern and southern sectors (where tidal currents are also at maximum) and a variety of fringing, ring and concentric platform reefs scattered across the landwards gently sloping ramp until a siliciclastic coastal zone is reached. In this coastal to inner shelf area 'turbid-zone' fringing *Galaxea*-reefs are found to be able to withstand high deposition rates from wave-induced mud suspension (*turbidophile* patch reefs are also known from the Neogene of Borneo). The history of the GBR revealed by a deep 210-m core reveals a shallowing-upwards succession, the younger part of which is punctuated by a series of erosion surfaces. Carbonate deposition, a series of debris flows, began about 770 ka in a relatively deepwater slope environment. Subsequent accretion involved downslope accumulation of grainstones and wackestones, sometimes cross-laminated, characterized by intervals with abundant rhodoliths and scattered, probably reworked, corals. Typical reefal associations of corals and calcareous algae developed unimpeded by erosion from about ca 500 ka. Periodic subaerial erosion surfaces between ca 300 ka and present indicate the reef tract had aggraded to the depth required for exposure during lowstands.

Framework reefs support not only a highly diverse coral community but also a flourishing often destructive reefal epifauna and infauna of molluscs, echinoids, coralline algae, microbial crusts and foraminifera. Integrated sediment budgets must compute (i) initial (gross) carbonate production by framework-building corals and encrusting coralline algae, (ii) loss of carbonate by bio- and physico-erosion and (iii) addition of carbonate through the biological activities of calcifying algae, molluscs and other organisms that coexist on the reef. A budget should also recognize the source and role of any outside sediment and analyse sediment storage flux. Results from the Kailua Bay, Oahu, Hawai reef tract (Tables 20.1–20.3), indicate an estimated 75% of total carbonate production here is retained in the reef, the remainder being resedimented landwards and seawards, the former slowly replenishing the lagoon beachface.

Sand fringes in the back-reef flat and fore-reef areas of Queensland, Florida and the Bahamas are often dominated by calcareous algal fragments, particularly *Halimeda*, coral not being a good sand-former. In the Floridan reef tract the extensive back-reef environment, with its patch reefs and sublittoral coral/algal sand spreads, grades into the back-reef lagoon. The seaward margins to the Belize barrier and atoll reefs comprise four facies belts passing seawards from the reef front. The reef front down to 70 m depth comprises coarse coral and *Halimeda* sands and conglomerates with grainstone fabrics. The reef wall (65–120 m) is made up of well-cemented coral-rich limestones, which yield ages in the range 8000–15 000 yr, spanning the period of most rapid sea-level rise. The fore-reef talus fans comprise muddy *Halimeda* sands showing packstone and wackstone fabrics. Cements in the reef wall include common high-Mg calcite and subordinate aragonite. Isotopic and trace-element analyses prove the marine origins of these cements. Accretionary models for the seaward margin to the Belize platform have erosion during lowstand, and rapid coral growth 'plastering' the reef wall during sea-level rise. Reef wall growth is thus envisaged as a discontinuous lateral accretion process, with submarine cementation occurring after each period of accretion. A similar pattern to Belize is found off the Florida shelf margin where relict reefs are now colonized by deeper-water communities and form a characteristic stepped profile offshore.

The passage of tropical storms leads to production of abundant coral debris and temporary cessation of growth (Fig. 20.28). Possible physical and biological reasons for more permanent reef demise, in addition to various coral diseases like whiteband, blackband and bleaching due to environmental stress, include (i) rapidity of sea-level rise, (ii) higher nutrient levels and (iii) enhanced oceanic swell energy.

Table 20.1 Summary of annual sediment production by source.

| Sediment source | Sediment production (mass) $\times 10^3$ kg yr$^{-1}$ | Sediment production (volume) m$^3$ yr$^{-1}$ | Unit sediment production kg m$^{-2}$ yr$^{-1}$ | Contribution by volume (%) |
|---|---|---|---|---|
| Framework bioerosion | $2828 \pm 646$ | $1911 \pm 436$ | | 47 |
| Mechanical erosion | ~441 | ~315 | | ~8 |
| *Total erosion* | *$3269 \pm 646$* | *$2226 \pm 436$* | *$0.33 \pm 0.13$* | *$55 \pm 19$* |
| *Halimeda* | $538 \pm 32$ | $769 \pm 46$ | | 19 |
| Branching cor-algal | $441 \pm 72$ | $283 \pm 46$ | | 7 |
| Forams | $482 \pm 97$ | $344 \pm 69$ | | 8 |
| Molluscs | $511 \pm 46$ | $426 \pm 39$ | | 11 |
| *Total bioflux* | *$1972 \pm 247$* | *$1822 \pm 200$* | *$0.20 \pm 0.06$* | *$45 \pm 12$* |
| **Total sediment production** | **$5242 \pm 892$** | **$4048 \pm 635$** | **$0.53 \pm 0.19$** | |

Volume conversion use densities specific to sediment origin (see text). Sediment production by framework bioerosion includes the contribution of grazing echinoids (given in parentheses). Normalized sediment production calculated using Eq. 13.

Table 20.2 Sediment contribution by volume.

| Carbonate source | Contribution (by volume) (%) | Sediment composition (%) |
|---|---|---|
| Coral | 35 | 12 |
| Coralline algae | 27 | 36 |
| (encrusing) | (20) | |
| (branching) | (7) | |
| Halimeda | 19 | 15 |
| Molluscs | 11 | 11 |
| foraminifera | 8 | 5 |

Early drilling results through Holocene reefs and use of independent regional sea-level time series suggested three possible reactions of reefs to sea level rise.

1 *Keep-up* reefs initiate as local substrate is submerged, subsequently accreting at the same rate of sea-level rise. However, detailed data on reef net accretion rates (rather than coral growth rates) relative to water depth during rising sea level through the Holocene reveals no statistically significant depth-related decrease in accretion rate as is commonly supposed—few reefs have built faster than the rate of sea-level rise in the early and mid-Holocene, but most reefs have been able to keep-up since about 5 ka.
2 *Catch-up* reefs show up to several thousand years lag in initiation and then accrete rapidly to rising sea level. It has been pointed out that the lag inherent in catch-up may be an artefact of patchy early colonization and inability of cores to intersect the earliest reef nucleus.
3 *Give-up* reefs fail to accrete normally for some reason and are killed by the rising sea level.

Great insight into the Holocene development of reefs is possible by careful coring and radiometric dating programmes. Studies in U.S. Virgin Islands reefs seem to indicate a definite lag of some 1500 years between initial transgression and reef development. Thus shelf flooding near Buck Island occurred as early as 9500 yr ago (all subsequent ages in Cal. yr BP) but preserved reefs lagged by as much as 1800 yr, initiating on platform currently at −13 to −16 m water depth. Earliest reef development was dominated by branching *Acropora palmate* (Fig. 20.29) near the shelf edge and massive corals closer inshore. By 7200 yr BP, *A. palmata* apparently declined near the platform margin and was absent until ca 5200 yr BP throughout the study area. Over subsequent time, the inshore reefs built upward (ca 16 m) and prograded seaward (ca 50 m) as rate of sea-level rise slowed and reef-front growth (*A. palmata* dominating) reached its upper limit. Branching coral apparently disappeared again between ca 3030 and 2005 yr BP for reasons that are not clear. This and the previous decline of *A. palmata* mimic patterns seen throughout the Caribbean. By 1000 yr BP, the inshore reefs had largely assumed their present character and continued to track slowly rising sea level until the present.

All reef development hinges upon the ability of framework-building corals to achieve growth rates

Table 20.3 Sediment storage in Kaihia Bay compared to 5000 years of potential sediment production.

| | ($\times 10^3 m^3$) | (%) Of modern storage | (%) Of holocene production |
|---|---|---|---|
| Potential sediment production (5 kyr) | | | |
| Framework erosion | 11,129 ± 2,178 | | |
| Direct production | 9,109 ± 999 | | |
| Total sediment productin | 20,239 ± 3,177 | | |
| Modern sediment storage | | | |
| Submarine reservoirs | | | |
| Channel | 2,220 ± 185 | 15 ± 4 | 11 ± 3 |
| Reef sand bodies (north) | 145 ± 15 | 1 ± 0.3 | 3 ± 1 |
| Reef sand bodies (south) | 604 ± 60 | 4 ± 1 | 3 ± 1 |
| Nearshore triangle | 285 ± 29 | 2 ± 1 | 1 ± 0.2 |
| Offshore mouth | 471 ± 47 | 3 ± 1 | 2 ± 1 |
| Total submarine storage | 3,726 ± 336 | 26 ± 6 | 19 ± 5 |
| Subaerial reservoirs | | | |
| Beach | 1,000 ± 1,000 | 7 ± 1 | 5 ± 1 |
| Coastal plain | 10,049 ± 1,809 | 68 ± 24 | 51 ± 17 |
| Total subaerial storage | 11,049 ± 1,909 | 75 ± 25 | 56 ± 18 |
| Total storage | 14,775 ± 2,244 | | 45 ± 23 |

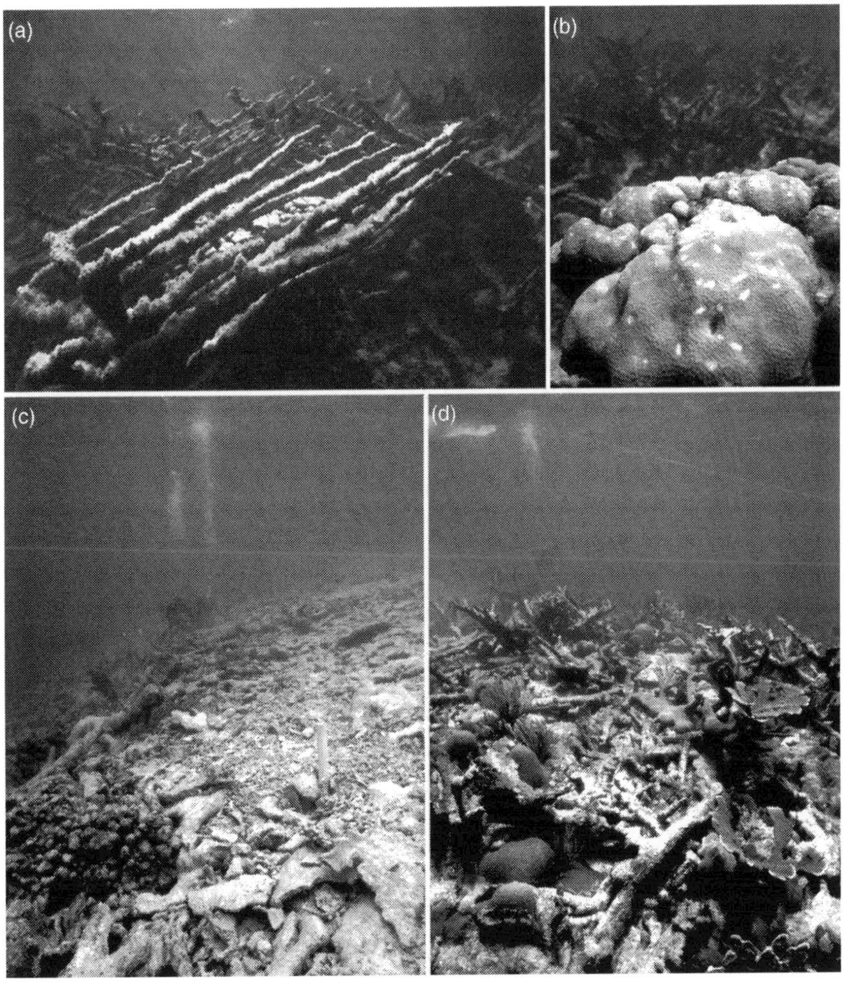


**Fig. 20.28** Virgin Islands coral reef scenes of the effects of disease and hurricanes on reef framework.(a) Underwater photograph along the eastern end of the northern forereef in the late 1970s. This area was dominated by large (2–3 m) colonies of branching *A. palmata* in excellent condition. (b) Photograph representative of the southern forereef in the late 1970s. Branching *A. palmata* dominates the reef crest in the background, while massive colonies of *Montastraea faveolata* and *M. annularis* can be seen in the deeper foreground. (c) Southern reef after the passage of Hurricane Hugo in 1989. Most of the devastated area is covered by standing but dead *A. palmata* that had been killed earlier by White Band Disease (WBD) plus live *Porites porites*. Note the surviving colony of *M. annularis* at lower left. (d) Buck Island Bar in April 1990. Whilst a few colonies of *A. palmata* are alive, most are dead, having been killed by WBD. Despite having succumbed to WBD over a decade before, numerous colonies were still standing after the passage of Hurricane Hugo in September of 1989. (After Hubbard *et al.*, 2005.)

that exceed the local bio-erosion rate. High nutrient availability affects bio-erosion, competition and coral growth, the former two by encouraging the growth of coral predators and competing 'fleshy' algae and the latter by making the corals excrete excess mucus, which encourages bacterial attack on the coral–algal community, so killing them. Since coastal upwelling often controls the amount of nutrient supplied, it follows that coral reefs should be largely absent amongst the carbonate platform margins of upwelling coastlines, a prediction confirmed by observation.

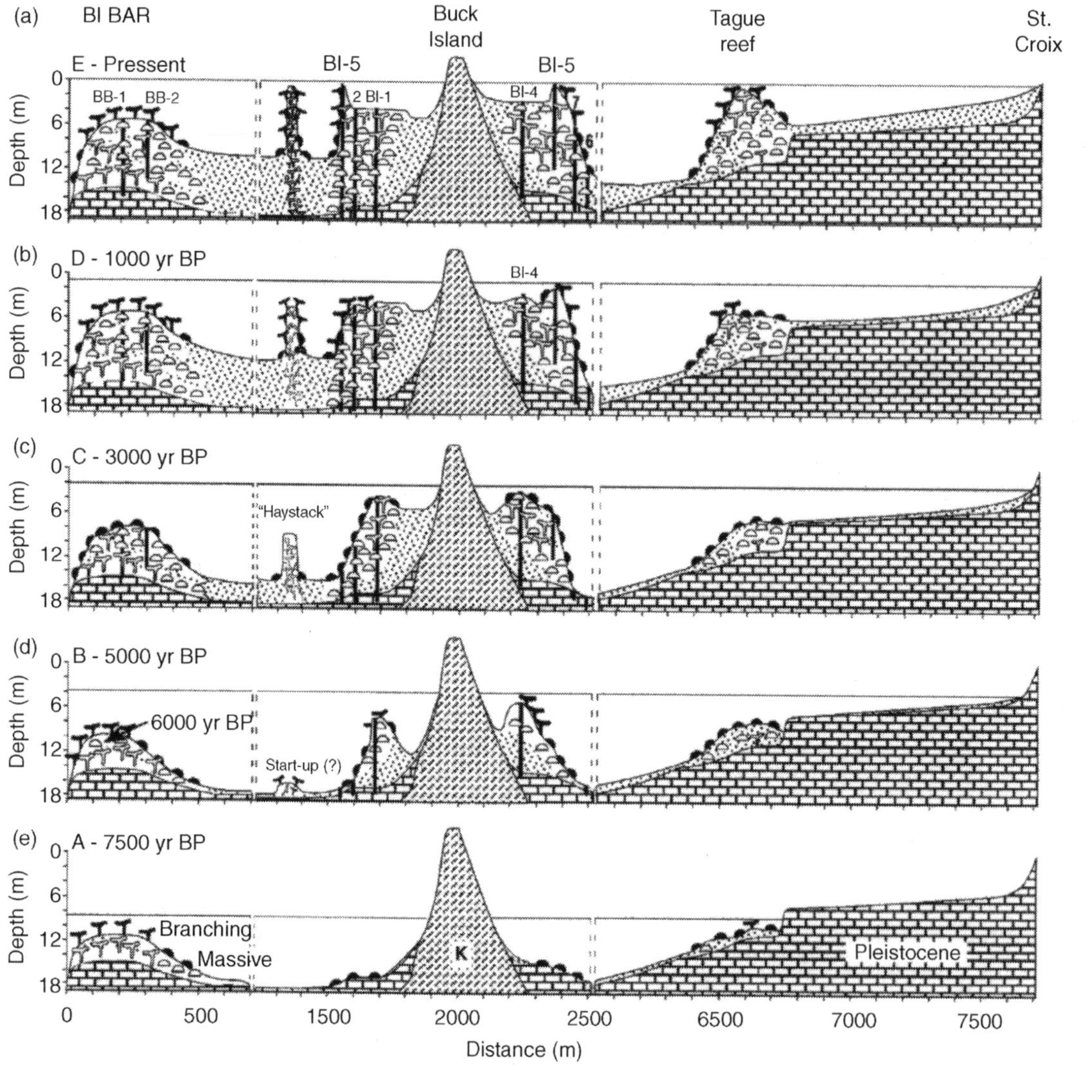


**Fig. 20.29** Changes in reef development across the northern St Croix Shelf from 7500 Cal yr BP to the present. The underlying Pleistocene carbonates are shown by the brick pattern. Buck Island is Cretaceous (K), volcaniclastic rock. Earliest reef development occurred along Buck Island Bar and Lang Bank, to the east (not shown). Buck Island reef built upward and seaward, steepening progressively over time. Reefs were dominated by massive corals at 6,000 Cal yr BP. The return to massive corals just before 3000 Cal yr BP lasted until ca 2100 Cal yr BP. (After Hubbard *et al.*, 2005.)

Coral growth can also occur only below some ambient energy threshold. Core studies of Hawaii Holocene reefs around Kauai, Oahu and Molokai show that in most cases vigourous north-fringing reef accretion during the early Holocene ended at the start of the mid-Holocene ca 5 ka, roughly at the time of highstand. Previous studies had assumed the termination was due to mid-Holocene sea-level fall but there is no independent evidence of this and a more likely correlation is with the onset of enhanced ENSO winter-season storm occurrence in the North Pacific, responsible for generating sustained large and damaging north-derived swell. (NB This mechanism is not to be confused with ENSO-related warming which leads to coral death by 'bleaching'.)

Quaternary sea-level oscillations have had major effects upon reefs developed on lowstand surfaces. The *antecedent karst theory* envisages lowstand subaerial exposure of a limestone platform surrounded by a relatively steep structural or depositional slope. $CaCO_3$ dissolution is concentrated in the middle of atoll-like offshore banks and on the landward flanks of barrier-like ramps. Karstic rims and tower karsts acted as nuclei for coral growth during rising sea level to produce the present-day atoll rims, barrier reefs and lagoonal pinnacle reefs. However, examples of the Belize/Honduras lagoons do not have karstic foundations but are built upon the undulating remains of a lowstand siliciclastic coastal plain. The behaviour of reef complexes during periods of sea-level change is epitomized by the well-exposed Miocene reefs of Mallorca. These remarkable rocks give evidence for prograding lowstand reef growth with erosion landwards, aggradational growth during sea-level rise, progradation once more at highstand, followed by offlapping and downlapping with landward erosion during forced regression (Fig. 20.30).

### Reefs as fine-scale tuners of the pleistocene sea-level curve

From the above discussions it will be obvious that the elevation of fossil reef crests, with their distinctive architecture, coral species and boulder gravels, will be a mean dipstick of use in calibrating the Pleistocene and older sea-level curve. This is primarily because of the present ability of mass spectrometers to date diagenetically unaltered coral by means of $^{230}$Th determinations, routinely obtaining ages from 130 ka accurate to within 2 kyr. A case history is outlined in section 23.17.

## 20.8 Platform margin slopes and basins

Much of our knowledge of platform margin slopes comes from integrated studies of the Bahamas Bank margin, in particular the results of submersible dives, deep-sea cores and seismic profiles (Figs 20.1 & 20.31). Interesting data also comes from the Belize platform. Much has also been inferred from exposures of ancient margins, in particular from magnificent exposures of Triassic carbonate platform margins in the Italian Dolomites (see further below), the Permian Capitan reef front of Texas and the Devonian reefs of the Canning Basin margin, Australia. These have shed much light on platform growth and form important field tests for computational depositional models. The following controls influence sedimentation at platform margins: (i) presence or absence of margin-bounding faults; (ii) direction and magnitude of off-platform sediment transport (modern platform green algal and chemical precipitation rates vastly exceed deposition rates); (iii) accretion by mass flows and pelagic deposits; (iv) occurrence of submarine landslides; (v) nature of the oceanic/thermohaline currents and accretion of basin-floor/distal-slope sediment wedges; (vi) degree of submarine cementation and microbial activity; and (vii) presence of deep-water organic build-ups. Item (viii) emerges as the major of these controls, with maximum export of bank-top carbonate sediment during highstands (when platform interior production was greatest) causing platform accretion to leeward. Platform margin deposits are thus important sources of information about sea-level changes and platform-top carbonate production. Highstands were often periods of high fore-reef production, these alternating with periods of relative sediment starvation (despite the influx of karstic detritus) and algal encrustation on the lower fore-reef slopes during lowstands.

The extensive scalloped margins of the smaller platforms of the southeast Bahamas suggest that periodic gravity collapse due to undercutting or faulting may have disrupted a once more extensive marginal platform, an inference borne out by dive-observations of the dominant role of lowstand collapse in forming periplatform breccias. However, despite these observations, seismic data (Figs 20.1 & 20.31) make clear that lateral bank growth and the infilling of original structural depressions like Bimini embayment and the Straits of Andros by lee-side accretion were predominant over the long term. Submersible observations emphasize the role of highstand lateral accretion of upper bank margins by outgrowing coral ledges, their periodic collapse down very steep (50°) bypass walls to the limit of Pleistocene sea-level lowstands, and the cementation of a residual fore-reef slope below about 150 m water depth. On leeward platform margins, exported bank-top aragonite sediment during highstands drapes the lowstand breccia wedge. Seismic and core data reveal that Belize platform evolution involved significant aggradation and slight progradation during the Quaternary of a reef-rimmed margin,

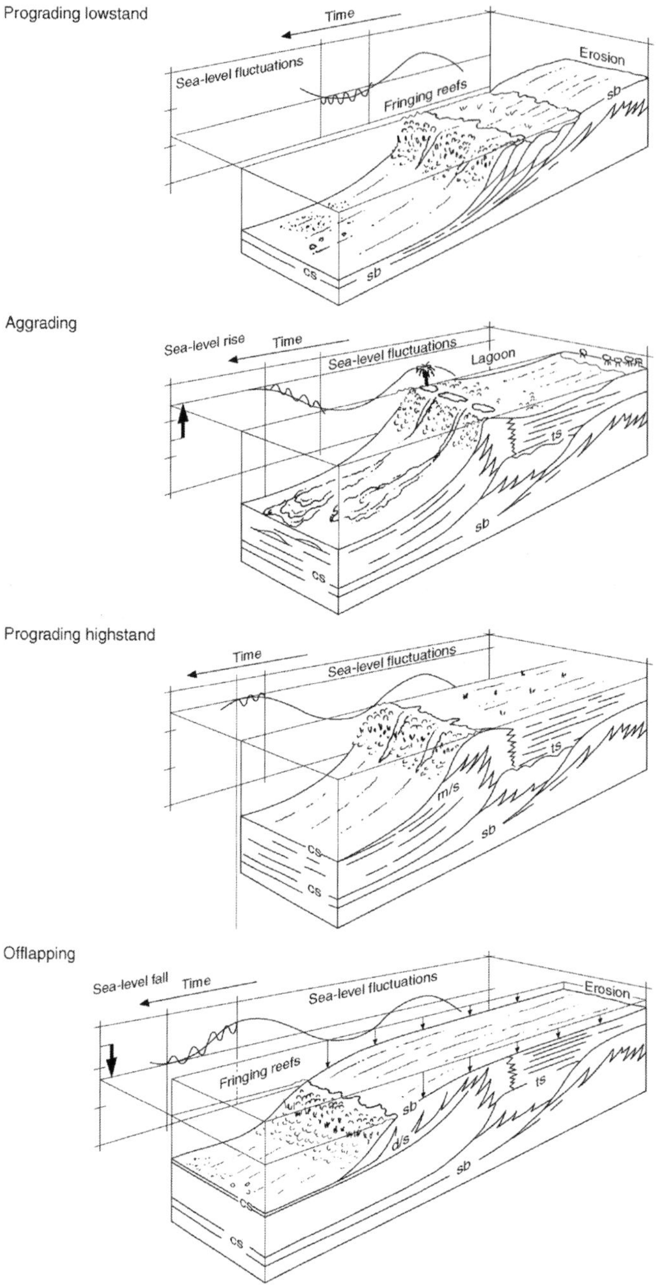


**Fig. 20.30** Schematic diagrams to show the response of late Miocene Mallorcan coral reefs to sea-level change and the types of sedimentary architecture produced. d/s, downlap surface; cs, condensed section; sb, sequence boundary; ts, transgressive surface. (After Pomar & Ward, 1994.)

in contrast to the unrimmed ramp that existed in Pliocene times.

Hemipelagic and gravity-flow processes dominate bank margin sedimentation. Suspension of fine- to sand-grade sediment during winter storms is the main export process, the dense plumes sinking laterally as underflows. Thick carbonate turbidites occur on the lower slopes around Little Bahamas Bank (LBB) and as thinner beds in the basins, where they are interbedded with pelagic oozes. Mass gravity flows are important where slopes are steep. Slope breccias of debris flow origin occur on the gentle muddy slopes, whilst grain flows occur at the base of very steep (~18°) slopes around the marginal escarpment. Pelagic carbonate deposition is important only when not winnowed by bottom currents, diluted by gravity flows or dissolved below the carbonate compensation depth. Cores through Middle Pleistocene to Holocene sediment (375 ka to present) in 460 m water depth off the northern slope of the LBB reveal two basic sediment types: (i) fine-grained periplatform aragonite and aragonite/calcite oozes with both shallow-water and pelagic particles and (ii) coarser intervals with cemented debris consisting of massive, poorly sorted, mud-supported or clast-supported deposits with an increased high-Mg calcite content (Fig. 20.32). During interglacial stages (marine isotope stages 1, 5, 7, 9 and 11) periplatform oozes are characterized by higher aragonite contents, finer grain-size and higher organic contents, whereas during glacial stages (marine isotope stages 2 to 4, 6, 8 and 10), increased low-magnesium and high-Mg calcite, coarser grain-size and lower organic contents occurred (Fig. 20.33). The coarser deposits (ii above) occur mainly at the transitions from glacial to interglacial and interglacial to glacial stages, and are interpreted as redeposition

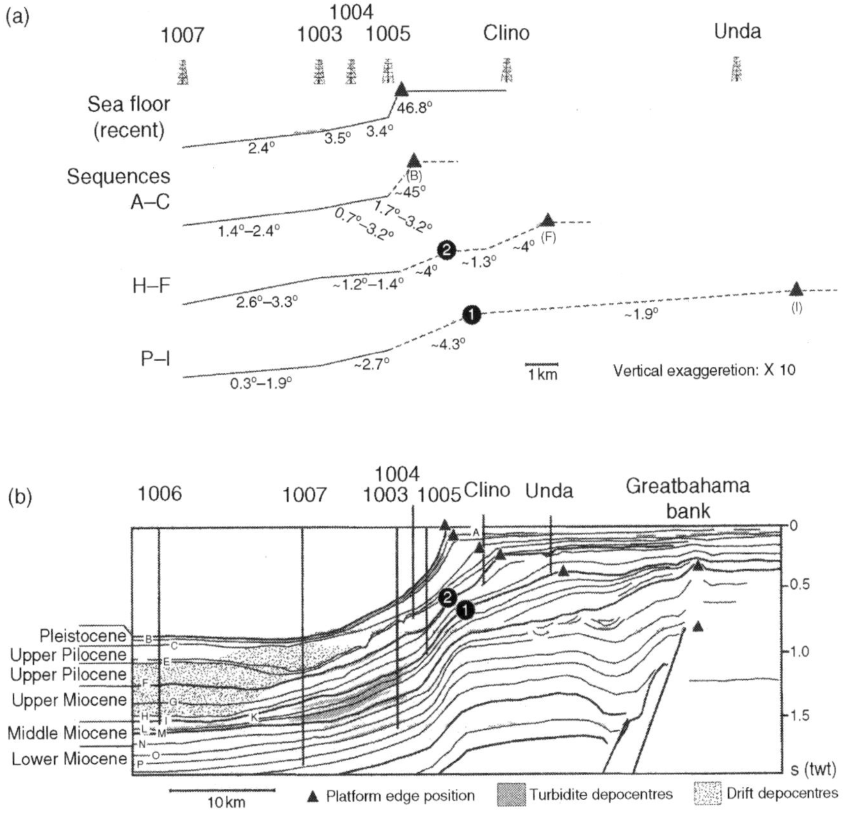


**Fig. 20.31** (a) Evolution of increasing slope-break gradients from ramp-like form on the leeward flank of Great Bahama Bank. (b) Location of turbidite and drift depocentres along the flank. (After Betzler *et al.*, 1999.)

events, indicating a direct link between sediment properties (changes in mineralogy, grain-size distribution, variations in organic contents) and sea-level fluctuations. Changes in hydrostatic pressure and the wave-base position during sea-level changes are proposed to have triggered these large-scale resedimentation events.

Cores through mid-Holocene sediments in the leeward accretionary wedge at a water depth 290 m off Bimini to the north of the Great Bahamas Bank platform show high sedimentation rates in the lower core section (the first 1600 yr) caused by rising sea level that switched on the platform-top carbonate factory. Sediment is rather uniform white to light grey ooze, a mixture of platform-derived components, mostly aragonite needles, and pelagic planktonic foraminifera and coccoliths: no evidence for resedimentation events was seen. Time series and spectral analysis of aragonite abundance shows a multimillennial signal and quasi-periodic oscillations of 1300–2000 yr, 500–600 yr, 380 yr, 260 yr, 200 yr, 100 yr, 88 yr and 60 yr period. Two major factors may control such aragonite variations: (i) rate of aragonite supply due to covarying production on the platform and (ii) offbank transport efficiency. The major impacts on the former are sea-level changes and the aragonite saturation state of the platform waters controlled by water depth/temperature, distance from platform margin and platform water residence time. Transport efficiency depends upon trade-wind vectors and winter cold-front activity. Regarding explanations for the aragonite variations, the origins of the multimillenial signal is unknown whereas the close link of the millennial-scale climatic fluctuations to proxies of solar forcing in the North Atlantic suggest a control via modification of the atmospheric circulation cells and surface winds possibly caused by changes in solar irradiance. The 200 yr and 100 yr signals may be attributed to solar forcing, whilst the 260 yr, 380 yr and 500–600 yr quasi-periodic signals are probably of climatic origin.

Cores at ~1000 m depth on the windward slope margin off Exhuma Cays into Exhuma Sound identify an anomalous negative excursion in foraminiferal $\delta^{13}C$ and $\delta^{18}O$ due to a change in coastal surface waters during or immediately after a period (4.0–3.8 ka) in which there was a regional net reduction in platform margin circulation, as well as local bank-margin erosion. The isotope anomaly is interpreted to record climate-controlled offbank transport of warmed, dilute waters and the formation of a brackish coastal zone, presumably reflecting increased precipitation onshore and possibly variations in trade-wind vectors or winter storms (the area is windward to Exhuma Cays but leeward of Cat Island Bank). Skeletal deposition has been dominant along this rimmed windward margin since 3.8–3.3 ka so that the reappearance and short-term dominance of non-skeletal allochems before 2.4 ka, then again in the past few hundred years is surprising, perhaps coinciding with periods of rejuvenated circulation due to climate change in this era of stable sea levels. Between 2.4 ka and the past few hundred years, aragonite and foraminiferal $\delta^{18}O$ stratigraphies indicate that the onshore carbonate factory of Lee Stocking Island was rimmed by shoal barriers. There was a reduced offshore flux of aragonite to the slope during this period with coastal surface water increasing in salinity, possibly because of seaward reflux of hypersaline waters through the shoal margin and/or cooled as a result of reduced offbank transport of warm bank waters relative to wind-generated upwelling.

Extensive areas of submarine cementation occur west of the northern Bahamas and in the Tongue of the Ocean down to depths >500 m. These lithified slopes are very stable and the cementation combined with strong current scour help to maintain the steep gradients observed in all the deep embayments separating the Bahamian platforms. In general the degree of cementation decreases downslope from well-lithified hardgrounds at depths < 375 m, to lithified nodules in a soft muddy matrix at depths from 375 to 500 m, and to soft oozes at depths greater than 500 m. The nodules are multigeneration intramicrites to intramicrudites cemented by high-Mg calcite in layers up to 1.5 m thick. The cementation is greatest along slopes where the Florida Current flows.

There are also widespread thick (>600 m) sediment drifts in the Straits of Florida and the Santaran Channel between Bahamas and Cuba at depths up to 800 m. These accumulations of mostly shallow-water-derived carbonate sand-grade sediment are reworked by energetic ocean-floor contour currents with velocities up to 0.6 m/s. Cementation and bottom currents also play a role in localizing the spectacular 'ribbon' of deep-sea build-ups, which extends over 200 km from the Blake Plateau along the western margin of the Little Bahama Bank to Bimini. These build-ups

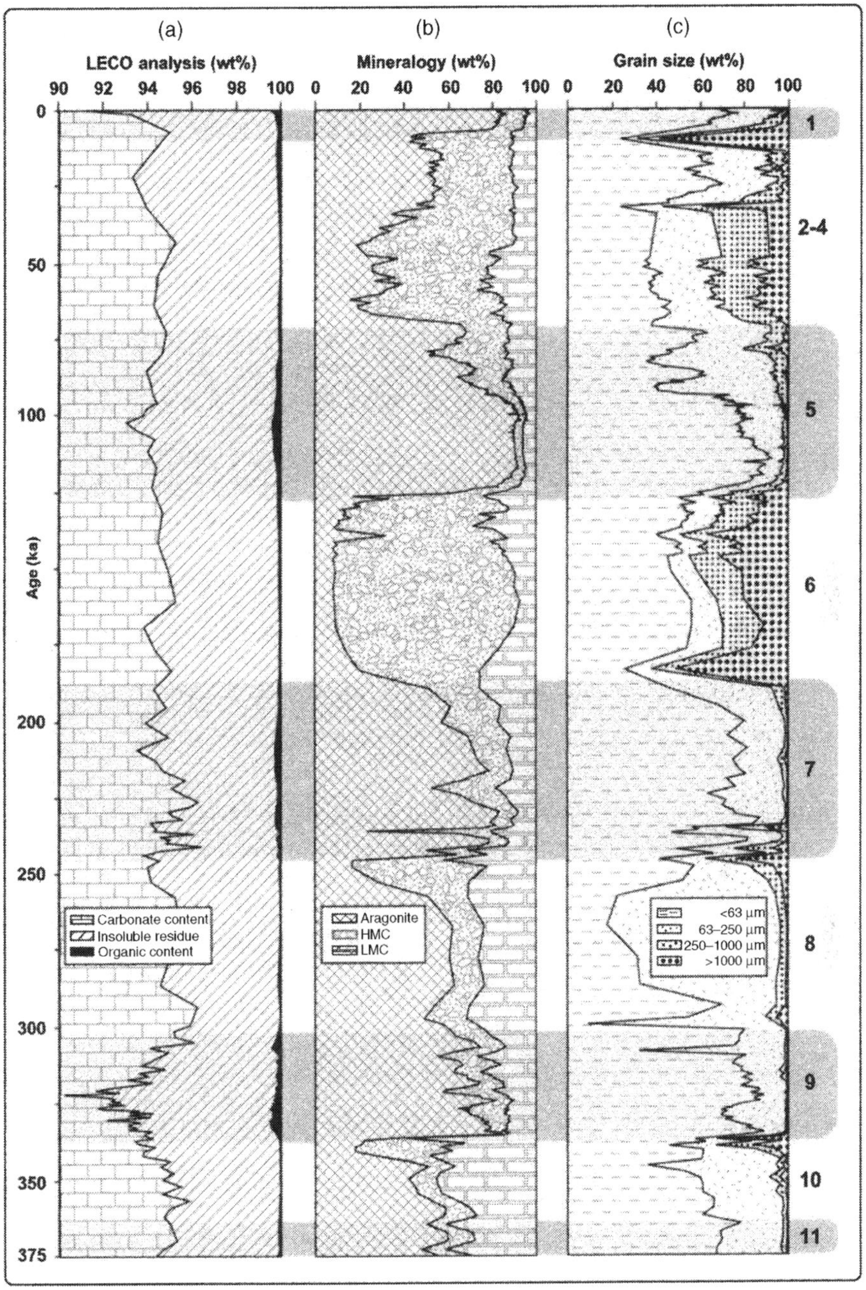


**Fig. 20.32** Data from piston core MD992202 taken from the north slope of Little Bahamas Bank (after Lantzsch *et al.*, 2007) to show glacial/interglacial fluctuations in (a) basic constituents, (b) mineralogy (HMC, high-Mg calcite; LMC, low-Mg calcite) and (c) grain size. Ordinate is age, abscissa is marine oxygen isotope stage (MIS): glacials, white; interglacials, grey.

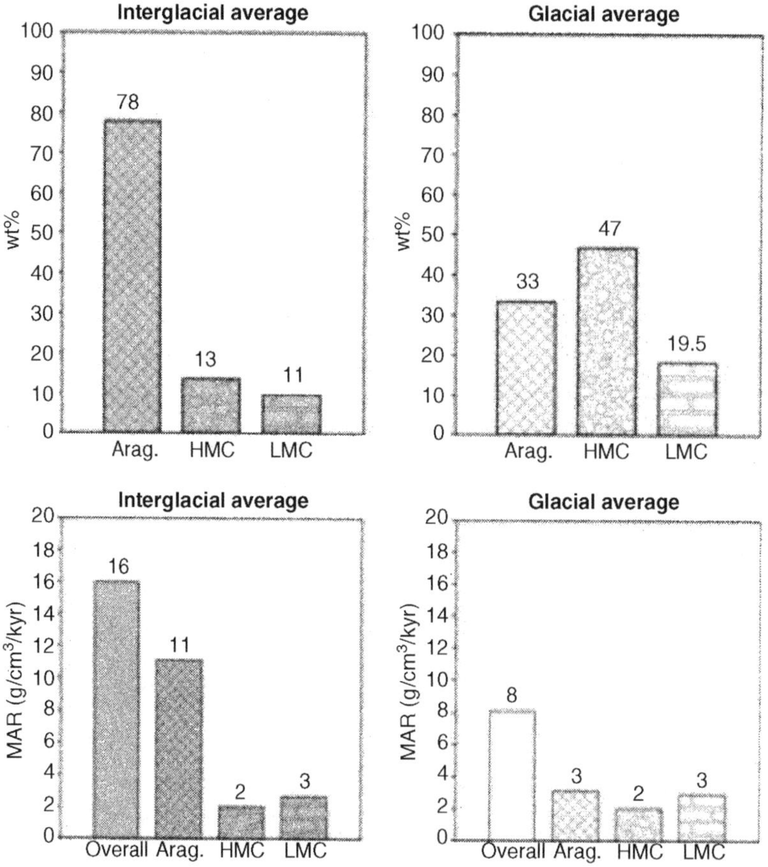


**Fig. 20.33** (a) Interglacial/glacial average values of mineralogy within core MD992202 of Fig. 20.32. (b) Interglacial/glacial average values of overall mass accumulation rates (MAR) and that separately of aragonite (Arag.), high-Mg calcite (HMC) and low-Mg calcite (LMC). (All after Lantzsch *et al.*, 2007.)

occupy a zone some 15 km wide in water depths of 600–700 m. They show up to 50 m relief and may be hundreds of metres long, orientated parallel to the northerly deep current flow. Observations from submersibles confirm their contemporary origin and reveal a dense and diverse benthic community of crinoids, ahermatypic corals and sponges, which baffle and trap sediment provided by the bottom currents. The build-ups are constructed *in situ* by lithification of successive layers of trapped sediment by micritic high-Mg calcite.

Particular insights into the evolution of carbonate platform margins have come from magnificent exposures across exhumed Triassic examples in the Latemaar Massif of the Italian Dolomites (Fig. 20.34). This is part of a set of atoll-like carbonate platforms and has a diameter of about 2.5 km. Clinoforms record episodic marginal progradation and at first sight might suggest Bahamian-style leeward platform accretion. However, the majority of steeply dipping clinoform sediment is more-or-less *in situ* microbial boundstone: it seems that the clinoforms were their own carbonate factory! Regardless of their origins, the nature of the lower and upper boundaries to the slope clinoforms yields important information on the processes and rates of aggradation and progradation. A strong case for the role of very slow gravitational collapse and shallow subsurface cementation in emphasizing clinoform surfaces is made for certain Triassic slope clinoforms (Fig. 20.35). More recent discoveries include a magnificent Moroccan Jurassic margin (Fig. 20.36) whose evolution from aggrading to retrograding, prograding, and retrograding was strongly influenced by syndepositional extensional tectonics superimposed on the general effects of eustasy.

## 20.9 Carbonate sediments, cycles and sea-level change

As we have seen throughout this Chapter, sea-level change has pronounced effects on marine carbonate sediments. Apart from general trends in sediment character caused by shoaling or deepening the effects may be briefly summarized thus:

1 Exposure to dessication and to freshwater percolation causes *tepee structures* to form by a combination of large-scale polygonal shrinkage-cracking and cementation due to dissolution of aragonitic/high-Mg calcite bioclasts and their reprecipitation as low-Mg calcite.
2 Small- to large-scale karstification occurs as meteoric/phreatic water tables descend through highstand sequences. The lower limit to karstification, cave dissolution and speleothem formation is found at the base of the freshwater lens that overlies saline phreatic seawater.
3 Distinctive supratidal soils and encrustations like *terra rossa* develop in humid carbonate environments whose origins owe much to root growth, *in situ* dissolution and deposition of wind-blown dust.

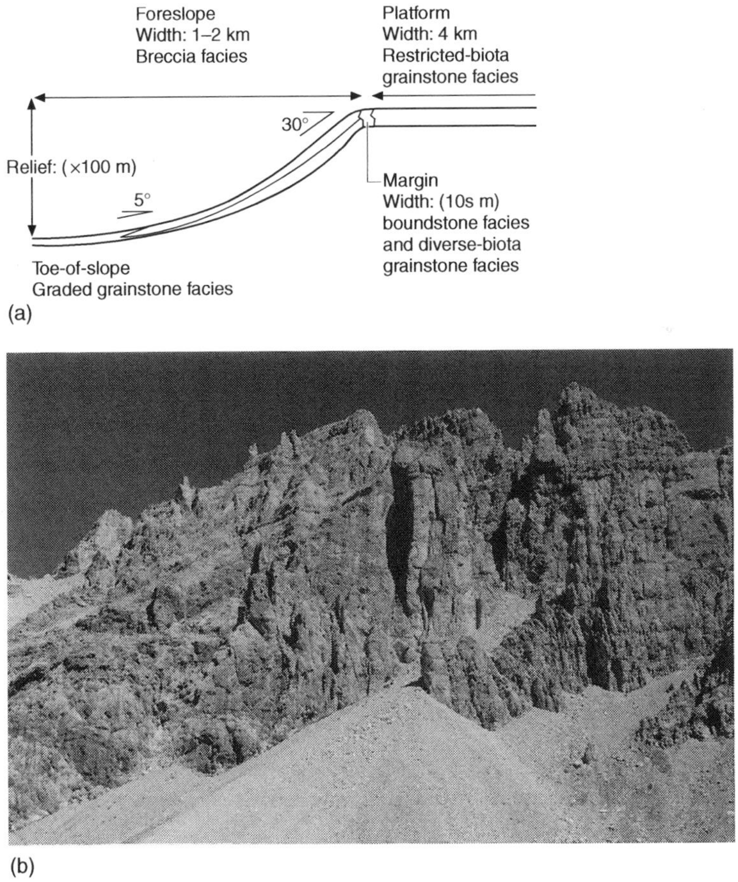


**Fig. 20.34** View of mountainside exposure of the transition from steeply dipping foreslope to horizontally bedded platform-top deposits of the Triassic Latemar platform, northern Italy. (From Harris, 1993.)

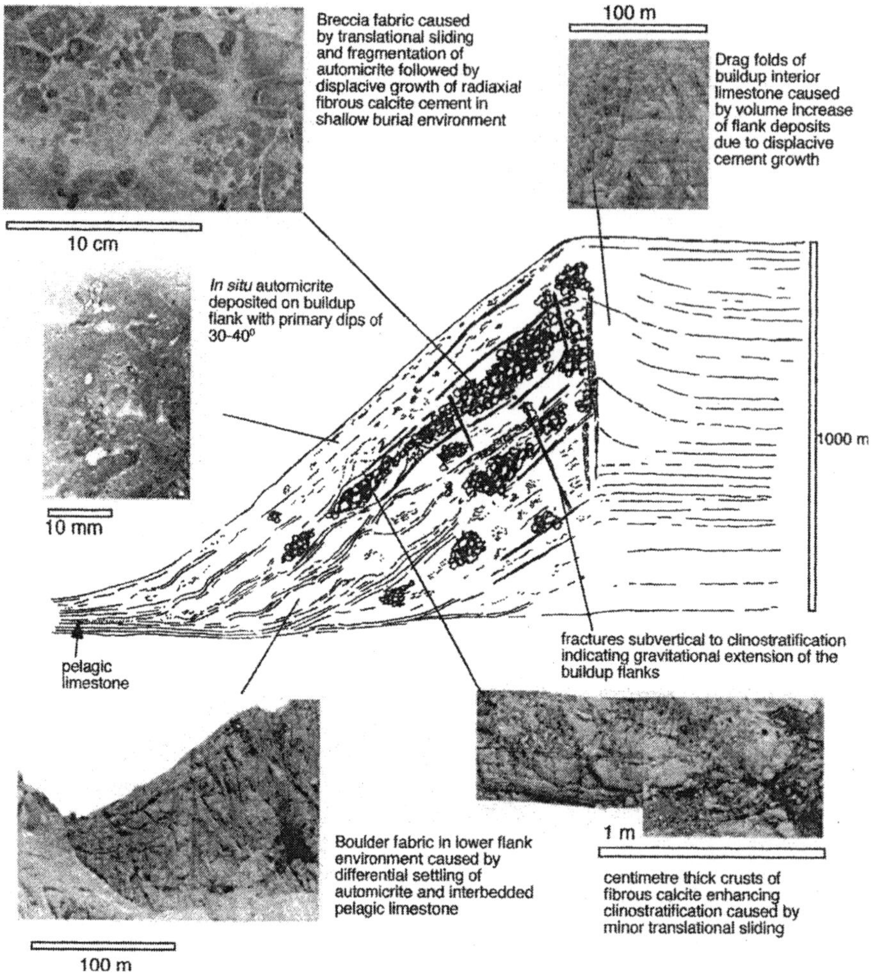


**Fig. 20.35** Diagram showing the suggested effects of marine early phreatic diagenesis on the rock fabric of clinoforms of Middle Triassic build-ups in the Dolomites. (After Blendinger, 2001.)

4 Development of significant spreads of wind-reworked carbonate sands, *aeolianites*, by onshore winds working on exposed carbonate sediment shoals. Meteoric cementation and soil development gives them high preservation potential.
5 Seaward spread of distinctive supratidal sabkha and salina evaporites during regression and lowstand in arid climates. Also, mixed saline and fresh conditions define *schizohaline* environments.
6 Seaward spread of siliciclastic sediments occurs over previous carbonate environments with evidence of incision by river and delta channels.
7 Solution notches produced at platform margins at the marine phreatic–meteoric interface.
8 Jointing and cracking of karstified and partly lithified build-up carbonates occurs along steep lowstand platform margins due to wave action with lowstand collapse of cracked and jointed blocks to form distinctive talus cones at the foot of periplatform slopes.

All these and other features due to exposure may be used to track the course of significant ancient sea-level change, either in core samples, seismic work or at outcrop. Lowstand features are often draped by con-

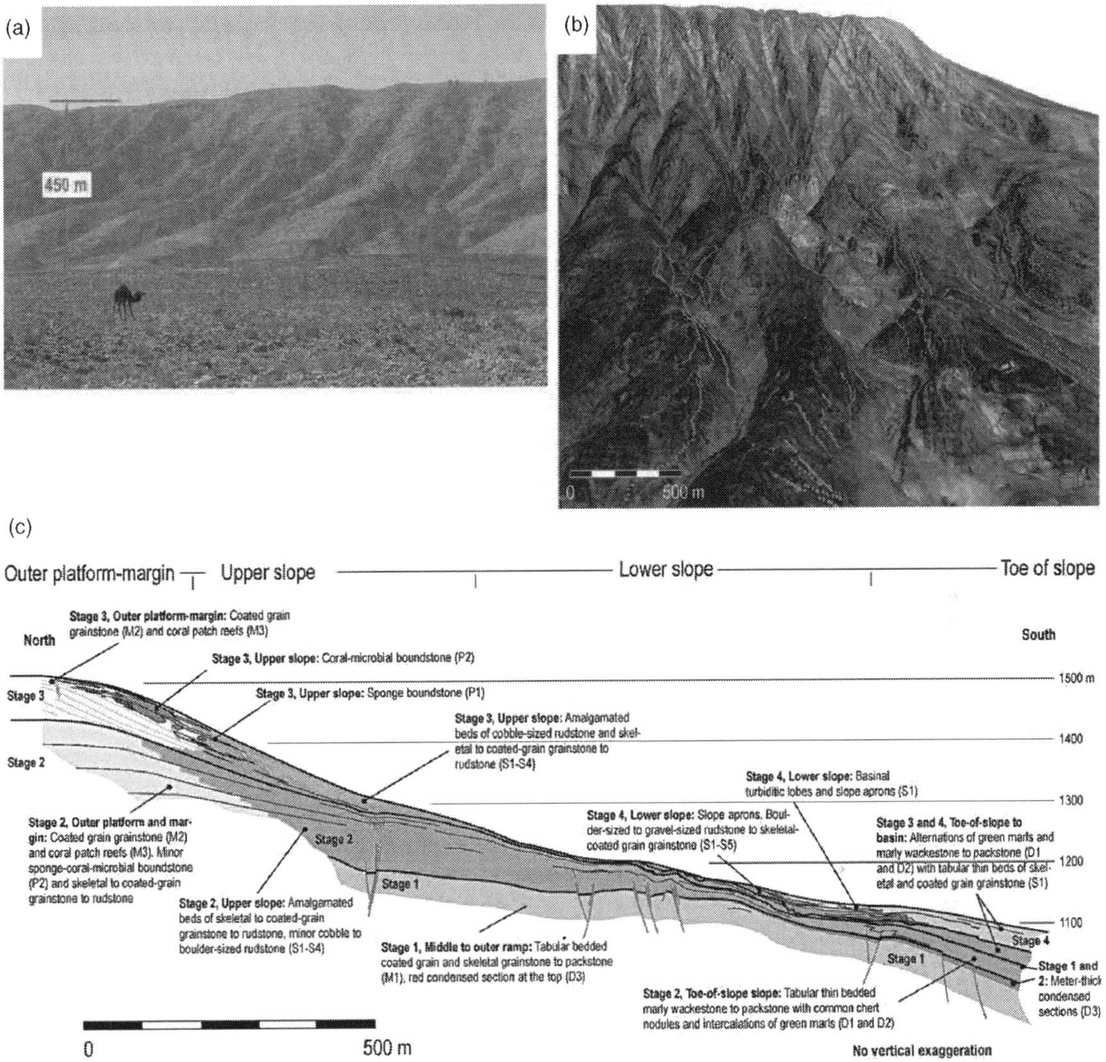


**Fig. 20.36** Moroccan Jurassic carbonate platform slope.(a) Outcrop photo showing the Lower Jurassic high-relief slope of the platform. (b) Oblique view of Quickbird satellite image draped over DEM, horizontal image resolution 0.7 m. (c) Summary cross-section illustrating lithofacies types, main stratigraphic units and traced stratal surfaces. Vertical scale is GPS height in metres. (After Verwer *et al.*, 2009.)

trasting transgressive and highstand sediments with a record of 'switching-on' of bank-top carbonate factories and the rapid burial of lowstand wedges by highstand finer-grained drapes resulting from renewed export of bank-top carbonate in response to prevailing wind directions.

It is commonplace to observe sedimentary cycles developed in carbonate sediments. Some are of shallow-water origins—witness those peritidal examples a few metres thick deposited on extensive ramp-type margins or in platform-top lagoons. Bed-by-bed correlation of such cycles establishes beyond doubt that they have been forced by factors external to the carbonate sediment factory, i.e. they are *allocyclic*, almost certainly the product of eustatic sea-level changes. However, some authors remain sceptical about the very existence of such cycles, pointing to the randomness of modern carbonate ramp sediment distributions. Though some ramp environments may show nearest-neighbour randomness this conclusion seems nugatory as a general principle. Spectral analysis of platform lagoon deposits like the famous Triassic Latemar Limestone (see Fig. 20.34) originally led to the proposal that shoaling-upwards cycles developed at orbital frequencies, but precise U–Pb radiometric dating of intercalated volcaniclastics and faunal correlation of the Latemar sequence reveal that depositional time was far too short (by about 50%) to allow such an interpretation for all 600 or so cycles. Thus the duration of the upper part of the Latemar succession is between 2–4.7 Myr. If this is correct, each cycle would have lasted about 5000–7000 yr, clearly not within any Milankovitch band—the origin of such multimillenial forcing seems obscure. Using the chronology, compacted carbonate accumulation rates ranged between 100 and 235 mm/kyr. In contrast to the Latemar, younger Triassic cycles of the Durrenstein Formation in the same general carbonate province indicate Milankovitch-driven cyclicity.

Another type of small-scale cyclic sequence common in the Mesozoic and Cenozoic record comprises relatively thin alternations of a few decimetres to metres between carbonates and marl intervals richer in siliciclastic clays. Pliocene examples in southern Italy have been ascribed to climatic fluctuations at precessional (~20 kyr) frequencies. The sedimentary origins of the cycles are somewhat controversial. They are clearly of ramp to basinal origins, mostly well below storm wave base. Some workers ascribe the marly layers to periods of increased freshwater runoff and higher clastic sediment yield; comparable cycles on the West Florida ramp result from periodic Mississippi-induced changes in shelf sediment plume activity. According to this idea the carbonate units were deposited during periods of more arid climate. However, evidence that the two components of the cycles have quite distinctive microfaunas and floras, with the carbonates richer in foraminiferae indicative of highly productive cool upwelling conditions, leads to an alternative explanation, *viz.* that the cycles represent alternation between high- (limestones) and low-intensity (marls) upwelling conditions related to the precession-dependent strength of Pliocene wind systems. It is likely that both ideas contribute, for changes in upwelling alone cannot cause clastic input to vary without concomittant changes in continental runoff.

## 20.10 Displacement and destruction of carbonate environments: siliciclastic input and eutrophication

It is very common in the stratigraphic record to find alternating carbonate and siliciclastic facies. It is also increasingly recognized that carbonate production in patch reefs can occur synchronous with deltaic clastic input. Well-defined alternations imply that carbonate depositional environments were periodically displaced by the introduction of siliciclastic sediment from rivers. This implies that a carbonate nursery and siliciclastic factory were adjacent at any one time, ready either to expand or contract in area in response to external forcing. But what kind of forcing? Climate change might dry the rivers up, but not completely, whereas sea-level change will cause transgression and at highstand the spread of carbonate sediments over former coastal plains. At falling stage and lowstand the situation reverses, with the shelfward spread of rivers and stranding of highstand carbonates above mean sea level and subject to the vagaries of cementation, karstification and soil development. Or, a relative sea-level change might be caused by tectonics, with subsidence causing the carbonate expansion shorewards on to the coastal plain and uplift the reverse. Clearly, each geologically ancient example of carbonate/clastic cyclicity must be approached on a case-by-case basis. However, there are clearly obvious constraints imposed when icehouse conditions pertained and it is reasonable to supppose that in such cases fast eustasy (Chapter 10) must have played a dominant role in cycle development.

The Queensland coast, northeast Australia (Fig. 20.37) is a haven for studying the effects of siliciclastic input, for here the coastal plain and inshore are dominated by detritus brought in by ferocious monsoon-fed rivers like the Burdekin (see Chapter 11). Only 15 km or so offshore lies the Great Barrier Reef complex (up to 100 km wide) in waters of 15–150 m depth so that the siliciclastic factory is coeval with the carbonate nursery and significant amounts of siliciclastics are shed on to the outer shelf and into the Queensland basin through gaps in the reefs. To a first order, the composition of surface sediment on the shelf reflects proximity to sediment sources. Along the North Queensland segment of the margin, modelling studies estimate that 15 rivers deliver about $12 \times 10^6$ tonnes of terrigenous sediment to the inner shelf each year, compared to $10^6$ tonnes/yr before land-use changes last century. Of these fluxes, Burdekin River supplies most of the sediment, with minimum discharge rates between 1.3 and $2.5 \times 10^6$ tonnes/yr averaged over the Holocene. Monsoon- and cyclone-influenced, high-discharge events deliver most of this river sediment to the GBR shelf. For example, during a 29 day discharge event in February and March of 2000, an estimated $3.7 \times 10^6$ tonnes of suspended sediment and $0.3 \times 10^6$ tonnes of bedload material were delivered through the mouth of Burdekin River alone. For comparison the production of carbonate from the GBR nursery is a staggering $68 \times 10^6$ tonnes/yr.

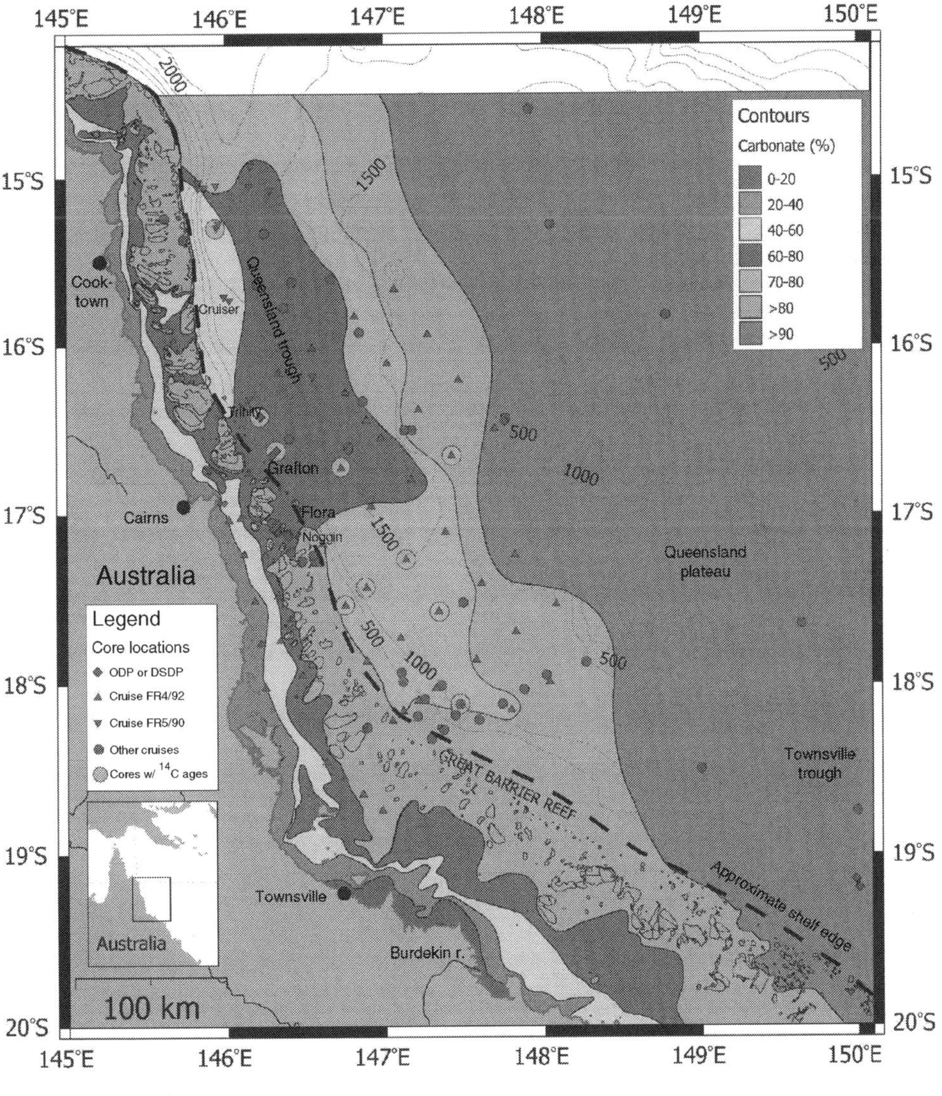


**Fig. 20.37** Map of North Queensland Margin to show coexisting coastal plain siliciclastic belt and Great Barrier Reef carbonate nursery, with contours of $CaCO_3$ content. (After Francis *et al.*, 2007, and sources cited therein).

With such an example as the Queensland margin, it is easy to imagine the seaward or landward shift in clastic and carbonate facies belts during periods of sea-level change in the geological record. A similar situation pertains in the Arabian Gulf where the Tigris–Euphrates delta and its extensive organic-rich wetlands are present adjacent to coastal sabkhas and shallow-water carbonate deposits. Further light is shed on the problem from shallow seismic and coring explorations in the Belize lagoons of Honduras, where it is evident that the modern highstand lagoonal reefs and carbonates behind the fringing barrier platform overlie lowstand clastic sediments deposited in fluvial and deltaic distributary channel complexes. The palaeomorphology (levees, etc.) and compactional relief over this siliciclastic foundation has strongly influenced the locus of subsequent reef formation. An example of a novel purely climatic model for control of mixed-sediment cycles is shown in Fig. 20.38, whilst Fig. 20.39 outlines the detailed sedimentary architecture of mixed-sediment cyclicity.

Particularly interesting mixed-sediment cycles and alternations characterize high-latitude margins of the geological past. During Permian icehouse times in the Australian sector of Gondwanaland the relatively isolated early and middle Permian carbonates in southern Queensland are important because they provide a window into regional palaeo-oceanography that is not obtainable from siliciclastic rocks. The carbonate nursery is principally a relatively low-productivity heterozoan association of pelmatozoans, bryozoans, bivalve molluscs and brachiopods. The factory lacks large benthic forams that characterized coeval warm-temperate neritic palaeoenvironments. A recurring deepening-upward pattern confirms that this high-latitude carbonate system was unable to keep pace with rising sea level brought about by large-scale deglaciation. Relative scarcity of limestones overall further demonstrates that because of slow growth rates in these cold waters, biogenic carbonates were easily overwhelmed by siliciclastic deposition and were confined to specific isolated environments.

A second reason for the demise of carbonate platforms involves the phenomena of eutrophication and anoxism, whereby an increase in the abundance of seawater nutrients (perhaps induced by upwelling or warming) causes shut down of the benthic carbonate nursery and reduced sedimentation so that coupled with ongoing subsidence the platforms effectively 'drown' and are subject to deposition of organic-rich black shales. In fact, there is a distinct absence of modern carbonate platforms along all eastern oceanic basin margins, precisely those that tend to be affected by upwelling. We have already noted the controls upon coral reefs provided by nutrient availability: many examples of carbonate successions succeeded by black shales are known in the sedimentary record. In particular the great carbonate platforms of the Jurassic Tethyan Ocean margin now exposed in southern Europe illustrate contrasting behaviour during the Early Toarcian *Oceanic Anoxic Event* (OAE). The western sector of the Trento Platform in the Southern Alps of Northern Italy shows evidence for deepening and the development of more clay-rich cherty facies, suggestive of eutrophic conditions due to increased nutrient availability. Although the platform recovered to some extent and did not drown definitely until Aalenian/Bajocian time, the oolitic and crinoidal facies that developed indicate a certain open-marine influence. By contrast, the carbonate platform in the Southern Apennines shows only minor facies changes recording the Early Toarcian OAE and these were registered only at its onset: during most of the OAE shallow-water oolitic sediments were deposited. A major difference in the two platforms is the reconstructed rate of subsidence during the Toarcian/Aalenian interval: 23–28 m/Myr for the Trento Platform versus 12–15 m/Myr for the Campania–Lucania Platform. The lesser subsidence rate in the case of the latter may have helped maintain environments as shallow as a few metres throughout the period of the OAE, hence shielding it from certain environmental factors that adversely affected the Trento Platform, which ultimately led to its definitive drowning during the Aalenian.

## 20.11 Subaqueous saltern evaporites

Until the importance of sabkha evaporites (section 6.2) was recognized in the mid-1960s, it was widely assumed that evaporitic salts in the geological record were precipitated subaqueously from bodies of standing brine, sometimes termed *salterns*, as generally outlined in Chapter 3. Standing brine bodies today form in a multitude of settings, including marine-flooded sabkha (e.g. Abu Dhabi, Trucial Coast),

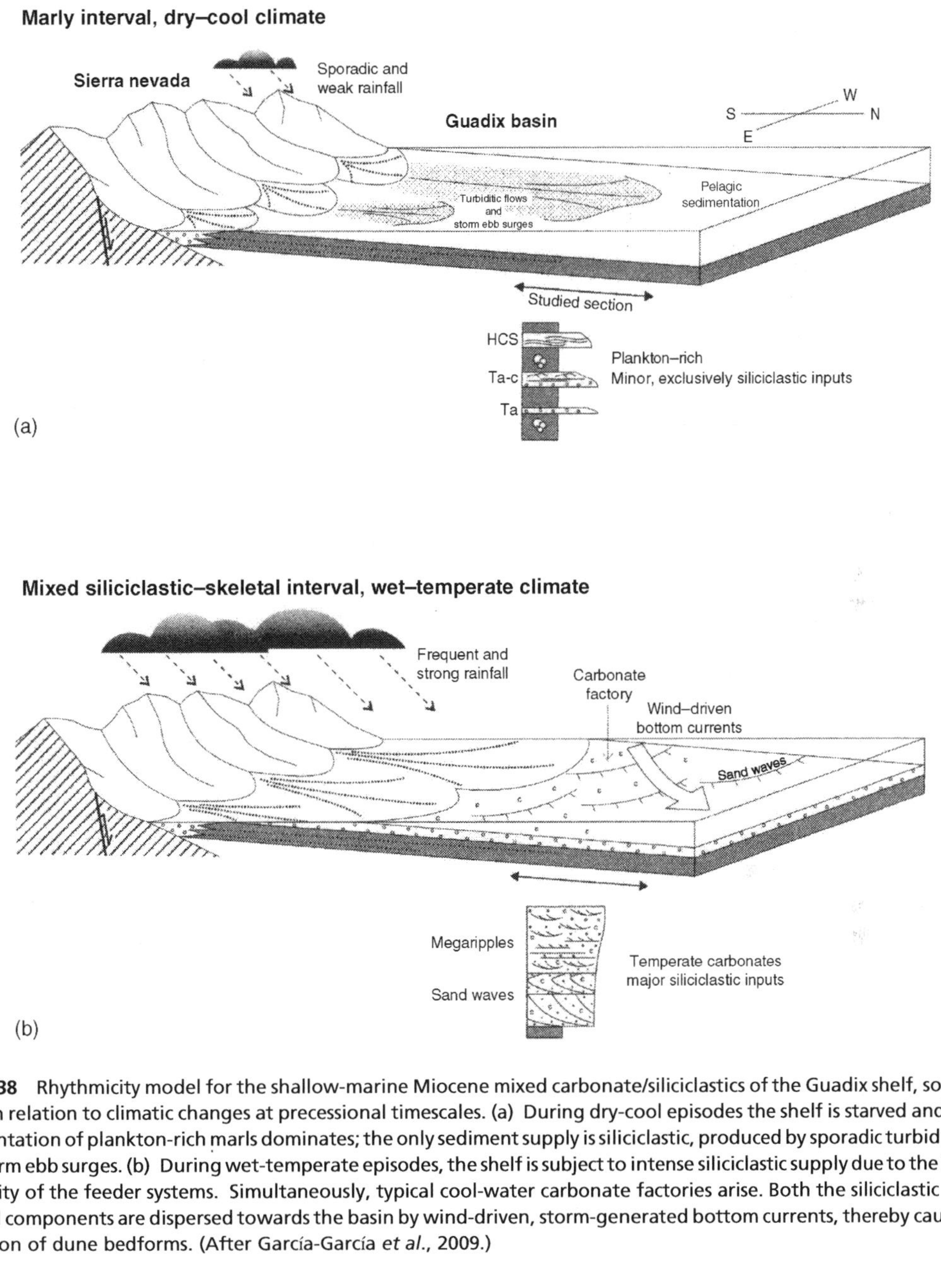


**Fig. 20.38** Rhythmicity model for the shallow-marine Miocene mixed carbonate/siliciclastics of the Guadix shelf, southwest Spain in relation to climatic changes at precessional timescales. (a) During dry-cool episodes the shelf is starved and pelagic sedimentation of plankton-rich marls dominates; the only sediment supply is siliciclastic, produced by sporadic turbidity flows and storm ebb surges. (b) During wet-temperate episodes, the shelf is subject to intense siliciclastic supply due to the increase in activity of the feeder systems. Simultaneously, typical cool-water carbonate factories arise. Both the siliciclastic and the skeletal components are dispersed towards the basin by wind-driven, storm-generated bottom currents, thereby causing the migration of dune bedforms. (After García-García *et al.*, 2009.)

abandoned coastal intradune depressions, desert rift basin playas (e.g. Death Valley, California), barred lagoons, deep perennial lake basins (e.g. the Dead Sea, Israel and Jordan), shallow perennial lakes (e.g. Great Salt Lake, Utah, USA) and non-marine saline pans and mudflats (e.g. Death Valley, California, USA, and Salar de Atacama, Chile). The crux of the dilemma regarding the origin of subaqueous evaporites hinges around: (i) the extremely large area over which some ancient evaporites may be traced, e.g. Silurian Michigan evaporite basin, USA (400 000 km$^2$), Permian Zechstein basin of Europe and whole Miocene extent

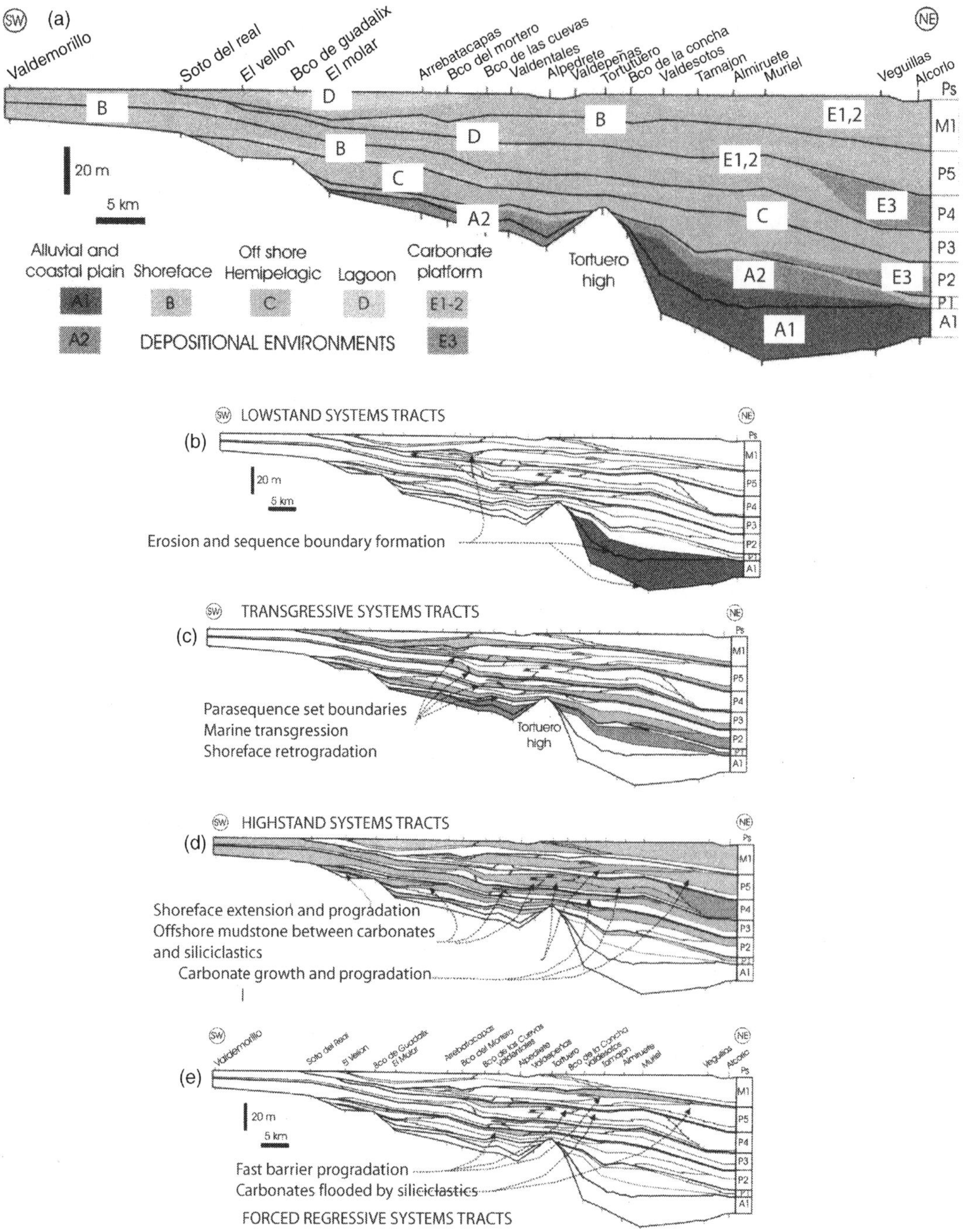


**Fig. 20.39** Carbonate–clastic cycles in the Cretaceous of Central Spain. (a) Cross-section showing depositional environments, sequences and parasequence set boundaries. (b)–(d) Facies and systems tracts recognized in the sequence and parasequence set. (b) Lowstand systems tracts (LST). (c) Transgressive systems tracts (TST). (d) Highstand systems tracts (HST) represent the change from retrogradational to progradational trends in shoreface deposits, marked by maximum flooding surfaces for the P2 sequence and parasequence sets. Carbonate growth and progradation occur in this part of the cycle, separated from siliciclastic shoreface by offshore mudstones. (e) Forced regressive systems tracts (FRST) are characterized by rapid barrier progradation with carbonate production stifled by siliciclastics. (After García-Hidalgo *et al.*, 2007.)

of the Mediterranean ocean; (ii) the scarcity of modern examples of such large evaporite basins. Ancient saltern basin tectonic settings include shelf thermal-sag basins (Zechstein), intracratonic carbonate bank-rimmed basins (Michigan) and reef-rimmed basins (Permian Castile evaporites, New Mexico/West Texas, USA). Lateral facies changes (Fig. 20.40) may allow basinal subaqueous evaporitic units to be traced into contemporary nearshore and saltern margin facies with evidence of shallow-water or continental deposition. In such cases a *prima facie* case for basin topography may emerge. Subaqueous nonevaporitic facies that overlie and underlie an evaporite unit may also be helpful in this respect. In the last resort it is the features of the evaporites themselves that may prove decisive. The following features would tend to indicate subaqueous deposition.

1 Shallow brine bodies are affected by waves and, perhaps, tides and so sedimenting evaporite crystals should exhibit primary bedforms and sedimentary structures. However, upon burial and particularly after structural deformation these may be partly or completely destroyed due to the weakness and recrystallization of evaporite minerals.
2 Gradual infill of brine bodies will lead to production of shoaling-upwards sequences.
3 Lateral gradations in brine salinity from basin centre to margin will lead to concentric or shore-parallel changes in evaporite mineralogy. Modern salinas are dominated by gypsum and less commonly halite precipitation (Fig. 20.40), in marked contrast to the anhydrites of Arabian sabkhas. It should be noted, however, that gypsum will always recrystallize to anhydrite with burial below about 500 m of overburden.
4 Should poorly lithified evaporites be uplifted and eroded then resedimentation may occur in adjacent basins as slumps, debris flows and turbidites. In this way some apparently deep-water evaporites may be derived from pre-existing shallow-water to supratidal evaporites that have been disrupted and uplifted by tectonic activity. Outstanding fine examples occur in the Messinian of Sicily, Italy.
5 Subaqueous evaporite deposition favours the production of widely traceable (tens to hundreds of kilometres) varve-like laminations. For example, the Castile laminae are particularly impressive 2-mm-thick alternations of calcite (containing organic matter) and anhydrite with the lamina of calcite forming about 6% of this value. The varves are generally remarkably regular in thickness, and are commonly repeated thousands of times without interruption. They formed at the brine surface or in the upper part of the brine and 'rained' on to the basin floor. Not only do the varves have great lateral persistence, the thicknesses of individual varves generally remain about the same over most of the basin. On the basis chiefly of thickness variations of anhydrite laminae, cores of alternating laminae of calcite and anhydrite have been correlated definitively over distances of up to 113 km.
6 Subaqueous evaporites should show evidence of crystal growth either at the brine–air interface or at the brine–sediment interface; using a term borrowed from igneous petrology such precipitates are termed *cumulates*—the Castile evaporitic anhydrite–calcite couplets were of this type. In Cl-brines, delicate rafts of hopper halite crystals form at the brine–air interface whilst upward-growing chevron halite and other growth forms occur at the brine– sediment interface as epitaxial growths from capsized hopper rafts. Layers of large, vertically standing, elongate to curved gypsum crystals (swallow-tail crystals) with internal horizontal inclusion trails and dissolution surfaces are also thought to indicate growth upwards from the sediment surface into brine. Intriguing and analogous features occur as halite growths in Dead Sea industrial salterns.
7 Crystal apices of primary, bottom-grown gypsum crystals (sabre-shaped selenites; Fig. 20.41) may be similarly aligned over broad areas and have a common azimuth due to a consistent synoptic direction of saltern brine flow during gypsum crystallization in water just a few metres deep.

The distinctive salt-pan evaporites that occupy many playas of the semiarid south western USA and Lake Eyre salt-pan of Australia show distinctive repeated cycles. The brackish lake phase occurs after flooding due to high runoff, followed by evaporative concentration in saline lakes and, finally, desiccation of the layered evaporites to give surface polygons and subsurface precipitation due to concentration of groundwater brine. The two most distinctive features are laterally extensive dissolution surfaces and thin clastic laminae, formed as the undersaturated brackish waters initially flood the dry pan.

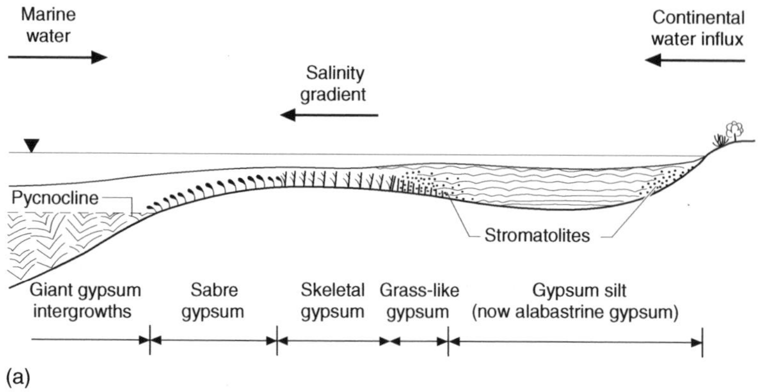


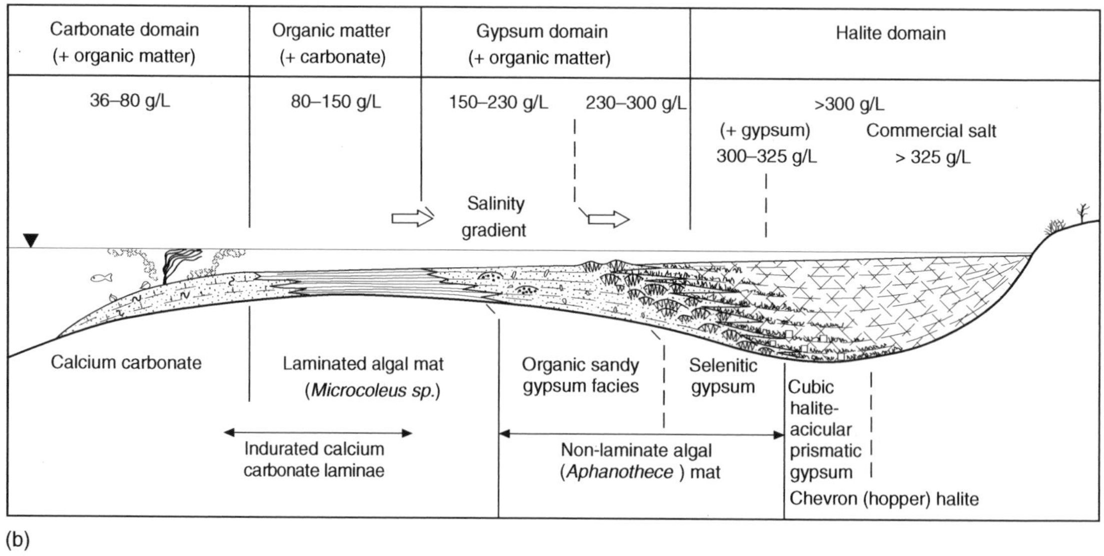


**Fig. 20.40** (a) Subaqueous evaporite facies reconstruction for the gypsiferous successions of the Badenian of western Ukraine. Note the flow-inclined (like sunflowers) sabre gypsum (see Fig. 22.41). (From Peryt, 1996.) (b) Subaqueous halite–gypsum environments of modern Spanish coastal Salinas. (From Orti-Cabo *et al.*, 1984.)

As a young proto-ocean is invaded by marine waters, initially poor circulation will encourage evaporite formation if the climatic regime is suitably arid. Tremendous thicknesses of evaporites, particularly halite, may form, e.g. in Red Sea shelf basins and the adjacent Danikil Depression of Ethiopia, up to 3 km of Miocene halite are recorded. Notable volumes of evaporites developed during the early history of the South Atlantic ocean and the Gulf of Mexico. In the former case continental break-up between South America and Africa led to the creation of a ca $10^6$ km$^2$ silled depocentre in which a staggering $1–4 \times 10^6$ km$^3$ of evaporites accumulated. Perhaps the most spectacular development of evaporites in a major ocean basin occurred in the Mediterranean during the late Miocene, an event termed the 'Messinian salinity crisis' when 5% of the dissolved salt of the oceans of the world was extracted in a fraction of a million years to form a deposit more than 1 million km$^3$ in volume. In the 1970s DSDP boreholes were drilled into seismically recognized halokinetic structures and cores revealed deepwater Pliocene pelagic and hemipelagic oceanic sediments overlying anhydrite and halite evaporites up to 1500 m thick. The evaporites provided evidence of alternating shallow-water playa to subaerial sabkha precipitation with desiccation cracks, stromatolites and chicken-mesh anhydrite texture. It was thus clear to early workers that the Mediterranean underwent cycles of complete desiccation and replenishment. Its isolation was mostly due to tectonic blockage at its former western opening into the central Atlantic in what is today Morocco. A deep basin, in places 3 km below Atlantic sea level, was formed, with streams and rivers like the Rhône and Nile deeply incising their valleys into the basin margins. The most recent revised model of evaporative concentration now has shallow-margin, shallow-water and deep-basin, deep-water precursors to desiccation. Evaporative drawdown began halfway through the salinity crisis when influx from the Atlantic no longer kept up with evaporation. Prior to that time, a million years passed as a sea of marine-fed brine concentrated towards halite precipitation, precipitating first cycle gypsum with clear marine $^{87}Sr/^{86}Sr$ signature in many onshore shallow basins. During the later part of this interval, more than 14 cyclic beds of gypsum accumulated along shallow margins, modulated by orbital forcing. The thick salt on the deep seabed precipitated in just the next few cycles when drawdown commenced and the brine volume shrank. Erosion surfaces and their detritus formed in response to the drop in base level during evaporative drawdown and may be traced down to the basin floor. Upon closure of the Atlantic spillway, the remnants of the briny sea transformed into salt pans and endorheic lakes fed from watersheds of Eurasia and Africa evidenced by the fact that the younger cycles of evaporites have clear nonmarine $^{87}Sr/^{86}Sr$ signature. Eventually deep marine sediments record an early Pliocene (5.33 Ma) reconnection of the sea to the Atlantic, possibly a catastrophic event termed the Zanclean Flood. Even today, the Messinian evaporites source deep-sea *brine pools* where they have been exposed to solution by tectonic uplift and erosion.

All of the above is not to suggest that the desiccating deep-basin model for Mediterranean Messinian evaporites is accepted by all sedimentologists. One re-examination of the evidence provided by the original DSDP cores concluded that the interpretations made to justify the original model were either equivocal or incorrect and that several features used as evidence for shallow-water are more compatible with deposition under deep-water (below wave base), whilst others can only be considered as of uncertain origin. Thus (i) the chicken-mesh anhydrite that figured large in the argument can originate by burial diagenesis, (ii) there is an absence of 'proper' sabkha cycles, (iii) only a single, and that equivocal, desiccation crack occurs and (iv) stromatolite laminae are unconvincing and could be of nonphotosynthetic microbial origins. In addition, coring has sampled only the upper few tens of metres of the evaporites, a factor that seriously limits any interpretation of the origin of the deposit as a whole. Some workers have even suggested that some of the evaporites may be resedimented from onshore salterns that were structurally uplifted during the severe thrust tectonics that typified the late Miocene period in the development of the Mediterranean area generally. However, none of these objections can overcome the existence of the Miocene deeply incised Nile (~2.5 km deep) and Rhône (~1 km) channels and fan complexes—completely unequivocal and independent evidence for ocean drawdown. Recent core and seismic data gained from the Europe–Africa tunnel project also lends very strong support to the notion of a catastrophic refill of the deep dessicated basin in the early Pliocene: this is evidence for the *Zanclean flood* very rapidly cutting a deep (250 m), long (150 km) submarine canyon from the Camarinal sill in the western Straits of Gibralter eastwards. Additionally, very recent evidence for extreme δD fractionation in organic Messinian biomarkers strongly supports the desiccated Mediterranean hypothesis.

## Further reading

### *General*

Bathurst's profoundly scholarly book (1975) was and is the 'bible' of modern carbonate sedimentology processes, though with the years it has become more of a much-loved old testament. Morse & McKenzie (1990) cover geochemical aspects, whilst Tucker & Wright (1990) is still a major source for all things carbonate. Kendall & Harwood (1996) and Wright & Burchette (1996) provide good summaries. See Harris & Kowalik (1994) for many superb satellite images of carbonate environments; three of these are reproduced by permission in the present text (Plates 15–17). Logan (1987) is the most impressive single study of an evaporite environment and repays close study. Statistical analysis of carbonate environments using

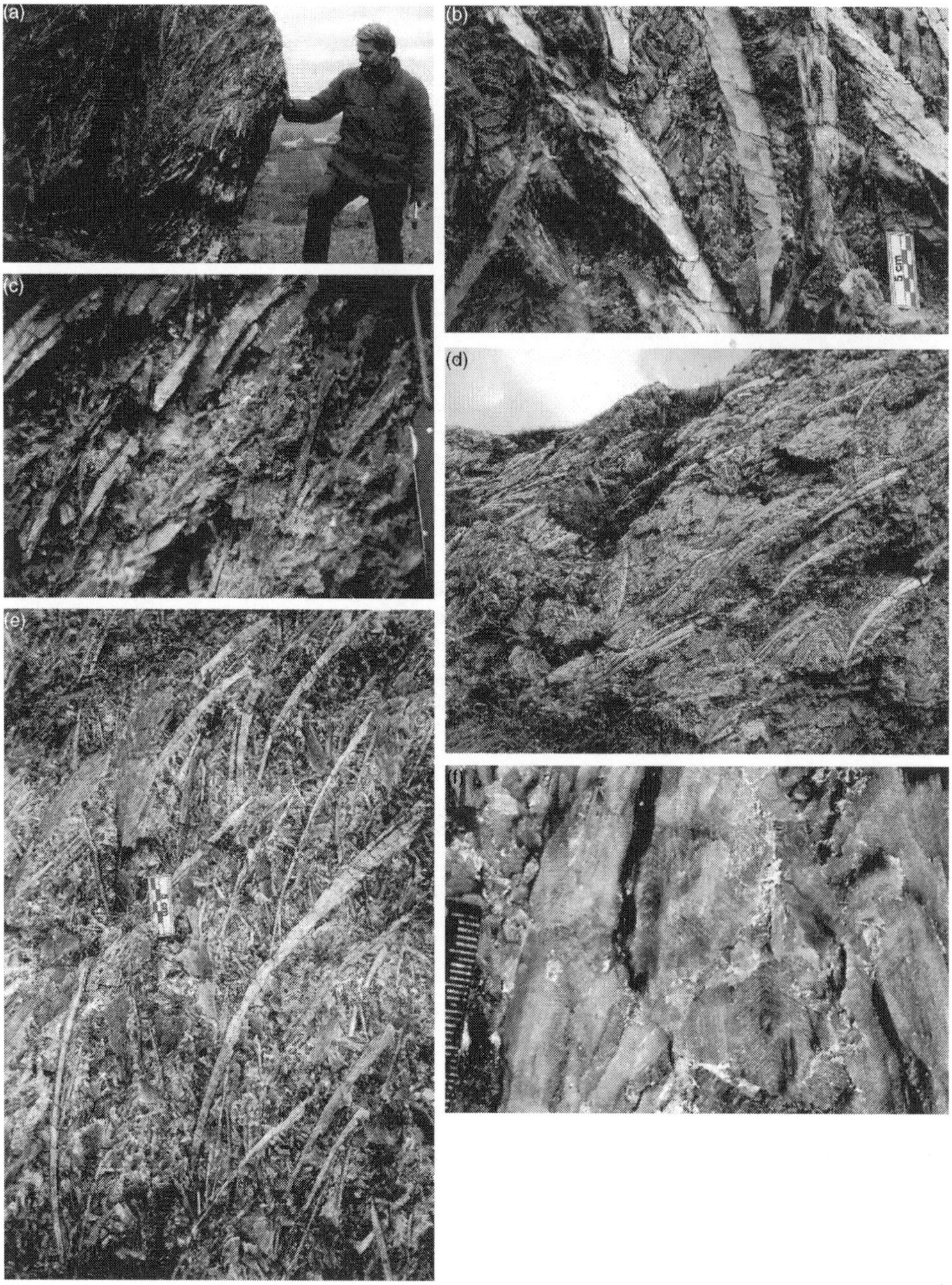


**Fig. 20.41** Oriented sabre gypsum facies from various parts of the Badenian northern Carpathian Foredeep basin of Poland, Ukraine, and the Czech Republic. (a) sabre gypsum with vertically oriented crystals. (b) Sabre gypsum with relatively thick crystals displaying competitive growth fabric. (c) Sabre gypsum with primary empty intercrystalline pores. (d) Sabre gypsum with subhorizontally oriented crystals. (e) Sabre gypsum with very narrow crystals showing competitive growth fabric. (f) Primary growth-zoning of sabre gypsum crystals related to advance of the {120} prism faces. (After Babel & Becker, 2006.)

satellite imagery and GIS-driven spatial statistics has appeared in recent years, see Rankey (2002, 2004) for examples. The Tribute volume to R.N. Ginsburg (Swart *et al.*, 2009), another carbonate pioneer has many interesting papers. Vacher & Quinn (1997) focus on the geology and hydrogeology of carbonate islands. Schreiber & Tabakh (2000) is a fine synthesis of depositional environments, structures and diagenetic fabrics in evaporites. Ryan (2009) is thrilling reading on the Mediterranean dry-up (but see notes in text above). Garcia-Castellanos *et al.* (2009) document their spectacular evidence for the Zanclean flood. Biomarker evidence for arid dessication is in Andersen *et al.* (2001).

*Specific*

For carbonate shelf margins (production rates, ramps, rimmed margins, etc.) see Maxwell & Swinchatt (1970), Ahr (1973), Hine *et al.* (1981b), Read (1985), Eberli & Ginsburg (1987), Bosence & Waltham (1990), Roof *et al.* (1991), Burchette & Wright (1992), Bosence *et al.* (1994), Aurell *et al.* (1995), Keene & Harris (1995), Whitaker *et al.* (1997) and Anastasa *et al.* (1997). A good collection of papers on ramps is in Wright & Burchette (1998), with a useful opening review of terminology. Demicco & Hardie (2002) cast a sceptical eye over simple depth-dependant sediment production rates. A collection of papers on carbonate platforms is provided by Insalaco *et al.* (2000).

For arid carbonate tidal flats and evaporite sabkhas: Arabian Gulf—Wells & Illing (1964), Illing *et al.* (1965), Kinsman (1966), Shearman (1966), Evans *et al.* (1969), Park (1976, 1977), Chafetz & Rush (1994), Kirkham (1998), Goodall *et al.* (2000), Purkis *et al.* (2005) and Lokier & Stobier (2008); Shark Bay—Logan *et al.* (1970, 1974), Hagan & Logan (1974) and Woods & Brown (1975).

Ancient sabkha facies are reported by Wood & Wolfe (1969), Shearman & Fuller (1969), Fuller & Porter (1969), Holliday & Shepard-Thorne (1974) and West (1975).

For humid carbonate tidal flats and marshes of Florida see Griffith *et al.* (1969), Enos & Perkins (1979), Purdy (1974) Wanless *et al.* (1988b) and Rankey (2004); for the Bahamas see Shinn *et al.* (1969), Hardie (1977), Shinn (1983), Hardie & Garret (1977), Hardie & Ginsberg (1977) and Rankey (2002).

Ancient Bahamian-type facies are discussed in Fischer (1964, 1975), Hardie (1977), Laporte (1971), Eriksson (1977), Francis (1984) and Strasser (1988).

For lagoons and bays of Honduras see Matthews (1966); for the Bahamas see Shinn *et al.* (1989), Milliman *et al.* (1993) and Loreau (1982); for Florida see Purdy (1974), Neumann & Land (1975), Enos & Perkins (1979), Wanless *et al.* (1988b), Rankey (2004) and Boardman & Neumann (1984); for Grand Cayman see MacKinnon & Jones (2002). For Polynesian atolls see Tribovillard *et al.* (1999), Gautret *et al.* (2004); for Yucatan/Belize see Gischler & Lomando (1999). For Maldive tsunamites see Nichol & Kench (2008).

For tidal delta and margin-spillover ooid tidal sands (mostly Bahamian) see Illing (1954), Purdy (1963), Ball (1967), Harris (1979), Hine *et al.* (1981b), Dill *et al.* (1986), Shinn *et al.* (1990), Shapiro *et al.* (1995) and Harris (1979). Modern integrated results are reported by Rankey *et al.* (2006) and Reeder & Rankey (2008).

Ancient oolitic complexes are well described from the Pleistocene of Florida (Halley *et al.*, 1977; Neal *et al.* 2008), the Mesozoic of northwest Europe (Sellwood & McKerrow, 1973; Purser, 1979) and the oil reservoirs of the Middle East (Davies *et al.* 2000; Handford *et al.*, 2002)

Open carbonate shelf ramps: for warm-water examples from Arabia see Pilkey & Noble (1967), Shinn (1969); Yucatan see Ginsburg & James (1974); northeast Brazil see Testa & Bosence (1999); for rhodoliths see Hetzinger *et al.* (2006); for cool water examples from New Zealand see Nelson *et al.* (1988); Lacapede see James *et al.* (1992); northwest Australia see Dix *et al.* (2005); South Australia see Rivers *et al.* (2007); southwest Australia see James *et al.* (1999).

Ancient carbonate ramp/shelf sediments (see also carbonate cycles) are discussed by Wilson (1975), Kerans *et al.* (1994), Wright & Burchette (1998), Drzewiecki & Simo (2000), Seguret *et al.* (2001), Wilmsen & Neuweiler (2008) and Phelps *et al.* (2008).

for ancient cool- to cool-warm transition water carbonate ramps see Randazzo *et al.* (1999), Lukasik *et al.* (2000), Stemmerik (2001), Pedley & Grasso (2002), Johnson *et al.* (2005), Rogala *et al.* (2007) and Bensing *et al.* (2008).

Platform margin reefs and carbonate build-ups: for definitions see Embrey & Klovan (1971), Wilson

(1975) and James & Gravestock (1990); role of microbials see Camoin & Montaggioni (1994) and Braga *et al.* (1996); mud mounds generally see Monty *et al.* (2005); mound carbonate production rates see Bosence *et al.* (1985); Algerian Devonian build-ups see Wendt *et al.* (1993, 1997); Canning basin Devonian build-ups see Playford (1980) and Stephens & Sumner (2003); Triassic Dolomites build-up see Blendinger *et al.* (2004); Capitan Reef see Tinker (1998); other ancient build-ups see Turner *et al.* (1993), Narbonne & James (1996), Savarese *et al.* (1993), Wood *et al.* (1993), Narbonne & James (1996), Harris (1993) and Lees & Miller (1985).

Modern reefs and build-ups: general papers are by James & Ginsburg (1979), Chappell (1980), Hallock & Schlager (1985) and Hallock (1988); Florida by Macintyre (1988); Grand Cayman by Blanchon & Jones (1997); Hawain reef budget by Harney & Fletcher (2003); Belize reefs by James & Ginsburg (1979), Gischler & Lomando (1999) and Mazzullo (2006); Great Barrier Reef by Maxwell & Swinchatt (1970), Braithwaite *et al.* (2004) and Francis *et al.* (2007). Turbid Great Barrier Reef conditions are discussed by Tudhope & Scoffin (1994), Amos *et al.* (2004), Larcombe *et al.* (2001), Dunbar & Dickens (2003), Fielding *et al.* (2003, 2006) Wilson (2005) and Francis *et al.* (2007). Virgin Islands reefs by Hubbard *et al.* (2005); Tahiti reefs by Cabioch *et al.* (1999); reef drowning, catch-up, etc. by Schlager (1981), Neumann & MacIntyre (1985), Blanchon & Blakeway (2003) and Hubbard (2009); reef death, wave stress and ENSO by Rooney *et al.* (2003), Engels *et al.* (2004) and Grossman & Fletcher (2004); Belize mud mounds by Mazzullo *et al.* (2003); slope bryozoan mounds by James *et al.* (2004) and Bjerager & Surlyk (2007); deep cold-water coral mounds by Ferdelman *et al.* (2006) and Wheeler *et al.* (2008); deep-water build-ups of the Bahamas by Neumann *et al.* (1977) and Mullins & Neumann (1979).

On sea-level changes and reef develoment see Purdy (1974), MacNeil (1954), Ginsburg *et al.* (1991), Choi & Ginsburg (1982), Blanchon & Shaw (1995), Schlager (1981), Neumann & MacIntyre (1985), Tinker (1998), Blanchon & Blakeway (2003), Hubbard (2009), Pomar & Ward (1994) and Hubbard *et al.* (2005).

For modern platform margin slopes and basins see Neumann & Land (1975), Schlager *et al.* (1976, 1994), Neumann *et al.* (1977), Mullins & Neumann (1979), Crevello & Schlager (1981), Hine *et al.* (1981a,b), Kendall & Schlager (1981), Boardman & Neumann (1984), Bosellini (1984), Mullins *et al.* (1980a,b, 1984), Schlager & Camber (1986), Eberli & Ginsburg (1987, 1989), Shinn *et al.* (1989), Bosence & Waltham (1990), Harris (1993), Melim & Scholle (1995), Bosence *et al.* (1994), Droxler & Schlager (1985), Mullins & Hine (1989), Wilber *et al.* (1990, 1993), Ginsburg *et al.* (1991), Wilson & Roberts (1992, 1995), Grammer *et al.* (1993), George *et al.* (1997), Betzler *et al.* (1999), Dix & Kyser (2000), Roth & Reijmer (2005), Mazzullo (2006) and Lantzsch *et al.* (2007).

Ancient platform margins: for examples from the Dolomites see Bosellini (1984)' Harris (1993) and Blendinger (1994, 2001); a lovely review with historical philosophy of the Dolomites record is provided by Schlager & Keim (200..); for the Permian Capitan see Melim & Scholle (1995) and Phelps & Kerans (2007); for the Devonian Canning see George *et al.* (1997); Moroccon Jurassic see Verwer *et al.* (2009) and Blomeier & Reijmer (2002); Cantabrian Carboniferous see Bahamonde *et al.* (2000) see Della Porta (2004); Oman Cretaceous see Immenhauser *et al.* (2001) and Hillgärtner *et al.* (2003).

On features associated with sea-level changes see Whitaker *et al.* (1997), Mylroie & Carew (1988, 1990), Vollbrecht & Meischner (1996), Boardman *et al.* (1995), Francis (1984), Strasser (1988), Joachimski (1994), Grammer *et al.* (1993) and Egenhoff *et al.* (1999).

On sedimentary cycles developed in carbonate sediments see Fischer (1964), Schwarzacher & Fischer (1982), Grotzinger (1986), Hardie *et al.* (1986), Read *et al.* (1986), Strasser (1988), Goldhammer *et al.* (1990), Hinnov & Goldhammer (1991), Drummond & Wilkinson (1993), Kerans *et al.* (1994), Brack *et al.* (1996), Tinker (1998), Egenhoff *et al.* (1999), Sattler *et al.* (2005), Anderson (2004), Colombié & Strasser (2005), Husinec & Jelaska (2006).

On whether peritidal cycles actually exist; pro—Spence & Tucker (1999, 2000, 2007), Spence *et al.* (2004) and Preto & Hinov 2003); con—Wilkinson *et al.* (1996, 1997a,b, 1999).

On carbonate/marl cycles see de Boer & Wonders (1984), DeVisser *et al.* (1989), Roof *et al.* (1991) and Thunnell *et al.* (1991).

Peritidal carbonate cycles forward modelled by Burgess (2001, 2006), Burgess *et al.* (2001), Burgess

& Wright (2003), Paterson *et al.* (2006), Castell *et al.* (2007, with field data).

On siliciclastic/carbonate cycles see Maxwell & Swinchatt (1970), Choi & Ginsburg (1982), Aqrawi & Evans (1994), Evans (1995), Rankey (1997), Soreghan (1994), García-García (2009), Bauer *et al.* (2003), Fielding *et al.* (2003, 2006), Wynn & Read (2008), Pufahl *et al.* (2004), García-Hidalgo *et al.* (2007), Francis *et al.* (2007).

On carbonates and eutrophication see Ziegler *et al.* (1984), Brasier (1995), Caplan *et al.* (1996) and Whalen (1995).

For Tethyan platforms see Woodfine *et al.* (2008).

On subaqueous evaporites see Warren (1982), Cathro *et al.* (1992), Warren & Kendall (1985), Lowenstein & Hardie (1985), Orti-Cabo *et al.* (1984), Schreiber *et al.* (1976), Rouchy *et al.* (1995), Shearman (1970), Arthurton (1973), Peryt (1996), Talbot *et al.* (1996), Hovorka (1987), Lowenstein (1988) and Cathro *et al.* (1992).

Evaporite basins in the Badenian of eastern Europe are considered by Peryt (1994, 1996, 2001) and Babel & Becker (2006); the Michigan basin by Satterfield *et al.* (2005); the Castile basin and Sr ratios by Kirkland *et al.* (2000); the Zechstein basin by Becker & Bechstädt (2006).

Oriented subaqueous gypsum growth is studied by Babel & Becker (2006); 1ry vs. 2ry evaporite textures by Schreiber & Helman (2005); hypersaline oceans by Kinsman (1975a,b).

Deep Miocene desiccation of Mediterranean: pro—Hsü (1972), Hsü *et al.* (1977), Barber (1981), Wallmann *et al.* (1997), Flecker & Ellam (2006) and Ryan (2009, highly reccomended review); con—Hardie & Lowenstein (2004, incisive critique, quite damning, but wrong in view of later new references). Garcia-Castellanos *et al.* (2009) document spectacular evidence for the Zanclean flood. Biomarker evidence for arid dessication is in Andersen *et al.* (2001).

# Chapter 21

# DEEP OCEAN

*Agus nan robh mi tráigh Mhùideart*
*Còmhla riut, a nodhachd ùidhe,*
*Chuirinn suas an cochur gaoil dhut*
*An cuan' s a' ghaineamh, bruan air bhraon dhiubh.*

Somhairle MacGill-Eain, '*Tráighean*', *Dain Chruinnichte* 1939–1941. Carcanet 1990

*And if I were on the shore of Moidart*
*With you, for whom my care is new,*
*I would put up a synthesis of love for you*
*The ocean and the sand, drop and grain.*

Sorley MacLean (transl. Somhairle MacGill-Eain) 'Shores', Collected Poems 1939–1941. Carcanet 1990

## 21.1 Introduction

Our understanding of deep-sea sedimentation, both its nature and physical/chemical/biological controls, has been revolutionized since the mid-twentieth century by: (i) massive intergovernmental investment in deep-sea coring; (ii) measurement of ocean water chemistry, deep sea currents and sediment fallout made from vessels and moored subsurface recorders; (iii) seafloor imaging by deep-tow sidescan sonar and multibeam bathymetry; (iv) deep-water hydrocarbon exploration, chiefly the advent of high-quality subsurface 3D seismic data and exploration drilling. Taking a broad view, the nature and preservation of deep-sea sediments are closely controlled (Fig. 21.1) by clastic sediment and nutrient fluxes, by variations in the depth of the oxygen minimum layer, and by the carbonate compensation depth (CCD). Clastic sedimentation around the ocean margins is closely related to:

1 tectonics, because this controls ocean-floor basement morphology, shelf width and onshore continental gradient;
2 magnitude and nature of river sediment and water input;
3 cycles of sea-level highstand and lowstand which lead to tendencies toward *shelf sink-mode* and *shelf bypass-mode* (Chapter 19) in controlling sediment flux to the deep ocean over time.

However, there has been some neglect of climate's role in all this because of an overreliance upon concepts inherent in **3**, the core of the sequence stratigraphy paradigm, a topic returned to below. Maximum production of plankton corresponds closely to areas of high organic primary productivity in the productive

*Sedimentology and Sedimentary Basins: From Turbulence to Tectonics*, 2nd edition. © Mike Leeder.
Published 2011 by Blackwell Publishing Ltd.

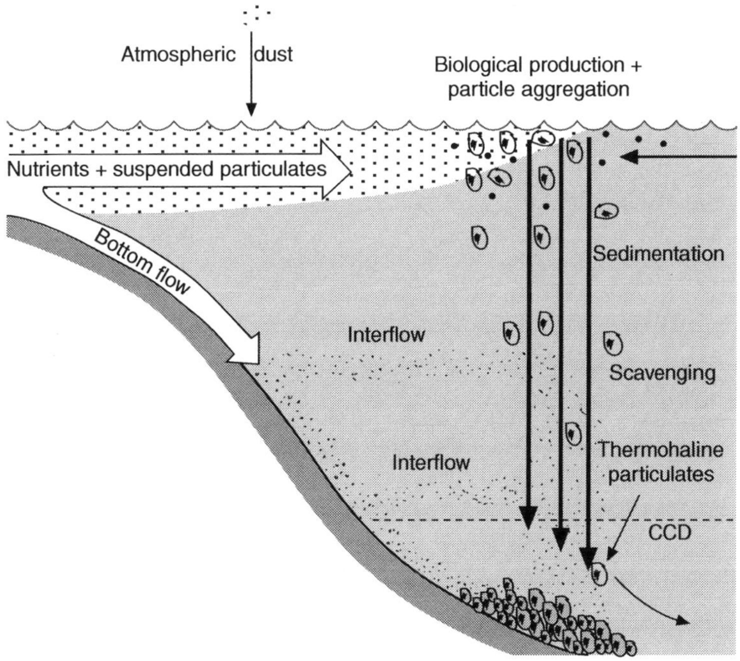


**Fig. 21.1** Particle fluxes to the oceans. CCD, carbonate compensation depth. (Mostly after Ittikot *et al.*, 1991.)

surface layer of the oceans, especially where mixing of water masses occurs. The magnitude of this biological contribution to abyssal ocean-floor sediment depends not only upon local dilution by clastic input but also on carbonate and silica saturation states of ocean-floor water. High-productivity ocean basin surface waters can also lead to deep hypoxia and production of carbon-rich muds whereas states of hypersalinity may be triggered by ocean basin isolation and evaporative sea-surface drawdown.

From the above generalities the history of deep-sea sedimentation over various time intervals at any location is likely to be highly complex. It is unlikely that any one scheme for the development of sedimentary deposits can be distilled from the intermixing of so many variables. Take two examples. First, plate tectonic evolution may drive a mid-ocean ridge coated with pelagic biogenic and chemical sediments closer and closer to a young continental margin of high onshore relief and high-magnitude clastic sediment influx. The magnitude of this flux will depend both upon continental drainage catchment geology and climate but also on cycles of sea-level change. Such a situation has been evolving along the Cascadia plate margin off western North America over tens of millions of years. Second, studies of ancient ocean basin sediment fills now preserved in mountain belts indicate that deep marine systems may change their architectural style as active tectonics cease and the tectonic basin 'passively' infills. In this scenario, topographic relief is gradually reduced, overall seafloor gradients decrease, shelves and coastal plains widen and potentially more coarse sediment is stored inboard the deep ocean. Lowered gradients may lead point-sourced, canyon-fed deep sea deposits to be replaced by broader, ramp-like sediment supply. Such scenarios are envisaged for evolution of former deep-marine Tertiary basins now emplaced as nappes by crustal shortening in the Pyrenees, northeast Spain.

## 21.2 Sculpturing and resedimentation: gullies, canyons and basin-floor channels

The continental slope, or more precisely the shelf–slope break, marks a fundamental division between shallow shelf and deep sea environments; tide and wave processes progressively (though not exclusively) give way to those of ocean currents, thermohaline currents and gravity flows. Below the break the continental rise and inner abyssal plains are the major environments where appreciable volumes of terrigenous clastic sediment are deposited. Average slope profiles from around the world are well represented by overall semi-Gaussian or logistic functions, with gradually increasing, then decreasing slopes on to abyssal plain or trench (Fig. 21.2). The feeding systems for base-of-slope environments—gullies and canyons cutting the continental slope, are sometimes sourced directly from incised channel systems on the continental shelf but many relying upon internal slope failures to povide sediment that feeds debris and turbidity flows downslope. The simple form of some slopes may be severely disrupted by the effects of salt intrusion (the salt originating during early rifting), with

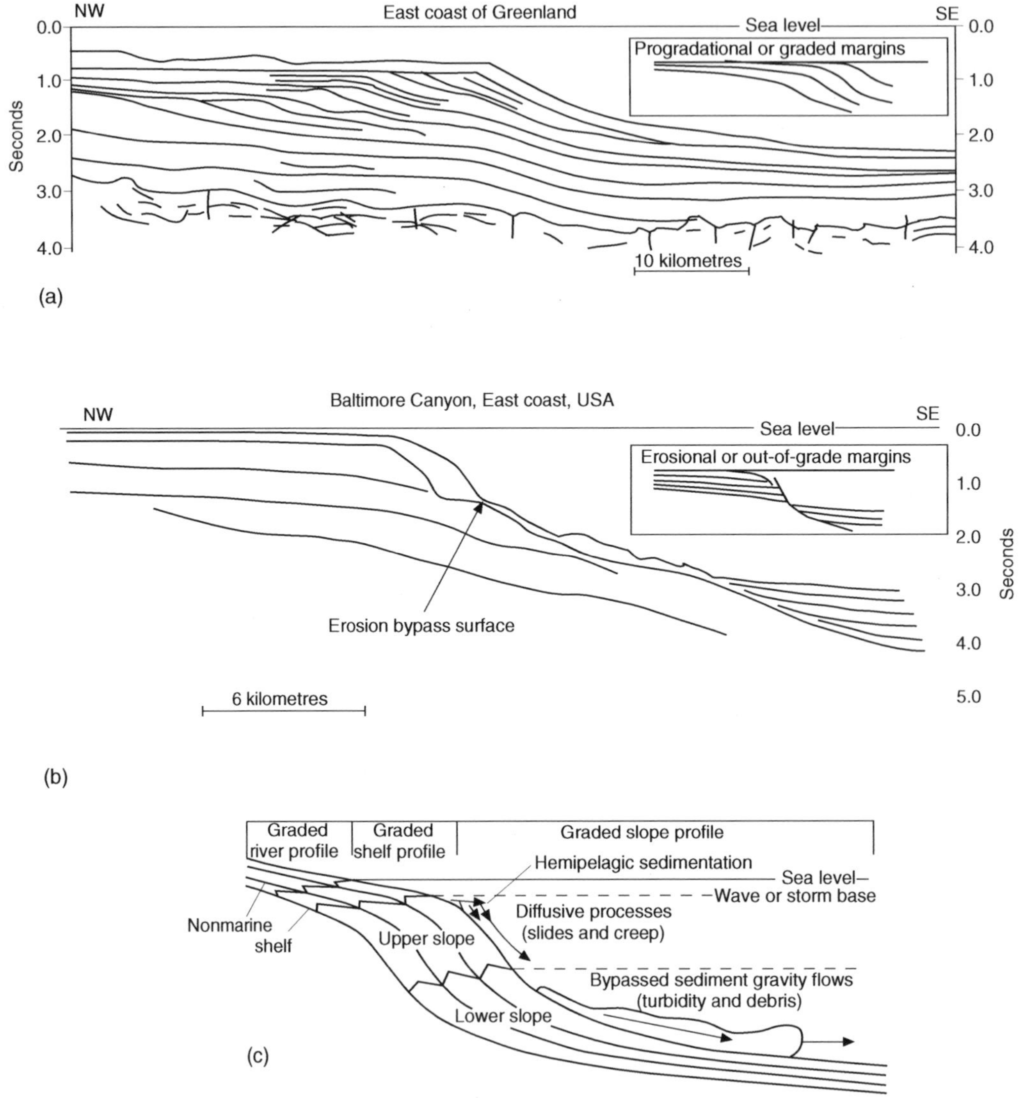


**Fig. 21.2** Progradational and erosional continental margins. (a) Progradational with outbuilding clinoform profile, periodically aggrading, margin. (b) Erosional margin. (c) Cartoon of prograding slope to show processes affecting the exact form of the 'graded' slope profile. (From Ross *et al.*, 1994, and sources cited therein.)

chaotic topography, as in the northern Gulf of Mexico. Faulted slopes with gradients up to 30° dominate young ocean rifted margins, many subducting margins and coastal extensional basins; they have been referred to as *slope aprons*. Characteristic base-of-slope-apron deposits include talus collapse breccias, slump mounds, debris-flow masses and steep-sided gully fills. Periodic small submarine fans are located at canyon exits through fault scarps. Point sources for sediment dispersal to the ocean floor are provided by canyons that were fed directly by major rivers during lowstand. They commonly have dimensions of hundreds of metres deep and kilometres wide, with a V-shaped cross-sectional form in their upper parts. Their precise course on the upper slope is governed by local topographic funnelling.

Continental slopes are thus dynamic surfaces, encouraged to prograde seawards by deposition, smoothed by slow downslope diffusional creep but at the same time ravaged by enigmatic gullies, seemingly generated at lowstand or falling stages, and slide scars due to downslope mass movements on scales of a few cubic metres to thousands of cubic kilometres (Fig. 21.2). During sea-level lowstands, much coarse sediment is delivered directly to the slope by rivers. During highstands, deposition is from powerful shelf-crossing hyperpycnal underflows and by more general shelf-edge fine-sediment plumes. The resulting alternation of relatively coarse- and fine-grained outer-shelf and continental-slope sediment is inherently unstable: generally high water contents and the development of gas hydrates encourage periodic mass failure after earthquake shocks and abnormal pressures produced by tsunami and internal waves. Lower values of shear strength in the vicinity of submarine canyons are related to a combination of increased concentrations of organic matter and fine-grained sediment. Passing surface waves during storms may induce pressure anomalies with wavelengths of 300 m and amplitudes of 70 kN/m$^2$ in water depths of 60 m. Although the amplitude of these wave-induced pressure anomalies decreases in deeper water, the effect is considered important on shelf-edge sediments always liable to failure.

Submarine canyons are arguably the greatest bypass-conduits for oceanwards coarse and fine sediment transfer. Currents in submarine canyons show a strong ocean tidal signature, with up and down water motions strongly modified by current reflections and refractions, the tendency for denser suspensions to move downslope, wave currents, storm surge (gradient) currents and internal waves. More persistent and strong downcanyon flows (over 1 m/s, sometimes over several days, but usually a few hours) record the onset of turbidity current events. Major submarine canyons such as Monterey Canyon, offshore central California (Fig. 21.3) are kilometres deep and tens of kilometres wide. Improved documentation of sedimentary processes and sediment distribution in canyons derives from examination of modern canyons, as well as the deposits of ancient canyon deposits in outcrop and the subsurface. Modern Monterey Canyon for example seems to be aggrading and has a narrow axial channel dominated by coarse-grained material with large sand waves recording the bypass of sediment through the uppermost reaches of the channel system. Channel bars in the axial channel are analogous in general shape to composite fluvial unit braid bars—cores reveal upward-fining, structureless sand beds sometimes > 2 m thick. Large-scale slumping is especially prevalent in submarine canyons, examples within Monterey Canyon system completely dammed canyon channels and large granitic blocks spalled from canyon walls into the channel axis during the 1989 Loma Prieta earthquake. Other seismically triggered examples of mass transport deposits up to ca 4 m thick are from the central Scotian Slope, identified from piston cores collected from canyon floors and intercanyon ridges. Stratigraphic evidence suggests four or five episodes of sediment failure within the past 17 ka; the most likely mechanism for triggering synchronous failures in separate canyons is seismicity. Sedimentary deposits resulted from local slides being overlain by mud-clast conglomerate deposits derived from failures farther upslope and finally by coarser-grained deposits resulting from retrogressive failure remobilizing upper slope sediments to form debris flows and turbidity currents.

Startling evidence from uplifted Paleogene canyon fills onshore from Monterey reveal thick (up to 75 m) lenticular, coarsening-upwards sandbodies comprising stacked coarse-grained turbidite beds deduced to have been deposited by high-density turbidity currents. Deposition is obviously due to flow deceleration but for what reason? The coarsening upwards may suggest the infill of a slump-generated minibasin within the palaeocanyon, perhaps dammed downstream. This would have led to 'ponding' and the gradual infill

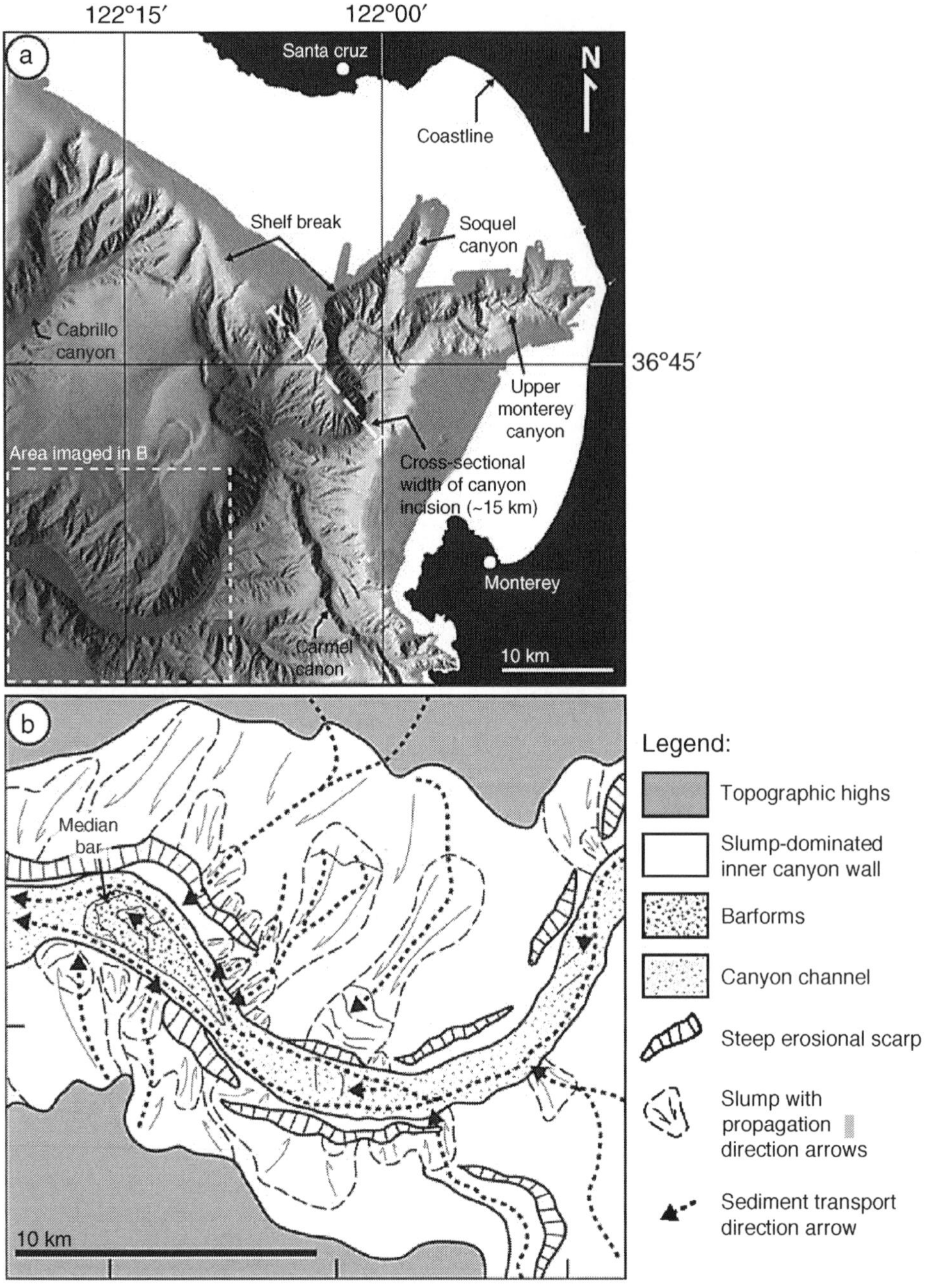


**Fig. 21.3** (a) Bathymetric image of the middle and upper parts of the Monterey Canyon system from multibeam sonar data. The line highlighting the cross-sectional width of canyon incision (15 km) represents the location where the canyon is comparable in scale to ancient canyons studied by the authors. The most distal (westward) extreme of the canyon is 20–25 km from the position where it passes into the upper fan valley. (From Anderson *et al.*, 2006; image by Monterey Bay Aquarium Research Institute.) (b) Geomorphologic map of the interior of the modern Monterey submarine canyon The channelized canyon floor is 1–4 km wide, filled with turbidite deposits, bar macroforms and landslide debris. (After Anderson *et al.*, 2006.)

of the depression, coarsening-upwards as successive turbidity currents prograded to renew an equilibrium gradient, whereupon the channels incise the ponded fill and pass the sediment downstream. Evidently the local palaeocanyon channel ceased operation as a conduit because of massive collapse of an adjacent canyon wall, the slump debris being the youngest part of the coarse active infill. The coarse lenticular channel fills occur repeatedly, stacked within a background of finer grained deposits possibly representing highstand/lowstand alternations along the Californian palaeoshelf—total reconstructed canyon dimensions were ca 2 km deep, ca 8 km wide. Other late Cretaceous–early Paleocene outcrops in Pacific coastal Baja California reveal evidence for repeated cycles of intracanyon wall channel stacking, attached channel lateral bar accretion and the generation of poorly sorted gravel levees. A remarkable ancient continental slope margin is preserved in the Cretaceous Tres Pasos Formation of the Magallanes Basin, southern Chile. The sedimentary succession here records both slope progradation and aggradation. Deposition in lower sections was dominated by poorly channelized to unconfined sand-laden flows and accumulation of mud-rich mass transport deposits, interpreted as base-of-slope to lower slope setting. Evidence for channelization and indicators of bypass of coarse-grained turbidity currents are more common in the upper part of the > 600 m thick succession, reflecting increasing slope gradient conditions.

But how do fan-feeding turbidity currents initiate from the shelf edge and evolve down canyon before reaching a depocentre like a submarine fan? Direct monitoring of natural processes on the slope have been rare. They may be inferred from seabed morphology, resulting deposits, and from sparsely monitored historical canyon turbidity currents. Experiments may also help. Three main end-member mechanisms are:

- *Sediment failure* caused by liquefaction of granular sediment and break-up of more cohesive sediment on pre-existing steep slopes. Retrogressive (upslope-propagating) failure may sustain flow for many hours. Such failure tends to produce broad, straight depressions in surficial slope sediment. Headless canyons (Figs 21.4 & 21.5) produced by this

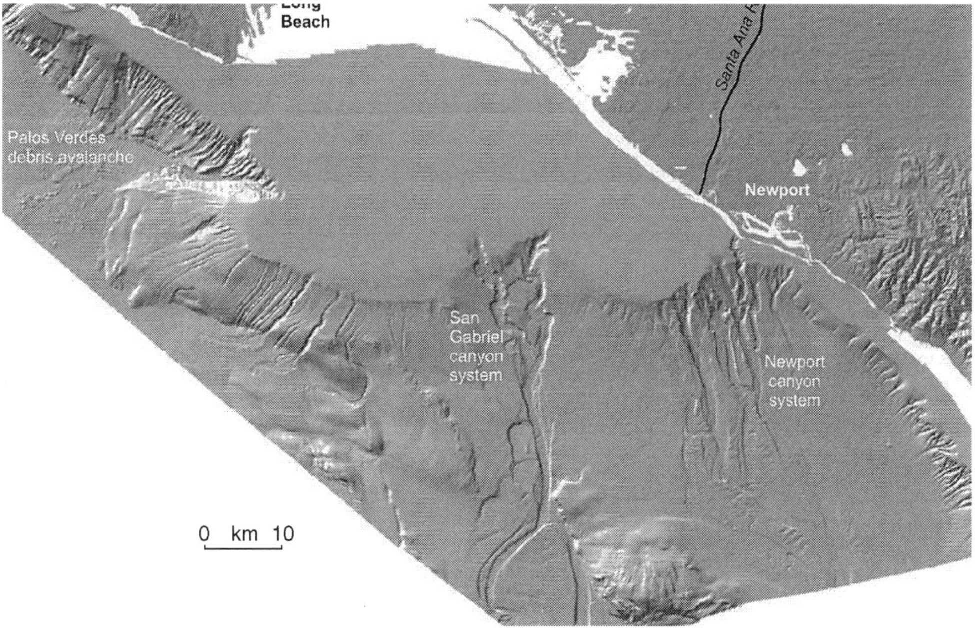


**Fig. 21.4** Multibeam bathymetry of part of the California continental borderland south of Los Angeles, showing morphology resulting from turbidity-current erosion. Most canyons do not cross the shelf slope break, with the notable exception of one in the eastern Newport system which may be the palaeocourse of the Santa Ana River. (After Piper & Normark, 2009.)

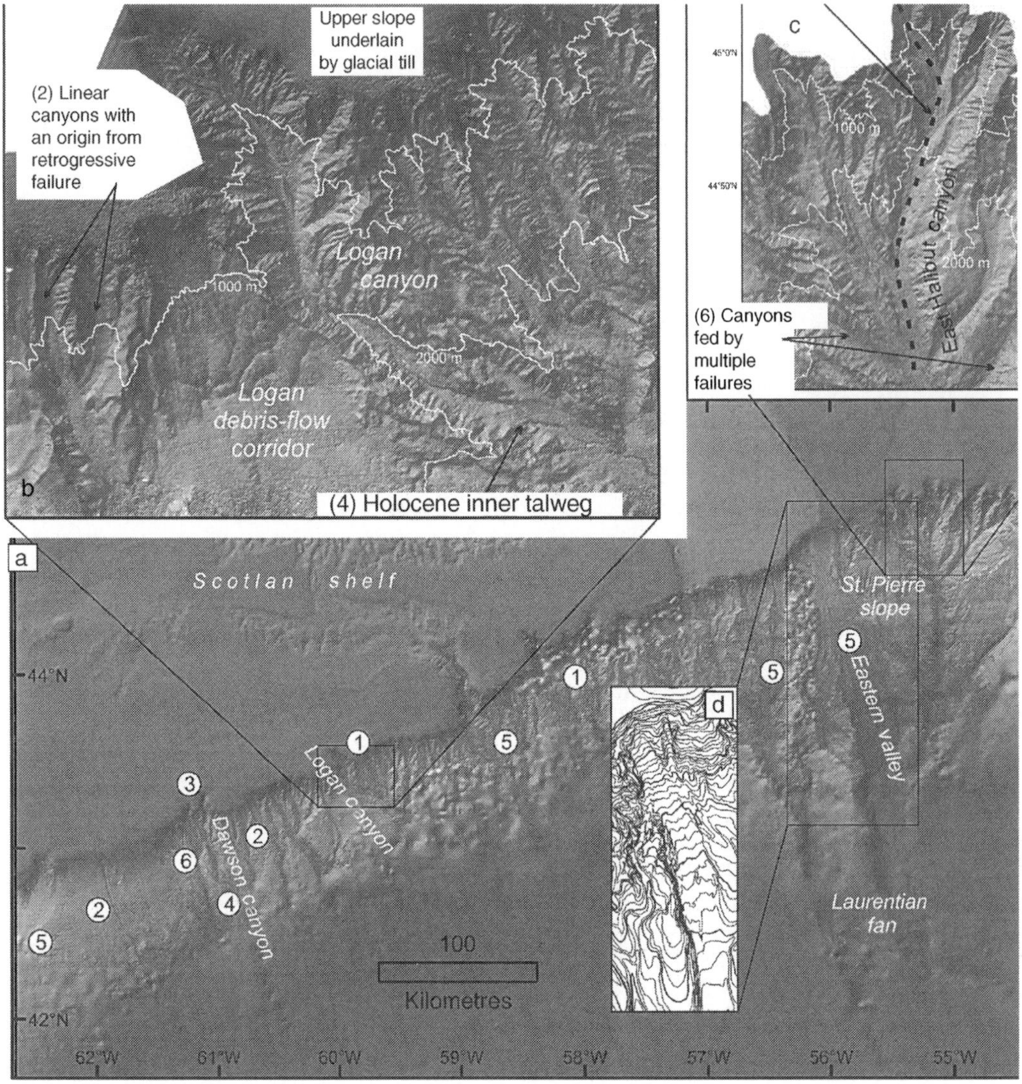


**Fig. 21.5** Multibeam bathymetry of part of the continental margin off Nova Scotia, Canada, showing morphology resulting from turbidity current erosion and six types of submarine canyon (a) Regional map: (1) canyons with dendritic tributary gullies probably resulting from plume fallout; (2) linear canyons originating from retrogressive failure; (3) shelf-indenting canyons, probably with multiple origins, that trap shelf sands; (4) Holocene inner thalweg; (5) partly filled Pleistocene ice-margin hyperpycnal flow canyons; (6) canyons fed by multiple small failures. (b) Detail of Logan Canyon area. (c) Detail of the Halibut Canyon area. (d) Detail of Eastern Valley of Laurentian Fan, in conventional bathymetry. (From Piper & Normark, 2009.)

mechanism may not extend on to the shelf. The competitive growth of along-slope scars leads to a fairly regular canyon spacing of a few kilometres or so. The orientation of canyons and gullies is sometimes strongly structurally controlled, by salt injection structures, faults and by sedimentary layering. Though headless, such canyons still act as conduits after retrogressive failure, serving to funnel ocean and internal tides, storm waves and shelf plumes into and out of shelf-slope margins.

- *Hyperpycnal underflow* flow of rivers is caused by extreme suspended load and can be maintained for the duration of the flood peak hydrograph. Large flows cut broad straight slope channels of relatively uniform width similar to that of the width of the source flow.
- *Storm wave processes* lead to fallout of mud from advective plumes and from shelf-edge resuspension that may lead to downcanyon flow on the slope. Waxing currents erode previous deposits in the canyon and may cause downslope widening of conduits as pick-up continues with ever increasing flow power.

Two key question remain. (i) What controls canyon infilling vs. canyon flushing, i.e. the ability of the three redistributive processes to cannabilize. The sheer volume of some gigantic turbidite beds in the sedimentary record seems to imply that conduit flushing is a major contributor. (ii) Can turbidite deposits be used to hindcast turbidity current genesis? These are left as interesting research and philosophical questions for the reader.

It is usually assumed that turbidity currents are totally confined within submarine canyons on continental slopes and that overtopping of canyon margins, termed *spillover*, contributes little to slope sedimentation or accretion. Such is the case in offshore Angola in the southeast Atlantic but 3D seismic data clearly shows an evolutionary trend here with initial low-sinuosity channels with levees and overspill features evolving into incised canyons bounded by canyon walls down which highly sinuous channels developed and with little evidence of spillover (Figs 21.6 & 21.7). Studies of MIS 6 lowstand channel-slope complexes (Bryant and Eastern canyons) off an ancestral Mississippi delta lobe reveal evidence for lowstand spillover in the form of levee morphology and mudwaves. It seems that turbidity current erosion occurred mainly in the thalweg of channels, with the remaining wide canyon floors supporting thick (up to 0.5 m) and

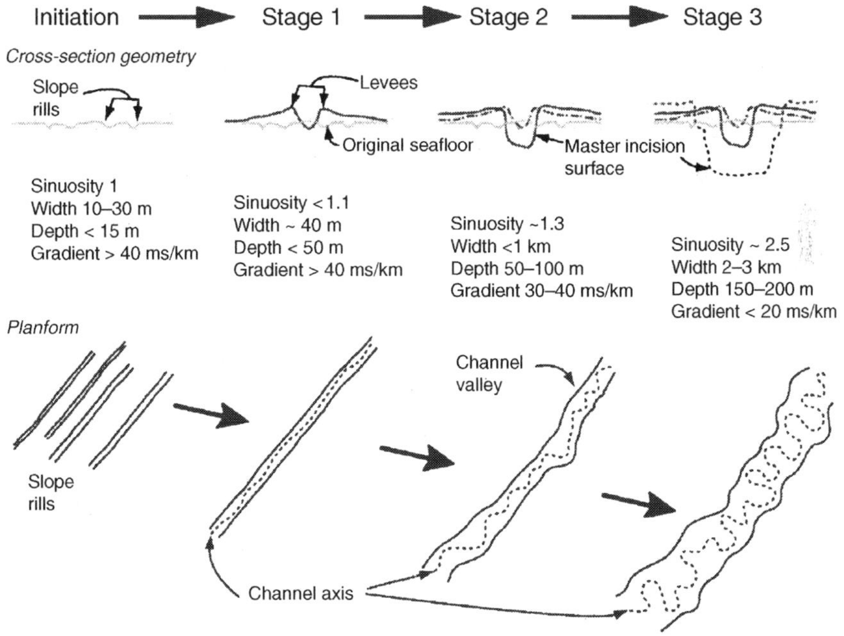


**Fig. 21.6** Scheme summarizing the initiation and evolution of erosional submarine channels. Channels evolve from rills on the open slope into high-gradient, low-sinuosity channels that are slightly incised into the slope and largely confined by levees. The authors think that as channels attempt to develop an equilibrium long profile they become more incised and increase sinuosity in order to lower long-profile channel-axis gradient. As a result they develop wider, deeply incised channel valleys, with lower gradient and a high-sinuosity channel axis. (After Gee *et al.*, 2007.)

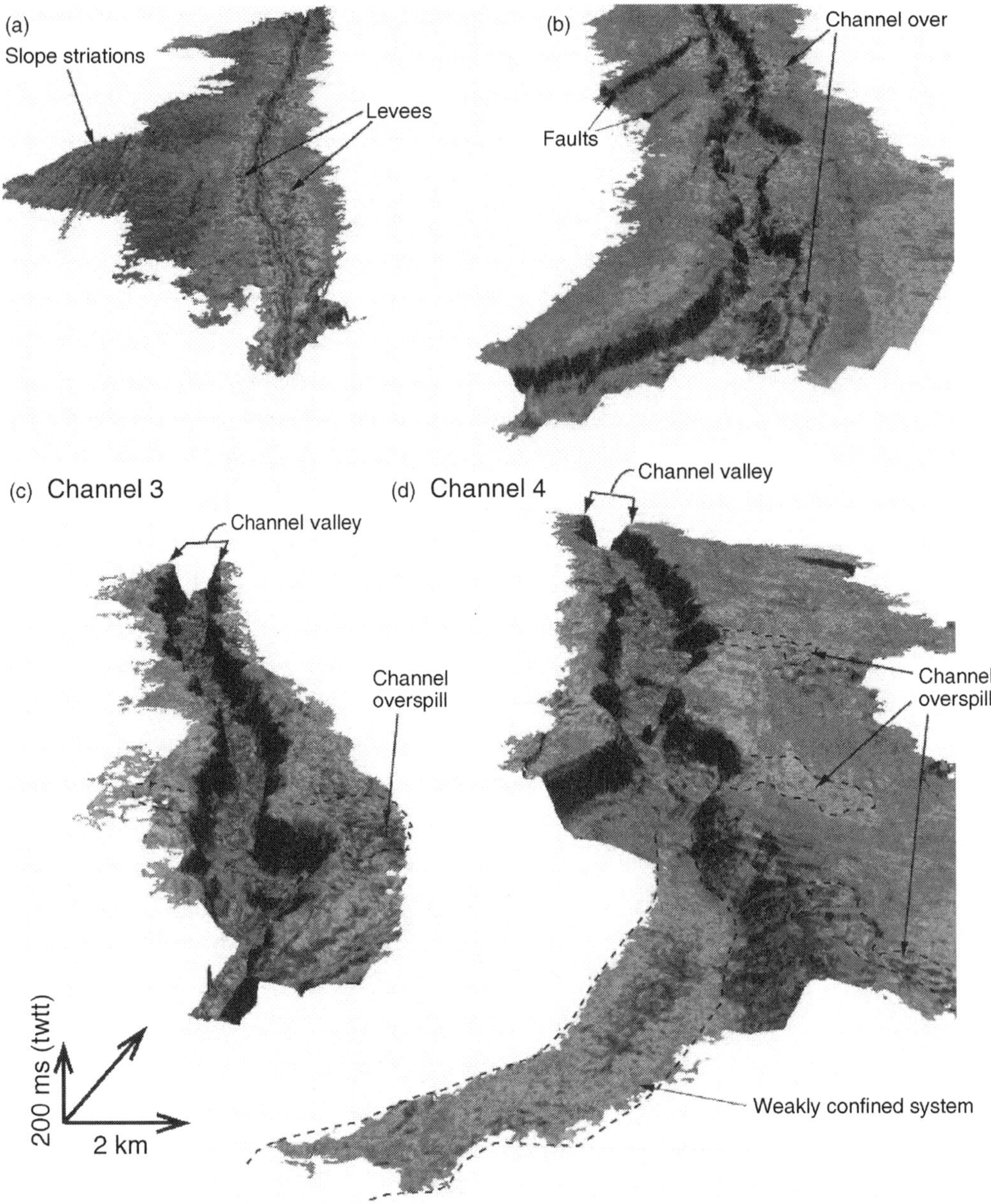


**Fig. 21.7** Oblique views of seismic amplitude draped on time–structure maps for four slope channels on the Angolan margin (Channels 1–4). Views look upslope, towards the northeast. The time–structure maps are for the bases of the channels; lighter, red and yellow colours are higher amplitude. See also Fig. 21.6. twtt, two-way travel time (After Gee *et al*., 2007.)

widespread (15–20 km) muddy turbidite deposits on inner levees and terraces. It is thought that these normally graded muddy turbidites resulted from sediment failures on the outer shelf and upper continental slope that fed the canyons.

The northern Gioia Basin of the southeast Tyrrhenian Sea (Fig. 21.8) is an interesting example of an upper slope dominated by fluvial input at a shelfless margin which generates hyperpycnal underflows that flow down canyon/channel systems. The channels are

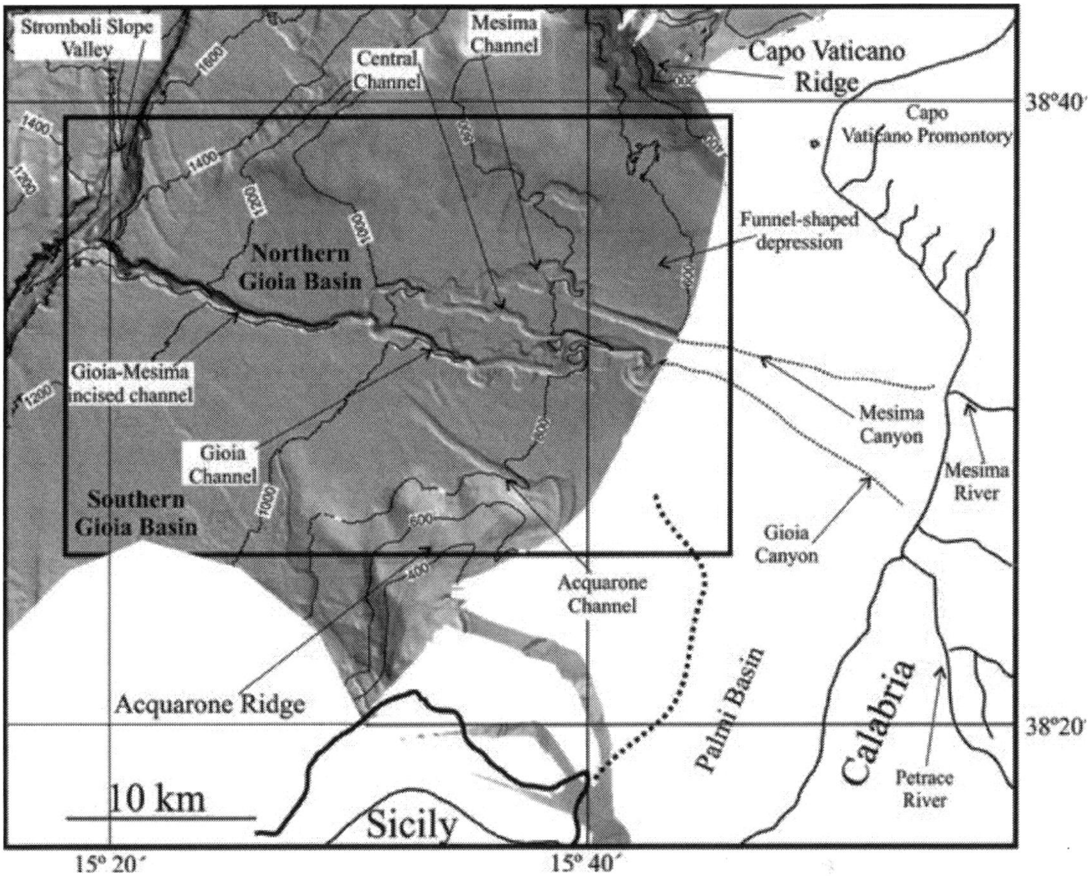


Fig. 21.8 Shaded relief map of a portion of the Gioia Basin, offshore southern Italy, based on multibeam bathymetric data. The inactive Acquarone Channel and fine-grained hemipelagic deposits in the southern part of the basin highlight the fact that the Sicilian margin now supplies little sediment to the basin, except for local slide/slump features. Coarse sediment is supplied from the Calabrian margin, where the shelf zone is extremely narrow or absent. Two major river deltas with hyperpycnal flows supply sediment directly to the head parts of the Gioia and Mesima bypass canyons, whereas an array of smaller coastal streams supplies sediment suspension across the very narrow shelf further to the north, where turbidity currents bypass the slope through a funnel-shaped scour zone with chutes. Note abandoned meander belts. (From Gamberi & Morani, 2008.)

a few hundred metres wide and a few tens of metres deep, with an interesting downslope change from straight to meandering planform as slope gradient decreases from 3.2% to 1.7%. The Mesima Channel has its lower segment abandoned because of avulsion and crevasse-splay formation at an upslope bend. The adjacent Gioia Channel has had its upper segment straightened and lower segment entrenched because of erosional deepening of the Stromboli Valley into which it debouches and which acts as the local base level. Overbank features include levees, coalescent splays and 'yazoo' channels; their nature and surface characteristics depend upon the magnitude and sediment grain-size of spillover flows.

Most slope canyons feed into the channelized apices of submarine fans (Fig. 21.9). However, some do not dissipate thus and feed truly remarkable *basin floor channels* that persist downflow for hundreds or thousands of kilometres. The most notable is the great 3800 km long Northwest Atlantic Mid-Ocean Channel (NAMOC) between Canada and Greenland that flows south to the Sohm abyssal plain. It is believed to

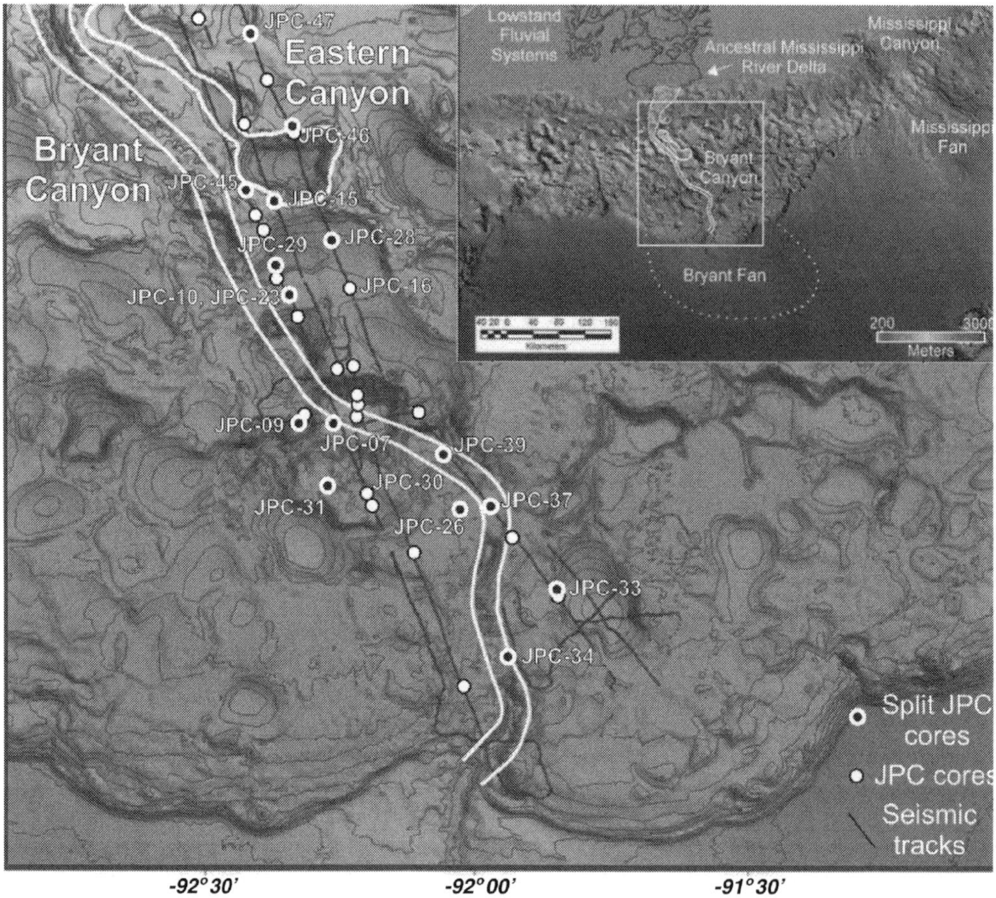


**Fig. 21.9** Shaded bathymetric maps of Bryant Canyon area and the northwest Gulf of Mexico displaying the locations of jumbo piston cores (JPC), acquisition tracks of the high-resolution geophysical information and locations of Bryant and Eastern canyon systems at MIS 6; along with an ancient lowstand fluvial system on the shelf and an ancestral Mississippi River Delta. The white lines indicate the pathways of Bryant and Eastern canyons. Contours represent isobaths at 100 m intervals. (Tripsanas *et al.*, 2008, and sources cited therein.)

have been cut by hyperpycnal glacial meltwater discharges from connected streams of the North American Laurentide ice-cap with the Labrador slope and abyssal plains. NAMOC shows spectacular depositional features within it, like meanders, channel bars and giant levees forced to grow asymmetrically westwards from overspill along the south-flowing channel. Local topographic forcing between seamounts in the lower reaches has caused channel incision. Impressive ancient examples are documented from deep marine foreland basins in outcrops from the late Cretaceous of southern Chile and the subsurface from the Miocene of Austria (Figs 21.10 & 21.11). The former channel belt was characterized by a low sinuosity planform architecture, as inferred from outcrop mapping and extensive palaeocurrent measurements. Internally the channel body architecture comprises (i) sandy matrix conglomerate with evidence of traction-dominated deposition and sedimentation from turbulent gravity flows, (ii) graded units of muddy matrix conglomerate interpreted as coarse-grained slurry-flow deposits and (iii) massive

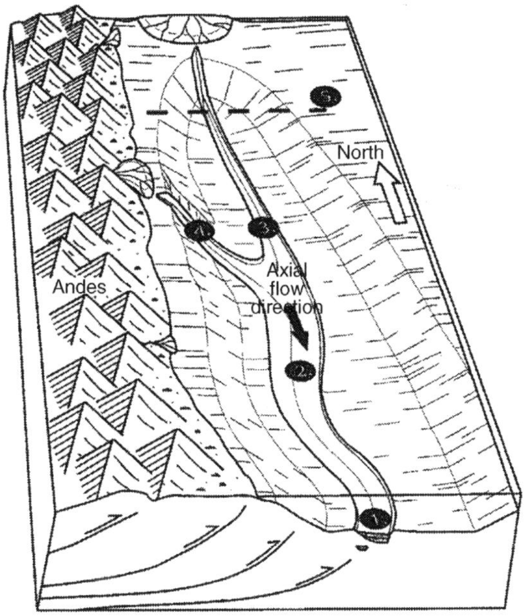


**Fig. 21.10** Palaeogeographical reconstruction of the Magallanes basin during deposition of the middle part of the Lago Sofia Member. The axial channel belt, hundreds of kilometres long, was fed coarse-grained sediment from a delta to the north and directly from an Andean catchment on the western margin of the basin. Note the presence of two tributaries in the north and the narrowing of the channel belt southwards. (From Hubbard *et al.*, 2008, not to scale.)

sandstone beds interpreted as high-density turbidity current deposits. Interbedded sandstone and mudstone intervals are present locally, interpreted as inner levee deposits. In laterally equivalent successions interbedded mudstone and sandstone sediments are interpreted from palaeocurrents, thinning trends and slump vergence as levees derived from gravity flows that spilled over the channel-belt margin. The Austrian example by way of contrast has a more sinuous planform (Fig. 21.11).

## 21.3 Well caught: intraslope basins

In addition to gullies and canyons, slopes often feature more-or-less enclosed areas of seafloor where downslope gradient is reversed and which act as local or regional sediment traps. These *intraslope basins* have received much attention in recent years because of their role as deep-sea hydrocarbon plays. There are many forms that such basins can take; the best known are from the hummocky, pock-marked northern Gulf of Mexico slope (Fig. 21.9), often crudely circular due to their origins as 3D subsurface salt withdrawal (called *halokinetic*) features. The salt is of Jurassic age, dating from early rifting of the Gulf of Mexico margin and is overlain by thick Cretaceous and Cenozoic sediments. There are over 90 such basins with a relief in excess of 150 m in the northwest Gulf of Mexico alone. Often the halokinesis has led to blockage of abandoned canyon systems. Bryant and Eastern canyons (Fig. 21.9) are a network of intraslope minibasins separated by salt-supported sills or plateaux. Bryant Canyon traverses the whole slope, linking basins by channels downslope to terminate in a submarine fan on the continental rise. Eastern Canyon is confined to the upper continental slope and terminates in a prominent intraslope basin ca 20 km across. Other intraslope basins may have more elongate forms when formed by thrust faulting (inner trench walls) or normal faulting. Whatever their origin, intraslope basins share the attribute that reversed gravity slopes, whether 2D or 3D in shape, cause downslope-advancing turbidity currents to decelerate (possibly forming internal hydraulic jumps) and deposit coarser sediment load as they encounter progressively lower slopes and then to lose energy, and lose more turbid water and sediment by *detrainment* as they traverse the basin plain and attempt to climb the negative slope. The kinetic energy degradation and sediment deposition reduces forward momentum as the current sloshes backwards as an internal bore, ever more quietly depositing its sediment load until exhausted; a process known as *ponding*. Or the current may have sufficient kinetic energy or height to rise or spill over the negative slope (possibly cutting an exit channel for itself or for subsequent flows as it does so) and carry on downslope to the next basin or finally on to the continental rise. Through time, slope basins infill, the deposits gradually onlapping basin margins, offering less and less of a hindrance to traversing turbidity currents—eventually all currents might be able to surmount their containing walls and escape down a channel or over a lip; the whole process memorably known as *fill-and-spill* (Fig. 21.12). If this is a gradual process, more-and-more coarser sediment should find its way out of the upslope basin into the downslope basin causing a coarsening upwards succession to form there. More 2D linear-to-curved ob-

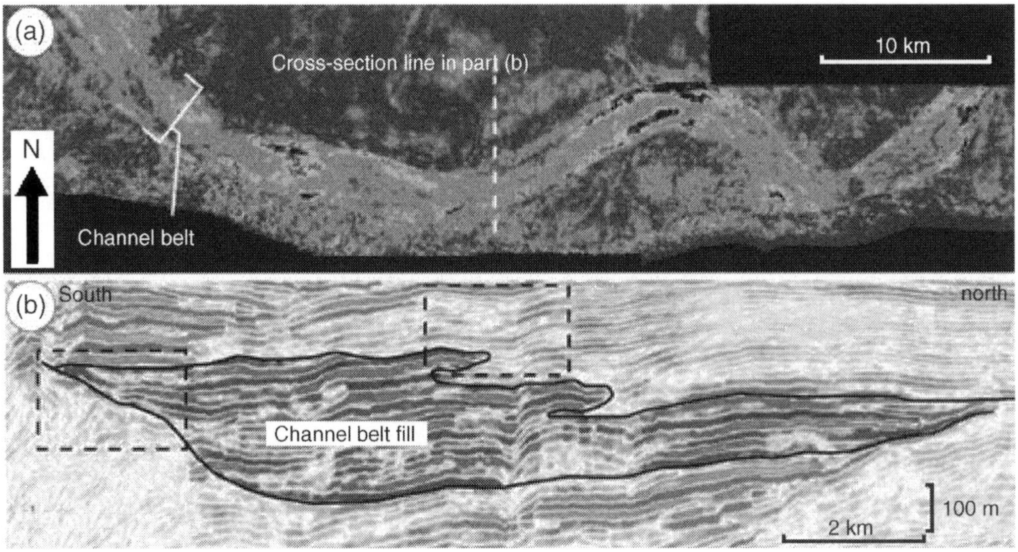


**Fig. 21.11** Seismic characterization of Oligocene–Miocene strata from the Molasse basin of Austria. (a) Plan view amplitude map showing a sinuous axial channel belt (characterized by bright amplitudes) surrounded by constructional, fine-grained overbank deposits (associated with darker amplitudes). (b) Seismic cross-section of the basin axial channel belt with outlined coarse-grained fill that is laterally juxtaposed against, finer-grained levee/overbank deposits (palaeoflow out of the page). The channel belt is 3 to 6 km wide and the channel fill succession is ca 250 m thick. The fill is coarse-grained (conglomeratic) and characterized by asymmetry. The two outlined boxes in are considered analogous to outcrop panels presented in figs 12A and 10B of Hubbard *et al.* (2008). (From Hubbard *et al.*, 2008, and sources cited therein.)

stacles formed by fault- or fold-induced topography will tend to deflect or obliquely reflect incident turbidity currents, rather in the manner of a skilled cricket batsman leg- or offside-glancing a low fast ball. Resulting turbidite successions will record all these processes by trends in grain size, sedimentary structures, bed thickness trends and palaeocurrents. Such features can lead to some pretty reconstructions of ancient basin evolution, including the role of local sediment sourcing from bounding slopes and the generation of fining- and coarsening-upwards successions. Given all this variability some rather neat experiments and satisfying theory have been concocted in recent years to help visualize and explain the physical controls upon sediment architecture of turbidites in intraslope basins.

## 21.4 Resedimentation: slides, slumps, linked debris/turbidity flows on the slope and basin plain

Detailed exploration of the ocean margins with sonar confirm earlier discoveries that large-volume submarine slides, avalanches and debris flows are very important depositional processes in a variety of deep marine environments (Fig. 21.13). Such mass flows are spectacularly illustrated by failure from the steep slopes (5–10°) associated with volcanic edifices like those of the Hawaiian chain in the Pacific and the Canary Islands in the central eastern Atlantic (Fig. 21.14). Volumes of individual events involved in the former area may exceed 5000 km$^3$ and occur as a halo of dispersed blocks (some of which are kilometres in extent) around the volcanic centres. It seems that gravity is the main destructive agent of igneous islands built up slowly over millions of years of volcanic eruptions. Passive margins are also common sites of large-scale mass failure and downslope flow transformations towards the basin plain. For example, the famous $M_s$ 7.2 Grand Banks earthquake of 1929 in the northwest North Atlantic triggered a series of upslope-propagating slide failures on the St Pierre slope south of Newfoundland that were able to source quasi-continuous debris flows from abundant late Quaternary glacial sediment deposits. These transformed rapidly downslope as they passed through

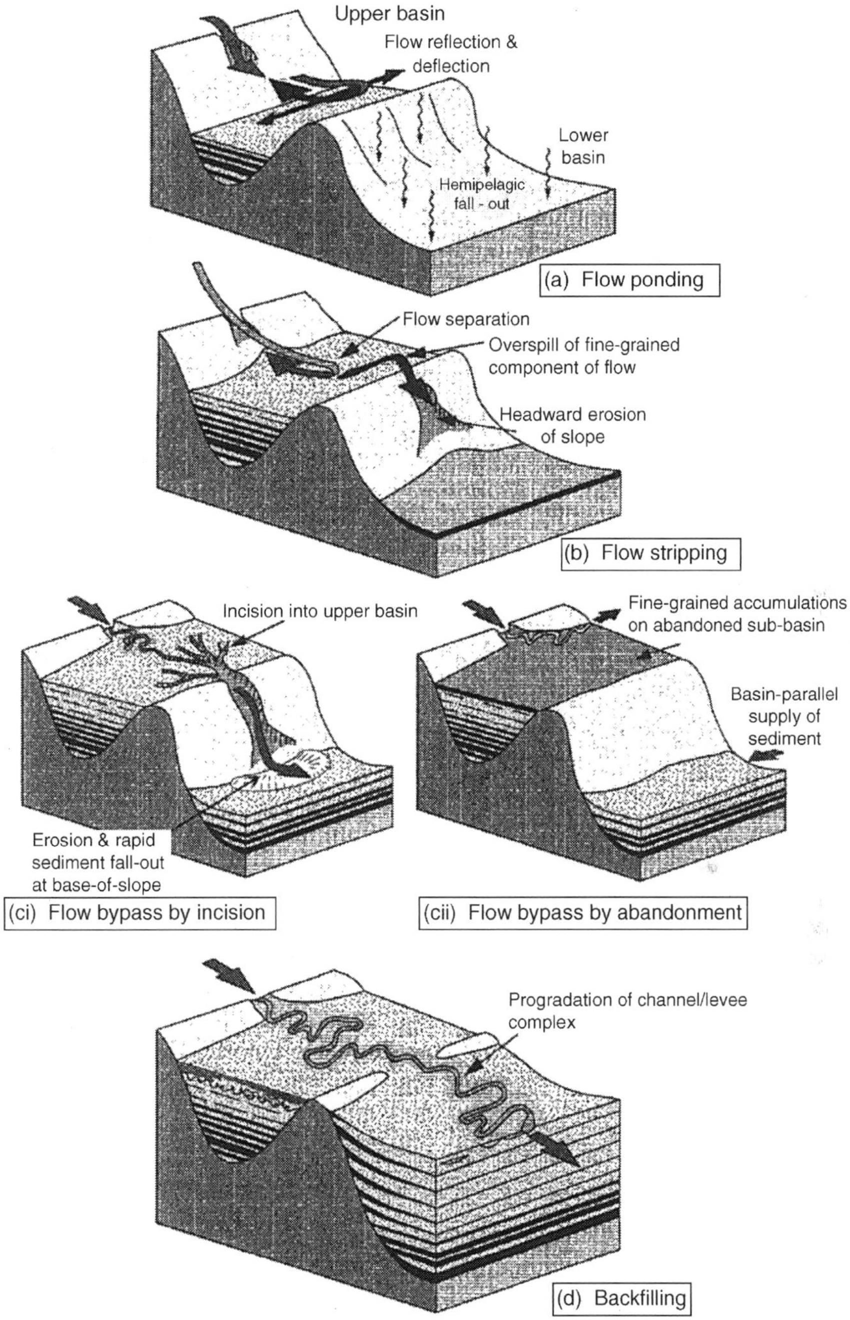


**Fig. 21.12** Depositional model for the progressive infill of a confined turbidite basin and associated deposits at the base of the slope of a lower basin. Based on the Annot Sandstone of southern France. (After Sinclair & Tomasso, 2002.)

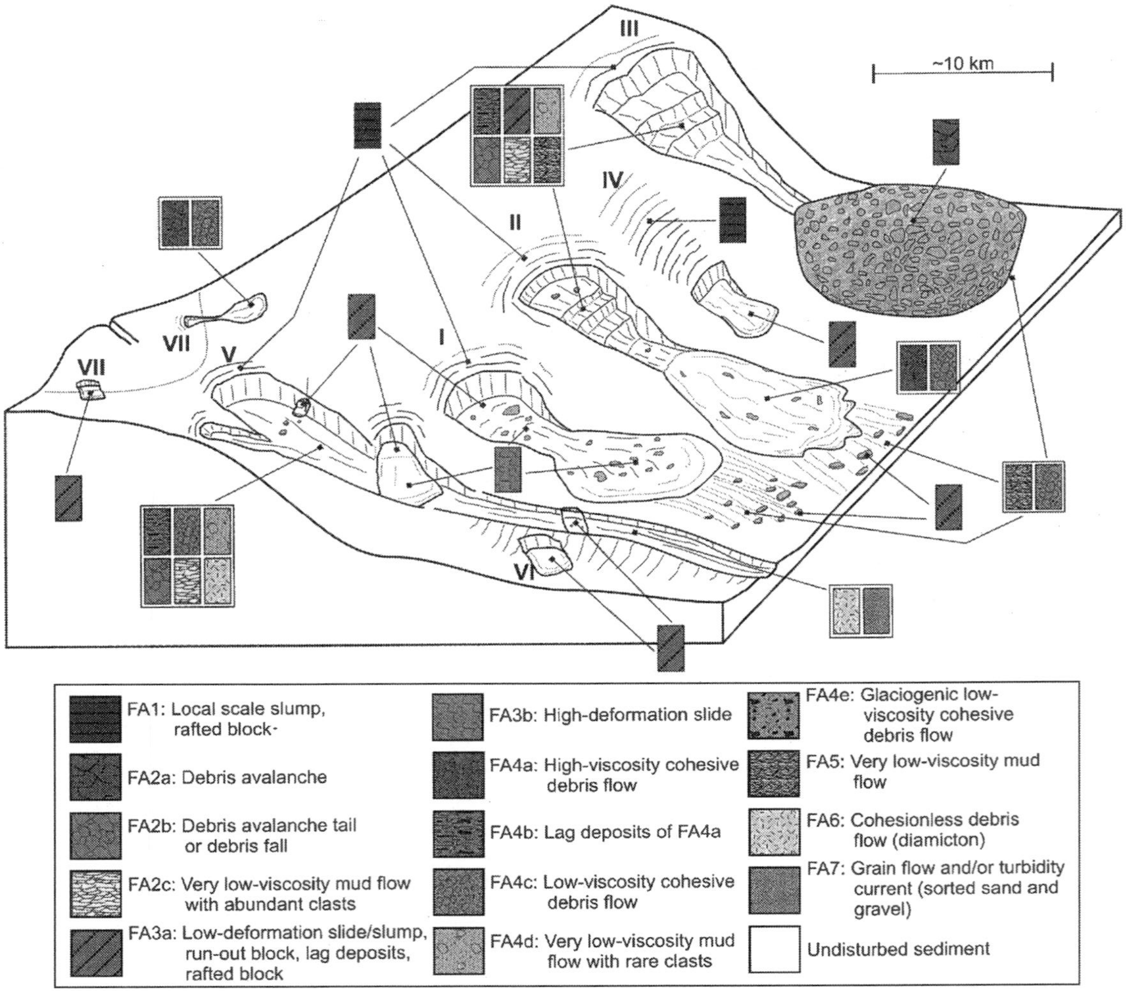


**Fig. 21.13** Cartoon displaying several different types of sediment failures and the spatial distribution of the mass transport deposits. (From Tripsanas *et al.*, 2008.)

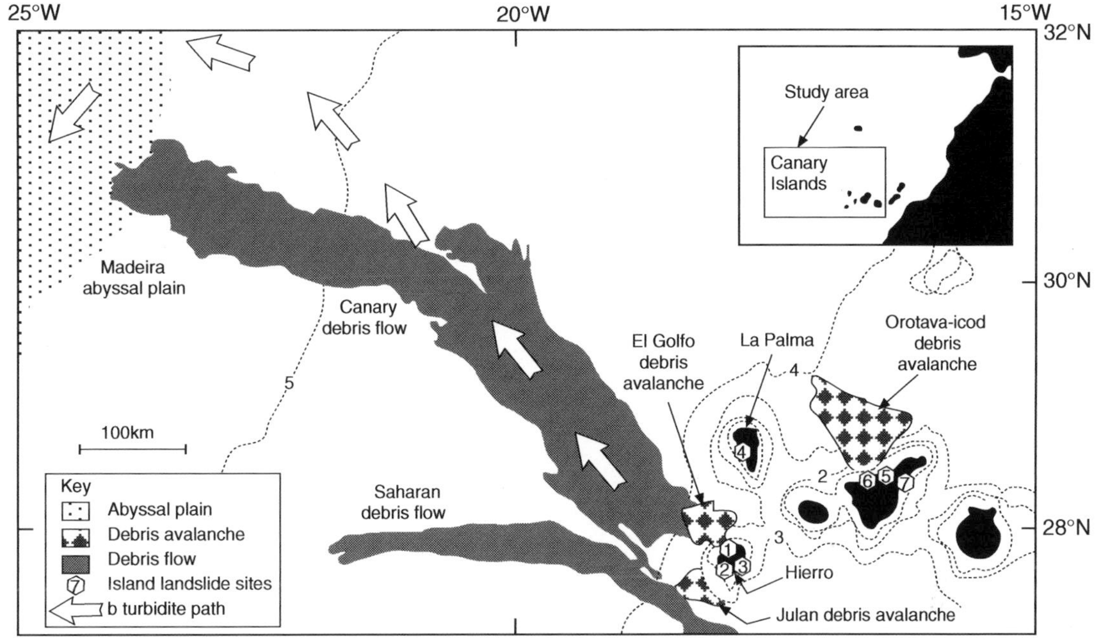


**Fig. 21.14** Location of debris flows and debris avalanches around the western Canary Islands, a group of plume-related volcanic edifices rising from the Atlantic floor to the west of the Sahara Desert. (After Masson, 1996.)

hydraulic jumps on slopes with changing curvature and were transformed into a turbidity current whose velocity peaked at more than 19 m/s, running out into the Sohm abyssal plain and depositing in total over 150 km$^3$ of sediment. The stupendous 5500 km$^3$ Storegga slide and debris flow ran out some 800 km into the Norwegian sea, giving rise to a basinwide megaturbidite and triggering a spectacular tsunami whose effects are widely recorded in Scotland and coastal Norway. Today the slide scar actually defines the continental margin and is some 300 km wide and 300 m deep.

Major slides and debris flows of enormous extent have been generated off the Spanish Sahara on the northwest African continental margin. About 60 ka, one Saharan debris flow travelled on slopes as low as 0.1° for a distance of several hundred kilometres. The deposits cover an area of about 30 000 km$^2$ and originated from a massive slump of volume 600 km$^3$ on the upper continental rise where a prominent slide scar several tens of metres deep now exists. Recognition of the debris flow deposits is based on a characteristic geometry, a distinctive acoustic character, a pebbly mudstone fabric and sharp angular contacts in cores, and an undulating surface morphology revealed by bottom photographs. The extreme mobility exhibited by the flow is thought to be due to it having overridden and loaded a porous and permeable pre-existing volcaniclastic deposit on the slope and blocking fluid escape due to debris flow matrix content. High pore pressures reduced friction facilitated shear within the volcaniclastic substrate. Sonar surveys identify numerous large volcanic avalanches and slides from the Canary Islands with downslope debris flow deposits and finally turbidity current deposition on the Madeira abyssal plain (Fig. 21.14). On one occasion a turbidity current and debris flow parted company close to the source area and followed separate paths. Eventually the slower debris flow caught up with the slowly depositing turbidity current on the abyssal plain, burying the lower deposits and then becoming buried itself by the later flow deposits. In other examples accelerated debris flow through submarine constrictions led to massive seafloor erosion and downslope transformation of turbidity current from the debris flow.

We featured previously (Chapter 9) the spectacular example of downflow transformation from turbidity flow to debris flow from the Agadir canyon and basin plain, a process thought to have been triggered by massive flow deceleration at a slope break. Such high-resolution studies of young deposits may shed light on the frequent occurrence of linked turbidites/debrites (*turdebrites*) in the older geological record. Amazingly detailed studies of over 100 sections through an easily recognizable 30 m interval in the Miocene Marnoso Arenacea Formation, Italy, established correlations for 120 km in the flow direction and 30 km across it. Extreme bed continuity and rare flow deflection suggests that seafloor relief was subtle. Thick beds are either ungraded mud-rich sandstones with ample evidence of en-masse *debrite* deposition or graded, mud-poor sandstones indicative of progressive settling and tractional turbidite deposition. Both types of deposit make up individual correlated beds, with transitions from turbidite to debrite common both streamwise and spanwise. The former thin and disappear abruptly, both down- and cross-flow whilst mud-poor sandstones define gradually tapering wedges in both down- and cross-flow directions. These relationships suggest either: (i) bypass of an initial debris flow past proximal sections, (ii) localized input of debris flows away from available sections, or (iii) generation of debris flows by transformation of turbidity currents on the basin plain because of seafloor erosion and/or abrupt flow deceleration, as seen in the Agadir basin. Another well-documented ancient example is from the lower Pleistocene Otadai Formation, Boso Peninsula, Japan, where 1–4 m thick turbidites are laterally equivalent to downflow debrites encased in turbidites. Such turbidites and encased debrites are thought to have been deposited from single flow events, perhaps due to intense erosion of muddy substrates in response to the increase in intensity of turbulence and consequent turbulence-damping in precursor turbidity currents at the mouths of middle-fan channels.

Little is known about the deep internal character and basal structure of slides which can help interpretation of the dynamics of runout and emplacement on a basal shearing surface. Such a surface has been imaged under the Saharan slides and it has been inferred that during runout a mysterious 'bulking' process led to transfer of volume from subshear layers into the moving slide. Potential sliding surfaces have been identified from 3D seismic profiling off Angola in the southeast Atlantic. Images of a large slide show slide scars, slide blocks, and debris flows involving ca 20 $km^3$ of material over an area ca 430 $km^2$. The internal structure of the slide shows evidence that seafloor remobilization occurred during initial failure, with spectacular basal striations that originate and extend downslope from a major growth fault. The slide deposit (*slidite*) consists of large blocks (1–5 km wide, 100–150 m thick) which have slid, rotated and deformed within a chaotic debris-flow matrix. Striations and furrows are thought to form by ploughing caused by detached sediment layers whose incorporation in the slide caused deposit volume to be double the volume of erosional slide scars within 5 km of source (Fig. 21.15).

## 21.5 Continental margin deposition: fans and aprons

Submarine fans are the largest sediment accumulations on Earth—millions of cubic kilometres of sediment have deposited on them for millions to tens of millions of years and the larger fans extend across the continental rise into the abyssal plain. These prodigious features represent the most invaluable archive for the investigation of climatic and tectonic processes that have shaped the surface of the Earth, not only active modern fans but also from ancient examples. Unravelling the sedimentary dynamics of such systems is essential in order to understand the processes responsible for their formation and to correctly interpret that vast archive. On submarine fans, channels are the arteries down which pass the all-nourishing sediment-charged currents that transport and deposit material eroded from the continents to the deep sea. The nourishment is derived from storm-driven plumes, surge-like and continuous underflowing turbidity currents, debris flows and slumps generated on the continental slope. In the past million years or so, fan activity has often, but not invariably, been at maximum during lowstands, with high deposition rates and the formation of lowstand 'wedges'. This trend was accentuated in fans like the Mississippi (see image of Fig. 21.9) where huge quantities of glacial meltwater and associated sediment were discharged to the shelf edge, subsequently to descend as underflows to the Mississippi submarine fan. Many fans (e.g. Indus, Amazon, Mississippi, Bengal) have become largely inactive during the Holocene highstand with very low deposition rates of fine-grained pelagic or hemipelagic sediment as their valley heads became far removed from

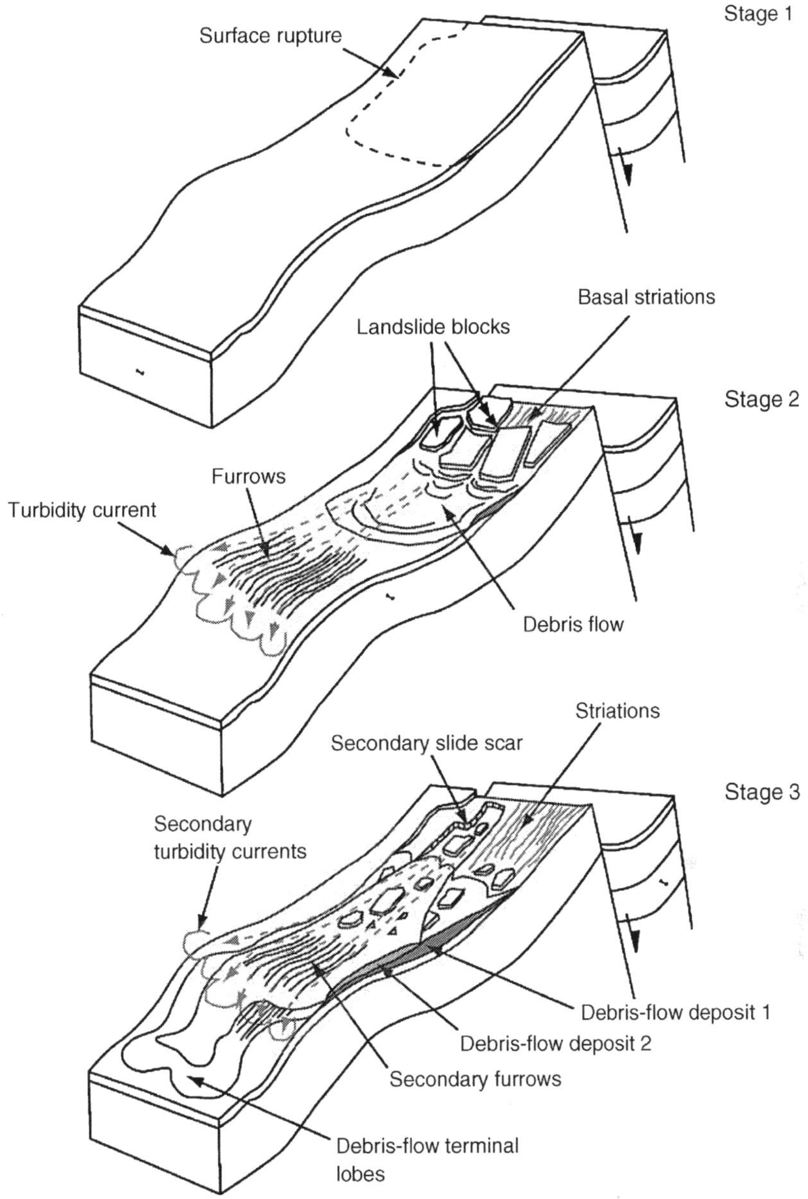


**Fig. 21.15** Scheme for landslide evolution based on offshore Angolan examples. Stage 1 shows seafloor rupture. Stage 2 shows tabular blocks, basal striations, debris flow and turbidity current generated in the headwall area. Downslope of the headwall, turbidity currents erode furrows in the seafloor. Stage 3 shows the development of secondary slide events within the headwall, triggering secondary debris flows and turbidity currents. (After Gee *et al*., 2006.)

direct riverine input of sediment, with the shelf acting as a sediment 'trap'. However, other fans, particularly those on tectonically active margins where coastal deposition is reduced and where the shelf is narrow and steep are nourished during highstand by storm-driven sediment plumes. These deposit sediment in canyon heads which are periodically flushed as turbidity currents. Also, the surface of fans and of travelling turbidity currents on western oceanic boundaries must be strongly affected both by ocean 'storms' descending from energetic upper ocean eddies and by thermohaline current activity, yet we know little about such influences at the present time.

High-resolution stratigraphic studies in older sediments pre-dating the 'eccentricity-driven' cycles of the upper Pleistocene also point to a periodic climatic control on turbidity current activity, driven perhaps in Mediterranean areas by precession-induced (~23 kyr) changes in continental water and sediment runoff rather than by direct sea-level controls. This is particularly proven for the Nile fan system (Fig. 21.16) where extensive coring and dating has led to one of the first 'source-to-sink' models for the evolution of a whole major deep-sea fan system. Six individual deep-sea fans have been active over the past 200 kyr, each comprising a canyon, channel system and terminal lobes. The turbidite system has been more active during periods of rising and high sea-level associated with wetter climates (pluvials) than during lowstands, and may rapidly become largely inactive, covered in hemipelagic muds during highstands coincident with arid periods. This study indicates that traditional sequence stratigraphic approaches to deep-sea sedimentary evolution should include a consideration of onshore catchment processes like climate and sediment yield.

Submarine fans are rarely regularly fan-shaped (or even fan-shaped at all) since their growth is often closely controlled by submarine topography, with which intruding turbidity currents and debris flows must interact. Thus not only is there a great range in fan area, reflecting sediment and water discharge from feeder canyons and slopes, but also great variability in shape. Yet only a pedant would argue against established usage since the term *fan* is a geomorphic one not generic and serves as a useful unifying concept, linking as it does all point-sourced submarine sediment bodies resulting from dispersal of sediment by a migrating and avulsive channel system separated by levees and interchannel areas. Basically the majority of a submarine fan deposit is made up of an amalgam of stacked and leveed channels (Fig. 21.17).

Fans are conveniently divided into the following sectors.

1 *Upper fan*. This contains the main active feeder channel issuing from a submarine canyon and rarely, as in the Crati fan, from a tributary network and usually bordered by major levees. Multibeam surveys on Redondo fan, California (Fig. 21.18) reveal nodal channel avulsion and the preservation within one inactive palaeochannel of gigantic (tens to a few hundred metres and depths of 20 m), partially filled flute-like scars and sand waves. Debris-flow lobes sourced from slope collapse may occur on the upper fan, spreading down on to the middle fan.
2 *Middle fan*. Here the main channel splits into leveed 'distributaries' of different ages. As in deltas and alluvial fans, usually only one channel is active at any given time and so use of the term distributary is misleading. Many fans show evidence of width reduction by internal levee construction during highstand. The channels show meandering patterns or braid between interchannel areas bounded by levees. Some channels terminate in the mid-fan to pass into sandy 'suprafan' lobes, which are elongate and up to 5 km long. Noteworthy examples occur on Hueneme, Dume, Delgada and Navy fans, California and the Corsican margin (Figs 21.19 & 21.20). Lobe switching and aggradation of lobes is controlled by random avulsions into topographic lows; the lobes are often 'nested' in the form of composite sandbodies and define aggrading offset cycles.
3 *Lower or outer fan*. This is often smooth, or with a myriad of smaller channels a few metres in depth, sometimes ending in well-defined terminal fan lobes; noteworthy examples occur on Navy and Mississippi fans. In the subsurface deposits consist of sheet-like alternations of sand and mud with shallow channels and lenses.

Submarine fan channels show decreasing channel slope and width with distance downslope. However, data from the Amazon and Zaire channels show zones of rapid change in width and zones of roughly constant width. The former mostly correspond to zones where channels narrow rapidly as channel slope falls, though occasionally a channel does the opposite by widening and steepening with distance. On the basis of these observations it has been suggested that there is a causal link between width and slope controlled by the Froude number of flows along a channel. For low Froude numbers, channel widths are approximately constant whilst rapid width-change segments support near-critical flows. Noting that critical-flow segments are terminated by abrupt widening suggests a simple sedimentological mechanism for generating channels of the correct width for maintaining near-critical flow:

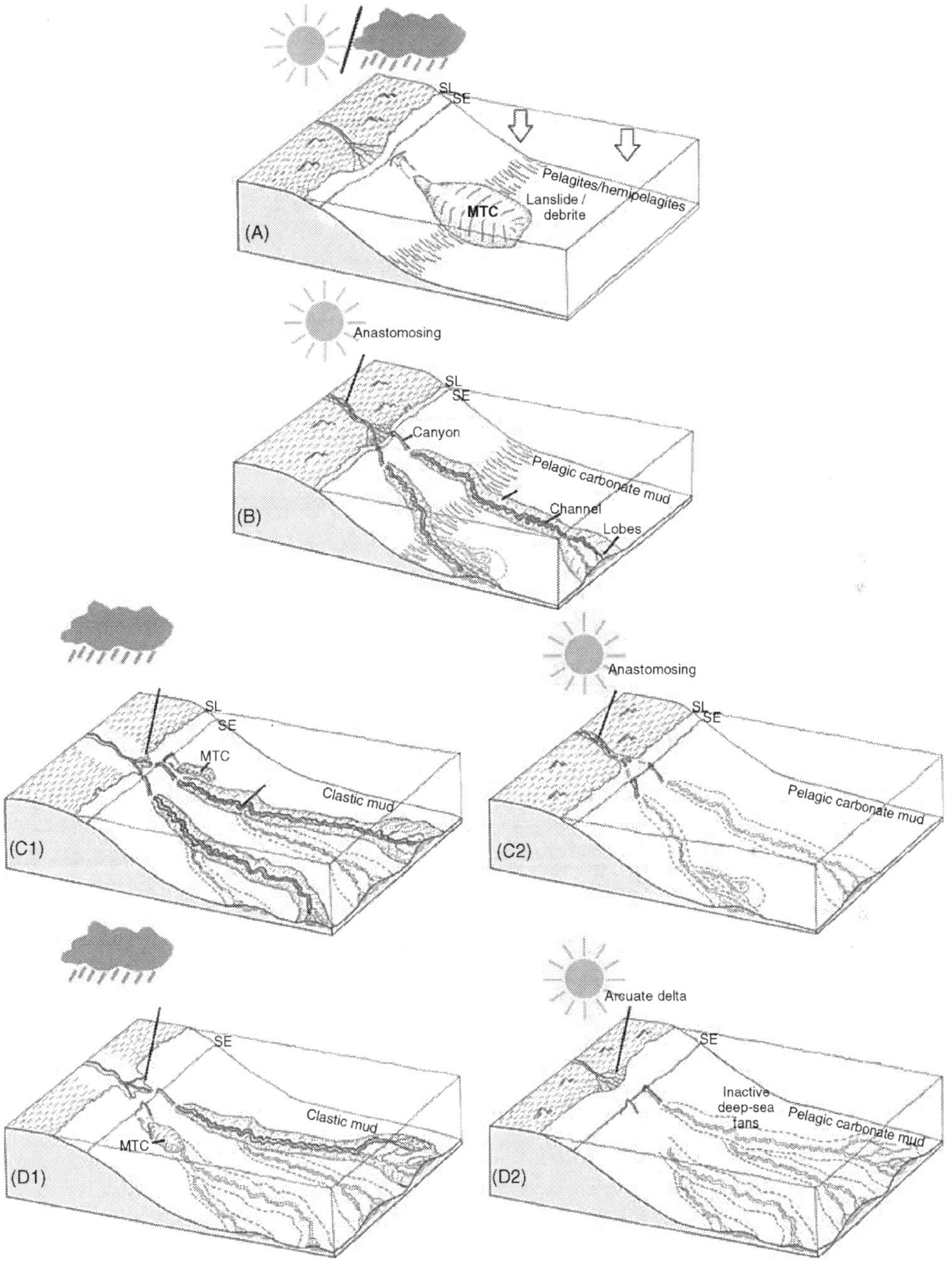


**Fig. 21.16** Interpretation of evolution of channel–levee systems and lobes fed by the Nile River during four stages of a sea-level cycle combined with climate conditions. Stage A: initial lowering of sea-level; a mass-transport complex was deposited in the basin. Stage B: continued lowering of sea-level and lowstand so that river valley and submarine canyons were connected; mainly turbidity current transport and deposition ensued. Stage C1: rise in sea-level under wet conditions; despite retreat of shoreline, significant turbidite deposition continued because of an increase in sediment flux following monsoon intensification in the source area. C2: rise in sea-level under arid conditions; retreat of shoreline. Stage D1: sea-level highstand under wet conditions. Stage D2: high sea-level stand under arid conditions. Not to scale. SL: shoreline; SE: shelf edge. (After Ducassu *et al.*, 2009.)

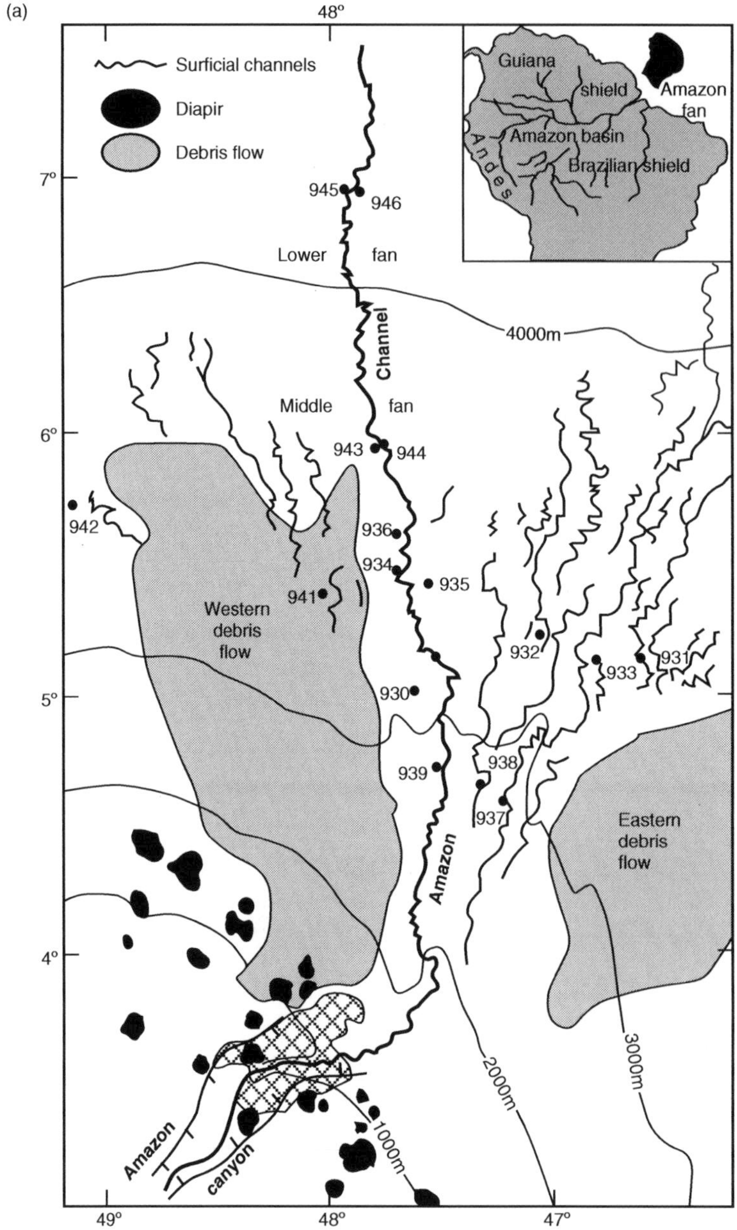


**Fig. 21.17** Amazon fan. (a) Map to show active submarine channel and numerous abandoned channels. Numbers refer to core locations. (b) Schematic cross section. Numbers are the sites shown in (a). Note the stacked and offset lowstand channel-levee complexes formed in response to channel switching, and the burial and onlap of debris flow deposits by fan sediment. MLC, middle levee complex; 30 ka Lake Mungo event. (After Cramp *et al.*, 1995.)

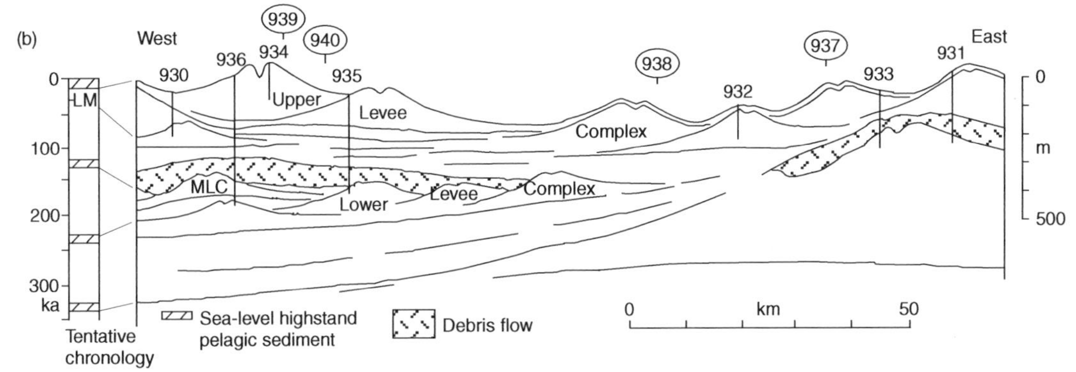


Fig. 21.17 *(Continued)*

flows will produce hydraulic jumps as they leave the segments of rapid width-change to enter the moderate Froude-number zones, and this will result in enhanced sedimentation and associated channel narrowing, thus allowing rapid-width-change segments to grow downstream.

In plan view submarine channels are morphologically analogous to coastal plain river channels but their slopes and cross-sectional areas may be orders of magnitude greater. For example, the main submarine canyon of the Zaire fan, offshore West Africa, decreases in slope from 1.5% to 0.2% downstream, reaches 15 km wide and 1300 m deep, with levees 50 km wide and 250 m relief upslope and 10 km wide with 10 m relief downslope. By way of contrast, the feeding river, with the second largest discharge on Earth, is a miserable 2 km wide and maximum 30 m deep. Abundant evidence exists for submarine channel meandering, lateral meander migration, lateral accretion, rarer meander loop cutoff and avulsion of entire channels (Figs. 21.8 & 21.17). Despite the similarity of these features and processes with subaerial river channels it is clear that the dynamics of channel flow may be different, chiefly in (i) the effect of gravity reduced by two orders of magnitude and thus (ii) high probability of supercritical flow, (iii) existence of a high-friction upper interface with ambient seawater and (iv) in the sense of rotation of the helical flow spiral in channel bends. Concerning the latter, some rather controversial experiments have identified helical flow with an opposite sense to that determined from river channels, i.e. inner-to-outer bend bottom flow rather than outer-to-inner (**Cookie 42**). Some channels end as distributaries with well-defined terminal lobes (Fig. 21.21), which mark the onset of rapid radial deceleration and turbidity current dissipation. It is possible that in fans dominated by underflow, terminal lobe sites mark the occurrence of flow lofting and rapid deposition of fines on the lower fan to form what have been termed *hemiturbidites*. Remarkable distal silty turbidites on the Toyoma fan, central Japan Sea are thought to be the result of depositional episodes from long-runout (700 km) sustained and quasi-steady hyperpycnal flows generated

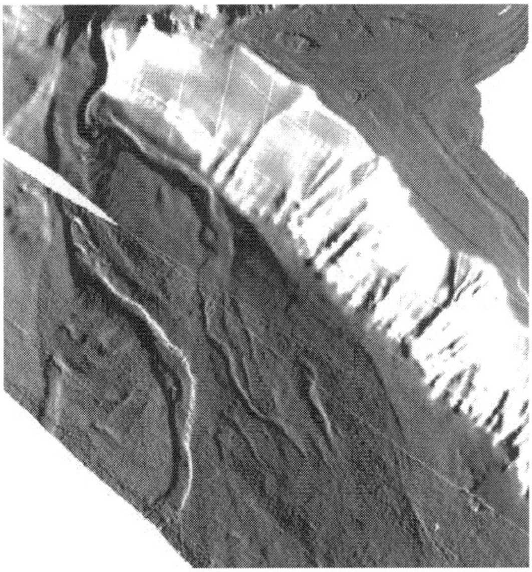

**Fig. 21.18** Redondo upper fan channel. Image based on multibeam data. (After Normark *et al*., 2009, and references cited therein.)

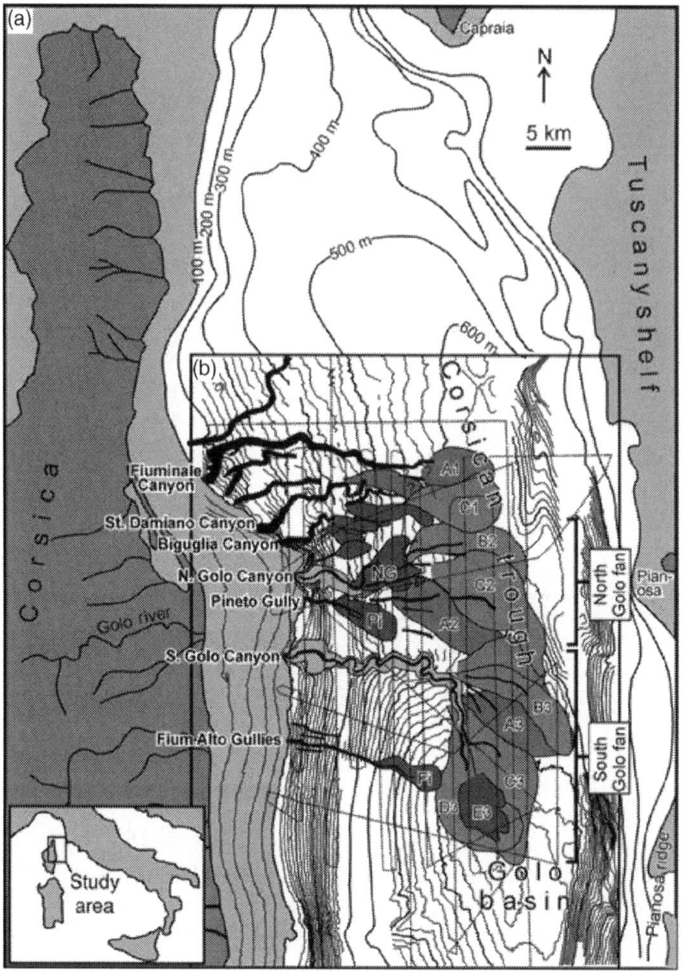


**Fig. 21.19** East Corsican margin. (a) Base map showing location of canyons, channels and 'lobes'. Dashed lines show location of Huntec deep-towed seismic boomer profiles. Fi, Fium Alto lobe; Pi, Pineto lobe; NG, proximal North Golo lobe. Inset shows regional setting. Lobes labelled A2, B3, C2, etc., are composite mid-fan lobes deposited outboard major fan valleys like the North and South Golo canyons. (After Deptuck *et al.*, 2007, and sources cited therein.)

in a mountainous terrain subject to high-magnitude precipitation. Seismic reflection profiles from the channel levees show large bedforms of climbing-dune type, attributed to the spillover of thick sustained turbidity currents. Distal silty turbidite beds are interbedded with bioturbated hemipelagites and show rhythmic layering indicative of flow-strength fluctuations during individual depositional episodes; some show a fining-upward internal trend (net flow-strength waning), whereas others show coarsening-upward (net flow-strength waxing) followed by fining.

Concerning timing of channel cutting, occupation and migration it seems that the majority of channels seen on or near the surface of modern fans were cut during lowstand or in the transition to highstand times, since Holocene highstand draping by fine hemipelagic sediment is common. However, some large fans are actively depositing, e.g. the Zaire fan discussed above. The very existence of fan channels means that at some point turbidity currents issuing from continental-slope feeder canyons must have had the excess power available locally in front of the exit

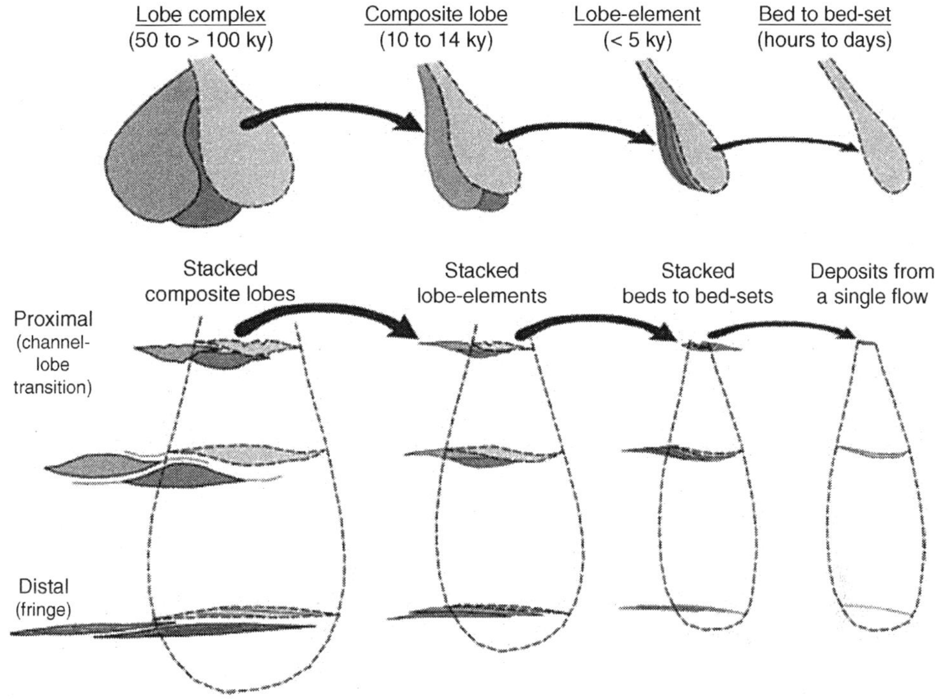


**Fig. 21.20** Diagram showing the hierarchy of compensation-stacking observed in composite mid-fan lobes. (After Deptuck *et al.*, 2007.)

point to cut such channels, otherwise the turbidity current would have spread radially, depositing uniformly as it did so. The fact that channels are still seen hundreds of kilometres downfan from the main feeder canyon means that point-concentrated power was almost always available—perhaps because the starting heads of exiting turbidity currents are *always* erosive, no matter how dilute the flow? One neat explanation is that channel formation is obligate, i.e. it must occur, perhaps due to initial erosion at the site of a hydraulic jump, by the conversion of horizontal to vertical turbulent momentum at the slope break or by development of a powerful roller vortex at the point of exit of a turbidity current from the confines of a slope canyon channel terminus. Concerning the latter possibility, experiments with continuous underflows exiting a sloping channel into a large tank identify a radial-spreading *ring vortex* developed rapidly at the back of the head of turbidity currents as the exiting head spreads and thins from the channel mouth to *circa* two channel widths downstream into the main tank (section 8.4). Such vortices develop to maintain angular momentum as velocity decreases in a spreading and lowering gravity current head. The strong vortex took clear ambient water nearly down to the bed, giving the appearance of head detachment, an effect seen most clearly in steeper channel runs (9° and 20° slopes). Other experiments show that given conditions of relatively fine substrate, high local shear stress and low angle of spreading, a dense bottom current is able to induce incision of a *self-formed channel* and that meandering can develop spontaneously for smaller values of current flux. Once cutting has begun, the channel then serves to guide subsequent currents, which may in turn enlarge the channel by erosion or infill the channel incrementally depending upon the state of the current locally. Another experimental condition for self-channelization is that the turbidity flow must be too thin to adequately cover the available downslope surface, it becomes unstable and concentrates into lanes. The relatively high flow velocity near the core of a lane inhibits deposition whilst toward the outer edges deposition and levee formation occurs.

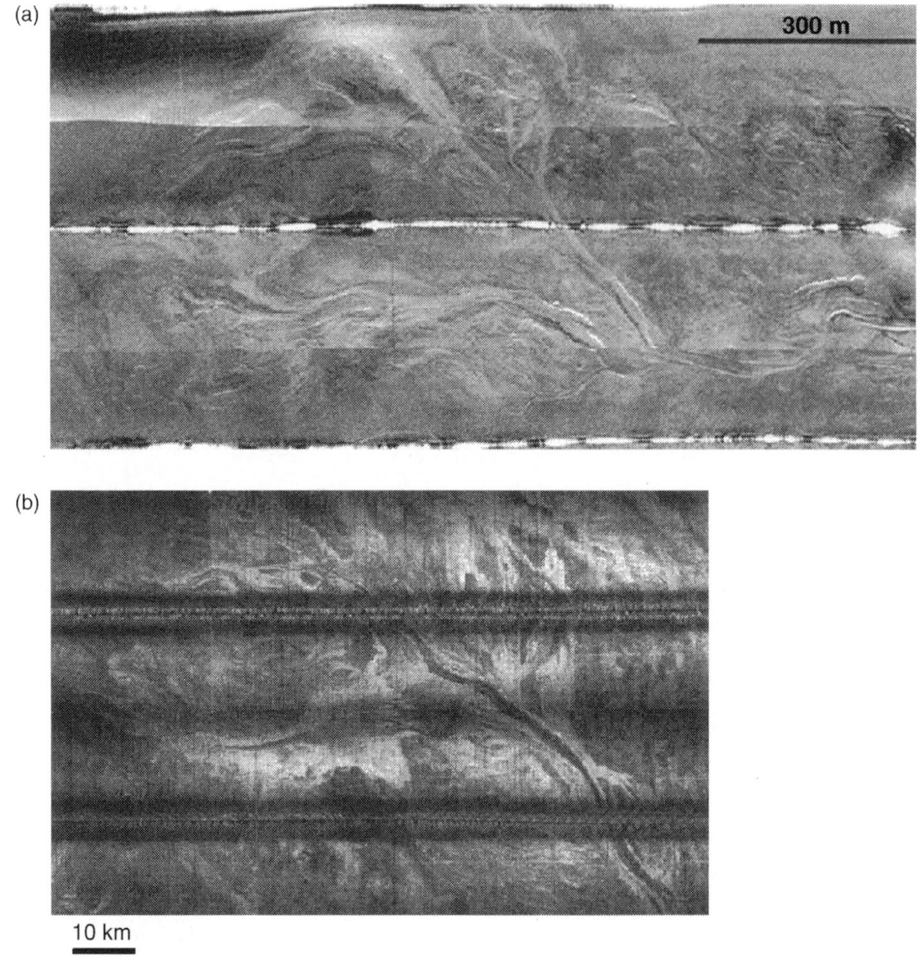


**Fig. 21.21** (a) Side-scan sonar image of the distal end of the Pochnoi submarine fan, Bering Sea, showing several mildly sinuous leveed channels. The flow direction is from right to left. (From Yu *et al.*, 2006, Image of N. Kenyon.) (b) Side-scan sonar image of the distal end of the fan complex of the Gulf of Cadiz submarine fan. Note the multitude of mildly sinuous, low-relief leveed channels. The flow direction is from right to left of the image. (From Yu *et al.*, 2006, image of A. Akmetzhanov.)

Natural channel-bounding levees are constructed as turbidity currents *overspill* channel margins and interact with the surrounding fan surface. Evidence for frequent, deep out-of-channel flow comes from acoustic imaging and sounding, direct measurement, lateral correlation of depositional beds, levee erosion and breeching, and widespread presence of fields of large-scale muddy or sandy *sediment waves*. Coriolis-turning of currents creates asymmetric levees in higher latitudes whereas symmetric levee pairs mark equatorial channels like those on Zaire and Amazon. The rates of growth of levees and of the deepening or shallowing of channels then control the dynamics of the whole fan system. In many upper and middle fan sectors it seems that some sort of depositional (aggradational) equilibrium between channel and levee was attained, since leveed channels now perch with their thalwegs (deepest parts) high above the surface of the surrounding fan. We know from evidence of modern rivers and modelling studies of rivers that such a topography leads to a system of offset stacking of successive channel–levee complexes as channel avulsion occurs with time. Such offset stacks have been nicely imaged by high-resolution seismic reflection studies (Fig. 21.17). Another process that may have great importance in controlling levee deposition, terminal lobe formation and avulsion is that of *flow stripping*. Here, large-magnitude flows negotiating a bend split into two, with the denser, sandier, lower portion carrying on round the bend and the upper portion leaving the channel confines to erode the levee and escape on to the surrounding fan surface, where it may lay down a lobe-shaped deposit during its radial deceleration. Under some circumstances of major flow and levee failure, the process may lead to complete channel avulsion. Studies on several fans show that levee sediments coarsen downfan as levees decrease in relief; presumably due to entrapment and deposition from progressively lower portions of grain-size stratified turbidity currents.

Concerning sediment waves, detailed studies on the Zaire fan channel (Fig. 21.22) reveal linear, channel-parallel forms with gentle slopes and amplitudes ranging from a few metres to 70 m and wavelengths

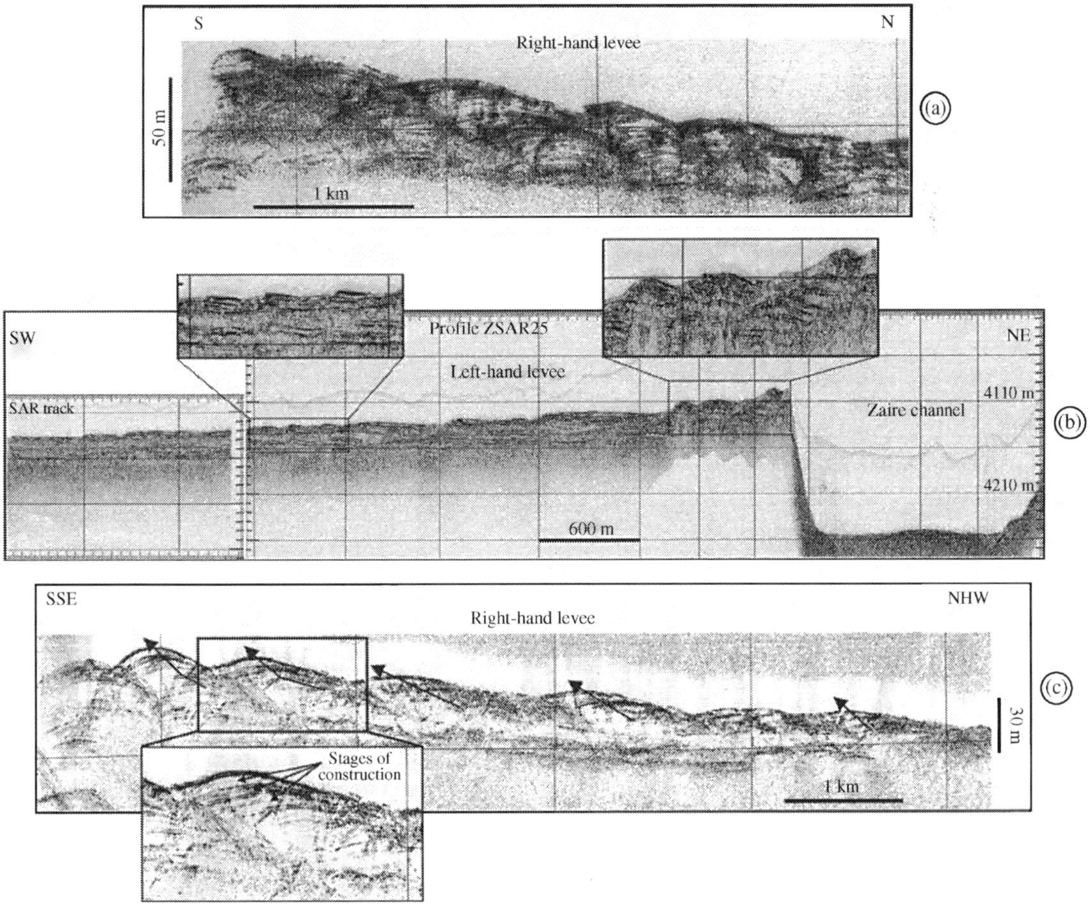


**Fig. 21.22** Zaire levee sediment waves revealed by 3.5 kHz shallow seismic profiling. (a) The upstream part of the modern channel–levee system shows sediment waves located near the crest of the levee which have migrated upslope. Downlevee, sediment waves are poorly developed and exhibit a symmetrical cross-section. (b) The middle part of the modern channel–levee system. Insets show detailed morphology and internal structures of sediment waves located near the levee crest and down levee. (c) The downstream part of the modern channel–levee system. Arrows indicate direction of sediment-wave migration. Inset shows detailed internal structure of a sediment wave located near the levee crest. (From Migeon *et al*., 2004.)

0.2 km–7 km. They show upslope and presumably upcurrent migration toward levee crestlines and channels. Most commonly sand wave fields are observed downstream from sharp bends of channel, on the concave sides of meanders, or along straighter reaches. The fields are larger and the number of sediment waves increases as levee relief decreases. Sediment-wave wavelengths increase gradually downchannel, from less than 400 m to 2000 m, and amplitudes decrease downchannel, from 15–20 m to 5 m. Within each field, both wavelengths and amplitudes decrease down levee (away from the channel). The basic flow dynamic processes creating such bedforms remain obscure, though they resemble well scaled-up versions of supercritical flow antidunes. Internal structures show evidence of both upflow migration and aggradation above subsurface irregularities that may have triggered wave growth.

Large-scale fan architecture is a response to both changing sea level and the shifting locus of sediment input at the continental shelf break. The Mississippi fan shows this latter effect *par excellence* as the changing Quaternary position of the Mississippi river has influenced the sites of successive shelf and slope canyons. The resulting spaghetti-like arrangement of about 17 Plio-Pleistocene channel–levee complexes in a lateral wedge-shaped amalgamation of fans is impressive. A similar sequence of lobe shifting following movement of feeder river channels is documented in many other fans, nicely shown by integrated surface and subsurface studies of the Nile fan (Fig. 21.16) and of the fans supplying the Santa Monica basin, offshore California. In the latter, coarser-grained mid-fan lobes prograded into the basin from the Hueneme, Mugu and Dume fans at times of rapid sea-level fall, sourced by river channel incision and delta cannibalization. Deposition rates of up to 13 mm $yr^{-1}$ are measured on the mid- and upper parts of Hueneme fan.

Submarine fans on high-latitude continental margins fed by continental ice-sheets at the shelf edge include the Upper Pleistocene lowstand-active Laurentian fan and Labrador Sea channels. The Laurentian fan also has large upperfan channels with huge relief from channel levee crest to channel floor (800 m). As an example of an active ice-sourced fan system, consider the Wilkes Land continental slope, Antarctica. Here upper fans also have humongous channels with depths up to 900 m, distances between levee crests up to 18 km, and channel-floor widths up to 6 km. Middle-fan channels have reduced but still substantial depths (ca 300 m) with local interchannel areas featuring both channel-overbank deposits from turbidity currents and mounded sediment of contour-current origins with relief ca 400 m. The lower fan has small, shallow channels (50–75 m relief). Such ice-sheet-fed fans and channels owe their prodigious size to the high magnitude of water/sediment discharge available right at the shelf margin.

Our all-too-brief synthesis of submarine fans must end with some attempt to guide geologists in their difficult attempts to make sense of ancient deposits thought to be of submarine fan origins. In the 70s and 80s it was popular to define 'cycles' of turbidite beds tens of metres thick displaying upward thinning/fining and upward thickening/coarsening, deposited in submarine fan channels and prograding submarine fan lobes respectively (Fig. 21.23). However, these claims have been shown to be statistically invalid and cannot provide criteria for identification of fan subenvironments. Observed disordered stacking of thicker and thinner beds are natural consequences of fan aggradation and are probably due to channel switching and avulsion causing irregular variations in flow volumes, concentrations and precise pathways of turbidity currents. Upward-fining trends that do occur in a minority of turbidite successions in channel deposits are interpreted as the result of channel filling or the stacking of onlapping deposits at a channel mouth.

When considering whether a succession of turbidites originated on a submarine fan, criteria such as specific deposit characteristics, large-scale geometry, and degree of sand-bed clustering provide the best tools for discrimination of submarine fan subenvironments (Fig. 21.24). Channel deposits are believed to fine downstream from source and to contain vertically or laterally accreted sequences of sand- and/or gravel-grade sediments depending on available sediment supply. Hydraulic considerations lead to the expectation of channel and point bars covered with dune bedforms and thus the production of laterally accreted cross-stratified deposits with frequent erosion surfaces. The magnitude of modern channels should prepare the geologist for major, kilometres-wide, hundreds of metres deep complexes of great internal complexity (Figs 21.19 & 21.20). Channel deposits may be divided into two end-members, *erosional* and *aggradational*. The former tend to occur on smaller fans and are usually low-sinuosity channels with low

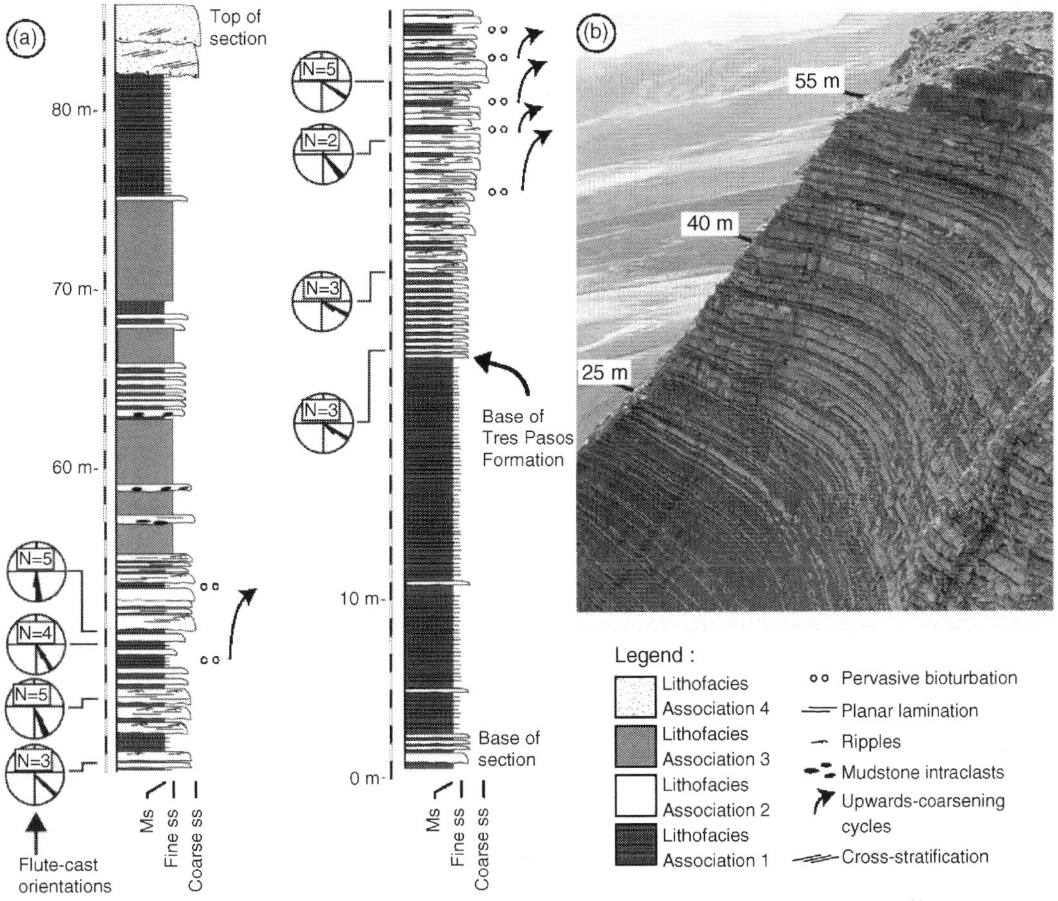


**Fig. 21.23** Cycles or not? (a) Detailed measured stratigraphic section through a pile of turbidites at El Chingue Bluff, southern Chile. The authors note the bundling of coarse-grained sandstone sedimentation units into cycles separated by pervasively bioturbated mudstone intervals. Deposits of these bundles coarsen upwards, and associated beds get thicker (and are more amalgamated) towards the tops of cycles. Rose diagrams indicate flute-cast orientations. (b) Overview of the succession showing the tabularity of individual sandstone beds. North is towards the top of the photo. (From Shultz & Hubbard 2005.)

levees transporting coarse sediment down relatively steep slopes; avulsions are frequent and a high degree of channel deposit density is achieved locally. The latter are usually highly sinuous channels with large levees on gentler slopes; avulsions are infrequent and depositional sequences are dominated by the vertical aggradation of the channel deposits. Cores from modern fan levees are dominated by thin-bedded fine sand beds with internal small-scale cross-laminations. The form of levees indicates that over kilometres or tens of kilometres all units should fine and thin away from the channels. Complete levee sequences might reach hundreds of metres thick for the largest submarine fans. Interchannel flats between channel levee complexes comprise bioturbated muds and thin silts. Terminal lobes like those of Navy and Delgada fans might be mapped out as such in an area of good exposure, showing downlobe diminution of channels, fining and thinning of turbidite laminae and beds.

The overall depositional pattern of a fan involves a downfan fining from thick coarse sand- or gravel-grade turbidites and debris flows in the upper fan

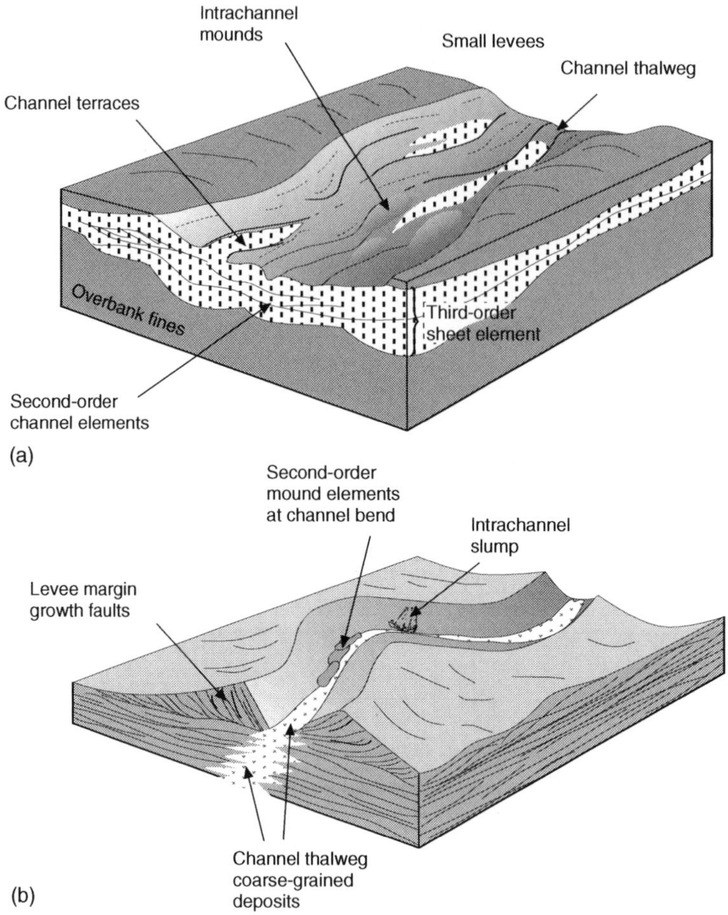


**Fig. 21.24** Block diagrams to contrast (a) low-sinuosity fan channels producing sheet sands by lateral migration, and (b) high-sinuosity channels with abundant spillover producing ribbon sands in an aggrading package of levee deposits. (After Clark & Pickering, 1996.)

channels to thin, very fine sand- or silt-grade turbidites on the lower fan apron. Sections normal to the fan axis on the mid-fan show channelized turbidites separated by fine interchannel areas. Fan progradation is postulated to cause a gross coarsening-upwards succession as progressively more proximal fan facies overlie the fan apron. On a basin scale, ancient submarine fans make a sedimentary architecture explicable by considering eustatic sea-level change. For example, icehouse conditions similar to the middle–late Pleistocene occurred in the Permo-Carboniferous and the effects of presumed Milankovitch cyclicity on fan waxing and waning is clearly set out from magnificent exposures in the Karoo basin, South Africa. Regional mapping reveals whole-fan cycles of lowstand activity whose commencement as regionally mappable erosion surfaces record sequence boundaries and subsequent progradation/aggradation. These periods alternated with highstand drapes of hemipelagic fines. Intrafan cycles record minor sea-level fluctuations, fan-channel spillover and fan-lobe shifting. These Karoo outcrops also reveal well-exposed mid-fan channel lobe deposits comprising many stacked finger-like elements comparable to those recorded from modern fan outlets and in experiments with quasi-continuous turbidity underflows.

As a coda it is important to mention that active tectonics, especially the growth and death of faults can lead to significant effects on deep-basinal fan turbidite evolution, chiefly by tilt-induced gradient changes to the seafloor and the production of barriers to flow. A fine example occurs in the Miocene Tabernas basin of southeast Spain.

## 21.6 Continental margin deposition: turbidite pathway systems connecting slopes and basin plains

On ocean margins with complex topography and structure (oceanic islands, sea mounts, etc.) point-sourced base-of-slope fan, or series of fans defining a submarine bajada, are not so well defined. Multiple sources contribute to basin filling and linkage takes place from canyon mouth across immense distances of ocean-floor basins and connecting channels. This type of deposition can hardly be said to represent a concentrated fan environment and has been termed a *turbidite pathway system*. The Moroccan Turbidite System, North Africa is a much-studied example that originates on the Moroccan margin, before running westwards between the Canary and Madeira archipelagos, and terminating on the Madeira Abyssal Plain (Fig. 21.25). It has a total length of 1500 km, and its morphology is largely controlled by the position of volcanic islands (Canaries, Madeira), seamounts, salt diapers and basin plains (Seine, Agadir). Frequent gradient changes and multiple sourcing leads to a complicated and linked deposit architecture. Large-volume ($> 100\,km^3$) turbidites sourced from the Morocco Shelf have a relatively simple architecture in the Madeira and Seine Abyssal Plains. They form distinct lobes or wedges that thin rapidly away from the basin margin and are overlain by basin-wide muds deposited by 'ponding' flows. In the Agadir Basin the turbidite fill is more complex owing to a combination of multiple source areas and large variations in turbidite volume. For example, a single, very large turbidity current (200–300 $km^3$ of sediment) deposited most of its sandy load within the Agadir Basin, but still had sufficient energy to carry most of the mud fraction 500 km further downslope to the Madeira Abyssal Plain. Large turbidity currents (100–150 $km^3$ of sediment) deposit most of their sand and mud fraction within the Agadir Basin, but also transport some of their load westwards to the plain. Small turbidity currents ($< 35\,km^3$ of sediment) are wholly confined within the Agadir Basin, and their deposits pinch out on the basin floor. Turbidity currents flowing beyond the Agadir Basin pass through a large distributary channel system. An interesting fact about the incidence of large-volume turbidity flows in the Morocco system is that they were deposited during periods of rapid sea-level change and do not appear to be specifically connected to sea-level lowstands. This contrasts with the classic fan model, which suggests that most turbidites are deposited during lowstands of sea level. This leads one to suspect a climatic control from the Moroccan onshore hinterland, perhaps analogous to that noted above for the Nile hinterland.

## 21.7 Continental margin deposition: thermohaline currents and contourite drifts

Deep ocean currents (aka *thermohaline currents* or *contour currents*) are particularly important on the western sides of the oceans; perhaps the best known example is the Western Boundary Undercurrent (WBU), that portion of North Atlantic Deepwater that flows along the western margin of the North Atlantic ocean. It flows parallel to and southwards over the continental rise at velocities of up to 0.25 m/s (Fig. 21.26). It is coincident with bottom *nepheloid layers*, increased suspended material in near-bottom water revealed by light-scattering techniques. These are attributed in part to the action of thermohaline currents and in part to the occurrence of deep-sea 'storms', i.e. deep downward transfer of ocean surface eddy current energy, in this case from the Gulf Stream. The source of the suspended sediment (Fig. 21.27) is variable: distant sourcing (from polar regions especially), local erosional resuspension of ocean-floor muds by 'storms' and enhanced thermohaline currents, and dilute distal turbidity current flows probably all have a role. Some nepheloid layers may be up to 2 km thick, although 1–200 m is a more usual figure. Sediment in nepheloid layers is usually $< 2\,\mu m$ although fine silt up to 12 μm may be suspended, normally at concentrations of up to 500 mg/L rising to 5000 mg/L a few metres off the bottom during deep-sea 'storms'. Nepheloid layers are also known in many areas from intermediate depths, often at the junction between different water masses. These are thought to arise through the erosion of bottom sediments by internal waves and tides amplified on certain critical

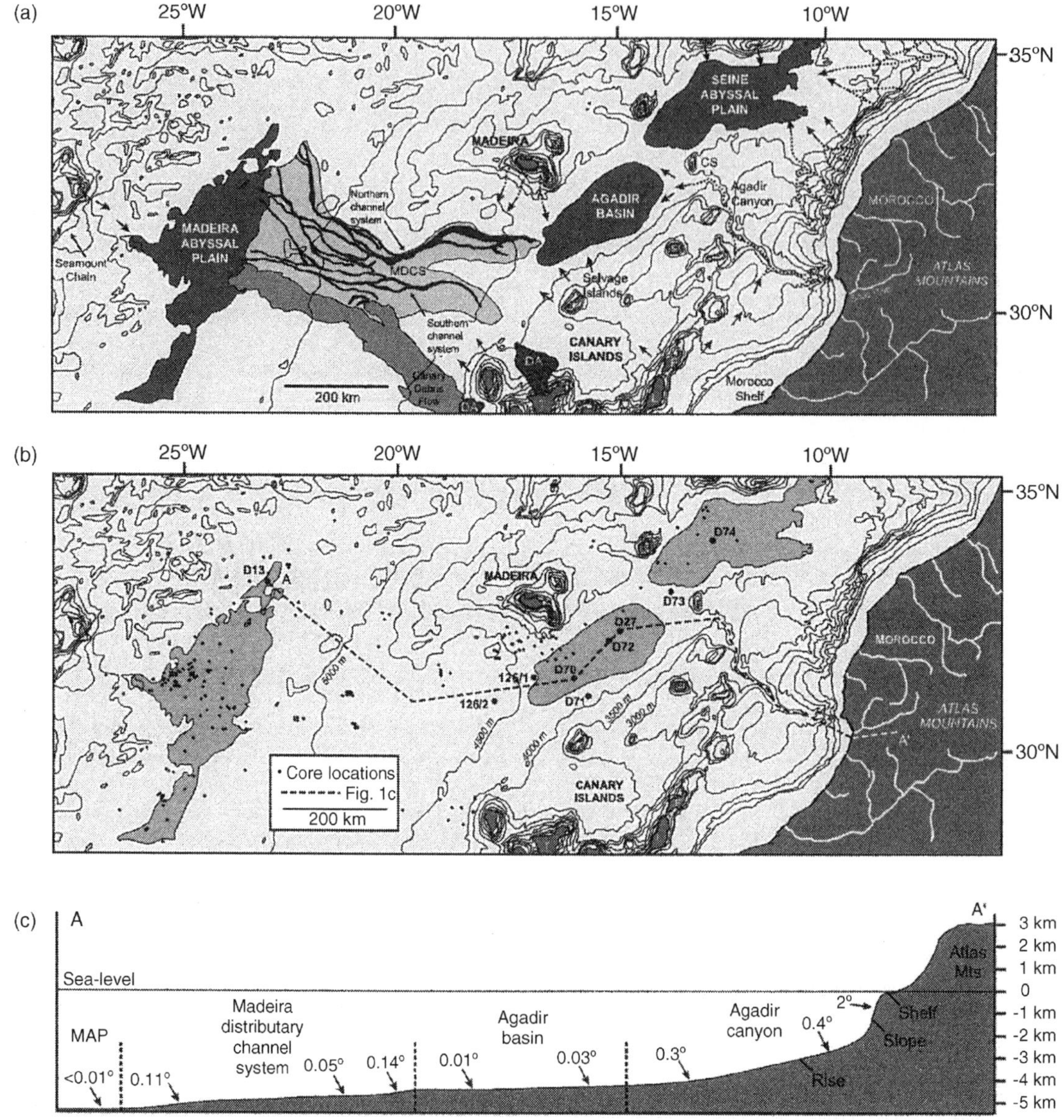


**Fig. 21.25** The prodigious Moroccan Turbidite System (MTS). (a) Location map with main features and principal transport directions for turbidity currents shown by arrows and dashed lines. CS, Casablanca Seamount; DA, debris avalanche; MDCS, Madeira Distributary Channel System; T, Tenerife; L, Lanzarote; F, Fuerteventura. (b) Map showing all core locations and the location of the line of cross-section shown in (c). Bathymetric contours spaced at 500-m intervals. (c) Cross-section showing variations in gradient across the MTS. Note the change in gradient between the almost flat Agadir Basin and the head of the MDCS. (After Wynn *et al.*, 2002.)

bottom slopes. The layers, once formed, intrude laterally into the adjacent open ocean as tongues many tens of metres thick (Fig. 21.28).

Seismic studies and seabed coring beneath WBU thermohaline currents reveal vast, thick (several 100s of metres to kilometres scale), relatively slowly

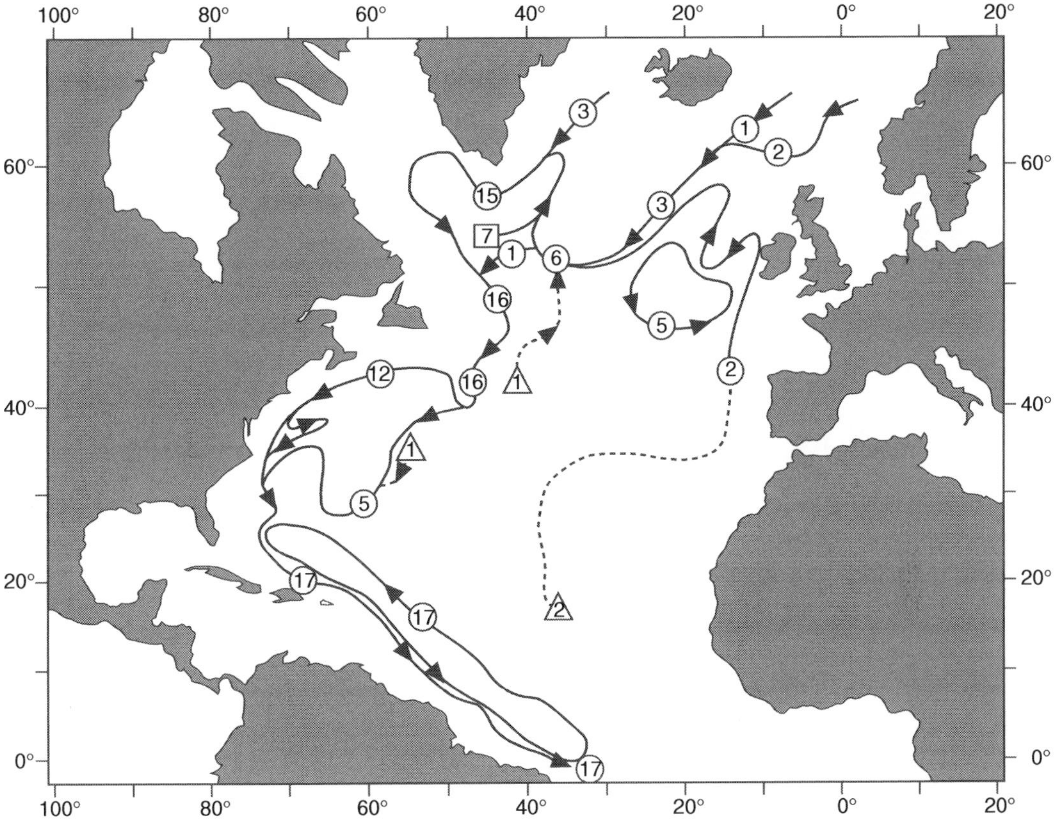


**Fig. 21.26** Map to show generalized paths and transport rates (numerals encircled are Sverdrups, Sv, $10^6$ m$^3$/s) of North Atlantic Deep Water (1.8–4 °C) circulation. Curved paths indicate that flow is following submarine topography, hence the older term for thermohaline flow used by marine geologists—contour currents. (After Schmitz & McCartney, 1993.)

deposited (typical rates, 50–100 mm per thousand years) *sediment drifts*, comprising alternations of thin, very fine sands, silts and bioturbated muds. The sands and silts are thinly bedded, ungraded, well sorted, and may contain heavy mineral placers in small-scale cross-laminations. Small ripple-like forms, other tractional features and current scour features are recorded during periodic intensification of the near-bottom flow during deep-sea storms. These muddy, silty and fine sandy deposits of boundary undercurrents are termed *contourites*; their good sorting, due to winnowing, distinguishes the coarser of them from unbioturbated thin distal turbidites, but pervasive bioturbation of contourite muddy silts is common. Erosional effects of cold undercurrents are also very important; many stratigraphic gaps in deep-sea sediment cores from Oligocene times onwards have been attributed to onset of contour-current erosion—it doesn't take long for erosive rates of a millimetre or so per year to remove deposits which have accumulated quite slowly. Drift formation and spatial development involves a complex interaction between thermohaline flow deceleration, seafloor lithology and sediment supply from submarine-fan sediment. The latter factor emphasizes that both along-slope and downslope processes operate together in deep-sea areas. Rapid gravity-current sediment delivery is followed by secondary reworking by bottom currents on a much longer timescale. This can result in modification of turbidite channel–levee architecture, creating elongate levees or submarine fans. It is also likely that turbidity currents travelling

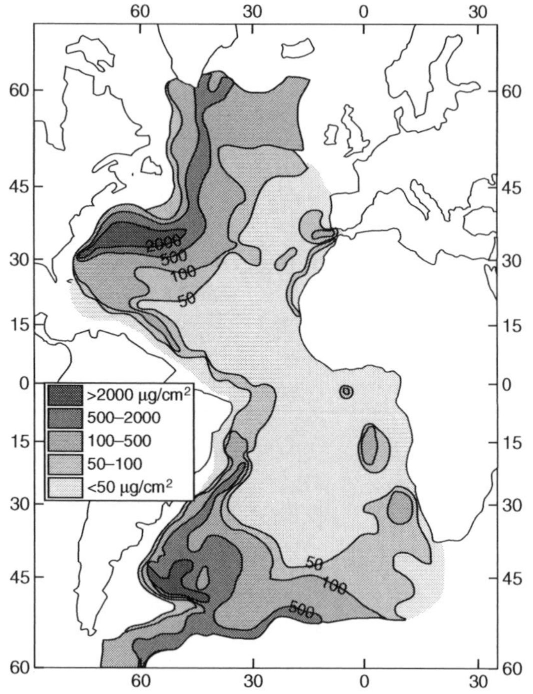


**Fig. 21.27** Biscaye and Eittreim's (1977) classic map to show the excess of suspended sediment in Atlantic Deep Water, defining a turbid nepheloid layer. Note that positive and negative spatial gradients in concentration along transport paths (see Fig 21.26) prove substantial local deposition on sediment drifts and erosion from glacigenic deposits and fans.

down slope in canyons and channels can intercept and divert bottom currents downslope.

Perhaps the most impressively energetic system of thermohaline currents occurs in the northeast Atlantic between Scotland and the Faroe Islands. Here, much of the deep cold-water flow (Norwegian Sea Deep Water) between the Norwegian Sea and the main North Atlantic basin passes southwest through the Faroe–Shetland Channel and is Coriolis-forced and focused into the northwest-trending, narrow (ca 30 km wide), silled Faroe Bank Channel, generating strong persistent bottom currents capable of eroding and transporting sediment up to and including gravel. A large variety of sedimentary bedforms, including scours, furrows, comet marks, barchan dunes, sand sheets and sediment drifts have been documented here using sidescan sonar images, seismic profiles, seabed photographs and sediment cores from the floor of the channel. Bottom current velocities $> 1.0$ m/s are measured, supported by the range of bedforms observed.

Another ocean current system, the west-to-east flowing Antarctic Circumpolar Current of the Southern Ocean, deserves special mention for not only is it the largest discharge of ocean water on Earth (110–144 Sv), but its influence extends from surface to deep ocean floor, down to below 4000 m. The shallower part of the current (1000–2000 m) is driven by the circumpolar westerlies whilst the deeper part below 2000 m, termed Circumpolar Deep Water (CPDW), has energy input from thermohaline flow generated by cool, dense Antarctic Bottom Water and North Atlantic Deep Water. The CPDW is responsible for depositing sediment in contourite drifts whose dynamics are closely controlled by rugged seafloor topography. For example in the northern Scotia Sea and Falklands Trough, CPDW flows north through a deep gap in the North Scotia Ridge before turning east into the Falkland Trough. South of the gap and offset from the main axis of the flow a pear-shaped sediment drift has developed since the early–middle Miocene opening of Drake Passage that allowed inception of the whole circum-Antarctic current system. The drift covers an area of 10 500 $km^2$ and forms a broadly asymmetrical mound up to 800 m thick with sediment thinning along the northwestern margin, due to accelerated flow around rough ocean-floor topography. Mean flow across the crest of the drift averages 11.6 cm/s, with intermittent benthic storm activity resuspending fine sediment fractions. Cores recovered from the crest and margins of the drift in water depths of 3900–4300 m reveal predominantly fine-grained contourites and diatom-rich hemipelagites, capped by sandy–silty contourites rich in planktonic forams. There is some evidence from subtle variations in silt grain size that the strength of the CPDW has fluctuated over the past 18.5 kyr since the LGM when ocean-floor conditions were periodically more energetic than seen at present.

An important class of deep ocean currents is generated by outflow or inflow of distinctive, denser, water masses between ocean basins. The best known are the anti-estuarine Mediterranean Outflow (MO) of warm but saline water into the central Atlantic slope at about 36 °N through the Straits of Gibralter and the estuarine Mediterranean Inflow (MI) into the brackish Black Sea through the Bosphorus Straits. Both are accompanied by compensatory less-dense surface in-

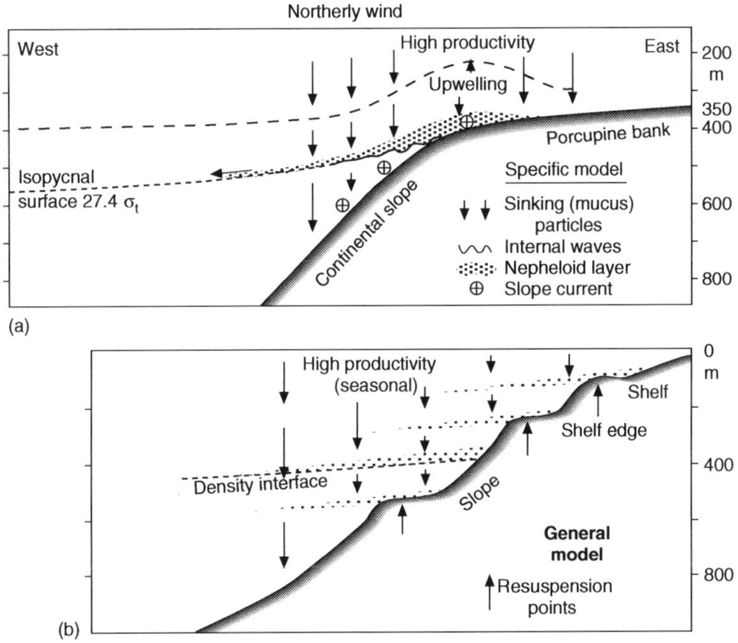


**Fig. 21.28** The origin of internal nepheloid layers at isopycnal surfaces which intersect shelf or slope ramps. The excess turbidity is due to a variety of factors, including bottom resuspension by internal waves, high local productivity and slope currents. (After Dickson & McCave, 1986.)

flows and outflows respectively; in the latter case once a major contributor to Eastern Mediterranean anoxism and sapropel formation (section ). The MO has long been known since ancient sailing ships cunningly deployed deep cables with drag-inducing devices to pull them through the Straits against the opposing inflow of cooler but less saline Atlantic water. Similarly, eastward-travelling submarines steam more efficiently at or near surface. As Figs 21.29 & 21.30 make clear, in the case of the MO a clear link occurs between slope-parallel and slope-normal flows. When the MO leaves the confined extension of the Strait of Gibraltar it is subject to competing Coriolis and gravity forces. The former causes an inertial turn to the north and contour-parallel geostrophic flow with generation of muddy sand waves on the upper slope above 900 m. But this flow is progressively opposed by the excess density of the MO; a continuous *ageostrophic* component eventually detaches into distributary jets which descend the slope as high-velocity flow through seafloor erosional channels that are floored by muddy sand waves and bordered by large levees. This strong downslope flow occurs down to 1300 m water depth, generating muddy waves. Deeper than this there are no longer any channels or downslope bottom flow; by this time the MO has mixed with the surrounding less dense water mass to a point where it has become neutrally buoyant and *lift-off* (aka *lofting*) occurs. Finally, gravity flow channels with terminal lobes on the basin-plain floor exist downslope from several of the bottom current channels indicating that these can be periodically scavenged by turbidity flows. Each gravity flow system has a narrow, slightly sinuous channel, up to 20 m deep, feeding a depositional lobe up to 7 km long. Cores from the lobes recovered up to 8.5 m of massive, well- sorted, fine sand, with occasional mud clasts.

## 21.8 Oceanic biological and chemical processes

The biological productivity of the *photic surface layer* of the oceans (depths $<$ 200 m) is due to the abundance of light, enabling the primary producers, *phytoplankton*, to thrive and photosynthesize, forming the basis of the oceanic food chain. Satellites now

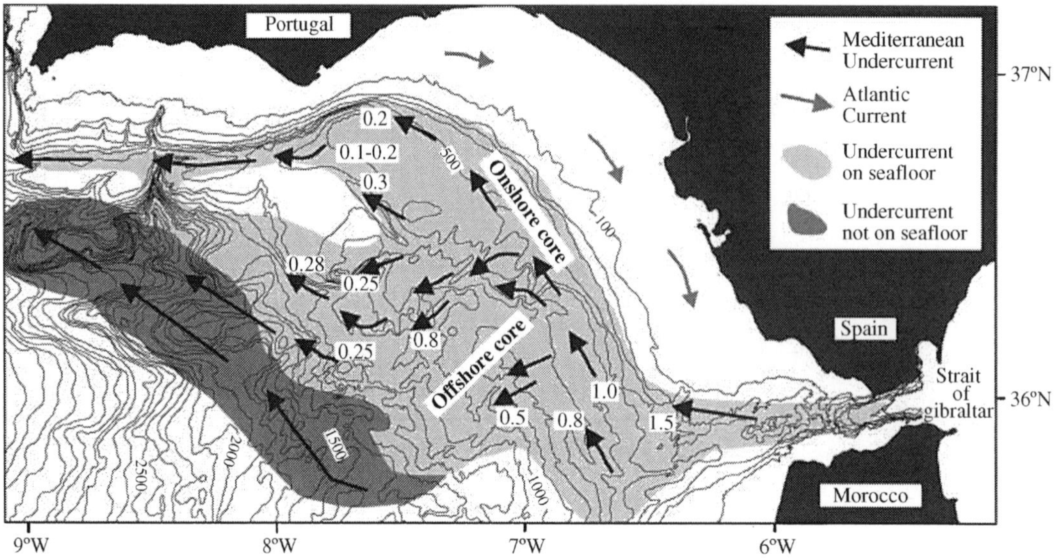


**Fig. 21.29** Map showing the main pathways of Mediterranean Outflow. Arrows with numbers show peak near-seafloor current speeds in m/s. Contour interval 100 m. (After Habgood *et al*., 2003, and references cited therein.)

regularly check levels of primary production by measuring levels of surface chlorophyll. The major phytoplanktonic groups are the siliceous diatoms, the calcareous nanoplanktonic coccoliths and the organic-walled dinoflagellates. *Primary production* is the term given to the daily amount of carbon fixed by photosynthetic reactions in the surface layers of the oceans, about half the planet's total. Across the oceans this production varies over two orders of magnitude, the amount of carbon fixed ranging up to about 200 g/m$^2$ yr (only ca 25% of the highest values for equatorial rainforest environments). About 1% of this carbon reaches the ocean floor as organic sediment, defining the mass sink from the surface layer *organic carbon pump*. The remainder is scavenged through the water column and utilized in bacterial respiration. Combined with the loss of carbon through $CaCO_3$ precipitation, the result is an unsteady and nonuniform rain of inorganic detritus from coastal plumes, organic tissue, siliceous opal and calcium carbonate to the ocean floor (Fig. 21.31). Unsteadiness is a seasonal, annual, decadal or longer-term response to water mass characteristics, including elemental supply, inorganic input, nutrient content, temperature, light levels and so on. Variation in these factors is responsible for near-surface planktonic 'blooms', increased probability of organic/inorganic aggregation and the production of sediment interlaminations. Low productivity occurs in the surface waters of the subtropical gyres. The highest production comes from areas within the influence of large-scale river-derived nutrient input and where forced upwelling of deep, cool, nutrient-rich waters occurs due to Ekman transport. In low-surface-productivity areas, vertical transport of nutrients by rising and descending diatom mats has been established as a dynamic process whereby some surface productivity still occurs in the absence of *in situ* nutrients. An interesting theory points out that in the Southern Ocean, even though upwelling occurs associated with the Circum-Antarctic Current, primary productivity is very low; high-nutrient/low-chlorophyll conditions are due to very low available iron contents due to the paucity of wind-blown dusts with their iron oxide coatings. Iron is the limiting element in primary productivity, a feature illustrated by experiments involving seeding the upper ocean waters of such areas with iron compounds and by observations of water mass characteristics at the great convergences of the Antarctic Polar Front. Additionally there is evidence for markedly increased primary productivity in glacial times when wind-blown dust supplies from Patagonia to the Southern ocean were much increased, general oceanic upwelling increased and nitrate loss by denitrification much reduced. It has been calculated

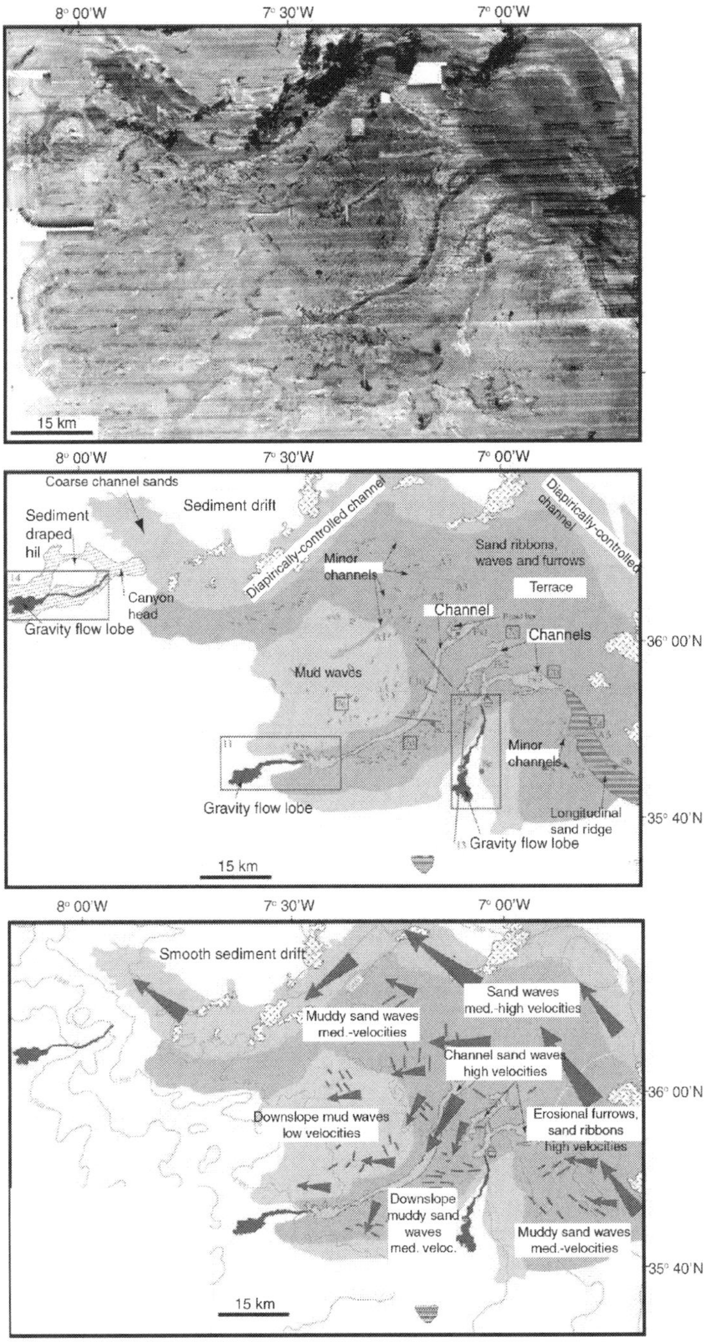


**Fig. 21.30** The sedimentary consequences of Mediterranean Outflow. (a) Sidescan sonar mosaic of the study area. High backscatter is shown by dark shades. (b) Interpretation of study area based on sidescan sonar, profile and core data. (c) Interpretative map identifying the major bedform zones and sediment transport paths with inferred relative current strength. Boundaries between the zones of sediment waves are often gradational. (After Habgood *et al.*, 2003.)

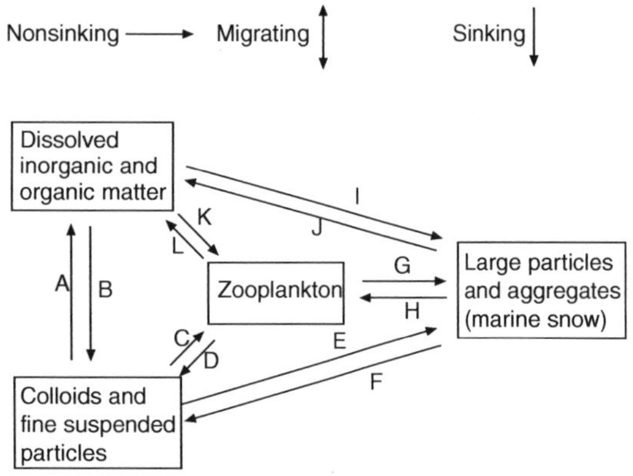


Fig. 21.31 The fundamental controls upon the transformation of organic matter, pelagic sediment and organisms into the aggregated and sedimenting particles known as 'marine snow'. The principle processes are labelled: (a) metabolic release of respiratory products, excretion and desorption; (b) assimilation by microorganisms, adsorption; (c) grazing on fine particles; (d) fine particle production via grazing; (e) aggregation (low turbulent shear); (f) fragmentation, disaggregation (high turbulent shear); (g) faecal packaging; (h) grazing on large particles; (i) assimilation of dissolved substances by microbes associated with large particles; (j) excretion and metabolic release of dissolved substances by microbes associated with large particles; (k) assimilation of dissolved substances by zooplankton; (l) excretion by zooplankton. '(After Bruland *et al.*, 1989; Jahnke, 1990.)

that this increased glacial productivity and the return of the sequestered carbon into sediments was more than ample to cause the reduced $CO_2$ content of the glacial atmosphere.

Two important chemical 'fences' relevant to sedimentation exist in ocean water; the junctions between carbonate sediment accumulation and dissolution (carbonate compensation depth (CCD), Chapter 2) and between oxygen enrichment and depletion. Modelling studies indicate that widespread anthropogenic ocean acidification and shallowing of the CCD, to the surface in the case of the Southern Ocean, will occur in the next 100 years. Oxygen in seawater is derived from the atmosphere by diffusion, wave-induced advective mixing, lateral intrusion of plumes and jets and as a byproduct of photosynthesis. Saturation is usually achieved in near-surface levels, but oxygen contents generally fall with increasing depth (50–100 m). *Anoxic* states represent longer-term low-oxygen content whilst *hypoxia* is a seasonal development (see also Chapter 19 for shelf hypoxia).

## 21.9 Oceanic pelagic sediments

In oceanic environments below the CCD and in areas of low surface-water productivity for siliceous plankton, the chief sediment type is 'red' clay (actually chocolate to red-brown silty clays). These accumulate at very slow rates (0.1–1.0 mm/kyr) and predominantly comprise clay minerals whose compositions (illite, chlorite, kaolinite) reflect continental climatic weathering regimes (Chapter 1) or intra-oceanic basic igneous source rocks (montmorillonite). Slow-growing manganese nodules are common in certain areas of red clay deposition. Adjacent to Trade Wind deserts such as the Sahara, appreciable amounts of wind-blown silt occur in red clay facies, much of it *wüstenquartz* with its characteristic iron oxide coating. As discussed previously, this wind-blown dust supplies biologically available iron to surface ocean waters. Studies of aeolian dust in North Atlantic cores (Chapter 13) have proved to be of great help in elucidating desert expansion and contraction during the Quaternary.

On the ocean floor above the CCD, biogenic calcareous oozes dominate, with the main culprits being planktonic coccolithophores, foraminifera and pteropods, all of which fall through the ocean column as part of 'marine snow' (Fig. 21.31) or in faecal pellets from predators higher in the food chain. Mapping the distribution and thickness of calcareous oozes in subsurface oceanic sediments provides critical evidence for the chemical dynamics of the ocean with time. Calcareous oozes predominate over the crests and flanks of the mid-ocean ridges where the resultant deposits may be strongly modified by bio-irrigation, local gravity flowage, density currents and bottom currents. Open-ocean calcareous plankton produce an estimated 0.72–1.4 Pg/C/yr, a mass that exceeds by a factor of two to four that produced in shallow-water coral reefs and shelf carbonate provinces; however, much of the open ocean production is returned to the water column by post-mortem solution, both in the water column below the CCD and in sediment porewaters. Other estimates of palaeoproductivity are the ratio of benthic to planktonic foraminiferae, the abundance of the former reflecting the amount of suitable organic matter supplied to the ocean floor.

The components of silicic oozes are the opaline skeletons of planktonic diatoms, silicoflagellates and the predatory protozoan radiolarians. Preservation of opal is largely independent of water depth, i.e. there is no silica compensation depth. Hence siliceous biogenic sediments may be a good indicator of ocean surface productivity *if* post-depositional dissolution can be estimated. Diatom oozes are typical high-latitude deposits at the present day, Antarctic waters accounting for over 50% of the world's opaline silica production. This has not always been the case in the past and diatoms are still abundant in certain lower-latitude areas today (e.g. Gulf of California), though radiolarians are usually overall more common in low latitudes. The distribution of opaliferous deposits reflects high-fertility areas of the ocean marked by either coastal upwelling, surface water divergence, as in equatorial regions like the eastern Pacific, or convergence, as at the Antarctic Polar Front. In such areas high phosphate and nitrate contents arise from annual thermocline breakdown and deep-water mixing processes. Delicately varved diatomaceous laminites are produced in this way, with important records of decadal to half-century cyclicity in the ocean current systems responsible in the eastern Pacific/Gulf of California. As noted previously, siliceous planktonic productivity is severely iron-limited in areas away from the influx of wind-blown dust.

## 21.10 Oceanic anoxic pelagic sediments

Below the photic zone the oxygen content of seawater is reduced by the oxidation of organic material. In deeper water oxygen content may rise again if intrusion occurs by shelf margin plumes or cool, oxygen-rich thermohaline currents. Vast areas of open ocean show marked oxygen deficiency at depths of 1000–2000 m. These oceanic *oxygen minimum layers* may intersect an ocean margin, oceanic plateaux or continental shelf margin. It is no coincidence that such areas are close to regions of upwelling and therefore of high organic productivity. The absence of anoxic waters in the open western ocean margins is attributed to more efficient flushing by ocean currents and to the general paucity of phosphate, which limits primary production levels. By way of contrast, periodic abnormal hypoxic episodes occur in the northeastern Pacific Oregon margin when saline, low-oxygen Subarctic North Pacific Deep Water upwells. The large area of anoxic ocean water found in the Arabian Sea is noteworthy because it is accompanied by abundant hydrogen sulphide in the bottom sediments. Anoxic water masses also exist in basins that have a narrow connection with the open ocean across a shallow sill. Intermittent anoxic conditions occur in many fjords and small bays, whilst permanent anoxic conditions occur in the Black Sea (see below).

Organic-rich sediments are termed *sapropels* (Gr. *sapros*, putrid) and are typically brown-black, frequently laminated with varves, and unbioturbated. They provide an analogue for *black shales* in the stratigraphic record which, in addition to their high C-content are also enriched in uranium, a feature that eases their recognition and correlation in uncored exploration and production wells, since the gamma-rays they emit are readily recorded. The reason for U-enrichment is that, in the porewaters of reducing sediments, dissolved uranium complexes are reduced to insoluble $U_4O_2$, the rate of accumulation being inversely proportional to the depth of the uranium redox boundary from the sediment–water interface. Many Mesozoic black shale horizons may be correlated over large areas; in fact some may mark episodes of globally enhanced organic carbon burial, termed *Oceanic Anoxic Events* (OAE). Possible triggers have been proposed for such OAE, including atmospheric-$CO_2$ increases caused by surges in magmatism, either by generation of large-igneous provinces or by phases of accelerated seafloor spreading. However, there is controversy over many claims that OAE were (i) global and (ii) forced by external controls; it seems that some, if not all, were regional in extent (RAE). It is therefore instructive to examine areas of modern oceanic anoxism to see what controls their genesis.

The Black Sea is perhaps the largest and best-known example of a sea that has been periodically anoxic. This silled basin, up to 2200 m deep, has an $O_2/H_2S$ interface (*chemocline*) at a mean depth of about 100 m. It is currently linked to the Mediterranean via the narrow Bosphorus channel through which passes saline Mediterranean water (denser) at depth, with compensatory surface Black Sea waters (fresher) at the surface. Surface salinities in the sea are 17.5–19‰, whilst the remainder is at about 22‰. Early Quaternary shallow-water deposits comprise 'megavarves' (10–100 mm thick), evaporites, chalky oozes (*seekreide*) and oil shales formed in a stratified water body fluctuating between fresh and saline. Upper

Pleistocene to Holocene deposits are, successively: (i) terrigenous turbidites and fines deposited in an oxic freshwater (ca 5‰) lake during lowstand and isolation from the Mediterranean (pre-7.5 ka); (ii) a sapropel representing marine anoxic conditions (beginning ca 7.5 ka); (iii) chalky pelagic oozes marking modern-day surface 'marine' oxic conditions since about 5 ka. The Holocene sapropel is a few centimetres to decimetres thick with ~10% organic matter. Well-developed varves are present, with dark microlaminae originating from seasonal mass mortality of planktonic bacteria. Associated pyrite evidently precipitated within the water column. The sapropel formed during times of warmer climate when quasi-continuous saline underflows from the Mediterranean sea intruded at the Bosphorus mouth and down a submarine channel complex on to the megalake bed to form a rising front that gradually moved up through the water mass. Termination of sapropel deposition occurred when the $O_2/H_2S$ interface was at constant depth, so that permanent density stratification produced a planktonic community adjusted to the new stable habitat. Thus the present stable conditions in the Black Sea have lasted for about 3000 yr, the youngest sapropel being overlain by annually varved coccolith oozes (with low organic contents) that continue to form today.

Periodic Quaternary anoxic events in the eastern Mediterranean have been responsible for centimetre- to decimetre-thick sapropel layers marked by absence of benthic microfossils and the presence of abnormal planktonic foraminifera containing a high proportion of salinity-sensitive forms. Carbonate with depleted $\delta^{18}O$ and organic matter with depleted $\delta D$ both support lowered salinities during sapropel events. The high (> 2%) organic-C contents are due almost entirely to well-preserved diatoms. Over the past 100 kyr, the deposition of sapropels has been synchronous in the eastern Mediterranean at a precessional periodicity of around 20 kyr. The youngest sapropel, S1, formed in the Holocene between 8 and 5 ka (the post-glacial climatic optimum) during a precessional orbital minimum that caused an insolation maximum in the northern hemisphere. This strengthened the African summer monsoonal circulation and led to increased precipitation, mixed-species forest growth around the Mediterranean basin and reduction of Saharan dust supply northwards. The climate change also forced the great Neolithic expansion of human society and agriculture in the Middle and Near East after the chilly rigours of the Younger Dryas. The adjacent continental pollen and speleothem record, occurrence of salinity-sensitive forams and stable isotope results indicate that the onset of anoxism was probably due to rapid attainment of oceanic stratification: enhanced precipitation over the ocean and higher summer runoff from North Africa, the Nile, Levant, Aegean and Black Sea (compensating for the inflow of Mediterranean saline water) generated a low-salinity oceanic surface layer. This shut off thermohaline circulation driven by high Eastern Mediterranean evaporation and prevented oxygenation at depth. Yet despite this potentially oligotrophic (low nutrient) system, high primary production was attained because of increased supply of riverborne nutrients. Varve studies of sapropels at submillimetre scales indicate that under this well-stratified system, highly productive diatom mats flourished during spring and summer. They were aided in this by energy sources made available by assocciated symbiotic N-fixing bacteria. The mats provided the primary food chain and were partly sedimented to the seafloor during the onset of autumnal seasonal mixing where they rapidly use up available oxygen.

It has long been an intrigueing question as to the degree of anoxism in the entire Mediterranean water column during periods of sapropel deposition—was the entire water column anoxic, or just a bottom layer? In the past decade molecular green sulphur-utilizing bacterial fossil pigments support the former case since they require a combination of light and hydrogen-sulphide-rich waters. This means that a chemocline separated oxic and anoxic waters and that this must have been at no great depth, near the base of the photic zone, perhaps as shallow as the 200 m water depth of the present-day Black Sea. Finally, sapropel S1 accumulated during the last phase of the MIS 1 eustatic transgression. Though coinciding with major regional climate change, its generation might be *entirely* due to another coincidence, major outflow of Black Sea freshwater as compensation was made for Mediterranean inflow into that sea, independently dated as occurring from about 7.5 ka to 5 ka.

Instructive examples of the periodic onset of anoxism and the role played by climate/oceanographic change occur in silled basins off the California Borderland and in Caraico basin, Venezuela. In the latter, $O_2$-deficient waters occur below sill depths at the present day. The low-$O_2$ concentrations and high mean annual surface productivity (400–500 gC/m$^2$/yr) induced by nutrient-rich deep oceanic upwelling encourage rapid seasonal deposition of sapropel (10–11 gC/m$^2$/yr). In the Santa Barbara basin, piston-cores and high-resolution $^{14}$C-dating record a superb upper Holocene sequence of varved sapropels with millimetre-scale annual organic-silt couplets, interrupted by centimetre–decimetre-scale (100–300 yr return periods typically) river flood and turbiditic deposits. Longer-scale sediment cores back to 60 ka reveal that anoxism has alternated with periods of oxic bottom waters, producing bioturbated sediments and assemblages of more oxygen-loving forams. These cycles (Fig. 21.32) correlate well with rapid climate excursions in Greenland ice-cores, giving evidence that changes in the outflow of thermohaline currents and related changes in Trade Wind strength and upwelling can alter the pattern of productivity and oxygenation worldwide in restricted basins over quite short timespans. By way of contrast, light–dark laminated sapropelic sediments in the suboxic Bay of La Paz, Gulf of California are due to periodic fluctuations in terrigenous clastic input, controlled by enhance river runoff at sunspot-recurrence intervals of ca 12 yr.

All of the above modern and Quaternary examples of anoxic events were/are regional in extent and it is to be doubted whether any older examples were global phenomena.

## 21.11 Palaeo-oceanography

Ocean sediments and their fossils provide the memory bank that can be used to trace the physical and chemical evolution of oceanic water masses; they are a proxy record of the ocean's ecology, bathymetry, currents sytems and productivity. The fact that oceanic circulation patterns have a major effect on world climate through their role as the major 'conveyor belt' of thermal energy transfer from the atmosphere means that palaeo-oceanography must also address and reconstruct this aspect. A prerequisite for oceanic reconstruction is some knowledge of the evolution of oceanic shape, size and depth. Linear ocean-crust magnetic anomalies enable shape and size to be reconstructed, while models for ocean-crust cooling and subsidence enable palaeodepths to be estimated. Data from offshore deep-water wells and multichannel seismic reflection profiles enable stratigraphic and sedimentary successions to be defined and delimited, and the nature of the sedimentary record to be related to ocean-basin evolution. Typically a low-latitude Mesozoic-aged opening ocean involved initial accumulation of evaporites (see below for an account of hypersaline oceans) followed by carbonate platforms and ramps and the subsequent development of submarine fans as continental-scale drainage systems developed. The onset of deep thermohaline circulation in the Cenozoic and the onset of rapid and high-magnitude sea-level variations in the Quaternary complete the major controls. The high-latitude Arctic Ocean has only recently revealed its secrets after deep cores taken from icebreaker-drillships on the Lomonosov Ridge. During the Paleocene–Eocene Thermal Maximum (PETA) the polar oceans were warm and balmy with temperatures of 18–23 °C and deposition of relatively shallow brackish and organic-C-rich sediments. Dropstones are recorded in sediments with cool-water microfloras about 45 Ma (middle Eocene), approximately the same time as frigidity took hold in the Antarctic. After that a remarkable hiatus or reduction in sedimentation occurred till ca 18 Ma when, after opening of the Fram Strait and connection to the Atlantic, sapropelic sediments gave way to 'normal' silty and clayey deposits containing an ever increasing ammount of ice-rafted debris upwards. This notable connection of the Arctic to the Atlantic enabled export of saline cool waters where they presently lie at 200–1500 m depth as Arctic Intermediate Water.

Estimates of the CCD in ancient oceans depend upon plots of $CaCO_3$ accumulation rates against palaeodepth. The position of the CCD can then be computed from a regression equation, being the intercept of depth as the accumulation rate tends to zero. Time series of CCD show strong similarities between all oceans, implying control by circulation of deep oceanic waters. Evidence for OAE/RAE comes from black-shale horizons (*palaeosapropels*), whilst upwelling is recorded by cherts, phosphorite and abundant fish debris. Very extensive cherts record high equatorial productivity of radiolarian species. Studies of biogenic silica in Pacific cores, for example, reveal that the maximum accumulation rates have occurred at the equator for the past 50 Myr, indicating the persistence of equatorial upwelling caused by divergence. A problem with this approach is that the opaline skeletons of plankton are vulnerable to early burial dissolution by porewaters, the silica diffusing

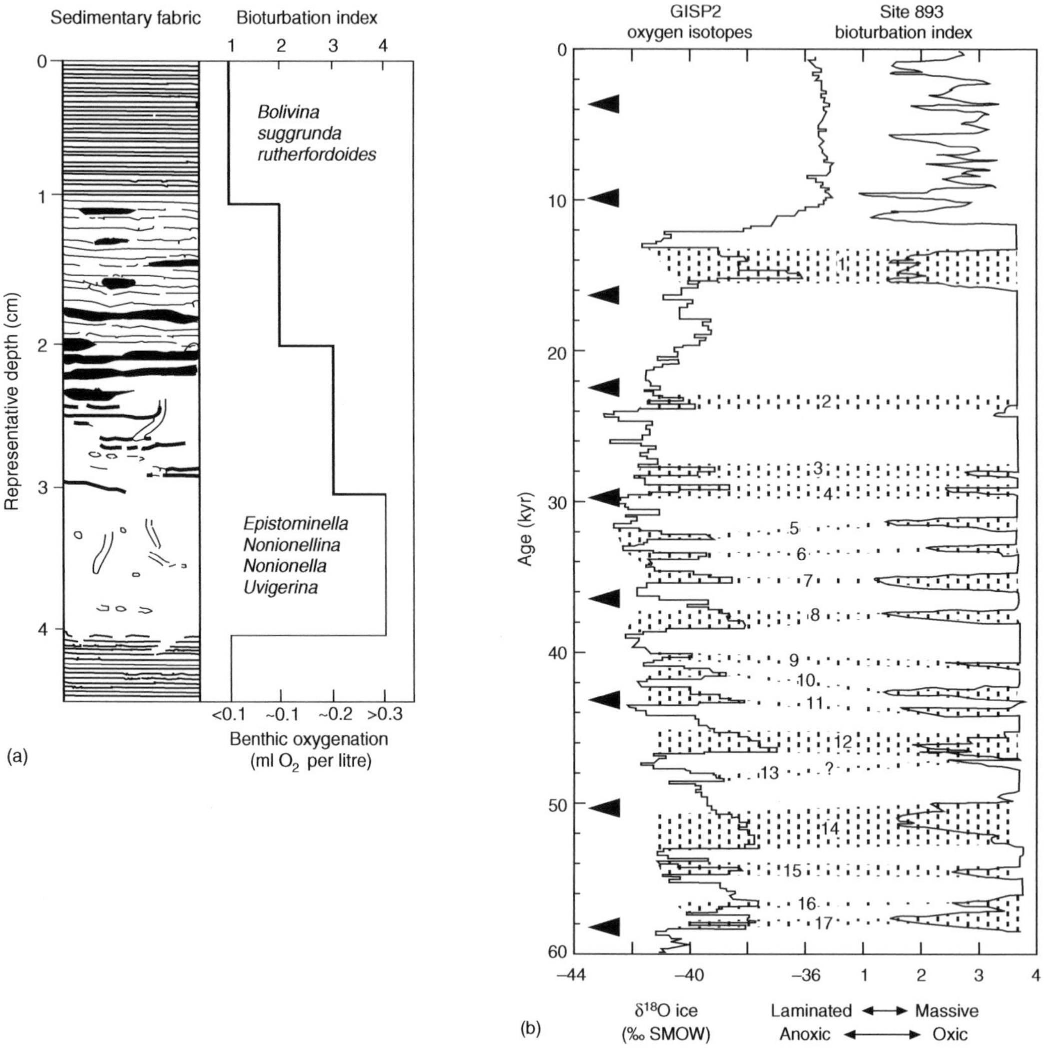


**Fig. 21.32** (a) Illustrative sedimentary log to show millimetre-scale varve laminations and centimetre-scale interlaminations of bioturbated (normal oxygen bottom-waters) and well-laminated, no-bioturbated pelagic sediments in Santa Barbara basin. (b) The stratigraphic record of anoxic laminated sediments in relation to periodic warm interstadial excursions in the late Pleistocene to Holocene oxygen isotope curve from the Greenland ice-cores. The good correlation indicates that the basin developed periodic anoxia in response to enhanced upwelling brought about by global ocean current fluctuations. (All after Behl & Kennett, 1996.)

back into ocean water. Studies in the Southern Ocean reveal a greatly increased diatom production, as seen in opal accumulation rates, but decreased carbonate accumulation rates (due to increased dissolution) in glacial times as compared with today. Generally it is found that the glacial equatorial oceans were characterized by a higher surface productivity and increases in the rate of burial of organic carbon ('carbon events'), probably due to an increase in rates of coastal upwelling. However, this increase in organic carbon burial does not seem to have given rise to more widespread oceanic anoxia even in more physically isolated basins such as the Panama basin. Contour-current activity is indicated by thick continental-rise 'contourites', whilst submarine fans of turbiditic origin witness the growth of terrestrial drainage systems on continental areas adjacent to the opening ocean.

As noted previously, two perturbations are responsible for changes in ocean-water dynamics—polar cooling to create cool dense waters, and equatorial heating and evaporation to give warm saline waters. Both have the potential to give rise to thermohaline currents, but the local oceanic topography must be capable of letting the water masses 'feed' into and circulate around the world's oceans. Oceanic circulation depends upon the particular topographic evolution of the ocean basins, the zonal position of the continents *and* the world climate. The present-day circulation results from thermohaline deep currents due to a strong zonal temperature gradient and the development of a circum-Antarctic seaway (see below). The former process provided the necessary dense water, whilst the latter allowed deep dispersal of this water into the world's oceans. Thus increased flux of warm equatorial water polewards causes a tendency for 'greenhouse' interglacial conditions, whilst increases in cool polar water equatorwards gives glacial conditions. OAE/RAE were also more easily triggered during longer-term greenhouse climate states (e.g. mid-Cretaceous, 120–80 Ma), when sluggish deep oceanic circulation led to stratification and low mixing rates.

The PETA noted above occurred when a major reorganization of oceanic circulation took place. It is probable that strong latitudinal gradients in sea-surface temperatures caused evaporation to exceed precipitation greatly in low-latitude semirestricted areas of the old and closing Tethys Ocean or the new, opening South Atlantic Ocean. Variation of the CCD with time is beautifully illustrated by events at the Eocene–Oligocene boundary (ca 34 Ma) revealed in deep-sea cores. Oxygen isotope measurements on deep benthic forams reveal a rapid temperature fall by ca 5 °C caused by the initiation of the deep Antarctic Bottom Current at this time as the 'gateway' between Antarctica and southernmost South America opened by seafloor spreading. Onset of deep circulation (see below) caused increased oceanic turnover, which in turn caused increased calcareous biogenic production in the central Pacific. This was coincident with onset of Antarctic glaciation and with major and apparently rapid deepening of the CCD. The reasons for this coincidence are hotly debated, four candidate hypotheses are: (i) increase in organic-C burial rate; (ii) increase in chemical weathering of silicate rocks; (iii) increase in global siliceous plankton export production; (iv) shift of global $CaCO_3$ sedimentation from shelf to deep ocean basins.

There is plentiful evidence from oxygen isotopes that a major uptake of ocean water into ice occurred in the upper Miocene to the Miocene–Pliocene boundary. This marks a global sea-level fall, which coincidentally made a major contribution to the subsequent isolation and desiccation of the entire Mediterranean Sea (see section 23.11). Data from ODP drilling in the Arctic Ocean indicates that outflow of NADW into the Atlantic did not occur before that late Miocene times. Prior to this, poorly ventilated bottom conditions dominated, with preservation of sapropels and a siliceous diatom record of high surface productivity. Abundant dropstones in lower Pliocene sediments attest to the development of ice, with a marked increase in their frequency at about 2.5 Ma.

Detailed isotopic and grain-size studies of ocean sediment and of ice-sheet cores reveal that, superimposed upon the shortest-term Milankovitch band timescale of 20 kyr, there are short-duration (~1–5 kyr) rapid climatic switches termed *Dansgaard–Oeschger events*. These record millennial-scale changes in both climate and deep-water production rates and give rise to cyclical variations in sediment type in high-latitude oceans (50–60°) because of the periodic increase in discharge of icebergs produced in glacial surges during colder snaps. They result in widespread dissemination of meltout siliciclastic sediments (*Heinrich layers*), both to the seafloor as *ice-rafted debris* and to deep thermohaline currents as the nepheloid layer intensifies (Fig. 21.33). The actual effects of massive iceberg discharge on thermohaline circulation are unknown, though theoretical studies suggest that the resultant freshening of the ocean surface waters (confirmed by oxygen data on planktonic forams) may temporarily stop the deeper circulation due to surface buoyancy increase. In this way the occurrence of climatic fluctuations has a major effect upon oceanic circulation and sedimentation patterns.

Spectacular results from DSDP and ODP cores in the equatorial Pacific ocean include the discovery of repeated episodes of increased equatorial primary production in the Neogene between 15 and 4 Ma (middle Miocene to lower Pliocene). These are

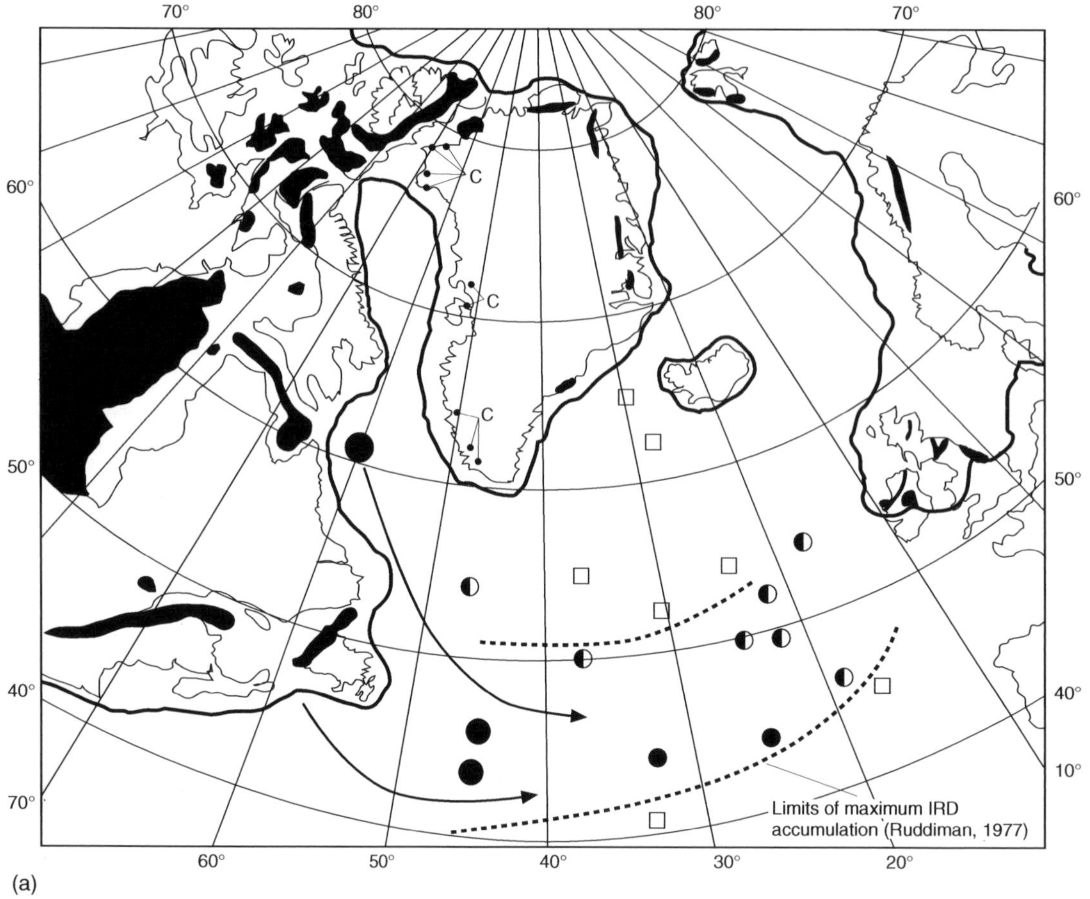


**Fig. 21.33** (a) Map of the North Atlantic to show the distribution of oceanic ice-rafted debris (IRD) sourced from carbonate-rich rocks (outcrops shaded; mainly North America) of glaciated catchments. Maximum limits of last glacial ice are outlined. The sizes of the filled circles indicate the relative abundance of IRD. Half-filled circles indicate IRD not present in all Heinrich layers. Squares have no limestone-rich IRD. Arrows indicate possible iceberg paths from the major calving area of Greenland–Labrador and St Lawrence ice-streams. (b) Graph to show relative abundance of all IRD and Heinrich layers with carbonate-rich IRD in DSDP Site 609. (All after Bond *et al.*, 1992.)

represented by laminated deposits rich in the strand-like diatom *Thalassiothrix* (Fig. 21.34) and interpreted to be the foundered remains of surface 'mats'. The laminated intervals, centimetres to decimetres in thickness, are distinctive and unusual in the sense that 'normal' oceanic sediment in the areas is bioturbated and unlaminated. The laminations are not due to development of open-ocean anoxia because the benthic foraminifera associated with the laminations are normal open-ocean forms. It is thought that the rapid descent of vast volumes of tough diatomaceous mat simply overwhelmed the local benthic organisms' ability to bioturbate the sediment. It may be that other intervals of laminated sediment may also be due to diatomaceous mat growth, rather than to the development of oxygen deficiency. The discoveries shed new light on variations in carbonate and silica production in ancient oceans, for the abundance of silica in this case is not due to preferential dissolution of carbonate. The periodic nature of the smaller-scale laminations, probably reflecting decadal timescales, suggests a mechanism such as the ENSO phenome-

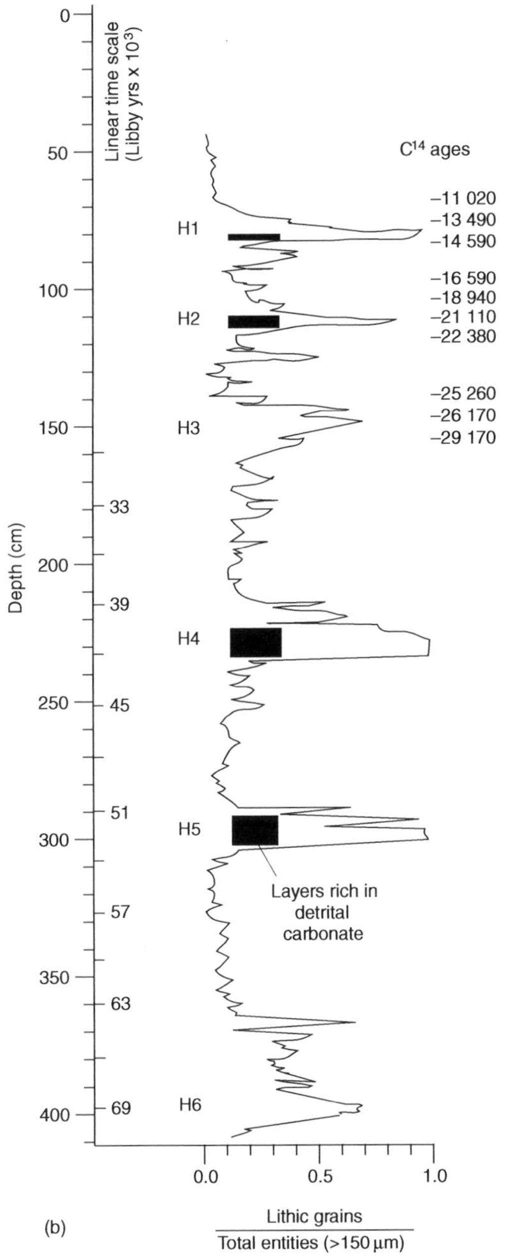


**Fig. 21.33** *(continued)*.

non. The decimetre-scale interbedding of the siliceous oozes and the 'normal' nanoplankton-rich sediment must be due to longer-scale processes. It is thought that the major phases of surface mat production coincide with major periods of cooling and reorganization of oceanic nutrient supplies. For example, an interval from 10.5 to 9.5 Ma is coeval with a major shift in silica production from the Atlantic to the Pacific, whilst the absence of mats in the upper Pliocene to recent correlates with the global shift of major silica production to the Southern Ocean.

## Further reading

### *General*

Oceanography and oceanic sedimentology is well-served with excellent introductory texts. Summerhayes & Thorpe (1996) is superb in every way. Thurman (1991) is deservedly popular. Pernetta (1994) has a mass of information. Pickering *et al.* (1989), though a little long in the tooth, still has much to offer in the way of review of oceanic clastic sediments. Summerhayes *et al.* (1995) has much research level stuff on ocean upwelling, whilst Lisitzin (1996) has an amazing amount of data. Numerous papers on the contourite versus turbidite debate are in the volume edited by Stow & Faugeres (1998). Results from the United States STRATAFORM research programme are in the impressive volume edited by Nittrouer *et al.* (2007). Publications devoted to deep-sea clastic systems are edited by Weimer *et al.* (2000), Lomas & Joseph (2004) and Hodgson & Flint (2005). Geological perspectives on deep-sea clastic-dominated margins, with beautiful 3D seismic images are in Posamentier & Kolla (2003). Meridional contrasts along northeast Atlantic margins are compared and contrasted by Weaver *et al.* (2000). Cascadia margin evolution is revealed by Underwood *et al.* (2005); Pyrenean former deep-sea basin evolution by Sutcliffe & Pickering (2009).

### *Specific*

Slope profiles are studied by Adams & Schlager (2000) and Mitchell & Huthnance (2008); canyons, slides, slumps, debris flows by Flood *et al.* (1979), Surlyk (1987), Ross *et al.* (1994), Keller *et al.* (1979), Watkins & Kraft (1978), Farre *et al.* (1983), Hesse *et al.* (1987, 1996), Hesse & Rakofsky (1992), Klaucke *et al.* (1997), Shepard *et al.* (1979), Moore *et al.* (1989), Mitchell (2004), Jenner *et al.* (2007) and Tripsanas *et al.* (2008); slope gulleys by Spinelli &

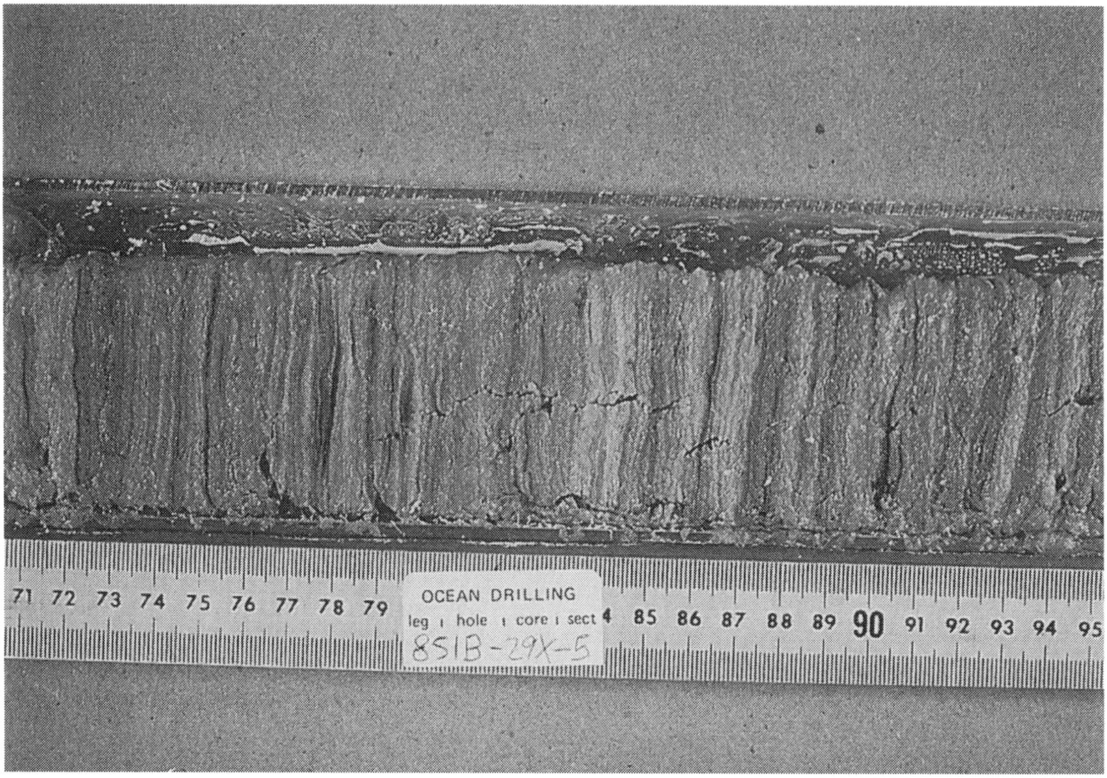


**Fig. 21.34** Core photo to show diatom-laminated ooze, Site 851, eastern Pacific. (From Kemp & Baldauf, 1993; photo courtesy of A. Kemp.)

Field (2001); ancient shelf-break gullies by Ricketts & Evenchick (1999).

Northeast Atlantic Saharan margin slides and flows are discussed by Embley (1976), Masson *et al.* (1992), Masson (1994, 1996), Gee *et al.* (1999, 2001) and Wynn *et al.* (2002); Storega slide by Bugge *et al.* (1988) and Dawson *et al.* (1993); Angola slide by Gee *et al.* (2006); Grand Banks slides and 1929 turbidity current by Piper *et al.* (1998, 1999a); Scotia Slope mass flow deposits by Jenner *et al.* (2007).

Slope–canyon–intraslope basin deposition are considered by Bryant *et al.* (1990), Bouma *et al.* (1990), Lee *et al.* (1996), Prather *et al.* (1998), Twichell *et al.* (2000), Underwood *et al.* (2003), Adeogba *et al.* (2005), Beaubouef & Friedmann (2000), Pirmez *et al.* (2000), Schultz *et al.* (2005), Paull *et al.* (2005) and D.P. Smith *et al.* (2005); ancient examples by Sinclair (2000), Sinclair & Tomasso (2002), Grecula *et al.* (2003), Schultz & Hubbard (2005), Hodgson & Flint (2005), Anderson *et al.* (2006), Tripsanas *et al.* (2006) and Gamberi & Marani (2008).

Experimental and theoretical studies of intraslope basin turbidites are reported by Brunt *et al.* (2004), Lamb *et al.* (2004, 2006), Violet *et al.* (2005), Toniolo *et al.* (2006a,b) and Khan & Imran (2008). Basin slope–floor channels are considered by Hubbard *et al.* (2008), De Ruig & Hubbard (2006), Gee *et al.* (2007) and Kane *et al.* (2009); NAMOC by Hesse *et al.* (1987, 1996), Hesse & Rakofsky (1992) and Klaucke *et al.* (1998); turbidity current initiation by Piper & Normark (2009); turbidity current to debris-flow transition by Talling *et al.* (2007a) and Ito (2008).

Submarine fans and sea level variations by Posamentier *et al.* (1991), Posamentier & Erskine (1991), Armentrout *et al.* (1991a,b); submarine fans generally by Normark & Piper (1972, 1984), Colella & Normark (1984), Colella & di Geronimo (1987),

Damuth *et al.* (1988), Flood *et al.* (1991), Twichell *et al.* (1991), Weimer (1991), O'Connell (1991), McHargue (1991), Piper *et al.* (1999a) and Peakall *et al.* (2000b).

Specific fans: Monterey Fan—Fildani & Normark (2004) and Fildani *et al.* (2006, 2008); Amazon fan—Damuth *et al.* (1988) and Cramp *et al.* (1995); Mississippi—Twichell *et al.* (1991), Weimer (1991) and Hay *et al.* (1982); Indus—McHargue (1991) and Prins *et al.* (2000); Nile—Ducassou *et al.* (2009); Delgada—Drake *et al.* (1989); Redondo—Normark *et al.* (2009); Petit-Rhône—Torres *et al.* (1997); Navy—Piper & Normark (1983) and Bowen *et al.* (1984); Crati—Colella & Normark (1984), Colella & di Geronimo (1987); Zaire—Khripounoff *et al.* (2003), Babonneau *et al.* (2002), Migeon *et al.* (2004); Toyama—Nakajima & Satoh (2001) and Nakajima (2006); Hueneme and Dume—Piper *et al.* (1999b) and Normark *et al.* (2006); Corsica—Deptuck *et al.* (2008); Antarctic—Escutia *et al.* (2000).

Sediment supply to submarine fans is considered by Postma *et al.* (1993a), Weltje & de Boer (1993) and Drexler *et al.* (2006); fan architecture by Walker & Mutti (1973), Walker (1976, 1978), Komar (1969), Klaucke *et al.* (1997), Clark *et al.* (1992) and Clark & Pickering (1996); turbidite lift off on fans by Sparks *et al.* (1993) and Stow & Wetzel (1990).

Experiments and numerical studies of fan dynamics are reported by Imran *et al.* (1998, 2008), Parsons *et al.* (2002), Al Ja'aidi *et al.* (2004), Alexander & Mulder (2002), Baas *et al.* (2004), Métivier *et al.* (2005), Yu *et al.* (2006), Alexander *et al.* (2008), Waltham (2008) and Corney *et al.* (2006, 2008). Ancient submarine fan interpretations are by Anderton (1995), Chen & Hiscott (1999), Johnson *et al.* (2001), Hickson & Lowe (2002), Lien *et al.* (2003) and Hodgson *et al.* (2006); ancient slope margin preservation by Romans *et al.* (2009).

Co-genetic turbidites-debrites (turdebrites) are discussed by Talling *et al.* (2004), Haughton *et al.* (2003), Sylvester & Lowe (2004), Amy & Talling (2006) and Butler & Tavarnelli (2006); turbidite pathway systems by Gee *et al.* (2001), Wynn *et al.* (2002) and Talling *et al.* (2004, 2007b); tectonic controls on turbidite architecture by Haughton (2000).

Contourites are considered by Heezen & Hollister (1963), Hollister & Heezen (1972), Faugères *et al.* (1985a,b, 1993, 1998, 1999), Gross *et al.* (1988), Stow & Lovell (1979), McCave *et al.* (1981), Shanmugam *et al.* (1993), Stow & Piper (1984), Hollister (1993), Carter & McCave (1997, 2002) and McCave *et al.* (2002); contour currents and channels in Mediterranean outflow by Habgood *et al.* (2003), northeast Atlantic examples by Schmitz & McCartney (1993) and Masson *et al.* (2004); Antarctic Circumpolar Current by Howe & Pudsey (1999); nepheloid layers by Eittreim *et al.* (1975), Biscaye & Eittreim (1977), Gross *et al.* (1988), Drake & Gorsline (1973), Pak *et al.* (1980), Dickson & McCave (1986) and Huthnance (1981).

Biological and chemical oceanic processes are studied by Villareal *et al.* (1993), Martin *et al.* (1994), de Baar *et al.* (1995), Kumar *et al.* (1995), Ganeshram *et al.* (1995), Kennett & Shackleton (1976), Demaison & Moore (1980), Wignall (1994), Lovely *et al.* (1991) and Behrenfield *et al.* (2006); rising CCDs by Orr *et al.* (2005); open-ocean CaCO3 budgets by Iglesias-Rodrigues (2002); pelagic diatomic oceanic sediments by Honjo (1976) and Pike & Kemp (1997).

Anoxic events of seas and oceans are considered by Thiede & van Andel (1977), Schlanger & Jenkyns (1976, originators of the term Oceanic Anoxic Event), Degens & Ross (1974), Degens & Stoffers (1980), Wignall (1994) and Negri *et al.* (2009).

Mediterranean sapropels are studied by Rohling (1994), Jones & Gagnon (1994), Lyons (1997), Calvert & Karlin (1998), Thunell *et al.* (1977), Cramp & O'Sullivan (1999), Hilgen (1991), Lourens *et al.* (1996), Sancetta (1994), Kemp *et al.* (1999), Yang *et al.* (1995), Krishnamurthy *et al.* (2000), Krom *et al.* (1999), Rossignol-Strick (1999), Negri *et al.* (2009), Capozzi & Negri (2009), Kotthoff *et al.* (2008), Meijer & Tuenter (2007), Marino *et al.* (2007), Bianchi *et al.* (2006) and Emeis & Weissert (2009); Oregon hypoxia by Grantham *et al.* (2004).

OEA triggers and verisimilitude are discussed by Jones & Jenkyns (2001), Cohen *et al.* (2004), Hesselbo *et al.* (2000), Jenkyns (1988), Beerling *et al.* (2002), Wilson & Norris (2001), McElwaine *et al.* (2005), van de Schootbrugge (2005), Turgeon & Creaser (2008) and Emeis & Weissert (2009); Black Sea by Degens & Ross (1974), Degens & Stoffers (1980), Jones & Gagnon (1994), Lyons (1997) and Flood *et al.* (2009); Cariaco by Hughen *et al.* (1996) and Muller-Karger *et al.* (2000) California Borderland by Behl &

Kennett (1996) and Schimmelmann *et al.* (2006); Gulf of California by Molina-Cruz *et al.* (2002). Palaeo-oceanography of modern oceans by Kennett *et al.* (1974), Kennett & Shackleton (1976), van Andel *et al.* (1977), Kennett (1977), Leinen (1979), Heinrich (1988), Bond *et al.* (1992), Berger & Herguera (1992), Kemp & Baldauf (1993), Paillard & Labeyrie (1994), Kumar *et al.* (1995), Yang *et al.* (1995), O'Connell *et al.* (1996), Hodell *et al.* (1986), McManus *et al.* (1999), Villareal *et al.* (1999), Merico *et al.* (2008) and Poli *et al.* (2000); Arctic Ocean by Moran *et al.* (2006) and Jakobsson *et al.* (2007).

# Part 7

# ARCHITECTURE OF SEDIMENTARY BASINS

*The meal continued with considerations on the art of war, the relative merits of Mahon cheese and Cheshire, and the surprising depth of the Mediterranean only a short way off the land...*

Patrick O'Brian, from *Master & Commander*, Collins 1970 (Stephen Maturin contemplative in the brig *Sophie* off Catalunya).

## Introduction

Sedimentary basins are topographic or bathymetric lows formed due to tectonically induced subsidence of the crust. The gradients produced induce gravity flow of water and sediment so that they may infill with both over time. During the course of the Wilson Cycle of plate tectonics there is sequential development of continental rifting, splitting and sagging during the evolution of passive continental margins. Ocean closure occurs along convergent margins, with subduction-related basins finally caught up in continent–continent collision as ocean destruction finally occurs and an orogenic belt is formed. Each stage of these events imprints a distinctive sedimentary pattern onto a distinctive first-order tectonic architecture. At the same time, climate and sea-level changes act upon basinal sedimentary processes to modify this large-scale architecture in a number of important ways. Thus the sedimentary fill of a basin provides unique evidence for the environmental conditions that pertained during the basin's lifetime. Sedimentology has a vital role here, for it is largely the evidence written in sedimentary rocks that provides data for reconstructing Earth evolution. This is particularly true in ancient orogenic mountain belts whose eroded stumps must be carefully mapped and examined for tell-tale signs of former evolution.

The lithostratigraphy of the basin fill and its relationship to structural deformation comprise what has been called *basin architecture*—the contents of a basin are, metaphorically speaking, carefully sieved, just as archaeologists do in an excavation (Part 7 Fig. 1). In fact, *basin analysis* means much more than just the examination of the sedimentary basin fill, for this can rarely be understood without careful consideration of wider global context (e.g. icehouse or greenhouse Earth), tectonics and structural evolution. The concept of a basin as a repository finds full meaning when the origin of economically important byproducts of the sedimentary process are considered. Thus it is virtually impossible nowadays to explore for, and produce, oil, gas, water and metals without a very detailed idea of basin evolution built up from outcrop, well and geophysical viewpoints. This is because the way a basin evolves determines the spatial relationships of sourcerock, migration, seal and reservoir.

The nature of a basin's sedimentary infill depends upon a number of sometimes interrelated variables, including local climatic regime, hinterland geology, continentality, pre-basinal elevation,

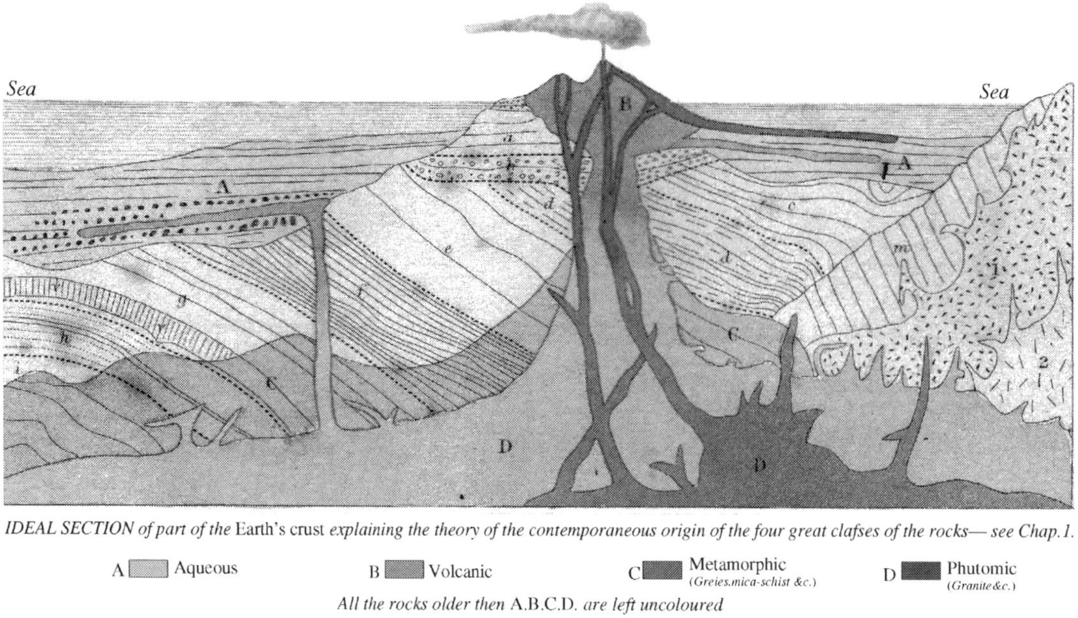
Sea
Sea
A
B
C
D
IDEAL SECTION of part of the Earth's crust explaining the theory of the contemporaneous origin of the four great clafses of the rocks— see Chap. 1.
A Aqueous
B Volcanic
C Metamorphic (Greies.mica-schist &c.)
D Phutomic (Granite&c.)
All the rocks older then A.B.C.D. are left uncoloured

**Part 7 Fig. 1** The frontispiece (originally hand-coloured) to Charles Lyell's 'Elements of Geology' published in 1838 is a schematic crustal cross-section to illustrate the main classes of rock and represents one of the earliest attempts to illustrate what we nowadays know as basin architecture.

topography, sediment flux, subsidence/uplift history, eustasy and volcanism. Thus any particular basin, whether active or deceased, will possess its own unique sedimentary architecture. A simple introduction to conceptual models for whole-basin architecture is presented in the following chapter of this book.

# Chapter 22

# SEDIMENT IN SEDIMENTARY BASINS: A USER'S GUIDE

*The word 'time' split its husk; poured its riches over him; and from his lips fell like shells, like shavings from a plane, without his making them, hard, white, imperishable, words, and flew to attach themselves to their places in an ode to Time; an immortal ode to Time.*

Virginia Woolf, *Mrs Dalloway*, p. 78, Penguin, 1992

## 22.1 Continental rift basins

The structural and sedimentary architecture of rift basins during the *syn-rift* phase of lithospheric extension is controlled by the length of normal fault segments (10–50 km), approximately equal to the thickness of the brittle crustal layer. Most rifts are typically half-grabens, with fault segments offset by *stepover* or separated by crossover in wider *transfer* or *accommodation zones* where fault polarity reverses and extension is taken up by numerous small normal faults (Fig. 22.1). This causes offset zones to stand at higher elevations than intervening rift floors, defining physical barriers to sediment transport between basins. Stepover and crossover accommodation zones are frequently the site of abundant volcanism in the East African and Rio Grande rift systems and in extensional back-arc basins. Multiple basin-margin faults occur where initially wide tiltblocks are fragmented into smaller blocks by new faults developed in the old hangingwall. Cessation of activity on the old bounding fault is followed by footwall uplift, erosional dissection and transport of the old basin fill into the newly subsiding basin. It should also be noted that asymmetric extensional basins occur associated with low-angle detachment normal faults and their associated mid-crustal *core complexes* in the western USA.

Over time a wedge-shaped sediment fill develops in a half graben (Fig. 21.1a), with the characteristic structural asymmetry exerting a fundamental control on the distribution of sedimentary environments. Footwall uplands adjacent to main basin-bounding faults increase relief as fault slip accumulates and fault segments merge (Figs. 22.2 & 22.3) and short, narrow, steep-gradient drainage basins gradually evolve. Hangingwall sourcelands are initially much wider, with gentler slopes normal to the tectonic gradient and larger drainage basins than those that drain footwall uplands. Closed basins with longitudinal tilt may develop short axial drainage channels fed by the transverse drainages. Basins become linked along-strike once crossover blocking zones are breached. Hangingwall sourcelands become progressively

*Sedimentology and Sedimentary Basins: From Turbulence to Tectonics*, 2nd edition. © Mike Leeder.
Published 2011 by Blackwell Publishing Ltd.

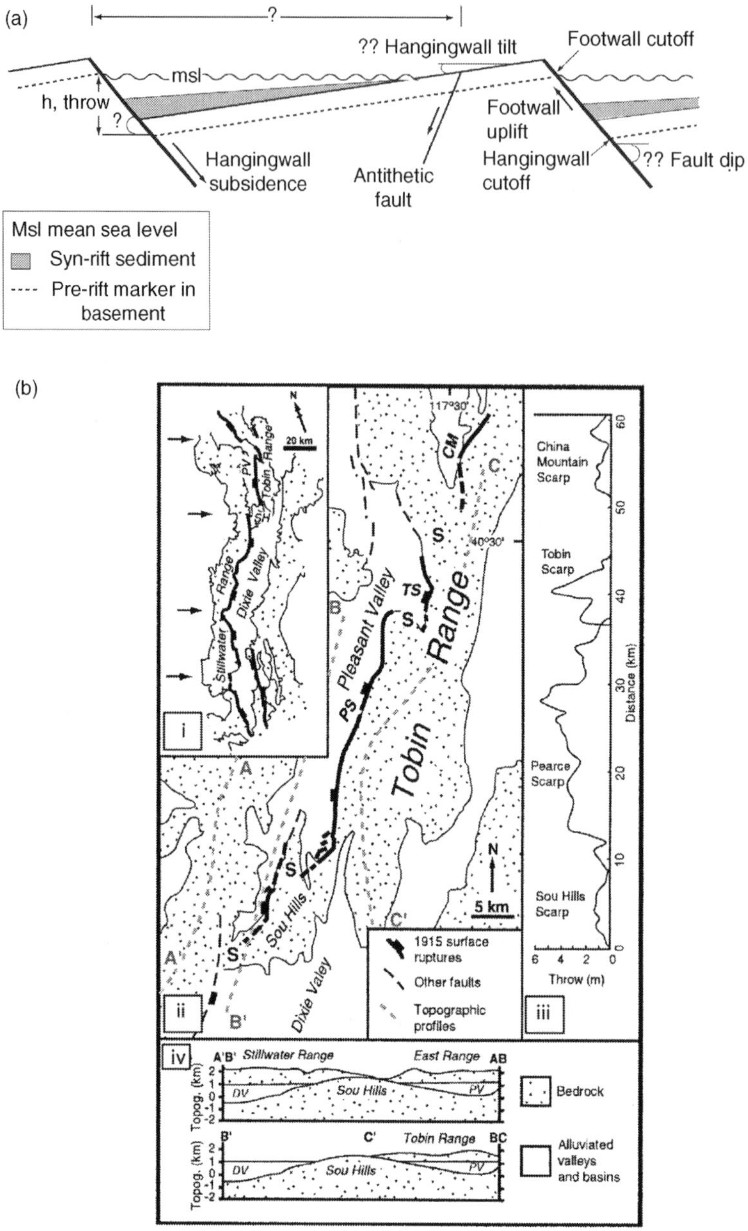


**Fig. 22.1** (a) Cross section through a typical half-graben with commonly used nomenclature. (b) Fault segments and slip and displacement profiles for the Dixie and Pleasant Valleys, Nevada, USA. (i) Major segment boundaries (arrows) and fault segments. (ii) Map of 1915 surface ruptures, showing rupture segment boundaries (S). Only some of the minor faults in the Sou Hills area are shown for clarity. (iii) Slip profile for the 1915 ruptures. (iv) Range crests and valley floor topography and estimated depth to basement profiles illustrating nature of long-term displacement profiles for the Dixie Valley and Pleasant Valley border faults. Note how the major segment boundary around Sou Hills is associated with high topography relative to the basins to the north and south. PS, Pearce scarp; TS, Tobin scarp; CM, China Mountain scarp; DV, Dixie Valley; PV, Pleasant Valley; SH, Sou Hills. (After Gawthorpe & Leeder, 2000, and sources cited therein.)

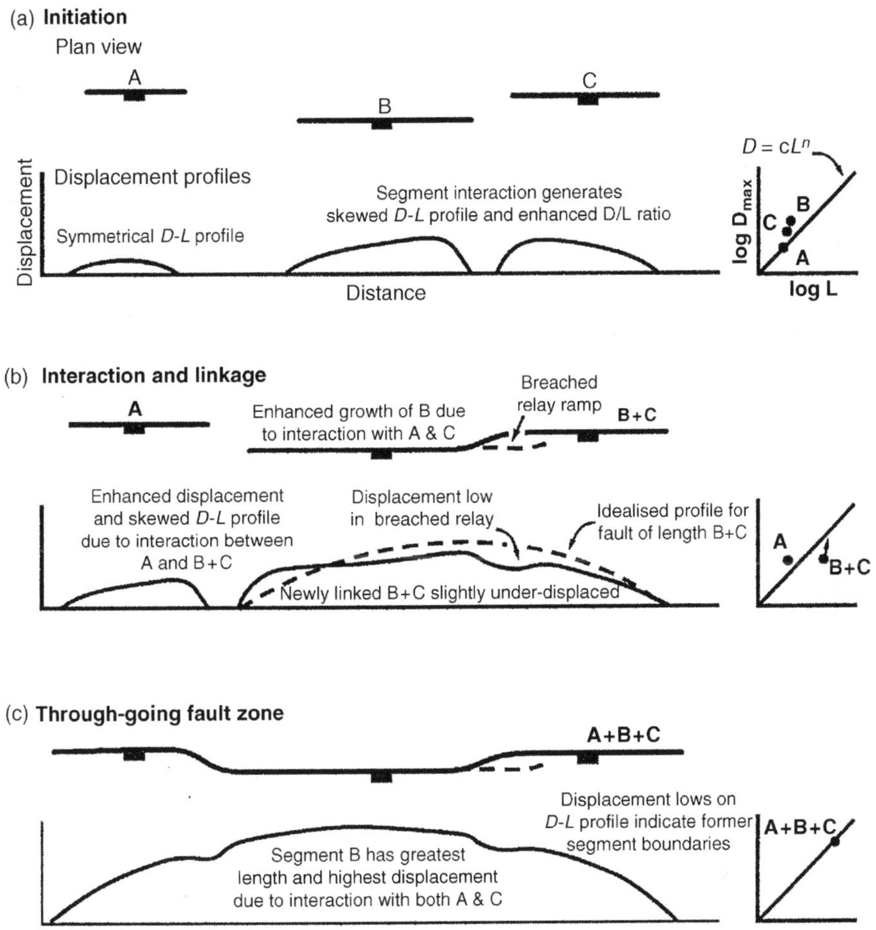


**Fig. 22.2** Evolution of three segments to produce a major border fault zone, illustrated by the evolution of segments A, B and C. (a) Fault initiation stage, (b) interaction and linkage stage and (c) through-going fault zone stage. Note how interaction between segments produces skewed displacement profiles and that the displacement (*D*) and length (*L*) characteristics evolve so that the through-going fault zone has similar characteristics to those of an isolated fault segment (e.g. fault A in (a)). (After Gawthorpe & Leeder, 2000, and sources cited therein.)

smaller as basin infill proceeds, as witnessed by progressive onlap of younger sedimentary strata (see below). Their drainage basins feed alluvial fans and fan deltas, larger and with gentler gradients than footwall-sourced fans. Channel and fan-surface gradients increase during each episode of fault slip, with local fan-surface incision resulting, though incision may also result from purely climatic or runoff changes. Coalescence of hangingwall fans creates prominent low-angle *bajadas*.

### Continental rift basins with interior drainage and/or lakes (fig. 22.4a & b)

Climate controls the degree of lake development—the prototype at the playa end of the spectrum is Death Valley and many other basins like it in the Basin and Range Province. During Pleistocene pluvial periods, widespread shallow lakes occupied such basins (e.g. Centennial Lake grabens of Idaho; Great Basin of Utah; Buffalo, Buena Vista and Dixie Valleys of Nevada). Their early Holocene shorelines may be

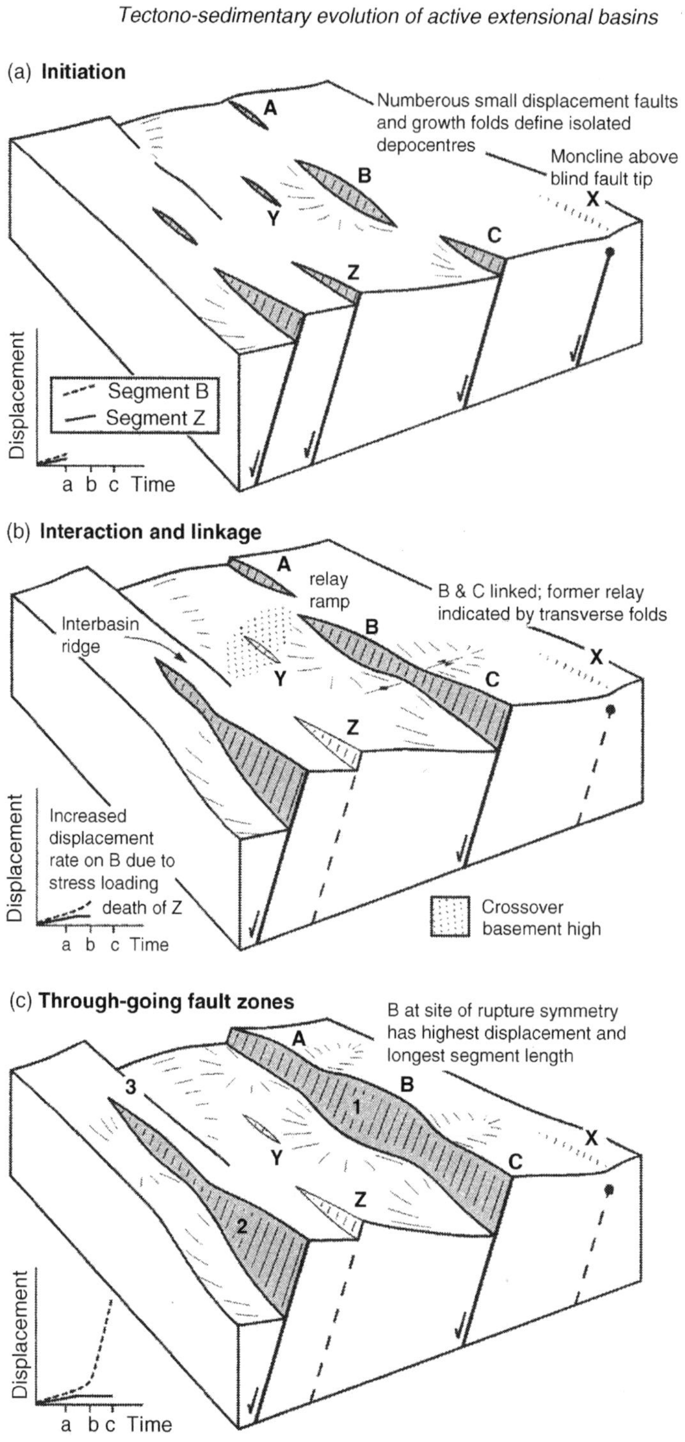
Tectono-sedimentary evolution of active extensional basins
(a) Initiation
Numberous small displacement faults and growth folds define isolated depocentres
Moncline above blind fault tip
A
B
Y
C
Z
X
Displacement
Segment B
Segment Z
a b c Time
(b) Interaction and linkage
relay ramp
B & C linked; former relay indicated by transverse folds
Interbasin ridge
A
B
Y
C
Z
X
Displacement
Increased displacement rate on B due to stress loading
death of Z
a b c Time
Crossover basement high
(c) Through-going fault zones
B at site of rupture symmetry has highest displacement and longest segment length
A
B
1
2
3
Y
C
Z
X
Displacement
a b c Time

easily mapped today using fan-delta remnants—these form important datum with which to measure post-highstand deformation by faulting and regional tilting. Lake Baikal (Chapter 14) and Lakes Malawi (Nyasa) and Tanganyika in the East African Rift Valley are good examples of deeper lakes in rift basins. The deep African lakes (water depths up to 1500 m and 500 m respectively) were not immune to climate-controlled fluctuations during the Pleistocene, with lake levels varying by several hundreds of metres and the resulting water dynamics fluctuating between shallow and saline during lowstand periods and deep and stratified during highstands (Chapter 14).

Basin architecture is controlled by interactions between basin-centre and encroaching alluvial fans and fan deltas sourced from footwall and hangingwall uplands. Periodic catastrophic denudation events are recorded by landslide megabreccias. In areas directly affected by Quaternary ice action, valley glaciers have left prominent terminal and side moraines that continue to influence sedimentation strongly. Climate-controlled cycles dominate the Pleistocene lake fills of many rift basins, with interesting analogues preserved in the Triassic grabens of the northeastern USA. Wind reworking of alluvial and lake shoreline sands is prominent in many arid and semiarid basins. The location of small ergs reflects the orientation of the basin relative to dominant prevailing winds. The aeolian sands interfinger and are partially reworked by fluviatile, axial and transverse fluvial systems. A fine example occurs in the present San Luis basin of Colorado, where the Great Sand Dunes National Monument occupies the eastern end of the graben, close by a mountainous footwall.

### Continental rift basins with axial through-drainage (fig. 22.4c & d)

These are widespread in the Aegean (e.g. Vardar, Lamia graben, Greece; Menderes graben, Turkey), Rhine (e.g. Erft graben), Basin and Range, USA (e.g. Lemhi, Beaverhead and Madison), southwest USA (Rio Grande) and Ethiopia (Awash). The position of the axial-stream channel reflects subtle controls exerted by periodic tilting, fault propagation, encroaching transverse fans, emplacement of landslide megabreccias, intrabasinal horst structures and mini-graben. Perhaps, the best example for these various scenarios is in the southern Rio Grande rift valley, where late Pleistocene and Holocene incision has revealed the deposits and architecture of a previous Plio-Pleistocene aggrading phase (Chapter 10).

### Coastal/marine siliciclastic gulfs (fig. 22.4e & f)

Fine examples occur in the Aegean (Gulfs of Corinth, Euboea, Thermaios) and along the Gulfs of Suez and Aquaba where axial-fluvial drainage passes into river- or wave-dominated deltas in virtually tideless environments. Steep, faulted footwall margins lead to development of talus cones, alluvial fans, fan deltas and submarine fans. Steep subaqueous slopes are mantled by hemipelagic sediment and cut by subaqueous channels that issue from both axial and transverse alluvial channels and fan deltas, the latter feeding base-of-slope submarine fans and axial submarine channels. The bypassed slopes and delta fronts are highly susceptible to mass failure during earthquakes so that side-scan sonar and shallow seismic profiling reveal a plethora of slump scars, slumps and debris-flow lobes. Some of these gravity flows are transformed to turbidity currents as they move downslope, the currents travelling over the basin floor to deposit parallel beds that build up smooth bottom topography. Opportunities for oblique and normal turbidity current reflection are plentiful in such basins. Direct access of fluviatile sediment in suspension as underflows is also possible. Basinal anoxic conditions lead to the development of

**Fig. 22.3** Three-dimensional evolution of a normal fault array, with graphs illustrating displacement history of fault segments B and Z (a) Fault initiation stage, characterized by a large number of small-displacement normal fault segments. Note the low surface topography influenced by fault propagation folds and surface-breaking normal fault scarps. (b) Fault interaction and linkage stage, where stress feedback between segments influences growth, and deformation in the fault array begins to become localized along major fault zones (A, B, C). Faults located in stress shadows begin to become inactive (X, Y, Z). (c) Through-going fault zone stage, where deformation is localized along major border fault zones (e.g. 1, 2 & 3) giving rise to major half-graben and graben depocentres. Note the death of segment Z and increase in the displacement rate on segment B as deformation becomes localized on major fault zones. (After Gawthorpe & Leeder, 2000. and sources cited therein.)

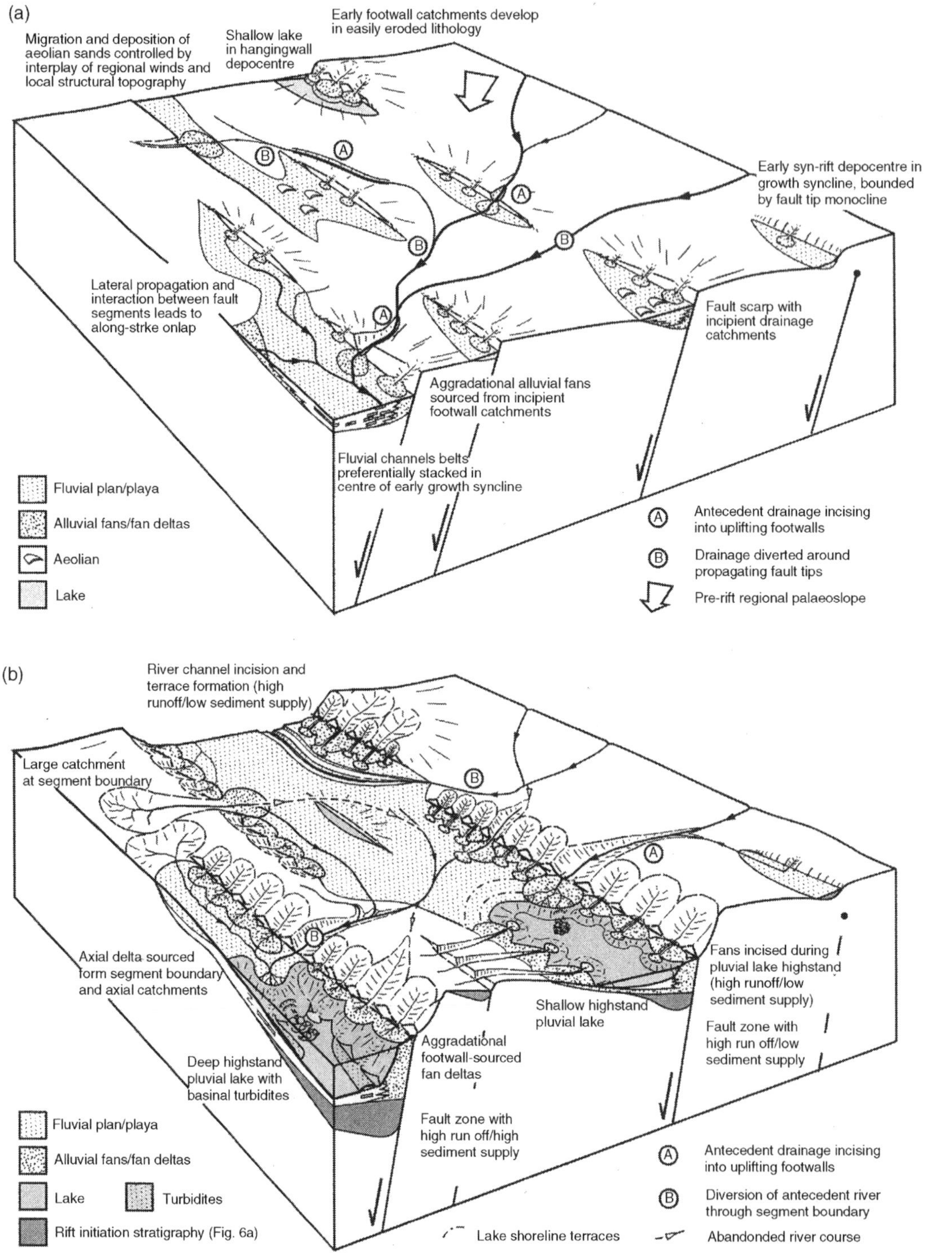
(a)
Migration and deposition of aeolian sands controlled by interplay of regional winds and local structural topography
Shallow lake in hangingwall depocentre
Early footwall catchments develop in easily eroded lithology
Early syn-rift depocentre in growth syncline, bounded by fault tip monocline
Lateral propagation and interaction between fault segments leads to along-strke onlap
Fault scarp with incipient drainage catchments
Aggradational alluvial fans sourced from incipient footwall catchments
Fluvial channels belts preferentially stacked in centre of early growth syncline
Fluvial plan/playa
Alluvial fans/fan deltas
Aeolian
Lake
A Antecedent drainage incising into uplifting footwalls
B Drainage diverted around propagating fault tips
Pre-rift regional palaeoslope
(b)
River channel incision and terrace formation (high runoff/low sediment supply)
Large catchment at segment boundary
Axial delta sourced form segment boundary and axial catchments
Fans incised during pluvial lake highstand (high runoff/low sediment supply)
Shallow highstand pluvial lake
Fault zone with high run off/low sediment supply
Aggradational footwall-sourced fan deltas
Deep highstand pluvial lake with basinal turbidites
Fault zone with high run off/high sediment supply
Fluvial plan/playa
Alluvial fans/fan deltas
Lake
Turbidites
Rift initiation stratigraphy (Fig. 6a)
Lake shoreline terraces
A Antecedent drainage incising into uplifting footwalls
B Diversion of antecedent river through segment boundary
Abandonded river course

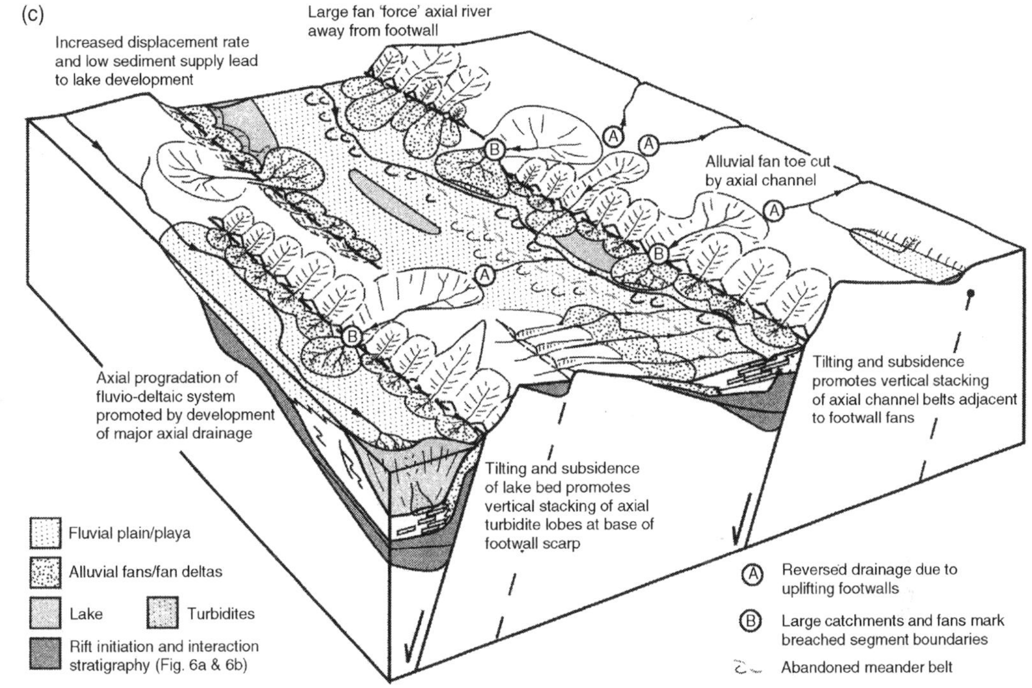
(c)
Large fan 'force' axial river away from footwall
Increased displacement rate and low sediment supply lead to lake development
Alluvial fan toe cut by axial channel
Axial progradation of fluvio-deltaic system promoted by development of major axial drainage
Tilting and subsidence promotes vertical stacking of axial channel belts adjacent to footwall fans
Tilting and subsidence of lake bed promotes vertical stacking of axial turbidite lobes at base of footwall scarp
Fluvial plain/playa
Alluvial fans/fan deltas
Lake
Turbidites
Rift initiation and interaction stratigraphy (Fig. 6a & 6b)
A Reversed drainage due to uplifting footwalls
B Large catchments and fans mark breached segment boundaries
Abandoned meander belt

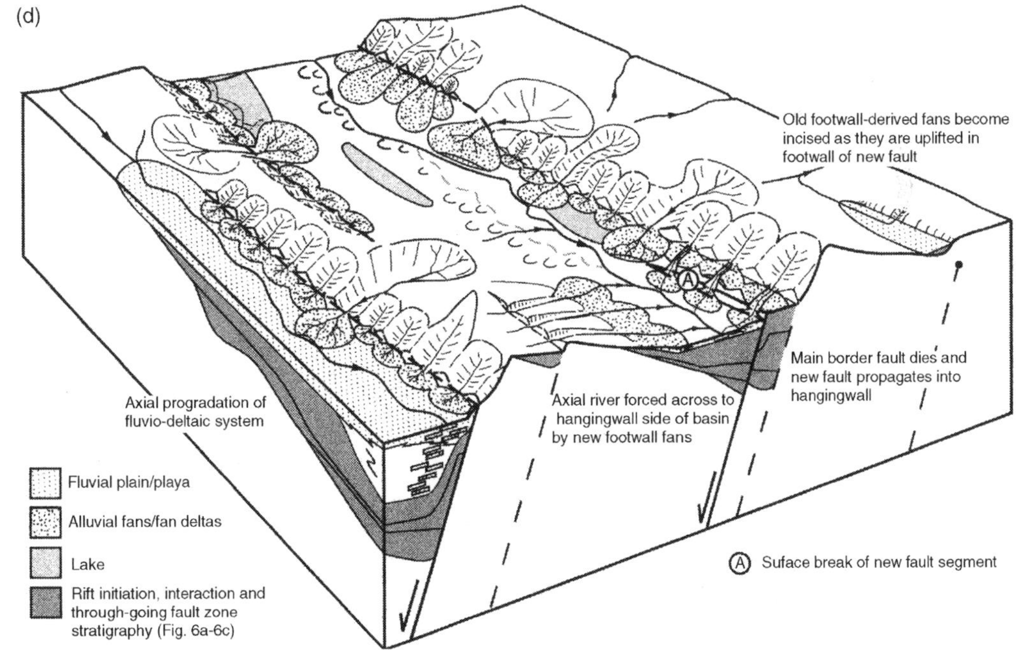
(d)
Old footwall-derived fans become incised as they are uplifted in footwall of new fault
Main border fault dies and new fault propagates into hangingwall
Axial progradation of fluvio-deltaic system
Axial river forced across to hangingwall side of basin by new footwall fans
Fluvial plain/playa
Alluvial fans/fan deltas
Lake
Rift initiation, interaction and through-going fault zone stratigraphy (Fig. 6a-6c)
A Suface break of new fault segment

**Fig. 22.4** *(Continued)*

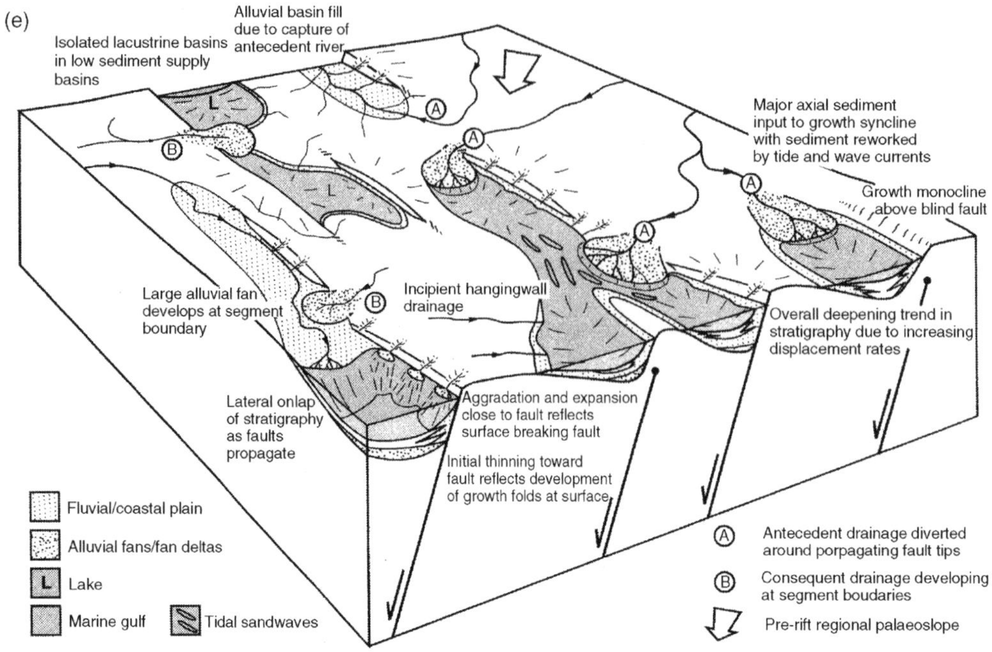
(e)
Alluvial basin fill due to capture of antecedent river
Isolated lacustrine basins in low sediment supply basins
Major axial sediment input to growth syncline with sediment reworked by tide and wave currents
Growth monocline above blind fault
Large alluvial fan develops at segment boundary
Incipient hangingwall drainage
Overall deepening trend in stratigraphy due to increasing displacement rates
Lateral onlap of stratigraphy as faults propagate
Aggradation and expansion close to fault reflects surface breaking fault
Initial thinning toward fault reflects development of growth folds at surface
Fluvial/coastal plain
Alluvial fans/fan deltas
Lake
Marine gulf
Tidal sandwaves
A Antecedent drainage diverted around porpagating fault tips
B Consequent drainage developing at segment boudaries
Pre-rift regional palaeoslope

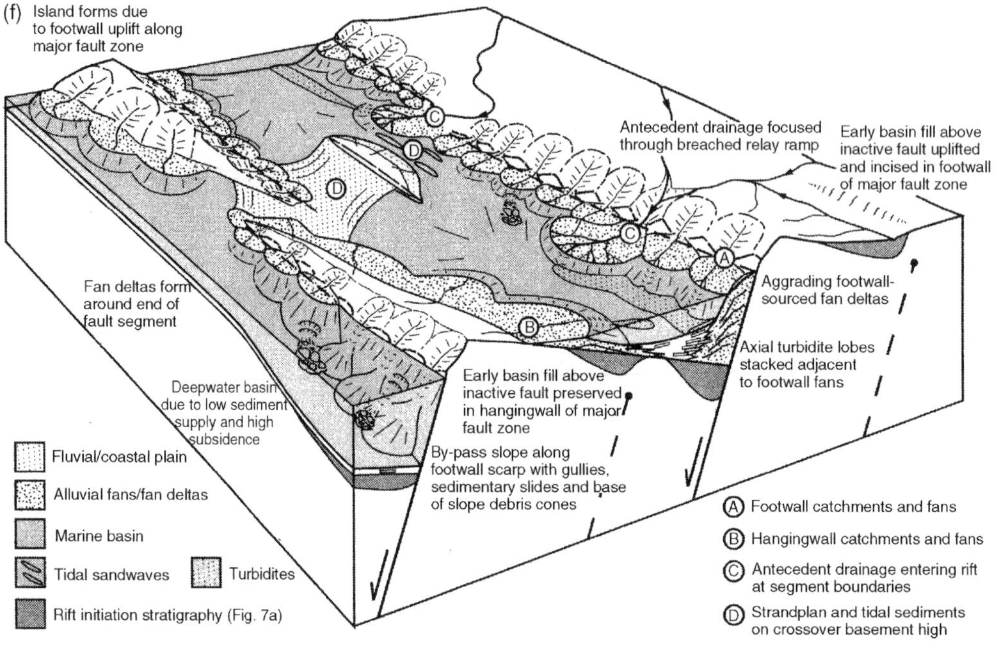
(f) Island forms due to footwall uplift along major fault zone
Antecedent drainage focused through breached relay ramp
Early basin fill above inactive fault uplifted and incised in footwall of major fault zone
Fan deltas form around end of fault segment
Aggrading footwall-sourced fan deltas
Axial turbidite lobes stacked adjacent to footwall fans
Deepwater basin due to low sediment supply and high subsidence
Early basin fill above inactive fault preserved in hangingwall of major fault zone
By-pass slope along footwall scarp with gullies, sedimentary slides and base of slope debris cones
Fluvial/coastal plain
Alluvial fans/fan deltas
Marine basin
Tidal sandwaves
Turbidites
Rift initiation stratigraphy (Fig. 7a)
A Footwall catchments and fans
B Hangingwall catchments and fans
C Antecedent drainage entering rift at segment boundaries
D Strandplan and tidal sediments on crossover basement high

**Fig. 22.4** *(Continued)*

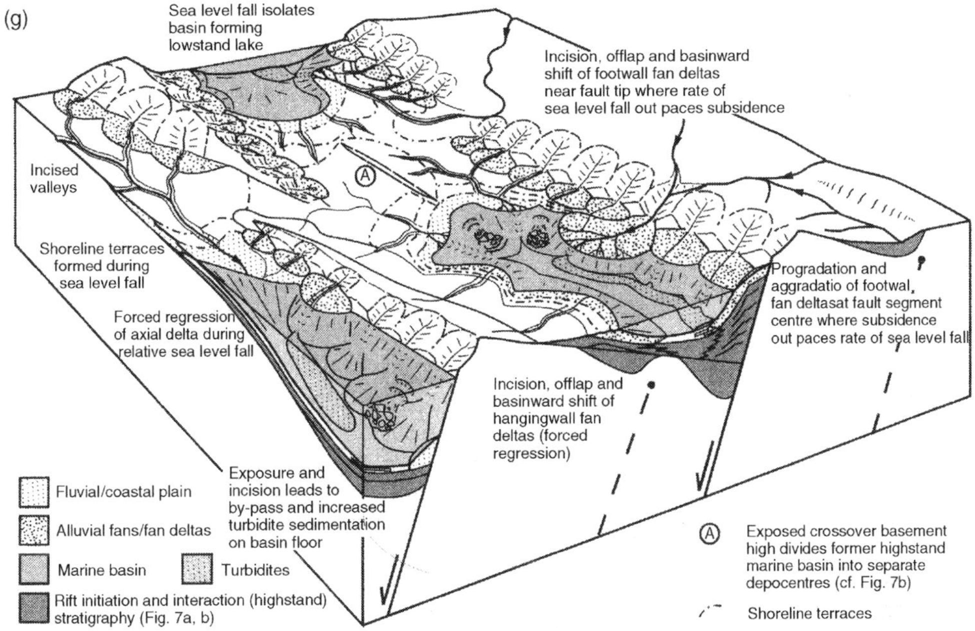


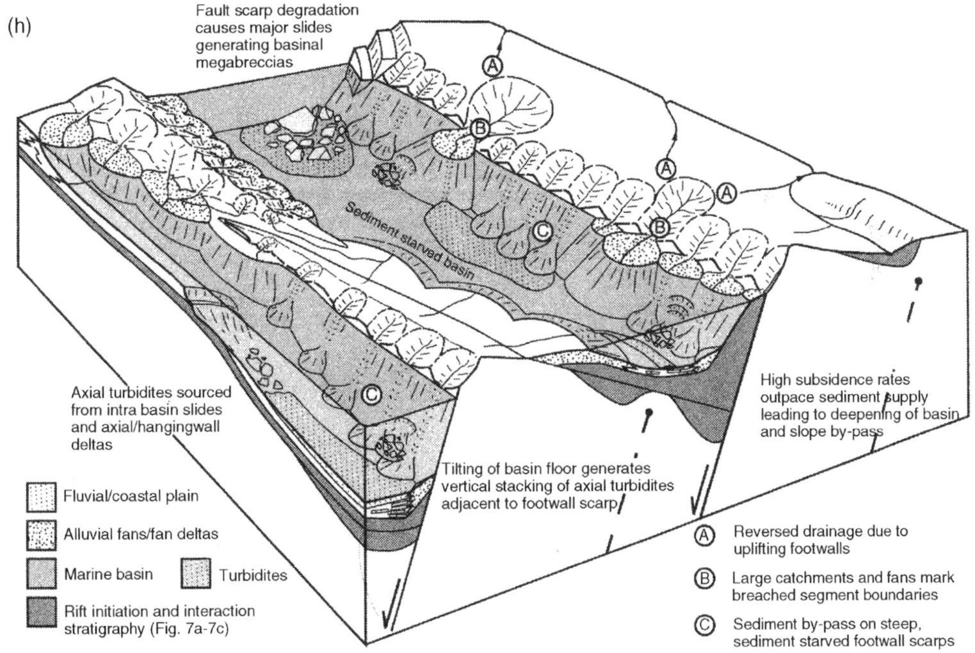


**Fig. 22.4** (*Continued*)

organic-rich basin-floor muds, with which transverse and axial clastic systems interdigitate. The best-exposed ancient analogues are in the Miocene of Suez (section 23.8) and the Mesozoic rifts of East Greenland, whilst the footwall-derived fan deltas of the Pliocene Loreto basin, Baja California, are

**Fig. 22.4** Tectono-sedimentary evolution of a normal fault array (continental environments). (After Gawthorpe & Leeder, 2000, and sources cited therein.) (a) Initiation stage: numerous isolated fluvio-lacustrine sub-basins in the hangingwalls of propagating normal fault segments. Major sediment transport pathways are dominated by antecedent drainage networks that are locally modified by surface topography associated with fault breaks and growth folds. Stratigraphic variability between individual basins is high, due to differences in sediment supply and whether surface deformation is associated with growth folds or faults. (b) Interaction and linkage stage: lateral propagation and interaction between fault segments leads to enlargement and coalescence of early fault depocentres, whilst other fault segments become inactive (dashed on front face). Basin fills adjacent to inactive faults are buried and preserved if located close to the hangingwall of a major fault, or are uplifted, incised and reworked if near footwall crest. Consequent drainage catchments continue to develop along faceted footwall scarps and hangingwall dip-slopes and act as transverse sediment sources to developing half graben depocentres. Note decrease in size of footwall catchments and associated fans towards fault tips. Location of isolated lakes is largely controlled by fault segmentation. Right-hand fault zones are shown for Basin and Range style interglacial with high runoff and low sediment yields. Lefthand fault zones are shown for interglacial with high runoff and high sediment yield (e.g. East African lakes). Tectono-sedimentary evolution of a normal fault array (continental environments). (After Gawthorpe & Leeder, 2000, and sources cited therein.) (c) Through-going fault stage: linkage of adjacent fault segments creates major linked fault zones defining half-graben basins. Displacement on linked faults reduces topography of former intrabasin highs, allowing axial river to flow between former isolated basin segments. Note asymmetric development of axial meander belt and interaction between meander belt and footwall fans. Localization of displacement causes increased displacement rates on active faults leading to the development of pronounced footwall topography and reversed antecedent drainage. (d) 'Fault death' stage: locus of active faulting migrates into hangingwall of right-hand fault zone causing uplift and incision of former footwall-derived fans and a shift of the axial river away from the rift shoulder. Tectono-sedimentary evolution of a normal fault array (coastal/marine environments). (After Gawthorpe & Leeder, 2000, and sources cited therein.) (e) Initiation stage: early formed fault segments and growth folds form low-lying topography and define numerous isolated depocentres partially linked at highstands of sea level to form shallow, elongate marine gulfs and lakes. Antecedent drainage, locally modified by the evolving fault and fold topography, forms major sediment transport pathways. These early syn-rift depocentres display a marked variation in sedimentary fill, depending on their position with respect to sea level, sediment sources and the relationship between topography and sea level. (f) Interaction and linkage stage during sea-level highstand: lateral propagation and interaction between fault segments leads to enlargement and coalescence of early fault depocentres, whilst other fault segments become inactive (dashed on front face). Development of drainage catchments along uplifting footwalls leads to the development of transverse sediment supply to footwall- and hangingwall-derived deltas. Right-hand fault zones along the rift shoulder are supplied by antecedent drainage that enters the rift through topographically low segment boundaries. Left-hand fault zones form isolated footwall islands. Limited transverse sediment supply from these islands leads to the development of starved, deep-water basins. Crossover basement highs at accommodation zones form shallow platforms along the axis of the rift that may become sites of shallow marine and tidal sedimentation. Tilting of the basin floor promotes axial transport of turbidites that stack and interfinger with the toes of footwall-derived deposits. Localization of deformation along major fault zones leads to increased subsidence rates that may outpace sediment supply and result in an overall deepening upward trend in the basin fill. Tectono-sedimentary evolution of a normal fault array (coastal/marine environments). (After Gawthorpe & Leeder, 2000, and sources cited therein.) (g) Interaction and linkage stage during sea-level lowstand: eustatic fall in sea level results in subaerial exposure and incision in low-subsidence-rate settings (hangingwall dipslope, around fault tips/segment boundaries and crossover basement highs) resulting in marked basinward facies shifts. Exposure of crossover basement highs may lead to isolation of depocentres. In the immediate hangingwall of fault segment centres, high subsidence rates may outpace eustatic sea-level falls, generating relative sea-level rise and resulting in depositional systems prograding and aggrading. (h) Through-going fault zone stage: continued fault growth and linkage leads to localization of deformation on to a limited number of major fault zones. Localization of deformation leads to high displacement rates on the active fault zones and to pronounced structural relief across the fault zones. Topographic highs associated with now breached segment boundaries along the linked fault zones become subdued and sub-basins become linked. In general, high displacement rates outpace sediment supply along the fault zones, leading to retrogradation of footwall-derived depositional systems and sediment starvation in basinal areas. Fault scarps become areas of sediment bypass characterized by chutes and slump scars with base of slope talus and debris flows. Major zones of footwall instability can lead to the generation of degradation complexes.

noteworthy because of their sedimentary record of short-term (a few thousand years) unsteadiness in fault movement (section 23.10).

### Marine carbonate coastal gulfs and pelagic tiltblocks (fig. 22.4g & h)

Since carbonate production and facies distributions are strongly depth- and slope-dependent, marked lateral changes in sediment type occur within the structural template provided by half-graben. The steep gradient caused by the main boundary fault ensures that the footwall slope transition is steep and abrupt, causing a structural bypass. The gentler hangingwall slope develops into a ramp-type margin, deepening towards the footwall scarp. Eventually, a rimmed shelf and a sedimentary-bypass margin develop. Axial basin environments may be starved of sediment and the facies developed are markedly condensed. Black shales will develop if the basin has a permanent anoxycline. The subsidence of shallow-marine carbonate-dominated tiltblocks to great depths leads to development of characteristic condensed pelagic and resedimented facies on the footwall and hangingwall slopes. Mixed clastic/carbonate environments occur in the late-Pleistocene Corinth rift in central Greece, with tiltblock topography locally controlling the magnitude of tidal currents and causing rapid lateral facies variations from ooid shoals to beach siliciclastics.

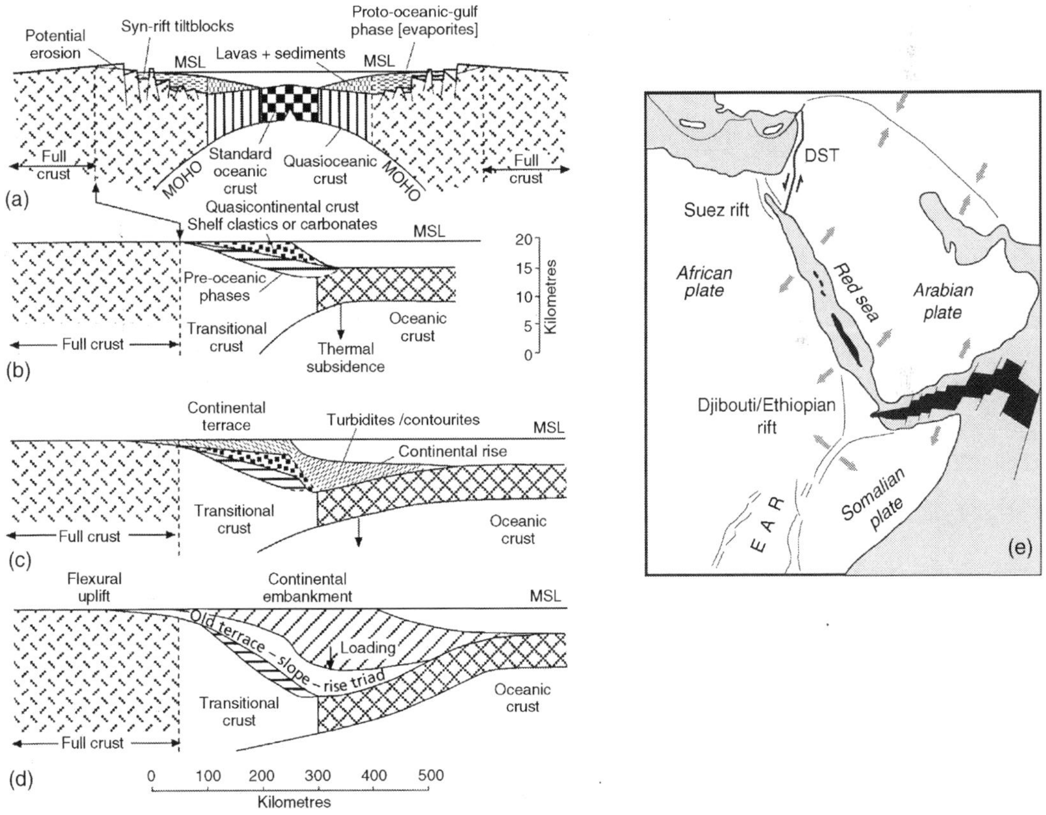


**Fig. 22.5** (a–d) Sketches to show the sequential development of a passive continental margin. MSL, mean sea level. (After Dickinson, 1976; Ingersoll, 1988.) (e) Setting of the Red Sea proto-oceanic basin and its connection southwards into the Djibouti/Ethiopian rift and East African Rift (EAR). Note also Suez rift and Dead Sea Transform fault (DST). Black indicates ocean crust in Red Sea and arrows are indicative of relative plate motions.

## 22.2 Proto-oceanic rifts

The Suez/Aquaba/Red Sea/Gulf of Aden/north Ethiopian continental rift-to-ocean basin province (Fig. 22.5) is the chief natural laboratory where the early evolution of continental extension and its active transition to oceanic crustal generation may be studied. The process is also documented in the back-arc Woodlark rift basin of Papua New Guinea and in the Gulf of California, where there is also significant and rapid transform fault motion causing transtensional tectonics.

The Red Sea province is flanked landwards by prominent marginal escarpments. Seawards, gently sloping shelves, up to 150 km across, flank the 50 km wide axial zone with its discontinuous rift valley and isolated axial deeps that descend to over 3 km depth. Only in the southern third (south of 20 °N) is the axial region floored by continuous young oceanic crust ( < 5 Myr old). Further north isolated deeps occur that become progressively less common. The shelves are underlain by stretched continental crust in central areas but by an abrupt ocean–continental crust transition in the north (particularly on the western margin). To the south, the western tip of the slow-spreading (16 mm/yr) Gulf of Aden rift comes onshore in Djibouti as the Asal rift. The Red Sea rift is also propagating westwards into the Afar depression towards Asal. Prior to formation of the initial rift, the Red Sea area had low relief and was part of a northward-deepening epicontinental area from the late Cretaceous to the early Oligocene. The widespread presence of marine conditions at many localities along the Red Sea and Gulf of Suez margins prior to and shortly after the first evidence for extensional faulting implies that pre-rift doming was not important. Fission-track dating of flank uplift indicates basement exhumation of up to 2.5–5 km since the middle Miocene. Syn-rift sedimentation was typified by marginal alluvial fans and fan deltas together with a variety of nearshore carbonate and clastic environments. The Gulf of Suez rift is noteworthy for its evidence of syn-rift fault growth and of intimate controls upon sedimentation by developing structures (section 23.8). During late Miocene times, enormous thicknesses of evaporites formed in the periodically isolated proto-oceanic trough. Normal marine salinities resumed during Pliocene–recent times. Holocene sedimentation in the central Red Sea has been dominated by pelagic foram–pteropod oozes with high sedimentation rates. A highly stratified water column developed during early deglaciation (14–8 ka) with low productivity in the upper mixed layer, very high bottom salinities and accumulation of sapropels. Today the situation is much changed, with generally good mixing and well-oxygenated conditions, though deep brine pools exist along the axial ridge.

## 22.3 Coastal plains, shelf terraces and continental rises

After active rifting has given way to seafloor spreading, the long process of thermal subsidence and continental margin construction begins (Figs. 22.5 & 22.6), the rate being greatest in the thinnest continental crust. Once syn-rift topography is eliminated by erosion and subsidence, construction of coastal plain, shelf terrace and rise begins (Fig. 22.7). The net result is a sigmoidal bathymetric profile that slowly increases in slope from the continent across the hinge-line with

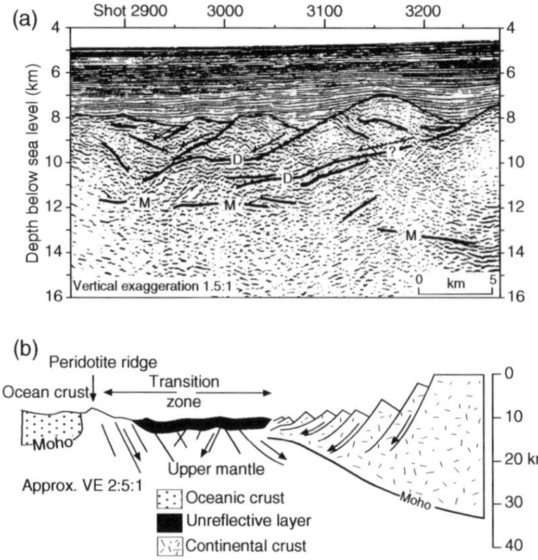


**Fig. 22.6** (a) Deep seismic reflection image (pre-stack, depth-migrated, partly interpreted) from the Iberian Atlantic passive continental margin to show highly extended continental crust as tiltblocks capped by syn-rift Mesozoic sediments and bounded by low-angle detachments (D). M is Moho. Note also draping by post-rift oceanward-dipping sediments. (b) Summary sketch of the highly extended Iberian continental margin; post-rift sediments not shown. (Both from Pickup *et al.*, 1996.)

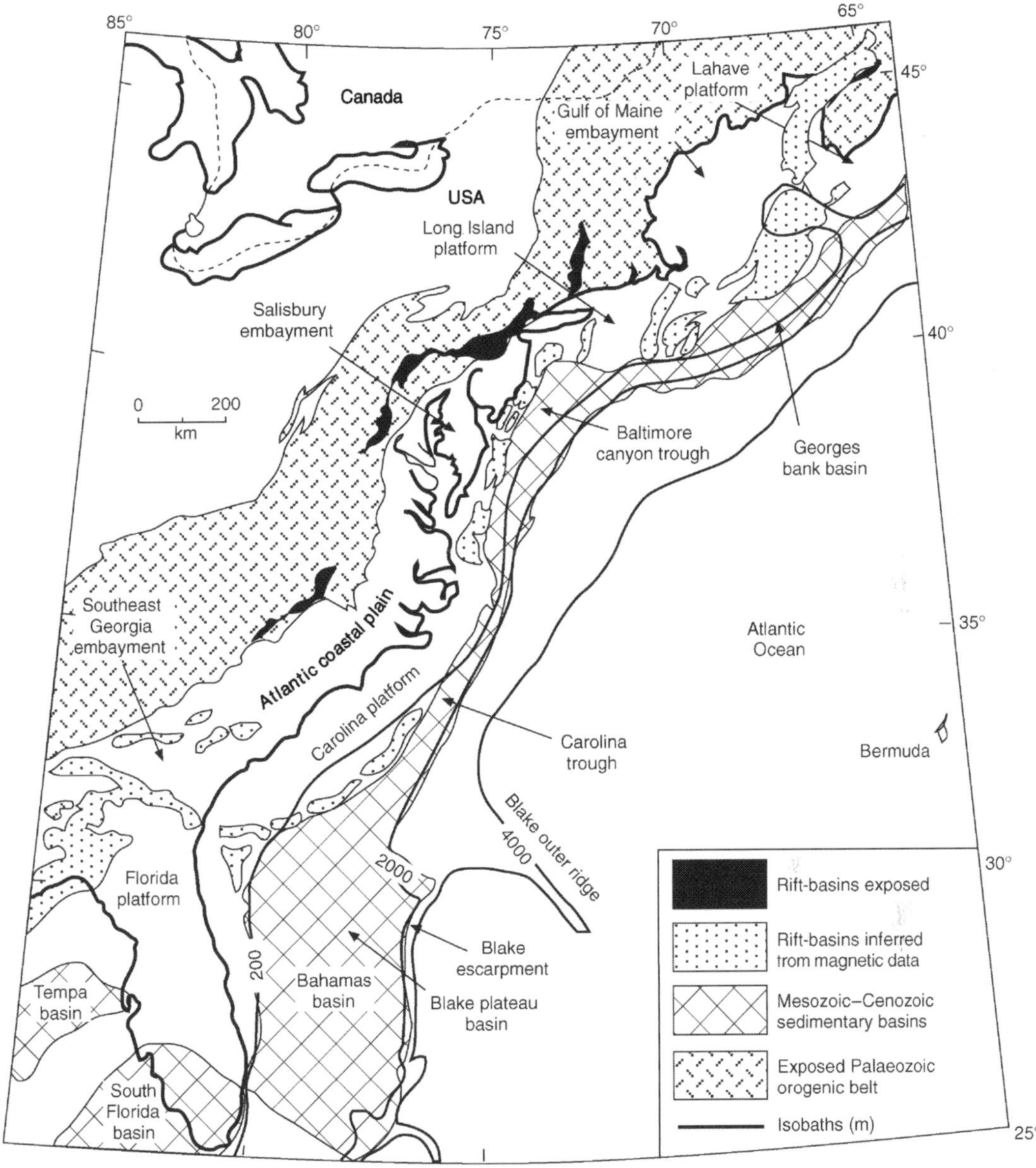


**Fig. 22.7** Map of the eastern United States passive continental margin, adjacent coastal plains and ocean floor. (From Sheridan & Grow, 1988.)

stretched continental crust, rapidly increases across the transition to ocean crust, and finally slowly decreases once more across the ocean basin proper. Construction of the shelf terrace (Chapter 19) is achieved by deposition from tides, waves and ocean currents, modified across the more steeply sloping shelf rise by mass failure and gravity flows. Lateral shelf growth is much affected by periodic sea-level changes, with lowstand lateral accretion at the shelf margin. The continued deposition of sediment over the terrace and rise takes the form of a seaward-expanding wedge. This exerts an additional load-induced flexure on the subsidence profile induced by thermal contraction. Flexure adds to subsidence across the coastal plain and terrace, causing the feather edge of subsidence to migrate continentwards. At the same time a flexural forebulge develops in the continental hinterland that provides continued uplift and provision of sediment to the coastline.

Where major continental drainage systems abut onto a shelf, huge prisms of deltaic sediment build out as embankments, sometimes directly onto abyssal ocean floor (Chapters 17, 19 & 21). The locus of deposition of such large rivers is strongly influenced by tectonic routing along syn-rift and earlier zones of drainage escape. Striking examples are afforded by the Mississippi drainage along the ancient rift of the New Madrid seismic zone, the Niger drainage along the failed rift arms of the Benue Trough, and the Amazon along the Amazonas and Solimos rift basins. In two of these cases the growth of huge subcontinental catchments has been a response to gradients set up along the distant active plate margins that formed new drainage sources—the Rocky Mountains for the Mississippi and the Andes for the Amazon.

The low gradient of shelf terraces and coastal plains on mature passive margins means that small changes in relative sea level can have marked effects on shoreline position, with lateral excursions up to 200 km. Slow thermal subsidence amplified by flexure allows preservation of transgressional and regressional cycles (Chapter 10) whose thickness and duration are controlled by local, regional or global sea-level changes. Nonmarine coastal plains (Fig. 22.7) are dominated by fluviatile and lacustrine environments, with the coastal zone more variable according to the predominance of wave, tide and river energy. Highstand and lowstand deposits may differ markedly. During lowstand, canyons are cut, shelf terraces exposed and a large flux of sediment is transported out to shelf-edge deltas and hence to the adjacent ocean floor as submarine fans. Highstand shelf terraces are swept by variable tidal currents and the effects of seasonal storm waves. On the slope, suspension fallout from nepheloid layers causes slope-parallel clinoforms to develop. Slopes are modified by failure, slumping and resedimentation processes that cause base-of-slope thickening in response to gravity and geostrophic flows. Since the middle Tertiary, thermohaline currents (Chapter 21) have dominated sedimentation along many continental rises. These are often strongly influenced by topography inherited from active rifting and early seafloor spreading. Burial of old syn-rift or proto-oceanic evaporites leads to progressive development of evaporitic salt domes, intrusions and salt flowage that can cause major bathymetric modifications.

## 22.4 Convergent/destructive margin basins: some general comments

All convergent (Fig. 22.8) destructive margins share a common feature—the consumption and/or shortening of lithosphere. The magnitude of the consumption would roughly balance that of plate creation at the mid-ocean ridges were it not for the accompanying production of voluminous subduction-related magmatic products and the offscraping of these and oceanic sediment against the overriding plate on to *accretionary subduction complexes* (Fig. 22.9). On a long timescale (> 4 Gyr) such accretion has given rise to growth of continental lithosphere but over the past 100 Myr or so there has been only a small net growth of continental crust. There are important differences in the nature of convergence that cause the morphology of convergent margins to differ markedly. These arise because of differences in relative plate motions and the degree of coupling between the subducting lower and the overriding upper plates as expressed in the angle of subduction. A paradoxical feature is that convergent margins may also be the site of significant extension, leading to preservation of characteristic back-arc extensional volcanic and sedimentary associations.

*Compressional arc–trench systems* have strongly coupled descending and overriding plates with shallow subduction and the whole overriding plate advancing faster and in the opposite or a strongly oblique direction relative to the descending plate. The majority of the overriding plate is in strong compression with many great earthquake epicentres in a broad band parallel to the shallow-dipping subduction zone. Such conditions tend to occur when the rate of plate convergence and the negative buoyancy of the subducting slab are high and the angle of plate descent is low. Morphologically, compressional margins are characterized by relatively shallow trenches,

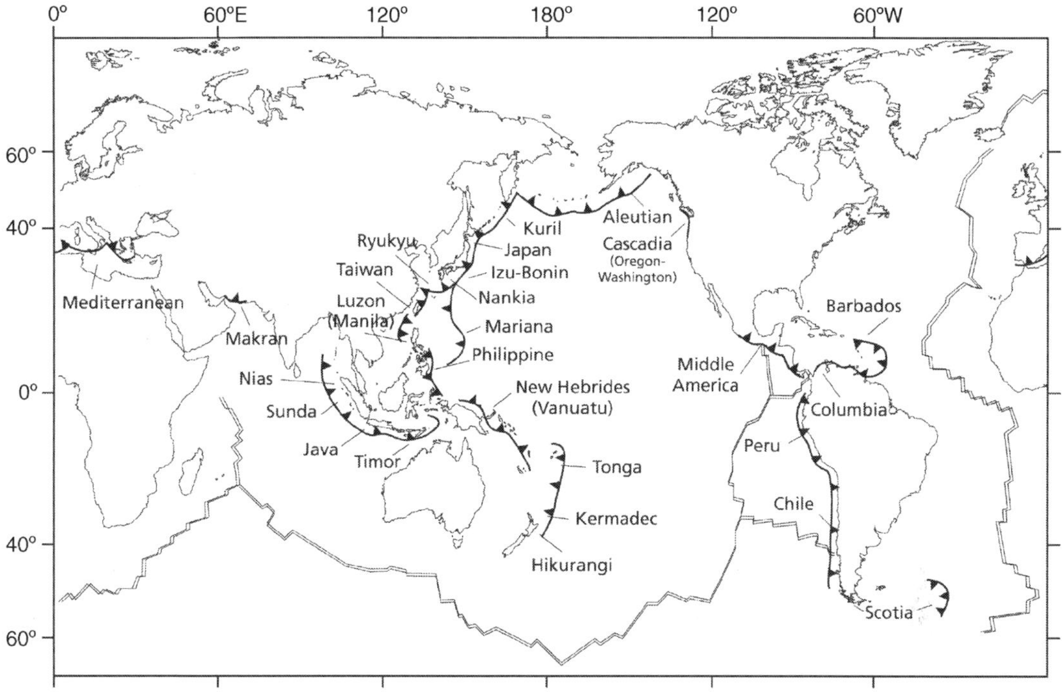


**Fig. 22.8** Global distribution of active oceanic subduction zones. (After Underwood & Moore, 1995).

voluminous calc-alkaline plutonism and volcanism, and high relief. The effects of compresssion are felt widely across the overriding plate to the extent that crustal-scale thrust tectonics may occur thousands of kilometres away from the trench behind the arc. This causes the formation of *retro-arc foreland basins* (see below).

*Extensional arc–trench systems* have weakly coupled descending and overriding plates with the whole overriding plate advancing slowly and/or in the same direction as the descending plate. The majority of the overriding plate is in extension with few strong earthquakes along a narrow seismic zone atop a steeply dipping subduction zone. Extensional margins tend to

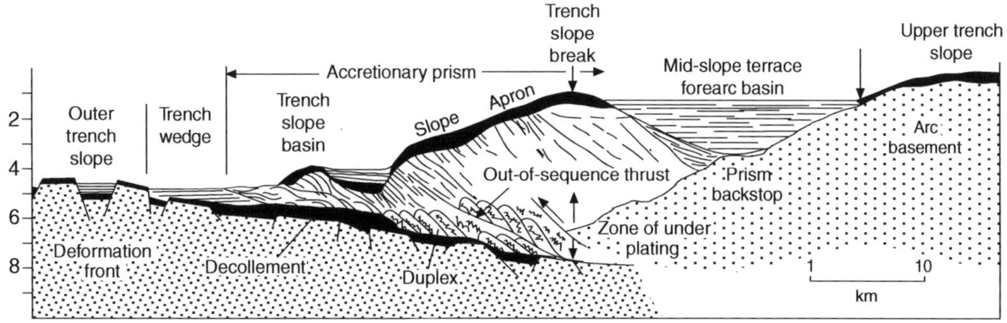


**Fig. 22.9** Section to show the major bathymetric features, structural deformation and depositional sites associated with subduction zones. (After Underwood & Moore, 1995.)

occur when the rate of plate convergence and the negative buoyancy of the subducting plate are both low and the angle of plate descent is high, a common situation along destructive intraoceanic convergent margins in the Pacific Ocean. Evidence from within the Rocky Mountain cordillera and Andes indicates that such margins were very typical of the early stages of convergence of the Pacific margins in Mesozoic times. Morphologically, extensional margins are characterized by deep trenches, voluminous calc-alkaline and basaltic volcanism. The effects of extension are felt widely across the overriding plate to the extent that crustal-scale extensional tectonics may occur thousands of kilometres away from the trench behind the arc. This results in the formation of a plethora of intra-arc and back-arc extensional basins with their characteristic block-faulted and tilted morphologies.

In conclusion, the majority of destructive margins show combinations of compressional, extensional or obliquely convergent features. This makes analysis of ancient convergent-margin sediments both difficult and interesting.

## 22.5 Subduction zones: trenches and trench-slope basins

During its journey in time and space from ocean ridge to subduction zone, oceanic lithosphere cools and slowly subsides. Some ocean-floor relief is obscured by sediment but still in evidence are a myriad of seamounts, remnant fracture zones and associated features. These all play vital roles as ocean boundary layer roughness elements (Chapter 21). As oceanic plate moves progressively closer to a subduction zone it comes under the influence of flexural forces caused by bending as the subducting lower plate turns into the

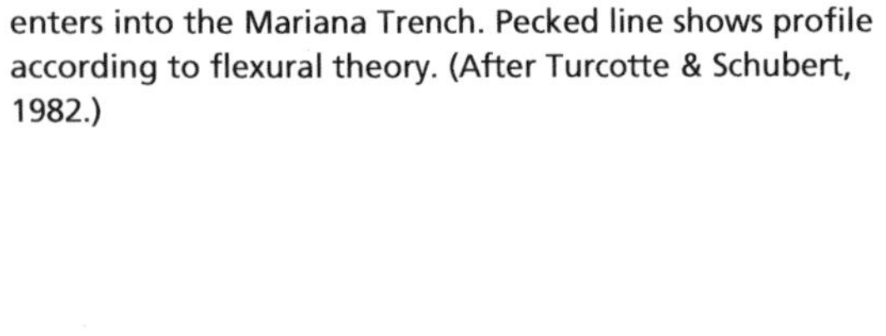

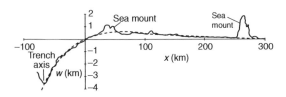


Fig. 22.10 Bending of the Pacific plate (solid line) as it enters into the Mariana Trench. Pecked line shows profile according to flexural theory. (After Turcotte & Schubert, 1982.)

subduction zone and by gravitational loading due to the weight of the overriding upper plate (Chapter 10; Fig. 22.10). These forces cause production of a peripheral bulge of 300–500 m relief up to 150 km oceanwards, the ocean floor then sloping gently towards the locus of maximum depth along the trench. The bending causes extensional stresses which reactivate basement weaknesses such as old fractures and normal faults. Renewed subsidence along normal faults occurs as the surface of the outer trench slope moves trenchwards, the perched basins receiving sediment from local submarine fans or turbidity flows sourced axially or landwards from the inner trench slope. The steeper inner trench slope is the surface expression of a landward wedge of offscraped sediment known as an *accretionary complex* or *prism* (Fig. 22.9).

Trench and trench-slope bathymetry and morphology are determined by:

- *style and rate of subduction*—shallow trenches have shallow, slow plate descent, vice versa for deep trenches
- *bathymetry and structure of lower (descending) plate*—seamounts, aseismic ridges, active ridges and fracture zones cause the trench-slope of the upper plate to be segmented axially;
- *flux of detrital sediment*—large fluxes causing trench infill, flat floors, gentle gradients and marginal onlap on to trench walls and axial segment boundaries, with low-flux trenches being deep, bathymetrically irregular and with steep local gradients;
- *flux of ocean-floor sediment*—efficiency and magnitude of the 'scrape-off' process as witnessed by the lateral and vertical extent of the accretionary prism.

The junction between trench floor and inner trench slope approximates to the position of the active thrust front (or *décollement*) bounding the accretionary prism. The position of this relative to height above oceanic crustal basement determines the preservation potential of trench sediment (Fig. 22.9): the higher the *décollement*, the more sediment is subducted and underplated. The position of the *décollement* depends upon a number of factors, but chiefly the internal shear resistance profile of the sedimentary sequence. It is usual for the majority of sediment supplied to the trench floor to be scraped off and accreted. At the same time oceanic pelagic sediment and ocean crust may also be thrust into the prism as *ophiolite*

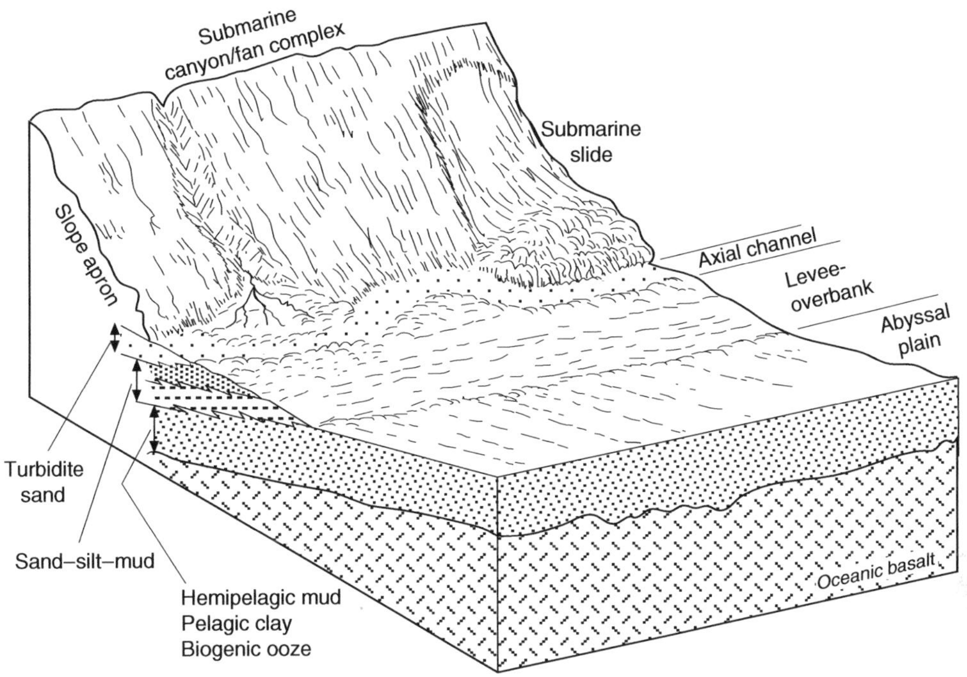


**Fig. 22.11** Simplest possible relationships for lateral submarine fan input to a subduction margin; a coarsening-upwards sequence is produced passing from hemipelagic deposits to fan-channel turbidite sands and gravels. In practice the slope apron has intraslope basins (see Fig. 22.12). (After Piper *et al*., 1973.)

slivers. Underplated material is subducted, perhaps eventually partly recycled as part of arc-derived magma.

The inner trench slope developed upon the ever-uplifting and seaward-expanding accretionary prism may show a relatively smooth surface draped by hemipelagic clays and silts or a very intricate bathymetric profile due to intersection of thrust and normal faults with the seafloor (Figs. 22.11 & 22.12). In general the major offscraping thrust faults become older towards the volcanic arc, though backthrusts, out-of-sequence thrusts and normal faults develop on the trench slope, with shallow extensional slides, slumps and debris-flow channels complicating the picture in detail. Many slopes have smooth-floored *intraslope basins* formed in depressions between active or once-active thrust surfaces. These act as efficient sediment traps for local turbidity currents. High-resolution seafloor mapping and multichannel seismic reflection data along the accretionary Sumatra trench system (Figs. 22.13 & 22.14) show that upper plate segmentation is reflected in varying modes of mass transfer, with active frontal erosion of the upper plate over large areas. The deformation front in the southern segment is characterized by neotectonic formation of a broad and shallow fold-and-thrust belt consistent with the resumption of frontal sediment accretion in the wake of oceanic relief subduction. Conversely, surface erosion by mass flow increasingly shapes the morphology of the lower slope and accretionary prism towards the north where significant relief along subducting fracture zones and an extinct spreading centre occurs. In general, increasing intensity of mass-wasting processes from south to north correlates with the extent of oversteepening of the lower slope (lower slope angle of 3.81° in the south compared with 7.61° in the north), probably in response to alternating phases of frontal accretion and sediment underthrusting. Accretionary mechanics along such segments thus pose a second-order factor in shaping upper plate morphology near the trench.

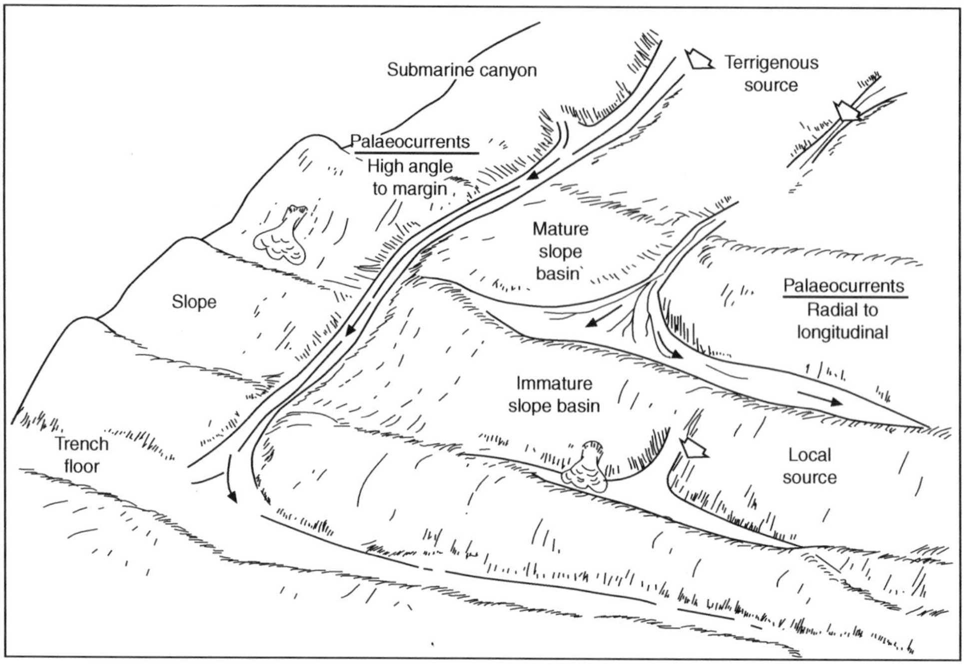


**Fig. 22.12** Sketch to show the nature of an inner trench slope on an accretionary prism, with slope basins forming at the intersection of thrust faults and associated monoclinal growth folds with the prism surface. (After Underwood & Bachman, 1982.)

## 22.6 Fore-arc basins

Along destructive margins, fore-arc refers to usually linear areas seawards of continental or oceanic volcanic arcs and landwards of any accretionary prism or trench (Fig. 22.15). A subsiding fore-arc is referred to as a *fore-arc basin*, rooted in either modified continental or oceanic crust. Fore-arc bathymetry is rather variable, usually shelf-like, with gentle slopes up to 100 km or more wide (e.g. Sunda arc; Fig. 22.16). Some are much more complex, like the Cretan fore-arc with its plethora of sub-basins and local uplifted highs associated with both thrust and normal faulting. These act as local sediment traps for often prodigious sediment fluxes issuing from the adjacent volcanic arc. Volcanic airfalls, submarine slumps and eruption-driven turbidity currents transfer such sediment downslope. Floating pumice rafts disperse more widely over the destructive margin. The efficiency of the fore-arc trap increases as ridge-like barriers form by accretionary offscraping at the trench-slope break. The tendency with time is for the initially shallow fore-arc, with its coarse-grained basal deposits, to deepen quickly and then to infill more gradually with a coarsening-upwards, predominantly turbiditic facies of arc volcanic provenance (Fig. 22.17). However, in many arcs there is a significant topographic barrier between the arc volcanoes and the trench slope (Fig. 22.18). Canyons that transfer sediment oceanwards rarely reach inboard as far as the arc volcanic front.

The distribution of sediment across inner-fore-arc regions, between sediment source and proximal sink, clearly reflects complex interactions between eustasy, climate change, rapid tectonic deformation, volcanism, and variations in sediment flux. For example, the rapidity of tectonic uplift (2–4 mm/yr) of the axial ranges and foothills of inner Hawke's Bay fore-arc basin, New Zealand, causes fluvial systems to deliver voluminous amounts of coarse-grained, largely

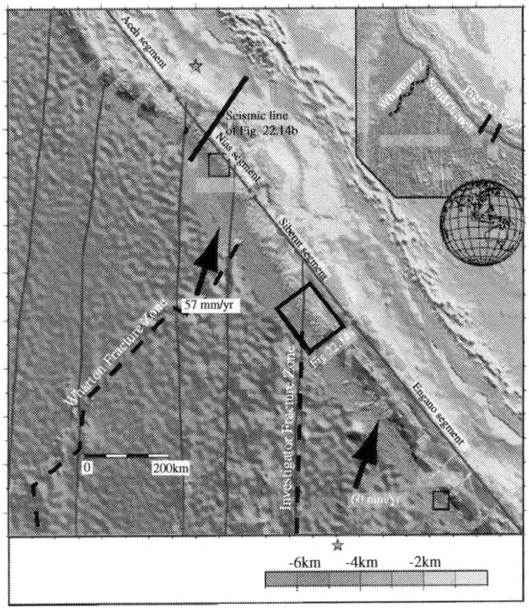


**Fig. 22.13** Composite bathymetric map of the Sumatra trench system with satellite altimetry overlain by high-resolution swath mapping data. The extent of the Enggano, Nias and Siberut segments is shown by red bars. Location of multichannel seismic lines presented in Fig. 22.14c & d is indicated. Black boxes show location of data presented in Fig. 22.14a–c. Fracture zones are from Cande *et al.* (1989). IFZ, Investigator Fracture Zone; FR, fossil ridge; FZ, fracture zone. (After Kopp *et al.*, 2008.)

nonvolcanic clastic sediment to intrabasin depocentres. Despite this, the rapidity of climate and sea-level change means that global glacial–interglacial climate cycles and eustatic change are the first-order drivers of sequence architecture on the convergent margin during the late Pleistocene. However, over sufficiently long periods of time within this interval ($\sim$100 kyr) the growth of structural ridges can be sufficiently fast to create syntectonic growth sequences, with the ridges forming localized barriers to sediment transport. Active deformation across the fore-arc controls the structure, location and geometry of the sedimentary depocentres, as well as the pathways available for drainage and sediment dispersal. Although the total width of the inner fore-arc between source (the crest of the axial ranges) and sink (the toe of the lowstand wedge) is narrow (ca 125 km), the sediment-dispersal pathway is a circuitous route along corridors of relative subsidence located between discontinuous, rising, thrust-faulted ridges associated with growing folds.

Ancient fore-arc basin fills may reach up to 10 km in thickness: the Great Basin of California, the Talara basin of northwest Peru and the Dras basin, Ladakh Himalaya, India are amongst the best-exposed and most-investigated examples.

## 22.7 Intra-arc basins

Although volcanic arcs have a constructional nature, ample opportunities exist for erosion as well as preservation of pyroclastic flows, ashfalls and volcanically derived sediments adjacent to volcanic centres in the often deep-water areas that separate individual volcanic centres. Many volcanic arcs are (or were) in a state of extension, with basin margins bounded by major normal faults allowing preservation of thick volcaniclastic sequences. In addition, rise and collapse of magmatic domes and caldera formation allow for shallow normal faulting and preservation of portions of the volcanic superstructure. Intra-arc basins may develop from back-arc or fore-arc basins as the style or polarity of subduction changes. Larger-scale forces are also at work in the arc environment such as when long-wavelength subsidence occurs as low-angle subduction changes to high-angle. Also, rising diapirs of calc-alkaline or alkaline magma may change a previously depositional submarine site into an uplifting terrestrial one.

Sediment deposited around oceanic volcanic arcs potentially provides the most complete record of the tectonic and geochemical evolution of active margins. Arc sedimentary sections in erosive plate margins can provide comprehensive records of volcanism and tectonism spanning $\sim$10 Myr. The use of such tectonic and geochemical records requires an accurate understanding of sedimentary dynamics in an arc setting. Evidence from the modern Mariana and Tonga arcs and a Jurassic arc crustal section from Alaska indicates all three have formed in tectonically erosive margin settings, resulting in long-term extension and subsidence. Debris aprons composed of turbidites and debris-flow deposits occur in the immediate vicinity of arc volcanoes, forming relatively continuous mass-wasted volcaniclastic records in abundant accommodation space. There is little erosion or reworking of old volcanic materials near the arc volcanic front.

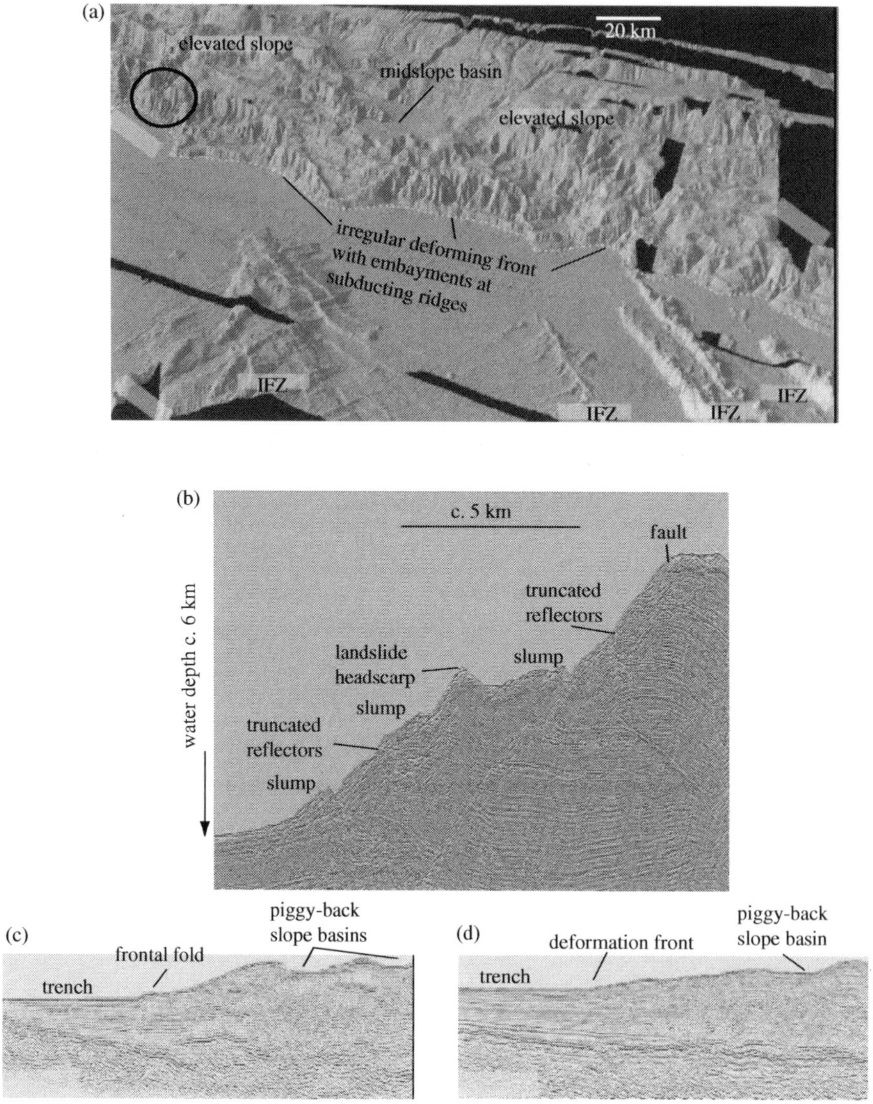


**Fig. 22.14** (a) Perspective view of the entry point of the Investigator Fracture Zone (IFZ) into the Sumatra trench (see Fig. 22.13 for location). The IFZ consists of four individual ridges and migrates northwards along the Sumatra trench. The geomorphology of the upper plate is profoundly affected by the subducted topography. To the south, margin healing after the passage of the IFZ is manifested in juvenile frontal thrust folds as frontal sediment accretion recommences. The frontal accretionary wedge has in parts been completely eroded, resulting in an irregular trend of the deformation front (stippled line) and local oversteepening of the lower slope (white ellipse). Alternating zones of uplift (local highs on an elevated slope) and subsidence are generated as the IFZ migrates along the margin. (b) Truncated reflectors in a time-migrated seismic section of the lower slope in the northern Nias segment indicate surface erosion and mass wasting processes resulting in slumps. Jumbled and chaotic reflectors characterize the slump masses on top of active folds, which are linked to landslide headscarps. (c)–(d) Reflection profiles across noneroding trench segments offshore southern Sumatra and off the Sunda Strait, respectively. Note absence of erosion, occurrence of relatively smooth frontal folds and a much shallower lower slope compared with the Nias segment (All after Kopp *et al.*, 2008.)

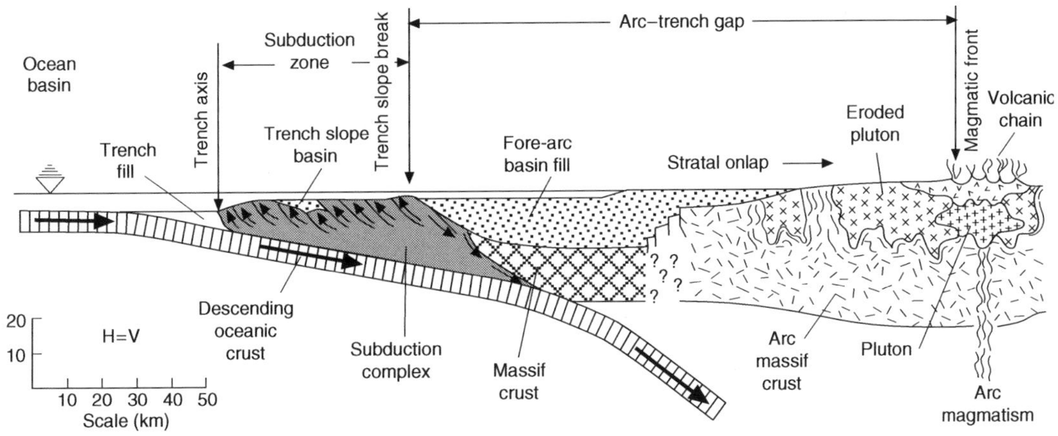


**Fig. 22.15** Section to show the tectonic setting of fore-arc basins. (From Dickinson, 1995.)

Tectonically generated topography in the fore-arc effectively blocks sediment flow from the volcanic front to the trench. Although some canyons deliver sediment to the trench slope, most volcaniclastic sedimentation is limited to the area immediately around volcanic centres. The chemical evolution of a limited section of an oceanic arc may be best reconstructed from sediments of the debris aprons for intervals up to ~20 Myr. For longer intervals, subduction erosion causes migration of the fore-arc basin crust and its sedimentary cover toward the trench where there is little volcaniclastic sedimentation and where older sediments are dissected and reworked along the trench slope.

## 22.8 Back-arc basins

The vast majority (> 70%) of back-arc basins occur associated with the destructive plate margins of the Pacific Ocean (Fig. 22.19). Other notable examples

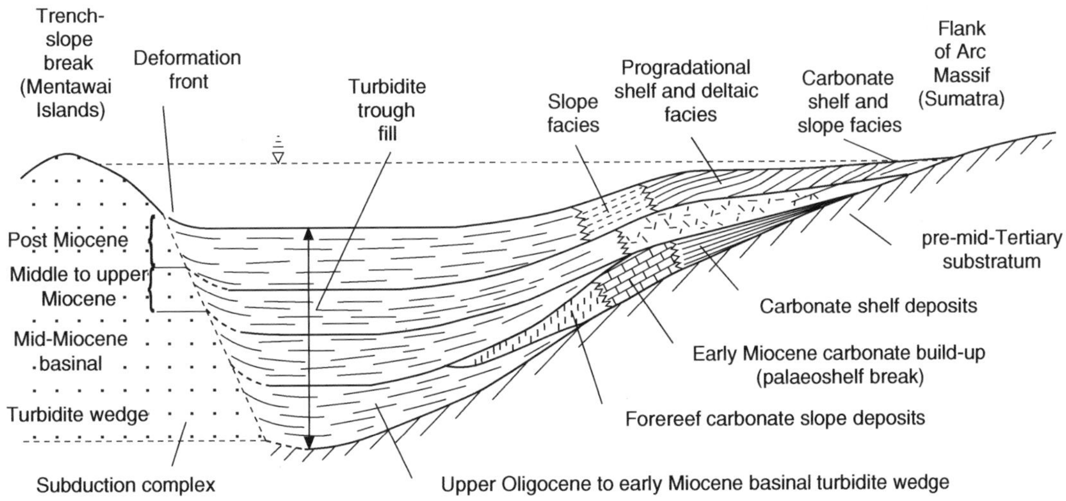


**Fig. 22.16** Sedimentary environments in the Sunda fore-arc basin (From Dickinson, 1995, and sources cited therein.)

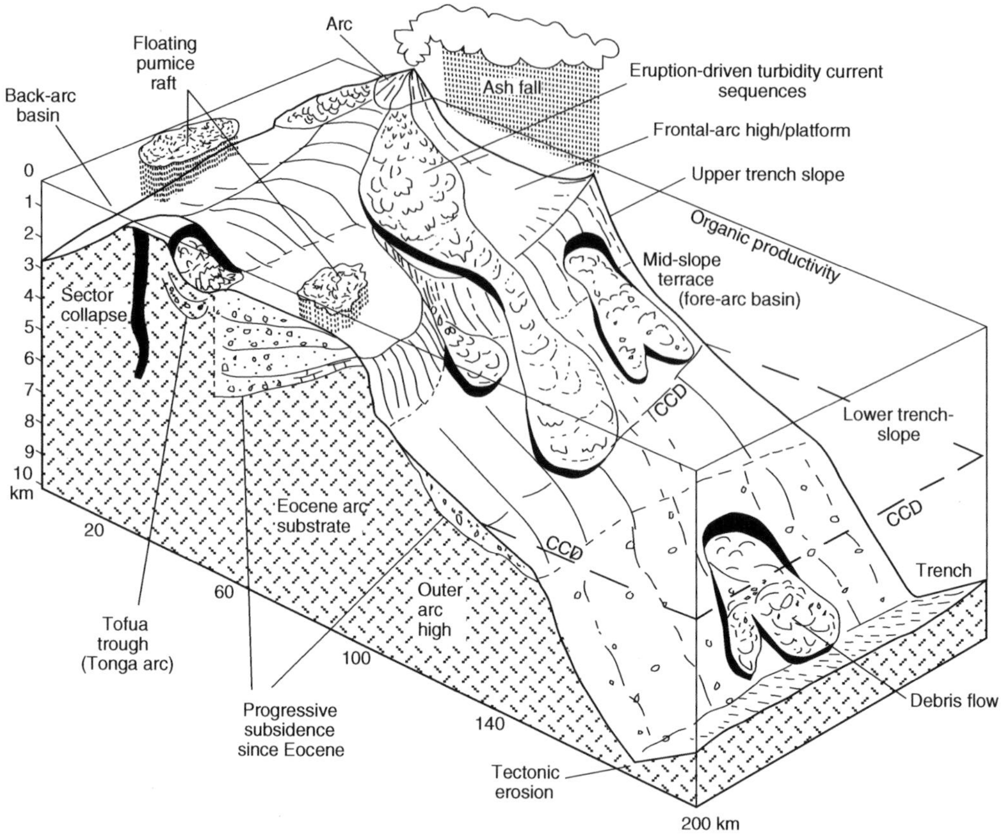


**Fig. 22.17** Context, sedimentary environments and processes along a typical volcanigenic oceanic fore-arc. CCD, carbonate compensation depth. (Based mainly on Tonga; after Underwood *et al.*, 1995.)

occur in the Scotia Sea and Mediterranean Ocean (Tyrrhenian, Aegean Seas). We have already noted how destructional plate margins may be brought under a state of extensional stress so that a formerly active volcanic arc or an area of continental crust may be converted into a syn-rift province accompanied perhaps by basaltic or calc-alkaline volcanism evolving into a site of new ocean crust formation. A second group of back-arc basins owes its origin to a reorganization of subduction whereby an oceanward shift in the site of subduction leaves a stranded and abandoned volcanic arc and an area of pre-existing ocean floor behind the new arc. Such passive or residual back-arc basins are typified by the Bering Sea and West Philippine basins.

Early stages of active back-arc rifting in oceanic settings are exemplified by the young ($<$ 5 Ma) Izu-Bonin arc, northern Iwo Jima ridge, Phillipine–Japan Sea (Fig. 22.20). This comprises half-grabens arranged in long (700 km) chains up to 40 km wide behind or sometimes between a line of arc volcanoes. The grabens, separated by volcanically active transfer zones, are bounded by active normal faults with up to 2.5 km throw. Fault development closely follows the pattern seen in major continental half-grabens, whereby a large half-graben develops against a master normal fault. Sediment fluxes are dominated by pyroclastic flows and by resedimented volcanically derived material in gravity flows down the slopes formed by tectonics. Localization of rifts occurs as seafloor spreading develops along a narrow axial zone and thermal subsidence dominates over the now inactive syn-rift remnant, volcanic arc and earlier spreading centres.

Back-arc basins forming in continental crust are typified by the largely submarine Okinawa Trough and the Aegean Sea where, in the latter case,

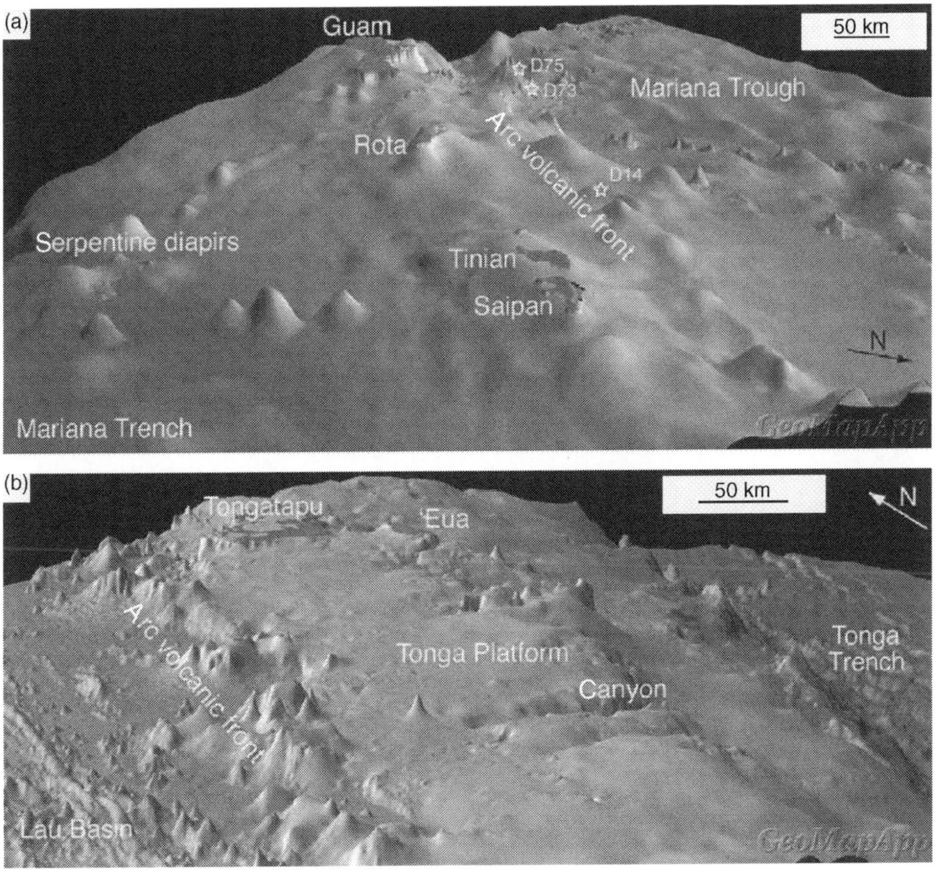


**Fig. 22.18** Perspective views of (a) the Mariana and (b) Tonga Arc and forearc regions, showing the great distance and topographic barrier between the arc volcanic front and the trench slope. Stars show the location of dredge samples. (After Draut & Clift, 2006. Images are from the GeoMapApp integrated mapping application developed by Lamont-Doherty Earth Observatory.)

well-defined half-grabens have horst flanks extending well above sea level, providing clastic sediment to the deep-water rift basins. There is a trend from lacustrine conditions early on in the extensional process to marine conditions with abundant influx of detrital sediment. Steep, fault-controlled continental and submarine slopes result in spectacular examples of sediment dispersal by gravity flows.

## 22.9 Foreland basins

Foreland basins (Fig. 22.21) develop on continental crust along the length of collisional plate margins or along compressional destructive margins. In both cases they are peripheral to a developing thrust-fold orogenic belt. 'Foreland' refers to the relatively undeformed (in the shortening sense) continental crust or continental margin over which major thrust faults transfer wedges of crust from the orogen. These thrust wedges play a dual role in the evolution of a foreland basin and its sedimentary infill:

- they load the foreland plate, which in turn responds by flexural bending (Chapter 10);
- they form the uplifted source areas for river catchments to develop and provide a sediment source for the developing basin.

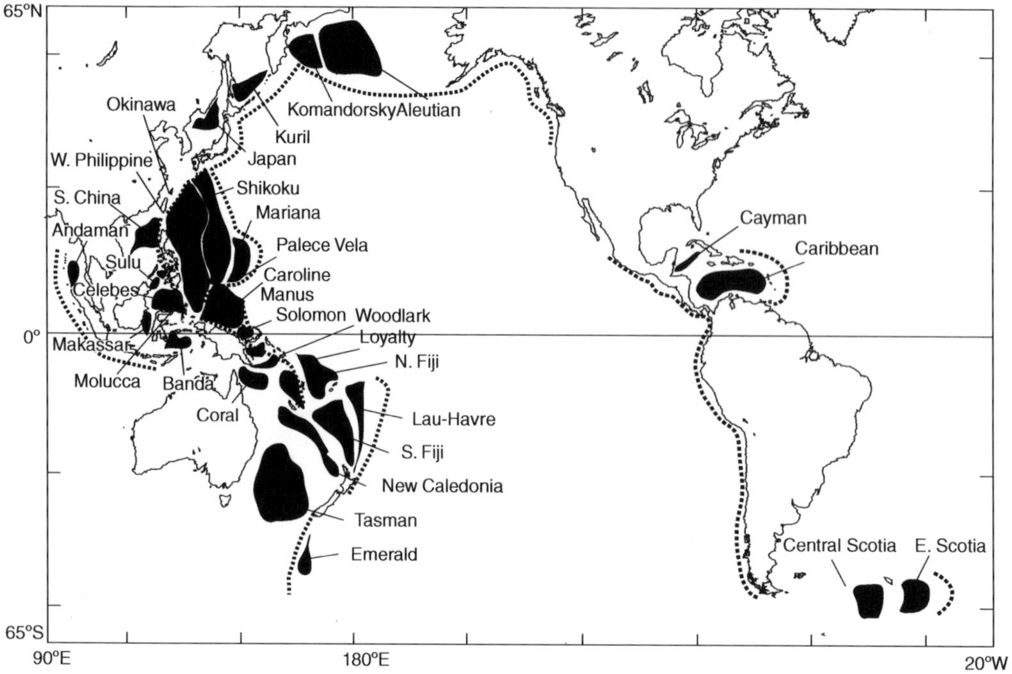


Fig. 22.19 Distribution of trenches and back-arcs in the circum-Pacific region. (After Tamaki & Honza, 1991.)

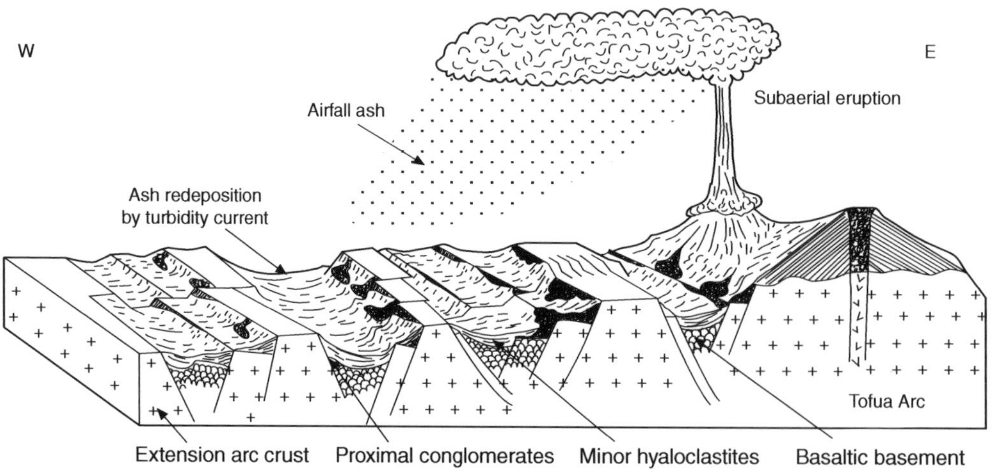


Fig. 22.20 Representation of a highly extended oceanic back-arc basin. Note the plethora of extensional tiltblocks in ocean crust and their role in trapping the proximal products of arc volcanism. (After Marsaglia *et al.*, 1995.)

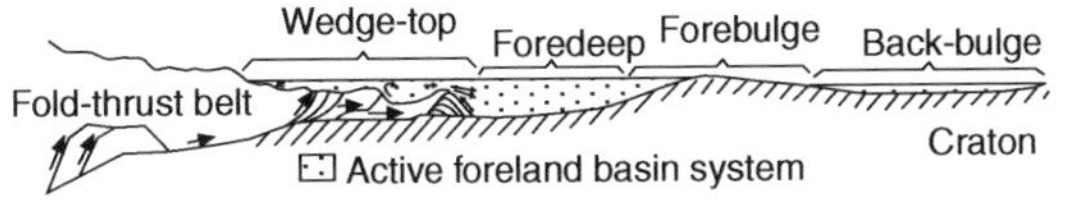


**Fig. 22.21** Cross-section of a composite foreland basin system, based on the central Andean foreland basin system. The origin of the thin saucer-shaped back-bulge depression is poorly known. For wedge-top basins, see Horton (1998). (After Horton & DeCelles, 1997.)

Foreland basins are certainly the most-studied type of sedimentary basin because they lie adjacent to Earth's past and current major mountain chains and contain the only real geological data with which the evolution of these *orogenic belts* can be established. Three-dimensional tectonic evolution is constrained by assembling many stratigraphic and restored structural sections whilst time–temperature–pressure paths for exhumed crust are gained from apatite fission-track and mineral closure temperature ages. A variety of syntectonic signals from foreland basin sediments are of use in tectonic reconstructions.

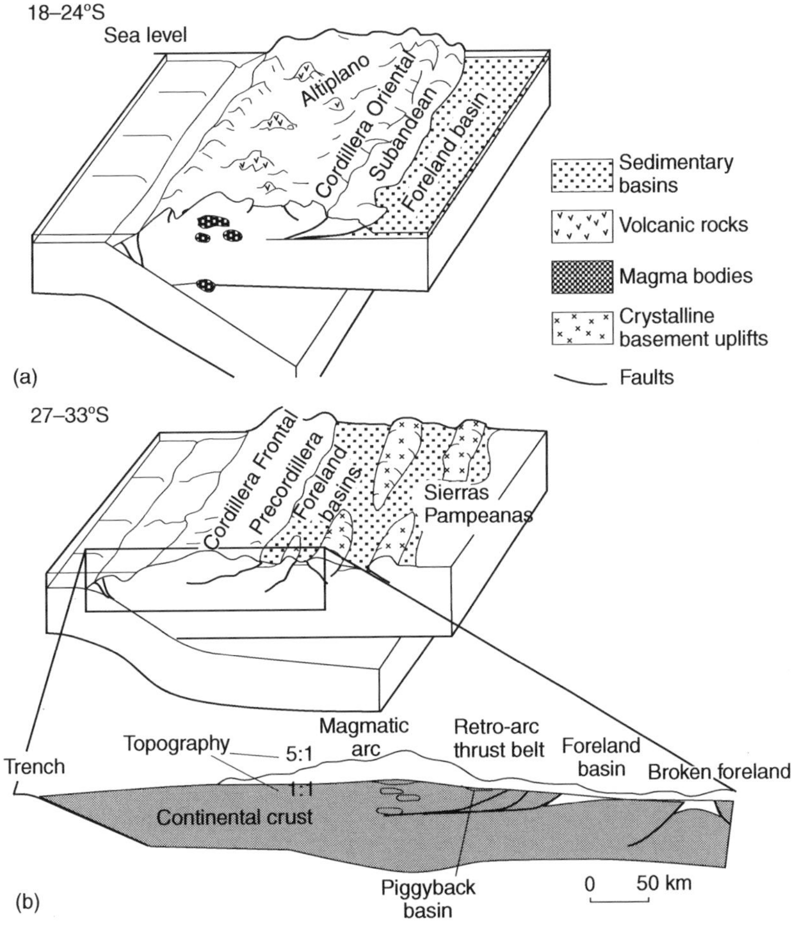


**Fig. 22.22** Diagrams to indicate longitudinal variations in subduction geometry and foreland basin development in the Andes. (a) Northern section with steeper subduction and thin-skinned sub-Andean foreland basin. (b) Central section has flat subduction causing strong plate interactions and shortening deformation far out into the sub-Andean broken foreland. (After Jordan, 1995, and sources cited therein.)

1 Compositional indicators of mountain evolution and exhumation come from age-dated mineral tracers and sediment mineralogical composition.
2 Vertical changes in sedimentation rates may reflect either changes in sediment flux from the uplifted source and/or changes in subsidence rate because of tectonic loadiñg.
3 An increase in the gravel sediment fraction may reflect either an increase in the relative or absolute production rate in the source area as uplifted units are exposed.
4 Changes in palaeotransport direction may reflect changes in slope resulting from uplift of adjacent source areas

However, differences in the influence of transport and storage processes mean that across most basins first occurrences of these four markers are not synchronous. Since an increase in grain size is easiest to observe, progradation of the gravel front ('*syntectonic conglomerate*') is often chosen as the best messenger of an approaching thrust front. But in addition, various external forces and internal feedback mechanisms, such as lithological state, climate, base-level change and nickpoint migration, can complicate the sedimentary response to even simple tectonic events. Thus in the Italian Venetian retro-arc foreland basin, glaciofluvial sediment fluxes during Pleistocene times overprinted a formerly clear tectonic increase seen in the Miocene. Also, playoff between increased sediment supply and increased basin subsidence rates resulting from tectonic uplift and loading can affect vertical aggradation rates. Finally, the interplay between timescales of various alluvial processes and the natural diffusional timescale that a basin needs in order to develop a steady-state response (*basin equilibrium time*) can produce widely divergent and nonlinear responses.

There are two major types of foreland basin, differentiated on the basis of position relative to the thrust belt (Fig. 22.22):

1 *Pro-foreland basins* lie on underriding crust at continental collision zones. They are typified by the active Indo-Gangetic foreland basins south of the Himalayan frontal thrusts, the inactive Molasse basins of the Alps and Spanish Pyrenees (Ebro basin), and the active Tigris–Euphrates–Arabian Gulf basins west of the Zagros Mountains in Iran and Iraq.
2 *Retro-foreland basins* are intracrustal structures of the overriding crust at destructive margins and at continental collision zones. They are typified by the mosaic of active basins to the east of the Andes in Argentina and by formerly active basins like the Cretaceous Western Interior basin in the eastern forelands of the Rocky Mountains, the Po basin, Italy and the Aquitaine basin, France.

The two types are distinctive in respect of tectonic position but share the common feature of

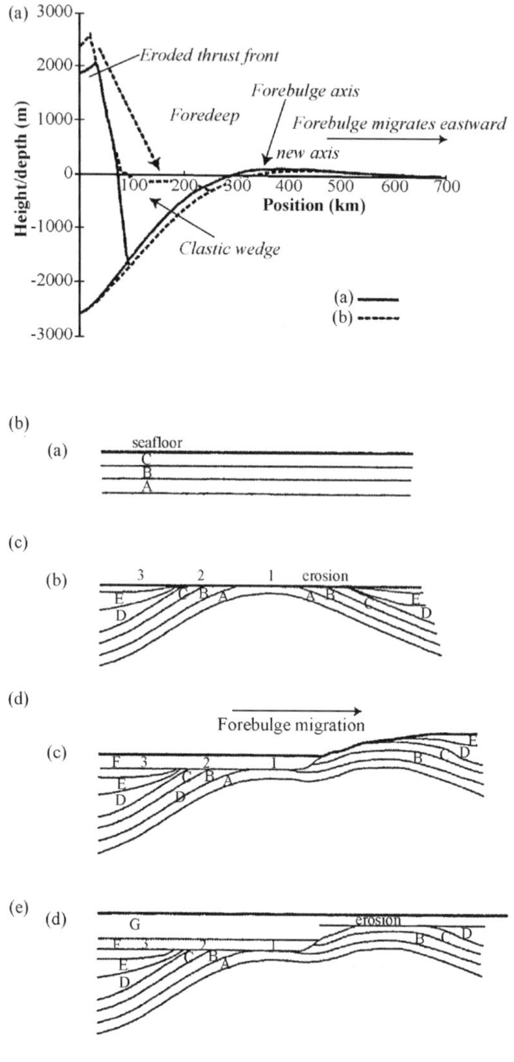


**Fig. 22.23** (a)–(e) Schematic forebulge migration, sedimentation, tilting and disconformity development. (After White *et al.*, 2002.)

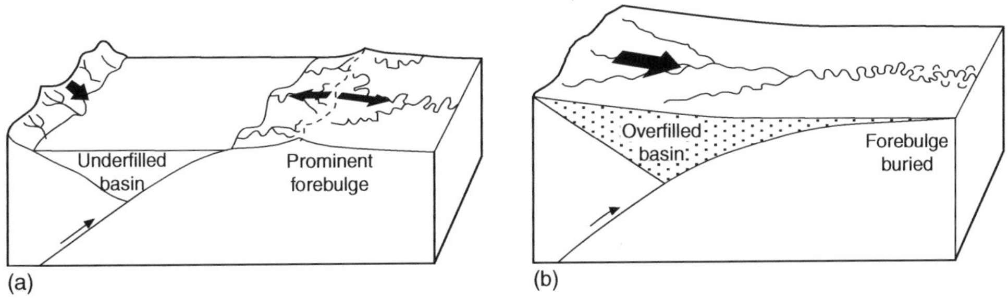


**Fig. 22.24** Cartoons to show contrasts between (a) underfilled (water-filled in this case) and (b) overfilled foreland basins. (After Crampton & Allen, 1995.)

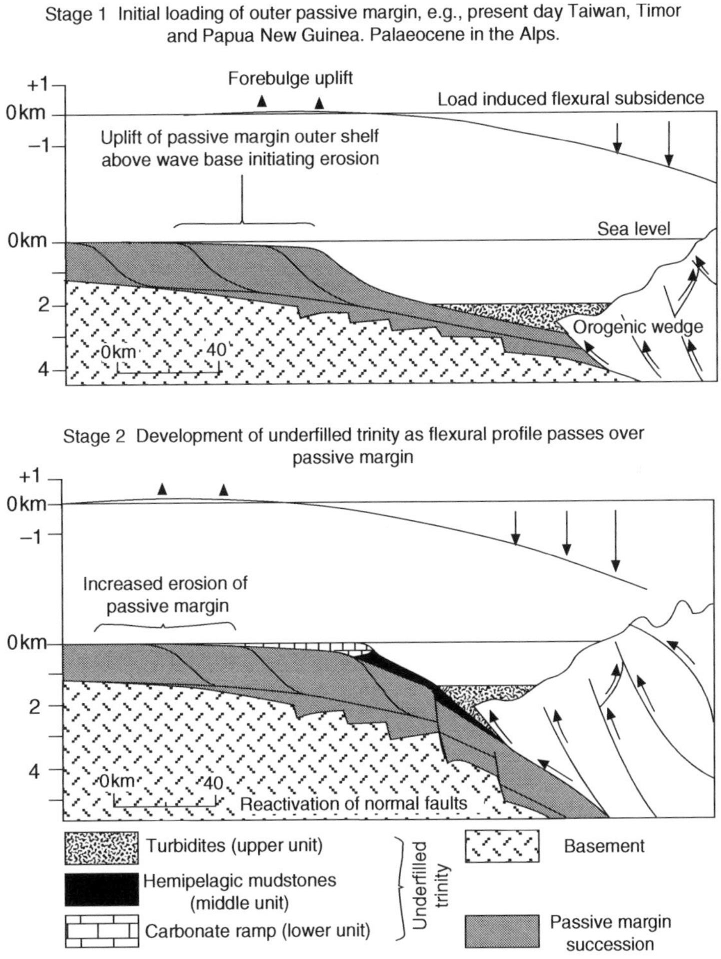


**Fig. 22.25** Sketches to show the sequential evolution of an initially marine foreland basin as it narrows and is affected by forebulge migration in response to advance of the orogenic wedge along a thrust front. (After Sinclair, 1997.)

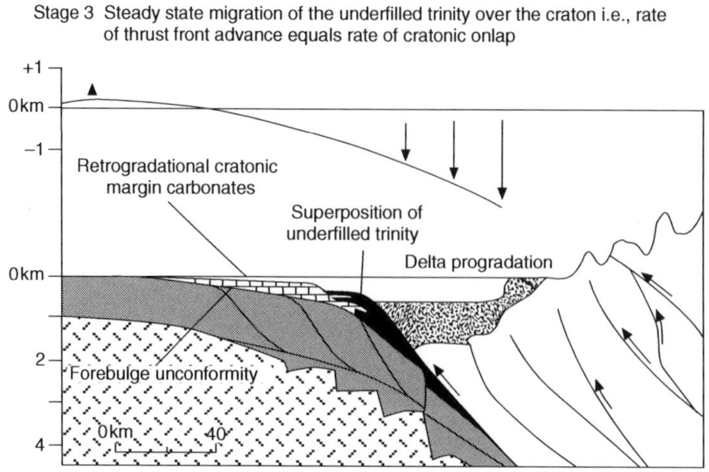
Stage 3 Steady state migration of the underfilled trinity over the craton i.e., rate of thrust front advance equals rate of cratonic onlap
+1
0km
−1
Retrogradational cratonic margin carbonates
Superposition of underfilled trinity
Delta progradation
0km
2
Forebulge unconformity
0km
40
4

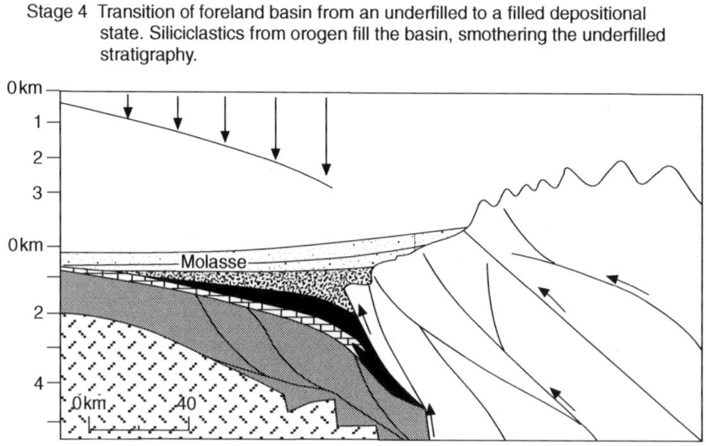
Stage 4 Transition of foreland basin from an underfilled to a filled depositional state. Siliciclastics from orogen fill the basin, smothering the underfilled stratigraphy.
0km
1
2
3
0km
Molasse
2
4
0km
40

**Fig. 22.25** *(continued)*

flexure-induced subsidence caused by thrust loading. Active thrusting produces geologically instantaneous subsidence in the footwall of the thrust around the periphery of the load. The magnitude and extent of the subsequent flexural subsidence depend upon those of the load and upon the elastic properties of the overridden plate. Loading is obviously a time-dependent feature since the total extent of thrusting is accumulative, but not necessarily steady. Not only will the loading mass increase with time, but the rate of removal of the load by erosion will also increase, giving an increase in sediment flux into the developing basin. Further out into the overthrust plate from the point of application of the thrusted load (several hundred kilometres for a crust of average flexural rigidity $\sim 3.10^{24}$ Nm), subsidence and uplift reflect the development of *peripheral bulging*, whose height and distance from the load closely depend upon the elastic properties of the underlying lithosphere (Fig. 22.23a).

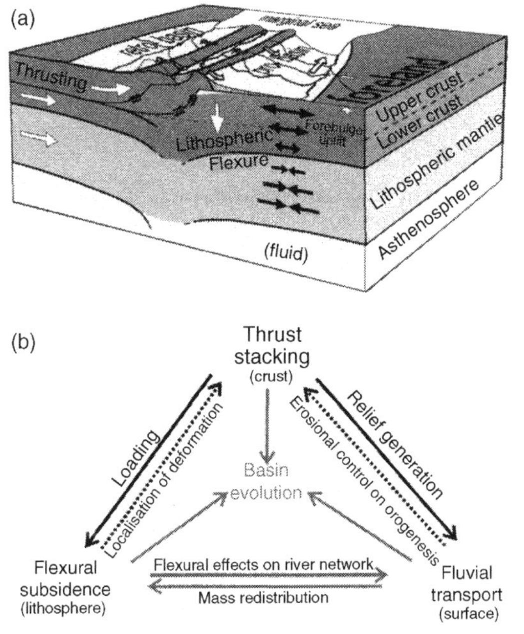


**Fig. 22.26** (a) Sketch to show the role of rivers in surface mass transport away from thrust-belt loading and across subsiding foreland basin. It can be seen that rivers drive the surface mass distribution. (b) Interplay between the main processes involved in foreland basin evolution. (After Garcia-Castellanos, 2002.)

A migrating, partly eroding load will increase in mass with time and drive a complete wave of subsidence and uplift across the loaded plate, increasing in magnitude as it goes. Any given point in space thus has a changing displacement vector over time. Superimposed will be the modifying effects of deposition. It seems that in the majority of cases, after an initial acceleration of subsidence as the load grows and advances from the far-field, high subsidence rates result in rapid water deepening close to the thrust front, with uplift of the peripheral bulge causing shallowing-upwards trends in any pre-existing deposits of the overridden plate. The shoaling upwards may eventually terminate with a phase of above wave-base or continental erosion. Deposition subsequently causes infill of the bathymetry and the production of coarsening-upwards sequences. The nature of any foreland basin thus depends critically upon the ability of the sediment depositional systems feeding the basin to fill it with sediment so that all slopes are changed from tectonic to depositional. Such considerations lead to concepts of *underfilled vs. overfilled basins* (Fig. 22.24).

Climate, sea level change and thrust propagation sequence play an extremely important role in the evolution of foreland basins:

1 Drainage sourcelands fed by orographic rainfall are prone to the effects of climatic change on runoff, vegetation and sediment supply.
2 Structural growth by thrust evolution can create rain-shadow effects on previously well-watered basins.
3 Glaciation in mid- to high-latitudes markedly increases sediment flux from established or growing orogenic sourcelands
4 Thrust fault/fold propagation causes unsteadiness and nonuniformity in tectonic slopes and may cause marked basinwards shifts in the locus of sediment outfall due to drainage growth by coalescence and the gradual replacement of initial smallish alluvial fans by a few *megafans* (Chapter 12).
5 Basinward fault propagation causes the uplift of portions of previously active thrust-front proximal basins into *thrust-/wedge-top basins* (Fig. 22.21) often called *piggyback basins* because they lie upon the back of low-angle thrust ramps.
6 Uplift and erosion of the sedimentary fill to piggyback basins may cause large-scale drainage reorganization and the rapid production of major alluvial fans.
7 Development of axial drainage from the coalescence of transverse drainages leads to a massive increase in throughflux of sediment (Fig. 22.26). Peripheral foreland basins are often characterized by axial drainage, classic examples being the Indo-Gangetic and Euphrates–Tigris basins, whilst much-segmented retro-arc basins exhibit local drainage or major cross-basin drainage normal to basin strike.
8 Regarding sea-level controls, these may be overshadowed by structural reorganization. For example, during early development of the North Alpine foreland basin there was slow coastal progradation controlled initially by eustatic sea-level fluctuations superimposed on flexural subsidence. After 37 Ma, a tectonically controlled coastline evolution and rapid progradation occurred in response to

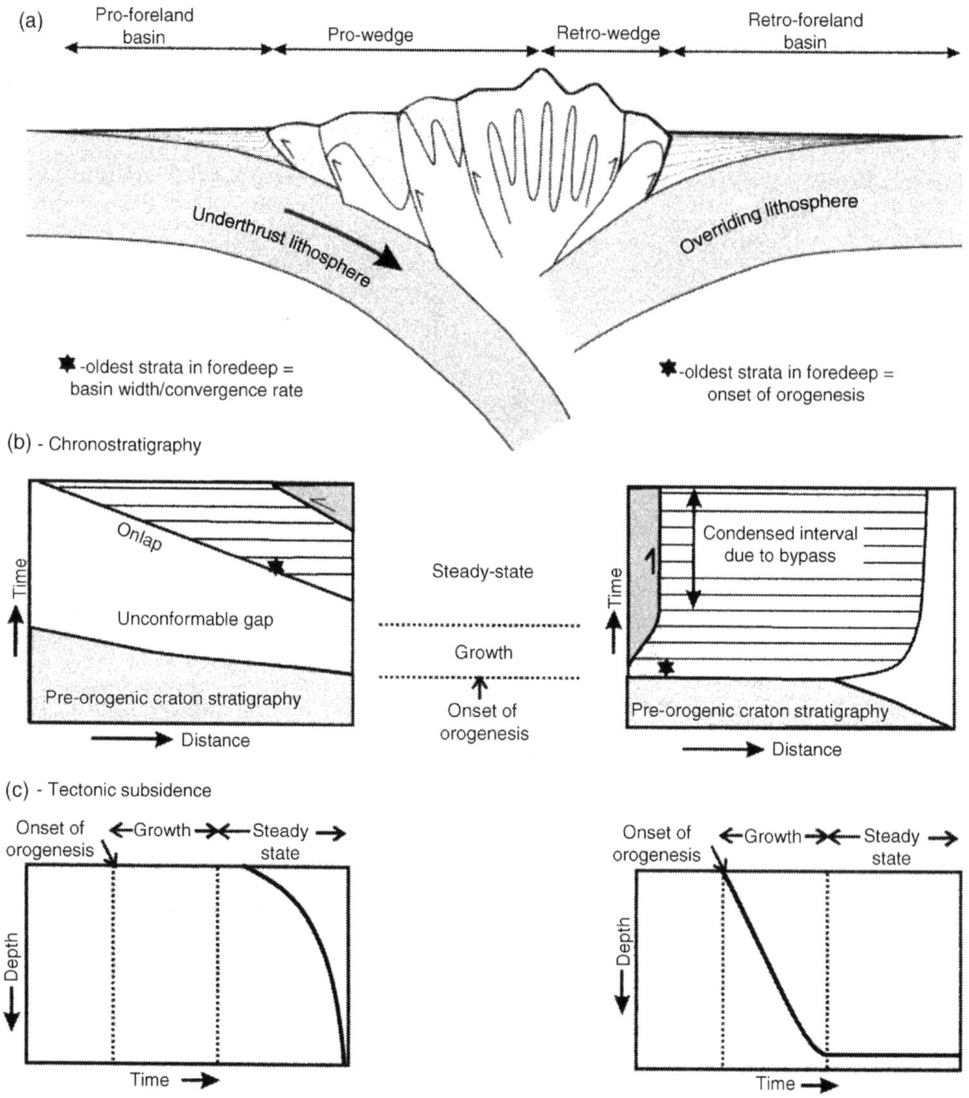


**Fig. 22.27** Contrasting the basin characteristics of pro-foreland (left) and retro-foreland (right) foreland basins. (a) The pro-foreland basin exhibits basin onlap of the cratonic margin at a rate greater or equal to the plate convergence rate, dependent upon whether the thrust wedge is in a growth or steady-state phase, respectively. The retro-foreland basin records little onlap except in the early stage of growth. (b) The aforementioned onlap pattern is clearly seen in the time equivalent, which also illustrates the relatively limited interval preserved in the pro-foreland basin. (c) The pro-foreland basin records accelerated subsidence over a relatively short interval of orogenesis. In contrast, the retro-foreland basin records the full history of the basin with initial uniform subsidence during growth of the mountain belt, and retro-thrust wedge, followed by zero subsidence during steady state when the retro-wedge no longer accretes new material. During this latter stage, the retro-foreland basin records a condensed stratigraphic succession which is likely to be dominated by bypass of sediment generated in the mountain belt and exported farther afield. (After Naylor & Sinclair, 2008.)

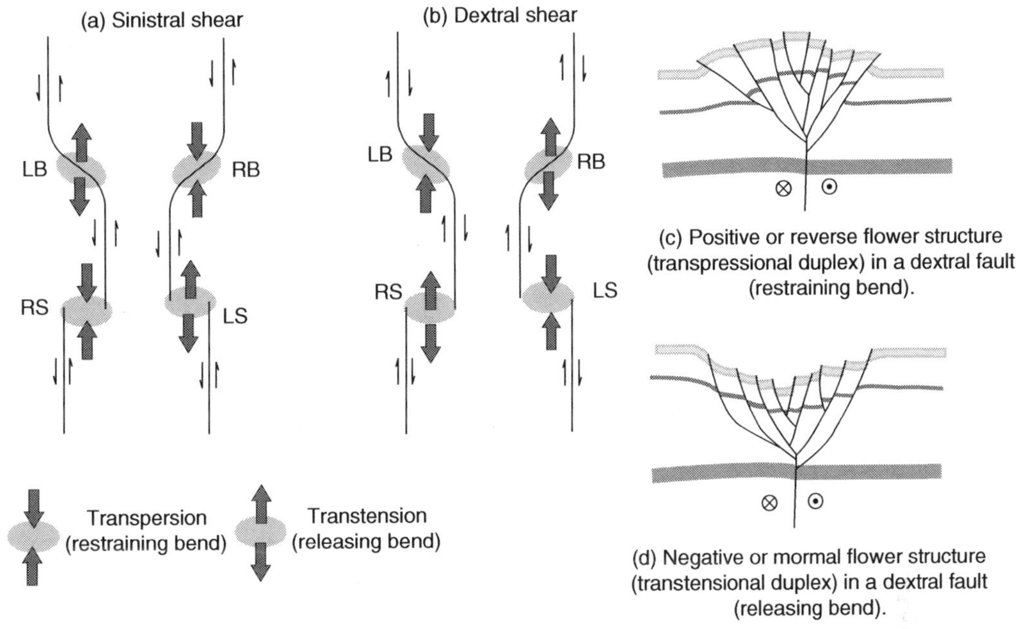


**Fig. 22.28** Bends and stepovers causing transpression and transtension in (a) sinistral and (b) dextral strike-slip faults. (c) & (d) Sectional views to show strike-slip duplex structures (aka *flower structures*) forming in areas subjected to compression or tension. RB, right bend; LF, left bend; RS, right stepover; LS, left stepover.

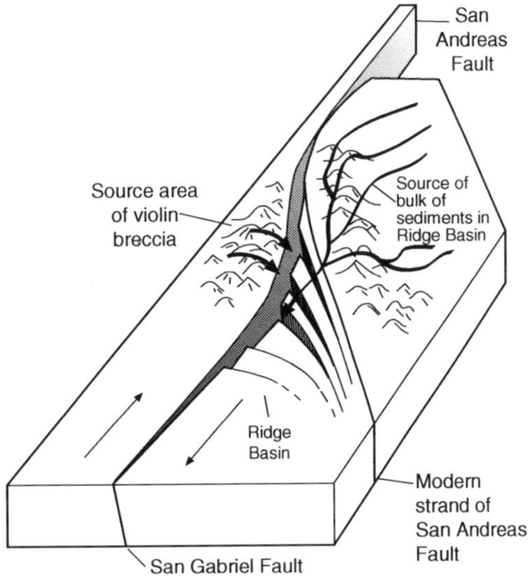


**Fig. 22.29** John Crowell's classic notion of a slide-apart basin formed along a bend in a major strike-slip fault, as gleaned from the famous Ridge Basin along the now defunct San Gabriel Fault, California. Drainage courses are indicative only. (From Crowell, 1974a.)

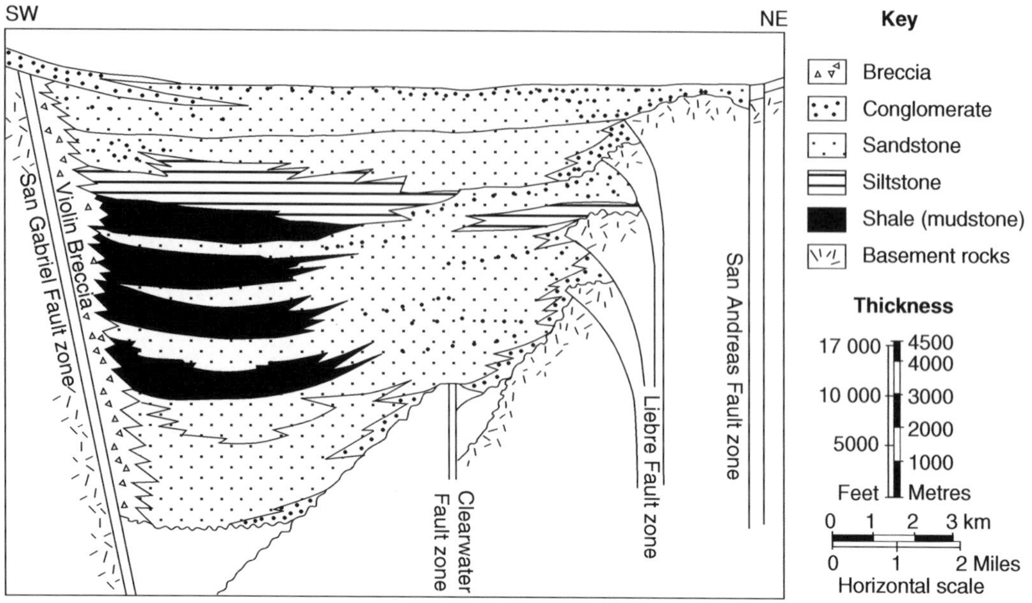


**Fig. 22.30** Cross-section through the Ridge Basin to show alternating cycles of central lacustrine fine-grained sediment and coarse alluvial fan/fan delta clastic wedges (e.g. Violin Breccia) sourced from narrow catchments cutting across the San Gabriel fault to the southwest and the wider sourcelands of the fragmented 'slide-apart uplands to the NE. (After Crowell & Link, 1982; see also Fig. 22.29.)

structural growth due to collision of the European and Adriatic margins.

One of the most characteristic features of retro-arc foreland basins is the occurrence of combined zones of thin-skinned tectonics and scattered basement uplifts originating along much deeper and steeper thrusts, which bring crystalline crustal basement to the surface in the hangingwall of *blind thrusts*, thus creating shallowing upwards trends in basinal facies over the gradually emerging thrust-related folds. Such foreland basins with their isolated uplifts and subsiding basins are termed *broken forelands*, as distinct from the more continuous peripheral foreland basin plains brought about by thin-skinned thrusts. The broken forelands of the eastern Andes, the Sierras Pampeanas of Argentina, have propagated far eastwards in an area coincident with the shallowest angle of the subducting plate (Fig. 22.22). They are characterized by both local basement sediment sources and more regional sourcing from the Andes thin-skinned thrust belt to the west. The identification of these sources is vital in the elucidation of the history of ancient broken foreland basins like the Mesozoic and early Tertiary Sevier and Laramide basins of the ancestral Rocky Mountains.

Modelling studies that incorporate suitable tectonic boundary conditions predict contrasting basin development in pro- and retro-basin types (Fig. 22.27). Pro-foreland basins are characterized by: (i) accelerating tectonic subsidence driven primarily by the translation of the basin fill towards the mountain belt at the convergence rate; (ii) stratigraphic onlap on to the cratonic margin at a rate at least equal to the plate convergence rate; (iii) a basin infill that records the most recent development of the mountain belt with a preserved interval determined by the width of the basin divided by the convergence rate. In contrast, retro-basins are relatively stable, are not translated into the mountain belt once steady state is achieved. They are characterised by: (i) constant tectonic subsidence rate during growth of the thrust wedge, with zero tectonic subsidence during the steady-state phase (i.e. ongoing accretion–erosion, but constant load); (ii) relatively little stratigraphic onlap driven only by the growth of the retro-wedge; (iii) a basin fill that records the entire growth phase of the mountain belt,

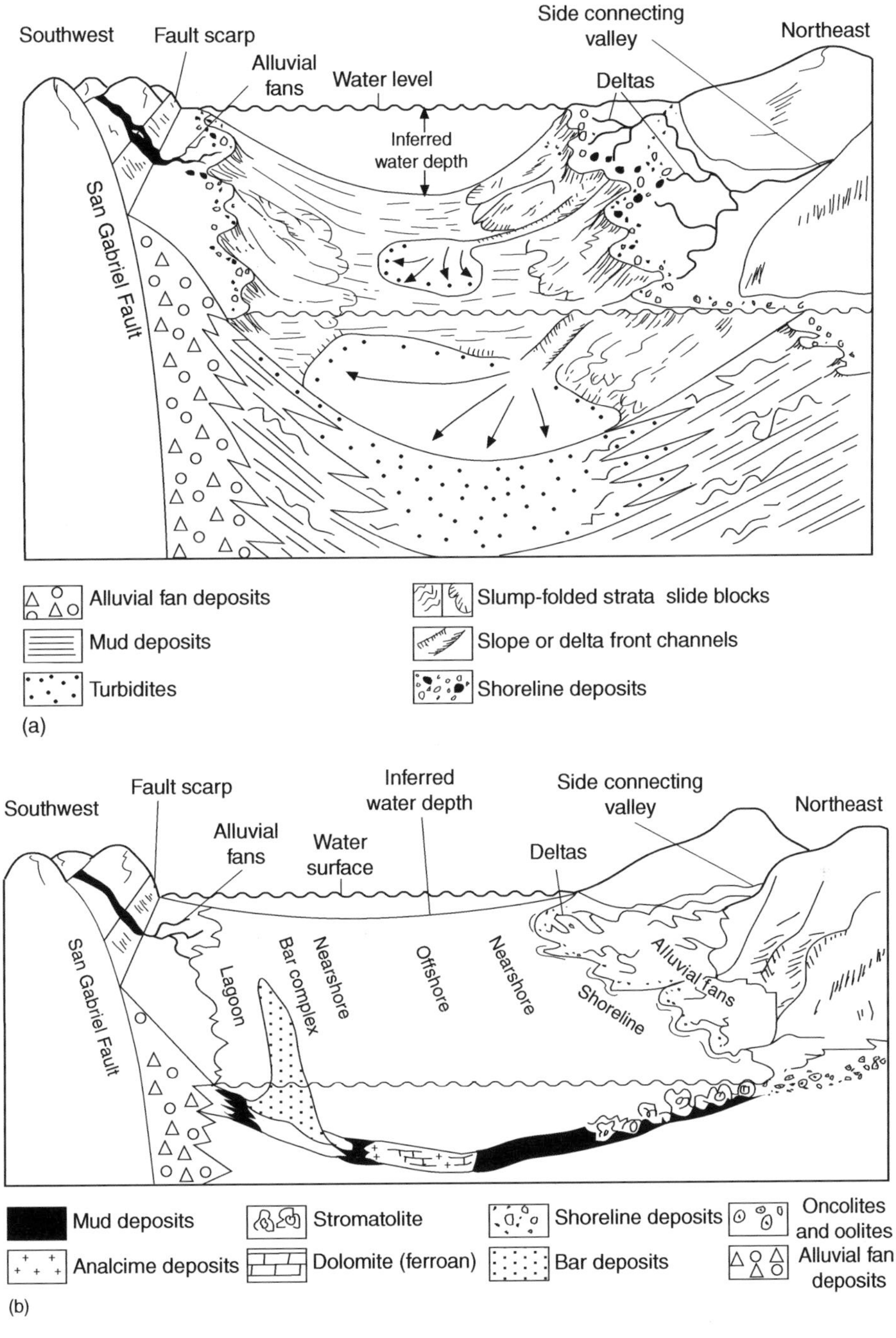


**Fig. 22.31** Palaeoenvironmental sketches to show depositional settings in the fluviolacustrine Ridge Basin during (a) lake highstands and (b) lake lowstands. (After Link & Osborne, 1982.)

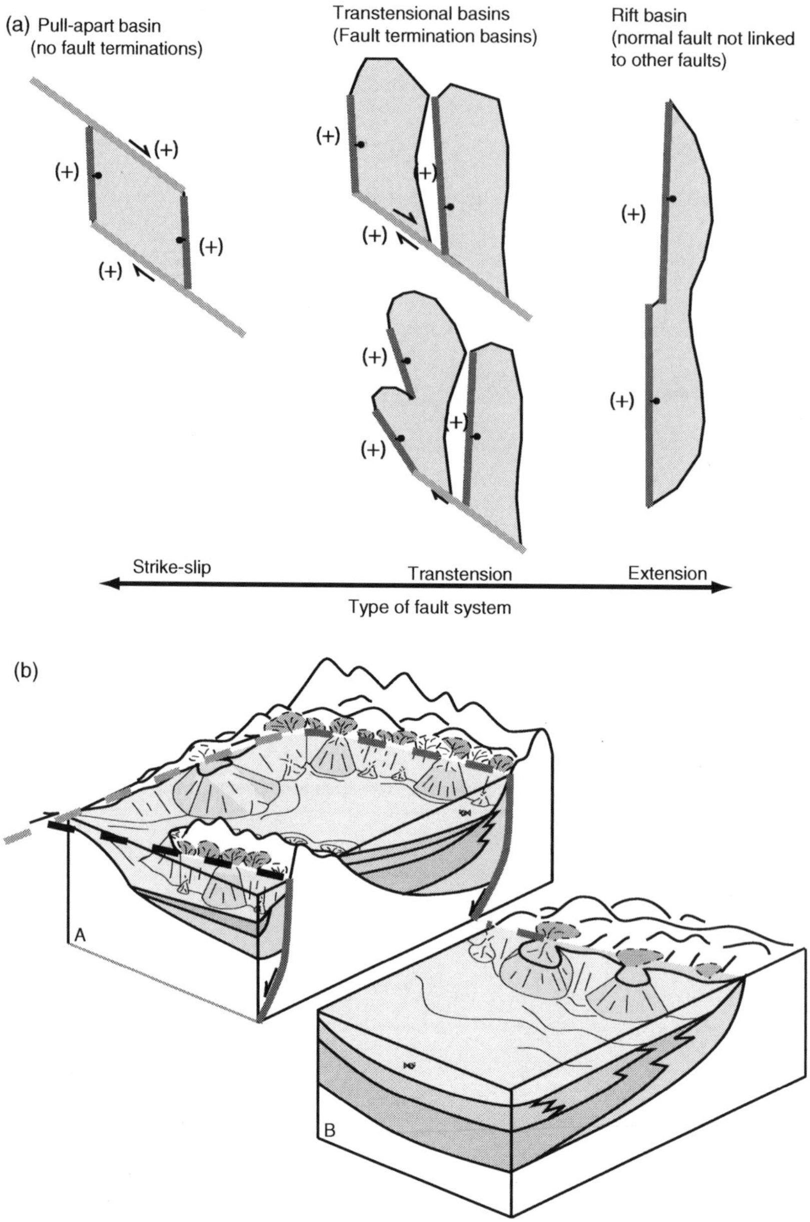


**Fig. 22.32** Terminator! (a) Simple scheme that depicts how transtensional fault-termination basins lie within the spectrum between pull-apart (aka strike-slip) basins and rift basins. Fault-termination basins have local fault bends and/or steps similar to pull-apart basins, but these latter have complete linkage across the fault step or bend so that the fault slip is through-going with no lateral loss of slip. By way of contrast, fault-termination basins end along the border of the basin and therefore the main strike-slip fault loses slip here so that unconformities are common. In cross- section, fault-termination basins are similar to rift basins because normal faults run parallel to the long axis of both types of basins. (b) Diagrams of idealized transtensional fault-termination basin and sedimentary architecture based on examples in southern Gulf of California. (After Umhoefer *et al.*, 2007.)

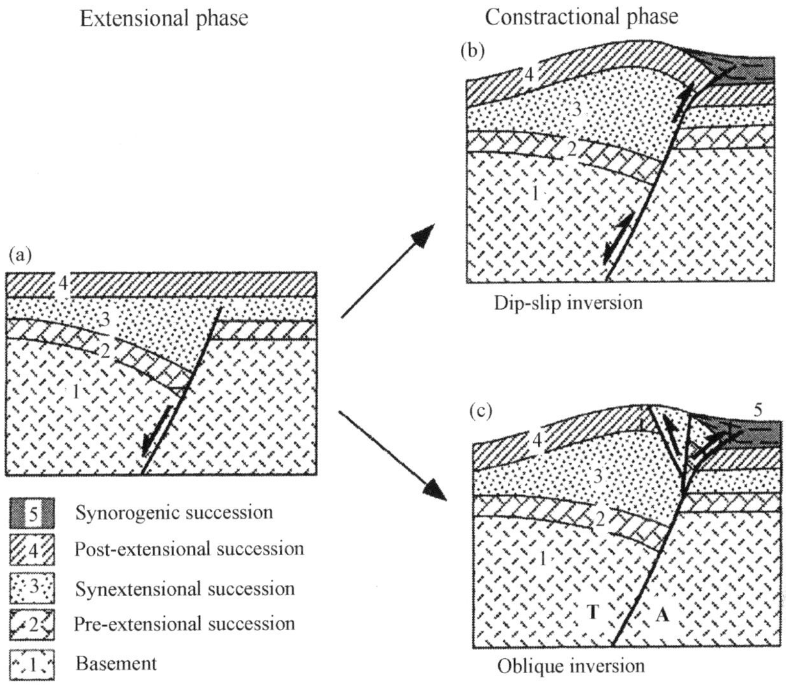


**Fig. 22.33** Scheme for fault inversion. (a) Initial stratal geometry in the syn-rift–post-rift extensional phase. (b) Location of syncontractural deposits during dip-slip inversion. (c) Ditto for oblique slip inversion with a positive flower structure. (After Bayona & Lawton, 2003.)

but only a condensed representation of steady-state conditions.

## 22.10 Strike-slip basins

Major strike-slip faults are broadly linear features. Some penetrate the whole lithosphere as plate-bounding transform faults like the San Andreas Fault of California. Others, the majority, penetrate the brittle crust as transcurrent boundaries to sliding and rotating crustal blocks driven by basal traction against a uniformly deforming lower lithosphere (section 10.3). These latter are common within impacting continental terrains, as in the North Anatolian Fault of Turkey or the Pannonian Basin margin to the Vienna Basin. They are often *en echelon*-segmented and may show marked departures from linearity along-strike. Major basins usually result from two 'pull-apart' mechanisms (Fig. 22.28).

- Lateral *en echelon stepping* or *jogging*, creating rhomboid-shaped areas of subsidence between the steps (*dilational jogs*) as in the Dead Sea basin of Jordan and Israel.
- Along-strike interaction between different parts of fault bends results in a vertical component of slip, leading on the one hand to thrust faults and upthrust source-blocks in restraining bends or to normal faults and basin opening along releasing bends. The latter basins feature fault-calving tendencies along normal to oblique-slip antithetic faults at major releasing bends (Fig. 22.28), as first proposed for the Ridge Basin of California. In vertical section, restraining bends show positive or reverse *flower structure* arrangement of minor faults that may sole down to the master strike-slip structure. Faults at releasing bends show negative or normal flower structures.

Because major strike-slip faults like the North Anatolian, Dead Sea and San Andreas often define plate or crustal block boundaries, associated basins show rapid (up to several millimetres per year) asymmetric subsidence and complementary uplifting highs with rapidly rotating antithetic faults. Sediment sourced

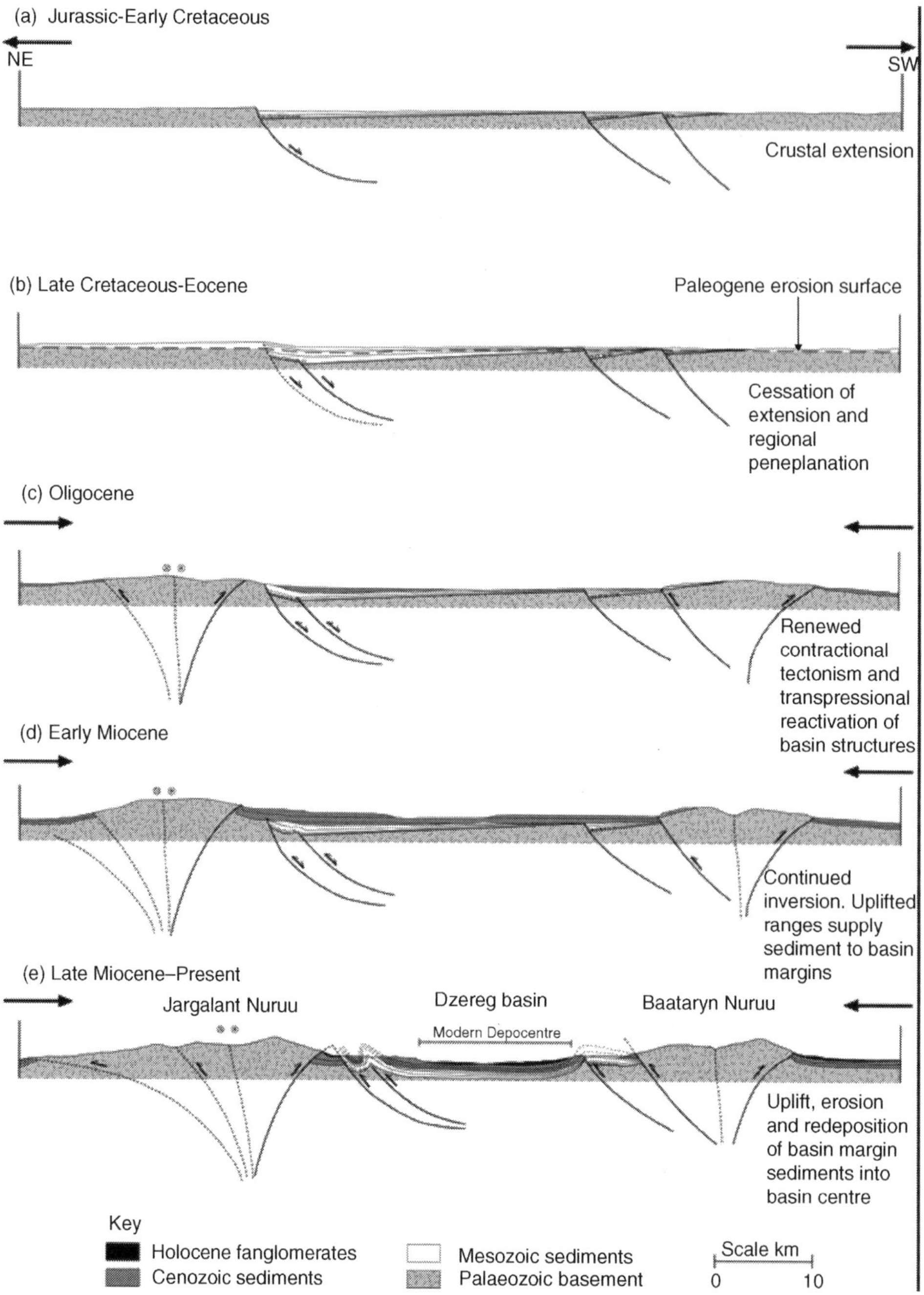


**Fig. 22.34** Cross-sections to show the interpreted evolution of the Dzereg basin, Mongolia, from the Jurassic to the present day. (a) Jurassic–early Cretaceous regional extension and syn-rift development. (b) Paleogene tectonic quiescence and regional erosion. (c) Oligocene onset of transpression and uplift to northeast. (d) & (e) Transpression continues to present day with deformation stepping well into the basin from both margins leading to uplift, erosion and recycling of Mesozoic and Cenozoic sediment downslope as the basin closes. (After Howard *et al*., 2003.)

from local transpressional highs infill basins by transverse fans feeding lakes or axial, fault-parallel rivers and submarine fans (Figs. 22.29-22.31). The strike-slip movement itself creates along-strike offset of source areas, changing sediment provenance with time at a point and the occurrence of laterally migrating centres of deposition along the length of the elongating basin.

*Transtensional fault-termination basins* occur at the ends of large strike-slip faults where normal- or oblique-slip faults splay off to terminate the deformation field. They have characteristics of both dip-slip half-graben and pull-apart basins along bends in strike-slip faults (Fig 22.32). Active examples are in the southern Gulf of California, associated with pre-spreading strike-slip faulting and in the northern Aegean Sea at the termination to the North Anatolian strike-slip fault. Basin tectonic architecture comprises faulting with various combinations of oblique-, strike- and normal-slip, with common steps and bends, buttress unconformities between fault steps and beyond the ends of faults. Sedimentary architecture in the Gulf of California involves terrestrial sediments passing up and away from faults into marine sediments and common Gilbert deltas (Fig. 22.32). Subsidiary intrabasinal fault blocks and rapid growth and death of fault-tip splays result in complex lateral facies relations, dramatic termination at basin margins as faults reorganize and propagate within an overall short basin history (few million years).

## 22.11 A note on basin inversion

Inversion is a term used in basin analysis for the process whereby a subsiding sedimentary basin changes its behaviour to one of uplift—the sediment sink becomes a sediment source. It is also a kinematic term used in structural geology for a change of slip vector from thrust to normal faulting, or vice versa. Of course it may be that the structural vector change is actually causing the basinal change, but this may not necessarily be the case, e.g. compression of a formerly thermal-sag cratonic basin or uplift due a magmatic underplating event. The fundamental reason for, and the timing of, changes in basin and/or fault behaviour is of great value to palaeotectonic reconstructions. Sedimentary evidence is obviously the key in reconstructing inversion history, because reactivated or newly initiated faults control erosion, stratigraphy and depositional environments in distinctive ways during tectonic evolution. However, field analysis of inversion may be rather difficult when contractual strains are large, as when extensional basins caught up in destructive margins or collision zones; patient and careful stratigraphic mapping and tectono-sedimentary insights are then called for.

Good examples of inverted extensional basins occur in the McCoy–Bisbee–Chihuahua–Sabinas basin of southern Arizona, southwest New Mexico and Mexico. This Jurassic–Cretaceous extensional province probably formed during a lengthy period of steep-slab subduction of Pacific plate under the southwest USA and Mexico that caused back-arc extension. As the slab shallowed in the late Cretaceous to Paleogene, compressional tectonics took over and the volcanic arc advanced eastwards into the area, both effects causing surface uplift, with inversion of normal to reverse faults (Fig. 22.33). Also recommended is a study of inverted rifts in the Columbian Andes. A particularly interesting scenario occurs when a basin changes from being influenced by purely extensional tectonics to a transpressive regime with time. Such an event is recorded from the Dzereg basin, Mongolia (Fig. 22.34).

## Further reading

### *General*

Texts and reference tomes: the excellent Allen & Allen (2004); the now-dated but fine Busby & Ingersoll (1995); many thematic issues of the journal *Basin Research*.

### *Specific*

Rifts: ronceptual sedimentary models—Gawthorpe & Leeder (2000); 3D all-singing numerical models—Gawthorpe *et al.* (2003) and Ritchie et al. (2004a,b, 1995, 1996); sequence stratigraphy—Dart *et al.* (1994) and Gawthorpe *et al.* (1994); fault growth and amalgamation by Peacock & Sanderson (1991, 1994), Jackson & Leeder (1994), Trudgill & Cartwright (1994), Cartwright *et al.* 1995, 1996), Gawthorpe *et al.* (1997), Gupta *et al.* 1998, 1999) and Sharp *et al.* (2000a,b). New Zealand example—Lamarche *et al.* (2000) and Taylor *et al.* (2004); accomodation zones and volcanism—Ebinger *et al.* (1987) and Mack & Seager (1995). rift margin case histories: Suez Rift—Jackson *et al.* (2002), Young *et al.* (2002) and Carr *et al.* (2003); Loreto basin—Dorsey *et al.* (1997) and Mortimer *et al.* (2005); Gulf of Corinth—Bell *et al.* (2009); Idaho—Janecke *et al.* (1999); Rio Grande, New Mexico— Mack & Seager (1990), Mack & Madoff (2005), Mack *et al.* (2006); Central Europe— Rajchl *et al.* (2009).

Supradetachment extensional basins are discussed by Friedmann & Burbank (1995), Dorsey & Martín-Barajas (1999) and Janecke *et al.* (1999); continental terraces and rises by Bond *et al.* (1995); trenches and trench-slope basins by Underwood & Moore (1995), Mountney & Westbrook (1996), Henstock *et al.* (2006), Bailleul *et al.* (2007) and Kopp *et al.* (2008).

Fore-arc basins are considered by Dickinson (1995), Underwood *et al.* (1995), Ingersoll (1979), Postma *et al.* (1993b), Clift *et al.* (2000), Fildani *et al.* (2008) and Paquet *et al.* (2009); intra-arc basins by Smith and Landis (1995), Underwood *et al.* (1995), Clift *et al.* 1998, 2005) and Draut & Clift (2006); back-arc basins by Carey & Sigurdsson (1984), Marsaglia *et al.* (1995), Clift (1995), Taylor *et al.* (1991) and Klaus *et al.* (1992); foreland/thrust-related basins, pro vs. retro by Dickinson (1975), Naylor & Sinclair (2008), Miall (1995) and Castle (2001).

Retro-arc (broken) forelands are discussed by Jordan (1995), DeCelles & Giles (1996) and Horton & DeCelles (1997); forebulges by Allen *et al.* (1991), Crampton & Allen (1995), Sinclair (1997) and White *et al.* (2002); piggybacks by Ori & Friend (1984).

Thrust and basin evolution are studied by Beaumont (1981)Beaumont *et al.* 1992, 2000, 2001), Lawton (1986), Talling *et al.* (1994), Burbank *et al.* (1992), Burbank & Raynolds (1988), Schlunegger *et al.* (1997, 1988), Kempf & Pfiffner (2004), DeCelles (2004), Naylor & Sinclair (2008), Clevis *et al.* (2004a,b) and Lu *et al.* (2009); unroofing signal, drainage and sediment flux by Koons (1989a,b), Garcia-Castellanos (2002), Jones *et al.* (2004), Clift & Bluszajn (2005), Clift *et al.* (2008) and Mancin *et al.* (2009).

Himalayan foreland basin drainage, sedimentation and thrust evolution are considered by Burbank & Beck (1991), Burbank *et al.* (1996), Gupta (1997), Horton & DeCelles (2001), DeCelles et al. (1998a,b, 2001), Najman *et al.* (2004) and Clift *et al.* (2008); growth and blind folds by Anadón *et al.* (1986), DeCelles (1988), De Boer *et al.* (1991), Jackson *et al.* (1996), Gupta (1997), Burbank *et al.* (1999) and Martín-Martín (2001); strike-slip basins by Crowell (1974a,b), Nilsen & Sylvester (1995), Dorsey *et al.* (1997), Umhoefer *et al.* (2007); basin inversion by Cooper & Williams (1989), Hayward & Graham (1989), Butler *et al.* (1997), Howard *et al.* (2003), Bayona & Lawton (2003) and Mora *et al.* (2009).

## Part 8

# TOPICS: SEDIMENT SOLUTIONS TO INTERDISCIPLINARY PROBLEMS

*The only sound was the wind over the dunes and the lapping water down there: it was in a way the world at the very beginning – the elements alone, and starlight.*

Patrick O'Brian, from *The Ionian Mission*, Collins 1981 (Stephen Maturin just landed in France)

### Introduction

The object of this final section is to make the interested reader aware of current research-led interdisciplinary problems which are being investigated using field and laboratory data gained from sediments and sedimentary rocks. The choice of topics is intentionally broad, many are controversial and most require further work. Though not all may suit all reader's areas of interest it is hoped that they at least show how an active and vigorous subject can adapt to problem-solving in cognate disciplines. It is hoped they may be of some use in seminar and class discussions. Some further reading for each topic is given at the end of the chapter. These should be studied carefully since the author's short précis is nothing more than an introductory (and sometimes probably inadequate) summary of complex arguments.

### Topic starters

#### *Evaporite fluid inclusions, global tectonics and seawater composition*

It is known from experiments that elevated Mg concentration in seawater inhibits calcite precipitation. In the present low-latitude global ocean this leads to predominant aragonite precipitation. However, periods of high-Mg and low-Mg calcite precipitation are known over the past 400 Myr. Is there a ready explanation for this?

*Links: Chapters 2, 10, 20.*

#### *Banded ironstone formations (BIFS), rise of cyanobacteria, Great Oxidation Event (GOE) and secular change in global tectonics*

Banded Ironstone Formations are 'nonuniformitarian' sediments, that is to say they cannot be seen forming today. Their abundance in the Archaean and Proterozoic points to different oxidation states then in both atmosphere and ocean, whilst their demise after 2.6 Ga indicates increasing atmospheric oxygen. But what happened at 2.6 Ga to trigger this change?

*Links: Chapters 2, 10.*

*Tibetan plateau uplift; palaeoaltimetry and monsoon intensity*

Timing is everything. Onset of Tibetan–Himalayan uplift is said to have led to Southeast Asian monsoon intensification, increased continental weathering and orogenic exhumation. Decisive application of stable isotope geochemistry to palaeoaltimetry and sediment proxy evidence to monsoon intensity yield revealing results. Recent (2010) modelling also suggests the Plateau has no influence on the monsoon. Just when was Tibet raised, and what were the consequences?

*Links: Chapters 1, 2, 10, 22.*

*Colorado plateau uplift and Grand Canyon incision dated by speleothem carbonate*

The Grand Canyon is probably the world's most famous landform—uplift of the Colorado Plateau is widely thought to have triggered deep incision by the Colorado River to form it. Upper mantle structure indicates that the uplift is due to upwelling warm asthenosphere under the plateau. Can the sedimentary record be used to accurately date the onset of this tectonic event?

*Links: Chapters 2, 10, 22.*

*River channels and large-scale regional tilting*

Rivers are sensitive spirit levels and may respond to tectonic forcing of gradient by systematic migration. Traces of migration in the form of abandoned channels are sometimes preserved on floodplains. Can this reveal coherent regional trends in crustal tilting, and how might such trends come about?

*Links: Chapters 10, 11, 22.*

*Regional drainage reversal*

Tectonic uplift can cause rivers to reverse course, but the timing and true origins of such reversals are often poorly constrained. Can syn-rift sedimentary and volcanic successions help constrain events that lead to reversal?

*Links: Chapters 10, 11, 14, 22.*

*Tectonic sedimentology of foreland basins*

Fold/thrust mountain belts (aka *orogens*) are made by tectonics, but erosion and sedimentation also play important roles in orogen evolution, as revealed by numerical simulations. The efficacy of erosion, and its concentration on windward slopes is especially important. Yet do sediment budgets from real-world basins support such theoretical results?

*Links: Chapters 10, 11, 12, 22.*

*Lengthwise fault growth and fault amalgamation*

Earthquake seismology has shed much light on the dynamics and kinematics of active surface faulting, but because it is essentially operating in a fixed time slot it cannot shed light on how faults may have evolved over timespans of $10^3$–$10^7$ years. Can the sedimentary record preserved in basins adjacent to fault structures help with this problem?

*Links: Chapters 10, 11, 12, 16, 17, 22.*

*Rivers, basement uplifts and fault growth*

Many rift basins feature axial drainage with a river flowing down the rift axis parallel to bounding fault systems. Such rivers deposit well sorted sands and gravels that form exceptional aquifers and reservoirs once preserved in the subsurface. How does fault spacing and fault growth affect position of axial sediments?

*Links: Chapters 10, 11, 23.*

*Unsteady strain and the sedimentary response*

Another major issue in tectonics is the possibility of unsteady, i.e. time-variable, strain. This is assumed unlikely when large, plate-wide strain is considered because of the inertia induced by large-scale requirements of the mantle driving-flow. But,

locally, do sedimentary systems respond to unsteady strain and the periodicity of its occurrence?
*Links: Chapters 10, 17, 22.*

*Tectonics versus climate as depositional controls*

It is clear from the world around us that active tectonics creates landforms which are modified, but not eliminated, by climatically induced surface erosion. How have Quaternary sediments deposited adjacent to tectonically active highlands reacted to high-frequency climate change?
*Links: Chapters 10, 12, 22.*

*River equilibrium, incision and aggradation—away from the knee-jerk of tectonic explanation*

The concept of equilibrium has a long history in geomorphology. In this rather ideal *stasis*, stream channels neither aggrade by deposition nor degrade by erosion. In view of the often extreme climatic changes experienced since the Pliocene, how have rivers flowing through sedimentary basins reacted to such events?
*Links: Chapters 10, 11, 12, 22.*

*Integrated sedimentary systems: modelling tectonics, sediment yield and sea-level change*

Environmental variables such as tectonic uplift, subsidence, sediment yield and sea-level change give rise to development of rich and complex sedimentary systems. Numerical modelling comes into its own when trying to assess the influence of these competing variables. How do river and deltaic sedimentary environments adjust to these?
*Links: Chapters 10, 11, 17, 22.*

*Extraterrestrial sedimentology—atmospheric and water flows on Mars*

An astonishing series of discoveries erupted into the literature in the mid-2000s concerning the role of water, ice and hydrocarbons in sediment transport and chemical weathering on the planets Mars and Venus and the satellite Titan. Unprecedented images by the small Mars landers were the highlight. Here we concentrate on Martian aspects and ask the question 'Can we apply Earth-bound sedimentary principles to explain extraterrestrial surface processes?'
*Links: Parts 1–3, 5–6.*

*Reefs and speleothem as suborbital-scale tuners of the Pleistocene sea-level curve*

A major new development over the past few years has been the recognition that eustatic sea level may suddenly change by up to 15 m at suborbital frequencies, i.e. $< 10^4$ years. The evidence for these changes comes from carbonate rock organic build-ups deposited at or close to previous highstand sea level. What are the origins of such changes, and will future human generations living in the present interglacial witness such events first hand?
*Links: Chapters 2, 10, 20, 22.*

*Speleothem: rosetta stones for past climate*

The growth of stalactite speleothem produces a calcitic structure whose internal layers may be sampled for oxygen, carbon and hydrogen stable isotope and U-series analysis, leading to a calibrated growth history and evidence for past climatic variations. Collection of samples world-wide enables teleconnections to be made between different areas and other proxies such as those preserved in ocean sediments and ice cores.
*Links: Chapters 2, 10, 20, 21.*

# Chapter 23

# SEDIMENTS SOLVE WIDER INTERDISCIPLINARY PROBLEMS

*. . .they climbed steadily mile after mile, sometimes so steeply that Stephen dismounted and walked beside the mule; and steadily the landscape grew more mineral.*
*'I wish I had paid more attention to geology,' said Stephen, for on his right hand on the far side of the gorge the bare mountainside showed a great band of red, brilliant in the declining sun against the grey rock below and the black above. 'Would that be porphyry, at all?'*

Patrick O'Brian, from *The Wine-Dark Sea*, Collins 1993 (Stephen Maturin with Joselito in the Peruvian Andes)

## 23.1 Sediments, global tectonics and seawater composition

The discovery that major element oceanic composition, most significantly Mg/Ca ratio and $SO_4^{2-}$ concentration, has varied over geological time came indirectly from the sedimentary record of oolite carbonate mineralogy, fossil skeletal composition and evaporite abundance. This led in the 1990s to a description of past oceans as either 'calcite seas' or 'aragonite seas', depending on whether Mg concentrations were high enough to inhibit low-Mg calcite precipitation and let aragonite dominate (**Cookie** 7). Over the past decade the case has been substantially strengthened by precise determination of Mg/Ca ratios from marine Phanerozoic echinoderm tests, detailed considerations of the mineralogy of other phyla and, most importantly, directly from compositional analysis of fluid inclusions in marine halite evaporite deposits (Fig. 23.1). There can be no doubting the reality of these temporal changes, though the time series is somewhat incomplete at the moment due to the vagaries of global evaporite production and preservation. Recent analysis of the fossil record somewhat complicates the calcite- vs. aragonite-sea idea because there is not a particularly clear correlation with shell and framework mineralogy over Phanerozoic time—this presumably comes about because periods of extinction are independent of ambient-ocean major ion composition.

The sedimentary evidence for oceanic change has been correlated to the beat of global tectonics over geological time, by direct and indirect reasoning. The basic link requires large-scale changes in global or regional mid-ocean ridge (MOR) activity so that during periods of high oceanic lithosphere production there is also more ridge-related hydrothermal (shallow subsurface hot water) activity, both by high-temperature

*Sedimentology and Sedimentary Basins: From Turbulence to Tectonics*, 2nd edition. © Mike Leeder.
Published 2011 by Blackwell Publishing Ltd.

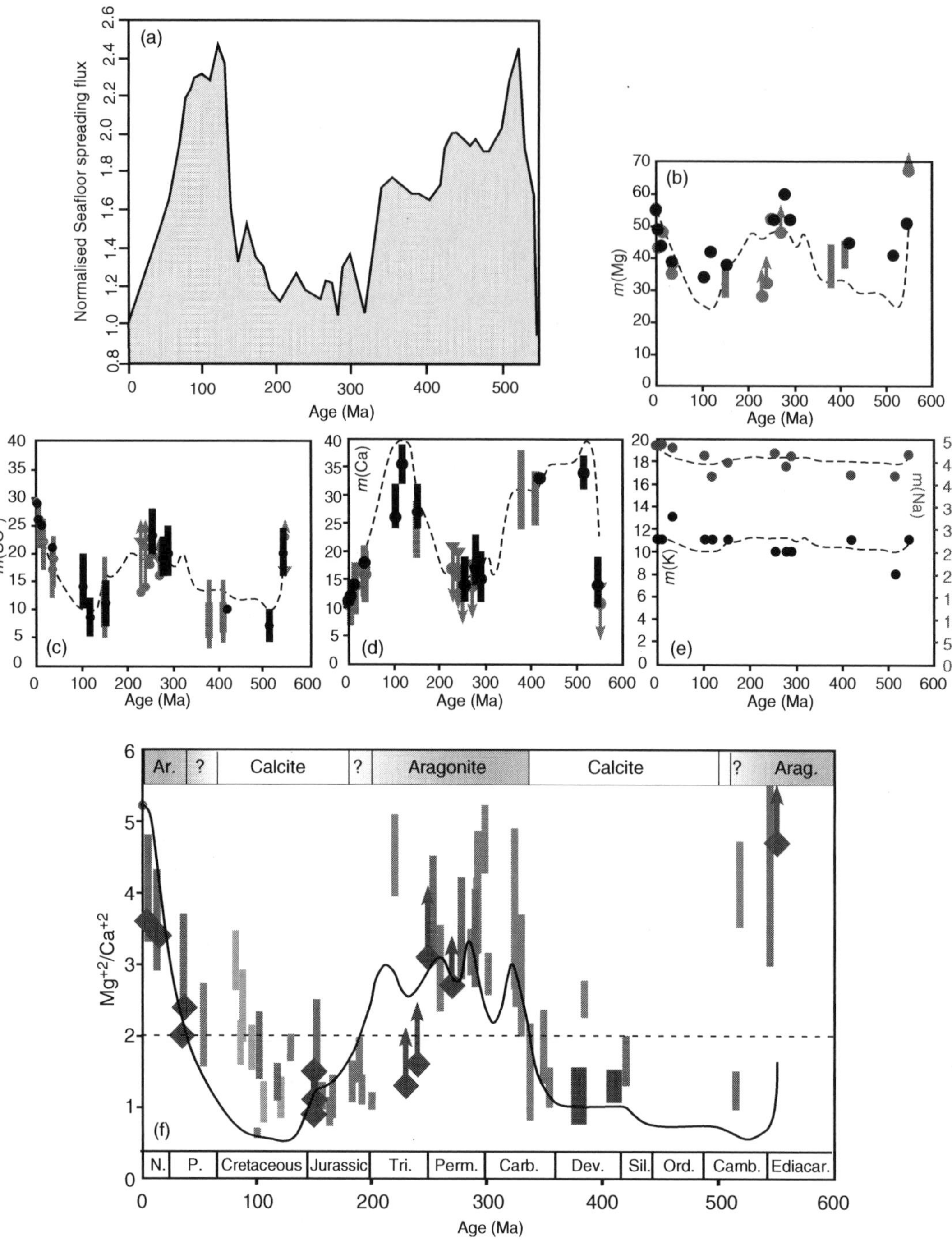
(a)
Normalised Seafloor spreading flux
Age (Ma)
(b)
m(Mg)
Age (Ma)
(c)
m(SO4)
Age (Ma)
(d)
m(Ca)
Age (Ma)
(e)
m(K)
m(Na)
Age (Ma)
(f)
Ar.
?
Calcite
?
Aragonite
Calcite
?
Arag.
Mg+2/Ca+2
N.
P.
Cretaceous
Jurassic
Tri.
Perm.
Carb.
Dev.
Sil.
Ord.
Camb.
Ediacar.
Age (Ma)

circulation at the ridge axis and low-temperature circulations at ridge flanks. This leads to more *sepiolite* (a Mg-rich clay mineral) formation as ocean crust basalts are transformed to altered clayey greenstone. The MOR thus acts as a huge *Mg-pump*, locking up more or less of the element as the rate of hydrothermal alteration varies, thus lowering or raising Mg/Ca ratios globally.

But is there independent evidence for secular changes in MOR activity over geological time? Computation of Mesozoic/Cenozoic seafloor spreading rates from oceanic crust magnetic anomalies and subduction rates (Fig. 23.1a) has proved controversial: previous and some recent evaluations find no clear trend over time whilst others support the notion. Either way, such inferences become impossible before the early Mesozoic because older ocean crust is not preserved. Instead, use is then made of the proxy sedimentary record of relative sea-level change, making the assumption that high MOR activity worldwide generates high global sea levels. Yet this seismic stratigraphic approach to the sedimentary record for *global* (eustatic) sea-level change has been fatally negligent in its technical analysis, chiefly in deconvolving eustatic from regional tectonic causes of the supposed relative changes.

The final step in this correlation of linked secular events comes from recognition that global spreading rate also correlates with alternating icehouse/greenhouse states, perhaps implying that more rapid mantle $CO_2$-degassing prompts the tipping point from state to state. However, there are other correlations to bear in mind here, notably the role of $CO_2$-drawdown due to enhanced silicate weathering during periods of mountain-building activity (**Cookie 5**). Either way there emerges a link between tectonics and the long-term C-cycle relevant to attempts to model past atmospheric $CO_2$ concentration.

The major role for the sedimentary record in the above arguments is revealing. It first provides *a priori* evidence from marine halite for global ocean compositional change over geological time, whose only rational explanation (the Mg-pump) implies fluctuating global rates of ocean crust formation. Seismic stratigraphy is then used to derive a global sea-level curve which is then used to independently infer spreading rates. The precariousness of some of the conclusions derived from such analyses seems obvious but that's as good as it gets in view of the myriad of assumptions made in the modelling.

## 23.2 Banded Iron Formations, rise of cyanobacteria and secular change in global tectonics

A great problem in sedimentology and sedimentary geochemistry has been the origin and significance of the most noteworthy nonuniformitarian rock type in the sedimentary record, Banded Ironstone Formations (BIFS). These Archaean to Proterozoic (ca 3.8–1.8 Ga) laminated chert–haematite sedimentary rocks (Fig. 23.2) have been the subject of much study and geochemical speculation in recent years as efforts have been made to understand the evolution of atmospheric composition. In particular, attention has focused on the rise of atmospheric $O_2$ content in what has been termed the 'Great Oxygenation Event' (GOE), thought by many to have begun around 2.4 Ga. Conventional explanations for the presence of haematite in BIFS so far back in the sedimentary record depend of the transport of ferrous iron from MOR hydrothermal

**Fig. 23.1** (a) Sea-floor production estimates normalized to a present-day value of unity for Phanerozoic time according to Gaffin (1987). Such curves are highly controversial (see text). (b)–(e) Filled circles with error bars indicate chemical composition of seawater (concentrations of Mg, $SO_4$, Ca, K + Na respectively) through the Phanerozoic back-calculated from measurements of fluid inclusions in marine halites (unpublished compilation courtesy of Bob Demicco, pers. comm., 2009). Symbols are data from Horita *et al.* (2002), Timofeeff (2000), Lowenstein *et al.* (2001, 2003, 2005), Timofeeff *et al.* (2000, 2006), Brennan, (2002), Brennan & Lowenstein (2002) and Brennan *et al.* (2004). Dashed lines show modelled results calculated from back estimates of mid-ocean ridge (MOR) seafloor creation rates, hydrothermal fluxes and constant continental runoff equal to present-day values (Demicco *et al.*, 2005). (f) Compilation of data for time series of *m*(Mg)/*m*(Ca) for the Phanerozoic (unpublished compilation courtesy of Bob Demicco, pers. comm., 2009) based on data from fluid inclusions in marine halite (Lowenstein *et al.*, 2001, Horita *et al.*, 2002) and skeletal compositions of marine invertebrates (data of Dickson, 2002, 2004, recalculated by Ries, 2004 and Steuber & Rauch, 2005). The line is from modelling by Demicco *et al.* (2005). An *m*(Mg)/*m*(Ca) of 2 (dashed line) is generally reckoned to separate so-called 'calcite seas' from 'aragonite seas'.

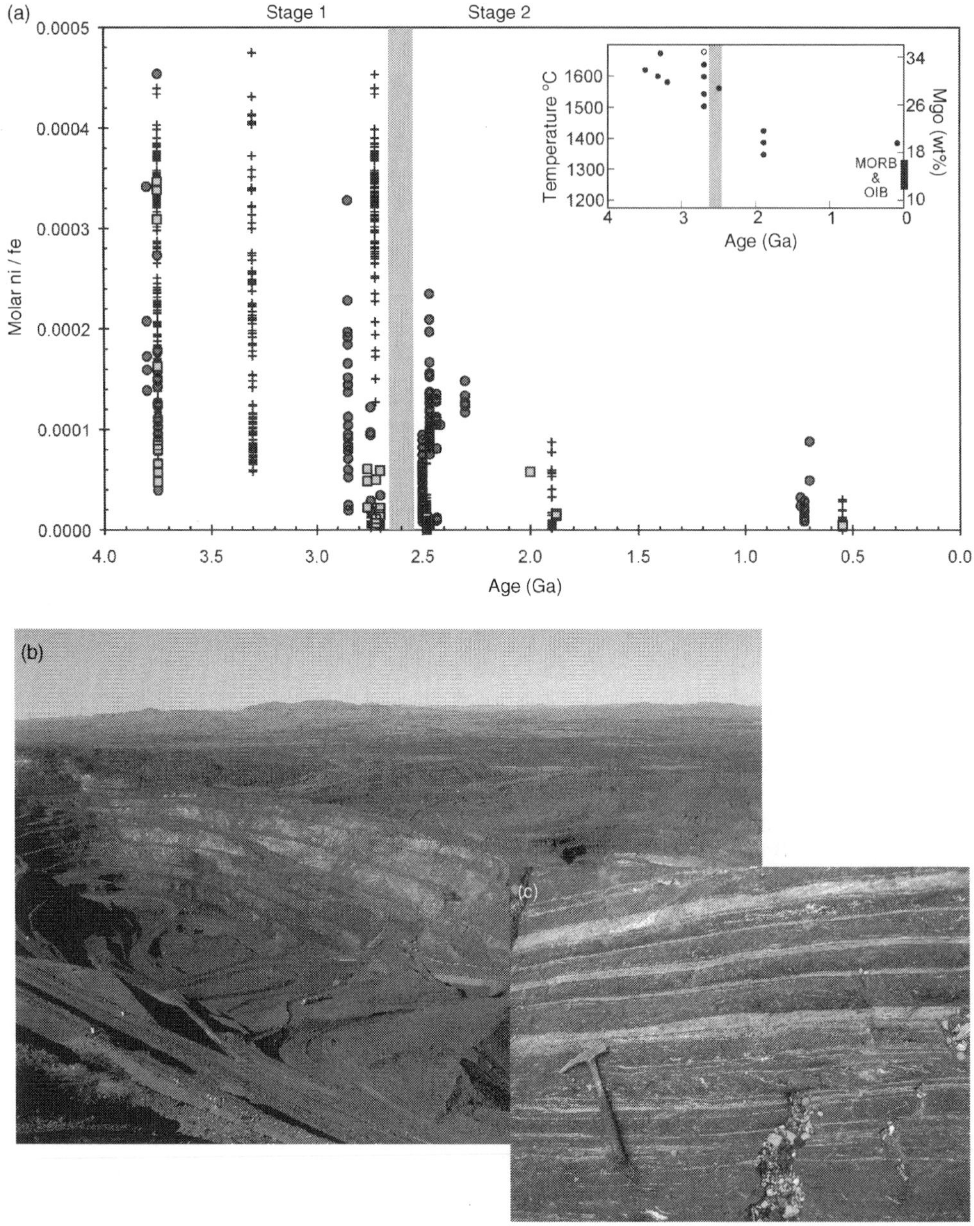


**Fig. 23.2** (a) Ranges of Ni/Fe mole ratios for 1214 measurements of worldwide BIF samples vs. age. (After Konhauser *et al.*, 2009; Figures courtesy of K. Konhauser.). Circles are data from literature, new analyses are squares (bulk analyses) and crosses (grain-by-grain laser ablation analyses). The greyscale bar marks the 2.4 Gyr transition from high to low Ni. Inset shows properties of parental komatiite liquids and denotes the change from high-temperature/high-Mg komatiites to low-temperature/low-Mg mid-ocean ridge basalts (MORB). (b) Banded Iron Formations being mined at Mount Tom Price, Western Australia. As sediments that salted directly out of primitive seawater, they captured a chemical record of the evolution of nutrients in Earth's early oceans. (Photo: Mark Barley, University of Western Australia.) (c) Close-up of BIF, Ontario, Canada. (Photo: Stefan Lalonde, University of Alberta.)

systems through an anoxic deep ocean to shallow waters bathed in ultraviolet light where it was oxidized by photochemical reactions aided by bacterial photosynthesis. The most popular theory for the demise of BIFs involves the rise of cyanobacteria, whose $O_2$-producing photosynthesic activities enabled, first, ocean surface water oxygenation, then atmospheric methane-oxidation and subsequent gradual atmospheric and deep ocean $O_2$-build-up. However, there is great controversy over the timing of such trends for there is also geochemical evidence from S-isotopes in sedimentary pyrite and the composition of palaeosols for appreciable oceanic and atmospheric oxygen well before the GOE.

A new aspect to the BIF/GOE story comes from recent evidence for the presence of primary haematite in thin (submillimetre) laminae within the 3.5 Gyr Marble Bar Chert Member of the Pilbara Craton in Western Australia. This unit is proposed as a relatively deep-water deposit formed in a volcanic depression in which reduced iron from hot hydrothermal fluids was oxidized by ambient seawater and precipitated. In support of this *early oxygen hypothesis* there is strong evidence from the sedimentary record for pre-3 Ga stromatolites and at one location for the existence of reef-like stromatolites as old as 3.4 Ga. Given the fragmentary nature of the Archaean record it would seem improbable that such communities were isolated and did not radiate widely into shallow marine niches worldwide where they may have facilitated the oxygen production necessary for primary haematite and BIF formation. However, a cautionary note is needed for it is not known whether the 3.4 Ga microbial buildups were constructed by oxygenic or anoxygenic cyanobacteria.

The latest twist to the fascinating BIFS/GOE problem makes an initially unexpected but nevertheless clear global tectonic link between MOR asthenospheric potential temperature, oceanic Ni-content and the rise of cyanobacteria at around 2.6 Ga. It adds more support to notions that Earth's oceanic and atmospheric elemental biogeochemical cycles are subject more to the vagaries of 'internal' solid earth controls rather than robust expressions of a Gaia-like driver. The argument comes from a temporal link between two distinct lines of evidence. First, analyses of BIFs reveal a steep decline in the molar Ni/Fe ratio about 2.6 Ga, the beginning of a 'nickel famine' that persists today (Fig. 23.2). Second, between about 2 and 2.6 Ga there was a marked decrease in Mg concentration in oceanic crustal lavas. This was due to secular mantle cooling: the resulting decreased eruption temperatures caused decreased production of Ni-rich oceanic ultrabasic komatiite MOR lavas at the expense of modern-type Ni-poor MOR tholeitic basalts. A consequence of cooling mantle was thus marked reduction in Ni-supply to the oceans from MOR hydrothermal alterations. Since abundant Ni is a key metal cofactor in several methanogenic bacterial enzymes, the subsequent demise of these at the expense of oxygen-producing cyanobacteria is explained.

## 23.3 Tibetan Plateau uplift; palaeoaltimetry and monsoon intensity

The first Sino-Western scientific cooperation in Tibet in the 1980s led to strenuous interdisciplinary earth science expeditions. The results of these suggested that the plateau accreted by docking of successive 'platelet' terranes during the Mesozoic, now seen conjoined along ophiolite-smeared sutures. The process culminated in the arrival of the fast-moving Indian plate in the Paleogene, ~50 Ma. Significantly, this major impact not only began to raise the Himalaya (though the exact timing of this is not known for certain) but also overprinted Mesozoic Tibetan tectonic architecture with thrust tectonics and associated thrust-front alluvial and lacustrine basins—high-level expressions of crustal shortening and thickening at depth. It was *felt* (but could not be proven) by structural geologists that such evidence for shortening and thickening could only explain 60% or so of > 5 km post-Cretaceous plateau uplift—thus the concept of post-thickening continental lithospheric *mantle delamination*, more recently horribly termed *mantle-dribbling*, was enthusiastically adopted. It was assumed that this remaining rapid uplift happened after shortening during Neogene times, causing a new E–W extensional tectonic regime to develop in south-central Tibet (Fig. 23.3).

During the 1990s the history of Tibetan–Himalayan uplift took on new and unexpected significance, with great interest in (i) the marked increase in secular marine $^{87}Sr/^{86}Sr$ ratio at about 40 Ma (Chapter 10) and (ii) whether the Urey–Walker silicate-weathering mechanism (**Cookie 5**) working on the crystalline rocks exhumed by Himalaya–Tibet denudation could have drawn down enough atmospheric $CO_2$ to initiate Cenozoic Icehouse Earth at about 34 Ma. Large-scale

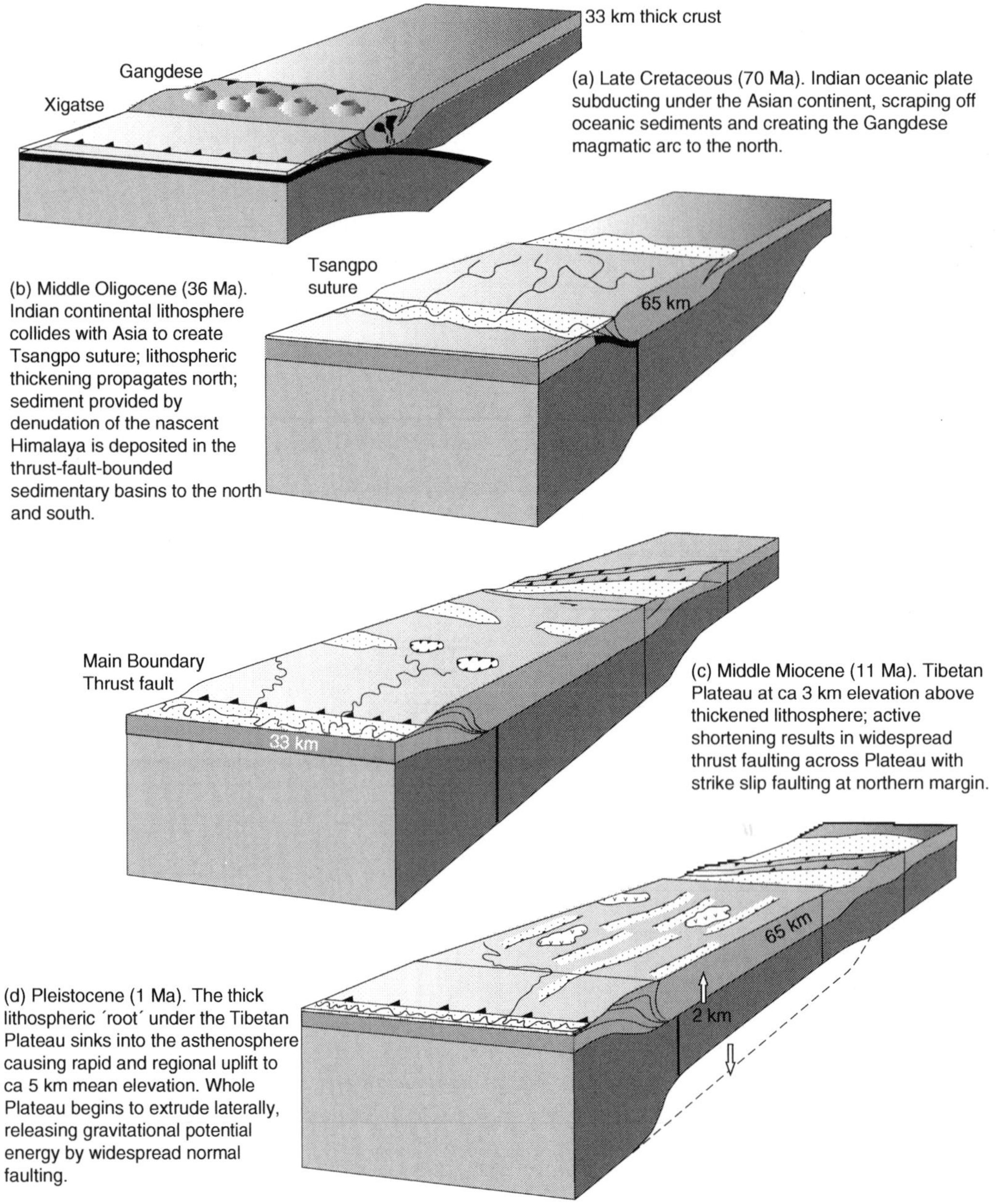

33 km thick crust
Gangdese
Xigatse
(a) Late Cretaceous (70 Ma). Indian oceanic plate subducting under the Asian continent, scraping off oceanic sediments and creating the Gangdese magmatic arc to the north.
Tsangpo suture
65 km
(b) Middle Oligocene (36 Ma). Indian continental lithosphere collides with Asia to create Tsangpo suture; lithospheric thickening propagates north; sediment provided by denudation of the nascent Himalaya is deposited in the thrust-fault-bounded sedimentary basins to the north and south.
Main Boundary Thrust fault
33 km
(c) Middle Miocene (11 Ma). Tibetan Plateau at ca 3 km elevation above thickened lithosphere; active shortening results in widespread thrust faulting across Plateau with strike slip faulting at northern margin.
65 km
2 km
(d) Pleistocene (1 Ma). The thick lithospheric 'root' under the Tibetan Plateau sinks into the asthenosphere causing rapid and regional uplift to ca 5 km mean elevation. Whole Plateau begins to extrude laterally, releasing gravitational potential energy by widespread normal faulting.


**Fig. 23.3** Continental shortening deformation on a grand scale. 'Traditional' view of the development of the collision of the Indian and Asian plates along the Himalayan mountain belt and the uplift and sideways collapse of the high Tibetan Plateau consequent upon lithospheric mantle thickening and subsequent thinning by delamination ('dripping'). (After Dewey *et al.*, 1988.)

climate changeologists preferred a young onset for Tibetan uplift ($< 10$ Ma) because it was considered responsible for amplifying or even initiating, the Indian monsoon. Many ingenious attempts were subsequently made to determine uplift timing, including timing of rock exhumation, time of initiation of normal faults and dyke swarms associated with southern Tibetan extension. However, these proxies were not necessarily reliable and/or precise indicators of the timing of uplift. Further, recent thermodynamic modelling studies (2010) reject any influence of the Tibetan Plateau on monsoon development or intensity. Rather it is the High Himalaya itself that acts to contain incoming moist monsoonal trade winds which are driven by intense heating over the Indo-Gangetic plains to convect deeply upwards, heating as condensation occurs.

Inspired use of the sedimentary archive in several Tibetan and Himalayan sedimentary basins has firmed-up conclusions regarding timing of uplift. Using calcic palaeosols and lake carbonates deposited in Paleogene thrust-front basins, new approaches to stable-isotope-based $O_2$ palaeoaltimetry in several independent studies have shown that late Eocene to early Oligocene sediments, independently dated by fossil evidence, show very strong depletion in $\delta^{18}O$ consistent with orographic fractionation at elevations $> 4$ km. This is supported by results from fossil leaf physiognomy in relation to atmospheric enthalpy which suggest that the elevation of one part of southern Tibet, the Namling basin at elevation 4.3–4.6 km, has changed little over the past 15 Myr. Changes in $\delta^{18}O$ from different sites across the plateau reveal a northward younging of uplift to $> 4$ km from about 40 Ma in the south to 20 Ma in the north (Fig. 23.4), results consistent with tectonic models for northward migration of lithospheric underthrusting and thickening causing the uplift rather than any involvement of plateau-wide lithospheric mantle root collapse. It has been objected that this conclusion fails to take account of a 500–1000 m decrease in mean surface elevation of the Lunpola basin, where one set of samples were taken, since extensional tectonics started to thin the Tibetan crust 8–15 Myr ago. However, these arguments ignore the facts that (i) the palaeoaltimetry data are minimum estimates, (ii) the magnitude of Tibetan extensional strain is *very* poorly constrained, (iii) there is very little seismological or geological evidence for neotectonic extension in central Tibet north of the Bangong suture line, placing Lunpola right on the northern margin of extension and (iv) the leaf physiognomy data noted above show little evidence for isostatic subsidence due to extension in southern Tibet. Additional support for early timing of uplift (perhaps too early) has come from pollen analysis in the northeast plateau which indicate high-altitude vegetation (*Picea*, spruce species) from about 38 Ma.

Early growth of the Tibetan Plateau also poses problems for tectonic models that require onset of significant Himalayan exhumation by erosionally focused denudation at the frontal Himalaya causing viscous crustal channel flow. This exhumation began in the early Miocene (ca 25 Ma), which leaves a time lag of ca 15 Myr between its onset and previous regional plateau uplift: there was obviously some decoupling between Tibetan plateau and Himalayan uplift. It is also noteworthy that explanations of northern plateau margin uplift and lateral growth due to flux of lower crustal mass by viscous channel flow are unnecessary following recognition that the great 2008 Wenchuan ($M = 7.9$) earthquake took place along a major thrust fault bounding the eastern extremity of the Tibetan Plateau along the Longmen Shan range. Here, crustal shortening, topographic relief and geological structure are all correlated along the thrusted range front that borders the Chengdu basin, in direct contradiction to predictions made by the viscous channel flow scheme.

Finally, an interesting take on the intriguing link between tectonic uplift, exhumation and monsoon intensity from the oceanic sedimentary record of marginal basins around Southeast Asia is provided by sedimentary analysis of deep drilling at the Pearl River site. This is outwith the area of high Himalayan–Tibetan relief and is located where the sediment source might be expected to track monsoonal development only. Chemical weathering rates were calculated from geochemical and mineralogical indices such as the Chemical Index of Alteration, the mineralogical ratio of chlorite/(chlorite + haematite + goethite) and the K/Al ratio. Results (Fig. 23.5) show that intensity of chemical weathering peaks markedly in the Middle Miocene (10–15 Ma) and declines steadily to about the late Pliocene before increasing into the Pleistocene. The Neogene trend is attributed to a middle Miocene peak in monsoonal intensity that subsequently declines. For sites like the Indus fan whose hinterland includes the Himalaya a parallel trend of post-middle Miocene decline in sedimentation rates is revealed,

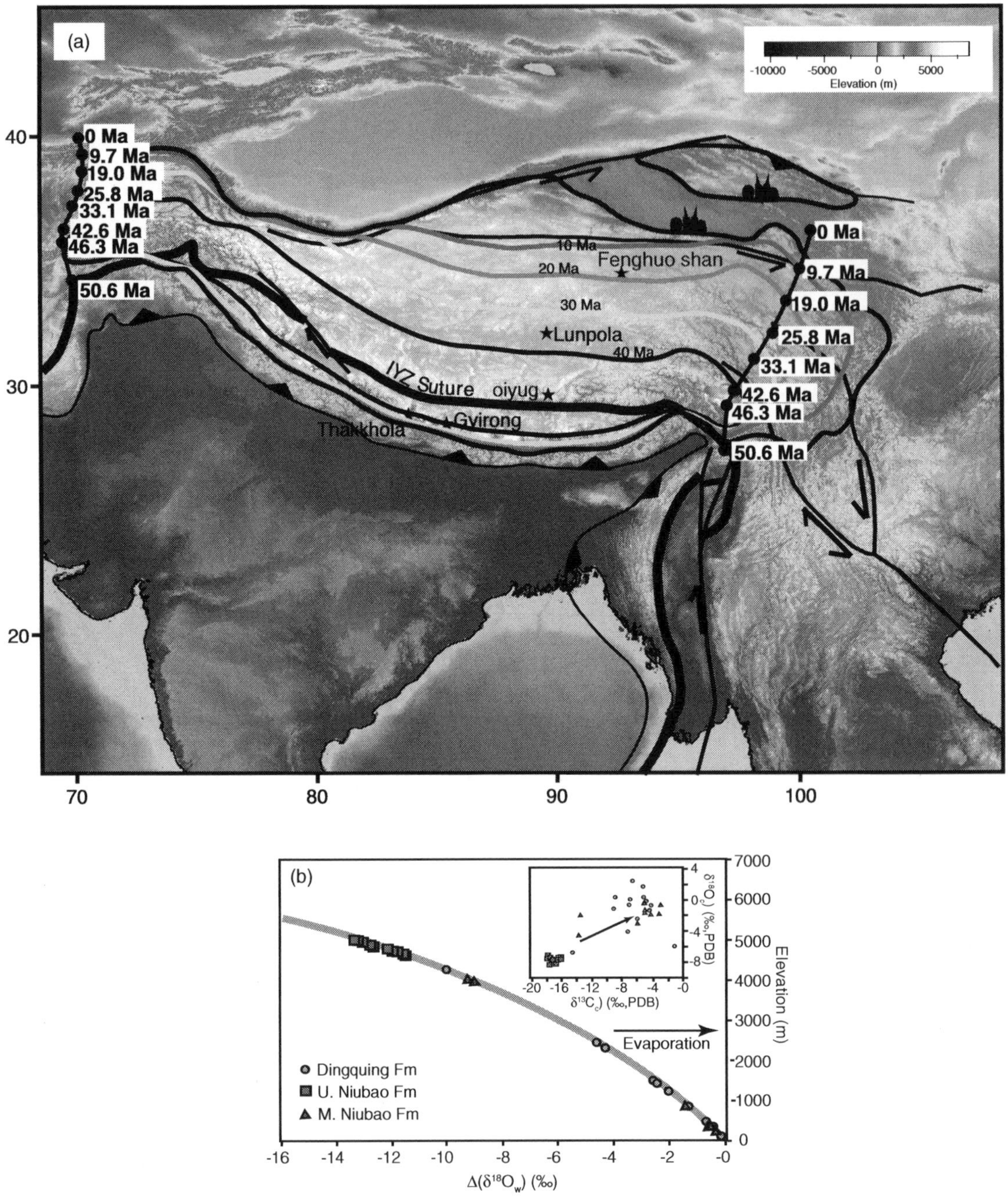


**Fig. 23.4** (a) Elevation history of the Tibetan Plateau established from oxygen isotope analysis in five sedimentary basins. Solid lines show the timing of northward growth of elevation to 4 km. (b) Graph to show $\Delta\delta^{18}O_w$ versus predicted elevations of carbonate rock-derived estimates of local surface and soil waters from the Lunpola basin, central Tibet. Model curve in greyscale. The arrow indicates the effects of evaporation on lake carbonates from the Dingquing Formation, with enrichment giving lower than expected elevations. (After Rowley & Currie, 2006, and references cited therein. Figures courtesy of D. Rowley.)

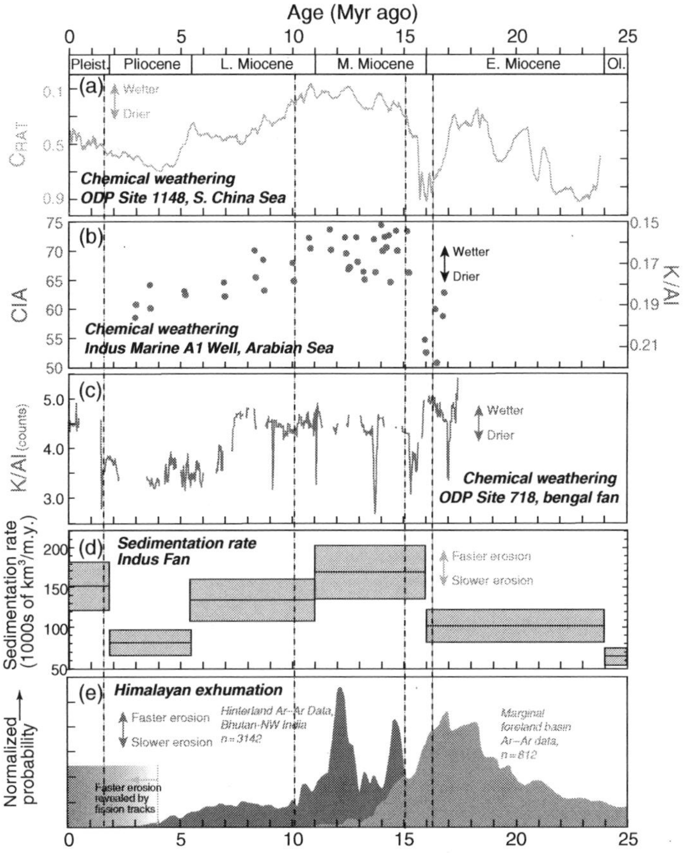


**Fig. 23.5** Correlation of Southeast Asian erosional and depositional histories with ODP Site 1148 monsoon intensity model. (a) The chemical weathering index $C_{RAT}$ (errors $\pm 0.1$). (b) K/Al ratios and CIA data from cuttings from Indus Marine well A-1. (c) K/Al ratios from whole-core XRF scanning from ODP Site 718. (d) Total sediment flux into the Indus fan—shaded bandwidths indicate compactional error limits. (e) Probability densities for $^{40}Ar/^{39}Ar$ muscovite dates from the Himalayan hinterland and proximal foreland basin. Vertical dashed lines across all figures indicate boundaries between periods of dominantly weak or strong summer monsoon. (After Clift *et al.*, 2008. Figure courtesy of P. Clift.)

until the well documented 'step-up' in Pleistocene rates. Having established a proxy trend in monsoonal intensity this is compared to thermochronometric data from muscovite $^{40}Ar/^{39}Ar$ dating of both detrital mica in foreland basin sediments and from bedrock micas from Himalayan hinterlands. The former suggest a gradual increase in Himalayan erosion from 24 to 18 Ma that seems to be independent of the middle Miocene peak in monsoonal intensity. This is not the case for the hinterland data set which seems to uncannily follow the monsoonal trends deduced from the various proxy methods noted above.

To conclude this section:

- beginning uplift of the vast Tibetan Plateau, and probably the Himalaya, began 10 Myr or so *before* the major Eocene–Oligocene transition to the state of Icehouse Earth, a result compatible with the Urey–Walker silicate-weathering hypothesis for global cooling;
- the South Asian monsoon developed to a peak intensity well *after* attainment of appreciable Tibetan elevation, precluding a causal link;
- thermodynamic climate models suggest the Tibetan Plateau has no effects on Southeast Asian monsoon intensity but topographic blocking by the High Himalaya does—it follows that timing of High

Himalayan uplift should correlate with monsoon amplification and increased sediment yields outboard.

## 23.4 Colorado Plateau uplift and Grand Canyon incision dated by speleothem carbonate

River incision through uplifting plateaux and the subsequent export of sediment to adjacent sedimentary basins is epitomized by the River Colorado and the Grand Canyon. This universal icon of natural grandeur is cut into the Colorado Plateau and the faulted margin of the adjacent Basin and Range to the west—its development led to progradation of the Colorado delta in the distant Gulf of California. Because erosion continually eliminates canyon palaeotopography, dateable *dipstick indicators* are needed that can indicate past valley-floor levels. An innovative approach is the use of cave *mammillary-type speleothem* deposits (Fig. 23.6) that mark former positions of the phreatic–meteoric water-table boundary in the Grand Canyon's walls as benchmarks of ancient levels of the river bed. Success of the approach depends entirely upon the key assumption that rates of groundwater table decline are equivalent to incision rates. Precise U–Pb dating of mammillary calcitic speleothem collected in the wider Grand Canyon area indicated that cutting may have began as long ago as 17 Ma and that a spatial migration of canyon formation and a change in incision rate subsequently occurred, with older, slower downcutting in the west and younger, faster downcutting in the east. Previously, incision rates were extrapolated from radiometric dating of young lava flows preserved on rock terraces throughout the canyon system. These rates were deemed too small to have cut the whole canyon in the ~6 million years required by indirect geological evidence for the initiation of the through-flowing Colorado River: either the rates were higher further back in time, or some canyon relief had been cut earlier—the discovery paper supported the latter alternative.

A later contribution has objected that only one speleothem radiometric age was as old as 17 Ma, with another at 7.5 Ma and the remaining eight dates < 3.8 Ma. Further, the two older dates were located some distance from the main canyon itself; the oldest from a Redwall Limestone cave where the water table is only 150–250 m below the dated sample and almost 1 km above the modern river and the other almost 40 km north of the canyon itself. It was also evident that the canyon acts as a seepage face for groundwater discharge from numerous perched water tables and that speleothem ages can only date incision where the trunk canyon intersects the main Redwall–Muav limestone karst system and aquifer. Two dates (3.9 Ma and 2.2 Ma) in the western canyon satisfy these conditions and give incision rates of 75 and 55 m/Myr respectively, much lower than rates in the eastern canyon of ca 250 m/Myr (Fig. 23.6). This supports one of the original workers chief conclusions. It was subsequently proposed that incision arose at ~6 Ma by a combination of (i) broad regional uplift due to asthenospheric upper mantle upwelling, (ii) initiation of the Grand Wash Fault along the western edge of the Colorado Plateau over which the River Colorado flowed causing a nickpoint to propagate eastwards and (iii) younger footwall uplift to the east of the west-throwing Hurrican–Torroweap normal fault system which served to increase incision by nickpoint development in the eastern Canyon.

## 23.5 River channels and large-scale regional tilting

Tectonic tilting can cause systematic migration of river channels and the formation of features such as asymmetric meander belts and preferred avulsion pathways. Although widely observed at individual rift-basin scales, recent digital-elevation-model (DEM) studies have found spectacular evidence for large-scale regional deflections. Meander-belt asymmetry and stepwise incision of unpaired terraces have been used to trace migration trends for three of the longest and highest discharge river systems on Earth: the Yenisei, Ob' and Irtysh rivers that traverse the West Siberian basin (Fig. 23.7). Lateral slopes, $S_L$, of tilted river valleys vary in gradient from 0.001 to 0.0001 whereas downstream valley gradients, $S_D$, are much lower. For example, the Ob has a typical $S_L/S_D$ ratio of 3. Regional incision and lateral shifts of the rivers demonstrate the effects of long wavelength surface tilting, directed away from the Urals and Central Asian mountains and towards the Siberian Craton. In the north of the basin, surface uplift of individual folds is recorded by local lateral drainage migration. The surface deformation is most plausibly

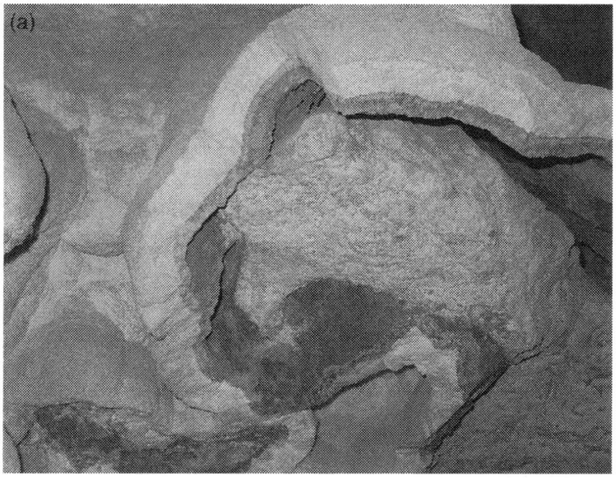

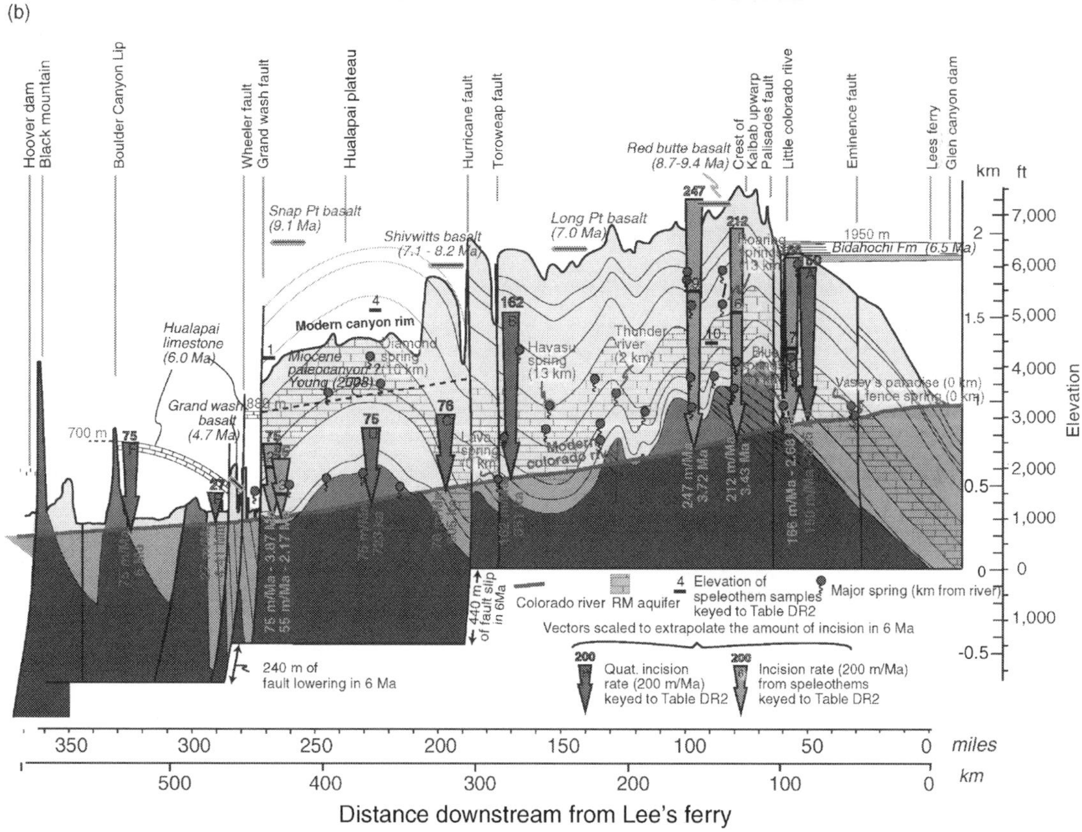

**Fig. 23.6** (a) Grand Canyon cave speleothem mammillaries as palaeowater table (dipstick) indicators from site 6 (Tsean Bida cave) dated by U–Pb to 3.42 Ma and indicating subsequent uplift at rates of ~0.2 mm/yr. (Polyak *et al.*, 2008. Photo courtesy of V. Polyak.) (b) Longitudinal Colorado River profile, modern canyon rim, stratigraphic units, spring elevations and incision points. (After Karlstrom *et al.*, 2008. Figure courtesy of K. Karlstrom.)

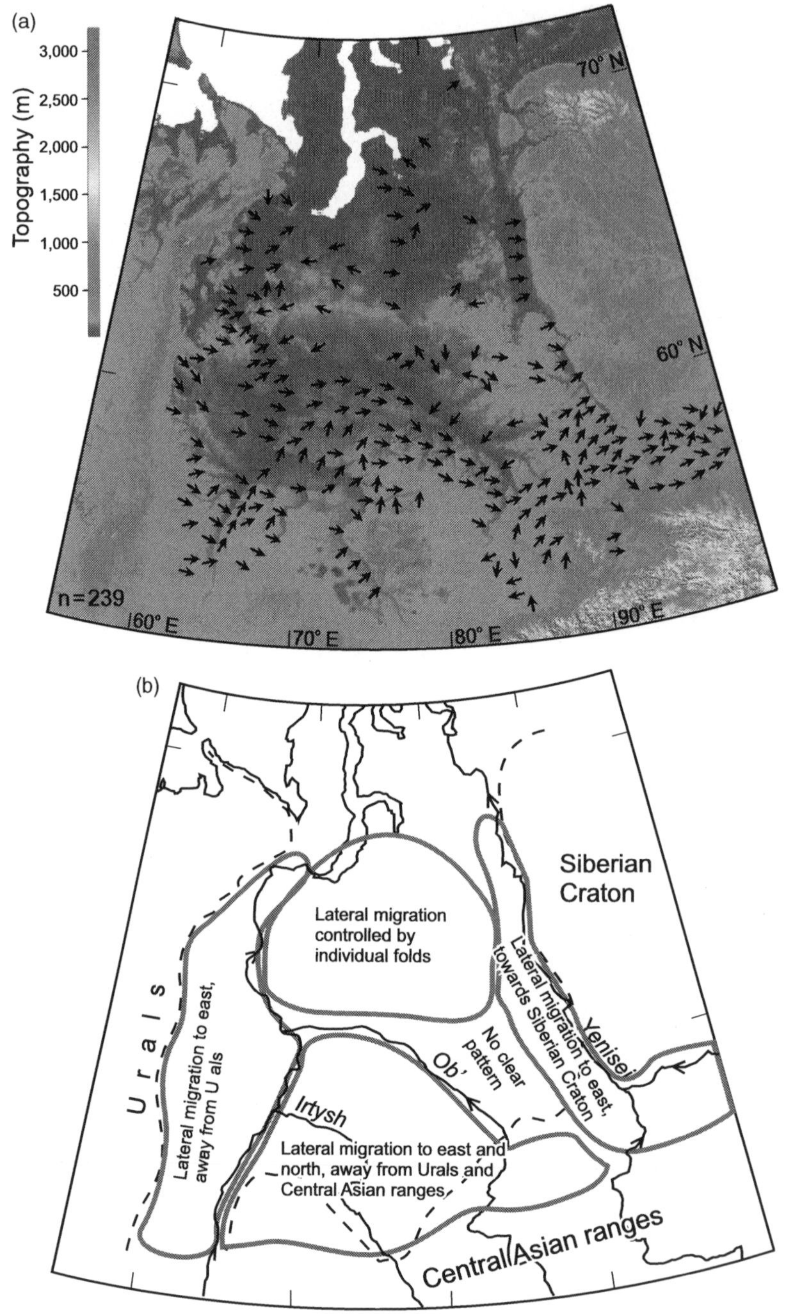


**Fig. 23.7** Summary lateral drainage migration in the West Siberian Basin. (a) Arrows give lateral channel migration directions based on geomorphic criteria within 10 000 $km^2$ bins, visible on satellite and/or digital-elevation-model (DEM) data and based on criteria discussed in the text. (b) Summary of tilt pattern in (a). (After Allen & Davies, 2007. Figures courtesy of M. Allen.)

caused by subtle faulting at depth, below the thick basin fill of Mesozoic and lower Cenozoic sediments. The active deformation may represent a far-field effect of the India-Eurasia collision, up to 1500 km north of the limit of major seismicity and mountain building. It may be an analogue for the formation of low-angle unconformities in terrestrial sedimentary basins on the periphery of other orogenic belts and whose origins have hitherto remained obscure.

## 23.6 Regional drainage reversal

A classic example of drainage reversal due to rift flank tilting was first highlighted by the legendary A. Holmes (Fig. 23.8) from the East African rift along the eastern side of the deformed horst containing Lake Victoria between the eastern and western branches of the rift in Uganda. It was envisaged by Holmes that late Pleistocene back-tilting occurred along Lake Albert boundary faults causing reversal of the drainage of the Rivers Katonga and Kafu and leading to the formation of Lake Kyoga.

A recent study illustrates the use of sedimentary environmental analysis to elucidate the timing and origin of drainage reversal in the Kivu–Nile portion of the rift in Rwanda. Here the modern sedimentary system involves fluvial sedimentation by the south-flowing River Nyabarongo that mysteriously cuts its way eastwards through gorges in the boundary fault footwall to join the Base River and then discharges into Lake Victoria (Fig. 23.9). The time evolution of the rift is revealed by detailed sedimentological studies of isolated Neogene sediments fortuitously preserved along the rift margin and in the adjacent footwall highlands. These show that a deep lake formed in the rift segment after extrusion of thick lavas dammed the northward flow of the palaeo-Nyabarongo. Subsequent overspill of the lake cut a spillpoint to allow the eastward flow seen today.

## 23.7 Sediment budgeting and modelling of foreland basins

Recent attempts to account for the influence of surface processes like erosion and deposition on the style of developing fold-thrust belts and foreland basins make use of a 2D (plane strain) elasto-visco-plastic rheology mechanical model incorporating fully dynamic coupling between mechanics and surface processes. The model is used to investigate how fold-thrust deformation and foreland basin development is influenced by the nondimensional parameter $k$, the ratio of surface process to deformation timescales. When $k \leq 1$, the rates of surface processes are so slow that a classic propagating fold-thrust belt develops, with wedge-top basins and a largely underfilled foreland flexural depression. Increasing $k$ causes deposition to shift progressively from the wedge top into the foreland basin centre, which deepens and may eventually become filled. Erosion causes widespread exhumation of the fold-thrust belt and reduced rates of frontal thrust propagation and possible attainment of a steady-state orogen width with change in the style and dynamics of deformation. Together, these effects indicate that erosion and sedimentation, rather than passively responding to tectonics, play an active and dynamic role in the development of fold-thrust belts and foreland basins. Because erosion and sediment flux is strongly dependent on climate the results of this study clear the way for a thorough attack on the role of climate and climate change on orogenic development.

Some modelling predictions can be checked when detailed sediment budgeting over geologic timescales can be attempted. Sediment sources and sinks are coupled by surface processes and their fluxes but there is no universal law of sediment yields to predict delivery rates to any basin. Modellers must either use a mean guesstimate or try to use a sensible empirical budget study with which to parameterize their machinations. Mass balance approaches have been used to quantify sediment budgets for some orogens but it can be difficult to estimate losses from the so-called oceanic sink in some of these, e.g. loss to the Makran accretionary prism from the Indus fan. So, sediment budgets provide some of the best constraints for inferring mountain palaeotopography and estimating denudation rates, but uncertainties are often large and/or not quantified.

The most recent and honest budgeting attempt applies to the Grande, Parapeti and Pilcomayo drainages of the central Andean fold-thrust belt and related deposits in the Chaco foreland of southern Bolivia (Fig. 23.10). By constraining source–sink dimensions, fluxes *and* their errors with topographic maps and satellite imagery, a hydrologically conditioned DEM was produced using reconstructions of a regional erosion surface, foreland sediment isopachs and estimated denudation rates. The budget rates are

(a) (b)

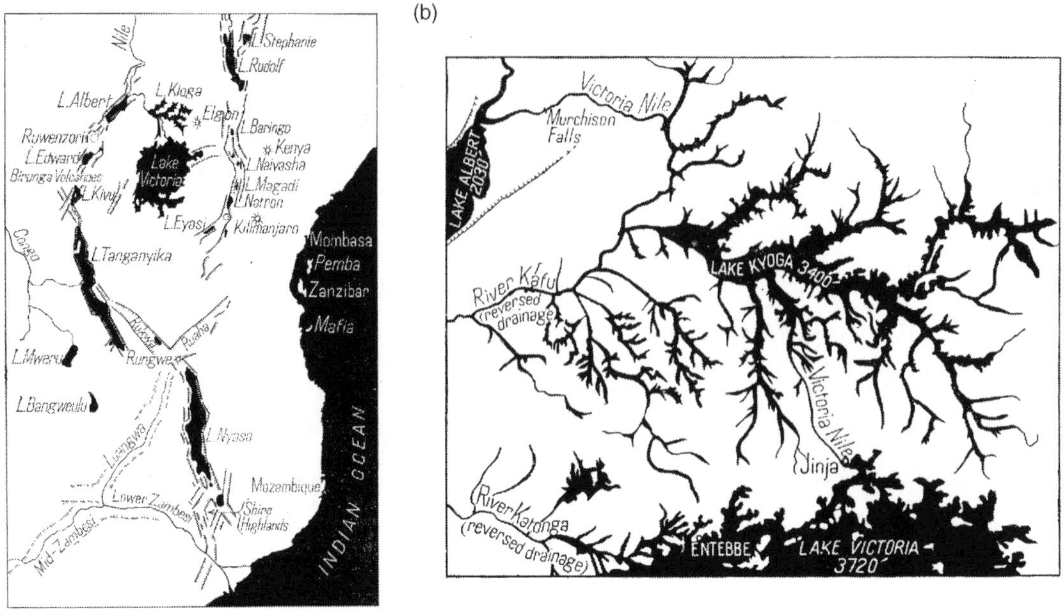


**Fig. 23.8** (a) Location map for Lake Kyoga in context of East African Rift. (b) Close-up of northern Lake Victoria and Lake Kyoga. (After Holmes, 1965.)

within the median range measured for the Neogene, but are up to two orders of magnitude higher than some observations. The predictions of the models discussed above are, to first-order, consistent with the central Andean fold-thrust belt, e.g. the arid, southern Chaco foredeep is basically underfilled and the wet, northern Beni foredeep is overfilled. It seems clear that more effort should be put into source-to-sink sediment budget analyses and associated interpretations must explicitly and quantitatively reconcile all available area, volume and rate observations because of their inherent imprecision and the potential for magnification when they are convolved.

## 23.8 Lengthwise growth and fault amalgamation

Quantification of fault growth rate can only come from carefully selected examples in which sedimentary 'event' horizons of known age can be traced from fault footwall to hangingwall. This is clearly impossible with most basement-bounded faults and so faults cutting sea or lake beds undergoing sedimentation must be used. The best example (Fig. 23.11) comes from offshore North Island, New Zealand, in the Whakatane Graben, Bay of Plenty, where an actively extending back-arc basin contains a thick sedimentary sequence broken by major active fault arrays. A dense network of seismic reflection data down to depths of almost 800 m collected across the Rangitaiki Fault indicated a linked and segmented normal fault as the dominant active structure in the graben. Total fault length is 20 km, with a displacement of up to $830 \pm 130$ m in the top 1.5 km of sediments. The fault has been actively growing for the past $1.34 \pm 0.51$ Myr and clearly developed from isolated fault segments with their own depocentres to a fully linked fault system. Initially, the dominant process of fault growth was tip propagation, with average and maximum displacement rates of $0.52 \pm 0.18$ and $0.72 \pm 0.23$ mm $yr^{-1}$ respectively. Interaction and linkage became more significant as the fault segments grew toward each other, resulting in the fault network becoming fully linked between 300 and 18 ka. Following fault segment linkages, the average displacement rate of the fault network increased by almost threefold to $1.41 \pm 0.31$ mm $yr^{-1}$, whilst the maximum displacement rate increased to $3.4 \pm 0.2$ mm $yr^{-1}$. This was the

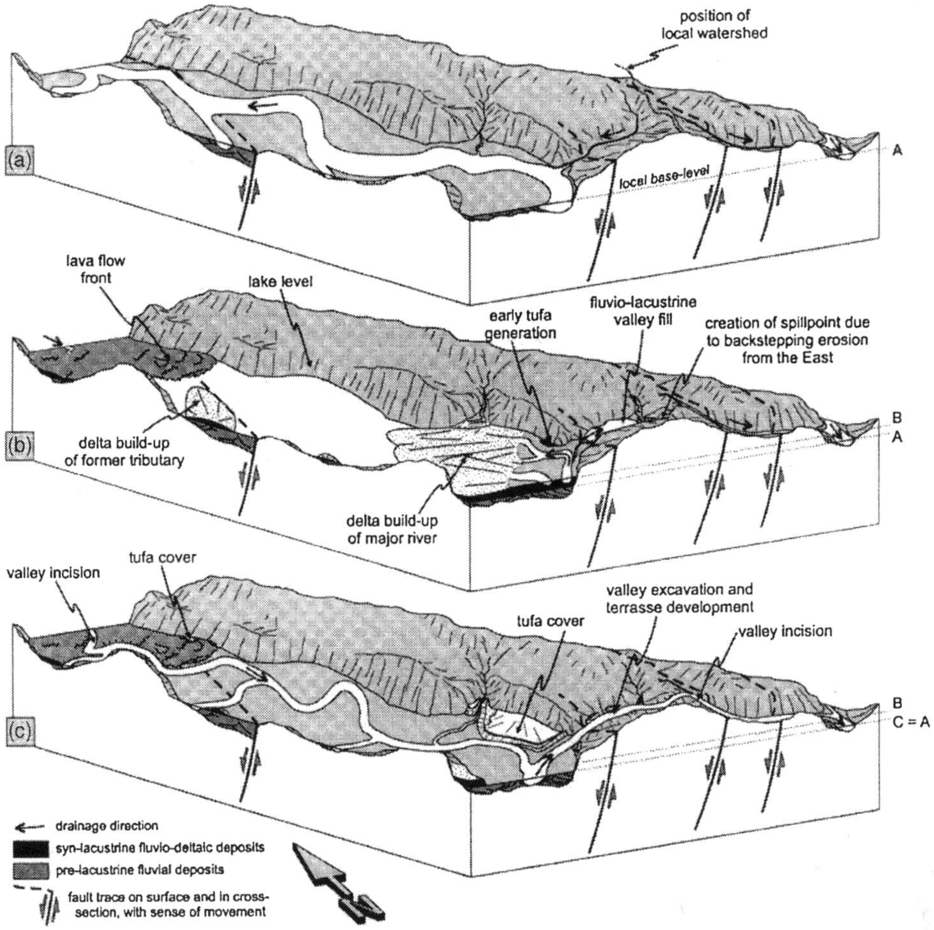


**Fig. 23.9** Reconstructions of rift axial drainage reversal and diversion, Kivu rift, East Africa. (a) The northward flowing palaeo-Nyabarongo and the southward flowing palaeo-Base rivers are largely fault-controlled pre-volcanic fluvial systems in low-gradient N–S-oriented valleys (marking the regional base level A). (b) Lava extrusion causes a thick flow to flood the palaeo-Nyabarongo River valley from downstream, damming the river flow. A major lacustrine delta develops from the upstream end of the lake, while former tributaries to the river develop small delta bodies. Contemporaneous backstepping erosion into the footwall to the east lowers the watershed, eventually resulting in a spillpoint (marking the local base level B for the western region) into the neighbouring palaeo-Base River system (still having a local base level A). (c) Readaption of the system to the pre-lacustrine local base-level (C = A) by enhanced erosion into the footwall and reworking of the lac ustrine deposits results in the deflection of the Nyabarongo River system to the east and a flow reversal within the former river valley north of Masangano. (After Holförster & Schmidt, 2007. Figures courtesy of F. Holförster.)

first time that the growth rate of unlinked fault segments had been resolved and shown to have been slower than in the subsequent linked fault system, exactly confirming previous conceptual, field and theoretical proposals.

Major information on fault growth and death and on the influence of structures on stratigraphic development has emerged from numerous studies in spectacular exposures of uplifted early syn-rift sediments in the Miocene of the Suez rift, Egypt. For example in the Hammam Faraun fault-block deposits of the Nukhul Formation are attributed to two linked depositional settings—offshore to shoreface and estuary. These were deposited during initial stages of

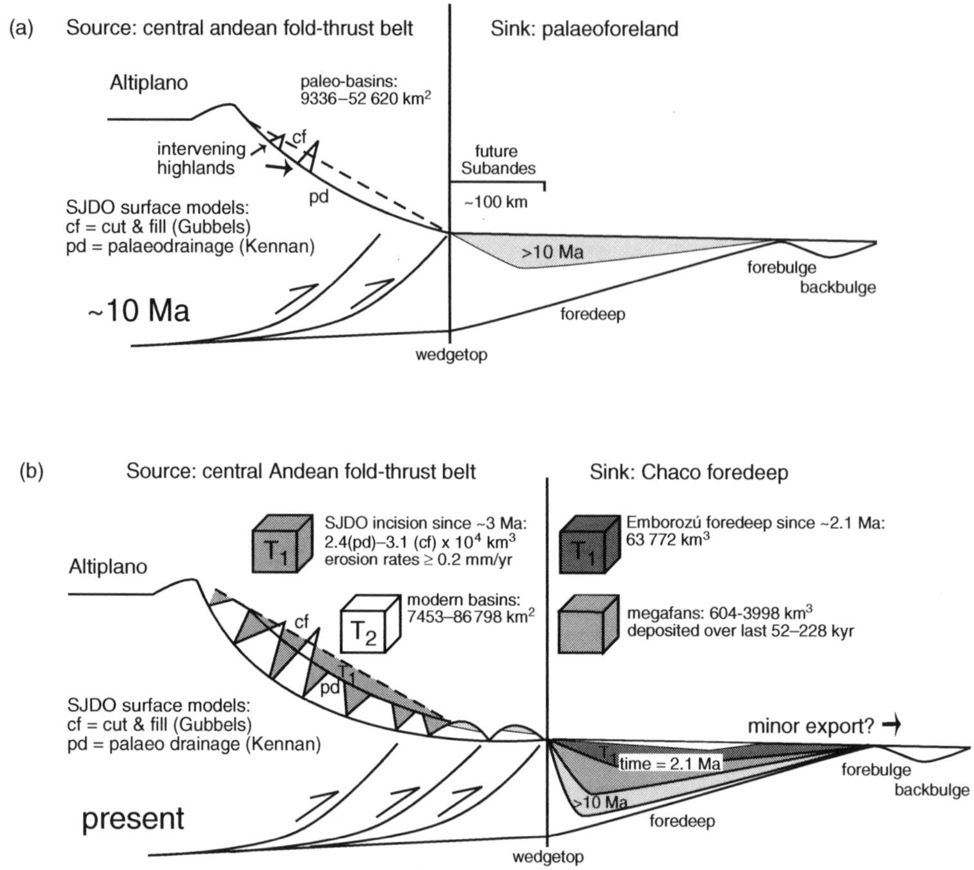


**Fig. 23.10** Schematic central Andean fold-thrust belt and Chaco foreland basin system sediment budget in cross-section at ~20°S. Symbols: cf, cut-and-fill model surface representation (dashed line in source); pd, palaeodrainage surface representation (solid concave up line in source). (a) Source and sink features during peak San Juan del Oro (SJDO) formation at ~10 Myr before incision. (b) Source and sink features after incision at present. Time slice $T_1$ is the Plio-Quaternary to recent (~2 or 3–0 Ma) represented by the volume eroded by incision into the SJDO surface in the source and deposited within the Emborozu Formation in the sink, respectively. Time slice $T_2$ is very recent time (230–0 ka) represented by the modern drainage areas and the megafan volumes in the foreland, respectively. Additional source region solid lines represent the modern maximum (jagged line) and minimum (lowest line) topography. (After Barnes & Heins, 2009. Figures courtesy of J. Barnes.)

rifting in hangingwall depocentres of early-formed propagating fault segments (Fig. 23.12). Stratigraphic development was strongly influenced by the evolving early-rift structure. Depocentres were narrow (2–5 km wide) and elongate (ca 10 km long) parallel to the strike of normal-fault segments. The elongate, fault-controlled geometry of the depocentres confined the bayhead delta and further enhanced tidal influence. Stratal geometry reflects deformation associated with low-relief growth folds and surface-breaking faults that, together, formed part of an evolving fault array. This basin configuration and associated Nukhul stratigraphy is markedly different to tectono-stratigraphic models for crustal-scale tilted fault blocks that are applicable from late stages of rifting.

Also in the Suez rift the sedimentological and architectural consequences of fault linkage across relay ramps can be clearly seen. The basin-bounding

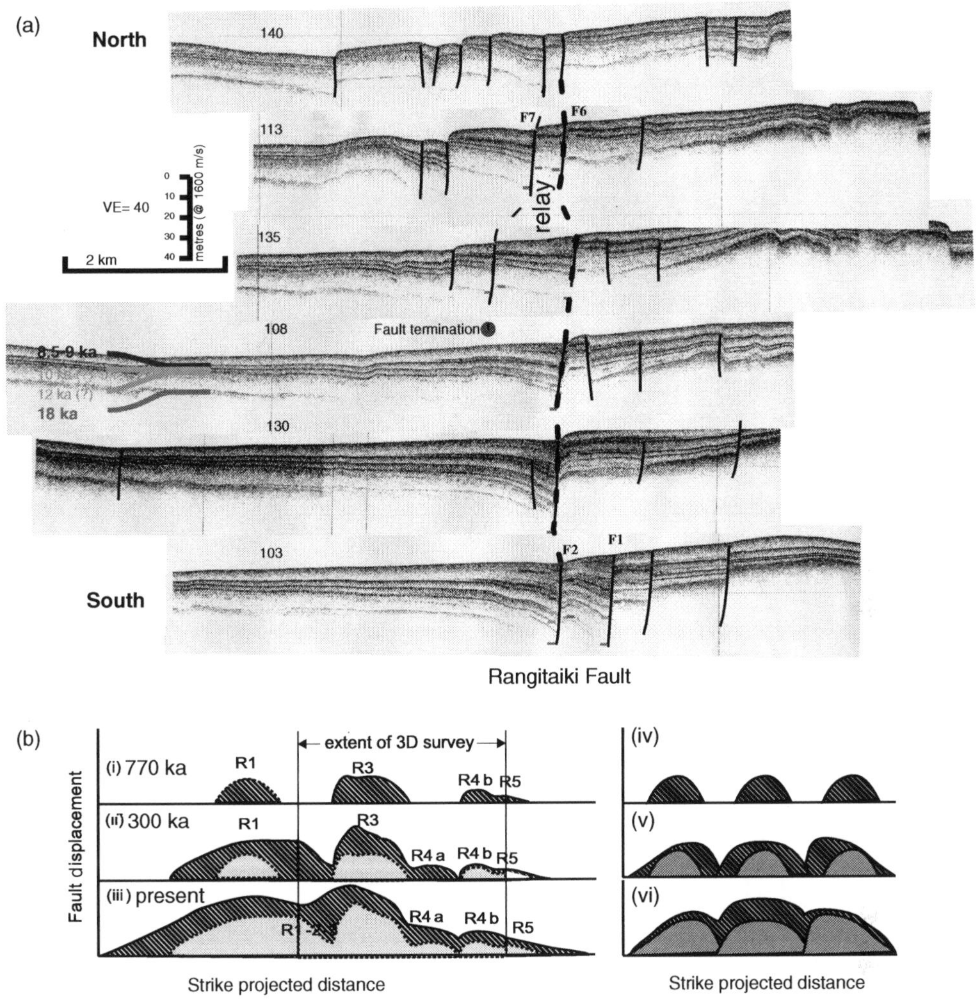


**Fig. 23.11** (a) 3.5-kHz records from parallel NE–SW sections across the Rangitaiki Fault to show a fault transfer across a relay zone. Two reflection horizons are correlated with an erosion surface dated as 18 ka and 10 ka. Fault activity can be determined from analysis of growth strata: active faults are identified either as blind faults that do not break surface and witnessed by monoclinal folds or as surface breaks marked by seabed displacement scarps. Sealed faults may be inactive. (After Lamarche *et al.*, 2000. Figures courtesy of G. Lamarche.) (b) Sketch showing the temporal development of displacement on the entire Rangitaiki Fault system. (i–iii) Strike projection of the oldest horizon (MCS3) within the detailed study area. The evolution from isolated faults to a fully linked system over 1.3 Myr is shown. The position of the southernmost segment of the Rangitaiki Fault, R1, is less well constrained for the earliest time period, and this is indicated by the dotted displacement profile. (iv–vi) Comparison with a model of tip propagation and linkage for normal faults proposed by Cartwright *et al.* (1995). Hatched areas indicate where new displacement is accumulated in comparison to previous time periods. (After Taylor *et al.*, 2004. Figures courtesy of J. Bull.)

fault is composed of two ca 6 km long precursor fault strands linked by a jog or transfer fault. Integration of structural and stratigraphic data indicates that the ramp between these two fault strands was a relative high across which drainage passed during much of the rift event, with hard-linkage and considerable displacement accumulation not occurring until at least ca 7.5 Myr after rift initiation. Separate depocentres

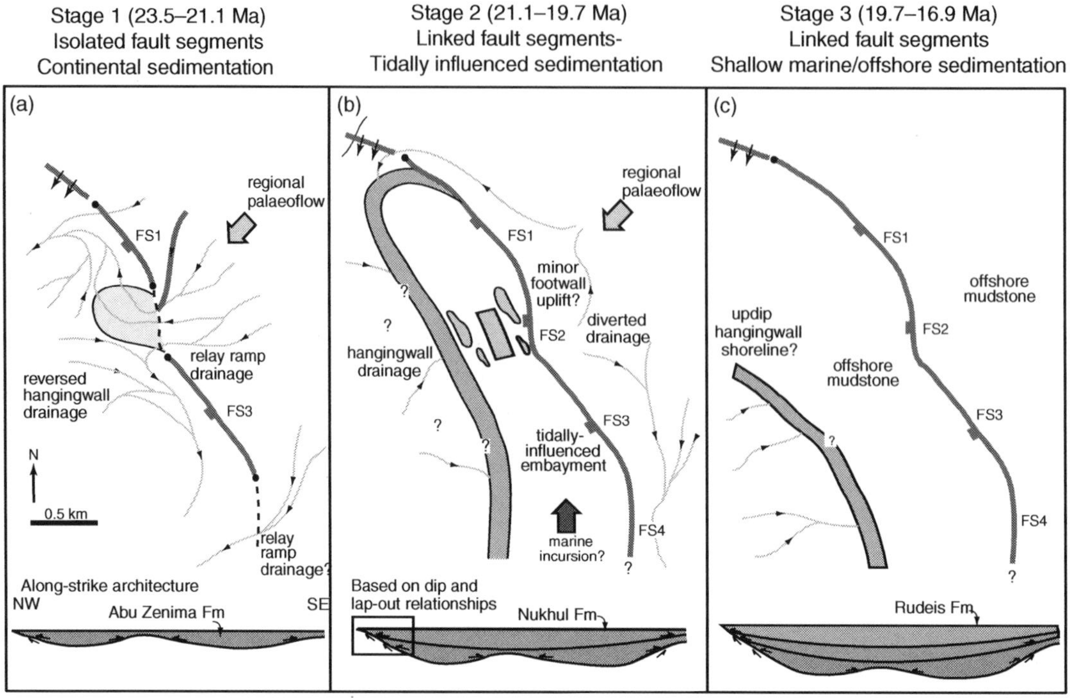


Fig. 23.12 Schematic tectono-stratigraphic evolution of the East Tanka fault zone, Suez rift, to show the consequences of fault segment linkage and propagation on local sedimentary environments. (After Jackson *et al.*, 2002. Figures courtesy of C. Jackson.)

developed adjacent to the long-lived basement high, with the latter propagating upsection as later post-linkage subsidence allowed burial of the structure and formation of linked hangingwall anticline–synclinal pairs.

## 23.9 Rivers, basement uplifts, tilting and fault growth

The extreme sensitivity of river channels to tectonically induced slope changes is illustrated from the southern Rio Grande rift, New Mexico, USA (Fig. 23.13). Careful large-scale mapping, geochronology of interbedded volcanic ash deposits and palaeomagnetic results have revealed that between ca 5 and ca 0.8 Ma the rift featured normal fault-related uplifts with half-grabens and grabens developing over time. The ancestral Rio Grande wound its way southwards as a structure-parallel axial channel through a maze of rift-floor depressions between structural footwall uplifts and aggraded up to ~150 m of sediment from its alluvial plains. In the half-graben (e.g. Engle, Palomas basins) successive axial channel sands were deposited in narrow belts near footwall uplifts in response to basin-floor hangingwall tilting. The river periodically interacted with alluvial fans issuing from relatively small catchment sources in adjacent footwall uplifted blocks. This interaction involved periodic erosion of fan margins ('toe-trimming'), sometimes involving lateral shifts of the river channel hundreds of metres towards active faults, a process probably triggered by tilting. In contrast the river traversed almost the entire width of full grabens (e.g. Hatch-Rincon, Mesilla basins), depositing interbedded channel and floodplain sediment with moderately mature palaeosols. As faults grew in length to cut formerly depositional areas, associated uplift and subsidence led to drainage diversions and very mature palaeosol development on long-lived geomorphic surfaces, best seen in the southward growth of the Robledo Mountains faults between Gauss and Matuyama times.

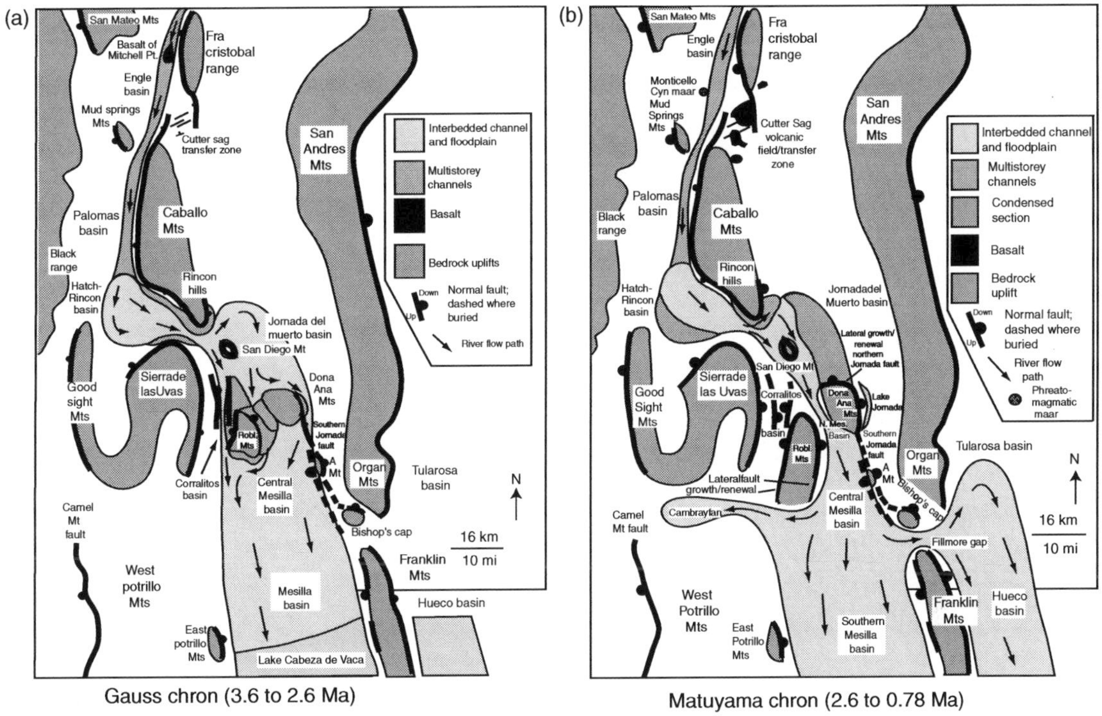


**Fig. 23.13** Palaeogeographical maps to show evolution of faulting, basement uplifts and river positions of the Rio Grande in the southern Rio Grande rift during (a) the Gauss and (b) the Matuyama palaeomagnetic chrons of the Plio-Pleistocene. (c) Sections to illustrate basin architecture and surface morphological forms (lobate fans, axial ancestral Rio Grande channels) across typical half-graben, graben and transfer zones. (All after Mack *et al.*, 2006.)

## 23.10 Unsteady strain and the sedimentary response

At regional or subregional scales, when fault arrays are competing to take up strain, for example as fault blocks rotate, it is possible that partitioning can lead to 'on–off' bursts of activity on adjacent structures. A neat study that illustrates such activity used the architecture of coarse-grained delta progradational units in the Pliocene Loreto basin (Baja California Sur, Mexico), a half-graben located on the western margin of the Gulf of California. Previous work indicated that delta progradation and transgression cycles in the basin were driven by episodic fault-controlled subsidence along the basin-bounding Loreto fault, with deltas prograding during periods of fault quiescence. A Milankovich-band control by fluctuating sediment supply or sea-level change was ruled out because radiometric data on duration of cycles pointed to a subprecessional mean recurrence interval of $\sim$10 kyr. Later investigations revealed that seven out of eleven deltaic units show a lateral transition from friction-dominated shallow water delta to deeper water Gilbert-delta morphology as they prograded (Fig. 23.14). The transition between the two types of deltas records increasing water depth through time during individual episodes of progradation. A mechanism that explains the transition is an accelerating rate of fault-controlled subsidence. During episodes of low slip rate, shoal-water deltas prograde across the submerged topography of the underlying delta unit. As displacement rate accelerates, increasing bathymetry at the delta front leads to steepening of foresets and initiation of Gilbert deltas which eventually terminate with sigmoidal foresets indicative of aggradation (build-up). A real puzzle is the cause of subsequent delta drowning and presence of *ravinement* shell lags that cap the delta

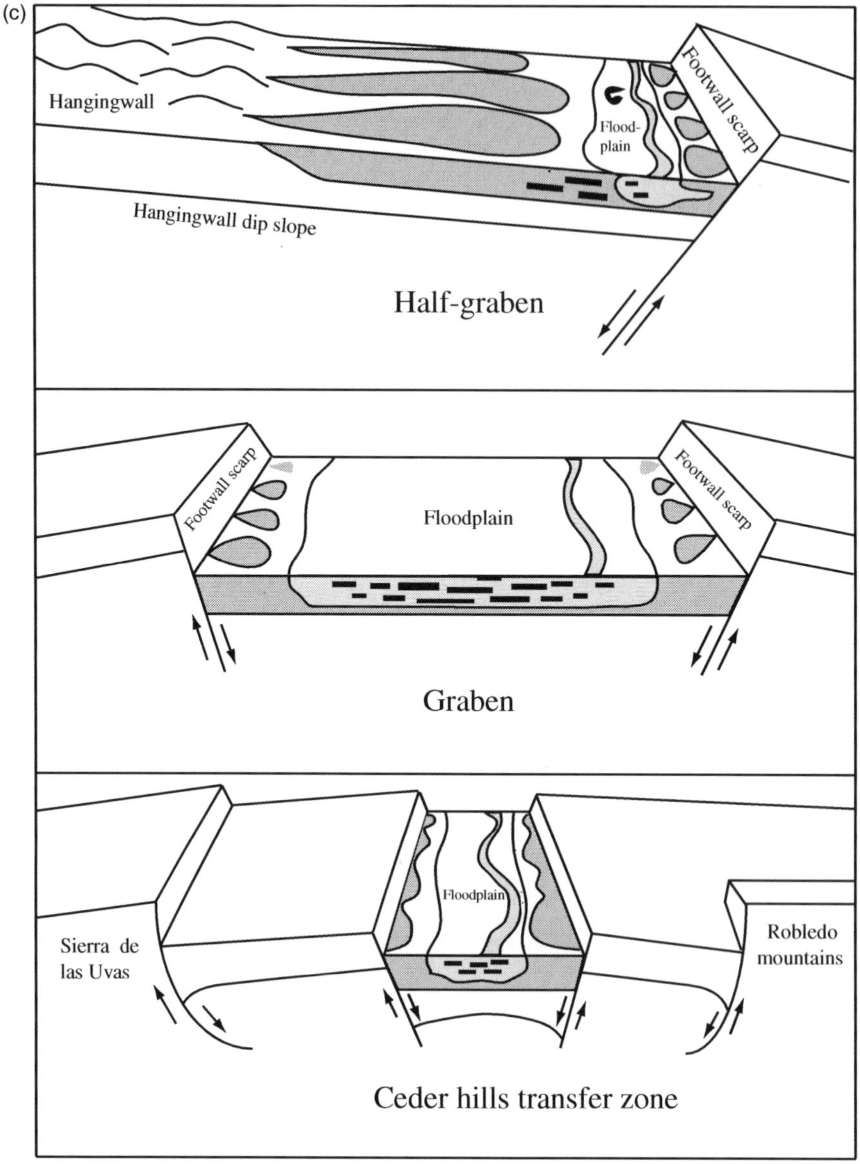


**Fig. 23.13** (*Continued*)

units. This is attributed to sediment starvation at the shoreline during periods of high slip rates because of sediment trapping upstream as the alluvial hinterlands tilt back into the active fault. The observed delta architecture suggests that the long-term (> 100 kyr) history of slip on the Loreto fault was characterized by repetitive episodes of accelerating displacement. Concerning rates, clinoform nucleation requires that displacement acceleration was up to two to five times the rate at the start of progradation, from < 0.5–2.2 to 1.7–4.8 mm/yr. Such episodic fault behaviour is most likely to be because of variations in temporal and

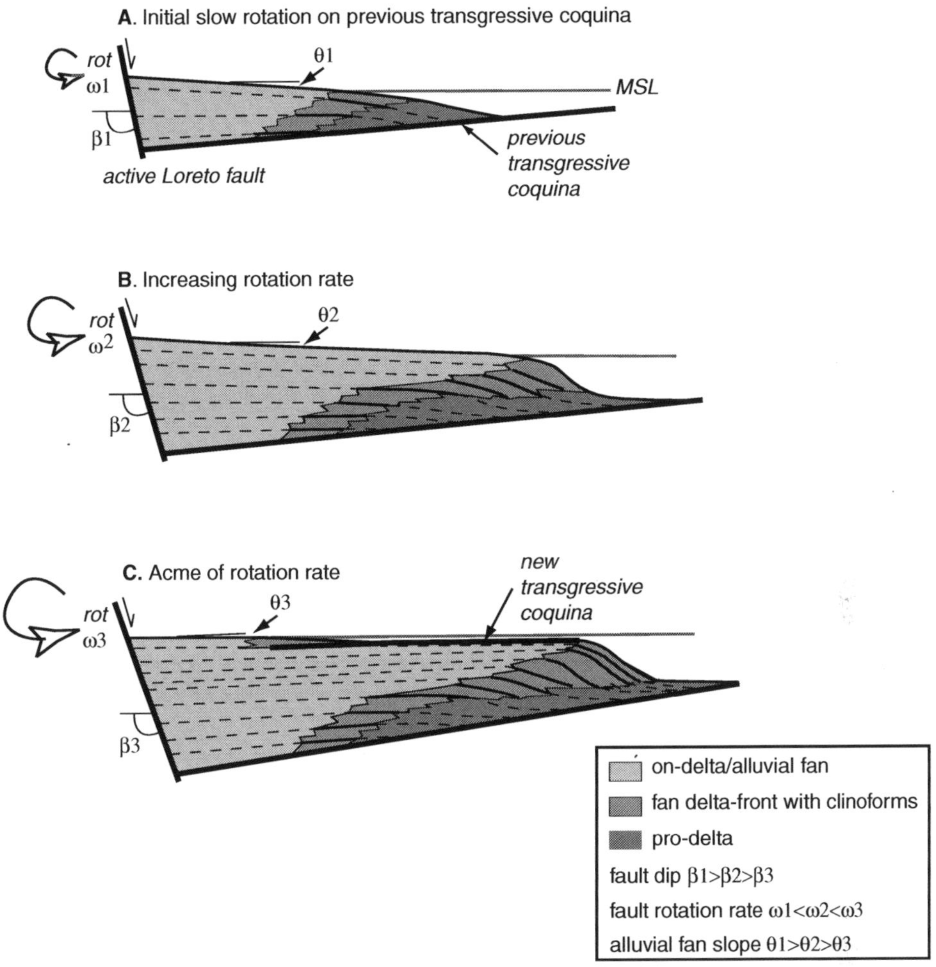


**Fig. 23.14** Schematics (simplified after Mortimer *et al.*, 2005) illustrating the development of observed fan-delta depositional architectures in the Loreto basin. The 2500m wide study area falls within a region 1500m basinward of the Loreto fault. The calculated amount of back-tilt rotation, w, in the most proximal region required to explain the full thickness of the measured field section (9 individual depositional cycles totalling 378 m) is 3.31° about a fulcrum 6.5 km distant. Cartoons (A-C) greatly exaggerate this backtilting in a single cycle.(A) Progradation is renewed onto an underlying marine transgressive depositional unit. The slope, q, of the alluvial section is initially low but increases up section as the alluvial fan system re-equilibriates from previous sediment storage and enhanced backtilting (see C.). During these initial stages, this leads to aggradation of the delta front and then progradation as sediment supply increases.(B) The rate of subsidence increases along the Loreto fault, generating an increase in water depth at the delta front. The height of the foresets increases as does their depositional angle in response to this increase in water depth. With continued increase in subsidence rate, the water depth at the delta front increases to a point where shoal-water delta avalanche foresets cannot be sustained, a greater angle of deposition is reached, and angle-of-repose clinoforms nucleate in the basin. During this time the alluvial slope reduces as the rate of subsidence increases. (C) Increasing subsidence rate increases delta front water depth, foreset heights increase and their geometry changes as they aggrade. Aggradation is enhanced by reduction in the slope of the alluvial system causing trapping and storage of sediment in proximal fan regions. Eventually, sediment storage in the proximal region, and the rate of increase in water depth at the delta front (both controlled by ever more rapid subsidence) exceed the amount of sediment being transported to the delta front, and the delta top is transgressed. During the initial period of transgression a depositional hiatus causes a shell bed to be deposited. Through subsequent time the sequence A-C is repeatedly duplicated.

spatial strain partitioning between the Loreto fault and other faults in the Gulf of California.

## 23.11 Tectonics and climate as depositional controls

A recent perspective on the relative influences of climate and tectonics is provided by studies of alluvial fan sequences in south-central Australia. The most voluminous alluvial fans in the Flinders Ranges region have developed adjacent to catchments uplifted by Plio-Quaternary reverse faults, implying that young tectonic activity has exerted a first-order control on long-term sediment accumulation rates along the range front. However, optically stimulated luminescence dating of the sediments making up alluvial fan sequences indicates that late Quaternary sedimentary environmental changes and trends towards sediment aggradation and fan dissection are not directly correlated with individual faulting events. The fan sequences record a transition from debris-flow deposition and soil formation to clast-supported conglomeratic sedimentation by ca 30 ka. This transition is interpreted to reflect a response to increasing climatic aridity, coupled with large flood events that episodically stripped previously weathered regolith from the landscape. Similar sequences have been documented from the prominent alluvial fans that drain the active Palomas basin in the Rio Grande rift of southern New Mexico, USA. Late Pleistocene to Holocene cycles of fan incision and aggradation in the Australian examples post-date the youngest-dated surface fault ruptures and are interpreted to reflect changes in the frequency and magnitude of large floods. These datasets indicate that tectonic activity controlled long-term sediment supply but that climate has governed the spatial and temporal patterns of range-front sedimentation.

As yet, no stratigraphic modelling studies have been able to simulate the detailed response of alluvial fans and their feeder catchments to high frequency (Milankovich-band) climate change and resultant variations in sediment/water runoff predicted by soil erosion models. The best that can be done is to determine sensitivity of first-order controls when rates of fault displacement or channelized runoff are step-changed from an equilibrium base model. This is the approach adopted in a one-dimensional model of a transport-limited footwall catchment feeding a fan whose surface slope is set by the balance between catchment sediment flux and tectonically generated basin accommodation. Rock uplift rate in the normal fault footwall is spatially variable across the model space. Results indicate that increasing the fault slip rate, or decreasing the precipitation rate, leads to an increase in fan slope, temporary back-stepping of the fan toe, and a pronounced angular unconformity. Conversely, a decrease in slip rate, or an increase in precipitation rate, results in a decrease in fan slope, and progradation and eventual stabilization of the fan toe. Once perturbed, the system evolves toward a new equilibrium state with time constants of 0.5–2 Myr; these response times are insensitive to slip rate but are strongly dependent on precipitation rate. Variations in fan slope are well described by a dimensionless parameter that expresses equilibrium slope as a function of slip rate, precipitation rate, system size and catchment lithology.

A fascinating study that shows a direct linkage between orogenic growth of relief, the incidence of orographic precipitation and erosion rates is from the arid Altiplano-Puna plateau of the southern Central Andes. Cenozoic distributed-shortening in the area formed intramontane basins between the Western and Eastern Cordillera that became isolated from the humid foreland to the east. Thick Tertiary and Quaternary sediment in Puna basins has reduced topographic contrasts between the compressional basins and ranges, leading to a typical low-relief plateau morphology. Structurally identical basins that are still externally drained straddle the eastern border of the Puna and document the eastward propagation of orographic barriers and ensuing aridification (Fig. 23.15). One of them, the Angastaco basin, is transitional between the highly compartmentalized Puna highlands and the undeformed Andean foreland. Sandstone petrography, structural and stratigraphic analysis are combined with detrital apatite fission-track thermochronology to analyse a 6200-m-thick Miocene to Pliocene stratigraphic section in the Angastaco basin. The data documents late Eocene to late Pliocene exhumation history of source regions along the eastern border of the Puna (Eastern Cordillera (EC)) as well as the construction of orographic barriers along the southeastern flank of the Central Andes. Onset of exhumation of a source in the EC in late Eocene time as well as a rapid exhumation of the Sierra de Luracatao (in the EC) at about 20 Ma are recorded

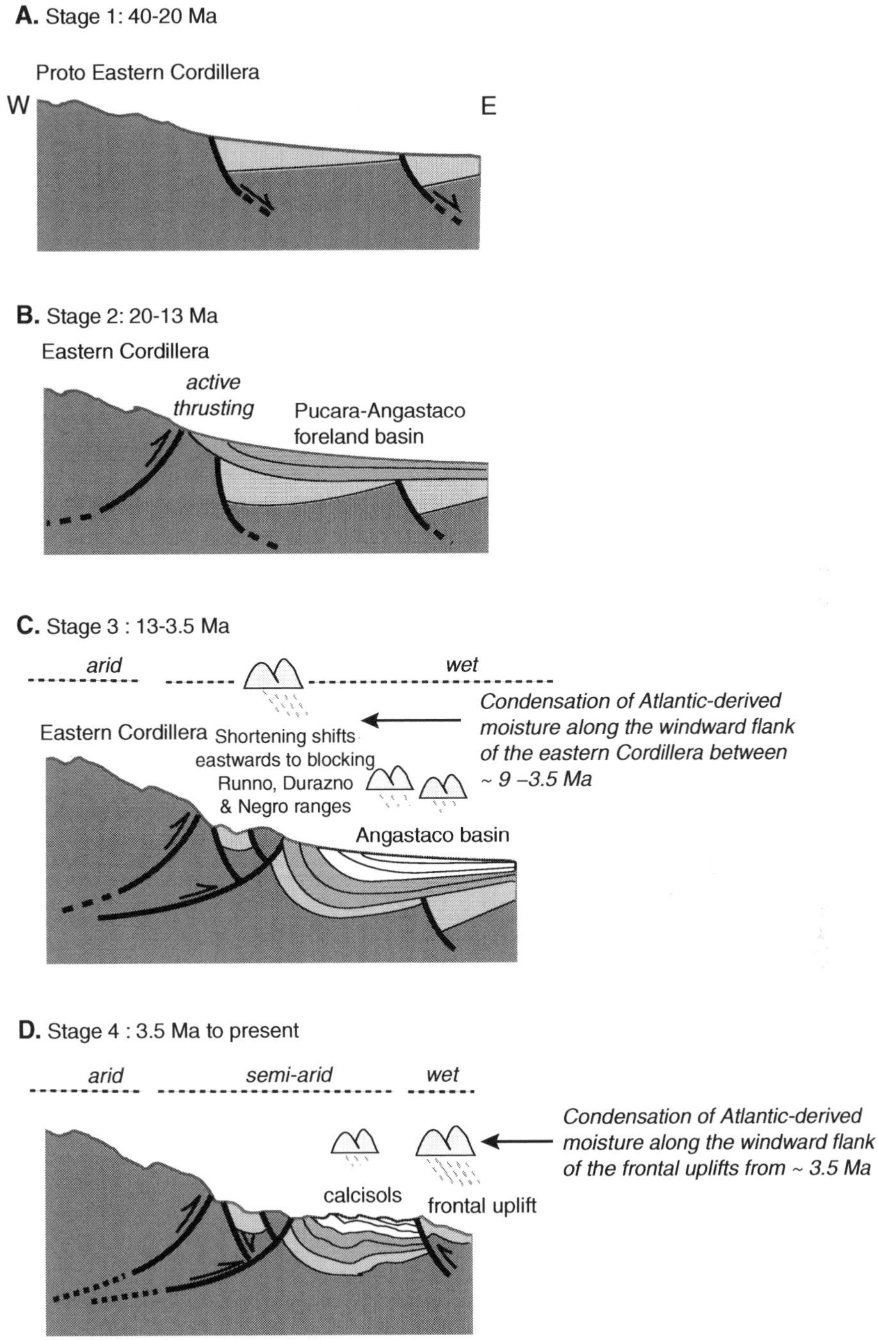


**Fig. 23.15** Cartoons illustrating the tectonic, topographic and climatic evolution of the Eastern Cordillera and Calchaqui valley at the latitude of the Angastaco basin, Argentina during the past 40 Myr. (After Coutand *et al.*, 2006. Figures courtesy of I. Coutand.)

in the detrital sediments of the Angastaco basin. Sediment accumulation in the basin began c. 15 Ma, a time at which the EC had already built sufficient topography to prevent Puna-sourced detritus from reaching the basin. After ca 13 Ma, shortening shifted eastward, and by 9 Ma the EC constituted an effective orographic barrier that prevented moisture penetration into the plateau. Between 3.4 and 2.4 Ma another orographic barrier was uplifted to the east, leading to further aridification and pronounced precipitation gradients along the mountain front.

## 23.12 River equilibrium, incision and aggradation—away from the knee-jerk of tectonic explanation

Disequilibrium in fluvial systems is commonplace due to spatial gradients in flow energy and mean sediment transport rate. A common climatic/hydrological reason for nonzero downstream transport gradient is excess water supply relative to sediment supply or vice versa. A second reason is change to gradients brought about by either tectonic tilting or base-level fall causing a river to run over a steeper reach. Gradient curvature induces excess bed shear stress and increased stream-power so that erosion at the point of curvature propagates upstream as an erosional knickpoint discontinuity at a rate determined by 2D diffusion. A key additional factor in determining whether long-term incision will occur is the requirement that enough turbulent kinetic energy be generated so that bedrock or older alluvial substrate can be directly scoured. Extreme flood events ($10^2$–$10^3$ year recurrences) are clearly implicated, with the extra energy for sediment transport and pickup coming from the nonuniform and unsteady contributions, together with a much increased efficiency of transport under flash flood conditions.

In the 1990s the concept of *equilibrium time* was applied to the study of river deposition in sedimentary basins. This is the time required for a basin to reach equilibrium, such that sedimentation rate balances basin subsidence rate. It also refers to the efficiency with which a river traversing any given basin can transmit depositional or erosional 'signals' in the form of changed grain size and/or sediment flux. The general concept had previously been used in computing the rate of upstream propagation of fluvial incision following a base-level fall. It assumes that sediment transport is linearly proportional to local bed slope, an assumption that leads to an estimate of a characteristic diffusion coefficient for the incision processes. Since equilibrium time is an analogue to relaxation time in conductively cooling thermal systems, the term *sediment relaxation time*, $T_s$, is an alternative. Generally, this is given by $T_s = l^2/\kappa$, where $l$ is a streamwise length scale and $\kappa$ is sediment diffusivity. Rivers with larger $T_s$ take longer to transmit sediment signals lengthwise.

The effects of climate on basin-fill evolution is well illustrated in the Rio Grande rift, southwest USA. A steady-state aggradational mode operated in the southern rift between $\sim$ 5 and 0.8 Ma, causing preservation of ancestral axial-channel and floodplain deposits due to relatively slow, long-term active rift subsidence and a state of mean equilibrium or supersaturation of sediment-transport. The onset of major climatic change around 0.8 Ma resulted in the axial river periodically incising $\sim$150 m, removing about 25% by volume of previously accumulated sediment, despite continued active faulting and fault-induced subsidence. This climatic mode is interpreted to be a periodic response to positive downstream gradients in sediment transport rate during glacial and glacial-transition periods, caused by low-level external sediment sourcing and a dominance of large magnitude, sediment-unsaturated, spring snowmelt floods from northern mountain valleys. $T_s$ for the Rio Grande is probably short despite the great length of the river system as a whole, since it is the balance between hydrological and sediment input from the many lateral tributaries that controls nonuniform transport capacity of the axial channel.

Field-based geomorphology and $^{10}$Be terrestrial cosmogenic nuclides have been used to investigate spectacular $>$ 300 m deep incision of High Atlas thrust-front derived pre-Pliocene alluvial fans and lake sediments in the formerly Miocene endorheic (hydrologically closed) Ouarzazate foreland basin of southern Morocco (Fig. 23.16). Alluvial fan and terrace incision in this active basin was controlled partially by a drop in base level during the Pliocene or early Pleistocene as an axial outlet channel, the Draa River, progressively cut through the earlier enclosed basin deposits into the Anti-Atlas to the south of the Ouarzazate foreland basin. The process may have been facilitated by overfilling and onlap of the older endorheic system onto the Anti-Atlas to the

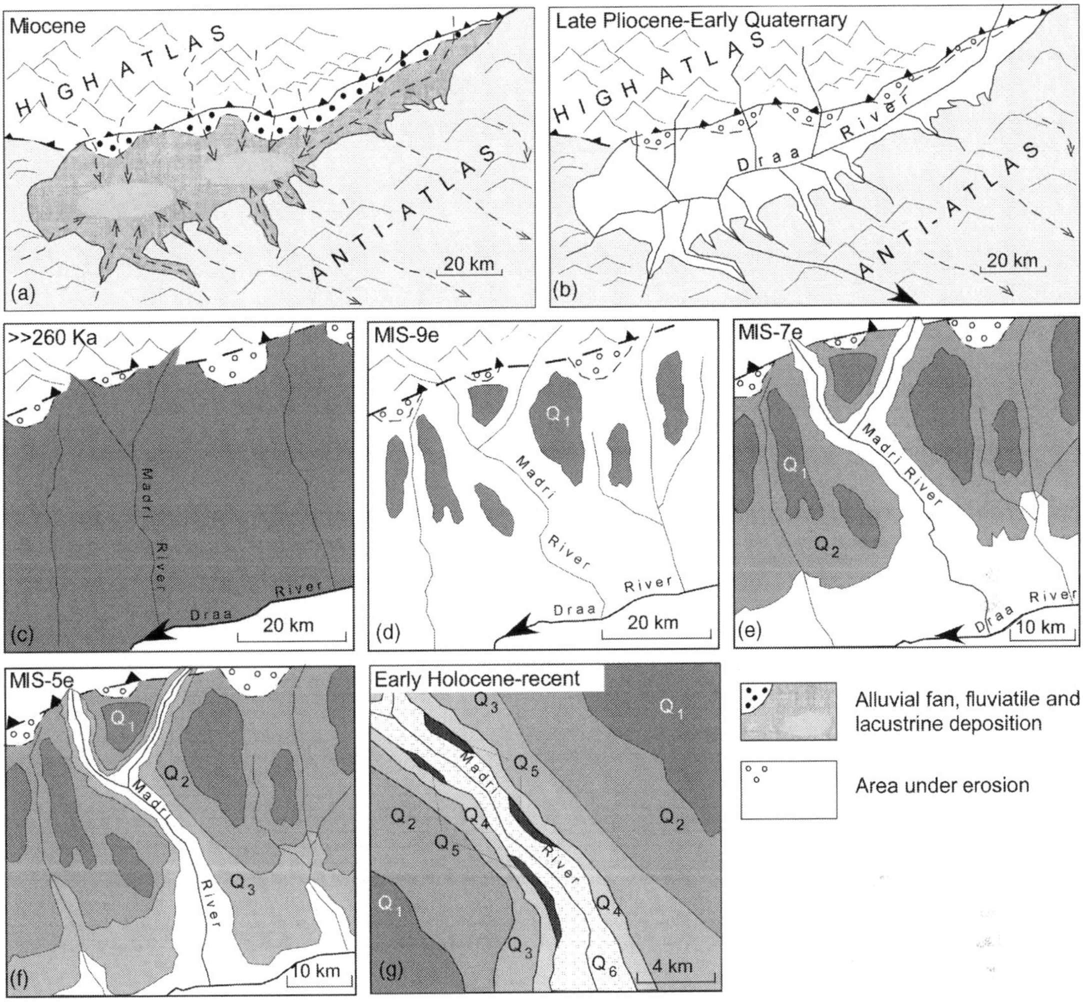


**Fig. 23.16** Schematic diagrams showing the evolution of the Ouarzazate basin highlighting the formation of terraces along the Madri River. (After Arboleya *et al.*, 2008.)

south. Alluvial fans and terrace surfaces have abandonment ages that correlate with the past four glacial cycles. Their formation was strongly modulated by climate on glacial–interglacial timescales as base level dropped. Mean rates of fluvial incision range between 0.3 and 1.0 mm/yr for the latter part of the Quaternary. Deformation of the fluvial terraces demonstrates that the transition from aggradation to erosion occurred while thrust loading was still active along the High Atlas margin, but despite this the modification of the drainage pattern from internal to external and oscillations of the Quaternary climate were the dominant factors controlling incision in the Ouarzazate basin at least during the past 300 ka.

## 23.13 Integrated sedimentary systems: modelling tectonics, sediment yield and sea level change

An instructive example comes from investigations of river and coastal delta development and sequence

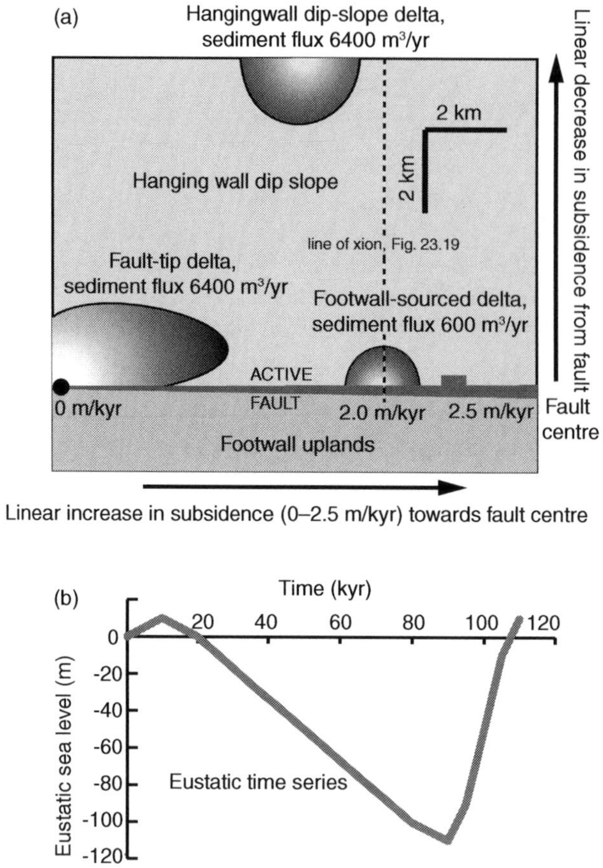


**Fig. 23.17** Definition diagram for modelling sedimentation adjacent to a fault segment tip with variable sediment flux, spatially variable displacement gradients and changing sea level. (After Gawthorpe *et al.*, 2003. Figures courtesy of R. Gawthorpe.)

variability (Figs. 23.17–23.19). Changing a single variable produces marked differences in three-dimensional morphology, cross-sectional stratal geometry and delta evolution during a cycle of sea-level change. Sediment supply strongly influences not only the timing of transgressive and maximum flooding surface development during sea-level rise, and hence the diachroneity of lowstand, transgressive and highstand systems tracts, but also the timing of onset of fluvial incision and the characteristics of incised valleys and forced regressive wedges. In contrast to other controls, tectonic subsidence (or uplift) leads to modification of accommodation, so that relative sea-level change may vary significantly along a basin margin. In high-subsidence settings during eustatic sea-level fall, local relative sea level may continue to rise. In such settings, deltas lack incised valleys, forced regressive delta lobes and prominent sequence boundaries. Furthermore, normal regression and even transgression in high-subsidence locations can be contemporaneous with forced regression and incised valley development in adjacent lower-subsidence locations. The results of this sensitivity analysis suggest that systems tracts and their bounding surfaces are probably diachronous along many basin margins and may locally be absent. More than one control can produce a similar stratigraphic response, making interpretation of controlling processes from the stratigraphic product equivocal. The models provide a framework for understanding the stratigraphy of natural basin fills, but further research on the interplay

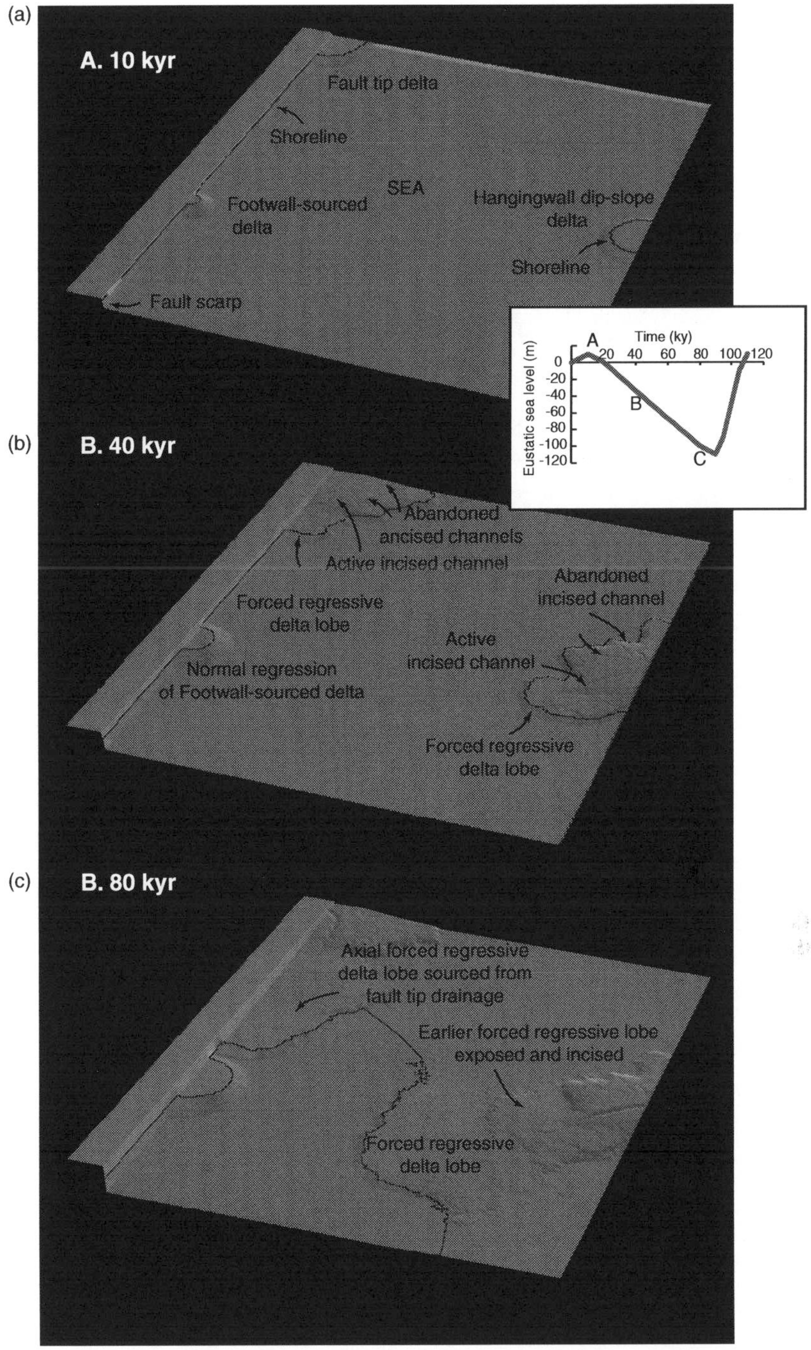


**Fig. 23.18** (a–f) Time snapshots of sedimentary response to eustatic sea-level variation within the context of spatially variable subsidence, sediment flux and time-varying sea level. See Fig. 23.17 for definition diagram. (After Gawthorpe *et al.*, 2003. Figures courtesy of Rob Gawthorpe.)

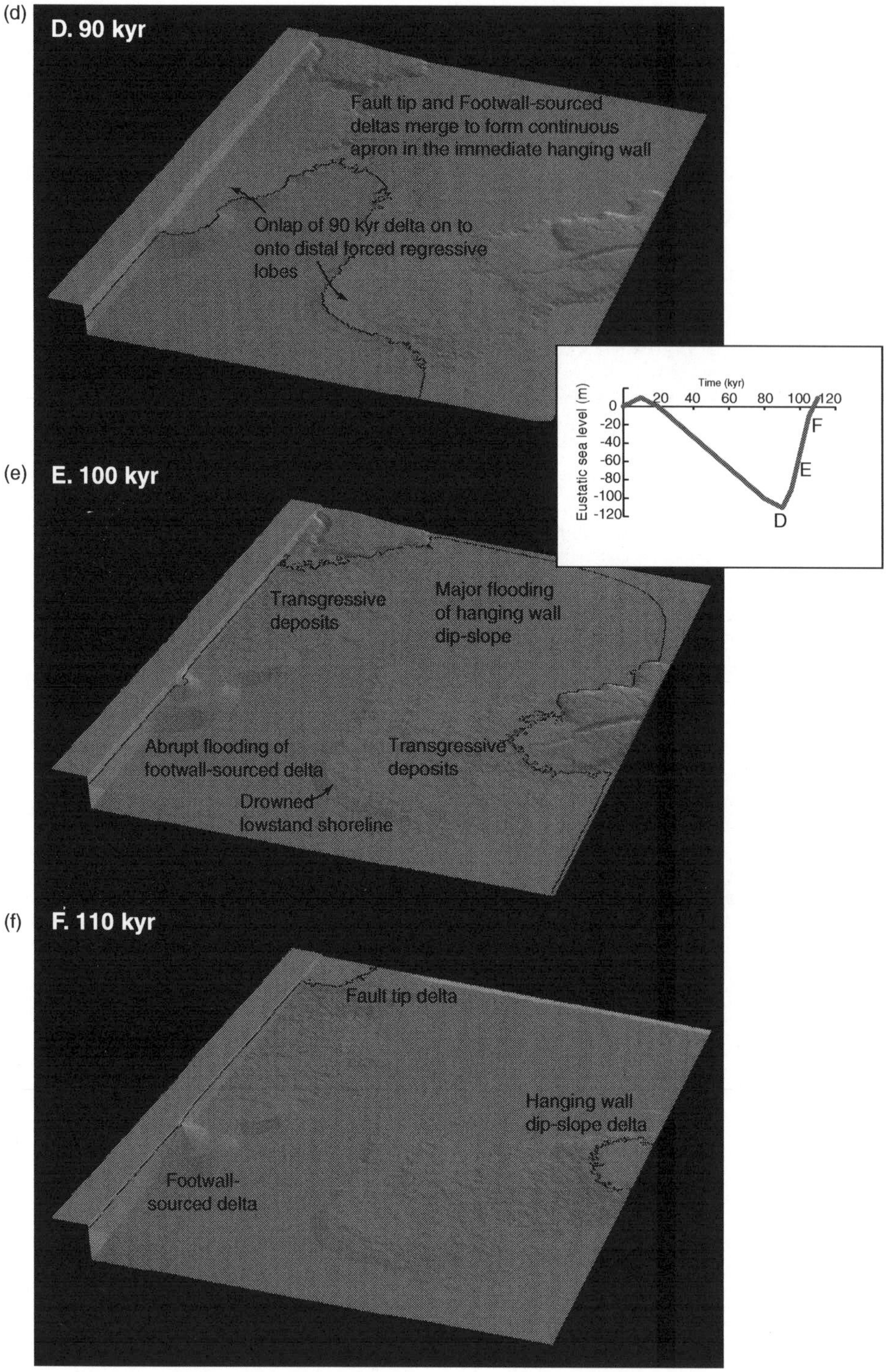


**Fig. 23.18** (*Continued*)

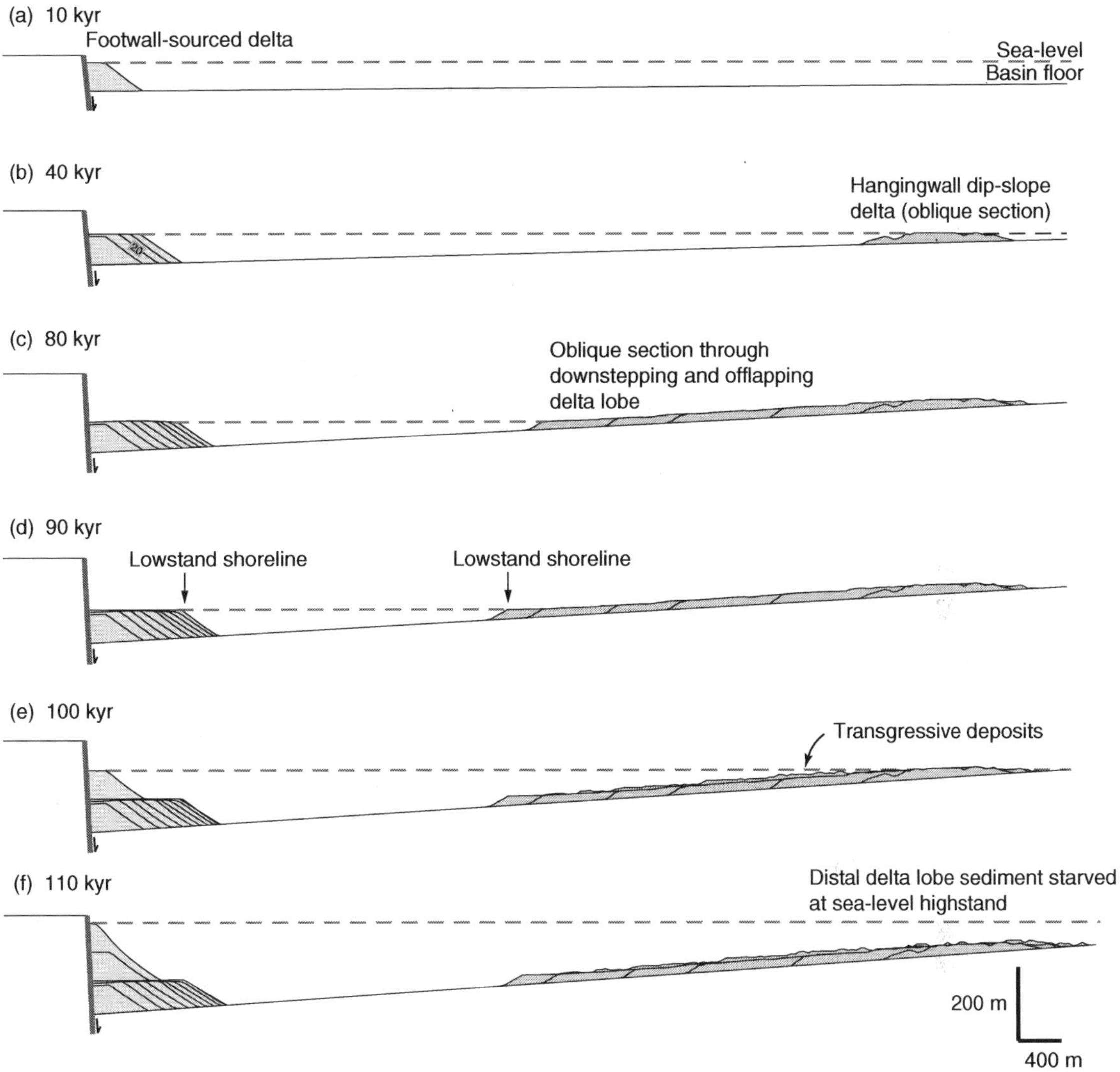


**Fig. 23.19** Sections to illustrate sedimentary architecture for the scenarios developed in model experiments of Fig. 23.17. See location of section line in Fig. 23.17. (After Gawthorpe *et al.*, 2003. Figures courtesy of Rob Gawthorpe.)

between the controls is required in order to understand the climatic, tectonic and sea-level signals concealed in the stratigraphic record.

Another example of informative 3D architectural modelling concerns the sedimentary evolution of a developing foreland basin in a 3D numerical model which explores the development of alluvial, deltaic and marine successions in foreland basins filled by transverse and axial depositional systems, under conditions of variable tectonism and eustatic sea-level change (Fig. 23.20). Note that sediment supply influenced by climate change was not included as a variable in these numerical experiments. The constant water and sediment supply rates to the model system were based on the time-averaged volumetric sedimentation rates of the Lower Montanyana Group in the Tremp Basin, with the supply ratio derived from the general notion that the average sediment/water discharge ratio for large rivers fluctuates around 1:10 000. The axial sediment entering the system is composed of

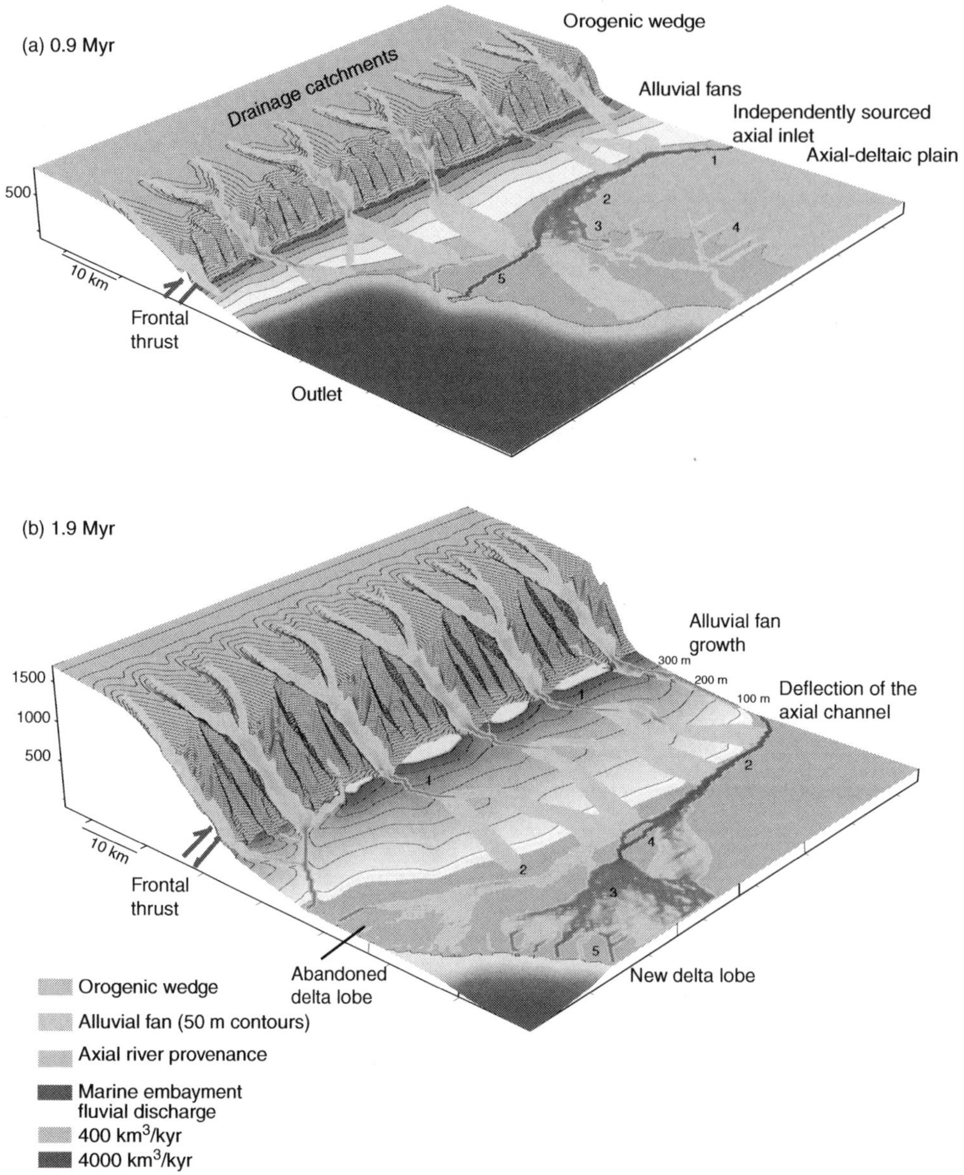


**Fig. 23.20** Two examples of a modelled foreland basin landscape, (a) at 0.9 Myr and (b) at 1.9 Myr, during two phases of tectonic quiescence and forced progradation. Both figures show the configuration of the main depositional elements filling the basin; transverse alluvial fans and an axial river discharging into the marine embayment as a delta. The main axial channel has a fixed inlet position but is drawn to the zone of high accommodation creation close to the fan fringes (a). Eventually, the fans deflect this channel basinward (b). Features in (a) are (i) incised axial channel, (ii) confined braidplain, (iii) deltaic distributary, (iv) incision into the upper deltaic plain and (v) incised channel triggered by a eustatic fall, funnelling the bulk of the discharge. Features discussed in (b) are (i) fan-head entrenchment, (ii) axial channel close to fan fringes, (iii) and (iv) locations of upstream avulsion leaving a downstream delta abandoned, and (v) new delta with juvenile incisions triggered by a eustatic fall. (After Clevis *et al*., 2004b.)

10% sand and 90% finer grained sediment. Results of the forward modelling experiments indicate that the onset of tectonic activity is reflected by rapid retrogradation of both depositional systems and by widespread marine transgression and onlap by marine sediments. Syntectonic architecture on the axial-deltaic plain is dominated by swiftly relocating channels in response to a general rise in relative sea level induced by flexural subsidence. The resulting surface morphology of the axial delta is convex upwards. Syntectonic eustatic sea-level fluctuations result in parasequence-scale packages of retrograding and prograding fan and delta sediments bounded by minor flooding surfaces and type 2 sequence boundaries. Incised channels are rare within the syntectonic parasequences and are formed only during phases of tectonic quiescence when eustatic falls are no longer compensated by the subsidence component in the rise in relative sea level. Suites of amalgamating, axial channels corresponding to multiple eustatic falls delineate the resulting type 1 unconformities. Coarse-grained, incised-channel fills are found in the zone between the alluvial fan fringes and the convex upward body of the axial delta since the axial streams tend to migrate towards this zone of maximum accommodation.

## 23.14 Extraterrestrial sedimentology—atmospheric and liquid flows on Mars

Sediment transport occurs widely in turbulent fluid flows on Earth but is also significant in the atmospheres and on the surfaces of other planets. Planetary sediment itself is diverse, comprising silicate minerals or ice whilst the fluid phase may be liquid water on palaeo-Mars, liquid methane on Titan or atmospheric wind on present-day Mars and Venus. Even the process of planetary formation involves primal sediment-like materials called interstellar dust that has been widely detected in the past decade by powerful radio arrays, new infrared and X-ray telescopes and detectors in orbit. The dust forms turbulent whorls and eddies that dominate the galactic centre in the constellation of Sagittarius. It is thought to play a major role in cosmogenesis through its role in granular accretion of protoplanetary discs. In this process clay-grade material ($< 0.1$ mm) rapidly coagulates in the disc around a young star to form larger millemetre-size grains that sediment into the disc plane over rather short timespans of a few million years. Grain size estimated by light-scattering sensors is remarkably similar to that of glassy chondrule particles in the carbonaceous chondrite class of meteorites.

A more directly observable extraterrestrial sedimentary phenomenon has been known for many years—the periodic red-brown hue to the Martian atmosphere due to suspended silt-grade sediment whipped up by vast atmospheric storms. The suspension process is thought to be aided by the saltation of bedload sediment on to dust carpets—the thin Martian atmosphere means that falling sand grains have a high kinetic energy. Also a role has been proposed for strongly rotating vortices in the atmosphere, visible to us in Earth deserts as *dust devils*, but also clearly photographed by NASA's Sprite rover. The strong vorticity associated with these and the enhanced lift pressures in the devil cores is thought to radically increase the level of applied turbulent stresses, so increasing the efficacy of suspension. Thus the 2001 dust storm of the early spring in the Martian Northern Hemisphere covered the whole planet for 3 months and was observed by Mars Global Surveyor and the Hubble Space Telescope. The storm began in and around the Hellas basin and propagated across the equator to move with the southern hemisphere jet stream which is much strengthened at this time because of large meridional temperature gradients. Estimates of the thickness and density of dust were made by thermal methods since dust cools the lower atmosphere whilst that at the top is heated up by Sun's shortwave radiation. The cooling allows the strong surface winds to decline and the dust to settle, though return of Sun's radiation may trigger further strong winds and set the whole process off again.

Sophisticated satellite images and remarkable imagery from NASA's Mars Exploration Rover lander missions has transformed our knowledge of Martian atmospheric loose-boundary hydraulics in recent years. These have imaged bedforms ranging in scale from vast dune fields to wind-blown ballistic ripples (Fig. 23.21). These first direct data on Martian sediment size and bedform characteristics have raised a serious problem concerning the Rouse criterion (**Cookie 33**) for turbulent suspension, which overestimates the size of sediment particles resisting suspension on the Martian surface. Thus observed particle diameter, $d$, of aeolian ripples formed

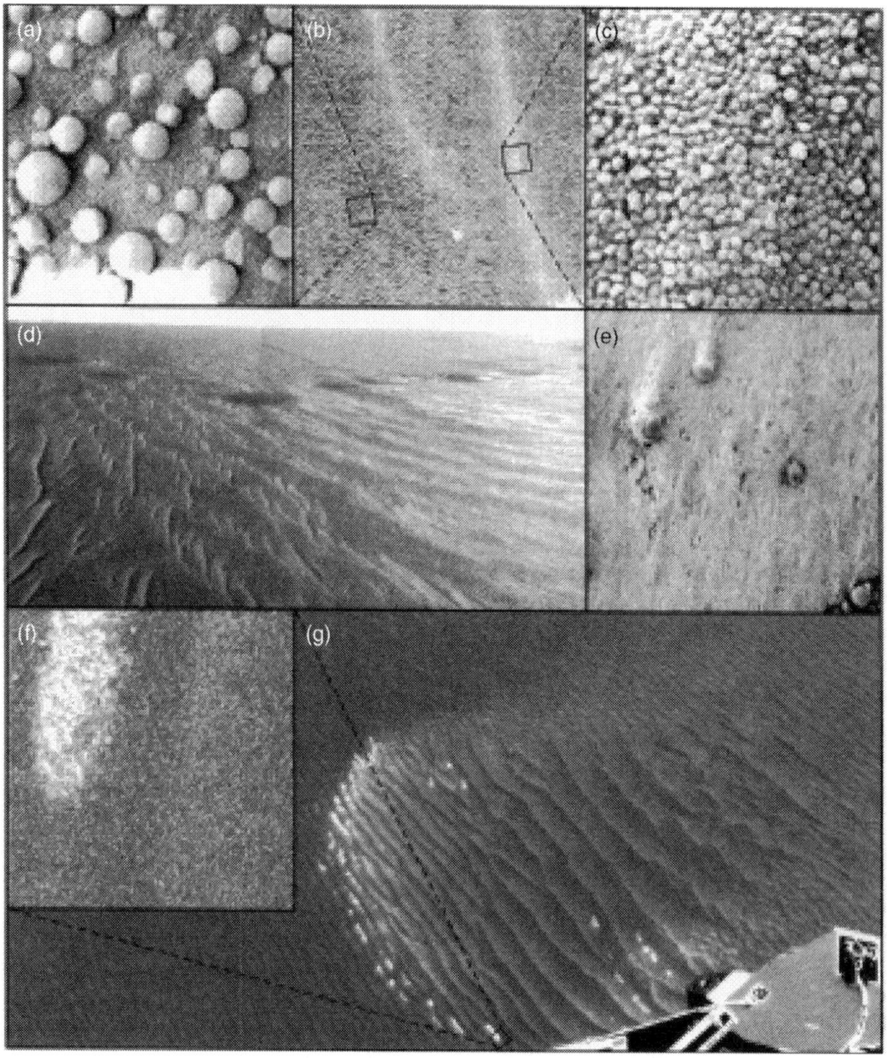


**Fig. 23.21** Martian sediments and bedforms from the pioneering and startling results reported by Sullivan *et al.* (2005). Microscopic Imager views at Meridiani Planum, Mars, are ca 31 mm across. (a)–(c) Locations of close-ups of a ballistic wind ripple crest and an area between wind ripples. (d) A 90° panorama of Plains wind ripple field organized into 1ry and 2ry orientations. (e) Ventifacts on sulphate-enriched outcrops at Fram crater. (f): Microscopic Imager view of basaltic sand wind ripples on Eagle crater floor. (g) Navcam view of the same ripples, longest wavelengths being ca 11 cm.

during bedload sediment transport range from $d = 50$–$125\,\mu m$. This is much smaller than estimates for suspension threshold gained from the Rouse criterion of $d = 240\,\mu m$. The Rouse criterion states that suspension occurs once the sediment particle fall velocity, $U_s$, is equalled or exceeded by the vertical component, $v'_{rms}$, of the root-mean-square turbulent velocity fluctuation. The disadvantages of the criterion arise from its lack of generality and dynamic content. The vertical component of the mean turbulent flow acceleration should be used to determine the net positive vertical force that may support a suspended load. To determine suspension thresholds the Shields approach is followed whereby for suspension of an originally stationary particle layer of fractional concentration, $C$, the vertical turbulent force component must exceed the force due to gravity acting on the sediment. The threshold criterion is derived as $v'_{rms\ max} \geq [dC(\sigma - \rho) g/\rho]^{0.5}$ (**Cookie 33**). Allowing for Martian gravity of about (3.73/9.81)Earth and a $\rho$ of (1/80)Earth we expect $v'_{rms\ max}$ on Mars to be [O] $[80/3]^{0.5}$ =~500% of Earth and for Venus, [O] $[1/90]^{0.5}$ = ~10% of Earth.

In order to estimate the shear velocity, $u_*$, and hence the surface velocity, $U$, of planetary winds, use is made of the experimental fact that $v'_{rms\ max} = 1.3u_*$ during turbulent flow over plane beds undergoing intense sediment transport. Detailed parameterization then yields the curves shown in Fig. 23.22. Active Martian ripples and dunes have $d = 50$–$125\ \mu m$. To avoid such sediment being carried away in suspension, this requires recent maximum wind shear velocities $\leq 5.2$ m/s. This corresponds to a maximum permitted streamwise wind velocity, $U_1$ ($U$ measured at 1 m height above a ground surface comprising aerodynamic roughness, $z_o$, of 125 μm diameter particles of 73 m/s. Experimental data on bedload transport threshold for 125 μm particles means that $u_*$ must have exceeded 2 m/s. Thus $u_*$ for recent Martian winds has been in the range 2–5.2 m/s, corresponding to $U_1 = 28$–$73$ m/s. Older ripples imaged by the Mars landers have $d = 1000$–$2000\ \mu m$. These coarse-grained deposits have two possible interpretations: either (i) particle sizes < 1 mm have been transported away in suspension from the observation site by hurricane-force palaeowinds of $u_* > 15$ m/s (corresponding to $U_1$ of > ~131 m/s) or (ii) the deposits are analogous to the granule ridges found on Earth that record particle transport by bedload creep due to the impact of finer saltating particles and are thus unrelated to palaeowind magnitude. For Venus, the sluggish atmospheric surface winds ($U_1 \approx 0.5$–$1.0$ m/s) recorded by the Venera 9 and 10 missions of 1972 are predicted by the new criterion to be capable of suspending 85 μm very fine sands at maximum shear velocities of ca 0.1 m/s.

Perhaps the most staggering discovery on Mars has been the evidence for past periods of running water

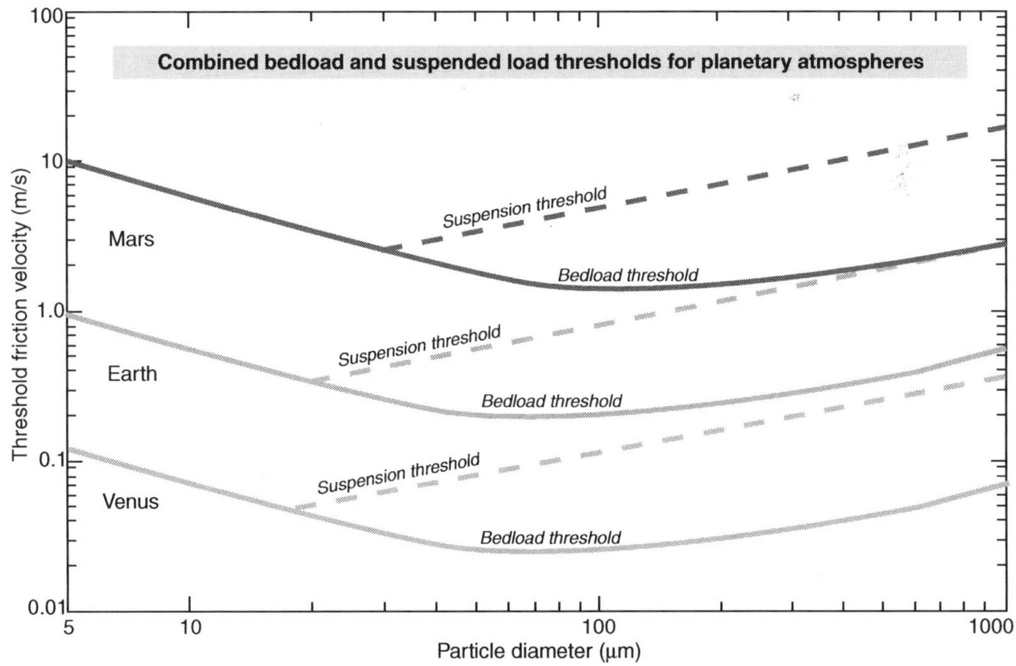


**Fig. 23.22** Combined bedload and suspended load thresholds for planetary atmospheres. (After Leeder, 2007).

that transported sediment by turbulent shear. It had been known for sometime that the polar regions supported both perennial and seasonal ice-caps but only recently have thermal and spectrometric data proved that the ice is largely water ice rather than $CO_2$ ice. The chief evidence for liquid water has come from the following observations.

- Spectacular river channels, gully headcuts, and stepped fan delta bodies that formerly were fed from stream systems that drained very large water discharges into crater lakes.
- Occurrence of bedded sedimentary rocks with clear evidence from trough-shaped cross laminations for subaqueous tractive transport of basaltic sand grains and haematite spherules in downcurrent-accreting current ripple bedforms. Alternative sub-aerial volcanic base-surge origins that have been proposed do not give rise to such fine-scale avalanche laminations.
- Widespread occurrence of water-bearing or water-derived minerals like jarosite and haematite derived from chemical weathering of basalt by groundwater.
- Occurrence of penetrative desiccation cracks and mud curl-up features formed by drying-out of water-saturated sediments.
- Young debris flows that have partially runout from crater walls into the floor

The great age of some of these water-produced features in relation to cratering episodes means that at ~4 Ga the young planet must have had a supergreenhouse climate warmed by copious $CO_2$, for with weak Noarchian luminosity and distance from the Sun it is difficult otherwise to produce liquid water. Of younger age are crater-infill flood deposits, crater ice-masses, remnant pack-ice pressure ridges and piedmont glacier remnants belonging to one of the several Ice Ages that have affected Mars over the past few millions of years. These features all indicate that substantial amounts of water must exist in the Martian subsurface, quite possibly as subsurface water ice which periodically escapes as short-lived flows. Surface water ice up to 200 m thick is also recorded in the shaded parts of high latitude craters. It seems most likely that, as on Earth, climatic forcing due to variations in planetary orbital parameters controls the Martian Ice Age rhythms responsible for the massive ice melting features noted above. Discovery of layered, dusty, water-ice mantling deposits many metres thick over extensive areas of the planetary surface from mid-latitudes to the poles formed between 2.1 and 0.4 Myr ago due to variations in obliquity reaching 30–35°. At present the planet is in interglacial mode and the ice-rich deposits are undergoing reworking, degradation and retreat. During glacial times, unlike Earth, warmer polar climates prevail during which sediment transport is enhanced.

## 23.15 Suborbital surprises: reefs and speleothem as fine-scale tuners of the Pleistocene sea-level curve

While sea-level records interpreted from oxygen isotope analyses of deep-sea sediment cores provide sufficient resolution to recognize interglacial–glacial sea-level changes and also less marked interstadial–stadial variations, this record only allows resolution of sea-level to $\pm 10$ m. Higher resolution records of sea-level variation are available only from a few emergent ocean island and continental records, which show, for example, that sea-level during the MIS 5e sustained highstand began sometime around 128–130 $\pm$ 1 ka and reached 3–6 m higher than present. The elevation of fossil reef crests, with their distinctive architecture, coral species and boulder gravels, is a convenient dipstick of use in calibrating the Pleistocene and older sea-level curve. This is primarily because of the present ability of mass spectrometers to date diagenetically unaltered coral by means of $^{230}$Th determinations, routinely obtaining ages from 130 ka accurate to within 2 kyr. However, in most tectonically stable areas, reef crests only reach an elevation of +3 m whereas other intertidal and quasi-sea-level indicators like *Lithophaga*-bored notches indicate clear highstand at 5–6 m. In the past decade it has become clear that uplifted and dated MIS 5e coral reef terraces record rapid sea-level fluctuations, both within the sustained highstand and in the very early part of MIS 5e around 135–130 ($\pm$2) ka. These sea-level changes have been corroborated and extended by a continuous, high-resolution oxygen isotope sediment record from the Red Sea, but that record itself is partly indirect. Extremely rapid rates of MIS 5e sea-level changes around 130 ka (~50 m/ka, sustained over 1000–2000 yr) inferred from Papua New Guinea coral data are much faster than those indicated by the marine oxygen isotope records. It is therefore important that these poorly constrained and still

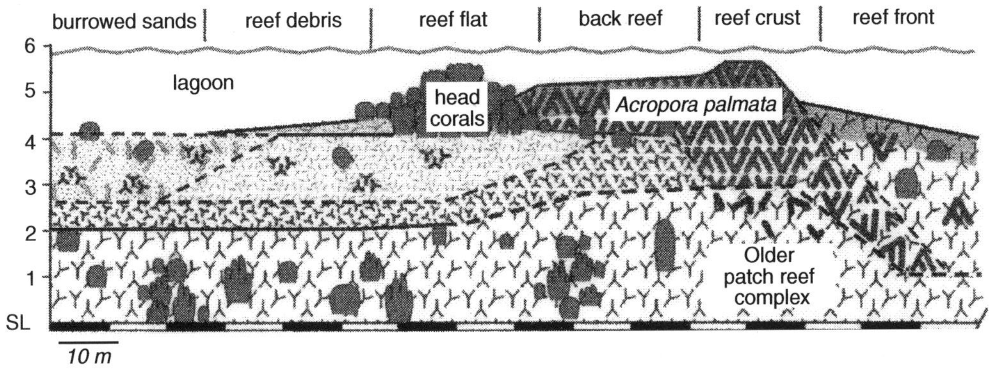


**Fig. 23.23** Reconstruction of upper and youngest MIS 5e reef development at Xcaret, Yucatan, Mexico during the last interglacial. (After Blanchon *et al.*, 2009). The older patch reef complex was a lagoonal assemblage associated with a slightly lower reef crest about 5ky before the date of the upper reef.

controversial findings are either supported or refuted by new observations made from records in other parts of the world.

In the tectonically stable Yucatan Peninsula, Mexico two distinct reef crests can be mapped (Fig. 23.23). The lower reef tract crops out for 500 m with the *Acropora palmata* colonized crest extending from below mean sea level to +3 m. The upper reef tract at maximum elevation +6 m crops out ~150 m inland and its crest parallels the lower one for ~400 m. It clearly developed in a retrogradational (backstepping) mode since mapping shows it grew on the substrate provided by the back-reef lagoon of the lower reef tract. That this backstepping was relatively sudden is indicated by an abrupt erosive contact between the upper reef and its lagoonal substrate. The eventual demise of both reef tracts is traced in a veneer of beach sediments that moved down the reefal topography as highstand gave way to falling sea level during stadial MIS 5d. From this field evidence it appears that the MIS 5e highstand was a pulsating affair, with an earlier stable +3 m sea level which relatively rapidly increased to +6 m. An intensive mapping and sampling programme involving an extremely parsimonious selection of the most accurate ages for groups of coral sampled from the two crests shows that the lower +3 m reef, together with additional data from other tectonically stable areas (Bahamas, Western Australia), was actively accreting between 126 and 121 ka before a rapid rise led to the landward relocation of the reef to its upper limit of +6 m for a further 3.5 kyr. Although the rate of the rapid rise cannot at present be computed it is inferred to have been due to a rapid ice-sheet melting event triggered by some unknown climatic or oceanographic instability. The consequences of such alarming, probably ecologically rapid, perturbations in interglacial sea level is of some relevance to present times.

A second example of suborbital sea-level change early in the last interglacial, this time sudden and startling falls, comes from carbonate sediments defining an extensive MIS 5e marine terrace level that has been tectonically uplifted along the shoreline of the Gulf of Corinth rift, central Greece (Fig. 23.24). The deposits demonstrate two rapid suborbitally forced sea-level oscillations in the early part of MIS 5e. The evidence comes from microbial bioherms with intergrown marine coralline algae interpreted as early highstand deposits. Presence of thin vadose flowstone (speleothem) coating interbioherm surfaces mark a shortlived regression of > 10 m, followed by sea-level recovery and re-establishment of the highstand, marked by coralline algae coating interbioherm cavity surfaces. These marine algae are then coated by a younger vadose flowstone with a $^{230}$Th date of 134.8 ± 2.0 ka. The dated flowstone is itself encrusted

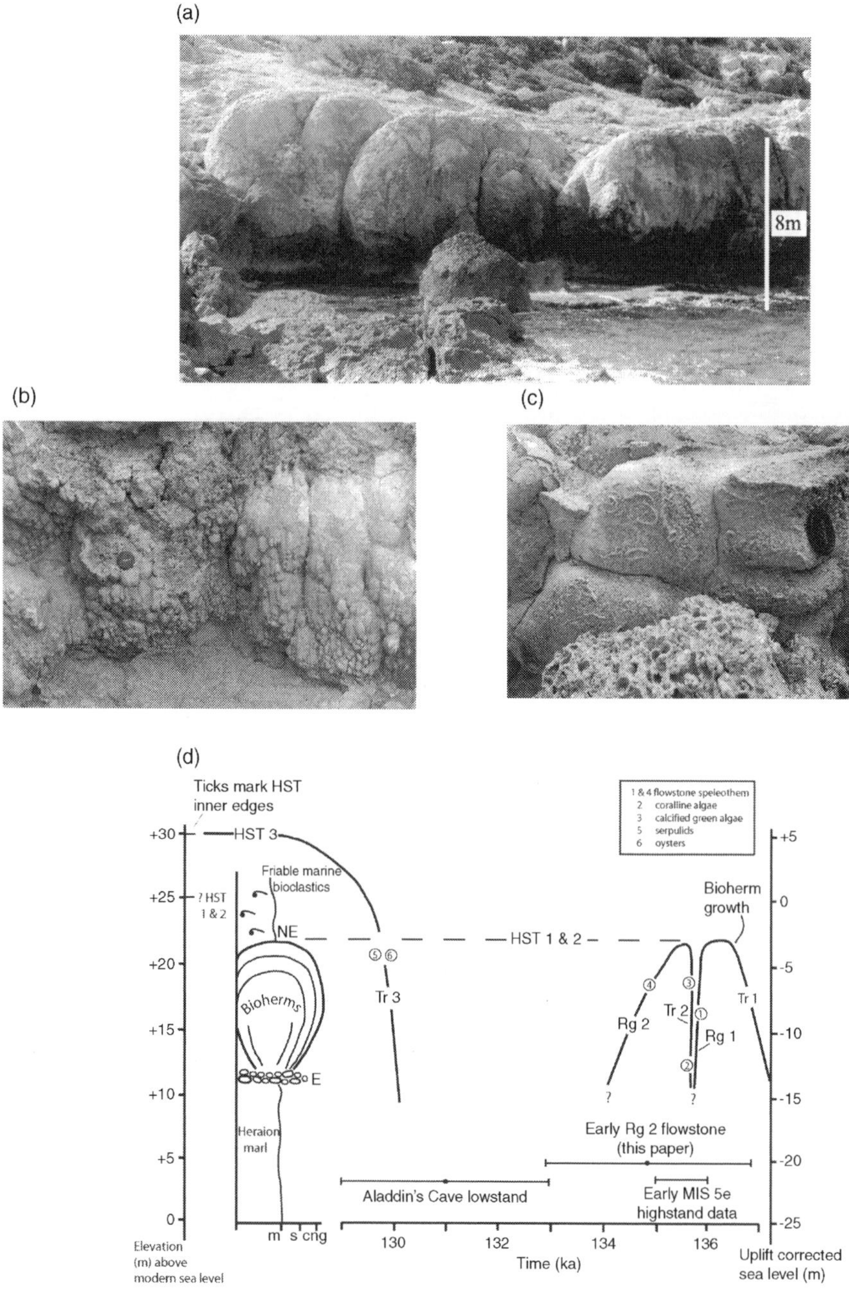


**Fig. 23.24** Evidence for suborbital sea-level changes from an uplifting coastline in central Greece. (a) Uplifted early MIS 5e algal bioherms. (b) Vadose flowstone dated by U–Th in cracks within bioherms. (c) Marine serpulid worms encrusting flowstone. (d) Deduced sea-level history. See text for discussion. (After Andrews *et al.*, 2007.)

by marine serpulid worm fauna and the entire sequence overlain by highstand marine sediments and marine aragonite cements dated to 114–118 ka, part of the sustained MIS 5e highstand. The age of the younger flowstone demonstrates that the early highstand occurred before $134.8 \pm 2.0$ ka, and uplift arguments suggest that the bioherms are unlikely to be older than ~136 ka. These data are consistent with the notion that most of termination II (TII) sea-level rise had occurred before 135 ka; indeed they suggest sea-level at this time reached about 2–4 ($\pm 4$) m below present sea-level. This early highstand was itself punctuated by a rapid sea-level oscillation of > 10 m (as yet undated), and this oscillation, supported by TII sea-level data from the Red Sea probably occurred in about 1000 yr. The flowstone dated at $134.8 \pm 2.0$ ka is interpreted to record the early part of the 'Aladdin's Cave' regressive event from Papua New Guinea, although in Greece only the first 16 m of the event is recorded.

## 23.16 Speleothem: Rosetta stone for past climate (provided by J.E. Mason)

Amongst the different types of speleothem (section 2.9), stalagmites are most suited to palaeoclimatic study because their internal structure and stratigraphy is relatively simple and the dynamics of their formation is reasonably well understood. Their internal structure comprises a series of growth layers, representing successive generations of stable low-Mg calcite precipitate. The apex of each lamina represents the point where the drip-water was received. An axial section through the stalagmite intersects these planar zones and provides a stratigraphic sequence through the stalagmite (Fig. 23.25). Geochemical analysis is made along samples taken from this stalagmite axis and the results can be related to palaeoclimatic parameters when chronologically constrained by uranium-series dating of the calcite.

### Uranium-series dating

Uranium-series methods provide the most reliable and precise method for dating speleothems that are less than ~500 ka old. The speleothem precipitated from meteoric waters contains dissolved components from bedrock. Thorium (Th) is commonly undectectable in these waters due to its very low solubility, whereas the waters may contain 0.1–3 ppm of U carried as a complex of uranyl$^{2+}$ ions with either carbonate or

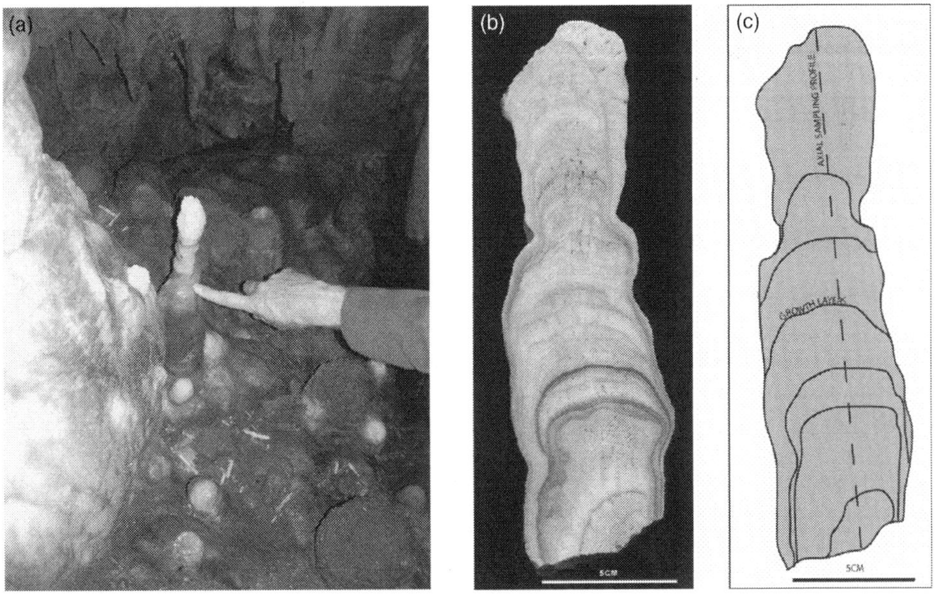


**Fig. 23.25** (a) Attractive stalagmite group with finger pointing to sampled example. (b) Sliced and scanned section through sampled stalagmite. (c) Sketch to show growth layers and trace of axial sampling transect.

sulphate ions or dissolved organic species. At the time of its deposition U is co-precipitated in the calcite lattice. The low content of Th in recharge waters hypothetically leads to correspondingly low (effectively absent) Th contents of freshly deposited speleothem. The ratio of daughter ($^{230}Th$) to parent ($^{234}U$) isotope then increases with time to a limit of unit activity ratio (secular equilibrium). These and other isotopes are measured in a calcite sample and because the relationship between $^{230}U/^{234}U$ in a sample and time since deposition is known, the age of the sample can then be calculated.

### Stable isotopes in speleothems

The stable isotopic composition of a stalagmite calcite sample is expressed as a departure of the ratio between the heavier and lighter isotope of an element from that of a standard, e.g. for oxygen:

$$\delta^{18}O‰ = \left( \frac{(^{18}O/^{16}O)_{\text{sample}} - (^{18}O/^{16}O)_{\text{standard}}}{(^{18}O/^{16}O)_{\text{standard}}} \right) \times 10^3$$

For carbon, the isotopes $^{13}C/^{12}C$ and for hydrogen $^{2}H/^{1}H$ are the isotope pairs used. The resulting values are expressed in the delta ($\delta$) notation in per mill (‰) units, more negative values representing lower ratios in the sample relative to the standard. For carbonates the standard used is the Vienna Pee Dee Belemnite (VPDB). For waters the standard used is the Vienna Standard Mean Ocean Water (VSMOW). The carbon and oxygen isotopic values are quoted relative to these standards.

The rationale for interpreting oxygen isotopes in speleothems is that, when deposited under equilibrium conditions: (i) $\delta^{18}O$ of the speleothem calcite reflects in part the $\delta^{18}O$ of the meteoric water from which it formed and this can be used to infer palaeoclimatic changes based on water composition and (ii) there is a temperature dependent water–calcite oxygen isotopic fractionation (partitioning of the isotopes of oxygen between water and calcite) which can be used to relate $\delta^{18}O$ calcite to formation temperature. Cave temperatures closely reflect the mean annual air temperature at the surface of the cave and so therefore this measurement is highly useful climatically. Although these principles are true, in natural systems there are potentially a large number of other fractionations and other effects that may influence the $\delta^{18}O$ of water vapour on its route from the ocean to cave-speleothem. Isolating different components of palaeoclimate such as temperature thus becomes difficult and requires a detailed understanding of all the different factors which may have affected $\delta^{18}O$. For example we have already seen how altitude affects fractionation (section 6.3).

The carbon isotopic composition of speleothem calcite ($\delta^{13}C$) may also be related to palaeoclimate, for example both the amount and type of vegetation above a cave can affect the $\delta^{13}C$ composition of cave drip-water. Changes in $\delta^{13}C$ values of speleothems have been interpreted to reflect climate-driven change in vegetation type (e.g. C3- to C4-dominated plant assemblages). Both thickness and productivity of the soil can also affect $\delta^{13}C$ composition. Incomplete equilibration with soil $CO_2$ and percolating water may also be responsible for elevated $\delta^{13}C$, and this may be associated with wetter periods.

It is also possible to isotopically analyse (for $\delta^{18}O$ and $\delta^{2}H$) palaeowater contained in sealed microscopic cavities called *fluid inclusions* within the speleothem calcite itself. These measurements provide direct information on rainfall provenance in the past. They may also yield information on temperature changes in the cave as the $\delta^{18}O$ of water in combination with the $\delta^{18}O$ composition of calcite can be used to determine the cave temperature at time of speleothem formation.

### Other climatic proxies in speleothems

Trace elements that substitute for Ca in the $CaCO_3$ lattice (e.g. Sr, Ba, U and Mg) can also be measured and used to yield palaeoclimatic information. For example increasing Mg/Ca and Sr/Ca ratios in cave waters have been related to relative aridity and this signal transferred into speleothem calcite. Different petrographic fabrics, morphologies and growth rates in speleothems can potentially be related to physical and chemical parameters in a cave, and hence they too are useful in palaeoclimatic study, as is visible and nonvisible growth banding (which may be annual).

### Case studies

Some of the most globally important speleothem records come from Hulu and Dongge caves, China, which span the past 224 kyr (Fig. 23.26). Fluctuations in oxygen isotopic composition in the Hulu

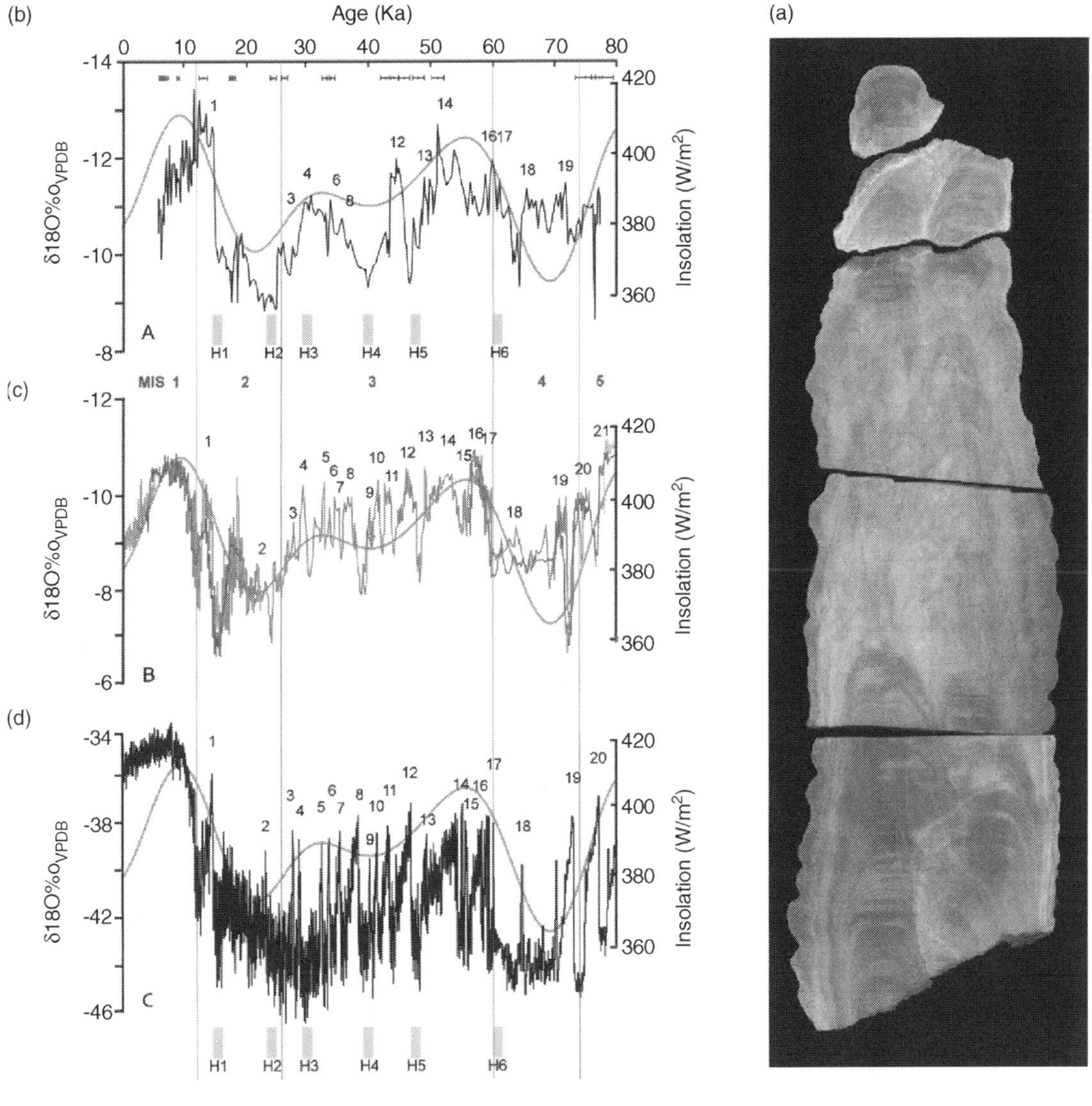


**Fig. 23.26** (a) Photo of sampled stalagmite (length ~60 cm) from Karaka cave, northeast Turkey. (b) Time series of $\delta^{18}$0 with insolation curve (greyscale) for Karaca cave. (After Mason, 2010.) (c) Time series of $\delta^{18}$0 with insolation curve (greyscale) for Hulu cave, China. (After Wang *et al.*, 2008.) (d) Time series of $\delta^{18}$0 with insolation curve (greyscale) for GRIP ice-core. Also shown are chief Heinrich events (H1, etc.) as defined from North Atlantic sediment cores.

speleothems have been interpreted as a measure of East Asian Monsoon intensity. Periods of enhanced monsoonal activity, termed Chinese Interstadials (CIS), have been correlated with large-scale temperature fluctuations (Dansgaard–Oeschger, DO, cycles) indentified in Greenland ice-cores, demonstrating large-scale climatic teleconnection. Oxygen isotope ratios in a stalagmite from Karaca Cave in northeast Turkey also exhibit similar temporal patterns to the Greenland ice-core oxygen isotope records and to the Chinese speleothem records of East Asian Monsoon fluctuations over the period 77 ka to 6 ka (Fig. 23.26). These variations have been related to seasonality of precipitation and/or switching of moisture source areas in response to (i) precessional-scale changes in insolation and (ii) stadial–interstadial climatic oscillations originating in the North Atlantic Basin.

Another important speleothem record spanning ~250 ka, comes from Soreq cave, Israel. Here oxygen isotope ratios have been related to temperature and rainfall amount on glacial–interglacial timescales. Good correlation has also been found between Soreq speleothems and oxygen isotopes from Eastern Mediterranean Sea core. A 30–85 ka record from Villars cave in southwest France shows evidence for rapid climate oscillations coincident with Dansgaard–Oeschger events in both the oxygen and carbon isotopic records. The oxygen isotopic signature is similar to that from Soreq cave and marine records; large, rapid shifts in temperature and vegetation in western Europe are inferred. There is also evidence for a long phase of extremely cold climate in southwest France between 61.2–67.4 ka.

## Further reading

*Evaporite fluid inclusions, global tectonics and seawater composition*

Discovery papers by Sandberg (1983), Hardie (1996) and Morse *et al.* (1997). Mg/Ca ratios by Dickson (2002, 2004), Stanley & Hardie (1998), Stanley *et al.* (2002), Ries (2004), Steuber & Rauch (2005) and Stanley (2006). Fluid inclusions in marine halite deposits by Lowenstein *et al.* (2001, 2003, 2005), Horita *et al.* (2002), Demicco *et al.* (2005) and Timofeef *et al.* (2006). Role of extinction rates by Kiessling *et al.* (2008). Mesozoic/Cenozoic seafloor spreading rates by Gaffin (1987), Engebretson *et al.* (1992), Heller *et al.* (1996), Rowley *et al.* (2001), Gaína *et al.* (2003) and Hays & Pitman (1973). Seismic stratigraphic approach by Vail *et al.* (1977) and Kominz (1984). Review critique by Miall & Miall (2001).

*Banded Ironstone Formations (bifs), rise of cyanobacteria, Great Oxidation Event (goe) and secular change in global tectonics*

General by Konhauser *et al.* (2007). Early oxygen by Ohmoto *et al.* (1993), Ohmoto (1996) and Hoashi *et al.* (2009). Oldest stromatolite reef by Allwood *et al.* (2006). Nickel story by Konhauser *et al.* (2009).

*Tibetan Plateau Uplift; palaeoaltimetry from carbonate stable isotopes, monsoon intensity and exhumation from sediment budgets*

Early Tibet expeditions and tectono-sedimentation by Allègre *et al.* (1984), Chang *et al.* (1986, 1988), Coward *et al.* (1988), Kidd *et al.* (1988) and Leeder *et al.* (1988). Mantle delamination by England & Houseman (1986). Marine 87Sr/86Sr and Tibetan-Himalayan uplift by Richter *et al.* (1992). Urey silicate-weathering mechanism by Raymo (1991) and Raymo & Ruddiman (1992). Young onset for Tibetan uplift by Molnar *et al.* (1993). Soil and lake carbonates and O2 stable isotope-based palaeoaltimetry by Rowley *et al.* (2001), Quade *et al.* (2007), Garzione *et al.* (2000), Cyr *et al.* (2005), Graham *et al.* (2005), Rowley & Currie (2006) and DeCelles *et al.* (2007). Fossil leaf physiognomy by Spicer *et al.* (2003). High-altitude vegetation by Dupont-Nivet *et al.* (2008). Northward younging of uplift by Rowley & Currie (2006). Discussion by Molnar *et al.* (2006). Northern plateau margin by Hubbard & Shaw (2009). Tibet uplift and Cenozoic cooling by Garzione (2008). Tectonic uplift, exhumation and monsoon intensity by Clift *et al.* (2008). Models for non-Tibetan involvement by Boos & Kuang (2010).

*Colorado Plateau uplift and Grand Canyon incision dated by speleothem carbonate*

Discovery paper by Polyak *et al.* (2008). Critique by Karlstrom *et al.* (2008).

*River channels and large-scale regional tilting*

Allen & Davies (2007).

*Regional drainage reversal*

Holmes (1965) and Holförster & Schmidt (2007).

*Tectonic sedimentology of foreland basins*

Pioneering coupled models by Koons (1989a,b) and Beaumont *et al.* (1992). Second generation coupled models by Whipple & Meade (2004) and Simpson (2006). Andean budgeting by Barnes & Heins (2009). Real world basins by Ford *et al.* (1999).

*Lengthwise fault growth and fault amalgamation*

New Zealand by Lamarche *et al.* (2000) and Taylor *et al.* (2004). Previous conceptual, field and theoretical proposals by Peacock & Sanderson (1991, 1994), Trudgill & Cartwright (1994) and Cartwright *et al.* (1995, 1996). Suez rift by Jackson *et al.* (2002), Young *et al.* (2002) and Carr *et al.* (2003).

*Rivers, basement uplifts and fault growth*

Southern Rio Grande rift by Mack *et al.* (2006).

*Unsteady strain and the sedimentary response*

Loreto basin by Dorsey *et al.* (1997) and Mortimer *et al.* (2005).

*Tectonics versus climate as depositional controls*

Flinders Ranges by Quigley *et al.* (2007). Palomas basin, Rio Grande rift USA by Mack & Leeder (1999). Altiplano-Puna plateau by Coutand *et al.* (2006).

*River equilibrium, incision and aggradation—away from the knee-jerk of tectonic explanation*

Equilibrium time by Paola *et al.* (1992a,b). Diffusional approach by Begin *et al.* (1981) and Begin (1988). Rio Grande rift by Mack *et al.* (2006) and Leeder & Mack (2007). Ouarzazate foreland basin by Arboleya *et al.* (2009a,b).

*Integrated sedimentary systems: modelling tectonics, sediment yield and sea level change*

Rift basin 3D all-singing models by Gawthorpe *et al.* (2003) and Ritchie *et al.* (2004a,b). Foreland basin all-dancing models by Clevis *et al.* (2004a,b).

*Extraterrestrial sedimentology—atmospheric and water flows on Mars*

Planetary accretion and sand grains by Herbst *et al.* (2008). Soil and diagenesis by Elwood Madden *et al.* (2004), Yen *et al.* (2005), Poulet *et al.* (2005), Haskin *et al.* (2005), Greenwood & Blake (2006), Andrews-Hanna *et al.* (2007), Chevrier *et al.* (2007), Ehlmann *et al.* (2008) and Mustard *et al.* (2008). Wind-blown sands and ripples by Jerolmack *et al.* (2006) and Sullivan *et al.* (2005). Dust transport by Fenton (2007). Dust devils by Balme *et al.* (2003), Ringrose *et al.* (2003), Whelley & Greeley (2008). Glacial features by Neukum *et al.* (2004), Bibring *et al.* (2004), Head *et al.* (2005), Vaniman *et al.* (2004). Ice Ages by Head *et al.* (2003), Tanaka (2005), Schorghofer (2007). Water flow features by Kraal *et al.* (2008), Christensen (2003), Grotzinger *et al.* (2006). Oceans Perron *et al.* (2007), Murray *et al.* (2005).

*Suborbital surprises: reefs and speleothem as fine-scale tuners of the Pleistocene sea-level curve*

General evidence for higher MIS 5e sea level by Chen *et al.* (1991), Zhu *et al.* (1993), Carew & Mylroie (1995), Hearty & Kindler (1995), Stirling *et al.* (1998), McCulloch & Esat (2000), Muhs *et al.* (2002) and Jedoui *et al.* (2003). Twin main MIS 5e peaks by Blanchon *et al.* (2009). Early MIS 5e sea-level fall by Andrews *et al.* (2007).

*Speleothem: Rosetta stone for past climate*

Hulu cave by Wang *et al.* (2008). Soreq cave by Almogi-Labin *et al.* (2009). Villars cave by Couchoud *et al.* (2009). Karaca cave by Mason (2010).

# COOKIES (BRIEF REFRESHER OR STUDY GUIDES AND DERIVATIONS)

*...Mr Dormer, ' he said to a young gentleman whose attention seemed to be wandering, 'pray define a logarithm.'*

*Dormer blushed, straightened himself, and said, ' A logarithm, sir, is when you raise ten to the power that gives the number you first thought of.'*

*After a few more answers of this kind Jack desired Mr Walkinshaw to return to his remarks on the principles of spherical trigonometry...*

Patrick O'Brian, from *The Yellow Admiral*, Harper Collins 1997 (Jack Aubrey and the mathematical education of his young gentlemen)

## Cookie 1: Properties of water: key player in rock weathering and mineral precipitation

Water molecules possess poles just like a magnet and similarly act as electric dipoles. So, when placed between plates of a charged capacitor, water molecules orientate themselves with the positive hydrogens towards the negative plate and the negative oxygens towards the positive plate. The effect is similar to the orientation of a magnet in a magnetic field. The strength of the dipole moment is determined by the product of the magnitude of the charges and the distance between them. Polar water molecules result from asymmetric covalent bonding where the strongly electronegative (electron-loving) oxygen atom takes a majority share of available electrons. This gives oxygen a partial negative charge at the expense of a partial positive charge on the less electronegative hydrogen atoms. The chemical bonds between the oxygen and the two hydrogen atoms arise from the overlap of the *p*-electron clouds of the oxygen atom with the electron clouds of the hydrogen atoms. The bonds should theoretically be at 90° to one another and consist of a molecular orbital occupied by a pair of electrons. In practice, because of repulsion between the hydrogen atoms, the bond angles meet at an angle of just over 105°. When the water molecules come together *en masse*, they interact loosely by hydrogen bonding because of the polar nature of the

*Sedimentology and Sedimentary Basins: From Turbulence to Tectonics*, 2nd edition. © Mike Leeder.
Published 2011 by Blackwell Publishing Ltd.

$H_2O$ covalent bonds. They form into tetrahedral groups of four by this process (Cookie 1 Fig. 1).

Many of the distinctive properties and reactive tendencies of water are explained by this molecular structure.

1 It is a highly effective solvent for ionically bonded compounds (NaCl is a familiar example) because of its polar nature. In this dissolution process the positive and negative ends of the water molecule attach themselves to the compound's negative and positive surface ions respectively, neutralizing their charges and enabling mechanical agitation to separate the constituent ions. Once in solution the ions are surrounded by water molecules whose combined electrostatic attraction may partly balance the ion's potential energy, thus stabilizing it by

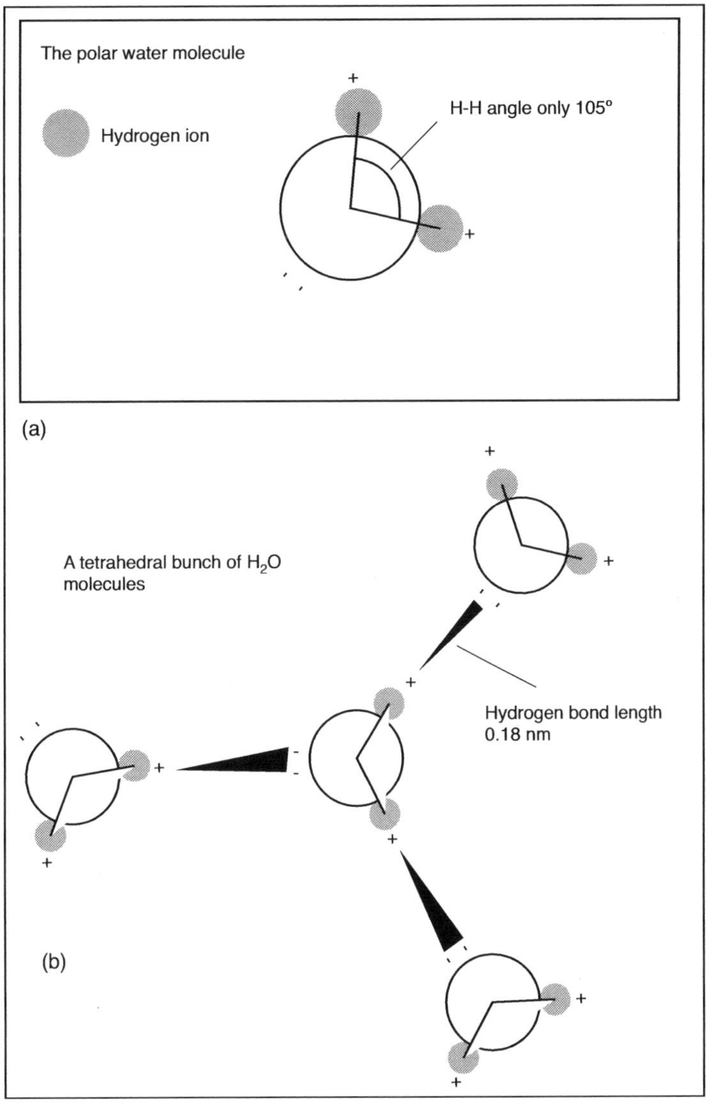


**Cookie 1 Fig. 1** (a) The polar water molecule. (b) Tetrahedral groupings of polar water molecules.

hydration, an ion–dipole interaction. The degree of stabilization by hydration is determined by the ionic potential, i.e. the ratio between the electrostatic charge and ionic radius (see Chapter 1).

2 The tetrahedral groupings of water molecules cause such properties as high surface tension and capillarity. High surface tension causes water to have a high saturation effect in porous materials, penetrating tiny cracks and crevices where it aids mineral solution, reprecipitation and mechanical failure.

3 Decreasing temperature is accompanied by an increase in density as a result of contraction caused by decreasing molecular thermal agitation. This effect is opposed by the production of more and more hydrogen bonds. The maximum density of freshwater is reached at 4 °C, well before the onset of freezing, but this temperature is depressed by increasing pressure (the *thermobaric* effect) and salinity. Expansion dominates below 4 °C as the tetrahedral molecular groups also begin to be taken up into hexagonal ice structures and so the density decreases. Expansion continues until −22 °C at which ice achieves its minimum density and maximum expansive pressure. Hexagonal ice crystals have a maximum growth rate normal to the basal plane, so that ice whisker crystals growing in microcrevices of rocks (see note above on water's high saturation) can exert high boundary stresses.

## Cookie 2: Metallic ions, electron transfer and Eh: chemical control of weathering processes

The loss of electrons from an element or ion is termed *oxidation* and leads to an increase in positive valency or a decrease in negative valency, and vice versa for *reduction*. For example, consider the oxidation of ferrous iron as might occur in water-mineral reactions involving silicate minerals containing $Fe^{2+}$, such as pyroxene, olivine, biotite or hornblende, or the ferrous disulphide, pyrite:

$$Fe^{2+} \leftrightarrow Fe^{3+} + e^{-} \qquad (C2.1)$$

Changes in oxidation state are important in chemical weathering and other sedimentary geochemical processes because they affect the solubility of metal species, particularly the oxides, hydroxides and sulphides of the transition-group metals. In the case of Fe, the $Fe^{2+}$ (ferrous) form is a mild reducing agent because of the relative ease with which it surrenders its electron. The $Fe^{3+}$ (ferric) form, by way of contrast, is a highly stable structural state because of its half-filled valence electron shell. The $Fe^{2+}$–O bond has a much lower site energy than the $Fe^{3+}$–O bond, and therefore $Fe^{2+}$ compounds are much more mobile (soluble). Highly important cycles of electron transfer occur between reduced and oxidized Fe and organic ligands (a ligand is an attaching compound, usually to a transition metal with its many vacant electron orbitals) at the boundary between oxygen-poor (anoxic) and oxygen-rich (oxic) zones. This produces *gley*-type soil horizons, blotched yellow-grey by mixtures of the two iron species (section 1.5). Dissolved oxygen in surface waters is the most important natural oxidizing agent on account of its very high electronegativity (second only to fluorine in its attraction for electrons).

It is possible to measure the oxidation–reduction potential (*redox potential*) by noting the potential difference ($E$) produced between an immersed inert electrode, usually platinum, and a hydrogen electrode of known potential. This redox potential for reactions involving protons, commonly termed *Eh* by geochemists, is compared to the arbitrary value of 0.00 mV for hydrogen in the reaction:

$$2H^{+} + 2e^{-} \leftrightarrow H_2 \qquad (C2.2)$$

at 25 °C and 1 atm pressure at a concentration of 1 mol/kg (pH = 0). Negative values of Eh indicate reducing conditions, and positive values oxidizing conditions, with respect to the arbitrary hydrogen scale. If we know the standard oxidation–reduction potential for a particular reaction from laboratory measurements, and we possess a field measurement of Eh from a particular weathering zone, then it is possible to predict the type of dissolved oxidation state for a particular ion. For example, the standard potential of the $Fe^{2+}$–$Fe^{3+}$ couple of Equation (C2.1) is 0.77 mV. If, for example, our field measurement shows a reading of 0.5 mV under acidic conditions (pH = 2), we would be able to predict that $Fe^{2+}$ is the stable phase. However, the reactions that determine Eh are often very slow in the natural environment and the aqueous environment does not quickly come to equilibrium with the measuring electrodes. Thus field redox measurements generally give only semiquantitative information and tell us little about reaction rates or the attainment of equilibrium.

## Cookie 3: Natural waters as proton donors: Ph, Eh–Ph diagrams, acid hydrolysis and calcium carbonate weathering and precipitation

Water molecules separate into $H^+$ (protons) and $OH^-$ (hydroxyl) ions at all temperatures. Protons are responsible for the acidity of an aqueous solution and we can thus consider weathering in terms of proton donors and proton budgets. The molar concentration of $H^+$ ions in dilute solutions is so small and variable that acidity is expressed on a logarithmic scale, as the negative logarithm of the free $H^+$ concentration. This defines pH, with units of mol/kg (note that the higher the pH, the lower the proton concentration). At room temperature there are only $10^{-7}$ mol/kg of protons (with an equal number of $OH^-$) in pure water, and this is used as a reference value for a neutral pH of 7. Larger values of pH are termed *alkaline* and smaller values *acidic*. These conditions are possible because $H^+$ or $OH^-$ ions may be provided by other reactions and reactants in aqueous solutions.

A particularly informative graph may be produced by plotting Eh against pH. The area emphasized in Cookie 3 Fig. 1 shows the usual limits of Eh and pH found in near-surface environments. The lower limit to pH of about 4 is produced by natural concentrations of $CO_2$ and organic acids dissolved in surface and soil waters. The upper limit to pH of about 9 is reached by waters in contact with carbonate rocks and still in contact with atmospheric $CO_2$. Local conditions may sometimes fall well outside the usual limits. Thus, oxidation of pyrite gives very acid conditions, as illustrated by the severe pollution caused by groundwaters issuing from many abandoned mines (*acid mine drainage*). Most elements in the weathering zone are oxidized. Exceptions exist in waterlogged soils or at the boundary between poorly and well-drained soils (gleys noted previously). Below the water table, conditions are oxygen-poor and anaerobic bacteria are abundant. Here the reaction in Equation C2.2 is reversed and insoluble ferric iron is reduced to soluble ferrous iron.

No natural waters are pure, all contain greater or lesser quantities of ions in solution (solutes) in addition to the ubiquitous $H^+$ ions. We would naturally like to be able to understand what concentrations of reactants are produced from particular reactions. But what about the state of saturation of a solution? How can we tell whether an aqueous solution like rainwater or streamwater will dissolve rock-forming minerals, i.e. a state of undersaturation exists, or precipitate minerals, i.e. state of oversaturation exists? Take a surface weathering example. A kilogram of natural streamwater is found by analysis to contain $2 \times 10^{-3}$ g of $CaCO_3$ in solution. A quantity such as this is known as the *Ion Activity Product* (IAP) or concentration. Since the relative molar mass of $CaCO_3$ is $40 + 12 + (3 \times 16) = 100$, we have what is known as the activity ($a$) of the solution as $0.002/100 = 2 \times 10^{-5}$. Such a solution is extremely undersaturated since laboratory experiments would show that it could dissolve appreciably more $CaCO_3$. In fact at 25 °C the water could hold $2.91 \times 10^{-2}$ g $CaCO_3$, an order of magnitude more than our natural sample. The water is thus said to be undersaturated and capable of dissolving more $CaCO_3$ locally as it moves around.

Since many natural aqueous solutions come into contact with the atmosphere, we must also express the concentration of soluble gaseous components in a solution ($CO_2$ is really the only significant example). This is done by noting that any gas exerts a pressure ($P$) across an interface with fluid and that this will be in direct proportion to the mole fraction ($x_i$) of component $i$ in the gas. Thus we speak of the *partial pressure* ($p_i$) of component $i$ as:

$$p_i = x_i P \tag{C3.1}$$

Many natural weathering and sediment-forming reactions are reversible in that they can move either 'forwards' or 'backwards' depending upon local conditions of availability of ions, dissolved $H^+$ and so on. Thus in pure water (no $H^+$ or $OH^-$ ions) calcite dissociates by the reversible reaction:

$$CaCO_3 \leftrightarrow Ca^{2+} + CO_3^{2-} \tag{C3.2}$$

A state of equilibrium exists for every such reaction in which the rate left to right (forwards) is exactly balanced by the rate right to left (backwards). At this equilibrium the activity of the reaction products on either side of the equation are balanced, so we may define a constant term $K$ (for given temperature) as:

$$K = \frac{aCa^{2+}\, aCO_3^{2-}}{aCaCO_3} \tag{C3.3}$$

$K$ is called the *equilibrium constant* for any particular reaction at a specified temperature. For this case

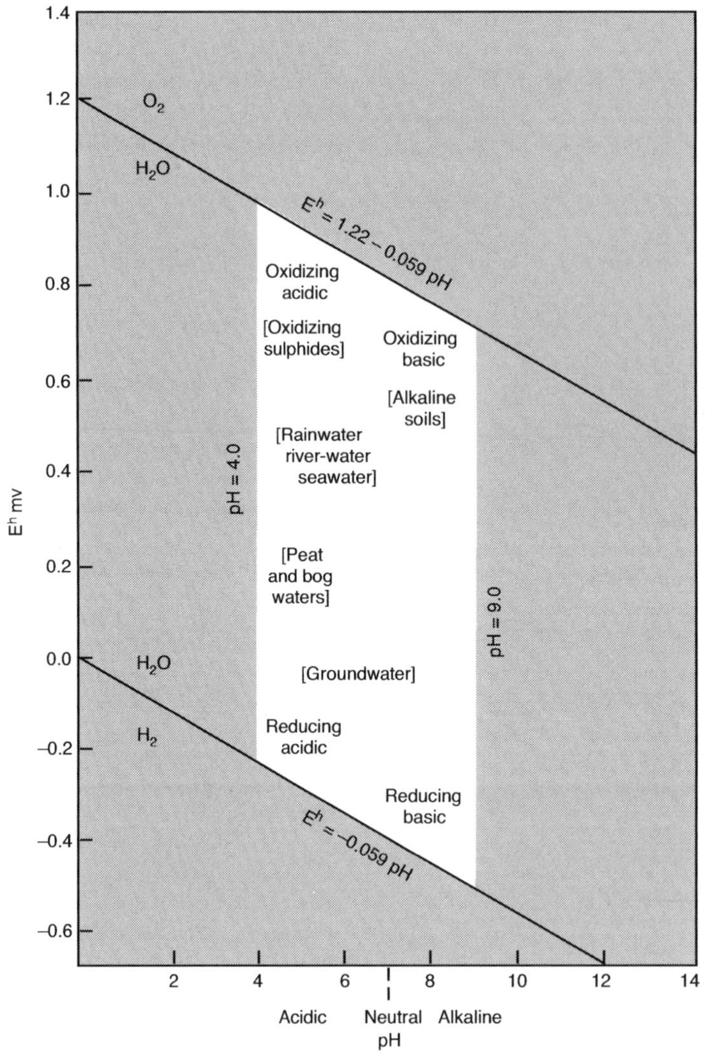


**Cookie 3 Fig. 1** An Eh–pH diagram to show the range expected in most earth-surface and near-surface aqueous environments. (After Andrews *et al.*, 2005.)

$CaCO_3$ is a pure solid and for all solids, activity $a = 1$. Thus:

$$K = aCa^{2+}\, aCO_3^{2-} \qquad (C3.4)$$

If we disturb the system, say by adding calcium ions, then the reaction will work to neutralize the change, preserving the value of $K$, in this case by moving into reverse to produce (precipitate) solid calcium carbonate. Generally for most salts whose dissolution in water is an endothermic (needing heat energy) process, an increase of temperature increases the value of $K$ and hence the solubility increases. This is not so for gases, for when temperature and/or pressure increase, solubility decreases. This makes the calcium carbonate weathering system in natural waters particularly complex, since dissolution is not controlled simply by the dissociation reaction above, but by reactions involving dissolved $CO_2$ whose source may be atmospheric or respiratory. In such cases $pCO_2$ is the critical quantity that controls pH and the rate of reaction.

Perhaps the most important source of $H^+$ ions is provided by atmospheric $CO_2$ dissolved in rain and by other humic and bacterially produced acids. Also, abundant biogenic $CO_2$ is produced by respiration of microorganisms in soil water. This is summarized by the following series of interlinked reactions for the combination of water and carbon dioxide and the two-stage ionization of carbonic acid:

$$H_2O + CO_2 \leftrightarrow H_2CO_3 \leftrightarrow H^+ + HCO_3^- \text{(1st stage)} \leftrightarrow 2H^+ + CO_3^{2-} \text{(2nd stage)} \quad (C3.5)$$

The dissociation of carbonic acid is strongly pH-dependent, the equilibrium pH of pure water in the $H_2O$–$CO_2$ system of Equation (C3.5) being about 5.7 at 25 °C. The equilibrium constants for the above reactions and for the whole carbonate system at various temperatures are given in Cookie 3 Table 1. It can be seen that:

1 $CO_2$ is less soluble at higher temperatures;
2 the first dissociation of carbonic acid ($K_1$) produces more hydrogen ions for acid reaction than the second ($K_2$).

These conclusions have many sedimentological applications, to which we return from time to time in the main text.

Finally, it has been found by field measurements and laboratory simulations that $pCO_2$ can reach large values (several times atmospheric) in the subsoil vadose zone because of the small rates of diffusion of respiratory $CO_2$ to the surface. This contribution of respiratory $CO_2$ in soils decreases the pH very substantially below that of normal rainwater, a factor of interest in cave and speleothem genesis (section 2.9). Such enhanced values of pH due to soil $CO_2$ are found to persist even in deep aquifers, such as the Florida aquifer, USA, where the chemical conditions are important in controlling mineral reactions.

**Cookie 3 Table 1** Various equilibrium constants ($K$) for the carbonate system of Equation C3.5. The equilibrium constants are: $K_{CO_2} = aH_2CO_3/pCO_2$, $K_1 = aH^+ aHCO_3^- / aH_2CO_3$, $K_2 = aH^+ aCO_3^{2-} / aHCO_3^-$, $K_{calcite}$ and $K_{aragonite}$ both $aCa^{2+} aCO_3^{2-}$

| $T$ (°C) | $K_{CO2}$ | $K_1$ | $K_2$ | $K_{calcite}$ | $K_{aragonite}$ |
|---|---|---|---|---|---|
| 5 | $10^{-1.19}$ | $10^{-6.52}$ | $10^{-10.55}$ | $10^{-8.39}$ | $10^{-8.24}$ |
| 25 | $10^{-1.47}$ | $10^{-6.35}$ | $10^{-10..33}$ | $10^{-8.48}$ | $10^{-8.34}$ |
| 60 | $10^{-1.78}$ | $10^{-6.29}$ | $10^{-10.14}$ | $10^{-8.76}$ | $10^{-8.64}$ |
| 90 | $10^{-1.94}$ | $10^{-6.38}$ | $10^{-10.14}$ | $10^{-9.12}$ | $10^{-9.02}$ |

## Cookie 4: The thermodynamics of weathering reactions that produce sediment

Chemical weathering is basically about breaking atomic bonds or changing the electron configuration of atoms in pre-existing minerals making up catchment bedrock. In order to do this, energy must be exchanged and the energy budget of the mineral reactions all balanced. In the first place, according to the scheme of Curtis (1976), the likelihood of a particular reaction occurring may be decided by reference to the change in free energy of a reaction ($\Delta G_r$). This the sum of the free energies of formation ($\Delta G_f°$) of all the reaction products minus those of the reactants. When negative, reactions will proceed spontaneously, although it must be said that the *rate* of reaction is *not* specified, though the more negative the value of $\Delta G_r$, the more likely a reaction will occur. In order to assess mineral stability in this way we write specific weathering reactions with the main silicate minerals of igneous rocks as the reactants. Consider the Ca-rich feldspar, anorthite:

$$\underset{\text{anorthite}}{CaAl_2Si_2O_8} + \underset{\text{in sol}^n}{2H^+} + H_2O \rightarrow \underset{\text{kaolinite}}{Al_2Si_2O_5(OH)_4} + \underset{\text{in sol}^n}{Ca^{2+}} \quad (C4.1)$$

In this reaction, $\Delta G_r$ is calculated as $-23.9$ kcal/mol, the negative energy change indicating that the feldspar will spontaneously react with hydrogen ions in aqueous solutions to form the clay mineral kaolinite plus calcium ions.

It is also possible to assess the stability of minerals directly from solubility data for IAP (see Cookie 3) and to compare it with the solubility product, $K$. When $IAP/K = 1$ then by definition $\Delta G$ is zero, and the various mineral phases are in equilibrium with the soil water. When $IAP/K < 1$ then $\Delta G_r$ is negative and the reactant mineral will dissolve. When the ratio is positive then the reactive mineral will precipitate. In all silicate weathering reactions it is found that IAP is directly dependant on $H^+$ activity (pH). (Cookie 4 Fig. 1)

Finally, despite the elegance of thermodynamics, mineral–rock–water reactions in the oxidative and organic-rich weathering zone are *highly complex.*

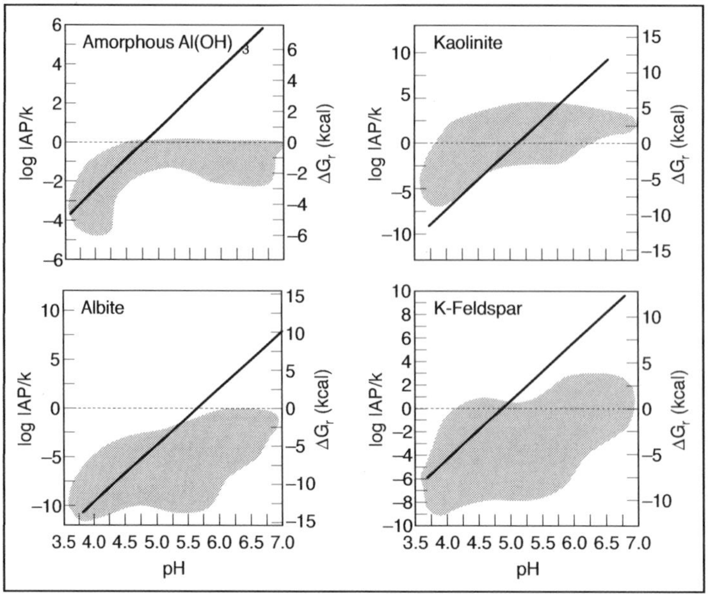


**Cookie 4 Fig. 1** The extent of dissolution, equilibrium (dotted lines) and precipitation achieved by the reaction of aqueous solutions on various minerals at different pH under standard earth-surface conditions. Shaded areas indicate envelope of data points. $\Delta G_r$ is the free energy of reaction. (After White, 1995.)

There is no easy answer as to rates and mechanisms because of the large number of variables involved (temperature, mineral surface properties, pH, solvent ionic strength, role of other ions, nonequilibrium thermodynamics). Successful computational schemes for given catchments use field data, a tribute to the persistent efforts of field geochemists.

## Cookie 5: Chemical weathering of silicate minerals as a geosink for global atmospheric $CO_2$

Buffer reactions use up any increased quantity of a reactant produced by other inputs or reactions, leading to no overall change in quantities. Such reactions are particularly relevant to maintenance or otherwise of pH. Equation (C5.1), written below in summary form, buffers the pH of the oceans in the face of increased atmospheric $CO_2$ concentrations, since extra $CO_2$ is used in the forward reaction:

$$CaCO_3 + CO_2 + H_2O \leftrightarrow Ca^{2+} + 2HCO_3^- \quad (C5.1)$$

Although this is true, note that the reaction is reversible so that in the long term a steady state is set up that will reissue the $CO_2$ to the ocean and atmosphere during inorganic and organic calcium-carbonate-precipitating reactions.

For a global sink to exist for $CO_2$ there must be longer term removal of $CO_2$. This is achieved by calcium-bearing silicate weathering systems, like that of anorthite, the Ca-feldspar considered in **Cookie 4**. A generalized form of this reaction, termed the *Urey equation*, written as:

$$CaSiCO_3 + 2CO_2 + 3H_2O \leftrightarrow Ca^{2+} + 2HCO_3^- + H_4SiO_4 \quad (C5.2)$$

Here, for every two carbon atoms used up in weathering, only one will be returned to the atmosphere by the ocean carbonate sink backward reaction of Equation C5.1. We see that any changes induced in the rate of silicate weathering can moderate atmospheric $CO_2$ composition. Further development of this idea leads to a global rock cycle control on Earth System dynamics (Chapter 10).

## Cookie 6: The rates and mechanisms of chemical weathering

Simply and generally we may view the loss of mass, $\Delta M$, in moles, resulting from chemical weathering over a time, $t$, as the product

$$\Delta M = kSt \quad (C6.1)$$

where $k$ is an overall rate constant (mol/m$^2$/s) and $S$ is the total reactive area of mineral surface (m$^2$). The temperature of weathering reactions is a key control over $k$. Like pure laboratory chemicals, natural minerals and the dilute acids that bathe them obey Arrhenius's law. This states that $k$ increases exponentially with increasing temperature according to:

$$k = Ae^{-[E_a/RT]} \quad (C6.2)$$

where $A$, the pre-exponential factor with units as $k$, and $E_a$, the activation energy with units kJ/mol, are known as the Arrhenius parameters with $R$ as the gas constant and $T$ is °K. Activation energies for the main silicate minerals fall broadly in the range 30–90 kJ/mol. The effect of the exponential form is highly important since weathering temperature causes reaction rates to vary about 25-fold from polar to tropical regions (Cookie 6 Fig. 1).

The reactive area term in Equation C6.1 is also important for it means that the rate of chemical weathering is also highly sensitive to processes of physical weathering, like glacial grinding, which 'prepares' a multitude of fresh mineral surfaces for chemical attack. Surface areas are conventionally measured in weathering horizons by dosing with $N_2$ and determining adsorbed monolayer volume, but the technique cannot reveal the awesome complexity of the internal pores present in many minerals, particularly feldspars with their turbid microporosity networks (see Fig. 1.7) that are entirely accessible to infiltrating waters.

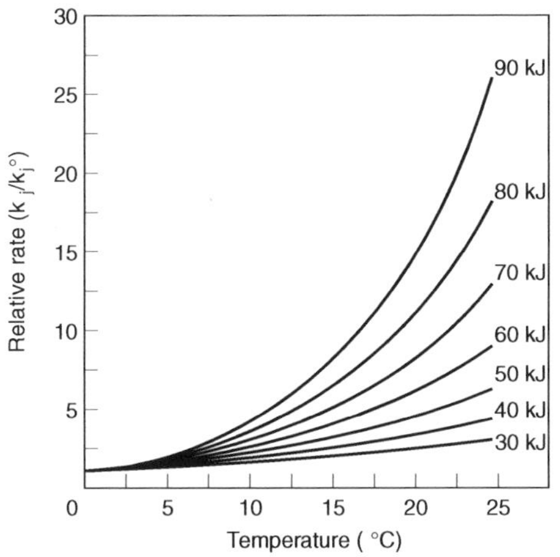


**Cookie 6 Fig. 1** The effect of the exponential Arrhenius temperature-driven term on weathering reaction rates for activation energies of reaction suitable for silicate minerals in the range 30–90 kJ/mol. (After White, 1995.)

In order to account for the role of water concentration we may rewrite Equation (C6.2) in the notional form:

$$k = f(P)Ae^{-E_a/RT} \quad (C6.3)$$

where $f(P)$ is a linear functional relationship involving some measure of effective monthly or yearly precipitation, $P$. The relationship is likely to be a complex one, if only because summer rains will be more effective agents of chemical attack than winter rains. A form of Equation C6.3 successfully accounts for Si and Na fluxes from many catchments, but is unsuccessful for other alkali and alkaline-earth elements.

The role of water composition has been emphasized in the main text and **Cookies 1 & 2** (in proton and electron transfers) and will not be repeated here, save to say that when the Eh and pH of weathering solutions extracted from soils and weathered bedrock are determined they usually (semiarid, alkaline environments excepted) fall in the stability fields for the ubiquitous weathering product, kaolinite. Concerning the rate of dissolution of silicate minerals with time in natural environments, there is much experimental work that suggests it is linear, controlled by surface reactions between mineral and aqueous phases (Cookie 6 Fig. 1). Surface in this case refers to fractured and cleaved mineral faces. There is also sound evidence from electron microscopy studies that weathering can occur along uncracked and uncleaved areas of minerals, the process occurring along submicroscopic diffusion paths at kink and step sites and other dislocations. The chemical attack is seen in the form of etch pits. The rates of surface reactions are usually very slow compared to diffusion rates, and the concentration of products, $c$, adjacent to the reacting mineral surface must be comparable to that of the weathering solution. In chemical language, a zero-order reaction rate law applies, so that:

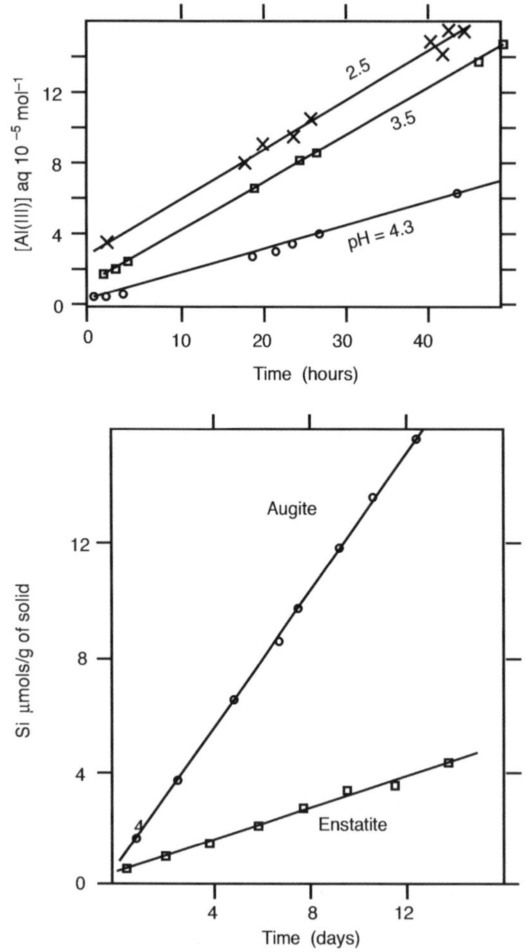


**Cookie 6 Fig. 2** Linear dissolution kinetics. (a) $Al_2O_3$; plotted as $Al^{3+}$ released vs. time. (b) Frame silicates; plotted as $Si^{4+}$ released vs. time for dissolution of etched pyroxenes at pH 6 and T = 20–50 °C. (After Schott & Berner, 1985; Furrer & Stumm, 1986; Stumm & Wollast, 1990.)

$$r = \frac{dc}{dt} = ka \qquad (C6.4)$$

where $r$ is the dissolution rate, seen to be directly linearly proportional to $a$, the surface area of mineral, and $k$, the reaction rate constant. Experimental dissolution (leach) rates for silicate minerals under acidic conditions are always low, being lowest for quartz (about $10^{-13}$ mol/m$^2$/s at pH $< 6$), compared with typical feldspar values up to 2–3 orders of magnitude higher.

How do surface reactions take place? We can envisage two steps (Cookie 6 Fig. 3).

1 A fast attachment of hydration and protonation reactants like $H_2O$, $H^+$, $OH^-$ and ligands for the transition metals like Fe and Mn to form surface species at the mineral-water interface. The reactants effectively polarize and weaken the bonds between metal atoms and their linked oxygens to form precursor compounds. The nature and structure of these latter is the subject of much current research. It has been suggested that clays and other amorphous products may form directly in the surface layer of reaction, perhaps without a solution phase being passed through.
2 A slow detachment of metallic ions from the surface into solution. This is the rate-determining step and is controlled, amongst other things, by the probability of occurrence of sites at the mineral surface suitable for formation of the precursor compound. Experimental evidence suggests that detachment reactions are nonstoichiometric, and that the order and ease of cation detachment is sensitive to pH. For example, under highly acidic conditions, Al is released from feldspar in preference to silicon, with the formation of cation- and Al-deficient layers up to 1200 Å thick.

## Cookie 7: Further information on basic carbonate minerals

Calcium carbonate exists as the mineral polymorphs *calcite* and *aragonite*. Both polymorphs may form as inorganic precipitates or as biological secretions in the hard parts of numerous organisms. Dolomite is a double Ca–Mg carbonate whose origins as primary or secondary precipitates in sediments and sedimentary rocks are still the subject of lively debate (see below); it has no primary role in skeletal build-ups. Recognition of carbonate minerals in thin-sections of limestones is greatly aided by staining techniques; calcite, Fe-calcite, high-Mg calcite, aragonite, dolomite and Fe-dolomite may all be distinguished in this way. Cathodoluminescence emissions in conjunction with electron microprobe analyses are also used as an aid to identification.

### Calcite

Calcite crystallizes in the trigonal system and its crystal growth habit may be fibrous, rhombic or the

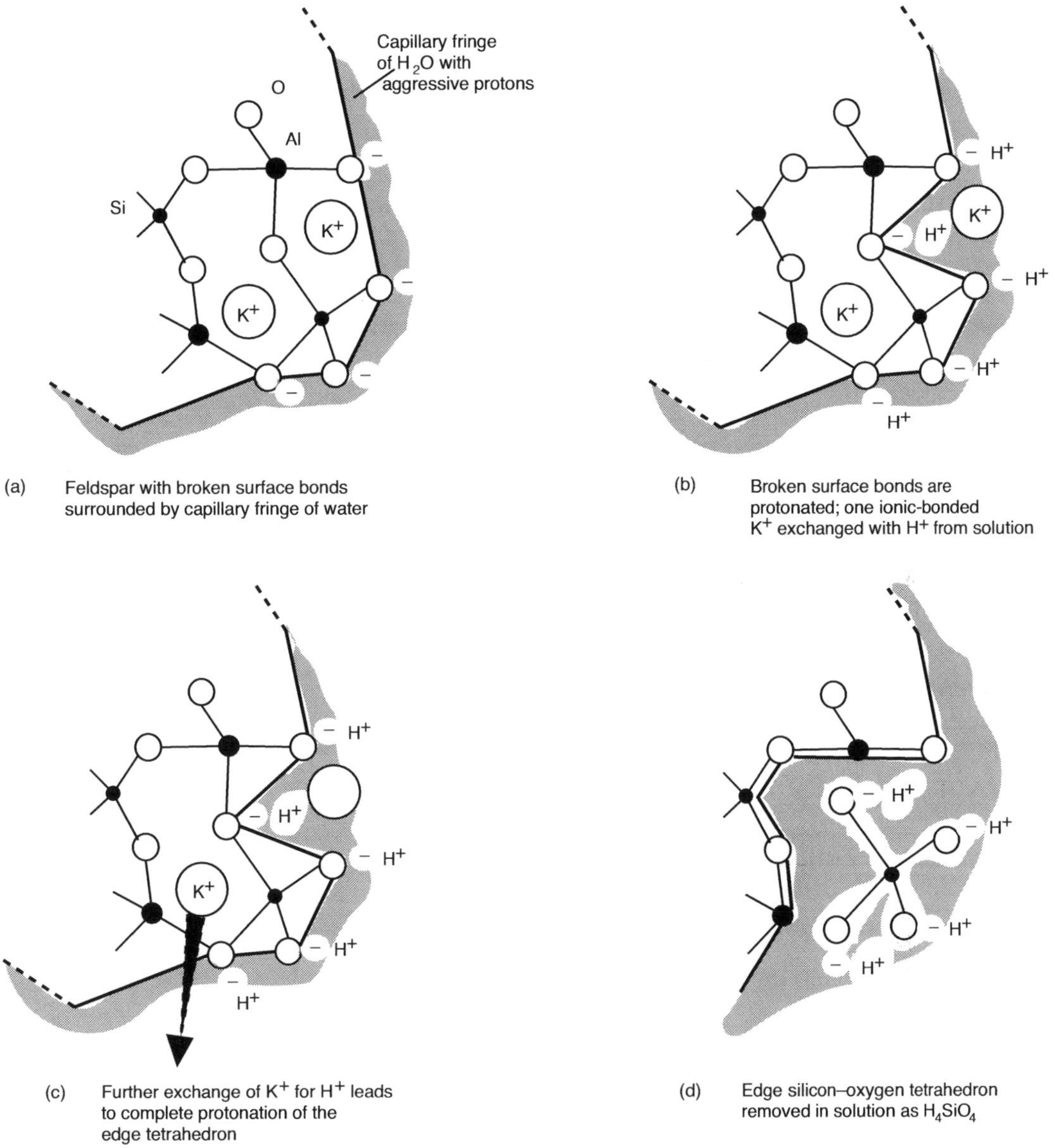

Capillary fringe of $H_2O$ with aggressive protons
O
Al
Si
K+
H+
(a) Feldspar with broken surface bonds surrounded by capillary fringe of water
(b) Broken surface bonds are protonated; one ionic-bonded K+ exchanged with H+ from solution
(c) Further exchange of K+ for H+ leads to complete protonation of the edge tetrahedron
(d) Edge silicon–oxygen tetrahedron removed in solution as $H_4SiO_4$

**Cookie 6 Fig. 3** Schematic view of highly simplified weathering mechanisms for the acid hydrolysis of a K-feldspar. (After Andrews *et al.*, 2005; see Blum & Stillings, 1995, for more detailed accounts of surface reactions.)

familiar 'dog's tooth' morphology. An important consequence of the charge similarity and ionic radius of of $Ca^{2+}$ and $Mg^{2+}$ ions and of the structure of the calcite lattice is that $Mg^{2+}$ may substitute extensively for $Ca^{2+}$ (see **Cookie 8**). These calcites with 5–40% $MgCO_3$ are known as *high-magnesian calcites*; they are the most reactive and soluble of all the calcium carbonate minerals. Of great importance is the fact that Mg content varies positively with temperature, leading to an important *palaeothermometry* technique for past oceanic temperatures using fossil calcitic foraminiferal tests obtained from sediment cores. Small amounts of $Fe^{2+}$ (up to a few thousand parts per million) may also substitute for $Ca^{2+}$ under low-Eh conditions, giving rise to *ferroan calcites*. The substitution of trace amounts of $Mn^{4+}$ causes calcite to luminesce under the influence of cathode ray bombardment. If the flux of $Mn^{4+}$ varied with time during calcite crystallization, then luminescence studies reveal tell-tale growth zones, which may often be mapped out in stratigraphic sections.

## Aragonite

Aragonite is metastable under aqueous earth-surface conditions—many of the interesting problems and features of carbonate sedimentology revolve around the timing and chemical constraints upon aragonite vs. calcite precipitation and aragonite dissolution in freshwaters (see **Cookie** 8). The crystal habit of aragonite in precipitated and some biogenic phases is usually of a fibrous type. Aragonite does not usually precipitate directly from freshwater and so is absent from cave speleothems, tufas and other freshwater calcium carbonate deposits. In contrast to calcite, the aragonite crystal lattice cannot take up much $Mg^{2+}$ when it is inorganically precipitated, e.g. corals may contain about 0.001% $Mg^{2+}$. By way of contrast, the larger strontium atom may substitute up to ~10 000 ppm.

## Calcite oceans vs. aragonite oceans

It is likely that there have been major secular changes in seawater chemistry through geological time. For example, Archaean and Proterozoic oceans are thought to have been deficient in $O_2$, with abundant $Fe^{2+}$ acting as a 'poison' to calcite precipitation. Under such conditions widespread seafloor precipitation of aragonite occurred, with characteristic morphologies (Cookie 7 Fig. 1). Data from the composition of marine authigenic carbonate minerals and evaporites (details in section 23.1) indicate that the oceans may have oscillated between two states, aragonite and calcite oceans. (Cookie 7 Fig. 2) Aragonite ocean states were marked by high Mg/Ca ratios (most notably in the Pennsylvanian/Permian and Cambro-Ordovician) and corresponded to 'icehouse Earth' conditions, as today, with widespread precipitation of aragonite rather than calcite in low-latitude oceanic

**Cookie 7 Fig. 1** Characteristic fan-shaped seafloor growths of primary aragonite, now pseudomorphed by calcite. Scale bar = 5 cm. From ca > 2 Ga, Revilo Formation, South Africa. (From Sumner & Grotzinger, 1996. Photo courtesy of D. Sumner.)

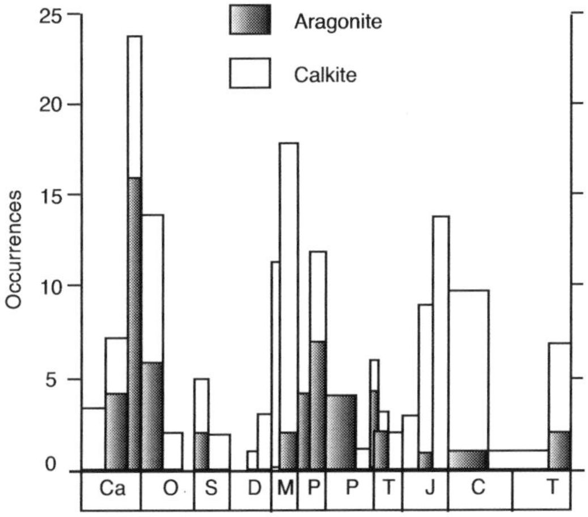


**Cookie 7 Fig. 2** Variation in calcium carbonate mineralogy over Phanerozoic time for chemically precipitated (authigenic/abiotic/non-skeletal) calcite and aragonite. (From Wilkinson *et al.*, 1985.)

waters. Calcite oceans dominated in early Palaeozoic and Mesozoic 'greenhouse Earth' times and were characterized by low Mg/Ca ratios.

### Dolomite

The double carbonate lattice is highly ordered, with alternating layers of cations and $CO_3^{2-}$ groups, in which the cation layers are alternatively $Ca^{2+}$ and $Mg^{2+}$. Dolomites fall into two groups; generally finely crystalline calcian forms (> 53% Ca) and near-stoichometric, fine-to-coarsely crystalline forms. Ferrous iron may substitute for up to 25% of $Mg^{2+}$ in the solid solution series dolomite–ankerite to define *ferroan dolomites*.

### The 'Dolomite problem'

The lack of dolomite precipitation in modern seawater must result from an inhibition mechanism, since from the known composition of seawater, dolomite should be the first mineral to precipitate (Table 2.2). The very high degree of ordering in the lattice seems to result in extremely slow crystal nucleation and growth rates. Attempts at laboratory precipitation of dolomite result in the formation of more poorly ordered and metastable magnesian calcites of dolomite composition, known as *protodolomites*. A surface poisoning effect, when hydrated $Mg^{2+}$ ions surround the $Ca^{2+}$ growth planes of tiny dolomite nuclei, may impede growth much as postulated for the inhibition of calcite growth. There is evidence that significant dolomite precipitation is microbially mediated and occurs under shallow subsurface oxic and anoxic conditions (Cookie 7 Fig. 3). Thus the presumed barriers to early dolomitization in normal seawater and in seawater-derived pore fluids at about standard $T$ and $P$ are overcome in organic-rich sediments as a result of bacterial sulphate reduction and methanogenesis. These processes promote early dolomitization by causing high pH, total alkalinity and $[CO_3^{2-}]$ in pore fluids at the same time as decreasing Mg-hydration.

Despite the evidence for lack of primary dolomite precipitation from the modern oceans, it has long been suspected that dolomite was precipitated from Precambrian seas. Not only are dolomitic rocks much commoner in the Precambrian, but petrographic and stable isotopic studies suggest that the pristine condition of dolomitic ooids, pisoliths and various cements could not have resulted from any diagenetic recrystallization. The 'dolomite problem' involves efforts to explain this phenomenon.

## Cookie 8: Calcium carbonate equilibria reactions: seawater pH, $CO_2$ buffering and carbonate mineral precipitation

Calcium carbonate reactions in the carbonate cycle play a key role in maintaining seawater pH. Seawater is well buffered and has a pH in the narrow range of 7.8–8.3. Any tendency to increase surface seawater acidity, caused for example by higher atmospheric $CO_2$, is opposed by the following reactions: first, the slow dissolution reaction of $CO_2$ with water to give the weak carbonic acid:

$$H_2O + CO_2 \leftrightarrow H_2CO_3 \qquad (C8.1)$$

$$H_2CO_3 + CaCO_3 \leftrightarrow Ca^{2+} + 2HCO^{3-} \qquad (C8.2)$$

and secondly, the rapid ionization of carbonic acid in two stages (see also Cookie 3 Table 1), the first more important than the second:

$$H_2CO_3 \leftrightarrow H^+ + HCO^{3-} \qquad (C8.3)$$

$$HCO_3^- \leftrightarrow H^+ + CO_3^{2-} \qquad (C8.4)$$

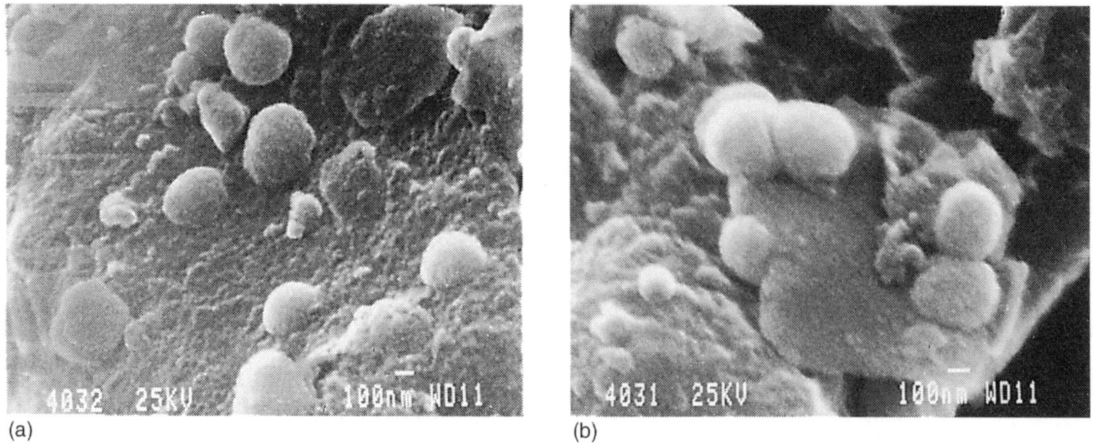


**Cookie 7 Fig. 3** Secondary electron images of microbially mediated ferroan dolomite coatings. (a) Nanobacterial coatings. (b) Coatings on possibly twinned nanobacteria caught *in flagrante delicto* during cell division. Note 100 nm scale bars. (From Vasconcelos & McKenzie, 1997. Photos courtesy of J. McKenzie.)

These two reactions provide alkalinity that balances the excess positive cation charge in seawater (neglecting the tiny contribution made by the $H^+$ ions). At normal oceanic pH of around 8, more than 80% of dissolved carbon is in the form of $HCO_3^-$, with $CO_3^{2-}$ making up the remainder (Cookie 8 Fig. 1). Not all of the bicarbonate is available for reaction because of the involvement of the anion with $Mg_2^+$ in the ion pair $MgHCO_3^+$. Any increase in alkalinity of seawater is opposed by the buffering reaction:

$$HCO^{3-} + OH^- \leftrightarrow H_2O + CO_3^{2-} \quad (C8.5)$$

From the above discussion it should be clear that the oceans are highly sensitive to changes in sources and sinks for $CaCO_3$. Any change in the global balance of

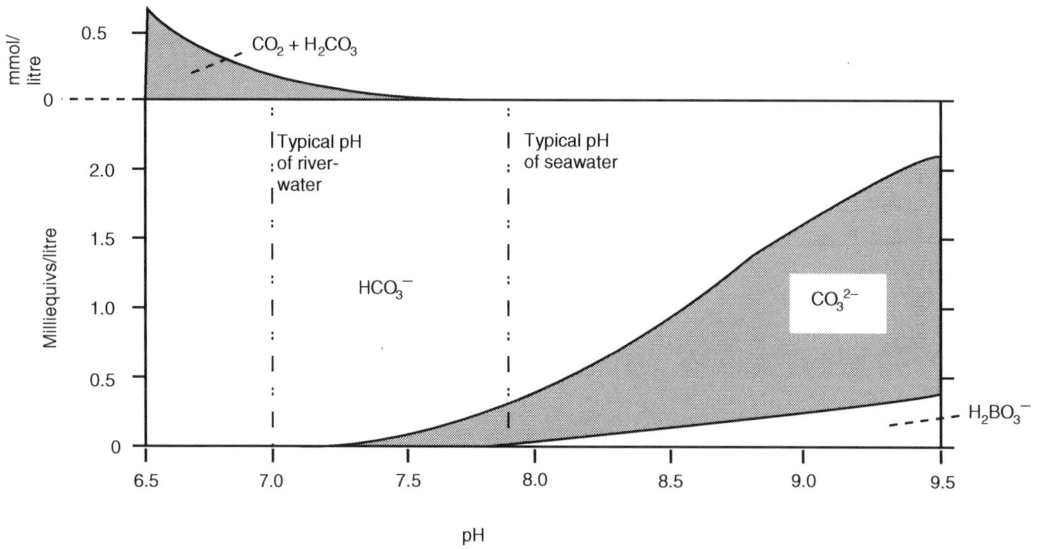


**Cookie 8 Fig. 1** Variation of alkalinity components of seawater with pH. (Data of Cloud, 1962.)

weathering and alteration (input) or shallow- and deep-water deposition (output) will change the oceanic carbonate-ion content so that a balance is restored (see **Cookie** 5). This is the $CaCO_3$ compensation mechanism and may be written as the following notional pH balance:

$$CO_2 + CO_3^{2-} + H_2O \leftrightarrow 2HCO^{3-} \quad (C8.6)$$

It is unlikely that oceanic pH has ever strayed far outside the range 6–9. Thus calcite is present in rocks as old as 3.8 Ga, and a pH of less than 6 is thus improbable (other Ca-bearing phases would have precipitated, for which there is no evidence) unless the $pCO_2$ value was unreasonably high (>€1 atm). Oceanic Ph > 9 is unlikely because massive sodium carbonate precipitation would have occurred, for which there is also no evidence.

In these years of increasing partial pressure of carbon dioxide ($pCO_2$) due to burning of fossil fuel the oceans are calculated to mop-up about 50% of the excess. Increase of atmospheric $pCO_2$ above the present world average of around 320 ppm can be accommodated but the global mean oceanic pH will obviously fall. This has implications for all marine organisms that construct calcareous skeletons and for accumulations of marine carbonate sediment, since carbonate will have to dissolve to compensate for the increased $CO_2$. This will initially favour the dissolution of skeletal high-Mg calcites, which are significantly more soluble than both pure calcite and, for those with > 12 mol% $MgCO_3$, aragonite.

The $pCO_2$ of modern surface ocean waters is controlled in large part by photosynthetic activities of planktonic plants. Any reduction in atmospheric $CO_2$ content by such uptake depends upon the efficiency with which nutrients like phosphates and nitrates are made available. The efficiency is high in most tropical and temperate seas, but low in the Antarctic and eastern Mediterranean. This is because of the lack of soluble iron, another essential cell-forming element, in these areas, the former due to lack of adjacent continental weathering, the latter to scavenging by sinking aeolian dust particles. It has been proposed that the addition of iron to both oceans would help to mop-up the excess $CO_2$ provided by future emissions; but dynamic calculations do not support the idea as significant.

Even though seawater is supersaturated with respect to aragonite, calcite and dolomite, only the

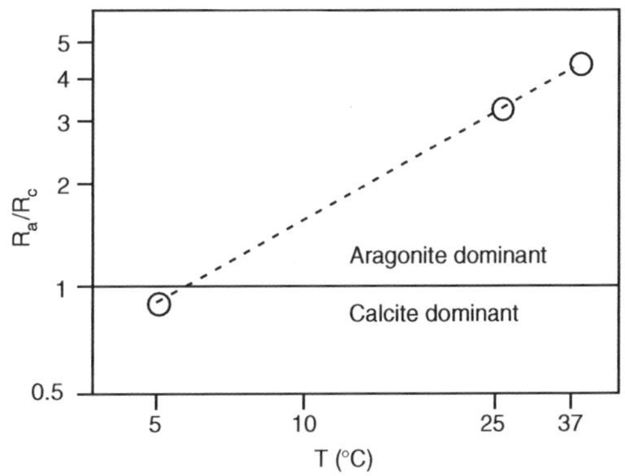


**Cookie 8 Fig. 2** Relative precipitation rates of aragonite versus calcite ($R_a/R_c$) as a function of temperature in seawater five times supersaturated with respect to calcite. (Data of Burton & Walter, 1987.)

former mineral seems to directly precipitate from seawater. Why should this be so?

First there is a very strong temperature control on polymorph selection. For any given ratio Mg/Ca > 1, aragonite is always favoured by higher seawater temperature (Cookie 8 Fig. 2).

Second, numerous experimental and theoretical studies indicate that, at low supersaturations, calcite and aragonite precipitation is inhibited by other dissolved ionic species.

Chief amongst these carbonate 'poisons' are $Mg^{2+}$, $Fe^{2+}$ and dissolved phosphate. Concerning $pCO_2$ and $Mg^{2+}$ ions:

- changing $pCO_2$ for fixed degrees of supersaturation has little effect upon the rate of calcite or aragonite precipitation;
- dissolved $Mg^{2+}$ in seawater has no effect upon the seeded precipitation of aragonite;
- dissolved $Mg^{2+}$ severely retards calcite precipitation;
- calcite precipitated from seawater on *pure* calcite seeds contains 7–10% $MgCO_3$ as an overgrowth of the more soluble high-Mg calcite.

Also illustrated in recent experimental results (Cookie 8 Fig. 3):

- in solutions with different Mg/Ca ratios, the transition between the aragonite and the calcite + aragonite precipitation fields is controlled by a combination of the saturation state of the solution with respect to $CaCO_3$ and the Mg/Ca ratio in solution;

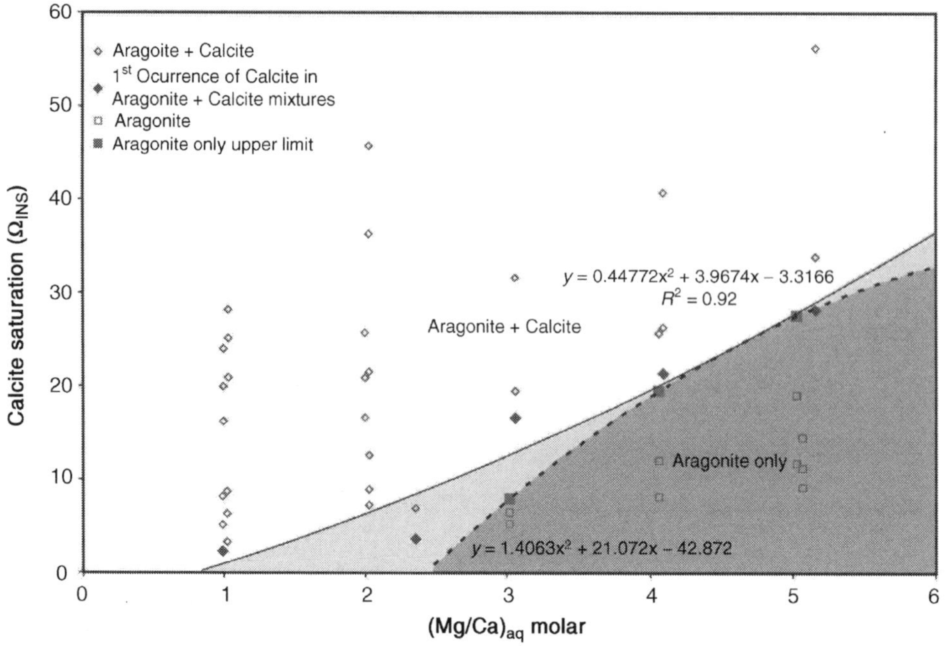


**Cookie 8 Fig. 3** Supersaturation curves defining the region in which calcite precipitation is first observed (solid curve) and the region where only aragonite precipitates (dashed curve). The lower boundary of the calcite + aragonite region is the best fit to the minimum instantaneous supersaturation values required for calcite precipitation (solid diamonds). The upper boundary of the aragonite region is defined by the maximum instantaneous supersaturation in which only aragonite precipitated (solid squares). Note that with increasing Mg/Ca ratio, higher supersaturation levels are required for calcite to precipitate. (After Choudens-Sanchez & Gonzalez, 2009.)

- as Mg/Ca ratios increase, a progressively higher supersaturation level is required for calcite precipitation;
- the predominance of aragonite at high solution Mg/Ca and low supersaturations is attributed to a relative decrease in calcite growth rates as a result of increasing incorporation of $Mg^{2+}$ in the calcite lattice at higher solution Mg/Ca ratios;
- increasing solution Mg/Ca decreases the growth rate of calcite while aragonite growth rates stay unaffected;
- the percentage of $MgCO_3$ in calcite increases with increasing solution Mg/Ca ratio—as calcite growth rates decrease, aragonite growth rates stay constant and it becomes the dominant mineral phase in solutions with high Mg/Ca ratio and low supersaturations.
- in $Mg^{2+}$-deficient 'seawater', with $< 5\%$ of the normal Mg content, $Mg^{2+}$ does not appreciably retard the seeded precipitation of calcite—low-magnesian calcite is thus stable in freshwater regimes of suitable pH.

The surface 'poison' effect of $Mg^{2+}$ upon calcite precipitation may be due to the fact that the smaller $Mg^{2+}$ ions are more firmly hydrated by polar water molecules than are $Ca^{2+}$ ions. More thermodynamic work must be done to dehydrate these $Mg^{2+}$ ions than to dehydrate $Ca^{2+}$ for the growth of aragonite lattices. $Mg^{2+}$ has no effect upon the formation of the orthorhombic aragonite lattice.

Both aragonite and calcite precipitation are inhibited by the trace quantities of dissolved phosphate ions present in most seawater. Calcite and aragonite precipitation rates depend upon both the saturation state of calcium carbonate and the phosphate concentration. With no phosphate around, aragonite precipitates much faster than calcite. In normal phosphate-bearing seawater, the relative rates depend upon pH through the balance between $PO_3{}^{3-}$ and $HPO_4{}^{2-}$. At normal seawater pH of around 8, the most stable phase to precipitate is calcite, but we have already seen above that this phase is actively discouraged by $Mg^{2+}$ poisoning. The pH control is probably more important in sediment porewaters where pH is more variable.

## Cookie 9: Carbonate mineral growth habits

Experimental techniques like atomic force microscopy (AFM) and scanning force microscopy (SFM) imaging, have shed light on the nature of crystal growth for the comparatively simple calcite lattice. It appears that growth occurs as advancing monomolecular steps on the calcite cleavage surfaces prepared for the experiments. Step nucleation occurs mostly at growth spirals, and the steps are fed by addition of material from the growth solution, not by surface diffusion as observed in the vapour growth of many artificial monatomic semiconductors and metals. SFM experiments on freshly prepared calcite growth surfaces, as distinct from cleavage surfaces, at saturations ($\Omega > 1$–2) appropriate for seawater reveal that precipitation begins with scattered surface nuclei. These nuclei are initially high and grow, spread and combine to form eventually beautifully layered growth patterns resembling spirals whose heights are only a few monolayers (Cookie 9 Fig. 1a). Growth rates decay from initial to layered stages, the longer-term equilibrium rates being linear (first order). A very important observation during the experiments was that exposure of growing crystals to air, even for short periods of time, caused re-establishment of rapid spot-nucleated growth. Thus those natural carbonates that form under conditions of frequent wetting–drying cycles (e.g. calcisols, tufas, beachrocks, etc.) experience rapid growth and the production of a highly porous fabric and low density. Rate laws for these environments will be of higher order than for the simple linear kinetics of a continuously wetted environment.

The AFM and SFM experiments shed light on the effects of phosphates upon precipitation. Addition of phosphate during the nucleation stage changes the nuclei to amorphous shapes, the phosphate perhaps blocking growth of the CaCO3 nuclei. The addition of phosphate during the layer stage of precipitation disrupts the growth of regular spiral steps and terraces,

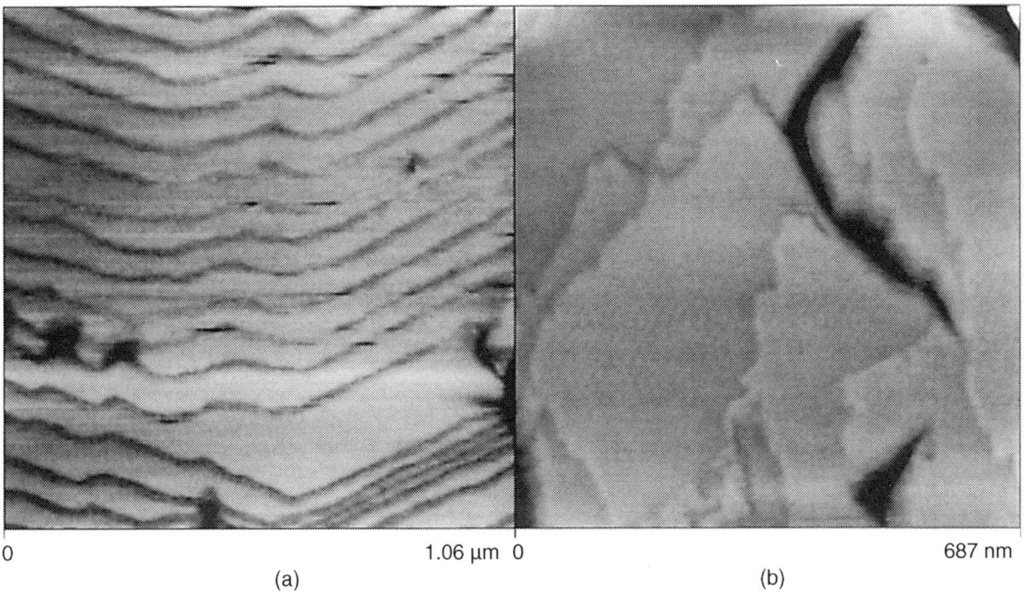


**Cookie 9 Fig. 1** Scanning force microscope photos showing: (a) calcite growth in smooth spiral steps, each 3 Å step being individual monolayers of sheets of carbonate and calcium groups on a growth surface; (b) the effect of adding phosphate in dilute solution to a similar stepped surface to that in (a)—the large, jagged, irregular steps coincide with inhibited and irregular growth. (After Dove & Holchella, 1993. Photos courtesy of P. Dove.)

perhaps because the phosphate ions block spiral growth sites like kinks, steps or holes (Cookie 9 Fig. 1b).

## Cookie 10: Further information on organic groups that calcify and contribute to carbonate production

### Molluscs

Many of this varied phyla (*bivalves*, *gastropods*, *cephalopods*) are marine benthic creatures, but pelagic forms are also common, e.g. vast numbers of pelagic gastropods known as *pteropods* inhabit high-productivity zones of the oceans associated with upwelling currents along shelf ramps. Terrestrial and freshwater gastropod molluscs are also very common. Their shells comprise wholly aragonite or calcite or complex alternations of the two minerals in layers within protein frameworks. The multilayers are arranged in various geometrical ways (useful in thin-section determinations) in different genera.

### Brachiopods

This phyla is now a pale shadow of its former self, for it dominated many benthic communties in the Palaeozoic era. Its place has been largely taken over by benthic molluscs, chiefly the bivalves. Brachiopod shells are wholly low-Mg calcitic with thin fibrous primary and thick secondary layers. A third prismatic layer may also occur. Canals may cut across the secondary layer in some forms and external spines feature in the common productid forms of the Upper Palaeozoic. The stability of brachiopod calcite in deep geological time means the phylum has been much sampled as a source for ancient seawater elemental composition, particularly Sr-isotope variations.

### Corals

Coral skeletons of both reef-builders and solitary forms include many architectural elements, including epithecal walls, internal horizontal partitions known as tabulae and vertical radial partitions known as septae. The extinct tabulate corals are entirely calcitic and have tabulae within the outer walls only. Rugose corals are also calcitic in mineralogy but have septae in addition. Modern hexacorals have only internal septae and are entirely aragonitic. Living coral reefs are prone to destructive 'bleaching' episodes due to pigment loss when attendant symbiotic algae are expelled as sea-surface temperatures in an ocean current 'hotspot' exceeds a colony's tolerance level. Aragonite from hexacorals contains trace amounts of uranium that makes the group much-sampled for $^{230}$Th-dating in uplifted late Quaternary marine terrace deposits worldwide.

### Foraminifera

The unicellular foraminifera may be either benthic or pelagic, making up a high proportion of oceanic calcareous oozes. The tests are usually of calcite or high-Mg calcite.

### Ostracods

These minute hermaphrodites have calcitic valves and an extraordinarily wide ecological range, from fresh to hypersaline waters on land and in the oceans. They are the only multicellular pelagic organisms to have left an abundant fossil record and provide an excellent palaeoproductivity indicator.

### Echinoderms

The echinoderms formerly dominated the benthic fauna of Upper Palaeozoic shallow seas. Their high-Mg calcite tests have always been secreted as single crystal plates, making them most distinctive in thin-section.

### Cyanobacteria

These *prokaryotes* (they have no cell nucleii) are also known as blue-green algae, blue-green bacteria or Cyanophyta. They are a phylum of filamantous or globular multicellular bacteria that obtain energy through photosynthesis and are thus an important primary producer in many areas of the ocean, and on land. *Stromatolites* (section 2.15) are fossilized oxygen-producing cyanobacteria that are hard mechanical structures caused by precipitated external carbonate around their mucilaginous filamentous sheaths and/or within trapped sediment laminae. *Thrombolites* are a variety of stromatolite produced by globular cyanobacteria and which do not show internal laminations, rather a sort of 'clotted' fabric, and which have been formed entirely by microbially induced precipitation of $CaCO_3$. Cyanobacteria go way back into the Archaean era and their photosynthetic activities gradually established an oxygenic atmosphere during early Proterozoic times. Chloroplasts in higher plants and eukaryotic algae (see below) have both evolved from cyanobacteria.

### Algae

These plants are *eukaryotes*, i.e. their cells have nucleii. Many calcareous forms show a cellular internal structure with calcification by aragonite or calcite within individual cells or in the cell walls. There are also those that may or may not calcify outside the cell walls, the choice depending upon external environmental factors such as water composition; these genera are known as *non-obligate calcifiers*. A good example is the genus *Rivularia* which only calcifies today in freshwater but has done so in the past in brackish-to-marine environments where they constructed substantial reef-like build-ups (section 23.17). Other marine examples are the filamentous calcareous algae *Girvanella*, *Garwoodia* and *Ortonella* that are common encrusting agents, forming *algaliths* in many ancient carbonate sediments.

Marine calcareous benthic algae which are obligate calcifiers show diverse morphologies. Erect branching forms such as the green alga *Halimeda* and the red alga *Lithothamnion* break down initially into gravel-sized segments. Red algae (rhodoliths) have recently been investigated for their potential as palaeotemperature indicators; their fine internal calcitic growth banding recording sea temperature by variations in Mg-content at bi-weekly resolution. After death, the delicate fronds of the codiacean alga *Penicillus* (Fig. 2.6) break down completely into their constituent aragonite needles that contribute to aragonite mud sediment production in modern semitropical lagoons and bays where they grow in vast numbers. Freshwater green alga, the *charophytes* or stoneworts, are dominated by the genus *Chara* which inhabit shallow lakewaters of high pH and high alkalinity. During photosynthesis the cells of the plant stems and fruiting bodies calcify externally, usually as low-Mg calcite. These break down post-mortem into silt-sized particles to form pure calcareous or mixed carbonate/quartz-silt and clay mineral deposits known as *marls*.

Marine calcareous planktonic algae of the phylum Chryophyta (golden plants) are abundantly represented in many oceanic pelagic deposits from Jurassic times onwards. They are referred to broadly as *calcareous nanoplankton*, important as primary producers in sunlit near-surface oceanic waters. *Coccoliths* (Fig. 2.8) are the individual discs or scales (sizes commonly 2–20 μm) that make up whole algal cells (diameter up to ca 80 μm).

## Cookie 11: Evaporite precipitation

The equilibrium constant for the reaction

$$\underset{\text{anhydrite}}{CaSO_4(\text{solid})} + 2H_2O \leftrightarrow \underset{\text{gypsum}}{CaSO_4 \cdot 2H_2O(\text{solid})} \qquad (C11.1)$$

is given by the activity of water, $a^2H_2O$, and enables stability fields for gypsum and anhydrite to be plotted from experimental results (Fig. 2.11).

## Cookie 12: Silica equilibrium

For the system:

$$SiO_2 + 2H_2O \rightarrow H_4SiO_4 \qquad (C12.1)$$

$K = aH_4SiO_4 = 2 \times 10^{-3}$ and IAP is $2 \times 10^{-4}$ to $1 \times 10^{-6}$. Most dissolved silicon occurs in the form $H_4SiO_4$, further ionization to $H_3SiO_4^-$ being very limited ($K = 10^{-9.9}$).

## Cookie 13: Iron reactions: bacterial ferric iron reduction and iron hydroxide (limonite/goethite) dehydration

The $Fe^{3+}$-reducing microorganism strain GS-15 (aka *Geobacter metallireducens*) metabolizes by using enzymes to link the oxidation of natural acetate to $CO_2$ with the reduction of $Fe^{3+}$ to $Fe^{2+}$. Thus:

$$CH_3COO^- + 8Fe^{3+} + 4H_2O \rightarrow 8Fe^{2+} + 2HCO^{3-} + 9H^+ \qquad (C13.1)$$

Interestingly, this same reaction, but now involving the reduction of $U^{6+}$ to $U^{4+}$, yields more energy to the bugs and has been proven to proceed independently, providing another method of mobilizing insoluble $U^{6+}$.

Once deposited, the iron hydroxide mineral known as limonite/goethite must reach an equilibrium with ferric oxide in the form of haematite by the dehydration reaction:

$$2HFeO_2 \rightarrow Fe_2O_3 + H_2O \qquad (C13.2)$$

$\Delta G^\circ$ (see **Cookie 4**) for this reaction is always negative, the exact value being dependent upon goethite crystallinity. Thus limonitic goethite is unstable relative to haematite + water under diagenetic conditions.

## Cookie 14: Notations and reference frames for fluid flow

1 Space is divided by $x$, $y$, $z$ rectangular (Cartesian) coordinates.
2 Local (at a point) instantaneous fluid velocity is $u$, comprising local instantaneous vectorial components $u_i$, $v_i$, $w_i$ corresponding to the $x$ (streamwise), $y$ (depthwise), $z$ (spanwise) axes respectively. In some texts and research papers the reader will often find $z$ (depthwise) and $y$ (spanwise), so be careful!
3 Local time-mean fluid velocity is $\bar{u}$.
4 Any local instantaneous turbulent fluctuation in velocity is written as $u'$, $v'$, or $w'$.
5 Any local time-mean fluctuation is a root-mean-square (*rms*) deviation, $u'_{rms}$, $v'_{rms}$, $w'_{rms}$. For example, to obtain a streamwise *rms* value, the measured instantaneous velocity obtained locally, $u_i$, is subtracted from the time-mean value, $\bar{u}$. The result is squared, then all the squared values taken over a period of time are summed and the mean of the squares obtained by dividing by the number of measurements used. The *rms* value is the square root of the mean square.
6 The velocity of any solid carried by a fluid flow is $U$.

## Cookie 15: Flow without physics: ideal (potential) flow

We consider some general aspects of fluid motion, but without going into the dynamics of the flow (in order to do dynamics, first read **Cookie 16**, on Newton's Laws). The discussion limits itself to the kinematic (Greek *kinema*—movement) property of velocity, excluding considerations of force. Such an approach may be exploratory and intuitive during the investigation of a particular flow and may be a useful first step in an eventual analysis of the forces involved.

Early on in the development of fluid mechanics, the effects of viscosity and turbulence were little understood and flows were analysed as if they had no viscosity, i.e. they were frictionless and showed no evidence for shearing stresses. In the absence of shearing stresses in an ideal fluid there can be no rotational motions (see below). Such flows are called *ideal* or *potential*. The formulation of the mathematical laws of hydrodynamics has been greatly aided by these simplifications, but, not surprisingly, there are many important fluid effects in sedimentology that cannot be explained or even predicted by such *inviscid theory*. It might thus seem perverse to spend any time on this nowadays, but the results are useful when considering flows well away from solid boundaries where the maximum effects of friction and viscosity are felt.

From the principles of the continuity equation discussed in the main text (section **4.3**) and further below, total fluid discharge between *streamlines* is constant. Thus it is possible to label streamlines according to the magnitude of the discharge that is carried past them (Cookie 15 Fig. 1). This discharge is known as the *streamfunction* ($\psi$) of a streamline. The magnitude of $\psi$ is obviously unique to any particular streamline and must be constant along the streamline. Velocity is higher when streamline spacing is closer and vice versa. Thus for any streamfunction in a 2D flow:

$$|u| = \psi/y \text{ and } |v| = \psi/x \qquad (C15.1)$$

Another useful concept is that of velocity potential lines ($\phi$). These are (imaginary) lines of equal velocity drawn normal to streamlines. They are best compared to, say, contour lines on a map where the direction of greatest rate of change of height with distance is along any local normal to the contours (i.e. gradient of the scalar height). The velocity is the gradient of $\phi$, i.e. the direction of largest rate of change of $\phi$, taken by definition normal to the *equipotential lines*. Thus, in a similar way to the relationships deduced for streamfunction,

$$|u| = \phi/x \quad \text{and} \quad |v| = \psi/y \qquad (C15.2)$$

From the 2D form of the continuity equation C. 15.3 we have:

$$\frac{\partial u}{\partial x} = \frac{\partial v}{\partial y} = 0 \qquad (C15.3)$$

Substituting for $u$ and $v$, from the potential equations we have:

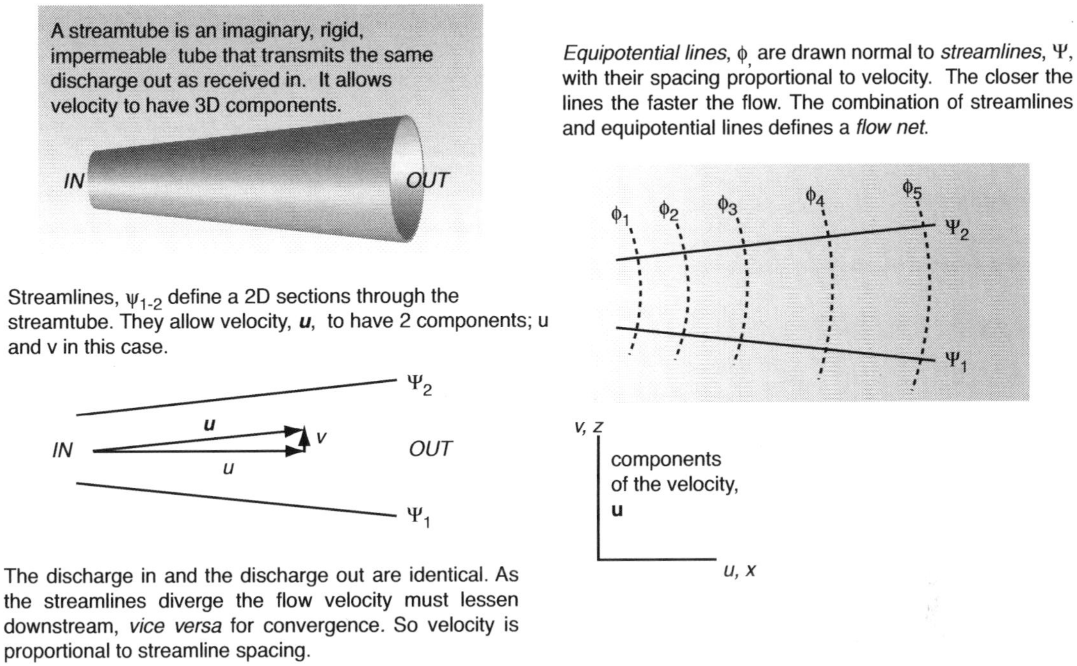
A streamtube is an imaginary, rigid, impermeable tube that transmits the same discharge out as received in. It allows velocity to have 3D components.
IN
OUT
Streamlines, $\psi_{1-2}$ define a 2D sections through the streamtube. They allow velocity, *u*, to have 2 components; u and v in this case.
$\Psi_2$
u
v
IN
OUT
u
$\Psi_1$
The discharge in and the discharge out are identical. As the streamlines diverge the flow velocity must lessen downstream, *vice versa* for convergence. So velocity is proportional to streamline spacing.
*Equipotential lines*, $\phi$, are drawn normal to *streamlines*, $\Psi$, with their spacing proportional to velocity. The closer the lines the faster the flow. The combination of streamlines and equipotential lines defines a *flow net*.
$\phi_1$
$\phi_2$
$\phi_3$
$\phi_4$
$\phi_5$
$\Psi_2$
$\Psi_1$
v, z
components of the velocity, **u**
u, x

**Cookie 15 Fig. 1** Streamtubes, streamlines and potentials.

$$\frac{\partial^2 \varphi}{\partial x^2} + \frac{\partial^2 \varphi}{\partial y^2} = 0 \qquad \text{(C15.4)}$$

This is the 2D form of *Laplace's equation* (2φ = 0), to which all ideal flows conform. Comparing the expressions for potential and streamfunction we get:

$$\frac{\partial \psi}{\partial y} = \frac{\partial \phi}{\partial x} \quad \text{and} \quad \frac{\partial \phi}{\partial y} = \frac{\partial \psi}{\partial x} \qquad \text{(C15.5)}$$

These are the *Cauchy–Riemann equations* enabling the velocity potential to be calculated from the streamfunction. They also cause the equipotential lines to be normal to streamlines.

Potential flow techniques enable some very useful flow constructions to be made. Thus if the distance between equipotential lines and streamlines is made close and equal, then the resultant pattern of small squares forms a flow grid. Construction of flow grids for flow through various 2D shapes may considerably aid physical analysis. The grid is built up by trial and error from an initial sketch of streamlines between the given boundaries. Then the equipotential lines are drawn so that their spacing is the same as the streamline spacing. Continuous adjustments are made until the grid is composed (as nearly as possible) of squares and the actual streamlines are then obtained.

An example of a flow net constructed for a shape like an advancing debris or turbidity flow is shown in Cookie 15 Fig. 2. From the streamline construction one may deduce velocity and, with a knowledge of Bernoulli's equation, pressure variations (**Cookie 16**). However, it will be obvious to the reader that flow nets are only a rather simple imitation of natural flow patterns. Experimental studies alone will reveal the true patterns of flow, in particular the phenomena of flow separation and turbulence.

## The 3D continuity equation

Continuity can be stated for a fixed point where the local velocity and/or density change is stipulated. This is relevant to the stationary observer recording the flow as it moves past. As noted in the main text, such a fixed observation site with respect to any moving fluid is termed the Eulerian reference frame. However, we also need to consider the changes as the flow moves from point to point, appropriate to a moving observer

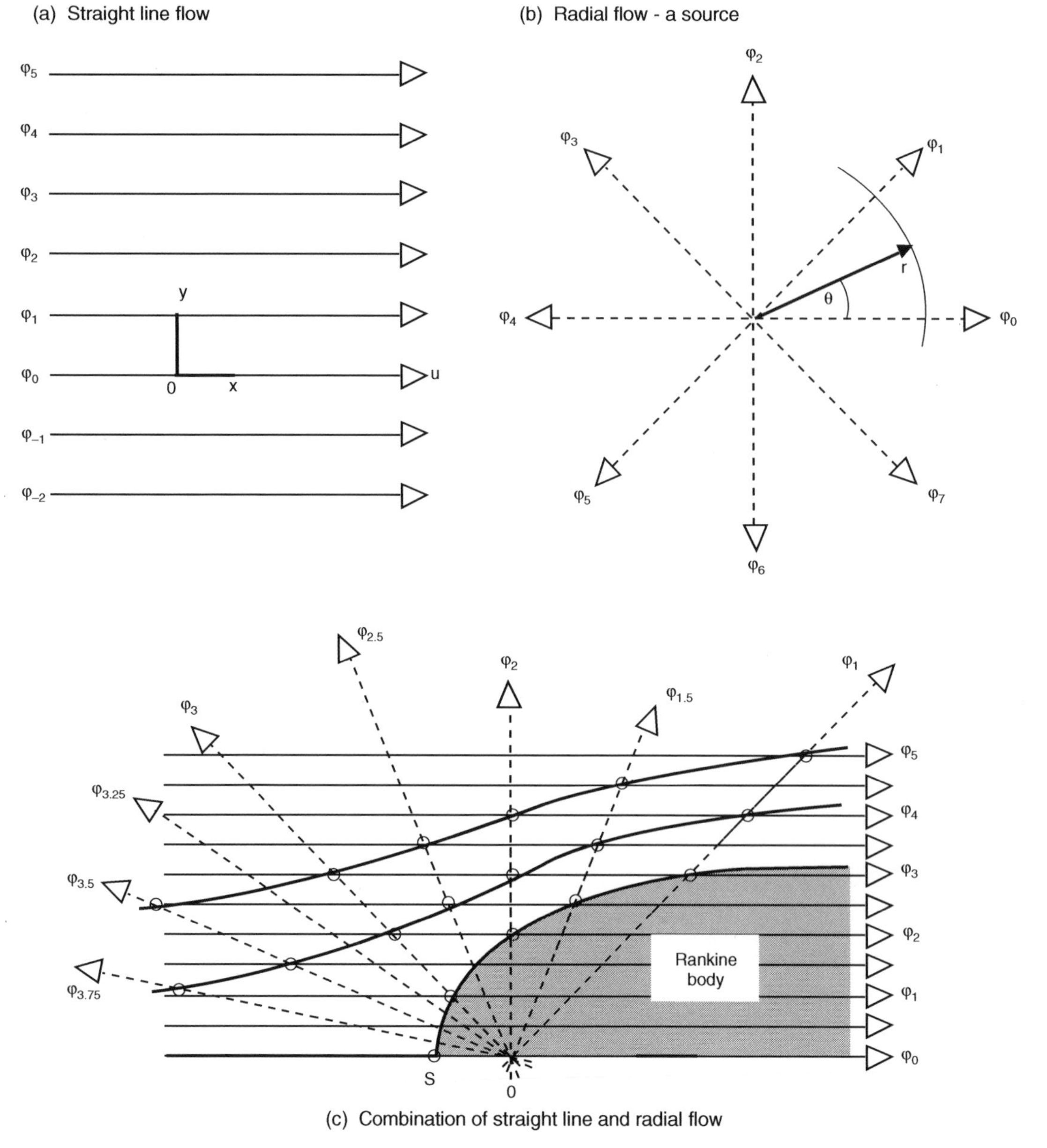


**Cookie 15 Fig. 2** Constructed patterns of potential flow for straight-line, radial-source and combined straight/radial as a *Rankine body*. In a sedimentological context the exercise produces a first-order account of flow over the surface of an intruding gravity current like a turbidity or density current or a cool thermal undercurrent. (After Massey, 1979.)

(you are surfing with the flow, measuring as you go). This is called the Lagrangian coordinate frame. It is important, for even if density doesn't vary at a point, it may well vary laterally or vertically, due to features like suspended sediment concentration, salinity stratification or temperature gradients. The convention adopted is that the Lagrangian operator $D/Dt$, termed the *substantive derivative*, sometimes known as the *total derivative,* expresses both the spatial and time gradients of any variable.

Using a control cube with sides d$x$, d$y$, d$z$ fixed in the fluid volume we may consider continuity at a point in space, in the Eulerian view. Let fluid pass through the cube and let us resolve the flow vector into its usual three components. Take the $x$-direction as an example. Mass flow IN over unit time is the product $\rho \boldsymbol{u}$ d$y$d$z$. You can check that this gives mass by dimensions. Mass flow OUT is mass flow in plus any change in $\rho \boldsymbol{u}$ along d$x$. The full workings to drive the algebraic expression for 3D continuity with constant density is given in Cookie 15 Fig. 3.

From the 2D form of the continuity equation we have:

$$\frac{\partial u}{\partial x} = \frac{\partial v}{\partial y} = 0 \qquad \text{(C15.3)}$$

Substituting for $u$ and $v$, from the potential equations we have:

$$\frac{\partial^2 \phi}{\partial x^2} + \frac{\partial^2 \phi}{\partial y^2} = 0 \qquad \text{(C15.4)}$$

This is the 2D form of *Laplace's equation* ($2\phi = 0$), to which all ideal flows conform. Comparing the expressions for potential and streamfunction we get:

$$\frac{\partial \psi}{\partial y} = \frac{\partial \phi}{\partial x} \quad \text{and} \quad \frac{\partial \phi}{\partial y} = \frac{\partial \psi}{\partial x} \qquad \text{(C15.5)}$$

These are the *Cauchy–Riemann equations* enabling the velocity potential to be calculated from the streamfunction. They also cause the equipotential lines to be normal to streamlines.

## Cookie 16: Newton's 1st, 2nd and 3rd laws

*Law 1. Every solid or fluid mass will remain stationary or continue in its moving path at constant speed unless it is acted upon by some external applied force*

The definition of force provided by Newton's 1st law is often known as the 'principle of inertia'. Inertia is simply the resistance of a mass to acceleration. Thus we see that movement of a sediment grain from rest or any acceleration or deceleration is due to the action of some force, $F$. Conversely, any forces acting on a grain at rest or in uniform motion must be in equilibrium.

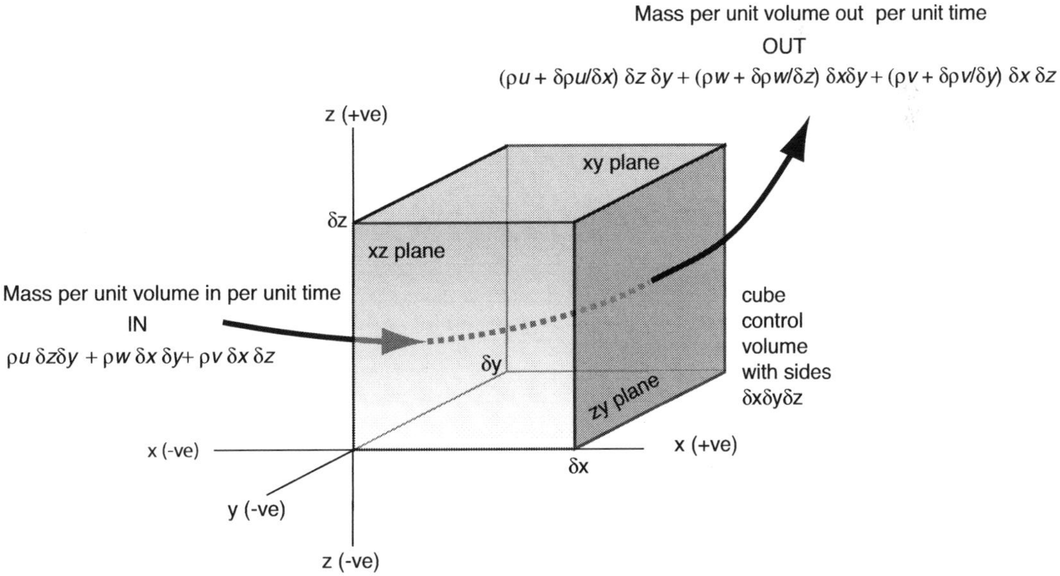


**Cookie 15 Fig. 3** To derive general mass continuity in 3D for flow through a fixed infinitesimal volume, d$x$d$y$d$z$. The accompanying working thereafter uses partial differential notation, for example, $\partial u/\partial x$, etc. (After Massey, 1979.)

*Law 2. Any change of momentum, either direction or strength, is proportional to the moving force applied, and takes place in the straight line in which that force is applied*

This gives information on exactly how to determine the inertial force—it is the most essential and general of the three laws and the basis of all dynamics. This is quite easy to imagine since we are all familiar with the forces that result when moving objects collide with stationary surfaces, the former sometimes losing momentum entirely. An appropriate sedimentological example is momentum transfer as desert rock surfaces are bombarded by bouncing sand grains. Loss, exchange or change of momentum over time is $d(mu)/dt$. Change of momentum or acceleration requires a force to produce it. We can say all this in simple terms using the 2nd law, $F = ma$.

*Law 3. An action is always opposed by an equal reaction; or, the mutual actions of two bodies are always equal and act in opposite directions*

When two masses interact with each other, like along surface contact due to friction, equal and opposite forces result. This means that natural forces come in opposing pairs. If substance of mass $m_1$ at position $x_1$ exerts force $F$ on mass $m_2$ at position $x_2$ the latter exerts an exactly equal force $-F$ on the former. This law can be applied to flows where there is a gradient of velocity, such as in a fluid boundary layer. Examples would be an atmospheric boundary layer shearing over a desert surface, exerting force $F_{ABL}$ and opposed by equal and opposite force from the ground of $-F_{GR}$. Or, in a fluid flowing down an inclined channel (Cookie 16 Fig. 1). In all such cases the total rate of change of momentum with respect to time is zero, and momentum is said to be conserved.

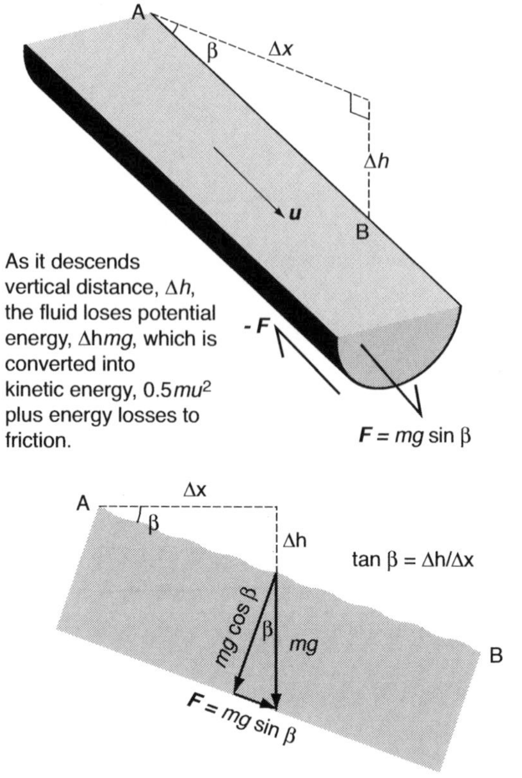


**Cookie 16 Fig. 1** A mass, $m$, of water, debris or turbidity current flowing downslope at angle, $\beta$, exerts a shearing force, $mg\sin\beta$, along its base.

## Cookie 17: Fluid forces and the equations of motion

In stationary fluids the *static forces* of hydrostatic pressure and buoyancy are due to gravity. These forces also exist in moving fluids but with additional *dynamic forces* present—viscous and inertial—due to gradients of velocity and accelerations affecting the flow. In order to understand the dynamics of such flows and to be able to calculate the resulting forces acting we need to understand the interactions between the dynamic and static forces that comprise $\Sigma F$, the total force. This will enable us to eventually solve some dynamic force equations, the Equations of Motion, for properties such as velocity, pressure and energy. Such a development will inform several chapters in this book, notably in the processes of sediment transport and bedform development.

### General momentum approach

To begin with we make simple use of Newton's 2nd law and consider the total force, $\Sigma F$, causing a change of momentum in a moving fluid, not inquiring into the various subdivisions of the force (Cookie 17 Fig. 1). To do this we take the simplest steady flow of constant density, incompressible fluid moving through an imaginary conic streamtube orientated parallel with a downstream flow unaffected by radial or rotational forces. From the continuity equation the discharges into and out of the tube are constant, but from the

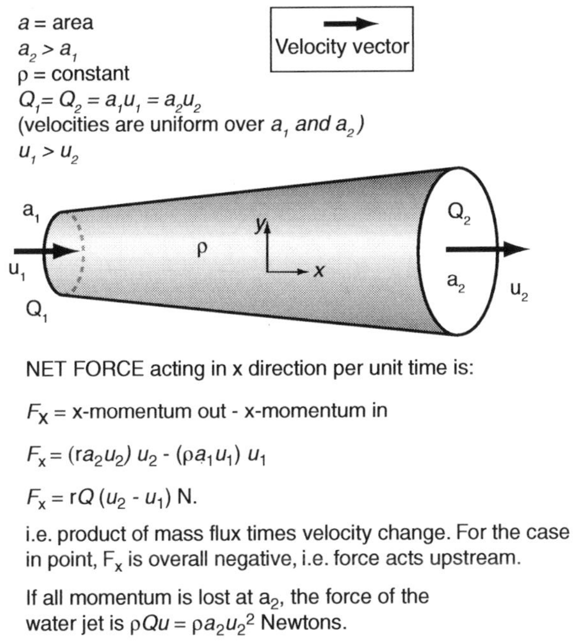

a = area
a2 > a1
ρ = constant
Q1= Q2 = a1u1 = a2u2
(velocities are uniform over a1 and a2)
u1 > u2
Velocity vector
a1
u1
Q1
ρ
y
x
Q2
a2
u2
NET FORCE acting in x direction per unit time is:
Fx = x-momentum out - x-momentum in
Fx = (ra2u2) u2 - (ρa1u1) u1
Fx = rQ (u2 - u1) N.
i.e. product of mass flux times velocity change. For the case in point, Fx is overall negative, i.e. force acts upstream.
If all momentum is lost at a2, the force of the water jet is ρQu = ρa2u2² Newtons.


**Cookie 17 Fig. 1** General momentum approach (After Massey, 1979.)

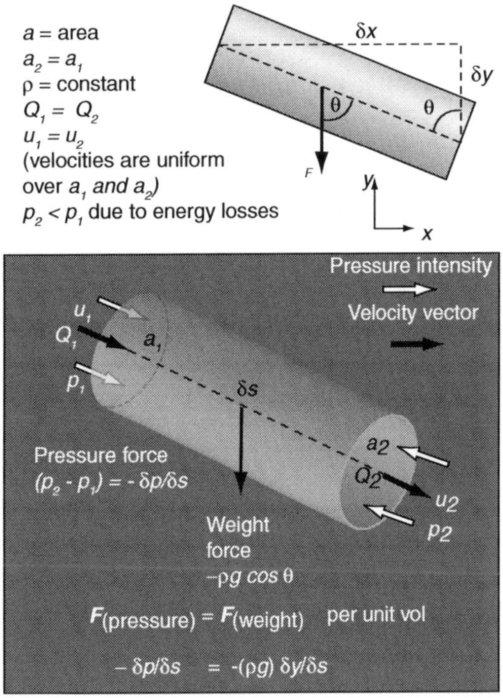


**Cookie 17 Fig. 2** Pressure-gravity approach for constant velocity (After Massey, 1979.)

principles of continuity a deceleration must be taking place along the tube, hence momentum must be changing and a net force acting. The net downstream force acts over the entire streamtube and comprises both pressure forces normal to the walls and ends of the tube and shear forces parallel to the walls. The approach also allows us to calculate the force exerted by fluid impacting onto solid surfaces and around bends.

## Momentum-gravity approach

In many cases we need to know more about the components of the total force in order to find relevant and interesting properties of environmental flows, such as velocity and pressure distributions. One major problem in the early development of fluid dynamics was what to do with Newton's discovery of viscosity and the existence of viscous stresses. This was because the origin and distribution of viscous forces was seen as an intractable problem. In a bold way, Euler, one of the pioneers of the subject, decided to ignore viscosity altogether, inventing *ideal* or *inviscid flow* (see Cookie 15). In fact, viscous friction can be relatively unimportant away from solid boundaries to a flow (e.g. away from channel walls, river or sea bed, desert surface, etc.) and the inviscid approach yields relevant and highly important results. In the interests of clarity, we again develop the approach for the simplest possible case (Cookie 17 Fig. 2), a steady and uniform flow through a cylindrical streamtube involving two forces, gravity and pressure, acting in a vectorially unresolved direction, $s$. The 2nd law tells us that:

$$\Sigma \boldsymbol{F} = \boldsymbol{F}(\text{pressure}) + \boldsymbol{F}(\text{gravity}) = \text{mass} \times \text{acceleration}$$

Since in this flow there is no acceleration:

$$\boldsymbol{F}(\text{pressure}) + \boldsymbol{F}(\text{gravity}) = 0 \text{ or } \boldsymbol{F}(\text{pressure}) = \boldsymbol{F}(\text{gravity})$$

But why should acceleration take place along some linear flow path? The principles involved may be illustrated by a simple but dramatic experiment (Cookie 17 Fig. 3). Water is fed into a length of horizontal tube which has a middle section of lesser diameter that leads smoothly and gradually to and from larger diameter end sections. Vertical tubes partially infilled with dyed dense fluid are let out from the bottom of the horizontal tube to measure the static pressures acting at

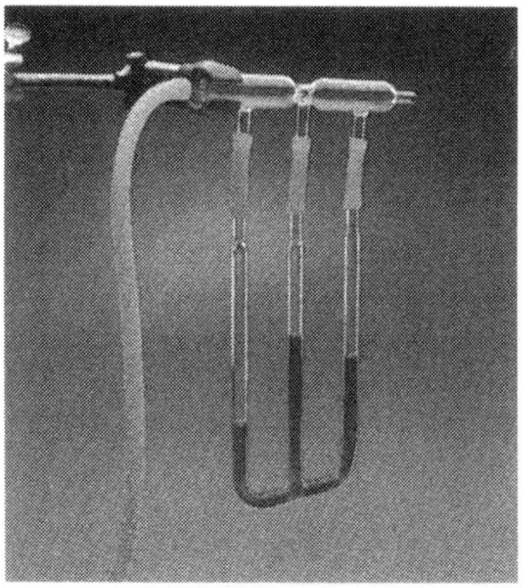

**Cookie 17 Fig. 3** Apparatus for illustrating the Bernouilli effect (see text for details).

the boundary. When the fluid is at rest, the pressures in each vertical tube are equal. The outlet valve is now opened and constant water discharge (i.e. steady flow conditions) is let into the inlet end of the tube to freely pass through the whole tube. A dramatic change occurs in the pressure, that in the narrow bore section being much reduced compared to that measured in the upstream and downstream wider bore sections.

How do we explain this startling result? As the flow passes into the narrow part of the tube, continuity tells us that the flow must accelerate (remember that water is incompressible under the experimental conditions) and that this must be caused by a net force. Since there is no change in the mean gravity force, the tube centreline being horizontal throughout, this net force must come about by the action of a pressure gradient in order that the force balance between inertia and pressure is maintained. We thus have:

$$\Sigma \boldsymbol{F} = \boldsymbol{F}(\text{pressure}) + \boldsymbol{F}(\text{gravity}) = \text{mass} \times \text{acceleration}$$

The result means that pressure gradient causes the acceleration and that since forces are balanced then energy must also be balanced, the increase in flow kinetic energy due to the acceleration being balanced by a decrease in the flow energy due to pressure.

By generalizing the approaches above we arrive at *Bernouilli's equation* (see **Cookie 18**).

### Real-world flows of increased complexity

For real-world flows of hydraulic, oceanographic and meteorological interest several additional terms are relevant, including those for friction (viscous and turbulent), buoyancy, radial and rotational forces. We sample just a few of the various possibilities here.

- *Frictionless oceanographic and meteorological flows.* In the open oceans and atmosphere, away from constraining boundaries to flow, currents have traditionally been viewed as uninfluenced by viscous or turbulent frictional forces. This is because in such regions there is thought to be very little in the way of spatial gradient to the velocity flow field and therefore not much in the way of viscous or turbulent forcing. Clearly this somewhat unrealistic scenario is inapplicable in regions of fast ocean surface and bottom current systems, where dominant turbulent mixing occurs. Nevertheless, the *geostrophic approximation* has enabled major progress in understanding large-scale oceanic circulation substantially affected by the Coriolis force:

$$\Sigma \boldsymbol{F} = \boldsymbol{F}(\text{pressure}) + \boldsymbol{F}(\text{gravity}) + \boldsymbol{F}(\text{Coriolis})$$
$$= \text{mass} \times \text{acceleration}$$

In situations of steady flow with no acceleration, where there are no density changes and where gravity is balanced in the hydrostatic condition this expression becomes an equality between the pressure and Coriolis forces:

$$\boldsymbol{F}(\text{pressure}) = \boldsymbol{F}(\text{Coriolis})$$

- *Viscous friction flows: Navier-Stokes approach.* The incorporation of frictional resistance via viscous forces into the Euler–Bernouilli versions of the Equations of Motion was a major triumph, attributed jointly to Navier and Stokes. (Refer forward to **Cookie 20** for an account of Reynold's derivation of the viscous stress and the net viscous force resulting in a flow boundary layer.) The simplest form of the *Navier–Stokes equation* may be written for an incompressible, non-rotating, straight-line flow, such as in a straight river channel or along a local wind, as:

$$\Sigma \boldsymbol{F} = \boldsymbol{F}(\text{pressure}) + \boldsymbol{F}(\text{gravity}) + \boldsymbol{F}(\text{viscous})$$
$$= \text{mass} \times \text{acceleration}$$

- *Turbulent friction flows: Reynold's approach.* As we see in Chapter 6, Reynolds neatly deconvolved turbulent flow velocities into mean and fluctuating components. The latter are responsible for a very large increase in the resisting forces to fluid motions on account of the immense accelerations produced in the flow. Thus through most of the flow thickness these fluctuating turbulent forces dominate over viscous frictional forces. However, there still remain strong residual viscous resisting forces close to any flow boundary and so we keep the viscous contribution in the Equation of Motion for turbulent flows, written here for a simple case of straight channel flow:

$$\Sigma \boldsymbol{F} = \boldsymbol{F}(\text{pressure}) + \boldsymbol{F}(\text{gravity}) + \boldsymbol{F}(\text{viscous}) + \boldsymbol{F}(\text{turbulent}) = \text{mass} \times \text{acceleration}$$

Natural accelerations in turbulent flows may be *extreme* compared to planetary gravity, $\boldsymbol{g}$, reaching instantaneous values [O] $10^3\boldsymbol{g}$.

### Cookie 18: A most useful expression: forms of the Bernoulli–Euler fluid energy conservation equation

The reader will be reassured to know that one can make a lot of progress by various simplifications of the Navier–Stokes equation. Ignoring viscosity enables us to develop the ideal (potential) flow approach touched on previously. The approach was first proposed by Euler, hence the equation is christened the Euler equation. Referring to Cookie 18 Fig. 1: from Newton's 2nd law, mass times acceleration (rate of change of momentum) equals the total force, $\Sigma F$, acting in the direction of motion down a streamtube of infinitessimal length, $\delta s$. Mass in the streamtube is $\rho(a + \delta a/2)\delta s$ and the acceleration due to nonuniformity in steady flow along the conic streamtube is $u\delta u/\delta s$. Neglecting

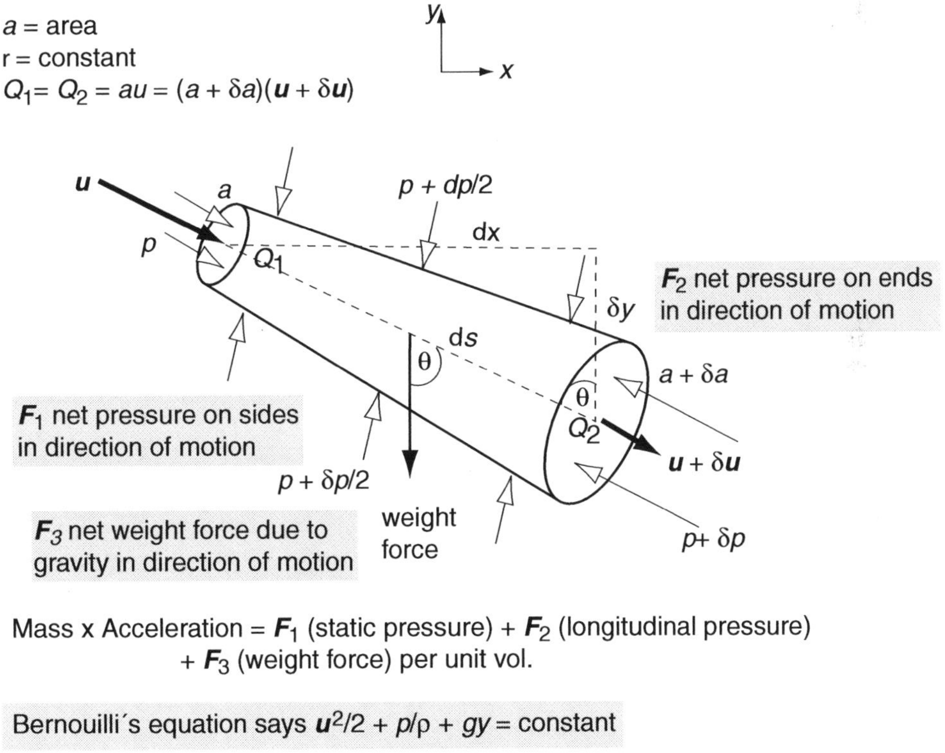


**Cookie 18 Fig. 1** Euler–Bernouilli energy approach for variable velocity. (After Massey, 1979.)

$\delta a/\delta s$ as very small, rate of change of momentum is thus $\rho a(u\delta u)$.

$\Sigma F$ has three components:

- $F_1$ is the result of varying pressure along the sides of the streamtube, given by the product of the mean pressure, $p + \delta p/2$ and the change in tube area, $\delta a$, over the small distance $\delta s$, i.e. $F_1 = p\delta a$, ignoring the second-order term since $\delta p/\delta s$ is small.
- $F_2$ results from the longitudinal change of pressure along $\delta s$ acting on opposing ends of the streamtube. In the direction of motion, $F_2 = -(p + \delta p)(a + \delta a) + pa = -p\delta a - a\delta p$.
- $F_3$ is the weight force component acting due to gravity, $g$, acting upon the mass of fluid, $\rho(a + \delta a/2)\delta s$ within the streamtube tilted at angle, $\theta$, so that $\cos\theta = \delta y/\delta s$ and $F_3 = \rho(a + \delta a/2)\delta s\, g\delta y/\delta s = -\rho ag\delta y$, ignoring the second order term as $\delta a/\delta s$ is small.

Since rate of change of momentum equals the sum of forces acting, we have $\rho a(u\delta u) = p\delta a + (-p\delta a - a\delta p) + (-\rho ag\delta y)$. Cancelling terms and changing to differential notation we have $\rho audu = -adp - \rho agdy$. Dividing through by $\rho ag$ we get:

$$\frac{u}{g}\mathrm{d}u + \frac{dp}{\rho g} + \mathrm{d}y = 0$$

or

$$\frac{1}{g}\int u\mathrm{d}u + \int\frac{\mathrm{d}p}{\rho g} + \int \mathrm{d}y = 0 \qquad \text{(C18.1)}$$

and for the condition of density not varying with pressure (i.e. incompressible liquid flows only) the integrals are solved to give the final Bernoulli–Euler equation for the conservation of energy

$$\frac{u^2}{2g} + \frac{p}{\rho g} + y = \text{constant}$$

Each term in the equation has units of length and should be regarded as the quantity of energy in a fluid volume of unit weight, i.e. J/N. The terms are recognizable as (i) kinetic energy, (ii) flow work energy due to pressure in the fluid and (iii) potential energy of position.

### Scope of application of Bernouilli's equation

The production of flow acceleration as a consequence of pressure change is a major feature of fluid dynamics which has major consequences. Despite its simplicity in ignoring the effects of frictional forces exerted by flow boundaries, application of the Bernouilli equation has enabled increased understanding of flight, wave generation, hydraulic jumps and erosion by wind and water, to name but a few.

## Cookie 19: Reynolds number, *Re*, from first principles

This is a simple way of deriving *Re*, following Reynolds own outline sketch in his classic 1885 paper. We seek to express the balance of inertial to viscous forces. Consider a cubic fluid volume of viscosity, $\mu$, density $\rho$, length $\delta$, cross-sectional area $\delta^2$ as shown in Cookie 19 Fig. 1. Let the upper surface of the volume move with speed $u$ relative to the lower surface, the gradient of speed being linear and due to viscous forces. The viscous force acting on the $xz$ plane is equal to the shear stress times area, $\tau\delta^2$. But we know that $\tau = \mu(\mathrm{d}u/\mathrm{d}y)$ by Newton's viscous equation. Since $\mathrm{d}u/\mathrm{d}y$ is the linear velocity gradient across the cubic element in the $xy$ plane, this is also simply $u/\delta$, so we have the viscous force as $\mu u\delta^2/\delta = \mu\delta u$. The inertial force is equal to mass $\times$ acceleration by Newton's 2nd law. Now mass = density $\times$ volume $= \rho\delta^3$ and acceleration $= u/t = u^2/\delta$ (since $u = \delta/t$ and $t = \delta/u$). Thus the inertial force $= \rho u^2\delta^3/\delta = \rho\delta^2u^2$. Arranging the inertial and viscous forces as a ratio gives *Re* as $\rho u^2\delta^2/\delta\mu u = \rho u\delta/\mu$.

## Cookie 20: Its a parabola! Equation of motion for a steady newtonian liquid viscous flow through a conduit

For steady Newtonian liquid flow along a horizontal streamtube there are no accelerations and no body force changes (see Cookie 17 Fig. 2). Under these conditions the net downstream-acting viscous force $\mu\mathrm{d}^2u/\mathrm{d}z^2$ is balanced only by the mean upstream acting pressure gradient force, $-\mathrm{d}p/\mathrm{d}x$, written for simplicity as $-p$. Using the coordinate system of Cookie 20 Fig. 1 with $p$ not varying *across* the flow (in the $z$-direction) and applying the boundary condition that $u = 0$ at $z = \pm a$, we have:

$\mu\dfrac{\mathrm{d}^2u}{\mathrm{d}z^2} = -p$ or $\dfrac{\mathrm{d}^2u}{\mathrm{d}z^2} = -\dfrac{p}{\mu} = -p\dfrac{1}{\mu}$ and so $\dfrac{\mathrm{d}u}{\mathrm{d}z} = \displaystyle\int -p\frac{1}{\mu}dz = -\frac{pz}{\mu} + c_1$. Since $\mathrm{d}u/\mathrm{d}z = 0$ at $z = 0$,

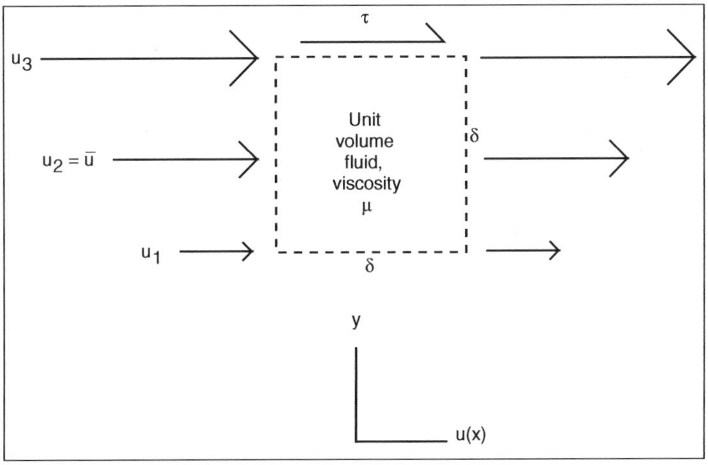


**Cookie 19 Fig. 1** Definition diagram for derivation of Reynolds number. (After Massey, 1979.)

then the integration constant $c_1 = 0$. We now have: $u = \int \frac{py}{\mu} dz + c_2 = -\frac{pz^2}{2\mu} + c_2$. To determine the second integration constant, $c_2$, since $u = 0$ at $z = \pm a$, $0 = c_2 + -\frac{pa^2}{2\mu}$ or $c_2 = \frac{pa^2}{2\mu}$. Therefore finally we have $u = -\frac{p}{2\mu}(z^2 - a^2)$ or

$$u = \frac{p}{2\mu}(a^2 - z^2) \qquad \text{(C20.1)}$$

One can see that the trajectory of $u$ across the boundary layer from $+a$ to $-a$ is a parabola, with $u_{max}$ at $z = 0$ and $u \to 0$ as $z \to \pm a$. Reynolds first derived this expression in 1895, though the type of parabolic flow between parallel plates is often called Couette flow. Further derivations are possible for non-Newtonian liquids, different shapes of conduits (pipes, channels, etc.) and to determine volumetric discharge through these.

## Cookie 21: Turbulent times: Reynold's accelerations, turbulent stresses and turbulent kinetic energy

Turbulence involves 3D eddy motions in a steady, uniform flow. If we measure velocity over time at a point in such a flow then the velocity signal will fluctuate about a steady time-mean. Although the mean fluctuation is always zero (because it is a change plus or minus about a mean value and these cancel out), the mean square of the fluctuations is not, because, as Maxwell developed, the square of the negative fluctuations becomes positive. Working in 1D for simplicity, the instantaneous flux of momentum in the $x$-direction is $u_i(mu_i)$, written per unit fluid volume as $u_i(\rho u_i)$, with $u_i = \bar{u} + u'$, where $\bar{u}$ is time mean velocity (remember the flow is steady) and $u'$ is the instantaneous fluctuation about the mean. So, we have $u_i(\rho u_i) = \rho(u_i^2) = \rho(\bar{u} + u')^2 = \rho(\bar{u}^2 + 2\bar{u}u' + u'^2)$. Now, the mean flux of momentum over a longer time interval is $\rho\bar{u}^2 = \rho\left(\overline{u^2} + \overline{2\bar{u}u'} + \overline{u'^2}\right)$ or simplified to $\rho\bar{u}^2 = \rho\left(\bar{u}^2 + \overline{u'^2}\right)$, since the central term in the triple term bracket involving the time mean of the fluctuation is zero. Turbulence thus involves time mean turbulent accelerations. These are caused by forces due to pressure fluctuations in the boundary layer. It has become customary, though dynamically misleading, to regard the mean turbulent accelerations as virtual stresses by taking the terms over to the sum of forces side of the Navier–Stokes equations of motion, $ma = \Sigma F$, where they are signed negative and termed *Reynold's stresses*. We can write similar expressions for the other two turbulent velocities ($y$ and $z$ components) and there are a total of six independent Reynolds stress terms in the tensor $\tau_{ij}$ (see Maths Appendix for information on tensors).

The total acceleration tensor for steady, uniform turbulent flow reduces to:

$$\frac{D(\bar{u}+u')}{Dt} \equiv \frac{\partial}{\partial x_j}\left(\overline{u'_i u'_j}\right) \qquad \text{(C21.1)}$$

The right-hand side represents the action of velocity fluctuations on the mean flow because of the existence of eddies transporting fluid of time-varying velocity, thus creating space-varying velocity—the expression states that a net mean acceleration is produced in steady, uniform turbulent flow due to gradients in space of turbulent fluctuations. So, going back to our basic point concerning accelerations and forces, net forces arise in unit fluid cells in steady uniform turbulent flows and produce rate of change of momentum. Or, more correctly, since we are viewing accelerations, we may say the turbulent acceleration requires a net force to produce it. This is provided by pressure drive, since viscous forces are far too small.

## Cookie 22: Skid row: transfer of turbulent momentum and the 'law of the wall' for steady, turbulent fluids

In the interval between the top of the viscous sublayer and about 0.4 flow depth, the rate of change of velocity with height, d$u$/d$z$, decreases. Neglecting the influence of viscosity we make the simplest possible guess as to the rate of change of this gradient, that it varies as 1/$z$. Then d$u$/d$z \propto 1/z$ or $u = k\int \frac{1}{z}\mathrm{d}z$, where $k$ is some constant, and integration gives:

$$u = \log_e z + c \qquad \text{(C22.1)}$$

For a fuller physical derivation, consider the spinning, rotary nature of fluid turbulence as an eddy passes through an observation volume (Cookie 22 Fig. 1). The instantaneous velocity at the mid-point of the eddy is $u$. The instantaneous velocity gradient across the length, $l$, of the eddy is d$u$/d$z$ and the velocity of the upper and lower boundaries to the eddy motions with respect to the mid-point velocity is $\pm 0.5l\mathrm{d}u/\mathrm{d}z$. This velocity transfers a mass of fluid, $m$, in unit time from fast to slow layers and vice versa. For the cylindrical volumes of cross-sectional area, $a$, this mass is $m = a\rho 0.5l\mathrm{d}u/\mathrm{d}z$ and each unit mass of fluid changes momentum by an amount $l\mathrm{d}u/\mathrm{d}z$. The total rate of change of momentum by both cylinders is $2\left(a\rho 0.5l\frac{\mathrm{d}u}{\mathrm{d}z}\right)\left(l\frac{\mathrm{d}u}{\mathrm{d}z}\right)$ and this is balanced by a shear stress acting over the ends of both cylinders of magnitude $\tau 2a$. Thus $\tau 2a = 2\left(a\rho 0.5l\frac{\mathrm{d}u}{\mathrm{d}z}\right)\left(l\frac{\mathrm{d}u}{\mathrm{d}z}\right)$, or $\tau = 0.5\rho\left(l\frac{\mathrm{d}u}{\mathrm{d}z}\right)^2$. Since eddies vary in size and strength over time and distance as they advect past an observer we must represent this expression with respect to mean velocity, mean velocity gradient and mean eddy length by means of a constant, $k$. Thus $\tau = k0.5\rho\left(l\frac{\mathrm{d}u}{\mathrm{d}z}\right)^2$, or $\frac{\mathrm{d}u}{\mathrm{d}z} = \left(\frac{\tau}{\rho}\right)^{0.5}\left(\frac{1}{(0.5k)^{0.5}l}\right)$. Experience and experiments show that the part $(0.5k)^{0.5}l$ varies directly with distance, $y$, from the solid

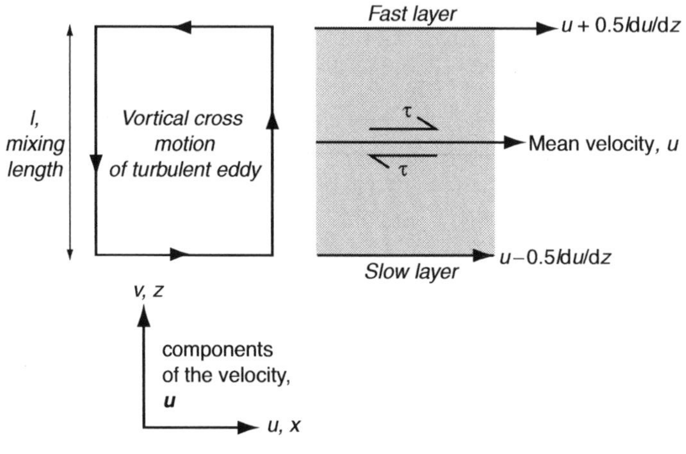


Cookie 22 Fig. 1 Definition diagram for derivation of Prandtl's 'law of the wall'. (After Francis, 1969.)

boundary to the flow (i.e. eddies grow in size as they move away from the bed of a river channel) where $\tau$ is designated a constant mean value at the boundary of $\tau_0$. In fact, $(0.5k)^{0.5}l = 0.4z$ exactly, with the integer termed the *Von Karman constant* and $l$ a characteristic *eddy mixing length*. The differential equation then becomes $\frac{du}{dz} = \left(\frac{\tau_0}{\rho}\right)^{0.5}\left(\frac{1}{0.4z}\right)$ or $du = \left(\frac{\tau_0}{\rho}\right)^{0.5} \int\left(\frac{1}{0.4z}\right)dz$. Integration gives $u = \left(\frac{\tau_0}{\rho}\right)^{0.5} 2.5[\log_e z] + c_1$. Setting $u = 0$ at positive height intercept $c$ we have $c_1 = -\left(\frac{\tau_0}{\rho}\right)^{0.5} 2.5\left[\log_e \frac{1}{c}\right]$ and $u = \left(\frac{\tau_0}{\rho}\right)^{0.5} 2.5\left[\log_e z - \log_e \frac{1}{c}\right]$, or $u = 2.5\left(\frac{\tau_0}{\rho}\right)^{0.5}\left[\log_e \frac{z}{c}\right]$. Switching to $\log_{10}$ units we have the final result for the Von Karman–Prandtl 'law of the wall' as:

$$u = 5.75\left(\frac{\tau_0}{\rho}\right)^{0.5}\left[\log_{10}\frac{z}{c}\right] \qquad \text{(C22.2)}$$

## Cookie 23: Touching matters: flow velocity in the viscous sublayer

Experiments show that there is a linear increase of velocity with height in the viscous sublayer of turbulent flow and that turbulent stresses are negligible. Assuming therefore that viscous stresses predominate we make use of Newton's lubricity equation, $\tau_0 = \mu du/dy$. Integrating

$$u = \int \tau_0 \frac{1}{\mu} dy \qquad \text{(C23.1)}$$

and

$$u = \tau_0 y/\mu. \qquad \text{(C23.2)}$$

Now, armed with the quantity $u* = \sqrt{(\tau/\rho)}$ we can develop the concept of flow boundary roughness by comparing the size of boundary irregularities or sediment grains to the thickness of the viscous sublayer of turbulent flows. First, since $\tau_0 = \rho(u_*)^2$, we can write Equation C23.2 as:

$$\frac{u}{u_*} = \frac{u_* y}{\nu} \qquad \text{(C23.3)}$$

This expression closely fits experimental data up to a value of $u/u_*$ of about 11.5 for a perfectly smooth boundary. Thus the viscous sublayer in a turbulent flow must have a thickness:

$$y = \delta = \frac{11.5\nu}{u_*} \qquad \text{(C23.4)}$$

## Cookie 24: Mean stresses in turbulent flows and palaeohydraulic analysis

The dimensionless friction term, $f$, is the ratio of the shearing stress, or drag, $\tau_0$, exerted by a surface on a flow to the mean kinetic energy per unit volume of the flow, $0.5\rho u^2$. That is:

$$\tau_0 = f(0.5\rho\bar{u}^2) \qquad \text{(C24.1)}$$

You may see this general expression referred to as the 'quadratic stress law'. The friction factor, $f$, takes the form of an experimentally determined drag coefficient, $C_D$, for objects falling through stationary flows, whilst for channelized flows the constant of proportionality becomes $0.25f$ in the expression:

$$\tau_0 = 0.25f(0.5\rho\bar{u}^2) = 0.125f(\rho\bar{u}^2) \qquad \text{(C24.2)}$$

$f$ is known in hydraulics as the Darcy–Weisbach friction coefficient. The quadratic term means that sediment transport rates are non-linear. In general terms, $f$ is strongly dependent on the Reynolds number for laminar to transitional flow regimes. For fully turbulent flows $f$ is more or less independent of Reynolds number, being constant for a particular relative roughness (ratio of roughness diameter to flow thickness) and increasing with relative roughness.

Now consider a steady, uniform, open-channel water flow of average depth $h$. The downslope shear component, $\tau_0$, of the normal stress, $\rho gh$, due to the water acting on unit area of the bed is simply:

$$\tau_0 = \rho ghS \qquad \text{(C24.3)}$$

where $S$ is the water slope (= bed slope for uniform flow). Equation (4.16) is known as the du Boys tractive stress equation. The Darcy–Weisbach and du Boys equations can be combined to give the useful expression for mean flow velocity:

$$u = \sqrt{8g/f} \times \sqrt{RS} \qquad \text{(C24.4)}$$

where $\sqrt{8g/f}$ is the Chezy coefficient and $R$ is the channel hydraulic radius, usually approximated as the mean depth in wide natural channels.

In everyday language we refer to flows as being fast or slow, strong or weak. In fact, there are a number of alternative parameters that can be used to express flow magnitude in a more exact way. We have already come across mean velocity, $\bar{u}$, and bed shear stress, $\tau_0$. The product of these two parameters gives us the stream power, $\omega$ (dimensions $MT^{-3}$) available to unit bed area of fluid:

$$\omega = \tau_0 \bar{u} \quad \text{(C24.5)}$$

In a channel the available power supply $\Omega$ to unit length is the time rate of liberation (in kinetic form) of the liquid's potential energy as it descends the gravity slope $S$. Thus:

$$\Omega = \rho g Q S \quad \text{(C24.6)}$$

where $Q$ is the whole discharge of the stream. The mean available power supply $\omega$ to the fluid column over unit bed area is thus:

$$\bar{\omega} = \frac{\Omega}{\text{flow width}} = \frac{\rho g Q S}{\text{flow width}} = \rho g d u = \tau_0 u \quad \text{(C24.7)}$$

The concept of available fluid power is an important one since Bagnold has made extensive use of it in his sediment transport theory (Chapter 6).

## Palaeohydraulics from river flood deposits

In the summer of 2006, very large localized storms dumped massive precipitation over many parts of the southwest USA during the summer Mexican monsoon, including on two active alluvial fans, Red Canyon and Palomas, positioned on opposite sides of the modern floodplain of the Rio Grande in the Palomas basin of southern New Mexico. A few months prior to the floods, the region had been mapped, which fortuitously provided the framework necessary to accurately assess the major erosional and depositional processes of the floods. The estimation of palaeohydraulic variables for the summer 2006 floods is made relatively straightforward since it was possible to measure accurately key post-flood parameters such as channel bed slope and flood depth. Three assumptions made in the analysis below are all consistent with field observations, namely:

1 Flood sediment deposits aggraded steadily during the flood events such that the elevation of the highest flood debris 'tidemark' above the final sediment bed represents effective flow depth throughout flood deposition. This was 1.5 m just upstream of the avulsion node in Red Canyon and 1.1 m for Palomas Canyon channel upstream of the channel mouth.
2 Mean sediment grain size was constant during the depositional process and is thus adequately represented by measurements taken on the final sediment surface.
3 Floods during peak flow involved turbulent streamflow sediment transport at local equilibrium with channel slope. This was indicated by the general consistency of observed bed sediment surface and 'tidemark' levels.

## Methodology

The DuBoys tractive stress equation, $\bar{\tau}_o = \rho g \bar{h} \bar{S}$, where $\rho$ = water density, $g$ = gravitational acceleration, $h$ = flow depth and $S$ = channel slope, is the fundamental palaeohydraulic expression whose solution is obtainable from field survey data, with an error constrained by the accuracy of slope measurements to ca 5%. The quadratic stress equation, $\bar{\tau}_o = 0.125 f \rho \bar{u}^2$, can then be solved for the flow velocity parameter, $u$, once values for $f$, the Darcy–Weisbach friction coefficient, are known. These may be found via computation of the Chézy coefficient, $C = \sqrt{(8g/f)}$, where $C = 32.6 \log h/k + 35.3$ and $k$ is the Nikuradse effective roughness height, taken as mean bed surface clast diameter, $\bar{d}$. Errors in $u$ come about because of natural variation of clast sizes calculated to $\pm 1$ standard deviation. Values for flood mean flow velocity may then be used to solve the standard expression for mean flood water discharge $\bar{Q}_w = \bar{u}\bar{h}\bar{w}$, in channels of measured bankfull width, $w$.

Concerning sediment transport rate, $i_b$, the most appropriate expression is based upon Bagnold's concept of the flow power immediately available to transport bedload (see above), $i_b = \frac{a}{\tan \alpha}(\tau_o - \tau_c)(u_* - u_*)$, where $a$ is a coefficient that has a mean value of 17 for high-stage flows most appropriate to the present situation, $\tan \alpha$ is the dynamic friction coefficient of magnitude ca 0.6, $\tau_c$ = critical bed shear stress for the threshold of bedload motion, $u_* =$ fluid shear velocity ($= \sqrt{(\tau/\rho)}$) and $u_{*c}$ is the critical shear velocity for threshold of bedload motion. Values of $\tau_c$ and $u_{*c}$ for gravel bed sediment transport by shallow flows are estimated from critical values of the Shields dimensionless stress (section 6.4), $\theta_c = \frac{\tau_c}{\sigma-\rho)gd}$. This has values in the range 0.03–0.07 for $d_{mean}/h \leq 0.2$ and we use the minimum value for measured mean grain diameters in our calculations. Of all palaeohydraulic calculations, $i_b$ is likely to have the most error, chiefly due to uncertainty in coefficient $a$ and tan $\alpha$ and inherent wide variation in experimentally determined $\theta_c$.

Finally we calculate flood duration time, $t_f$, from estimates of flood deposit immersed mass corrected for porosity ($m_d$) and width-integrated bedload transport rate according to the expression, $t_f = m_d/i_b \cdot w$. For the reasons noted above our methodology is expected to yield a minimum estimate.

### Discussion of palaeohydraulic results

For Red Canyon channel just upstream of the avulsion node, $d_{mean} = 19.4 \pm 7.9$ cm. Flood depth here is 1.5 m, channel width 75 m and slope $0.69° \pm 0.03°$. For these parameters values of mean flood flow velocity of some 8.6 m/s and a water discharge of 971 $m^3$/s are derived. The total volume of Red Canyon flood lobe deposits downstream from the avulsion node is $3.19 \times 10^5$ $m^3$. This represents an immersed mass of $3.35 \times 10^8$ kg, assuming fractional porosity of 0.4. For the minimum transport rate calculated across the 75 m wide channel upstream from the avulsion node a sediment delivery rate of approximately $1.7 \times 10^3$ kg/s is estimated. Deposition of the total sediment volume would thus have required a minimum flood transport duration time of around 55 h.

For Palomas Canyon channel, parameters are calculated for mean flow depth = 1.1 m, channel slope = 0.65°, mean channel width = 22 m and $d_{mean} = 12 \pm 3$ cm. For these parameters hydraulic variables are a mean flood flow velocity of some 7.4 m/s and a water discharge of 179 $m^3$/s. The total volume of coarse flood deposits (neglecting the sandy overspill lobe) is $1.72 \times 10^4$ $m^3$. This represents an immersed mass of $1.8 \times 10^7$ kg, assuming fractional porosity of 0.4. At the minimum transport rate calculated for the 22 m wide channel a sediment delivery rate of $4.61 \times 10^2$ kg/s is estimated. Deposition of the whole sediment volume would thus require a minimum flood sediment transport flow time of around 37 h.

## Cookie 25: Simplest possible derivation of wave theory for deep water waves

Here we regard the wave surface as a streamline and apply Bernoulli's theory to the difference of water pressure across the trough and crest of a travelling wave. Take the harmonic water wave of amplitude, $H/2$, travelling at velocity, $c$, shown in Cookie 25 Fig. 1. The speed of water flow, $u$, in the wave orbital motion is the distance round the wave orbit, the circumference $\pi H$, travelled over the wave period $T$, i.e. $u = \pi H/T$. In a thought experiment in which you are the stationary observer, bring the travelling waveform to a halt by imposing an opposite water flow, $-c$, then $u_1 = u\text{-}c = \pi H/T - c$ at the wave crest and $u_2 = -\pi H/T - c$ in the trough. Now, we use the notion that the wave surface is a streamline and that although the wave form no longer travels due to the counter current, there is still orbital velocity along the streamline. We use Bernoulli's equation (**Cookie 18**) to conserve energy between crestal position 1 and trough position 2, with the trough level $H/2$ taken

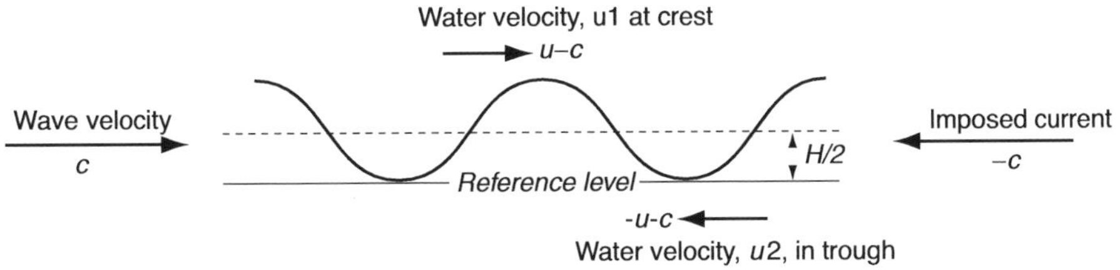


**Cookie 25 Fig. 1** Derivation diagram for the artifice of bringing a travelling deep water wave of velocity, c, to a halt by applying an equal and opposite current. (After Denny, 1993.)

arbitrarily, but conveniently, as the reference height with which to compare potential energy i.e.

$$\frac{\rho u_1^2}{2} + p_1 + \rho g H = \frac{\rho u_2^2}{2} + p_2 \qquad \text{(C25.1)}$$

Neglecting the tiny variation of atmospheric pressure between trough and crest, $p_1 = p_2$ and dividing by $\rho$ we get $0.5u_1^2 + gH = 0.5\,u_2^2$. Substituting $u_1 = \pi H/T - c$ and $u_2 = -\pi H/T - c$ we can expand and cancel terms to yield $c = gT/2\pi$. We see that the velocity of deep-water waves depends only upon wave period or wavelength, since wavelength, $\lambda$, is the product of wave velocity, $c$, and period, $T$, i.e. $\lambda = cT$, then $\lambda = gT^2/2\pi$. Alternatively in terms of celerity, since $T = \lambda/c$, $c = \sqrt{(g\lambda/2\pi)}$.

## Cookie 26: Group activities: wave group velocity and energy flux in deep-water waves

Because deep-water waves travel at speeds determined by wavelength it is common that waves interact as they pass into each other from some source. This is a distinctive interaction rather different from that discussed in the main text for interacting shallow-water waves and solitary waves. For the two wave trains illustrated in Cookie 26 Fig. 1 the small difference in wavelength causes remarkable patterns of constructive and destructive interference, with peaks and troughs in a larger waveform, the group wave, where the individual crests either reinforce each other's wave height or cancel out to zero. This group wave is also called the *beat*. The simple case considered is an example of *linear wave interaction*, where the effect is achieved by wave combination (more complicated schemes of *nonlinear wave interaction* involving wave transformations into smaller and larger frequencies occurs for high frequency ($< 0.3$ Hz) waves). A wave group has lower frequency and travels more slowly than the individual waves which simply appear to pass into the group at the rear and out again at the front. A nice analogy is with the *kinematic wave* phenomena of traffic concentrations along single carriageway roads; the slow-moving queue or tiresomely stationary jam receives your vehicle joining from behind and after some time interval you (hopefully) pass through to the front where you resume your journey at your chosen speed. But another wave group

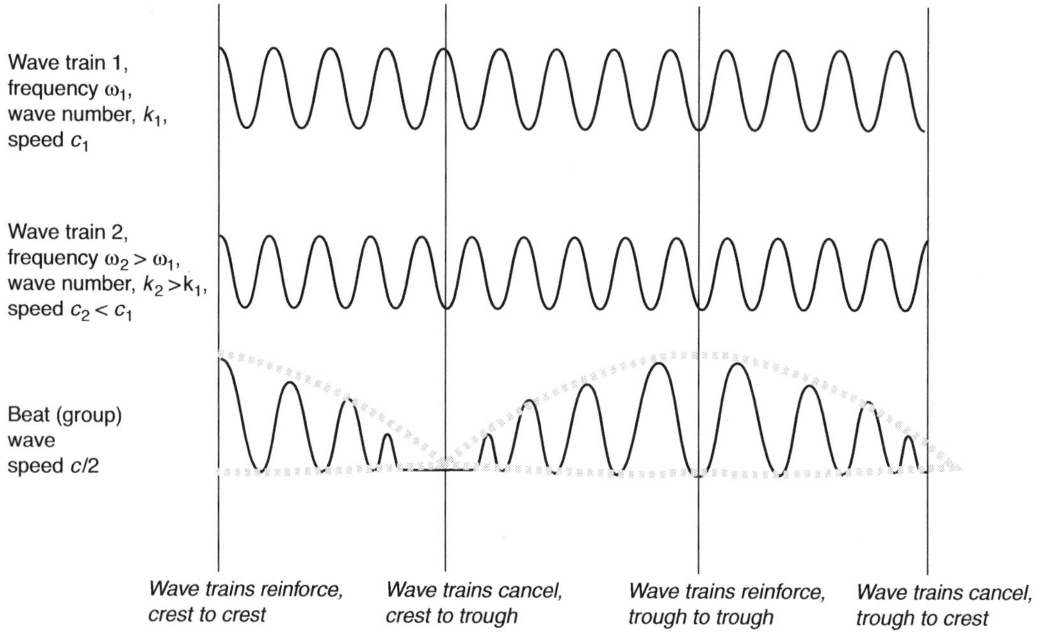


**Cookie 26 Fig. 1** To show how the linear interference between wave trains produces slower-travelling beat (group) waves.

soon occurs, demonstrating both the pointlessness of overtaking and the dependence of individuals on group velocity.

The slower speed of a group wave compared to the waves that pass through it is derived as follows. For realistic situations of group wave generation, the angular frequency, $\omega$, and wave number, $k$, differ little between two wave trains so that the $\delta\omega = (\omega_1 - \omega_2)$ and $\delta k = (k_1 - k_2)$. The wave group speed, $c_{group}$ is then the ratio $\delta\omega/\delta k$ since generally for any wave, $c = \omega/k$, we can derive $c_{group}$ as $c_1c_2/(c_1 - c_2)$. Since $c_1$ and $c_2$ differ little this reduces to $c^2/2c$, or $c/2$, and we have the simple result that a group or beat wave travels half as fast as the waves that pass through it. In such situations the energy of the waves is all carried in the group wave, because there can be no energy at the troughs of such waves where the constituent waves cancel as they arrive or leave.

## Cookie 27: Simplest possible derivation of Airy wave theory for shallow-water waves, dimensionless wave parameters and radiation stress

We utilize a similar approach to that in **Cookie 25** to determine the speed of a shallow-water wave or a solitary wave like a bore. Here, again with the water surface representing a streamline, we place $u_1$ at the wave crest and $u_2$ in front of the wave at stillwater level so that wave height is $H$ (Cookie 27 Fig. 1). Assuming dynamic pressures are the same along the streamline and ignoring atmospheric and capillary pressure, Bernoulli's theorem gives:

$$\frac{\rho u_1^2}{2} + \rho g H = \frac{\rho u_2^2}{2} \qquad \text{(C27.1)}$$

Now, the water at the crest has velocity $u_1 = (u - c)$ and in front of the wave has velocity, $-c$. Substituting gives $c^2 = u^2 - 2uc + c^2 + 2gH$. Since $u$ is always small relative to $c$ we ignore it and have simply $c = gh/u$. To find $u$ we use the continuity principle which states that the discharge of fluid through sections 1 and 2 must be equal. Since discharge is the product of flow speed and cross-sectional area, i.e. $-ch = (u - c)(h - H)$ or $u = c - (ch/(\mathrm{h} + \mathrm{H}))$, where $h$ is water depth. Substituting for $u$ in $c = gh/u$ we get $c = \sqrt{(g(h + H))}$ or more simply still, when wave height is much less than water depth

$$c = \sqrt{gh} \qquad \text{(C27.2)}$$

Full development of linear wave theory gives a continuous function for the influence of water depth on wave speed through the expression

$$c = \sqrt{g\lambda/2\pi}\sqrt{\tanh(2\pi h/\lambda)} \qquad \text{(C27.3)}$$

in which the tan$h$ term is the hyperbolic tangent which approaches value 1 for deep-water waves and tends to $\sqrt{(2\pi h/\lambda)}$ for shallow water, the latter giving the expression developed previously, i.e. $c = \sqrt{(gh)}$.

### Dimensionless wave parameters

Five useful dimensionless parameters used for prediction of sediment transport and bedform development under water waves include:

1 a dimensionless oscillation period, $Tv/d_{50}^2$, where $T$ is wave period, $v$ is kinematic viscosity and $d_{50}$ is median grain diameter;

2 a form of Reynolds number which can be understood as a dimensionless maximum oscillation

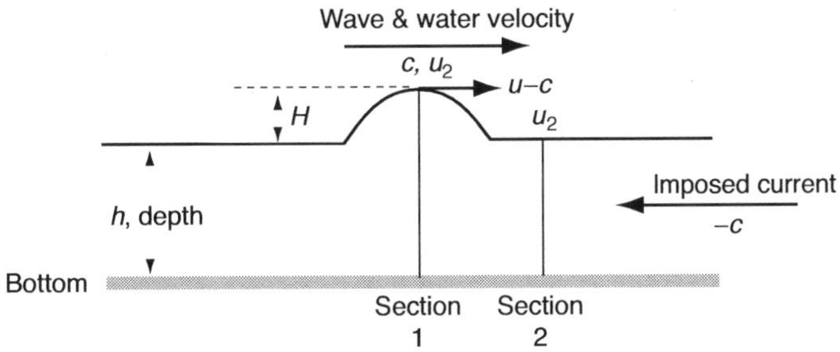


**Cookie 27 Fig. 1** Derivation diagram for the artifice of bringing a deep travelling shallow water wave of velocity, $c$, to a halt by applying an equal and opposite current. (After Denny, 1993.)

velocity, $u_{max}d_{50}/\nu$, where $u_{max}$ is maximum near-bed oscillation velocity;

3 a form of grain Reynolds number or dimensionless grain size, $\sqrt{(gRd_{50})} \cdot d_{50}/\nu$, where $R$ is immersed specific sediment density, $(\sigma/\rho) - \rho$, about 1.65 for quartz mineral grains;

4 a wave mobility parameter, $\psi = u_{max}{}^2/Rgd_{50}$;

5 a form of the Shields number suitable for waves, $\theta = \tau_w/(\sigma - \rho)gd_{50}$, where $\tau_w$ is the wave shear stress exerted on bed grains of magnitude $0.5\rho f_w u_{max}{}^2$, where $f_w$ is a friction factor.

### Practical wave hydraulics

The interested reader is referred to Soulsby (1997) for accounts of how to calculate hydraulic parameters for wave and combined tide/wave situations.

### Radiation stress

Concerning radiation stress, as waves arrive at coasts the forward energy flux or power is, $Ecn$, where $E$ is the wave energy per unit area, $c$ is the local wave velocity and $n = 0.5$ in deep water and 1 in shallow water. Because of this forward energy flux there exists a shoreward-directed momentum flux or *radiation stress* outside the zone of breaking waves. This radiation stress is the excess shoreward flux of momentum due to the presence of groups of water waves, the waves outside the breaker zone exerting a thrust on the water inside the breaker zone. This thrust arises because the forward velocity associated with the arrival of groups of shallow-water waves gives rise to a net flux of wave momentum (Cookie 27 Fig. 2). For wave crests advancing towards a beach there are two relevant components of the stress, $\tau_{ij}$. One is $\tau_{xx}$, with the $x$-axis in the direction of wave advance and the other, $\tau_{yy}$, with the $y$-axis parallel to the wave crest. These components are $\tau_{xx} = E/2$ for deep water or $3E/2$ for shallow water, and $\tau_{yy} = 0$ for deep water or $E/2$ for shallow water. Radiation stress plays an important role in the origin of a number of coastal processes, including wave set-up and set-down, generation of longshore currents and the origin of rip currents.

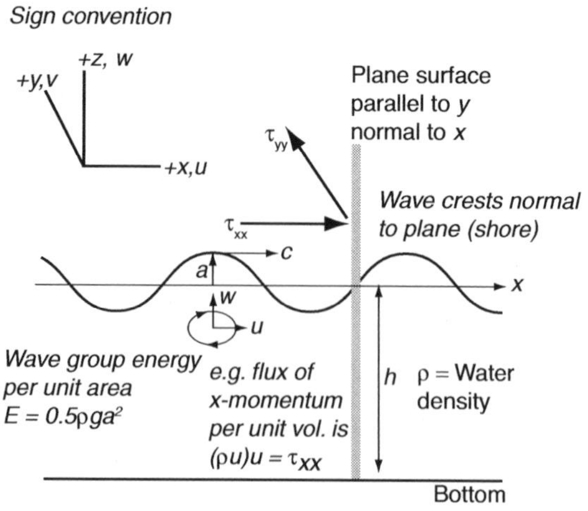


**Cookie 27 Fig. 2** Definition diagram for the radiation stress, $\tau$, exerted on the $+$ve side of the $xy$ plane by wave groups approaching from the left-hand side The radiation stress is the momentum flux (i.e. pressure) due to the waves. (After Longuet-Higgins & Stewart, 1964.)

## Cookie 28: In and out: origin of tides

The tides are explained by Newton's well-known inverse square law of gravitation which states that:

$$F = \frac{Gm_1m_2}{r^2}$$

where $F$ is the force of attraction, $G$ is the universal gravitational constant, $m_1$ and $m_2$ are the masses of two objects (planetary bodies in this case) and $r$ is the distance between the masses. In addition to gravitational forces acting on it from Moon and Sun, each water particle in the oceans undergoes a centripetal acceleration of equal strength and direction (see **Cookie 42** for an explanation of accelerations in curved flows) as the Earth–Moon system revolves in space. Since $F$ varies as the square of $r$, and since water particles vary in their distance from the Moon, a net residual force exists, given by the difference between $F$ and the constant centripetal acceleration (Cookie 28 Fig. 1). The tide-raising force is small but unopposed until the pressure gradient of the sloping tidal bulge balances it exactly. The so-called equilibrium tide causes two areas of high water, one directly under the Moon and another on the opposite side of the Earth. As Earth completely rotates daily, any point on its surface will pass under the position of the tidal bulges twice, giving two times of high water and two times of low water every 24 hours—a semidiurnal tide. The magnitude of the tide-producing force varies with latitude; it is zero immediately underneath the moon and also normal to this position. The maximum tidal amplitude is predicted half-way

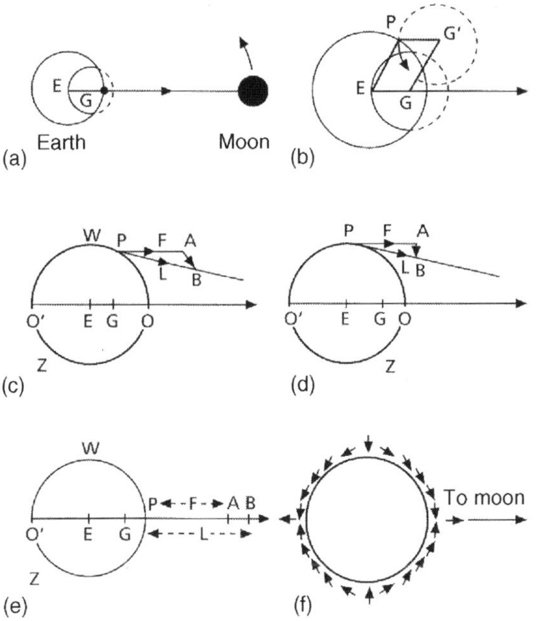


**Cookie 28 Fig. 1** Definition diagrams to illustrate the origin of tide-raising forces. (a) The Moon revolves about *G*, the common centre of gravity of the Earth and Moon, in an anticlockwise direction as viewed from the Pole Star. (b) The centre of the Earth, *E*, and any point, *P*, on its surface, describe circles of constant radius *PG′EG* as the centre rotates about *G*. (c)–(e) *PA* is drawn parallel to *EG* and represents the centripetal force required to make the particle, *P*, move in its circle. *PA* is constant over the Earth's surface. The line *PB* represents the force of the Moon's attraction and by Newton's inverse square law of gravitation must decrease from equator to pole. The force *PB* is equivalent to the resultant of the forces *PA* + *AB*. The force *AB* is thus the tide-raising force. The direction of *AB* varies from equator to pole as shown. (f) Summary of the directions of the tide-raising forces over the Earth's surface; the horizontal component gives rise to the equilibrium tide. (After Tricker, 1964.)

between. The tide is in fact a shallow-water wave of great speed (20–200 m/s) and long wavelength.

The effect of the Sun's large mass upon Earth tides is overcompensated by its longer distance from the Earth compared to the less massive but closer Moon. Even so the Sun exerts about 50% of the total tide-rasing force. Important effects arise when the Sun and Moon act together on the oceans to raise extremely high tides (spring tides) and act in opposition to raise extremely low tides (neap tides) in a two-weekly rhythm. Variations in these tides come about because of the eccentricity of the lunar orbit, the very highest springs forming when the perigee falls at either new or full moon, and because of variations in the Earth–Sun distance in summer and winter. A number of further complications arise because of the declination of the Earth's axis at 23.5° to the plane of orbit about the Sun. Declination with respect to both Moon and Sun introduces a diurnal inequality into the tides, so that high tides will be successively larger and smaller. Other longer-term periodic components of the Earth–Moon–Sun behaviour also affect the tides.

## Cookie 29: Derivation of Stokes law for the fall velocity of solid grains through stationary liquids

There are two ways of approaching this. We could group together likely variables that might control fall velocity, $U_s$, using scientific logic. Thus velocity should vary directly with effective grain density $(\sigma - \rho)$, diameter, $d$, or radius, $r$, and gravity, $g$. It would also seem logical that it should vary inversely with fluid molecular viscosity, $\mu$. Arranging the parameters we get a proportionality relationship of the form $U_s = k(\sigma - \rho)dg/\mu$. This leaves the determination of $k$ and any nonlinear behaviour to experiment. The approach of the mathematical physicist (originally G.G. Stokes) is to solve the problem theoretically by developing arguments from the Navier–Stokes equations of motion and potential flow theory, both topics that we have touched upon lightly in previous Cookies. Stokes considered viscous fluid resistance forces only and assumed that during steady fall, all acceleration terms in the equations of motion vanish, leaving a balance between the grain's immersed weight force $F_w$ and the viscous drag force, $F_v$ acting over the grain surface, i.e. $F_v = F_g$. Derivation of $F_v$ from first principles (a lengthy procedure) gives $6\pi\mu r U_s$ and the force balance therefore as $4/3\pi r^3(\sigma - \rho)g = -6\pi\mu r U_s$, whereupon it can be easily solved for $U_s = 0.22[r^2(\sigma - \rho)g/\mu]$. Note that $U_s$ changes sign for negatively buoyant systems when $\sigma < \rho$.

As noted in the main text, Stokes' law accurately predicts the fall velocity only when the grain Reynolds number is so very small ($< 0.5$; corresponding to silt-sized quartz grains and finer in water) that flow separation effects around the grain (Cookie 29 Fig. 1) do not increase the drag force. Under such conditions recourse has to be made to experimental data on rates of fall and drag coefficients. For the Stokes range only we may equate the two forms of expression for the

**Cookie 29 Fig. 1** Smoke visualisation of the onset of turbulent flow separation around a sphere held motionless by a vertical spigot in an air stream moving upwards (equivalent to a sphere falling down through still fluid). Under such conditions Stokes law no longer applies since the grain Reynolds number allows non-viscous drag forces due to the separation to slow the sphere down somewhat. (After van Dyke, 1982.)

surface drag force and get $6\pi\mu r U_s = C_D \pi r^2 \rho (U_s^2/2)$ or $C_D = 24/Re$ for the drag coefficient. Outside the Stokes range $C_D$ must be obtained from experiment, but it seems to be constant for high *Re* systems.

## Cookie 30: Collisions: Bagnold's derivation of intraparticle stresses in shearing Newtonian fluid-particle suspensions

In 1906, Einstein extended Newton's analysis of viscous shear resistance to include the effects of very dilute suspensions of solid particles; very dilute meaning that the particles had no effect on each other. In nature grains are commonly sufficiently close that interaction or physical collision occurs. Making use of Reynold's concept of dilitancy (Fig. 8.2) and general ideas of kinetic theory of molecular gases, Bagnold imagined a dispersion of rigid spheres of uniform diameter, $D$, and density, $\sigma$, previously arranged in canonball-style (rhombohedral) packing. Here, during shear, the mean distance between particle centres is increased from $D$ to $bD$ and the resulting free distance between individual particle rims is denoted, $s$, we have $b = s/D + 1 = 1/\lambda + 1$, where $\lambda = D/s$, is *linear concentration*. $L$ is $\infty$ for $s = 0$, 22.5 for rhombohedral packing and 8.3 for the loosest cubic packing. The lower the value of $\lambda$ below 8.3 the less likely that intraparticle interactions will take place. For dispersions where there is no slip between fluid and solid, the fluid and solid densities are equal, i.e. neutral buoyancy occurs. The system is maintained in a state of uniform, constant shear strain, $\mathrm{d}U/\mathrm{d}z$, where $U$ is the particle velocity, the kinetic energy per unit volume of the system is maintained by frictional losses involving either intraparticle collisons and/or a general 3D oscillation due to pressure effects at particle near-approaches and recessions. For the case where particle collisions dominate (grain inertia conditions) the grains of an upper layer, $B$, are sheared over a lower, $A$, at mean relative velocity $\mathrm{d}U = kb\mathrm{d}U/\mathrm{d}z$ where $k$ has some value between $\surd 0.66$ (rhombic packing) and $(0.5)\surd 2$ (cubic packing). Each grain in layer $B$ makes $f(\lambda)\mathrm{d}U/s$ collisions with an $A$ grain in unit time, where $f(\lambda)$ is unknown. Since the number of grains in unit area in the $xz$ plane (spanwise plane) in each layer is $1/b^2D^2$, and at each collision each $B$ grain experiences a total momentum change of $2m\delta U\cos\alpha$, where $\alpha$ is an unknown collision angle, the repulsive pressure in the $z$-direction is:

$$P_z = \frac{1}{b^2D^2}\frac{f(\lambda)\delta \boldsymbol{U}}{s}2m\delta \boldsymbol{U}\cos\alpha \qquad \text{(C30.1)}$$

or

$$P_z = k\sigma\lambda f(y)D^2\left(\frac{\mathrm{d}U}{\mathrm{d}z}\right)^2\cos\alpha, \qquad \text{(C30.2)}$$

where $k$ is some proportionality constant. There is also an intraparticle shear stress (in addition to any fluid shear stress) of magnitude $T_{xz} = P_z \tan\alpha$.

From the discussion above it is implicit that, once grains are in motion as bedload, impacting grains will transfer momentum from flow to the bed and give rise to additional solid transmitted stresses. For example, consider a saltating grain of mass $m$ impacting on to a bed with velocity $U$ and making an angle $\zeta$ with the bed surface. If elastic rebound occurs, and neglecting frictional effects, then momentum of magnitude $2mU\sin\zeta$ is transferred normal to the bed surface and momentum $2mU\cos\zeta$ is transferred tangential to the bed surface. If no rebound occurs and a period of rolling ensues after impact, then this momentum transfer is halved.

It is difficult to directly measure the magnitude of the solid stresses $T$ and $P$. The results of the experiment reveal that there is a great increase in the shear resistance of the solid–fluid mixture when compared to the plain fluid alone. Two regions of behaviour were definable in terms of a dimensionless number now known as the *Bagnold number*. A viscous region of behaviour at low strain rates and/or low grain concentrations showed the ratio $T/P$ to be constant at around 0.75. In this region grain–grain effects were caused by near approaches causing a repulsion before solid collision could occur. An inertial region of behaviour at high strain rates and/or high grain concentrations showed the ratio $T/P$ to be constant at around 0.32. Grain–-grain collisions dominated in this region. Applying these results to natural flows, Bagnold postulated that:

1 in air, owing to the 'chain reaction' process described above at threshold, so much fluid momentum is transferred to the saltation that virtually the whole of the applied shear stress is resisted by it;
2 in water, where the density ratio of solid to fluid is low, the contribution of $T$ gradually increases with transport stage—when one whole, formerly stationary grain layer is in motion as bedload at high transport stage, Bagnold argues that $\tau = T$;
3 the normal stress, $P$, also termed the dispersive stress, should be in equilibrium with the normal stress due to the bedload weight over unit bed area.

## Cookie 31: Bagnold's subaqueous bedload sediment transport approach

During bedload transport the immersed mass of transported grains present over unit bed area must be in equilibrium with an equal and opposite stress due to momentum transfer between grains and stationary bed. Further, in order to maintain transport in the face of frictional resistance it is necessary to apply an external fluid impulse to the grains from the fluid. The magnitude of all forces involved depends upon the applied fluid bed shear stress, $\tau$. The bedload immersed weight, $w_b$, is $m'_b g$, where $m'_b$ is the bedload immersed mass per unit area. This must be in equilibrium with a solid transmitted normal reaction stress $P$. From basic considerations of friction, $P \tan\alpha = T$, where $\tan\alpha$ is a dynamic friction coefficient and $T$ is the transverse shear component of $P$. $T$ is that stress directly balanced by the applied fluid shear stress, $\tau$, minus that shear stress needed to initiate motion, $\tau_{crit}$. For equilibrium we have $\tau - \tau_{crit} = m'_b \tan\alpha$.

Introducing Bagnold's concept of flowing fluid as a transporting machine, where *work rate* = *available power* × *efficiency*, the work rate is given by $(m'_b g U_b) \tan\alpha$, where the bracketed term gives the transport rate, $i_b$, and $U_b$ is mean bedload velocity. The available power is given by $[k(\tau - \tau_{crit})u_b]/\tan\alpha$, where $k$ is a coefficient ($<1$) determining that part of $\tau$ that is transferred to the bed indirectly via the saltating solids and $u_b$ is the mean fluid velocity in the centre of gravity of the bedload layer. The efficiency is given by $U_b/u_b$ and is a function of the 'slip' or relative velocity of solid grains and fluid. Finally we have the complete bedload transport function as $[k(\tau - \tau_{crit})U_b]/\tan\alpha$. The expression can be solved only with an experimental knowledge of $k$ for different transport stages and bed states, a parameter that has proven difficult to determine. Experimental values for $\tan\alpha$ and $U_b$ are better established. Further, the expression can be solved only where there is no relief on the transporting bed greater than a few grain diameters, establishing that there is no stress contribution arising from energy dissipation in friction due to flow separation over bed features (i.e. $k \approx 1$). These somewhat restrictive conditions prevent widespread application of the theory, but tests whose parameterization fall within the experimental conditions are in good agreement with Bagnold's basic theory.

## Cookie 32: Bagnold's bedload transport prediction for wind-driven sediment

Concerning wind-blown bedload transport, from experiments and field measurements the rate is found to be a cube function of the shear velocity. Above threshold for motion, wind accelerates each sediment grain in transport from rest until splashdown. Solid momentum is thus gained from the air and subsequently released to the bed on collision. Each particle of mass, $m$, gains velocity, $U$, over unit length, $l$ and so the air loses momentum $mU/l$. For a total mass of particles, $i_b$, moving over unit area in unit time the total loss of momentum equals the drag (resisting) stress, $\tau$, exerted by the bedload on the air, so that $i_b U/l = \tau$. Since $\tau = \rho u_*$, where $u_*$ is shear velocity, and both $U$ and $l$ are proportional to $u_*$ we have $i_b \sim k\rho u_*^3$ where $k$ is an experimental constant.

## Cookie 33: A dynamic sediment suspension theory

To suspend something is literally to hold it up. No suspension of grains picked up from the bed occurs in laminar flow, so a mass of sediment grains must be held above a bed in a state of suspension by the eddies of fluid turbulence. Applied to individual grains, the concept of suspension is statistical, because of the continuous exchange of grains between bedload and the overlying turbulent flow. However, a steady state exists with respect to a suspended mass in a steady, uniform flow. Thus over a sufficiently long period of time the measured mass will itself be constant whereas the constituent grains that define the mass may be exchanged continuously between bed, bedload and suspended load. A state of dynamic equilibrium thus exists. Two questions follow.

- What mechanism acts to keep up the suspended grains?
- What dynamic quantity controls the magnitude of the suspended mass?

Bagnold's answers to these questions are, as usual, startlingly original and incisive. Concerning the turbulent stress needed to transport suspended sediment, the vertical turbulent force required has to be balanced by a corresponding vertical pressure gradient. Vertical gradients in the vertical turbulent velocity, $v'$, occur throughout turbulent boundary layers, with marked positive gradients close to the bed. It is these upward gradients in turbulent acceleration that provide the motive force for sediment suspension. These conclusions fall like pearls out of the Navier–Stokes–Reynolds equations for turbulent fluid motion (**Cookie 21**).

So, a vertical turbulent force must support any suspended load. To determine suspension threshold we follow the same general approach as originally developed by Shields for the initiation of general bedload particle motion (Cookie 33 Fig. 1). For suspension of an originally stationary particle layer comprising grains of diameter, $d$, immersed density, $(\sigma - \rho)$, and fractional concentration, $C$, the vertical turbulent force component must exceed the force due to gravity acting on the sediment. The upward turbulent force is the gradient of mean vertical turbulent stress from the wall ($y = 0$) up to the level of $y = v'_{rmsmax}$, where $v'_{rmsmax}$ is the maximum value of $v'_{rms}$ measured in the wall layer. The upward turbulent stress is derived as $\rho(v'_{rmsmax})^2$, where $\rho$ is fluid density. The

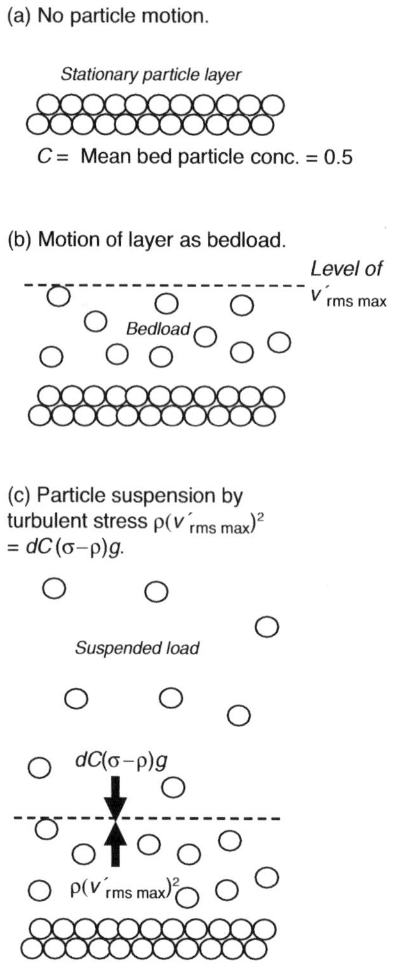


Cookie 33 Fig. 1 Sketches to illustrate (a) an originally stationary particle layer, (b) the layer undergoing bedload transport and (c) the layer in suspension transport in response to applied turbulent stresses. (After Leeder, 2007.)

stress due to gravity acting on the sediment layer over unit bed area is $dC(\sigma - \rho)g$ and therefore the stress balance required for suspension is: $\rho(v'_{rms\max})^2 \geq dC(\sigma - \rho)g$. Written in terms of the turbulent velocity fluctuation alone, this becomes the threshold criterion: $v'_{rms\max} \geq [dC(\sigma - \rho)g/\rho]^{0.5}$.

## Cookie 34: The sediment continuity equation (SCE)

Let $i$ be the quantity of sediment transported in the $x$-direction. Let $q$ be the amount added from outside the system, say by wind deposition or contour currents. Both $i$ and $q$ can vary with both incremental space, $\delta x$, and time, $\delta t$. The definition diagram (Cookie 34 Fig. 1) sketches a concave surface that has grown by deposition in time $\delta t$, defining an elemental area ABCD. For this area:

(mass flux in)−(mass flux out)

= increase/decrease of mass in ABCD

or

$$(i_x - i_{x+\delta x})\delta t + q\delta x\delta t = \delta z\delta x.$$

Dividing by $\delta x\delta t$ gives

$$(i_x - i_{x+\delta x})/\delta x + q = \delta x/\delta t.$$

Now, as $\delta x$ and $\delta t$ tend to zero we get the exact partial differential SCE equation that has universal validity

$$-\partial i/\partial x + q = \partial z/\partial t.$$

We see that the rate of erosion (negative right-hand side) or deposition (positive right-hand side) depends on the sign of the input term, $q$, and of the horizontal rate of change of the transport term, $i$. For example, a negative downstream change in $i$ alone means that deposition will occur, vice versa for positive change. Note that the SCE as written assumes no internal density changes due to deposition and compaction and also no internal sinks or sources such as mineral dissolution/precipitation. More involved versions of the SCE can cater for these, and other complications such as tectonic subsidence, etc.

## Cookie 35: Bedform 'lag' effects

Many of the remarks made in Chapter 7 assume the presence of an 'equilibrium' bed state adjusted to a steady flow. However, natural flows of water and wind are unsteady on a variety of scales, e.g. tidal flows for hours, and river and air flows for weeks or months. Subaqueous dunes that form in response to a steady flow may persist for a considerable time as the flow decays into the ripple stability field. Within the

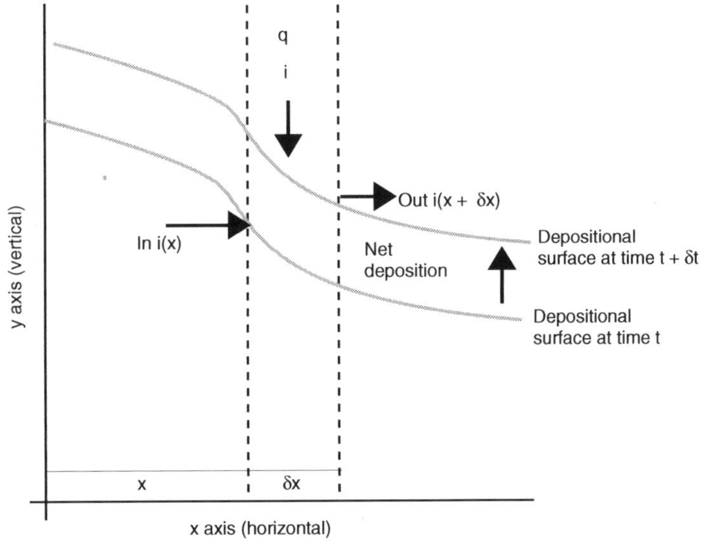


**Cookie 34 Fig. 1** Definition diagram for derivation of the simplest version of the Sediment Continuity Equation. (After Wilson & Kirkby, 1975.)

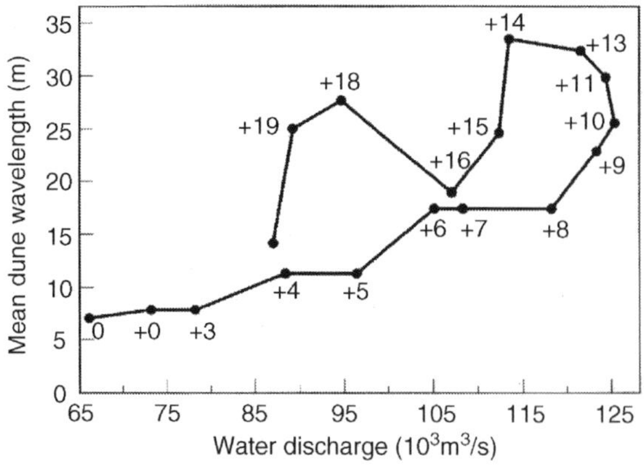


**Cookie 35 Fig. 1** Variation of mean dune wavelength with water discharge over a period of 19 days (points 0–19) in the Fraser River, British Columbia, Canada. Note the well developed lag as the dunes continue to grow despite falling discharge. The reasons for this behaviour are not well understood. (After Allen, 1973.)

dune stability field, changes in water depth during rising or falling river stage may cause changes in the dune wavelength or height to lag behind the flow (Cookie 35 Fig. 1). To illustrate this phase lag, graphs of $y$ and $x$ are plotted so that both variables are also functions of time; $y$ could be dune wavelength whilst $x$ could be discharge. If no lag exists, then a straight-line plot results. Lag reaches extreme values of 90°.

Bedform lag in natural environments means that we should be very careful in interpreting field measurements of bedform sizes in relation to existing flow conditions. Criteria for bedform equilibrium should be investigated. One relevant example might be the equations relating dune wavelength and height to water depth, in which very considerable scatter was evident. Much of this scatter could have resulted from lag effects.

## Cookie 36: Bedform theory for water flows

Ripples are stable at low values of applied fluid shear stress in fine sediment. The spanwise sequence of fluid bursts and sweeps in the viscous sublayer plays an important role here. Entrainment of grains during a sweep is followed by deposition as the sweep fluid decelerates. All over a sand bed, made artificially plane prior to experiment, tiny discontinuities are formed from heaps of deposited grains two to three diameters high. Certain heaps now begin to influence flow structure as flow separation and reattachment occur on their lee sides. These heaps become magnified and propagate downstream since the turbulent stresses at attachment erode too much material to be held in transport. More heaps are thus formed downstream, which in turn give rise to flow separation, and so on. Tiny current ripples now form, coalesce and interact all over the bed. An equilibrium ripple assemblage evolves after an hour or so. The ripples grow to an equilibrium height and wavelength determined by the size of the lee-side separation eddy. They are not related to outer flow structures since they have no relationship of magnitude to flow depth. The curved crest equilibrium form suggests some control by longitudinal vortices.

The well-known inability of coarser sands and gravels of grain size 0.6 mm to form ripple-like bedforms has been explained by changes brought about in the flow by such coarse bed grains. It can be no coincidence that the threshold for motion of coarse sands of diameter ~0.6 mm is the point at which the viscous sublayer begins to be broken and disrupted by the grain roughness. It is thus expected that at this point the near-bed boundary layer over small bed defects produced by turbulent streaks should be more intensely mixed due to the production of eddies shed off from the lee sides of both bed grains and those in saltation transport. This increase of mixing is seen in the increased values of the vertical component of flow turbulent stresses. The effect of enhanced mixing is to steepen the velocity gradient and decrease the pressure rise at the bed in the lee of defects. This causes separation to be inhibited and the defects are unable to amplify to form ripples. An alternative view is that the 0.6 mm limit represents the onset of grain saltations whose wavelength matches that of the bed defects produced by turbulence. Thus grains saltate their way out of the lee-side separation eddies and defects are unable to amplify.

The positive relationship between dune height, wavelength and flow depth indicates that the magnitude of dunes is somehow related to thickness of the boundary layer or flow depth. In view of the prominent occurrence of large-scale turbulent eddies that are generated in the separation-zone free shear layer, it is likely that dunes grow to an equilibrium determined by the effects of such large-scale eddies on the transport of suspended sediment. The instability of a ripple or lower plane bed field as flow strength is increased may be due to the random growth of very large ripples or dunoids (Cookie 36 Fig. 1) that are able to trigger large-scale eddy formation in their downstream free shear layers. These statistical freaks ('rogue ripples') somehow trigger the appearance of like forms downstream and so a dune field gradually equilibrates. The change from dune to upper plane bed states occurs at large values of applied bed shear stress. It is thought that the change may somehow be due to the effects of increasing suspended load and bedload sediment transport upon the structure of the turbulent flow close to the bed. It is possible that turbulence may be suppressed by increasing concentrations of sediment in transport and hence the likelihood that turbulent flow separation will occur becomes less. This reasoning is supported by the fact that the dune stability field is much reduced or absent in sediments below 0.1 mm diameter, and also by the discovery of natural symmetrical duneforms without lee-side flow separation at high bed shear stresses.

Antidunes occur as stable forms when the flow Froude number, $Fr$, is $0.7 < Fr < 1.8$, approximately indicative of rapid (supercritical) flow and the occurrence of stationary or migrating water-surface sinusoidal waves that are in-phase with the bedform. They are thus common in fast, shallow free-surface flows, but may also occur in two-layer flows. Antidune wavelength of free-surface flows is related to the square of the mean flow velocity by $\lambda = 2\pi u^2/g$. The corresponding equation for two-layer flows is a little more complex and will not be discussed here. The rare occurrence of antidunes preserved on bedding planes facilitates direct determination of palaeoflow velocity from the expression if $\lambda$ can be measured. Because antidune height/wavelength ratios are constant at ~0.014, the longer wavelength examples formed under faster flows are more subdued. Also, the height, $H$, of water surface waves in supercritical flows is a function of water depth, $h$, such that $H \approx 0.7h$.

## Cookie 37: Measurement of palaeocurrents and problems arising from trough-shaped sets of cross-stratification

It is commonplace for students (and professionals) to misunderstand the very principles of palaeocurrent measurement in the field or in core by an insufficient appreciation of the influence of the 3D shape of cross-bedding. It is absolutely necessary to establish the 3D geometry of any set before a palaeocurrent measurement is used to reconstruct palaeoflow (Cookie 37 Fig. 34). Only with tabular cross-sets can mean flow direction be firmly established from field observations in well-exposed 3D sections, when the angle of maximum dip may be directly measured or reconstructed from the stereographic projection of two apparent dips. Trough-shaped sets formed by curved crested dunes cannot be reconstructed from two measurements of apparent dip—they must be measured when the trough shape may be seen in plan view, as a measurement along the trough axis, or from numerous measurements around a trough assembled on a stereographic projection.

## Cookie 38: Dispersive pressure in granular flow

Granular flows consist of a multitude of grains kept aloft above a basal shear plane. Equilibrium demands that the weight force, $W$, of the grains be balanced by an equal and opposite force, $P$, arising from the transfer of grain momentum from all colliding grains onto the shear plane. Using the experimental results and notation outlined in Cookie 30, we have $W = Cg(\sigma - \rho)(Y - y)\cos\beta$ and $P = k\sigma\lambda^2 D^2 (dU/dY)^2 \cos\alpha$, where $k$ is an experimental constant, $C$ is mean grain concentration by volume above point $y$ in the flow, $\rho$ is fluid density, $Y$ is flow thickness and $\beta$ is local bed slope. Since $P = W$ at equilibrium, we may solve for $dU/dy$ by integration, the *Bagnold–Lowe derivation*, with boundary conditions that $U = 0$ at $y = 0$ and the

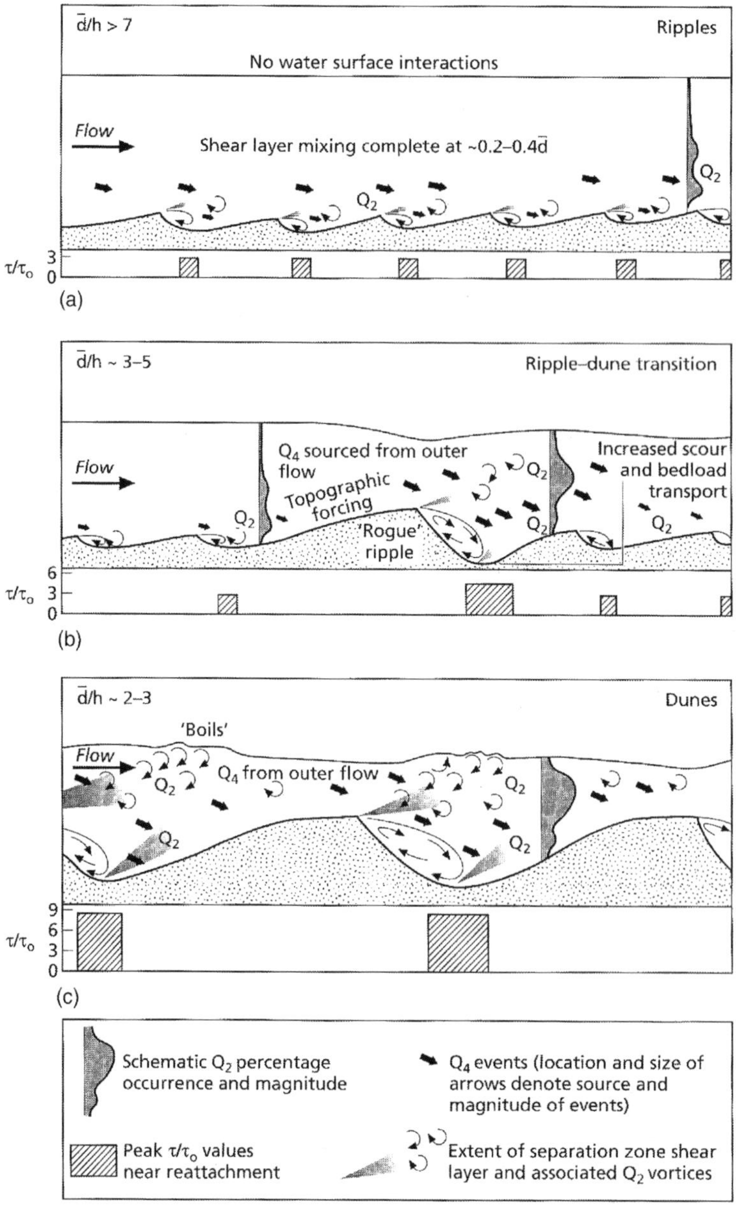


**Cookie 36 Fig. 1** Sketches of bedform changes across the ripple-to-dune transition, involving the growth of 'rogue' current ripples of larger-than-average size that trigger changes in the response of the sediment bed and turbulence, chiefly the increased involvement of the outer flow and the production of large Kelvin–Helmholtz vortices along free shear layers in the lee of the bedform. (after Bennett & Best 1996.)

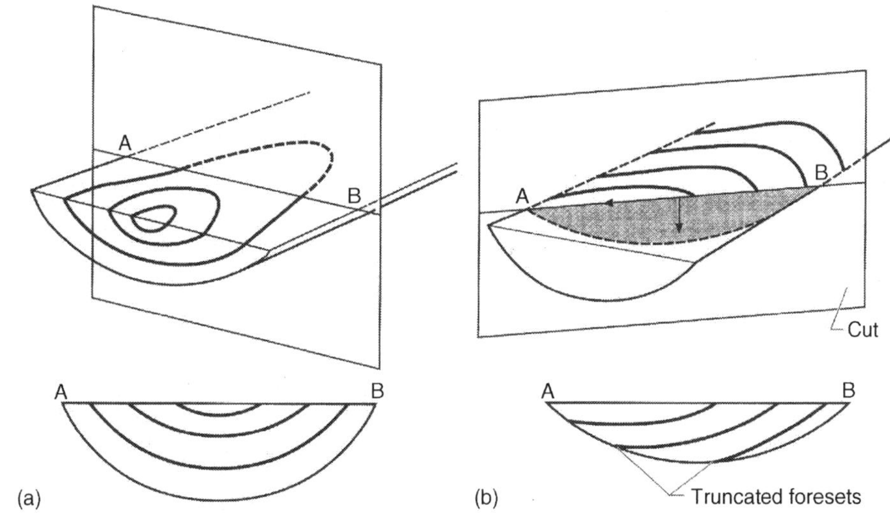


**Cookie 37 Fig. 1** To illustrate the effects of orientation on the appearance and in the measurement of trough cross-stratification. (After DeCelles *et al.*, 1983.)

assumption that C, λ and α do not vary with height. This gives:

$$U = \frac{2}{3}\left[\frac{Cg(\sigma-\rho)\cos\beta}{\rho k\cos a}\right]^{0.5}\frac{1}{\lambda D}\left[Y^{1.5}-(Y-y)^{1.5}\right] \quad \text{(C38.1)}$$

an expression that yields a velocity profile showing a thick shearing zone overlain by a surficial 'plug' of nonshearing granular flow. More recent work shows that the shearing zone is much thinner than predicted and that the assumption of constant C, λ and α with height is not realized. Further, the assumption that variations in dispersive pressure are due to the quadratic term in $D$ for $P$ from Bagnold's work is mistaken. A recent theoretical analysis suggests that dispersive pressure in a grain flow, modified grain flow, or traction carpet cannot account for the upward segregation of large particles, as sometimes assumed. The analysis shows that rapid granular flows are self-organized in such a way that dispersive pressure at any level in a shearing mass of grains is always equal to the applied normal stress, as noted above. An increase in dispersive pressure causes an immediate expansion of the flow and a consequent decrease in dispersive pressure until it equals the applied normal stress again. The gradient of dispersive pressure is therefore 'hydrostatic', and only particles lighter than the bulk density of the flow are pushed upward. The inverse grading of heavy particles observed in a great variety of deposits is not caused by dispersive pressure and must be explained by other mechanisms, chiefly *kinetic sieving* (see Chapter 8).

## Cookie 39: Flow runout

Gravity flow downslope exchanges potential energy $mgh$, where $m$ is flow mass and $h$ is the distance of fall, for kinetic energy which is then totally lost to the effects of frictional resistance, $f$, as flow ceases. The energy lost over runout length, λ, is $mgf\lambda$. For energy balance, $mgh = mgf\lambda$ and therefore flow efficiency, $1/f$, is given by $\lambda/h$, an attractively simple expression since parameters $h$ and λ are relatively easy to measure.

## Cookie 40: Debris flow rheology and run-up

Debris flows are non-Newtonian in that they possess a yield strength that increases as the amount of cohesive clay-grade material is added to an initially cohesionless granular flow. Shear strength, $T$, then depends upon three parameters, the matrix property of cohesion, $c$, granular friction caused by grain–grain interactions, $F_N \tan\theta$ (where $F_N$ is normal weight stress and $\theta$ is the angle of dynamic friction) and bulk Newtonian viscous resistance, $\mu du/dy$. For viscous flow to begin then, $T \geq c + F_N \tan\theta$. Recognizing that this means a bulk yield stress, $k$, must be exceeded before flow begins we have the *Johnson model* for debris flow as deforming like a Bingham plastic, $T = k + \mu_B\, du/dy$ with $\mu_B$ as a Bingham viscosity.

Concerning the palaeohydraulics of debris flows, in situations where run-up has occurred on local adverse slopes and the flow has come to rest, we assume that all kinetic energy is converted to potential energy with no loss due to friction. Then, $0.5mu^2 = mgh$ and $u = \sqrt{(2gh)}$. For debris flows that have travelled round channel bends then centrifugal forces drive the flow slightly higher on the outside of the bend (Cookie 42). This superelevation, $\Delta h$, may be measured from debris marks and used to solve for mean flow velocity in the expression $u = \sqrt{(g\Delta h r_c/w)}$, where $r_c$ is radius of channel centreline curvature and $w$ is channel width.

## Cookie 41: Autosuspension in turbidity currents

Suspended sediment is held up by the eddies of fluid turbulence generated at flow boundaries. In normal river flow, gravity causes fluid movement down a slope, which then causes turbulence, which in turn supports a given mass of suspension. Now, a turbidity current owes its kinetic energy to a suspension giving it excess density, but motion also engenders turbulence and therefore more sediment suspension. A feedback effect thus arises, originally termed autosuspension by Bagnold in the 1960s. In terms of flow power we can say that the available power in the current exceeds that needed for support of the suspension and so sediment pick-up from the bed (assuming it to be granular) results. This is expressed simply as $e_x\beta U/V_g \geq 1$, where $e_x$ is an efficiency factor, $\beta$ is bed slope, $U$ is the velocity of the suspension and $V_g$ is the sediment fall velocity. So autosuspension is favoured by high efficiency, large slope, high flow velocity and fine sediment.

## Cookie 42: Flow in curved channels

In our discussion of speed and velocity (Chapter 4) we saw that fluid travels at a certain speed or velocity in straight lines *or* in curved paths, the latter expressed as *angular* speed, velocity or acceleration. Many physical environments on land and in the ocean and atmosphere allow motion in curved space, with substance moving from point to point along circular arcs, like the river bend illustrated in Cookie 42 Fig. 1a. In many cases, where the radius of the arc of curvature is very large relative to the path travelled, it is possible to ignore the effects of curvature and to still assume linear velocity. But in many flows the angular velocity of slow-moving flows gives rise to secondary flows. Assume a bend to have a constant discharge and an unchanging morphology and identical cross-sectional area throughout, the latter a rather unlikely scenario in nature, but a necessary restriction for our present purposes. From continuity for unchanging (steady) discharge, the magnitude of the velocity at any given depth is constant. Consider surface velocity: although there is no change in the length of the velocity vector as water flows around the bend, i.e. the magnitude is unchanged, the velocity is in fact changing—in direction. This kind of spatial acceleration is termed a *radial acceleration* and it occurs in every curved flow. The curved flow of water is the result of a net force being set up. This is familiar to us during motorized travel as we negotiate a sharp bend in the road slightly too fast; the car heaves outwards on its suspension as the tyres (hopefully) grip the road surface and set up a frictional force that opposes the acceleration. The existence of this radial force follows directly from Newton's 2nd law, since, although the speed of motion, $u$, is steady, the direction of the motion is constantly changing, inwards all the time, around the bend and hence an inwards angular acceleration is set up. This inward-acting acceleration acts centripetally towards the virtual centre of radius of the bend. To demonstrate this, refer to the definition diagrams (Cookie 42 Fig. 1b). Water moves uniformly and steadily at speed $\boldsymbol{u}$ around the centreline at 90° to lines $OA$ and $OB$ drawn from position points $A$ and $B$. In going from $A$ to $B$ over time $\delta t$ the water changes direction and thus velocity by an amount $\delta\boldsymbol{u} = \boldsymbol{u}_B - \boldsymbol{u}_A$ with an inwards acceleration, $\boldsymbol{a} = \delta\boldsymbol{u}/\delta t$. A little algebra gives the instantaneous acceleration inwards along $r$ as equal to $\boldsymbol{u}^2/r$. The centripetal acceleration increases more than linearly with velocity, but decreases with increasing radius of bend curvature. For the case of the River Wabash channel illustrated, the upstream bend has a very large radius of curvature, ca 2350 m, compared to the downstream bend, ca 575 m. For a typical surface flood velocity at

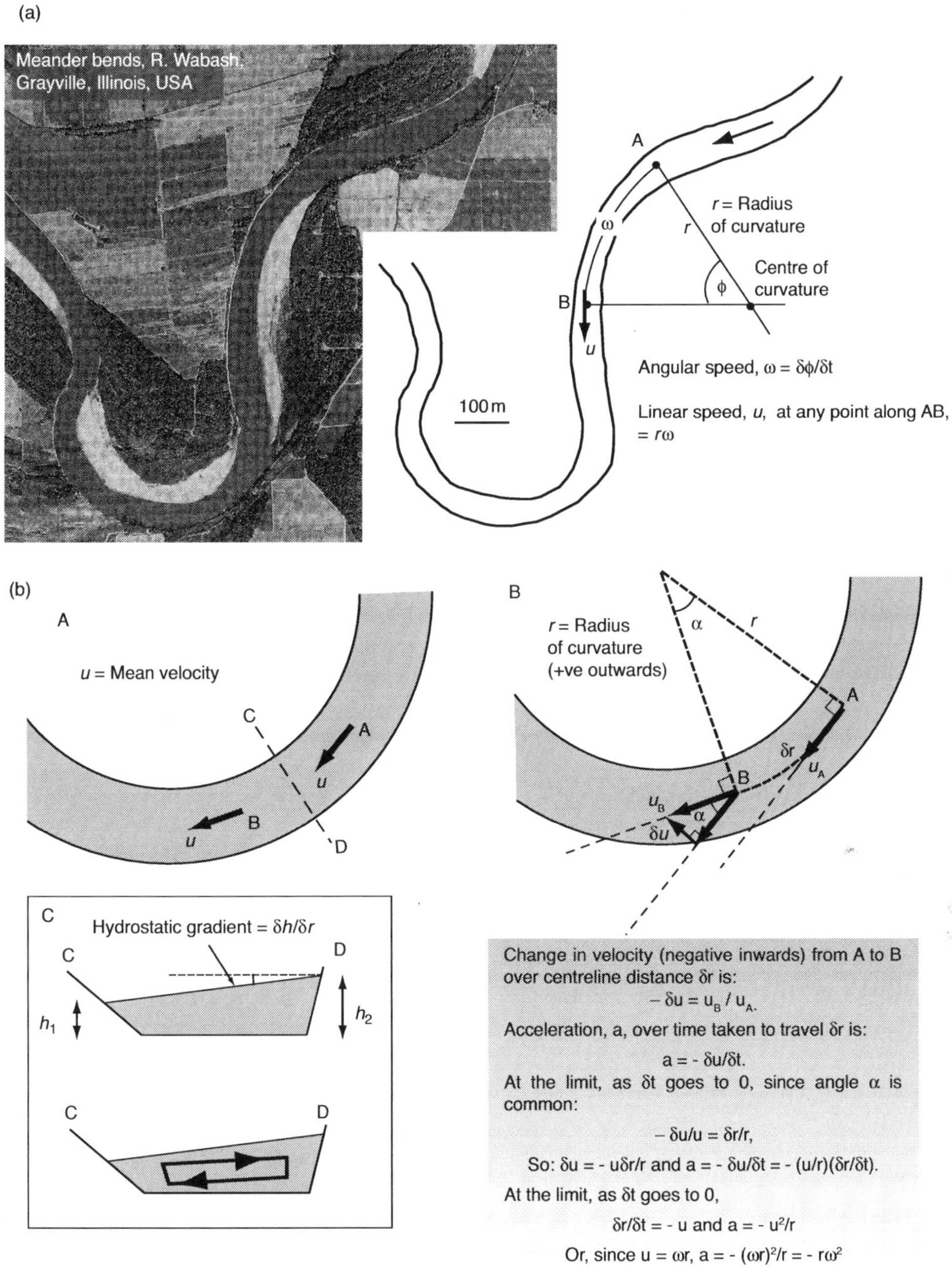


**Cookie 42 Fig. 1** (a) The speed of flow in channel bends. (b) To define the radial acceleration acting in curved flow around channel bends. Note superelevation of water surface on the outside of any bend and the resulting helical secondary flow cell.

channel centreline of $u = \sim 1.5$ m/s, the inward accelerations are $9.6 \times 10^{-4}$ and $4.5 \times 10^{-3}$ m/s$^2$ respectively.

Although the computed inward accelerations illustrated from the River Wabash bends are small, they create a flow pattern of great interest. The mean centripetal acceleration must be caused by a centripetal force. From Newton's 3rd law we know this will be opposed by an equal and opposite centrifugal (outward-acting) force. This tends to push water outwards to the outside of the bend, causing a linear water slope inwards and therefore a constant lateral hydrostatic pressure gradient that balances the mean centrifugal force (Cookie 42 Fig. 1b). Although the *mean* radial force is hydrostatically balanced, the value of the radial force due to the faster flowing surface water exceeds the hydrostatic contribution whilst that of the slow-moving deeper water is less. This inequality drives a secondary circulation of water, outwards at the surface and inwards at the bottom, that spirals around the channel bend and is partly responsible for predictable areas of channel outer bank erosion and inner bank point-bar deposition as it progresses. The principle of this is familiar to us whilst stirring a cup of black or green tea with tea leaves in the bottom. Visible signs of the force balance involved are the inwards motion of tea leaves to the centre in the bottom of the cup as the flow spirals outwards at the surface and down the sides of the cup towards the cup centre point.

## Cookie 43: Twister: the Ekman spiral

Turbulent wind shear mixes surface lake or ocean layers and drives water along in response to the momentum exchanged. In small-scale experiments (blowing surface floats) or field situations the water is driven leewards with the wind. On a large scale, in the open ocean, Coriolis accelerations make the water flow direction diverge from the windflow, right in the northern hemisphere and left in the southern. This was first systematically noted by Nansen in the Arctic seas where he measured icebergs moving 20–40° to the right of the prevailing wind. Ekman solved this problem theoretically by making use of the equations of motion for a rotating fluid system with the effects of constant Coriolis acceleration, $f = 2\Omega \sin\phi$, and upper boundary friction, $F$, taken into account (**Cookie 17**). The very small effect of pressure gradients due to water surface slope is ignored (i.e. geostrophic component of velocity is zero), as is bottom friction (i.e. very deep water). Thus the relevant equations of motion simplify to:

Coriolis force + friction force = mass × acceleration

For steady-state (no accelerations) conditions involving horizontal turbulent water flow in the $xy$ plane, constant Reynolds turbulent accelerations, $a_y$, and vertical ($y$-axis) velocity gradients only, the horizontal equations of motion per unit mass of water (the *Ekman equations*) are then:

$$fv + a_y \frac{\partial^2 u}{\partial y^2} = 0$$

and

$$-fu + a_y \frac{\partial^2 v}{\partial y^2} = 0$$

The Ekman equations can be solved (not developed here) to demonstrate that water velocity is at maximum at the surface in a direction 45° clockwise to wind velocity in the northern hemisphere (anticlockwise in the southern hemisphere). The magnitude of the velocity decreases downwards but increasing in clockwise turning (anticlockwise turning in the southern hemisphere) until at a certain depth a vanishingly small velocity points upstream relative to the surface Ekman velocity. This depth defines the thickness of the wind-driven surface current known as the *Ekman layer.*

## Cookie 44: More about tides in estuaries

The incoming progressive tidal wave is modified as it travels along a funnel-shaped estuary whose width and depth steadily decrease upstream. For a 2D wave that suffers little energy loss due to friction or reflection (a simplification), the wave energy flux will remain constant, causing the wave to amplify and shorten as it passes upstream into narrower reaches. This is the *convergence effect.* Thus for wave energy, $E$, per unit length of an estuary, $Eb$ is the energy per unit length, where $b$ is total estuary width. Multiplying by the wave speed, $c$, gives the energy flux up the estuary as $Ebc = \text{constant}$. Writing $E = (\rho g a^2)/2$ and the wave equation for shallow water waves as $c = (gh)^{0.5}$, we have $0.5(\rho g a^2) b \sqrt{gh} = \text{constant}$, or, $a \propto b^{-0.5} h^{-0.25}$. We can see that narrowing has more effect on changing wave amplitude than shallowing.

Shallowing causes wave speed to decrease and, since wave frequency is constant, the wavelength must decrease by the argument $c = f\lambda$. Since $\lambda = c/f = \sqrt{gh}/f$, we have $\lambda \propto h^{0.5}$. Thus tidal waves increase in amplitude and decrease in wavelength as they travel up estuaries, the *hypsosynchronous effect*. But we cannot ignore frictional retardation of the tidal wave in this discussion; this causes a reduction in amplitude of the tide upstream and is greatest when channel depth decreases rapidly. In some estuaries the tidal wave changes little in amplitude since the convergence effect is balanced by frictional retardation. Resonant effects with tide or wave may also affect currents in estuaries (**Cookie 45**).

## Cookie 45: Resonant effects on shelf tides

In the oceans the twice-daily tidal wavelength, $\lambda$, is very large (about $10^4$ km) compared with water depth, $h$ (say 5 km) and is thus still of shallow-water (long-wave) type (i.e. $h/\lambda \ll 0.1$). The maximum tidal wave velocity in the open oceans is given approximately by $u = \xi(gh)^{0.5}$, where $\xi$ is tidal amplitude, about 0.5 m. The open ocean tidal wave travelling at $> 200$ m/s decelerates as it crosses the shallowing waters of the shelf edge. This causes wave refraction of obliquely incident waves into parallelism with the shelf break and partial reflection of normally incident waves. At the same time the wave amplitude, $\zeta$, of the transmitted tidal wave is enhanced—this follows from the energy equation for gravity waves $E = 0.5\rho g\xi^2(gh)^{0.5}$; the supremacy of the square versus the square root terms means that as water depth shallows and the tidal wave slows down, then wave amplitude must increase to conserve energy. The tidal current velocity of a water particle (as distinct from the tidal wavelength) also increases because, as noted above, this depends upon the instantaneous amplitude of the wave.

A chief cause of spatially varying tidal strength is the *resonant* effect of the shelf acting upon the open oceanic tide (Cookie 45 Fig. 1) which creates *standing waves*. Resonance greatly increases the oceanic tidal range in nearshore environments and leads to the establishment of very strong tidal currents. In the limiting scenario of a closed basin with a resonant period, $T$, for a standing wave with a node in the middle and antinodes at the ends, the wavelength, $\lambda$, is twice the length of the basin. The speed of the wave is thus $2L/T$ and, treating the tidal wave as a shallow water wave ($h/\lambda < 0.1$), we may write *Merian's formula* as $T = 2L/\sqrt{gh}$. Most shelfs are too narrow and deep to show significant resonance across them, i.e. shelf basin width, $L < 0.25\lambda$, where $\lambda$ is tidal wavelength. In such cases, for example the shelf of the eastern USA, a simple slow linear increase of tidal

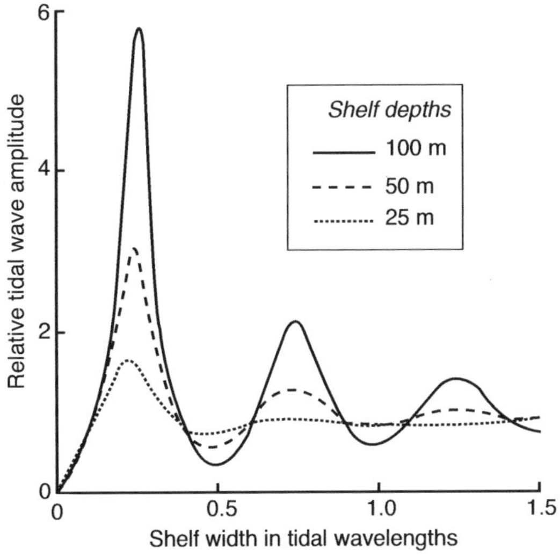


**Cookie 45 Fig. 1** Tidal wave resonance across shelves of different width and water depth. (After Haworth, 1982.)

amplitude and currents occurs across the shelf. Open coastal basins like estuaries, bays and lagoons must receive the 12-hourly oceanic tidal wave and a standing wave (of period 12 h) may be set up, with a node at the mouth and an antinode at the end (by no means the only resonant possibility). In this scenario $L$ is only $0.25\lambda$ and we have $T = 4L/\sqrt{gh}$. The Bay of Fundy, Maritime Canada, is the world's most spectacular example of a gulf that resonates with the ca 12-h period of the semidiurnal ocean tide. The gulf has a length of about 270 km (calculated from the gulf head to the major change of slope at the shelf edge) and is about 70 m deep on average, giving the required ~12 h characteristic resonant period. The standing resonant oscillation has a node at its entrance, which causes the tidal range to increase from 3 m to a spring maximum of some 15.6 m along its length to the antinode.

## Cookie 46: Rotary tidal (Kelvin) waves

Why should such rotary tidal motions occur? The answer is that the tidal gravity wave, unlike normal surface gravity waves due to wind shear or swell, has a sufficiently long period that it must be deflected by the Coriolis force. Since water on continental shelf embayments like the North Sea is largely bounded by solid coastlines, the deflected tide rotates against the sides (Cookie 46 Figs 1 & 2) as a *boundary wave*.

Such waves of rotation against solid boundaries are termed *Kelvin waves*, the propagating wave being forced against the solid boundaries by the effects of the Coriolis parameter, $f$. The water builds up as a wave whose radial slope exerts a pressure gradient that exactly balances the Coriolis effect at equilibrium. Tidal currents due to the wave are coast parallel at the coast (Cookie 46 Fig. 2) with velocities at maximum in the crest or trough (reverse) and minimum at the half wave height. The wave decays in height exponentially seawards towards an *amphidromic node* of zero displacement. The resonant period in the North Sea for example is around 40 h, a figure large enough to support three multinodal standing waves. The crest of the tidal Kelvin wave is a radius of the roughly circular basin and is also a *cotidal line* along which tidal minima and maxima coincide. Concentric circles drawn about the node are lines of equal tidal displacement. Tidal range is thus increased outwards from the amphidromic node by the rotary action. Further resonant and funnelling amplification

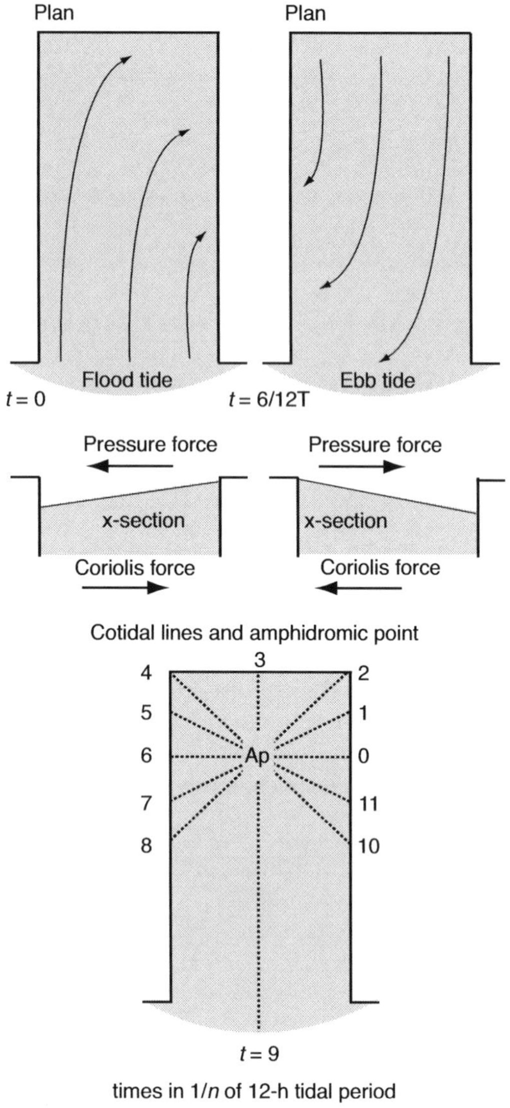


**Cookie 46 Fig. 1** The development of amphidromic circulation within a partly enclosed shelf sea by Coriolis-turning of the tidal wave into a Kelvin wave of circulation. (After Wells, 1986.)

may of course take place at the coastline, particularly in estuaries. Not all basins can develop a rotary tidal wave: there must be sufficient width, since the wave decays away exponentially with distance. The critical width is termed the *Rossby radius of deformation*, $R$, given by the ratio of the velocity of a shallow-water wave to the magnitude of the Coriolis parameter, i.e. $R = \sqrt{gh}/f$. At this distance the amplitude of any

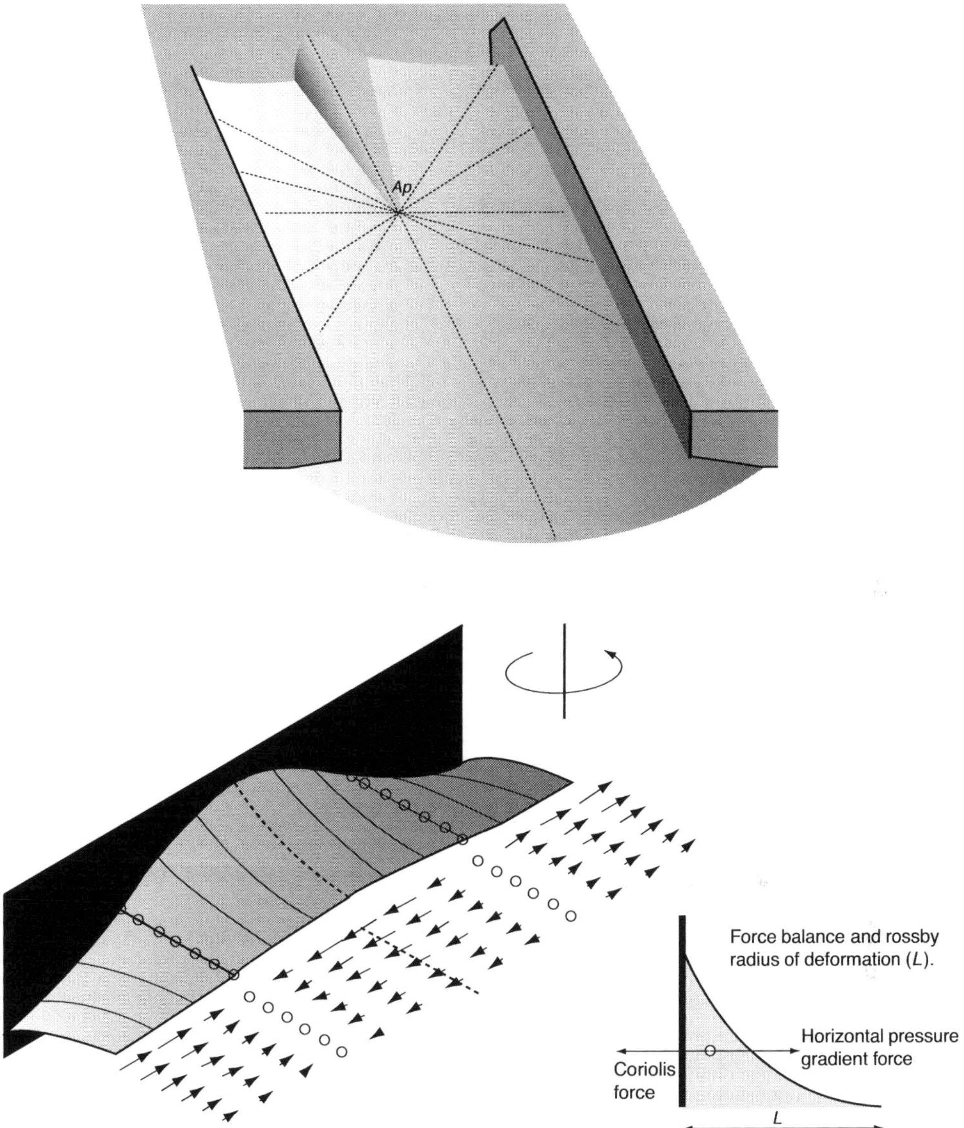


**Cookie 46 Fig. 2** (a) The Kelvin rotating tidal wave travels anticlockwise in the northern hemisphere, decreasing in amplitude inwards towards the amphodromic point, Ap, of zero displacement. (b) Topography and bottom flow associated with the edge of an anticlockwise-rotating Kelvin tidal wave. The rotary component is neglected for clarity. (All after Wells, 1986.)

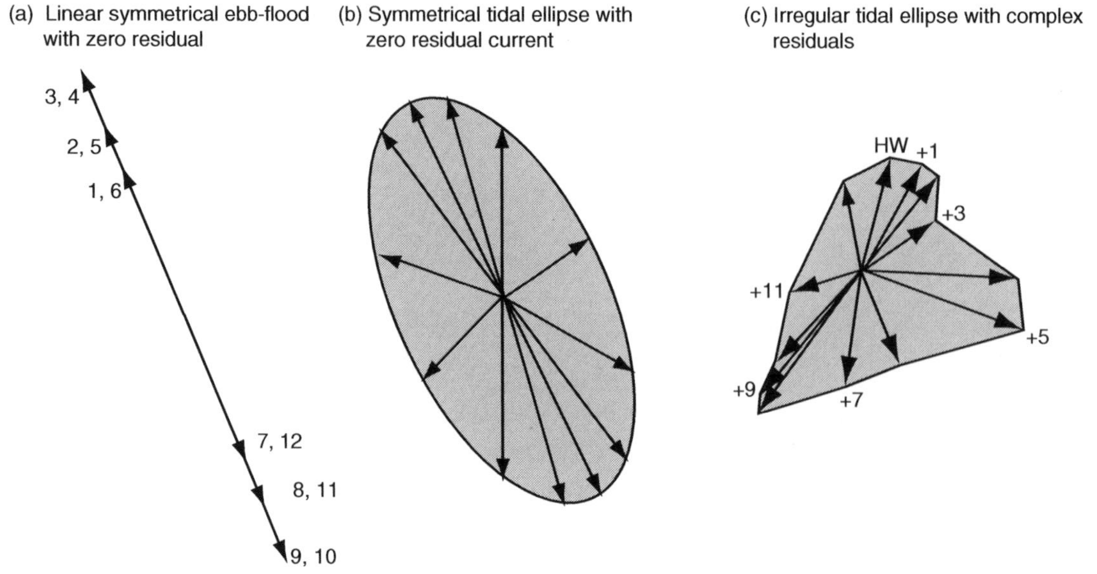


**Cookie 46 Fig. 3** Notional tidal current velocity changes over time shown as vectorial variation over a complete tidal cycle of 12 hours duration. Computation of residual vector defines net sediment transport direction (remember sediment transport is a quadratic function of current velocity).

Kelvin wave has reduced to 1/*e*, 0.37 of its initial value.

Finally, we may usefully summarize the vector variation of tidal currents by means of *tidal current ellipses* whose ellipticity is a direct function of tidal current type and vector asymmetry (Cookie 46 Fig. 3). For example, the inequality between ebb and flood on the northwest European continental shelf is largely determined by a harmonic of the main lunar tide. Since sediment transport is a cubic function of current velocity it can be appreciated that quite small residual tidal currents can cause appreciable net sediment transport in the direction of the residual current. The turbulent stresses of the residual currents will be further enhanced should there be a superimposed wave oscillatory flow close to the bed. A further consideration arises from the fact that turbulence intensities are higher during decelerating tidal flow than during accelerating tidal flow, due to unfavourable pressure gradients. Increased bed shear stress during deceleration thus causes increased sediment transport compared to that during acceleration, so that the net transport direction of sediment will lie at an angle to the long axis of the tidal ellipse.

# MATHS APPENDIX

## Power or exponent

A power or exponent is a number which raises another number to the value given by the product of the number with itself as many times as the value of the power indicates. Thus $x^3$ is $x$ raised to the power 3, or $(x.x.x)$ and $x^2$ is $(x.x)$. For negative powers we use the reciprocal of the positive equivalent, e.g. $x^{-3} = 1/x^3$.

## Logarithm

Referring back to the Cookies' motto, Mr Dormer should have replied to Captain Aubrey as follows: 'A logarithm is a number that is the power of a certain base integer to which the integer must be raised to get the number.' Thus the logarithm to base 2 of the number 8 is $\log_2 8 = 2^3 = 3$. This is also the logarithm to the base 10 of 1000, $\log_{10} 1000 = 10^3 = 3$. Thus we must always quote the base integer used.

## The exponential, e

e, the exponential function is given by the infinite series

$$e = 1 + \frac{1}{1!} + \frac{1}{2!} + \frac{1}{3!} + \frac{1}{4!} + \ldots.$$

where the exclamation mark means 'factorial' or an instruction to multiply out the number from 1 to its value, e.g. 3! means $3 \times 2 \times 1 = 6$. The series quickly converges to the value 2.718. It is used as the base for the natural or Napierian logarithm, $\log_e$ or ln e.

## Functions

A functional relationship is when one variable depends upon the value of another variable. Thus if we are travelling at constant speed then our travel distance (displacement, $s$) depends upon time, $t$, passed. We write $s = f(t)$. For any case in general, $y = f(x)$, where $y$ is the dependent variable and $x$ the independent variable; this means that $y$ depends upon the value of $x$. For example we saw in section 1.10 that velocity may be a function of both position in space and of time, i.e. $\boldsymbol{u} = f(x, y, z, t)$.

Functional relationships can be linear ($y = mx + c$), power ($y = x^n$), logarithmic ($y = \log_a x$), exponential ($y = e^x$, or $y = \exp x$), trigonometric ($y = \sin^{-1} x$), etc. A wider range of functions includes polynomial and rational functions of the second ($y = x^2 + 2x + 1$) or third ($y = 3x^3 + 2x^2 + x + 1$) degrees, often termed quadratic or cubic respectively.

More complicated-looking functions may also be constructed, for example hyperbolic functions from exponential functions, like $y = \sinh x = (e^x - e^{-x})/2$ or the equation for a catenary curve, $y = \cosh x = (e^x + e^{-x})/2$, both used in wave theory. The Gaussian normal function is obtained by taking the exponential of the quadratic $y = -ax^2$ so that $y = e^{-ax^2}$.

## Differential calculus

The change in the $y$ part of a function like $y = f(x)$ may be quite simple if, say, $y = 2x$. Then for every unit change in $x$ there is a proportional change in $y$ of 2. In other words the gradient of $y$ with respect to $x$ is a constant, 2. But for any other function of $x$ that is not

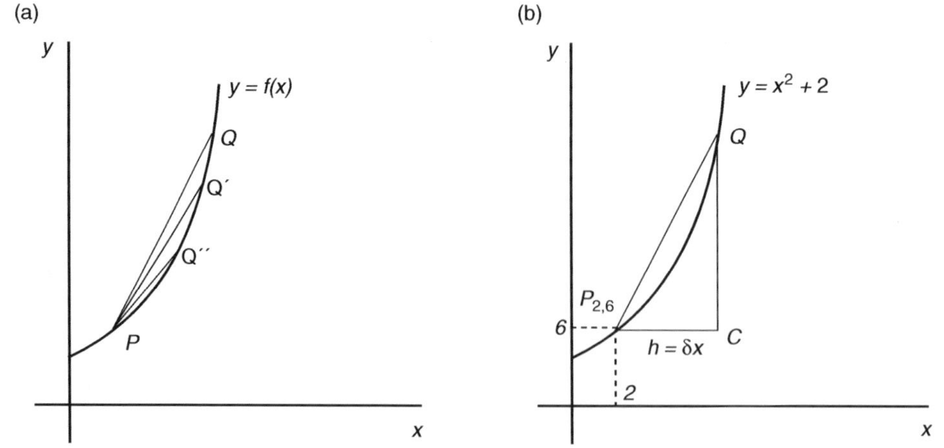


**Fig. 1** To illustrate the derivation of the principles of the differential calculus. (a) For any function $y = f(x)$, as *Q* approaches fixed point, *P*, the chord *PQ* becomes, in the limit, the tangent to the function curve at *P*. This is best visualized by imagining *PQ* shrinking gradually via *PQ′* and *PQ″* to fixed point *P*. (b). To illustrate the argument (see text) that the derivative of the function $y = x^2 + 2$ for an infinitesimal gradient at point *P* is $2x$ (= 4 for this case).

linear then the change in *y* is not immediately obtainable without recourse to the logic of *differential calculus*. To generalize this we graph up a nonlinear $y = f(x)$ (Appendix Fig. 1a). The differential calculus enables us to estimate the gradient of the function at a point, making use of an initial chord *PQ* on the function which we shrink to the point *P* so that we can say that the chord *PQ* becomes, at the limit, as *Q* diminishes to *P*, the tangent to the curve of the function at *P*.

Specifically, with reference to the graphed example (Appendix Fig. 1b ) function $y = x^2 + 2$, point *P* is at position $x = 2, y = 6$, or $P_{2,6}$. Now form a right-angled triangle *PQC*, placing *Q* as another point on the function curve with C at the normal apex. Let position C be $C_{2+h,6}$, where $PC = h = \delta x$ for short. Point *Q* is also at *x*-position $(2 + h)$, with its *y*-coordinate as $y = x^2 + 2 = (2 + h)^2 + 2 = h^2 + 4h + 6$. The side *QC* of the triangle *PQC* is thus $h^2 + 4h$. Simple trigonometry of the right-angled triangle gives the gradient of the chord *PQ* as $QC/PC = (4h + h^2)/h = 4 + h^2$. Now, shrinking the chord *PQ* to *P* (i.e. $h$ or $\delta x \rightarrow 0$) the gradient at the limit at *P* (where $x = 2$) becomes 4 or $2x$. Thus at $P_{2,6}$, $y = x^2 + 2$ and $dy/dx = 2x$. Note that the differential gradient does not depend on the integer at the end of the function $y = x^2 + 2$.

Generally, the gradient *QC/PC* is expressed as $\delta y/\delta x$ for the infinitesimal gradient. The lower case deltas signify tiny (infinitesimal) increments. At the limit, for the gradient at point, *P* we say $dy/dx$, the differential rate of change of *y* with respect to *x*, also written in various forms, including $f'(x)$ or $y'$. Formally, $\frac{dy}{dx} = \text{limit}(\delta x \rightarrow 0)\frac{\delta y}{\delta x}$.

Switching from a generalized functional form to a physical expression, the function in question may be the rate of change of distance, *y*, with time, *t*, in other words, velocity, $\boldsymbol{u} = dy/dt$. This is termed a first differential. Or it could be the rate of change of velocity with time, $du/dt$, an acceleration. Now in this case, since $\boldsymbol{u} = dy/dt$ the acceleration is also $d/dt(dy/dt)$ or $d^2y/dt^2$. This is termed the second differential coefficient. Similarly the rate of change of acceleration with time is $d/dt(d^2y/dt^2)$ or $d^3y/dt^3$, the third differential.

There are some common standard first derivatives that are tempting to learn because we come across them quite often in physical processes. Our example used above is the first in the list and ought to be remembered if nothing else is! Try it out for $y = x^3$.

| | | | |
|---|---|---|---|
| $y =$ | $f(x)$ | $dy/dx = f'(x)$ | or, equivalently |
| $y =$ | $x^n$ | $nx^{n-1}$ | |
| $y =$ | $\log_e x$ | $1/x$ | |
| $y =$ | $e^x$ | $e^x$ | |
| $y =$ | $\sin x$ | $\cos x$ | |
| $y =$ | $\cos x$ | $-\sin x$ | |
| $y =$ | $\tan x$ | $\sec^2 x$ | $1/\cos^2 x$ |

| | | | |
|---|---|---|---|
| $y =$ | $\cosh x$ | $\sinh x$ | $(e^x - e^{-x})/2$ |
| $y =$ | $\sinh x$ | $\cosh x$ | $(e^x + e^{-x})/2$ |
| $y =$ | $\tanh x$ | $\text{sech}^2 x$ | $1/(\cosh x)^2$ |
| $y =$ | $\sin^{-1} x$ | $1/(1 - x^2)^{0.5}$ | |
| $y =$ | $\cos^{-1} x$ | $-1/(1 - x^2)^{0.5}$ | |
| $y =$ | $\tan^{-1} x$ | $1/(1 + x^2)$ | |
| $y =$ | $\sinh^{-1} x$ | $1/(1 + x^2)^{0.5}$ | |
| $y =$ | $\text{osh}^{-1} x$ | $1/(x^2 - 1)^{0.5}$ | |
| $y =$ | $\tanh^{-1} x$ | $1/(1 - x^2)$ | |

### Partial differential symbol, ∂

When $y$ depends only upon $x$, i.e. the variation of $y$ with $x$ is $dy/dx$, a curve is defined in 2D. In 3D $y$ also depends upon $z$. In such cases, $y = f(x,z)$, and a surface is defined and we say that $y$ partially depends upon $x$ and $z$, or $\partial y/\partial x + \partial y/\partial z$.

### Integral calculus

Whilst differential calculus is concerned with the gradient of a curve (function) at a point, the integral calculus calculates the area under the curve (function), like area $UPQV$ in Appendix Fig. 2. Integration, symbol $\int$, reverses the process of differentiation so that if we are given $d\alpha/dx = y$, then $\alpha = \int y dx + c = \int f(x)dx + c$. The parameter $c$ that has mysteriously crept in is known as the constant of integration and must generally be present since the function may have an unknown integer that drops out during differentiation (like the integer 2, in our function $y = x^2 + 2$ used previously).

It is easiest to think about integration physically using velocity as an example. We know velocity is the rate of change of distance, $s$, over time, $u = ds/dt$. If we want to determine $s$, this is the integral $s = \int u dt + c$. Also we know that acceleration is the rate of change of velocity over time, $a = du/dt$. If we want to determine velocity we need the integral $u = \int a dt + c$ We now go on to prove that all this is true and not just some mathematical gibberish.

With reference to Appendix Fig. 2 we first need to find any infinitesimal area under the curve of the generalized function, $y = f(x)$. We first define a tiny area under the function curve, $UVQP$ by letting $P$ and $Q$ be any two points on the curve. The position of $P$ is defined as $P_{x,y}$ and the position of $Q$ as $Q_{x+\delta x, y+\delta y}$. Then $ABPU$ is the area under the curve from $x = a$ to $x = x$; $ABQV$ is that area from $x = a$ to $x = x + \delta x$; the increment in area, $\delta A$ is $UPQV$. Note that $UVSP \leq \delta A \leq UVQR$:

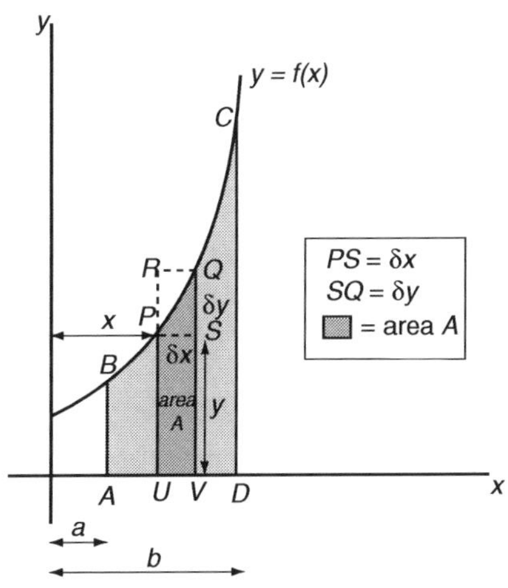


**Fig. 2** To illustrate the derivation of the principles of the integral calculus. For any function $y = f(x)$, the integration provides the area under the functional curve. This total area, for example $ABCD$, can be regarded as comprising an infinite number of small strip areas like $UVQP$ shown. So, for such a strip, as point $Q$ approaches position $P$ then the shaded strip of area, $A$, vanishes as $dx$ goes to zero and $dy$ also goes to zero. At this point $dA/dx = y$ and $A = \int y\,dx$. This is the indefinite integral. For the definite integral, like the total area $ABCD$ under the curve between $x = a$ and $x = b$, see text.

$$y\delta x \leq \delta A \leq (y + \delta y)dx,$$

dividing by $\delta x$

$$y \leq \delta A/\delta x \leq y + \delta y.$$

Now let $\delta x$ go to zero, then $\delta y$ also goes to zero and the lower and upper bounds of $\delta A/\delta x$ also go to zero. Switching to infinitesimals we have the defining expression for the infinitesimal limit, as $\delta x$ goes to zero,

$$\delta A/\delta x = dA/dx$$

then,

$$dA/dx = y$$

or

$$A = \int y\,dx = \int f(x)dx + c.$$

This is the *indefinite integral* and is the general integral solution to a given function. For the given function

$y = x^n$ we have the general rule

$$\int y\mathrm{d}x = \int x^n \mathrm{d}x = \frac{x^{n+1}}{n+1}$$

A few common integrands are

| $y =$ | $f(x)$ | $\int y\mathrm{d}x$ |
|---|---|---|
| | $x^n$ | $x^{n+1}/(n+1)$ |
| | $\cos x$ | $\sin x + c$ |
| | $\sin x$ | $-\cos x + c$ |
| | $e^x$ | $e^x + c$ |
| | $1/x$ | $\ln x + c$ |

But what about a specific defined area under the functional curve, like *ABCD*? This is taken between the limits $x = a$ and $x = b$ (Appendix Fig. 2), defined by the definite integral

$$A = \int_a^b y\mathrm{d}x = \int_a^b f(x)dx$$

Substituting for $x = b$ we subtract this from the value of $x = a$ Taking our example function, $y = x^2 + 2$, we thus have as its integral

$$A = \int_a^b (x^2+2)\mathrm{d}x = \left[\frac{1}{3}x^3 + 2x\right]_a^b = \frac{1}{3}b^3 + 2b - \frac{1}{3}a^3 - 2a$$

## Scalars

A scalar quantity has only dimensions of its own magnitude, with no direction additionally specified. Easy examples are temperature (as on a weather chart), speed (as on a speeding ticket, the direction is irrelevant), energy or power (my old souped-up car engine had 142 horsepower, but obviously no direction is specified). Here is a list of scalars: *length, area, volume, linear speed, angular speed, work done, electrical resistance, power, energy, mass, density, temperature*.

## Vectors—scalars with attitude

A vector quantity has both magnitude and direction, though the latter is not specified in the units involved, only in the direction taken with respect to given $x$, $y$, $z$ coordinates. Vectors add crucial information to any physical description. With Cartesian coordinate axes fixed at right angles, three components to any vector are possible, in the $x$, $y$ and $z$ directions; all vectors may be resolved into such 3D components. Easy examples are velocity (speed in a given direction; $u_x$, $u_y$, $u_z$) and force (mass times acceleration, positive in the direction of the acceleration). For the common case of velocity, the three different components are usually given different symbols, though these vary between disciplines; thus in this text we may state the overall velocity, $\boldsymbol{u}$, with components $u$ in the streamwise ($x$) direction, $w$ in the vertical ($z$) direction and $v$ in the spanwise ($y$) direction. The reader should be careful with notation as some workers switch notation of the $y$ and $z$ axes and we do sometimes in this text. There is no problem with this since the axes are arbitrary anyway (Descartes is supposed to have invented orthogonal coordinate axes whilst watching a fly crawling across his rectangular bedroom ceiling!), but it's important always to check for consistency and usage. Here is a list of common vectors; *displacement, velocity, vorticity, force, acceleration, momentum, electrostatic force, magnetic force, electric current*.

## The meaning of vector operator ∇

Consider elevation or temperature. In order to discuss possible variations in topography or thermal energy in space (a common situation in Nature) it is necessary to be able to succinctly state the variation of these scalars in 3D, e.g. the variation of height or temperature in space. We did this with vectors above with the implicit assumption that 3D components are an essential part of the definition of vectors. We have seen repeatedly that 3D components are defined with respect to three coordinate axes. These define a *scalar field*, defined as a region of space where a scalar quantity can be associated with every point. So, temperature or elevation in a steady scalar field is a function of axes $x$, $y$, $z$ only, i.e. $T(x,y,z)$ or $h(x,y,z)$. Now a field defines a geographical spread of values, a space rate of change in fact, for the field variable; it may be greater or lesser depending on direction. We would like to be able to state this succinctly; like as your eye would seek out the steepest topographic slope for a toboggan run or a gentler slope for a more cautious ski run. We have seen already that rates of change are a concept used in differential calculus; we can simply denote the space rate of change of $T$ or $h$ as $dT/ds$ or $dh/ds$, in 3D ($\partial T/\partial x + \partial T/\partial y + \partial T/\partial z$) or ($\partial h/\partial x + \partial h/\partial y + \partial h/\partial z$). The partial differentials are there because the scalar can vary in the three independent coordinate directions. This is rather clumsy to write out so we refer to *gradients* in a scalar quantity as Grad $T$ or Grad $h$, or by the symbol, $\nabla$, denoted *DEL* or *NABLA*), thus we have $\nabla T$ or $\nabla h$. Note that in vector geometry a gradient is the *steepest* slope possible in a given scalar field. Now we can simply say that

$$\text{Grad}\, h = \nabla h = (\partial h/\partial x + \partial h/\partial y + \partial h/\partial z)$$

the quantity $\nabla = (\partial/\partial x + \partial/\partial y + \partial/\partial z)$ being the *vector operator*. In our examples, $\nabla T$ gives the heat flow direction and $\nabla h$ the topographic gradient at any point.

## Tensors

Tensors are vector quantities whose vectorial components are each resolvable into three further components, giving a total of nine in all. Stress and strain are tensors (Chapters 24 & 25; Cookie 22), as is vorticity. The concept is best imagined for stress, $\tau$, with each main vectorial component $\tau_x$, $\tau_y$, $\tau_z$ further divisible into every possible combination, viz. ($\tau_{xx}$, $\tau_{xy}$, $\tau_{xz}$), ($\tau_{yx}$, $\tau_{yy}$, $\tau_{yz}$) and ($\tau_{zx}$, $\tau_{zy}$, $\tau_{zz}$). Early workers in vectorial physics recognized the clumsiness of the equation sets released by such tensor notation and refer simply to the matrix $\tau_{ij}$, with $i = 1$–3 for the vectorial compoments and $j = 1$–3 for the subordinate tensorial components, the repeated summation $ij$ indicating the sum of the various combinations.

## Grain size statistics

1 Mean grain size, the statistical first moment of any distribution, is calculated as:

$$\mu = \sum_{i=1}^{n} x_i/n$$

where $x_i$ is the value of the $i$th measurement and the $\sum$ term means 'add together all the $n$ values of $x$ from 1 to $n$'.

2 Standard deviation of grain size, the statistical second moment, $\sigma$, is calculated as the square root of the variance, $\sigma^2$:

$$\sigma^2 = \sum_{i=1}^{n} (x_i - \mu)^2/n$$

3 The skewness of a distribution is the statistical third moment: $sk = \sum_{i=1}^{n} (x_i - \mu)^3/n$

## Further reading for maths

The best introductory text for relatively innumerate earth and environmental scientists which is seriously good at bringing your maths up to a decent level is Dave Waltham's *Mathematics for Geologists*.

# REFERENCES


Aalto, R. *et al.* 2003. *Nature*, **425**, 493–497.
Abbott, J.E. & Francis, J.R.D. 1977. *Philosophical Transactions of the Royal Society, London (A)*, **284**, 225–254.
Abers, G.A. *et al.* 1977. *Journal of Geophysical Research*, **102**, 15301–15317.
Abrahams, A.D. & Gao, P. 2006. *Earth Surface Processes & Landforms*, **31**, 910–928.
Abrahams, A.D. & Parsons, A.J. 1994. *Geomorphology of Desert Environments*, Chapman & Hall.
Acarlar, M.S. & Smith, C.R. 1987. *Journal of Fluid Mechanics*, **175**, 1–41.
Acheson, D.J. 1990. *Elementary Fluid Dynamics*, Oxford University Press.
Adams, E.W. & Schlager, W. 2000. *Journal of Sedimentary Research*, **70**, 814–828.
Adams, E.W. *et al.* 2001. *Sedimentology*, **48**, 661–679.
Adams, K.D. 2003. *Sedimentology*, **50**, 565–577.
Adamson, D.A. *et al.* 1980. *Nature*, **288**, 50–55.
Adeogba, A.A. *et al.* 2005. *Bulletin of the American Association of Petroleum Geologists*, **89**, 627–643.
Adrian, R.J. *et al.* 2000. *Journal of Fluid Mechanics*, **422**, 1–54.
Ahnert, F. 1970. *American Journal of Science*, **268**, 243–263.
Ahr, W.M. 1973. *American Association of Petroleum Geologists Bulletin*, **57**, 1826–1839.
Aigner, T. & Reineck, H.E. 1982. *Senckenbergiana Maritima*, **14**, 183–215.
Aitken, J.F. & Flint, S.S. 1995. *Sedimentology*, **42**, 3–30.
Albarède, F. 2003. *Geochemistry: An Introduction*. Cambridge University Press.
Alexander, J. 2007. *Sedimentology*, **55**, 845–868.
Alexander, J. & Fielding, C. 1997. *Sedimentology*, **44**, 327–337.
Alexander, J. & Fielding, C.R. 2006, *Journal of Sedimentary Research*, **76**, 539–556.
Alexander, J. & Leeder, M.R. 1987. In: *Recent Developments in Fluvial Sedimentology* (eds Ethridge, F.G. *et al.*), pp. 243–252. Special Publication 39, Society of Economic Paleontologists and Mineralogists.
Alexander, J. & Morris, S. 1994. *Journal of Sedimentary Research*, **64**, 899–909.
Alexander, J. & Mulder, T. 2002. *Marine Geology*, **186**, 195–210.
Alexander, J. *et al.* 1999. In: *Floodplains: Interdisciplinary Approaches* (eds Marriott, S. & Alexander, J.), pp. 27–40. Special Publication 163, Geological Society of London.
Alexander, J. *et al.* 2001. *Sedimentology*, **48**, 133–152.
Alexander, J. *et al.* 2008. *Sedimentology*, **55**, 845–868.
Al Ja'aidi, O.S. *et al.* 2004. In: *Confined Turbidite Systems* (eds Lomas, S.A. & Joseph, P.), pp. 45–58. Special Publication 222, Geological Society of London.
Allègre, C.J. *et al.* 1984. *Nature*, **307**, 17–22.
Allen, G.P. 1991. In: *Clastic Tidal Sedimentology* (eds Smith, D.G. *et al.*), v. 16, p. 29–40. Canadian Society of Petroleum Geologists.
Allen, G.P. & Posamentier, H.W. 1993. *Journal of Sedimentary petrology*, **63**, 378–391.
Allen, G.P. *et al.* 1975. *Estuarine Processes 2*, Academic Press, New York, 63–79.
Allen, G.P. *et al.* 1980. *Sedimentary Geology*, **26**, 69–90.
Allen, J.R.L. 1964. *Sedimentology*, **3**, 89–108.
Allen, J.R.L. 1969. *Proceedings of the Geologists Association*, **80**, 1–42.
Allen, J.R.L. 1970. *Physical Processes of Sedimentation*. George Allen and Unwin.
Allen, J.R.L. 1972. *Geografiska Annaler*, **53A**, 157–187.
Allen, J.R.L. 1973. *Sedimentology*, **20**, 323–329.
Allen, J.R.L. 1978. *Sedimentary Geology*, **21**, 129–147.
Allen, J.R.L. 1982. *Sedimentary Structures: Their Character and Physical Basis*. Elsevier, Amsterdam.
Allen, J.R.L. 1985. *Principles of Physical Sedimentology*. Allen and Unwin.
Allen, J.R.L. & Banks, N.L. 1972. *Sedimentology*, **19**, 257–283.



*Sedimentology and Sedimentary Basins: From Turbulence to Tectonics*, 2nd edition. © Mike Leeder.
Published 2011 by Blackwell Publishing Ltd.

Allen, M.B. & Davies, C.E. 2007. *Basin Research*, **19**, 379–392.

Allen, P.A. 1984. *Marine Geology*, **60**, 455–473.

Allen, P.A. 1997. *Earth Surface Processes*, Blackwell Science.

Allen, P.A.& Homewood, P. (eds) 1986. *Foreland Basins*. Special Publication 8, International Association of Sedimentologists.

Allen, P.A. & Allen, J.R. 1990. *Basin Analysis*. Blackwell Scientific Publications, Oxford.

Allen, P.A. & Allen, J.R. 2004. *Basin Analysis: Principles and Applications*, 2nd edn. Blackwell.

Allen, P.A. *et al.* 1991. *Basin Research*, **3**, 143–163.

Allen, P.A. *et al.* (eds) 1992. Thematic set on foreland basins. *Basin Research*, **4**, 169–352.

Alley, R.B. *et al.* 1986. *Nature*, **322**, 57–59.

Alley, R.B. *et al.* 2005. *Science*, **310**, 456–461.

Allison, M.A. *et al.* 1998. *Geology*, **26**, 175–178.

Allwood, A.C. *et al.* 2006. *Nature*, **441**, 714–718.

Almedeij, J. & Diplas, P. 2005. *Eos, Transactions American Geophysical Union*, **86**, 44.

Almogi-Labin, A. *et al.* 2009. *Quaternary Science Reviews*, **28**, 2882–2896.

Alonso-Zarza, A.M. *et al.* 1992. *Sedimentology*, **39**, 17–35.

Amiotte Suchet, P. & Probst, J.L. 1993. *Comptes Rendus, Academie des Sciences, Paris (II)*, **317**, 615–622.

Amorosi, A. *et al.* 1996. *Sedimentary Geology*, **102**, 275–295.

Amorosi, A. *et al.* 2005. In: *Mediterranean Prodelta Systems: Proceedings of the International ComDelta Symposia, Aix-en-Provence: France* (eds Trincardi, F. & Syvitski, J.), pp. 7–18. Marine Geology, **222–223**.

Amos, K.J. *et al.* 2004. *Sedimentology*, **51**, 145–162.

Amy, L.A. & Talling, P.J. 2006. *Sedimentology*, **53**, 161–212.

Amy, L.A. *et al.* 2005. *Sedimentary Geology*, **179**, 163–174.

Anadón, P. *et al.* 1986. In: *Foreland Basins* (eds Allen, P.A. & Homewood, P), pp. 259–271. Special Publication 8, International Association of Sedimentologists.

Anastasa, A.S. *et al.* 1997. *Sedimentology*, **44**, 869–891.

7–3– *et al*1– Anderson, E.J. 2004. *Sedimentology*, **51**, 455–477.

Anderson J.A.R. 1964. *Journal of Tropical Geography*, **18**, 7–16.

Anderson, J.B. 2005. In: *River Deltas: Concepts, Models, and Examples* (eds Giosan, L. & Bhattacharya, J.P.), pp. 257–278. Special Publication 83, Society of Economic Paleontologists and Mineralogists.

Anderson, J.B.& Ashley, G.M. (eds) 1991. *Glacial Marine Sedimentation: Palaeoclimatic Significance*. Special Publication 261, Geological Society of America.

Anderson, J.B. *et al.* 1991. In: *Glacial Marine Sedimentation: Palaeoclimatic Significance* (eds Anderson, J.B. & Ashley, G.M.), pp. 1–26. Special Publication 261, Geological Society of America.

Anderson, J.B. *et al.* 2004. In: *Late Quaternary Stratigraphic Evolution of the Northern Gulf of Mexico Margin* (eds Anderson, J.B. & Fillon, R.H.), pp. 1–23. Special Publication 79, Society of Economic Paleontologists and Mineralogists.

Anderson, J.G. *et al.* 2001. *Geology Today*, 4–9.

Anderson, K.S. *et al.* 2006. *Journal of Sedimentary Research*, **76**, 819–838.

Anderson, M.G. *et al.* 1996. *Floodplain Processes*. Wiley, Chichester.

Anderson, R.S. & Bunas, K.L. 1993. *Nature*, **365**, 740–743.

Anderson, S.P. *et al.* 2000. *Geochimica et Cosmochimica Acta*, **64**, 1173–1189.

Andersen, N. *et al.* 2001. *Geology*, **29**, 799–802.

Anderton, R. 1976. *Sedimentology*, **23**, 429–458.

Anderton, R. 1995. In: *Characterization of Deep Marine Clastic Systems* (eds Hartley, A.J. & Prosser, D.J.), pp. 5–11. Special Publication 94, Geological Society of London.

Andrews, J.E. *et al.* 1998. *Quaternary Research*, **50**, 240–251.

Andrews, J.E. *et al.* 2004a. *Palaeogeography. Palaeoclimatology. Palaeoecology*, **204**, 101–114.

Andrews, J.E. *et al.* 2004b. *An Introduction to Environmental Chemistry*, Blackwell.

Andrews, J.E. *et al.* 2005. *An Introduction to Environmental Chemistry*. Blackwell. 3rd edn.

*et al* Andrews, J.E. *et al.* 2007. *Earth and Planetary Science Letters*, **259**, 457–468.

Andrews, J.T. 2000. *Journal of Sedimentary Research*, **70**, 782–787.

Andrews-Hanna, J.C. *et al.* 2007. *Nature*, **446**, 163–166.

Anthony, E.J. *et al.* 2002. *Sedimentology*, **49**, 1095–1112.

Aqrawi, A.A.M. & Evans, G. 1994. *Sedimentology*, **41**, 755–776.

Arboleya, M.-L. *et al.* 2008. *Journal of the Geological Society, London*, **165**, 1059–1073.

Archer, A.W. 1994. *Mathematical Geology*, **26**, 47–65.

Archer, A.W. & Johnson, T.W. 1997. *Sedimentology*, **44**, 991–1010.

Archer, A.W. *et al.* 1995. *Journal of Sedimentary Research*, **65**, 408–416.

Archer, D. 1991. *Journal of Geophysical Research*, **96**, 17037–17050.

Archer, D. & Maier-Reimer, E. 1994. *Nature*, **367**, 260–263.

Armentrout, J.M. 1991a. In: *Seismic Facies and Sedimentary Processes of Submarine Fans and Turbidite Systems* (eds Weimer, P. & Link, M.H.), pp. 137–170. Springer.

Armentrout, J.M. *et al.* 1991b. In: *Seismic Facies and Sedimentary Processes of Submarine Fans and Turbidite Systems* (eds Weimer, P. & Link, M.H.), pp. 223–240. Springer.

Arnott, R.W.C. & Hand, B.M. 1989. *Journal of Sedimentary Petrology*, **59**, 1062–1069.

Arp, G. *et al.* 2002. *Geology*, **30**, 579.

Arthurton, R.S. 1973. *Sedimentology*, **20**, 145–160.

Ashley, G.M. 1975. In: *Glaciofluvial and Gaciolacustrine Sediments* (eds Jopling, A.V. & McDonald, B.C.), pp. 304–320. Special Publication 23, Society of Economic Paleontologists and Mineralogists.

Ashley, G.M. 1990. *Journal of Sedimentary Petrology*, **60**, 160–172.

Ashley, G. 1995. In: *Glacial Environments Volume 1* pp. 417–444. Butterworth-Heinemann.

Ashley, G.M. *et al.* 1991. *Bulletin of the Geological Society of America*, **103**, 1607–1621.

Ashley, G.M. & Sheridan, R.E. 1994. In: *Incised-Valley Systems: Origin and Sedimentary Sequences* (eds Dalrymple, R.W. *et al.*), pp. 285–301. Special Publication 51, Society of Economic Paleontologists and Mineralogists.

Ashley, G.M. & Smith, N.D. 2000. *Bulletin of the Gelogical Society of America*, **112**, 657–667.

Ashworth, P.J. *et al.* 1996. *Coherent Flow Structures in Open Channels*. Wiley, Chichester.

Ashworth, P.J. *et al.* 1999. In: *Fluvial Sedimentology VI* (eds Smith, N.D. & Rogers, J.), pp. 333–346. Special Publication 28, International Association of Sedimentologists.

Ashworth, P.J. *et al.* 2000. *Sedimentology*, **47**, 533–555.

Ashworth, P.J. *et al.* 2004. *Geology*, **32**, 21–24.

Ashworth, P.J. *et al.* 2007. *Sedimentology*, **54**, 497–513.

Aslan, A. & Autin, W.J. 1998. *Bulletin of the Geological Society of America*, **110**, 433–449.

Aslan, A. & Autin, W.J. 1999. *Journal of Sedimentary Research*, **69**, 800–815.

Aslan, A. *et al.* 2003. *Bulletin of the Geological Society of America*, **115**, 479–498.

Aslan, A. *et al.* 2005. *Journal of Sedimentary Research*, **75**, 650–664.

Asselman, N.E.M. & Middelkoop, H. 1995. *Earth Surface Processes and Landforms*, **20**, 481–499.

Atkins, P.W. 1992. *The Elements of Physical Chemistry*. Oxford University Press.

Atkinson, B.W. & Zhang, J. Wu. 1996. *Reviews of Geophysics*, **34**, 403–431.

Atwater, B.F. *et al.* 2001. *Bulletin of the Geological Society of America* **113**, 1193–1204.

Aubrey, J. & Geise, F.J. 1993. *Formation and Evolution of Multiple Tidal Inlets*, Monograph, American Geophysical Union.

Aurell, M. *et al.* 1995. *Sedimentology*, **42**, 75–94.

Autin, W.A. *et al.* 1991. In: *The Geology of North America*; Vol. K2, *Quaternary Nonglacial Geology; Conterminous US* (ed. Morrison, R.B). Geological Society of America.

Awadallah, S.A.M. & Hiscott, R.N. 2004. *Canadian Journal of Earth Sciences*, **41**, 1299–1317.

Baas, J.H. *et al.* 1994. *Sedimentology*, **41**, 185–209.

Baas, J.H. 1999. *Sedimentology*, **46**, 123–138.

Baas, J.H. & Best, J.L. 2002. *Journal of Sedimentary Research*, **72**, 336–340.

Baas, J.H. & Best, J.L. 2008. *Sedimentology*, **55**, 635–666.

Baas, J.H. *et al.* 1993. *Terra Nova*, **5**, 29–35.

Baas, J.H. *et al.* 2004. *Sedimentology*, **51**, 1053–1088.

Baas, J. *et al.* 2009. *Journal of Sedimentary Research*, **79**, 162–183.

Babel, M. & Becker, A. 2006. *Journal of Sedimentary Research*, **76**, 996–1011.

Babonnneau, N. *et al.* 2002. *Marine and Petroleum Geology*, **19**, 445–467.

Bagnold, R.A. 1936. *Proceedings of the Royal Society of London*, **A157**, 594–620.

Bagnold, R.A. 1940. *Journal of the Institute of Civil Engineers*, **15**, 27–52.

Bagnold, R.A. 1941. *The Physics of Blown Sand and Desert Dunes*. Chapman & Hall, London.

Bagnold, R.A. 1946. *Philosophical Transactions of the Royal Society, London (A)*, **187**, 1–18.

Bagnold, R.A. 1954a. *Philosophical Transactions of the Royal Society, London (A)*, **225**, 49–63.

Bagnold, R.A. 1954b. *The Physics of Blown Sand and Desert Dunes*, 2nd edn. Chapman & Hall, London.

Bagnold, R.A. 1955. *Proceedings of the Institute of Civil Engineers*, **4**, 174–205.

Bagnold, R.A. 1956. *Philosophical Transactions of the Royal Society, London (A)*, **249**, 335–397.

Bagnold, R.A. 1962. *Philosophical Transactions of the Royal Society, London (A)*, **265**, 315–319.

Bagnold, R.A. 1963. In: *The Sea* (ed. Hill, M.N.), pp. 507–523. Wiley, New York.

Bagnold, R.A. 1966a. *Philosophical Transactions of the Royal Society, London (A)*, **295**, 219–232.

Bagnold, R.A. 1966b. Professional Paper 422–I, US Geological Survey.

Bagnold, R.A. 1973. *Philosophical Transactions of the Royal Society, London (A)*, **332**, 473–504.

Bagnold, R.A. 1991. *Sand, Wind and War: Memoirs of a Desert Explorer*. University of Arizona Press.

Bagnold, R.A. & Barndorff-Nielsen, O. 1980. *Sedimentology* **27**, 199–207.

Bahamonde, J.R. *et al.* 2000. *Sedimentology*, **47**, 645–664.

Baikal Drilling Project Members. 1997. Project-96 (Leg 2). *Eos, Transactions American Geophysical Union*, December, 597–604.

Bailleul, J. *et al.* 2007. *Journal of Sedimentary Research*, 77, 263–283.

Baker, E.K. *et al.* 1995. In: *Tidal Signatures in Modern and Ancient Sediments* (eds Flemming, B.W. & Bartholoma, A.), pp. 193–211. Special Publication 24, International Association of Sedimentologists.

Baker, E.T. & Hickey, B.M. 1986. *Marine Geology*, **71**, 15–34.

Baker, V.R. 1973. Special Papers **144**, Geological Society of America.

Baker, V.R. 1990. *Bulletin of the Geological Society of America*, **100**, 1157–1167.

Baker, V.R. 1994. *Geomorphology*, **10**, 139–56.

Ball, M.M. 1967. *Journal of Sedimentary Petrology*, **37**, 556–591.

Balme, M. *et al.* 2003. *Geophysical Research Letters*, **30**, doi: 1029/2003GL:017493.

Balson, P.S.& Collins, M.B. (eds). 2007. *Coastal and Shelf Sediment Transport*. Special Publication 274, Geological Society of London.

Banks, N.L. 1973. *Journal of Sedimentary Petrology*, **43**, 423–427.

Bannerjee, I. & McDonald, B.C. 1975. In: *Glaciofluvial and Glaciolacustrine Sediments* (eds Jopling, A.V. & McDonald, B.C), pp. 132–154. Special Publication 23, Society of Economic Paleontologists and Mineralogists.

Barber, P.M. 1981. *Marine Geology*, **44**, 253–272.

Barbosa, C.F. & Suguio, K. 1999. *Journal of Sedimentary ResearChapter* **69**, 576–587.

Bardou, E. *et al.* 2007. *Sedimentology*, **54**, 469–480.

Barker, G. & Grimson, M. 1990. *New Scientist*, May.

Barnard, P.L. *et al.* 2006. *Eos*, Transactions American Geophysical Union, 87/29 321, 285, 289.

Barnes, J.B. & Heins, W.A. 2009. *Basin Research*, **21**, 91–109.

Barnett, A.J. *et al.* 2002. *Basin Research*, **14**, 417–438.

Barrett, P.J. 2007. In: *Glacial Sedimentary Processes and Products* (eds Hambrey, M. *et al.*), 7– pp. 259–288. Special Publication 39, International Association of Sedimentologists.

Barros, A.P. & Lettenmaier, D.P. 1994. *Reviews of Geophysics*, **32**, 265–284.

Barshad, I. 1966. In: *Proceedings of the 1st International Clay Conference*, Jerusalem, pp. 167–173.

Bartholomaus, T.C. *et al.* 2008. *Nature Geoscience*, **1**, 33–37.

Bartov, Y. *et al.* 2007. *Journal of Sedimentary Research*, **77**, 680–692.

Bastos, A.C. *et al.* 2003. *Sedimentology*, **50**, 1105–1122.

Bates, C.C. 1953. *Bulletin of the American Association of Petroleum Geologists*, **37**, 2119–2162.

Bathurst, R.G.C. 1966. *Geological Journal*, **5**, 15–32.

Bathurst, R.G.C. 1968. In: *Recent Developments in Carbonate Sedimentology in Central Europe* (eds Muller, G. & Friedman, G.M.), pp. 1–10. Springer, Berlin.

Bathurst, R.G.C. 1975. *Carbonate Sediments and their Diagenesis*. Elsevier, Amsterdam.

Bauer, J. *et al.* 2003. *Sedimentology*, **50**, 387–414.

Bayona, G. & Lawton, T.F. 2003. *Basin Research* **15**, 251–270.

Bazylinski, D.A. *et al.* 1993. *Nature*, **366**, 218.

Beard, D.C. & Weyl, P.K. 1973. *Bulletin of the American Association of Petroleum Geologists*, **51**, 349–369.

Beaty, C.B. 1970. *American Journal of Science*, **268**, 50–77.

Beaty, C.B. 1990. In: *Alluvial Fans:A Field Approach* (eds Rachocki, A.H. & Church, M), pp. 69–89. Wiley.

Beaubouef, R.T. & Friedman, S.J. 2000. In: Weimer, P. *et al.* (eds) Deep-Water Reservoirs of the World: Gulf Coast Geological Society.

Beaumont, C. 1981. *Geophysical Journal of the Royal Astronomical Society*, **65**, 291–329.

Beaumont, C. *et al.* 1992. In: *Thrust Tectonics* (Ed. K. McClay). Chapman & Hall, London, 1–18.

Beaumont, C. *et al.* 2000. In: *Geomorphology and Global tectonics*. (ed. M. Summerfield), pp. 29–55, Wiley.

Beaumont, C. *et al.* 2001. *Nature*, **414**, 738–742.

Beavais, A. 1999. *Geochimica et Cosmochimica Acta*, **63**, 3939–3957.

Becker, F. & Bechstädt. 2006. *Sedimentology*, **53**, 1083–1120.

Becker, P. 1994. In: *AGU Fall Meeting Abstract Volume*, p. 317.

Beerling, D.J. *et al.* 2002. *American Journal of Science*, **302**, 28–49.

Beghin, P. *et al.* 1981. *Journal of Fluid Mechanics*, **107**, 407–422.

Begin, Z.B. 1988. *Earth Surface Processes and Landforms*, **13**, 487–500.

Begin, Z.B. *et al.* 1981. *Earth Surface Processes and Landform*, **6**, 49–68.

Behl, R.J. & Kennett, J.P. 1996. *Nature*, **379**, 243–246.

Behrenfield, M.J. *et al.* 2006. *Nature*. **442**, 1025–1028.

Belderson, R.H. *et al.* 1978. *Marine Geology*, **28**, 65–75.

Belderson, R.H. *et al.* 1982. In: *Offshore Tidal Sands: Processes and Deposits* (ed. Stride, A.H.), pp. 27–57. Chapman & Hall, London.

Beletsky, D. *et al.* (1999) *Journal of Great Lakes Research*, **25**, 78–93.

Bell, R.E. 2008. *Nature Geoscience*, **1**, 297–304.

Bell, R.E. *et al.* 2007. *Nature*, **445**, 904–907.

Bell, R.E. *et al.* 2009. *Basin Research*, **21**, 824 DOI: 10.1111/j.1365-2117.2009.00401.x

Belleck, B.W. *et al.* 2007. *Journal of Sedimentary Research*, **77**, 980–991.

Bellotti, P. *et al.* 1994. *Journal of Sedimentary Research*, **64**, 416–432.

Benan. C. & Kocurek, G. 2000. *Sedimentology*, **47**, 1069–1080.

Benison, K.C. *et al.* 2007. *Journal of Sedimentary Research*, **77**, 366–388.

Benn, D.I. 1994. *Sedimentology*, **41**, 279–292.

Benn, D.I. 1995. *Sedimentology*, **42**, 735–747.

Benn, D.I. & Evans, D.J.A. 1998. *Glaciers and Glaciation*. Arnold, London.

Bennett, M.R. 2007. In: *Glacial Sedimentary Processes and Products* (eds Hambrey, M. *et al.*), pp. 177–202. Special Publication 39, International Association of Sedimentologists.

Bennett, S.J. & Best, J.L. 1995. *Sedimentology*, **42**, 491–514.

Bennett, S.J. & Best, J.L. 1996. In: *Coherent Flow Structures in Open Channels* ( Ashworth, P.J. *et al.*). pp. 281–304, Wiley.

Bennett, S.J. & Bridge, J.S. 1995. *Journal of Sedimentary Research*, **65**, 29–39.

Bensing, J.P. *et al.* 2008. *Journal of Sedimentary Research*, **78**, 2–15.

Benson, L. & Thompson, R.S. 1987. In: *The Geology of North America*, Vol. K3, *North America and Adjacent Oceans During the Last Deglaciation* (eds W.F. Ruddiman & H.E. Wright), pp. 241–260. Geological Society of America, Boulder, CO.

Berger, W.H. 1982. *Naturwissenschaften*, **69**, 87–88.

Berger, W.H. & Herquera, J.C. 1992. *Primary Productivity and Biogeochemical Cycles in the Sea* (eds Falkowski, P.G & Woodhead, A.D.), pp. 455–486. Plenum.

Berné, S. *et al.* 1993. *Journal of Sedimentary Petrology*, **63**, 780–793.

Berner, R.A. 1969. *Geochimica et Cosmochimica Acta*, **33**, 267–273.

Berner, R.A. 1970. *American Journal of Science*, **208**, 1–23.

Berner, R.A. 1971. *Principles of Chemical Sedimentology*. McGraw-Hill, New York.

Berner, R.A. 1975. *Geochimica et Cosmochimica Acta*, **39**, 489–504.

Berner, R.A. 1976. *American Journal of Science*, **276**, 713–30.

Berner, R.A. 1992. *Geochimica et Cosmochimica Acta*, **56**, 3225–3232.

Berner, R.A. 1994. *American Journal of Science*, **294**, 56–91.

Berner, R.A. 1995. In: *Chemical Weathering Rates of Silicate Minerals* (eds White, A.F. & Branfield, S.L.). American Mineralogical Society.

Berner, R.A. *et al.* 1978. *American Journal of Science*, **278**, 816–837.

Berner, R.A. *et al.* 1983. *American Journal of Science*, **283**, 641–683.

Berner, R.A. & Lasaga, A.C. 1989. *Scientific American*, **260**, 54–61.

Best, J.L. 1987. In: *Recent Developments in Fluvial Sedimentology* (eds Ethridge, F.G. *et al.*), pp. 27–35. Special Publication 39, Society of Economic Paleontologists and Mineralogists.

Best, J.L. 1988. *Sedimentology*, **35**, 481–498.

Best, J.L. 1992a. *Sedimentology*, **39**, 797–811.

Best, J.L. 1992b. *Bulletin of Volcanology*, **54**, 299–318.

Best, J.L. 1993. In: *Turbulence: Perspectives on Flow and Sediment Transport* (eds Clifford, N.J. *et al.*), pp. 62–91. Wiley.

Best, J.L. 2005. *Journal of Geophysical Research*, **110**, F04S02, doi: 10.1029/2004JF000218.

Best, J.L. & Ashworth, P.J. 1997. *Nature*, **387**, 275–277.

Best, J.L. & Bridge, J.S. 1992. *Sedimentology*, **39**, 737–752.

Best, J.L.& Bristow, C.S. (eds) 1993. *Braided Rivers*. Special Publication 75, Geological Society of London.

Best, J.L. & Leeder, M.R. 1993b. *Sedimentology*, **40**, 1129–1137.

Best, J.L. & Roy, A.G. 1991. *Nature*, **350**, 411–413.

Best, J.L. *et al.* 2005. *Geology*, **33**, 765–768.

Beveridge, C. *et al.* 2006. *Sedimentology*, **53**, 1391–1409.

Betzler, C. *et al.* 1999. *Sedimentology*, **46**, 1127–1143.

Bhattacharya, J.P. & Giosan, L. 2003. *Sedimentology*, **50**, 187–210.

Bhattacharya, J. P. & MacEeachern. 2009. *Journal of Sedimentary Research*, **79**, 184–209.

Bianchi, D. *et al.* 2006. *Palaeogeography Palaeoclimatology Palaeoecology*, **235**, 265–287.

Bianchi, G.G. 1999. *Sedimentology*, **46**, 1001–1014.

Bianchi, G.G. & McCave, I.N. 1999. *Nature*, **397**, 515–517.

Bibring, J.-P. *et al.* 2004. *Nature*, **428**, 627–630.

Bickle, M.J. *et al.* 1995. *Contributions to Mineralogy and Petrology*, **121**, 400–413.

Bidle, K.D. & Azam, F. 1999. *Nature*, **397**, 508–512.

Biegel, E. & Hoekstra, P. 1995. In: *Tidal Signatures in Modern and Ancient Sediments* (eds Flemming, B.W. & Bartholoma, A.), pp. 85–99. Special Publication 24, International Association of Sedimentologists.

Bigarella, J.J. 1973. *Geologische Rundschau*, **62**, 447–477.

Biscaye, P.E. & Eittreim, S.L. 1977. *Marine Geology*, **23**, 155–172.

Bjerager, M. & Surlyk, F. 2007. *Journal of Sedimentary Research*, **77**, 634–660.

Black, K.S. *et al.* (eds). 1998. *Sedimentary Processes in the Intertidal Zone*. Special Publication 139, Geological Society of London.

Blair, T.C. 1999a. *Sedimentology*, **46**, 913–940.

Blair, T.C. 1999b. *Sedimentology*, **46**, 941–965.

Blair, T.C. 1999c. *Sedimentology*, **46**, 1015–1028.

Blair, T.C. & McPherson, J.G. 1994a. *Journal of Sedimentary Research*, **64**, 450–489.

Blair, T.C. & McPherson, J.G. 1994b. In: *Geomorphology of Desert Environments* (eds Abrahams, A.D & Parsons, A. J), pp. 354–402. Chapman & Hall.

Blakey, R.C. *et al.* 1996. *Journal of Sedimentary Research*, **66**, 324–342.

Blanchon, P. & Blakeway, D. 2003. *Sedimentology*, **50**, 1271–1282.

Blanchon, P. & Jones, B. 1997. *Sedimentology*, **44**, 479–506.

Blanchon, P. & Shaw, J. 1995. *Geology*, **23**, 4–8.

Blanchon, P. *et al.* 2009. *Nature*, **458**, 881–886.

Blankenship, D.D. 1993. *Eos, Transactions American Geophysical Union, Abstracts Spring Meeting*, pp. 309.

Blankenship, D.D. *et al.* 1986. *Nature*, **322**, 54–57.

Blas, L. *et al.* 2000. *Sedimentology* **47**, 1135–1156.

Blatt, H. *et al.* (1980). *Origin of Sedimentary Rocks*, 2nd edn. Prentice Hall, New Jersey.

Blendinger, W. 1994. *Sedimentology*, **41**, 1147–1159.

Blendinger, W. 2001. *Sedimentology*, **48**, 919–933.

Blendinger, W. *et al.* 2004. *Sedimentology*, **51**, 297–314.

Blikra, L.H. & Nemec, W. 1998. *Sedimentology*, **45**, 909–959.

Blomeier, D.P.G. & Reijmer, J.J.G. 2002. *Journal of Sedimentary Research*, **72**, 462–475.

Blott, S.J. & Pye, K. 2008. *Sedimentology* **55**, 31–63.

Blount, G. & Lancaster, N. 1990. *Geology*, **18**, 724–728.

Bluck, B.J. 1967. *Journal of Sedimentary Petrology*, **37**, 128–156.

Blum, A.E. & Stillings, L.L. 1995. In: *Chemical Weathering Rates of Silicate Minerals* (eds A.F. White & S.L. Branfield). American Mineralogical Society.

Blum, J.D. *et al.* 1998. *Geology*, **26**, 411–414.

Blum, J.D. *et al.* 1999. *Geochimica, Cosmochimica Acta*, **63**, 1905–1925.

Blum, M.D. 1990. *GCSSEPM Foundation 11th Annual Research Conference*, pp. 71–83.

Blum, M.D. 1993. In: *Siliciclastic Sequence Stratigraphy: Recent Developments and Applications* (eds P. Weimer & H.W. Posamentier), pp. 259–83. Memoir 58, American Association of Petroleum Geologists.

Blum, M. & Roberts, H. 2009. *Nature geosciences*, **2**, 488–491.

Blum, M.D. & Törnqvist, T.E. 2000. *Sedimentology*, **47** (**Suppl. 1**), 2–48.

Blumel, W.D. *et al.* 1998. *Palaeogeography, Palaeoclimatology, Palaeoecology*, **138**, 139–149.

Boardman, M.R. & Neuman, A.C. 1984. *Journal of Sedimentary Petrology*, **54**, 1110–1123.

Boardman, M.R. *et al.* 1995. In: *Terrestrial and Shallow Marine Geology of the Bahamas and Bermuda* (eds A.A. Curran & B. White), pp. 33–51. Special Paper 300, Geological Society of America.

Boersma, J.R. 1967. *Geologie Mijnbouw*, **46**, 217–235.

Boersma, J.R. & Terwindt, J.H.J. 1981. *Sedimentology*, **28**, 151–170.

Boesch, D.F. *et al.* 2009. *Eos, Transactions American Geophysical Union*, **90/14**, 117–118.

Boettcher, S.S. & Milliken, K.L. 1994. *Journal of Geology*, **102**, 655–663.

Bogaart, P. W. & van Balen, R. T. 2002. *Global Planetary Change*, **27**, 147–163.

Bogaart, P. W. *et al.* 2002. In: *Sediment Flux to Basins: Causes, Controls and Consequences* (eds Jones S.J. & Frostick, L. E.) pp. 187–198. Special Publication 191, Geological Society of London.

Bogaart, P.W. *et al.* 2003. *Quaternary Science Reviews*, **22**, 2097–2110.

Bond, G. *et al.* 1992. *Nature*, **360**, 245–249.

Bond, G. *et al.* 1993. *Nature*, **365**, 143–147.

Bond, G. *et al.* 1995. In: *Tectonics of Sedimentary Basins* (eds C. Busby & R. Ingersoll), pp. 149–178, Blackwell Science, Oxford.

Bond, G. *et al.* 1997. *Science*, **278**, 1257–1266.

Bondevik, S. 2008. *Nature*, **455**, 1183–1184.

Boos, W.F. & Kuang, Z. 2010. *Nature*, **463**, 218–222.

Boothroyd, J.C. & Ashley, G.M. 1975. In: *Glaciofluvial and Glaciolacustrine Sediments* (eds A.V. Jopling & B.C. McDonald), pp. 193–222. Special Publication 23, Society of Economic Paleontologists and Mineralogists.

Borchert, H. & Muir, R.O. 1964. *Salt Deposits*. Van Nostrand Reinhold, London.

Borrego, J. *et al.* 1995. In: *Tidal Signatures in Modern and Ancient Sediments* (eds Flemming, B.W. & Bartholoma, A.), pp. 151–170. Special Publication 24, International Association of Sedimentologists.

Bosak, T. & Newman, D.K. 2005. *Journal of Sedimentary Research*, **75**, 190–199.

Bosellini, A. 1984. *Sedimentology*, **31**, 1–24.

Bosence, D.W.J. 1973. *Geologie Mijnbouw*, **52**, 63–67.

Bosence, D.W.J. 1983a. In: *Coated Grains* (ed. Peryt, T.M.), pp. 217–24. Springer, Berlin.

Bosence, D.W.J. 1983b. In: *Coated Grains* (ed. Peryt, T.M.), pp. 225–42. Springer, Berlin.

Bosence, D.W.J. & Waltham, D.A. 1990. *Geology*, **18**, 26–30.

Bosence, D.W.J. *et al.* 1985. *Sedimentology*, **32**, 317–343.

Bosence, D.W.J. *et al.* 1994. *Bulletin of the American Association of Petroleum Geologists*, **78**, 247–266.

Boulton, G.S. 1972. *Quarterly Journal of the Geological Society of London*, **128**, 361–393.

Boulton, G.S. 1976. *Journal of Glaciology*, **17**, 287–309.

Boulton, G.S. 1978. *Sedimentology*, **25**, 773–799.

Boulton, G.S. 1979. *Journal of Glaciology*, **23**, 15–38.

Boulton, G.S. 1987. In: Menzies, J. & Rose, J. (eds) 25–80, *Drumlin Symposium*. Balkema.

Boulton, G.S. 1990. In: *Glacimarine Environments* (eds Dowdeswell, J.A. & Scourse, J.D.), pp. 15–52. Special Publication 53, Geological Society of London.

Boulton, G.S. & Hindmarsh, R.C.A. 1987. *Journal of Geophysical Research*, **92B**, 9059–9082.

Bouma, A.H. 1962. *Sedimentology of Some Flysch Deposits: A Graphic Approach to Facies Interpretations*. Elsevier, Amsterdam.

Bouma, A.H. & Hollister, C.D. 1973. In: *Turbidites and Deep Water Sedimentation*, pp. 79–118. Society of Economic Paleontologists and Mineralogists Short Course, Anaheim.

Bouma, A.H. *et al.* 1990. *Geo-Marine Letters*, **10**, 200–208.

Bourgeois, J. 1980. *Journal of Sedimentary Petrology*, **50**, 681–702.

Bourne, J.K. 2004. *National Geographic Oct*, 87–106.

Bourne, S.J. *et al.* 1998. *Nature*, **391**, 655–659.

Bowen, A.J. 1969. *Journal of Geophysical Research*, **74**, 5467–5478.

Bowen, A.J. & Inman, D.L. 1969. *Journal of Geophysical Research*, **74**, 5479–5490.

Bowen, A.J. *et al.* 1968. *Journal of Geophysical Research*, **73**, 256–277.

Bowen, A.J. *et al.* 1984. *Sedimentology*, **31**, 169–185.

Bowen, G.J. *et al.* 2008. *Journal of Sedimentary Research*, **78**, 162–174.

Bowler, J.M. 1977. *Earth Science Reviews*, **12**, 279–310.

Bowman, D. 1978. *Earth Surface Processes*, **3**, 265–276.

Boyd, R. *et al.* 1992. *Sedimentary Geology*, **80**, *p.* 139–150.

Boyer, J. *et al.* 2005. In: *Mediterranean Prodelta Systems: Proceedings of the International ComDelta Symposia, Aix-en-Provence, France* (eds Trincardi, F. & Syvitski, J.). *Marine Geology*, **222–223**, 267–298.

Boylan, A.L. *et al.* 2002. *Basin Research*, **14**, 401–415.

Braaksma, H. *et al.* 2006. *Journal of Sedimentary Research*, **76**, 175–199.

Brack, P. *et al.* 1996. *Geology*, **24**, 371–375.

Bradshaw, M.J. *et al.* 1980. *Journal of Sedimentary Petrology*, **50**, 295–299.

Bradshaw, P. 1971. *Introduction to Turbulence and its Measurement*. Elsevier

Brady, P.V. *et al.* 1999. *Geochimica et Cosmochimica Acta*, **63**, 3293–3300.

Braga, J.C. *et al.* 1996. *Geology*, **24**, 35–38.

Braissant, O. *et al.* 2003. *Journal of Sedimentary Research*, **73**, 485–490.

Braithwaite, C.J. *et al.* 2004. *Journal of Sedimentary Research*, **74**, 298–310.

Brantley, S.L. & Chen, Y. 1995. In: *Chemical Weathering Rates of Silicate Minerals* (eds White, A.F. & Branfield, S. L.), American Society of Mineralogy, Washington, DC.

Brasier, M.D. 1995. In: *Marine Palaeoenvironmental Analysis from Fossils* (eds Bosence, D.J.W. & Allison, P.A.), pp. 113–132. Special Publication 83, Geological Society of London.

Brayshaw, A.C. 1984. In: *Sedimentology of Gravels and Conglomerates* (eds Koster, E.H. & Steel, R.J.), pp. 77–85. Memoir 10, Canadian Society of Petroleum Geologists.

Brayshaw, A.C. *et al.* 1983. *Sedimentology*, **30**, 137–143.

Breed, C.S. 1977. *Icarus*, **30**, 326–340.

Bréhéret, J-G. 2008. *Sedimentology*, **55**, 557–578.

Brenchley, P.J. *et al.* 1993. *Sedimentology*, **40**, 359–382.

Brennan, S.T. 2002. The major-ion evolution of seawater: Fluid inclusion evidence from the terminal Proterozoic, Early Cambrian, Silurian and Tertiary marine halites. Unpublished PhD thesis, University of Binghamton, New York, 186 pp.

Brennan, S.T. & Lowenstein, T.K. 2002. *Geochimica Cosmochimica Acta*, **66**, 2683–2700.

Brennan, S.T. *et al.* 2004. *Geology*, **32**, 473–476.

Brice, J.C. 1984. In: *River Meandering, Proceedings Rivers '83 Conference*, pp. 1–15. American Society of Civil Engineers, New York.

Bridge, J.S. 1976. *Sedimentology*, **23**, 407–414.

Bridge, J.S. 1977. *Earth Surface Processes*, **2**, 401–416.

Bridge, J.S. 1981. *Journal of Sedimentary Petrology*, **51**, 1109–1124.

Bridge, J.S. 1985. *Journal of Sedimentary Petrology*, **55**, 579–589.

Bridge, J.S. 1993. In: *Braided Rivers* (eds Best, J.L. & Bristow, C.S.), pp. 13–71. Special Publication 75, Geological Society of London.

Bridge, J.S. 2003. *Rivers and Floodplains: Forms, Processes and Sedimentary Record.* Blackwell Science.

Bridge, J.S. & Bennett, S.J. 1992. *Water Resources Research*, **28**, 337–363.

Bridge, J.S. & Best, J.L. 1997. *Sedimentology*, **44**, 253–262.

Bridge, J.S. & Gabel, S.L. 1992. *Sedimentology*, **39**, 125–142.

Bridge, J.S. & Jarvis, J. 1982. *Sedimentology*, **29**, 499–541.

Bridge, J.S. & Leeder, M.R. 1979. *Sedimentology*, **26**, 617–644.

Bridge, J.S. & Mackey, S.D. 1993. In: *The geological Modelling of Hydrocarbon Reservoirs and Outcrop Analogues* (eds Flint, S.S. & Bryant, I.D.), pp. 213–236. Special Publication 15, International Association of Sedimentologists.

Bridge, J.S. *et al.* 1986. *Sedimentology*, **33**, 851–870.

Bridge, J.S. *et al.* 1995. *Sedimentology*, **42**, 839–852.

Bridges, P.H. 1982. In: *Offshore Tidal Sands* (ed. Stride, A. H.), pp. 172–192. Chapman & Hall, London.

Bridges, P.H. & Leeder, M.R. 1976. *Sedimentology*, **23**, 533–552.

Bridgman, P.W. 1911. *Proceedings of the American Academy of Arts and Sciences*, **47**, 439–558.

Brimblecombe, P. & Davies, T.D. 1982. In: *The Cambridge Encyclopaedia of Earth Sciences* (ed. Smith, D.G.). Cambridge University Press, Cambridge.

Bristow, C.S. 1993. In: *Braided Rivers* (eds Best, J.L. & Bristow, C.S.), pp. 277–290. Special Publication 75, Geological Society of London.

Bristow, C.S. & Pucillo, K. 2006. *Sedimentology*, **53**, 769–788.

Bristow, C.S. *et al.* 1993. In: *Alluvial Sedimentation* (eds Marzo, M. & Puigdefabregas, C.), pp. 91–100. Special Publication 17, International Association of Sedimentologists.

Bristow, C.S. *et al.* 1996. *Sedimentology*, **43**, 995–1003.

Bristow, C.S. *et al.* 1999. *Sedimentology*, **46**, 1029–1047.

Bristow, C.S. *et al.* 2000. *Sedimentology*, **47**, 923–944.

Bristow, C.S. *et al.* 2007. *Geology*. **35**, 555–558.

Britter, R.E. & Linden, P.F. 1980. *Journal of Fluid Mechanics*, **99**, 531–543.

Broecker, W.S. 1974. *Chemical Oceanography*. Harcourt Brace Jovanovich, New York.

Broecker, W.S. & Takahashi, T. 1966. *Journal of Geophysical Research*, **71**, 1575–1602.

Brookfield, M. 1970. *Zeitschrift fur Geomorphologie, Suppl.* **10**, 121–153.

Brookfield, M.E. 1977. *Sedimentology*, **24**, 303–330.

Brooks, G.R. 2003. *Sedimentology*, **50**, 441–458.

Brown, D.A. *et al.* 1997. *Geochimica et Cosmochimica Acta*, **61**, 3341–3348.

Brown, D.J. *et al.* 2003. *Geochimica et Cosmochimica Acta*, **67**, 2711–2723.

Brown, M.A. *et al.* 1990. *Journal of Sedimentary Petrology*, **60**, 152–159.

Brown, P. *et al.* 1999. *Paleoceanography*, **14**, 498–510.

Bruckschen, P. *et al.* 1999. *Chemical Geology*, **161**, 127–163.

Bruland, K.W. *et al.* 1989. In: *Productivity of the Oceans: Present and Past* (eds Berger, W.H. *et al.*), pp. 193–216. Wiley, New York.

Brunner, C.A. *et al.* 1999. *Geology*, **27**, 463–466.

Brunt, R.L. *et al.* 2004. *Journal of Sedimentary Research*, **74**, 438–446.

Brutsaert, W. 1999. *Reviews of Geophysics*, **37**, 439–451.

Bryant, M. *et al.* 1995. *Geology*, **23**, 365–368.

Bryant, W.R. *et al.* 1990. *Geo-Marine Letters*, **10**, 182–199.

Buchardt, B. *et al.* 1997. *Nature*, **390**, 129–130.

Buchardt, B. *et al.* 2001. *Journal of Sedimentary Research*, **71**, 176–189.

Buffington, J.M. & Montgomery, D.R. 1997. *Water Resources Research*, **33**, 1993–2029.

Bugge, T. *et al.* 1988. *Philosophical Transactions of the Royal Society, London (A)*, **325**, 357–388.

Bull, W.B. 1979. *Bulletin of the Geological Society of America*, **90**, 453–464.

Bull, W.B. 1991. *Geomorphic Response to Climate Change.* Oxford University Press, New York.

Bullimore, S. & Helland-Hansen, W. 2004. In: *Shoreline and Shelf-Edge Trajectories: Effects upon Facies, Environments and Stratigraphic Architecture in Wave-Dominated Regressive Systems* (ed. Bullimore, S.) University of Bergen, Bergen, 31 pp.

Burbank, D.W. & Raynolds, R.G.H. 1988. In: *New Perspectives in Basin Analysis* (eds Kleinspehn, K.L. & Paola, C.), pp. 331–351. Springer, New York.

Burbank, D.W. & Beck, R.A. 1991. *Geologische Rundschau.* **80**, 623–638.

Burbank, D.W. *et al.* 1992. *Bulletin of the Geological Society of America*, **104**, 1101–1120.

Burbank, D.W. *et al.* 1996. In: *The Tectonic Evolution of Asia* (ed. Yin, A. & Harrison, T.M.), pp. 149–188. Cambridge University Press, Cambridge.

Burbank, D.W. *et al.* 1999. *Basin Research*, **11**, 75–92.

Burchette, T.P. & Wright, V.P. 1992. *Sedimentary Geology*, **79**, 3–57.

Burchfiel, B.C. 1966. *Bulletin of the Geological Society of America*, **77**, 95–100.

Burgess, P.M. 2001. *Geology*, **29**, 1127–1130.

Burgess, P.M. 2006. *Journal of Sedimentary Research*, **76**, 962–977.

Burgess, P.M. & Gurnis, M. 1995. *Earth and Planetary Science Letters*, **136**, 647–663.

Burgess, P.M. & Wright, V.P. 2003. *Journal of Sedimentary Research*, **73**, 637–652.

Burgess, P.M. *et al.* 1997. *Bulletin of the Geological Society of America*, **108**, 1515–1535.

Burgess, P.M. *et al.* 2001. *Basin Research*, **13**, 1–16.

Burkham, D.E. 1972. *Channel Changes of the Gila River in Safford Valley, Arizona, 1846–1970.* Professional Paper 655-G, US Geological Survey.

Burt, N. *et al.* (eds). 1997. *Cohesive Sediments.* Wiley, Chichester.

Burton, E.A. & Walter, L.M. 1987. *Geology*, **15**, 111–114.

Burton, E.A. & Walter, L.M. 1990. *Geochimica et Cosmochimica Acta*, **54**, 797–808.

Busby, C.J.& Ingersoll, R.V. (eds). 1995. *Tectonics of Sedimentary Basins.* Blackwell Science, Oxford.

Butler, G.P. 1970. In: *Third Salt Symposium* (Eds Rau, J.L. & Dellwig, L.F.), pp. 120–152. Northern Ohio Geological Society, Cleveland.

Butler, R.W.H. *et al.* 1997. *Journal of the Geological Society, London*, **154**, 69–71.

Butler, R.W.H. & Tavarnelli, E. 2006. *Sedimentology*, **53**, 655–670.

Cabioch, G. *et al.* 1999. *Sedimentology*, **46**, 985–1000.

Cacchione, D.A. & Southard, J.B. 1974. *Journal of Geophysical Research*, **79**, 2237–2242.

Calvert, S.E. & Karlin, R.E. 1998. *Geology*, **26**, 107–110.

Calvo, J.P. *et al.* 1989. *Palaeogeography, Palaeoclimatology, Palaeoecology*, **70**, 199–214.

Camoin, G.F. & Montaggioni, L.F. 1994. *Sedimentology*, **41**, 655–676.

Campbell, C.S. 1989. *Journal of Geology*, **97**, 653–665.

Campbell, C.S. *et al.* 1995. *Journal of Geophysical Research*, **100**, 8267–8283.

Campbell, C.V. & Oakes, R.Q. 1973. *Journal of Sedimentary Petrology*, **43**, 765–778.

Campbell, K.A. *et al.* 2006. *Sedimentology* **53**, 945–969.

Campbell, S.E. 1979. *Origins of Life*, **9**, 335–348.

Caplan, M.L. *et al.* 1996. *Geology*, **24**, 715–718.

Capozzi, R. & Negri, A. 2009. *Palaeogeography, Palaeoclimatology, Palaeoecology*, **273**, 249–257.

Carew, J.L. & Mylroie, J.E. 1995. In: *Terrestrial and Shallow Marine Geology of the Bahamas and Bermuda* (eds Curran, A.A. & White, B.), pp. 5–32. Special Paper 300, Geological Society of America.

Carew, J.L. & Mylroie, J.E. 1997. In: *Geology and Hydrogeology of Carbonate Islands* (eds Vacher, H. L. & Quinn, T. M.), pp. 91–139. Developments in Sedimentology 54. Elsevier, Amsterdam.

Carey, S.N. & Sigurdsson, H. 1984. In: *Marginal Basin Geology* (eds Kokelaar, B.P. & Howell, M.F.), pp. 37–58. Special Publication 16, Geological Society of London.

Carling, P.A. 1981. In: *Holocene Marine Sedimentation in the North Sea Basin* (eds Nio, S.D. *et al.*), pp. 65–80. Special Publication 5, International Association of Sedimentologists.

Carling, P.A. 1999. *Journal of Sedimentary Research*, **69**, 534–545.

Carling, P.A. & Glaister, M.S. 1987. *Journal of the Geological Society of London*, **44**, 543–551.

Carling, P.A. & Shvidchenko, A.B. 2002. *Sedimentology*, **49**, 1269–1282.

Carling, P.A. *et al.* 2000. *Sedimentology*, **47**, 227–252.

Carr, I.D. *et al.* 2003. *Journal of Sedimentary Research*, **73**, 407–420.

Carroll, A.R. & Bohacs, K.M. 1999. *Geology*, **27**, 99–102.

Carroll, D. 1958. *Geochimica et Cosmochima Acta*, **14**, 1–27.

Carson, M.A. 1971. *The Mechanics of Erosion.* Pion, London.

Carson, M.A. & Kirkby, M.J. 1972. *Hillslope Form and Process.* Cambridge University Press, Cambridge.

Carter, L. & McCave, I.N. 1997. *Journal of Sedimentary Research*, **67**, 1005–1017.

Carter, L. & McCave, I.N. 2002. In: *Deep-Water Contourite Systems: Modern Drifts and Ancient Series, Seismic and Sedimentary Characteristics* (eds Stow, D.A.V. *et al.*), pp. 385–497. Memoir 22, Geological Society of London.

Carter, R.W.G.& Woodroffe, C.D. (eds). 1994. *Coastal Evolution: Late Quaternary Shoreline Morphodynamics*. Cambridge University Press, Cambridge.

Cartwright, J. A. *et al.* 1995. *Journal of Structural Geology*, **17**, 1319–1326.

Cartwright, J. A. *et al.* 1996. In: *Modern Developments in Structural Interpretation, Validation and Modelling* (eds Buchanan, P. G. & Nieuwland, D. A.), pp. 163–177. Special Publication 99, Geological Society of London.

Casey, W.H. & Bunker, B. 1991. *Reviews of Mineralogy*, **23**, 397–426.

Casey, W.H. *et al.* 1993. *Nature*, **366**, 253–255.

Castell, J.M.C. *et al.* 2007. *Sedimentology*, **54**, 423–441.

Castle, J.W. 2001. *Basin Research*, **13**, 397–418.

Cathro, D.L. *et al.* 1992. *Sedimentology*, **39**, 983–1002.

Causse, C. *et al.* 1989. *Geology*, **17**, 922–925.

Chadwick, O.A. *et al.* 1999. *Nature*, **397**, 491–497.

Chafetz, H.S. & Rush, P.F. 1994. *Sedimentology*, **41**, 409–421.

Chahine, M.T. 1992. *Nature*, **359**, 373–380.

Chamberlain, C.P. 1998. *Geology*, **26**, 411–414.

Chan, M.A. *et al.* 1994. *Geology*, **22**, 791–794.

Chan, M.A. *et al.* 2004. *Nature*, **429**, 731–734.

Chandler, D. *et al.* 2008. *Geophysical Research Letters*, **35**, Article Number L12502.

Chang, C. *et al.* 1986. *Nature*, **323**, 501–507.

Chang, C. *et al.* 1988. *Philosophical Transactions of the Royal Society of London*, **327A**, 1–413.

Chang, T.S. & Flemming, B.W. 2006. *Sedimentology*, **53**, 687–691.

Chapman, M.R. & Shackleton, N.J. 2000. *Holocene*, **10**, 287–291.

Chappell, J. 1980. *Nature*, **286**, 249–252.

Chappell, J. & Shackleton, N.J. 1986. *Nature*, **324**, 137–140.

Charles, C.D. & Fairbanks, R.G. 1992. *Nature*, **355**, 416–419.

Charnock, H. 1996. In: *Oceanography, An Illustrated Guide* (eds Summerhayes, C.P. & Thorpe, S.A.), Manson Press, London.

Chave, K.E. & Suess, E. 1970. *Limnology and Oceanography*, **15**, 633–637.

Chen, C. & Hiscott, R.N. 1999. *Journal of Sedimentary Research*, **69**, 486–504.

Chen, J.H. *et al.* 1991. *Geological Society of American Bulletin*, **103**, 82–97.

Cheng, N-S. 1997. *American Society of Civil Engineers*, **123**, 149–152.

Chepil, W.S. 1961. *Proceedings of the Soil Science Society of America*, **25**, 343–345.

Chesworth, W. 1992. In: *Weathering, Soils and Palaeosols* (eds Martini, I.P. & Chesworth, W.), Elsevier, Amsterdam.

Chevrier, V. *et al.* 2007. *Nature*, **448**, 60–63.

Chidsey, T.C. Jr. 2001. Utah Geological Survey report.

Chikita, K.A. *et al.* 1996. *Sedimentology*, **43**, 865–875.

Choi, D.R. & Ginsburg, R.N. 1982. *Bulletin of the Geological Society of America*, **93**, 116–126.

Choi, K.S. *et al.* 2004. *Journal of Sedimentary Research*, **74**, 677–689.

Choudens-Sanchez, V. & Gonzalez, L.A. 2009. *Journal of Sedimentary Research*, 79, 363–376.

Chough, S.K. *et al.* 1990. *Journal of Sedimentary Petrology*, **60**, 445–455.

Chrintz, T. & Clemmenson, L.B. 1993. In: *Aeolian Sediments Ancient and Modern* (eds Pye, K. & Lancaster, N.), pp. 151–161. Special Publication 16, International Association of Sedimentologists.

Christensen, P.R. 2003. *Nature* **422**, 45–48.

Christiansen, C. *et al.* 1984. *Geological Magazine*, **121**, 47–51.

Church, M. 2006. *Annual Reviews of Earth and Planetary Science*, **34**, 325–354.

Clark, J.D. & Pickering, K.T. 1996. *Bulletin of the American Association of Petroleum Geologists*, **80**, 194–221.

Clark, J.D. *et al.* 1992. *Geology*, **20**, 633–636.

Cleary, P.W. & Campbell, C.S. 1993. *Journal of Geophysical Research*, **98**, 21911–21924.

Clemente, P. & Pérez-Arlucea, M. 1993. *Journal of Sedimentary Petrology*, **63**, 437–452.

Clemmenson, L.B. 1987. In: *Desert Sediments: Ancient and Modern* (eds Frostick, L. & Reid, I.), pp. 213–31. Special Publication 35, Geological Society of London.

Clemmenson, L.B. 1989. *Sedimentary Geology*, **65**, 139–151.

Clemmenson, L.B. *et al.* 1989. *Bulletin of the Geological Society of America*, **101**, 759–773.

Clevis, Q. *et al.* 2004a. *Sedimentology*, **51**, 809–835.

Clevis, Q. *et al.* 2004b. *Basin Research*, **16**, 145–163.

Clifford, N.J. *et al.* (eds). 1993. *Turbulence: Perspectives on Flow and Sediment Transport*. Wiley, Chichester.

Clift, P.D. 1995. In: *Active Margins and Marginal Basins of the Western Pacific*, pp. 67–96. Geophysical Monograph 88, American Geophysical Union.

Clift, P.D. & Bluszajn, J. 2005. *Nature*, **438**, 1001–1003.

Clift, P.D. *et al.* 1998. *Bulletin of the Geological Society of America*, **110**, 483–496.

Clift, P.D. *et al.* 2000. *Bulletin of the Geological Society of America*, **112**, 450–466.

Clift, P.D. *et al.* 2005. *Bulletin of the Geological Society of America*, **117**, 902–925.

Clift, P.D. *et al.* 2008. *Nature Geoscience*, **1**, 875–880.

Clifton, H.E. & Dingler, J.R. 1984. *Marine Geology*, **60**, 165–198.

Clifton, H.E. *et al.* 1971. *Journal of Sedimentary Petrology*, **41**, 651–670.

Cloud, P.E. 1962. *Environment of Calcium Carbonate Deposition West of Andros Island, Bahamas.* Professional Paper 350, US Geological Survey.

Cody, R.D. & Hull, A.B. 1980. *Geology*, **8**, 505–509.

Coe, A.L. (ed.) 2003. *The Sedimentary Record of Sea-Level Change.* Cambridge University Press.

Cohen, A.S. *et al.* 1997. *Bulletin of the Geological Society of America*, **109**, 444–460.

Cohen, A.S. *et al.* 2004. *Geology*, **32**, 157–160.

Cohen, K.M. 2005. In: *River Deltas: Concepts, Models, and Examples* (eds Giosan, L. & Bhattacharya, J.P.), pp. 341–364. Special Publication 83, Society of Economic Paleontologists and Mineralogists.

Cohen, K.M. *et al.* 2002. *Netherlands Journal of GeoSciences*, **81**, 389–405.

Colbeck, S.C. (ed.). 1980. *Dynamics of Snow and Ice Masses.* Academic Press, New York.

Colella, A. 1988. In: *Fan Deltas: Sedimentology and Tectonic Settings* (eds Nemec, W. & Steel, R.J.), pp. 50–74. Blackie, London.

Colella, A. & di Geronimo, I. 1987. *Sedimentary Geology*, **51**, 257–277.

Colella, A. & Normark, W.R. 1984. *Memoir of the Geological Society of Italy*, **27**, 381–90.

Colella, A.& Prior, D.B. (eds). 1990. *Coarse-Grained Deltas.* Special Publication 10, International Association of Sedimentologists.

Coleman, J.M. 1969. *Sedimentary Geology*, **3**, 129–239.

Coleman, J.M. 1976. *Deltas: Processes of Deposition and Models for Exploration.* Continuing Education Publishing, Champaign, IL.

Coleman, J.M. & Gagliano, S.M. 1964. *Transactions of the Gulf Coast Association of Geological Societies*, **14**, 67–80.

Coleman, J.M. *et al.* 1964. *Marine Geology*, **1**, 240–258.

Coleman, J.M. *et al.* 1967. *Deltas*, pp. 185–197. Special Publication 15, Society of Economic Paleontologists and Mineralogists.

Coleman, N.L. 1981. *Journal of Hydraulic Research*, **19**, 211–229.

Collier, R.E.Ll. *et al.* 1990. *Geological Magazine*, **127**, 117–128.

Collier, R.E.Ll. & Thompson, J. 1991. *Sedimentology*, **38**, 1021–1040.

Collier, R.E.Ll. *et al.* 1995. In: *Mediterranean Quaternary River Environments* (eds Lewin, J. *et al.*), pp. 31–44. Balkema, Rotterdam.

Collin, P.Y. *et al.* 2005. *Sedimentology*, **52**, 969–985.

Collins, J.I. 1976. In: *Beach and Nearshore Sedimentation* (eds Davis, K.S. & Ethington, R.L.), pp. 54–68. Special Publication 24, Society of Economic Paleontologists and Mineralogists.

Collins, M.B. *et al.* 1981. In: *Holocene Marine Sedimentation in the North Sea Basin* (eds Nio, S.-D. *et al.*), pp. 81–98. Special Publication 5, International Association of Sedimentologists.

Collinson, J.D. 1970. *Geografiska Annaler*, **52A**, 31–56.

Collinson, J.D. & Thompson, D.B. 1982. *Sedimentary Structures.* Allen and Unwin, London.

Collinson, J.D. *et al.* 2006. *Sedimentary Structures*, 3rd edn. Terra Publishing.

Colman, S.M. 1981. *Quaternary Research*, **15**, 250–264.

Colman, S.M. & Pierce, K.L. 1981. *Weathering Rinds on Andesitic and Basaltic Stones as a Quaternary Age Indicator, Western United States.* Professional Paper 1210, US Geological Survey.

Colman, S.M. *et al.* 1995. *Nature*, **378**, 769–771.

Colman, S.M. *et al.* 2003. *Journal of Sedimentary Research*, **73**, 941–956.

Colombié, C. & Strasser, A. 2005. *Sedimentology*, **52**, 1207–1227.

Colombini, M. 1993. *Journal of Fluid Mechanics*, **254**, 701–719.

Colombo, F. 2005. In: *Alluvial Fans: Geomorphology, Sedimentology, Dynamics* (eds Harvey, A.M. *et al.*), pp. 69–84. Special Publication 251, Geological Society of London.

Colson, J. & Cojan, I. 1996. *Sedimentology*, **43**, 175–188.

Constantine, J.A. & Dunne, T. 2008. *Geology*, **36**, 23–26.

Cook, P.J. *et al.* 1977. *Bulletin of the Bureau of Mineral Resources of Australia*, **2**, 81–88.

Cooke, R.U. 1979. *Earth Surface Processes*, **4**, 347–359.

Cooke, R.U. & Warren, A. 1973. *Geomorphology in Deserts.* Batsford, London.

Cooke, R.U. *et al.* 1993. *Desert Geomorphology.* UCL Press, London.

Cooper, J.A.G. 1993. *Sedimentology*, **40**, 979–1017.

Cooper, M.A.& Williams, G.D. (eds) 1989. *Inversion Tectonics.* Special Publication 44, Geological Society of London.

Copeland, P. & Harrison, T.M. 1990. *Geology*, **18**, 354–357.

Corbett, I. 1993. In: *Aeolian Sediments Ancient and Modern* (eds Pye, K. & Lancaster, N), pp. 45–60. Special Publication 16, International Association of Sedimentologists.

Cornelis, W.M. & Gabriels, D. 2004. *Sedimentology*, **51**, 39–52.

Corney, R.K.T. *et al.* 2006. *Sedimentology*, **53**, 249–257.

Corney, R.K.T. *et al.* 2008. *Sedimentology*, **55**, 241–247.

Correggiari, A. *et al.* 2001. *Journal of Sedimentary Research*, **71**, 218–236.

Correggiari, A. *et al.* 2005a. In: *Mediterranean Prodelta Systems, Proceedings of the International ComDelta Symposia, Aix-en-Provence: France* (eds Trincardi, F. & Syvitski, J.). *Marine Geology*, **222-223**, 49–74.

Correggiari, A. *et al.* 2005b. In: *River Deltas: Concepts, Models, and Examples* (eds Giosan, L. & Bhattacharya, J.P.), pp. 365–392. Special Publication 83, Society of Economic Paleontologists and Mineralogists.

Cortijo, E. *et al.* 1994. *Nature*, **372**, 446–448.

Costas, S. *et al.* 2006. *Journal of Sedimentary Research*, **76**, 1077–1092.

Couchoud, I. *et al.* 2009. *Quaternary Science Reviews*, **28**, 3263–3274.

Coudé-Gaussen, G. 1984. *Bulletin du Centre Recherches et Exploration-Production Elf-Aquitaine*, **8**, 167–182.

Coussot, P. & Proust, S. 1996. *Journal of Geophysical Research*, **101**, 25217–25229.

Coutand, I. *et al.* 2006. *Basin Research*, **18**, 1–26.

Cowan, E.A. & Powell, R.D. 1990. In: *Glacimarine Environments* (eds Dowdeswell, J.A. & Scourse, J.D.), pp. 75–89. Special Publication 53, Geological Society of London.

Cowan, E.A. & Powell, R.D. 1991. In: *Glacial Marine Sedimentation: Palaeoclimatic Significance* (eds Anderson, J.B. & Ashley, G.M.), pp. 61–73. Special Publication 261, Geological Soeiety of America.

Cowan, E.A. *et al.* 1998. *Geo-Marine Letters*, **18**, 40–48.

Coward, M.P. *et al.* 1988. *Philosophical Transactions of the Royal Society*, **327A**, 307–336.

Crabaugh, M. & Kocurek, G. 1993. In: *The Dynamics and Environmental Context of Aeolian Sedimentary Systems* (ed. Pye, K.), pp. 103–26. Special Publication 72, Geological Society of London.

Craig, H. 1994. *American Geophysical Union Fall Meeting Abstracts*, p. 388.

Craik, A.D.D. 2005. *Annual Reviews Fluid Mechanics*, **37**, 23–42.

Cramp, A. & O'Sullivan, G. 1999. *Marine Geology*, **153**, 11–28.

Cramp, A. *et al.* 1995. *Geoscientist*, **5**, 23–25.

Crampton, S.L. & Allen, P.A. 1995. *Bulletin of the American Association of Petroleum Geologists*, **79**, 1495–1514.

Crane, K. *et al.* 1991. *Eos, Transactions American Geophysical Union*, **72**, 585.

Crans, W. *et al.* 1980. *Journal of Petroleum Geology*, **2**, 265–307.

Crevello, P.D. & Schlager, W. 1980. *Journal of Sedimentary Petrology*, **50**, 1121–1148.

Crews, S.G. & Ethridge, F.G. 1993. *Journal of Sedimentary Petrology*, **63**, 420–436.

Crockett, J.S. & Nittrouer, C.A. 2004. *Continental Shelf Research*, **24**, 55–73.

Cronin, S.J. *et al.* 1996. *Geology*, **24**, 1107–1110.

Cronin, T.M. & Raymo, M.E. 1997. *Nature*, **385**, 624–627.

Crowell, J.C. 1974a. In: *Modern and Ancient Geosynclinal Sedimentation* (eds Dott, R.H. & Shaver, R.H.), pp. 292–303. Special Publication 19, Society of Economic Paleontologists and Mineralogists.

Crowell, J.C. 1974b. pp. 190–204. Special Publication 22, Society of Economic Paleontologists and Mineralogists.

Crowell, J.C. 1983. *Transactions of the Geological Society of South Africa*, **86**, 230–261.

Crowell, J.C. & Link, M.H. 1982. *Geological History of the Ridge Basin, Southern California (Book 22)*. Pacific Section of the Society of Economic Paleontologists and Mineralogists, Los Angeles.

Csanady, G.T. 1978. In: *Lakes: Chemistry, Geology, Physics* (ed. Lerman, A.), pp. 21–64. Springer, New York.

Csato, I. *et al.* 1997. *Bulletin of the Geological Society of America*, **108**, 1485–1501.

Cuffey, K.M. & Marshall, S.J. 2000. *Nature*, **404**, 591–594.

Cummings, D.I. *et al.* 2009. *Journal of Sedimentary Research*, **79**, 83–93.

Curtis, C.D. 1976. *Earth Surface Processes*, **1**, 63–70.

Curtis, C.D. & Spears, D.A. 1968. *Economic Geology*, **63**, 257–270.

Cyr, A. *et al.* 2005. *Journal of Geology*, **113**, 517–533.

Daams, R. *et al.* 1996. *Sedimentary Geology*, **102**, 187–209.

Dabrio, C.J. & Polo, M.D. 1988. In: *Fan Deltas: Sedimentology and Tectonic Settings* (eds Nemec, W. & Steel, R. J.), pp. 354–367. Blackie, London.

Dabrio, C.J. *et al.* (eds). 1991a. *The Dynamics of Coarse-Grained Deltas*. Cuadernos de Geologia Iberica 15, Universidad Complutense, Madrid.

Dabrio, C.J. *et al.* 1991b. In: *The Dynamics of Coarse-Grained Deltas* (eds Dabrio, C.J. *et al.*), pp. 103–137. Cuadernos de Geologia Iberica 15, Universidad Complutense, Madrid.

Dade, W.B. & Friend, P.F. 1998. *Journal of Geology*, **106**, 661–675.

Dade, W.B. *et al.* 1994. *Journal of Sedimentary Research*, **64**, 423–432.

Dadson, S.J. *et al.* 2003. *Nature*, **426**, 648–650.

Dail, M.B. *et al.* 2007. *Continental Shelf Research*, **27**, 1857–1874.

Dalai, T.K. *et al.* 2002. *Geochimica et Cosmochimica Acta*, **66**, 3397–3416.

Dalrymple, R.W. & Zaitlin, B.A. 1994. *Sedimentology*, **41**, 1069–1091.

Dalrymple, R.W. *et al.* 1990. *Sedimentology*, **37**, 577–612.

Dalrymple, R.W. *et al.* 1991. In: *Clastic Tidal Sedimentology* (eds Smith, D.G. *et al.*), pp. 137–160. Memoir 16, Canadian Society of Petroleum Geologists.

Dalrymple, R.W. *et al.* 1992. *Journal of Sedimentary Petrology*, **62**, 1130–1146.

Dalrymple, R.W. *et al.* (eds) 1994. *Incised-Valley Systems: Origin and Sedimentary Sequences*. Special Publication 51, Society of Economic Paleontologists and Mineralogists.

Dalrymple, R.W. *et al.* 2006. *Sedimentology*, **53**, 693–696.

Damuth, J.E. *et al.* 1988. *Bulletin of the American Association of Petroleum Geologists*, **72**, 885–911.

Darmadi, Y. *et al.* 2007. *Journal of Sedimentary Research*, **77**, 225–238.

Darrah, P.R. 1993. *Plant Soil*, **155**, 1–20.

Dart, C. *et al.* 1994. *Marine and Petroleum Geology*, **11**, 545–560.

Das, A. *et al.* 2005. *Geochimica et Cosmochimica Acta*, **69**, 2067–2084.

Dashtgard, S.E. *et al.* 2006. *Sedimentology*, **53**, 279–296.

Davidson, J.F. *et al.* 1977. *Annual Reviews of Fluid Mechanics*, **9**, 55–86.

Davidson-Arnott, R.G.D. & Greenwood, B. 1974. *Journal of Sedimentary Petrology*, **44**, 698–704.

Davidson-Arnott, R.G.D. & Greenwood, B. 1976. In: *Beach and Nearshore Sedimentation* (eds Davis, K.S. & Ethington, R.L.), pp. 149–68. Special Publication 24, Society of Economic Paleontologists and Mineralogists.

Davidson-Arnott, R.G.D. & Pember, G.F. 1980. In: *The Coastline of Canada* (ed. McCann, S.B.), pp. 417–428. Paper 80-10, Geological Society of Canada.

Davies, C.E. *et al.* 2005. In: *River Deltas: Concepts, Models, and Examples* (eds Giosan, L. & Bhattacharya, J.P.), pp. 207–230. Special Publication 83, Society of Economic Paleontologists and Mineralogists.

Davies, G.R. 1970. *Memoir of the American Association of Petroleum Geologists*, **13**, 169–205.

Davies, P.J. *et al.* 1978. *Sedimentology*, **25**, 703–730.

Davies, R. *et al.* 2000. In: *Middle East Models of Jurassic/Cretaceous Carbonate Systems* (eds Alsharan, A. & Scott, R.), pp. 273–286. Special Publication 69, Society of Economic Paleontologists and Mineralogists.

Davies-Vollum, K-S & Smith, N.D. 2008. *Journal of Sedimentary Research*, **78**, 683–692.

Davis, K.S.& Ethington, R.L. (eds) 1976. *Beach and Nearshore Sedimentation.* Special Publication 24, Society of Economic Paleontologists and Mineralogists.

Davis, R.A. (ed.) 1985. *Coastal Sedimentary Environments*, 2nd edn. Springer, New York.

Davis, R.A. & Flemming, B.W. 1995. In: *Tidal Signatures in Modern and Ancient Sediments* (eds Flemming B.C. & Bartoloma, A.), pp. 121–132. Special Publication 24, International Association of Sedimentologists.

Dawson, A.G. 1992. *Ice Age Earth.* Routledge, London.

Dawson, A.G. *et al.* 1993. In: *Tsunamis in the World* (ed. Tinti, S.), pp. 31–42. Kluwer, Dordrecht.

Day, J.W. *et al.* 2007. *Science*, **315**, 1679–1684.

Dean, W. *et al.* 2002. *Eos, Transactions American Geophysical Union*, **83/9**, 86–91.

De Baar, H.J.W. *et al.* 1995. *Nature*, **373**, 412–415.

De Batist, M.& Jacobs, P. (eds). 1996. *Geology of Siliciclastic Shelf Seas.* Special Publication 117, Geological Society of London.

De Boer, P.L.& Smith, D.G. (eds) 1994a. *Orbital Forcing and Cyclic Sequences.* Special Publication 19, International Association of Sedimentologists.

De Boer, P.L. & Smith, D.G. 1994b. In: *Orbital Forcing and Cyclic Sequences* (eds de Boer, P.L. & Smith, D.G.), pp. 1–14. Special Publication 19, International Association of Sedimentologists.

De Boer, P.L. & Wonders, A.A.H. 1984. In: *Milankovich and Climate* (eds Berger, A. *et al.*), Part 1, pp. 177–190. Reidel, Dordrecht.

De Boer, P.L. *et al.* 1989. *Journal of Sedimentary Petrology*, **59**, 912–921.

De Boer, P.L. *et al.* 1991. *Basin Research*, **3**, 623–678.

DeCelles, P.G. 1987. *Journal of Sedimentary Petrology*, **57**, 250–264.

DeCelles, P.G. 1988. *Geology*, **16**, 1039–1043.

DeCelles, P.G. 2004. *American Journal of Science*, **304**, 105–168.

DeCelles, P.G. & Cavazza, W. 1999. *Bulletin of the Geological Society of America Bulletin*, **111**, 1315–1334.

DeCelles, P.G. & Giles, K.A. 1996. *Basin Research*, **8**, 105–123.

DeCelles, P.G. *et al.* 1983. *Journal of Sedimentary Petrology*, **53**, 629–642.

DeCelles, P.G. *et al.* 1991. *Bulletin of the Geological Society of America*, **103**, 1458–1475.

DeCelles, P.G. *et al.* 1998a. *Tectonics*, **17**, 741–765.

DeCelles, P.G. *et al.* 1998b. *Bulletin of the Geological Society of America*, **110**, 2–21.

DeCelles, P.G. *et al.* 2001. *Tectonics*, **20**, 487–509.

DeCelles, P.G. *et al.* 2007. *Earth and Planetary Science Letters*, **253**, 389–401.

DeDekker, P. 1988. In: *Lacustrine Petroleum Source Rocks* (eds Fleet, A.J. *et al.*), pp. 45–58. Special Publication 40, Geological Society of London.

Deelman, J.C. 1978. *Journal of Sedimentary Petrology*, **48**, 471–472.

Degens, E.T.& Ross, D.A. (eds) 1974. *The Black Sea Geology, Chemistry and Biology.* Memoir 20, American Association of Petroleum Geologists.

Degens, E.T. & Stoffers, P. 1980. *Journal of the Geological Society of London*, **137**, 131–138.

Deibert, J.E. *et al.* 2003. *Journal of Sedimentary Research*, **73**, 546–558.

Delaney, C. 2007. In: *Glacial Sedimentary Processes and Products* (eds Hambrey, M. *et al.*), pp. 149–163. Special Publication 39, International Association of Sedimentologists.

Della Porta, G. *et al.* 2004. *Sedimentology*, **51**, 267–295.

Deloffre, J. *et al.* 2006. *Marine Geology*, **235**, 151–164.

De Lurio, J.L. & Frakes, L.A. 1999. *Geochimica et Cosmochimica Acta*, **63**, 1039–1048.

Demaison, G.J. & Moore, G.T. 1980. *Bulletin of the American Association of Petroleum Geologists*, **64**, 1179–1209.

Demicco, R.V. & Hardie, L.A. 2002. *Journal of Sedimentary Research*, **72**, 849–857.

Demicco, R.V. *et al.* 2005. *Geology*, **33**, 877–880.

De Mowbray, T. & Visser, M.J. 1984. *Journal of Sedimentary Petrology*, **54**, 811–824.

Denny, M.W. 1993. *Air and Water: The Biology and Physics of Life's Media.* Princeton University Press, Princeton, NJ.

DePloey, J. *et al.* 1991. *Earth Surface Processes and Landforms*, **16**, 339–409.

Deptuck, M.E. *et al.* 2008. *Sedimentology*, **55**, 869–898.

De Putter, T. 1998. *Geophysical Research Letters*, **25**, 3193–3196.

De Raaf, J.F.M. *et al.* 1977. *Sedimentology*, **24**, 451–483.

Derbyshire, E. & Owen, L.A. 1990. In: *Alluvial Fans: A Field Approach* (eds Rachocki, A.H. & Church, M.), pp. 27–53. Wiley, Chichester.

De Ruig, M.J. & Hubbard, S.M. 2006. *Bulletin of the American Association of Petroleum Geologists*, **90**, 735–752.

De Visser, J.P. *et al.* 1989. *Palaeogeography, Palaeoclimatology, Palaeoecology*, **69**, 45–66.

Dewey, J.F. 1982. *Journal of the Geological Society of London*, **139**, 371–414.

Dewey, J.F. & Pitman, W.C. 1998. In: *Eustasy and the Tectonostratigraphic Evolution of Northern South America*. Special Publication 58, Society of Economic Paleontologists and Mineralogists.

Dewey, J. F. *et al.* 1988. *Philosophical Transactions of the Royal Society of London*, Series A, **327**, 379–413.

Diaz-Molina, M. 1993. In: *Alluvial Sedimentation* (eds Marzo, M. & Puigdefabregas, C.), pp. 115–131. Special Publication 17, International Association of Sedimentologists.

Dickinson, J.A. & Wallace, M.W. 2009. *Sedimentology*, **56**, 547–565.

Dickinson, W.R. 1976. *Plate Tectonic Evolution of Sedimentary Basins*. Continuing Education Course Notes Series 1, American Association of Petroleum Geologists.

Dickinson, W.R. 1995. In: *Tectonics of Sedimentary Basins* (eds Busby, C. & Ingersoll, R.), pp. 221–262. Blackwell Science, Oxford.

Dickman, M. 1985. *Sedimentology*, **32**, 109–118.

Dickson, J.A.D. 1995. *Geology*, **23**, 535–538.

Dickson, J.A.D. 2002. *Science*, **298**, 1222–1224.

Dickson, J.A.D. 2004. *Journal of Sedimentary Research*, **74**, 355–365.

Dickson, R.R. & McCave, I.N. 1986. *Deep Sea Research*, **33**, 791–818.

Diem, B. 1985. *Sedimentology*, 32, 685–704.

Dietrich, W.E. 1982. *Water Resources Research*, **18**, 1615–1626.

Dill, R.F. *et al.* 1986. *Nature*, 324, 55–58.

Dix, G.R. & Kyser, K. 2000. *Sedimentology*, **47**, 421–434.

Dix, G.R. *et al.* 2005. *Journal of Sedimentary Research*, **75**, 665–678.

Dixon, T.H. *et al.* 2006. *Nature*, **441**, 587–588.

Doake, C.S.M. & Wolff, E.W. 1985. *Nature*, **314**, 255–257.

Domack, E.W. & Ishman, S. 1993. *Bulletin of the Geological Society of America*, **105**, 1175–1189.

Domingues, C.M. *et al.* 2008. *Nature*, **453**, 1090–1093.

Donaldson, W.S. *et al.* 1999. *Sedimentology*, **46**, 1159–1182.

Donovan, R.N. & Foster, R.J. 1972. *Journal of Sedimentary Petrology*, **42**, 309–317.

Donselaar, M.E. & Schmidt, J.M. 2005. *Sedimentology*, **52**, 1021–1042.

Dorsey R. & Martín-Barajas. 1999. *Basin Research*, **11**, 205–221.

Dorsey, R.J. *et al.* 1997. *Geology*, **25**, 679–682.

Dott, R. & Bourgeois, J. 1982. *Bulletin of the Geological Society of America*, **93**, 663–680.

Dove, P.M. 1995. In: *Chemical Weathering Rates of Silicate Minerals* (eds White, A.F. & Branfield, S.L.), Chapter 6. American Mineralogical Society, Washington, DC.

Dove, P.M. & Elston, S.F. 1992. *Geochimica et Cosmochimica Acta*, **56**, 4147–4156.

Dove, P.M. & Hochella, M.F. 1993. *Geochimica et Cosmochimica Acta*, **57**, 705–714.

Dowdeswell, J.A. & Sharp, M.J. 1986. *Sedimentology*, **33**, 699–710.

Dowdeswell, J.A.& Scourse, J.D. (eds). 1990a. *Glacimarine Environments*. Special Publication 53, Geological Society of London.

Dowdeswell, J.A. & Scourse, J.D. 1990b. In: *Glacimarine Environments* (eds Dowdeswell, J.A. & Scourse, J.D.), pp. 1–13. Special Publication 53, Geological Society of London.

Dowdeswell, J.A. *et al.* 1994. *Sedimentology*, **41**, 21–35.

Dowdeswell, J.A. *et al.* 2000. *Sedimentology*, **47**, 557–576.

Drake, D.E. & Gorsline, D.S. 1973. *Bulletin of the Geological Society of America*, **84**, 3949–3968.

Drake, D.E. *et al.* 1972. In: *Shelf Sediment Transport* (eds Swift, D.J.P. *et al.*), pp. 307–31. Dowden, Hutchinson and Ross, Stroudsberg.

Drake, D.E. *et al.* 1989. *Journal of Geophysical Research*, **94**, 3139–3158.

Draut, A.E. & Clift, P.D. 2006. *Journal of Sedimentary Research*, **76**, 493–514.

Dreimanis, A. 1989. In: *Genetic Classification of Glacigenic Deposits* (eds Goldthwait, R.P. & Matsch, C.L.), pp. 17–83. Balkema, Rotterdam.

Drever, J.I. 1988. *The Geochemistry of Natural Waters*. Prentice Hall, New Jersey.

Drever, J.I. & Clow, D.W. 1995. In: *Chemical Weathering Rates of Silicate Minerals* (eds White, A.F. & Branfield, S. L.), Chapter 10. American Mineralogical Society, Washington, DC.

Drever, J.I. *et al.* 1988. In: *Chemical Cycles in the Evolution of the Earth* (eds Gregor, C. B. et al.), pp. 17–54. Wiley, New York.

Drexler, T. M. *et al.* 2006. *Journal of Sedimentary Research*, **76**, 839–853.

Droxler, A.W. & Schlager, W. 1985. *Geology*, **13**, 799–802.

Drummond, C. & Sheets, H. 2001, *Journal of Sedimentary Research*, **71**, 621–627.

Drummond, C.N. & Wilkinson, B.H. 1993. *Journal of Sedimentary Petrology*, **63**, 369–377.

Drummond, C.N. *et al.* 1996. *Sedimentology*, **43**, 677–689.

Drzewiecki, P.A. & Simo, J.A. 2000. *Sedimentology*, **47**, 471–495.

Ducassou, E. *et al.* 2009. *Sedimentology*, **56**, 2061–2090.

Duke, W.L. 1990. *Journal of Sedimentary Petrology*, **60**, 870–883.

Duller, R.A. *et al.* 2008. *Sedimentology*, **55**, 939–964.

Dumas, S. *et al.* 2005. *Journal of Sedimentary Research*, **75**, 501–513.

Dunbar, D.B. & Barrett, P.J. 2005. *Sedimentology*, **52**, 253–269.

Dunbar, G.B. & Dickens, G.R. 2003. *Sedimentology*, **50**, 1061–1077.

Dupont-Nivet, G. *et al.* 2008. *Geology*, **36**, 987–990.

Dyer, K.R. 1989. *Journal of Geophysical Research*, **94C**, 14327–14339.

Dyer, K.R. 2000. *Continental Shelf Research*, **20**, 1039–1060.

Dyer, K.R. *et al.* 2004. *Estuarine, Coastal and Shelf Science*, **59**, 237–248.

Dyer, K.T. *et al.* 2000. *Estuarine, Coastal and Shelf Science*, **50**, 607–625.

Eberli, G.P. & Ginsburg, R.N. 1987. *Geology*, **15**, 75–79.

Eberli, G.P. & Ginsburg, R.N. 1989. In: *Controls on Carbonate Platform and Basin Development* (eds Crevello, P. D. *et al.*) pp. 339–351. Special Publication 44, Society of Economic Paleontologists and Mineralogists.

Ebinger, C.J. *et al.* 1987. *Tectonophysics*, **141**, 215–235.

Eden, D.J. & Eyles, N. 2001. *Sedimentology*, **48**, 1079–1102.

Edmond, J.M. 1992. *Science*, **258**, 1594–1597.

Edwards, D.A. *et al.* 1994. *Sedimentology*, **41**, 437–461.

Egenhoff, S.O. *et al.* 1999. *Sedimentology*, **46**, 893–912.

Eggleton, R.A. 1986. In: *Rates of Chemical Weathering of Rocks and Minerals* (eds Colman, S.M. & Dethier, D.P.), pp. 21–40. Academic Press, New York.

Ehlmann, B.L. *et al.* 2008. *Nature Geoscience*, **1**, 355–358.

Einarsson, P. *et al.* 1997. *Eos, Transactions American Geophysical Union*, **78**, 369–375.

Einsele, G. 1992. *Sedimentary Basins: Evolution, Facies and Sediment Budget*. Springer, Berlin.

Eisma, D. *et al.* 1997. In: *Cohesive Sediments* (eds Burt, N. *et al.*), pp. 17–44. Wiley, New York.

Eitel, B. 1997. *Zeitschrift fur Geomorphologie*, **111**, 73–95.

Eittreim, S. *et al.* 1975. *Journal of Geophysical Research*, **80**, 5061–5067.

Elbelrhiti, H. *et al.* 2005. *Nature*, **437**, 720–723.

Ellegaard, C. *et al.* 1998. *Nature*, **392**, 767–768.

El-Tabakh, M. *et al.* 1997. *Sedimentology*, **44**, 767–790.

Elwood Madden, M.E. *et al.* 2004. *Nature*, **431**, 821–823.

Ely, L.L. *et al.* 1996. *Bulletin of the Geological Society of America*, **108**, 1134–1148.

Embley, R.W. 1976. *Geology*, **4**, 371–374.

Embrey, A.F. & Klovan. 1971. *Bulletin of Canadian Petroleum Geology*, **19**, 730–781.

Emeis, K-C. & Weissert, H. 2009. *Sedimentology*, **56**, 247–266.

Emery, D. & Myers, K.J. 1996. *Sequence Stratigraphy*. Blackwell Science, Oxford.

Engebretson, D.C. *et al.* 1992. *GSA Today*, **2**, 93.

Engels, M. *et al.* 2004. *Journal of Sedimentary Research*, **74**, 255–269.

Engelstadter, S. *et al.* 2006. *Earth Science Reviews*, **22**, 1835–1857.

England, P.C. & Houseman, G.A. 1986. *Journal of Geophysical Research*, **91**, 3664–3676.

Enos, P. & Perkins, R.D. 1979. *Bulletin of the Geological Society of America*, **90**, 59–83.

*Eos*. 1992. Andrew shortens lifetime of Louisiana barrier islands. *Eos, Transactions American Geophysical Union*, **73/47**, 505.

EPICA community members. 2004. *Nature*, **429**, 623–628.

Eriksson, K.A. 1977. *Sedimentary Geology*, **18**, 223–244.

Escutia, C. *et al.* 2000. *Journal of Sedimentary Research*, **70**, 84–93.

Ethridge, F.G. & Westcott, W.A. 1984. In: *Sedimentology of Gravels and Conglomerates* (eds Koster, E.H. & Steel, R. J.), pp. 217–235. Memoir 10, Canadian Society of Petroleum Geologists.

Ethridge, F.G. *et al.* 1998. In: pp. 17–29. Special Publication 59, Society of Economic Paleontologists & Mineralogists.

Eugster, H.P. & Hardie, L.A. 1975. *Bulletin of the Geological Society of America*, **86**, 319–334.

Eugster, H.P. & Hardie, L.A. 1978. In: *Lakes: Chemistry, Geology, Physics* (ed. Lerman, A.), pp. 237–334. Springer, New York.

Evans, G. 1965. *Quarterly Journal of the Geological Society of London*, **121**, 209–245.

Evans, G. 1995. *Cuadernos de Geologia Iberica*, **19**, 61–96.

Evans, G. *et al.* 1969. *Sedimentology*, **12**, 145–159.

Evans, G. *et al.* 1995. *Marine Geology*, **128**, 127–136.

Everts, A.J.W. *et al.* 1999. *Sedimentology*, **46**, 261–278.

Everts, C.H. 1987. In: *Sea Level Fluctuation and Coastal Evolution* (eds Nummedal, D. *et al.*), pp. 49–57. Special Publication 41, Society of Economic Paleontologists and Mineralogists.

Eyles, C.H. *et al.* 1993. *Sedimentology*, **40**, 1–25.

Eyles, C.H. *et al.* 1998. *Sedimentology*, **45**, 121–161.

Eyles, N. & Eyles, C.H. 1989. *Sedimentology*, **36**, 601–620.

Eyles, N. & McCabe, A.M. 1989. *Sedimentology*, **36**, 431–448.

Eyles, N. *et al.* 1985. *Palaeogeography, Palaeoclimatology, Palaeoecology*, **51**, 15–84.

Eyles, N. *et al.* 1988. *Sedimentology*, **35**, 465–480.

Faber, T.E. 1995. *Fluid Dynamics for Physicists*. Cambridge University Press, Cambridge.

Falco, R.E. 1977. *Physics of Fluids*, **20**, 124–132.

Falk, P.D. & Dorsey, R.J. 1998. *Sedimentology*, **45**, 331–349.

Fan, H. *et al.* 2006. *Geomorphology*, **74**, 124–136.

Farre, J.A. *et al.* 1983. In: *The Shelfbreak: Critical Interface on Continental Margin* (eds Stanley, D.J. & Moore, G.T.), pp. 25–39. Special Publication 33, Society of Economic Paleontologists and Mineralogists.

Faugères, J.-C. *et al.* 1984. *Geology*, **12**, 296–300.

Faugères, J.-C. *et al.* 1985a. *Bull. Inst. Geol. Bassin Aquitaine*, **37**, 229–258.

Faugères, J.-C. *et al.* 1985b. *Bull. Inst. Geol. Bassin Aquitaine*, **37**, 259–287.

Faugères, J.-C. *et al.* 1993. *Sedimentary Geology*, **82**, 189–203.
Faugères, J.-C. *et al.* 1998. *Sedimentary Geology*, **115**, 81–110.
Faugères, J.-C. *et al.* 1999. *Marine Geology*, **162**, 1–38.
Felix, M. 2002. *Sedimentology*, **49**, 397–419.
Felix, M. & Peakall, J. 2006. *Sedimentology*, **53**, 107–123.
Felix, M. *et al.* 2005. *Sedimentary Geology*, **179**, 31–47.
Felix, M. *et al.* 2006. *Journal of Sedimentary Research*, **76**, 382–387.
Fenton, L.K. *et al.* 2007. *Nature*, **446**, 646–649.
Ferdelman, T.G. *et al.* 2006. *Scientific Drilling*, **2**, 11–16.
Ferentinos, G. *et al.* 1988. *Marine Geology*, **83**, 43–61.
Ferguson, J. *et al.* 1978. *Chemical Geology*, **22**, 285–308.
Ferguson, R. 1987. In: *River Channels: Environments and Process* (ed. Richards, K.S.), pp. 125–158. Blackwell Scientific, Oxford.
Ferguson, R. & Brierley, G.J. 1999. *Sedimentology*, **46**, 627–648.
Ferguson, R.I. & Church, M. 2004. *Journal of Sedimentary Research*, **74**, 933–937.
Ferguson, R. *et al.* 1996. *Geology*, **24**, 179–182.
Fernando, H.J.S. 1991. *Annual Reviews of Fluid Mechanics*, **23**, 455–493.
Ferrier, K.L. & Kirchner, J.W. 2008. *Earth and Planetary Science Letters*, **272**, 591–599.
Field, M.E. 1980. *Journal of Sedimentary Petrology*, **50**, 505–528.
Fielding, C.R. *et al.* 1997. *Palaeogeography, Palaeoclimatology, Palaeoecology*, **135**, 123–144.
Fielding, C.R. *et al.* 1999. In: *Fluvial Sedimentology VI* (eds Smith ND & Rogers, J.), pp. 347–362. Special Publication 28, International Association of Sedimentologists.
Fielding, C.R. *et al.* 2003. *Sedimentary Geology*, **157**, 291–301.
Fielding, C.R. *et al.* 2005a. In: *River Deltas: Concepts, Models, and Examples* (eds Giosan, L. & Bhattacharya, J.P.), pp. 467–496. Special Publication 83, Society for Economic Palaeontologists and Mineralogists.
Fielding, C.R. *et al.* 2005b. *Journal of Sedimentary Research*, **75**, 55–66.
Fielding, C.R. *et al.* 2006. *Journal of Sedimentary Research*, **76**, 411–428.
Fildani, A. & Normark, W.R. 2004. *Marine Geology*, **206**, 199–223.
Fildani, A. *et al.* 2006. *Sedimentology*, **53**, 1265–1287.
Fildani, A. *et al.* 2008. *Basin Research*, **20**, 305–331.
Filgueira-Rivera, M. *et al.* 2007. *Sedimentology*, **54**, 905–919.
Fineberg, J. 1996. *Nature*, **382**, 763–764.
Finkelstein, K. & Ferland, M.A. 1987. In: *Sea Level Fluctuation and Coastal Evolution* (eds Nummedal, D. *et al.*), pp. 145–155. Special Publication 41, Society of Economic Paleontologists and Mineralogists.
Fischer, A.G. 1964. In: *Symposium on Cyclic Sedimentation* (ed. Merriam, D.F.), pp. 107–149. Bulletin 169, Geological Survey of Kansas.
Fischer, A.G. 1975. In: *Tidal Deposits* (ed. Ginsburg, R.N.), pp. 235–242. Springer, New York.
Fischer, U.H. & Clarke, G.K.C. 1994. *Journal of Glaciology*, **40**, 97–106.
Fishbane, P.M. *et al.* 1993. *Physics for Scientists and Engineers Extended Version*. Prentice-Hall.
Fisk, H.N. 1944. *Geological Investigations of the Alluvial Valley of the Lower Mississippi River*. Mississippi River Commission, Vicksberg, MS.
Fisk, H.N. 1947. *Fine-grained Alluvial Deposits and their Effects on Mississippi River Activity*. Mississippi River Commission, Vicksburg, MS.
Fisk, H.N. 1959. Padre Island and the Laguna Madre flats, coastal South Texas. *National Academy of Science-National Research Council 2nd Coastal Geography Conference*, 103-SI.
Fisk, H.N. *et al.* 1954. *Journal of Sedimentary Petrology*, **24**, 76–99.
Fitzgerald, D.M. 1988. In: *Hydrodynamics and Sediment Dynamics of Tidal Inlets* (eds Aubrey, D.G. & Weishar, L.), pp. 186–225. Springer, Berlin.
Fitzgerald, D.M. 1993. In: *Formation and Evolution of Multiple Tidal Inlets* (eds Aubrey, J. & Geise, F.J.), pp. 1–61. Monograph, American Geophysical Union.
Fitzgerald, D.M. & van Heteren, S. 1999. *Sedimentology*, **46**, 1083–1108.
Fitzgerald, D.M. *et al.* 2004. *Sedimentology*, **51**, 1157–1178.
Fitzgerald, D.M. *et al.* 2008. *Annual Reviews of Earth and Planetary Science*, **36**, 601–647.
Flament, P.J. *et al.* 1996. *Nature*, **383**, 610–613.
Flather, R.A. 1984. *Quarterly Journal of the Royal Meteoreological Society*, **110**, 591–612.
Flecker, R. & Ellam, R.M. 2006. *Sedimentary Geology*, **188**, 189–203.
Fleet, A.J. *et al.* (eds) 1988. *Lacustrine Petroleum Source Rocks*. Special Publication 40, Geological Society of London.
Flemings, P.B. & Jordan, T.E. 1990. *Geology*, **18**, 430–435.
Flemming, B.W. 2007. *Sedimentary Geology*, **202**, 425–435.
Flemming, B.W.& Bartholema, A. (eds) 1995. *Tidal Signatures in Modern and Ancient Sediments*. Special Publication 24, International Association of Sedimentologists.
Fletcher, C.H. *et al.* 1990. *Bulletin of the Geological Society of America*, **102**, 283–297.
Flocks, J.G. *et al.* 2006. *Journal of Sedimentary Research*, **76**, 429–443.
Flood, R.D. *et al.* 1979. *Marine Geology*, **32**, 311–334.
Flood, R.D. *et al.* 1991. In: *Seismic Facies and Sedimentary Processes of Submarine Fans and Turbidite Systems* (eds Weimer, P. & Link, M.H.), pp. 415–434. Springer, Berlin.
Flood, R.D. *et al.* 2009. *Sedimentology*, **56**, 807–839.

Flugel, E. 2004. *Microfacies of Carbonate Rocks*. Springer.

Folk, R.L. 1955. *American Mineralogist*, **40**, 356–357.

Folk, R.L. 1962. In: *Classification of Carbonate Rocks* (ed. Ham, W.E.), pp. 62–84. Memoir, American Association of Petroleum Geologists.

Folk, R.L. 1971. *Sedimentology*, **16**, 5–54.

Folk, R.L. 1974. *Journal of Sedimentary Petrology*, **44**, 40–53.

Folk, R.L. & Patton, E.B. 1982. *Zeitschrift für Geomorphologie*, **26**, 17–32.

Follmi, K.B. 1995. *Geology*, **23**, 859–862.

Follmi, K.B. 2008. *Sedimentology*, **55**, 1029–1051.

Ford, M. *et al.* 1999. *Basin Research*, **11**, 315–336.

Forristall, G.Z. *et al.* 1977. *Journal of Physical Oceanography*, **7**, 532–546.

Forterre, Y. & Pouliquen, O. 2008. *Annual Reviews of Fluid Mechanics*, **40**, 1–24.

Foster, G.L. & Vance, D. 2006. *Nature*, **444**, 918–921.

Fountain, A.G. & Walder, J.S. 1998. *Reviews of Geophysics*, **36**, 299–328.

Fountain, A.G. *et al.* 2005. *Nature*, **433**, 618–621.

Fournier, F. 1960. *Climat et Erosion: la Relation entre l'Erosion du Sol par l'Eau et les Precipitations Atmospheriques*. Presses Universitaire de France, Paris.

Fowler, A.C. 2000. In: *Deformation of Glacial materials* (eds Maltman, A.J.), pp. 307–319. Special Publications 176, Geological Society, London.

Frakes, L.A. *et al.* 1992. *Climate Modes of the Phanerozoic*. Cambridge University Press, Cambridge.

Francis, J.E. 1984. *Palaeogeography, Palaeoclimatology, Palaeoecology*, **48**, 285–307.

Francis, J.M. *et al.* 2007. *Journal of Sedimentary Research*, **77**, 572–586.

Francis, J.R.D. 1969. *A Textbook of Fluid Mechanics*. Edward Arnold, London.

Francis, J.R.D. 1973. *Philosophical Transactions of the Royal Society, London (A)*, **332**, 443–471.

Fraser, C. *et al.* 2005. *Sedimentology*, **52**, 141–160.

Fraticelli, C.M. 2006. *Journal of Sedimentary Research*, **76**, 1067–1076.

Frazier, D.E. 1967. *Transactions of the Gulf Coast Association of Geological Societies*, **17**, 287–315.

Fredsoe, J. & Deigaard, R. 1992. *Mechanics of Coastal Sediment Transport*. World Scientific, Singapore.

Frenz, M. & Henrich, R. 2007. *Sedimentology*, **54**, 391–404.

Friedmann, S.J. 1998. *Journal of Sedimentary Research*, **67**, 792–804.

Friedmann, S.J. & Burbank, D.W. 1995. *Basin Research*, **7**, 109–127.

Friend, P.F. 1978. In: *Fluvial Sedimentology* (ed. Miall, A. D.), pp. 531–542. Memoir 5, Canadian Society of Petroleum Geologists.

Frings, R.M. & Kleinhans, M.G. 2008. *Sedimentology*, **55**, 1145–1171.

Frisia, S. *et al.* 2000. *Journal of Sedimentary Research*, **70**, 1183–1196.

Frostick, L.E. & McCave, I.N. 1979. *Estuarine and Coastal Marine Science*, **9**, 569–576.

Frostick, L.E. & Reid, I. 1986. In: *Sedimentation in the African Rifts* (ed. Frostick, L.E.), pp. 113–125. Special Publication 25, Geological Society of London.

Fryberger, S.G. 1993. In: *Characterisation of Fluvial and Aeolian Reservoirs* (eds North, C.P. & Prosser, D.J.), pp. 167–197. Special Publication 73, Geological Society of London.

Fryberger, S.G. & Ahlbrandt, T.S. 1979. *Zeitschrift fur Geomorphologie*, **23**, 440–450.

Fryberger, S.G. *et al.* 1979. *Journal of Sedimentary Petrology*, **49**, 733–746.

Fryberger, S.G. & Schenk, C.J. 1988. *Sedimentary Geology*, **55**, 1–15.

Fuller, J.G.C.M. & Porter, J.W. 1969. *Bulletin of the American Association of Petroleum Geologists*, **53**, 909–926.

Furbish, D.J. 1997. *Fluid Physics in Geology*. Oxford University Press.

Furrer, G. & Stumm, W. 1986. *Geochimica et Cosmochimica Acta*, **50**, 1847–1860.

Gabel, S.L. 1993. *Sedimentology*, **40**, 237–269.

Gábris, G. & Nagy, B. 2005. In: *Alluvial Fans: Geomorphology, Sedimentology, Dynamics* (eds Harvey, A.M. *et al.*), pp. 61–67. Special Publication 251, Geological Society of London.

Gadow, S. & Reineck, H.E. 1969. *Senckenbergiana Maritima*, **3**, 103–133.

Gaffin, S. 1987. *American Journal of Science*, **287**, 596–611.

Gaína, C. *et al.* 2003. *Geophysical Research Abstracts*, **5**: 04842.

Gale, S.J. & Hoare, P.G. 2007. In: *Glacial Sedimentary Processes and Products* (eds Hambrey, M. *et al.*), pp. 203–234. Special Publication 39, International Association of Sedimentologists.

Galloway, W.E. 1975. In: *Deltas, Models for Exploration* (ed. Broussard, M.L.), pp. 87–98. Houston Geological Society.

Galloway, W.E. 1980. pp. 59–69. Memoir 38, New Mexico Bureau of Mines and Mineral Resources.

Galloway, W.E. 2002. *Journal of Sedimentary Research*, **72**, 476–490.

Galvin, C.J. 1968. *Journal of Geophysical Research*, **73**, 3651–3659.

Galy, A. *et al.* 1999. *Geochimica et Cosmochimica Acta*, **63**, 1905–1925.

Gamberi, F. & Marani, M. 2008. *Sedimentology*, **55**, 1889–1903.

Ganeshram, R.S. *et al.* 1995. *Nature*, **376**, 755–758.

Gani, M.R. & Bhattacharya, J.P. 2005. In: *River Deltas: Concepts, Models, and Examples* (eds Giosan, L. & Bhattacharya, J.P.), pp. 31–48. Special Publication

38, Society of Economic Paleontologists and Mineralogists.
Gani, M.R. & Bhattacharya, J.P. 2007. *Journal of Sedimentary Research*, **77**, 284–302.
Garcia-Castellanos, D. 2002. *Basin Research*, **14**, 89–104.
Garcia-Castellanos, D. *et al.* 2009. *Nature*, **462**, 778–781.
García del Cura, M. 2001. *Sedimentology*, **48**, 897–915.
García-García, F. *et al.* 2009. *Journal of Sedimentary Research*, **79**, 302–315.
García-Hidalgo *et al.* 2007. *Sedimentology*, **54**, 1245–1271.
Gardulski, A.F. *et al.* 1990. *Sedimentology*, **37**, 727–743.
Garrels, R.M. & Christ, C.L. 1965. *Solutions, Minerals and Equilibrium*. Harper and Row, New York.
Garvie, L.A.J. *et al.* 2008. *Geology*, **36**, 215–218.
Garzione, C.N. 2008. *Geology*, **36**, 1003–1004.
Garzione, C.N. *et al.* 2000. *Geology*, **28**, 339–342.
Gasse, F. *et al.* 1990. *Nature*, **346**, 141–146.
Gastaldo, R.A. *et al.* 1995. In: *Coarse-Grained Deltas* (eds Colella, A. & Prior, D.B.), pp. 193–211. Special Publication 10, International Association of Sedimentologists.
Gautret, P. *et al.* 2004. *Journal of Sedimentary Research*, **74**, 462–478.
Gawthorpe, R.L. & Colella, A. 1990. In: *Coarse-Grained Deltas* (eds Colella, A. & Prior, D.B.), pp. 113–128. Special Publication 10, International Association of Sedimentologists.
Gawthorpe, R.L. & Leeder, M.R. 2000. *Basin Research*, **12**, 195–218.
Gawthorpe, R.L. *et al.* 1993. In: *Characterization of Fluvial and Aeolian Reservoirs* (eds North, C.P. & Prosser, D.J.), pp. 421–432. Geological Society of London, Special Publication 73.
Gawthorpe, R.L. *et al.* 1994. *Marine and Petroleum Geology*, **11**, 642–658.
Gawthorpe, R.L. *et al.* 1997. *Geology*, **25**, 795–798.
Gawthorpe, R.L. *et al.* 2003. *Sedimentology*, **50**, 169–185.
Gee, M.J.R. *et al.* 1999. *Sedimentology*, **46**, 317–335.
Gee, M.J.R. *et al.* 2001. *Sedimentology*, **48**, 1389–1411.
Gee, M.J.R. *et al.* 2006. *Journal of Sedimentary Research*, **76**, 9–19.
Gee, M.J.R. *et al.* 2007. *Journal of Sedimentary Research*, **77**, 433–446.
Geissler, P.E. & Haberle, R.M. 2007. *Nature*, **446**, 646–649.
3–George, A.D. *et al.* 1997. *Sedimentology*, **44**, 843–867.
George, G. & Berry, J.K. 1993. In: *Characterisation of Fluvial and Aeolian Reservoirs* (eds North, C.P. & Prosser, D.J.), pp. 291–320. Special Publication 73, Geological Society of London.
Gerard, R. 1978. Journal of the Hydraulic Division, American Society of Civil Engineers, **104**, 755–773.
Gerdes, G. *et al.* 2000. *Sedimentology*, **47**, 279–308.
Gibbard, P. 1980. *Boreas*, **9**, 71–85.
Gibbard, P.L. 1988. Philosophical Transactions of the Royal Society, London, **318**, 559–602.
Gibbs, R.J. *et al.* 1971. *Journal of Sedimentary Petrology*, **41**, 7–18.
Gibling, M.R. 2006. *Journal of Sedimentary Research*, **76**, 731–770.
Gibling, M.R. & Bird, D.J. 1994. *Bulletin of the Geological Society of America*, **106**, 105–117.
Gibling, M.R. *et al.* 1998. *Sedimentology*, **45**, 595–619.
Gienapp, H. 1973. *Senckenbergiana Maritima*, **5**, 135–151.
Gile, L.H. *et al.* 1981. *Soils and Geomorphology in the Basin and Range Area of Southern New Mexico—Guidebook to the Desert Project*. Memoir 39, New Mexico Bureau of Mines and Mineral Resources, Socorro, NM.
Gill, R. 1989. *Chemical Fundamentals of Geology*. Chapman & Hall, London.
Gill, W.D. & Kuenen, P.H. 1958. *Quarterly Journal of the Geological Society of London*, **113**, 441–460.
Gilli, A. *et al.* 2005. *Sedimentology*, **52**, 1–23.
Ginsburg, R.N. & James, N.P. 1974. In: *The Geology of Continental Margins* (eds Burk, C.A. & Drake, C.L.), pp. 137–155. Springer, Berlin.
Ginsburg, R.N. *et al.* 1991. *Journal of Sedimentary Petrology*, **61**, 976–987.
Giosan, L.& Bhattacharya, J.P. (eds). 2005. *River Deltas: Concepts, Models, and Examples*. Special Publication 83, Society of Economic Paleontologists and Mineralogists.
Giosan, L. *et al.* 2005. In: *River Deltas: Concepts, Models, and Examples* (eds Giosan, L. & Bhattacharya, J.P.), pp. 393–412. Special Publication 83, Society of Economic Paleontologists and Mineralogists.
Girard, J.-P. *et al.* 2000. *Geochimica et Cosmochimica Acta*, **64**, 409–426.
Giraudi, C. 2005. *Palaeogeography, Palaeoclimatology, Palaeoecology*, **218**, 161–173.
Gischler, E. & Lomando, A.J. 1999. *Journal of Sedimentary Research*, **69**, 747–763.
Gladstone, C. 1998. *Sedimentology*, **45**, 833–843.
Gleadow, A.J.W. *et al.* 1986. *Contributions to Mineralogy and Petrology*, **94**, 405–415.
Glenn, C.R. *et al.* 1994. *Eclogae Geologica Helvetica*, **87**, 747–788.
Glennie, K.W. 1970. *Desert Sedimentary Environments*. Elsevier, Amsterdam.
Glennie, K.W. 1982. *Sedimentary Geology*, **34**, 245–265.
Glennie, K.W. 1986. In: *Introduction to the Petroleum Geology of the North Sea*. (ed. Glennie, K.W.), pp. 63–86, Blackwell Science, Oxford.
Glennie, K.W. & Buller, A.T. 1983. *Sedimentary Geology*, **35**, 43–81.
Gloor, M. *et al.* 1994. *Hydrobiologia*, **284**, 59–68.
Goddéris, Y. *et al.* 2006. *Geochimica et Cosmochimica Acta*, **70**, 1128–1147.
Gohain, K. & Parkash, B. 1990. In: *Alluvial Fans: A Field Approach* (eds Rachocki, A.H. & Church, M.), pp. 151–178. Wiley, New York.
Goldhammer, R.K. *et al.* 1990. *Bulletin of the Geological Society of America*, **102**, 535–562.

Gole, C.V. & Chitale, S.V. 1966. *Journal of the Hydraulics Division, American Society of Civil Engineers*, **92**, 111–126.

Gomez, B. *et al.* 1995. *Geology*, **23**, 963–966.

Gomez, E.A. & Amos, C.L. 2005. *Sedimetology*, **52**, 183–189.

González-Muñoz, M.T. *et al.* 2000. *Journal of Sedimentary Research*, **70**, 559–564.

Goodall, T.M. *et al.* 2000. *Sedimentology*, **47**, 99–118.

Goolsby, D.A. 2000. *Eos, Transactions American Geophysical Union*, **81/29321**, 326–327.

Gooseff, M.N. *et al.* 2009. *Eos, Transactions American Geophysical Union*, **90/4**, 29–30.

Goudie, A.S. 1985. *Salt Weathering*. Research Paper 33, School of Geography, University of Oxford.

Goudie, A.S. 2008. *Annual Reviews of Earth and Planetary Science*, **36**, 97–119.

Goudie, A.S. & Middleton, N.J. 2006. *Desert Dust in the Global System*. Springer.

Goudie, A.S. *et al.* 1970. *Area*, **4**, 42–48.

Goudie, A.S. *et al.* 2000. *Sedimentology*, **47**, 1011–1021.

Gouw, M. 2007. *Alluvial architecture of the Holocene Rhine–Meuse delta (The Netherlands) and the Lower Mississippi Valley (USA)*. Nederlandse Geografische Studies, 364. Utrecht University.

Gouw, M.J.P. 2008. *Sedimentology*, **55**, 1487–1516.

Gouw, M.J.P. & Berendson, H.J.A. 2007. *Journal of Sedimentary Research*, **77**, 124–138.

Grabemann, I. & Krause, G. 1989. *Journal of Geophysical Research*, **94C**, 14373–14379.

Graf, J.B. *et al.* 1991. *Bulletin of the Geological Society of America*, **103**, 1405–1415.

Graham, D.J. & Hambrey, M.J. 2007. In: *Glacial Sedimentary Processes and Products* (eds Hambrey, M. *et al.*), pp. 235–256. Special Publication 39, International Association of Sedimentologists.

Graham, S.A. *et al.* 2005. *American Journal of Science*, **305**, 101–118.

Grammer, G.M. *et al.* 1993. In: *Carbonate Sequence Stratigraphy* (eds Loucks, R.G. & Sarg, J.F.), pp. 107–132. Memoir 57, American Association of Petroleum Geologists.

Grant, W.D. & Williams, A.J. 1984. *Journal of Physical Oceanography*, **14**, 506–527.

Grantham, B.A. *et al.* 2004. *Nature*, **429**, 749–754.

Grass, A.J. 1970. *Journal of the Hydraulics Division, American Society of Civil Engineers*, **96**, 619–632.

Grass, A.J. 1971. *Journal of Fluid Mechanics*, **50**, 233–255.

Gratz, A.J. *et al.* 1993. *Geochimica et Cosmochimica Acta*, **57**, 491–495.

Gray, T.E. *et al.* 2005. *Sedimentology*, **52**, 467–488.

Gray, W.A. 1968. *The Packing of Solid Particles*. Chapman & Hall, London.

Greb, S.F. & Archer, A.W. 1995. *Journal of Sedimentary Research*, **65**, 96–106.

Grecula, M. *et al.* 2003. *Journal Sedimentary Research*, **73**, 603–620.

Greeley, R. 2002. *Planetary & Space Science*, **50**, 151–155.

Greeley, R. & Iverson, J.D. 1985. *Wind as a Geological Process: on Earth, Mars, Venus and Titan*. Cambridge University Press, Cambridge.

Green, M. 1999. *Sedimentology*, **46**, 427–441.

Green, M.O. *et al.* 1990. *Journal of Geophysical Research*, **C95** 9629–9244.

Greene, D.L. *et al.* 2007. *Journal of Sedimentary Research*, **77**, 139–158.

Greenwood, J.P. & Blake, R.E. 2006. *Geology*, **34**, 953–956.

Greenwood, B. & Sherman, D.J. 1993. *Marine Geology*, **60**, 31–61.

Gregory, K.J. & Walling, D.E. 1973. *Drainage Basin Form and Process*. Edward Arnold, London.

Gretener, B. & Stromquist, L. 1987. *Geografiska Annaler*, **69A**, 139–146.

Griffin, R.A. & Jurinak, J.J. 1973. *Proceedings of the Soil Science Society of America*, **37**, 847–850.

Griffith, L.S. *et al.* 1969. In: *Depositional Environments in Carbonate Rocks* (ed. Friedman, G.H.), pp. 120–137. Special Publication, Society of Economic Paleontologists and Mineralogists.

Gross, T.F. *et al.* 1988. *Nature*, **331**, 518–520.

Grossman, E.E. & Fletcher, C.H. III. 2004. *Journal of Sedimentary Research*, **74**, 49–63.

Grotzinger, J.P. 1986. *Bulletin of the Geological Society of America*, **97**, 1208–1131.

Grotzinger, J. P. *et al.* 2006. *Geology* **34**, 1085–1088.

Gruszczynski, M. *et al.* 1993. *Sedimentology*, **40**, 217–236.

Guidry, M.W. & MacKenzie, F.T. 2003. *Geochimica et Cosmochimica Acta*, **67**, 2949–2963.

Gunatilaka, A. 1976. *Journal of Sedimentary Petrology*, **46**, 548–554.

Gupta, S. 1997. *Geology*, **25**, 11–14.

Gupta, S. *et al.* 1998. *Geology*, **26**, 595–598.

Gupta, S. *et al.* 1999. *Basin Research*, **11**, 167–189.

Gupta, S. *et al.* 2007. *Nature*, **448**, 342–346.

Gurnis, M. 1992. *Science*, **255**, 1556–1558.

Gurnis, M. 1993. *Nature*, **364**, 589–593.

Gust, G. 1976. *Journal of Fluid Mechanics*, **75**, 29–47.

Guzzetti, F. *et al.* 1997. *Geomorphology*, **18**, 119–136.

Haase-Schram, A. *et al.* 2004. *Geochimica and Cosmochimica Acta*, **68**, p. 985–1005.

Habgood, E.L. *et al.* 2003. *Sedimentology*, **50**, 483–510.

Hack, J.T. 1957. *US Geological Survey Professional Paper* 294B.

Hagan, G.M. & Logan, B.W. 1974. In: *Evolution and Diagenesisof Quaternary Carbonate Sequences, Shark Bay, W. Australia* (eds Logan, B.W. *et al.*), pp. 61–139. Memoir 22, American Association of Petroleum Geologists.

Hails, J.& Carr, A. (eds) 1975. *Nearshore Sediment Dynamics and Sedimentation*. Wiley, London.

Halfar, J. *et al.* 2008. *Geology*, **36**, 463–466.

Hall, I.R. *et al.* 1998. *Earth and Planetary Science Letters*, **164**, 15–21.

Hall, I.R. *et al.* 2001. *Nature*, **412**, 809–812.

Hallam, A. 1997. *Journal of the Geological Society of London*, **154**, 773–779.

Halley, R.B. 1977. *Journal of Sedimentary Petrology*, **47**, 1099–1020.

Halley, R.B. *et al.* 1977. *Bulletin of the American Association of Petroleum Geologists*, **61**, 519–526.

Hallock, P. 1988. *Palaeogeography, Palaeoclimatology, Palaeoecology*, **63**, 275–291.

Hallock, P. & Schlager, W. 1985. *Palios*, **1**, 389–398.

Halsey, S.D. 1979. In: *Barrier Islands* (ed. Leatherman, S.P.), pp. 185–210. Academic Press, New York.

Hambrey, M. 1994. *Glacial Environments*. UCL Press, London.

Hambrey, M.& Harland, W.B. (eds) 1981. *Earth's pre-Pleistocene Glacial Record*. Cambridge University Press, Cambridge.

Hambrey, M.J. & McElvey, B. 2000. *Sedimentology*, **47**, 577–607.

Hambrey, M.J. *et al.* (eds) 2007. *Glacial Sedimentary Processes and Products*. Special Publication 39, International Association of Sedimentologists.

Hampson, G.J. 2000. *Journal of Sedimentary Research*, **70**, 325–340.

Hampson, G.J. & Storms, J.E.A. 2003. *Sedimentology*, **50**, 667–701.

Hampson, G.J. & Howell, J.A. 2005. In: *River Deltas: Concepts, Models, and Examples* (eds Giosan, L. & Bhattacharya, J.P.), pp. 133–145. Special Publication 83, Society of Economic Paleontologists and Mineralogists.

Hampton, B.A. & Horton, B.K. 2007. *Sedimentology*, **54**, 1121–1147.

Hampton, M.A. 1972. *Journal of Sedimentary Petrology*, **42**, 775–793.

Handford, C.R. *et al.* 2002. *Bob F. Perkins 22nd Annual Research Conference*, pp. 539–564. Gulf Coast Section, Society of Economic Paleonotologists and Mineralogists.

Häner, G. & Spencer, N. 1998. *Physics Today*, September issue, 22–27.

Hanna, S.R. 1969. *Journal of Applied Meteorology*, **8**, 874–883.

Hansen, J.P.V. & Rasmussen, E.S. 2008. *Journal of Sedimentary Research*, **78**, 130–146.

Harbor, J. *et al.* 1997. *Geology*, **25**, 739–742.

Hardie, L.A. 1967. In: *Holocene Marine Sedimentation in the North Sea Basin* (eds Nio, S.D. *et al.*), pp. 187–210. Special Publication 5, International Association of Sedimentologists.

Hardie, L.A. (ed.) 1977. *Sedimentation on the Modern Carbonate Tidal Flats of NW Andros Island, Bahamas*. Johns Hopkins Press, Baltimore.

Hardie, L.A. 1996. *Geology*, **24**, 279–283.

Hardie, L.A. & Garrett, P. 1977. In: *Sedimentation on the Modern Carbonate Tidal Flats of NW Andros Island, Bahamas* (ed. Hardie, L. A.), pp. 12–49. John Hopkins Press, Baltimore.

Hardie, L.A. & Ginsburg, R.N. 1977. In: *Sedimentation on the Modern Carbonate Tidal Flats of NW Andros Island, Bahamas* (ed. Hardie, L. A.), pp. 50–123. John Hopkins Press, Baltimore.

Hardie, L.A. & Lowenstein, T.K. 2004. *Journal of Sedimentary Research*, **74**, 453–461.

Hardie, L.A. *et al.* 1978. In: *Modern and Ancient Lake Sediments* (eds Matter, A. &. Tucker, M.E.), pp. 7–42. Special Publication 2, International Association of Sedimentologists.

Hardie, L.A. *et al.* 1986. *Palaeooceanography*, **1**, 447–457.

Hardisty, J. 1986. *Earth Surface Processes and Landforms*, **11**, 327–333.

Hardisty, J. & Whitehouse, R.J.S. 1988. *Nature*, **332**, 532–534.

Hardy, S. *et al.* 1994. *Marine and Petroleum Geology*, **11**, 561–574.

Harland, W.B. *et al.* 1990. *A Geological Time Scale*. Cambridge University Press, Cambridge.

Harms, J.C. *et al.* 1975. *Depositional Environments as Interpreted from Primary Sedimentary Structures and Stratification Sequences*. SEPM Short Course Notes 2.

Harney, J.N. & Fletcher, C.H. III. 2003. *Journal of Sedimentary Research*, **73**, 856–868.

Harris, M.T. 1993. *Sedimentology*, **40**, 383–401.

Harris, P.M. 1979. *Facies Anatomy and Diagenesis of a Bahamian Ooid Shoal*. Sedimenta 7, Comparative Sedimentology Laboratory, University of Miami, Florida.

Harris, P.M. & Kowalik, W.S. 1994. *Satellite Images of Carbonate Depositional Settings. Examples of Reservoir and Exploration Scale Geological Facies Variation*. Methods in Exploration Series 11, American Association of Petroleum Geologists.

Harris, P.T. & Collins, M.B. 1991. *Marine Geology*, **101**, 209–216.

Harris, P.T. *et al.* 1995. In: *Tidal Signatures in Modern and Ancient Sediments* (eds Flemming, B.C. & Bartoloma, A.), pp. 3–18. Special Publication 24, International Association of Sedimentologists.

Harris, P.T. *et al.* 2002. *Journal of Sedimentary Research*, **72**, 858–870.

Hart, B.S. & Plint, A.G. 1989. *Sedimentology*, **36**, 551–557.

Hart, J.K. 2006. *Sedimentology*, **53**, 125–146.

Hart, J.K. & Boulton, G.S. 1991. In: *The Glacial Deposits of Britain* (eds Rose, J. *et al.*), pp. 233–244. Balkema, Rotterdam.

Harvey, A.M. 1990. In: *Alluvial Fans: A Field Approach* (eds Rachocki, A.H. & Church, M.), pp. 247–269. Wiley, Chichester.

Harvey, A.M. 1997. In: *Arid Zone Geomorphology* (ed. Thomas, D.G.), pp. 231–259. Wiley, Chichester.

Harvey, A.M. 2005. In: *Alluvial Fans: Geomorphology, Sedimentology, Dynamics* (eds Harvey, A.M. *et al.*), pp. 117–131. Special Publication 251, Geological Society of London.

Harvey, J.G. 1976. *Atmosphere and Ocean: Our Fluid Environments*. Artemis Press, Sussex.

Harvie, C.E. *et al.* 1982. *Geochimica et Cosmochimica Acta*, **46**, 1603–1618.

Harvie, C.E. *et al.* 1984. *Geochimica et Cosmochimica Acta*, **48**, 723–751.

Harwood, D. *et al.* 2009. *Eos, Transactions American Geophysical Union*, **90** /11, 90–91.

Haskin, L.A. *et al.* 2005. *Nature*, **436**, 66–69.

Hassan, F.A. 2007. *Quaternary International*, **173–174**, 101–112.

Haughton, P.D.W. 1994. *Journal of Sedimentary Research*, **A64**, 233–246.

Haughton, P.D.W. 2000. *Sedimentology*, **47**, 497–518.

Haughton, P.D.W. *et al.* 2003. *Sedimentology*, **50**, 459–482.

Havholm, K.G. *et al.* 1993. In: *Aeolian Sediments Ancient and Modern* (ed. Pye, K. & Lancaster, N.), pp. 87–107. Special Publication 16, International Association of Sedimentologists.

Hawley, N. & Lee, C.-H. 1999. *Sedimentology*, **46**, 791–805.

Haworth, M. 1982. In: *Offshore Tidal Sands: Processes and Deposits* (ed. Stride, A.H.), pp. 8–40. Chapman & Hall, London.

Hay, A.E. *et al.* 1982. *Science*, **217**, 833–835.

Hayes, M.O. 1971. In: *The Estuarine Environment* (ed. Schubel, J.R.), pp. 1–71. American Geological Institute, Washington, DC.

Hayes, M.O. 1975. In: *Estuarine Research* (ed. Cronin, L.E.), pp. 3–22. Academic Press, New York.

Hayes, M.O. 1979. In: *Barrier Islands* (ed. Leatherman, S.P.), pp. 1–27. Academic Press, New York.

Haynes, C.V. 1989. *Quaternary Research*, **32**, 153–167.

Hays, J.D. & Pittman, W.C. 1973. *Nature*, **246**, 18–22.

Hays, J.D. *et al.* 1976. *Science*, **194**, 1121–1132.

Hayward, A.B. 1982. *Coral Reefs*, **1**, 109–114.

Hayward, A.B. 1985. *Sedimentary Geology*, **53**, 241–260.

Hayward, A.B. & Graham, R.H. 1989. In: *Inversion Tectonics* (eds Cooper, M.A. & Williams, G.D.), pp. 17–39. Special Publication 44, Geological Society of London.

Head, J.W. *et al.* 2003. *Nature*, **426**, 797–782.

Head, J.W. *et al.* 2005. *Nature*, **434**, 346–351.

Head, M.J.& Gibbard, P.L. (eds) 2005. *Early–Middle Pleistocene Transitions: The Land–Ocean Evidence*. Special Publication 247, Geological Society of London.

Head, M.R. & Bandyopadhyay, P. 1981. *Journal of Fluid Mechanics*, **107**, 297–338.

Hearty, P.J. & Kindler, P. 1995. *Journal of Coastal Research*, **11**, 675–689.

Heath, G.R. 1974. In: *Studies in Paleo-oceanography* (ed. Hay, W.W.), pp. 77–93. Society of Economic Paleontologists and Mineralogists.

Heathershaw, A.D. 1985. *Continental Shelf Research*, **4**, 485–493.

Heckel, P.H. 1986. *Geology*, **14**, 330–334.

Heezen, B.C. & Hollister, C.D. 1963. *International Union Geodesy and Geophysics*, **6**, 111.

Heikoop, J.M. *et al.* 1996. *Geology*, **24**, 759–762.

Heinrich, H. 1988. *Quaternary Research*, **29**, 143–152.

Helland-Hansen, W. & Gjelberg, J.G. 1994. *Sedimentary Geology*, **92**, 31–52.

Helland-Hansen, W. & Martinsen, O.J. 1996. *Journal of Sedimentary Research*, **66**, 670–688.

Heller, P.L. & Paola, C. 1996. *Journal of Sedimentary Research*, **66**, 297–306.

Heller, P.L. *et al.* 1996. *Geology*, **24**, 491–494.

Hellmann, R. & Tisserand, D. 2006. *Geochimica et Cosmochimica Acta*, **70**, 364–383.

Henderson, G.M. *et al.* 1994. *Earth and Planetary Letters*, **128**, 643–651.

Henstock, T.J. *et al.* 2006. *Geology*, **34**, 485–488.

Hequette, A. & Hill, P.R. 1993. *Marine Geology*, **113**, 283–304.

Herbst, W. *et al.* 2008. *Nature*, **452**, 194–197.

Herd, D.G. 1986. *Eos, Transactions American Geophysical Union*, 13 May, 457–460.

Hereford, R. 1993. *Entrenchment and Widening of the Upper San Pedro River, Arizona*. Special Paper 282, Geological Society of America.

Hernández-Molina, F.J. *et al.* 2000. In: *Sedimentary responses to Forced Regression*. (eds Hunt, D. & Gawthorpe, R.L.), pp. 329–362. Special Publication 172, Geological Society of London.

Herries, R.D. 1993. In: *Characterisation of Fluvial and Aeolian Reservoirs* (eds North, C.P. & Prosser, D.J.), pp. 199–218. Special Publication 73, Geological Society of London.

Hervouet, J.-M. 2000. *Hydrological Processes*, **14**, 2209–2210.

Hesse, R. & Khodabakhsh, S. 1998. *Geology*, **26**, 103–106.

Hesse, R. & Rakofsky, A. 1992. *Bulletin of the American Association of Petroleum Geologists*, **104**, 680–707.

Hesse, R. *et al.* 1987. *Canadian Journal of Earth Sciences*, **24**, 1595–1624.

Hesse, R. *et al.* 1996. *GSA Today*, September, 3–8.

Hesse, R. *et al.* 2004. *Geology*, **32**, 449–452.

Hesselbo, S.P. *et al.* 2000. *Nature*, **406**, 392–395.

Hesselbo, S. & Huggett, J.M. 2001. *Journal of Sedimentary Research*, **71**, 599–607.

Hetzinger, S. *et al.* 2006. *Journal of Sedimentary Research*, **76**, 670–682.

3–Heward, A.P. 1978. In: *Fluvial Sedimentology* (ed. Miall, A.D.), pp. 669–702, Memoir 5, Canadian Society of Petroleum Geologists.

Hickin, A.S. *et al.* 2009. *Journal of Sedimentary Research*, **79**, 457–477.

Hickin, E. 1974. *American Journal of Science*, **274**, 414–442.

Hickson, T.A. & Lowe, D.R. 2002. *Sedimentology*, **49**, 335–362.

Hicock, S.R. *et al.* 1981. *Canadian Journal of Earth Sciences*, **18**, 71–80.

Hilgen, F.J. 1991. *Earth Planetary Science Letters*, **104**, 226–244.

Hilgen, F.J. 1994. In: *Orbital Forcing and Cyclic Sequences* (eds de Boer, P.L. & Smith, D.G.), pp. 109–116. Special Publication 19, International Association of Sedimentologists.

Hill, P.R. 2001. *Sedimentology*, **48**, 1047–1078.

Hillgartner, H. *et al.* 2003. *Journal of Sedimentary Research*, **73**, 756–773.

Hine, A.C. *et al.* 1981a. *Marine Geology*, **42**, 327–348.

Hine, A.C. *et al.* 1981b. *Bulletin of the American Association of Petroleum Geologists*, **64**, 261–290.

Hinnov, L.A. & Goldhammer, R.K. 1991. *Journal of Sedimentary Petrology*, **61**, 1173–1193.

Hinsinger, P. *et al.* 2001. *Geochimica et Cosmochimica Acta*, **65**, 137–152.

Hiroki, Y. & Masuda F. 2000. *Sedimentology*, **47**, 135–149.

Hiroki, Y. & Terasaka, T. 2005. *Sedimentology*, **52**, 65–75.

Hirst, J.P.P. & Nichols, G.J. 1986. In: pp. 153–164. Special Publication 8, International Association of Sedimentologists.

Hiscott, R.N. 1994. *Journal of Sedimentary Research*, **64**, 204–208.

Hoashi, M. *et al.* 2009. *Nature Geoscience*, **2**, 301–306.

Hochella, M.F. & Banfield, J.F. 1995. In: *Chemical Weathering Rates of Silicate Minerals* (eds White, A.F. & Branfield, S.L.), Chapter 8. American Mineralogical Society, Washington, DC.

Hodell, D.A. *et al.* 1986. *Nature*, **320**, 411–414.

Hodgson, D.M. *et al.* 2006. *Journal of Sedimentary Research*, **76**, 20–40.

Hodgson, D.M.& Flint, S.S. (eds) 2005. *Submarine Slope Systems: Processes and Products*. Special Publication 244, Geological Society of London.

Holdren, G.R. & Speyer, P.M. 1985. *Geochimica et Cosmochimica Acta*, **49**, 675–681.

Holförster, F. & Schmidt, U. 2007. *Quaternary Science Reviews*, **26**, 1771–1789.

Holland, H.D. 1978. *The Chemistry of the Atmospheres and Oceans*. Wiley, New York.

Holliday, D.W. & Shephard-Thorne, E.R. 1974. *Basal –Purbeck Evaporites of the Fairlight Borehole, Sussex*. Report 74/4, Institute of Geological Sciences.

Hollister, C.D. 1993. *Sedimentary Geology*, **82**, 5–15.

Hollister, C.D. & Heezen, B.C. 1972. In: *Studies in Physical Oceanography* (ed. Gordon, A.L.), pp. 37–66. Gordon and Breach, New York.

Hollister, C.D. & McCave, I.N. 1984. *Nature*, **309**, 220–225.

Holmes, A. 1965. *Principles of Physical Geology*. Nelson, London.

Holzhausen, G.R. 1989. *Engineering Geology*, **27**, 225–278.

Homsy, G.M. *et al.* 2007. *Multimedia Fluid Mechanics*. Cambridge University Press.

Honji, H. *et al.* 1980. *Sedimentology*, **27**, 225–229.

Honjo, S. 1976. *Marine Micropalaeontology*, **1**, 65.

Honjo, S. & Doherty, K.W. 1988. *Deep Sea Research*, **35**, 133–149.

Hooke, R.LeB. 1967. *Journal of Geology*, **75**, 438–460.

Hooke, R.LeB. 1972. *Bulletin of the Geological Society of America*, **83**, 2073–2098.

Hooke, R.LeB. 1998. *Principles of Glacier Mechanics*. Prentice Hall, Englewood Cliffs, NJ.

Hopfinger, E.J. 1983. *Annual Review of Fluid Mechanics*, **15**, 47–76.

Hopkins, M. *et al.* 2008. *Nature*, **456**, 493–496.

Hori, K. *et al.* 2002. *Journal of Sedimentary Research*, **72**, 884–897.

Horita, J. *et al.* 2002. *Geochimica et Cosmochimica Acta*, **66**, 3733–3756.

Horppila, J. & Niemistö, J. 2008. *Sedimentology*, **55**, 1135–1144.

Horton, B.K. 1998. *Bulletin of the Geological Society America*, **110**, 1174–1192.

Horton, B.K. & Schmitt, J.G. 1996. *Sedimentology*, **43**, 133–155.

Horton, B.K. & DeCelles, P.G. 1997. *Geology*, **25**, 895–898.

Horton B.K. & DeCelles P.G. 2001. *Basin Research*, **13**, 43–63.

Houboult, J.J.H.C. 1968. *Geologie en Mijnbouw*, **47**, 245–273.

Hovikoski, J. *et al.* 2008. *Sedimentology*, **55**, 499–530.

Hovius, N. 1997. In: *Relative Role of Eustasy, Climate and Tectonics in Continental Rocks* (eds Shanley, K.W. & McCabe, P.J.) Special Publication, Society of Economic Paleontologists and Mineralogists.

Hovius, N. & Leeder, M.R. (eds) 1998. Thematic Set on Sediment Supply to Basins. *Basin Research*, **10**, 1–174.

Hovius, N. *et al.* 1997. *Geology*, **25**, 231–234.

Hovius, N. *et al.* 1998. *Geology*, **26**, 1071–1074.

Hovorka, S. 1987. *Sedimentology*, **34**, 1029–1054.

Howard, J.P. *et al.* 2003. *Basin Research*, **15**, 45–72.

Howe, J.A. & Pudsey, C.J. 1999. *Journal of Sedimentary Research*, **69**, 847–861.

Hoyle, F. 1981. *Ice*. Hutchinson, London.

Hsü, K.J. 1966. *Mineralum Depositum*, **2**, 133–138.

Hsü, K.J. 1972. *Earth Science Reviews*, **8**, 371–196.

Hsü, K.J. 1975. *Bulletin of the Geological Society of America*, **80**, 129–140.

Hsü, K.J. *et al.* 1977. *Nature*, **267**, 399–403.

Hubbard, D.K. 2009. In: *Perspectives in Carbonate Geology* (eds Swart, P.K. *et al.*), pp. 1–18. Special Publication 41, International Association of Sedimentologists.

Hubbard, D.K. *et al.* 2005. *Journal of Sedimentary Research*, 75, 97–113.

Hubbard, J. & Shaw, J.H. 2009. *Nature*, **458**, 194–197.

Hubbard, S.M. *et al.* 2008. *Sedimentology*, **55**, 1333–1359.

Huckleberry, G. 1994. *Geology*, **22**, 1083–1086.

Hudson, J.D. 1963. *Palaeontology*, **6**, 318–326.

Hughen, K.A. *et al.* 1996. *Nature*, **380**, 51–54.

Hughes, M.G. *et al.* 1997. *Marine Geology*, **138**, 91–103.

Hunter, R.E. 1977. *Sedimentology*, **24**, 361–387.

Hunter, R.E. & Clifton, H.E. 1982. *Journal of Sedimentary Petrology*, **52**, 127–143.

Huntley, D.A. & Bowen, A.J. 1975a. In: *Nearshore Sediment Dynamics and Sedimentation* (eds Hails, J. & Carr, A.). Wiley, New York.

Huntley, D.A. & Bowen, A.J. 1975b. *Journal of the Geological Society of London*, **131**, 69–81.

Huntington, K. *et al.* 2007. *Eos, Transactions American Geophysical Union*, **88/52**, 577–578.

Husinec, A. & Jelaska, V. 2006. *Journal of Sedimentary Research*, **76**, 1120–1136.

Hutchinson, G.E. 1957. *A Treatize on Limnology, 1: Geography, Physics and Chemsitry*. Wiley, New York.

Hutchinson, D.R. *et al.* 1992. *Geology*, **20**, 589–593.

Huthnance, J.M. 1981. *Progress in Oceanography*, **10**, 193–226.

Hutter, K. & Pudasaini, S.P. (eds) 2005. *Philosophical Transactions of the Royal Society*, **363A**, 1493–1700.

Hwang, I.-G. & Heller, P.L. 2002. *Sedimentology*, **49**, 977–999.

Ibbeken, H. & Schleyer, R. 1991. *Source and Sediment: A Case Study of Provenance and Mass Balance at an Active Plate Margin (Calabri, Southern Italy)*. Springer, Berlin.

Iglesias-Rodrigues, M.D. *et al.* 2002. *Eos, Transactions American Geophysical Union*, **83/34365**, 374–375.

Illing, L.V. 1954. *Bulletin of the American Association of Petroleum Geologists*, **38**, 1–95.

Illing, L.V. *et al.* 1965. In: *Dolomitisation and Limestone Diagenesis: A Symposium* (eds Pray, L.C. & Murray, R. C.), pp. 89–111. Special Publication 13, Society of Economic Paleontologists and Mineralogists.

Imbrie, J. & Imbrie, K.P. 1979. *Ice Ages: Solving the Mystery*. Macmillan, New York.

Imbrie, J. *et al.* 1984. In: *Milankovitch and Climate*, Part 1 (eds Berger, A. *et al.*), pp. 269–305. Reidel, Dordrecht.

Immenhauser, A. *et al.* 2001. *Sedimentology*, **48**, 1187–1207.

Imran, J. *et al.* 1998. *Journal of Geophysical ResesarChapter* **103C**, 1219–1238.

Imran, J. *et al.* 2008. *Sedimentology*, **55**, 235–239.

Ingersoll, R.V. 1979. *Bulletin of the Geological Society of America*, **90**, 813–826.

Ingersoll, R.V. 1988. *Geological Society of America Bulletin*, **100**, 1704–1719.

Ingersoll, R.V. & Eastmond, D.J. 2007. *Journal of Sedimentary Research*, **77**, 784–796.

Inman, D.L. & Bagnold, R.A. 1963. In: *The Sea*, Vol. 3 (ed. Hill, M.N.), pp. 529–583. Wiley, New York.

Inman, D.L. & Bowen, A.J. 1963. *Proceedings of the 8th Conference on Coast Engineering*, pp. 137–150.

Insalaco, E. *et al.* (eds) 2000. *Carbonate Platform Systems*. Special Publication **178**, Geological Society of London.

Iseya, F. & Ikuda, H. 1989. In: *Sedimentary Facies in the Active Plate Margin*, pp. 81–112. Terra Scientific Publishing, Tokyo.

Ito, M. 2008. *Journal of Sedimentary Research*, **78**, 668–682.

Ittekot, V. *et al.* 1991. *Nature*, **351**, 385.

Iverson, R.M. 1997. *Reviews of Geophysics*, **35**, 245–296.

Iverson, R.M. & Denlinger, R.P. 2001. *Journal of Geophysical Research*, **106**, 537–552.

Iverson, R.M. & Vallance, J.W. 2001. *Geology*, **29**, 115–118.

Iverson, R.M. *et al.* 1997. *Annual Review of Earth & Planetary Sciences*, **25**, 85–138.

Ivey, G.N. *et al.* 2008. *Annual Reviews of Fluid Mechanics*, **40**, 169–184.

Jackson, C.A-L. *et al.* 2002. *Journal of the Geological Society of London*, **159**, 175–187.

Jackson, D. & Launder, B. 2007. *Annual Reviews of Fluid Mechanics*, **39**, 19–35.

Jackson, J.A. & Leeder, M.R. 1994. *Journal of Structural Geology*, **16**, 1041–1059.

Jackson, J.A. *et al.* 1988. *Journal of Structural Geology*, **10**, 155–170.

Jackson, J.A. *et al.* 1996. *Journal of Structural Geology*, **18**, 217–234.

Jackson, R.G. 1975. *Bulletin of the Geological Society of America*, **86**, 1511–1522.

Jackson, R.G. 1976. *Journal of Sedimentary Petrology*, **46**, 579–594.

Jackson, T.A. & Keller, W.D. 1970. *American Journal of Science*, **269**, 446–466.

Jacobs, G.A. *et al.* 1994. *Nature*, **370**, 360–363.

Jacobsen, S.B. & Kaufman, A.J. 1999. *Chemical Geology*, **161**, 37–57.

Jaeger, H.M. *et al.* 1996. *Physics Today*, **49**, 32–36.

Jahnke, R.A. 1990. *Reviews of Geophysics*, **28**, 381–98.

Jakobsson, M. *et al.* 2007. *Nature*, **447**, 986.

James, N.P. 1978a. In: *Facies Models* (ed. Walker, R.G.), pp. 105–8. Geological Society of Canada.

James, N.P. 1978b. In: *Facies Models* (ed. Walker, R.G.), pp. 121–132. Geological Society of Canada.

James, N.P. 2001. *Journal of Sedimentary Research*, **71**, 549–567.

James, N.P. & Ginsburg, R.N. 1979. *The Seaward Margin of Belize Barrier and Atoll Reefs* (eds James, N.P. & Ginsberg, R.N.). Special Publication 3, International Association of Sedimentologists.

James, N.P. & Gravestock, D.I. 1990. *Sedimentology*, **37**, 455–489.

James, N.P. *et al.* 1992. *Sedimentology*, **39**, 877–903.

James, N.P. *et al.* 1999. *Journal of Sedimentary Research*, **69**, 1297–1321.

James, N.P. *et al.* 2000. *Geology*, **28**, 647–650.

James, N.P. *et al.* 2004. *Journal of Sedimentary Research*, **74**, 20–48.

James, N.P. *et al.* 2005a. *Journal of Sedimentary Research*, **75**, 454–463.

James, N.P. *et al.* 2005b. *Geology*, **33**, 9–12.

Janecke, S. *et al.* 1999. *Basin Research*, **11**, 143–165.

Jankaew, K. *et al.* 2008. *Nature*, **455**, 1228–1234.

Jansen, J.D. & Brierley, G.J. 2004. *Sedimentology*, **51**, 901–925.

Jansson, M.B. 1988. *Geografiska Annaler*, **70A**, 81–98.

Jedoui, Y. *et al.* 2003. *Quaternary Science Reviews*, **22**, 343–351.

Jeffreys, D.J. 1982. *Advances in Colloid and Interface Science*, **17**, 213–218.

Jenkyns, H.C. 1988. *American Journal of Science*, **288**, 101–151.

Jenner, K.A. *et al.* 2007. *Sedimentology*, **54**, 19–38.

Jerolmack, D.J. *et al.* 2006. *Journal of Geophysical Research*, **111**, E12S02.

Jiongxin, X. 2004. *Geografiska Annaler*, **86A**, 349–366.

Joachimski, M.M. 1994. *Sedimentology*, **41**, 805–824.

Johannessen, P.N. & Nielsen, L.H. 1986. *IAS 7th Regional Meeting Abstracts Volume*, Cracow. International Association of Sedimentologists.

Johansen, C. *et al.* 1997. In: *Cohesive Sediments* (eds Burt, N. *et al.*), pp. 305–314. Wiley, New York.

Johnson, A.M. 1970. *Physical Processes in Geology*. Freeman, Cooper, San Francisco.

Johnson, C.L. *et al.* 2005. *Sedimentology*, **52**, 513–536.

Johnson, M.A. *et al.* 1981. In: *Holocene Marine Sedimentation in the North Sea Basin* (eds Nio, S.-D. *et al.*), pp. 247–256. Special Publication 5, International Association of Sedimentologists.

Johnson, S.D. *et al.* 2001. *Sedimentology*, **48**, 987–1023.

Johnson, T.C. *et al.* 1987. *Bulletin of the Geological Society of America*, **98**, 439–447.

Johnson, T.C. *et al.* 1995. *Bulletins of the Geological Society of America*, **107**, 812–829.

Jones, C.E. & Jenkyns, H.C. 2001. *American Journal of Science*, **301**, 112–149.

Jones, C.M. & McCabe, P.J. 1980. *Journal of Sedimentary Petrology*, **50**, 613–620.

Jones, G.A. & Gagnon, A.R. 1994. *Deep-Sea Research*, **41**, 531–557.

Jones, K.P.N. *et al.* 1992. *Marine Geology*, **107**, 149–173.

Jones, M. *et al.* 2004. *Basin Research*, **16**, 467–488.

Jongmans, A.G. *et al.* 1997. *Nature*, **389**, 682–683.

Jopling, A.V. & McDonald, B.C. (eds) 1975. *Glaciofluvial and Glaciolacustrine Sediments*. Special Publication 23, Society of Economic Paleontologists and Mineralogists.

Jordan, T.E. 1995. In: *Tectonics of Sedimentary Basins* (eds Busby, C.J. & Ingersoll, R.V.), pp. 331–362. Blackwell Science, Boston.

Jullien, R. *et al.* 1992. *Physical Review Letters*, **69**, 640–643.

Juyal, N. *et al.* 2006. *Quaternary Science Reviews*, **25**, 2632–2650.

Kahle, C.F. 1974. *Journal of Sedimentary Petrology*, **44**, 30–39.

Kamenos, N.A. 2008. *Geochimica et Cosmochimica Acta*, **72**, 771–779.

Kane, I.A. *et al.* 2009. *Sedimentology*, **56**, 2207–2234.

Kaneps, A.G. 1979. *Science*, **204**, 297–301.

Kapdasli, M.S. 1991. In: *Sand Transport in Rivers, Estuaries and the Sea*, Eromecht 262 (eds Soulsby, R.L. & Bettess, R.), pp. 31–36. Balkema, Rotterdam.

Kapitsa, A.P. *et al.* 1996. *Nature*, **381**, 684–686.

Karlstrom, K.E. *et al.* 2008. *Geology*, **36**, 835–838.

Karssenberg, D. & Bridge, J.S. 2008, *Sedimentology*, **55**, 1717–1745.

Kassem, A. & Imram, J. 2001. *Geology*, **29**, 655–658.

Ke, X. *et al.* 1996. *Sedimentology*, **43**, 157–174.

Keene, J.B. & Harris, P.T. 1995. In: *Tidal Signatures in Modern and Ancient Sediments* (eds Flemming, B.C. & Bartoloma, A.), pp. 225–236. Special Publication 24, International Association of Sedimentologists.

Keighley, D. *et al.* 2003. *Journal of Sedimentary Research*, **73**, 987–1006.

Keller, C.K. & Wood, B.D. 1993. *Nature*, **364**, 223–225.

Keller, G.H. *et al.* 1979. In: *Geology of Continental Slopes* (eds Doyle, L.J. & Pilkey, O.H.), pp. 131–151. Special Publication 27, Society of Economic Paleontologists and Mineralogists.

Kelly, J.C. *et al.* 2001. *Sedimentolog*, **48**, 325–338.

Kelts, K. & Hsu, K.J. 1978. In: *Lakes: Physics, Chemistry and Geology* (ed. Lerman, A.), pp. 295–321. Springer, New York.

Kemp, A.E.S. & Baldauf, J.G. 1993. *Nature*, **362**, 141–144.

Kemp, A.E.S. *et al.* 1999. *Nature*, **398**, 57–61.

Kemp, P.H. 1975. In: *Nearshore Sediment Dynamics and Sedimentation* (eds Hails, J. & Carr, A.), pp. 47–68. Wiley, New York.

Kemp, P.H. & Simons, R.R. 1982. *Journal of Fluid Mechanics*, **116**, 227–250.

Kempf, O. & Pfiffner, O.A. 2004. *Basin Research*, **16**, 549–567.

Kendall, A.C. & Harwood, G.M. 1996. In: *Sedimentary Environments: Processes, Facies and Stratigraphy* (ed. Reading, H.G.), pp. 281–324. Blackwell Science, Oxford.

Kendall, G.S.C. & Schlager, W. 1981. *Marine Geology*, **44**, 180–212.

Kennedy, J.F. 1963. *Journal of Fluid Mechanics*, **16**, 521–544.

Kennett, J.P. 1977. *Journal of Geophysical Research*, **82**, 3843–3860.

Kennett, J.P. & Shackleton, N.J. 1976. *Nature*, **260**, 513–515.
Kennett, J.P. *et al.* 1974. *Science*, **186**, 144–147.
Kent, P.E. 1966. *Journal of Geology*, **74**, 79–83.
Kenyon, N.H. & Stride, A.H. 1970. *Sedimentology*, **14**, 159–173.
Kenyon, N.H. *et al.* 1981. In: *Holocene Marine Sedimentation in the North Sea Basin*, (eds Nio, S.-D. *et al.*), pp. 257–268. Special Publication 5, International Association of Sedimentologists.
Kenyon, R.M. & Turcotte, D.L. 1985. *Bulletin of the Geological Society of America*, **96**, 1457–1465.
Kerans, C. *et al.* 1994. *Bulletin of the American Association of Petroleum Geologists*, **78**, 181–216.
Kersey, D.G. & Hsu, K.J. 1976. *Sedimentology*, **23**, 761–90.
Keulegan, G.H. 1957. *Thirteenth Progress Report on Model Laws for Density Currents: An Experimental Study of the Motion of Saline Water from Locks into Freshwater Channels*. Report 5168, US National Bureau of Standards.
Khan, I.A. *et al.* 1997. *Sedimentology*, **44**, 221–251.
Khan, S.A. & Imran, J. 2008. *Journal of Sedimentary Research*, **78**, 245–257.
Khripounoff, A. *et al.* 2003. *Marine Geology*, **194**, 151–158.
Kidd, W.S.F. *et al.* 1988. *Philosophical Transactions of the Royal Society*, **327A**, 287–305.
Kiehl, J.T. 1994. *Physics Today*, November, 36–42.
Kiessling, W. *et al.* 2008. *Nature geosciences*, **1**, 527–530.
Kim, S.B. 1995. *Journal of Sedimentary Research*, **65**, 706–708.
Kimberley, M.M. 1979. *Journal of Sedimentary Petrology*, **49**, 111–132.
Kinsman, D.J.J. 1966. In: *Second Symposium on Salt* (ed. Rau, J.L.), pp. 302–326. Northern Ohio Geological Society, Cleveland, OH.
Kinsman, D.J.J. 1975a. *Nature*, **255**, 375–378.
Kinsman, D.J.J. 1975b. In: *Petroleum and Global Tectonics* (eds Fischer, A.G. & Judson, S.), pp. 83–126. Princeton University Press, Princeton, NJ.
Kinsman, D.J.J. 1976. *Journal of Sedimentary Petrology*, **46**, 273–279.
Kirby, R. & Parker, W.R. 1983. *Canadian Journal of Fisheries and Aquatic Sciences*, **40**, 83–95.
Kirkbride, A. 1993. In: *Turbulence: Perspectives on Flow and Sediment Transport* (eds Clifford, N. J. *et al.*), pp. 185–196. Wiley, Chichester.
Kirkby, M.J. 1995. *Geomorphology*, **13**, 319–335.
Kirkby, M.J. 1999. In: *Process Modelling and Landform Evolution* (eds Hergarten, S. & Neugebauer, H. J.), pp. 189–203. Lecture Notes in Earth Sciences 78, Springer Verlag, Berlin, Heidelberg.
Kirkby, M.J. & Cox, N.J. 1995. *Catena*, **25**, 333–352.
Kirkby, M.J. & Neale, R.H. 1987. In: *International Geomorphology 1986, Part 2* (ed. Gardiner, V.), pp. 189–210. Wiley, New York.
Kirkby, M.J. *et al.* 1998. *Geomorphology*, **24**, 35–49.
Kirkham, A. 1998. In: *Carbonate Ramps* (eds Wright, V.P. & Burchette, T. P.), pp. 15–42. Special Publication 149, Geological Society of London.
Kirkland, D.W. *et al.* 2000. *Journal of Sedimentary Research*, **70**, 749–761.
Kjær, K.H. & Kröger, J. 2001. *Sedimentology*, **48**, 935–952.
Klaucke, I. *et al.* 1997. *Sedimentology*, **44**, 1093–1102.
Klaucke, I. *et al.* 1998. *Bulletin of the Geological Society of America*, **110**, 22–34.
Klaus, A. *et al.* 1992. *Proceedings of the Ocean Drilling Program, Scientific Results*, **126**, 555–574.
Kleinhans, M.G. 2005. *Sedimentology*, **52**, 291–311.
Kline, S.J. *et al.* 1967. *Journal of Fluid Mechanics*, **30**, 741–773.
Kneller, B.C. 1996. In: *Characterisation of Deep Marine Clastic Systems* (eds Hartley, A. & Prosser, D. J.), pp. 29–46. Special Publication 94, Geological Society of London.
Kneller, B.C. & Branney, M.J. 1995. *Sedimentology*, **42**, 607–616.
Kneller, B.C. & Buckee, C. 2000. *Sedimentology*, **47** (Suppl. 1), 62–94.
Kneller, B.C. & McCaffrey, W. 1999. *Journal of Sedimentary Research*, **69**, 980–991.
Kneller, B.C. *et al.* 1991. *Geology*, **19**, 250–252.
Kneller, B.C. *et al.* 1999. *Journal of Geophysical Research*, **104**, 5381–5391.
Knight, J.& Harrison, J. (eds). 2009. *Periglacial and Paraglacial Processes and Environments*. Special Publication 320, Geological Society of London.
Knight, J.B. *et al.* 1993. *Physical Review Letters*, **70**, 3728–3731.
Knight, J. *et al.* 2002. *Sedimentology*, **49**, 1229–1252.
Knighton, D. 1998. *Fluvial Forms and Processes: a New Perspective*. Arnold, London.
Knox, J.C. 1983. In: Late Quaternary Environments of the United States, Vol. 2, *The Holocene* (ed. Wright, H.E. Jr), pp. 26–41. University of Minnesota Press, Minneapolis.
Knox, J.C. 2006. *Geomorphology*, **79**, 286–310.
Knuepfer, P.L.K. 1988. *Bulletin of the Geological Society of America*, **100**, 1224–1236.
Kobluk, D.R. & Risk, M.J. 1977. *Journal of Sedimentary Petrology*, **47**, 517–528.
Kochel, R.C. 1990. In: *Alluvial Fans: A Field Approach* (eds Rachocki, A.H. & Church, M.), pp. 109–130. Wiley, New York.
Kocurek, G. 1981. *Sedimentology*, **28**, 753–780.
Kocurek, G. 1988. *Sedimentary Geology*, **56**, 193–206.
Kocurek, G. 1996. In: *Sedimentary Environments: Processes, Facies and Stratigraphy* (ed. Reading, H.G.), pp. 125–155. Blackwell Science, Oxford.
Kocurek, G. & Havholm, K.G. 1993. In: *Siciliclastic Sequence Stratigraphy* (eds Weimer, P. & Posamentier, H. W.), pp. 393–409. Memoir 58, American Association of Petroleum Geologists.

Kocurek, G. & Lancaster, N. 1999. *Sedimentology*, **46**, 505–515.

Kocurek, G. & Nielson, J. 1986. *Sedimentology*, **33**, 795–816.

Kocurek, G. *et al.* 1991. *Sedimentology*, **38**, 751–772.

Kocurek, G. *et al.* 1992. *Journal of Sedimentary Petrology*, **62**, 622–635.

Kolb, C.R. & van Lopik, J.R. 1958. Geology of the Mississippi River Deltaic Plain. Technical Reports 3483 and 3484, US Corps of Engineers Waterways Experimental Station.

Komar, P.D. 1969. *Journal of Geophysical Research*, **74**, 4544–4558.

Komar, P.D. 1971. *Journal of Geophysical Research*, **76**, 713–721.

Komar, P.D. 1974. *Journal of Sedimentary Petrology*, **44**, 169–180.

Komar, P.D. 1975. In: *Nearshore Sediment Dynamics and Sedimentation* (eds Hails, J. & Carr, A.), pp. 17–46. Wiley, New York.

Komar, P.D. 1998. *Beach Processes and Sedimentation.* Prentice Hall, Englewood Cliffs, NJ.

Komar, P.D. & Inman, D.L. 1970. *Journal of Geophysical Research*, **75**, 5914–5927.

Komar, P.D. *et al.* 1972. In: *Shelf Sediment Transport: Process and Pattern* (eds Swift, D. J. P. *et al.*), pp. 601–619. Dowden, Hutchinson and Ross, Stroudsberg, PA.

Komar, P.D. & Miller, M. 1973. *Journal of Sedimentary Petrology*, **43**, 1111–1113.

Kominz, M.A. 1984. In: *Interregional Unconformities and Hydrocarbon Accumulation* (ed. Schlee, J.S.), pp. 109–127. American Association of Petroleum Geologists Memoir.

Konhauser, K.O. 2007. *Introduction to Geomicrobiology.* Blackwell Publishing.

Konhauser, K.O. *et al.* 1992. *Geology*, **20**, 227–230.

Konhauser, K.O. *et al.* 2007. *Earth and Planetary Science Letters*, **258**, 87–100.

Konhauser, K.O. *et al.* 2009. *Nature*, **458**, 750–753.

Kontrovitz, M. *et al.* 1979. *Journal of Foraminiferal Research*, **9**, 228–232.

Koons, P.O. 1989a. *Geology*, **18**, 679–682.

Koons, P.O. 1989b. *Annual Review of Earth and Planetary Sciences*, **23**, 375–408.

Kopp, H. *et al.* 2008. *Basin Research*, **20**, 519–529.

Kostaschuk, R. 2000. *Sedimentology*, **47**, 519–532.

Kostaschuk, R.A. & Church, M.A. 1993. *Sedimentary Geology*, **85**, 25–37.

Kostaschuk, R.A. & Villard, P. 1996. *Sedimentology*, **43**, 849–863.

Kostaschuk, R.A. *et al.* 1992. *Sedimentology*, **39**, 305–317.

Koster, E.H. 1978. In: *Fluvial Sedimentology* (ed. Miall, A. D.), pp. 161–186. Memoir 5, Canadian Society of Petroleum Geologists.

Kosters, E.C. 1989. *Journal of Sedimentary Petrology*, **59**, 98–113.

Kostick, B. & Aigner, T. 2007. *Sedimentology*, **54**, 789–808.

Kotthoff, U. *et al.* 2008. *Quaternary Science Reviews*, **27**, 832–845.

Kraal, E.R. 2008. *Nature*, **451**, 973–976.

Kraft, J.C. 1971. *Bulletin of the Geological Society of America*, **82**, 2131–2158.

Kraft, J.C. & John, C.J. 1979. *Bulletin of the American Association of Petroleum Geologists*, **63**, 2145–2163.

Kraft, J.C. *et al.* 1987. In: *Sea Level Fluctuation and Coastal Evolution* (eds Nummedal, D. *et al.*), pp. 129–144. Special Publication 41, Society of Economic Paleontologists and Mineralogists.

Kranck, K. 1975. *Sedimentology*, **22**, 111–123.

Kranck, K. 1981. *Sedimentology*, **28**, 107–114.

Kraus, M.J. 1996. *Journal of Sedimentary Research*, **66**, 354–363.

Kraus, M.J. & Aslan, A. 1993. *Journal of Sedimentary Petrology*, **63**, 453–463.

Krause, N.C. & Horikawa, K. 1990. In: *The Sea*, Vol. 9, *Ocean Engineering Science* (eds LeMehaute, B. & Hanes, D. M.), pp. 775–814. Wiley, New York.

Krauskopf, K.B. 1979. *Introduction to Geochemistry*, 2nd edn. McGraw-Hill, New York.

Krishnamurthy, R.V. *et al.* 2000. *Geology*, **28**, 263–266.

Krishnaswami, S. *et al.* 1992. *Earth and Planetary Science Letters*, **109**, 243–253.

Krom, M.D. *et al.* 1999. *Marine Geology*, **160**, 45–61.

Kroonenberg, S.B. *et al.* 2005. In: *River Deltas: Concepts, Models, and Examples* (eds Giosan, L. & Bhattacharya, J. P.), pp. 231–256. Special Publication 83, Society of Economic Paleontologists and Mineralogists.

Kuecher, G.J. *et al.* 1990. *Sedimentary Geology*, **68**, 211–221.

Kuehl, S.A. *et al.* 2005. In: *River Deltas: Concepts, Models, and Examples* (eds Giosan, L. & Bhattacharya, J. P.), pp. 413–434. Special Publication 83, Society of Economic Paleontologists and Mineralogists.

Kuhnle, R.A. *et al.* 2006. *Sedimentology*, **53**, 631–654.

Kulm, L.D. *et al.* 1975. *Journal of Geology*, **83**, 145–176.

Kulp, M. *et al.* 2005. In: *River Deltas: Concepts, Models, and Examples* (eds Giosan, L. & Bhattacharya, J. P.), pp. 279–294. Special Publication 83, Society of Economic Paleontologists and Mineralogists.

Kumar, N. & Sanders, J.E. 1974. *Sedimentology*, **21**, 491–532.

Kumar, N. *et al.* 1995. *Nature*, **378**, 675–680.

Kuriyama, Y. *et al.* 2005. *Sedimentology*, **52**, 1123–1132.

Kurtz, A.C. *et al.* 2001. *Geochimica et Cosmochimica Acta*, **65**, 1971–1983.

Kusuda, T. *et al.* 1985. *Water Science Technology*, **17**, 891–901.

Kutzbach, J.E. & Street-Perrott, F.A. 1985. *Nature*, **317**, 130–134.

Kvale, E.P. *et al.* 1994. *Geology*, **22**, 331–334.

Kvale, E.P. *et al.* 1999. *Journal of Sedimentary Research*, **69**, 1154–1168.

Laban, C. & Schüttenhelm, R.T.E. 1981. In: *Holocene Marine Sedimentation in the North Sea Basin* (eds Nio, S. -D. *et al.*), pp. 239–245. Special Publication 5, International Association of Sedimentologists.

Lagmay *et al.* 2006. *Eos, Transactions American Geophysical Union*, **87** (12), 121, 124.

Lamarche, G. *et al.* 2000. *Eos, Transactions American Geophysical Union*, **81 481**, 485–486.

Lamb, H. 1945. *Hydrodynamics*, 6th edn. Cambridge University Press, Cambridge.

Lamb, H.H. 1948. *Hydrodynamics*. Cambridge University Press, Cambridge.

Lamb, H.H. 1995. *Climate, History and the Modern World*, 2nd edn. Routledge, London.

Lamb, M.P. *et al.* 2004. *Journal of Sedimentary Research*, **74**, 148–155.

Lamb, M. *et al.* 2006. *Sedimentology*, **53**, 147–160.

Lamb, M.P. *et al.* 2008. *Journal of Sedimentary Research*, **78**, 480–498.

Lambe, T.W. & Whitman, R.V. 1969. *Soil Mechanics*. Wiley, New York.

Lancaster, N. 1989. *Sedimentology*, **36**, 273–289.

Lancaster, N. 1992. *Sedimentology*, **39**, 631–644.

Lancaster, N. 1995. *Geomorphology of Desert Dunes*. Routledge, London.

Lancaster, N. & Baas, A. 1998. *Earth Surface Processes & Landforms*, **23**, 69–82.

Langbein, W.B. & Schumm, S.A. 1958. *Transactions of the American Geophysical Union*, **39**, 1076–1084.

Langford, R.P. 1989. *Sedimentology*, **36**, 1023–1035.

Langford, R.P. & Chan, M.A. 1989. *Sedimentology*, **36**, 1037–1051.

Langford, R.P. & Chan, M.A. 1993. In: *Aeolian Sediments Ancient and Modern* (eds Pye, K. & Lancaster, N.), pp. 109–126. Special Publication 16, International Association of Sedimentologists.

Langford, R.P. *et al.* 2008. *Journal of Sedimentary Research*, **78**, 410–422.

Langhorne, D.N. & Read, A.A. 1986. *Journal of the Geological Society of London*, **143**, 957–962.

Lanier, W.P. *et al.* 1993. *Journal of Sedimentary Petrology*, **63**, 860–873.

Lantzsch, H. *et al.* 2007. *Sedimentology*, **54**, 1307–1322.

Laporte, L.F. 1971. *Journal of Sedimentary Petrology*, **41**, 724–740.

Larcombe, P. & Jago, C.F. 1996. *Sedimentology*, **43**, 541–559.

Larcombe, P. *et al.* 2001. *Sedimentology*, **48**, 811–835.

Larronne, J.B. & Reid, I. 1993. *Nature*, **36**, 148–150.

Larronne, J.B. & Shlomi, Y. 2007. *Sedimentary Geology*, **195**, 21–37.

Larsen, M.C. *et al.* 2001. *Eos, Transactions American Geophysical Union*, November 20 issue, 572–573.

Lasaga, A.C. 1995. In: *Chemical Weathering Rates of Silicate Minerals* (eds White, A.F. & Branfield, S.L.), Chapter 2. American Mineralogical Society, Washington, DC.

Lastras, G. *et al.* 2005. *Journal of Sedimentary Research*, **75**, 784–797.

Lautridou, J.-P. & Ozouf, J.C. 1982. *Progress in Physical Geography*, **6**, 215–232.

Lautridou, J.-P. & Seppala, M. 1986. *Geografiska Annaler*, **68A**, 89–100.

Lavoie, D. 1995. *Sedimentology*, **42**, 95–116.

Lawton, T.F. 1986. pp. 423–42. Memoir 41, American Association of Petroleum Geologists.

Lawson, M.P. *et al.* 2002. *Quaternary Science Reviews*, **21**, 825–836.

Leatherman, S.P. (ed.) 1979. *Barrier Islands*. Academic Press, New York.

Leckie, D.A. 1988, *Journal of Sedimentary Petrology*, **58**, 607–622.

Leckie, D.A. & Krystink, L.F. 1989. *Journal of Sedimentary Petrology*, **59**, 862–870.

Leclair, S.F. 2002. *Sedimentology*, **49**, 1157–1180.

Leclair, S.F. & Arnott, R.W.C. 2005. *Journal of Sedimentary Research*, **75**, 1–5.

Leclair, S.F. & Bridge, J.S. 2001. *Journal of Sedimentary Research*, **71**, 713–716.

Leddy, J.O. *et al.* 1993. In: *Braided Rivers* (eds Best, J.L. & Bristow, C. S.), pp. 119–127. Special Publication 75, Geological Society of London.

Ledwell, J.R. *et al.* 1993. *Nature*, **364**, 701–703.

Lee, G.H. *et al.* 1996. *Bulletin of the American Association of Petroleum Geologists*, **80**, 340–358.

Lee, H.J. & Chough, S.K. 1989. *Marine Geology*, **87**, 195–205.

Lee, H.J. & Chu, Y.S. 2001. *Journal of Sedimentary Research*, **71**, 144–154.

Lee, H.J. *et al.* 2006 *Journal of Sedimentary Research*, **76**, 284–291.

Lee, K. *et al.* 2007. *Journal of Sedimentary Research*, **77**, 303–323.

Lee, M.R. & Parsons, I. 1995. *Geochimica et Cosmochimica Acta*, **59**, 4465–4488.

Lee, M.R. & Parsons, I. 1998. *Journal of Sedimentary Research*, **68**, 198–211.

Lee, M.R. *et al.* 2005. *Journal of Sedimentary Research*, **75**, 313–322.

Lee, T. *et al.* 1993. *Geology*, 423–426.

Leeder, M.R. 1973. *Geological Magazine*, **110**, 265–276.

Leeder, M.R. 1975. *Geological Magazine*, **112**, 257–270.

Leeder, M.R. 1978. In: Fluvial Sedimentology (ed. Miall, A. D.), pp. 585–596. Memoir 5, Canadian Society Petroleum Geologists.

Leeder, M.R. 1983. *Sedimentology*, **30**, 485–491.

Leeder, M.R. 1993. In: *Characterization of Fluvial and Aeolian Reservoirs* (eds North, C. P. & Prosser, D. J.), pp. 7–22. Special Publication 73, Geological Society of London.

Leeder, M.R. 1995. In: *Tectonics of Sedimentary Basins* (eds Busby, C. & Ingersoll, R.), pp. 119–148, Blackwell Science, Boston.

Leeder, M.R. 1996. *Earth Surface Processes and Landforms*, 22, 229–237.

Leeder, M.R. 1998. In: *Lyell: the Past is the Key to the Present* (eds Blundell, D.J. & Scott, A. C.), pp. 97–110. Special Publication 143, Geological Society of London.

Leeder, M.R. 2007. *Geophysical Research Letters*, **34**, Article Number: **L01201**.

Leeder, M.R. & Gawthorpe, R.L. 1987. In: Continental Extensional Tectonics (eds Coward, M. P. *et al.*), pp. 139–152. Special Publication 28, Geological Society of London.

Leeder, M.R. & Jackson, J.A. 1993. *Basin Research*, **5**, 79–102.

Leeder, M.R. & Mack, G.H. 2001. *Journal of the Geological Sciety of London*, **158**, 885–893.

Leeder, M.R. & Mack, G.H. 2007. In: *Sedimentary Processes, Environments and Basins: A Tribute to Peter Friend* (eds Nichols G. *et al.*), pp. 9–28. Special Publication 38, International Association of Sedimentologists.

Leeder, M.R. & Pérez-Arlucea, M. 2005. *Physical Processes in Earth and Environmental Sciences*. Blackwell Publishing.

Leeder, M.R. & Stewart, M.D. 1996. In: *Sequence Stratigraphy in British Geology* (eds Hesselbo, S.P. & Parkinson, D. N.), pp. 25–39. Special Publication 103, Geological Society of London.

Leeder, M.R. *et al.* 1988. *Philosophical Transactions of the Royal Society*, **327A**, 107–143.

Leeder, M.R. *et al.* 1998. *Basin Research*, **10**, 7–18.

Leeder, M.R. *et al.* 2005. *Sedimentology*, **52**, 683–691.

Lees, A. 1975. *Marine Geology*, **19**, 159–198.

Lees, A. & Miller, J. 1985. *Geological Journal*, **20**, 159–180.

Leg 155 Shipboard Scientific Party. 1994. *Eos, Transactions American Geophysical Union*, September, 435–437.

Legros, F. 2002. *Journal of Sedimentary Research*, **72**, 166–170.

Leinen, M. 1979. *Bulletin Geological Society America*, **90**, 801–803.

Leithold, E.L. & Hope, R.S. 1999. *Marine Geology*, **154**, 183–195.

Leopold, A. 1991. *The River of the Mother of God and Other Essays* (eds Flacer, S.L. & Callicott, J. B.). University of Wisconsin Press, Madison.

Leopold, L.B. & Bull, W.B. 1979. *Proceedings of the American Philosophical Society*, **123**, 168–202.

Leopold, L.B. & Wolman, M.G. 1957. *US Geological Survey Professional Paper* **282B**, 39–85.

Leopold, L.B. *et al.* 1964. *Fluvial Processes in Geomorphology*. Freeman, *San Francisco*.

Leprince, S. *et al.* 2008. *Eos, Transactions American Geophysical Union*, **89/1** 1–2.

Lerman, A. (ed.) 1978. *Lakes: Physics, Chemistry and Geology*. Springer, New York.

Le Roux, J.P. 2004. *Sedimentology*, **51**, 669–670.

Le Roux, J.P. 2007. *Sedimentology*, **54**, 1447–1448.

Lesueur, P. *et al.* 2002. *Sedimentology*, **49**, 1299–1320.

Levell, B.K. *et al.* 1988. *Bulletin of the American Association of Petroleum Geologists*, **72**, 775–796.

Lever, A. & McCave, I.N. 1983. *Journal of Sedimentary Petrology*, **53**, 811–832.

Levin, D.R. 1995. In: *Tidal Signatures in Modern and Ancient Sediments* (eds Flemming, B.C. & Bartoloma, A.), pp. 71–84. Special Publication 24, International Association of Sedimentologists.

Levin, N. *et al.* 2008. *Sedimentology*, **55**, 751–772.

Lewis, C.F.M. *et al.* 2008a. *Eos, Transactions American Geophysical Union*, **89/52** 541–542.

Lewis, C.F.M. *et al.* 2008. *Aquatic Ecosystem Health and Management*, **11**, 127–136.

Lezzar, K.E. *et al.* 1996. *Basin Research*, **8**, 1–28.

Li, G. *et al.* 2001. *Continental Shelf Research*, **21**, 607–625.

Li, J. *et al.* 1997. *Bulletin of the Geological Society of America*, **109**, 1361–1371.

Li, M. & O'Connor, B.A. 2007. *Sedimentology*, **54**, 1345–1363.

Li, M.Z. & Gust, G. 2000. *Sedimentology*, **47**, 71–86.

Li, M.Z. & Komar, P.D. 1992. *Journal of Sedimentary Research*, **62**, 584–590.

Li, X. *et al.* 1996. *Nature*, **380**, 416–419.

Li, Y.-H. 1976. *Geology*, **4**, 105–107.

Lien, T. *et al.* 2003. *Sedimentology*, **50**, 113–148.

Lighthill, J. 1978. *Waves in Fluids*. Cambridge University Press, Cambridge.

Ligrani, P.M. & Moffat, R.J. 1986. *Journal of Fluid Mechanics*, **162**, 69–98.

Linacre, E. & Geerts, B. 1997. *Climates and Weather Explained*. Routledge, London.

Link, M.H. & Osborne, R.H. 1982. In: *Geologic History of Ridge Basin, Southern California* (eds Crowell, J.C. & Link, M. H.), pp. 63–78. Pacific Section, Society of Economic Paleontologists and Mineralogists.

Lintern, G. & Sills, G. 2006. *Journal of Sedimentary Research*, **76**, 1183–1195.

Lippmann, F. 1973. *Sedimentary Carbonate Minerals*. Springer, New York.

Lisitzin, A.P. 1996. *Oceanic Sedimentation*. American Geophysical Union, Washington, DC.

Liu, T. & Broecker, W.S. 2008. *Geology*, **36**, 403–406.

Liu, J.P. *et al.* 2009. *The Sedimentary Record*, **7**, 4–9.

Liutkus, C.M. & Wright, J.D. 2008. *Sedimentology*, **55**, 965–978.

Livingstone, D.A. 1963. *Chemical Composition of Rivers and Lakes*. Professional Paper 440G, US Geological Survey.

Livingstone, I. & Thomas, D.S.G. 1993. In: *The Dynamics and Environmental Context of Aeolian Sedimentary Systems* (ed. Pye, K.), pp. 91–101. Special Publication 72, Geological Society of London.

Lobo, F.J. *et al.* 2000. *Marine Geology*, **164**, 91–117.

Lobo, F.J. *et al.* 2003. *Journal of Sedimentary Research*, **73**, 973–986.

Lockwood, J.G. 1979. *Causes of Climate*. Edward Arnold, London.

Logan, B.W. 1987. *The MacLeod Evaporite basin, Western Australia*. Memoir 44, American Association of Petroleum Geologists.

Logan, B.W. *et al.* (eds) 1970. *Carbonate Sedimentation and Environments, Shark Bay, Western Australia*. Memoir 13, American Association of Petroleum Geologists.

Logan, B.W. *et al.* (eds) 1974. *Evolution and Diagenesis of Quaternary Carbonate Sequences, Shark Bay, W. Australia*. Memoir 22, American Association of Petroleum Geologists.

Lokier, S. & Stobier, T. 2008. *Journal of Sedimentary Research*, **78**, 423–431.

Lomas S.A. & Joseph P. (eds) 2004. *Confined Turbidite Systems*, Special Publication 222, Geological Society of London.

Longuet-Higgins, M.S. 1953. *Philosophical Transactions of the Royal Society, London (A)*, **245**, 535–581.

Longuet-Higgins, M.S. 1970. *Journal of Geophysical Research*, **75**, 6778–6801.

Longuet-Higgins, M.S. & Stewart, R.W. 1964. *Deep-Sea Research*, **11**, 529–563.

Lønne, I. *et al.* 2001. *Journal of Sedimentary Research*, **71**, 922–943.

Loope, D.B. 1984. *Sedimentology*, **31**, 123–132.

Loope, D.B. *et al.* 2004. *Sedimentology*, **51**, 315–322.

Lorang, M.S. & Komar, P.D. 1990. *Nature*, 433–434.

Loreau, J.-P. 1982. *Sédiments Aragonitique et leur Genese*. Memoires, Serie C, 47, Museum d'Histoire Naturelle, Paris.

Loreau, J.-P. & Purser, B.H. 1973. In: *The Persian Gulf Holocene Carbonate Sedimentation and Diagenesis in a Shallow Epicontinental Sea* (ed. Purser, B.H.), pp. 279–328. Springer, Heidelberg.

Løseth, T.M. *et al.* 2006. *Sedimentology*, **53**, 735–767.

Loucks, R.G. & Sarg, J.F. 1993. *Carbonate Sequence Stratigraphy*, Memoir 57, American Association of Petroleum Geologists.

Lourens, L.J.A. *et al.* 1996. *Palaeoceanography*, **11**, 391–431.

Lovelock, J.E. 1979. *Gaia: A New Look at Life on Earth*. Oxford University Press, Oxford.

Lovely, D.R. *et al.* 1987. *Nature*, **330**, 252–254.

Lovely, D.R. *et al.* 1991. Microbial reduction of uranium. *Nature*, **350**, 413–15.

Lowe, D.R. 1975. *Sedimentology*, **22**, 157–204.

Lowe, D.R. 1976. *Journal of Sedimentary Petrology*, **46**, 188–199.

Lowe, D.R. 1982. *Journal of Sedimentary Petrology*, **52**, 279–297.

Lowe, D.R. & Lopiccolo, R.D. 1974. *Journal of Sedimentary Petrology*, **44**, 484–501.

Lowenstam, H.A. 1963. In: *The Earth Sciences—Problems and Progress in Current Research* (ed. Donnelly, T.W.). pp. 137–195. University of Chicago Press, Chicago.

Lowenstein, T.K. 1988. *Bulletin of the Geological Society of America*, **100**, 592–608.

Lowenstein, T.K. & Hardie, L.A. 1985. *Sedimentology*, **32**, 627–644.

Lowenstein, T.K. *et al.* 2001. *Science*, **294**, 1086–1088.

Lowenstein, T.K. *et al.* 2003. *Geology*, **31**, 857–860.

Lowenstein, T.K. *et al.* 2005. *Geochimica Cosmochimica Acta*, **69**, 1701–1719.

Lu, H. *et al.* 2009. *Basin Research*, doi: 10.1111/j.1365-2117.2009.00412.x

Lucchi, R. & Camerlenghi, A. 1993. *Bollettino di Oceanologia Teorica ed Applicata*, **11**, 3–25.

Ludwig, W. & Probst, P. 1996. *Global BioGeoChemical Cycles*, **10**, 23–41.

Lukasik, J.J. *et al.* 2000. *Sedimentology*, **47**, 851–881.

Lunt, I.A. *et al.* 2004. *Sedimentology*, **51**, 377–414.

Lunt, I.A. *et al.* 2007. *Sedimentology*, **54**, 71–87.

Lustig, L.K. 1965. *Clastic sedimentation in Deep Springs Valley, California*. pp. 131–192. Professional Paper 352F, US Geological Survey.

Lynch, H.D. & Morgan, P. 1987. In: *Continental Extensional Tectonics* (eds Coward, M. P. *et al.*), pp. 53–65. Special Publication 28, Geological Society of London.

Lyons, P.C. & Alpern, B. 1989. *International Journal of Coal Geology*, **12**, 1–4.

Lyons, T.W. 1997. *Geochimica et Cosmochimica Acta*, **61**, 3367–3382.

Macdonald, R.G. *et al.* 2009, *Sedimentology*, **56**, 1346–1367.

Macintyre, I.G. 1988. *Bulletin of the American Association of Petroleum Geologists*, **72**, 1360–1369.

Mack, G.H. & James, W.C. 1994. *Journal of Geology*, **102**, 360–366.

Mack, G.H. & Leeder, M.R. 1998. *Sedimentary Geology*, **117**, 207–219.

Mack, G.H. & Leeder, M.R. 1999. *Journal of Sedimentary Research*, **69**, 635–652.

Mack, G. H. & Madoff, R. D. 2005. *Sedimentology*, **52**, 191–211.

Mack, G.H. & Seager, W.R. 1990. *Bulletin of the Geological Society of America*, **102**, 45–53.

Mack, G.H. & Seager, W.R. 1995. *Journal of the Geological Society*, **152**, 551–560.

Mack, G.H. *et al.* 1993. *Bulletin of the Geological Society of America*, **105**, 129–136.

Mack, G.H. *et al.* 1994. *American Journal of Science*, **294**, 621–640.

Mack, G.H. *et al.* 2003. *Palios*, **18**, 403–420.

Mack, G.H. *et al.* 2006. *Earth-Science Reviews*, **79**, 141–162.

Mack, G.H. *et al.* 2008. *Journal of Sedimentary Research*, **78**, 432–442.

Mackenzie, F.T. & Morse, J.W. 1992. *Geochimica et Cosmochimica Acta*, **56**, 3281–3295.

Mackey, S.D. & Bridge, J.S. 1995. *Journal of Sedimentary Research*, **65**, 7–31.

MacKinnon, L. & Jones, B. 2001. *Journal of Sedimentary Research*, **71**, 568–580.

Macklin, M.G. *et al.* 1992. *Bulletin of the Geological Society of America*, **104**, 631–643.

MacNeil, F.S. 1954. *American Journal of Science*, **252**, 385–401.

Macquaker, J.H.S. *et al.* 1996. In: *Sequence Stratigraphy in British Geology* (eds Hesselbo, S.P. & Parkinson, D. N.), pp. 25–39. Special Publication 103, Geological Society of London.

McArthur, J.M. *et al.* 1986. *Earth Planetary Science Letters*, **77**, 20–34.

McArthur, J.M. *et al.* 2001. *Journal of Geology*, **109**, 155–170.

McCabe, M. & Eyles, N. 1988. *Sedimentary Geology*, **59**, 1–14.

McCabe, P.J. 1984. In: *Sedimentology of Coal and Coal-bearing Sequences* (eds Rahmani, R.A. & Flores, R. M.), pp. 13–42. Special Publication 7, International Association of Sedimentologists.

McCabe, P.J. & Shanley, K.W. 1992. *Geology*, **20**, 741–744.

McCaffrey, W. & Kneller, B. 2001. *American Association of Petroleum Geologists Bulletin*, **85**, 971–988.

McCaffrey, W. *et al.* (eds) 2001. *Particulate Gravity Currents*. Special Publication 31, International Association of Sedimentologists.

McCarthy, P.J. & Plint, A.G. 2003. *Sedimentology*, **50**, 1187–1220.

McCarthy, P.J. *et al.* 1997. *Sedimentology*, **44**, 197–220.

McCarthy, P.J. *et al.* 1999. *Sedimentology*, **46**, 861–891.

McCarthy, T.S. & Candle, A.B. 1995. *Journal of Sedimentary Research*, **65**, 581–583.

McCarthy, T.S. *et al.* 1991. *Sedimentology*, **38**, 471–487.

McCauley, S.E. & DePaola, D.J. 1997. In: *Tectonic Uplift and Climate Change* (ed. Ruddiman W.F.), pp. 428–470. Plenum, New York.

McCave, I.N. 1971. *Marine Geology*, **10**, 199–225.

McCave, I.N. 1972. In: *Shelf Sediment Transport: Process and Pattern* (eds Swift, D. P. *et al.*), pp. 225–248. Dowden, Hutchinson and Ross, Stroudsberg, PA.

McCave, I.N. 1979. *Marine Geology*, **31**, 101–114.

McCave, I.N. 1985. *Marine Geology*, **66**, 169–188.

McCave, I.N. 1995. *Philosophical Transactions of the Royal Society of London*, **B348**, 229–241.

McCave, I.N. & Jones, K.P.N. 1988. *Nature*, **333**, 250–252.

McCave, I.N. *et al.* 1981. *Journal of Sedimentary Petrology*, **50**, 1049–1062.

McCave, I.N. *et al.* 1995. *Paleooceanography*, **10**, 593–610.

McCave, I.N. *et al.* 2002. In: *Deep-water Contourite Systems: Modern Drifts and Ancient Series, Seismic and Sedimentary Characteristics* (eds Stow, D. A. V. *et al.*), pp. 21–38. Memoir 22, Geological Society of London.

McCave, I.N. *et al.* 2006. *Sedimentology*, **53**, 919–928.

McClennan, S.M. 1993. *Journal of Geology*, **101**, 295–303.

McClung, D.M. 2001. *Journal of Geophysical Research*, **106**, 16489–16498.

McCulloch, M.T. & Esat, T. 2000. *Chemical Geology*, **169**, 107–129.

McElwaine, J. & Nishimura, K. 2001. In: *Particulate Gravity Currents* (eds McCaffrey, W. *et al.*), pp. 135–148. Special Publication 31, International Association of Sedimentologists.

McElwaine, J.C. *et al.* 2005. *Nature*, **435**, 479–482.

McEwan, I.K. & Willetts, B.B. 1993. *Journal of Fluid Mechanics*, **252**, 99–115.

McEwan, I.K. & Willetts, B.B. 1994. *Sedimentology*, **41**, 1241–1251.

McEwan, I.K. *et al.* 1999. *Sedimentology*, **46**, 407–416.

McGillycuddy, D.J. *et al.* 1998. *Nature*, **394**, 263–266.

McHargue, T.R. 1991. In: *Seismic Facies and Sedimentary Processes of Submarine Fans and Turbidite Systems* (eds Weimer, P. & Link, M. H.), pp. 403–414. Springer, Berlin.

McHugh, C.M.G. *et al.* 2004. *Geology*, **32**, 169–172.

McKean, J.A. *et al.* 1993. *Geology*, **21**, 343–346.

McKee, E.D. 1966. *Sedimentology*, **7**, 1–61.

McKee, E.D. (ed.) 1978. *A Study of Global Sand Seas*. Professional Paper 1052, US Geological Survey.

McKenna-Neumann, C. 2004. *Sedimentology*, **51**, 1–17.

McKenna-Neumann, C. *et al.* 2000. *Sedimentology*, **47**, 211–226.

McKenzie, D.P. 1978. *Earth and Planetary Science Letters*, **40**, 25–32.

McLelland, S.J. *et al.* 1999. In: *Fluvial Sedimentology VI* (eds Smith, N.D. & Rodgers, J.), pp. 43–57. Special Publication 28, International Association of Sedimentologists.

McNamara, K.J. & Awramik, S.M. 1994. *Science Progress*, **77**, 1–20.

McLellan, H. 1965. *Elements of Physical Oceanography*. Pergamon, Oxford.

McManus, J.F. *et al.* 1999. *Science*, **283**, 971–975.

McTigue, D.F. 1981. *Journal of the Hydraulics Division, American Society of Civil Engineers*, **107**, HY6, 659–673.

Maddox, J. 1990. *Nature*, **347**, 225.

Magirl, C.S. *et al.* 2007. *Eos, Transactions American Geophysical Union*, **88/17** 191–193.

Maillet, G.M. *et al.* 2006. *Marine Geology*, **234**, 159–177.

Mainguet, M. 1978. *Geoforum*, **9**, 17–28.

Mainguet, M. 1983. In: *Mega-geomorphology* (eds Gardner, R. & Scoging, H.), pp. 113–133. Clarendon Press, Oxford.

Mainguet, M. & Canon, L. 1976. *Revue Geographie Physique et de Geologie Dynamique*, **18**, 241–250.

Maizels, J. 1989. *Journal of Sedimentary Petrology*, **59**, 204–223.

Maizels, J. 1990. In: *Alluvial Fans: A Field Approach* (eds Rachocki, A.H. & Church, M.), pp. 271–304. Wiley, New York.

Major, J.J. 1997. *Journal of Geology*, **105**, 345–366.

Major, J.J. 2000. *Journal of Sedimentary Research*, **70**, 64–83.

Major, J.J. & Iverson, R.M. 1999. *Geological Society of America Bulletin*, **111**, 1424–1434.

Mäkinen, J. 2003. *Sedimentology*, **50**, 327–360.

Makaske, B. 1998. *Anastomosing Rivers: Forms, Processes and Sediments*. Nederlandse Geografische Studies 249, Utrecht University.

Makaske, B. *et al.* 2002. *Sedimentology*, **49**, 1049–1071.

Makaske, B. *et al.* 2007. *Journal of Sedimentary Research*, **77**, 110–123.

Makse, H.A. *et al.* 1997. *Nature*, **386**, 379–382.

Makse, H.A. *et al.* 1998. *Philosophical Magazine*, **77B**, 1341–1351.

Malakoff, D. 1998. *Science*, **281**, 190–192.

Malmström, M. & Banwart, S. 1997. *Geochimica et Cosmochimica Acta*, **61**, 2779–2799.

Maltman, A.J. *et al.* (eds). 2000. *Deformation of Glacial materials*. Special Publication 176, Geological Society of London.

Mancin, N. *et al.* 2009. *Basin Research*, **21**, 799–823.

Mann, P. 1997. *Geology*, **25**, 211–214.

Manning, A.J. & Dyer, K. R. 1999. *Marine Geology*, **160**, 147–170.

Mantz, P.A. 1978. *Sedimentology*, **25**, 83–103.

*et al.* 5–Marino, G.G. *et al.* 2007. *Geology*, **35**, 675–678.

Marriott, S.B. 1996. In: *Floodplain Processes* (eds Anderson, M. G. *et al.*), pp. 63–93. Wiley, New York.

Marsaglia, K.M. 1995. In: *Tectonics of Sedimentary Basins* (eds Busby, C. & Ingersoll, R.), pp. 299–330, Blackwell Science, Oxford.

Marsaglia, K.M. *et al.* 1995. In: *Active Margins and Marginal Basinsof the Western Pacific*, pp. 291–314. Geophysical Monograph 88, American Geophysical Union.

Martin, J.H. *et al.* 1990. *Nature*, **345**, 156–158.

Martin, J.H. *et al.* 1994. *Nature*, **371**, 123–129.

Martín-Martín, M. *et al.* 2001. *Basin Research*, **13**, 419–433.

Martino, R.L. & Sanderson, D.D. 1993. *Journal of Sedimentary Petrology*, **63**, 105–119.

Marzo, M. & Anadon, P. 1988. In: *Fan Deltas: Sedimentology and Tectonic Settings* (eds Nemec, W. & Steel, R. J.), pp. 318–341. Blackie, London.

Marzo, M. *et al.* 1988. *Sedimentology*, **35**, 719–738.

Mason, J.E. 2010. PhD thesis, University of East Anglia, England.

Massari, F. & Colella, A. 1988. In: *Fan Deltas: Sedimentology and Tectonic Settings* (eds Nemec, W. & Steel, R. J.), pp. 103–122. Blackie, London.

Massari, F. & Parea, G.C. 1988. *Sedimentology*, **35**, 881–913.

Massari, F. *et al.* 2007. *Journal of Sedimentary Research*, **77**, 461–468.

Masselink, G. 2008. *Sedimentology*, **55**, 667–687.

Masselink, G. & Hughes, M.G. 2003. *Introduction to Coastal Processes and Geomorphology*. New York, Oxford University Press.

Masselink, G. *et al.* 2007. *Sedimentology*, **54**, 39–53.

Massey, B.S. 1979. *Mechanics of Fluids*, 4th edn. Van Nostrand Reinhold, New York.

Masson, D.G. 1994. *Basin Research*, **6**, 17–33.

Masson, D.G. 1996. *Geology*, **24**, 231–234.

Masson, D.G. *et al.* 1992. In: *Geologic Evolution of Atlantic Continental Rises* (eds Poag, C.W. & de Graciansky, P. C.), pp. 327–343. Van Nostrand Reinhold, New York.

Masson, D.G. *et al.* 2004. *Sedimentology*, **51**, 1207–1241.

Mastandrea, A. *et al.* 2006. *Sedimentology*, **53**, 465–480.

Mather, A.E. & Hartley, A. 2005. In: *Alluvial Fans: Geomorphology, Sedimentology, Dynamics* (eds Harvey, A. M. *et al.*), pp. 9–29. Special Publication 251, Geological Society of London.

Matter, A. & Tucker, M.E. (eds) 1978. *Modern and Ancient Lake Sediments*. Special Publication 2, International Association of Sedimentologists.

Mattheus, C.R. *et al.* 2007. *Journal of Sedimentary Research*, **77**, 213–224.

Matthews, R.K. 1966. *Journal of Sedimentary Petrology*, **36**, 428–454.

Matyas, E.L. 1984. *Canadian Journal of Earth Sciences*, **21**, 1156–1160.

Maxwell, W.G.H. & Swinchatt, J.P. 1970. *Bulletin of the Geological Society of America*, **81**, 691–724.

Mayer, L. *et al.* 1984. *Catena Supplement*, **5**, 137–151.

Maynard, J.R. 2006. *Sedimentology*, **53**, 515–536.

Maynard, J.R. & Leeder, M.R. 1992. *Journal of the Geological Society of London*, **149**, 303–311.

Mazzullo, S.J. 2000. *Journal of Sedimentary Research*, **70**, 10–23.

Mazzullo, S.J. 2006. *Sedimentology*, **53**, 1015–1047.

Mazzullo, S.J. *et al.* 1995. *Geology*, **23**, 341–344.

Mazzullo, S.J. *et al.* 2003. *Sedimentology*, **50**, 743–770.

Meckel, T.A. *et al.* 2007. *Basin Research*, **19**, 19–31.

Medwedeff, D.A. & Wilkinson, B.H. 1983. In: *Coated Grains* (ed. Peryt, T.M.), pp. 109–115. Springer, Berlin.

Meehl, G.A. 1992. *Annual Review of Earth and Planetary Sciences*, **20**, 85–112.

Mehta, A.J. 1989. *Journal of Geophysical Research*, **94**, 14303–14314.

Mehta, A.J. (ed.) 1993. *Nearshore and Estuarine Cohesive Sediment Transport*. American Geophysical Union.

Mehta, A. & Barker, G.C. 1991. *Physical Review Letters*, **67**, 394–397.

Meijer, P.T. & Tuenter, E. 2007. *Journal of Marine Systems*, **68**, 349–365.

Melchor, R.N. 2007. *Sedimentology*, **54**, 1417–1446.

Melim, L.A. & Scholle, P.A. 1995. *Journal of Sedimentary Research*, **65** 107–118.

Mellere, D. & Steel, R.J. 1995. *Sedimentology*, **42**, 551–574.

Melosh, H.J. 1979. *Journal of Geophysical Research*, **84**, 7513–7520.

Melosh, H.J. 1987. In: *Debris Flows/Avalanches: Processes, Recognition, and Mitigation* (eds Costa, J.E. & Wieczorek, G. F.), pp. 41–49. Reviews in Engineering Geology 7, Geological Society of America.

Melton, M.A. 1965. *Journal of Geology*, **73**, 1–38.

Menzies, J. 1995. *Modern Glacial Environments*. Butterworth Heinemann, Oxford.

Menzies, J. & Rose, J.(eds) 1989. Sedimentary Geology, **62**, 2–4.

Merico, A. *et al.* 2008. *Nature*, **452**, 979–982.

Mertes, L.A.K. 1994. *Geology*, **22**, 171–174.

Mertes, L.A.K. *et al.* 1996. *Bulletin of the Geological Society of America*, **108**, 1089–1107.

Métivier, F. *et al.* 2005. *Journal of Sedimentary Research*, **75**, 6–11.

Meyer, G.A. *et al.* 1992. *Nature*, **357**, 147–150.

Meyer, H.J. 1984. *Journal of Crystal Growth*, **66**, 639–646.

Miall, A.D. 1985. *Sedimentology*, **32**, 763–788.

Miall, A.D. 1991. *Journal of Sedimentary Petrology*, **61**, 497–505.

Miall, A.D. 1994. *Journal of Sedimentary Research*, **64**, 146–158.

Miall, A.D. 1995. In: *Tectonics of Sedimentary Basins* (eds Busby, C. & Ingersoll, R.), pp. 393–424, Blackwell Science, Oxford.

Miall, A.D. 1996. *The Geology of Fluvial Deposits*. Springer, Berlin.

Miall, A.D. 1997. *The Geology of Stratigraphic Sequences*. Springer, Berlin.

Miall, A.D. 2001. *Sedimentology*, **48**, 971–985.

Miall, A.D. 2006. *Sedimentary Geology*, **186**, 39–50.

Miall, A.D. & Miall, C.E. 2001. *Earth-Science Reviews*, **54**, 321–348.

Middleton, G.V. 1965. *Journal of Sedimentary Petrology*, **35**, 922–927.

Middleton, G.V. 1966a. *Canadian Journal of Earth Sciences*, **3**, 523–546.

Middleton, G.V. 1966b. *Canadian Journal of Earth Sciences*, **3**, 627–637.

Middleton, G.V. 1966c. *Canadian Journal of Earth Sciences*, **4**, 475–505.

Middleton, G.V. 1970. In: *Flysch Sedimentology in North America* (ed. Lajoie, J.), pp. 253–272. Report 7, Geological Association of Canada.

Middleton, G.V. & Wilcock, P.R. 1994. *Mechanics in the Earth and Environmental Sciences*. Cambridge University Press, Cambridge.

Middleton, N.J. *et al.* 1986. In: *Aeolian Geomorphology* (ed. Nickling, W.G.), pp. 237–260. Allen and Unwin, Boston.

Midgley, N.G. *et al.* 2007. In: *Glacial Sedimentary Processes and Products* (eds Hambrey, M. *et al.*), pp. 11–22. Special Publication 39, International Association of Sedimentologists.

Midtgaard, H.H. 1996. *Journal of Sedimentary Research*, **66**, 343–353.

Migeon, S. *et al.* 2004. *Journal of Sedimentary Research*, **74**, 580–598.

Milhous, R.T. 1973. Sediment transport in a gravelbottomed stream. PhD thesis, Oregon State University, Corvallis.

Miller, D.J. & Eriksson, K.A. 1997. *Journal of Sedimentary Research*, **67**, 653–660.

Miller, E.K. *et al.* 1993. *Nature*, **362**, 438–441.

Miller, M.C. *et al.* 1977. *Sedimentology*, **24**, 507–528.

Milligan, T.G. *et al.* 2007. *Continental Shelf Research*, **27**, 309–321.

Milliman, J.D. & Mead, R.H. 1983. *Journal of Geology*, **91**, 1–21.

Milliman, J.D. & Syvitski, J.P.M. 1992. *Journal of Geology*, **100**, 525–544.

Milliman, J.D. *et al.* 1993. *Journal of Sedimentary Petrology*, **63**, 589–595.

Millot, R. *et al.* 2002. *Earth and Planetary Letters*, **196**, 83–98.

Milner, C. & Hughes, R.E. 1968. *Methods for Estimating the Primary Production of Grasslands*, IBP Handbook 6. Blackwell Scientific Publications, Oxford.

Mitchell. N.C. 2004. *American Journal of Science*, **304**, 590–611.

Mitchell, N.C. & Huthnance, J.M. 2008. *Journal of Sedimentary Research*, **78**, 29–44.

Mohrig, D. *et al.* 1998. *Bulletin Geological Society of America*, **110**, 387–394.

Mohrig, D. *et al.* 2000. *Bulletin of the Geological Society of America*, **112**, 1787–1803.

Molina-Cruz, L. *et al.* 2002. *Sedimentology*, **49**, 1401–1410.

Molnar, P. & England, P.C. 1990. *Nature*, **346**, 29–34.

Molnar, P. *et al.* 1993. *Reviews of Geophysics*, **31**, 357–396.

Molnar, P. *et al.* 2006. *Nature*, **444**, doi: 10.1038/nature053682006

Molnar, P. *et al.* 2007. *Journal of Geophysical Research*, **112**, F03014.

Molnia, B.F. (ed.) 1983a. *Glacial-Marine Sedimentation*. Plenum, New York.

Molnia, B.F. 1983b. In: *Glacial-Marine Sedimentation* (ed. Molnia, B.F.), pp. 95–144. Plenum, New York.

Monaghan, J.J. 2007. *Annual Reviews of Fluid Mechanics*, **39**, 245–261.

Monty, C.L.V. *et al.* (eds) 1995. *Carbonate Mud Mounds—Their Origin and Evolution*. Special Publication 23, International Association of Sedimentologists

Moore, G.T. 1979. *Bulletin of the American Association of Petroleum Geologists*, **63**, 660–667.

Moore, J.G. & Mark, R.K. 1986. *Eos, Transactions American Geophysical Union*, **2** December, 1353–1362.

Moore, J.G. *et al.* 1989. *Journal of Geophysical Research*, **94**, 17465–17484.

Moore, J.N. *et al.* 1984. *Journal of Sedimentary Petrology*, **54**, 615–625.

Moore, L.J. *et al.* 2004. *Journal of Sedimentary Research*, **74**, 690–696.

Mora, A. *et al.* 2009. *Basin Research*, **21**, 111–137.

Moran, K. *et al.* 2006. *Nature*, **441**, 601–605.

Morozova, G.S. & Smith, N.D. 2000. *Sedimentary Geology*, **130**, 81–105.

Morris, J.D. 1991. *Annual Review of Earth and Planetary Sciences*, **19**, 313–350.

Morris, J.E. *et al.* 2006. *Sedimentology*, **53**, 1229–1263.

Morris, S.A. *et al.* 1998. *Sedimentology*, **45**, 365–377.

Morse, J.W. 1974. *American Journal of Science*, **274**, 97–107.
Morse, J.W. & Berner, R.A. 1972. *American Journal of Science*, **272**, 840–851.
Morse, J.W. & Mackenzie, F.T. 1990. *Geochemistry of Sedimentary Carbonates*. Developments in Sedimentology 48. Elsevier, Amsterdam.
Morse, J.W. *et al.* 1984. *Journal of Geophysical Research*, **89**, 3604–3614.
Morse, J.W. *et al.* 1997. *Geology*, **25**, 85–87.
Morse, J.W. *et al.* 2006. *Geochimica et Cosmochimica Acta*, **70**, 5814–5830.
Mortimer, R.G.J. & Coleman, M.L. 1997. *Geochimica et Cosmochimica Acta*, **61**, 1705–1711.
Mortimer, E. *et al.* 2005. *Basin Research*, **17**, 337–359.
Morton, R.A. 1981. In: *Holocene Marine Sedimentation in the North Sea Basin* (eds Nio, S.-D. *et al.*), pp. 385–396. Special Publication 5, International Association of Sedimentologists.
Morton, R.A. *et al.* 2008. *Journal of Sedimentary Research*, **78**, 624–637.
Mossop, G. *et al.* 1983. *Sedimentology*, **30**, 493–509.
Moum, J.N. & Caldwell, D.R. 1994. *Eos, Transactions American Geophysical Union*, **75**, 489–490.
Mountjoy, E.W. *et al.* 1972. *Proceedings of the 24th International Geological Congress*, Vol. 6, pp. 172–89.
Mountney, N.P. 2006. *Sedimentology*, **53**, 789–823.
Mountney, N.P. & Howell, J. 2000. *Sedimentology*, **47**, 825–849.
Mountney, N.P. & Jagger, A. 2004. *Sedimentology*, **51**, 713–743.
Mountney, N.P. & Thompson, D.B. 2002. *Sedimentology*, **49**, 805–833.
Mountney, N.P. & Westbrook, G.K. 1996. *Basin Research*, **8**, 85–101.
Muck, M.T. & Underwood, M.B. 1990. *Geology*, **18**, 54–57.
Muhs, D.R. *et al.* 2002. *Quaternary Science Reviews*, **21**, 1355–1383.
Muir, I.J. *et al.* 1990. *Geochimica et Cosmochimica Acta*, **54**, 2247–2256.
Mukerji, A.B. 1990. In: *Alluvial Fans: A Field Approach* (eds Rachocki, A.H. & Church, M.), pp. 131–149. Wiley, New York.
Mulder, T. & Alexander, J. 2001. *Sedimentology*, **48**, 269–299.
Mulder, T. & Syvitski, J.P.M. 1995. *Journal of Geology*, **103**, 285–299.
Mulder, T. *et al.* 1997. *Sedimentology*, **44**, 305–326.
Müller, A. & Gyr, A. 1982. In: *Mechanics of Sediment Transport*, Euromech 156 (eds Sumer, B.M. & Müller, A.), pp. 41–45. Balkema, Rotterdam.
Müller, A. & Gyr, A. 1986. *Journal of Hydraulic Research*, **24**, 359–275.
Muller-Karger, F. *et al.* 2000. *Eos, Transactions American Geophysical Union*, **81/45** 534–535.
Mullins, H.T. & Hine, A.C. 1989. *Geology*, **17**, 30–33.
Mullins, H.T. & Neumann, A.C. 1979. In: *Geology of Continental Slopes* (eds Doyle, L.J. & Pilkey, O. H.). Special Publication 27, Society of Economic Paleontologists and Mineralogists.
Mullins, H.T. *et al.* 1980a. *Journal of Sedimentary Petrology*, **50**, 117–131.
Mullins, H.T. *et al.* 1980b. *Bulletin of the American Association of Petroleum Geologists*, **64**, 1701–1717.
Mullins, H.T. *et al.* 1984. *Sedimentology*, **31**, 141–168.
Munro-Stasiuk, M.J. 2000. *Journal of Sedimentary et al. Research*, **70**, 94–106.
Murray, A.B. & Paola, C. 1994. *Nature*, **371**, 54–57.
Murray, J.B. *et al.* 2005. *Nature*, **434**, 353–356.
Murray, P.B. *et al.* 1991. In: *Sand Transport in Rivers, Estuaries and the Sea*, Euromech 262 (eds Soulsby, R. L. & Bettess, R.), pp. 37–44. Balkema, Rotterdam.
Murray, S.P. 1970. *Journal of Geophysical Research*, **75**, 4579–4582.
Murray, T. & Clarke, G.K.C. 1995. *Journal of Geophysical Research*, **100**, 10231–10245.
Murray, T. *et al.* 2000. *Journal of Geophysical Research*, **105**, 13491–13507.
Murray, T. *et al.* 2008. *Geophysical Research Letters*, **35**, 12504, doi: 10.1029/2008GL033681
Mustard, J.F. *et al.* 2008. *Nature*, **454**, 305–309.
Muto, T. 1989. *Journal of Geology*, **97**, 640–645.
Mutti, E. & Normark, W.R. 1991. In: *Seismic Facies and Sedimentary Processes of Submarine Fans and Turbidite Systems* (eds Weimer, P. & Link, M. H.), pp. 75–106. Springer, Berlin.
Myers, K.J. & Milton, N.J. 1996. In: *Sequence Stratigraphy* (eds Emery, D. & Myers, K. J.), pp. 11–44. Blackwell Science, Oxford.
Mylroie, J.E. & Carew, J.L. 1988. *Quaternary Science et al. Reviews*, **7**, 55–64.
Mylroie, J.E. & Carew, J.L. 1990. *Earth Surface Processes and Landforms*, **15**, 413–424.
Myrow, P.M. & Southard, J.B. 1991. *Journal of Sedimentary Petrology*, **61**, 202–210.
Nadon, G.C. 1998. *Geology*, **26**, 727–730.
Nageswara Rao, K. *et al.* 2005. In: *River Deltas: Concepts, Models, and Examples* (eds Giosan, L. & Bhattacharya, J. P.), pp. 435–451. Special Publication 83, Society of Economic Paleontologists and Mineralogists.
Nagy, K.L. 1995. In: *Chemical Weathering Rates of Silicate Minerals* (eds White, A.F. & Branfield, S.L.), Chapter 5. American Mineralogical Society, Washington, DC.
Naish, T. 1997. *Geology*, **25**, 1139–1142.
Naish, T. *et al.* 2007. *Eos, Transactions American Geophysical Union*, **88/50** 557–558.
Naish, T. *et al.* 2009. *Nature*, **458**, 322–328.
Najman, Y. *et al.* 2004. *Basin Research*, **16**, 1–24.
Nakajima, T. 2006. *Journal of Sedimentary Research*, **76**, 60–73.

Nakajima, T. & Satoh, M. 2001. *Sedimentology*, **48**, 435–463.

Nanayama, F. *et al.* 2003. *Nature*, **424**, 660–662.

Nanson, G.C. 1980. *Sedimentology*, **27**, 3–29.

Narbonne, G.M. & James, N.P. 1996. *Sedimentology*, **43**, 827–848.

Naruse, H. & Masuda, F. 2006. *Journal of Sedimentary Research*, **76**, 854–865.

Naylor, N. & Sinclair, H.D. 2008. *Basin Research*, **20**, 285–303.

Neal, A. *et al.* 2002. *Sedimentology*, **49**, 789–804.

Neal, A. *et al.* 2008. *Journal of Sedimentary Research*, **78**, 638–653.

Needham, R.S. 1978. *Sedimentology*, **25**, 285–296.

Neev, D. 1978. *Abstracts Volume 10th International Association of Sedimentologists Meeting*, Jerusalem, Vol. 2, p. 459.

Neev, D. & Emery, K.O. 1967. *Israel Geological Survey Bulletin*, **41**, 1–147.

Negri, A. *et al.* 2009. *Palaeogeography, Palaeoclimatology, Palaeoecology*, **273**, 213–227.

Nelson, C.S. *et al.* 1988. *Sedimentary Geology*, **60**, 71–94.

Nemec, W. & Kazanci, N. 1999. *Sedimentology*, **46**, 139–170.

Nemec, W. & Postma, G. 1993. In: *Alluvial Sedimentation* (eds Marzo, M. & Puigdefabregas, C.), pp. 235–276. Special Publication 17, International Association of Sedimentologists.

Nemec, W. & Steel, R.J. (eds) 1988. *Fan Deltas: Sedimentology and Tectonic Settings.* Blackie, London.

Nesbitt, H.W. & Young, G.M. 1982. *Nature*, **299**, 715–717.

Nesbitt, H.W. & Young, G.M. 1989. *Journal of Geology*, **97**, 129–147.

Nesbitt, H.W. *et al.* 1997. *Journal of Geology*, **105**, 173–191.

Neukum, G. *et al.* 2004. *Nature*, **432**, 971–979.

Neumann, A.C. & Land, L.S. 1975. *Journal of Sedimentary Petrology*, **45**, 763–786.

Neumann, A.C. & MacIntyre, I. 1985. *Proceedings 5th International Coral Reef Symposium*, **3**, 105–110.

Neumann, A.C. *et al.* 1977. *Geology*, **5**, 4–10.

Neumeier, U. & Amos, C.L. 2006. *Sedimentology*, **53**, 259–277.

Newbould, P.J. 1967. *Methods for Estimating the Primary Production of Forests.* IBP Handbook 2. Blackwell Scientific Publications, Oxford.

Newton, M.S. 1994. In: *Sedimentology and Geochemistry of Modern and Ancient Saline Lakes* (eds Renaut, R.W. & Last, W. M.), pp. 143–157. Special Publication 50, Society of Economic Paleontologists and Mineralogists.

Nguyen, Q.D. & Boger, D.V. 1992. *Annual Review of Fluid Mechanics*, **24**, 47–88.

Nichol, S.L. & Kench, P.S. 2008. *Sedimentology*, **55**, 1173–1187.

Nichol, S.L. *et al.* 1997. *Sedimentology*, **44**, 263–286.

Nicholas, A.P. *et al.* 2006. *Journal of Hydrology*, **329**, 577–594.

Nichols, G.J. 1987. In: *Recent Developments in Fluvial Sedimentology* (eds Ethridge, F. G. *et al.*), pp. 269–278. Special Publication 39, Society of Economic Paleontologists and Mineralogists.

Nichols, G.J. 2005. In: *Alluvial Fans: Geomorphology, Sedimentology, Dynamics* (eds Harvey, A. M. *et al.*), pp. 187–206. Special Publication 251, Geological Society of London.

Nichols, G.J. 2008. In: *Sedimentary Processes, Environments and Basins* (eds Nichols, G. *et al.*), pp. 569–590. Special Publication 38, International Association of Sedimentologists.

Nichols, G.J. & Fisher, J.A. 2008. *Sedimentary Geology*, **195**, 75–90.

Nichols, G.J. & Hirst, J.P.P. 1998. *Journal of Sedimentary Research*, **68**, 879–889.

Nichols, G.J. & Thompson, B. 2005. *Sedimentology*, **52**, 571–585.

Nichols, R.J. 1995. In: *Characterisation of Deep Marine Clastic Systems* (eds Hartley, A.J. & Prosser, D. J.), pp. 63–76. Special Publication 94, Geological Society of London.

Nichols, R.J. *et al.* 1994. *Sedimentology*, **41**, 233–253.

Nickling, W.G. *et al.* 2002. *Sedimentology*, **49**, 171–190.

Nielsen, L.H. & Johannessen, P.N. 2009. *Sedimentology*, **56**, 935–968.

Nielsen, L.H. *et al.* 1988. *Sedimentology*, **35**, 915–937.

Nielsen, P. 1992. *Coastal Bottom Boundary Layers and Sediment Transport.* World Scientific, Singapore.

Nijman, W. 1998. In: *Cenozoic Foreland Basins of Western Europe* (eds Mascle, A. *et al.*), pp. 135–162. Special Publication 134, Geological Society of London.

Nijman, W. *et al.* 1998. *PreCambrian Research*, **88**, 83–108.

Nilsen, T.H. & Sylvester, A.G. 1995. In: *Tectonics of Sedimentary Basins* (eds Busby, C. & Ingersoll, R.), pp. 425–458. Blackwell Science, Oxford.

Nittrouer, C.A. & Kuehl, S.A.(eds) 1995. *Marine Geology*, **125**, 175–399.

Nittrouer, C.A. & Wright, L.D. 1994. *Reviews of Geophysics*, **32**, 85–113.

Nittrouer, C.A. *et al.* (eds) 2007. *Continental Margin Sedimentation: from Sediment Transport to Sequence Stratigraphy.* Special Publication 37, International Association of Sedimentologists.

Noffke, N. *et al.* 2006. *Geology*, **34**, 253–256.

Nordfjord, S. *et al.* 2006. *Journal of Sedimentary Research*, **76**, 1284–1303.

Normark, W.R. & Piper, D.J.W. 1972. *Journal of Geology*, **80**, 198–223.

Normark, W.R. & Piper, D.J.W. 1984. *Geo-Marine Letters*, **3**, 101–108.

Normark, W.R. *et al.* 1979. *Sedimentology*, **26**, 749–774.

Normark, W.R. *et al.* 1993. *Reviews of Geophysics*, **31**, 91–116.

Normark, W.R. *et al.* 2006. *Sedimentology*, **53**, 867–897.

Normark, W.R. *et al.* 2009. *Sedimentology*, doi: 10.1111/j.1365-3091.2009.01052.x.

North, C.P. & Prosser, D.J. (eds) 1993. *Characterisation of Fluvial and Aeolian Reservoirs*. Special Publication 73, Geological Society of London.

North, C.P. & Warwick, G.L. 2007. *Journal of Sedimentary Research*, **77**, 693–701.

Nott, J. & Roberts, R.G. 1996. *Geology*, **24**, 883–887.

Nummedal, D. & Penland, S. 1981. In: *Holocene Marine Sedimentation in the North Sea Basin* (eds Nio, S. D. *et al.*), pp. 187–210. Special Publication 5, International Association of Sedimentologists.

Nummedal, D. & Swift, D.J.P. 1987. In: *Sea Level Fluctuation and Coastal Evolution* (eds Nummedal, D. *et al.*), pp. 241–260. Special Publication 41, Society of Economic Paleontologists and Mineralogists.

O'Connell, S. *et al.* 1991. In: *Seismic Facies and Sedimentary Processes of Submarine Fans and Turbidite Systems* (eds Weimer, P. & Link, M.H.), pp. 365–382. Springer, Berlin.

O'Connell, S. *et al.* 1996. *Bulletin Geological Society America*, **108**, 270–284.

O'Connor, J.E. *et al.* 1994. *Geology*, **102**, 1–9.

Odin, G.S. & Dodson, M.H. 1982. In: *Numerical Dating in Stratigraphy* (ed. Odin, G.S.), pp. 277–305. Wiley, UK.

Odin, G.S. & Matter, A. 1981. *Sedimentology*, **28**, 611–641.

Oertel, G.F. 1979. In: *Barrier Islands* (ed. Leatherman, S.P.), pp. 273–290. Academic Press, New York.

Oertel, G.F. 1988. In: *Hydrodynamics and Sediment Dynamics of Tidal Inlets* (eds Aubrey, D.G. & Weishar, L.), pp. 297–318. Springer, Berlin.

Ohmoto, H. 1996. *Geology*, **24**, 1135–1138.

Ohmoto, H. *et al.* 1993. *Science*, **262**, 555–557.

Okazaki, H. & Masuda, F. 1995. In: *Tidal Signatures in Modern and Ancient Sediments* (eds Flemming, B.C. & Bartoloma, A.), pp. 85–99. Special Publication 24, International Association of Sedimentologists.

Olariu, C. & Bhattacharya, J.P. 2006. *Journal of Sedimentary Research*, **76**, 212–233.

Olariu, C. *et al.* 2005. In: *River Deltas: Concepts, Models, and Examples* (eds Giosan, L. & Bhattacharya, J.P.), pp. 155–178. Special Publication 83, Society of Economic Paleontologists and Mineralogists.

Olsen, P.E. 1986. *Science*, **234**, 842–848.

Ono, Y. 1990. In: *Alluvial Fans: A Field Approach* (eds Rachocki, A.H. & Church, M.), pp. 91–107. Wiley, New York.

Oomkens, E. 1974. *Sedimentology*, **21**, 195–222.

Oost, A.P. & de Boer, P.L. 1994. *Senckenbergiana Maritima*, **24**, 65–115.

Opdyke, B.N. & Walker, J.C.G. 1992. *Geology*, **20**, 733–736.

Opdyke, B.N. & Wilkinson, B.H. 1993. *American Journal of Science*, **293**, 217–234.

Orange, D.L. *et al.* 1994. *GSA Today*, **4**, 36–39.

Orford, J.D. 1975. *Sedimentology*, **22**, 441–463.

Ori, G.-G. 1989. *Geology*, **17**, 918–921.

Ori, G.-G. & Friend, P.F. 1984. *Geology*, **12**, 475–478.

Ori, G.-G. *et al.* 2008. In: *Sedimentary Processes, Environments and Basins* (eds Nichols, G. *et al.*), pp. 519–533. Special Publication 38, International Association of Sedimentologists.

Orme, G.R. *et al.* 1978. *Philosophical Transactions of the Royal Society, London (A)*, **291**, 85–99.

Orr, J.C. *et al.* 2005. *Nature*, **437**, 681–686.

Orti, F. *et al.* 2003. *Sedimentology*, **50**, 361–386.

Orti-Cabo, F. *et al.* 1984. *Rev. Inst. Inv. Geol.*, **38/39**, 169–220.

Orton, G.J. & Reading, H.G. 1993. *Sedimentology*, **40**, 475–512.

Osborne, P.D. & Greenwood, B. 1992. *Marine Geology*, **106**, 25–51.

Osborne, P.D. & Greenwood, B. 1993. *Sedimentology*, **40**, 599–622.

Ovviatt, C.G. *et al.* 1994. *Bulletin of the Geological Society of America*, **106**, 133–144.

Owen, G. 1996. *Sedimentology*, **43**, 279–293.

Owen, L.A. 1988. In: *Glaciotectonics: Forms and Process* (ed. Croot, D.G.), pp. 123–147.

Owen, R.B. *et al.* 1982. *Nature*, **298**, 523–529.

Page, K.J. & Nanson, G.C. 1996. *Sedimentology*, **43**, 927–945.

Page, K.J. *et al.* 2003. *Journal of Sedimentary Research*, **73**, 5–14.

Paillard, D. & Labeyrie, L. 1994. *Nature*, **372**, 162–164.

Paine, M.D. *et al.* 2005. *Journal of Sedimentary Research*, **75**, 742–759.

Pak, H. *et al.* 1980. *Journal of Geophysical Research*, **85**, 6697–6708.

Palmer, M.R. & Elderfield, H. 1985. *Nature*, **314**, 526–528.

Panagiotopoulos, I. *et al.* 1994. *Sedimentology*, **41**, 951–962.

Pantin, H.M.P. 1979. *Marine Geology*, **31**, 59–99.

Pantin, H.M.P. & Leeder, M.R. 1987. *Sedimentology*, **34**, 1143–1155.

Pantin, H.M. *et al.* 1981. *Geo-Marine Letters*, **1**, 255–260.

Pantin, H.M.P. *et al.* 1987. *Marine Geology*, **76**, 163–167.

Paola, C. & Borgman, L. 1991. *Sedimentology*, **38**, 553–565.

Paola, C. & Voller, V.R. 2005. *Journal of Geophysical Research*, **110**, F04014.

Paola, C. *et al.* 1992a. *Basin ResearChapter* **4**, 73–90.

Paola, C. *et al.* 1992b. *Science*, **258**, 1757–1760.

Paphitis, D. *et al.* 2001. *Sedimentology*, **48**, 645–659.

Paphitis, D. *et al.* 2002. *Sedimentology*, **49**, 211–225.

Paquet, F. *et al.* 2009. *Journal of Sedimentary Research*, **79**, 97–124.

Park, H.-B. & Vincent, C.E. 2007. *Journal of Coastal Research*, Special Issue **50**, 868–873.

Park, R.K. 1976. *Sedimentology*, **23**, 379–393.

Park, R.K. 1977. *Sedimentology*, **24**, 485–506.

Park, S.-C. *et al.* 2006. *Journal of Sedimentary Research*, **76**, 1093–1105.

Parker, A.G. *et al.* 2006. *Quaternary Research*, **66**, 465–476.

Parker, G. 1976. *Journal of Fluid Mechanics*, **76**, 457–480.

Parker, G. *et al.* 1982. *Journal of the Hydraulics Division, American Society of Civil Engineers*, **108**, 544–571.

Parker, G. *et al.* 1986. *Journal of Fluid Mechanics*, **171**, 145–181.

Parrish, J.T. & Peterson, F. 1988. *Sedimentary Geology*, **56**, 261–282.

Parsons, B. & Sclater, J.G. 1977. *Journal of Geophysical Research*, **82**, 803–827.

Parsons, D.R. *et al.* 2005. *Journal of Geophysical Research*, **110**, F04S03, doi: 10.1029/2004JF000231.

Parsons, I. *et al.* 2005. *Journal of Sedimentary Research*, **75**, 921–942.

Parsons, J.D. *et al.* 2001. *Sedimentology*, **48**, 465–478.

Parsons, J.D. *et al.* 2002. *Journal of Sedimentary Research*, **72**, 619–628.

Partheniades, E. 1993. In: *Nearshore and Estuarine Cohesive Sediment Transport* (ed. Mehta, A.J.), pp. 40–59. Monograph, American Geophysical Union.

Pascucci, V. *et al.* 2009. *Sedimentology*, **56**, 529–545.

Paterson, D.M. 1997. In: *Cohesive Sediments* (eds Burt, N. *et al.*), pp. 215–229. Wiley, New York.

Paterson, R.J. *et al.* 2006. *Journal of Sedimentary Research*, **76**, 1162–1182.

Paterson, W.S.B. 1994. *The Physics of Glaciers*, 3rd edn. Pergamon, Oxford.

Pattison, S.A.J. 2005. *Journal of Sedimentary Research*, **75**, 420–439.

Pattison, S.A.J. & Walker, R.G. 1992. *Journal of Sedimentary Petrology*, **62**, 292–309.

Pattison, S.A.J. *et al.* 2007. *Sedimentology*, **54**, 1033–1063.

Paull, C.K. *et al.* 2005. *Bulletin of the Geological Society of America*, **117**, 1134–1145.

Pauly, H. 1963. *Arctic*, **16**, 263–264.

Pavich, M.J. 1985. In: *Tectonic Geomorphology* (eds Morisawa, M. & Hack, J.T.), pp. 299–320. Allen and Unwin, Boston.

Pavich, M.J. 1986. *In: Rates of Chemical Weathering of Rocks and Minerals* (eds Colman, S.M. & Dethier, D.P.), pp. 552–590. Academic Press, Orlando.

Pazzaglia, F.J. & Brandon, M.T. 1996. *Basin Research*, **8**, 255–278.

Peacock, D.C.P. & Sanderson, D.J. 1991. *Journal of Structural Geology*, **13**, 721–733.

Peacock, D.C.P. & Sanderson, D.J. 1994. *Bulletin of the American Association of Petroleum Geologists*, **78**, 147–165.

Peakall, J. 1998. *Journal of Sedimentary Research*, **68**, 788–799.

Peakall, J. *et al.* 1996. In: *The Scientific Nature of Geomorphology* (eds Rhoads, B.L. & Thorn, C.R.), pp. 221–253. Wiley, New York.

Peakall, J. *et al.* 2000a. *Basin Research*, **12**, 413–424.

Peakall, J. *et al.* 2000b. *Journal of Sedimentary Research*, **70**, 434–448.

Peakall, J. *et al.* 2007. *Journal of Sedimentary Research*, **77**, 197–212.

Pedocchi, F. & Garcia, M.H. 2009a. *Journal of Geophysical Research Oceans*, **114**, **C12014**.

Pedocchi, F. & Garcia, M.H. 2009b. *Journal of Geophysical Research-Oceans*, **114**, C12015.

Pedley, M. & Grasso, M. 2002. *Sedimentology*, **49**, 533–553.

Peizhen, Z. *et al.* 2001. *Nature*, **410**, 891–897.

Peng, T.-H. & Broecker, W.S. 1991. *Nature*, **349**, 227–229.

Penland, S. *et al.* 1988. *Journal of Sedimentary Research*, **58**, 932–949.

Pérez-Arlucea, M. & Smith, N.D. 1999. *Journal of Sedimentary Research*, **69**, 62–73.

Perlmutter, M.A. & Mathews, M.D. 1989. In: *Quantitative Dynamic Stratigraphy* (ed. Cross, T.A.), pp. 233–260. Prentice Hall, Englewood Cliffs, NJ.

Pernetta, J. 1994. *Atlas of the Oceans*. Mitchell Beazly, London.

Perri, E. & Tucker, M. 2007. *Geology*, **35**, 207–210.

Perron, J.T. *et al.* 2007. *Nature*, **447**, 840–843.

Perry, C.T. *et al.* 2008. *Journal of Sedimentary Research*, **78**, 77–97.

Peryt, T.M. 1994. *Sedimentology*, **41**, 83–113.

Peryt, T.M. 1996. *Sedimentology*, **43**, 571–588.

Peterson, C.D. *et al.* 2008. *Journal of Sedimentary Research*, **78**, 390–409.

Peterson, F. 1988. *Sedimentary Geology*, **56**, 207–260.

*et al.* Phelps, R.M. & Kerans, C. 2007. *Journal of Sedimentary Research*, **77**, 939–964.

Phelps, R.M. *et al.* 2008. *Sedimentology*, **55**, 1777–1813.

Phillips, A.C. *et al.* 1991. In: *Glacial Marine Sedimentation: Palaeoclimatic Significance* (eds Anderson, J.B. & Ashley, G.M.), pp. 1–26. Special Publication 261, Geological Society of America.

Pickering, K.T. & Hiscott, R.N. 1985. *Sedimentology*, **34**, 1143–1155.

Pickering, K.T. & Hiscott, R.N. 1995. In: *Atlas of Architectural Styles in Turbidite Systems* (Ed: Pickering. K.T. *et al.*), 310–316. Chapman & Hall, London.

Pickering, K.T. *et al.* 1989. *Deep Marine Clastic Environments: Clastic Sedimentation and Tectonics*. Unwin Hyman, London.

Pickering, K.T. *et al.* 1992. *Geology*, **20**, 1099–1102.

Pickering, K.T. *et al.* (eds) 1999. *Atlas of Deep Water Environments: Architectural Style in Turbidite Systems*. Chapman & Hall, London.

Pickrill, R.A. & Irwin, J. 1983. *Sedimentology*, **30**, 63–75.

Pickup, S.L.B. *et al.* 1996. *Geology*, **24**, 1079–1082.

Pierson, T.C. 1981. *Sedimentology*, **28**, 49–60.

Pierson, T.C. 1985. *Bulletin of the Geological Society of America*, **96**, 1056–1069.

Pierson, T.C. 1995. *Journal of Volcanology and Geothermal Research*, **66**, 283–294.

Pierson, T.C. & Scott, K.M. 1985. *Water Resources Research*, **21**, 1511–1524.

Pietras, J.T. & Carroll, A.R. 2006. *Journal of Sedimentary Research*, **76**, 1197–1214.

Pigott, J.D. & Mackenzie, F.T. 1979. *Geological Society of America Abstracts*, **11**, 495–496.

Pike, J. & Kemp, A.E.S. 1997. *Paleooceanography*, **12**, 227–238.

Pilkey, O.H. & Davis, T.W. 1987. In: *Sea Level Fluctuation and Coastal Evolution* (eds Nummedal, D. *et al.*), pp. 59–70. Special Publication 41, Society of Economic Paleontologists and Mineralogists.

Pilkey, O.H. & Noble, D. 1967. *Deep Sea Research*, **13**, 1–16.

Pinet, P. & Souriau, M. 1988. *Tectonics*, **7**, 563–582.

Pingree, R.D. & Griffiths, D.K. 1979. *Journal of the Marine and Biological Association, UK*, **59**, 497–513.

Piovano, E.L. 2002. *Sedimentology*, **49**, 1371–1384.

Piper D.J.W. & Normark, W.R. 1983. *Sedimentology*, **30**, 681–694.

Piper, D.J.W. & Normark, W.R. 2009. *Journal of Sedimentary Research*, **79**, 347–362.

Piper, D.J.W. *et al.* 1973. *Geology*, **1**, 19–22.

Piper, D.J.W. *et al.* 1985. *Geology*, **13**, 538–541.

Piper, D.J.W. *et al.* 1999a. *Sedimentology*, **46**, 79–97.

Piper, D.J.W. *et al.* 1999b. *Sedimentology*, **46**, 47–78.

Pirmez, C. *et al.* 2000. In: Deep-Water Reservoirs of the World (eds Weimer, P. *et al.*), pp. 782–805. *20th Annual Research Conference, Gulf Coast Society and Society of Economic Paleontologists and Mineralogists.*

Pirazzoli, P.A. 2005. *Quaternary Science Reviews*, **24**, 1989–2001.

Pitman, W.C. 1978. *Bulletin of the Geological Society of America*, **89**, 1389–1403.

Pizzuto, J.E. 1987. *Sedimentology*, **34**, 301–317.

Pizzuto, J.E. *et al.* 2008. *Journal of Sedimentary Research*, **78**, 16–28.

Plafker, G. & Ericksen, G.E. 1978. In: *Rockslides and Avalanches*, Vol. **1**, *Natural Phenomena* (ed. Voight, B.), pp. 277–314. Developments in Geotechnical Engineering 14A.

Platt, N.H. 1989. *Sedimentology*, **36**, 665–684.

Playford, P.E. 1980. *Bulletin of the American Association of Petroleum Geologists*, **62**, 814–841.

Plink-Björklund, P. 2005. *Sedimentology*, **52**, 391–428.

Plink-Björkland, P. & Steel, R. 2005. In: *River Deltas: Concepts, Models, and Examples* (eds Giosan, L. & Bhattacharya, J.P.), pp. 179–206. Special Publication 83, Society of Economic Paleontologists and Mineralogists.

Plint, A.G. 1988. In: *Sea Level Changes: An Integrated Approach* (eds Wilgus, C.K. *et al.*), pp. 357–370. Special Publication 42, Society of Economic Paleontologists and Mineralogists.

Plint, A.G. & Wadsworth, J. A. 2003. *Sedimentology*, **50**, 1147–1186.

Plint, A.G. *et al.* 1986. *Bulletin of Canadian Petroleum Geologists*, **34**, 313–325.

Plummer, L.N. 1977. *Water Resources Research*, **13**, 801–812.

Poli, M.S. *et al.* 2000. *Geology*, **28**, 807–810.

Polyak, V. *et al.* 2008. *Science*, **319**, 1377–1380.

Polzin, K.L. *et al.* 1996. *Nature*, **380**, 54–57.

Pomar, L. & Ward, W.C. 1994. *Geology*, **22**, 131–134.

Pond, S. & Pickard, G.L. 1983. *Introductory Dynamical Oceanography*, 2nd edn. Pergamon, London.

Pontén, A. & Plink-Björkland, P. 2007. *Sedimentology*, **54**, 969–1006.

Porebski, S.J. 1995. *Studia Geologica Polonica*, **107**, 7–97.

Porebski, S.J. & Steel, R.J. 2003. *Earth Science Reviews*, **62**, 283–326.

Porebski, S.J. & Steel, R.J. 2006. *Journal of Sedimentary Research*, **76**, 390–403.

Porebski, S.J. *et al.* 1991. *Sedimentology*, **38**, 691–715.

Posamentier, H.W. & Erskine, R.D. 1991. In: *Seismic Facies and Sedimentary Processes of Submarine Fans and Turbidite Systems* (eds Weimer, P. & Link, M. H.), pp. 197–222. Springer, Berlin.

Posamentier, H.B. & Kolla, V. 2003. *Journal of Sedimentary Research*, **73**, 367–388.

Posamentier, H.W. *et al.* 1991. In: *Seismic Facies and Sedimentary Processes of Submarine Fans and Turbidite Systems* (eds Weimer, P. & Link, M.H), pp. 127–136. Springer, Berlin.

Postma, G. 1990. *Terra Nova*, **2**, 124–130.

Postma, G. & Roep, T.B. 1985. *Journal of Sedimentary Petrology*, **55**, 874–885.

Postma, G. *et al.* 1988. In: *Fan Deltas: Sedimentology and Tectonic Settings* (eds Nemec, W. & Steel, R.J.), pp. 91–102. Blackie, London.

Postma, G. *et al.* 1993a. *Terra Nova*, **5**, 438–444.

Postma, G. *et al.* 1993b. In: *Special Publication 20, International Association of Sedimentologists*, pp. 335–62.

Potter, P.E. 1978. *Journal of Geology*, **86**, 13–33.

Potter, P.E. & Hamblin, W.K. *Big Rivers*. 2006. Brigham Young University Geology Studies 48, 1–45.

Poulet, F. *et al.* 2005. *Nature*, **438**, 623–627.

Pouliquen, O. *et al.* 1997. *Nature*, **386**, 816–817.

Poulos, S.E. & Collins, M.B. 2002. In: *Sediment Flux To Basins: Causes, Controls and Consequences* (eds Jens, S.J. & Frostick, L.E.), pp. 227–245. Special Publication 191, Geological Society of London.

Poulsen, C.J. *et al.* 1998. *Bulletin Geological Society of America*, **110**, 1105–1122.

Powell, R.D. 1990. In: *Glacimarine Environments* (eds Dowdeswell, J.A. & Scourse, J.D.), pp. 53–73. Special Publication 53, Geological Society of London.

Pozzobon, J.G. & Walker, R.G. 1990. *Bulletin of the American Association of Petroleum Geologists*, **74**, 1212–1227.

Prather, B.E. *et al.* 1998. *Bulletin of the American Association of Petroleum Geologists*, **82**, 701–728.

Pratson, L.F. *et al.* 1994. *Bulletin of the Geological Society of America*, **106**, 395–412.

Pratt, B.R. & Bordonaro, O.L. 2007. *Journal of Sedimentary Research*, 77, 256–262.

Prave, A.R. 1990. *Sedimentology*, **37**, 1049–1052.

Prave, A.R. *et al.* 1996. *Sedimentology*, **43**, 611–629.

Prélat, A. *et al.* 2009. *Sedimentology*, **56**, 2132–2154.

Prentice, I.C. *et al.* 1992. *Nature*, **360**, 658–660.

Prentice, J.E. *et al.* 1968. *Nature*, **218**, 1207–1210.

Preto, N. & Hinov, L.A. 2003. *Journal of Sedimentary Research*, **73**, 774–789.

Priestley, C.H.B. & Taylor, R.J. 1972. *Monthly Weather Review*, **100**, 81–92.

Prins, M.A. *et al.* 2000. *Marine Geology*, **169**, 327–349.

Prior, D.B. & Bornhold, B.D. 1988. In: *Fan Deltas: Sedimentology and Tectonic Settings* (eds Nemec, W. & Steel, R.J.), pp. 125–143. Blackie, London.

Prior, D.B. & Bornhold, B.D. 1990. In: *Coarse-Grained Deltas* (eds Colella, A. & Prior, D.B.), pp. 75–90. Special Publication 10, International Association of Sedimentologists.

Prior, D.B. *et al.* 1987. *Science*, **237**, 1330–1333.

Pritchard, D.W. 1955. *Proceedings, American Society of Civil Engineers*, **81**, 1–11.

Pritchard, D.W. 1967. In: *Estuaries* (ed. Lauff, G.H.), pp. 1–10. American Association for the Advancement of Science, Washington, DC.

Pritchard, D.W. & Carter, H.H. 1971. In: *The Estuarine Environment* (ed. Schubel, J.R.), pp. 1–17. American Geological Institute, Washington, DC.

Pritchard, H.D. *et al.* 2009. *Nature*, **461**, 971–975.

Prokoph, A. & Veizer, J. 1999. *Chemical Geology*, **161**, 225–240.

Pufahl, P.K. *et al.* 2004. *Sedimentology*, **51**, 997–1027.

Pugh, D.T. 1987. *Tides, Surges and Mean Sea Level*. Wiley, London.

Purdy, E.G. 1963. *Journal of Geology*, **71**, 472–497.

Purdy, E.G. 1974. In: *Reefs in Time and Space* (ed. Laporte, L.F.), pp. 9–76. Special Publication 18, Society of Economic Paleontologists and Mineralogists.

Purkis, S.J. *et al.* 2005. *Journal of Sedimentary Research*, **75**, 861–876.

Purseglove, J. 1989. *Taming the Flood: A History and Natural History of Rivers and Wetlands*. Oxford University Press, Oxford.

Purser, B.H. 1979. *Symp. Sed. Jurass. W. Europe*, pp. 75–84. Publication 1.

Pye, K. 1982. *Geografiska Annaler*, **A64**, 213–227.

Pye, K. 1993. In: *Aeolian Sediments: Ancient and Modern* (eds Pye, K. & Lancaster, N.), pp. 23–44. Special Publication 16, International Association of Sedimentologists.

Pye, K. & Allen, J.R.L. (eds) 2000 *Coastal and Estuarine Environments*. Special Publication 175, Geological Society of London.

Pye, K.& Lancaster, N. (eds) 1993. *Aeolian Sediments: Ancient and Modern*. Special Publication 16, International Association of Sedimentologists.

Pye, K. & Tsoar, H. 1990. *Aeolian Sand and Sand Dunes*. Chapman & Hall, London.

Pyrce, R.S. & Ashmore, P.E. 2005. *Sedimentology*, **52**, 839–857.

Pysklywec, R.N. & Mitrovica, J.X. 1998. *Geology*, **26**, 687–690.

Pytkowicz, R.M. 1965. *Journal of Geology*, **73**, 196–199.

Qian, N. 1990. *Journal of Sedimentary Research*, **5**, 1–13.

Quade, J. *et al.* 1997. *Science*, **276**, 1828–1831.

Quade, J. *et al.* 2007. *Reviews in Mineralogy & Geochemistry*, **66**, 53–87.

Quigley, M. *et al.* 2007. *Basin Research*, **19**, 491–505.

Quin, J.G. 2008. *Sedimentology*, **55**, 1053–1082.

Quinlan, G.M. & Beaumont, C. 1984. *Canadian Journal of Earth Sciences*, **21**, 973–996.

Rabineau, M. *et al.* 2006. *Earth and Planetary Science Letters*, **252**, 119–137.

Rahmani, R.A.& Flores, R.M. (eds) 1984. *Sedimentology of Coal and Coal-bearing Sequences*, Special Publication 7, International Association of Sedimentologists.

Rai, H. & Hill, G. 1980. *Hydrobiologia*, **72**, 85–99.

Rao, K.N. *et al.* 2005. In: *River Deltas: Concepts, Models, and Examples* (eds Giosan, L.& Bhattacharya, J.P.), pp. 435–452. Special Publication 83, Society of Economic Paleontologists and Mineralogists.

Radies, D. *et al.* 2004. *Sedimentology*, **51**, 1359–1385.

Rainbird, R.H. *et al.* 1997. *Journal of Geology*, **105**, 1–17.

Raiswell, R.W. *et al.* 1980. *Environmental Chemistry*. Edward Arnold.

Rajchl, M. & Uličný, D. 2005. *Sedimentology*, **52**, 601–625.

Rajchl, M. *et al.* 2009. *Basin Research*, **21**, 269–294.

Ramos, A. *et al.* 1986. *Journal of Sedimentary Petrology*, **56**, 862–875.

Rampino, M.R. & Sanders, J.E. 1981. *Sedimentology*, **28**, 37–48.

Randazzo, A.F. *et al.* 1999. *Journal of Sedimentary Research*, **69**, 283–293.

Rankey, E.C. 1997. *Bulletin of the Geological Society of America*, **109** 1089–1100.

Rankey, E.C. 2002. *Journal of Sedimentary Research*, **72**, 591–601.

Rankey, E.C. 2004. *Journal of Sedimentary Research*, **74**, 2–6.

Rankey, E.C. *et al.* 2006. *Sedimentology*, **53**, 1191–1210.

Rannie, W.F. 1990. In: *Alluvial Fans: A Field Approach* (eds Rachocki, A.H.& Church, M.), pp. 179–193, Wiley, New York.

Rannie, W.F. *et al.* 1989. *Canadian Journal of Earth Sciences*, **26**, 1834–1841.

Raudkivi, A.J. 1976. *Loose Boundary Hydraulics*. Pergamon, Oxford.

Raudkivi, A.J. & Hutchinson, D.L. 1974. *Proceedings of the Royal Society of London*, **A337**, 537–554.

Raymo, M.E. 1991. *Geology*, **19**, 344–347.

Raymo, M.E. & Ruddiman, W.F. 1992. *Nature*, **359**, 117–122.

Raymond, P.A. *et al.* 2008. *Nature*, **451**, 449–452.

Raynaud, D. *et al.* 2005. *Nature*, **436**, 39–40.

Rea, D.K. 1993. *GSA Today*, **3**, 205–210.

Read, J.F. 1982. *Tectonophysics*, **81**, 195–212.

Read, J.F. 1985. *Bulletin of the American Association of Petroleum Geologists*, **69**, 1–21.

Read, J.F. *et al.* 1986. *Geology*, **14**, 107–110.

Reddy, M.M. 1977. *Journal of Crystal Growth*, **41**, 287–295.

Reeder, M.S. *et al.* 2000. *Marine & Petroleum Geology*, **17**, 199–218.

Reeder, S.L. & Rankey, E.C. 2008. *Journal of Sedimentary Research*, **78**, 175–186.

Reid, I. & Frostick, L.E. 1987. *Hydrological Processes*, **1**, 239–253.

Reid, R.P. *et al.* 2000. *Nature*, **406**, 989–992.

Reid, R.P. *et al.* 2003. *Facies*, **49**, 45–53.

Reineck, H.E. 1958. *Geologische Rundschau*, **47**, 73–82.

Reineck, H.E. 1963. *Sedimentgefuge in Bereich der Sudlichen Nordsee*. Gesellschaft Nr. 505, Abteilung Senckenbergiana Naturforschung.

Reineck, H.E. 1967. In: *Estuaries* (ed. Louff, G.D.), pp. 191–206. American Association for the Advancement of Science, Washington, DC.

Reineck, H.E. 1972. In: *Recognition of Ancient Sedimentary Environments* (eds Rigby, J.K.& Hamblin, W.K.), pp. 146–159. Special Publication 16, Society of Economic Paleontologists and Mineralogists.

Reineck, H.E. & Singh, I.B. 1973. *Sedimentology*, **18**, 123–128.

Reineck, H.E. & Singh, I.B. 1980. *Depositional Sedimentary Environments*, 2nd edn. Springer, Berlin.

Reineck, H.E. & Wunderlich, F. 1968. *Lethaia*, **49**, 321–345.

Renaut, R.W.& Last, W.M. (eds) 1994. *Sedimentology and Geochemistry of Modern and Ancient Saline Lakes*. Special Publication 50, Society of Economic Paleontologists and Mineralogists.

Renaut, R.W. & Tiercelin, J.-J. 1994. In: *Sedimentology and Geochemistry of Modern and Ancient Saline Lakes* (eds Renaut, R.W.& Last, W.M.), pp. 101–123. Special Publication 50, Society of Economic Paleontologists and Mineralogists.

Reneau, S.L. & Dietrich, W.E. 1991. *Earth Surface Processes and Landforms*, **16**, 307–322.

Retallack, G.J. 1990. *Soils of the Past: An Introduction to Palaeopedology*. Unwin Hyman, London.

Reymer, A. & Schubert, G. 1984. *Tectonics*, **3**, 63–77.

Reynaud, J.-Y. 1999. *Sedimentology*, **46**, 703–721.

Reynaud, J.-Y. *et al.* 1999. *Journal of Sedimentary Research*, **69**, 351–364.

Rice, S. 1999. *Journal of Sedimentary Research*, **69**, 32–39.

Richards, M.T. 1994. *Sedimentology*, **41**, 55–82.

Richards, P.L. & Kump, L.R. 2003. *Geochimica et Cosmochimica Acta*, **67**, 3803–3815.

Richardson, J.F. & Zaki, W.N. 1954. *Chemical Engineering Science*, **3**, 65–73.

Richter, D.K. 1983. In: *Coated Grains* (ed. Peryt, T.M.), pp. 71–99. Springer, Berlin.

Richter, F.M. *et al.* 1992. *Earth and Planetary Science Letters*, **109**, 11–23.

Richter-Bernberg, G. 1955. *Deutsche Geologische Gesellschaft*, **105**, 593–596.

Ricketts, B.D. & Evenchick, C.A. 1999. *Journal of Sedimentary Research*, **69**, 1232–1240.

Ridge, M.J.H. & Carson, B. 1987. *Continental Shelf Research*, **7**, 759–772.

Ridgeway, K.D. & DeCelles, P.G. 1993. *Sedimentology*, **40**, 645–666.

Riding, R. 1983. In: *Coated Grains* (ed. Peryt, T.M.), pp. 277–283. Springer, Berlin.

Riding, R. 2000. *Sedimentology*, **47 Suppl. 1**, 179–214.

Riding, R. 2002. *Geology*, **30**, 31–34.

Riebe, C.S. *et al.* 2003. *Geochimica et Cosmochimica Acta*, **67**, 4411–4427.

Riech, V. & von Rad, U. 1979. In: *M. Ewing Series Monograph*, **3**, 315–340. American Geophysical Union.

Ries, J.B. 2004. *Geology*, **32**, 981–984.

Ries, J.B. *et al.* 2006a. *Journal of Sedimentary Research*, **76**, 515–523.

Ries, J.B. *et al.* 2006b. *Geology*, **34**, 525–528.

Ringrose, T.J. *et al.* 2003. *Icarus*, **163**, 78–87.

Ritchie, B.D. *et al.* 2004a. *Journal of Sedimentary Research*, **74**, 203–220.

Ritchie, B.D. *et al.* 2004b. *Journal of Sedimentary Research*, **74**, 221–238.

Rivers, J.M. *et al.* 2007. *Journal of Sedimentary Research*, **77**, 480–494.

Roberts, H.H. *et al.* 1980. *Bulletin of the American Association of Petroleum Geologists*, **64**, 264–79.

Roden, J.E. 1998. *Sedimentology and dynamics of mega-sand dunes, Jamuna River, Bangladesh*. PhD thesis, University of Leeds.

Rodrigues, S. *et al.* 2006. *Sedimentary Geology*, **186**, 89–109.

Rodriguez, A.B. *et al.* 2000. *Journal of Sedimentary Research*, **70**, 283–295.

Rodriguez, A.B. *et al.* 2001. *Sedimentology*, **48**, 837–853.

Rodriguez, A.B. *et al.* 2004. *Journal of Sedimentary Research*, **74**, 405–421.

Rodriguez, A.B. *et al.* 2005. *Journal of Sedimentary Research*, **75**, 608–620.

Rodriguez, A.B. & Meyer, C.T. 2006. *Journal of Sedimentary Research*, **76**, 257–269.

Rogala, B. *et al.* 2007. *Journal of Sedimentary Research*, **77**, 587–606.

Rohling, E.J. 1994. *Marine Geology*, **122**, 1–28.

Rohling, E.J. & Pälike, H. 2005. *Nature*, **434**, 975–979.

Romans, B.W. *et al.* 2009. *Sedimentology*, **56**, 737–764.

Roof, S.R. *et al.* 1991. *Journal of Sedimentary Petrology*, **61**, 1070–1088.

Rooney, J.J.R. *et al.* 2003. *Science*, **58**, 305–324.

Rosato, A. *et al.* 1987. *Physical Review Letters*, **58**, 1038–1040.

Rosen, M.R. 1994. In: *Paleoclimate and Basin Evolution of Playa Systems* (ed Rosen, M.R.). Special Paper 289, Geological Society of America.

Ross, W.C. *et al.* 1994. *Geology*, **22**, 511–514.

Rossignol-Strick, M. 1999. *Quaternary Science Reviews*, **18**, 515–530.

Roth, S. & Reijmers, J.J.G. 2005. *Sedimentology*, **52**, 161–181.

Rothwell, R.G. *et al.* 1992. *Basin Research*, **4**, 103–131.

Rothwell, R.G. *et al.* 1998. *Nature*, **392**, 377–380.

Rothwell, R.G. *et al.* 2000. *Sedimentary Geology*, **135**, 75–88.

Rottman, J.W. & Simpson, J.E. 1989. *Quarterly Journal of the Royal Meteorological Society*, **115**, 941–963.

Rouchy, J.-M. *et al.* 1995. *Sedimentology*, **42**, 267–282.

Rowe, P.W. 1962. *Proceedings of the Royal Society, London (A)*, **269**, 500–527.

Rowley, D.B. 2002. *Geological Society of America Bulletin*, **114**, 927–933.

Rowley, D.B. & Currie, B.S. 2006. *Nature*, **439**, 677–681.

Rowley, D.B. *et al.* 2001. *Earth and Planetary Science Letters*, **188**, 253–268.

Roy, P.S. 1994. In: *Incised-Valley Systems: Origin and Sedimentary Sequences* (eds Dalrymple, R.W. *et al.*), pp. 241–263. Special Publication 51, Society of Economic Paleontologists and Mineralogists.

Roy, A.G. & Bergeron, N. 1988. *Earth Surface Processes and Landforms*, **13**, 583–598.

Roy, M. *et al.* 2004. *Earth and Planetary Letters*, **227**, 281–296.

Royden, L. 1988. In: pp. 27–48. Memoir 45, American Association of Petroleum Geologists.

Rozovskii, I.L. 1961. *Flow of Water in Bends of Open Channels*. Israel Programme for Scientific Translations, Jerusalem.

Rubin, D.M. 2004. *Journal of Sedimentary Research*, **74**, 160–165.

Ruddiman, W.F. & Kutzbach, J.E. 1991. *Scientific American*, March issue.

Ruegg, G.H.J. 1991. *Mededelingen Rijks Geologische Dienst*, **46**, 3–25.

Rusnak, G.A. 1960. In: *Recent Sediments of the NW Gulf of Mexico* (eds Shepard, F.P. *et al.*), pp. 153–196. American Association of Petroleum Geologists.

Russell, H.A.J. *et al.* 2007. In: *Glacial Sedimentary Processes and Products* (eds Hambrey, M. *et al.*), pp. 85–108. Special Publication 39, International Association of Sedimentologists.

Russell, H.A.J. & Arnott, R.W.C. 2003. *Journal of Sedimentary Research*, **73**, 887–905.

Rust, B.R. 1975. In: *Glaciofluvial and Glaciolacustrine Sediments* (eds Jopling, A.V. & McDonald, B.G.), pp. 238–248. Special Publication 23, Society of Economic Paleontologists and Mineralogists.

Rust, B.R. & Romanelli, R. 1975. In: pp. 177–192. Special Publication 23, Society of Economic Paleontologists and Mineralogists.

Ruszczyńa-Skaszenajch, H. 2001. *Sedimentology*, **48**, 585–597.

Ryan, W.B.F. 2009. *Sedimentology*, **56**, 95–136.

Sack, D. 1994. In: *Geomorphology of Desert Environments* (eds Abrahams, A.D. & Parsons, A.J.), pp. 616–630. Chapman & Hall, London.

Saez, A. & Cabrera, L. 2002. *Sedimentology*, **49**, 1073–1094.

Saez, A. *et al.* 2007. *Sedimentology*, **54**, 1191–1222.

Said, R. 1981. *The Geological Evolution of the River Nile*. Springer, New York.

Sak, P.B. *et al.* 2004. *Geochimica et Cosmochimica Acta*, **68**, 1453–1472.

Sallenger, A.H. 1979. *Journal of Sedimentary Petrology*, **49**, 553–562.

Sambrook-Smith, G. & Ferguson, R.I. 1995. *Journal of Sedimentary Research*, **A65**, 423–430.

Sambrook-Smith, G. *et al.* 2006. *Sedimentology*, **53**, 413–434.

Samimy, M. *et al.* (eds) 2003. *A Gallery of Fluid Motion*. Cambridge University Press.

Sancetta, C. 1994. *Palaeoceanography*, **9**, 195–196.

Sandberg, P.A. 1975. *Sedimentology*, **22**, 497–537.

Sandberg, P.A. 1983. *Nature*, **305**, 19–22.

Sandiford, M. & Cupper, M.L. 2007. *Basin Research*, **19**, 491–505.

Santantonio, M. 1993. *Sedimentology*, **40**, 1039–1067.

Sanz, M.E. *et al.* 1995. *Sedimentology*, **42**, 437–452.

Sarg, J.F. 2001. *Sedimentary Geology*, **140**, 9–34.

Satterfield, C.L. *et al.* 2005. *Journal of Sedimentary Research*, **75**, 534–546.

Sattler, U. *et al.* 2005. *Sedimentology*, **52**, 339–361.

Saunderson, H.C. & Lockett, F.P. 1983. In: *Modern and Ancient Fluvial Sediments* (eds Collinson, J.D. & Lewin, J.), pp. 49–58. Special Publication 6, International Association of Sedimentologists.

Savage, S.B. 1979. *Journal of Fluid Mechanics*, **92**, 53–96.

Savarese, M. *et al.* 1993. *Geology*, **21**, 917–920.

Schimmelmann, A. *et al.* 2006. *Journal of Sedimentary Research*, **76**, 74–80.

Scheihing, M.H. & Gaynor, G.C. 1991. *Sedimentology*, **38**, 433–444.

Schieber, J. *et al.* 2000. *Nature*, **406**, 981–985.

Scheiber, J. *et al.* 2007. *Science*, **318**, 1760–1763.

Schlager, W. 1981. *Bulletin Geological Society of America*, **92**, 197–211.

Schlager, W. & Camber, O. 1986. *Geology*, **14**, 762–765.

Schlager, W. & Keim, L. 2009. *Sedimentology*, **56**, 191–204.

Schlager, W. *et al.* 1976. *Bulletin of the Geological Society of America*, **87**, 1115–1118.

Schlager, W. *et al.* 1994. *Journal of Sedimentary Research*, **64**, 270–281.

Schlanger, S.O. & Jenkyns, H.C. 1976. *Geologieen Mijnbouw*, **55**, 79–84.

Schlische, R.W. 1992. *Bulletin of the Geological Society of America*, **104**, 1246–1263.

Schlische, R.W. & Olsen, P.E. 1990. *Journal of Geology*, **98**, 135–155.

Schlunegger, F. *et al.* 1997. *Bulletin of the American Association of Petroleum Geologists*, **81**, 1185–1207.

Schlunegger, F. *et al.* 1998. *Basin Research*, **10**, 197–212.

Schmid, M. *et al.* 2008. *Geophysical Research Letters*, **35**, 09605, doi: 10.1029/2008GL033223.

Schminke, H.V. *et al.* 1975. *Sedimentology*, **20**, 553–574.

Schmitz, W.J. 1995. *Reviews of Geophysics*, **33**, 151–174.

Schmitz, W.J. & McCartney, M.S. 1993. *Reviews of Geophysics*, **31**, 29–49.

Scholz, C.A. & Finney, B.P. 1994. *Sedimentology*, **41**, 163–179.

Scholz, C.A. *et al.* 1990. *Geology*, **18**, 140–144.

Scholz, C.A. *et al.* 1993. *Eos, Transactions American Geophysical Union*, **74**, 465.

Schorghofer, N. 2007. *Nature*, **449**, 192–194.

Schott, J. & Berner, R.A. 1985. In: *The Chemistry of Weathering* (ed. Drever, J.I.), pp. 35–53. Reidel, Hingham, MA.

Schreiber, B.C. & Helman, M.L. 2005. *Journal of Sedimentary Research*, **75**, 525–533.

Schreiber, B.C. & Tabakh, M.El. 2000. *Sedimentology*, **47 Suppl 1**, 215–238.

Schreiber, B.C. *et al.* 1976. *Sedimentology*, **23**, 729–760.

Schubel, J.R. 1971. In: *The Estuarine Environment* (ed. Schubel, J.R.). American Geological Institute, Washington, DC.

Schubel, J.R. & Okabo, A. 1972. In: *Shelf Sediment Transport: Process and Pattern* (eds Swift, D.J.P. *et al.*), pp. 333–346. Dowden, Hutchinson & Ross, Stroudsberg, PA.

Schumm, S.A. 1960. *American Journal of Science*, **258**, 177–184.

Schumm, S.A. 1963. *Geological Society of America Bulletin*, **74**, 1089–1100.

Schumm, S.A. 1968a. *Bulletin of the Geological Society of America*, **79**, 1573–1588.

Schumm, S.A. 1968b. *River Adjustment to Altered Hydrologic Regimen—Murrumbidgee River and paleochannels, Australia*. Professional Paper 598, US Geological Survey.

Schumm, S.A. 1977. *The Fluvial System*. Wiley, New York.

Schumm, S.A. 1993. *Journal of Geology*, **101**, 279–294.

Schumm, S.A. & Brackenridge, G.R. 1987. In: *The Geology of North America*, Vol. **K3**, *North America and Adjacent Oceans during the Last Deglaciation* (eds Ruddimann, W.F. & Wright, H.E. Jr), pp. 221–240. Geological Society of America, Boulder, CO.

Schwartz, R.K. 1975. *Nature and Genesis of Some Washover Deposits*. Technical Memo 61, US Army Corps of Engineers, Coastal Engineering Research Centre.

Schwartzman, D.W. 1993. *Geochimica et Cosmochimica Acta*, **57**, 2145–2146.

Schwartzman, D.W. & Volk, T. 1989. *Nature*, **340**, 457–460.

Schwarz, T. 1996. In: *Proceedings of the 4th International Symposium on the Geochemistry of the Earth's Surface* (ed. Bottrell, S.H.), pp. 216–220. University of Leeds.

Schwarzacher, W. & Fischer, A.G. 1982. In: *Cyclic and Event Stratification* (eds Einsele, G. & Seilacher, A.), pp. 72–95. Springer, Berlin.

Schwimmer, R.A. & Pizzuto, J.E. 2000. *Journal of Sedimentary Research*, **70**, 1026–1035.

Scoffin, T.P. 1970. *Journal of Sedimentary Petrology*, **40**, 249–273.

Scoffin, T.P. 1987. *An Introduction to Carbonate Sediments and Rocks*. Blackie, London.

Scorer, R.S. 1997. *Dynamics of Meteorology and Climate*. Wiley, New York.

Scott, K.M. & Williams, R.P. 1978. *Erosion and Sediment Yields in the Transverse Ranges, Southern California*. Professional Paper 1030, US Geological Survey.

Seeber, L. & Gornitz, V. 1983. *Tectonophysics*, **92**, 335–367.

Seelos, K. & Sirocko, F. 2005. *Sedimentology* **52**, 669–681.

Seguret, M. *et al.* 2001. *Sedimentology* **48**, 231–254.

Selby, M.J. 1993. *Hillslope Materials and Processes*. Oxford University Press, Oxford.

Selleck, B.W. *et al.* 2007. *Journal of Sedimentary Research*, **77**, 980–991.

Sellin, R.H.J. 1964. *Houille Blanche*, **19**, 793–801.

Sellwood, B.W. 1968. *Journal of Sedimentary Petrology*, **38**, 854–858.

Sellwood, B.W. & McKerrow, W.S. 1973. *Proceedings of the Geologists Association*, **85**, 189–210.

Seltzer, G.O. *et al.* 1998. *Geology*, **26**, 167–170.

Seppälä, M. & Linde, K. 1978. *Geografiska Annaler*, **60**, 29–42.

Sestini, G. 1989. In: *Deltas: Sites and Traps for Fossil Fuels* (eds Whateley, M.K.G. & Pickering, K.T.), pp. 99–128. Special Publication 41, Geological Society of London.

Sha, L.P. & de Boer, P. L. 1991. In: *Clastic Tidal Sedimentology* (eds Smith, D.G. *et al.*), pp. 199–218. Memoir 16, Canadian Society of Petroleum Geologists.

Shanley K.W. & McCabe, P.J. 1994. *Bulletin of the American Assoiciation of Petroleum Geologists*, **78**, 544–568.

Shanley, K.W. *et al.* 1992. *Sedimentology*, **39**, 905–930.

Shanmugam, G. 2000. *Marine & Petroleum Geology*, **17**, 285–342.

Shanmugam, G. 2006. *Journal of Sedimentary Research*, **76**, 718–730.

Shanmugam, G. *et al.* 1993. *Geology*, **21**, 929–932.

Shapiro, A.H. 1961. *Shape and Flow: The Fluid Dynamics of Drag*. Doubleday, New York; Heinemann, London.

Shapiro, R.S. *et al.* 1995. In: *Terrestrial and Shallow Marine Geology of the Bahamas and Bermuda* (eds Curran, A.A. & White, B.), pp. 139–155. Special Paper 300, Geological Society of America.

Sharma, G.D. *et al.* 1972. *Bulletin of the American Association of Petroleum Geologists*, **56**, 2000–2012.

Sharp, I.R. 2000a. *Bulletin of the Geological Society of America*, **112**, 1877–1899.

Sharp, I.R. 2000b. *Basin Research*, **12**, 285–305.

Sharp, M. *et al.* 1994. *Journal of Glaciology*, **40**, 327–340.

Shearman, D.J. 1966. *Transactions of the Institute of Mining and Metallurgy*, **75B**, 208–215.

Shearman, D.J. 1970. *Transactions of the Institute of Mining and Metallurgy*, **79B**, 155–162.

Shearman, D.J. & Fuller, J.G.C.M. 1969. *Bulletin of the Canadian Society of Petroleum Geologists*, **17**, 496–525.

Shearman, D.J. *et al.* 1970. *Proceedings of the Geologists Association*, **81**, 561–575.

Shearman, D.J.J. *et al.* 1989. *Bulletin of the Geological Society of America*, **101**, 913–917.

Shepard, F.P. & Inman, D.L. 1950. In: *Proceedings of the 1st Conference on Coastal Engineering*, pp. 50–59. Council on Wave Research, Berkeley, CA.

Shepard, F.P. *et al.* 1979. *Currents in Submarine Canyons and Other Sea Valleys*. Studies in Geology 8, American Association of Petroleum Geologists.

Sheridan, R.E.& Grow, J.A. (eds) 1988. *The Geology of North America*, Vol. I2, *The Atlantic Margin: US*. Geological Society of America.

Shi, Z. 1991. *Sedimentary Geology*, **73**, 43–58.

Shideler, G.L. 1978. *Marine Geology*, **26**, 284–313.

Shinn, E.A. 1969. *Sedimentology*, **12**, 109–144.

Shinn, E.A. 1983. *Journal of Sedimentary Petrology*, **53**, 619–628.

Shinn, E.A. *et al.* 1969. *Journal of Sedimentary Petrology*, **39**, 1202–1228.

Shinn, E.A. *et al.* 1989. *Journal of Sedimentary Petrology*, **59**, 147–161.

Shinn, E.A. *et al.* 1990. *Journal of Sedimentary Petrology*, **60**, 952–67.

Shotyk, W. & Metson, J.B. 1994. *Reviews of Geophysics*, **32**, 197–220.

Shreve, R.L. 1968. *The Blackhawk Landslide*. Special Paper 108, Geological Society of America.

Shreve, R.L. 1985. *Bulletin of the Geological Society of America*, **96**, 639–646.

Shukla, U.K. *et al.* 2001. *Sedimentary Geology*, **144**, 243–262.

Shultz, M.R. & Hubbard, S.M. *Journal of Sedimentary Research*, **75**, 440–453.

Shultz, M.R. *et al.* 2005. In: *Submarine Slope Systems: Processes and Products* (eds Hodgson, D.M. & Flint, S. S.), pp. 27–50. Special Publication 244, Geological Society of London.

Sibson, R.H. 1985. *Nature*, **316**, 248–251.

Sidle, R.C. & Chigira, M. 2004. *Eos, Transactions American Geophysical Union*, **85** /15, 145, 151.

Siegenthaler, C. & Huggenberger, P. 1993. In:, pp. 147–162. Special Publicaton 75, Geological Society of London.

Siegenthaler, U. & Sarmiento, J.L. 1993. *Nature*, **365**, 119–125.

Siegert, M.J. 2007. In: *Glacial Sedimentary Processes and Products* (eds Hambrey, M. *et al.*), pp. 53–64. Special Publication 39, International Association of Sedimentologists.

Siegert, M.J. *et al.* 2007. In: *Glacial Sedimentary Processes and Products* (eds Hambrey, M. *et al.*), pp. 3–10. Special Publication 39, International Association of Sedimentologists.

Siever, R. *et al.* 1992. *Geochimica et Cosmochimica Acta*, **56**, 3265–3272.

Simkiss, K. 1964. *Biological Review*, **39**, 487–505.

Simons, D.B. *et al.* 1965. In: *Primary Sedimentary Structures and Their Hydrodynamic Interpretation* (ed. Middleton, G.V.), pp. 34–52. Special Publication, Society of Economic Paleontologists and Mineralogists.

Simpson, G.D.H. 2004. *Journal of Geophysics Research*, **109**, F03007, doi: 10.1029/2003JF000111.

Simpson, G.D.H. 2006. *Basin Research*, **18**, 125–143.

Simpson, J.E. 1972. *Journal of Fluid Mechanics*, **53**, 759–768.

Simpson, J.E. 1987 (1997, 2nd edn). *Gravity Currents: In the Environment and the, Laboratory*. Ellis Horwood/Wiley, Chichester.

Simpson, J.E. & Britter, R.E. 1979. *Journal of Fluid Mechanics*, **94**, 477–495.

Sinclair, H.D. 1993. *Sedimentology*, **40**, 955–978.

Sinclair, H.D. 1994. *Journal of Sedimentary Research*, **64**, 42–54.

Sinclair, H.D. 1997. *Bulletin of the Geological Society of America*, **109**, 324–346.

Sinclair, H.D. 2000. *Journal of Sedimentary Research*, **70**, 504–519.

Sinclair, H.D. & Tomasso, M. 2002. *Journal of Sedimentary Research*, **72**, 451–456.

Singh, H. *et al.* 1993. *Sedimentary Geology*, **85**, 87–113.

Siringan, F.P. & Anderson, J.B. 1993. *Journal of Sedimentary Petrology*, **63**, 794–808.

Siringan, F.P. & Anderson, J.B. 1994. *Journal of Sedimentary Research*, **64**, 99–110.

Sleath, J.F.A. 1984. *Sea Bed Mechanics*. Wiley, New York.

Slingerland, R. & Smith, N.D. 2004. *Annual reviews of Earth and Planetary Science*, **32**, 257–285.

Slingerland, R.L. *et al.* 1996. *Bulletin Geological Society America*, **108**, 941–952.

Smith, A.M. 1997. *Journal of Geophysical Research*, **102**, 543–552.

Smith, C.R. & Walker, J.D.A. 1990. In: *Proceedings of the NASA Langley Boundary Layer Workshop* (ed. Robinson, S.).

Smith, D.G. 1983. In: *Modern and Ancient Fluvial Systems* (eds Collinson, J.D. & Lewin, J.), pp. 155–168. Special Publication 6, International Association of Sedimentologists.

Smith, D.G. 1986. *Sedimentary Geology*, **46**, 177–196.

Smith, D.G. & Smith, N.D. 1980. *Journal of Sedimentary Petrology*, **50**, 157–164.

Smith, D.G. *et al.* 1999. *Journal of Sedimentary Research*, **69**, 1290–1296.

Smith, D.G. *et al.* 2005. In: *River Deltas: Concepts, Models, and Examples* (eds Giosan, L.& Bhattacharya, J.P.), pp. 295–318. Special Publication 83, Society of Economic Paleontologists and Mineralogists.

Smith, D.P. *et al.* 2005. *Bulletin of the Geological Society of America*, **117**, 1123–1133.

Smith, G.A. & Landis, C.A. 1995. In: *Tectonics of Sedimentary Basins* (eds Busby, C. & Ingersoll, R.), pp. 263–298, Blackwell Science, Oxford.

Smith, J.D. & McLean, S.R. 1977. *Journal of Geophysical Research*, **82**, 1735–1746.

Smith, N.D. & Ashley, G.M. 1985. In: *Glacial Sedimentary Environments.*, pp. 135–215. Short Course 16, Society of Economic Paleontologists and Mineralogists.

Smith, N.D. & Pérez-Arlucea, M. 1994. *Journal of Sedimentary Research*, **64**, 159–168.

Smith, N.D. *et al.* 1989. *Sedimentology*, **36**, 1–23.

Smith, N.D. *et al.* 1998. *Canadian Journal of Earth Sciences*, **35**, 453–466.

Snedden, G.A. *et al.* 2007. *Estuarine, Coastal and Shelf Science*, **71**, 181–193.

Sohn, M.F. 2007. *Quaternary International*, **166**, 49–60.

Somoza, L. & Rey, J. 1991. In: *The Dynamics of Coarse-Grained Deltas* (eds Dabrio, C.J. *et al.*), pp. 37–48. Cuadernos de Geologia Iberica 15, Universidad Complutense, Madrid.

Somoza, L. *et al.* 1997. *Geo-Marine Letters*, **17**, 133–139.

Sondi, I. *et al.* 1995. *Sedimentology*, **42**, 769–782.

Sonett, C.P. *et al.* 1988. *Nature*, **335**, 806–808.

Soreghan, G.S. 1994. In: (eds Embry, A.F. *et al.*), pp. 523–543. Memoir 17, Canadian Society of Petroleum Geologists.

Soulsby, R.L. 1997. *Dynamics of Marine Sands*. Thomas Telford Publications.

Southard, J.B. 1971. *Journal of Sedimentary Petrology*, **41**, 903–915.

Southard, J.B. 1991. *Annual Review of Earth and Planetary Sciences*, **19**, 423–455.

Southard, J.B. & Boguchwal, L.A. 1990a. *Journal of Sedimentary Petrology*, **60**, 658–679.

Southard, J.B. & Boguchwal, L.A. 1990b. *Journal of Sedimentary Petrology*, **60**, 680–686.

Southard, J.B. *et al.* 1990. *Journal of Sedimentary Petrology*, **60**, 1–17.

Sparks, R.S.J. *et al.* 1993. *Earth and Planetary Science Letters*, **114**, 243–257.

Spence, G.H. & Tucker, M.E. 1997. *Sedimentary Geology*, **112**, 163–193.

Spence, G.H. & Tucker, M.E. 1999. *Journal of Sedimentary Research*, **69**, 947–961.

Spence, G.H. & Tucker, M.E. 2000. *Journal of Sedimentary Research*, **70**, 1335–1336.

Spence, G.H. & Tucker, M.E. 2007. *Journal of Sedimentary Research*, **77**, 797–808.

Spence, G.H. *et al.* 2004. *Sedimentology*, **51**, 1243–1271.

Spencer, R.J. *et al.* 1984. *Contributions to Mineralogy and Petrology*, **86**, 321–334.

Sperling, C.H.B. & Cooke, R.U. 1985. *Earth Surface Processes and Landforms*, **10**, 541–555.

Spicer, R.A. *et al.* 2003. *Nature*, **421**, 622–624.

Spinelli, G.A. & Field, M.E. *Journal of Sedimentary Research*, **71**, 237–245.

Sridhar, V. *et al.* 2006. *Science*, **313**, 345–347.

Stanistreet, I.G. & McCarthy, T.S. 1993. *Sedimentary Geology*, **85**, 115–133.

Stanistreet, I.G. & Stollhofen, H. 2002. *Sedimentology*, **49**, 719–736.

Stanistreet, I.G. *et al.* 1993. *Sedimentary Geology*, **85**, 135–156.

Stanley, D.J. & Warne, A.G. 1993. *Nature*, **363**, 435–438.

Stanley, S.M. 2006. *Palaeogeography, Palaeoclimatology, Palaeoecology*, **232**, 214–236.

Stanley, S.M. & Hardie, L.A. 1998. *Palaeogeography, Palaeoclimatology, Palaeoecology*, **144**, 3–19.

Stanley, S.M. & Hardie, L.A. 1999. *GSA Today*, **9/2**, 1–7.

Stanley, S.M. *et al.* 2002. *Proceedings of the National Academy of Sciences of the USA*, **99**, 15323–15326.

Starkel, L. 1983. In: *Background to Paleohydrology* (ed. Gregory, K.J.), pp. 213–235. Wiley, New York.

Staub, J.R. & Esterle, J.S. 1993. *Sedimentary Geology*, **85**, 191–201.

Stearns, L. *et al.* 2008. *Nature Geosciences*, **1**, 827–831.

Steele, R.P. (1983) In: *Eolian Sediments and Processes* (eds Brookfield M.E. & Ahlbrandt, T.S.), pp. 543–550. Developments in Sedimentology 38. Elsevier, Amsterdam.

Stefani, M. & Vincenz, S. 2005. In: *Mediterranean Prodelta Systems, Proceedings of the International ComDelta Symposia, Aix-en-Provence, France* (eds Trincardi, F. & Syvitski, J.), pp. 19–48. Marine Geology, **222–223**.

Steiger, J. *et al.* 2005. *River Research and Applications*, **21**, 719–737.

Steinberg, T. 1995. *Slide Mountain, or the Folly of Owning Nature*. University of California Press, Berkeley, CA.

Stemmerik, L. 2001. *Sedimentology*, **48**, 79–97.

Stephens, N.P. & Sumner, D.Y. 2003. *Sedimentology*, **50**, 1283–1302.

Steuber, T. & Rauch, M. 2005. *Marine Geology*, **217**, 199–213.

Stewart, B.W. *et al.* 2001. *Geochimica et Cosmochimica Acta*, **65**, 1087–1099.

Stirling, C.H. *et al.* 1998. *Earth and Planetary Science Letters*, **160**, 745–762.

Stockman, K.W. *et al.* 1967. *Journal of Sedimentary Petrology*, **37**, 633–648.

Stokes, C.R. & Clark, C.D. 2001. *Quaternary Science Reviews*, **20**, 1437–1457.

Stokes, D. *et al.* 1997. *Geomorphology*, **20**, 81–93.

Stokes, W.L. 1968. *Journal of Sedimentary Petrology*, **38**, 510–515.

Stommel, H. 1948. *Transactions of the American Geophysical Union*, **29**, 202–206.

Stommel, H. 1957. *Deep-Sea Research*, **41**, 49–84.

Stone, G.W. *et al.* 2005. *Eos, Transactions American Geophysical Union*, **86**, 497–501.

Storms, J.E.A. & Hampson, G.J. 2005. *Journal of Sedimentary Research*, **75**, 67–81.

Storms, J.E.A. *et al.* 1999. *Sedimentology*, **46**, 189–200.

Stouthamer, E. 2005. In: *River Deltas: Concepts, Models, and Examples* ( Giosan, L. & Bhattacharya, J.P.), pp. 319–340. Special Publication 83, Society of Economic Paleontologists and Mineralogists.

Stouthamer, E. & Berendson, H.J.A. 2000. *Journal of Sedimentary Research*, **70**, 1051–1064.

Stouthamer, E. & Berendson, H.J.A. 2001. *Journal of Sedimentary Research*, **71**, 589–598.

Stow, D.A.V. & Bowen, A.J. 1980. *Sedimentology*, **27**, 31–46.

Stow, D.A.V. & Faugeres, J.C. 1998. *Sedimentary Geology*, **115**, 1–386.

Stow, D.A.V. & Lovell, J.P.B. 1979. *Earth Science Reviews*, **14**, 251–291.

Stow, D.A.V. & Piper, D.J.W. 1984. In: *Fine-Grained Sediments: Deep-Water Processes and Facies* (eds Stow, D.A. V. & Piper, D.J.W.), pp. 611–646. Special Publication 15, Geological Society of London.

Stow, D.A.V. & Shanmugam, G. 1980. *Sedimentary Geology*, **25**, 23–42.

Stow, D.A.V. & Wetzel, A. 1990. *Proceedings of the Ocean Drilling Program Scientific Results*, **116**, 25–34.

Stow, D.A.V. *et al.* (eds) 2002. *Deep-Water Contourite Systems: Modern Drifts and Ancient Series, Seismic and Sedimentary Characteristics.* Memoir 22, Geological Society of London.

Strasser, A. 1988. *Sedimentology*, **35**, 369–383.

Strachan, L.J. 2002. *Sedimentology*, **49**, 25–41.

Strachan, L.J. 2008. *Sedimentology*, **55**, 1311–1332.

Straub, S. 1997. *Geologische Rundschau*, **86**, 415–425.

Straub, S. 2001. In: *Particulate Gravity Currents* (eds McCaffrey W. *et al.*), pp. 91–109. Special Publication 31, International Association of Sedimentologists.

Stride, A.H. 1963. *Quarterly Journal of the Geological Society of London*, **119**, 175–199.

Stride, A.H. *et al.* 1982. In: *Offshore Tidal Sands: Processes and Deposits* (ed. Stride, A.H.), pp. 95–125. Chapman & Hall, London.

Strom, K.B. & Papanicolaou, A.N. 2008. *Sedimentology*, **55**, 137–153.

Stumm, W. 1992. *Chemistry of the Solid–Water Interface.* Wiley, New York.

Stumm, W. & Wollast, R. 1990. *Reviews of Geophysics*, **28**, 153–169.

Sturesson, U. *et al.* 2000. *Sedimentary Geology*, **136**, 137–146.

Sturm, M. & Matter, A. 1978. In: *Modern and Ancient Lake Sediments* (eds Matter, A. & Tucker, M.E.), pp. 145–166. Special Publication 2, International Association of Sedimentologists.

Suess, E. & Futterer, D. 1972. *Sedimentology*, **19**, 129–139.

Sugden, D.E. *et al.* 1995. *Nature*, **376**, 412–414.

Sullivan, R. *et al.* 2005. *Nature*, **436**, 58–61.

Sumer, B.M. & Deigaard, R. 1979. *Experimental Investigation of Motion of Suspended Heavy Particles and the Bursting Process.* Series Publication 23, Institute of Hydrodynamics and Hydraulic Engineering, Technical University of Denmark.

Sumer, B.M. & Oguz, B. 1978. *Journal of Fluid Mechanics*, **86**, 109–127.

Summerfield, M.A. & Hulton, N.J. 1994. *Journal of Geophysical Research*, **99B**, 13871–13885.

Summerhayes, C.P. & Thorpe, S.A. 1996. *Oceanography: an Illustrated Guide.* Manson Publishing, London.

Summerhayes, C.P. *et al.* (eds) 1995. *Upwelling in the Ocean.* Wiley, Chichester.

Sumner, E.J. *et al.* 2008. *Journal of Sedimentary Research*, **78**, 529–547.

Sumner, E.J. *et al.* 2009. *Geology*, **37**, 991–994.

Sumner, D.Y. & Grotzinger, J.P. 1996. *Geology*, **24**, 119–122.

Sumner, D.Y. & Grotzinger, J.P. 2004. *Sedimentology*, **51**, 1273–1299.

Surdam, R.C. & Stanley, K.O. 1979. *Bulletin of the Geological Society of America*, **90**, 93–110.

Suresh, N. *et al.* 2007. *Sedimentology*, **54**, 809–833.

Surlyk, F. 1978. *Submarine Fan Sedimentation Along Fault Scarps on Tilted Fault Blocks (Jurassic–Cretaceous Boundary, East Greenland).* Bulletin 128, Gronlands Geologiske Undersogelse.

Surlyk, F. 1987. *Bulletin of the American Association of Petroleum Geologists*, **71**, 464–675.

Surlyk, F. 1990. In: *Correlation in Hydrocarbon Production* (ed. Collinson, J.D.), pp. 231–241. Graham and Trotman, London.

Sutcliffe, C. & Pickering, K.T. 2009. *Sedimentology*, doi: 10.1111/j.1365-3091.2009.01051.x.

Suter, J.R. *et al.* 1987. In: *Siliciclastic Shelf Sediments* (eds Tillman, R.W. & Siemans, C.T.), pp. 199–219. Special Publication 34, Society of Economic Paleontologists and Mineralogists.

Suttner, L.J. *et al.* 1981. *Journal of Sedimentary Petrology*, **51**, 1235–1246.

Sverdrup, H. & Warfvinge, P. 1995. In: *Chemical Weathering Rates of Silicate Minerals* (eds White, A.F. & Branfield, S.L.), Chapter 11. American Mineralogical Society, Washington, DC.

Swart, P.K. *et al.* (eds) 2009. *Perspectives in Carbonate Geology*. Special Publication 41, International Association of Sedimentologists.

Sweet, M.L. 1999. *Sedimentology*, **46**, 171–187.

Swift, D.J.P. 1968. *Journal of Geology*, **76**, 444–456.

Swift, D.J.P. 1972. In: *Shelf Sediment Transport: Process and Pattern* (eds Swift, D.J.P. *et al.*), pp. 363–371. Dowden, Hutchinson & Ross, Stroudsberg, PA.

Swift, D.J.P. 1974. In: *The Geology of Continental Margins* (eds Burk, C.A. & Drake, C.L.), pp. 117–135. Springer, Berlin.

Swift, D.J.P. *et al.* 1973. *Marine Geology*, **15**, 227–247.

Swift, D.J.P. *et al.* 1981. In: *Holocene Marine Sedimentation in the North Sea Basin* (eds Nio, S.-D. *et al.*), pp. 361–383. Special Publication 5, International Association of Sedimentologists.

Swift, D.J.P. *et al.* (eds) 1991. *Shelf Sand and Sandstone Bodies*. Special Publication 14, International Association of Sedimentologists.

Swift, D.J.P. *et al.* 2003. *Sedimentology*, **50**, 81–111.

Sylvester, Z. & Lowe, D.R. 2004. *Sedimentology*, **51**, 945–972.

Syvitski, J.P.M. 1989. *Marine Geology*, **85**, 301–329.

Syvitsky, J.P.M. 1991. *Principles, Methods and Applications of Particle Size Analysis*. Cambridge University Press.

Syvitski, J.P.M. & Farrow, G.E. 1989. In: *Deltas: Sites and Traps for Fossil Fuels* (eds Whateley, M.K.G. & Pickering, K.T.), pp. 21–43. Special Publication 41, Geological Society of London.

Syvitski, J.P.M. *et al.* 1987. *Fjords: Processes and Products*. Springer, New York.

Sztano, O. & de Boer, P.L. 1995. *Sedimentology*, **42**, 665–682.

Ta, T.K.O. *et al.* 2005. In: *River Deltas: Concepts, Models, and Examples* (eds Giosan, L. & Bhattacharya, J.P.), pp. 453–467. Special Publication 83, Society of Economic Paleontologists and Mineralogists.

Takahashi, T. 1975. In: pp. 11–26, Special Publication 13. Cushman Foundation for Foraminiferal Researh.

Takahashi, T. 1978. *Journal of the Hydraulics Division, American Society of Civil Engineers*, **104**, 1153–1169.

Talbot, C.J. *et al.* 1996. *Sedimentology*, **43**, 1025–1047.

Talbot, M.R. 1973. *Palaeogeography, Palaeoclimatology, Palaeoecology*, **14**, 293–317.

Talbot, M.R. 1980. In: *The Sahara and the Nile* (eds Williams, M.A.J. & Faure, H.), pp. 37–62. Balkema, Rotterdam.

Talbot, M.R. 1985. *Sedimentology*, **32**, 257–265.

Talbot, M.R. 1990. *Chemical Geology*, **80**, 261–279.

Talbot, M.R. & Williams, M.A.J. 1979. *Catena*, **6**, 43–62.

Talbot, M.R. & Allen, P.A. 1996. In: *Sedimentary Environments: Processes, Facies and Stratigraphy* (ed. Reading, H. G.), pp. 83–124. Blackwell Science, Oxford.

Talling, P.J. 2007. *Sedimentology*, **54**, 737–769.

Talling, J.F. & Talling, I.B. 1965. *International Review of Hydrobiology*, **50**, 421–463.

Talling, P.J. *et al.* 1994. *Journal of Geology*, **102**, 181–196.

Talling, P.J. *et al.* 2004. *Sedimentology*, **51**, 163–194.

Talling, P.J. *et al.* 2007a. *Nature*, **450**, 541–544.

Talling, P.J. *et al.* 2007b. *Journal of Sedimentary Research*, **77**, 172–196.

Tamaki, K. & Honza, E. 1991. *Episodes*, **14**, 224–230.

Tamura, T. *et al.* 2007. *Sedimentology*, **54**, 1149–1162.

Tanaka, K.L. 2005. *Nature*, **437**, 991–994.

Tandon, S.K. & Gibling, M.R. 1997. *Sedimentary Geology*, **112**, 43–67.

Tanner, W.F. 1971. *Sedimentology*, **16**, 71–88.

Taylor, A. & Blum, J.D. 1995. *Geology*, **23**, 979–982.

Taylor, B. *et al.* 1991. *Journal of Geophysical Research*, **96**, 16113–16129.

Taylor, G. & Woodyer, K.D. 1978. In: *Fluvial Sedimentology* (ed. Miall, A.D.), pp. 257–275. Memoir 5, Canadian Society of Petroleum Geologists.

Taylor, K.G. & Curtis, C.D. 1995. *Journal of Sedimentary Research*, **65**, 358–368.

Taylor, K.G. *et al.* 2002. *Journal of Sedimentary Research*, **72**, 316–327.

Taylor, S.K. *et al.* 2004. *Journal of Geophysics Research*, **109**, B02408. doi: 10.1029/2003JB002412.

Taylor, S.R. & McLennan, S.M. 1995. *Reviews of Geophysics*, **33**, 241–265.

Teal, C.S. *et al.* 2000. *Journal of Sedimentary Research*, **70**, 649–663.

Tegen, I. *et al.* 1996. *Nature*, **380**, 419–421.

Ten Brinke, W.B.M. *et al.* 1998. *Earth Surface Processes and Landforms*, **23**, 809–824.

Teng, H.H. *et al.* 2001. *Geochimica et Cosmochimica Acta*, **65**, 3459–3474.

Tessier, B. 1993. *Marine Geology*, **110**, 355–367.

Tessier, B. & Gigot, P.A. 1989. *Sedimentology*, **36**, 767–776.

Testa, V. & Bosence, D.W.J. 1999. *Sedimentology*, **46**, 279–301.

Tewes, D.W. & Loope, D.B. 1992. *Sedimentology*, **39**, 251–261.

Tharp, T.M. 1987. *Bulletin of the Geological Society of America*, **99**, 94–102.

Thiede, J. & van Andel, T. H. 1977. *Earth and Planetary Science Letters*, **33**, 301–309.

Thomas, D.S.G. (ed.) 1989. *Arid Zone Geomorphology*. Belhaven Halsted, New York.

Thomas, D.G.S. *et al.* 1997. *The Holocene*, **7**, 273–281.

Thomas, D.S.G. *et al.* 2005. *Nature*, **435**, 1218–1221.

Thorne, C.R. *et al.* 1987. *Sediment Transport in Gravel Bed Rivers*. Wiley, Chichester.

Thorne, C.R. *et al.* (eds). 1988. *The Physics of Sediment Transport by Wind and Water: a Collection of Hallmark Papers by R.A. Bagnold.* American Society of Civil Engineers, New York.

Thunell, R.C. 1982. *Marine Geology*, **47**, 165–180.

Thunell, R.C. *et al.* 1977. *Marine Micropalaeontology*, **2**, 371–388.

Thunnell, R. *et al.* 1991. *Journal of Sedimentary Petrology*, **61**, 1109–1122.

Thurman, H.V. 1991. *Introductory Oceanography*, 7th edn. Macmillan, New York.

Tiedermann, W.G. *et al.* 1985. *Journal of Fluid Mechanics*, **156**, 419–437.

Till, R. 1978. In: *Sedimentary Environments and Facies* (ed. Reading, H.G.), Blackwell Science, Oxford.

Timofeeff. M.N. 2000. *Secular variations in seawater chemistry: A new technique for analysing fluid inclusions in marine halites from the modern, Permian, and Cretaceous.* Unpublished PhD thesis, University of Binghamton New York, 104 pp.

Timofeeff, M. N. *et al.* 2006. *Geochimica et Cosmochimica Acta*, **70**, 1977–1994.

Tinker, S.W. 1998. *Journal of Sedimentary Research*, **68**, 1146–1174.

Tipper, E.T. *et al.* 2006. *Geochimica et Cosmochimica Acta*, **70**, 2737–2754.

Tischer, M. *et al.* 2001. *Journal of Sedimentary Research*, **71**, 355–364.

Tolhurst, T.J. *et al.* 2006. *Aquatic Ecology*, **40**, 533–541.

Tolhurst, T.J. *et al.* 2009. *Journal of Hydraulic Engineering*, **135**, 73–87.

Tomkins, M.R. *et al.* 2005. *Sedimentology*, **52**, 1425–1432.

Toniolo, H. *et al.* 2006a. *Journal of Sedimentary Research*, **76**, 783–797.

Toniolo, H. *et al.* 2006b. *Journal of Sedimentary Research*, **76**, 798–818.

Törnqvist, T.E. 1993. *Journal of Sedimentary Petrology*, **63**, 683–693.

Törnqvist, T.E. & Bridge, J.S. 2002. *Sedimentology*, **49**, 891–905.

Törnqvist, T.E. & Bridge, J.S. 2006. *Journal of Sedimentary Research*, **76**, 959.

Törnqvist, T.E. *et al.* 1993. *Sedimentary Geology*, **85**, 203–219.

Törnqvist, T.E. *et al.* 1996. *Science*, **273**, 1693–1696.

Törnqvist, T.E. *et al.* 2008. *Nature Geoscience*, **1**, 173–176.

Torres, J. *et al.* 1997. *Sedimentology*, **44**, 457–477.

Trewin, N.H. 1986. *Transactions of the Royal Society of Edinburgh, Earth Sciences*, **77**, 21–46.

Tribovillard, N. *et al.* 1999. *Sedimentology*, **46**, 1183–1197.

Tricker, R.A.R. 1964. *Bores, Breakers, Waves and Wakes.* Mills and Boon, London; Elsevier, New York.

Tripati, A. *et al.* 2005. *Nature*, **436**, 341–346.

Tripsanas, E.K. *et al.* 2006. *Journal of Sedimentary Research*, **76**, 1012–1034.

Tripsanas, E.K. *et al.* 2008. *Sedimentology*, **55**, 97–136.

Tritton, D.J. 1988. *Physical Fluid Dynamics.* Oxford University Press, Oxford.

Trouwborst, R.E. 2006. *Geochimica et Cosmochimica Acta*, **71**, 4629–4643.

Trudgill, B. D. & Cartwright, J. A. 1994. *Bulletin of the Geological Society of America*, **106**, 1143–1157.

Tsoar, H. 1978. *The Dynamics of Longitudinal Dunes.* Final technical report, European Research Office, US Army, London.

Tsoar, H. 1982. *Journal of Sedimentary Petrology*, **52**, 823–831.

Tsoar, H. 1983. *Sedimentology*, **30**, 567–578.

Tucker, M.E. 1982. *Geology*, **10**, 7–12.

Tucker, M.E. & Wright, V.P. 1990. *Carbonate Sedimentology.* Blackwell Scientific, Oxford.

Tucker, R.M. & Cann, J.R. 1986. *Sedimentology*, **33**, 401–412.

Tudhope, A.W. & Scoffin, T.P. 1994. *Journal of Sedimentary Research*, **64**, 752–764.

Turcotte, D.L. & Schubert, G. 1982. *Geodynamics: Applications of Continuum Physics to Geological Problems.* Wiley, New York.

Turgeon, S.C. & Creaser, R.A. 2008. *Nature*, **454**, 323–329.

Turner, E.C. *et al.* 1993. *Geology*, **21**, 259–262.

Turner, R.E. *et al.* 2006. *Science*, **314**, 449–452.

Tuschall, J.R. & Brezonik, P.L. 1980. *Limnology and Oceanography*, **25**, 495–504.

Twichell, D.C. *et al.* 1991. In: *Seismic Facies and Sedimentary Processes of Submarine Fans and Turbidite Systems* (eds Weimer, P. & Link, M.H.), pp. 349–364. Springer, Berlin.

Twichell, D.C. *et al.* 2000. In: *Deep-Water Reservoirs of the World* (eds Weimer, P. *et al.*), pp. 1032–1044. 20th Annual Research Conference. CD-ROM, Gulf Coast Society and Society of Economic Paleontologists and Mineralogists.

Tzedakis, P.C. 1993. *Nature*, **364**, 437–440.

Umbanhowar, P.B. *et al.* 1996. *Nature*, **382**, 793–796.

Umhoefer, P.J. *et al.* 2007. *Basin Research*, **19**, 297–322.

Uncles, R.J. & Stephens, J.A. 1989. *Journal of Geophysical Research*, **94**, 14395–14405.

Uncles, R.J. *et al.* 1998a. In: *Sedimentary Processes in the Intertidal Zone* (eds Black, K.S. *et al.*), pp. 211–219. Special Publication 139. Geological Society of London.

Uncles, R.J. *et al.* 1998b. *Marine Pollution Bulletin*, **37**, 206–215.

Uncles, R.J. *et al.* 2002a. *Continental Shelf Research*, **22**, 1835–1856.

Uncles, R.J. *et al.* 2002b. *Estuarine Coastal and Shelf Science*, **55**, 829–856.

Underhill, J.R. 1991. *Basin Research*, **3**, 79–98.

Underwood, M.B. & Bachman, S.B. 1982. In: *Trench-Forearc Geology* (ed. Leggett, J.K.), pp. 537–550. Special Publication 10, Geological Society of London.

Underwood, M.B. & Moore, G.F. 1995. In: *Tectonics of Sedimentary Basins* (eds Busby, C. & Ingersoll, R.), pp. 179–220. Blackwell Science, Oxford.

Underwood, M.B. *et al.* 1995. In: *Active Margins and Marginal Basins of the Western Pacific*, pp. 315–353. Geophysical Monograph 88, American Geophysical Union.
Underwood, M.B. *et al.* 2003. *Journal of Sedimentary Research*, 73, 589–602.
Underwood, M.B. *et al.* 2005. *Journal of Sedimentary Research*, 75, 149–164.
UNESCO. 1977. *World Distribution of Arid Regions*. CNRS, Paris.
Vacher, H. L.& Quinn, T. M. (eds) 1997. *Geology and Hydrogeology of Carbonate Islands*. Developments in Sedimentology 54, Elsevier.
Vago, R. *et al.* 1997. *Nature*, **386**, 30–31.
Vail, P.R. *et al.* 1977. In: *Seismic Stratigraphy-Applications to Hydrocarbon Exploration* (ed. Payton, C.E.), pp. 49–205. Memoir 26, American Association of Petroleum Geologists.
Valeton, I. 1996. In: *Proceedings of the 4th International Symposium on the Geochemistry of the Earth's Surface* (ed. Bottrell, S.H.), pp. 234–240. University of Leeds.
Vallance, J.W. & Scott, K.M. 1997. *Bulletinof the Geological Society of America*, **109**, 143–163.
Valyashko, M.G. 1972. In: *Geology of Saline deposits* (ed. Richter-Bernberg, G.), pp. 41–51. UNESCO, Paris.
Van Andel, T.H. 1975. *Earth and Planetary Science Letters*, **26**, 187–194.
Van Andel, T.H. & Komar, P.D. 1969. *Bulletin of the Geological Society of America*, **80**, 1163–1190.
Van Andel, T.H. *et al.* 1977. *Journal of Geology*, **85**, 651–698.
Vance, D. *et al.* 2003. *Earth and Planetary Science Letters*, **206**, 273–288.
Vance, D. *et al.* 2009. *Nature*, **458**, 493–496.
Van den Berg, J.H. 1995. *Geomorphology*, **12**, 259–279.
Van den Berg, J.H. & van Gelder, A. 1993. In: *Alluvial Sedimentation* (eds M. Marzo & C. Puigdefabregas), Special Publication 17, pp. 11–21. International Association of Sedimentologists.
Van der Beek, P. *et al.* 1994. *Earth and Planetary Science Letters*, **121**, 417–433.
Vanderburghe, J. 1995. *Quaternary Science Reviews*, **14**, 631–638.
Van de Schootbrugge, B. *et al.* 2005. *Paleoceanography*, **20**, PA3008, doi: 1029/2004PA001102, 2005.
Van Dyke, M. (1982) *An Album of Fluid Motion*. Parabolic Press, Stanford, CA.
Van Gelder, A. *et al.* 1994. *Sedimentary Geology*, **90**, 293–305.
Van Houten, F.B. 1962. *American Journal of Science*, **260**, 561–576.
Van Houten, F.B. 1964. *Bulletin of the Geological Survey of Kansas*, **169**, 497–531.
Vaniman, D.T. *et al.* 2004. *Nature*, **431**, 663–665.
Van Leussen, W. 1997. In: *Cohesive Sediments* (eds Burt, N. *et al.*), pp. 45–62. Wiley, New York.
Van Maren, D.S. 2005. *Marine Geology*, **224**, 123–143.
Van Maren, D.S. *et al.* 2009. *Sedimentology*, **56**, 785–806.
Van Waggoner, J.C. *et al.* 1988. In: *Sea Level Changes—An Integrated Approach* (eds Wilgus, C.K. *et al.*), pp. 71–108. Special Publication 42, Society of Economic Paleontologists and Mineralogists.
Varban, B.L. & Plint, A.G. 2008. *Sedimentology*, **55**, 395–421.
Vardy, A. 1990. *Fluid Principles*. McGraw-Hill, New York.
Vasconcelos, C. & McKenzie, J.A. 1997. *Journal of Sedimentary Research*, **67**, 378–390.
Vasconcelos, C. *et al.* 1995. *Nature*, **377**, 220–222.
Veizer, J. *et al.* 1999. *Chemical Geology*, **161**, 59–88.
Velbel, M.A. 1993. *Geology*, **21**, 1059–1062.
Verwer, K. *et al.* 2009. *Journal of Sedimentary Research*, **79**, 416–439.
Vilas, F. *et al.* 1991. *Marine Geology*, **97**, 391–404.
Villard, P.V. & ChurChapter M. 2005. *Sedimentology*, **52**, 737–756.
Villard, P.V. & Osborne, P.D. 2002. *Sedimentology*, **49**, 363–378.
Villareal, T.A. *et al.* 1993. *Nature*, **363**, 709–12.
Villareal, T.A. *et al.* 1999. *Nature*, **397**, 423–425.
Vincent, C.E. & Green, M.O. 1990. *Journal of Geophysical Research*, **95**, 11591–11601.
Vincent, P. 1996. *Sedimentary Geology*, **103**, 273–280.
Vincent, S.J. 2001. *Sedimentology*, **48**, 1235–1276.
Violet, J. *et al.* 2005. *Journal of Sedimentary Research*, **75**, 820–843.
Visscher, P.T. *et al.* 2000. *Geology*, **28**, 919–922.
Visser, M.J. 1980. *Geology*, **8**, 543–546.
Vitousek, P.M. *et al.* 1997. *GSA Today*, **7**, 1–8.
Vogel, S. 1994. *Life in Moving Fluids*. Princeton University Press, Princeton, NJ.
Vogt, P.R. *et al.* 1993. *Eos, Transactions American Geophysical Union*, October 5 issue.
Vollbrecht, R. & Meischner, D. 1996. *Journal of Sedimentary Research*, **66**, 243–258.
Von Heune, R. & Scholl, D.W. 1991. *Reviews of Geophysics*, **29**, 279–316.
Von Heune, R. *et al.* 1989. *Journal of Geophysical Research*, **94**, 1703–1714.
Wagner, N.J. & Brady, J.F. 2009. *Physics Today*, October Issue, 27–32.
Wagreich, M. & Strauss, P.E. 2005. In: *Alluvial Fans: Geomorphology, Sedimentology, Dynamics* (eds Harvey, A. M. *et al.*), pp. 207–216. Special Publication 251, Geological Society of London.
Wahlstrom, E.E. 1948. *Bulletin of the Geological Society of America*, **59**, 1173–1190.
Wakelin-King, G.A. & Webb, J.A. 2007. *Journal of Sedimentary Research*, **77**, 702–712.
Walder, J.S. & Fowler, A. 1994. *Journal of Glaciology*, **40**, 3–15.
Walder, J. & Hallet, B. 1985. *Bulletin of the Geological Society of America*, **96**, 336–346.

Walker, J.C.G. 1977. *Evolution of the Atmosphere.* Macmillan, New York.
Walker, J.C.G. 1980. In: *Handbook of Environmental Chemistry* (ed. Hutzinger, O.), pp. 87–104. Springer, Heidelberg.
Walker, J.C.G. & Drever, J.I. 1988. In: *Chemical Cycles in the Evolution of the Earth* (eds Gregor, C.B. *et al.*), pp. 55–76. Wiley, New York.
Walker, R.G. 1965. *Proceedings of the Yorkshire Geological Society*, **35**, 1–32.
Walker, R.G. 1976. *Geosciences Canada*, **3**, 25–36.
Walker, R.G. 1978. *Bulletin of the American Association of Petroleum Geologists*, **62**, 932–966.
Walker, R.G. & Mutti, E. 1973. In: *Turbidites and Deep Water Sedimentation* (eds Middleton, G.V. & Bouma, A.H.), pp. 119–157. Pacific Section Short Course, Society of Economic Paleontologists and Mineralogists.
Walker, F.D.L. *et al.* 1995. *Mineralogical Magazine*, **59**, 507–536.
Walker, H.J. *et al.* 1987. *Geografiska Annaler*, **69A**, 189–200.
Walker, J.G.C. *et al.* 1981. *Journal of Geophysical Research*, **86**, 9776–9782.
Wallbridge, S. *et al.* 1999. *Sedimentology*, **46**, 17–32.
Wallmann, K. *et al.* 1997. *Nature*, **387**, 31–32.
Walsh, J.P. *et al.* 2006. *Eos, Transactions American Geophysical Union*, **87** /44, 477–478.
Waltham, D. 2004. *Journal of Sedimentary Research*, **74**, 129–134.
Waltham, D. 2008. *Journal of Sedimentary Research*, **78**, 317–322.
Wan, Z. 1982. *Bed Material Movement in Hyperconcentrated Flows.* Series Paper 31, Institute of Hydrodynamics and Hydraulic Engineering, Technical University of Denmark.
Wan, Z. & Wang, Z. 1994. *Hyperconcentrated Flow.* Balkema, Rotterdam.
Wang, D. *et al.* 2008. *Sedimentology*, **55**, 461–470.
Wang, P. & Horwitz, M.H. 2007. *Sedimentology*, **54**, 545–564.
Wang, T. *et al.* 2005. *Sedimentology*, **52**, 429–440.
Wang, Y.J. *et al.* 2008. *Nature*, **451**, 1090–1093.
Wang, Z.Y. *et al.* 1994. *Journal of Hydraulic Research*, **32**, 495–516.
Wanless, H.R. *et al.* 1988a. *Journal of Sedimentary Petrology*, **58**, 724–738.
Wanless, H.R. *et al.* 1988b. *Journal of Sedimentary Petrology*, **58**, 739–750.
Ward, W.C. & Brady, M.J. 1979. *Bulletin of the American Association of Petroleum Geologists*, **63**, 362–369.
Warren, J.K. 1982. *Journal of Sedimentary Petrology*, **52**, 1171–1201.
Warren, J.K. & Kendall, C.G.S.C. 1985. *Bulletin of the American Association of Petroleum Geologists*, **69**, 1013–1023.
Warren, W.P. & Ashley, G.M. 1994. *Journal of Sedimentary Research*, **64**, 433–449.
Wasson, R.J. & Hyde, R. 1983. *Nature*, **304**, 337–339.
Watkins, D.J. & Kraft, L.M. 1978. In: *Framework Facies and Oil-Trapping Characteristics of the Upper Continental Margin* (eds Bouma, A.H. *et al.*)pp. 267–286. Studies in Geology 7, American Association of Petroleum Geologists.
Weaver, P.P.E. & Thomson, J. 1993. *Nature*, **364**, 136–138.
Weaver, P.P.E. *et al.* 2000. *Sedimentology*, **47 Suppl 1**, 239–256.
Weber, K.J. & Daukoru, E. 1975. In: *Proceedings of the 9th World Petroleum Congress* Vol. 2, pp. 209–21. Applied Science, London.
Webster, P.J. 1994. *Reviews of Geophysics*, **32**, 427–476.
Weertman, J. 1957. *Journal of Glaciology*, **3**, 33–38.
Wei, T. & Willmarth, W.W. 1991. *Journal of Fluid Mechanics*, **223**, 241–252.
Weimer, P. 1991. In: *Seismic Facies and Sedimentary Processes of Submarine Fans and Turbidite Systems* (eds Weimer, P. & Link, M.H.), pp. 323–348. Springer, Berlin.
Weimer, P.& Posamentier, H.W. (eds) 1993. *Siliciclastic Sequence Stratigraphy.* Memoir 58, American Association of Petroleum Geologists.
Weimer, P. *et al.* (eds) 2000. *Deep-Water Reservoirs of the World.* 20th Annual Research Conference, Gulf Coast Society and Society of Economic Paleontologists and Mineralogists.
Weirich, F.H. *et al.* 1986. *Sedimentology*, **33**, 261–377.
Weiss, R. & Bahlburg, H. 2006. *Journal of Sedimentary Research*, **76**, 1267–1273.
Weissman, G.S. *et al.* 2005. In: *Alluvial Fans: Geomorphology, Sedimentology, Dynamics* (eds Harvey, A.M. *et al.*) pp. 169–186. Special Publication 251, Geological Society of London.
Welch, S.A. & Banfield, J.F. 2002. *Geochimica et Cosmochimica Acta*, **66**, 213–221.
Wellner, R.W. & Bartek, L.R. 2003. *Journal of Sedimentary Research*, **73**, 926–940.
Wellner, R.W. *et al.* 1993. *Geology*, **21**, 109–12.
Wells, A.J. & Illing, L.V. 1964. In: *Deltaic and Shallow Marine Deposits* (ed. Van Straten, L.M.J.U.), pp. 429–435. Elsevier, Amsterdam.
Wells, J.T. (ed.) 1987. *Effects of Sea-Level Rise on Deltaic Sedimentation in South-Central Louisiana.* Special Publication, Society of Economic Paleontologists and Mineralogists.
Wells, M.R. *et al.* 2005. *Sedimentology*, **52**, 715–735.
Wells, M.R. *et al.* 2007. *Sedimentary Research*, **77**, p. 843–865.
Wells, N. 1986. *The Atmosphere and Ocean: A Physical Introduction.* Taylor and Francis, London.
Wells, N.A. & Dore, J.A. 1987. *Geology*, **15**, 204–207.
Wells, N.C. *et al.* 1996. In: *Oceanography: an illustrated guide* (eds Summerhayes, C.P. & Thorpe, S.A.), Chapter 3. Manson Press, London.
Weltje, G.J. & de Boer, P.L. 1993. *Geology*, **21**, 307–310.

Weltje, G.J. & Prins, M.A. 2007. *Sedimentary Geology*, **202**, 409–424.
Weltje, G.J. & von Eynatten, H. 2004. *Sedimentary Geology*, **171**, 1–11.
Wendt, J. *et al.* 1993. *Geology*, **21**, 723–726.
Wendt, J. *et al.* 1997. *Journal of Sedimentary Research*, **67**, 424–436.
Werner, B.T. 1995. *Geology*, **23**, 1107–1110.
West, I.M. 1975. *Proceedings of the Geological Association*, **86**, 205–225.
Westbroek, P. 1991. *Life as a Geological Force*. Norton, New York.
Wetzel, R.G. 1983. *Limnology*, 2nd edn. Saunders College Publishing, Philadelphia.
Whalen, M.T. 1995. *Geology*, **23**, 625–628.
Whateley, M.K.G.& Pickering, K.T. (eds) 1989. *Deltas: Sites and Traps for Fossil Fuels*. Special Publication 41, Geological Society of London.
Wheatcroft, R.A. 2000. *Continental Shelf Research*, **20**, 2059–2294.
Wheatcroft, R.A. & Borgeld, J.C. 2000. *Continental Shelf Research*, **20**, 2163–2190.
Wheatcroft, R.A. & Drake, D.E. 2003. *Marine Geology*, **199**, 123–137.
Wheatcroft, R.A. & Sommerfield, C.K. 2005. *Continental Shelf Research*, **24**, 311–332.
Wheeler, A.J. *et al.* 2008. *Sedimentology*, **55**, 1875–1887.
Whelley, P. & Greeley, R. 2008. *Journal of Geophysics Research*, doi: 10.1029/2007JE002966.
Whipple, K.X. 1997. *Journal of Geology*, **105**, 243–262.
Whipple, K.X. & Dunne, T. 1992. *Bulletin of the Geological Society of America*, **104**, 887–900.
Whipple, K. X. & Meade, B. J. 2004. *Journal of Geophysical Research*, **109**, F01011, doi: 10.1029/2003JF000019.
Whitaker, F. *et al.* 1997. *Geology*, **25**, 175–178.
Whitaker, J.H.McD. 1973. *Norsk Geologisk Tidsskrift*, **53**, 403–417.
White, A.F. 1995. In: *Chemical Weathering Rates of Silicate Minerals* (eds White, A.F. & Brantley, S.L.), Chapter 9. American Mineralogical Society, Washington, DC.
White, A.F.& Brantley, S.L. (eds) 1995a. *Chemical Weathering Rates of Silicate Minerals*. American Mineralogical Society, Washington, DC.
White, A.F. & Brantley, S.L. 1995b. In: *Chemical Weathering Rates of Silicate Minerals* (eds White, A.F. & Brantley, S.L.), Chapter 1. American Mineralogical Society, Washington, DC.
White, A.F. *et al.* 1999. *Geochimica et Cosmochimica Acta*, **63**, 3277–3291.
White, A.F. *et al.* 2001. *Geochimica et Cosmochimica Acta*, **65**, 847–869.
White, A.F. *et al.* 2005. *Geochimica et Cosmochimica Acta*, **69**, 1455–1471.
White, T. *et al.* 2002. *Basin Research*, **14**, 43–54.
Whiting, P.J. *et al.* 1988. *Geology*, **16**, 105–108.
Widdel, F. *et al.* 1993. *Nature*, **362**, 834–836.
Wiegand, R.C. & Chamberlain, V. 1987. *Limnology and Oceanography*, **32**, 29–42.
Weins, D.A. *et al.* 2008. *Nature*, **453**, 770–774.
Wiggs, G.F.S. *et al.* 1995. *Earth Surface Processes & Landforms*, **20**, 515–529.
Wiggs, G.F.S. *et al.* 2004. *Sedimentology*, **51**, 95–108.
Wignall, P.B. 1994. *Black Shales*. Oxford University Press, Oxford.
Wignall, P.B. & Best, J.L. 2004. *Sedimentology*, **51**, 1343–1358.
Wilber, R.J. *et al.* 1990. *Geology*, **18**, 970–974.
Wilber, R.J. *et al.* 1993. *Geology*, **21**, 667–669.
Wilcock, P.R. & McArdell, B.W. 1993. *Water Resources Research*, **29**, 1297–1312.
Wilcock, P.R. & Southard, J.B. 1989. *Water Resources Research*, **25**, 1629–1641.
Wilde, S. A. *et al.* 2001. *Nature*, **409**, 175–178.
Wilkinson, B.H. *et al.* 1985. *Journal of Sedimentary Petrology*, **55**, 171–183.
Wilkinson, B.H. *et al.* 1996. *Journal of Sedimentary Research*, **66**, 1065–1078.
Wilkinson, B.H. *et al.* 1997a. *Journal of Sedimentary Research*, **67**, 1068–1082.
Wilkinson, B.H. *et al.* 1997b. *Geology*, **25**, 847–850.
Wilkinson, B.H. *et al.* 1999. *Journal of Sedimentary Research*, **69**, 338–350.
Williams, G.E. 1989. *Journal of the Geological Society of London*, **146**, 97–111.
Williams, G.E. 2001. *Geological Magazine*, **138**, 161–184.
Williams, M.A.J. 1994. In: *Geomorphology of Desert Environments* (eds Abrahams, A.D. & Parsons, A.J.), pp. 644–670. Chapman & Hall, London.
Williams, M.A.J. *et al.* 1993. *Quaternary Environments*. Arnold, London.
Williams, P.B. & Kemp, P.H. 1971. *Journal of the Hydraulics Division, American Society of Civil Engineers*, **97**, 505–522.
Williams, T. *et al.* 2006. *Eos, Transactions American Geophysical Union*, **87**, 525–526.
Willis, B.J. 1993. In: *Alluvial Sedimentation* (eds Marzo, M. & Puigdefabregas, C.), pp. 101–114. Special Publication 17, International Association of Sedimentologists.
Willis, B.J. 1997. *Sedimentology*, **44**, 735–757.
Willis, B.J. 2005. In: *River Deltas: Concepts, Models, and Examples* (eds Giosan, L. & Bhattacharya, J.P.), pp. 87–132. Special Publication 83, Society of Economic Paleontologists and Mineralogists.
Willis, B.J. & Behrensmeyer, A.K. 1994. *Journal of Sedimentary Research*, **64**, 60–67.
Willis, B.J. & Gabel, S.L. 2001. *Sedimentology*, **48**, 479–506.
Willis, B.J. & Gabel, S.L. 2003. *Journal of Sedimentary Research*, **73**, 246–263.
Willis, B.J. *et al.* 1999. *Sedimentology*, **46**, 667–688.
Willmott, V. *et al.* 2007. In: *Glacial Sedimentary Processes and Products* (eds Hambrey, M. *et al.*), pp. 67–84.

Special Publication 39, International Association of Sedimentologists.

Wilmsen, M. & Neuweiler, F. 2008. *Sedimentology*, **55**, 773–807.

Wilson, A.G. & Kirkby, M.J. 1975. *Mathematics for Geographers and Planners*. Oxford University Press, Oxford.

Wilson, I.G. 1971. *Geographical Journal*, **137**, 180–99.

Wilson, I.G. 1972a. *Sedimentology*, **19**, 173–210.

Wilson, I.G. 1972b. *Journal of Sedimentary Petrology*, **42**, 667–669.

Wilson, I.G. 1973. *Sedimentary Geology*, **10**, 77–106.

Wilson, J.B. 1982. In: *Offshore Tidal Sands: Processes and Deposits* (ed. Stride, A.H.), pp. 126–171. Chapman & Hall, London.

Wilson, J.L. 1975. *Carbonate Facies in Geologic History*. Springer, Berlin.

Wilson, M.E.J. 2005. *Journal of Sedimentary Research*, **75**, 114–133.

Wilson, P.A. & Norris, R.D. 2001. *Nature*, **412**, 425–430.

Wilson, P.A. & Roberts, H.H. 1992. *Geology*, **20**, 713–716.

Wilson, P.A. & Roberts, H.H. 1995. *Journal of Sedimentary Research*, **A65**, 45–56.

Wingham, D.J. *et al.* 2006. *Nature*, **440**, 1033–1036.

Winn, R.D. *et al.* 1995. *Bulletin of the Geological Society of America*, **107**, 851–866.

Winsemann, J. *et al.* 2007, In: *Glacial Sedimentary Processes and Products* (eds Hambrey, M. *et al.*), pp. 121–148. Special Publication 39, International Association of Sedimentologists.

Witten, T.A. 1990. *Physics Today*, July, 21–28.

Wolfe, S.A. & Nickling, W.G. 1996. *Earth Surface Processes & Landforms*, **21**, 607–619.

Wollast, R. & Mackenzie, F.T. 1983. In: *Silicon Geochemistry and Biochemistry* (ed. Ashton, S.R.), pp. 39–76. Academic Press, San Diego.

Wollast, R. *et al.* 1980. *American Journal of Science*, **280**, 831–848.

Wood, J.M. 1989. *Bulletin of the Canadian Society of Petroleum Geologists*, **37**, 169–181.

Wood, G.V. & Wolfe, M.J. 1969. *Sedimentology*, **12**, 165–191.

Wood, L.J. 2007. *Journal of Sedimentary Research*, **77**, 713–730.

Wood, R. *et al.* 1993. *Sedimentology*, **40**, 829–858.

Woodcock, N.H. 1986. *Philosophical Transactions of the Royal Society, London (A)*, **317**, 13–29.

Woodfine, R.G. *et al.* 2008. *Sedimentology*, **55**, 1011–1028.

Woodroffe, C.D. *et al.* 1985. *Nature*, **317**, 711–713.

Woods, P.J. & Brown, R.G. 1975. In: *Tidal Deposits* (ed. Ginsburg, R.N.), pp. 223–233. Springer, New York.

Wopfner, H. & Twidale, C.R. 1988. *Geologische Rundschau*, **77**, 815–834.

Worden, R.H. *et al.* 1990. *Contributions to Mineralogy and Petrology*, **104**, 507–515.

Wright, D.T. & Wacey, D.W. 2005. *Sedimentology*, **52**, 987–1008.

Wright, H.E. *et al.* (eds) 1993. *Global Climates Since the Last Glacial Maximum*. University of Minnesota Press, Minneapolis.

Wright, L.D. 1977. *Bulletin of the Geological Society of America*, **88**, 857–868.

Wright, L.D. 1985. In: *Coastal Sedimentary Environments* (ed. Davis, R.A.), pp. 1–76. Springer, New York.

Wright, L.D. & Coleman, J.M. 1973. *Bulletin of the American Association of Petroleum Geologists*, **57**, 370–398.

Wright, L.D. *et al.* 1986. *Geo-Marine Letters*, **6**, 97–105.

Wright, L.D. *et al.* 1991. *Marine Geology*, **96**, 19–51.

Wright, V.P. & Burchette, T.P. 1996. In: *Sedimentary Environments: Processes, Facies and Stratigraphy* (ed. Reading, H.G.), pp. 325–394. Blackwell Science, Oxford.

Wright, V.P.& Burchette, T.P. (eds). 1998. *Carbonate Ramps*. Special Publication 149, Geological Society of London.

Wright, V.P. & Marriott, S.B. 1993. *Sedimentary Geology*, **86**, 203–210.

Wunderlich, F. 1972. *Senckenbergiana Maritima*, **4**, 15–45.

Wunderlich, W.O. 1971. In: *Reservoir Fisheries and Limnology* (ed. Hall, G.E.), pp. 219–231. Special Publication 8, American Fisheries Society, Washington, DC.

Wyngaard, J.C. 1992. *Annual Reviews of Fluid Mechanics*, **24**, 205–233.

Wynn, R.B. *et al.* 2002. *Sedimentology*, **49**, 669–695.

Wynn, T.C. & Read J.F. 2008. *Sedimentology*, **55**, 357–394.

Yan, Y. *et al.* 2007. *Sedimentology*, **54**, 1–17.

Yang, B.C. *et al.* 2005. *Sedimentology*, **52**, 235–252.

Yang, B.C. *et al.* 2007. *Journal of Sedimentary Research*, **77**, 757–771.

Yang, B.C. *et al.* 2008. *Geology*, **36**, 39–42.

Yang, Y.-L. *et al.* 1995. *Geology*, **23**, 1115–1118.

Yarnold, J.C. 1993. *Bulletin of the Geological Society of America*, **105**, 345–360.

Yaylor & Chavetz. 2004. *Journal of Sedimentary Research*, **74**, 328–341.

Yechieli, Y. *et al.* 1998. *Geology*, **26**, 755–758.

Yen, A.S. *et al.* 2005. *Nature*, **436**, 49–54.

Yoo, K. & Mudd, S.M. 2008. *Geology*, **36**, 35–38.

Yordanova, E.K. & Hohenegger, J. 2007. *Sedimentology*, **54**, 1273–1306.

Yoshida, F. 1994. *Sedimentary Geology*, **92**, 97–115.

Yoshida, S. 2007. *Journal of Sedimentary Research*, **77**, 447–460.

You, Z.-Y. & Yin, B. 2006. *Sedimentology*, **53**, 1181–1190.

Young, M. J. *et al.* 2002. *Basin Research*, **14**, 1–23.

Yu, B. *et al.* 2006. *Journal of Sedimentary Research*, **76**, 889–902.

Yu, E.-F. *et al.* 1996. *Nature*, **379**, 689–694.

Yu, S-Y. 2007. *Geology*, **35**, 891–894.

Yuretich, R.F. 1979. *Sedimentology*, **26**, 313–332.

Yuretich, R.F. & Cerling, T.E. 1983. *Geochimica et Cosmochimica Acta*, **47**, 1099–1109.

Zakaznova-Herzog, V.P. *et al.* 2008. *Geochimica et Cosmochimica Acta*, **72**, 69–86.

Zaleha, M. 1997a. *Sedimentology*, **44**, 369–390.

Zaleha, M. 1997b. *Sedimentology*, **44**, 349–368.

Zavala, C. *et al.* 2006. *Journal of Sedimentary Research*, **76**, 41–59.

Zecchin, M. 2005. *Journal of Sedimentary Research*, **75**, 300–312.

Zhang, D.D. *et al.* 2001. *Journal of Sedimentary Research*, **71**, 205–216.

Zhu, Z.R. *et al.* 1993. *Earth Planetary Science Letters*, **118**, 281–293.

Ziegler, A.M. *et al.* 1984. In: *Fossils and Climate* (ed. Brenchley, P.J.), pp. 3–25. Wiley, New York.

Zieliński, T. & van Loon, A.J. 2000. *Geologie en Mijbouw*, **79**, 93–107.

# INDEX

Note: page numbers in italics refer to figures; page numbers in bold refer to tables



*Sedimentology and Sedimentary Basins: From Turbulence to Tectonics*, 2nd edition. © Mike Leeder.
Published 2011 by Blackwell Publishing Ltd.